Dating, 544
Debts, 337, 339
Decibel, 641
Degree requirements, 170
Demand, 117–119, 122, 123, 868, 875, 920, 1017,
    1087, 1098, 1171–1175
Dental plans, 76, 94
Depletion of natural gas reserves, 1155
Depreciation, 574, 577, 659
Designing a puzzle, 331, 382–385
Dice, 341, 380, 481
Diet, 121
Diffusion of new technology, 793
Disease, 524, 527
Dividends, 597, 598
DNA, 363
Doggie puddles, 911
Donations, 161
Drug tests, 457–461
$E = mc^2$, 980
Earthquakes, 641
Ecology, 76, 777
Econometrics, 160, 161, 171
Education fund, 582, 585–586
Education, 454–456, 863, 880, 912, 913
Effective interest rate, 570, 571, 600
Effective yield, 571, 620
Einstein's energy equation ($E = mc^2$), 980
Elasticity of demand, 916–922, 927
Electric current, 160, 1057
Electrostatic repulsion, 1121
Emission control, 725
Employment, 456, 466, 493, 513, 533, 626, 652, 803,
    912
Energy, 76, 219–220
English spelling, 390
Entertainment, 351, 352
Environmental Protection Agency, 741
Epidemics, 77, 577, 645, 650, 651, 653–656, 749, 786,
    793, 819, 879
Equilibrium price, 117–119, 1039, 1089
Estimating mortgage rates, 813–819
  estimating revenue from, 1015,
    1078–1083
Exchange rates, 185, 202
Exports, 419, 964, 978, 1045, 1047
Factoring, A-19–A-23
Faculty salaries, 1094
Fahrenheit, 72
Family values, 465
Farm population, 512
Fast cars, 926, 966
Fast food, 1042, 1053

Federal deficit, U.S., 15
Feeding sched
Fences, 859
File names,
Finance, 531
Financing lo
Fitness tests
Floppy disks, 379
Football, 515
Foreign investments, 465
Foreign trade, 1044
Frogs, 749
Fuel economy, 726, 762, 818, 979
Fuel efficiency, 88
Future value, 555, 610
Gambling, 420
Gas mileage, 926
Gasoline stations, 273
Germany, 236
GMAT, 99, 100, 121, 122, 169, 363, 379, 439–440,
    657, 661
Gold coins, 480
Gold stocks, 625
Government bonds, 650
Grades, 463–464, 476, 803, 913, 1120
GRE economics exam, 40, 78, 122, 160, 161, 171, 473,
    657, 726, 727, 763, 778, 803, 812, 864, 921
Greek life, 428
Half-life, 646–649
Harvesting forests, 864, 881
Health care, 492–493
Health insurance, 593, 513, 533
Health spending, 979
HIV testing, 420, 464, 473
HMOs, 793
HMOs, growth of, 793
Horse races, 403
Hospital finance, 400
Hospital staffing, 320
Hotel rooms, 926
Housing costs, 531, 725, 803, 997, 1112, 1121
Income, 22, 73, 103, 105, 484–485, 494, 509, 514,
    528, 533, 656, 921, 1079
Indonesia, 230–231
Industrial output, 879
Inelastic demand, 917
Infants, 16, 549, 1155
Inflation, 40, 569–570, 575–576, 600, 625, 657
Information highway, 674, 690, 777, 793
Inheritance, 169
Insurance, 75, 158, 504, 514, 546
Intelligence, alien, 1099
Interest, 555, 564, 570, 571, 576, 597, 598

FINITE MATHEMATICS & CALCULUS APPLIED TO THE REAL WORLD

FINITE MATHEMATICS & CALCULUS APPLIED TO THE REAL WORLD

STEFAN WANER
Hofstra University

STEVEN R. COSTENOBLE
Hofstra University

HarperCollins*CollegePublishers*

Sponsoring Editor: Kevin Connors
Developmental Editors: Louise Howe and Lynn Mooney
Project Coordination: Elm Street Publishing Services, Inc.
Design Administrator: Jess Schaal
Text and Cover Design: Lesiak/Crampton Design, Inc: Cindy Crampton
Cover Photo: Russell Phillips
Production Administrator: Randee Wire
Compositor: Interactive Composition Corporation
Printer and Binder: R. R. Donnelley & Sons Company
Cover Printer: Phoenix Color Corporation

Finite Mathematics and Calculus Applied to the Real World

Copyright © 1996 by HarperCollins College Publishers

HarperCollins® and ■® are registered trademarks of HarperCollins Publishers Inc.

All rights reserved. Printed in the United States of America. No part of this book may be used or reproduced in any manner whatsoever without written permission, except in the case of brief quotations embodied in critical articles and reviews. For information address HarperCollins College Publishers, 10 East 53rd Street, New York, NY 10022.

Library of Congress Cataloging-in-Publication Data

Waner, Stefan, 1949–
   Finite mathematics and calculus applied to the real world / Stefan Waner, Steven R. Costenoble.
      p.   cm.
   Includes index.
   ISBN 0-06-501816-8
    1. Mathematics. 2. Calculus. I. Costenoble, Steven R., 1961– . II. Title.
QA37.2.W34 1996
510—dc20   95–18448

95 96 97 98   9 8 7 6 5 4 3 2 1

To my parents, Ben and Mary Waner—Stefan Waner

To my parents, Earl and Kayla Costenoble, and to my wife, Nancy, and my son, Alexander—Steven R. Costenoble

Contents

Preface xiii

▶ **CHAPTER 1** FUNCTIONS AND GRAPHS 2

    1.1 COORDINATES AND GRAPHS 5
    1.2 FUNCTIONS AND THEIR GRAPHS 18
    1.3 LINEAR FUNCTIONS 41
    1.4 LINEAR MODELS 58
    1.5 QUADRATIC FUNCTIONS AND MODELS 79
    1.6 SOLVING EQUATIONS USING GRAPHING CALCULATORS OR COMPUTERS 89

    You're the Expert—Modeling the Demand for Poultry 95

▶ **CHAPTER 2** SYSTEMS OF LINEAR EQUATIONS AND MATRICES 102

    2.1 SYSTEMS OF TWO LINEAR EQUATIONS IN TWO UNKNOWNS 105
    2.2 USING MATRICES TO SOLVE SYSTEMS WITH TWO UNKNOWNS 124
    2.3 USING MATRICES TO SOLVE SYSTEMS WITH THREE OR MORE UNKNOWNS 138
    2.4 APPLICATIONS OF SYSTEMS OF LINEAR EQUATIONS 150

    You're the Expert—The Impact of Regulating Sulfur Emissions 164

▶ **CHAPTER 3** MATRIX ALGEBRA AND APPLICATIONS 172

    3.1 MATRIX ADDITION AND SCALAR MULTIPLICATION 174
    3.2 MATRIX MULTIPLICATION 187
    3.3 MATRIX INVERSION 204
    3.4 INPUT-OUTPUT MODELS 216

    You're the Expert—The Japanese Economy 232

▶ CHAPTER 4   LINEAR PROGRAMMING   238

    4.1   GRAPHING LINEAR INEQUALITIES   241

    4.2   SOLVING LINEAR PROGRAMMING PROBLEMS GRAPHICALLY   254

    4.3   SOLVING STANDARD LINEAR PROGRAMMING PROBLEMS USING THE SIMPLEX METHOD   276

    4.4   SOLVING NONSTANDARD LINEAR PROGRAMMING PROBLEMS   300

You're the Expert—Airline Scheduling   322

▶ CHAPTER 5   SETS AND COUNTING   330

    5.1   SETS   332

    5.2   THE NUMBER OF ELEMENTS IN A SET   344

    5.3   THE MULTIPLICATION PRINCIPLE; PERMUTATIONS   352

    5.4   THE ADDITION PRINCIPLE; COMBINATIONS   366

You're the Expert—Designing a Puzzle   382

▶ CHAPTER 6   PROBABILITY   388

    6.1   SAMPLE SPACES AND EVENTS   390

    6.2   INTRODUCTION TO PROBABILITY   404

    6.3   PRINCIPLES OF PROBABILITY   421

    6.4   COUNTING TECHNIQUES AND PROBABILITY   429

    6.5   CONDITIONAL PROBABILITY AND INDEPENDENCE   441

    6.6   BAYES' THEOREM AND APPLICATIONS   458

    6.7   BERNOULLI TRIALS   467

You're the Expert—The Monty Hall Problem   474

▶ CHAPTER 7   STATISTICS   478

    7.1   RANDOM VARIABLES AND DISTRIBUTIONS   480

    7.2   MEAN, MEDIAN, AND MODE   495

    7.3   VARIANCE AND STANDARD DEVIATION   516

    7.4   NORMAL DISTRIBUTIONS   534

You're the Expert—Lies, Damned Lies, and Statistics   548

Contents

▶ **CHAPTER 8**   THE MATHEMATICS OF FINANCE   552

     8.1   SIMPLE INTEREST   554
     8.2   COMPOUND INTEREST   562
     8.3   ANNUITIES AND LOANS   579
     You're the Expert—Pricing Auto Loans   594

▶ **CHAPTER 9**   EXPONENTIAL AND LOGARITHMIC FUNCTIONS   600

     9.1   EXPONENTIAL FUNCTIONS AND APPLICATIONS   602
     9.2   CONTINUOUS GROWTH AND DECAY AND THE NUMBER e 615
     9.3   LOGARITHMIC FUNCTIONS   627
     9.4   APPLICATIONS OF LOGARITHMS   642
     You're the Expert—Epidemics   653

▶ **CHAPTER 10**   INTRODUCTION TO THE DERIVATIVE   658

     10.1   RATE OF CHANGE AND THE DERIVATIVE   662
     10.2   GEOMETRIC INTERPRETATION OF THE DERIVATIVE   674
     10.3   LIMITS AND CONTINUITY   690
     10.4   DERIVATIVES OF POWERS AND POLYNOMIALS   704
     10.5   MARGINAL ANALYSIS   716
     10.6   MORE ON LIMITS, CONTINUITY, AND DIFFERENTIABILITY   727
     You're the Expert—Reducing Sulfur Emissions   741

▶ **CHAPTER 11**   TECHNIQUES OF DIFFERENTIATION   750

     11.1   THE PRODUCT AND QUOTIENT RULES   752
     11.2   THE CHAIN RULE   764
     11.3   DERIVATIVES OF LOGARITHMIC AND EXPONENTIAL FUNCTIONS   779
     11.4   IMPLICIT DIFFERENTIATION   794
     11.5   LINEAR APPROXIMATION AND ERROR ESTIMATION   804
     You're the Expert—Estimating Mortgage Rates   813

▶ **CHAPTER 12**   APPLICATIONS OF THE DERIVATIVE   820

     12.1   MAXIMA AND MINIMA   823
     12.2   APPLICATIONS OF MAXIMA AND MINIMA   845
     12.3   THE SECOND DERIVATIVE: ACCELERATION AND CONCAVITY   865

- **12.4** CURVE SKETCHING   882
- **12.5** RELATED RATES   900
- **12.6** ELASTICITY OF DEMAND   914

You're the Expert—Production Lot Size Management   922

▶ CHAPTER 13   THE INTEGRAL   928

- **13.1** THE INDEFINITE INTEGRAL   930
- **13.2** SUBSTITUTION   941
- **13.3** APPLICATIONS OF THE INDEFINITE INTEGRAL   951
- **13.4** GEOMETRIC DEFINITION OF THE DEFINITE INTEGRAL   967
- **13.5** ALGEBRAIC DEFINITION OF THE DEFINITE INTEGRAL   981
- **13.6** THE FUNDAMENTAL THEOREM OF CALCULUS   990
- **13.7** NUMERICAL INTEGRATION   996

You're the Expert—The Cost of Issuing a Warranty   1008

▶ CHAPTER 14   FURTHER INTEGRATION TECHNIQUES AND APPLICATIONS OF THE INTEGRAL   1014

- **14.1** INTEGRATION BY PARTS   1016
- **14.2** INTEGRATION USING TABLES   1025
- **14.3** AREA BETWEEN TWO CURVES AND APPLICATIONS   1029
- **14.4** AVERAGES AND MOVING AVERAGES   1045
- **14.5** IMPROPER INTEGRALS AND APPLICATIONS   1057
- **14.6** DIFFERENTIAL EQUATIONS AND APPLICATIONS   1069

You're the Expert—Estimating Tax Revenues   1078

▶ CHAPTER 15   FUNCTIONS OF SEVERAL VARIABLES   1086

- **15.1** FUNCTIONS OF TWO OR MORE VARIABLES   1088
- **15.2** THREE DIMENSIONAL SPACE AND THE GRAPH OF A FUNCTION OF TWO VARIABLES   1100
- **15.3** PARTIAL DERIVATIVES   1112
- **15.4** MAXIMA AND MINIMA   1122
- **15.5** CONSTRAINED MAXIMA AND MINIMA AND APPLICATIONS   1132
- **15.6** LEAST-SQUARES FIT   1145
- **15.7** DOUBLE INTEGRALS   1159

You're the Expert—Constructing a Best-Fit Demand Curve   1171

▶ **APPENDIX A**   ALGEBRA REVIEW   A-1

         A-1   OPERATIONS ON THE REAL NUMBERS   A-2
         A-2   INTEGER EXPONENTS   A-6
         A-3   RADICALS AND RATIONAL EXPONENTS   A-10
         A-4   THE DISTRIBUTIVE LAW: MULTIPLYING ALGEBRAIC EQUATIONS   A-16
         A-5   FACTORING ALGEBRAIC EXPRESSIONS   A-19
         A-6   RATIONAL EXPRESSIONS   A-23
         A-7   EQUATIONS   A-25
         A-8   SOLVING POLYNOMIAL EQUATIONS   A-29
         A-9   SOLVING MISCELLANEOUS EQUATIONS   A-40

▶ **APPENDIX B**   USING A GRAPHING CALCULATOR   A-49

▶ **APPENDIX C**   USING A COMPUTER SPREADSHEET   A-69

▶ **APPENDIX D**   AREA UNDER A NORMAL CURVE   A-79

▶ **APPENDIX E**   A TABLE OF INTEGRALS   A-81

         Answers to Selected Exercises   A-85
         Subject Index   A-146
         Company Index   A-160

Preface

This book is intended for a two-semester course in finite mathematics and basic calculus for students majoring in business, the social sciences, or the liberal arts. It is designed to address the considerable challenge of generating enthusiasm and developing mathematical sophistication in an audience that is often ill-prepared for and disaffected by the traditional applied mathematics courses offered on many college campuses.

This text is ambitious; we take the positive view that interested and motivated students can overcome any deficiencies in their mathematical background and attain a surprising degree of mathematical sophistication. The strong emphasis in this book on developing mathematical concepts and the abundance of relevant applications are dictated by this view. No less importantly, this book is one that a student whose primary interest is not mathematics can relate to and enjoy reading.

OUR APPROACH AND PHILOSOPHY

Our approach has been influenced by the current calculus reform movement. Within a framework of fairly traditional topics, we incorporate important features of the various calculus reform projects, including a thorough integration of graphing technology, a focus on real applications, and an emphasis on mathematical concepts through the extensive use of conceptual exercises and pedagogical techniques, such as the Rule of Three (numerical, geometric, and algebraic approach to concepts). In fact, we include a fourth rule: verbal communication of mathematical concepts. We implement this element through verbalizing mathematical concepts, rephrasing sentences into forms that easily go over to mathematical statements, and writing exercises.

At the same time, we retain the strong features of more traditional texts, including an abundance of practice and drill exercises where appropriate, large numbers of applications to choose from, and inclusion of the standard topics in applied calculus.

In addition to combining what we view as the strongest features of reform and traditional texts, we have worked to create a unique and fresh approach to pedagogy and style.

EMPHASIS ON CONCEPTS

As we develop each mathematical concept, we steer the student directly toward the most important ideas with as few obstacles as possible. We do this by equipping the student with relevant skills and a working knowledge of a topic before considering its abstract foundations in detail. We avoid prematurely side-tracking the student with rigor and abstract subtleties; the appropriate time for abstraction and rigor is later, once the student has a firm grasp of the underlying concepts. Our approach enables students to learn mathematics in the most natural way—by example. Just as one learns to speak before mastering grammar or to play a musical instrument before studying harmony, so the understanding of new mathematical concepts is best established on a solid foundation of relevant skills rather than upon abstraction devoid of context.

So that learning of concepts by example will be effective, discussion and explanation of key concepts within each example must not obscure the mathematical simplicity of the solution. We address this issue in many examples with a separate *Before we go on* discussion following the solution. Here we explain, discuss, and explore the example's solution. We sometimes remind the student that an example is not always finished when a solution is found—we should check the solution and examine its implications.

More globally, the organization of material within each chapter and section has also been planned with conceptual development in mind. We present a new concept or an application of an old one as directly as possible, with many worked examples and references to actual data, even if it means postponement of some of the underlying theory. Once the topic has progressed to the point where the student is sufficiently comfortable with the material and can relate it to the real world, we return for a retroactive in-depth look at the foundations of what we have been doing and then develop the concept further, if need be. In this way, the theory reinforces the student's knowledge and view of the world and is not seen as meaningless abstraction. For example, we introduce the student to the derivative as the rate of change, along with formulas and real applications, before we formally discuss limits.

THOROUGH INTEGRATION OF GRAPHING TECHNOLOGY

The use of graphing calculators and computer software has been thoroughly integrated throughout the discussion, examples, and exercise sets, beginning with the first example of the graph of an equation in Chapter 1. In many examples we discuss how to use either graphing technology or computer spreadsheet software to solve the example; the sections are marked with the ▦ symbol. Groups of exercises for which the use of graphing calculators or computers is suggested or required are also fully integrated in the exercise sets and carry the ▦ symbol.

This focus on graphing technology plays an important conceptual and pedagogical role in the presentation of many topics. For example, Chapter 1 includes a section on the numerical solution of equations using a graphing calculator. Chapter 2's discussion of row reduction includes examples and exercises written specifically for the use of technology. Chapter 12's treatment of curve sketching was written with the graphing calculator in mind. It also follows the increasingly popular approach of using graphing calculators to draw the graphs and then using calculus to explain the results. As a result, our approach to curve sketching is more concise, less rigid, and also less long-winded than the standard treatments.

Appendix B: Using a Graphing Calculator introduces the student to a graphing calculator and provides programs and interesting applications. It also includes several programs referred to in the text. Appendix C: Using a Computer Spreadsheet shows how to use spreadsheet software. Discussion and exercises on proper entry of functions into graphing calculators and computers are included in the first section of Appendix A: Algebra Review.

Although we emphasize graphing technology throughout the text, we are mindful of the varying degree of emphasis on and use of graphing calculators in college courses. The text is not dependent on this technology. Students who are equipped with nothing more than a scientific calculator will not find themselves at a disadvantage.

FOCUS ON REAL APPLICATIONS

We are particularly proud of the diversity, breadth, and sheer abundance of examples and exercises that are based on real, referenced data from business, economics, the life sciences, and the social sciences. This focus on real data has helped create a text that students in diverse fields can relate to and that instructors can use to demonstrate the importance and relevance of calculus in the real world.

Our coverage of real applications begins with the very first example of Chapter 1, where the Dow Jones Average is used to introduce the discussion on coordinates and graphs. It continues uniformly throughout the text, which includes innumerable examples and exercises based on real data.

At the same time, we have been careful to strike a pedagogically sound balance between applications based on real data and more traditional generic applications. The density and selection of real data-based applications have been tailored to the pedagogical goals and appropriate difficulty level for each section.

STYLE

It is a common complaint that many students do not actually read mathematics texts but simply search through them for examples that match the assigned exercises. We would like students to read this book. We would like students to *enjoy* reading this book. Thus we have written this book in a conversational and student-oriented style. We make frequent use of a question-and-answer dialogue format (indicated with the Q/A symbol) in order to encourage the development of the student's mathematical curiosity and intuition. We hope that this text will give the student insight into how a mathematician develops and thinks about mathematical ideas and their applications.

EXERCISE SETS

We regard the strength of our exercise sets as one of the best features of our text. More than 5,000 exercises provide a wealth of material that can be used to challenge students at almost every level of preparation. The exercises include everything from straightforward drill exercises to interesting and rather challenging applications. We also include in virtually every section of every chapter, applications based on real data (including data from approximately 100 major corporations and government agencies); conceptual and discussion exercises useful for writing assignments; graphing calculator exercises; and what we hope are amusing exercises. In addition, every chapter contains a collection of chapter review exercises. Communication and Reasoning Exercises appear at appropriate places in the text.

Many of the scenarios used in application examples and exercises are revisited several times throughout the book. Thus, for instance, students will find themselves using a variety of techniques, from graphing through the use of derivatives to elasticity of demand, to maximize revenue in the same application. The Cobb-Douglas production function is used in several different contexts throughout the text, including applications of derivatives, implicit differentiation, related rates, maxima and minima, and, in the chapter on calculus of several variables, linear regression, where we show the student how to obtain a best-fit Cobb-Douglas function. Our treatment of the logistic function is similar. Reusing scenarios and important functions this way provides unifying threads and shows students the complex texture of real-life problems.

PEDAGOGICAL FEATURES

- **The Rule of Four** Many of the central concepts, such as functions, limits, derivatives, and integrals, are discussed *numerically, graphically, and algebraically*. We have gone to some lengths to draw the student's attention to these distinctions. (See, for instance, the section headings in Chapter 13 in the Contents.) In addition, we verbally communicate mathematical concepts as a fourth element. We do so through our conceptual exercises at the end of each section, our emphasis on verbalizing mathematical concepts, and our discussions of rephrasing sentences into forms that easily go over to mathematical statements.

- ***Q/A*** An important pedagogical tool in this text is the frequent use of informal question-and-answer dialogues. These often anticipate the kind of questions that may occur to the student and also guide the student through the development of new concepts.

- **You're the Expert** Each chapter begins with an application—an interesting problem—and returns to it at the end of that chapter in a section titled *You're the Expert!* The themes of these applications are varied, and they are designed to be as unintimidating as possible. For example, we do not pull complicated formulas out of thin air but focus instead on the development of mathematical models appropriate to the topics. Among the more notable of the *You're the Expert* applications are an early example of modeling poultry demand based on actual data and an example using marginal analysis to design a strategy for regulating sulfur emissions. These applications are ideal for assignment as projects, and it is to this end that we have included groups of exercises at the end of each.
- **Before We Go On** Most examples are followed by supplementary discussions under the heading *Before we go on*. These discussions may include a check on the solution to a problem, a discussion of the feasibility and significance of a solution, or an in-depth look at what the solution means.
- **Systematic and Careful Treatment of Word Problems** We take an organized approach to the analysis and discussion of word problems, where the student is shown how to zero in on the unknown in a word problem and how to reword pertinent phrases into forms readily translatable into mathematical expressions. For example, such a potentially confusing phrase as "there are at least twice as many goats as sheep" is transformed into a clearer form, such as: "the number of goats is at least twice the number of sheep" (which translates directly into $x \geq 2y$).
- **Communication and Reasoning Exercises** These exercises are designed to broaden the student's grasp of the mathematical concepts. The student might be asked to provide examples, to illustrate a point, to design an application with a given solution, to fill in the blank, or to discuss and debate. These exercises often have no single correct answer.
- **Conceptual and Computational Devices** The text features a wide variety of novel devices to assist the student in overcoming hurdles. These include a calculation thought experiment for the analysis and differentiation of complicated functions, the use of the remarkably quick and efficient *column integration* method for integration by parts, and a *template* method for evaluating definite integrals that avoids the common errors in signs. We also use computational techniques for matrix reduction and simplex method pivoting that are easiest for hand calculation, including a technique for avoiding manipulating fractions.
- **Cautions, Hints, and Notes** Most sections include suggestions to assist students in avoiding common errors and in tracking down their source when they occur in a calculation.
- **Footnotes** Footnotes throughout the text provide sources, interesting background, extended discussion, and various asides.

RELATED BOOKS IN THE SERIES

This is one of three books by the authors on mathematics for business, the social sciences, and the liberal arts. *Finite Mathematics Applied to the Real World* contains the material of the first half of this book with the addition of matrix systems and games theory. *Calculus Applied to the Real World* contains the material of the second half of this book.

ORGANIZATION AND COURSE OPTIONS

Care has been taken to provide the greatest possible flexibility in course design.

- The chapter on linear programming does not require the material in Chapter 3 (Matrix Algebra).
- The Mathematics of Finance chapter can be covered at any time after Chapter 1.
- The material on counting arguments (Sections 3 and 4 of Chapter 5) is not required for the chapter on probability theory (optional sections 4 and 7 excluded).
- The Exponential and Logarithmic Functions chapter can be studied at any time in the course prior to Section 3 in Chapter 11 (on derivatives of exponential and logarithmic functions).
- The chapter on functions of several variables can be studied at any time after the chapter on applications of the integral.

SUPPLEMENTS

For the Instructor

HarperCollins Test Generator/Editor for Mathematics with QuizMaster is fully networkable and is available in both IBM and Macintosh versions. The system features printed graphics and accurate mathematical symbols. The program allows the instructor to choose problems either randomly from a section or problem type or manually while viewing them on the screen, with the options to regenerate variables or scramble the order of questions while printing if desired. The editing feature allows instructors to customize the chapter data disks by adding their own problems. The Test Generator comes free to adopters.

Instructor's Resource Manuals for *Finite Mathematics Applied to the Real World* **and** *Calculus Applied to the Real World* provide detailed discussion of the material in each section, complete solutions to the *Communication and Reasoning Exercises* and to the *You're the Expert* exercise sets, teaching tips, and a large collection of sample tests.

Instructor's Complete Solution Manuals for *Finite Mathematics Applied to the Real World* **and** *Calculus Applied to the Real World* contain solutions to every exercise in the texts.

Steven's Software is a set of tools for the Macintosh that can be used either by students for independent exploration or by instructors for demonstration of graphing the Gauss-Jordan method, matrix algebra, and the simplex method.

For the Student

Interactive Tutorial Software with Management System is available in IBM and Macintosh versions and is fully networkable. As with the Test Generator/Editor, this innovative software is algorithm driven, automatically regenerating constants so that a student will not see values repeated if he or she revisits any particular problem for additional practice. The tutorial is self-paced and provides unlimited opportunities to review lessons and to practice problem solving. If a student gives a wrong answer, he or she can request to see the problem worked out and get a textbook page reference. The program is menu-driven for ease of use, and on-screen help can be obtained at any time with a single keystroke. Students' scores are automatically recorded and can be printed for a permanent record. The optional **Management System** lets instructors record student scores on disk and print diagnostic reports for individual students or classes. This software is free to adopters but also may be purchased by students for home use. (Macintosh version ISBN 0-06-502585-7; IBM version 0-06-502584-9)

GraphExplorer provides students and instructors with a comprehensive graphing utility able to graph rectangular, conic, polar, and parametric equations; zoom; transform functions; and experiment with families of equations quickly and easily. It is available in IBM and Macintosh formats.

Explorations in Finite Mathematics (IBM format only), by David Schneider, University of Maryland, (ISBN 0-673-46932-8) contains on one disk a wider selection of routines than in any similar software supplement. Included are utilities for Gaussian elimination, matrix operations, graphical and simplex methods for linear programming problems, probability, binomial distribution, simple and compound interest, loan and annuity analysis, finance table, difference equations, and more. Refined monitor display for fractions, color capabilities, choice of exact or approximate calculations with matrices, and refined printing capabilities further set this apart from other programs.

Matrix with Linear Programming (IBM format only), by Maylin Dittmore (ISBN 0-06-501266-6) is designed to assist the student in any course of study that involves the use of matrices. MATRIX was not only created to help the student with tedious calculations associated with matrices but also to help them gain understanding of and appreciation for real work problems that can be analyzed and solved using matrices.

Student's Solution Manual (ISBN 0-06-501817-6) provides complete, worked-out solutions to the odd-numbered exercises in the text. The manual also includes comprehensive chapter summaries and true/false quizzes for each chapter that help students both review and test their understanding

Topics in Finite Mathematics: An Introduction to the Electronic Spreadsheet, by Sam Spero, Cuyahoga Community College (ISBN 0-06-500300-4), is a user-friendly guide designed to introduce students to the various ways one can approach problem solving with spreadsheets. Knowledge of spreadsheets is not assumed, and the approach is adaptable to all spreadsheet programs.

The Electronic Spreadsheet and Elementary Calculus, by Sam Spero, Cuyahoga Community College (ISBN 0-673-46595-0), is a companion to the above-mentioned *Topics in Finite Mathematics.* This guide helps students get started with graphing and problem solving by means of the spreadsheet. As with the companion volume, knowledge of spreadsheets is not assumed, and the approach is adaptable to all spreadsheet programs.

Graphing Calculator Lessons for Finite Mathematics by Paula G. Young (ISBN 0-06-501330-1) contains activities for utilizing the graphing calculator in the course.

CLASS TESTED

This book was used in manuscript form by many of our colleagues at Hofstra University. Their reactions and input were crucial to the development of this book. We thank them for all of their help; their names appear in the Acknowledgments section on the following page.

ACCURACY

Accuracy checking was carried out at every stage of the production process. In particular, each chapter was checked by at least two different mathematicians. We gratefully acknowledge their help in this critical aspect of the project: Steven Blasberg, West Valley College; Patricia Blus, National Lewis; Susan Boyer, University of Maryland—College Park; Richard L. Conlon, University of Wisconsin—Stevens Point; Pamela G. Coxson, Lawrence Berkeley National Laboratory; Carol DeVille, Louisiana Technical University; Vivian Freund; Richard Leedy, Polk Community College; Ron Netzel; Kathleen R. Pirtle; Franklin Pirtle; Richard Porter, Northeastern University; Jane Rood, Eastern Illinois University; James Wooland, Florida State University; Earl J. Zwick, Indiana State University.

ACKNOWLEDGMENTS

This project would not have been possible without the contributions and suggestions of numerous colleagues, students, and friends. We are particularly grateful to our many colleagues who class tested the various preliminary editions of this book, and to our editors at HarperCollins for their encouragement and guidance throughout the project. Specifically, we would like to thank George Duda, acquisitions editor, for his enthusiasm and for believing this would work; Kevin Connors, acquisitions editor, for vision in the final phases of the project; Louise Howe, developmental editor, for pushing us in the right directions; Lynn Mooney, developmental editor, for keeping us in check and seeing it through to the end; Connie Low for her work with the Game Theory chapter; David Knee for his detailed critiques of the text; Bill McKeough for his careful criticism and detailed suggestions for the logic, sets, counting and probability chapters; Aileen Michaels, Marysia Weiss, Nick

Frangos, and Robert M. Bumcrot for pioneering the class testing of the earliest drafts of the finite math portion of the book; Safwan Akbik and Michael Steiner for numerous helpful suggestions; Daniel Rosen for his careful analysis of the material and numerous suggestions and especially for his version of tabular integration by parts; Joan Arndt, Theresa Vechiarelli, and Michelle Lisi for class testing several versions of the finite math material and for many innovative ideas; Edward Ostling for his encouragement and for class testing several versions of the calculus portion; and Stan Kertzner for his careful analysis of the logic and set theory portions of the book.

We would also like to thank the numerous reviewers who read carefully and commented on successive drafts, providing many helpful suggestions that have shaped the development of this book:

Mary Kay Abbey, Montgomery College
Jeffrey S. Albritten, Middle Tennessee State University
William L. Armacost, California State University—Dominquez Hills
Bill Aslan, East Texas State University
Stephen A. Bacon, Central Connecticut State University
Louise B. Bernauer, City College of Morris
Steven Blasberg, West Valley College
Chris Boldt, Eastfield College—Dallas
Bob Bradsaw, Ohlone College
Barbara M. Brook, Camden City College
Mark Carpenter, Sam Houston State University
Mitzi Chaffer, Central Michigan University
Charles E. Cleaver, The Citadel
Richard L. Conlon, University of Wisconsin—Stevens Point
B. Jan Davis, University of Southern Missouri
Kenneth A. Dodaro, Florida State University
William Drezden, Oakton Community College
William L. Etheridge, University of North Carolina—Wilmington
Joe S. Evans, Middle Tennessee State University
Elise Fischer, Johnson County Community College
Carol J. Flakus, Lower Columbia College
Donald R. Goral, Northern Virginia Community College
John Gregory, Southern Illinois University
Joan F. Guetti, Seton Hall
Frances F. Gulick, University of Maryland
Dianne Hendrickson, Becker College

Edwin M. Klein, University of Wisconsin—Whitewater
Martin Kotler, Pace University
Richard Leedy, Polk Community College
James T. Loats, Metropolitan State College of Denver
Kelly M. Locke, Hartnell College
Vicky Lymbery, Stephen F. Austin State University
Randall B. Maddox, Pepperdine University
Don Mahaffey, Missouri Western State College
Steven E. Martin, Richard Bland College
James McCullough, Arapahoe Community College
Thomas McCullough, California State University—Long Beach
Charles J. Miller, Foothills College
Donald F. Myers, University of Arizona
Gertrude Okhuysen, Mississippi State University
Kevin O'Neil, Kankakee Community College
James H. Parker, Husson College
Gregory B. Passty, Southwest Texas State University
Kathleen R. Pirtle, Franklin University
Georgia B. Pyrros, University of Delaware
Ken Reeves, San Antonio College
Mohamad Riaza, Fort Hays State University
John A. Roberts, University of Louisville
Jane M. Rood, Eastern Illinois University
Arthur Rosenthal, Salem State University
Judith F. Ross, San Diego Mesa College
Derald D. Rothman, Moorhead State University
Daniel E. Scanlon, Orange Coast College
Arnold L. Schroeder, Long Beach City College
Richard H. Schroeder, Ball State University
Richard Semmler, Northern Virginia Community College
Sally Sestini, Cerritos College
Hari Shankar, Ohio University
Robert E. Sharpton, Miami-Dade Community College
Gordon Shilling, University of Texas, Arlington
Mahendra P. Singhal, University of Wisconsin—Parkside
Joseph D. Sloan, Lander University

Clifford W. Sloyer, University of Delaware
Samuel W. Spero, Cuyahoga Community College
Stanley L. Stephens, Anderson University
Lowell Stultz, Kalamazoo Valley Community College
Anthony E. Vance, Austin Community College
Lyndon C. Weberg, University of Wisconsin—River Falls
Patrick Webster, Marymount Palo Verde
Donald D. Weddington, San Jose State University
Lynn J. Wolfmeyer, Western Illinois University
Ellen Wood, Stephen F. Austin University
James Wooland, Florida State University
Anne L. Young, Loyola College
Fredric Zerla, University of South Florida
Earl J. Zwick, Indiana State University

Stefan Waner
Steven R. Costenoble
November 1995

CHAPTER 1

Boston Chicken Wields Midas Touch

By FLOYD NORRIS

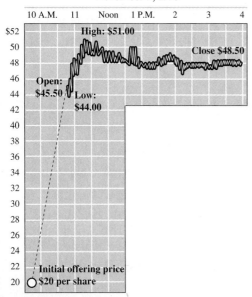

Nasdaq trading in Boston Chicken stock
November 9, 1993

Source: Bloomberg Financial Markets
The New York Times, Nov. 10, 1993 p. D1

In the most successful initial public offering of stock in years, Boston Chicken, a fast-food chain that has yet to earn its first annual profit, went public and promptly more than doubled in price.

At yesterday's closing price of $48.50 a share, Boston Chicken had a market value of $839 million, or almost 27 times revenues over the last 12 months. While the offering had been tipped for weeks as this year's hottest, the amount of the increase stunned both company officials and Wall Street traders.

The appetite for the stock reflected a strong growth story for the chain. Based in the Chicago suburb of Naperville, Ill., Boston Chicken claims to be the largest franchise operation specializing in rotisserie chicken, a segment that many think will grow rapidly as consumers become more health-conscious.

The top officers of Boston Chicken include several early investors in Blockbuster Video, including Scott A. Beck, the 35-year-old chairman and chief executive of Boston Chicken who "retired" last year as Blockbuster vice chairman. The connection gave Boston Chicken a following on Wall Street. In a highly unusual move, those officers yesterday bought stock in a companion offering, a fact that helped to persuade investors to buy.

Boston Chicken sold 2.06 million shares to the public at $20, in an offering whose lead underwriters were Merrill Lynch and Alex Brown & Sons.

The shares began trading in the public market at $45.50—a more than twofold gain—and traded as high as $51 before closing at $48.50. Volume was 8.4 million shares, meaning the average share in the offering was traded more than four times during the day.

The price of $20, while well above the range of $15 to $17 the company said it would seek when it filed to go public, proved to be far less than the demand. Leaving all that money on the table meant the company had less money than it might have had to finance its ambitious expansion plans, but Mark W. Stephens, the company's chief financial officer and a former investment banker, said the company was happy with the price.

Source: From Floyd Norris, "Boston Chicken Wields Midas Touch," *The New York Times*, November 10, 1993, p. D1.

Functions and Graphs

SECTIONS

1. Coordinates and Graphs
2. Functions and Their Graphs
3. Linear Functions
4. Linear Models
5. Quadratic Functions and Models
6. Solving Equations Using Graphing Calculators or Computers

You're the Expert
Modeling the Demand for Poultry

APPLICATION ▶ A government agency plans to regulate the price of poultry with the goal of increasing revenue to poultry producers. Economists have determined that the amount of poultry people will buy depends on both the price of poultry and the price of beef. They have devised an equation relating the demand for poultry (measured by annual consumption) to the price of poultry and the price of beef. The agency has hired you to develop a pricing policy and to analyze the effect of fluctuations in the beef price on revenue to poultry producers under your proposed policy. How do you go about developing the policy?

INTRODUCTION ▶ To advise the government agency, we need to identify, given the price of beef, the price of poultry that will maximize revenue to the poultry industry. This will tell us how the price of poultry should depend on the price of beef. By the time we get to the end of this chapter we will be able to give the government agency some good advice.

But where do we start? In order to model real-life situations, we need to build on a solid understanding of basic concepts and simple examples. Most problems that we try to solve using mathematics involve a relationship between two or more quantities, such as the price and the annual sales of poultry or the length and width of a rectangle. When we look at two related quantities, the relationship usually can be expressed as an equation in which the two quantities are the only unknown numbers. Such an equation is called an "equation in two unknowns." A very useful way of picturing such a relationship is to draw the *graph* of the equation. The simplest relationships, with the simplest graphs, are the linear and quadratic equations in two unknowns, and we shall look at these in some detail (in Sections 3 through 5). In our discussion of linear equations, we shall meet the concepts of *slope* and *rate of change,* which are the starting point of calculus.

Mathematics once revolved around the concept of an equation, but a more sophisticated notion evolved during the seventeenth and eighteenth centuries: a *function.* Briefly, if one quantity (the demand for poultry, for instance) depends on another (the price of poultry), then we say that the first quantity is a function of the second. In Section 2, we shall discuss the concept of a function and see again that graphs are a very useful tool.

You are urged to spend as much time as possible visualizing the functions you come across by means of their graphs—especially if you are using a graphing calculator. In particular, you should learn the graphs of the standard functions discussed in Section 2. Much of the intuition behind calculus comes from such pictures. You are also encouraged to glance through the Algebra Review in Appendix A, particularly Sections A.6 through A.9, before you begin your study of this chapter, and to refer to the Review as needed.

The material in the first part of this chapter is too simple to require the use of a graphing calculator or computer program. If you plan on using one of these devices later in the course, however, now is a good time to start practicing, so we have put comments throughout the chapter pointing out places where you can do so. A graphing calculator is handy for Section 2, and Section 6 discusses one particularly good use for these devices.

1.1 COORDINATES AND GRAPHS

PLOTTING DATA IN THE COORDINATE PLANE

The following table lists the daily closing value of the Dow Jones Industrial Average over the course of the week from June 21–25, 1993.*

Day	June 21	22	23	24	25
Dow	3511	3498	3467	3491	3491

Although this table tells you everything you need to know about the behavior of the closing values that week, it is much more striking to see the *graph* of the Dow, as in Figure 1. This is a graph you might see published in the newspaper.

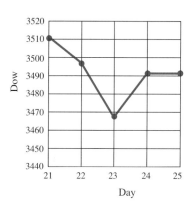

FIGURE 1

The movements up and down, and their relative sizes, are easier to see in the graph than in the tabulated data. This is the purpose of a graph: it is a way of *visualizing information*. Throughout this book we will be using graphs to visualize the relationships between quantities (here, time and the Dow), and so we need to recall some basic facts about the Cartesian plane, or "xy-plane." Although you probably think you know what the xy-plane is, you have probably never actually tried to *define* what it is. Thus, we pose the following question.

Q Just what is the xy-plane?

A The xy-plane is an infinite flat surface with two perpendicular lines, usually labeled the ***x*-axis** and ***y*-axis.** These axes are calibrated as shown in Figure 2.

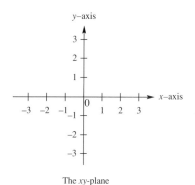

FIGURE 2

Thus the xy-plane is nothing more than a very large—in fact, infinitely large—flat surface. The purpose of the axes is to allow us to locate specific positions, or **points,** on the plane, with the use of **coordinates.** (If Captain Picard wants to have himself beamed to a specific location, he must supply its coordinates, or he's in trouble.)

▼ *Source: *Newsday,* issues of June 22–26, 1993.

Q So how do we use coordinates to locate points?

A The rule is simple. Each point in the plane has two coordinates, an **x-coordinate** and a **y-coordinate**. These can be determined in two ways:

1. The x-coordinate measures a point's distance to the right or left of the y-axis. It is positive if the point is to the right of the axis, negative if it is to the left of the axis, and 0 if it is on the axis. The y-coordinate measures a point's distance above or below the x-axis. It is positive if the point is above the axis, negative if it is below the axis, and 0 if it is on the axis. Briefly, the x-coordinate tells us the *horizontal* position (distance left or right), and the y-coordinate tells us the *vertical* position (height).

2. Given a point P, we get its x-coordinate by drawing a vertical line from P and seeing where it intersects the x-axis. Similarly, we get the y-coordinate by extending a horizontal line from P and seeing where it intersects the y-axis.

Here are a few examples to help you review coordinates.

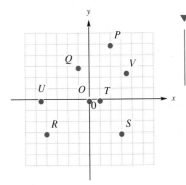

FIGURE 3

▼ **EXAMPLE 1**

Find the coordinates of the indicated points. (See Figure 3. The grid lines are placed at intervals of one unit.)

SOLUTION Taking them in alphabetical order, we start with the origin O. This point has height zero and is also zero units to the right of the y-axis, so its coordinates are $(0, 0)$. Turning to P, dropping a vertical line gives $x = 2$ and extending a horizontal line gives $y = 5$. Thus, P has coordinates $(2, 5)$. For practice, determine the coordinates of the remaining points, and check your work against the list that follows.

$$Q(-1, 3), R(-4, -3), S(3, -3), T(1, 0),$$
$$U(-\tfrac{9}{2}, 0), V(\tfrac{7}{2}, \tfrac{5}{2})$$

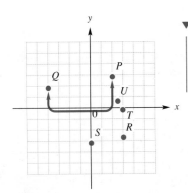

FIGURE 4

▼ **EXAMPLE 2**

Locate the following points in the xy-plane.

$$P(2, 3), Q(-4, 2), R(3, -\tfrac{5}{2}), S(0, -3), T(3, 0),$$
$$U(\tfrac{5}{2}, \tfrac{2}{3}).$$

SOLUTION In order to locate each of these points, we start at the origin $(0, 0)$, and proceed as follows (see Figure 4):

To locate P, we move 2 units to the right and 3 up, as shown.

To locate Q, we move -4 units to the right (i.e., 4 to the *left*) and 2 up, as shown.

To locate R, we move 3 units right and $\tfrac{5}{2}$ down, and so on.

▶ **NOTES**

1. The correspondence between a *point* on the xy-plane and the *pair of numbers* representing its coordinates is a one-to-one correspondence. That is,
   (i) to every point in the xy-plane, there corresponds one and only one pair of numbers (as in Example 1);
   (ii) to every pair of numbers, there corresponds one and only one point (as in Example 2).

   As a result, we usually think of a point as "being" a pair of numbers and vice-versa. Analogously, points in *three*-dimensional space may be thought of as *triples of numbers*.

2. One drawback of graphs is that it is difficult to plot points with fractional coordinates (such as R and U) with perfect accuracy. For instance, imagine trying to graph the point with coordinates (1.0000000001, 3.999998889997659).

3. This way of assigning coordinates to points in the plane is often called the system of **Cartesian** coordinates, in honor of the mathematician and philosopher René Descartes (1596–1650), who was the first to use them extensively. There are other useful coordinate systems that we shall not discuss. ◀

It is sometimes convenient to think of the xy-plane as being divided by the axes into four **quadrants** (first, second, third, and fourth). These consist of the following sets of points. (See Figure 5.)

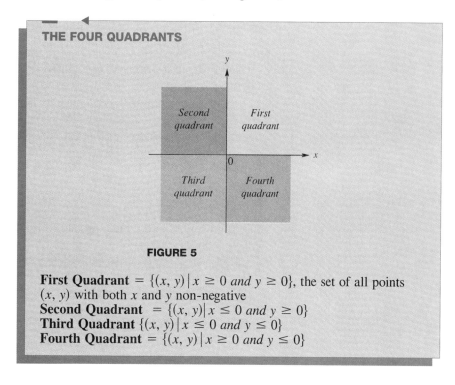

FIGURE 5

First Quadrant $= \{(x, y) \mid x \geq 0 \text{ and } y \geq 0\}$, the set of all points (x, y) with both x and y non-negative
Second Quadrant $= \{(x, y) \mid x \leq 0 \text{ and } y \geq 0\}$
Third Quadrant $\{(x, y) \mid x \leq 0 \text{ and } y \leq 0\}$
Fourth Quadrant $= \{(x, y) \mid x \geq 0 \text{ and } y \leq 0\}$

THE GRAPH OF AN EQUATION

One of the more surprising developments of mathematics was the realization that equations, which are algebraic objects, can be represented by graphs, which are geometric objects. The kinds of equations that we have in mind are equations in two variables, such as

$$y = 4x - 1,$$
$$2x^2 - y = 0,$$
$$q = 3p^2 + 1,$$
$$y = \sqrt{x - 1}.$$

In order to describe the graph of an equation, we must first say something about the *solutions* of an equation.

> ### SOLUTION TO AN EQUATION
> A **solution** to an equation in one or more unknowns is an assignment of numerical values to each of the unknowns so that when these values are substituted for the unknowns, the equation becomes a *true statement about numbers*. We say that a solution **satisfies** the equation.

Thus, for example, one solution to the equation

$$x + y^2 = 5$$

is $x = 1, y = -2$, since substituting these values for the unknowns gives

$$1 + (-2)^2 = 5,$$

which is a true statement about numbers. There are infinitely many solutions to this equation. (Two more are $x = 5, y = 0$, and $x = 3, y = \sqrt{2}$.) We can represent each solution by a single point in the coordinate plane. For instance, we can represent the solution $x = 1, y = -2$ by the point $(1, -2)$. With most equations, if we string together all the points representing solutions, we get an elegant curve, called the *graph* of the equation.

> ### GRAPH OF AN EQUATION
> The **graph** of an equation in the two variables x and y consists of all points (x, y) in the plane whose coordinates are solutions of the equation.

▼ **EXAMPLE 3**

Sketch the graph of the equation $x + y = 4$.

SOLUTION The graph consists of all points representing solutions of the equation, so we must first find the solutions. Since a solution of $x + y = 4$

consists of values for x and y that add up to 4, possible solutions are $(1, 3)$, $(0.5, 3.5)$, and $(100, -96)$.

To be more systematic about this, we proceed as follows. Notice that no matter what value we choose for x, we can always find a corresponding value for y so that the two numbers add up to 4. So we do the following.

1. Solve the equation $x + y = 4$ for y, getting $y = 4 - x$.
2. Choose values for x and substitute them in this formula to get the corresponding values for y.

This procedure gives a table of solutions.

x	-4	-3	-2	-1	0	1	2	3	4	5	...	411	...
$y = 4 - x$	8	7	6	5	4	3	2	1	0	-1	...	-407	...

The pairs (x, y) in this table, such as $(-4, 8)$, $(-3, 7)$, $(-2, 6)$, ... are solutions to the equation. Of course, there are infinitely many "in-between" solutions, such as $(-4.1, 8.1)$, $(-4.11, 8.11)$, and so on, that are not shown in this table. For now, you need only be aware that these other solutions exist. Reading the table from left to right gives the points

$$(-4, 8), (-3, 7), (-2, 6), (-1, 5), (0, 4), (1, 3), (2, 2),$$
$$(3, 1), (4, 0), (5, -1), \ldots, (411, -407), \ldots$$

on the graph of $x + y = 4$. Plotting several of them gives the picture on the left in Figure 6.

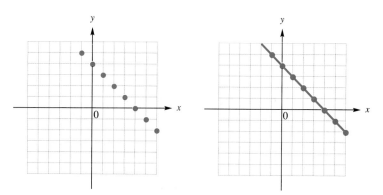

FIGURE 6

Now this figure is very suggestive of a straight line. If we decided to plot "in-between" points, such as $(0.5, 3.5)$, $(0.25, 3.75)$ and so on, we would find that they all lie precisely on the line shown on the right in Figure 6. This line is the graph of the equation $x + y = 4$.

Note that the line segment sketched gives only a *range* of solutions; the entire solution set would encompass an infinitely long line extending the segment shown in both directions. We can imagine what this may look like, but we can never actually see it.

Before we go on... Our first step, solving the equation for *y*, gives us *y* **as a function of** *x*. By this phrase we mean that the equation $y = 4 - x$ gives us a *rule* for calculating the value of *y* if we are given any value of *x*, as shown in the table. Through this rule, we think of *y* as **depending on** *x*. We shall discuss functions in the next section.

The graphing programs used by calculators and computers operate by simply plotting hundreds of solutions to a given equation and then joining them, usually with straight lines, creating the effect of a smooth curve. If you have a graphing calculator at hand, you can experiment by using it to plot the graph of $y = -x + 4$.* (On a TI-82, for example, you press [Y=] to get the display "$Y_1 =$" and enter

$$Y_1 = -X + 4$$

then press [GRAPH].) In order to view different portions of the graph, you must set the **viewing window coordinates** accordingly. These are usually denoted by *xMin*, *xMax*, *yMin*, and *yMax*, and they designate the portion of the *xy*-plane you will see. On the left in Figure 7 you see a rectangle drawn around the portion of the *xy*-plane defined by $xMin = -1$, $xMax = 5$, $yMin = -2$, and $yMax = 3$. On the right you see what the resulting graph would look like on a calculator.

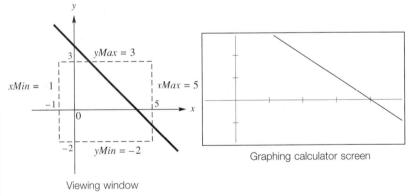

Viewing window

Graphing calculator screen

FIGURE 7

Once you have graphed the equation, you can use the trace feature to move around on the graph and find the coordinates of several points. Some of these points may correspond to the ones we calculated by hand in the table. (On the TI-82, you can obtain such a table automatically by pressing [2nd] [TABLE].)

Be warned, however, that the trace feature will not show all the points on the graph, only the finite number that are plotted on the screen. Also, the accuracy of the coordinates shown may be limited. (Check the

▼ * See the first section of Appendix G for more details on using a calculator to graph.

1.1 Coordinates and Graphs

device you are using: Do the x- and y-coordinates always add up to *exactly* 4?) You can use your calculator's zoom feature to see more of the "in-between" points on the graph.

Note finally that most graphing calculators and computer programs will only graph functions. Graphing a general equation (not solved for y) is difficult to do mechanically.

▼ **EXAMPLE 4**

Sketch the curve $xy = 1$.

SOLUTION We start, as before, by solving for y to get $y = \frac{1}{x}$. We then make a table of values by choosing values for x and calculating the corresponding y-values from the equation. Notice that this means we cannot choose $x = 0$, since there is no such number as $\frac{1}{0}$. But that shouldn't stop us from getting as close to zero as we like. For example, we can choose $x = \pm\frac{1}{2}$, $x = \pm\frac{1}{100}$ or $x = \pm\frac{1}{100,000}$. Think of it as a dangerous exploration, very close to the "forbidden" zone. Here are some values:

x	-3	-2	-1	$-\frac{1}{2}$	$-\frac{1}{3}$	$-\frac{1}{100,000}$	0	$\frac{1}{100,000}$	$\frac{1}{3}$	$\frac{1}{2}$	1	2	3
$y = \frac{1}{x}$	$-\frac{1}{3}$	$-\frac{1}{2}$	-1	-2	-3	$-100,000$	✘	$100,000$	3	2	1	$\frac{1}{2}$	$\frac{1}{3}$

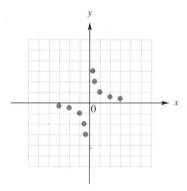

FIGURE 8

Plotting those points we can manage gives us Figure 8. Before joining up the points, note that the curve is not permitted to cross the y-axis.

Q Why not?

A Because if it did, then you would have a point on the curve with $x = 0$. But, as the "✘" reminds you, there is no possible y-value when $x = 0$. Thus, the y-axis should be regarded here as a wall that cannot be breached.

Q So how can the curve possibly get from the left-hand portion to the right-hand portion without crossing the y-axis?

A It can't, so there are two separate curves instead of one, as shown in Figure 9.

This curve is a **hyperbola**. Notice that it gets closer and closer to the axes as it extends outward in all directions, but never quite touches them.

As we've mentioned, graphing calculators and computer graphing programs draw graphs by plotting a large number of points (and possibly joining them up) to get an approximation to a curve. If you use a graphing calculator or computer to graph the equation $y = \frac{1}{x}$, then, depending on the type of graphing calculator or program you are using, you may not get the exact picture shown in Figure 9. Instead, you may see a near-vertical

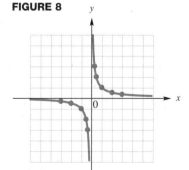

FIGURE 9

line down the y-axis joining the left-hand portion of the curve to the right-hand portion. The reason for this is that graphing software is not very good at spotting **singularities**—isolated points where the function of x is not defined. In the case we are looking at, there is a singularity at $x = 0$. When you instruct your graphing calculator to draw the graph, it may simply skip over the singularity and join the two portions of the curve with a near-vertical line.

INTERVALS ON AN AXIS

There are some sets of numbers that arise often enough to have their own notation. Suppose a and b are real numbers with $a < b$. Then by $[a, b]$ we mean the set of all real numbers x with $a \leq x \leq b$. For instance, $[1, 3]$ is the set of all real numbers between 1 and 3, inclusive. We can think of $[1, 3]$ graphically as a segment of the real line, or an axis.

Here, the segment is shown as a heavy line. The solid dots at the ends indicate that we are including the two points 1 and 3. We call $[a, b]$ a **closed interval.**

By (a, b) we mean the set of all real numbers x with $a < x < b$. Thus, for example, the set $(1, 3)$ is the set of all real numbers *strictly between* 1 and 3. We represent $(1, 3)$ graphically as follows.

Here, the hollow circles at 1 and 3 indicate that these two numbers are left out. We call (a, b) an **open interval.**

▶ CAUTION   The notation (a, b) is ambiguous. For instance, $(1, 3)$ could mean the open interval as shown above, or it could refer to a point in the xy-plane with coordinates 1 and 3. Mathematical notation can sometimes be ambiguous. You are expected to infer its meaning from the context of the discussion. Here we are talking about intervals, not points in the plane. ◀

It sometimes happens that we wish to consider an infinitely long interval, such as the set of all real numbers (the whole real line), or the set of positive real numbers or real numbers less than 6 (half-lines). For these, we use the notations $(-\infty, +\infty)$, $(0, +\infty)$, and $(-\infty, 6)$ respectively. This is consistent with our previous use of the notation.

Formally:

$(a, +\infty)$ is the set of all real numbers x with $a < x < +\infty$, or $x > a$.*

$(-\infty, b)$ is the set of all real numbers with $-\infty < x < b$, or $x < b$.

$(-\infty, +\infty)$ is just the whole real line, sometimes written as $\mathbb{R}$.

▼ *Saying $x < +\infty$ is redundant; *all* real numbers are less than infinity, so we could leave this out, and simply say $a < x$, or equivalently, $x > a$. There is a similar redundancy in the description of $(-\infty, b)$.

For example, $(-\frac{1}{4}, +\infty)$ is the set of all real numbers $x > -\frac{1}{4}$, as shown below.

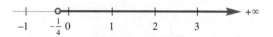

Similarly, $(-\infty, \frac{1}{4})$ is the set of all real numbers $x < -\frac{1}{4}$.

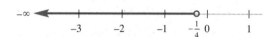

We also have four kinds of *half-open intervals*. Rather than defining each type formally, we'll simply give examples using diagrams.

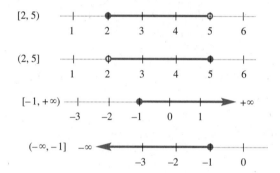

Just when you thought it was over, we mention one more thing: we'll also be considering *unions* of intervals of the various types just discussed. Again, we'll simply illustrate this with several examples.

This is the set of all points that are either in $(-\infty, 0]$ *or* in $(2, 3)$.

Here, the hollow dot at zero indicates that 0 is missing, as though we have "punctured" the line there. Had we wanted $(-3, 0) \cup [0, +\infty)$ instead, the diagram would have been exactly the same, except for the fact that 0 would not have been missing. This would, of course, be identical to the interval $(-3, +\infty)$.

▶ **NOTE** When using a graphing calculator or computer program, we can use the interval notation

$$[xMin, xMax]$$

and

$$[yMin, yMax]$$

1.1 EXERCISES

1. Referring to the following figure, determine the coordinates of the indicated points as accurately as you can. (The grid lines are placed at intervals of one unit.)

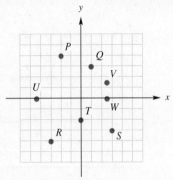

2. Referring to the following figure, determine the coordinates of the indicated points as accurately as you can. (The grid lines are placed at intervals of one unit.)

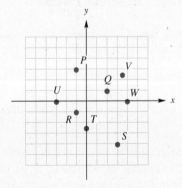

3. Graph the following points.

$P(4, 4)$, $Q(6, -5)$, $R(3, 0)$, $S(4, \frac{1}{4})$, $T(\frac{1}{2}, \frac{3}{4})$, $U(-2, 0)$, $V(-100, 0)$

4. Graph the following points.

$P(4, -2)$, $Q(2, -5)$, $R(1, -3)$, $S(-4, \frac{1}{4})$, $T(\frac{1}{4}, -1)$, $U(-\frac{1}{4}, 0)$, $V(0, 100)$

APPLICATIONS

Stock Market Index Exercises 5 and 6 are based on the following table, which shows the monthly highs and lows of the Dow Jones Industrial Average for the year 1929.*

	Jan	Feb	Mar	Apr	May	Jun	Jul	Aug	Sep	Oct	Nov	Dec
High	318	322	321	319	327	334	348	380	381	353	258	264
Low	297	296	297	299	293	299	335	338	344	230	199	231

5. Graph the monthly highs. Choose the scales on your axes carefully so that you can clearly see the upward and downward moves of the Dow.

6. Repeat the previous exercise for the monthly lows.

7. *Deficits* The monthly U.S. trade balance for the months from March 1992 to March 1993 is shown in the following graph.† Use this graph to estimate the approximate values (to the nearest $0.5 billion) of the trade deficit for these months.

8. *Deficits* The U.S. federal deficit for the years 1982–1993 is shown in the following graph.‡ Use this graph to estimate the values of the deficit (to the nearest $10 billion) for these years.

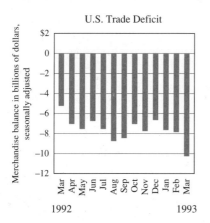

Source: U.S. Department of Commerce

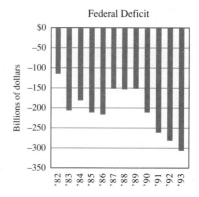

Source: Federal Reserve Board, Congressional Budget Office

9. *Demand* Market research for Fullcourt Press makes the following estimates of sales of a new book at various possible prices:

Price ($)	20	30	40	50	60
Sales (1,000)	110	90	70	50	30

(a) Graph these data.
(b) What effect does a price increase of $10 per book have on sales?

10. *Demand* For another book, Fullcourt Press's market research projects the following sales at various possible prices:

Price ($)	30	40	60	80	120
Sales (1,000)	150	110	75	55	37

(a) Graph these data.
(b) What is the effect on sales of doubling the price per book?

▼ *Source: Standard & Poor's *Security Price Index Record,* Statistical Service, 1992 Edition.
† As printed in the *Chicago Tribune* business section, May 20, 1993.
‡ As printed in the *Chicago Tribune* business section, March 7, 1993.

11. Cannons A cannon is fired horizontally from the top of a 500-ft cliff. The height of the cannonball after t seconds is shown in the following graph. Use this graph to estimate the cannonball's height at 0, 1, 2, 3, and 4 seconds.

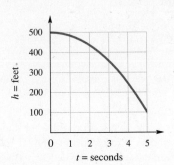

12. Archery An arrow is shot straight up at a speed of 100 ft/s. Its height after t seconds is shown in the following graph. Use this graph to estimate the arrow's height at 0, 1, 2, 3, 4, 5, and 6 seconds.

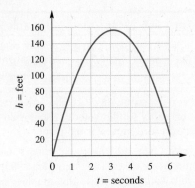

13. Infant Height Alexander Costenoble's height during his first year of life was measured as follows:

Age (months)	0	2	3	4	5	6	9	12
Height (inches)	21	23	24	25	26	27	28	29

(a) Graph these data.
(b) Use the graph to estimate Alexander's approximate growth during his ninth month.

14. Infant Weight Alexander Costenoble's weight during his first year of life was measured as follows:

Age (months)	0	2	3	4	5	6	9	12
Weight (pounds)	8	9	13	14	16	17	18	19

(a) Graph these data.
(b) Use the graph to estimate Alexander's approximate gain in weight during his tenth month.

Carefully sketch the graphs of each of the equations in Exercises 15–30.

15. $2x + y = 2$
16. $x - y = 4$
17. $2x - 3y = 4$
18. $2x + 3y = 4$
19. $y = -x^2 + 2$
20. $y = -2x^2 + 1$
21. $y = \dfrac{1}{x^2}$
22. $y = \dfrac{1}{x^3}$
23. $y = \dfrac{1}{x - 2}$
24. $y = \dfrac{1}{x + 1}$
25. $y = \dfrac{1}{(x + 1)^2}$
26. $y = \dfrac{1}{(x - 1)^2}$
27. $y(x^2 + 2x + 1) = x$
28. $y(x^2 + 3x + 2) = -x$
29. $y = \sqrt{x}(x - 1)$
30. $y = \sqrt{x}(x + 1)$

Exercises 31–36 require the use of a graphing calculator or computer software.

31. (a) Graph the equation $y = x^3 - 2x - 5$ with $-5 \leq x \leq 5$ and $-10 \leq y \leq 10$.
(b) What happens if you change the range to $-1 \leq x \leq 1$ and $-1 \leq y \leq 1$? Explain the result.

32. (a) Graph the equation $y = 1/(x^3 - 2x - 5)$ with $-5 \leq x \leq 5$ and $-10 \leq y \leq 10$.
(b) Find viewing window coordinates that show the interesting features of the graph more clearly.

Use a graphing calculator (or a graphing computer program) to display the graphs in Exercises 33–36. Answer the accompanying questions.

33. (a) $y = x^{1/2}(x - 1)$ (b) $y - 1 = x^{1/2}(x - 1)$
(c) $y - 2 = x^{1/2}(x - 1)$ (d) $y + 1 = x^{1/2}(x - 1)$

What is the effect on the graph of an equation if you replace y with the quantity $(y - c)$?

34. (a) $y = x^3 + x^2 + 4$ (b) $y - 1 = x^3 + x^2 + 4$
(c) $y - 2 = x^3 + x^2 + 4$ (d) $y + 1 = x^3 + x^2 + 4$

What is the effect on the graph of an equation if you replace y with the quantity $(y - c)$?

35. (a) $y = x^3 - 2x^2 + 4$ (b) $y = (x - 1)^3 - 2(x - 1)^2 + 4$
(c) $y = (x + 1)^3 - 2(x + 1)^2 + 4$ (d) $y = (x - 2)^3 - 2(x - 2)^2 + 4$

What is the effect on the graph of an equation if you replace x with the quantity $(x - c)$?

36. (a) $y = x^{1/2}(x - 1)$ (b) $y = (x - 1)^{1/2}(x - 2)$
(c) $y = (x + 1)^{1/2}(x)$ (d) $y = (x - 2)^{1/2}(x - 3)$

What is the effect on the graph of an equation if you replace x with the quantity $(x - c)$?

In Exercises 37–52, draw the given subsets of the real line.

37. $(-3, 4)$
38. $(-\infty, 4)$
39. $(-\infty, 0]$
40. $[4, 8]$
41. $[-1, +\infty)$
42. $(-\infty, +\infty)$
43. $(-4, 3) \cup (3, 4)$
44. $(-\infty, 1) \cup (1, +\infty)$
45. $(-4, 3) \cup [3, 4)$
46. $(-1, 0) \cup (1, 2) \cup (2, 3)$
47. $(-1, 0) \cup [1, 2] \cup (3, 4)$
48. $(-1, 1) \cup [1, 2] \cup (2, 4)$
49. $[2, 2]$
50. $[6.1, 6.1]$
51. $(2, 2)$
52. $(6, 6)$

In Exercises 53–58, represent the given line segments in interval notation.

53.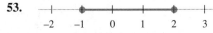

54.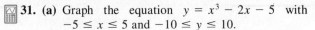

55.

56.

57.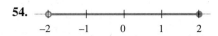

58.

1.2 FUNCTIONS AND THEIR GRAPHS

FUNCTIONS

Simply speaking, a **function** is a *rule* or *set of instructions* for manufacturing a new object from an old one. The functions we deal with in this text manufacture new *real numbers* from old ones, and we call them *real-valued functions of a real variable*. The rule or set of instructions can be viewed in a number of ways, including *algebraically* when the rule is given by an algebraic formula, *numerically* using a table of values, and *geometrically* using a graph. We shall use all three perspectives to help us understand functions.

First, a formal definition. We shall explore the notation and terminology used in this definition in the examples below.

> **FUNCTION**
>
> A **real-valued function f of a real variable** is a rule that assigns to each real number x in a specified set of numbers a single real number $f(x)$. The set of numbers x for which $f(x)$ is defined is called the **domain** of f.

It is customary to give functions single-letter names, such as "f" or "g." In applications, we generally use a letter related to the quantities that we are studying.

▶ **NOTE** Since a function f assigns a *single* number $f(x)$ to each x in the domain, it follows that $f(x)$ cannot have more than one value. For instance, we would not write $f(0) = \pm 1$, since $f(0)$ cannot equal both $+1$ and -1. ◀

A simple example of a function is the *squaring* function, by which we mean the rule that *squares* numbers. Thus, given an *input* number, say 4, this function produces the *output* 4^2, or 16. Similarly, if the input is -1, then the output is $(-1)^2 = 1$. If we call this function f, we would write

$$f(4) = 16,$$

meaning that "the function f, when applied to the input 4, yields the output 16," or, more concisely,

"f, when evaluated at 4, is 16,"

or even more concisely,

"f of 4 is 16."

Similarly, $f(-1) = 1, f(3) = 9, f(-11) = 121$, and so on. More succinctly, we can write the algebraic formula

$$f(x) = x^2.$$

This formula tells us that no matter what number x we use as the input, the output will be the square of that number, x^2. We can use this formula to evaluate f at any number we like. For example, if we wish to evaluate f at 6, we just take the formula and *replace x with the quantity* 6 *everywhere x occurs,* getting

$$f(6) = 6^2 = 36.$$

If we wish to evaluate f at -707, we replace x with the quantity -707 everywhere it occurs, getting

$$f(-707) = (-707)^2 = 499,849.$$

We say that we **substitute** the value -707 for x in the formula. Note that when we replace x with -707, we get $(-707)^2$ (which is positive), not -707^2 (which would be negative), on the right. Similarly, evaluating f at the unknown quantity a gives

$$f(a) = a^2.$$

This substitution rule also works for algebraic expressions. For instance, if we want to evaluate f at $x + h$ (which we shall want to when we start discussing calculus), then we substitute the quantity $x + h$ for x, getting

$$f(x + h) = (x + h)^2 = x^2 + 2xh + h^2.$$

In other words, we substitute the quantity $(x + h)$ for the quantity x. (We got the expression $x^2 + 2xh + h^2$ by expanding* the binomial $(x + h)^2$.)

There is a nice way of picturing the concept of a function: we can think of a function as a "black box," as shown in Figure 1. When the number x is fed into this particular function, out pops the square of that number, x^2. Thus, when the number -3 is fed in, we get the number 9 as the output.

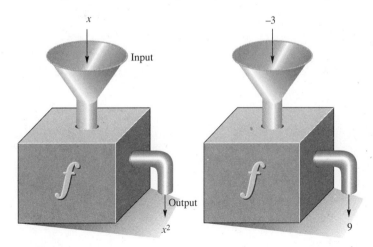

FIGURE 1

▼ *See the Algebra Review (Appendix A).

EXAMPLE 1

(a) Given that $f(x) = 3x^2 + 4x - 11$, find $f(0)$, $f(-1)$, $f(\frac{1}{2})$, $f(\pi)$, and $f(x + h)$.

(b) With f as in part (a), calculate and simplify the quantity
$$\frac{f(x+h) - f(x)}{h}.$$

SOLUTION

(a) Substituting,
$$f(0) = 3(0)^2 + 4(0) - 11 = -11.$$
$$f(-1) = 3(-1)^2 + 4(-1) - 11 = -12.$$
$$f\left(\frac{1}{2}\right) = 3\left(\frac{1}{2}\right)^2 + 4\left(\frac{1}{2}\right) - 11 = \frac{3}{4} + 2 - 11 = -\frac{33}{4}.$$
$$f(\pi) = 3\pi^2 + 4\pi - 11,$$

which, as good mathematicians, we'll simply leave like that. (Attempting decimal approximations will ruin the exactness of the answer!)

Finally,
$$f(x + h) = 3(x + h)^2 + 4(x + h) - 11$$

We've replaced every occurrence of x with the quantity $x + h$.

$$= 3(x^2 + 2xh + h^2) + 4x + 4h - 11$$
$$= 3x^2 + 6xh + 3h^2 + 4x + 4h - 11.$$

(b) We have already calculated
$$f(x + h) = 3x^2 + 6xh + 3h^2 + 4x + 4h - 11.$$

Thus,

Note that we are subtracting the **whole quantity** $f(x)$ as required.

$$\frac{f(x+h) - f(x)}{h} = \frac{[3x^2 + 6xh + 3h^2 + 4x + 4h - 11] - [3x^2 + 4x - 11]}{h}$$
$$= \frac{3x^2 + 6xh + 3h^2 + 4x + 4h - 11 - 3x^2 - 4x + 11}{h}$$
$$= \frac{6xh + 3h^2 + 4h}{h} \quad \text{All the other terms cancel.}$$
$$= \frac{h(6x + 3h + 4)}{h}$$
$$= 6x + 3h + 4.$$

You can enter a formula for a function into a graphing calculator (as well as certain other calculators and computer software) and have it evaluate the function at several numbers. For instance, you could have used a graphing calculator to find the first three answers above. You could also *approximate* the fourth answer (because a calculator or computer uses an approximation to π, not the exact value). On the other hand, it would require a more sophisticated calculator or computer software to evaluate $f(x + h)$ or the expression in part (b) since the calculation of $f(x + h)$ above is *symbolic,* and not *numerical.*

For details about using a graphing calculator to evaluate a function, see the corresponding section in Appendix B.

▼ EXAMPLE 2

If $f(x) = \frac{1}{x}$, find $f(-1), f(\frac{1}{2}), f(\pi), f(x + h),$ and $f(x + h) - f(x)$.

SOLUTION Substituting, we get

$$f(-1) = \frac{1}{-1} = -1,$$

$$f\left(\frac{1}{2}\right) = \frac{1}{1/2} = 2,$$

$$f(\pi) = \frac{1}{\pi},$$

$$f(x + h) = \frac{1}{x + h}.$$

Finally,

$$f(x + h) - f(x) = \frac{1}{x + h} - \frac{1}{x}$$

$$= \frac{x - (x + h)}{x(x + h)} \quad \text{by the rule for subtracting rational expressions*}$$

$$= \frac{x - x - h}{x(x + h)}$$

$$= \frac{-h}{x(x + h)}.$$

▼ *Here is the rule, as given in the Algebra Review (Appendix A):

$$\frac{A}{B} \pm \frac{C}{D} = \frac{AD \pm BC}{BD}.$$

▶ NOTES

1. When we write, say, $f(x) = \frac{1}{x}$, we call x the **independent variable,** since we think of it as a freely varying quantity. If we have an equation $y = f(x)$, we call y the **dependent variable,** and also say that y **depends on x** via the function f.
2. There is nothing magical about the letter x. We might just as well say $f(t) = \frac{1}{t}$ which means *exactly the same thing* as $f(x) = \frac{1}{x}$. For example, if we are told that $f(t) = \frac{1}{t}$, then, to get $f(3)$, we would substitute 3 for t, getting $f(3) = \frac{1}{3}$. This is exactly the same thing we would do, and exactly the same answer we would get, if we were told that $f(x) = \frac{1}{x}$. We replace x by t (or some other letter) if the independent variable stands for a particular quantity, such as time, and we want to use an appropriate letter for the quantity. ◀

▼ **EXAMPLE 3** Income Tax

Each year millions of citizens of the United States are required to compute a function. The income tax they owe is a function of the taxable income (gross income less deductions) that they earned. If we write I for the taxable income, then we can write the income tax owed as $T(I)$.

It is interesting how the function T is defined. There is a formula for T, but at some point the government decided that people could more easily and accurately read a table than use the formula, so they publish *tax tables* in which you can look up the tax owed on a given taxable income. (Figure 2 shows a portion of the 1993 tax table.)

Curiously, if your income is high enough, the government trusts you to use the formulas, for the tax tables stop at a taxable income of $99,999. The government does publish the formula for T as well as the tax tables. For 1993, the formula appeared as in Figure 3.

1993 Tax Table

If your taxable income is		And you are
At least	But less than	Single
		Your tax is
23,000	23,050	3,574
23,050	23,100	3,588
23,100	23,150	3,602
23,150	23,200	3,616

FIGURE 2

Schedule X – Use if your filing status is **single**

If your taxable income is		Your tax is	
Over	But not over		of the amount over
$0	$22,100	15%	$0
22,100	53,500	$3,315.00 + 28%	22,100
53,500	115,000	12,107.00 + 31%	53,500
115,000	250,000	31,172.00 + 36%	115,000
250,000	....	79,772.0 + 39.6%	250,000

FIGURE 3

Before we go on... The tax table is an example of a **numerical definition** of a function. That is, a function can be defined by simply listing its values at all possible values of the independent variable, instead of giving a formula. On the other hand, even if we are given a formula for a function it may be useful to list some of its values (we shall see this when we discuss graphs below).

There are many functions, such as the value of the Dow Jones industrial average as a function of time, for which it is difficult or impossible to write down a formula, but for which at least some values may be easily listed. One of the goals of mathematical modeling is to find algebraic formulas for numerically specified functions, since a formula is easier to work with, and more amenable to the use of mathematical tools, than a table of values. We shall see examples of this in Section 4 and scattered throughout the book.

THE DOMAIN OF A FUNCTION

It sometimes happens that you can't evaluate a function at a specified real number. For example, the income tax $T(I)$ is not defined for a negative value of I. Similar difficulties may occur if a function is specified algebraically: if $f(x) = \sqrt{x}$, for instance, then what is the value of $f(-3)$? If you try to calculate the square root of -3 on a calculator, it will give you an error message (unless it is a sophisticated calculator that does complex number arithmetic). In the realm of real numbers, we just don't have square roots of negative numbers, so we must restrict the possible inputs to prevent them from being negative.* This is why we mentioned the *domain* of a function in the definition at the beginning of this section. The domain is the set of all numbers that we permit as inputs to the function.

For example, the square root function should be defined as

$$f(x) = \sqrt{x}, \text{ with domain } [0, +\infty).$$

Similarly, if $g(x) = \frac{1}{x}$, then we dare not attempt to evaluate $g(0)$. Thus, we must restrict the domain of g to consist of real numbers *other than* 0. This we can do by defining g as

$$g(x) = \frac{1}{x}, \text{ with domain } (-\infty, 0) \cup (0, +\infty).$$

If, for some reason, we did not want to consider negative numbers, we could restrict this function further, by defining

$$h(x) = \frac{1}{x}, \text{ with domain } (0, +\infty).$$

▼ *Some calculators will also give an error message if you try to take the *cube* root of a negative number. This is not because there is no such real number—cube roots of negative numbers always exist (for instance, the cube root of -8 is -2). This result is caused by the way the calculator has been programmed. An *ideal* calculator would always give an answer, but most calculators are not ideal.

Q What is the difference between g and h? After all, they are defined by the same formula.

A The two functions have different domains. For example, whereas $g(-11)$ is defined and is $-\frac{1}{11}$, $h(-11)$ is *not* defined. Also notice that g has the *largest possible domain* given its formula $\frac{1}{x}$, whereas h does not. No matter, there is no requirement that all functions must be given their largest possible domain! In fact, we shall see that it is often useful to consider functions with restricted domains. Remember that *the domain of a function is part of its definition.*

▼ **EXAMPLE 4** Cost

Your electronics plant is equipped to manufacture up to 40 large-screen television sets per day. The cost in dollars for producing the sets is given by

$$C(x) = 400x + 10,000,$$

where x represents the number of large-screen sets manufactured in a day. How should you completely specify the function C?

SOLUTION To completely specify a function, we need to know its domain as well as the rule that tells us how to evaluate it. But all we have been given is the rule, so we need to come up with a domain. Since the plant can make at most 40 large-screen television sets per day, we must restrict x to be less than or equal to 40 ($x \leq 40$). Furthermore, it is meaningless to talk about manufacturing a *negative* number of sets, so we must also restrict x to be greater than or equal to zero ($x \geq 0$).

Thus, the domain is $[0, 40]$, and we can completely specify the function C by

$$C(x) = 400x + 10,000, \text{ with domain } [0, 40].$$

Before we go on... Actually, it does not make sense to evaluate C on any number other than an integer (unless you can make sense of what it means to manufacture, say, $\frac{1}{3}$ of a television in a day). Strictly speaking, then, the domain of C should be the set of integers $\{0, 1, 2, \ldots, 40\}$. There is, however, a great *mathematical* advantage to using a whole interval as the domain, and in particular the techniques of calculus require it. In practice, this will not be a problem.

You may have seen exercises such as this: "Find the domain of the function $g(x) = \sqrt{x+3}$." This is not worded well. What is meant is that you should find the *largest possible domain* of g. Now, this is a convention we all agree to adopt: if a domain is not specified, we mean that it should be as large as possible. While we won't devote too much time to the calculation of largest possible domains, an example or two might be illuminating.

EXAMPLE 5

Find the largest possible domain of the function $g(x) = \sqrt{x + 3}$.

SOLUTION We are looking for all the values of x for which the given formula makes sense. For example, the formula clearly *doesn't* make sense when $x = -479$, for then we would be trying to evaluate the square root of -476. If we want the function to make sense, we had better make sure that *the quantity under the radical is at least* 0. Writing this phrase as a formula,

$$x + 3 \geq 0.$$

We solve this inequality by adding -3 to both sides.

$$x \geq -3$$

Thus, we must have $x \geq -3$ for the formula to make sense. In other words, x must be in the interval $[-3, +\infty)$. It follows that the largest possible domain for this function is $[-3, +\infty)$. Further, we can now specify the function precisely, as follows:

$$g(x) = \sqrt{x + 3}, \text{ with domain } [-3, +\infty).$$

EXAMPLE 6

Find the largest possible domain of the function

$$H(t) = \frac{-3}{\sqrt{1 + t}}.$$

SOLUTION We must find all values of t for which $H(t)$ makes sense. Since there is a denominator, a warning bell should be going off in our heads:

- *denominators can never be zero!*

Another restriction is imposed by the square root sign:

- *the quantity under the radical can never be negative.*

Writing these two restrictions mathematically,

$1 + t \neq 0$ (or else its square root will be zero),
$1 + t \geq 0$ (or else we can't take its square root in the first place).

These restrictions must hold simultaneously. Together, they say that $1 + t > 0$, or, adding -1 to both sides, that $t > -1$. Thus, t must be in the interval $(-1, +\infty)$, and this is the largest possible domain for the function H.

GRAPHS OF FUNCTIONS

It is very useful to visualize a function by means of its **graph.** Given a function f, we get an equation in x and y by setting $y = f(x)$. For instance, if the function f is specified by

$$f(x) = x^2 + 1,$$

then we get the corresponding equation by setting $y = f(x)$, getting

$$y = x^2 + 1.$$

The graph of f is then the graph of the equation $y = x^2 + 1$, as discussed in the previous section.

> **THE GRAPH OF A FUNCTION**
>
> The **graph of the function** f is the graph of the equation $y = f(x)$, where we restrict the values of x to lie in the domain of f. In other words, the graph of f is the set of all points $(x, f(x))$, where x lies in the domain of f.

▼ **EXAMPLE 7**

Let $f(x) = -x + 4$, with domain the set $\mathbb{R}$ of all real numbers. Then the graph of f is just the graph of the equation $y = -x + 4$, with no restriction on x since the domain is the set of all real numbers. As we know from Section 1, the graph of this equation is the straight line shown in Figure 4.

Before we go on... Recall how we obtained this graph in Section 1. We first constructed a table of values of the function:

x	−4	−3	−2	−1	0	1	2	3	4	5
$f(x) = 4 - x$	8	7	6	5	4	3	2	1	0	−1

We then plotted the points $(-4, 8)$, $(-3, 7)$, and so on. We then connected up these points in the most reasonable way possible, which looked like a straight line.

We now have three different ways of thinking about this function: algebraically, from the formula $f(x) = -x + 4$, numerically from the table above, and graphically from the graph above.

Q Of what use is the table of values in the last example?

A First, we needed it to draw the graph. But there are features of the function that show up most clearly in the table. For example, it is obvious from the table that the value of f decreases by 1 whenever x increases by 1.

Q Of what use is the graph?

A Again, there are features of the function that show up very clearly in the graph. For example, we can see once again the steady decrease by 1 in the value of f for every increase by 1 in x. Also, we can see clearly that $f(x)$ is positive precisely when $x < 4$ (this is where the graph lies above the x-axis) and that $f(x)$ is negative when $x > 4$ (this is where the graph lies below the x-axis).

Also, we can use the graph to find $f(x)$ for any value of x in the domain. For example, suppose we want to find $f(1)$, looking only at the graph and not at the formula. We can reason as follows: If $(1, y)$ is a point on the graph, then y is given by the equation $y = f(1)$. Therefore, we look for the y-value of the point on the graph above $x = 1$ on the x-axis. (See Figure 5.)

The point above $x = 1$ has y-coordinate 3, so $f(1) = 3$. In this way, the graph can be used as another way to *define* a function.

In general, the graph is a visual way of "taking in the whole function." A picture is worth a thousand words.

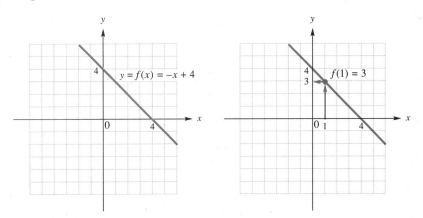

FIGURE 4   **FIGURE 5**

▼ **EXAMPLE 8**

Let $f(x) = x + \frac{1}{x}$, with domain $(0, +\infty)$. Its graph is the curve $y = x + \frac{1}{x}$, but with x restricted to be strictly positive. Figure 6 shows what the graph of f looks like.

Notice that as x increases, $\frac{1}{x}$ is small, so $x + \frac{1}{x}$ is very close to x. Thus, the graph of $y = x + \frac{1}{x}$ gets close to the line $y = x$. For values of x very close to 0, x is small, so the graph is close to the curve $y = \frac{1}{x}$. These dotted graphs $y = x$ and $y = \frac{1}{x}$ are called **asymptotes** of the function f. An *asymptote* is a line that the graph of a function approaches arbitrarily closely. Note that the function f also has the *vertical asymptote* $x = 0$ (why?).

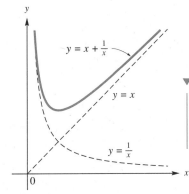

FIGURE 6
Graph of $f(x) = x + \frac{1}{x}$, with domain $(0, +\infty)$

Before we go on... It is obvious from the graph (but not from the formula) that this function has a minimum value at some smallish x (actually, at $x = 1$). Notice that the domain of f is not the largest possible domain. What would the graph look like if the domain were extended to $(-\infty, 0) \cup (0, +\infty)$?

 How did we get this curve? We could have plotted several points by hand by using a table of x- and y-values, or we could have instructed a graphing calculator to plot the curve $y = x + \frac{1}{x}$. (Actually, we used a computer graphing package something like the one available with this book.) If we use a graphing calculator, then we can "trace" the graph to obtain approximate coordinates of the lowest point of the curve.

Let us restrict the domain further.

▼ **EXAMPLE 9**

Let $g(x) = x + \frac{1}{x}$, with domain $(0, 2]$. Then its graph is the curve $y = x + \frac{1}{x}$, but this time x is restricted to be larger than 0 but less than or equal to 2. The graph is shown in Figure 7. Here, we get only a small segment of the graph described in Example 8.

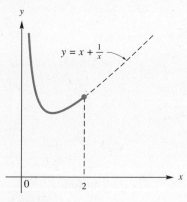

FIGURE 7

Graph of $g(x) = x + \frac{1}{x}$, with domain $(0, 2]$

▼ **EXAMPLE 10**

Let f be given by

$$f(x) = \begin{cases} -1 & \text{if } -4 \leq x < -1 \\ x & \text{if } -1 \leq x \leq 1 \\ x^2 - 1 & \text{if } 1 < x \leq 2 \end{cases}$$

with domain $[-4, 2]$. Sketch its graph, and use the graph to evaluate $f(1)$, $f(-1)$, and $f(-3)$.

SOLUTION   This is an example of a **piecewise-defined function:** it is defined using different formulas for different intervals of its domain (another example is the income tax function in Example 3). To sketch its graph, we need to sketch the three graphs $y = -1$, $y = x$, and $y = x^2 - 1$, using the appropriate domain for each (Figure 8). From the graph, we see that $f(1) = 1$ (and not 0, since there is a hollow dot there), $f(-1) = -1$, and $f(-3) = -1$.

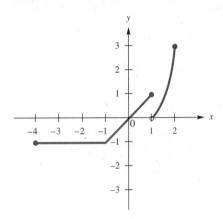

FIGURE 8

Before we go on... We didn't really need to specify that the domain of f is $[-4, 2]$, as that is implied by the description of $f(x)$ in the displayed formula. We could have simply let f be given by

$$f(x) = \begin{cases} -1 & \text{if } -4 \leq x < -1 \\ x & \text{if } -1 \leq x \leq 1 \\ x^2 - 1 & \text{if } 1 < x \leq 2. \end{cases}$$

The only values of x for which $f(x)$ is defined by this formula are those between -4 and 2, so the domain is $[-4, 2]$.

Some graphing calculators will not allow you to draw a piecewise-defined function very easily, since they apply the same x-range to all the graphs. Thus, you could plot the following three graphs with $-4 \leq x \leq 2$:

$$y_1 = -1,$$
$$y_2 = x,$$
$$y_3 = x^2 - 1.$$

You would then want to pay attention to the first graph only when $-4 \leq x < -1$, the second one when $-1 \leq x \leq 1$, and the third when $1 < x \leq 2$. (See Appendix B for other methods of graphing piecewise-defined functions.)

We have graphed functions by graphing corresponding equations. This raises the question, is the graph of every equation the graph of some function? To answer this, recall that if f is a function, there can be only a single value $f(x)$ assigned to each value of x. It follows that in the graph of a function, there should be only one y corresponding to any value of x, namely, $y = f(x)$. In other words, the graph of a function cannot contain two or more points with the same x-coordinate—that is, two or more points on the same vertical line. This gives us the following rule.

VERTICAL LINE TEST

For a graph to be the graph of a function, each vertical line must intersect the graph in *at most* one point.

▼ **EXAMPLE 11**

Which of the graphs shown in Figure 9 are the graphs of functions?

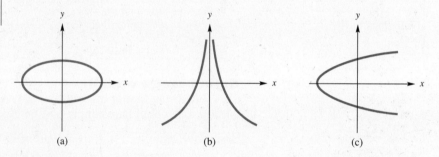

FIGURE 9

SOLUTION As illustrated in Figure 10, only graph (b) passes the vertical line test and is the graph of a function.

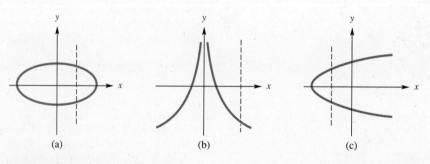

FIGURE 10

1.2 Functions and Their Graphs

Following is a table listing some important functions with their graphs. Knowing these graphs will be *immensely* useful in the rest of this book.

Function	Equation for Graph	Comments	Graph
$f(x) = x$, with domain $\mathbb{R}$	$y = x$	The graph is a straight line passing through the origin and inclined at 45°.	
$f(x) = \|x\|$, with domain $\mathbb{R}$	$y = \|x\|$	For x positive or zero, this agrees with $y = x$. For x negative or zero, it agrees with $y = -x$.	
$f(x) = x^2$, with domain $\mathbb{R}$	$y = x^2$	This is a **parabola** passing through the origin.	

Continued

Function	Equation for Graph	Comments	Graph
$f(x) = \sqrt{x}$, with domain $[0, +\infty)$	$y = \sqrt{x}$	The domain must be restricted to the nonnegative numbers, since the square root of a negative number is not real. Also note that the graph is the top half of a horizontally oriented parabola, since squaring both sides of the equation gives $x = y^2$.	
$f(x) = \frac{1}{x}$, with domain $(-\infty, 0) \cup (0, +\infty)$	$y = \frac{1}{x}$	This is a **hyperbola.** The domain excludes zero since there is no such thing as the reciprocal of 0.	

▶ ## 1.2 EXERCISES

In Exercises 1–6, perform the given evaluations, simplifying the answer where appropriate.

1. Given $f(x) = x^2 + 2x + 3$, find
   (a) $f(0)$ (b) $f(1)$ (c) $f(-1)$ (d) $f(-3)$
   (e) $f(a)$,   (f) $f(x + h)$ (g) $\dfrac{f(x + h) - f(x)}{h}$ (if $h \neq 0$).

2. Given $g(x) = 2x^2 - x + 1$, find
   (a) $g(0)$ (b) $g(-1)$ (c) $g(r)$ (d) $g(x + h)$
   (e) $\dfrac{g(x + h) - g(x)}{h}$ (if $h \neq 0$).

3. Given $g(s) = s^2 + \frac{1}{s}$, find
   (a) $g(1)$ (b) $g(-1)$ (c) $g(4)$ (d) $g(x)$
   (e) $g(s + h)$ (f) $g(s + h) - g(s)$.

4. Given $h(r) = \dfrac{1}{r + 4}$, find
   (a) $h(0)$ (b) $h(-3)$ (c) $h(-5)$,   (d) $h(x^2)$
   (e) $h(x^2 + 1)$ (f) $h(x^2) + 1$.

5. Given $\phi(x) = \sqrt{x^2 + 3}$, find
   (a) $\phi(0)$ (b) $\phi(-2)$ (c) $\phi(x + h)$ (d) $\phi(x) + h$.

6. Given $\alpha(\gamma) = \gamma^{3/2} - \gamma$, find
   (a) $\alpha(9)$ (b) $\alpha(16)$ (c) $\alpha(\sigma + h)$ (d) $\alpha(\sigma) + h$.

In Exercises 7–14, calculate and simplify the quotient $\dfrac{f(x + h) - f(x)}{h}$ $(h \neq 0)$.*

7. $f(x) = -x^2 - 2x - 1$

8. $f(x) = 3x^2 - 2x - 1$

9. $f(x) = \dfrac{2}{x + 1}$

10. $f(x) = \dfrac{1}{2 - x}$

11. $f(x) = x + \dfrac{1}{x}$

12. $f(x) = x^2 - \dfrac{1}{x}$

13. $f(x) = \dfrac{1}{x^2}$

14. $f(x) = \dfrac{1}{x^2 + 1}$

In Exercises 15–20, say whether $f(x)$ is defined for the given values of x. If it is defined, give its value.

15. $f(x) = x - \dfrac{1}{x^2}$, with domain $(0, +\infty)$
    (a) $x = 4$ (b) $x = 0$ (c) $x = -1$

16. $f(x) = \dfrac{2}{x} - x^2$, with domain $[2, +\infty)$
    (a) $x = 4$ (b) $x = 0$ (c) $x = 1$

17. $(x) = \sqrt{x + 10}$, with domain $[-10, 0)$
    (a) $x = 0$ (b) $x = 9$ (c) $x = -10$

18. $f(x) = \sqrt{9 - x^2}$, with domain $(-3, 3)$
    (a) $x = 0$ (b) $x = 3$ (c) $x = -3$

19. $f(x) = \sqrt{1 - x}$
    (a) $x = 0$ (b) $x = 2$ (c) $x = -3$

20. $f(x) = \dfrac{1}{\sqrt{x - 3}}$
    (a) $x = 0$ (b) $x = 3$ (c) $x = 4$

Calculate the largest possible domain of each of the functions in Exercises 21–32.

21. $f(x) = x^2 - 1$

22. $f(x) = \sqrt{x}$

23. $g(x) = \sqrt{3x}$

24. $g(x) = \sqrt{x^2}$

25. $h(x) = \sqrt{x - 1}$

26. $h(x) = \sqrt{x + 1}$

27. $f(x) = \dfrac{1}{x}$

28. $f(x) = 4 - \dfrac{1}{x}$

29. $g(x) = 4 - \dfrac{1}{x^2}$

30. $g(x) = \dfrac{1}{\sqrt{x}}$

31. $h(x) = \dfrac{1}{x - 2}$

32. $h(x) = \dfrac{3}{x - 1}$

▼ *This quotient, known as the *difference quotient*, measures how fast the value of the function changes as x changes. We shall study the difference quotient in detail when we study calculus.

34 Chapter 1 Functions and Graphs

In Exercises 33–40, use the graph of the function f to find approximations of the given values.

33.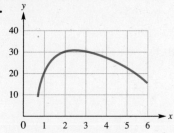
(a) $f(1)$ (b) $f(2)$
(c) $f(3)$ (d) $f(5)$
(e) $f(3) - f(2)$

34.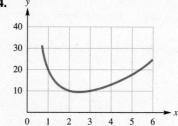
(a) $f(1)$ (b) $f(2)$
(c) $f(3)$ (d) $f(5)$
(e) $f(3) - f(2)$

35.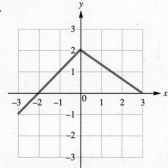
(a) $f(1)$ (b) $f(-2)$
(c) $f(0)$ (d) $f(3)$
(e) $f(3) - f(2)$

36.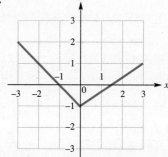
(a) $f(1)$ (b) $f(-2)$
(c) $f(0)$ (d) $f(3)$
(e) $f(3) - f(2)$

37.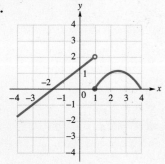
(a) $f(-3)$ (b) $f(0)$
(c) $f(1)$ (d) $f(2)$
(e) $\dfrac{f(3) - f(2)}{3 - 2}$

38.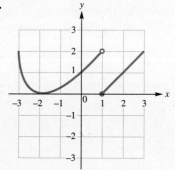
(a) $f(-2)$ (b) $f(0)$
(c) $f(1)$ (d) $f(3)$
(e) $\dfrac{f(3) - f(1)}{3 - 1}$

39.

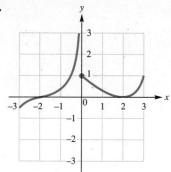

(a) $f(-3)$ (b) $f(-2)$
(c) $f(0)$ (d) $f(2)$
(e) $\dfrac{f(2) - f(0)}{2 - 0}$

40.

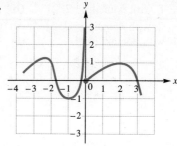

(a) $f(-2)$ (b) $f(0)$
(c) $f(1)$ (d) $f(3)$
(e) $\dfrac{f(2) - f(-1)}{2 - (-1)}$

In Exercises 41 and 42, match the functions to the graphs

41. (a) $f(x) = x$, with domain $[-1, 1]$ (b) $f(x) = -x$, with domain $[-1, 1]$
  (c) $f(x) = \sqrt{x}$, with domain $[0, 4]$ (d) $f(x) = x + \frac{1}{x} - 2$, with domain $(0, 4)$
  (e) $f(x) = |x - 1|$, with domain $(-2, 2)$ (f) $f(x) = x - 1$, with domain $(-1, 2)$

(I)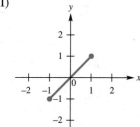

(II)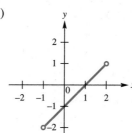

(III)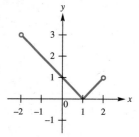

(IV)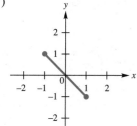

(V)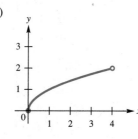

(VI)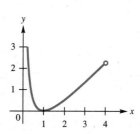

42. (a) $f(x) = -x + 1$, with domain $(-2, 2]$ (b) $f(x) = 2 - |x|$, with domain $[-2, 1)$
(c) $f(x) = \sqrt{x + 2}$, with domain $(-2, 2]$ (d) $f(x) = -x^2 + 3$, with domain $(-2, 2]$
(e) $f(x) = \frac{1}{x} - 1$, with domain $(0, 4)$ (f) $f(x) = x^2 - 1$, with domain $(-2, 2]$

(I)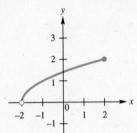

(II)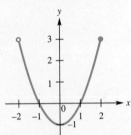

(III)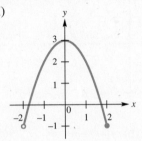

(IV)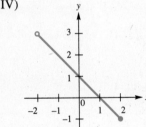

(V)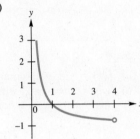

(VI)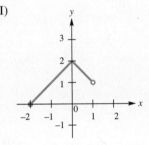

Graph the functions given in Exercises 43–48. We suggest that you become familiar with these graphs in addition to those in the chart at the end of the section.

43. $f(x) = x^3$, with domain $\mathbb{R}$

44. $f(x) = x^3$, with domain $[0, +\infty)$

45. $f(x) = x^4$, with domain $\mathbb{R}$

46. $f(x) = \sqrt[3]{x}$, with domain $\mathbb{R}$

47. $f(x) = \dfrac{1}{x^2}$, with domain $(-\infty, 0) \cup (0, +\infty)$

48. $f(x) = x + \dfrac{1}{x}$, with domain $(-\infty, 0) \cup (0, +\infty)$

In each of Exercises 49–54, sketch the graph of the given function. (The domain is implied by the formula. See the "Before we go on" discussion in Example 10.)

49. $f(x) = \begin{cases} x & \text{if } -4 \leq x < 0 \\ 2 & \text{if } 0 \leq x \leq 4 \end{cases}$

50. $f(x) = \begin{cases} -1 & \text{if } -4 \leq x \leq 0 \\ x & \text{if } 0 < x \leq 4 \end{cases}$

51. $f(x) = \begin{cases} x & \text{if } -1 < x \leq 0 \\ x + 1 & \text{if } 0 < x \leq 2 \\ x & \text{if } 2 < x \leq 4 \end{cases}$

52. $f(x) = \begin{cases} -x & \text{if } -1 < x < 0 \\ x - 2 & \text{if } 0 \leq x \leq 2 \\ -x & \text{if } 2 < x \leq 4 \end{cases}$

53. $f(x) = \begin{cases} x^2 & \text{if } -2 < x \leq 0 \\ \frac{1}{x} & \text{if } 0 < x \leq 4 \end{cases}$

54. $f(x) = \begin{cases} -x^2 & \text{if } -2 < x \leq 0 \\ \sqrt{x} & \text{if } 0 < x < 4 \end{cases}$

55. The **greatest integer function** I is defined by taking $I(x)$ to be the greatest integer $\leq x$. For example, $I(3.2) = 3$; $I(3.999) = 3$; $I(5) = 5$; $I(-4.1) = -5$ (since -5 is ≤ -4.1); $I(-4) = -4$.* Sketch the graph of this function. The domain of I is all of $\mathbb{R}$.

▼ * In other words, it drops the decimal part of positive numbers and rounds negative numbers down.

56. The **rounding** function R is defined by taking $R(x)$ to be the integer nearest to x, with the convention that a number midway between two integers is rounded up in magnitude. Thus, for example, $R(3.499) = 3$; $R(3.5) = 4$; $R(-2.4999) = -2$; $R(-2.5) = -3$. Sketch the graph of this function. The domain of R is all of $\mathbb{R}$.

In Exercises 57–70, use either a graphing calculator or computer graphing software. Sketch the graphs of the indicated functions, and answer the additional questions.

57. $f(x) = (x - 1)(x - 2)(x - 3)(x - 4)$, with domain $\mathbb{R}$

58. $f(x) = (x - 1)(x - 2)(x - 3)(x - 4)(x - 5)$, with domain $\mathbb{R}$

59. $f(x) = \dfrac{1}{(x - 1)(x - 2)(x - 3)(x - 4)}$
What is the largest possible domain of f?

60. $f(x) = \dfrac{1}{(x - 1)(x - 2)(x - 3)(x - 4)(x - 5)}$
What is the largest possible domain of f?

61. $f(x) = \begin{cases} \dfrac{1}{x^2 + 1} & \text{if } -2 \leq x \leq 0 \\ \dfrac{x}{x^2 + 1} & \text{if } 0 < x \leq 6 \end{cases}$

62. $f(x) = \begin{cases} \sqrt{x^2 + x} & \text{if } -10 < x \leq -1 \\ \sqrt{x^2 + 2x + 4} & \text{if } -1 < x < 10 \end{cases}$

63. $f(x) = \dfrac{x^2}{(x - 1)(x - 2)(x - 3)}$, $-5 \leq x \leq 5$
Find the largest possible domain of f.

64. $f(x) = \dfrac{2x - 5}{(x^2 - 9)(x+1)}$, $-5 \leq x \leq 5$
Find the largest possible domain of f.

65. $f(x) = (x - 1)(x + 2)\sqrt{x} - \dfrac{1}{(x - 1)(x - 2)}$, $0 \leq x \leq 5$
Find the largest possible domain of f.

66. $f(x) = \dfrac{(x - 1)(x + 2)}{x} - \dfrac{1}{(x - 1)(x - 2)}$, $-2 \leq x \leq 5$
Find the largest possible domain of f.

67. $f(x) = \sqrt{x}(x - 1)$, with domain $(0, 5]$. What is the lowest point on this graph?

68. $f(x) = \sqrt{x}(x + 1)$, with domain $(0, 5]$. Is this function increasing (getting larger) or decreasing (getting smaller)?

69. $f(x) = \dfrac{\sqrt{x}}{x - 1}$ Use your graphing calculator to find the largest possible domain. Is there any region in which this function is increasing?

70. $f(x) = \dfrac{x}{\sqrt{x - 1}}$ Use your graphing calculator to find the largest possible domain. Is there any region in which this function is increasing?

APPLICATIONS

71. *Demand* The demand for Sigma Mu Fraternity plastic brownie dishes is
$$q(p) = 361{,}201 - (p+1)^2,$$
where q represents the number of brownie dishes Sigma Mu can sell per month at a price of $p¢$ each. Use this function to determine
(a) the number of brownie dishes Sigma Mu can sell per month if the price is set at 50¢,
(b) the number of brownie dishes they can unload per month if they give them away,
(c) the price at which Sigma Mu will be unable to sell any dishes.

72. *Revenue* The total weekly revenue earned at Royal Ruby Retailers is given by
$$R(p) = -\frac{4}{3}p^2 + 80p,$$
where p was the price (in dollars) RRR charges per ruby. Use this function to determine
(a) the weekly revenue to the nearest dollar when the price is set at $20 per ruby,
(b) the weekly revenue to the nearest dollar when the price is set at $200 per ruby (interpret your result),
(c) the price RRR should charge in order to obtain a weekly revenue of $1200.

73. *Investments in South Africa* The number of U.S. companies that invested in South Africa from 1986 through 1994 closely followed the function
$$n(t) = 5t^2 - 49t + 232.$$
Here, t is the number of years since 1986, and $n(t)$ is the number of U.S. companies that own at least 50% of their South African subsidiaries and employ 1,000 or more people.*
(a) Find the appropriate domain of n.
(b) Is $[0, +\infty)$ the appropriate domain? Give reasons for your answer.

74. *Sony Net Income* The annual net income for *Sony Corp.* from 1989 through 1994 can be approximated by the function
$$I(t) = -77t^2 + 301t + 524.$$
Here, t is the number of years since 1989 and $I(t)$ is Sony Corp.'s net income in millions of dollars for the corresponding fiscal year.†
(a) Find the appropriate domain of t.
(b) Is $[0, +\infty)$ the appropriate domain? Give reasons for your answer.

75. *Toxic Waste Treatment* The cost of treating waste by removing PCPs goes up rapidly as the quantity of PCPs removed goes up. Here is a possible model:
$$C(q) = 2{,}000 + 100q^2,$$
where q is the reduction in toxicity (in pounds of PCPs removed per day) and $C(q)$ is the daily cost (in dollars) of this reduction.
(a) Find the cost of removing 10 pounds of PCPs per day.
(b) Government subsidies for toxic waste cleanup amount to
$$S(q) = 500q,$$
where q is as above and $S(q)$ is the daily dollar subsidy. Calculate the net cost function (the cost after the subsidy is taken into account) $N(q)$, given the cost function and subsidy above, and find the net cost of removing 20 pounds of PCPs per day.

76. *Dental Plans* A company pays for its employees' dental coverage at an annual cost C given by
$$C(q) = 1{,}000 + 100\sqrt{q},$$
where q is the number of employees covered, and $C(q)$ is the annual cost in dollars.
(a) If the company has 100 employees, find its annual outlay for dental coverage.
(b) Assuming that the government subsidizes cover-

▼ *The model is the authors' (least squares quadratic regression with coefficients rounded to nearest integer). Source for raw data: Investor Responsibility Research Center Inc., Fleming Martin/New York Times, *The New York Times*, June 7, 1994, p. D1.

† The model is the authors' (least squares quadratic regression with coefficients rounded to nearest integer). Source for raw data: Sony Corporation/New York Times, *The New York Times*, May 20, 1994, p. D1.

age by an annual dollar amount of
$$S(q) = 200q,$$
calculate the net cost function, $N(q)$, to the company, and calculate the net cost of subsidizing its 100 employees. Comment on your answer.

77. **Cost** I want to fence in a 20-square-foot rectangular vegetable patch. For reasons too complicated to explain, the fencing for the east and west sides costs $4 per foot, while the fencing for the north and south sides costs only $2 per foot.
   (a) Express the total cost of the project as a function of the length x of the east and west sides stretch of fencing.
   (b) Graph this function, and, if you are using graphing technology, determine the approximate value of x that leads to the lowest total cost.

78. **Cost** Professor Gaunce Lewis is a keen gardener, specializing in the rarest exotic orchids. He is planning to start a new garden at the far end of the university's property, and he wishes to keep stray animals and students out. He decides to buy razor wire fencing at cost of $100 per yard and decides that he can afford to lay out enough money for 20 yards of fencing. His garden is planned as shown in the figure, although he can't quite decide on a value of x for its length.

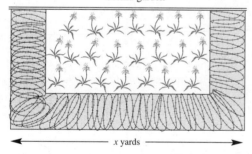

Orchid garden

← x yards →

   (a) Express the area of the garden as a function A of x, being careful to state the domain.
   (b) Graph this function, and, if you are using graphing technology, determine the approximate value of x that leads to the largest area.

79. **Volume** The volume of a cone is given by the formula $V = \frac{1}{3}\pi r^2 h$, where r is the radius of the base, and h is the height.
   (a) If the ratio of h to r is given by $\frac{h}{r} = 3$, express V as a function of h only, and graph this function.

   (b) Given the proportions of the cone in part (a), express h as a function of V, and graph this function.

80. **Surface Area** The surface area of a hollow cylinder of cross-sectional radius r and height h is given by $S = 2\pi rh$. Further, the surface area of a disc of radius r is given by $A = \pi r^2$.

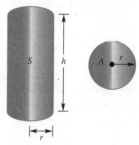

You are the design manager for Pebbles and Blips gourmet cat food, and you wish to design the shape of the cans that are to hold Pebbles and Blips Lobster Delight. Your expert consultant decides (after reading this book) that it would be most economical to design the cans with height equal to the diameter. Further, your cost analysis reveals that the gold/silver alloy you plan to use for the cans will cost the company $40 per square inch. Use this data to express the cost of a single can as a function of the height h. Graph this function.

81. **Biology—Reproduction** The Verhulst Model for population growth specifies the reproductive rate of an organism as a function of the total population according to the following formula:
$$R(p) = \frac{r}{1 + kp}.$$

Here, p is the total population, r and k are constants that depend on the particular circumstances and the organism being studied, and $R(p)$ is the reproduction

rate in new organisms per time period. Suppose that $r = 45$ and $k = 1/8000$ for a particular population.
(a) What is an appropriate domain for this function?
(b) Predict the reproduction rate when the population is 4,000.
(c) Graph R with the domain $0 < p \leq 10,000$.

82. **Biology—Reproduction** Another model, the Predator Satiation Model for population growth, specifies the reproductive rate of an organism as a function of the total population, according to the following formula:

$$R(p) = \frac{rp}{1 + kp}.$$

Here, p is the total population, r and k are constants that depend on the particular circumstances and the organism being studied, and $R(p)$ is the reproduction rate in new organisms per time period.* Suppose $r = 1/125$ and $k = 1/5,000$ for a particular population.
(a) What is an appropriate domain for this function?
(b) Predict the reproduction rate when the population is 3,000.
(c) Graph R with the domain $0 \leq p \leq 50,000$.

83. **Inflation Rates (GRE economics exam)**†
Suppose the inflation rate in an economy is given by i (where, for example, a 20% inflation rate corresponds to $i = 0.20$). According to an economic model, inflation will cause the Gross National Product (GNP) Y to differ from the potential GNP Y_P according to the formula

$$\frac{Y - Y_P}{Y_P} = -9i.$$

Assuming a current inflation rate of 5%, use the formula to obtain a function G that specifies the GNP (Y) in terms of the potential GNP (Y_P), and use your function to calculate the GNP in the event that the potential GNP is $2 trillion. How does inflation affect GNP?

84. **Inflation Rates (GRE economics exam)**†
Suppose the inflation rate in an economy is given by i (where, for example, a 20% inflation rate corresponds to $i = 0.20$). According to an economic model, if an economy has a potential Gross National Product (GNP) of Y_P then the inflation rate j the following year is related to the present year's inflation rate i by the formula

$$Y_P(j - i) = 0.9(Y - Y_P),$$

where Y is this year's GNP. Assuming a current GNP of $2 trillion and a current inflation rate of 2.5%, specify j as a function of Y_P and use your function to estimate next year's inflation rate given a potential GNP of $2.2 trillion. Interpret your answer.

85. **Acquisition of Language** The percentage $p(t)$ of children who are able to speak in at least single words by the age of t months can be approximated by the equation‡

$$p(t) = 100\left(1 - \frac{12{,}196}{t^{4.478}}\right) \qquad (t \geq 8.5).$$

(a) Graph p for $9 \leq t \leq 20$ and $0 \leq p \leq 100$.
(b) What percentage of children are able to speak in at least single words by the age of 12 months?
(c) By what age are 90% of children speaking in at least single words?

86. **Acquisition of Language** The percentage $p(t)$ of children who are able to speak in sentences of five or more words by the age of t months can be approximated by the equation‡

$$p(t) = 100\left(1 - \frac{5.2665 \times 10^{17}}{t^{12}}\right) \qquad (t \geq 30).$$

(a) Graph p for $30 \leq t \leq 45$ and $0 \leq p \leq 100$.
(b) What percentage of children are able to speak in sentences of five or more words by the age of 36 months?
(c) By what age are 75% of children speaking in sentences of five or more words?

▼ *Source: *Mathematics in Medicine and the Life Sciences* by F. C. Hoppenstaedt and C. S. Peskin (Springer-Verlag, New York, 1992) pp. 20–22.

†Based on a sample question in a GRE Economics exam. Source: *GRE Economics* by G. G. Gallagher, G. E. Pollock, W. J. Simeone, and G. Yohe, p. 48. (Research and Education Association, Piscataway, N.J., 1989).

‡The model is the authors' and is based on data presented in the article *The Emergence of Intelligence* by William H. Calvin, *Scientific American*, October, 1994, pp. 101–7.

87. *The Theory of Relativity* In science fiction terminology, a speed of *warp 1* is the speed of light—about 3×10^8 meters per second. (Thus, for instance, a speed of warp 0.8 corresponds to 80% of the speed of light—about 2.4×10^8 meters per second.) According to Einstein's Special Theory of Relativity, a moving object appears to get shorter to a stationary observer as its speed approaches that of light. If a rocket ship whose length is l_0 meters at rest travels at a speed of warp p, its length in meters, as measured by a stationary observer, will be given by

$$l(p) = l_0\sqrt{1 - p^2}, \text{ with domain } [0, 1]$$

(a) Assuming that a rocket is 100 meters long at rest ($l_0 = 100$), estimate $l(0.95)$. What does this tell you?
(b) At what speed should a rocket ship be traveling in order that it appears to be one-half as long as when it is at rest?

88. *Newton's Law of Gravity* The gravitational force exerted on a particle with mass m by another particle with mass M is given by the following function of distance:

$$F(r) = G\frac{Mm}{r^2}, \text{ with domain } (0, +\infty).$$

Here, r is the distance between the two particles in meters, the masses M and m are given in kilograms, $G \approx 0.0000000000667$, or 6.67×10^{-11}, and the resulting force is given in newtons.
(a) Given that M and m are both 1,000 kilograms, estimate $F(10)$. What does the answer tell you?
(b) How far apart should two 1,000-kilogram masses be in order that they experience an attractive force of 1 newton?

COMMUNICATION AND REASONING EXERCISES

89. Give a real-life scenario leading to a function with domain [2, 100].

90. Give a real-life scenario leading to a function with domain $[-100, +\infty)$.

91. Why is the following assertion false?
"If $f(x) = x^2 - 1$, then $f(x + h) = x^2 + h - 1$."

92. Why is the following assertion false?
"If $f(x) = \sqrt{x}$, then $f(x + h) - f(x) = \sqrt{x + h} - \sqrt{x} = \sqrt{h}$."

93. How do the graphs of two functions differ if they are specified by the same formula but have different domains?

94. How do the graphs of two functions $f(x)$ and $g(x)$ differ if $g(x) = f(x) + 10$? (Try an example.)

95. How do the graphs of two functions $f(x)$ and $g(x)$ differ if $g(x) = f(x - 5)$? (Try an example.)

96. How do the graphs of two functions $f(x)$ and $g(x)$ differ if $g(x) = f(-x)$? (Try an example.)

1.3 LINEAR FUNCTIONS

The simplest interesting equations are the **linear equations.** The simplest interesting functions are the **linear functions.** Understanding linear functions and equations is crucial to understanding calculus, and we shall see some of the basic ideas of calculus in this and the next section.

> **LINEAR EQUATION**
>
> A **linear equation in the two unknowns** (or **variables**) x and y is an equation of the form
>
> $$ax + by + c = 0$$
>
> with a, b, and c being fixed numbers, and a and b not both zero.

▼ **EXAMPLE 1**

The following are linear equations.

(a) $2x + 3y + 4 = 0$ ($a = 2$, $b = 3$, $c = 4$)
(b) $x - y + 1 = 0$ ($a = 1$, $b = -1$, $c = 1$)
(c) $-2x + 1 = 0$ ($a = -2$, $b = 0$, $c = 1$)
(d) $\frac{3}{4}y = 0$ ($a = 0$, $b = \frac{3}{4}$, $c = 0$)
(e) $4 = 0$ is *not* a linear equation in two unknowns, since both a and b are zero.
(f) $x^2 + 2xy^2 + xy = 0$, while a perfectly respectable equation in x and y, is not a *linear* equation in x and y, since the definition does not allow expressions such as x^2, y^2, or xy in linear equations.

In the case of equations **(a)**, **(b)**, and **(d)** of the above example, we can solve for y as function of x and write

(a) $y = -\frac{2}{3}x - \frac{4}{3}$
(b) $y = x + 1$
(d) $y = 0$.

These are examples of **linear functions.**

> **LINEAR FUNCTION**
>
> A **linear function** is a function that can be written in the form
>
> $$f(x) = mx + b$$
>
> with m and b being fixed numbers (the names m and b are traditional).

Note that a linear function $f(x) = mx + b$ leads to the linear equation $y = mx + b$, so the graph of a linear function is the graph of a linear equation. Therefore, all that we say about the graph of linear equations below will also apply to linear functions.

1.3 Linear Functions

In Section 1 we sketched the graph of the linear equation $x + y = 4$ and got a straight line. In fact, the graph of *any* linear equation is a straight line (hence the name *linear*). Before we discuss easy methods of drawing the graph of any linear equation, we need to introduce an extremely important concept, the **slope** of a straight line. The slope of a straight line is a way of measuring its steepness, similar to the "gradient" used by civil engineers and surveyors to measure the steepness of roads or railway tracks. Figure 1 shows a car proceeding from left to right up or down various inclines.

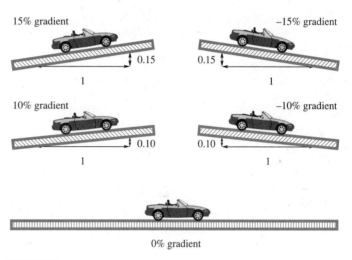

FIGURE 1

Look first at the 15% gradient at the top left. For every one unit across to the right, the road goes up 0.15 units, or 15% of the distance across. The -15% gradient at the top right goes *down* 0.15 units for every one unit across to the right. The $\pm 10\%$ gradients are similar: the road goes up or down 0.10 units for every unit across. The 0% gradient is horizontal: it goes up 0 units for every unit across.

We measure the steepness of a line in the same way, but instead of saying that the road at the top left of Figure 1 has a gradient of 15%, we shall say that, as a straight line, it has a **slope of 0.15.** Similarly, the line with a -10% gradient has a slope of -0.10. In other words, the **slope** of a line is the number of units it goes up for every unit across to the right. If it goes down instead, the line has negative slope. Consider the line of slope 2 shown in Figure 2. This line goes up two units for every unit across to the right. Thus, it goes up 6 units for every 3 units across, and it goes up 8 units for every 4 units across, and so on. We say that we have a **rise** of 2 for a **run** of 1, a rise of 6 for a run of 3, and so on.

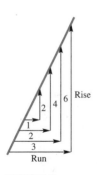

FIGURE 2

Referring to Figure 2, we see that the ratios *rise/run* all reduce to the same answer, 2 (since the right triangles are all similar). Thus, another way of defining the slope is to say that the slope is the ratio *rise/run*.

44 Chapter 1 Functions and Graphs

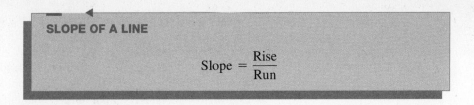

SLOPE OF A LINE

$$\text{Slope} = \frac{\text{Rise}}{\text{Run}}$$

Figure 3 gives several more examples. Notice that the larger the numerical value of the slope, the steeper the line.

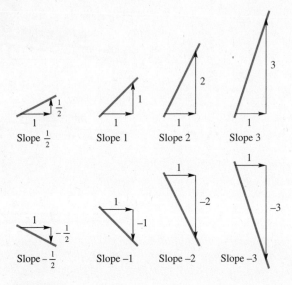

FIGURE 3

Q Suppose we have a straight line in the xy-plane, passing through two given points (x_1, y_1) and (x_2, y_2). Can we compute its slope in terms of these coordinates?

A According to Figure 4, the rise is $y_2 - y_1$, while the run is $x_2 - x_1$.

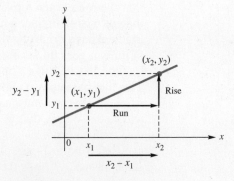

FIGURE 4

Thus, the slope, which is traditionally called* m, is given by the following formula.

> **SLOPE m OF THE LINE THROUGH THE POINTS (x_1, y_1) AND (x_2, y_2)**
>
> $$m = \frac{y_2 - y_1}{x_2 - x_1}$$

▶ **NOTE** Here is a good way of thinking about the above formula. The quantity $y_2 - y_1$ in the numerator is the change in the y-coordinate as we go from the first point (x_1, y_1) to the second point (x_2, y_2). (Refer to Figure 4.) Similarly, the quantity $x_2 - x_1$ in the denominator is the change in the x-coordinate as we go from the first point to the second. Thus, we can write

$$m = \frac{y_2 - y_1}{x_2 - x_1}$$
$$= \frac{\text{Change in } y}{\text{Change in } x} = \frac{\Delta y}{\Delta x},$$

where we use the shorthand Δy and Δx (Δ is the Greek letter *delta*) for "change in y" and "change in x." ◀

So to get the slope of a line in the xy-plane, all we need to know are the coordinates of any two distinct points on that line.

▼ **EXAMPLE 2**

Find the slope of the line through $(1, 3)$ and $(5, 11)$.

SOLUTION Take $(x_1, y_1) = (1, 3)$ and $(x_2, y_2) = (5, 11)$ and use the formula to get

$$m = \frac{\Delta y}{\Delta x} = \frac{y_2 - y_1}{x_2 - x_1} = \frac{11 - 3}{5 - 1} = \frac{8}{4} = 2.$$

Before we go on... What if we had chosen to list the two points in reverse order? That is, suppose we had taken $(x_1, y_1) = (5, 11)$ and $(x_2, y_2) = (1, 3)$. Then we would have obtained

$$m = \frac{\Delta y}{\Delta x} = \frac{y_2 - y_1}{x_2 - x_1} = \frac{3 - 11}{1 - 5} = \frac{-8}{-4} = 2,$$

the same answer. The order in which we take the points is not important, *as long as we use the same order on the top and the bottom.*

▼ * This is done for reasons that no one knows. There has actually been some research into this question lately, but still no one has found why the letter m is used.

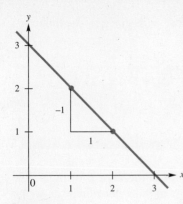

FIGURE 5

> ### EXAMPLE 3
>
> Find the slope of the line through (1, 2) and (2, 1).
>
> **SOLUTION**
>
> $$m = \frac{\Delta y}{\Delta x} = \frac{y_2 - y_1}{x_2 - x_1} = \frac{1 - 2}{2 - 1} = \frac{-1}{1} = -1$$
>
> ***Before we go on...*** Recall that lines with negative slope go downhill as you move from left to right. The line in this example is shown in Figure 5.

> ### EXAMPLE 4
>
> Find the slope of the line through the points (2, 3) and (−5, 3).
>
> **SOLUTION**
>
> $$m = \frac{\Delta y}{\Delta x} = \frac{3 - 3}{-5 - 2} = \frac{0}{-7} = 0$$
>
> ***Before we go on...*** A line of slope 0 has a 0 rise, so it is a *horizontal* line.

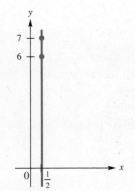

FIGURE 6

> ### EXAMPLE 5
>
> Finally, find the slope of the line through $(\frac{1}{2}, 6)$ and $(\frac{1}{2}, 7)$.
>
> **SOLUTION**
>
> $$m = \frac{\Delta y}{\Delta x} = \frac{7 - 6}{\frac{1}{2} - \frac{1}{2}} = \frac{1}{0} \; ✘$$
>
> We have used the symbol ✘ to remind you that there is no such number as $\frac{1}{0}$. This line has an **undefined** or **infinite** slope. (Although there is no such number as $\frac{1}{0}$, we sometimes think of it as being infinite.) If we plot the two points in question, we see that the line passing through them is *vertical* (Figure 6).

Learn to recognize the approximate slope of a line by looking at it. For example, recognize a line of slope 0 as being horizontal, a line of slope 1 as going up at an angle of 45°, and a line of slope −1 as going downward at the same angle. A line rising at an angle steeper than 45° would have a positive slope larger than 1, a line falling at an angle shallower than 45° would have a slope between −1 and 0, and so on.

PARALLEL AND PERPENDICULAR LINES

From Figure 7, you can conclude the following.

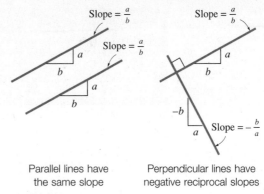

Parallel lines have the same slope

Perpendicular lines have negative reciprocal slopes

FIGURE 7

1. Parallel lines have the same slope. For example, if a line has slope -3, then any line parallel to it also will have slope -3.
2. Perpendicular lines (i.e., lines at right angles to each other) have negative reciprocal slopes. For example, if a line has slope 3, then any line perpendicular to it will have slope $-\frac{1}{3}$.

THE EQUATION OF THE STRAIGHT LINE WITH A GIVEN SLOPE THROUGH A GIVEN POINT

The slope by itself is not enough to completely specify a line. We need two pieces of information:

1. the *slope m* (which specifies the direction of the line), and
2. a *point* (x_0, y_0) on the line (which pins down its location in the plane).

Q What is the equation of the line through the point (x_0, y_0) with slope m?

A Before stating the answer in general, let us look at a specific example: suppose we want to find the equation of the line through (1, 2) with slope 3. If (x, y) is a point on this line other than (1, 2), then, since the line through the points (1, 2) and (x, y) has slope 3, the slope formula tells us that

$$m = 3 = \frac{y - 2}{x - 1}.$$

Multiplying both sides by the quantity $(x - 1)$ gives

$$y - 2 = 3(x - 1),$$

which is an equation of this line, since it must be satisfied by any point (x, y) on the line. Thus, an equation of the line through (1, 2) with slope 3 is $y - 2 = 3(x - 1)$.

We can easily generalize this derivation to get the following formula.

> **THE POINT-SLOPE FORMULA**
> An equation of the line through the point (x_0, y_0) with slope m is given by
> $$y - y_0 = m(x - x_0).$$

▶ **CAUTION** The point (x_0, y_0) represents a given point, so the subscripted variables x_0 and y_0 will always be replaced by actual numbers, just as in the formula for the slope of a line passing through two given points: $m = (y_2 - y_1)/(x_2 - x_1)$. The terms x and y, on the other hand, remain as x and y, since they are the variables in the equation of the line. ◀

▼ **EXAMPLE 6**

Find an equation of the line through (1, 3) with slope 2.

SOLUTION In order to apply the point-slope formula, we need:

- a point—given here by $(x_0, y_0) = (1, 3)$;
- the slope—given here by $m = 2$.

An equation of the line is therefore
$$y - y_0 = m(x - x_0).$$

In other words,
$$y - 3 = 2(x - 1)$$

and
$$y - 3 = 2x - 2, \quad \text{or} \quad y = 2x + 1.$$

Before we go on... It is a good idea to check our answers whenever possible. Here, the answer is $y = 2x + 1$. We can check that it is correct by making sure that it does pass through (1, 3) and has slope 2. To check that it passes through (1, 3), we substitute $x = 1$ and $y = 3$ into our equation, getting
$$3 = 2(1) + 1. \checkmark$$

To make certain that the line has slope 2, notice first that the formula $y - y_0 = m(x - x_0)$ tells us that the coefficient of x in the equation should be the slope m (since m is the only term multiplied by x in the point-slope formula). Thus, to check that the line has slope 2, we need only check that the coefficient of x in our answer is 2.

Another way you could check the answer is by using a graphing calculator or computer. Use it to plot the graph of $y = 2x + 1$, and then use the trace feature to check that it does in fact pass through the point (1, 3) and that it rises 2 units for every unit to the right.

▶ **CAUTION** If the axes are scaled differently on your graphing calculator, then lines won't appear to have the correct slope. Figure 8 shows the graph of $y = 2x + 1$ twice: on the xy-plane, with axes scaled the same, and on a graphing calculator, with axes scaled differently.

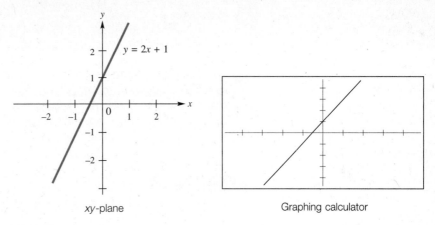

FIGURE 8

The graphing calculator line appears to have a slope slightly less than 2. This distortion is caused by the fact that (as with most calculators) the screen is not square. If, as is done above, you draw the graph for $-5 \leq x \leq 5$ and $-5 \leq y \leq 5$, the units on the x-axis will be different from the units on the y-axis, and lines will not appear to have their correct slopes. ◀

▼ **EXAMPLE 7**

Find an equation of the line through the points $(1, 2)$ and $(3, -1)$.

SOLUTION We need the following:

- a point—we have two to choose from, so we take the first, $(x_0, y_0) = (1, 2)$;
- the slope—not given *directly*, but we do have enough information to calculate it. Since we are given two points on the line, we can use the slope formula:

$$m = \frac{y_2 - y_1}{x_2 - x_1} = \frac{-1 - 2}{3 - 1} = \frac{-3}{2}.$$

An equation of the line is therefore

$$y - y_0 = m(x - x_0).$$

Thus, $$y - 2 = -\frac{3}{2}(x - 1)$$

and $$y - 2 = -\frac{3}{2}x + \frac{3}{2}, \quad \text{or} \quad y = -\frac{3}{2}x + \frac{7}{2}.$$

Before we go on... Once again we check the answer. All we need to verify is that it passes through the given points $(1, 2)$ and $(3, -1)$. Substituting each of these into the equation gives

$$2 = -\frac{3}{2}(1) + \frac{7}{2} \quad \checkmark$$

$$-1 = -\frac{3}{2}(3) + \frac{7}{2}. \quad \checkmark$$

As in the preceding example, you could check the answer by using a graphing calculator or computer to plot $y = -\frac{3}{2}x + \frac{7}{2}$, and then use the trace feature to check that it passes through the two points $(1, 2)$ and $(3, -1)$. Always bear in mind that the trace feature has limited accuracy (even if you use it in conjunction with the zoom feature).

EXAMPLE 8

Find the points where the line in Example 7 crosses the x- and y-axes.

SOLUTION The line is shown in Figure 9. The point where it crosses the x-axis has y-coordinate 0, so we substitute $y = 0$ in the equation and solve for x.

$$0 = -\frac{3}{2}x + \frac{7}{2}$$

$$\frac{3}{2}x = \frac{7}{2}$$

$$x = \frac{7}{3}$$

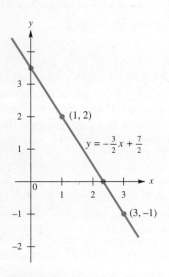

FIGURE 9

Therefore, the line crosses the x-axis at $(\frac{7}{3}, 0)$. The point where it crosses the y-axis has x-coordinate 0, so we substitute $x = 0$ in the equation.

$$y = -\frac{3}{2} \cdot 0 + \frac{7}{2} = \frac{7}{2}$$

Therefore, the line crosses the y-axis at $(0, \frac{7}{2})$.

Before we go on... These numbers $\frac{7}{3}$ and $\frac{7}{2}$ are called the x-intercept and the y-intercept of the line. In general, the x-coordinate of the point where a line crosses the x-axis is called the ***x*-intercept** of the line. Similarly, the y-coordinate of the point where it crosses the y-axis is the ***y*-intercept** of the line. Given an equation of a line, one easy way of sketching the line is to find its intercepts and draw the line connecting the corresponding points. This technique does not always work: try it with $2x - y = 0$. This equation has both x- and y-intercepts zero and thus gives only a single point—the origin—to plot. To draw the graph of such an equation, you could either find the coordinates of an additional point on the line or use the technique we are about to discuss.

EXAMPLE 9

Find the equation of the line with slope m and y-intercept b.

SOLUTION Although we have probably seen this answer somewhere before, let's pretend this is entirely new to us. We need two things:

- a point—not given directly, but the point b on the y-axis has coordinates $(0, b)$, so take $(x_0, y_0) = (0, b)$;
- the slope—given here simply as m.

The equation of the line is therefore

$$y - y_0 = m(x - x_0).$$

Thus,

$$y - b = m(x - 0)$$

and

$$y - b = mx, \quad \text{or} \quad y = mx + b.$$

Before we go on... To check that the graph of $y = mx + b$ does in fact pass through $(0, b)$, substitute this into the equation to get

$$b = m(0) + b. \checkmark$$

THE SLOPE-INTERCEPT FORMULA

An equation of the line with slope m and y-intercept b is

$$y = mx + b.$$

Figure 10 illustrates the graph of the line $y = mx + b$. Notice that the line crosses the y-axis at $y = b$, and that, since the slope is $m = \frac{m}{1}$, it goes up m units for every 1 unit in the x-direction.

Q What use is this formula if we already have the point-slope formula to give us an equation for any line?

A Suppose you are given the equation of a straight line in this form: for example, $y = \frac{3}{4}x - 1$. Then you can tell at a glance that the slope is $\frac{3}{4}$ and the y-intercept is -1.

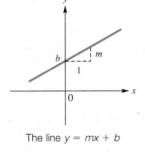

The line $y = mx + b$

FIGURE 10

FINDING THE SLOPE AND y-INTERCEPT FROM THE EQUATION OF A LINE

In the equation $y = mx + b$, the coefficient m of x is the slope, and the constant term b is the y-intercept of the line.

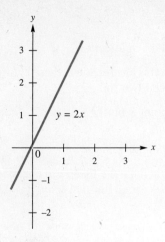

FIGURE 11

EXAMPLE 10

Sketch the graph of $2x - y = 0$.

SOLUTION Rewriting this equation in the form $y = mx + b$, we get

$$y = 2x.$$

We can now see that the slope of the line is 2 and the y-intercept is 0. Figure 11 shows the line.

Although you could use a graphing calculator or computer to sketch this line, it would really be overkill to do so. It may take you longer to use the device to draw the plot than it would take you to draw it yourself.

Note that everything we say about the graph of the equation $y = mx + b$ applies equally well to the graph of the general linear *function* $f(x) = mx + b$.

EXAMPLE 11

Find the linear function whose graph is the straight line through $(2, -2)$ that is parallel to the line $3x + 4y = 5$.

SOLUTION We begin by finding an equation of this line. As usual, we need

- a point—given here as $(2, -2)$;
- the slope. Since the required line is parallel to $3x + 4y = 5$, it must have the same slope. We can find the slope of this line by solving for y and then looking at the coefficient of x. Solving for y gives

$$y = -\frac{3}{4}x + \frac{5}{4},$$

so the slope is $-\frac{3}{4}$, since this is the coefficient of x.

Thus, an equation for the line is

$$y - y_0 = m(x - x_0)$$

$$y + 2 = -\frac{3}{4}(x - 2)$$

or $\quad y + 2 = -\frac{3}{4}x + \frac{3}{2}, \quad$ or $\quad y = -\frac{3}{4}x - \frac{1}{2}.$

Notice that we have put the answer in $y = mx + b$ form by solving for y. The corresponding function is therefore

$$f(x) = -\frac{3}{4}x - \frac{1}{2}.$$

Before we go on... To check the solution, $f(x) = -\frac{3}{4}x - \frac{1}{2}$, we first test that the graph of this function passes through the point $(2, -2)$.

$$f(2) = -\frac{3}{4}(2) - \frac{1}{2} = -2 \checkmark$$

To test that it is indeed parallel to the given line, all we need to do is make sure that the slope is $-\frac{3}{4}$. Since this is the coefficient of x, all is well.

With a graphing calculator or computer you can plot both $y = -\frac{3}{4}x + \frac{5}{4}$ and $y = -\frac{3}{4}x - \frac{1}{2}$ and look at the graphs to check that they do in fact appear to be parallel. As usual, you can use the trace feature to check (to within the accuracy of the display) that the second line does pass through the point $(2, -2)$.

▼ **EXAMPLE 12**

Find the linear function whose graph is the straight line through $(2, -2)$ that is perpendicular to the line $3x + 4y = 5$.

SOLUTION Again, we start by finding an equation of the line. As usual, we need

- a point—given here as $(2, -2)$;
- the slope. Since the required line is perpendicular to $3x + 4y = 5$, its slope is the negative reciprocal of the slope of $3x + 4y = 5$. As in Example 11, we solve for y to get $y = -\frac{3}{4}x + \frac{5}{4}$. Thus, the slope we want is the negative reciprocal of $-\frac{3}{4}$, namely, $\frac{4}{3}$.

So the desired equation is

$$y - y_0 = m(x - x_0)$$

$$y + 2 = \frac{4}{3}(x - 2)$$

or

$$y + 2 = \frac{4}{3}x - \frac{8}{3}, \quad \text{or} \quad y = \frac{4}{3}x - \frac{14}{3}.$$

The corresponding function is

$$f(x) = \frac{4}{3}x - \frac{14}{3}.$$

Before we go on... First, we test that the graph passes through $(2, -2)$.

$$f(2) = \frac{4}{3}(2) - \frac{14}{3} = -2 \checkmark$$

Finally, the coefficient of x tells us that the slope is $\frac{4}{3}$, which is what we wanted.

With a graphing calculator or computer, you can plot both $y = -\frac{3}{4}x + \frac{5}{4}$ and $y = \frac{4}{3}x - \frac{14}{3}$. Do they appear to cross at right angles? If not, why not? (See the caution after Example 6.)

▼ EXAMPLE 13

Sketch the lines found in Examples 11 and 12 above.

SOLUTION The line in Example 11 has equation $y = -\frac{3}{4}x - \frac{1}{2}$. Thus, its slope is $-\frac{3}{4}$ and its y-intercept is $-\frac{1}{2}$. Since the slope is $-\frac{3}{4}$, it goes down 3 units for every 4 units to the right, so we can sketch it as in Figure 12. For the line in Example 12, the equation is $y = \frac{4}{3}x - \frac{14}{3}$, so we can sketch it as in Figure 13.

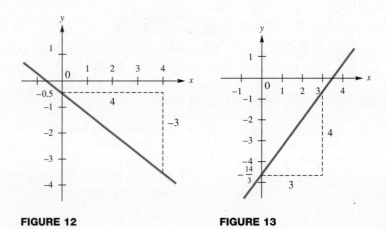

FIGURE 12   FIGURE 13

▼ EXAMPLE 14

Find the equation of **(a)** the horizontal line passing through $(-9, 5)$, and **(b)** the vertical line passing through the same point.

SOLUTION

(a) Here the point is $(-9, 5)$ and the slope is 0, yielding $y = 5$. (Check this for yourself using the point-slope formula.)

(b) Here, the point remains $(-9, 5)$, but the slope is undefined! So we can't use the point-slope formula, since there isn't a well-defined slope. (That formula only makes sense for a slope m that is a real value.) What can we do? Well, here are some points on the required line:

$$(-9, 1), (-9, 2), (-9, 3), \ldots,$$

so $x = -9$, and $y = $ *anything*. If we simply say that $x = -9$, then these points are all solutions, so the equation is $x = -9$.

Before we go on... The horizontal line is the graph of the function $f(x) = 5$. However, the vertical line is not the graph of a function (why not?).

SUMMARY: SURE-FIRE METHOD OF FINDING AN EQUATION OF A STRAIGHT LINE

1. All we need are two pieces of information: **(a)** a *point* on the line and **(b)** the *slope* of the line. We can then use the point-slope formula $y - y_0 = m(x - x_0)$.

2. Sometimes the problems are not posed so simply, but there must *always* be enough information to get these two quantities. Here are some instances that can occur:

   (a) We are given two points (x_1, y_1) and (x_2, y_2): use $m = (y_2 - y_1)/(x_2 - x_1)$ to get the slope. If the slope is undefined, see (d).

   (b) We are given a point and the equation of a parallel line: obtain the slope of the other line by solving its equation for y. This value is the slope to use. However, if the parallel line is vertical, and hence has no slope, see (d).

   (c) We are given a point and the equation of a perpendicular line: obtain the slope of the other line by solving its equation for y, and then take its negative reciprocal. This value is the slope to use. However, if the slope of the given line is zero, then the required line, being perpendicular to a horizontal line, is vertical, so see (d).

   (d) An equation of the vertical line passing through (x_0, y_0) is $x = x_0$.

3. The equations we will be using are these, and we strongly suggest you commit them to memory.

   (a) Point-Slope Formula An equation of the line through (x_0, y_0) with slope m is
   $$y - y_0 = m(x - x_0).$$

   (b) Slope-Intercept Formula An equation of the line with y-intercept b and slope m is
   $$y = mx + b.$$

   (c) Horizontal and Vertical Lines An equation of the horizontal line through (x_0, y_0) is
   $$y = y_0.$$
   An equation of the vertical line through (x_0, y_0) is
   $$x = x_0.$$

1.3 EXERCISES

In each of Exercises 1–18, find the slope of the straight line through the given pair of points. Try to do as many as you can without writing anything down.

1. $(0, 0)$ and $(1, 2)$
2. $(0, 0)$ and $(-1, 2)$
3. $(-1, -2)$ and $(0, 0)$
4. $(2, 1)$ and $(0, 0)$
5. $(4, 3)$ and $(5, 1)$
6. $(4, 3)$ and $(-1, -5)$
7. $(1, -1)$ and $(2, -2)$
8. $(-2, 2)$ and $(-1, -1)$
9. $(0, 1)$ and $(-\frac{1}{2}, \frac{3}{4})$
10. $(\frac{1}{2}, 1)$ and $(-\frac{1}{2}, \frac{3}{4})$
11. $(4, \sqrt{2})$ and $(5, \sqrt{2})$
12. $(1, 1)$ and $(\sqrt{2}, \sqrt{2})$
13. $(4, \sqrt{2})$ and $(5, 2\sqrt{2})$
14. $(4\sqrt{2}, 2\sqrt{2})$ and $(0, -2\sqrt{2})$
15. (a, a) and $(a, 3a)$ $\quad(a \neq 0)$
16. $(a, 1)$ and $(a, 2)$
17. (a, b) and (c, d) $\quad(a \neq c)$
18. (a, b) and (a, d) $\quad(b \neq d)$

Referring to Exercises 1–18, say whether each of the following pairs of lines are parallel, perpendicular, or neither.

19. 1 and 3
20. 2 and 4
21. 5 and 7
22. 6 and 8
23. 11 and 15
24. 12 and 14
25. 5 and 9
26. 16 and 18

27. Referring to the following figure, match the lines with their slopes.
    (I) $-\frac{1}{3}$; (II) 3; (III) $-\frac{1}{4}$; (IV) 1; (V) -1; (VI) undefined; (VII) $\frac{1}{2}$; (VIII) -2; (IX) 0.

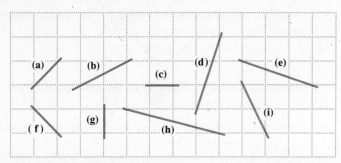

28. Referring to the following figure, match the lines with their slopes.
    (I) $-\frac{1}{3}$; (II) 3; (III) 2; (IV) 1; (V) -1; (VI) undefined; (VII) $-\frac{1}{2}$; (VIII) -2; (IX) 0.

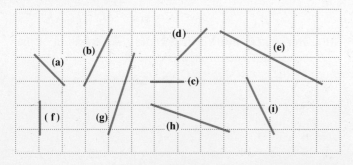

1.3 Linear Functions

Find the linear functions whose graphs are the straight lines in Exercises 29–44.

29. Through $(1, 3)$ with slope 3
30. Through $(2, 1)$ with slope 2
31. Through $(1, -\frac{3}{4})$ with slope $\frac{1}{4}$
32. Through $(0, -\frac{1}{3})$ with slope $\frac{1}{3}$
33. Through $(2, -4)$ and $(1, 1)$
34. Through $(1, -4)$ and $(-1, -1)$
35. Through $(1, -\frac{3}{4})$ and $(\frac{1}{2}, \frac{3}{4})$
36. Through $(\frac{1}{2}, -\frac{3}{4})$ and $(\frac{1}{4}, \frac{3}{4})$
37. Through $(6, 6)$ and parallel to the line $x + y = 4$
38. Through $(\frac{1}{3}, -1)$ and parallel to the line $3x - 4y = 8$
39. Through $(\frac{1}{2}, 5)$ and parallel to the line $4x - 2y = 11$
40. Through $(\frac{1}{3}, 0)$ and parallel to the line $6x - 2y = 11$
41. Through $(0, 2)$ and perpendicular to the line $x + y = 4$
42. Through $(0, 4)$ and perpendicular to the line $2x + 4y = 4$
43. Through $(3, -2)$ and perpendicular to the line $y = -2$
44. Through $(3, -2)$ and perpendicular to the line $x = 3$

Sketch the straight lines with equations given in Exercises 45–56.

45. $2x + 3y = 6$
46. $x + 2y = 6$
47. $y + \frac{1}{4}x = -4$
48. $y - \frac{1}{4}x = -4$
49. $7x - 3y = 5$
50. $2x - 3y = 1$
51. $3x = 8$
52. $2x = -7$
53. $6y = 9$
54. $3y = 4$
55. $2x = 3y$
56. $3x = -2y$

Sketch each of the lines in Exercises 57–66 on a graphing calculator. In each case, use your graph to find the approximate value of the x-intercept (if any) of the given line. Check by finding the exact value of the intercept.

57. $y = 4.1x - 5.4$
58. $y = 2.3x + 5.5$
59. $y = -10.4x + 10$
60. $y = 20.3x - 31.2$
61. $y = 10{,}050x + 4{,}323$
62. $y = -5{,}300x - 2{,}000$
63. $13x - 15y = 23$
64. $7x + 19y = 43$
65. $100x + 50y = -1{,}020$
66. $1{,}000x - 3y = 66$

COMMUNICATION AND REASONING EXERCISES

67. To what linear function of x does the linear equation $ax + by = c$ $(b \neq 0)$ correspond?
68. Why did we specify $b \neq 0$ in Exercise 67?
69. Which linear equations $ax + by = c$ have graphs that are *not* the graphs of linear functions?
70. In terms of a, b, and c, what are the slope, x-intercept, and y-intercept of the line with equation $ax + by = c$ $(a \neq 0, b \neq 0)$?
71. Complete the following. The slope of the line with equation $y = mx + b$ is the number of units that _____ increases per unit increase in _____ .

72. Complete the following. If, in a straight line, y is increasing three times as fast as x, then its _____ is _____ .

73. A friend tells you that the line through the points (3, 5) and (4, 5) has no slope. Is she correct? (Explain your answer.)

74. Another friend tells you that the line through the points (3, 5) and (3, 6) has no slope. Is he correct? (Explain your answer.)

1.4 LINEAR MODELS

Linear functions can be used to represent a wide variety of situations in everyday life. Using linear functions to represent situations in real life is called **linear modeling,** and we illustrate linear modeling with several examples.

EXAMPLE 1 Cost Function

The Yellow Cab Company charges $1 on entering the cab, plus an additional $2 per mile.*

(a) Find the cost C of an x-mile trip.
(b) Use your answer to calculate the cost of a 40-mile trip.
(c) What is the cost of the second mile?
(d) What is the cost of the tenth mile?
(e) Graph C as a function of x.

SOLUTION

(a) We are being asked to find how the cost C depends on the length x of the trip, or to find C as a function of x. Look at the cost in a few instances.

$$\text{When } x = 1, \text{ the cost is } C = 1 + 2(1) = 3.$$
$$\text{When } x = 2, \text{ the cost is } C = 1 + 2(2) = 5.$$
$$\text{When } x = 3, \text{ the cost is } C = 1 + 2(3) = 7.$$
$$\ldots$$
$$\text{When } x = x, \text{ the cost is } C = 1 + 2x.$$

Thus, the cost of an x-mile trip is given by the linear function

$$C(x) = 2x + 1.$$

---

* This is equivalent to charging $3 for the first mile and $2 for each subsequent mile. In 1993, cab fare in Chicago cost $2.40 for the first mile and $1.40 for each subsequent mile. (We are allowing for inflation.)

Notice that the slope 2 is equal to the incremental cost per mile, which we call **marginal cost,** while the varying quantity $2x$ we call the **variable cost.** The y-intercept is equal to the basic fee, which we call the **fixed cost.** In general, a linear cost function has the following form.

$$C(x) = \overbrace{mx}^{\text{Variable cost}} + b$$

$$\underset{\text{Marginal cost}}{\uparrow} \qquad \underset{\text{Fixed cost}}{\uparrow}$$

(b) We can now use the cost function to calculate the cost of a 40-mile trip as

$$C(40) = 2(40) + 1 = \$81.$$

(c) To calculate the cost of the second mile, we *could* proceed as follows.

1. Find the cost of a one-mile trip: $C(1) = 2(1) + 1 = \$3$.
2. Find the cost of a two-mile trip: $C(2) = 2(2) + 1 = \$5$.
3. Therefore, the cost of the second mile is $\$5 - \$3 = \$2$.

But notice that this is just the marginal cost. In fact, the marginal cost is the cost of each additional mile, so we could have done this more simply as follows.

$$\text{Cost of second mile} = \text{Marginal cost} = \$2.$$

(d) Since the marginal cost is the cost of each additional mile, the answer once again is \$2.

(e) Figure 1 shows the graph of the cost function, which we can interpret as a *cost vs. miles* graph. The fixed cost is the starting height on the left, while the marginal cost is the slope of the line.

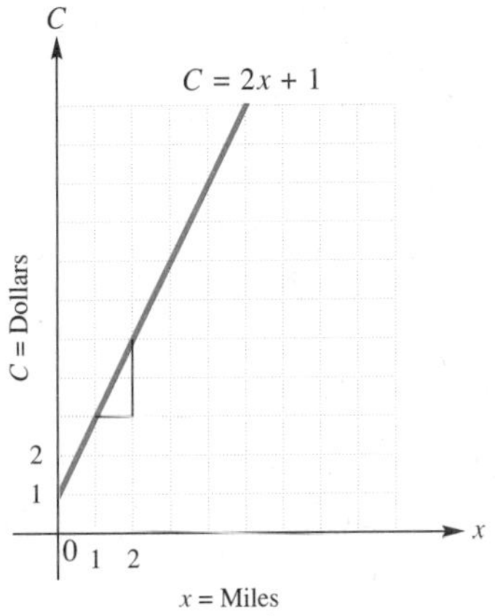

FIGURE 1

LINEAR COST FUNCTION

A **cost function** specifies the cost C as a function of the number of items x. Thus, $C(x)$ is the cost of x items. A cost function of the form

$$C(x) = mx + b$$

is called a **linear cost function.** The quantity mx is called the **variable cost,** and the intercept b is called the **fixed cost.** The slope m, which represents the **marginal cost,** measures the incremental cost per item.

EXAMPLE 2 Cost Function

The manager of the FrozenAir Refrigerator factory notices that on Monday it cost them a total of $25,000 to build 30 refrigerators, and on Tuesday it cost them $30,000 to build 40 refrigerators. Assuming a linear cost function, what is their daily fixed cost, and what is the marginal cost?

SOLUTION The secret in this kind of problem is to realize that we have been given two points on a line whose equation we want to find. The assumption is that the cost C will depend on x, the number of refrigerators, by an equation of the form

$$C = mx + b,$$

where m is the marginal cost and b is the fixed cost. We are told that $C = 25,000$ when $x = 30$, and this amounts to being told that $(30, 25,000)$ is a point on the line we get by graphing $C = mx+b$. Similarly, $(40, 30,000)$ is another point on the line. The graph of C vs. x is shown in Figure 2. To find the equation of this line we recall that you need two items of information: a point on the line, and the slope.

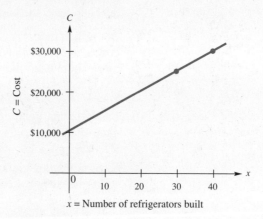

FIGURE 2

Point: We are given two of them, so you can choose the first: $(30, 25,000)$.
Slope: We use the slope equation (with C playing the role of y).

$$m = \frac{C_2 - C_1}{x_2 - x_1}$$
$$= \frac{30,000 - 25,000}{40 - 30}$$
$$= \frac{5,000}{10} = 500$$

In other words, the marginal cost is $500 per refrigerator. Put another way, each refrigerator adds $500 to the total cost. To complete the problem, we use

the point-slope formula:
$$C - C_0 = m(x - x_0)$$
$$C - 25{,}000 = 500(x - 30) = 500x - 15{,}000$$
$$C = 500x + 10{,}000$$

or

$$C(x) = 500x + 10{,}000.$$

Since $b = 10{,}000$, the factory's fixed cost is $10,000 each day.

Before we go on... We can check our equation by making sure that it gives the known costs of making 30 and 40 refrigerators:

$$C(30) = 500 \cdot 30 + 10{,}000 = 15{,}000 + 10{,}000$$
$$= 25{,}000 \quad \checkmark$$
$$C(40) = 500 \cdot 40 + 10{,}000 = 20{,}000 + 10{,}000$$
$$= 30{,}000. \quad \checkmark$$

We should also think about the domain of C. Certainly it makes no sense to speak of a negative number of refrigerators here, so $x \geq 0$. On the other hand, there likely is a largest number of refrigerators that this factory can churn out. If, for example, they could make no more than 70 refrigerators in one day, we would say that the domain of C is $[0, 70]$.

We also wish to consider revenue and profit.

REVENUE AND PROFIT

The **revenue** resulting from one or more business transactions is the total payment received, sometimes called the gross proceeds. The **profit,** on the other hand, is the *net* proceeds, or what remains of the revenue when costs are subtracted. Profit, revenue, and cost are related by the following formulas.

$$\text{Profit} = \text{Revenue} - \text{Cost}$$
$$P = R - C$$

If the profit is negative, say $-$500$, we refer to a **loss** (of $500 in this case).

▼ **EXAMPLE 3**   Revenue, Profit, and Break-Even Analysis

Lumber futures prices at the Chicago Mercantile Exchange reached a record high of $363.50 per lot of 1,000 board feet in February 1993.*

▼ *Source: *The Chicago Tribune*, February 10, 1993, Business Section, p. 1.

(a) Find the revenue that a seller of lumber at the Merc would earn if she sold x lots of lumber at the above price.

(b) Assuming that she has fixed costs of $1,000 per day and can buy lumber at $200 per lot, express her profit as a linear function of the number of lots she trades (buys and sells) per day.

(c) How many lots should she trade per day to break even?

SOLUTION

(a) Here, the revenue the seller obtains from the sale of a single lot is $363.50. To obtain the revenue from the sale of x lots, let us do a little accounting.

$$x \text{ lots @ } \$363.50 \text{ per lot gives } x \cdot 363.50 = 363.5x$$

Thus, the revenue function is

$$R(x) = 363.5x.$$

(b) We are now asked for the *profit*. In order to calculate profit, we use the formula

$$\text{Profit} = \text{Revenue} - \text{Cost}.$$

We already have the revenue function, so we must now find the cost function. Since the fixed daily cost is $1,000 per day and the marginal cost is $200 per lot, we can simply write down the total cost function.

$$C(x) = 200x + 1{,}000$$

We can now write the profit function.

$$P(x) = R(x) - C(x)$$
$$= 363.5x - (200x + 1{,}000)$$
$$= 163.5x - 1{,}000$$

Here, $P(x)$ is the daily profit the seller can make by trading x lots per day (assuming prices do not vary).

(c) To "break even" means to make zero profit. Thus, the question can be rephrased as

$$\text{Find } x \text{ so that } P(x) = 0.$$

So, all we have to do is set $P(x) = 0$ and solve for x.

$$163.5x - 1{,}000 = 0$$
$$163.5x = 1{,}000$$
$$x = \frac{1{,}000}{163.5} \approx 6.1$$

Thus, in order to break even, the seller should trade approximately 6.1 lots per day.

Q How do we interpret this answer? In other words, how is it possible to trade 6.1 lots per day?

A We can interpret it as follows. To break even, she should trade an *average* of 6.1 lots per day. For instance, if she trades 61 lots in 10 days, she will have traded an average of $\frac{61}{10} = 6.1$ lots per day, and is thus breaking even.

Before we go on... As we saw, break-even occurs when the profit $P(x) = 0$. We can also look at it as the point where Revenue = Cost: $R(x) = C(x)$. We can interpret this by graphing the revenue and cost functions on the same axes (Figure 3).

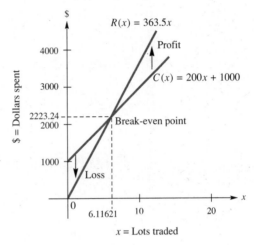

Break-even occurs at the point of intersection of the graphs of revenue and cost.

FIGURE 3

There are several things we can see from the figure.

1. Break-even occurs when the revenue and cost graphs intersect. The x-coordinate of the break-even point gives the number of lots that must be traded to break even: 6.1 per day. The y-coordinate gives the amount of money traded: the daily costs and revenue will each amount to $2,223 at break-even. (This figure was obtained from a calculator by taking $x = 1,000/163.5$ rather than the approximation $x \approx 6.1$.)
2. If she trades fewer than 6.1 lots per day, then she is to the left of 6.1 on the x-axis, where the graph of the cost function is above the graph of the revenue function, so she is making a loss amounting to the vertical distance between the graphs. Similarly, if she trades more than 6.1 lots per day, the revenue graph is above the cost graph, telling us that she is making a profit.

With a graphing calculator or computer, you can graph $y = 363.5x$ and $y = 200x + 100$ on the same axes. You can then use the trace and zoom features to locate the coordinates of the break-even point. Alternatively, you could graph the profit $P = 163.5x - 1000$ and locate the point where the profit changes from negative (a loss) to positive.

We summarize what we have gleaned from this example.

> **PROFIT FUNCTION AND BREAK-EVEN**
>
> The **profit function** P is given by
> $$P(x) = R(x) - C(x)$$
> (Profit = Revenue − Cost).
>
> **Break-even** occurs when *the revenue equals the cost*, or
> $$R(x) = C(x).$$
> Equivalently, it occurs when *the profit equals zero*, or
> $$P(x) = 0.$$
> The **break-even quantity** is the number of items at which break-even occurs.

> **OBTAINING THE BREAK-EVEN QUANTITY**
>
> We can obtain the break-even quantity *graphically* by finding the coordinates of the point of intersection of the graphs of revenue and cost. We can calculate the break-even quantity *algebraically* by solving one of the equations $R(x) = C(x)$ or $P(x) = 0$ for x. The revenue (and cost) corresponding to the break-even quantity is then the value of either $R(x)$ or $C(x)$ for this value of x.

▼ Velocity

You are driving down the Ohio Turnpike watching the mileage markers to stay awake. Measuring time in hours after you see the 20-mile marker, you see the following markers each half hour:

Time (hrs)	0	0.5	1	1.5	2
Marker (mi)	20	47	74	101	128

Find your location s as a function of t, the number of hours you have been driving. (The number s is called your **displacement,** or **position**).

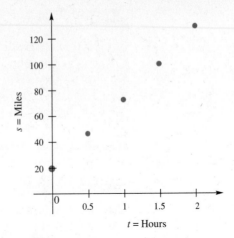

FIGURE 4

SOLUTION If we plot the location s versus the time t, the 5 markers listed give us the graph in Figure 4.

These points appear to lie along a straight line. We can verify this by calculating how far you traveled in each half-hour. In the first half-hour you traveled $47 - 20 = 27$ miles. In the second half-hour you traveled $74 - 47 = 27$ miles also. In fact, you traveled exactly 27 miles each half-hour. The points we plotted lie on a straight line that rises 27 for every $\frac{1}{2}$ unit we go to the right, for a slope of $27 \div (\frac{1}{2}) = 54$.

To get the equation of that line, we notice that we have the s-intercept, which is the starting marker of 20. From the slope-intercept form (using s in place of y and t in place of x) we get

$$s(t) = 54t + 20.$$

Before we go on... Notice the significance of the slope: for every hour you travel, you drive a distance of 54 miles. In other words, you are traveling at a constant speed of 54 mph. We have uncovered one of the most important principles in applied mathematics:

In the graph of displacement vs. time, velocity is given by the slope.

In Example 1, the slope was the marginal cost (in dollars per mile). Here, the slope is velocity (in miles per hour). In both instances, the slope can be interpreted as *the speed or rate at which a quantity y (or s) changes per unit of x (or t)*. Be sure to spend some time thinking about this idea. It is one of the central concepts underlying calculus.

A computer (particularly a spreadsheet program) is very useful in analyzing tables of data like this. By taking the differences in successive s values, we were able to notice a regularity: you traveled the same distance in each half-hour. This regularity indicated a linear relationship between s and t. You can analyze more complicated relations this way, too, and we shall occasionally return to this.

> **VELOCITY AND POSITION**
>
> If $s(t)$ is the position (or displacement) of a moving object at time t, and s is given by
>
> $$s(t) = mt + b,$$
>
> then the slope m is the **velocity** of the object, and b is its **starting position** or **initial position**.

▼ **EXAMPLE 5**   Fish Tanks

My 36-gallon tropical fish tank is leaking at a rate of 6 gallons per day. Assuming that it starts out full, how long will it take to empty?

SOLUTION   Before attempting to answer the question, let us first get an equation for the amount of water y left in the tank after t days, assuming that it is full when $t = 0$.

When $t = 1$, the amount of water left is $36 - 6(1)$.

When $t = 2$, the amount of water left is $36 - 6(2)$.

. . .

When $t = t$, the amount of water left is $36 - 6t$.

Thus,

$$y = -6t + 36.$$

Notice that the slope is negative. The reason: since the quantity y of water is *decreasing* at 6 gal/day, the rate at which y is changing is -6 gal/day. Thus, the slope is again measuring the rate of change of y, as it did in the previous examples. As usual, 36 is the initial value of y.

Now we can answer the question, "How long will it take to empty?" Mathematically, the question reads:

"Find that t for which y will equal 0."

We take the equation $y = -6t + 36$, put $y = 0$, and solve for t.

$$0 = -6t + 36$$
$$6t = 36, \text{ so } t = 6$$

Thus, my tropical fish tank will empty in exactly 6 days.

Before we go on... You might argue that you could have answered this by using "common sense" (36 gal $\div$ 6 gal/day = 6 days). Well, yes, but this is because the numbers were simple to work with. What is important here is the interpretation of slope as the rate of change.

It is commonplace to observe that demand for a commodity goes down as its price goes up. It is traditional to use the letter q for the (quantity of) demand. Consider the following example.

▼ **EXAMPLE 6** Setting Up a Demand Equation

You are the owner of the upscale Workout Fever Health Club and have been charging an annual membership fee of $600.* You are disappointed with the response: the club has been averaging only 10 new members per month. In order to remedy this, you decide to lower the fee to $500, and you notice that this boosts new membership to an average of 16 per month. Taking the demand q to be the average number of new members per month, express q as a linear function of the annual membership fee p.

SOLUTION A **demand equation** or **demand function** expresses demand q (in this case, the number of new memberships sold per month) as a function of the unit price p (in this case, membership fees). Since we are asking for a *linear* demand function, we are looking for an equation of the form

$$q = mp + b,$$

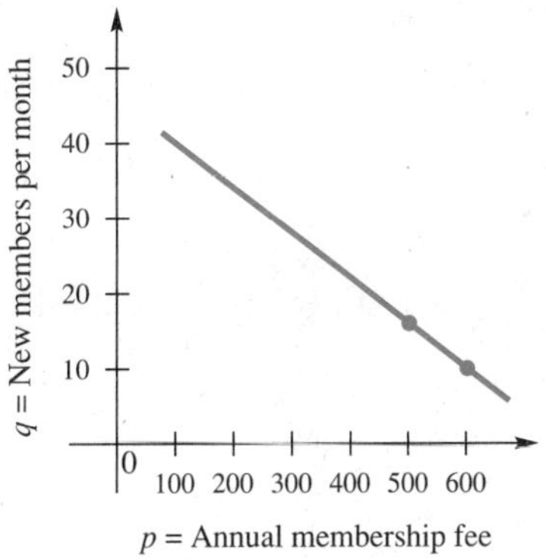

where m and b are constants to be determined. (Thus, we take p to play the role of x and q to play the role of y.) In order to find these constants, notice that, just as in the case of Example 2, we are given two points on the graph of q versus p: (600, 10) and (500, 16) (see Figure 5).

We can therefore use this information to obtain the equation of the line.
Point: (600, 10) (As usual, we can choose either of the points we are given.)

Slope:
$$m = \frac{q_2 - q_1}{p_2 - p_1}$$
$$= \frac{16 - 10}{500 - 600}$$
$$= -\frac{6}{100} = -\frac{3}{50}$$

Thus, the equation we want is

$$q - q_0 = m(p - p_0)$$
$$q - 10 = -\frac{3}{50}(p - 600)$$

or
$$q = -\frac{3}{50}p + 46.$$

Before we go on... We should check that this equation gives us the correct membership figures for the prices of $600 and $500, but we leave this to you as an exercise. What we shall do instead is use the equation to make some predictions about membership.

▼ *This is $1 more than what the East Bank Club in Chicago was charging for individual memberships in 1993, according to the *Chicago Tribune*. ("What the Clubs Cost," Section 6, May 10, 1993, p. 15)

FIGURE 5
q = New members per month vs. p = Annual membership fee

Q Suppose you decide to lower the annual fee to $100. How will this affect new membership?

A Here is where the demand equation is useful. Simply substitute $p = 100$ in the equation to obtain

$$q(100) = -\frac{3}{50}(100) + 46$$
$$= 40,$$

so you expect your new membership to increase to 40 new members per month.

Q Suppose, in a fit of greed, you decided to *increase* the price to $2,000. How would this affect membership?

A Our linear demand equation tells us that

$$q(2,000) = -\frac{3}{50}(2,000) + 46$$
$$= -74.$$

Our linear model seems to be telling us that -74 new members will join each month! In other words, the model has failed to give a meaningful answer.

This answer suggests the following further question.

Q Just how reliable is the linear model?

A Look at it this way: the *actual* demand graph could, in principle, be obtained by tabulating new membership figures for a large number of different annual fees. If the resulting points were plotted on the pq plane, they would probably suggest a curve and not a straight line. However, if you looked at a small enough portion of this curve, you could closely *approximate* it by a straight line. In other words, *over a small range of values of p, the linear model is accurate.* Linear models of real-life situations are generally reliable only for small ranges of the variables. (This point will come up again in some of the exercises.)

DEMAND EQUATION

A **demand equation** or **demand function** expresses demand q (the number of items demanded) as a function of the unit price p (the price per item). A **linear demand equation** has the form

$$q = mp + b.$$

It is usually the case that demand decreases as the unit price increases, so m is usually negative.

INTERPRETATION OF m
The (usually negative) slope m measures the change in demand per unit change in price. Thus for instance, if $m = -400$, p is measured in dollars and q in monthly sales, then each \$1 increase in the price per item will result in a drop in sales of 400 items per month.

INTERPRETATION OF b
The y-intercept b gives the demand if the items were given away.*

▼ **EXAMPLE 7** Demand Equation

The demand for rubies at Royal Ruby Retailers (RRR) is given by the equation

$$q = -\frac{4}{3}p + 80,$$

where p is the price RRR charges (in dollars) and q is the number of rubies sold per week. If we plot demand q vs. price p, we get a graph with q-intercept 80 and slope $-\frac{4}{3}$. (See Figure 6.)

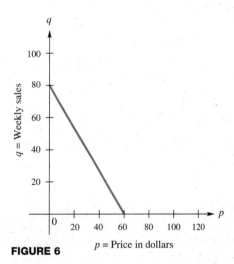

FIGURE 6

How should we interpret this? Since the slope is $-\frac{4}{3}$, we know that q, the quantity of rubies sold weekly, is decreasing by $\frac{4}{3}$ of a ruby per \$1 increase in price. In other words, a rise in price of \$3 results in a drop in sales of 4 rubies

---

▼ *This demand is not unlimited. For instance, campus newspapers are sometimes given away, and yet piles of them are often left unclaimed.

per week. Since the q-intercept is 80, RRR would hand out only 80 rubies per week if it were to give them away! (We see again what can go wrong with a linear model if we push it to extremes.)

Q At what price would sales stop completely?

A Translating into mathematics, "Find p when $q = 0$." We put $q = 0$ and solve for p, getting

$$0 = -\frac{4}{3}p + 80,$$

or

$$\frac{4}{3}p = 80,$$

and

$$p = \frac{3}{4}(80) = 60.$$

Thus, if RRR were to raise the price to $60, it would be stuck with them! Since $60 is ridiculously cheap for a good ruby, these stones must be very poor in quality!

▼ **EXAMPLE 8**   Using a Demand Function to Calculate Revenue

Referring to the demand equation $q = -\frac{4}{3}p + 80$ for rubies in Example 7, calculate the weekly revenue as a function of price.

SOLUTION   The weekly revenue is the total amount of money Royal Ruby Retailers will make from their sales of rubies. To calculate this as a function of p, we assume that they are selling them at $\$p$, and then do a little accounting.

q rubies @ $\$p$ per ruby gives a total revenue of $q \cdot p = pq$

In other words,

$$\text{Revenue} = \text{Price} \cdot \text{Quantity}$$
$$R = pq.$$

This formula gives R, but not as a function of p, since there is still the term q on the right-hand side. To get R in terms of p alone, we substitute the formula for q into this equation.

$$R = pq$$
$$= p(-\frac{4}{3}p + 80)$$

Thus,

$$R(p) = -\frac{4}{3}p^2 + 80p$$

is the function we are after.

Before we go on... Notice that R is not a *linear* function of p, it is a *quadratic* function of p because of the p^2 term. (We shall be studying quadratic expressions in the next section.)

Q If RRR gives rubies away, it makes no money. We saw in the last example that, if it sells them at $60, it still makes no money, because it doesn't sell any. Therefore, there should be a price somewhere *in between* $0 and $60 at which RRR makes the *largest* revenue. What is that price, and how much money does it bring in?

A We just saw that the weekly revenue is given by

$$R(p) = -\frac{4}{3}p^2 + 80p.$$

We now need to know: *What value of p gives the largest value of R(p)?* We'll see how to find the answer quickly in the next section. Calculus also provides an easy way to find it. In the meantime, we can make a pretty good guess at the answer by drawing the graph of R vs. p and then locating the highest point. We start by making a table.

p	0	10	20	30	40	50	60
$R(p) = -\frac{4}{3}p^2 + 80p$	0	667	1067	1200	1067	667	0

We suspect from this table that the best price is $30, which will bring RRR a total revenue of $1,200 per week. To support this claim, take a look at the graph in Figure 7.

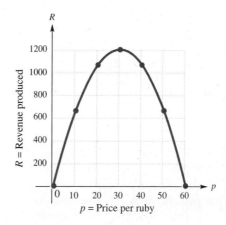

FIGURE 7

By the way, this curve is a parabola. (The next section is devoted to the study of parabolas.)

With a calculator or computer you can easily generate a table of revenues corresponding to lots of different prices, to get more points on this graph. Or, you can just use the calculator or computer to graph the equation $R = -\frac{4}{3}p^2 + 80p$ directly (whereupon it will generate its own very large table and plot large numbers of points to draw the graph). You will need to enter the equation as

$$Y_1 = -(4/3) X^2 + 80X$$

Notice that graphing calculators usually expect y to be given as a function of x, rather than p as a function of q. (Also, why have we inserted the parentheses, and what would happen if we omitted them?) You can use the ranges $0 \leq x \leq 60$ and $0 \leq y \leq 1300$. (You know that y will be between 0 and 1200, since we have already computed the values in the above table. In general, these ranges will not be given to you, and you will need to experiment with different ranges for x and y to obtain a suitable graph.)

Once you have plotted the curve, you can then find the price that gives the largest revenue by using the trace feature to move to the highest point of the graph—zooming in if you want more accuracy—and reading the x-coordinate from the display. (The y-coordinate gives the corresponding largest revenue.)

All the preceding examples illustrate the following principle.

GENERAL LINEAR MODELS

If $y = mx + b$ is a linear model of changing quantities x and y, then the slope m is the rate at which y is increasing per unit increase in x, while the y-intercept b is the value of y that corresponds to $x = 0$.

▶ **1.4 EXERCISES**

APPLICATIONS

1. *Cost* A piano manufacturer has a daily fixed cost of $1,200 and a marginal cost of $1,500 per piano. Find the daily cost $C(x)$ of manufacturing x pianos per day.

2. *Cost* A soft-drink manufacturer has a weekly fixed cost of $1,000 and a marginal cost of $5 per case of soda. Find the weekly cost $C(x)$ if the company produces x cases of soda per week.

3. *Demand* Sales figures show that your company sold 1,960 pen sets per week for $1 per pen set, and 1,800 pen sets per week for $5 per pen set. What is the linear demand function for your pen sets?

4. *Demand* A large department store is prepared to buy 3,850 of your neon-colored shower curtains per month for $5 apiece, but only 3,700 shower curtains per month for $10 apiece. What is the linear demand function for your neon-colored shower curtains?

5. *Revenue* The annual revenue of *United Airlines* increased from $8.50 billion in 1988 by approximately $0.95 billion per year for several years.*
   (a) Use these data to express *United's* annual revenue R (in billions of dollars) as a linear function of the number of years t since 1988.
   (b) Use your model to predict *United's* revenue in the year 2000.
   (c) Comment on the limitations of this model (if any).

6. *Profit* The annual profit of *United Airlines* decreased from $1,124 million in 1988 by approximately $520 million per year for several years.*
   (a) Use these data to express *United's* annual profit P (in millions of dollars) as a linear function of the number of years t since 1988.
   (b) Use your model to predict *United's* profit (or loss) in the year 2000.
   (c) Comment on the limitations of this model (if any).

7. *Fahrenheit and Celsius* In the Fahrenheit temperature scale, water freezes at 32°F and boils at 212°F. In the Celsius (or Centigrade) scale, water freezes at 0°C and boils at 100°C. Assuming that the Fahrenheit temperature f and the Celsius temperature c are related by a linear equation, find f in terms of c. Use your equation to find the Fahrenheit temperatures corresponding to 30°C, 22°C, −10°C, and −14°C.

8. *Fahrenheit and Celsius* Use the relationship between degrees Celsius and Fahrenheit you obtained in the previous exercise to obtain a linear equation for c in terms of f, and use your equation to find the Celsius temperatures corresponding to 100°F, 70°F, 10°F, and −40°F.

9. *Cost* The cost of renting tuxes for the Choral Society's concert is $20 down, plus $88 per tux. Express the cost C as a function of x, the number of tuxes rented, and graph $C(x)$. Use your function to answer the following questions.
   (a) What is the cost of renting 2 tuxes?
   (b) What is the cost of the second tux?
   (c) What is the cost of the 4,098th tux?
   (d) What is the marginal cost per tux?

10. *Income* The Enormous State University (ESU) Information Office pays student aides $3.50 per hour in addition to a daily meal allowance of $5.00 (even though the cafeteria charges $6.50 for a rather tasteless hamburger). Express a student aide's daily earnings q in terms of the number of hours h of work per day, and graph it. Use your equation to answer the following questions:
    (a) How much does a student aide earn in an eight-hour work day?
    (b) How much does a student aide earn in the eighth hour of an eight-hour work day?
    (c) What is the marginal earning rate for a student aide?
    (d) How long (to the nearest minute) must a student aide work in order to afford two cafeteria hamburgers?

11. *Income* The well-known romance novelist, Celestine A. Lafleur (a.k.a. Bertha Snodgrass), has decided to sell the screen rights to her latest book, *Henrietta's Heaving Heart*, to Boxoffice Success Productions, Inc. for $50,000. In addition, the contract assures Ms. Lafleur royalties of 5% of the net profits.[†] Express her income I as a function of the net profit N, and determine the net profit necessary to bring her an income of $100,000. What is her marginal income (share of each dollar of net profit)?

12. *Income* Due to the enormous success of the movie *Henrietta's Heaving Heart* based on a novel by Celestine A. Lafleur (see the preceding exercise), Boxoffice Success Productions Inc. decides to film the sequel, "Henrietta, Oh Henrietta." At this point, Bertha Snodgrass (whose novels now top the best-seller lists) feels she is in a position to demand $100,000 for the screen rights, and royalties of 8% of

---

▼ * Based on data published in *The New York Times*, December 24, 1993, p. D1. Source: Company Reports.
[†] Percentages of net profit are commonly called "monkey points." Few movies ever make a net profit on paper, and anyone with any clout in the business gets a share of the *gross*, not the net.

the net profits. Express her income I as a function of the net profit N, and determine the net profit necessary to bring her an income of $1,000,000. What is her marginal income (share of each dollar of net profit)?

13. *Cost* The RideEm Bicycles company can produce 100 bicycles in a day at a total cost of $10,500, and it can produce 120 bicycles in a day at a total cost of $11,000. What are the company's daily fixed costs, and what is the marginal cost to build a bicycle?

14. *Biology* The Snowtree cricket behaves in a rather interesting way: the rate at which it chirps depends linearly on the temperature. One summer evening you hear a cricket chirping at a rate of 140 chirps per minute, and you notice that the temperature is 80°F. Later in the evening the cricket has slowed down to 120 chirps per minute, and you notice that the temperature has dropped to 75°F. Express the temperature, T, as a function of the cricket's rate of chirping, r. What is the temperature if the cricket is chirping at a rate of 100 chirps per minute?

15. *Tax Depreciation* In calculating the value of its assets for tax purposes, a large law firm calculates that its office equipment, which cost them $60,000 new, is decreasing in value at a rate of 5% of the original price per year.
    (a) Express the total value v of the equipment as a function of n, the numbers of years after purchase.
    (b) How long will it take for the law firm's office equipment to depreciate to $1,000?
    (c) How long will it take for the law firm's office equipment to be worthless?
    (d) When does your linear model cease to be meaningful?

16. *Cost* Joe Silly's overdue library books (which he can never remember to return) are costing him a small fortune. For each overdue book, the Enormous State University (ESU) Library charges a $5 late fee, plus an additional $10 for each week a book remains overdue. Joe had 47 overdue books and his parents had $50,000 in the bank when the fines started mounting.
    (a) Express the total fine f for a *single* book as a function of the number w of overdue weeks.
    (b) Now express the total fine F on Joe Silly's 47 books as a function of w.
    (c) How long does Joe have before his overdue library fees wipe out his family's entire life savings?

17. *Muscle Recovery Time* Most workout enthusiasts will tell you that muscle recovery time is about 48 hours. But it is not quite as simple as that; the recovery time ought to depend on the number of sets you do involving the muscle group in question. For example, if you do no sets of biceps exercises, then the recovery time for your biceps is (of course) zero. To take a compromise position, let us assume that, if you do three sets of exercises on a muscle group, then its recovery time is 48 hours. Use this data to write a linear function that gives the recovery time (in hours) in terms of the number of sets affecting a particular muscle. Use this model to calculate how long it would take your biceps to recover if you did 15 sets of curls. Comment on your answer with reference to the usefulness of a linear model.

18. *Oil Reserves* The total estimated crude oil reserves (i.e., oil in the ground) in Saudi Arabia on January 1, 1989, were 255 billion barrels.[*] Further, Saudi Arabia produces on the order of 10 million (= .01 billion) barrels of crude oil per day. Use these data to give a function that estimates the crude oil reserves n years after the January 1, 1989 date, and use your equation to predict the Saudi Arabia crude oil reserves at the start of the year 2000. When (to the nearest year) will Saudi Arabia run out of crude?

19. *Profit Analysis—Aviation* The operating cost of a Boeing 747-100, which seats up to 405 passengers, is estimated to be $5,132 per hour.[†] If an airline charges each passenger a fare of $100 per hour of flight, find the hourly profit P it earns operating a 747-100 as a function of the number of passengers x (be sure to specify the domain). What is the least number of passengers it must carry in order to make a profit?

20. *Profit Analysis—Aviation* The operating cost of a McDonnell Douglas DC 10-10, which seats up to 295 passengers, is estimated to be $3,885 per hour.[†] If an airline charges each passenger a fare of $100 per

▼ [*] Source: *Oil and Gas Journal* (December 26, 1989).
[†] In 1992. Source: Air Transportation Association of America

hour of flight, find the hourly profit P it earns operating a DC 10-10 as a function of the number of passengers x (be sure to specify the domain). What is the least number of passengers it must carry in order to make a profit?

21. **Insurance Losses** *Allstate* Insurance Company charged an average of $360 per year for household insurance in Florida in 1992 (the year Hurricane Andrew struck the East Coast), and it reported that it had covered 1.1 million homes during that year.* Assume that *Allstate* paid an average of $50,000 per damaged home.
    (a) Find the profit $P(x)$ giving the total profit if x homes were damaged.
    (b) Given that its total losses were $2 billion that year, how many of the homes it covered were damaged?[†] What percentage is this of the total number of homes it insured?

22. **Insurance Losses** After Hurricane Andrew, *Allstate* applied for a Florida state increase of 30% for household insurance premiums and was seeking to reduce the total number of Florida homes covered to 750,000.*Repeat the preceding exercise to see what would have happened if these changes had gone into effect before 1992, but *Allstate* still lost $2 billion.

23. **Break-Even Analysis (based on a question from a CPA exam)** The Oliver Company plans to market a new product. Based on its market studies, Oliver estimates that it can sell up to 5,500 units in 1992. The selling price will be $2 per unit. Variable costs are estimated to be 40% of total revenue. Fixed costs are estimated to be $6,000 for 1992. How many units should the company sell to break even?

24. **Break-Even Analysis (based on a question from a CPA exam)** The Metropolitan Company sells its latest product at a unit price of $5. Variable costs are estimated to be 30% of the total revenue, while fixed costs amount to $7,000 per month. How many units should the company sell per month in order to break even, assuming that it can sell at up to 5,000 units per month at the planned price?

25. **Break-Even Analysis (from a CPA exam)** Given the following notations, write a formula for the break-even sales level.

    SP = Selling price per unit
    FC = Total fixed cost
    VC = Variable cost per unit

26. **Break-Even Analysis (based on a question from a CPA exam)** Given the following notation, write a formula for the total fixed cost.

    SP = Selling price per unit
    VC = Variable cost per unit
    BE = Break-even sales level in units

27. **Profit Analysis—Wireless Communications** You are the CEO of a new communications company that is interested in boosting subscriptions. Your monthly operating costs are calculated as $20 per customer and $10,000 overheads (fixed monthly costs). Your monthly income averages $50 per customer each month.
    (a) Find the profit function $P(x)$, where x is the total number of subscribers.
    (b) The U.S. Government hopes to auction a large amount of the nation's airwaves to cellular telephone companies for about $10 billion.[‡] This translates into about $40 per potential subscriber—or "pop" in the industry jargon. Assuming that your company has 1,000 subscribers at present, how many new "pops" could you afford to buy from the government at the start of next month at the estimated price of $40 in order to break even by the end of the month? (Assume that due to your large waiting list you will be able to begin immediate service to up to 4,000 new customers.)

28. **Profit Analysis—Automobile Manufacturing** You are the CEO of an automobile company that has just spent $6 billion developing a new "world car" planned for worldwide sales at a wholesale price of $14,000 per car.[§] Manufacturing costs amount to $12,000 per car, plus monthly overheads of $200,000.

▼ * Source: *The Chicago Tribune*, May 28, 1993, Section 3, p. 3.

† Allstate actually reported a $2.7 billion loss resulting from Hurricane Andrew.

‡ Such a plan was announced by the Federal Communications Commission in September, 1993. (Source: *The New York Times*, September 27, 1993, p. D1.)

§ This was the actual manufacturing cost of the Ford "world car" known variously as the Ford Mondeo, the Ford Contour and the Mercury Mistique. The planned sticker price was $16,000–$17,000 in the U.S. (Source: *The New York Times*, September 27, 1993, p. D1.)

(a) Ignoring development costs, find the profit function $P(x)$, where x is the total number of automobiles you manufacture each month (assuming that you can sell every car your company manufactures).

(b) How many "world cars" should your company manufacture in the coming year in order to cover the development costs by the end of the year?

29. *Population Growth* According to the U.S. Bureau of the Census, population growth in the U.S. has followed the declining pattern of most of the developed world, dropping from 55.2 live births per thousand in 1820 to 15.7 in 1987. Use these data to give a linear equation showing the annual number of live births b per thousand in terms of the number of years n since 1820, and use your model to predict when the rate will drop to zero (assuming the trend continues).

30. *Motion* On the day I got my learner's driving permit, I wanted to impress my friends with the speed at which I could maneuver a car in reverse along an 800-ft country road (at the end of which was a large mud pool). Assuming that at time $t = 0$ seconds I was at the 10-ft mark and that I subsequently maintained a speed of 30 mph (= 44 ft/sec), express my position s as a function of the time t in seconds. At what point in time did I wind up in the mud pool?

31. *Pollution* Radioactive pollutants are seeping into a lake at a rate of 3,000 gallons per year. At the start of 1992, it was estimated that the lake contained 2,500 gallons of pollutant. Find a function that gives the number of gallons p of pollutant in the lake n years since the start of 1992, and use your function to estimate the amount of pollutant at the start of 2001.

32. *Recycling* In Sweden, the problem of abandoned cars was addressed by an innovative policy introduced in 1976: Swedish car owners got a government bonus of $60 for selling their jalopies to a registered scrap dealer. The cost of this bonus was subsidized by a $60 surcharge on new cars. Since not every car ultimately wound up at a registered scrap yard, the government stood to make a profit from this arrangement. Here is a test scenario: let n be the number of new cars sold in Sweden per year, and let us assume that half that number of cars were scrapped per year. Express the annual surplus s accruing to the government as a function of n, and estimate the annual surplus if 50,000 new cars were sold per year.

33. *Ecology—Logging* In 1990, the U.S. Forest Service unveiled a plan calling for a reduction of logging in national forests from the then-current annual cut of 12.2 billion board feet to 10.8 billion board feet by the end of 1995.* Assuming that logging decreased linearly over this period, specify a function L, where $L(n)$ is the total cut (in billions of board feet) n years after 1990. Graph this function, and give the largest possible domain for which the model makes sense.

34. *Energy—Oil Reserves* The U.S. Strategic Oil Reserve in 1990 amounted to 590 million barrels of oil.* Assuming that it was tapped at a rate of 25 million barrels per year, specify a function R, where $R(n)$ is the total oil reserve (in millions of barrels) n years after 1990. Graph this function, and give the largest possible domain for which the model makes sense.

35. *Break-even Analysis—Organized Crime* The organized crime boss and perfume king Butch (Stinky) Rose has daily overheads (bribes to corrupt officials, motel photographers, wages for hitmen, explosives, etc.) amounting to $20,000 per day. On the other hand, he has a substantial income from his counterfeit perfume racket: he buys imitation French perfume (Chanel No. $22\frac{1}{2}$) at $20 per gram, pays an additional $30 per 100 grams for transportation, and sells it via his street thugs for $600 per gram. Specify Stinky's profit function, $P(x)$, where x is the quantity (in grams) of perfume he buys and sells, and use your answer to calculate how much perfume should pass through his hands per day in order that he break even.

36. *Break-even Analysis—Disorganized Crime* Butch (Stinky) Rose's counterfeit Chanel No. $22\frac{1}{2}$ racket has run into difficulties; it seems that the *authentic* Chanel No. $22\frac{1}{2}$ perfume is selling at only $500 per gram, whereas his street thugs have been charging $600 per gram, and his costs amount to $400 per gram plus $3,000 per 100 grams transportation costs. (The perfume's smell is easily detected by specially trained Chanel Hounds, and this necessitates elaborate packaging measures.) He therefore decides to price it at $420 per gram in order to undercut the competition. Specify Stinky's profit function, $P(x)$, where x is the quantity (in grams) of perfume he buys and sells, and use your answer to calculate how much perfume should pass through his hands per day in order that he break even. Interpret your answer.

---

* Source: Feb.-Mar. 1991 issue of *National Wildlife Magazine*.

37. *Demand and Revenue—Poultry* A linear model of the demand for chicken in the U.S. predicts that if the average per capita disposable income is $30,000, then

$$q = 65.4 - 0.45p + 0.12b,$$

where q is the per capita demand for chicken in pounds per year, p is the wholesale price of chicken in cents per pound, and b is the wholesale price of beef in cents per pound.*

(a) If the wholesale price of beef is fixed at 45¢ per pound, recast the above formula as a demand function for chicken in terms of the wholesale price per pound.

(b) Use your demand function to predict the demand for chicken (to the nearest pound) if the wholesale price is set at 20¢ per pound.

(c) At what price (to the nearest cent) does the model predict that the demand will drop to zero? Use this result to specify a domain for the demand function so that the demand can never be negative.

(d) Use your demand function from parts (a) and (c) to calculate the revenue function R, which specifies annual per capita revenue in terms of the wholesale price of chicken per pound.

38. *Demand and Revenue—Poultry* Repeat the preceding exercise, assuming that the wholesale price of beef is fixed at 60¢ per pound.

39. *Epidemics* The following chart shows the number of new cases of tuberculosis per 100,000 people in New York City since the 1920s.†

(a) Use the data from 1920 and 1980 to model the incidence of tuberculosis per 100,000 New York City residents as a linear function of time since 1920.

(b) What year gives the greatest discrepancy between your model and the actual data?

(c) Comment on the reliability of the linear model.

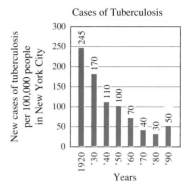

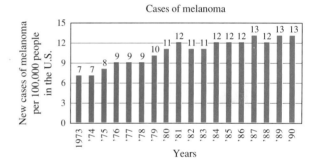

40. *Incidence of Melanoma* The following chart shows the number of new cases of melanoma per 100,000 people in the U.S. since 1973.‡

(a) Use the data from 1975 and 1990 to model the incidence of melanoma per 100,000 U.S. residents as a linear function of time since 1975.

(b) What year gives the greatest discrepancy between your model and the actual data?

(c) Comment on the reliability of the linear model.

41. *Demand and Revenue* You have been hired as a marketing consultant to Johannesburg Burger Supply, Inc., and you wish to come up with a demand equation for its hamburgers. In order to make life as simple as possible, you assume that the demand equation for Johannesburg hamburgers has the linear form

---

▼ * This equation is based on actual data from poultry sales in the period 1950–1984. In the "You're the Expert" section at the end of this chapter, we shall discuss this equation further. (Source: A. H. Studenmund, *Using Econometrics*, Second Edition (HarperCollins, 1992), pp. 180–181.)

† Sources: State Health Department, New York City Department of Health, Centers for Disease Control (New York Times, January 24, 1994, p. B1.)

‡ Source: American Cancer Society/New York Times (*The New York Times*, January 25, 1994, p. C3.)

$q = mp + b$, where p is the price per hamburger, q is the demand in weekly sales, and m and b are certain constants you'll have to figure out.

(a) Your market studies reveal the following sales figures: when the price is set at $2.00 per hamburger, the sales amount to 3,000 per week, but when the price is set at $4.00 per hamburger, the sales drop to zero. Use these data to calculate m and b and, hence, the actual demand equation.

(b) Use the demand equation obtained in (a) to estimate the number of hamburgers Johannesburg Burgers can unload per week if it gives them away.

(c) Use the equation to estimate Johannesburg's weekly revenue R if it were to sell the hamburgers at p dollars apiece.

(d) By tabulating values or otherwise, determine the price at which you would advise Johannesburg to sell its their hamburgers in order to maximize weekly revenue. What would this maximum revenue be?

42. *Demand and Revenue* Johannesburg Burger Supply, Inc. decides to market a new product: the "Fatfree Slim Thin" pork lard patty. Once again, you assume that the demand equation for Johannesburg patties should have the linear form $q = mp + b$, where p is the price per meat patty, q is the demand in weekly sales, and m and b are unknown constants.

(a) Your market studies reveal the following sales figures: when the price is set at $2.00 per pork lard patty, Johannesburg sells 6,000 per week, but when the price is set at $10.00 per pork lard patty, the sales plummet to zero. Use these data to calculate m and b and, hence, the actual demand equation.

(b) Use the demand equation obtained in (a) to estimate the number of pork lard patties Johannesburg can unload per week if it gives them away.

(c) Use the equation to estimate Johannesburg Burger's weekly revenue R if it were to sell the pork lard patties at p dollars apiece.

(d) By tabulating values or otherwise, determine the price at which you would advise Johannesburg to sell its patties in order to maximize weekly revenue. What would this maximum revenue be?

Exercises 43–46 use the idea of the **average cost**. If $C(x)$ is the total cost for x items, then the average cost per item for x items is

$$\overline{C}(x) = \frac{C(x)}{x}.$$

43. *Cost and Average Cost* A firm has monthly fixed costs of $30,000, variable (marginal) costs of $400 per item, and a manufacturing capacity of 1,000 items per month.

(a) Write the cost function $C(x)$, where x is the number of items produced per month.

(b) Write the average cost function (be sure to specify the domain).

(c) Find the production level (number of items produced per month) that gives an average cost of $450 per item.

44. *Cost and Average Cost* The RideEm Bicycles company can produce 100 bicycles in a day at a total cost of $10,500, and it can produce 120 bicycles in a day at a total cost of $11,000. Its maximum production capacity is 200 bicycles per day.

(a) Construct a linear cost function for RideEm Bicycles (be sure to specify the domain).

(b) Write the average cost function.

(c) How many bicycles should RideEm manufacture per day in order to meet an average cost goal of $75 per bicycle?

45. *Break-Even Analysis (based on a GRE economics exam question*)* A firm's average cost function is presented as

$$\text{Average cost} = 350 + \frac{9000}{Q}.$$

The revenue from one unit of output is $500. How many units of output must this firm sell to break even?

46. *Break-Even Analysis (based on a GRE economics exam question*)* A firm's average cost function is presented as

$$\text{Average cost} = 600 + \frac{900}{Q}.$$

The revenue from one unit of output is $700. How many units of output must this firm sell to make a profit of $100?

▼ * Source: *GRE Economics* by G. G. Gallagher, G. E. Pollock, W. J. Simeone and G. Yohe (Research and Education Association, Piscataway, N.J., 1989), p. 254.

COMMUNICATION AND REASONING EXERCISES

47. Describe some of the limitations of a linear demand model $q = mp + b$.

48. Describe some of the limitations of a linear cost function $C(x) = mx + b$.

49. If y and x are related by the linear expression $y = mx + b$, when should the quantity m be positive, when should it be negative, and when should it be zero?

50. If cost and revenue are expressed as linear expressions of the number of items x, what does the break-even quantity signify?

51. Suppose the cost function is $C(x) = mx + b$ (with m and b positive), the revenue function is $R(x) = kx$ ($k > m$) and the number of items is increased from the break-even quantity. Does this result in a loss, a profit, or is it impossible to say? Explain your answer.

52. You have been constructing a demand equation, and you obtained a (correct) expression of the form $p = mq + b$, whereas you would have preferred one of the form $q = mp + b$. Should you simply switch p and q in the answer, should you start again from scratch, using p in the role of x and q in the role of y, or should you solve your demand equation for q? Give reasons for your decision.

53. Come up with an interesting application leading to the linear model $s = -100k + 200$.

54. Come up with an interesting application leading to the linear model $s = 100k - 200$.

55. Explain intuitively why there is always a unit price that results in the maximum revenue, given a linear demand equation with negative slope.

56. Explain why there are always at least two unit prices that result in zero revenue, given a linear demand equation with negative slope.

▶ 1.5 QUADRATIC FUNCTIONS AND MODELS

We saw at the end of the preceding section that a linear demand function gives rise to a *quadratic* revenue function. Quadratic functions are the next most complicated functions after the linear ones, and they will be very useful as examples as we study calculus.

> **QUADRATIC FUNCTION**
> A **quadratic function** is a function of the form
> $$f(x) = ax^2 + bx + c \quad \text{(where } a \neq 0\text{)}.$$

▼ **EXAMPLE 1**

Sketch the graph of $y = x^2$.

SOLUTION We first construct a table with sufficiently many points to give us an idea of what the graph will look like.

x	−3	−2	−1	0	1	2	3
$y = x^2$	9	4	1	0	1	4	9

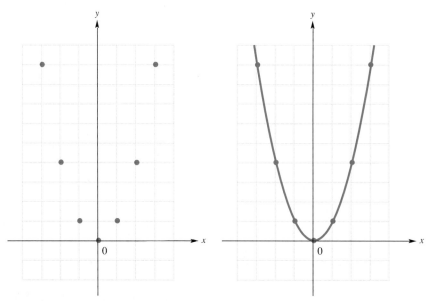

FIGURE 1

Plotting these points gives the picture on the left in Figure 1, suggesting the curve on the right.

This curve is called a **parabola,** and its lowest point, at the origin, is called its **vertex.**

 With a calculator or computer, you can generate this graph using the format

$$Y_1 = X\wedge 2.$$

(The caret symbol "^" is the standard graphing calculator and computer symbol for raising to a power.)

Any quadratic function $f(x) = ax^2 + bx + c$ ($a \neq 0$) has a parabola as its graph (possibly upside-down, and with its vertex not necessarily at the origin). These curves have the general shape shown in Figure 2.

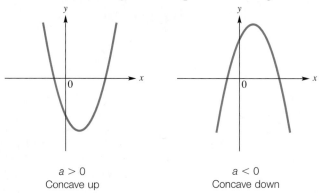

FIGURE 2

1.5 Quadratic Functions and Models

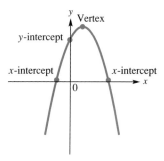

FIGURE 3

In this section, we present an easy method of sketching by hand the graph of any quadratic function, based on the features shown in Figure 3. If you are using a graphing calculator or computer program, think about how the following summary explains the picture the device gives you.

DETERMINING THE FEATURES OF A PARABOLA

The graph of $f(x) = ax^2 + bx + c$ is a parabola with the following features.

Vertex The x-coordinate of the vertex is $-b/(2a)$. The y-coordinate is $f(-b/(2a))$.

x-Intercepts (if any) These occur when $f(x) = 0$, or when
$$0 = ax^2 + bx + c.$$
We can solve this equation for x using the quadratic formula. Thus, the x-intercepts are
$$x = \frac{-b \pm \sqrt{b^2 - 4ac}}{2a}.$$
If the discriminant $\Delta = b^2 - 4ac$ is positive, there are two x-intercepts. If it is zero, there is a single x-intercept (at the vertex). If it is negative, there are no x-intercepts (so the parabola doesn't touch the x-axis at all).

y-Intercept This occurs when $x = 0$. So
$$y = a \cdot 0^2 + b \cdot 0 + c = c.$$

Symmetry The parabola is symmetric with respect to the vertical line through the vertex, which is the line $x = -b/(2a)$.

We shall not fully justify the formula for the vertex (and the axis of symmetry), but notice the following. If there are two x-intercepts, they are
$$x = -\frac{b}{2a} - \frac{\sqrt{b^2 - 4ac}}{2a}$$

and
$$x = -\frac{b}{2a} + \frac{\sqrt{b^2 - 4ac}}{2a}.$$

These should be symmetric around the axis of symmetry, but the formulas make it clear that they are symmetric around the line $x = -b/(2a)$. So this must be the axis of symmetry, and the vertex is located on this line as well. You can see the symmetry clearly in the next example. (We shall give a better justification after we have studied some calculus.)

▼ **EXAMPLE 2**

Sketch the parabola $f(x) = x^2 + 2x - 8$.

SOLUTION Here, $a = 1, b = 2$, and $c = -8$. Since $a > 0$, the parabola is concave up (Figure 4).

Vertex The x-coordinate of the vertex is $x = -b/(2a) = -2/2 = -1$. To get its y-coordinate, we substitute this value back into $f(x)$ to get $y = f(-1) = (-1)^2 + 2(-1) - 8 = 1 - 2 - 8 = -9$. Thus, the coordinates of the vertex are $(-1, -9)$.

x-Intercepts To calculate the x-intercepts (if any), we solve the equation $x^2 + 2x - 8 = 0$. Luckily, this factors as $(x + 4)(x - 2) = 0$. Thus, the solutions are $x = -4$ and $x = 2$, giving these values as the x-intercepts.

y-Intercept The y-intercept is given by $c = -8$.

Symmetry The graph is symmetric around the vertical line $x = -1$.

Now we can sketch the curve as in Figure 5.

FIGURE 4

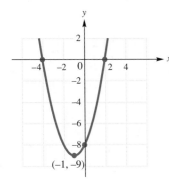

FIGURE 5

▼ **EXAMPLE 3**

Sketch the parabola $f(x) = -x^2 - 2x + 1$.

SOLUTION Here, $a = -1, b = -2$, and $c = 1$. Since $a < 0$, the parabola is concave down (Figure 6).

Vertex The x-coordinate of the vertex is $x = -b/(2a) = 2/-2 = -1$, and substitution gives its y-coordinate as $y = f(-1) = -(-1)^2 - 2(-1) + 1 = 2$. Thus, the vertex is the point $(-1, 2)$.

x-Intercepts The x-intercepts are the solutions to $-x^2 - 2x + 1 = 0$. This quadratic does not factor nicely, so we use the quadratic formula to obtain

$$x = -\frac{1}{2}(2 \pm \sqrt{8}) = -\frac{1}{2}(2 \pm 2\sqrt{2}) = -1 \pm \sqrt{2}.$$

Thus, the x-intercepts are $-1 + \sqrt{2}$ and $-1 - \sqrt{2}$. In order to plot them, we approximate $\sqrt{2}$ by 1.4. (We don't need much accuracy when we sketch

FIGURE 6

this curve, as we only want its general features.) Thus, the *x*-intercepts are approximately 0.4 and −2.4.

***y*-Intercept** The *y*-intercept is $c = 1$.

Symmetry The graph is symmetric around the vertical line $x = -1$.

The graph is shown in Figure 7.

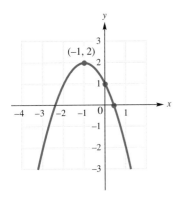

FIGURE 7

EXAMPLE 4

Sketch the graph of $f(x) = 4x^2 - 12x + 9$.

SOLUTION We have $a = 4$, $b = -12$, and $c = 9$. Since $a > 0$, this parabola is concave up.

Vertex The *x*-coordinate of the vertex is given by $x = -b/(2a) = \frac{12}{8} = \frac{3}{2}$. Substituting to get the *y*-coordinate gives $y = f(\frac{3}{2}) = 4(\frac{3}{2})^2 - 12(\frac{3}{2}) + 9 = 0$. Thus, the vertex is at the point $(\frac{3}{2}, 0)$.

x*-Intercepts To get the *x*-intercepts, we must solve $4x^2 - 12x + 9 = 0$. This quadratic factors as $(2x - 3)^2 = 0$, so the only solution is $2x - 3 = 0$, or $x = \frac{3}{2}$. Note that this coincides with the vertex, which also lies on the *x*-axis.

y*-Intercept The *y*-intercept is $c = 9$.

Symmetry The graph is symmetric around the vertical line $x = \frac{3}{2}$.

The graph is the very narrow parabola shown in Figure 8.

Why is the parabola so narrow? One way you can find out experimentally is by varying the coefficients of the equation $y = 4x^2 - 12x + 9$ and plotting them on the same set of axes. For instance, you could vary the coefficient of x^2 by plotting

$$Y_1 = 4X^2 - 12X + 9$$
$$Y_2 = 3X^2 - 12X + 9$$
$$Y_3 = 2X^2 - 12X + 9$$

to examine the effect of this coefficient on the shape of the graph. You could then vary the coefficient of *x*, and finally, the constant.

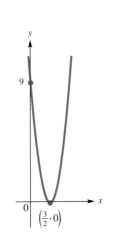

FIGURE 8

EXAMPLE 5

Sketch the graph of $f(x) = -\frac{1}{2}x^2 + 4x - 12$.

SOLUTION Here, $a = -\frac{1}{2}$, $b = 4$, and $c = -12$. Since $a < 0$, the parabola is concave down.

Vertex The vertex has *x*-coordinate $-b/(2a) = -4/-1 = 4$, with corresponding *y*-coordinate $f(4) = -\frac{1}{2}(4)^2 + 4(4) - 12 = -4$. Thus, the vertex is at $(4, -4)$.

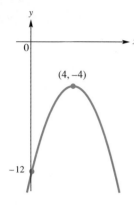

FIGURE 9

x-Intercepts For the x-intercepts, we must solve $-\frac{1}{2}x^2 + 4x - 12 = 0$. But in doing so, we discover that the discriminant Δ is $b^2 - 4ac = 16 - 24 = -8$. Since this is negative, there are no solutions of the equation, so there are no x-intercepts.

y-Intercept The y-intercept is given by $c = -12$.

Symmetry The graph is symmetric around the vertical line $x = 4$.

Since there are no x-intercepts, the graph lies entirely below the x-axis, as shown in Figure 9.

We can now put our study of parabolas to use.

▼ **EXAMPLE 6** Demand and Revenue

A publishing company predicts that the demand equation for the sale of its latest romance novel is

$$q = -2{,}000p + 150{,}000,$$

where q is the number of books it can sell per year at a price of $\$p$ per book. What price should it charge in order to obtain the maximum annual revenue?

SOLUTION The total revenue depends on the price, as follows.

$$\begin{aligned} R &= pq \\ &= p(-2{,}000p + 150{,}000) \\ &= -2{,}000p^2 + 150{,}000p \end{aligned}$$

We are after the price p that gives the largest possible revenue. Notice that what we have is a quadratic function of the form $R(p) = ap^2 + bp + c$, where $a = -2{,}000$, $b = 150{,}000$, and $c = 0$. Since a is negative, the graph of the function is a parabola, concave down, and so its vertex is its highest point. The p-coordinate of the vertex is

$$p = -\frac{b}{2a} = -\frac{150{,}000}{-4{,}000} = 37.5.$$

This value of p gives the highest point on the graph, and thus gives the largest value of $R(p)$. We may conclude that the company should charge $\$37.50$ per book to maximize its annual revenue.

Before we go on... You might ask what the maximum annual revenue is. The answer is supplied for us by the revenue function $R(p) = -2{,}000p^2 + 150{,}000p$. Since we have $p = 37.5$, we can substitute this value into the equation and obtain $R(37.5) = -2{,}000(37.5)^2 + 150{,}000(37.5) = 2{,}812{,}500$. In other words, the company will earn total annual revenues from this book amounting to $\$2{,}812{,}500$. Not bad!

1.5 Quadratic Functions and Models

EXAMPLE 7 Break-Even Analysis

As the operator of Workout Fever Health Club (see Example 6 in the preceding section) you calculate your demand equation to be

$$q = -\frac{3}{50}p + 46,$$

where q is the number of new members who join the club per month, and p is the annual membership fee you charge.

(a) Since you are running a shoestring operation, your annual operating costs amount to only $5,000 per year. At what price should you set annual memberships in order to break even?

(b) How would the situation change if your operating costs went up to $10,000 per year?

SOLUTION

(a) First recall from the preceding section that break-even occurs when total revenue equals total cost. The annual revenue is given by

$$R = pq$$
$$= p\left(-\frac{3}{50}p + 46\right)$$
$$= -\frac{3}{50}p^2 + 46p$$

while the annual cost C is fixed at $5,000. Thus, for break-even,

$$R = C,$$

or

$$-\frac{3}{50}p^2 + 46p = 5,000$$

or

$$-\frac{3}{50}p^2 + 46p - 5,000 = 0.$$

Note that the second equation above has the form

$$R - C = 0.$$

In other words,

$$\text{Profit} = 0.$$

We now have a quadratic equation in p, and we must solve it for the break-even price p. To simplify the equation, we can first multiply both sides by the bothersome 50 in the denominator, getting

$$-3p^2 + 2{,}300p - 250{,}000 = 0.$$

Now solve, using the quadratic formula to obtain

$$p = 635.55 \quad \text{or} \quad 131.12.$$

Although you may be tempted to choose the larger figure on the grounds that you will be getting more in membership fees, remember that both these options will result in your operation breaking even—there is no advantage to one or the other. Thus, to break even, you should charge either $635.55 or $131.12 per year.

(b) If we repeat the above analysis with the $5,000 fixed cost replaced by $10,000, we wind up with the quadratic equation

$$-3p^2 + 2{,}300p - 500{,}000 = 0.$$

This equation has no real solutions since the discriminant, $b^2 - 4ac$, is negative. You simply cannot break even with these costs, so you might as well close down the operation.

You can answer both parts of the question by using a graphing calculator or computer program to sketch the revenue and cost functions. We had $R = -\frac{3}{50}p^2 + 46p$, and $C = 5{,}000$ for part (a) and $C = 10{,}000$ for part (b). If you plot all three functions on the same coordinate system, you get the picture shown in Figure 10.

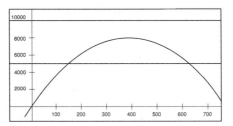

FIGURE 10

The horizontal lines represent the two fixed annual costs, while the parabola is the graph of the revenue function. The two intersection points are the break-even points we calculated in part (a). You can zoom and trace to obtain the coordinates of the break-even point and confirm the calculations we did. The higher cost line fails to touch the revenue curve, confirming our conclusion in part (b) that the health club cannot break even with annual costs of $10,000.

The graph also shows the price you should charge to obtain the largest profit. Trace to the highest point of the parabola (which has a p-coordinate midway between the break-even points) and you will find that in order to obtain the largest revenue, and hence profit, you should charge an annual fee of $383.33. You can then find the annual revenue predicted by the demand equation as the y-coordinate.

1.5 EXERCISES

Sketch the graphs of the quadratic functions given in Exercises 1–12, indicating the coordinates of the vertex, the y-intercept and the x-intercepts (if any).

1. $f(x) = x^2 + 3x + 2$
2. $f(x) = -x^2 - x - 12$
3. $f(x) = x^2 + x - 1$
4. $f(x) = x^2 + \sqrt{2}x + 1$
5. $f(x) = \frac{1}{4}x^2 + \sqrt{2}x - 1$
6. $f(x) = -\frac{1}{3}x^2 + 3x - 1$
7. $f(x) = x^2 + 2x + 1$
8. $f(x) = -x(3x + 2)$
9. $f(x) = x^2$
10. $f(x) = -x^2$
11. $f(x) = x^2 + 1$
12. $f(x) = -x^2 + 5$

For each of the demand equations in Exercises 13–16, express the total revenue R as a function of the price p per item, sketch the graph of the resulting function, and determine the price p that maximizes total revenue in each case.

13. $q = -4p + 100$
14. $q = -3p + 300$
15. $q = -2p + 400$
16. $q = -5p + 1200$

APPLICATIONS

17. *Revenue* The market research department of the Better Baby Buggy Co. notices the following. When its buggies are priced at $80 each, it can sell 100 each month. However, when the price is raised to $100, it can only sell 90 each month. Assuming that the demand is linear, at what price should it sell the buggies to get the largest revenue? What is the largest monthly revenue?

18. *Revenue* The Better Baby Buggy Co. has just come out with a new model, the Turbo. The market research department now estimates that the company can sell 200 Turbos per month at $60, but only 120 per month at $100. Assuming that the demand is linear, at what price should it sell its buggies to get the largest revenue? What is the largest monthly revenue?

19. *Revenue* Pack-Em-In Real Estate is building a new housing development. The more houses it builds, the lower the price people will be willing to pay, due to the crowding and smaller lot sizes. In fact, if the company builds 40 houses in this particular development, it can sell them for $200,000 each, but if it builds 60 houses, it will only be able to get $160,000 each. Assuming that the demand is linear, how many houses should the company build in order to get the largest revenue? What is the largest possible revenue?

20. *Revenue* Pack-Em-In has another development in the works. If it builds 50 houses in this development, it will be able to sell them at $190,000 each, but if it builds 70 houses, it will only get $170,000 each. Assuming that the demand is linear, how many houses should it build in order to get the largest revenue? What is the largest possible revenue?

21. *Revenue (computer or graphing calculator recommended)* The wholesale price for chicken in the U.S. fell from 25¢ per pound in 1951 to 14¢ per pound in 1958. At the same time, per capita chicken consumption rose from 21.7 pounds per year to 28.1 pounds per year.*
    (a) Use these data to set up a linear demand equation for poultry. (Round all decimals to four places.)
    (b) Use your demand equation to express the wholesale revenue per capita as a function of the price p of poultry per pound.
    (c) Calculate, to the nearest cent, the price per pound that should result in the largest per capita wholesale revenue.

---

*Source: U.S. Department of Agriculture, *Agricultural Statistics*. Also see the next footnote.

22. *Revenue (computer or graphing calculator recommended)* Repeat Exercise 21, given that the wholesale price for chicken in the U.S. fell from 10¢ per pound in 1962 to 9¢ per pound in 1968. At the same time, per capita chicken consumption rose from 30.8 pounds per year to 41.8 pounds per year.*

23. *Fuel Efficiency* The fuel consumption of an automobile engine increases with speed. If the latest model Guzzler is driven at 30 mph, it burns a gallon of gas in $\frac{1}{2}$ hour. If it is driven at 90 mph it burns a gallon in $\frac{1}{6}$ hour. Assuming that the time it takes to burn a gallon of gas depends linearly on the speed, find the speed at which the fuel efficiency (in miles/gallon) is highest. What is the highest fuel efficiency that the Guzzler can reach?

24. *Fuel Efficiency* If the latest model Sipper is driven at 25 mph, it takes $1\frac{1}{2}$ hours to burn a gallon of gas. If it is driven at 50 mph, it takes 1 hour to burn a gallon. Assuming that the time it takes to burn a gallon of gas depends linearly on the speed, find the speed at which the fuel efficiency (in miles/gallon) is highest. What is the highest fuel efficiency that the Sipper can reach?

25. *Break-Even Analysis* You are the only supplier of beef in a town where the demand equation for beef is given by

$$q = 500 - 40p.$$

Here, q is the number of pounds of beef consumed per month in your town, and p is the price of beef in dollars per pound.† Assume that you have stockpiled a large amount of frozen beef and wish to set the price in order to break even. Your fixed costs for storage and refrigeration amount to $1,000 per month. What is the most you could charge per pound?

26. *Break-Even Analysis* Repeat Exercise 25 using the demand equation $q = 400 - 30p$.

27. *Motion Under Gravity* If a ball is tossed straight up from ground level ($h = 0$) at a velocity of v_0 feet per second, its height h (in feet) after t seconds will be

$$h = v_0 t - 16t^2.$$

(This formula neglects the effect of air resistance.)
(a) With $v_0 = 64$, sketch the graph of h as a function of t, and use your graph to determine how long it will take the ball to reach the ground.
(b) True or false: If v_0 is doubled, then the time the ball is airborne is also doubled. Justify your answer.

28. *Motion Under Gravity* If a stone is thrown down a shaft from ground level ($d = 0$) at a velocity of v_0 feet per second, its depth d (in feet) after t seconds will be

$$d = v_0 t + 16t^2.$$

(This formula neglects the effect of air resistance.)
(a) With $v_0 = 2$, sketch the graph of h as a function of t, and use your graph to determine how long it will take the ball to reach a depth of 100 feet.
(b) True or false: If v_0 is doubled, then the time it takes the ball to reach a depth of 100 feet is halved. Justify your answer.

COMMUNICATION AND REASONING EXERCISES

29. Suppose the graph of revenue as a function of unit price is a parabola that is concave down. What are the significance of the coordinates of the vertex, the (possible) x-intercepts, and the y-intercept?

30. Suppose the height of a stone thrown vertically upward is given by a quadratic function of the time. What are the significance of the coordinates of the vertex, the (possible) x-intercepts, and the y-intercept?

▼ *Source: U.S. Department of Agriculture, *Agricultural Statistics*. If you happen to do both Exercises 21 and 22, you will notice that entirely different demand equations result. This is a further illustration of the limitations of applying linear modeling to every pair of data points in sight. It turns out that the case study from which these figures are quoted had to take into account not only the effect of the price of poultry, but also the effect of the price of *beef* on the demand for poultry. It also had to take into account the U.S. per capita disposable income! See "You're the Expert" at the end of this chapter.

† Source: *Using Econometrics: A Practical Guide,* by A. H. Studenmund (New York: HarperCollins 1992). The equation was obtained from actual statistics compiled by the U.S. Department of Agriculture. We have adapted the figures to match the scenario of the exercise.

31. Explain why, if demand is a linear function of unit price p (with negative slope) then there must be a *single value* of p that results in a maximum revenue.

32. Explain why, if the average cost of a commodity is given by $y = 0.1x^2 - 4x - 2$, where x is the number of units sold, there is a single choice of x that results in the lowest possible average cost.

33. If the revenue function for a particular commodity is $R(p) = -50p^2 + 60p$, what is the (linear) demand function? Give a reason for your answer.

34. If the revenue function for a particular commodity is $R(p) = -50p^2 + 60p + 50$, can the demand function be linear? What is the associated demand function?

1.6 SOLVING EQUATIONS USING GRAPHING CALCULATORS OR COMPUTERS

In this section we'll have a short discussion on the use of graphing calculators to solve equations in a single unknown (such as $x^2 - 4\sqrt{x} = 0$, or $x^3 - 4x + 1 = 0$). In the language of functions, this amounts to **finding zeros of functions:** finding values of x for which $f(x) = 0$. (For instance, solving $x^3 - 4x + 1 = 0$ is the same as finding the zeros of $f(x) = x^3 - 4x + 1$.) If you will not be using a graphing calculator or similar computer software, you can safely skip this section.

Before we start, we should point out that there are two methods of solving an equation: analytical and numerical. To solve an equation **analytically** means to obtain exact solutions using algebraic techniques. (The Algebra Review in Appendix A has several sections dealing with the analytic solution of equations.) To solve an equation **numerically** means to use a graphing calculator or computer program to obtain *approximate* solutions. Although numerical solutions are only approximations of true solutions, we can calculate them as accurately as we want. Further, some equations can be solved analytically only with great difficulty, and some cannot be solved analytically at all. Often, numerical solution is the best we can do.

Most standard graphing calculators come equipped with "trace" and "zoom" features. The trace feature allows you to move a cursor along the displayed graph and gives you the coordinates of the points as you go. The zoom feature lets you magnify a portion of the curve. Now, in any process of approximation it is important to have some idea of how accurate your answer is. In this regard the trace feature is misleading—it can fool you into thinking that your answer is more accurate than it is. We shall rely more on the zoom feature to keep track of the accuracy of our answers.

▼ **EXAMPLE 1**

Use a graphing calculator to solve the equation $3x^3 - x + 1 = 0$. The solution(s) should be accurate to within ± 0.05 (that is, accurate to one decimal place).

SOLUTION Begin by using your calculator to graph the equation $y = 3x^3 - x + 1$. If you need to specify x- and y-ranges, start by specifying x

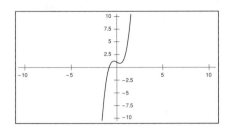

FIGURE 1

between -10 and 10, and do the same for y. Figure 1 shows the kind of picture you should get on your calculator display.

You are looking for a solution of $3x^3 - x + 1 = 0$. Since you have graphed $y = 3x^3 - x + 1$, you are looking for a point on the graph where $y = 0$. That is, you are looking for a point where the graph crosses the x-axis. Looking at the graph, notice that it crosses the x-axis at exactly one point, somewhere between -2 and 0. This observation tells you two things. First, there is only one real solution to the equation $3x^3 - x + 1 = 0$. (If there were another solution, you would see the curve crossing the x-axis again.) Second, this solution is somewhere between -2 and 0.

As your first estimate of the solution, take the point midway in this range $[-2, 0]$, that is, $x = -1$. Since you are not sure exactly where in this range the solution x lies, all you can say with certainty is that $x = -1$ with a possible error of 1 unit in either direction, that is, $x = -1 \pm 1$. (See Figure 2.)

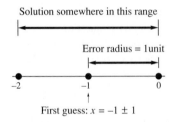

FIGURE 2

To get a more accurate estimate of where this solution occurs, "zoom in" by using your calculator's "zoom" feature or by redrawing the graph specifying the x-range as $-2 \le x \le 0$, since this is where you know the solution lies. As for the y-range, you can use any range that includes zero, say $-1 \le y \le 1$. Figure 3 shows what the output should look like.

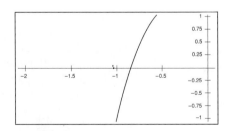

FIGURE 3

Now you see that the solution lies somewhere between -1 and -0.5, so as your next estimate, take the midpoint, -0.75, of this range. Since the width of the interval $[-1, -0.5]$ is 0.5 units, your estimate is accurate to within plus

or minus half that, or 0.25. Thus, your second estimate is
$$x = -0.75 \pm 0.25.$$

Q Wait! My graphing calculator does not put numbers on the axes as shown in the diagram. How can I tell that the curve crosses the x-axis between -1 and -0.5?

A You can use the scale feature to place tick marks on the axes as shown in the figures. Alternatively, you can use the trace feature as follows.

1. Place the cursor on any point of the curve to the left of the point of intersection with the x-axis, and read the value of x from the display. Call this value a. If—as often occurs when you use trace—the point a is a messy decimal such as $a = -0.9874512$, you can choose any convenient *smaller* number as a, for instance, $a = -1$. (Why smaller?)

2. Now trace to any point of the curve to the *right* of the point of intersection with the x-axis, and again read the value of x. Call this value b. If the point b is a messy decimal such as $b = -0.0821348$, you can choose any convenient *larger* number as b, for instance, $b = -0.08$. (Why larger?)

You can now see that the solution lies somewhere in the interval $[a, b]$. In fact, this technique can get you to the desired accuracy rather quickly.

The error is still too big, so you can't stop here. (Remember that you are looking for an accuracy of ± 0.05 or less.) So "zoom in" once again, using the x-range $-1 \leq x \leq -0.5$ and a smaller y-range if you like, say $-0.5 \leq y \leq 0.5$. (Figure 4)

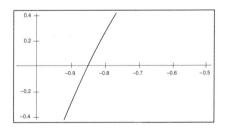

FIGURE 4

Now we are getting somewhere. According to the graph, the solution is somewhere between -0.9 and -0.8. Thus, as your next estimate, choose the midpoint -0.85, with a possible error of ± 0.05, since this is half the length of the interval $[-0.9, -0.8]$. This is the accuracy you needed, so you are done: the solution of $3x^3 - x + 1 = 0$ is $x = -0.85$, to within ± 0.05.

EXAMPLE 2

Using a graphing calculator, find all zeros of the function $f(x) = x^5 - 2x^2 + 1$, to within ± 0.02.

SOLUTION Recall that to find the zeros of $f(x)$ means to solve the equation $f(x) = 0$. Begin by having the calculator draw the graph of $y = x^5 - 2x^2 + 1$ in the range $-10 \leq x \leq 10$, and do the same for y. The output is shown in Figure 5.

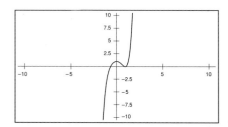

FIGURE 5

This time, it looks as though there are several solutions, one between -2 and 0, and one or two between 0 and 2. To get a better view, zoom in to $-2 \leq x \leq 2$, and do the same for y. (Figure 6)

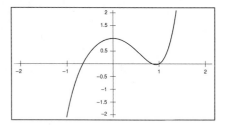

FIGURE 6

You can see now that there are three solutions. You will have to zoom in on each of the zeros separately to get the desired accuracy. We shall find the middle one, and leave the others to you! Looking at the last figure, notice that the solution second from the right is somewhere between 0.75 and 1. So zoom in there, getting Figure 7.

This figure shows the solution as slightly to the left of 0.85, and definitely in the range $0.8 \leq x \leq 0.9$. Thus, you get the estimate $x = 0.85 \pm 0.05$. Although you suspect it to be accurate to within ± 0.02 (why?) zoom in once more to confirm this (Figure 8).

1.6 Solving Equations using Graphing Calculators or Computers

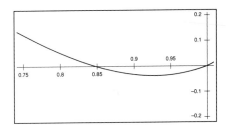

FIGURE 7

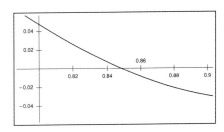

FIGURE 8

Now you can be absolutely certain that the solution is 0.85 ± 0.01, which is even more accurate than what was required!

Q We know how to locate the zeros of a function f—values of x for which $f(x) = 0$. What if we need to locate values of x for which $f(x) = g(x)$, where g is some other function?

A There are two ways to do this, and we illustrate with the example $f(x) = x^5 - x^2 + 2$, and $g(x) = \sqrt{4 - x^2}$.
1. Notice that solving
    $$x^5 - x^2 + 2 = \sqrt{4 - x^2}$$
    is the same as solving
    $$x^5 - x^2 + 2 - \sqrt{4 - x^2} = 0,$$
    and we already know how to locate zeros of functions such as this.
2. Alternatively, graph both $y = x^5 - x^2 + 2$ and $y = \sqrt{4 - x^2}$ on the same set of axes, and find values of x where the graphs cross. (This is what we did in our discussion of break-even analysis.)

1.6 EXERCISES

Use a graphing calculator to solve the equations in Exercises 1–8 to the specified accuracy.

1. $x^2 + 2x - 5 = 0$, to within ± 0.05
2. $-x^2 + 5x + 12 = 0$, to within ± 0.05
3. $-x^3 - 2x^2 + x - 1 = 0$, to within ± 0.01
4. $x^3 - 2x^2 + x - 1 = 0$, to within ± 0.01
5. $x^5 - 10x + 5 = 0$, to within ± 0.001
6. $x^5 - 16x - 1 = 0$, to within ± 0.001
7. $x^7 - x^5 + x - 2 = 0$, to within ± 0.02
8. $3x^7 - 2x^3 + x = 0$, to within ± 0.02

Use a graphing calculator to locate all zeros of each of the functions in Exercises 9–12 to the specified accuracy.

9. $f(x) = x^2 + 1/x - 4x$, to within ± 0.001
10. $f(x) = x^2 - 1/x - 5$, to within ± 0.001
11. $f(x) = x^5 - x - 3$, to within ± 0.05
12. $f(x) = x^4 - 3x^2 - x$, to within ± 0.05

Locate all values of x that satisfy the equations in Exercises 13–16.

13. $2^x = x$, to within ± 0.05
14. $3^x = 4x^2$, to within ± 0.05
15. $\dfrac{x^2 + 1}{x^2 - 1} = 2x - \sqrt{x}$, to within ± 0.05
16. $(x - 1)^{2/3} = 2^x$, to within ± 0.05

APPLICATIONS

17. *Dental Plans* A company pays for its employees' dental coverage at an annual cost C given by
$$C(q) = 1{,}000 + 100\sqrt{q},$$
where q is the number of employees covered, and $C(q)$ is the annual cost in dollars. If the government subsidizes coverage by an annual dollar amount of
$$S(q) = 50q,$$
at what number of employees does the company actually start making a profit on dental coverage?

18. *Surface Area* The surface area of a hollow cylinder of cross-sectional radius r and height h is given by $S = 2\pi rh$. Further, the surface area of a disc of radius r is given by $A = \pi r^2$. (See the diagram on page 39 for this exercise.)

You are the design manager for Pebbles and Blips gourmet cat food, and you wish to design the shape of the cans that are to hold Pebbles and Blips Lobster Delight. Your can must hold 20 cubic inches (the volume of a can is given by $V = \pi r^2 h$). Further, the alloy you plan to use to manufacture the cans will cost 4¢ per square inch. Express the cost of a single can as a function of the radius r, and determine at what radius the can will use $2.50 worth of metal to build. (Give your answer to the nearest 0.01 in.)

19. *Investments* If you invest $2,000 at interest rate r compounded monthly, the amount of money you will have after 10 years is given by the function
$$A(r) = 2{,}000\left(1 + \frac{r}{12}\right)^{120}.$$
What must the interest rate be in order for you to double your money in 10 years? (Give your answer to the nearest 0.01%.)

20. *Radioactive Decay* Carbon 14 is radioactive and decays over time to nitrogen. If you start with 10 g of Carbon 14, after t years you will still have
$$A(t) = 10\left(\frac{1}{2}\right)^{t/5{,}730}$$
grams. How long will it take your original 10 g to decay to 1 g? (Give your answer to the nearest year.)

You're the Expert

MODELING THE DEMAND FOR POULTRY

A government agency plans to regulate the price of poultry with the goal of increasing revenue to poultry producers. Economists developed the demand equation

$$q = 56.9 - 0.45p + 0.12b$$

in which q is the annual per capita demand for poultry in pounds per year, p is the wholesale price of poultry in cents per pound, and b is the wholesale price of beef in cents per pound.*

The agency has hired you to develop a pricing policy and to analyze the effect that fluctuations in the beef price would have on revenue to poultry producers under your proposed policy.

You observe immediately that the demand for poultry depends on the price of beef as well as the price of poultry. You decide that the pricing policy should "tie" the poultry price to the beef price in some way. You must do the following things.

1. Find the function f such that if b is the price of beef, then $p = f(b)$ is the price of poultry that will maximize revenue to poultry producers. (The price of beef is expected to be between $0.50 and $1.00 per pound.)
2. Estimate the revenue to poultry producers for the range of beef prices given.
3. Determine the increase in per capita annual revenue for poultry production for each one-cent increase in the price of beef, and find the largest value of this quantity over the range of beef prices given.

You then get to work. The first part of the project reminds you of the kind of calculation done in Sections 3 and 4. First, you recall that

$$\text{Revenue} = \text{Price per pound} \times \text{Number of pounds},$$

so you obtain

$$R(p) = p(56.9 - 0.45p + 0.12b).$$

You express this as a quadratic function in p, getting

$$R(p) = -0.45p^2 + (56.9 + 0.12b)p.$$

▼ *Source: A. H. Studenmund, *Using Econometrics*, Second Edition (HarperCollins, 1992), pp. 180–81. This equation was calculated using the "least squares" method, from data collected from 1950 through 1984 (see the chapter on functions of several variables). The equation originally involved a parameter for per capita disposable income, which we have assumed to be $15,000.

for the annual per capita revenue (in cents) to the poultry industry. You now calculate the value of p that gives the largest revenue.

$$f(b) = \frac{56.9 + 0.12b}{0.90} = 63.2 + 0.13b$$

Since you are interested in values of b between 50¢ and 100¢, you take the domain to be [50, 100], and you discover that you have completed the requirements for the first part of the assignment (and you take the rest of the day off).

Upon returning to your desk, you turn to the calculation of the total revenue. You notice that you already have the formula

$$R(p) = -0.45p^2 + (56.9 + 0.12b)p$$

and since $p = f(b)$, you calculate R as a function of b by substituting.

$$R(f(b)) = -0.45(63.2 + 0.13b)^2 \\ + (56.9 + 0.12b)(63.2 + 0.13b)$$

You give this new function of b a name, S, and simplify by expanding and combining terms.

$$S(b) = -0.45(3994.24 + 16.432b + 0.0169b^2) \\ + (3596.08 + 14.981b + 0.0156b^2) \\ \approx 1800 + 7.6b + 0.008b^2$$

with domain [50, 100] as before. This *almost* completes the second part of your assignment, but since the agency wants to *see* the revenue, you decide to draw the graph of S (Figure 1). (This is a tiny piece of a parabola, so small that you can hardly see the curvature.) You can now see at a glance, for example, that if the price of beef is set at 60¢ per pound, you can expect the average person to consume about $23 worth of poultry per year.

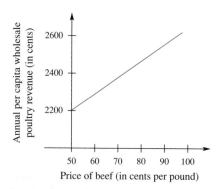

FIGURE 1

Q As the price of beef goes up, so does the price of poultry. Why?

A The higher the price of beef, the greater the demand for poultry. This allows the government to increase the price of poultry, and at the same time increase revenue from poultry.

Finally, the agency wants to know how much the revenue from poultry will go up for each 1¢ increase in the price of beef under your pricing policy. Now the revenue from poultry at a beef price of b is given by

$$S(b) = 1800 + 7.6b + 0.008b^2.$$

An increase of 1¢ in this price gives

$$S(b+1) = 1800 + 7.6(b+1) + 0.008(b+1)^2.$$

Thus, the increase in the revenue from poultry is

$$\begin{aligned}S(b+1) - S(b) &= [1800 + 7.6(b+1) + 0.008(b+1)^2] \\ &\quad - [1800 + 7.6b + 0.008b^2] \\ &= [1800 + 7.6b + 7.6 + 0.008b^2 + 0.016b \\ &\quad + 0.008] - [1800 + 7.6b + 0.008b^2] \\ &\approx 7.6 + 0.016b.\end{aligned}$$

The change in S goes up as the beef price goes up. Thus, the largest value $S(b+1) - S(b)$ can have occurs at a beef price of $b = 100$. So, if beef is priced at $1.00 per pound, a 1¢ increase would result in an increase in poultry revenue of

$$7.6 + 0.016 \cdot 100 \approx 9.2¢ \text{ per capita per year.}$$

Exercises

1. If the wholesale price of beef is 82¢ per pound, what, according to your proposed policy, should the wholesale price of poultry be?

2. Under your proposal, by how much will the price of poultry rise under your proposal for each 1¢ increase in the price of beef?

3. Explain the significance of each of the constants in the demand equation given by the economists.

4. Explain in general terms why one expects the revenue for poultry producers to increase as the price of beef increases.

5. Repeat the analysis assuming a per capita disposable income of $30,000, which makes the demand equation look like this:

$$q = 65.4 - 0.45p + 0.12b.$$

6. Just as you are about to submit your proposal to the agency, you get an urgent phone call: The term b in the demand equation should have been b^2. (The exponent got lost during faxing.) What adjustments will you need to make to your report?

7. Given the demand equation used in the discussion, assume that it costs poultry producers 35¢ to produce one pound of poultry. Repeat the first part of your analysis, but this time with a view to maximizing poultry producers' profit rather than revenue.

Review Exercises

Find the equations of the lines described in Exercises 1–10, in the form $ax + by + c = 0$.

1. Through the origin with slope 3
2. Through the origin with slope $-\frac{1}{2}$
3. Through the point $(1, -1)$ with slope 3
4. Through the point $(-1, -2)$ with slope -1
5. Through the points $(-3, -6)$ and $(1, -1)$
6. Through the points $(1, -2)$ and $(-1, 0)$
7. Through the point $(1, -2)$ and parallel to the line $x + 3y = 1$
8. Through the point $(1, -1)$ and parallel to the line $2x - 3y = 11$
9. Through the point $(1, -2)$ and perpendicular to the line $2x - y = 1$
10. Through the point $(0, -1)$ and perpendicular to the line $2x + 2y = 3$

Sketch the lines whose equations are given in Exercises 11–16.

11. $2x + y = 6$
12. $x - 2y = 8$
13. $2y = -3$
14. $3y = 0$
15. $2x + 1 = 0$
16. $4x - 3 = 0$

Sketch the parabolas whose equations are given in Exercises 17–22.

17. $y = x^2 - 3x + 2$
18. $y = -x^2 - 11x - 30$
19. $y = -5x^2 + x - 2$
20. $y = 6x^2 - 4x - 2$
21. $y = x^2 - x - 1$
22. $y = -x^2 + x - 1$

In Exercises 23–28, sketch the graphs of the given functions without plotting points, by referring to the graphs of the standard functions you have learned. (Unless otherwise stated, the domain is the largest possible.)

23. $f(x) = x^3$, with domain $[-\infty, 0)$
24. $f(x) = x^4$, with domain $[-1, 1]$
25. $f(x) = \sqrt{x}$, with domain $[0, 9]$
26. $f(x) = \sqrt[4]{x}$, with domain $[0, 16]$
27. $f(x) = 1/|x|$
28. $f(x) = |x| + 1/|x|$

Sketch the graph of each of the functions in Exercises 29–34.

29. $f(x) = x^2 + x + 2$
30. $f(x) = x^2 + \dfrac{1}{x}$
31. $f(x) = \dfrac{1}{(x+1)^2}$
32. $f(x) = \sqrt{x} + \dfrac{1}{x^2}$
33. $f(x) = \sqrt{1 - x^2} - x$
34. $f(x) = \sqrt{1 - x^2 + 2x}$

Use a graphing calculator to find at least one solution for each of the equations in Exercises 35–40 to within ± 0.005.

35. $x^5 - 4 = 0$
36. $x^5 - 2x = 0$
37. $x^2 - 5\sqrt{x} = 0$
38. $x^3 - 4\sqrt{x} = 0$
39. $4 = \sqrt{x} + \dfrac{1}{x^3}$
40. $\dfrac{1}{(x+1)^2} = 7x - 1$

41. Given $f(r) = \dfrac{1}{r+2}$, find
(a) $f(0)$ (b) $f(1)$ (c) $f(-1)$ (d) $f(x-2)$ (e) $f(x^2+x)$ (f) $f(x^2)+x$

42. Given $f(x) = x - \dfrac{1}{x}$, find
(a) $f(1)$ (b) $f(-1)$ (c) $f(\pi)$ (d) $f(x+h) - f(x)$ (e) $f(x) + h$ (f) $f(x) + h - f(x)$

43. Given $g(x) = \sqrt{x^2 - 1}$, find
(a) $g(1)$ (b) $g(-1)$ (c) $g(\sqrt{x+h})$ (d) $g(\sqrt{x}) + h$.

44. Given $g(y) = \sqrt[3]{y}$ find
(a) $g(8)$ (b) $g(16)$ (c) $g((y+h)^3)$ (d) $g(y^3) + h$.

45. Given $f(x) = \dfrac{x^2 + 1}{x}$, with domain $(0, +\infty)$, find
(a) $f(1)$ (b) $f(a^2)$ (c) $f(x+h) - h$ (d) $f(\sqrt{x}) + h \ (x > 0)$

46. Given $f(x) = \dfrac{6}{\sqrt{x}} - \sqrt{x}$, with domain $(0, +\infty)$, find
(a) $f(x^2)$ (b) $(f(x))^2$ (c) $\sqrt{x} f(x)$ (d) $f(x\sqrt{x})$

In Exercises 47–50, calculate and simplify the quotient $\dfrac{f(x+h) - f(x)}{h} \ (h \neq 0)$.

47. $f(x) = x^2 + x - 1$

48. $f(x) = -x^2 - 2x - 1$

49. $f(x) = \dfrac{2}{2x - 1}$

50. $f(x) = \dfrac{1}{3 - 2x}$

APPLICATIONS

51. *Cost* Rock Solid Insurance, Inc.'s premium for a $100,000 life insurance policy is $96 for a 25-year-old nonsmoker, and $186 for a 45-year-old nonsmoker. If n represents the number of years since a policyholder was 25, and p the premium, express p as a linear function of n, and use your model to predict what the premium will be for a 90-year-old nonsmoker.

52. *Demand* The market research department for Ultrafast Computers, Inc., finds that it can sell 10,000 computers at $5,000 apiece, but only 6,000 computers priced at $7,000. Express the demand q as a linear function of the price p.

53. *Motion* As you start your cross-country drive, your odometer reads 45,000 miles. If you maintain a constant speed of 55 mph, find a linear function giving your odometer reading s after t hours.

54. *Motion* Your brand new Corvette can accelerate from zero to 60 mph in 6 seconds. Write a linear function giving its speed v (mph) in terms of time t (seconds) What is your average acceleration over this period (that is, at what rate is your speed increasing)? Assuming you continue this rate of acceleration, how fast would your Corvette be going after 1 minute?

55. *Free Fall* If you drop a cannonball from the top of the leaning tower of Pisa, it will start with a velocity of 0, but after 5 seconds will have a velocity of 160 ft/s. Assuming that the velocity increases linearly, find an equation for the velocity v in terms of the time t. What is the acceleration of the cannonball?

56. *Cost Equation* Dirty Dudley's coin-operated clothes dryers require 50¢ to get started, and $1 for each fifteen minutes of drying time. Express the cost of drying a load of laundry as a function of the time t in hours. What is the marginal cost per hour? Because these dryers operate at close to room temperature, a typical load requires two hours of drying time. How much does this cost?

57. *Sales Commission (from the GMAT)* An employee is paid a salary of $300 per month and earns a 6 percent commission on all her sales. What must her annual sales be in order for her to have a gross annual salary of exactly $21,600?

58. *Break Even Analysis (from the GMAT)* Ken left a job paying $75,000 per year to accept a sales job paying $45,000 per year plus 15 percent commission. If each of his sales is for $750, what is the least number of sales he must make per year if he is not to lose money because of the change?

59. *Getting Ahead (from the GMAT)* In 1980 John's salary was $15,000 a year and Don's salary was $20,000 a year. If every year thereafter John receives a raise of $2,450 and Don receives a raise of $2,000, what is the first year in which John's salary will be more than Don's salary?

60. *Cost Equations (from the GMAT)* Marion rented a car for $18.00 plus $0.10 per mile driven. Craig rented a car for $25.00 plus $0.05 per mile driven. If each drove d miles and each was charged exactly the same amount for the rental, then what was d?

61. *Demand and Revenue* Your underground used book business is doing a booming trade. Your policy is to sell all used versions of *Calculus and You* at the same price (regardless of condition). When you set the price at $10, sales amounted to 120 volumes during the first week of classes. What was your total revenue that week? The following semester, you set the price at $30, and sold not a single book. Use these data to set up a demand equation, and use your equation to express the total revenue as a function of the price per book. What price gives you the maximum revenue, and what does that revenue amount to?

62. *Demand and Revenue* Banana Computers has just introduced a new model. The company estimates that it can sell 12,000 units at a price of $2,000, but only 11,000 units at a price of $2,250. Assuming a linear demand function, at what price will it get the largest revenue?

63. *Break-even Analysis (from the CPA exam)* At a break-even point of 400 units sold, the variable costs were $400 and the fixed costs were $200. What will the 401st unit contribute to profit before income taxes?

64. *Break-even Analysis (adapted from a CPA exam question)* At a break-even point of 200 units sold, the variable costs were $800 and the fixed costs were $100. What will the 401st unit contribute to profit before income taxes?

65. *Demand* The demand function for a commodity is
$$q(p) = 40 - 18p - p^2,$$
where q represents the number of items the manufacturer can sell per month at a price of p dollars each. Use this function to determine
(a) the number of items the manufacturer can sell per month if the price is set at $1;
(b) the price at which all sales will stop.

66. *Demand* The demand function for a commodity is
$$q(p) = 200 - 10p - p^2,$$
where q represents the number of items the manufacturer can sell per month at a price of p dollars each. Use this function to determine
(a) the number of items the manufacturer can sell per month if the price is set at $5;
(b) the price at which all sales will stop.

*Exercises 67 and 68 use the idea of the **average cost**. If $C(x)$ is the total cost for x items, then the average cost per item for x items is*
$$\overline{C}(x) = \frac{C(x)}{x}$$

67. *Cost and Average Cost* Lite Up My Life, Inc. can produce 1,000 light fixtures in a month at a total cost of $20,000, and it can produce 2,000 fixtures in a month at a total cost of $30,000.
(a) Construct a linear cost function for Lite Up My Life, Inc.
(b) Write the average cost function.
(c) How many light fixtures should Lite Up My Life, Inc. manufacture per month in order to meet an average cost goal of $12.50 per fixture?

68. *Cost and Average Cost* A firm has monthly fixed costs of $50,000 and variable (marginal) costs of $100 per item.
(a) Write the cost function $C(x)$, where x is the number of items produced per month.
(b) Write the average cost function.
(c) Find the production level (number of items produced per month) that gives an average cost of $200 per item.

69. *The Theory of Relativity* In science fiction terminology, a speed of *warp 1* is the speed of light—about 3×10^8 meters per second. (Thus, for instance, a speed of warp 0.8 corresponds to 80% of the speed of light—about 2.4×10^8 meters per second.) According to Einstein's Special Theory of Relativity, a moving object appears to get shorter to a stationary observer as its speed approaches that of light. If a rocket ship whose length is l_0 meters at rest travels at a speed of warp p, its length in meters, as measured by a stationary observer, will be given by
$$l(p) = l_0\sqrt{1 - p^2}, \text{ with domain } [0, 1].$$
(a) Assuming that a 100-meter rocket ship is traveling at warp 0.9, what will be its length as measured by a stationary observer?

(b) At what speed will it need to travel in order that it appears to be squashed to a length of 1 meter?
(c) What would happen at the speed of light (warp 1)?

70. *Newton's Law of Gravity* The gravitational force exerted on a particle with mass m by another particle with mass M is given by the following function of distance:

$$F(r) = G\frac{Mm}{r^2}, \text{ with domain } (0, +\infty).$$

Here, r is the distance between the two particles in meters, the masses M and m are given in kilograms, $G \approx 0.0000000000667$, or 6.67×10^{-11}, and the resulting force is given in newtons.
(a) Given that M and m are both 1,000 kilograms, find $F(1)$ and $F(10)$.
(b) How much would a battleship have to weigh in order to attract a 1-kg mass situated 1,000 meters away with a force of 1 newton?

71. *Sales Commissions* The Bigger the Better Publishing Company hires students to sell its 100-volume *Encyclopedia Galactica* (which also includes graphing software) for $3,025 per set. The sales staff are each paid a commission of 5% of the square root of total sales, plus a basic wage of $100 per month.
(a) Write a function that expresses the student's earnings on sales of x sets per month.
(b) Use your formula to calculate the earnings from the sale of 1 set per month and 100 sets per month.
(c) Approximately how many sets should a student sell in order to earn $200 per month? Comment on the company's policy.

72. *Sales Commissions* The Smaller the Better Publishing Company hires students to sell its 100-page *Encyclopedia Miniscula* (which includes graphing hardware) for $25 per volume. The sales staff are each paid a commission of 5% of the square of total sales, plus a basic wage of $100 per month.
(a) Write down a function giving a student's earnings on sales of x volumes per month.
(b) Use your formula to calculate the earnings from 1 volume per month and 100 volumes per month.
(c) Approximately how many sets should a student sell in order to earn $2,000 per month? Comment on the company's policy.

73. *Salary Scales in Japan* According to data in a *New York Times* article, the average annual salary for college graduates in Japan is approximately a linear function of the age, rising from an average of $30,000 per year for a 28-year-old to $65,000 per year for a 48-year-old worker.*
(a) Use this information to express the average salary S of a Japanese college graduate in thousands of dollars as a function of age x. (Also give the domain.)
(b) The same data also show the average salary increasing more slowly, at about $1,000 per year, from age 48 to 54, and then leveling off at $71,000 per year. Use these data to express annual salary as a piecewise-defined function of age x for $28 \leq x \leq 58$.

74. *Salary Scales in Japan* According to data in the *New York Times* article mentioned in Exercise 73, the average annual salary for high school graduates in Japan is approximately a linear function of the age, and rises from an average of $28,000 per year for a 28-year-old to $55,000 per year for a 48-year-old worker.*
(a) Use this information to express the average salary S of a Japanese high school graduate in thousands of dollars as a function of age x. (Also give the domain.)
(b) The same data also show the average salary increasing more slowly, at about $1,000 per year, from age 48 to 54, and then leveling off at $61,000 per year. Use these data to express annual salary as a piecewise-defined function of age x for $28 \leq x \leq 58$.

75. *Demand* The demand equation for donuts is

$$p = \frac{500}{q + 100},$$

where p is the price per donut and q is the quantity of demand. Solve for q as a function of p. What does this function signify?

76. *Demand* The demand equation for bagels is

$$p\sqrt{q} = 400,$$

where p is the price per bagel and q is the quantity of demand. Find q as a function of p, and also find p as a function of q. What do these functions signify?

---

*Source: *The New York Times*, Oct. 2, 1993, p. 6, and the Japan Federation of Employee Associations.

CHAPTER 2

EPA Chief Ties Ecology to Economy

By Casey Bukro

The nation's chief environmental enforcer says that industry must adopt practices that prevent pollution and preserve natural ecosystems.

That means manufacturers must find ways to produce products without ecological harm, said Carol Browner, new administrator of the U.S. Environmental Protection Agency.

"Pollution prevention and source reduction are going to be very important," she said, adding that she intends to avoid traditional command and control methods that set limits on pollution.

Browner said she has suggested several job-creating ideas to President Clinton, including watershed restoration and wastewater treatment facilities construction.

For every $1 billion spent on wastewater projects, she said 50,000 jobs are created.

The ideas she has presented are intended to help jump-start the economy while saving the environment.

Carl Pope, the Sierra Club's executive director, adds that it's no longer pie-in-the sky to expect industry to operate in environmentally friendly ways while becoming more competitive.

"We are doing it in virtually every industry," said Pope. He offered these examples:

- Pacific Gas & Electric Co. in San Francisco announced in 1991 that 75 percent of the expected increase in energy demand over the next decade will be met by energy efficiency, not new power plants. The utility promotes energy-conservation programs for residential and commercial customers.

- Minnesota Mining & Manufacturing Co. saved $570 million since 1975 by preventing pollution at its plants and reducing costs for waste disposal or pollution-control equipment.

- Monsanto Co. in St. Louis believes it has met its pledge to cut toxic air emissions by 90 percent by the end of 1992 through technological improvements, recycling and shutting down polluting equipment.

Environmentalists are calling attention to such examples as President Clinton's administration embarks on a program to stimulate the economy and increase the global competitiveness of U.S. industry. Before becoming president, Clinton said it was time to recognize that "Adam Smith's invisible hand can have a green thumb."

"At long last, we have an administration that recognizes the link between a healthy environment and a healthy economy," said Pope. "The two areas we probably will spend most time talking to the Clinton administration about are clean energy and transportation and, in the West, better management and restoration of public lands."

Source: From Casey Bukro, "EPA Chief Ties Ecology to Economy," *Chicago Tribune*, February 13, 1993, p. 1, Section 2.

Systems of Linear Equations and Matrices

APPLICATION ▶ You have been hired as a consultant to the Environmental Protection Agency. The agency is considering legislation to require a 15-million-ton reduction in sulfur emissions in an effort to curb the effects of acid rain on the ecosystem. You have been asked to estimate the cost of the proposed regulation to the major utility companies and also the effect on jobs in the coal-mining industry. The only data that are available show the annual cost to utilities and the cost in jobs for emission reductions of up to 12 million tons. Your assignment is to use these figures to compute projections for a 15-million-ton reduction.

SECTIONS

1. Systems of Two Linear Equations in Two Unknowns
2. Using Matrices to Solve Systems with Two Unknowns
3. Using Matrices to Solve Systems with Three or More Unknowns
4. Applications of Systems of Linear Equations

You're the Expert
The Impact of Regulating Sulfur Emissions

INTRODUCTION ▶ In Chapter 1 we considered equations in two unknowns. In this chapter we seek solutions to **systems** of two or more equations. Here is a simple example: *Find two numbers whose sum is 2 and whose product is $\frac{3}{4}$.* We are being asked to find two numbers x and y such that $x + y = 2$ and $x \cdot y = \frac{3}{4}$. It turns out that (the only) two solutions are: $x = \frac{1}{2}, y = 1\frac{1}{2}$ and $x = 1\frac{1}{2}, y = \frac{1}{2}$. You may wonder why you should be interested in this. Many problems boil down to solving a system of equations, and we'll see a few as we go along. In particular, the last section of this chapter is devoted entirely to applications.

As the title of this chapter suggests, we shall restrict ourselves to systems of *linear* equations: that is, systems of fairly simple equations such as $2x + y = 4$ or $3x + 4y - 2z = 1$. One reason for this is that there is a very elegant method for finding the solutions to such systems. Another reason is that scores of applications give rise to just such linear equations.

In the first section, we study systems of two linear equations in two unknowns. In Sections 2 and 3, we introduce a powerful matrix method, called "row reduction," for solving systems of linear equations in any number of unknowns. Finally, we explore a large number of applications in Section 4. The method we use virtually eliminates the need for extensive calculations with fractions or decimals. The reason for using this method is twofold. First, for hand calculations our method is a good deal easier than the standard technique using fractions. Second, there is a mathematical elegance to working with integer matrices. Since almost all of the reduction of integer matrices can be done using integers, why bother with fractions until we need to?

You should also be aware of the role of technology in solving systems of linear equations. For the large systems actually used in practice, computers have been used for many years to do the computations we shall teach you in this chapter. You probably already have access to devices that will do the row operations we shall talk about. Many graphing calculators can do them, as can spreadsheets and special-purpose computer programs. Using such a program makes the calculations quicker and helps to prevent arithmetic mistakes. Then there are programs (and calculators) where we simply feed in the system of equations and the program finds the solutions. We can think of what we do in this chapter as looking inside the "black box" of such a program. More importantly, we shall talk about how, starting from a real problem, we come up with the system of equations to solve in the first place. This is something no computer will yet do for us.

2.1 SYSTEMS OF TWO LINEAR EQUATIONS IN TWO UNKNOWNS

In Chapter 1, we said that a linear equation in two unknowns is one that can be written in the form

$$ax + by = c$$

with a, b, and c being real numbers. An example of such an equation is $3x - 5y = 15$. We saw that these equations may have infinitely many solutions. For example, in the equation just given, we could solve for $y = \frac{3}{5}x - 3$, and then for every value of x we choose we can easily get the corresponding value of y, giving a solution (x, y). Graphically, these solutions are the points on a straight line, the *graph* of the equation.

Now we are concerned with pairs (x, y) that are solutions to *two* given linear equations at the same time. You'll see what we mean in the next few examples.

EXAMPLE 1

Find the solution(s) of the system

$$x + y = 1$$
$$x - y = 0.$$

SOLUTION We shall show how to obtain the solution(s) in three ways: by "common sense," algebraically, and graphically. Remember that a solution is a pair (x, y) that satisfies *both* equations at the same time.

*Method 1: Common-sense** We look at the equations and try to reason our way to a solution. We see from the first equation that the numbers x and y must add up to 1. Of course, many such pairs are possible. For example, $(x, y) = (0, 1)$, $(\frac{1}{4}, \frac{3}{4})$, and so on. On the other hand, (x, y) must also be a solution to the second equation, so the difference of the two numbers must be zero. In other words, the two numbers x and y must be the same. Thus, the problem in a nutshell is the following: *find two numbers that add up to 1 and are the same*. Well, now we realize that the only possibility is that they are both $\frac{1}{2}$. In other words, $x = \frac{1}{2}$ and $y = \frac{1}{2}$. Thus, the only solution is $(\frac{1}{2}, \frac{1}{2})$.

---

* This approach only works well when the equations are very simple, as in this example, so we'll use this method sparingly. On the other hand, it's always helpful to think about problems this way; it should help sharpen your mathematical intuition.

Method 2: Algebraic The approach here consists of trying to combine the equations in such a way as to eliminate one of the variables. In this case, notice that if we add the left-hand sides of the equations, the ys are eliminated. Thus, we add the first equation to the second, getting

$$2x + 0 = 1$$

or

$$2x = 1,$$

giving

$$x = \tfrac{1}{2}.$$

Now that we know that x has to be $\tfrac{1}{2}$, we can substitute it back into either equation to get y. Choosing the first equation for this (it doesn't matter which we choose), we have

$$\tfrac{1}{2} + y = 1$$

giving

$$y = 1 - \tfrac{1}{2} = \tfrac{1}{2}.$$

Thus, we have found that the only possible solution is $x = \tfrac{1}{2}$ and $y = \tfrac{1}{2}$, or

$$(x, y) = (\tfrac{1}{2}, \tfrac{1}{2}).$$

Method 3: Graphical We already know that the solutions to a single linear equation are the points on its graph, which is a straight line. For a point to represent a solution to two linear equations, it must lie simultaneously on *both* of the corresponding lines. In other words, it must be the point where the two lines cross, or intersect. A look at Figure 1 should convince us that this is the point $(\tfrac{1}{2}, \tfrac{1}{2})$, so this is the only possible solution.

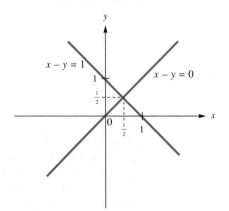

FIGURE 1

Before we go on... Having obtained the same solution in three different ways, we suspect that it is correct. Still, we should check our solution, $x = \tfrac{1}{2}, y = \tfrac{1}{2}$, by substituting it back into the original equations. Remember that a solution to an equation must yield a true statement when substituted into that equation. Substituting, we get

$$\frac{1}{2} + \frac{1}{2} = 1 \quad ✔$$
$$\frac{1}{2} - \frac{1}{2} = 0. \quad ✔$$

Thus, the solution is correct!

 You can reproduce the graphical approach by using your graphing calculator or computer to draw the graphs of $y = -x + 1$ and $y = x$ on the same set of axes. You can then find the intersection point by using the trace and/or zoom features. You can use this method to check the solutions to all of the examples that follow.

▼ **EXAMPLE 2**

Solve the system

$$3x + 5y = 0$$
$$2x + 3y = 1.$$

SOLUTION This example is too complicated for the "common-sense" approach. Instead, we rely on more mechanical methods.

Algebraic We first notice that adding the equations is not going to eliminate either the xs or the ys. Notice, however, that the coefficients of x are 3 and 2. If we multiply the first equation by 2 and the second by -3, the coefficients will become 6 and -6. *Then* if we add them, the xs will be eliminated. So we proceed as follows:

$$(3x + 5y = 0) \times (2)$$
$$(2x + 3y = 1) \times (-3)$$

giving

$$6x + 10y = 0$$
$$-6x - 9y = -3.$$

Adding them, we get

$$y = -3.$$

Substituting this value in the first equation gives

$$3x + 5(-3) = 0,$$

so

$$3x = 15; \text{ therefore, } x = 5.$$

Thus, the solution is $x = 5$ and $y = -3$, or $(x, y) = (5, -3)$.

Graphical We sketch the two lines in question in Figure 2.

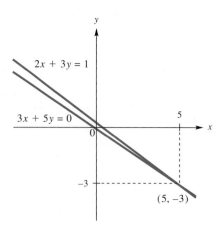

FIGURE 2

Again, we see that the solution is $(5, -3)$ by noting the point of intersection. How do we know that the point of intersection is *exactly* $(5, -3)$? From the algebraic method! It is difficult graphically to find the exact solution, but we *can* use the graphical method to confirm the algebraic solution.

Before we go on... We check the solution $(5, -3)$ by substitution.
$$3(5) + 5(-3) = 15 - 15 = 0 \quad ✔$$
$$2(5) + 3(-3) = 10 - 9 = 1 \quad ✔$$

 Even with a graphing calculator or computer it may be difficult to locate the exact point of intersection from the graph. Try it, and you'll see why.

▼ **EXAMPLE 3**

Solve the system
$$2x + 3y = 1$$
$$-3x + 2y = 1.$$

SOLUTION Using the algebraic method, we can eliminate the xs by multiplying the first equation by 3 and the second by 2, giving
$$6x + 9y = 3$$
$$-6x + 4y = 2.$$

Adding them together gives
$$13y = 5, \quad \text{so} \quad y = \frac{5}{13}.$$

Substituting in the first equation gives

$$2x + 3\left(\frac{5}{13}\right) = 1$$

or

$$2x + \frac{15}{13} = 1,$$

so

$$2x = 1 - \frac{15}{13} = -\frac{2}{13}.$$

Thus,

$$x = \frac{1}{2}\left(-\frac{2}{13}\right) = -\frac{1}{13}.$$

The solution is therefore $(x, y) = \left(-\frac{1}{13}, \frac{5}{13}\right)$.

Now we already can see that the graphical method is going to cause a bit of a problem. The intersection of the two lines will be the point $\left(-\frac{1}{13}, \frac{5}{13}\right)$, and it is by no means an easy task to get these coordinates precisely! However, we can tell from Figure 3 that the answer we just obtained is reasonable.

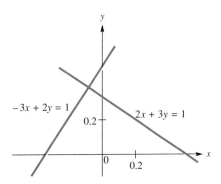

FIGURE 3

Before we go on... Substituting this solution into the original equations gives

$$2\left(-\frac{1}{13}\right) + 3\left(\frac{5}{13}\right) = -\frac{2}{13} + \frac{15}{13} = \frac{13}{13} = 1 \checkmark$$

$$-3\left(-\frac{1}{13}\right) + 2\left(\frac{5}{13}\right) = \frac{3}{13} + \frac{10}{13} = \frac{13}{13} = 1. \checkmark$$

You will not be able to write the *exact* answer if you use a graphing calculator or computer.

Q Why not?

A If you write the solution in decimal form, you will get repeating decimals: $(-0.0769230769230\ldots, 0.384615384615\ldots)$. Since a graphing calculator only gives the coordinates of points to several decimal places, it can show only an approximate answer.

Q How accurate is the answer shown on a graphing calculator?

A That depends. You can increase the accuracy up to a point by "zooming in" on the point of intersection of the two graphs. Most graphing calculators will be capable of giving an answer correct to at least eight or so decimal places.

Q According to my calculator (I used the window $-2 \leq x \leq 2$, $-2 \leq y \leq 2$, and the "trace" function) the point of intersection is $(-0.0851064, 0.37234043)$. Why is my answer wrong?

A The coordinates you have written provide an approximation to the point of intersection, and not a very accurate one at that. To get an estimate of how accurate your answer is, move your cursor location one step to the left and one step to the right of the point of intersection. This gives you two more values for x and y. For instance, you might obtain these values:

Point	Left	Intersection	Right
x	-0.11775796	-0.0851064	-0.0425532
y	0.32336306	0.37234043	0.43617021

Looking along the x-row, all we can say for sure is that the point of intersection is somewhere between -0.1276596 and -0.0425532. The difference between these numbers is 0.07520476, and a good estimate of the accuracy is half that difference, ± 0.0376, which we round to one significant digit, ± 0.04. In other words, the answer of $x = -0.0851064$ is only accurate to within ± 0.04, so all the decimal places beyond the second are quite meaningless, since the second decimal place could be off by as much as 4! Thus, we round the x-coordinate of the midpoint to two decimal places and write $x = -0.09 \pm 0.04$.

Similarly, the y-coordinate is somewhere between 0.32336306 and 0.43617021. If we again take half the difference and round to one significant digit, we get ± 0.06, which we use to estimate the accuracy of the y-coordinate. This says that we should round the y-coordinate to two decimal places as well (the second decimal place may be off by as much as 6). Thus,

$$(x, y) = (-0.09 \pm 0.04, 0.37 \pm 0.06).$$

The exact answer we got using algebra does lie in this range. You can get a more accurate approximation by zooming in.

▼ **EXAMPLE 4**

Solve the system

$$x - 3y = 5$$
$$-2x + 6y = 8.$$

2.1 Systems of Two Linear Equations in Two Unknowns

SOLUTION To eliminate the xs, all we need do is multiply the first equation by 2 and then add.

$$2x - 6y = 10$$
$$-2x + 6y = 8$$

Adding gives

$$0 = 18.$$

But this is absurd! This calculation shows that, if we had two numbers x and y that satisfied both equations, it would be true that $0 = 18$. Since 0 is *not* equal to 18, there can be no such numbers x and y. In other words, *the system has no solutions*.

To see what is going on graphically, plot the two lines. In slope-intercept form they are $y = \frac{1}{3}x - \frac{5}{3}$ and $y = \frac{1}{3}x + \frac{4}{3}$. Plotting them gives the two lines shown in Figure 4.

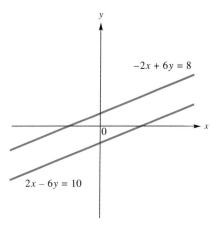

FIGURE 4

Since they both have slope $\frac{1}{3}$, they are parallel and thus do not intersect. A solution must be a point of intersection, so we conclude that there is no solution.

▼ **EXAMPLE 5**

Solve the system

$$x + y = 2$$
$$2x + 2y = 4.$$

SOLUTION Multiplying the first equation by -2 gives

$$-2x - 2y = -4$$
$$2x + 2y = 4.$$

Adding gives the not-very-enlightening result

$$0 = 0.$$

Now what has happened? To answer this, note that the second equation is really the first equation in disguise. (It is the first equation multiplied by 2.) Therefore we really have only one equation in two unknowns. From Chapter 1, we know that there are infinitely many solutions, one for each value of x. (Recall that to get the corresponding solution for y, we solve the equation for y and substitute the x-value.) The entire set of solutions can be summarized as follows.

x is arbitrary.

$y = 2 - x$ (Solve the first equation for y.)

Now, what about the graphical method? Well, the first line has x-intercept 2 and y-intercept 2. Similarly, the second line has x-intercept $\frac{4}{2} = 2$ and y-intercept $\frac{4}{2} = 2$. Since both lines have the same x-intercept and the same y-intercept, they are the same line! (Of course we knew this, since the two equations were really the same.) Looked at another way, since the two lines intersect at every point, there is a solution for each point on the common line.

We summarize the three possible outcomes we have encountered.

POSSIBLE OUTCOMES FOR A SYSTEM OF TWO LINEAR EQUATIONS IN TWO UNKNOWNS

1. **A single (or *unique*) solution**   This happens when the lines corresponding to the two equations are not parallel, so that they intersect at a single point.
2. **No solution**   This happens when the two lines are parallel and different.
3. **An infinite number of solutions**   This occurs when the two equations represent the same straight line. In this case, you represent the solutions by choosing one variable arbitrarily and solving for the other, as in Example 5 above.

You should think about straight lines and convince yourself that these are the only three possibilities. Now here are some applications.

▼ **EXAMPLE 6**   Blending

Acme Babyfoods mixes two strengths of apple juice. One quart of Beginner's juice is made from 30 fluid ounces of water and 2 fluid ounces of apple juice concentrate. One quart of Advanced juice is made from 28 fluid ounces of water and 4 fluid ounces of concentrate. Every day Acme has available 215 gallons (27,520 fluid ounces) of water and 25 gallons (3,200 fluid ounces) of concentrate. If Acme wants to use up all of the water and concentrate, how many quarts of each type of juice should it mix?

SOLUTION   In solving problems like this, we shall follow a general strategy.

Step 1: *Identify and label the unknowns.*   In other words, what are we asked to find? In answering this question, it is a common error to respond by saying, "The unknowns are Beginner's juice and Advanced juice." Quite frankly, this is a baffling statement. Just what is unknown about juice? We need to be more precise, saying

> The unknowns are (1) the **number of quarts** of Beginner's juice, and (2) the **number of quarts** of Advanced juice made each day.

We now label the unknowns as follows.

> Let x = the number of quarts of Beginner's juice, and let y = the number of quarts of Advanced juice made each day.

Step 2: *Use the information given to set up equations in the unknowns.* This step is trickier, and the strategy varies from problem to problem. Here, the amount of juice the company can make is constrained by the fact that it has only so much water and so much concentrate. This example shows a kind of problem we will often see, and it is helpful in these problems to use a table to record the amounts of the resources used. When setting up such a table, it is best to list the products associated with the unknowns across the top and the resources on the side, with an extra column for the amounts of the resources available.

	Beginner's (x qt)	**Advanced (y qt)**	**Available**
Water (fl oz)	30	28	27,520
Concentrate (fl oz)	2	4	3,200

We can now set up an equation for each of the items listed down the side.

Water: Read across the first row. If Acme mixes x quarts of Beginner's juice, each using 30 fluid ounces of water, and y quarts of Advanced juice, each using 28 fluid ounces of water, it will use a total of $30x + 28y$ fluid ounces of water. But we are told that the total has to be 27,520 fluid ounces. Thus, $30x + 28y = 27,520$. This is our first equation.

Concentrate: Read across the second row. If Acme mixes x quarts of Beginner's juice, each using 2 fluid ounces of concentrate, and y quarts of Advanced juice, each using 4 fluid ounces of concentrate, it will use a total of $2x + 4y$ fluid ounces of concentrate. But we are told that the total has to be 3,200 fluid ounces. Thus, $2x + 4y = 3,200$.

Now we have two equations:

$$30x + 28y = 27,520$$
$$2x + 4y = 3,200.$$

Notice that the coefficients of the first equation are divisible by 2, as are those of the second. We can divide both equations by 2 to make the numbers smaller and easier to work with.

$$15x + 14y = 13{,}760$$
$$x + 2y = 1{,}600$$

We can now eliminate x by multiplying the second equation by -15 and adding.

$$15x + 14y = 13{,}760$$
$$-15x - 30y = -24{,}000$$

Adding gives $-16y = -10{,}240$, and so $y = 640$. Substituting this into the equation $x + 2y = 1{,}600$, we get $x + 1{,}280 = 1{,}600$, and so $x = 320$. Thus, the solution is $(x, y) = (320, 640)$. In other words, the company should mix 320 quarts of Beginner's juice and 640 quarts of Advanced juice.

Before we go on... Let us check that our answer fits the conditions of the problem. The 320 quarts of Beginner's juice will require $320 \cdot 30 = 9{,}600$ fluid ounces of water and $320 \cdot 2 = 640$ fluid ounces of concentrate. The 640 quarts of Advanced juice will require $640 \cdot 28 = 17{,}920$ fluid ounces of water and $640 \cdot 4 = 2{,}560$ fluid ounces of concentrate. Thus, Acme will use a total of

$$9{,}600 + 17{,}920 = 27{,}520 \text{ fluid ounces of water}$$

and

$$640 + 2{,}560 = 3{,}200 \text{ fluid ounces of concentrate,}$$

which is precisely the amount available of each. What we have just done amounts to checking that the system of equations is satisfied, since we can rewrite the calculation as follows.

$$30(320) + 28(640) = 27{,}520 \quad ✔$$
$$2(320) + 4(640) = 3{,}200 \quad ✔$$

▼ **EXAMPLE 7** Nutrition

According to the nutrition information on a package of Honey Nut Cheerios® brand cereal, each 1-ounce serving of Cheerios contains 3 grams of protein and 24 grams of carbohydrates.* Each half-cup serving of enriched skim milk contains 4 grams of protein and 6 grams of carbohydrates. Your athletics coach has recommended that each of your meals provide 39 grams of protein (approximately half the U.S. RDA) and twice as much carbohydrates as

▼ *Actually, it is 23 grams of carbohydrates. We made it 24 grams in order to simplify the calculation.

2.1 Systems of Two Linear Equations in Two Unknowns

protein by weight. Assuming you want to breakfast on Cheerios and milk, how should you prepare your meal to satisfy your coach?

SOLUTION

Step 1: *Identify and label the unknowns.* To find the unknowns, we focus on the question at the end of the problem: "How should you prepare your meal . . . ?" In other words, we are asking for the amount of Cheerios and milk your breakfast should contain. Thus, the unknowns may be specified as follows:

Let x = the number of 1-oz servings of Honey Nut Cheerios, and
y = the number of half-cup servings of skim milk.

Step 2: *Use the information given to set up equations in the unknowns.* Again, start with a table showing the information given.

	Cheerios (x servings)	Milk (y servings)	Total Required
Protein (g)	3	4	39
Carbohydrates (g)	24	6	78

We can now set up an equation for each of the items listed down the side.

Protein: Read across the first row. Since you are using x servings of Cheerios, each supplying 3 grams of protein, and y servings of milk, each supplying 4 grams of protein, your total protein intake will be $3x + 4y$. But you are told that the total has to be 39. Thus, $3x + 4y = 39$.

Carbohydrates: Read across the second row. Since you are using x servings of Cheerios, each supplying 24 grams of carbohydrates, and y servings of milk, each supplying 6 grams of carbohydrates, your total carbohydrate intake will be $24x + 6y$. But you are told that the total has to be 78. Thus, $24x + 6y = 78$.

Now we have two equations.

$$3x + 4y = 39$$
$$24x + 6y = 78$$

The second equation is divisible by 6, so divide it by 6 to obtain the simpler system

$$3x + 4y = 39$$
$$4x + y = 13.$$

We can now eliminate y by multiplying the second equation by -4 and adding.

$$3x + 4y = 39$$
$$-16x - 4y = -52$$

Adding gives
$$-13x = -13,$$
so $x = 1$. Substituting this value of x in the first equation gives
$$3 + 4y = 39,$$
or
$$4y = 36,$$
so $y = 9$. The solution is as follows: you should prepare your breakfast with a single ounce of Cheerios and 9 half-cups of skim milk. Not the most exciting combination!

Before we go on... Notice that the large carbohydrates-to-protein ratio in Cheerios suggests that this answer is in the right ball park. As usual, we can check our answers by direct substitution.
$$3(1) + 4(9) = 39 \quad \checkmark$$
$$24(1) + 6(9) = 78 \quad \checkmark$$

▼ **EXAMPLE 8**   Blending

A medieval alchemist's love potion calls for a number of eyes of newt and toes of frog, the total being 20, but with twice as many newt eyes as frog toes. How many of each are required?

SOLUTION   As in the previous examples, the first step is to identify and label the unknowns. Thus,

Let x = the number of newt eyes, and let y = the number of frog toes.

As for the second step—that of setting up the equations—a table is less appropriate here than in the preceding example. Instead, we translate each phrase of the problem into an equation. The first sentence tells us that the total number of eyes and toes is 20. Thus,
$$x + y = 20.$$

The rest of the first sentence gives us more information, but the phrase "twice as many newt eyes as frog toes" is not in the simple form we desire, and seems a little ambiguous: does it mean that $x = 2y$ or that $y = 2x$? The trick is to *rephrase the information using the phrases* "the number of newt eyes," which is x and "the number of frog toes," which is y. Rephrased, the statement reads:

The number of newt eyes is twice the number of frog toes.

With this rephrasing, we can translate directly into algebra, giving
$$x = 2y.$$

In standard form, this equation reads
$$x - 2y = 0.$$
Thus, we have the two equations
$$x + y = 20$$
$$x - 2y = 0.$$
To eliminate x, we multiply the second equation by -1 and then add.
$$x + y = 20$$
$$-x + 2y = 0$$
Adding the two together gives
$$3y = 20,$$
so
$$y = \tfrac{20}{3} = 6\tfrac{2}{3}.$$
Substituting for y in the first equation, $x + y = 20$, now gives
$$x + \tfrac{20}{3} = 20;$$
therefore,
$$x = 20 - \tfrac{20}{3} = \tfrac{40}{3} = 13\tfrac{1}{3}.$$

So the recipe calls for exactly $13\tfrac{1}{3}$ eyes of newt and $6\tfrac{2}{3}$ toes of frog. Thus, the alchemist needed a very sharp scalpel and a very accurate balance (not to mention a strong stomach).

Before we go on... This time, we'll let *you* check that the solution fits the equations!

 This is another example in which your graphing calculator won't give the exact answer because of repeating decimals. (See Example 3.)

▼ **EXAMPLE 9**   Equilibrium Price

The **equilibrium price** of a commodity occurs when supply equals demand. The market is said to **clear** at this point. The demand for refrigerators in West Podunk is given by $q = -p/10 + 100$, where q is the number of refrigerators that the citizens will buy each year if they are priced at p dollars each. The supply is $q = p/20 + 25$, where now q is the number of refrigerators the manufacturers will be willing to ship into town each year if they are priced at p dollars each.

(a) Find the equilibrium price and the number of refrigerators that will be sold at that price.

(b) What would happen if the price were set at $400? At $600?

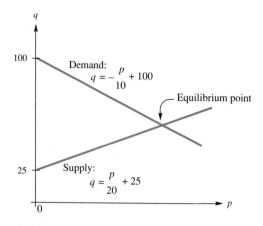

FIGURE 5

SOLUTION

(a) Figure 5 shows the demand and supply curves: $q = -p/10 + 100$ and $q = p/20 + 25$. The equilibrium price occurs at the point where these two lines cross, which is where demand equals supply.

The two equations can be written

$$q + \tfrac{1}{10}p = 100$$
$$q - \tfrac{1}{20}p = 25.$$

When fractions occur, it is always a good idea to multiply each equation by a number that will clear the fractions. Thus, we multiply the first equation by 10 and the second equation by 20, getting

$$10q + p = 1000$$
$$20q - p = 500.$$

If we now add the two equations together we can eliminate p, getting $30q = 1500$, so $q = 50$. Substituting into the first equation gives $500 + p = 1000$, so $p = 500$. So the equilibrium price is $500, and 50 refrigerators will be sold at that price.

(b) If the refrigerators were priced at $400, then the demand would be $-400/10 + 100 = 60$, but the supply would be $400/20 + 25 = 45$. Thus, demand would outstrip supply.

On the figure, the demand curve is above the supply curve when $p = 400$. The difference in heights gives the excess of demand over supply. On the other hand, if the refrigerators were priced at $600, the demand would be $-600/10 + 100 = 40$, and the supply would be $600/20 + 25 = 55$. In this case, the supply would be larger than the demand. Figure 6 shows the two situations.

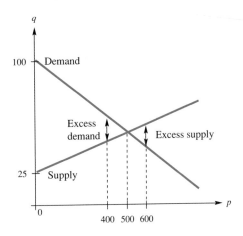

FIGURE 6

Before we go on... The demand equation represents the consumers' willingness to buy refrigerators at a given price. For example, at a price of $400 consumers would buy 60 refrigerators, whereas at a price of $600 they would buy only 40. Since demand goes down as price goes up,* the slope of the demand curve is negative. On the other hand, the supply equation represents the manufacturers' willingness to supply refrigerators at a given price. At a price of $400 they are willing to ship only 45 refrigerators, but at a price of $600 they will ship 55. Since their willingness to provide increases with increasing price, the slope of the supply curve is positive.

Notice that the manufacturers could safely raise the price above $400, since they are guaranteed to sell more refrigerators than they are supplying at $400. On the other hand, they should set the price at something less than $600 because they do not want unsold inventory on their hands. The optimum price is the equilibrium price, and there the price will tend to settle.

By the way, economists traditionally draw supply and demand curves on a graph with p on the *vertical* axis and q on the *horizontal* axis. We shall generally stick with the mathematical tradition, which dictates the axes that we have used here. But you need to be aware of the difference if you should run into an economist.

 As usual, you can locate the equilibrium point using the trace and zoom features on your calculator or graphing program. Another useful exercise is to use a spreadsheet or other program to calculate supply and demand figures for a number of different prices, starting with zero. You can see the region in which supply is lower than demand. As the price increases, you can see the numbers get closer together, and then you see the numbers cross so that supply becomes larger than demand. This procedure allows you to see numerically where the equilibrium point is.

▼ *This is true for most goods, but for luxury goods the opposite can be true!

2.1 EXERCISES

Solve the systems of linear equations in Exercises 1–6 graphically.

1. $x - y = 0$
   $x + y = 4$

2. $x - y = 0$
   $x + y = -6$

3. $x + y = 0$
   $2x + y = -1$

4. $x + y = 0$
   $x + 2y = 3$

5. $x + y = 4$
   $x - y = 2$

6. $2x + y = 2$
   $-2x + y = 2$

Find all solutions of the systems of linear equations in Exercises 7–20.

7. $3x - 2y = 6$
   $2x - 3y = -6$

8. $2x + 3y = 5$
   $3x + 2y = 5$

9. $0.5x + 0.1y = 0.7$
   $0.2x - 0.2y = 0.6$

10. $-0.3x + 0.5y = 0.1$
    $0.1x - 0.1y = 0.4$

11. $\dfrac{x}{3} - \dfrac{y}{2} = 1$
    $\dfrac{x}{4} + y = -2$

12. $-\dfrac{2x}{3} + \dfrac{y}{2} = -\dfrac{1}{6}$
    $\dfrac{x}{4} - y = -\dfrac{3}{4}$

13. $2x + 3y = 1$
    $-x - \dfrac{3y}{2} = -\dfrac{1}{2}$

14. $2x - 3y = 1$
    $6x - 9y = 3$

15. $2x + 3y = 2$
    $-x - \dfrac{3y}{2} = -\dfrac{1}{2}$

16. $2x - 3y = 2$
    $6x - 9y = 3$

17. $x = 2y$
    $y = x - 10$

18. $x = 2y$
    $y = x + 10$

19. $2x + 10 = 3y$
    $3x + 10 = 4y$

20. $2x + 10 = 4y$
    $3x + 15 = 6y$

APPLICATIONS

21. *Voting* The U.S. House of Representatives has 435 members. If an appropriations bill passes the House with 49 more members voting in favor than against, how many voted in favor and how many against?

22. *Voting* The U.S. Senate has 100 members. For a bill to pass with a supermajority, at least twice as many senators must vote for the bill as vote against it. If all 100 senators vote, how many must vote for a bill for it to pass with a supermajority?

23. *Resource Allocation* You manage an ice cream factory that makes two flavors: Creamy Vanilla and Continental Mocha. Into each quart of Creamy Vanilla go two eggs and three cups of cream. Into each quart of Continental Mocha go one egg and three cups of cream. You have in stock 500 eggs and 900 cups of cream. How many quarts of each flavor should you make in order to use up all the eggs and cream?

24. *Class Scheduling* The Enormous State University's Math Department offers two courses: Finite Math and Calculus. Each section of Finite Math has 60 students, while each section of Calculus has 50. The department will offer a total of 110 sections in a semester, and there are 6000 students who would like to take a math course. How many sections of each course should the department offer in order to accommodate all of the students?

25. *Nutrition* *Gerber* Mixed Cereal for Baby contains, in each serving, 60 calories and 11 grams of carbohydrates.* *Gerber* Mango Tropical Fruit Dessert contains, in each serving, 80 calories and 21 grams of carbohydrates.[†] If you want to provide your child with 200 calories and 43 grams of carbohydrates, how many servings of each should you use?

▼ *Source: nutrition information printed on the box.
  [†] Source: nutrition information printed on the jar.

26. *Nutrition* Anthony Latino is mixing food for his young daughter, and he is concerned that the meal supply 1 gram of protein and 5 milligrams of iron. He is mixing together cereal, with 0.5 gram of protein and 1 milligrams of iron per ounce, and fruit, with 0.2 gram of protein and 2 milligram of iron per ounce. What mixture will provide the desired nutrition?

27. *Investments* Index funds try to duplicate the performance of a set of stocks, such as the Standard & Poor's 500. Two of the top composite stock index funds in the 1992–1993 financial year were Fidelity Market Index Fund, which yielded a 10% return on investments (to the nearest percent) and Vanguard Small Capitalization Fund, which was up 13% (to the nearest percent).* If you invested a total of $10,000 in the two funds and your investment grew by 12% for the year, how much did you invest in each fund?

28. *Investments* Referring to the preceding exercise, how much of your total investment of $10,000 would you have had to invest in each fund in order to wind up with $11,100 at the end of the 1992–1993 financial year?

29. *Intramural Sports* The best sports dorm on campus, Lombardi House, has won a total of 12 games this semester. Some of these games were soccer games, and the others were football games. According to the rules of the university, each win in a soccer game earns the winning house two points, while each win in a football game earns them four points. If the total number of points Lombardi House earned was 38, how many of each type of game did they win?

30. *Law* Enormous State University's campus publication, *The Campus Inquirer,* ran a total of 10 exposés dealing with alleged recruiting violations by the football team and theft by the student treasurer of the film society. Each exposé dealing with recruiting violations resulted in a $4 million libel suit, and the treasurer of the film society sued the paper for $3 million as a result of each exposé concerning his alleged theft. Unfortunately for *The Campus Inquirer,* all the lawsuits were successful, and the paper wound up being ordered to pay $37 million in damages. (It closed shortly thereafter.) How many of each type of exposé did the paper run?

31. *Purchasing (from the GMAT)* Elena purchased Brand X pens for $4.00 apiece and Brand Y pens for $2.80 apiece. If Elena purchased a total of 12 of these pens for $42.00, how many Brand X pens did she purchase?

32. *Purchasing (based on a question from the GMAT)* Earl is ordering supplies. Yellow paper costs $5.00 per ream, while white paper costs $6.50 per ream. He would like to order 100 reams total and has a budget of $560. How many reams of each color should he order?

33. *Pollution* According to the results of a 10-year study funded by Congress in 1980,† the New York Adirondacks and Florida are the two regions in the U.S. with the highest percentage of acidified lakes. These two regions account for a total of 664 acidified lakes, with Florida accounting for 302 more of these polluted lakes than the Adirondacks.
  (a) How many acidified lakes did the study find in each of the two regions?
  (b) Given that there are a total of 1,290 lakes in the Adirondacks and 2,098 lakes in Florida, which of the two regions has the worse pollution record?

34. *Pollution* Chlorofluorocarbons (CFCs) have been implicated in the depletion of the earth's ozone shield in the stratosphere. The two principal CFCs are referred to as CFC-11 and CFC-12. It has been estimated‡ that by 1987 a total of 16.570 million tons of CFCs had been released into the atmosphere. The amount of CFC-11 is only $\frac{2}{3}$ the amount of CFC-12. How many million tons of each variety of CFC had been released into the atmosphere?

35. *Pollution* Joe Slo, a college sophomore, neglected to wash his dirty laundry for six weeks. By the end of that time, his roommate had had enough, and he tossed Joe's dirty socks and T-shirts into the trash, counting a total of 44 items. He noticed that there were three times as many dirty socks as T-shirts. How many of each item did he throw out?

36. *Diet* The local sushi bar serves 1-ounce pieces of raw salmon (consisting of 50% protein) and $1\frac{1}{4}$ ounce pieces of raw tuna (40% protein). Assuming a customer's total intake of protein amounts to 1.5 ounces after consuming a total of three pieces, how

---

* Source: "No-brainer funds gaining in appeal," *The Chicago Tribune,* March 4, 1993, Section 3, p. 3.
† The National Acid Rain Precipitation Assessment Program
‡ Source: OECD Environmental Data (Paris: OECD, 1985, 1989.)

many of each did the customer consume? (Fractions of pieces are permitted.)

37. *Budgeting* Radio station WBAH is planning a year-long promotional prize give-away. The prize budget is $18,000 for the year, and the prizes are trips for two to the Bahamas and VCRs. Each trip for two to the Bahamas costs the station $800, and each VCR costs $200. The station manager decides that twice as many VCRs as trips to the Bahamas should be awarded. How should the station allocate the prizes?

38. *Test Taking* Your class has just taken a standardized test. Fifty of the questions were true-false items and the remaining 50 questions were multiple-choice items. Each true-false question was worth 10 points, and each multiple-choice question was worth 20 points. At the conclusion of the test, your friend said that she had managed to answer a total of 90 questions in the time allotted and had answered 6 more multiple-choice questions than true-false questions. If she got all the questions she answered right, what was her score?

39. *Management (from the GMAT)* A manager has $6,000 budgeted for raises for 4 full-time and 2 part-time employees. Each of the full-time employees receives the same raise, which is twice the raise that each of the part-time employees receives. What is the amount of the raise that each full-time employee receives?

40. *Publishing (from the GMAT)* There were 36,000 hardback copies of a certain novel sold before the paperback version was issued. From the time the first paperback copy was sold until the last copy of the novel was sold, 9 times as many paperback copies as hardback copies were sold. If a total of 441,000 copies of the novel were sold in all, how many paperback copies were sold?

41. *Publishing* The demand per year for Finite Math books is given by $q = -1000p + 140,000$, where $p is the price per book. The supply is given by $q = 2000p + 20,000$. Find the price at which supply and demand will balance.

42. *Airplane Manufacture* The demand per year for jumbo jets is given by $q = -2p + 18$, where p is the price per jet in millions of dollars. The supply is given by $q = 3p + 3$. Find the price at which supply and demand will balance.

43. *Supply and Demand (from the GRE economics test)* The demand curve for widgets is given by $D = 85 - 5P$, and the supply curve is given by $S = 25 + 5P$, where P is the price of widgets. When the widget market is in equilibrium, how many widgets are bought and sold?

44. *Supply and Demand (from the GRE economics test)* In the market for soybeans, the demand and supply functions are $Q_D = 100 - 10P$, and $Q_S = 20 + 5P$, where Q_D is quantity demanded, Q_S is quantity supplied, and P is price in dollars. If the government sets a price floor of $7, what will be the resulting surplus or shortage?

45. *Nutrition* A 4-ounce serving of *Campbell's* Pork & Beans® contains 5 grams of protein and 21 grams of carbohydrates.* A typical slice of "lite" rye bread contains 4 grams of protein and 12 grams of carbohydrates. The U.S. RDA is 60 grams of protein per day.
  (a) I am planning a meal of "beans-on-toast" and wish to have it supply one third of the U.S. RDA for protein, and 80 grams of carbohydrates. How should I prepare my meal?
  (b) Is it possible to have my meal supply the same amount of protein as in part (a) but only 60 grams of carbohydrates?

46. *Nutrition* A 4-ounce serving of *Campbell's* Pork & Beans® contains 5 grams of protein and 21 grams of carbohydrates. A typical slice of white bread provides 2 grams of protein and 11 grams of carbohydrates per slice.
  (a) I am planning a meal of "beans-on-toast" and wish to have it supply one half of the U.S. RDA for protein (see Exercise 45), and 139 grams of carbohydrates. How should I prepare my meal?
  (b) Is it possible to have my meal supply the same amount of protein as in part (a) but only 100 grams of carbohydrates?

47. *Supply and Demand* Having recently been hired as Chief Financial Consultant to Gourmet Pet Foods Inc., I attempted to impress my boss by revising the company's pricing policies for the company's latest product, Filet Mignon Tidbits (FMT).
  (a) In order to come up with a reasonable model for the demand equation, I decided to use a linear form, $q = mp + b$. I studied the company's sales records and discovered the following: In June of

▼ *Label information on a 16-ounce can

1992, the company obtained orders for 5,500 cans of FMT at $4 per can. At the start of the following month, the company raised the price to $10 per can. Unfortunately, the demand dropped dramatically, and the company obtained orders for only 100 cans that month. What was the demand equation I came up with?

(b) Once I had produced the demand equation for Filet Mignon Tidbits, I spoke with the production manager, who informed me that as a result of resource allocation restrictions and cost restraints, the company could afford to produce 1,900 cans per month if they sold for $2 per can, and 3,300 cans per month if they sold for $4 per can. Based on these figures, what should I have used as a linear model, $q = mp + b$, for the supply equation?

(c) Use your answers to parts (a) and (b) to determine how much Gourmet Pet Foods Inc. should charge per can of Filet Mignon Tidbits in order to balance supply and demand.

48. **Supply and Demand** Gourmet Pet Foods's latest product is Kitty Kaviar. As the company's Chief Financial Consultant, I decided (flushed with the success of solving the Filet Mignon Tidbits pricing problem—see the preceding exercise) to turn my attention to this troublesome product.

(a) It appears that the company was having problems with demand. It had priced Kitty Kaviar at $10 per can and sold only 1,000 cans per month. Then, for some inexplicable reason, it raised the price to $11 per can, and monthly sales dropped to 200 cans. What linear demand equation models these data?

(b) The production manager informed me that the company could afford to manufacture 7,400 cans per month to sell for $10 per can and only 3,900 cans per month to sell for $5 per can. What linear supply equation models these data?

(c) Use your answers to parts (a) and (b) to determine how much Gourmet Pet Foods Inc. should charge per can of Kitty Kaviar in order to balance supply and demand.

COMMUNICATION AND REASONING EXERCISES

49. A system of three equations in two unknowns corresponds to three lines in the plane. Describe several ways that these lines might be positioned if the system has a unique solution.

50. A system of three equations in two unknowns corresponds to three lines in the plane. Describe several ways that these lines might be positioned if the system has no solutions.

51. Both the supply and demand equations for a certain product have negative slope. Can there be an equilibrium price? Explain.

52. Referring back to Exercise 23, suppose (for the sake of argument) that the correct answer to the corresponding system of equations was 198.7 gallons of vanilla and 100.89 gallons of chocolate. If your factory can produce only whole numbers of gallons, would you recommend rounding the answers to the nearest whole number? Explain.

53. Invent an interesting application leading to a system of two equations in two unknowns with a unique solution.

54. Invent an interesting application leading to a system of two equations in two unknowns with no solution.

2.2 USING MATRICES TO SOLVE SYSTEMS WITH TWO UNKNOWNS

In this section, we describe another method for solving systems of equations with two unknowns. While this method may seem a little cumbersome at first, it will prove *immensely* useful in this and the next several chapters.

First of all, notice that a linear equation with two unknowns (e.g. $2x - y = 3$) is entirely determined by its *coefficients* (here, the numbers 2 and -1) and its *constant term* or *right-hand side* (here, 3). In other words, if you were simply given the row of numbers

$$[2 \quad -1 \quad 3],$$

you could easily reconstruct the original linear equation by multiplying the first number by x, the second by y, and inserting a plus sign and an equals sign, as follows.

$$2 \cdot x + (-1) \cdot y = 3$$

or

$$2x - y = 3$$

Similarly, the equation

$$-4x + 2y = 0$$

is represented by the row

$$[-4 \quad 2 \quad 0],$$

and

$$-3y = \tfrac{1}{4}$$

by

$$[0 \quad -3 \quad \tfrac{1}{4}].$$

As the last example shows, the first number is always the coefficient of x and the second the coefficient of y. If an x or a y is missing, then we write down a zero. We shall call such a row the **coefficient row** of an equation.

Q What is the purpose of representing linear equations in this way?

A Recall, first, what happens when we multiply an equation by a number. For example, consider multiplying the equation $2x - y = 3$ by -2 to get $-4x + 2y = -6$. All we've done is multiply the coefficients and the right-hand side by -2. This corresponds to *multiplying the row* $[2 \quad -1 \quad 3]$ *by* -2, i.e., multiplying every number in the row by -2. We shall see that any manipulation we would do with the equations can be done instead with rows.

Here is the same operation in the language of equations and in the language of rows. (We refer to the equation here as *Equation* 1, or simply E_1 for short, and to the row as *Row* 1, or R_1.)

	Equation	Row
	E_1: $\quad 2x - y = 3$	$[2 \quad -1 \quad 3] \quad R_1$
Multiply by -2:	$(-2)E_1$: $\quad -4x + 2y = -6$	$[-4 \quad 2 \quad -6] \quad (-2)R_1$

2.2 Using Matrices to Solve Systems with Two Unknowns

> Multiplying an equation by the number a corresponds to multiplying the coefficient row by a.

Now look at what happens if we add two equations:

	Equation	Row
	E_1: $\ 2x - 3y = \ \ \ 3$	$\begin{bmatrix} 2 & -3 & 3 \end{bmatrix}\ R_1$
	E_2: $-x + 2y = -4$	$\begin{bmatrix} -1 & 2 & -4 \end{bmatrix}\ R_2$
Add:	$E_1 + E_2$: $\ \ x - \ \ y = -1$	$\begin{bmatrix} 1 & -1 & -1 \end{bmatrix}\ R_1 + R_2$

All we are really doing is *adding the corresponding entries in the rows*, or *adding the rows*. In other words,

> Adding two equations corresponds to adding their coefficient rows.

In short, the manipulations of equations that we saw in the last section can be done more easily with rows, since we don't have to carry the xs and ys along with us. The xs and ys can always be inserted at the end if desired.

You might also believe that it is easier to get a computer to manipulate rows than equations, and this is true. The technique that we develop in this section is the basis for most computer and calculator programs written to solve systems of linear equations. (If the truth be told, computers use a variation of this method optimized for speed and memory use, but the method we discuss in this book seems the easiest for hand calculations.)

Now, besides adding equations or multiplying equations by nonzero numbers, we have seen that it is also useful to combine these, as in

$$\text{Compute } 2E_1 + 4E_2,$$

or

$$\text{Compute } E_2 - 7E_1.$$

Here is an example.

	Equation	Row
	E_1: $\ 2x - 3y = \ \ \ 3$	$\begin{bmatrix} 2 & -3 & 3 \end{bmatrix}\ R_1$
	E_2: $-x + 2y = -4$	$\begin{bmatrix} -1 & 2 & -4 \end{bmatrix}\ R_2$
	$2E_1 + 3E_2$: $\ \ x \ \ \ \ \ \ \ \ \ = -6$	$\begin{bmatrix} 1 & 0 & -6 \end{bmatrix}\ 2R_1 + 3R_2$

On the Row side, all we are doing is taking 2(*first row*) + 3(*second row*) resulting in the answers

$$2(2) + 3(-1) = 1, \text{ going down the first column;}$$
$$2(-3) + 3(2) = 0, \text{ going down the second column;}$$
$$2(3) + 3(-4) = -6, \text{ going down the third column.}$$

(Just why we wish to do such things with equations or rows will become apparent shortly, if it is not already.)

Instead of systems of equations, we can now work with collections of rows. We shall label the rows R_1, R_2, and so on. The manipulations we are talking about are known as **elementary row operations,** and there are really three kinds.

ELEMENTARY ROW OPERATIONS*

Type 1: Replacing R_i by aR_i (where $a \neq 0$)

(in words: multiplying or dividing a row by a nonzero number)

Type 2: Replacing R_i by $aR_i \pm bR_j$ (where $a \neq 0$)

(multiplying a row by a nonzero number and adding or subtracting a multiple of another row)

Type 3: Switching the order of the rows

(This corresponds to switching the order in which we write the equations, and occasionally this will be convenient.)

Q Why are we not allowed to multiply a row by 0?

A Multiplying a row or an equation by 0 gives us the not-very-surprising result $0 = 0$. In fact, we lose any information that the equation provided, which usually means that the resulting system has more solutions than the original system. The elementary row operations as given guarantee that the new system will have exactly the same solutions as the original system.

SOLUTION OF SYSTEMS OF EQUATIONS BY ROW OPERATIONS

Now we put rows to work for us in solving systems of equations. First of all, let's start with a complicated-looking system of equations, such as

$$-\frac{2x}{3} + \frac{y}{2} = -3$$
$$\frac{x}{4} - y = \frac{11}{4}.$$

---

* We are using the term "elementary row operations" a little more freely than most books. Some mathematicians insist that $a = 1$ in an operation of Type 2, but the less restrictive version is very useful.

This system corresponds to the following two rows.

$$\begin{bmatrix} -\frac{2}{3} & \frac{1}{2} & -3 \\ \frac{1}{4} & -1 & \frac{11}{4} \end{bmatrix}$$

We call this the **augmented matrix** of the system of equations. (The term "augmented" means that we have included the right-hand sides -3 and $\frac{11}{4}$. We shall often drop the word "augmented" and simply refer to the matrix of the system.) A **matrix** (plural: **matrices**) is nothing more than a rectangular array of numbers as above. (We'll be studying matrices in their own right more carefully in Chapter 3.) Now what do we do with this matrix?

Step 1

Clear the fractions and/or decimals (if any) using operations of Type 1. Note that this is exactly what we did with equations. To clear the fractions, we multiply the first row by 6 and the second row by 4. We record the operations by writing the symbolic form of an operation next to the row we want to change, as follows.

$$\begin{bmatrix} -\frac{2}{3} & \frac{1}{2} & -3 \\ \frac{1}{4} & -1 & \frac{11}{4} \end{bmatrix} \begin{matrix} 6R_1 \\ 4R_2 \end{matrix}$$

By this we mean that we will replace the first row by $6R_1$ and the second by $4R_2$. Doing these operations gives

$$\begin{bmatrix} -4 & 3 & -18 \\ 1 & -4 & 11 \end{bmatrix}.$$

Step 2

*Designate the first nonzero entry in the first row as the **pivot**.* All we are doing here is designating the entry -4 in the first row as the "pivot." We designate it thus by highlighting it in color. You might want to draw a circle around it.

$$\begin{bmatrix} -4 & 3 & -18 \\ 1 & -4 & 11 \end{bmatrix} \leftarrow \text{Pivot row}$$
↑
Pivot column

Step 3

Use the pivot to clear its column using operations of Type 2. By **clearing a column**, we mean producing a matrix in which the pivot is the only nonzero number in its column.

$$\begin{bmatrix} -4 & 3 & -18 \\ 0 & \# & \# \end{bmatrix} \leftarrow \text{Desired row 2 (the ``\#''s stand for as yet unknown numbers)}$$
↑
Cleared pivot column

To get this, we replace R_2 by something else of the form $aR_2 \pm bR_1$. We need to choose a and b so that we get the desired cancellation. If you cannot see right away what numbers will work, there is a simple mechanical rule-of-thumb to do this.

1. Write the row you need to change on the left, and the pivot row on the right.

$$\underset{\underset{\text{Row to change}}{\uparrow}}{R_2} \quad \underset{\underset{\text{Pivot row}}{\uparrow}}{R_1}$$

2. *Ignoring signs,* decide what *positive whole number* to multiply each row by so that the row you want to change and the pivot row have *the same numerical values in the pivot column*. In the situation at hand, the pivot column is $\begin{bmatrix} -4 \\ 1 \end{bmatrix}$, so you can accomplish this by multiplying the second row by 4 and the first by 1. Doing so gives

$$4R_2 \quad 1R_1.$$

3. If the two entries in the pivot column have the same sign, insert a minus ($-$), but if they have different signs, insert a plus ($+$). Here you get the following instruction:

$$4R_2 + 1R_1.$$

4. Write this instruction next to the row you want to change, and then replace that row using the instruction.

$$\begin{bmatrix} -4 & 3 & -18 \\ 1 & -4 & 11 \end{bmatrix} {}_{4R_2 + 1R_1} \rightarrow \begin{bmatrix} -4 & 13 & -18 \\ 0 & -13 & 26 \end{bmatrix}$$

Thus, we have completed Step 3 and cleared the pivot column. The next step is one that can be performed at any time.

Simplification Step

If at any stage of the process, all the numbers in a row are multiples of an integer, divide by that integer—a Type 1 operation.

This is an optional, but extremely helpful step, as it makes the numbers smaller and thus easier to work with. Here, we notice that the entries in R_2 are divisible by the lucky number 13, so we divide that row by 13.

$$\begin{bmatrix} -4 & 13 & -18 \\ 0 & -13 & 26 \end{bmatrix} {}_{\frac{1}{13}R_2} \rightarrow \begin{bmatrix} -4 & 3 & -18 \\ 0 & -1 & 2 \end{bmatrix}$$

Step 4

Select the first nonzero number in the second row as pivot, and clear its column. Here we have combined two steps in one: selecting the new pivot and clearing the column. The pivot is shown below, as well as the desired result when the column has been cleared.

$$\begin{bmatrix} -4 & 3 & -18 \\ 0 & -1 & 2 \end{bmatrix} \rightarrow \begin{bmatrix} \# & 0 & \# \\ 0 & -1 & 2 \end{bmatrix} \leftarrow \text{Desired row}$$

$\qquad\qquad\uparrow \qquad\qquad\qquad\uparrow$
$\qquad$ Pivot column $\quad$ Cleared pivot column

We now wish to get a zero in place of the 3 in the pivot column. Thus, we need to change R_1. We again go through the steps required to come up with the instruction.

1. Write the row you need to change on the left and the pivot row on the right.

$$\begin{array}{cc} R_1 & R_2 \\ \uparrow & \uparrow \\ \text{Row to change} & \text{Pivot row} \end{array}$$

2. *Ignoring signs,* decide what to multiply each row by so that the row you want to change and the pivot row have *the same numerical values in the pivot column.* In the situation at hand, the pivot column is $\begin{bmatrix} 3 \\ -1 \end{bmatrix}$, so you can accomplish this by multiplying the first row by 1 and the second by 3. Doing so gives

$$1R_1 \qquad 3R_2.$$

3. If the two entries in the pivot column have the same sign, insert a minus $(-)$, but if they have different signs, insert a plus $(+)$. Here you get the instruction

$$1R_1 + 3R_2.$$

4. Write this instruction next to the row you want to change, and then replace that row using the instruction.

$$\begin{bmatrix} -4 & 3 & -18 \\ 0 & -1 & 2 \end{bmatrix} \begin{array}{c} 1R_1 + 3R_2 \\ \end{array} \rightarrow \begin{bmatrix} -4 & 0 & -12 \\ 0 & -1 & 2 \end{bmatrix}$$

Now we are essentially done, except for one last step.

Final Step

Using operations of Type 1, turn each pivot (i.e., the first nonzero entry in each row) into a 1.

We can accomplish this by multiplying the first row by $-\frac{1}{4}$ and the second row by -1.

$$\begin{bmatrix} -4 & 0 & -12 \\ 0 & -1 & 2 \end{bmatrix} \begin{array}{c} -\frac{1}{4}R_1 \\ -R_2 \end{array} \rightarrow \begin{bmatrix} 1 & 0 & 3 \\ 0 & 1 & -2 \end{bmatrix}$$

Now we can stop. Notice that the final matrix has the following nice form.

$$\begin{bmatrix} 1 & 0 & \# \\ 0 & 1 & \# \end{bmatrix}$$

(This is the form we will always obtain when there is a unique solution.) Translating back into equations gives

$$1x + 0y = 3$$
$$0x + 1y = -2.$$

In other words,

$$x = 3 \quad \text{and} \quad y = -2,$$

and so we have found the solution, which we can also write as $(x, y) = (3, -2)$.

You may have the impression that this was a laboriously long and tedious process. However, it is possible to repeat the whole thing very quickly in the following six steps.

$$\begin{bmatrix} -\frac{2}{3} & \frac{1}{2} & -3 \\ \frac{1}{4} & -1 & \frac{11}{4} \end{bmatrix} \begin{matrix} 6R_1 \\ 4R_2 \end{matrix} \rightarrow \begin{bmatrix} -4 & 3 & -18 \\ 1 & -4 & 11 \end{bmatrix} \begin{matrix} \\ 4R_2 + 1R_1 \end{matrix} \rightarrow$$

$$\begin{bmatrix} -4 & 3 & -18 \\ 0 & -13 & 26 \end{bmatrix} \begin{matrix} \\ \frac{1}{13}R_2 \end{matrix} \rightarrow \begin{bmatrix} -4 & 3 & -18 \\ 0 & -1 & 2 \end{bmatrix} \begin{matrix} 1R_1 + 3R_2 \\ \end{matrix} \rightarrow$$

$$\begin{bmatrix} -4 & 0 & -12 \\ 0 & -1 & 2 \end{bmatrix} \begin{matrix} -\frac{1}{4}R_1 \\ -R_2 \end{matrix} \rightarrow \begin{bmatrix} 1 & 0 & 3 \\ 0 & 1 & -2 \end{bmatrix}$$

This procedure is called **row reduction**.

Here is a summary of the procedure we used.

PROCEDURE FOR ROW REDUCING A MATRIX WITH TWO ROWS

Step 1 Clear the fractions and/or decimals (if any) using operations of Type 1.

Step 2 Designate the first nonzero entry in the first row as the "pivot."

Step 3 Use the pivot to clear its column using operations of Type 2.

Step 4 Now designate the first nonzero entry in the second row as the pivot and apply Step 3.

Simplification Step If at any stage of the process all the numbers in a row can be divided by an integer, do it.

Final Step Using operations of Type 1, turn each pivot (i.e., the first nonzero entry in each row) into a 1.

USING TECHNOLOGY

Calculators

Since many calculators can now store matrices and do row operations, you can use one to help you with row reduction, as long as the calculator can hold a matrix large enough for the problem you are doing (some calculators are severely limited in the size of matrix they can hold). We have included programs for the TI-81 and TI-82 in Appendix B. The TI-82 program allows you to completely reduce any matrix with dimensions up to 99×99 according to the above procedure. You can also consult *Graphing Calculator Lessons for Finite Mathematics* by Paula Young (published by HarperCollins College Publishers in 1993), which focuses on the TI-81.

Special-Purpose Computer Software

There are many computer software packages capable of doing row operations and/or pivoting (including the software package that is available with this book). There are also software packages that perform the whole row reduction automatically. Their value for learning the workings of row reduction is nil, though, since they do *all* the work for you. On the other hand, these are the programs that are used to solve the really huge systems of linear equations that arise in practice.

Spreadsheet Software

Electronic spreadsheets are proving to be useful tools for doing matrix operations. The text *Topics in Finite Mathematics—An Introduction to the Electronic Spreadsheet* by Samuel W. Spero (published by HarperCollins College Publishers in 1993) will show you how to program your spreadsheet for row reduction as well as for other procedures discussed in this text.

▶ NOTE Since all of the above technologies make the handling of decimals easy, you might find the following convenient when using one of them.

1. Don't bother to get rid of decimals—calculators can handle them with ease.
2. After selecting your pivot, and prior to clearing the pivot column, *divide the pivot row by the value of the pivot, thereby turning the pivot into a* 1. Although doing so might well result in decimals, it makes the row operations easy to specify. It also makes the simplification step unnecessary. ◀

We now go through several more examples, but this time in accelerated fashion. It would be good practice to try to go through the steps first on your own, and then check your solution against the one given.

▼ **EXAMPLE 1**

By row reduction, solve the system

$$0.5x + 0.3y = 0.2$$
$$4x + y = 1.$$

SOLUTION

$$\begin{bmatrix} 0.5 & 0.3 & 0.2 \\ 4 & 1 & 1 \end{bmatrix} \begin{matrix} 10R_1 \\ \, \end{matrix} \to \begin{bmatrix} 5 & 3 & 2 \\ 4 & 1 & 1 \end{bmatrix} \begin{matrix} \, \\ 5R_2 - 4R_1 \end{matrix} \to$$

$$\begin{bmatrix} 5 & 3 & 2 \\ 0 & -7 & -3 \end{bmatrix} \begin{matrix} 7R_1 + 3R_2 \\ \, \end{matrix} \to \begin{bmatrix} 35 & 0 & 5 \\ 0 & -7 & -3 \end{bmatrix} \begin{matrix} \frac{1}{35}R_1 \\ -\frac{1}{7}R_2 \end{matrix} \to$$

$$\begin{bmatrix} 1 & 0 & \frac{1}{7} \\ 0 & 1 & \frac{3}{7} \end{bmatrix}$$

Thus, $x = \frac{1}{7}$ and $y = \frac{3}{7}$.

Before we go on... We always end by checking our solution:

$$0.5\left(\frac{1}{7}\right) + 0.3\left(\frac{3}{7}\right) = \frac{5}{70} + \frac{9}{70} = \frac{14}{70} = 0.2 \checkmark$$

$$4\left(\frac{1}{7}\right) + \frac{3}{7} = \frac{7}{7} = 1. \checkmark$$

If your graphing calculator or computer is operating in decimal mode, you will not obtain the exact answers after the last step—turning the pivots into 1s. (You will get a decimal approximation instead.) You can avoid this by using the calculator to get to the step immediately before that, where all the entries in the matrix are whole numbers, and then do the easy last step by hand.

You might also try following our advice on using technology—don't bother clearing fractions or decimals, and always convert pivots into 1s *before* pivoting—and proceed as follows.

$$\begin{bmatrix} 0.5 & 0.3 & 0.2 \\ 4 & 1 & 1 \end{bmatrix} \begin{matrix} \frac{1}{0.5}R_1 \\ \, \end{matrix} \to \begin{bmatrix} 1 & 0.6 & 0.4 \\ 4 & 1 & 1 \end{bmatrix} \begin{matrix} \, \\ R_2 - 4R_1 \end{matrix} \to$$

$$\begin{bmatrix} 1 & 0.6 & 0.4 \\ 0 & -1.4 & -0.6 \end{bmatrix} \begin{matrix} \, \\ -\frac{1}{1.4}R_2 \end{matrix} \to \begin{bmatrix} 1 & 0.6 & 0.4 \\ 0 & 1 & 0.428571\ldots \end{bmatrix} \begin{matrix} R_1 - 0.6R_2 \\ \, \end{matrix} \to$$

$$\begin{bmatrix} 1 & 0 & 0.1428571\ldots \\ 0 & 1 & 0.428571\ldots \end{bmatrix}$$

Notice that the procedure required the same number of steps as the "by hand" procedure, but it differed in that
1. the row operations were easier to specify (you specify these), and
2. the calculations were more difficult (the computer or calculator does these).

This makes it ideal for technology.

In the next example, we use row reduction to solve a system of three equations in two unknowns. (Incidentally, systems with one more equation than the number of unknowns arise in the study of Markov chains, a subject that combines linear equations and probability.)

▼ **EXAMPLE 2**

Solve the system

$$x + y = 1$$
$$13x - 26y = -11$$
$$26x - 13y = 2.$$

SOLUTION We shall use row reduction, and once more explain the steps. First, we start with the augmented matrix, which now has three rows.

$$\begin{bmatrix} 1 & 1 & 1 \\ 13 & -26 & -11 \\ 26 & -13 & 2 \end{bmatrix}$$

Since there is no need to eliminate fractions or decimals, select the pivot and choose the row operations to use.

$$\begin{bmatrix} 1 & 1 & 1 \\ 13 & -26 & -11 \\ 26 & -13 & 2 \end{bmatrix}$$
↑
Pivot column

Note that the pivot column now has two numbers we wish to clear: the 13 in R_2 and the 26 in R_3. Thus, we must replace both the second and third rows. The instruction for replacing each of these rows is composed in the same way as above.

$$\begin{bmatrix} 1 & 1 & 1 \\ 13 & -26 & -11 \\ 26 & -13 & 2 \end{bmatrix} \begin{matrix} \\ R_2 - 13R_1 \\ R_3 - 26R_1 \end{matrix}$$

(Row to change — Pivot row)

We can do both these row operations at once, since the rows we are changing don't affect each other.

$$\begin{bmatrix} 1 & 1 & 1 \\ 13 & -26 & -11 \\ 26 & -13 & 2 \end{bmatrix} \begin{matrix} \\ R_2 - 13R_1 \\ R_3 - 26R_1 \end{matrix} \rightarrow \begin{bmatrix} 1 & 1 & 1 \\ 0 & -39 & -24 \\ 0 & -39 & -24 \end{bmatrix}$$

Before going on, notice two things. First, the second and third rows are both divisible by 3, so we can divide these rows by 3 in order to make the numbers smaller. (Again, this step is optional.) The next thing to notice is that the second and third rows are the *same*, so the corresponding equations are the same, and we can throw one out without changing the solutions. The process of row reduction will do this automatically for us by changing one of the rows into a row of zeros. Anyway, we now divide the rows by 3, select the next pivot, and clear its column.

$$\begin{bmatrix} 1 & 1 & 1 \\ 0 & -39 & -24 \\ 0 & -39 & -24 \end{bmatrix} \begin{matrix} \\ \frac{1}{3}R_2 \\ \frac{1}{3}R_3 \end{matrix} \rightarrow \begin{bmatrix} 1 & 1 & 1 \\ 0 & -13 & -8 \\ 0 & -13 & -8 \end{bmatrix} \begin{matrix} 13R_1 + R_2 \\ \\ R_3 - R_2 \end{matrix} \rightarrow \begin{bmatrix} 13 & 0 & 5 \\ 0 & -13 & -8 \\ 0 & 0 & 0 \end{bmatrix}$$

As promised, we wound up with a row of zeros. If we translate a row of zeros into an equation, we get $0 = 0$, which tells us precisely nothing! Thus we can safely discard the row of zeros if we so choose. We shall, however, continue to show it. The remaining step is to turn the pivots into 1s.

$$\begin{bmatrix} 13 & 10 & 5 \\ 0 & -13 & -8 \\ 0 & 0 & 0 \end{bmatrix} \begin{matrix} \frac{1}{13}R_1 \\ -\frac{1}{13}R_2 \end{matrix} \rightarrow \begin{bmatrix} 1 & 0 & \frac{5}{13} \\ 0 & 1 & \frac{8}{13} \\ 0 & 0 & 0 \end{bmatrix}$$

Thus, the solution is $x = \frac{5}{13}$, $y = \frac{8}{13}$.

Row-Reduced Echelon Form

Before we go on to the next example, we pause for some terminology. Look at the matrices we wound up with in the preceding two examples.

$$\begin{bmatrix} 1 & 0 & \frac{1}{7} \\ 0 & 1 & \frac{3}{7} \end{bmatrix} \text{ and } \begin{bmatrix} 1 & 0 & \frac{5}{13} \\ 0 & 1 & \frac{8}{13} \\ 0 & 0 & 0 \end{bmatrix}$$

Both these matrices are said to be *row-reduced*. In general, row-reduced matrices have the properties specified by the following definition.

> **DEFINITION OF A ROW-REDUCED MATRIX**
>
> A matrix is said to be **row-reduced** or is referred to as being in **reduced row echelon form** if it satisfies the following properties.
>
> **P1.** The first nonzero entry in each row (called the **leading entry** or **pivot** of that row) is a 1. (These were our pivots as we did the row reduction.)
>
> **P2.** The columns of the leading entries are **clear** (i.e., they contain all zeros in all positions other than those of the leading entries).
>
> **P3.** The leading entry in each row is to the right of the leading entry in the row above, and the rows of zeros (if any) are down at the bottom.

Q Is it possible to convert any matrix whatsoever into reduced row echelon form by using row operations?

A Yes. We can accomplish P2 by clearing columns just as we have been doing. We can then satisfy P1 by using the "final step." Finally, we are allowed to shuffle the rows around to accomplish P3. Again, we call this procedure **row reduction**. It is also known as **Gauss-Jordan reduction**.*

Row reduction is the method we shall use to solve *all* systems of equations from now on, and with it we will do a lot more besides!

Now we'll continue with several more examples.

▼ **EXAMPLE 3**

Solve the system

$$-\frac{2x}{3} + \frac{y}{2} = -3$$

$$\frac{x}{3} - \frac{y}{4} = \frac{3}{2}.$$

SOLUTION We proceed as usual.

$$\begin{bmatrix} -\frac{2}{3} & \frac{1}{2} & -3 \\ \frac{1}{3} & -\frac{1}{4} & \frac{3}{2} \end{bmatrix} \begin{matrix} 6R_1 \\ 12R_2 \end{matrix} \rightarrow \begin{bmatrix} -4 & 3 & -18 \\ 4 & -3 & 18 \end{bmatrix} R_2 + R_1 \rightarrow \begin{bmatrix} -4 & 3 & -18 \\ 0 & 0 & 0 \end{bmatrix}$$

The second row just disappeared! Don't panic. Just complete the row reduction as usual. If you check the requirements P1 through P3, you will notice that the only one that fails to apply to the present matrix is P1, since the leading entry in the first row is not a 1. We can remedy that.

$$\begin{bmatrix} -4 & 3 & -18 \\ 0 & 0 & 0 \end{bmatrix} -\frac{1}{4}R_1 \rightarrow \begin{bmatrix} 1 & -\frac{3}{4} & \frac{9}{2} \\ 0 & 0 & 0 \end{bmatrix}$$

The matrix is now row-reduced. Since this is as simple as we can get the matrix to be,† we abandon the matrix and go back to the realm of equations to see what we can do there. We obtain the single equation

$$x - \tfrac{3}{4}y = \tfrac{9}{2}.$$

Instead of solving for *y*, it seems much easier to solve for *x* from the way the equation is set up. Thus, we choose *y* to be arbitrary, and get $x = \tfrac{3}{4}y + \tfrac{9}{2}$. The general solution is

$$x = \tfrac{3}{4}y + \tfrac{9}{2}, \quad y \text{ being arbitrary.}$$

▼ * Carl Friedrich Gauss (1777–1855) was one of the great mathematicians, making fundamental contributions to number theory, analysis, probability and statistics, as well as many fields of science. He developed a method of solving systems of linear equations by row reduction, which was then refined by Wilhelm Jordan (1842–1899) into the form we are showing you here.

† If you don't believe this, just try to make it simpler by doing row operations, and see what happens!

Before we go on... As in the previous section, what we mean by this is that there are infinitely many solutions. We can get any particular solution by choosing a value (*any* value) for *y* and computing the corresponding value of *x* using the formula above. For example, we have the following particular solutions.

$$y = 0, x = \tfrac{9}{2}$$
$$y = 1, x = \tfrac{21}{4}$$
$$y = 2, x = 6$$
$$y = -100, x = -\tfrac{141}{2}$$

You should check that each of these is a solution to the original system of two equations.

 The last steps of the solution—converting back to equations and solving—must still be done by hand (unless you are using a very fancy software package). Remember that all the graphing calculator will do for you is the arithmetic for the matrix row operations. Once a matrix is reduced, it is up to you to interpret the result.

▼ **EXAMPLE 4**

Solve the system

$$\frac{x}{3} - \frac{y}{4} = -\frac{5}{2}$$

$$\frac{2x}{3} - \frac{y}{2} = -3$$

SOLUTION Proceeding as usual,

$$\begin{bmatrix} \tfrac{1}{3} & -\tfrac{1}{4} & -\tfrac{5}{2} \\ \tfrac{2}{3} & -\tfrac{1}{2} & -3 \end{bmatrix} \begin{matrix} 12R_1 \\ 6R_2 \end{matrix} \rightarrow \begin{bmatrix} 4 & -3 & -30 \\ 4 & -3 & -18 \end{bmatrix} R_2 - R_1 \rightarrow \begin{bmatrix} 4 & -3 & -30 \\ 0 & 0 & 12 \end{bmatrix}.$$

Now, the pivot seems too far to the right in the second row. Before proceeding with the reduction, look at the equation form of that suspicious second row:

$$0x + 0y = 12,$$

i.e., $\qquad\qquad\qquad 0 = 12.$

But this is impossible! Thus, we conclude (as we did in Example 4 in Section 1) that there are no solutions.

▶ **CAUTION** The following row operations are *not valid*:

1. multiplying a row by 0 (this removes an equation and changes the set of solutions);
2. multiplying one row by another (this makes no sense when you look at the corresponding equations);

3. dividing one row by another [just as nonsensical as (2)];
4. adding a constant to each element of a row (think about what this would mean for the corresponding equation). ◄

> **HINTS AND GENERAL GUIDELINES FOR ROW-REDUCING A MATRIX**
>
> 1. If you end up with fractions or decimals before the last step, you are not following the procedure correctly. A possible problem is that you may have multiplied a row by a fraction at some point.
> 2. If, when clearing a column, you *unclear* a column previously cleared, you have done something wrong. Most likely you have chosen a pivot in a row that already had one.
> 3. If, as a result of a row operation, a leading entry that was a 1 is changed to something else, don't worry. It will be changed back to a 1 in the last step.
> 4. Work *extremely* carefully. A simple error, even in a single sign, will cause your answer to be wrong. It is better to work slowly and accurately and get the right answer the first time than to have to redo the whole problem when your answer does not check. Your speed will pick up with practice.
> 5. *Always* check your answers in all of the equations of the original system. If even one equation fails, you've done something wrong.

▶ 2.2 EXERCISES

In Exercises 1 through 6, use row operations to reduce the given matrices to reduced row echelon form.

1. $\begin{bmatrix} -\frac{2}{3} & \frac{1}{3} & -3 \\ \frac{1}{3} & -\frac{2}{3} & \frac{11}{4} \end{bmatrix}$

2. $\begin{bmatrix} -\frac{2}{3} & \frac{2}{3} & -\frac{2}{3} \\ \frac{1}{3} & -\frac{2}{3} & \frac{11}{4} \end{bmatrix}$

3. $\begin{bmatrix} -2 & 1 & -3 \\ 1 & 3 & 1 \\ -1 & 4 & -2 \end{bmatrix}$

4. $\begin{bmatrix} 0 & 1 & -3 \\ -3 & 5 & -2 \\ -9 & 14 & -3 \end{bmatrix}$

5. $\begin{bmatrix} -2 & 1 & -3 \\ 1 & 4 & 1 \\ -1 & 4 & -2 \end{bmatrix}$

6. $\begin{bmatrix} 0 & 1 & 3 \\ -3 & 1 & -2 \\ -9 & 14 & -3 \end{bmatrix}$

Exercises 7–28 include some of the exercises from Section 2.1. The only difference is that here you should solve the given systems using row reduction.

7. $x + y = 4$
   $x - y = 2$

8. $2x + y = 2$
   $2x - 3y = 2$

9. $3x - 2y = 5$
   $2x - 3y = -5$

10. $2x + 3y = 5$
    $3x + 2y = 5$

11. $0.5x + 0.1y = 0.7$
    $0.2x - 0.2y = 0.6$

12. $-0.3x + 0.5y = 0.1$
    $0.1x - 0.1y = 0.4$

13. $\dfrac{x}{3} - \dfrac{y}{2} = 1$
    $\dfrac{x}{4} + y = -2$

14. $-\dfrac{2x}{3} + \dfrac{y}{2} = -\dfrac{1}{6}$
    $\dfrac{x}{4} - y = -\dfrac{3}{4}$

15. $2x + 3y = 1$
    $-x - \dfrac{3y}{2} = -\dfrac{1}{2}$

16. $2x - 3y = 1$
    $6x - 9y = 3$

17. $2x + 3y = 2$
    $-x - \dfrac{3y}{2} = -\dfrac{1}{2}$

18. $2x - 3y = 2$
    $6x - 9y = 3$

19. $x = 2y$
    $y = x - 10$

20. $x = 2y$
    $y = x + 10$

21. $2x + 10 = 3y$
    $3x + 10 = 4y$

22. $2x + 10 = 4y$
    $3x + 15 = 6y$

23. $x + y = 1$
    $3x - y = 0$
    $x - 3y = -2$

24. $x + y = 1$
    $3x - 2y = -1$
    $5x - y = \tfrac{1}{5}$

25. $x + y = 0$
    $3x - y = 1$
    $x - y = -1$

26. $x + 2y = 1$
    $3x - 2y = -2$
    $5x - y = \tfrac{1}{5}$

27. $0.5x + 0.1y = 1.7$
    $0.1x - 0.1y = 0.3$
    $x + y = \tfrac{11}{3}$

28. $-0.3x + 0.5y = 0.1$
    $x - y = 4$
    $\dfrac{x}{17} + \dfrac{y}{17} = 1$

In Exercises 29–34, use a graphing calculator or appropriate computer software. Round your solutions to four decimal places.

29. $2.1x - 3.9y = 4.3$
    $4.4x + 6.7y = 6.6$

30. $2.4x + 7.5y = 5.2$
    $1.3x - 4.7y = 6.3$

31. $3.12x + 1.11y = 26.9$
    $4.43x - 26.7y = 0$

32. $2.13x - 2.14y = 4.23$
    $31.3x - 1.71y = 1.3$

33. $31.3x - 1.71y = 1.3$
    $-12.52x + 0.684y = -0.52$

34. $4.4x + 6.7y = 6.6$
    $-1.76x - 2.68y = -2.64$

COMMUNICATION AND REASONING EXERCISES

35. By referring to an example (either from the text or your own example), explain why it is often simpler to pivot if the pivot entry happens to be a 1.

36. By referring to an example (either from the text or your own example), explain why it is often simpler to pivot if you first eliminate fractions and decimals (as we do in this text).

37. Why have we suggested that when using technology, it may be easier to (1) ignore the step about removing decimals and fractions, and (2) to change the pivot into 1 before clearing the column?

38. Give an example of a matrix representing a system of equations for which the "fraction-free" method of the text requires fewer steps than the "technology" method of first converting the pivots into 1s before pivoting.

2.3 USING MATRICES TO SOLVE SYSTEMS WITH THREE OR MORE UNKNOWNS

Next, we turn to the solution of systems of linear equations with three or more unknowns. The technique of row reduction explained in the preceding section works to solve any number of equations with any number of unknowns! We will begin, however, with equations with three unknowns.

2.3 Using Matrices to Solve Systems with Three or More Unknowns

> **LINEAR EQUATION IN THREE UNKNOWNS**
>
> A **linear equation in three unknowns** is an equation that can be written in the form
>
> $$ax + by + cz = d,$$
>
> where a, b, c, and d are numbers and x, y, and z* are the unknowns (as usual).

Examples of such equations are given below.

$2x + 3y - z = 0$ $\quad a = 2, b = 3, c = -1, d = 0$

$\dfrac{3}{4}x + 3z = -11$ $\quad a = \dfrac{3}{4}, b = 0, c = 3, d = -11$

$x = y - 4z + 1$ $\quad$ This can be written as
$\quad\quad\quad\quad\quad\quad\quad\quad x - y + 4z = 1.$
$\quad\quad\quad\quad\quad\quad\quad\quad a = 1, b = -1, c = 4, d = 1$

$0 = 0$ $\quad a = b = c = d = 0$

In order to solve systems involving three or more unknowns, we first translate linear equations into rows, just as we did in the last section.

▼ **EXAMPLE 1**

The equation $2x - 3y + z = 7$ may be represented by the row $[2 \ -3 \ 1 \ 7]$.

The equation $2x - z = 7$ may be represented by the row $[2 \ 0 \ -1 \ 7]$, since the coefficient of y is zero.

The equation $-x - 3y + z + 0.5w = 7$ may be represented by the row $[-1 \ -3 \ 1 \ 0.5 \ 7]$.

The row $[-1 \ \frac{1}{3} \ 4 \ 0 \ 0]$ represents the equation $-x + \frac{1}{3}y + 4z = 0$ in the *four* unknowns x, y, z and w.

The row $[0 \ 0 \ 0 \ 0 \ 1]$ represents the absurd equation $0 = 1$.

▼ * If we wanted to solve an equation with 27 unknowns, we would not have enough letters of the alphabet to name them all. Thus, you will sometimes find equations written

$$a_1 x_1 + a_2 x_2 + a_3 x_3 = b.$$

Here, a_1, a_2, a_3, and b are numbers, while the unknowns are x_1, x_2, and x_3. Since we shall rarely encounter more than four or five unknowns, we shall seldom use the subscripts, and we shall instead use the letters x, y, z, w, and so on, for the unknowns.

As was the case with equations in two unknowns, adding rows still corresponds to adding equations, and multiplying a row by a constant still corresponds to multiplying an equation by that constant. Therefore, it still makes sense to use the row operations discussed in the previous section, and the procedure of row reduction gives us a systematic way of solving any system.

▼ EXAMPLE 2

Solve the system

$$x - y + 5z = -6$$
$$3x + 3y - z = 10$$
$$x + 3y + 2z = 5.$$

SOLUTION The augmented matrix for the system is

$$\begin{bmatrix} 1 & -1 & 5 & -6 \\ 3 & 3 & -1 & 10 \\ 1 & 3 & 2 & 5 \end{bmatrix}.$$

Proceeding as before, we select the pivot in the first row and clear its column.

$$\begin{bmatrix} 1 & -1 & 5 & -6 \\ 3 & 3 & -1 & 10 \\ 1 & 3 & 2 & 5 \end{bmatrix} \begin{matrix} \\ R_2 - 3R_1 \\ R_3 - R_1 \end{matrix} \rightarrow \begin{bmatrix} 1 & -1 & 5 & -6 \\ 0 & 6 & -16 & 28 \\ 0 & 4 & -3 & 11 \end{bmatrix}$$

Now we use the optional simplification step to divide R_2 by 2.

$$\begin{bmatrix} 1 & -1 & 5 & -6 \\ 0 & 6 & -16 & 28 \\ 0 & 4 & -3 & 11 \end{bmatrix} \tfrac{1}{2}R_2 \rightarrow \begin{bmatrix} 1 & -1 & 5 & -6 \\ 0 & 3 & -8 & 14 \\ 0 & 4 & -3 & 11 \end{bmatrix}$$

Next, we select the pivot in the second row and clear its column.

$$\begin{bmatrix} 1 & -1 & 5 & -6 \\ 0 & 3 & -8 & 14 \\ 0 & 4 & -3 & 11 \end{bmatrix} \begin{matrix} 3R_1 + R_2 \\ \\ 3R_3 - 4R_2 \end{matrix} \rightarrow \begin{bmatrix} 3 & 0 & 7 & -4 \\ 0 & 3 & -8 & 14 \\ 0 & 0 & 23 & -23 \end{bmatrix}$$

We simplify R_3.

$$\begin{bmatrix} 3 & 0 & 7 & -4 \\ 0 & 3 & -8 & 14 \\ 0 & 0 & 23 & -23 \end{bmatrix} \tfrac{1}{23}R_3 \rightarrow \begin{bmatrix} 3 & 0 & 7 & -4 \\ 0 & 3 & -8 & 14 \\ 0 & 0 & 1 & -1 \end{bmatrix}$$

Now we have a pivot in the third row. We select it and clear its column in the usual way.

$$\begin{bmatrix} 3 & 0 & 7 & -4 \\ 0 & 3 & -8 & 14 \\ 0 & 0 & 1 & -1 \end{bmatrix} \begin{matrix} R_1 - 7R_3 \\ R_2 + 8R_3 \\ \end{matrix} \rightarrow \begin{bmatrix} 3 & 0 & 0 & 3 \\ 0 & 3 & 0 & 6 \\ 0 & 0 & 1 & -1 \end{bmatrix}$$

2.3 Using Matrices to Solve Systems with Three or More Unknowns

Finally, we turn all the pivots into 1s.

$$\begin{bmatrix} 3 & 0 & 0 & 3 \\ 0 & 3 & 0 & 6 \\ 0 & 0 & 1 & -1 \end{bmatrix} \begin{matrix} \frac{1}{3}R_1 \\ \frac{1}{3}R_2 \\ \end{matrix} \rightarrow \begin{bmatrix} 1 & 0 & 0 & 1 \\ 0 & 1 & 0 & 2 \\ 0 & 0 & 1 & -1 \end{bmatrix}$$

Now the matrix is reduced, so we translate back into equations to obtain the solution.

$$x = 1, y = 2, z = -1 \quad \text{or} \quad (x, y, z) = (1, 2, -1)$$

Before we go on... Just as in the case with equations in two unknowns, we check the solution by substitution.

$$1 - 2 + 5(-1) = -6 \quad \checkmark$$
$$3(1) + 3(2) - (-1) = 10 \quad \checkmark$$
$$1 + 3(2) + 2(-1) = 5 \quad \checkmark$$

You can use a graphing calculator or computer to help with row reduction, as explained in the previous section. While there are limitations on the size of matrix that a graphing calculator can handle, these limits are usually large enough to handle the examples and exercises in this text. A further point: You could try the above example on a graphing calculator without bothering to do the two simplification steps we performed in the calculation. Remember that simplification steps are only a device for making the numbers more manageable when you work by hand.

In the next example, we can skip steps. For instance, if a column is already cleared when we get to it, we can skip that step. If a column is partially cleared already, then we need only complete the process.

▼ **EXAMPLE 3**

Solve the system

$$\begin{aligned} x \quad\quad\, - z &= -3 \\ y/3 + z/3 &= 3 \\ x \quad\quad\, + z &= 7. \end{aligned}$$

(Notice how we've grouped the xs, ys and zs neatly underneath each other. This arrangement makes setting up the matrix easy.)

SOLUTION Here is the sequence of row operations required to reduce the matrix.

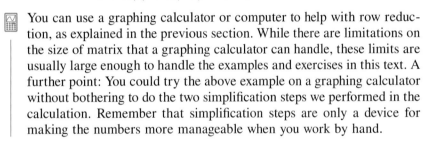

$$\begin{bmatrix} 1 & 0 & -1 & -3 \\ 0 & 1 & 1 & 9 \\ 0 & 0 & 2 & 10 \end{bmatrix} \tfrac{1}{2}R_3 \rightarrow \begin{bmatrix} 1 & 0 & -1 & -3 \\ 0 & 1 & 1 & 9 \\ 0 & 0 & 1 & 5 \end{bmatrix} \begin{matrix} R_1 + R_3 \\ R_2 - R_3 \\ {} \end{matrix} \rightarrow$$

$$\begin{bmatrix} 1 & 0 & 0 & 2 \\ 0 & 1 & 0 & 4 \\ 0 & 0 & 1 & 5 \end{bmatrix}$$

Thus, the solution is $x = 2$, $y = 4$, and $z = 5$, or $(x, y, z) = (2, 4, 5)$.

Before we go on... We check our answers.

$$2 \quad - 5 = -3 \ \checkmark$$
$$\tfrac{4}{3} + \tfrac{5}{3} = 3 \ \checkmark$$
$$2 \quad + 5 = 7 \ \checkmark$$

To show how easy it really is, let's solve a system of equations in *four* unknowns.

▼ **EXAMPLE 4**

Solve the system

$$\begin{aligned} x + y - z &= 3 \\ 2y + 2z - w &= 6 \\ x - 2z + 3w &= -5 \\ x - 4z &= 2. \end{aligned}$$

SOLUTION

$$\begin{bmatrix} 1 & 1 & -1 & 0 & 3 \\ 0 & 2 & 2 & -1 & 6 \\ 1 & 0 & -2 & 3 & -5 \\ 1 & 0 & -4 & 0 & 2 \end{bmatrix} \begin{matrix} {} \\ {} \\ R_3 - R_1 \\ R_4 - R_1 \end{matrix} \rightarrow \begin{bmatrix} 1 & 1 & -1 & 0 & 3 \\ 0 & 2 & 2 & -1 & 6 \\ 0 & -1 & -1 & 3 & -8 \\ 0 & -1 & -3 & 0 & -1 \end{bmatrix} \begin{matrix} 2R_1 - R_2 \\ {} \\ 2R_3 + R_2 \\ 2R_4 + R_2 \end{matrix} \rightarrow$$

$$\begin{bmatrix} 2 & 0 & -4 & 1 & 0 \\ 0 & 2 & 2 & -1 & 6 \\ 0 & 0 & 0 & 5 & -10 \\ 0 & 0 & -4 & -1 & 4 \end{bmatrix} \tfrac{1}{5}R_3 \rightarrow \begin{bmatrix} 2 & 0 & -4 & 1 & 0 \\ 0 & 2 & 2 & -1 & 6 \\ 0 & 0 & 0 & 1 & -2 \\ 0 & 0 & -4 & -1 & 4 \end{bmatrix} \begin{matrix} R_1 - R_3 \\ R_2 + R_3 \\ {} \\ R_4 + R_3 \end{matrix} \rightarrow$$

↑
Pivot column

Note that the pivot in the third row has been selected, since we work with the rows in order. The fact that the pivot is farther to the right than we expect needn't perturb us. If we wanted, we could always interchange Rows 3 and 4 first.

2.3 Using Matrices to Solve Systems with Three or More Unknowns

$$\begin{bmatrix} 2 & 0 & -4 & 0 & 2 \\ 0 & 2 & 2 & 0 & 4 \\ 0 & 0 & 0 & 1 & -2 \\ 0 & 0 & -4 & 0 & 2 \end{bmatrix} \begin{matrix} \frac{1}{2}R_1 \\ \frac{1}{2}R_2 \\ \\ \frac{1}{2}R_4 \end{matrix} \rightarrow \begin{bmatrix} 1 & 0 & -2 & 0 & 1 \\ 0 & 1 & 1 & 0 & 2 \\ 0 & 0 & 0 & 1 & -2 \\ 0 & 0 & -2 & 0 & 1 \end{bmatrix} \begin{matrix} R_1 - R_4 \\ 2R_2 + R_4 \\ \\ \end{matrix} \rightarrow$$

$$\underset{\text{Pivot column}}{\uparrow}$$

$$\begin{bmatrix} 1 & 0 & 0 & 0 & 0 \\ 0 & 2 & 0 & 0 & 5 \\ 0 & 0 & 0 & 1 & -2 \\ 0 & 0 & -2 & 0 & 1 \end{bmatrix} \begin{matrix} \\ \frac{1}{2}R_2 \\ \\ -\frac{1}{2}R_4 \end{matrix} \rightarrow \begin{bmatrix} 1 & 0 & 0 & 0 & 0 \\ 0 & 1 & 0 & 0 & \frac{5}{2} \\ 0 & 0 & 0 & 1 & -2 \\ 0 & 0 & 1 & 0 & -\frac{1}{2} \end{bmatrix} \begin{matrix} \\ \\ R_3 \leftrightarrow R_4 \end{matrix} \rightarrow$$

$$\begin{bmatrix} 1 & 0 & 0 & 0 & 0 \\ 0 & 1 & 0 & 0 & \frac{5}{2} \\ 0 & 0 & 1 & 0 & -\frac{1}{2} \\ 0 & 0 & 0 & 1 & -2 \end{bmatrix}$$

Thus, $x = 0$, $y = \frac{5}{2}$, $z = -\frac{1}{2}$, and $w = -2$, or $(x, y, z, w) = (0, \frac{5}{2}, -\frac{1}{2}, -2)$.

This calculation took eight steps. Solving systems with four or more unknowns can be very time-consuming (and error-prone) without a handy graphing calculator or computer software!

EXAMPLE 5

Solve the system

$$x + y + z = 1$$
$$\tfrac{1}{4}x - \tfrac{1}{2}y + \tfrac{3}{4}z = 0$$
$$x + 7y - 3z = 3.$$

SOLUTON Proceeding as usual,

$$\begin{bmatrix} 1 & 1 & 1 & 1 \\ \frac{1}{4} & -\frac{1}{2} & \frac{3}{4} & 0 \\ 1 & 7 & -3 & 3 \end{bmatrix} 4R_2 \rightarrow \begin{bmatrix} 1 & 1 & 1 & 1 \\ 1 & -2 & 3 & 0 \\ 1 & 7 & -3 & 3 \end{bmatrix} \begin{matrix} R_2 - R_1 \\ R_3 - R_1 \end{matrix} \rightarrow$$

$$\begin{bmatrix} 1 & 1 & 1 & 1 \\ 0 & -3 & 2 & -1 \\ 0 & 6 & -4 & 2 \end{bmatrix} \tfrac{1}{2}R_3 \rightarrow \begin{bmatrix} 1 & 1 & 1 & 1 \\ 0 & -3 & 2 & -1 \\ 0 & 3 & -2 & 1 \end{bmatrix} \begin{matrix} 3R_1 + R_2 \\ \\ R_3 + R_2 \end{matrix} \rightarrow$$

$$\begin{bmatrix} 3 & 0 & 5 & 2 \\ 0 & -3 & 2 & -1 \\ 0 & 0 & 0 & 0 \end{bmatrix}.$$

There are no leading entries in the third row, so there will be no pivot there. Thus, the matrix is almost reduced, except that the leading entries are not 1s. We finish the job with the final step.

$$\begin{bmatrix} 3 & 0 & 5 & 2 \\ 0 & -3 & 2 & -1 \\ 0 & 0 & 0 & 0 \end{bmatrix} \begin{matrix} \frac{1}{3}R_1 \\ -\frac{1}{3}R_2 \\ \, \end{matrix} \rightarrow \begin{bmatrix} 1 & 0 & \frac{5}{3} & \frac{2}{3} \\ 0 & 1 & -\frac{2}{3} & \frac{1}{3} \\ 0 & 0 & 0 & 0 \end{bmatrix}$$

Since that is as far as we can go with the matrix, we translate back into equations and obtain

$$x \quad\quad + \tfrac{5}{3}z = \tfrac{2}{3}$$
$$y - \tfrac{2}{3}z = \tfrac{1}{3}.$$

Now how do we obtain a solution? The thing to notice is that we can easily solve the first equation for x and the second for y, obtaining

$$x \quad\quad = \tfrac{2}{3} - \tfrac{5}{3}z$$
$$y = \tfrac{1}{3} + \tfrac{2}{3}z.$$

This is in fact the solution! We can choose z to be any number, and then get corresponding values for x and y according to the formulas, to get infinitely many solutions. Thus, the **general solution** is

$$x = \tfrac{2}{3} - \tfrac{5}{3}z$$
$$y = \tfrac{1}{3} + \tfrac{2}{3}z$$

z is arbitrary.

When we write the general solution this way, we say that we have a **parametrized solution,** and that z is the **parameter.** Specific choices of values for the parameter z give **particular solutions.** For example, the choice $z = 6$ gives the particular solution

$$x = \tfrac{2}{3} - \tfrac{5}{3}(6) = -\tfrac{28}{3}$$
$$y = \tfrac{1}{3} + \tfrac{2}{3}(6) = \tfrac{13}{3}$$
$$z = 6,$$

while the choice $z = 0$ gives the particular solution $(x, y, z) = (\tfrac{2}{3}, \tfrac{1}{3}, 0)$.

Before we go on... Why were there infinitely many solutions to this example? Because one of the equations "dropped out" during the row reduction (by turning into a row of zeros). The reason for this was that the third equation was really a combination of the first and second equations to begin with, so you really had only two equations in three unknowns.* Choosing a specific value for z (say, $z = 6$) had the effect of supplying the "missing"

▼ *In fact, you can check that the third equation, E_3, is equal to $3E_1 - 8E_2$. Thus, the third equation could have been left out, since it conveys no more information than the first two. The process of row reduction always eliminates such a redundancy by creating a row of zeros.

equation. How we can check this solution? We do what we always do: substitute the solution back into the original equations.

$$(\tfrac{2}{3} - \tfrac{5}{3}z) + (\tfrac{1}{3} + \tfrac{2}{3}z) + z = 1 - z + z = 1 \checkmark$$
$$\tfrac{1}{4}(\tfrac{2}{3} - \tfrac{5}{3}z) - \tfrac{1}{2}(\tfrac{1}{3} + \tfrac{2}{3}z) + \tfrac{3}{4}z = \tfrac{1}{6} - \tfrac{5}{12}z - \tfrac{1}{6} - \tfrac{1}{3}z + \tfrac{3}{4}z = 0 \checkmark$$
$$(\tfrac{2}{3} - \tfrac{5}{3}z) + 7(\tfrac{1}{3} + \tfrac{2}{3}z) - 3z = \tfrac{2}{3} - \tfrac{5}{3}z + \tfrac{7}{3} + \tfrac{14}{3}z - 3z = 3 \checkmark$$

Notice that we must leave z unknown; in particular, we do not substitute anything for the z that is already in each equation, but leave it as z.

▼ **EXAMPLE 6**

Solve the system

$$\begin{aligned} x + y - z &= 3 \\ 2y + 2z - w &= 6 \\ x - y - 3z + w &= -3. \end{aligned}$$

SOLUTION

$$\begin{bmatrix} 1 & 1 & -1 & 0 & 3 \\ 0 & 2 & 2 & -1 & 6 \\ 1 & -1 & -3 & 1 & -3 \end{bmatrix} R_3 - R_1 \rightarrow \begin{bmatrix} 1 & 1 & -1 & 0 & 3 \\ 0 & 2 & 2 & -1 & 6 \\ 0 & -2 & -2 & 1 & -6 \end{bmatrix} \begin{matrix} 2R_1 - R_2 \\ \\ R_3 + R_2 \end{matrix} \rightarrow$$

$$\begin{bmatrix} 2 & 0 & -4 & 1 & 0 \\ 0 & 2 & 2 & -1 & 6 \\ 0 & 0 & 0 & 0 & 0 \end{bmatrix} \begin{matrix} \tfrac{1}{2} R_1 \\ \tfrac{1}{2} R_2 \end{matrix} \rightarrow \begin{bmatrix} 1 & 0 & -2 & \tfrac{1}{2} & 0 \\ 0 & 1 & 1 & -\tfrac{1}{2} & 3 \\ 0 & 0 & 0 & 0 & 0 \end{bmatrix}$$

As the matrix is now reduced, we translate back into equations.

$$\begin{aligned} x - 2z + \tfrac{1}{2}w &= 0 \\ y + z - \tfrac{1}{2}w &= 3 \end{aligned}$$

Just as in the last example, we solve for the first variable in each equation.

$$\begin{aligned} x &= 2z - \tfrac{1}{2}w \\ y &= 3 - z + \tfrac{1}{2}w \end{aligned}$$

Here, we can get a solution after making *two* arbitrary choices, one for z and one for w. For example, we could choose $z = 0$ and $w = -2$ to obtain the solution $(1, 2, 0, -2)$. Alternatively, we could choose $z = 1$ and $w = 1$ and obtain a second solution, $(\tfrac{3}{2}, \tfrac{5}{2}, 1, 1)$. In general, the solution is

$$\begin{aligned} x &= 2z - \tfrac{1}{2}w \\ y &= 3 - z + \tfrac{1}{2}w \end{aligned}$$
z is arbitrary
w is arbitrary.

Here, our parametrized solution has two parameters, z and w.

Before we go on... We should check this solution by substituting it into the original equations, as in the previous example. Notice that, since both z and w are arbitrary, they will both be left as is (nothing will be substituted in their places).

$$(2z - \tfrac{1}{2}w) + (3 - z + \tfrac{1}{2}w) - z = 3 \checkmark$$
$$2(3 - z + \tfrac{1}{2}w) + 2z - w = 6 \checkmark$$
$$(2z - \tfrac{1}{2}w) - (3 - z + \tfrac{1}{2}w) - 3z + w = -3 \checkmark$$

▼ **EXAMPLE 7**

Solve the system

$$x + y + z = 1$$
$$2x - y + z = 0$$
$$4x + y + 3z = 3.$$

SOLUTION

$$\begin{bmatrix} 1 & 1 & 1 & 1 \\ 2 & -1 & 1 & 0 \\ 4 & 1 & 3 & 3 \end{bmatrix} \begin{matrix} \\ R_2 - 2R_1 \\ R_3 - 4R_1 \end{matrix} \rightarrow \begin{bmatrix} 1 & 1 & 1 & 1 \\ 0 & -3 & -1 & -2 \\ 0 & -3 & -1 & -1 \end{bmatrix} \begin{matrix} 3R_1 + R_2 \\ \\ R_3 - R_2 \end{matrix} \rightarrow$$

$$\begin{bmatrix} 3 & 0 & 2 & 1 \\ 0 & -3 & -1 & -2 \\ 0 & 0 & 0 & 1 \end{bmatrix}$$

Stop. That last row translates into $0 = 1$, which is a contradiction, and so this system has no solutions.

▶ **NOTES**

1. One of three things will happen in any system of linear equations.
   a. **There are no solutions** (as in Example 7). This occurs when you end up with a row of the form

   $$[0 \ \ 0 \ \ 0 \ \ \ldots \ \ 0 \ \ \#]$$

   where # is a nonzero number. As soon as you spot such a row, *check to see that you haven't made an error* and then stop. The given system has no solution, so there is no point in continuing any further.
   b. **There is exactly one solution (unique solution)** (as in Example 4). This occurs if, at the conclusion of the row reduction, translation back into equations yields a single value for *each* of the unknowns.
   c. **There are infinitely many solutions** (as in Example 5). This occurs if, at the conclusion of the row reduction, translation

back into equations does not yield a single value for *each* of the unknowns. In this case you can easily solve for the unknowns corresponding to the pivots, which will be the first unknowns in each equation. The other unknowns can be assigned arbitrary values. The arbitrary unknowns are called **parameters**, and the general solution as we have written it is called a **parametrized solution.**

These are the only things that can happen. It is impossible, for example, to have a system of linear equations with exactly eleven solutions. If it has more than one solution, it must have infinitely many solutions.*

2. If you end up with one or more rows of zeros, this means that one of the equations is redundant and can thus be ignored.

3. If a system of linear equations has no solution, it is said to be **inconsistent.** The reason for this term is that the equations actually contradict one another. Here is an obvious example of such a contradiction.

$$x + y + z = 0$$
$$x + y + z = 44$$

A reasonable person could hardly expect x, y, and z to add up to 0 and 44 at the same time!

4. A system of linear equations where there are fewer equations than unknowns is said to be **underdetermined.** These are the systems that usually give infinitely many solutions (as in Example 6) but may also result in no solutions if they are contradictory (as in the case immediately above). Such a system can never have a unique solution.

5. A system of equations in which the number of equations exceeds the number of unknowns is said to be **overdetermined.** In an overdetermined system, anything can happen, but such a system will usually be inconsistent. ◄

2.3 EXERCISES

In Exercises 1–42, solve each system.

1. $-x + 2y - z = 0$
$-x - y + 2z = 0$
$2x - z = 4$

2. $x + 2y = 4$
$y - z = 0$
$x + 3y - 2z = 5$

3. $x + y + 6z = -1$
$\frac{1}{3}x - \frac{1}{3}y + \frac{2}{3}z = 1$
$\frac{1}{2}x + z = 0$

4. $x - \frac{1}{2}y = 0$
$\frac{1}{3}x + \frac{1}{3}y + \frac{1}{3}z = 2$
$\frac{1}{2}x - \frac{1}{2}z = -1$

---

* The reason is roughly this: if you have any two solutions, say u and v, then the *average* of the two solutions, $\frac{1}{2}(u + v)$, is again a solution, so we get a third solution in this way. Now take the average of the first and third, giving a fourth solution, and so on, and so on . . . .

5. $-x + 2y - z = 0$
   $-x - y + 2z = 0$
   $2x - y - z = 0$

6. $x - \frac{1}{2}y = 0$
   $\frac{1}{2}x - \frac{1}{2}z = -1$
   $3x - y - z = -2$

7. $x + y + 2z = -1$
   $2x + 2y + 2z = 2$
   $\frac{3}{4}x + \frac{3}{4}y + z = \frac{1}{4}$
   $-x - 2z = 21$

8. $x + y - z = -2$
   $x - y - 7z = 0$
   $\frac{3}{4}x - \frac{1}{2}y + \frac{1}{4}z = 14$
   $x + y + z = 4$

9. $2x - y + z = 4$
   $3x - y + z = 5$

10. $3x - y - z = 0$
    $x + y + z = 4$

11. $0.75x - 0.75y - z = 4$
    $x - y + 4z = 0$

12. $2x - y + z = 4$
    $-x + 0.5y - 0.5z = 1.5$

13. $3x + y - z = 12$

14. $x + y - 3z = 21$

15. $x + y + 5z = 1$
    $y + 2z + w = 1$
    $x + 3y + 7z + 2w = 2$
    $x + y + 5z + w = 1$

16. $x + y + 4w = 1$
    $2x - 2y - 3z + 2w = -1$
    $4y + 6z + w = 4$
    $2x + 4y + 9z = 6$

17. $-\frac{3}{2}x + 2y - \frac{3}{4}z = 0$
    $-x - y + 2z = 0$
    $x - \frac{1}{3}y - z = 0$

18. $x - \frac{1}{2}y = 0$
    $\frac{1}{2}x - \frac{1}{2}z = -1$
    $\frac{3}{2}x - \frac{1}{2}y - \frac{1}{2}z = -2$

19. $x + y + 2z = -1$
    $2x + 2y + 2z = 2$
    $3x + 3y + 3z = 2$

20. $x + y - z = -2$
    $x - y - 7z = 0$
    $\frac{2}{7}x - \frac{8}{7}z = 14$

21. $-0.5x + 0.5y + 0.5z = 1.5$
    $4.2x + 2.1y + 2.1z = 0$
    $0.2x + 0.2z = 0$

22. $0.25x - 0.5y = 0$
    $0.2x + 0.2y - 0.2z = -0.6$
    $0.5x - 1.5y + z = 0.5$

23. $0.2x + 0.2y - z = 0$
    $x - 0.5y - 0.5z = 0$
    $0.4x + 0.1y - 1.1z = 0$

24. $0.2x + y = 0$
    $0.25x + y - 0.25z = 0$
    $0.3x + 1.3y - 0.2z = 0$

25. $x + y + 5z = 1$
    $y + 2z + w = 1$
    $x + y + 5z + w = 1$
    $x + 2y + 7z + 2w = 2$

26. $x + y + 4w = 1$
    $2x - 2y - 3z + 2w = -1$
    $4y + 6z + w = 4$
    $3x + 3y + 3z + 7w = 4$

27. $x_1 - 2x_2 + x_3 - 4x_4 = 1$
    $x_1 + 3x_2 + 7x_3 + 2x_4 = 2$
    $2x_1 + x_2 + 8x_3 - 2x_4 = 3$

28. $x_1 - 3x_2 - 2x_3 - x_4 = 1$
    $x_1 + 3x_2 + x_3 + 2x_4 = 2$
    $2x_1 - x_3 + x_4 = 3$

29. $z = 4(x - y)$
    $z = x - 2$
    $0.1x + 0.1y + 0.1z = 1.5$

30. $x = y + z$
    $3(y - z) = x$
    $0.1x + 0.2y + 0.1z = 0.8$

31. $p = -x + y + z$
    $y = x - z$
    $y = 2x$
    $p = 2x - 2y + 2$
    (x, y, z, and p are the unknowns.)

32. $c = -x - y + 2z$
    $x = -2y$
    $c + y = -x$
    $c = 2y + z - 1$
    (x, y, z and c are the unknowns.)

33. $2x + y + s = 300$
$x + 2y + t = 300$
$c = x + y$

(The unknowns are $x, y, c, s,$ and t.
Find a solution using s and t as parameters.)

34. $x + y + s = 100$
$4x + 2y + t = 200$
$p = 2x + 3y$

(The unknowns are $x, y, p, s,$ and t.
Find a solution using s and t as parameters.)

35. $-0.9x + 0.3z = 0$
$0.2x - 0.4y + 0.1z = 0$
$0.7x + 0.4y - 0.4z = 0$
$x + y + z = 1$

36. $-0.5x + 0.2y + 0.2z = 0$
$0.2x - 0.5y + 0.3z = 0$
$0.3x + 0.3y - 0.5z = 0$
$x + y + z = 1$

37. $\dfrac{x}{y} = -\dfrac{1}{2}$
$\dfrac{z-y}{x} = 4$
$\dfrac{z+10}{2x+y+z} = \dfrac{3}{2}$

38. $\dfrac{y}{z} = -\dfrac{2}{5}$
$\dfrac{x+2y}{z} = -1$
$\dfrac{3x+2y+100}{x+y-z} = 1$

39. $x + y + z + u + v = 15$
$y - z + u - v = -2$
$z + u + v = 12$
$u - v = -1$
$v = 5$

40. $x - y + z - u + v = 1$
$y + z + u + v = 2$
$z - u + v = 1$
$u + v = 1$
$v = 1$

41. $x - y + z - u + v = 0$
$y - z + u - v = -2$
$x - 2v = -2$
$2x - y + z - u - 3v = -2$
$4x - y + z - u - 7v = -6$

42. $x + y + z + u + v = 15$
$y + z + u + v = 3$
$x + 2y + 2z + 2u + 2v = 18$
$x - y - z - u - v = 9$
$x = 12$

In Exercises 43–46, use a graphing calculator or appropriate computer software to solve the systems of equations. Round your solutions to one decimal place.

43. $1.6x + 2.4y - 3.2z = 4.4$
$5.1x - 6.3y + 0.6z = -3.2$
$4.2x + 3.5y + 4.9z = 10.1$

44. $2.1x + 0.7y - 1.4z = -2.3$
$3.5x - 4.2y - 4.9z = 3.3$
$1.1x + 2.2y - 3.3z = -10.2$

45. $-0.2x + 0.3y + 0.4z - t = 4.5$
$2.2x + 1.1y - 4.7z + 2t = 8.3$
$3.4x + 0.5z - 3.4t = 0.1$
$9.2y - 1.3t = 0$

46. $1.2x - 0.3y + 0.4z - 2t = 4.5$
$1.9x - 0.5z - 3.4t = 0.2$
$12.1y - 1.3t = 0$
$3x + 2y - 1.1z = 9$

In Exercises 47–50, use a graphing calculator or appropriate computer software to solve the systems of equations. Express all solutions as fractions.

47. $x + 2y - z + w = 30$
$2x - z + 2w = 30$
$x + 3y + 3z - 4w = 2$
$2x - 9y + w = 4$

48. $4x - 2y + z + w = 20$
$3y + 3z - 4w = 2$
$2x + 4y - w = 4$
$x + 3y + 3z = 2$

49. $x + 2y + 3z + 4w + 5t = 6$
$2x + 3y + 4z + 5w + t = 5$
$3x + 4y + 5z + w + 2t = 4$
$4x + 5y + z + 2w + 3t = 3$
$5x + y + 2z + 3w + 4t = 2$

50. $x - 2y + 3z - 4w = 0$
$-2x + 3y - 4z + t = 0$
$3x - 4y + w - 2t = 0$
$-4x + z - 2w + 3t = 0$
$y - 2z + 3w - 4t = 1$

COMMUNICATION AND REASONING EXERCISES

51. Suppose that a system of equations has a unique solution. What must be true of the number of pivots in the reduced matrix of the system? Why?
52. Suppose that a system has infinitely many solutions. What must be true of the number of pivots in the reduced matrix of the system? Why?
53. Can you think of any practical reasons for wanting to consider systems of equations in which the number of equations is not equal to the number of unknowns?
54. Come up with a realistic situation represented by a single equation in three unknowns. What is the significance of the fact that there are infinitely many solutions?
55. Give an example of a system of three linear equations with the general solution $x = 1$, $y = 1 + z$, z abitrary. (Check your system by solving it.)
56. Give an example of a system of three linear equations with the general solution $x = y - 1$, y arbitrary, $z = y$. (Check your system by solving it.)
57. How likely do you think it is that a "random" system of two equations in two unknowns has a unique solution? Give some justification for your answer.
58. How likely do you think it is that a "random" system of three equations in two unknowns has a unique solution? Give some justification for your answer.

2.4 APPLICATIONS OF SYSTEMS OF LINEAR EQUATIONS

In this section, we consider real-life problems that lead to systems of linear equations in three or more unknowns. We can view many of these examples as further uses of the linear models considered in Chapter 1. In each of these examples, we'll content ourselves with setting up the problem as a linear system and then giving the solution. For practice, you should do the row reduction to get the solution.

▼ EXAMPLE 1   Blending

The Arctic Juice Company makes three juice blends: PineOrange, using 2 quarts of pineapple juice and 2 quarts of orange juice per gallon; PineKiwi, using 3 quarts of pineapple juice and 1 quart of kiwi juice per gallon; and OrangeKiwi, using 3 quarts of orange juice and 1 quart of kiwi juice per gallon. Each day they have 800 quarts of pineapple juice, 650 quarts of orange juice, and 350 quarts of kiwi juice available. How many gallons of each blend should they make each day if they want to use up all of their supplies?

SOLUTION   We follow the same procedure for setting up the problem as we did in Section 2, so the first step is to identify and label the unknowns. Looking at the question, we see that we should label the unknowns like this.

Let x be the number of gallons of PineOrange.

Let y be the number of gallons of PineKiwi.

Let z be the number of gallons of OrangeKiwi made each day.

Next, we notice that we can organize the information we are given into a table.

	PineOrange (x gallons)	PineKiwi (y gallons)	OrangeKiwi (z gallons)	Total Available
Pineapple Juice (quarts)	2	3	0	800
Orange Juice (quarts)	2	0	3	650
Kiwi Juice (quarts)	0	1	1	350

Notice how we have arranged the table: we have placed headings corresponding to the unknowns along the top, rather than down the side, and we have added a heading for the available totals. We shall always arrange tables in this manner, as this corresponds to the way we organize a matrix, with a column for each unknown. Reading across each row, the limitations on the amount of each juice lead to the following three equations.

$$2x + 3y = 800$$
$$2x + 3z = 650$$
$$ y + z = 350$$

The solution to this system is $(x, y, z) = (100, 200, 150)$, so Arctic Juice should make 100 gallons of PineOrange, 200 gallons of PineKiwi, and 150 gallons of OrangeKiwi each day.

Before we go on... Notice that when we can set up a table as above, the table is really the matrix of the system. We do not recommend relying on this coincidence—it is too easy to set up the table "sideways" and get the wrong matrix. We should always write down the system of equations *and be sure we know why each equation is true.* For example, the equation $2x + 3y = 800$ in this example indicates that the number of quarts of pineapple juice that will be used ($2x + 3y$) is equal to the amount available (800 quarts). By thinking of the reason for each equation, we can check that we have the correct system. If we have the wrong system of equations to begin with, solving it won't help us.

▼ **EXAMPLE 2**   Aircraft Purchases

An airline is considering the purchase of aircraft to meet an estimated demand for 2,000 seats. The airline has decided to buy Boeing 747s seating 400 passengers and priced at $150 million each, Boeing 777s seating 300 passengers and priced at $115 million, and Airbus A321s seating 200 passengers and priced at $60 million.* Assuming the airline wishes to buy as many 747s

▼ *Japan Air System actually placed an order for seven Boeing 777s at the price quoted above in June, 1993 (*The New York Times*, June 30, 1993, p. D4), while Airbus launched the A321 in Hamburg, Germany in March 1993. (*The Chicago Tribune*, March 4, 1993, Section 3, p. 1).

as 777s, how many of each should it order to meet the demand for seats, given a $710 million purchasing budget?

SOLUTION We label the unknowns this way.

Let x be the number of 747s ordered from Boeing.

Let y be the number of 777s ordered from Boeing.

Let z be the number of A321s ordered from Airbus.

We must now set up the equations. We can organize some (but not all) of the information given in a table.

	Boeing 747s (x)	Boeing 777s (y)	Airbus A321s (z)	Total
Passengers	400	300	200	2,000
Cost ($ millions)	150	115	60	710

Reading across gives us two equations.

$$400x + 300y + 200z = 2{,}000$$
$$150x + 115y + 60z = 710$$

An additional piece of information has not yet been used: the airline wishes to purchase as many 747s as 777s. This can be rephrased as: "the number of 747s ordered is equal to the number of 777s ordered," or

$$x = y,$$

or

$$x - y = 0.$$

We now have a system of three equations in three unknowns.

$$400x + 300y + 200z = 2{,}000$$
$$150x + 115y + 60z = 710$$
$$x - y = 0$$

Solving the system, we get the solution $(x, y, z) = (2, 2, 3)$. Thus, the airline should order two 747s, two 777s and three A321s.

▼ **EXAMPLE 3** Traffic Flow

Traffic through downtown Urbanville flows through the one-way system shown in Figure 1.

Traffic counting devices installed in the road (see Figure 1) count 200 cars entering town from the west every hour and 100 cars leaving town on each road to the east every hour. Using this information, is it possible to

2.4 Applications of Systems of Linear Equations

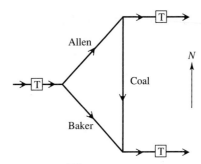

The symbols $\boxed{T}$ represent traffic counters.

FIGURE 1

determine how many cars drive along Allen, Baker, and Coal Streets every hour?

SOLUTION Our unknowns are

x = the number of cars per hour on Allen Street
y = the number of cars per hour on Baker Street
z = the number of cars per hour on Coal Street.

Assuming that, at each intersection, cars do not fall into a pit or materialize out of thin air, the number of cars entering each intersection has to equal the number exiting. For example, at the intersection of Allen and Baker Streets there are 200 cars entering, and $x + y$ cars exiting, giving the equation

$$x + y = 200.$$

At the intersection of Allen and Coal Streets we get

$$x = z + 100,$$

and at the intersection of Baker and Coal Streets we get

$$y + z = 100.$$

We now have the following system of equations.

$$x + y = 200$$
$$x - z = 100$$
$$y + z = 100$$

This system has many solutions, which can be written in the following way.

$$x = z + 100$$
$$y = -z + 100$$
z is arbitrary.

In other words, since we do not have a unique solution, it is not possible to determine how many cars drive along Allen, Baker, and Coal Streets every hour.

Before we go on... How arbitrary is z? It makes no sense for any of the variables x, y, or z to be negative in this problem, so certainly $z \geq 0$. Further, in order to keep y positive we must have $z \leq 100$. So, z is not completely arbitrary, but we can say that it has to satisfy $0 \leq z \leq 100$ in order to get a realistic answer. Here is a final question to think about: If you wanted to nail down x, y, and z to see where the cars are really going, how would you do it with only one more traffic counter?

EXAMPLE 4 Investments

You run a small accounting firm, and you are preparing an annual statement for one of your clients, Professor Goodbrain. Unfortunately, Professor Goodbrain has misplaced many of the documents pertaining to his investments over the past year. You do have the following information: Professor Goodbrain invested a total of $8,000 at the start of the year in one-year certificates at various Canadian institutions: Canadian Western Bank, which paid 4% interest on the certificates; Montreal Trust, which also paid 4%; Maritime Life, which paid 3.5%; and Bayshore Trust, which paid 4.5%.* At the end of the year, his investments had appreciated by $320, and all he could remember was that he had invested the same amount in Maritime Life as in Bayshore Trust, and 10 times as much in Canadian Western as in Maritime Life.

(a) Do you have enough information to determine how much he invested in each institution?

(b) The next day, Professor Goodbrain marches into your office and declares that he now remembers investing $1,000 in Bayshore Trust. Could he be correct?

SOLUTION

(a) We first identify the unknowns.

x = the amount invested in Canadian Western Bank
y = the amount invested in Montreal Trust
z = the amount invested in Maritime Life
w = the amount invested in Bayshore Trust

The first piece of information we are given is that Professor Goodbrain invested a total of $8,000. Thus,

$$x + y + z + w = 8,000.$$

▼ *The rates given were quoted by these institutions on December 9, 1993. (Source: Commercial Paper and Treasury Bill Rates, RBC Dominion Securities, as quoted in the *Canadian Financial Post*, Dec. 11–13, 1993, p. 14.)

Second, the total amount of interest he earned was $320. This amount was made up of the following.

$x in Canadian Western Bank at 4% interest: $0.04x$
$y in Montreal Trust at 4% interest: $0.04y$
$z in Maritime Life at 3.5% interest: $0.035z$
$w in Bayshore Trust at 4.5% interest: $0.045w$

Thus,
$$0.04x + 0.04y + 0.035z + 0.045w = 320.$$

Next, we are told that he invested the same amount in Maritime Life as in Bayshore Trust, so
$$z = w,$$
or
$$z - w = 0.$$

Finally, we are told that he invested 10 times as much in Canadian Western as in Maritime Life. Rephrasing this to use the phrases "the amount invested in Canadian Western Bank" and "the amount invested in Maritime Life," we get the following.

The amount invested in Canadian Western Bank is 10 times the amount invested in Maritime Life.

In other words,
$$x = 10z,$$
or
$$x - 10z = 0.$$

Putting these equations together gives us the following system of four equations in four unknowns.
$$x + y + z + w = 8,000$$
$$0.04x + 0.04y + 0.035z + 0.045w = 320$$
$$z - w = 0$$
$$x - 10z = 0$$

When we reduce the associated augmented matrix, we find that there are infinitely many solutions.
$$x = 10w$$
$$y = 8,000 - 12w$$
$$z = w$$
w is arbitrary.

Thus, we do not have enough information to piece together his original investments.

(b) Professor Goodbrain now seems to recall that $w = 1,000$. But substituting this value for w in the above solution gives $y = 8,000 - 12,000 = -4,000$, which is impossible. Thus, Professor Goodbrain must surely be mistaken!

▶ 2.4 EXERCISES

APPLICATIONS

1. **Resource Allocation** You manage an ice cream factory that makes three flavors: Creamy Vanilla, Continental Mocha, and Succulent Strawberry. Into each batch of Creamy Vanilla go two eggs, one cup of milk, and two cups of cream. Into each batch of Continental Mocha go one egg, one cup of milk, and two cups of cream, while into each batch of Succulent Strawberry go one egg, two cups of milk, and one cup of cream. You have in stock 350 eggs, 350 cups of milk, and 400 cups of cream. How many batches of each flavor should you make in order to use up all of your ingredients?

2. **Resource Allocation** You own a hamburger franchise and are planning to shut down operations for the day, but you are left with 13 bread rolls, 19 defrosted beef patties, and 16 opened cheese slices. Rather than throw them out, you decide to use them to make burgers which you will sell at a discount. Plain burgers each require one beef patty and one bread roll, double cheeseburgers each require two beef patties, one bread roll and two slices of cheese, and regular cheeseburgers each require one beef patty, one bread roll and one slice of cheese. How many of each should you make?

3. **Purchasing** In Example 2, we saw that *Boeing* 747s seat 400 passengers and are priced at $150 million each, *Boeing* 777s seat 300 passengers and are priced at $115 million, and the European *Airbus* A321s seat 200 passengers and are priced at $60 million. You are the purchasing manager of an airline company, and have a $1.625 billion budget to purchase new aircraft to seat a total of 4,500 passengers. Your company has a policy of supporting U.S. industries, and you have been instructed to buy twice as many U.S.-manufactured aircraft as foreign aircraft (*Boeing* is a U.S. company). Given the above selection of aircraft, how many of each should you order?

4. **Purchasing** Repeat Exercise 3, but with the following data: Your purchasing budget is $4.560 billion, you must seat a total of 13,600 passengers, and are required to spend twice as much on U.S.-manufactured aircraft as on foreign aircraft.

5. **Feeding Schedules** Your 36-gallon tropical fish tank contains three types of carnivorous creatures: baby sharks, piranha, and squids, and you feed them three types of delicacies: goldfish, angelfish, and butterfly fish. Each baby shark can consume 1 goldfish, 2 angelfish, and 2 butterfly fish per day; each piranha can consume 1 goldfish and 3 butterfly fish per day (the piranha are rather large as a result of their diet), and each squid can consume 1 goldfish and 1 angelfish per day. After a trip to the local pet store, you were able to feed your creatures to capacity, and you noticed that 21 goldfish, 21 angelfish, and 35 butterfly fish were eaten. How many of each type of creature do you have?

6. **Resource Allocation** The Enormous State University Choral Society is planning its annual Song Festival, at which it will serve three kinds of delicacies: granola treats, nutty granola treats, and nuttiest granola treats. The following table shows the ingredients required for a single serving of each delicacy, as well as the total amount of each ingredient available.

	Granola	Nutty Granola	Nuttiest Granola	Total Available (oz)
Toasted Oats	1	1	5	1,500
Almonds	4	8	8	10,000
Raisins	2	4	8	4,000

The Song Festival planners would like to use up all the ingredients. Is this possible? If so, how many servings of each kind of delicacy can they make?

7. Resource Allocation Urban Community College is planning to offer courses in Finite Math, Business Calculus, and Computer Methods. Each section of Finite Math holds 40 students and earns the college $1,000 in revenue per student. Each section of Business Calculus holds 40 students and earns the college $1,500 per student, while each section of Computer Methods holds 10 students and earns the college $2,000 per student. Assuming the college can offer a total of six sections, wishes to accommodate 210 students, and wants to bring in $260,000 in revenues, how many sections each course should it offer?

8. Resource Allocation The Enormous State University History Department offers three courses, Ancient, Medieval, and Modern History, and the department chair is trying to decide how many sections of each to offer this semester. The department is allowed to offer 45 sections total, there are 5000 students who would like to take a course, and there are 60 professors to teach them (each professor teaches only one section). Sections of Ancient History hold 100 students each, sections of Medieval History hold 50 students each, and sections of Modern History hold 200 students each. Modern History sections are taught by a team of two professors, while Ancient and Medieval History need only one professor per section. How many sections of each course should the chair schedule in order to offer all the sections the department is allowed to offer, accommodate all of the students, and give teaching assignments to all of the professors?

9. Asset Appreciation *Renault* owns 10 percent of *Volvo A.B.*, 25 percent of the *Volvo Car Corporation*, and 45 percent of the *Volvo Truck Corporation*.* Consider the following (fictitious) scenario: *Renault* finds at the end of a financial year (in which it controlled the above percentages of *Volvo* assets) that the total value of *Volvo* has increased by $2 million, whereas the total value of *Renault's Volvo* holdings appreciated by $400,000, and the combined value of its holdings in *Volvo A. B.* and the *Volvo Truck*

Corporation appreciated by $650,000. By how much did the value of each *Volvo* division increase or decrease that year?

10. Asset Appreciation Referring to Exercise 9, assume that the market value of *Renault's* total *Volvo* holdings increased by $550,000, the value of its holdings in *Volvo Truck Corporation.* increased by twice as much as its holdings in Volvo A.B., and that the total value of Volvo increased by $1 million. By how much did the total market value of each *Volvo* division increase or decrease that year?

11. Pension Plans Many small firms are discontinuing their pension plans. Statistics for the three years 1980, 1985, and 1987 show that a total of 187,210 U.S. companies with fewer than 100 employees scrapped their pension plans.† In 1985, the worst year for pension funds, only 14,932 fewer pension plans were scrapped than in 1980 and 1987 combined. The least damaging year in this regard was 1980, when the total number of pension plans scrapped was 45,815 less than the total for 1987. How many pension plans were scrapped in each of the three years?

12. Investments Many life insurance issuers of guaranteed investment contracts had a considerable investment in junk bonds in the 1980s. According to Dec. 31, 1988 figures, the three such companies with the highest percentage of junk bonds relative to total assets (cash plus investment assets) were *Executive Life of NY* (45% of total assets), *Executive Life* (45%), and *United Pacific* (22%).‡ These companies invested a total of about $8.08 billion in junk bonds, and had assets totaling approximately 20 billion dollars. If Executive Life's total assets were three times that of United Pacific, calculate the approximate assets of each of the three companies named.

13. Supply A bagel store orders cream cheese from 3 suppliers, Cheesey Cream Corp. (CCC), Super Smooth & Sons (SSS), and Bagel's Best Friend Co. (BBF). One month, the total order of cheese came to

---

▼ *Statistics were current as of June 30, 1993. (Source: *The New York Times,* June 30, 1993, p. D4.) Renault and Volvo were increasing their investments in each other as part of a merger plan. (The planned merger was subsequently scrapped in December 1993.)

† Source: "Many Small Firms Junking Pensions: Complex, Ever-Changing Federal Regulations Blamed," *Lincoln Journal,* March 13, 1990, p. 14.

‡ Source: Earl C. Gottschalk, Jr., "The Risk in 'Guaranteed' Investments," *The Wall Street Journal,* November 14, 1989. Figures are rounded to the nearest percentage point.

100 tons (the store does a booming trade). The costs were $80, $50, and $65 per ton from the three suppliers, respectively, with total cost amounting to $5,990. Given that the store ordered the same amount from CCC and BBF, how many tons of cream cheese were ordered from each supplier?

14. *Supply* Referring to Exercise 13, the bagel store's outlay for cream cheese the following month was $2,310 for a total of 36 tons. Two more tons of cream cheese came from Bagel's Best Friend Co. than from Super Smooth & Sons. How many tons of cream cheese came from each supplier?

15. *Pest Control* Conan the Great has boasted to his hordes of followers that many a notorious villain has fallen to his awesome sword: his total of 560 victims consists of evil sorcerers, warriors, and orcs. These he has slain with a total of 620 mighty thrusts of his sword; evil sorcerers and warriors each require two thrusts (to the chest) and orcs each require one thrust (to the neck). When asked about the number of warriors he has slain, he replies, "The number of warriors I, the mighty Conan, have slain is five times the number of evil sorcerers that have fallen to my sword!" How many of each type of villain has he slain?

16. *Manufacturing* The Fancy French Perfume Company recently had its secret formula divulged. It turned out that it was using three ingredients: rose oil, oil of fermented prunes, and alcohol. Moreover, each 22-ounce econo-size bottle contained 4 more ounces of alcohol than oil of fermented prunes, while the amount of alcohol was equal to the combined volume of the other two ingredients. How much of each ingredient did the company put into each econo-size bottle?*

17. *Mortality Rates* The Commissioners 1980 Standard Ordinary Mortality Table yields the following mortality figures for males in their 32nd, 50th, and 60th years.† The mortality rates in 1980 were respectively 0.2%, 0.7%, and 1.5%, and this accounted for a total of 202,000 deaths in a total population of 26.5 million males of the designated ages. Given that there were 500,000 more 32-year-old males than 50-year-old males, how many males of each designated age where there?

18. *Insurance Premiums* Consider the following data regarding auto insurance premiums on comparable policies in Adams County, Pennsylvania.‡ Of the three companies (*Aetna Casualty, Prudential Property & Casualty,* and *State Farm Mutual*), Prudential's annual premium is the highest—$64 higher than Aetna's, which in turn is $37 higher than State Farm's. Given that the Aetna's annual premium is $251 less than the sum of the premiums of the other two, find the annual premium charged by each of the three companies.

19. *Voting* In the 75th Congress (1937–39) there were in the U.S. House of Representatives 333 Democrats, 89 Republicans, and 13 members of other parties. Suppose a bill passed the House with 31 more votes in favor than against, with 10 times as many Democrats voting for the bill as Republicans, and with 36 more non-Democrats voting against the bill than for it. How many members of each party voted in favor of the bill?

20. *Voting* In the 75th Congress (1937–39) there were in the Senate 75 Democrats, 17 Republicans, and 4 members of other parties. Suppose a bill passed the Senate with 16 more votes in favor than against, with three times as many Democrats voting in favor as non-Democrats voting in favor, and 32 more Democrats voting in favor than Republicans. How many senators of each party voted in favor of the bill?

21. *Inventory Control* Big Red Bookstore wants to ship books from its warehouses in Brooklyn and Queens to its stores: one in Long Island and one in Manhattan. Its warehouse in Brooklyn has 1,000 books, and its warehouse in Queens has 2,000. Each store orders 1,500 books. It costs $1 to ship each book from Brooklyn to Manhattan, and $2 to ship each book from Queens to Manhattan. It costs $5 to ship each book from Brooklyn to Long Island, and $4 to ship each book from Queens to Long Island.
   (a) If Big Red has a transportation budget of $9,000 and is willing to spend all of it, how many books should it ship from each warehouse to each store in order to fill all orders?
   (b) Is there an obvious way to do this for less money?

▼ * Most perfumes consist of 10 to 20% perfume oils dissolved in alcohol. This may or may not be reflected in this company's formula!

† Source: Black/Skipper, *Life Insurance,* eleventh edition, (Englewood Cliffs, N.J.: Prentice Hall, Inc., 1987 p. 314). We have adjusted these figures a little for computational ease.

‡ Source: *Your Guide to Auto Premiums,* Pennsylvania Insurance Department, July 5, 1989.

22. *Inventory Control* The Tubular Ride Boogie Board Company has manufacturing plants in Tucson, Arizona, and Toronto, Ontario. You have been given the job of coordinating distribution of its latest model, the Gladiator, to its outlets in Honolulu and Venice Beach. The Tucson plant, when operating at full capacity, can manufacture 620 Gladiator boards per week, while the Toronto plant, beset by labor disputes, can produce only 410 boards per week. The outlet in Honolulu orders 500 Gladiator boards per week, and Venice Beach orders 530 boards per week. Transportation costs are as follows:

Tucson to Honolulu: $10 per board;
Tucson to Venice Beach: $5 per board;
Toronto to Honolulu: $20 per board;
Toronto to Venice Beach: $10 per board.

(a) Assuming that you wish to fill all orders and assure full-capacity production at both plants, is it possible to meet a total transportation budget of $10,200? If so, how many Gladiator boards are shipped from each manufacturing plant to each distribution outlet?

(b) Is there a way to do this for less money?

23. *Investments* Things have not been going too well here at Accurate Accounting Inc. since we hired Todd Smiley. He has a tendency to lose important documents, especially around April, when tax returns of our business clients are due. Today Smiley accidentally shredded Colossal Conglomerate Corp.'s investment records and must therefore reconstruct them based on the information he can gather. He recalls that the company earned an $8,000,000 return on investments totaling $65,000,000 last year. After a few frantic telephone calls to sources in Colossal, he learned that Colossal had made investments in four companies last year: X, Y, Z, and W. (For reasons of confidentiality, we are withholding their names.) Investments in Company X earned 15% last year, investments in Y depreciated by 20% last year, investments in Z neither appreciated nor depreciated last year, and investments in W earned 20% last year. Smiley has also been told that Colossal invested twice as much in Company X as in Company Z, and three times as much in Company W as in Company Z. Does Smiley have sufficient information to piece together Colossal's investment portfolio before its tax return is due next week? If so, what does the investment portfolio look like?

24. *Investments* Things are going from bad to worse here at Accurate Accounting Inc.! Colossal Conglomerate Corp.'s tax return is due tomorrow, and accountant Todd Smiley seems to have no idea of how Colossal Conglomerate Corp. earned a return of $8 million on a $65-million investment last year. It appears that although the returns from companies X, Y, Z, and W were as listed in Exercise 23, the rest of the information there was wrong. What Smiley is now being told is that Colossal only invested in Companies X, Y, and Z, and that the investment in X amounted to $30 million. His sources in Colossal still maintain that twice as much was invested in Company X as in Company Z. What should Smiley do?

25. *Traffic Flow* One-way traffic through Enormous State University is shown in the accompanying figure, where the numbers indicate daily counts of vehicles.

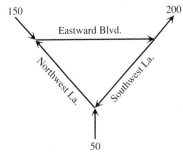

(a) Is it possible to determine the daily flow of traffic along each of the three streets from the information given? If your answer is "yes," what is the traffic flow along each street? If your answer is "no," what additional information would suffice?

(b) Is a traffic flow of 60 vehicles per day along Southwest Lane consistent with the information given?

26. *Traffic Management* The Outer Village Town Council has decided to convert its (rather quiet) main street, Broadway, to a one-way street, but is not sure of the direction of most of the traffic. The accompanying diagram illustrates the downtown area of Outer Village, as well as the *net* traffic flow along the intersecting streets (in vehicles per day). (There are no one-way streets; a net traffic flow in a certain direction is defined as the traffic flow in that direction minus the flow in the opposite direction.)

(a) Is the given information sufficient to determine the net traffic flow along the three portions of Broadway shown? If your answer is "yes," give the traffic flow along each stretch. If your answer is "no," what additional information would suffice? (*Hint:* For the direction of net traffic flow, choose either east or west. If a corresponding value is negative, it indicates net flow in the opposite direction.)

(b) Assuming that there is very little traffic (less than 30 vehicles per day) east of Fleet Street, in what direction is the net flow of traffic along Broadway?

27. *Traffic Flow* The traffic through downtown East Podunk flows through the one-way system shown below.

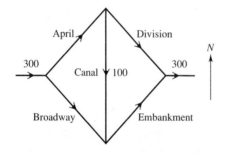

Traffic counters find that 300 vehicles enter town from the west each hour, and 300 leave town to the east each hour. Also, 100 cars drive down Canal Street each hour.

(a) Find the general solution to the associated system of linear equations.

(b) On which street would you put another traffic counter in order to determine the flow completely?

28. *Electric Current* Electric current measures (in **amperes** or **amps**) the flow of electrons through wires. Like traffic flow, the current entering an intersection of wires must equal the current leaving it.*

Here is an electrical circuit known as a **Wheatstone bridge.**

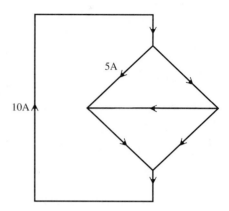

(a) If the currents in two of the wires are 10 amps and 5 amps as shown, determine the currents in the unlabeled wires in terms of suitable parameters.

(b) In which wire should you measure the current in order to know all of the currents exactly?

29. *Econometrics* (*from the GRE economics test*) This and the next exercise are based on the following simplified model of the determination of the money stock.

$$M = C + D$$
$$C = 0.2D$$
$$R = 0.1D$$
$$H = R + C$$

where

M = money stock
C = currency in circulation
R = bank reserves
D = deposits of the public
H = high-powered money

If the money stock were $120 billion, what would bank reserves have to be?

---

* This is known as **Kirchoff's current law,** named after Gustav Robert Kirchoff (1824–1887). Kirchoff made important contributions to the fields of geometric optics, electromagnetic radiation, and electrical network theory.

30. *Econometrics (from the GRE economics test)* With the model in the previous exercise, if H were equal to $42 billion, what would M equal?

31. *Donations* The Enormous State University Good Works Society recently raised funds for three worthwhile causes: the Math Professors' Benevolent Fund (MPBF), the Society of Computer Nerds (SCN), and the NY Jets. As they are closet jocks, they donated twice as much to the NY Jets as to the MPBF, and equal amounts to the first two funds (they are unable to distinguish between mathematicians and nerds). Further, for every $1 they gave to the MPBF, they decided to keep $1 for themselves; for every $1 they gave to the SCN, they kept $2, and for every $1 to the Jets, they also kept $2. The treasurer of the Society, Johnny Treasure, was required to itemize all donations for the Dean of Students, but discovered to his consternation that he had lost the receipts! The only information available to him was that the Society's bank account had swelled by $4,200. How much did it donate to each cause?

32. *Tenure* Professor Walt is up for tenure, and he wishes to submit a portfolio of written student evaluations as evidence of his good teaching. He begins by grouping all the evaluations into four categories: good reviews, bad reviews—a typical one being "GET RID OF WALT! THE MAN CAN'T TEACH!"—mediocre reviews—such as "I suppose he's OK, given the general quality of teaching here at this university" and reviews left blank. When he tallies up the piles, he gets a little worried: there are 280 more bad reviews than good ones and only half as many blank reviews as bad ones. The good reviews and blank reviews together total 170. On an impulse, he decides to even up the piles a little by removing 280 of the bad reviews (including the memorable "GET RID OF WALT" review), and this leaves him with a total of 400 reviews of all types. How many of each category of reviews were there originally?

CAT Scans CAT scans (Computerized Axial Tomographic Scanners) are used to map the exact location of interior features of the human body. CAT scan technology is based on the following principles: (1) Different components of the human body (water, grey matter, bone, etc.) absorb X rays to different extents. (2) To measure the X ray absorption by a specific region of, say, the brain, it suffices to pass a number of line-shaped pencil beams of X rays through the brain at different angles and measure the total absorption for each beam. The accompanying diagram illustrates a simple example. (The number in each region shows its absorption, and the number on each X ray beam shows the total absorption for that beam.)*

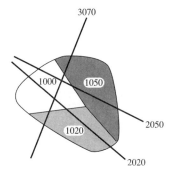

▼ * Based on a COMAP article, *Geometry: New Tools for New Technologies*, by J. Malkevitch, Video Applications Library, COMAP, 1992. The absorptions are actually calibrated on a logarithmic scale. In real applications, the size of the regions is very small, and very large numbers of beams must be used.

162 Chapter 2 Systems of Linear Equations and Matrices

In Exercises 33–38, use the table below and the given X ray absorption diagrams to identify the composition of each of the regions marked by a letter.

Type	Air	Water	Grey matter	Tumor	Blood	Bone
Absorption	0	1,000	1,020	1,030	1,050	2,000

33.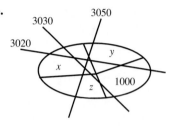

34.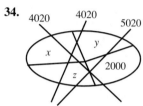

35.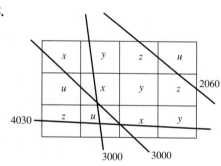

36.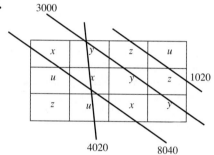

37. Identify the composition of site x.

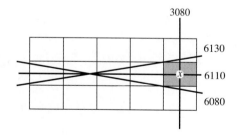

38. Identify the composition of site x.

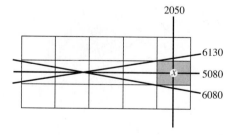

(The horizontal and slanted beams each pass through five regions.)

2.4 Applications of Systems of Linear Equations

Airline Costs In Exercises 39 and 40, use a graphing calculator or appropriate computer software. Exercises 39 and 40 are based on the following chart, which shows the amount spent by five major U.S. carriers to fly one available seat one mile in the second quarter of 1994.*

Carrier	TWA	Northwest	American	Delta	USAir
Cost	9.2¢	9.7¢	9.8¢	10.7¢	11.7¢

39. Suppose that on a 3,000-mile New York to Los Angeles flight, TWA, Northwest, and American flew a total of 160 empty seats, costing them a total of $46,230. If Northwest had three times as many empty seats as TWA, how many empty seats did each airline carry on its flight?

40. Suppose that on a 2,000-mile Miami to Memphis flight, American, Delta, and USAir flew a total of 180 empty seats, costing them a total of $37,320. If American had twice as many empty seats as Delta, how many empty seats did each airline carry on its flight?

COMMUNICATION AND REASONING EXERICISES

In Exercises 41–46, x, y, and z represent the weights of the three ingredients X, Y, and Z in a gasoline blend. Decide whether each of the following is represented by a linear equation in x, y, and z, and if it is, write a form of the equation.

41. The blend contains 30% ingredient Y by weight.

42. The weight of ingredient X is the product of the weights of ingredients Y and Z.

43. The blend consists of 100 pounds of ingredient X.

44. The blend is free of ingredient X.

45. There is twice as much ingredient X by weight as Y and Z combined.

46. There is at least 30% ingredient Y by weight.

47. Make up an entertaining word problem leading to the following system of equations.

$$10x + 20y + 10z = 100$$
$$5x + 15y = 50$$
$$x + y + z = 10$$

48. Make up an entertaining word problem leading to the following system of equations.

$$10x + 20y = 300$$
$$ 10z + 20w = 400$$
$$20x + 10z = 400$$
$$ 10y + 20w = 300$$

▼ *Source: Avitas Incorporated/New York Times, *The New York Times,* September 23, 1994, p. D3.

You're the Expert

THE IMPACT OF REGULATING SULFUR EMISSIONS

You have been hired as a consultant to the Environmental Protection Agency. The agency is considering regulations requiring a 15-million-ton rollback in sulfur emissions in an effort to curb the effects of acid rain on the ecosystem, and it would like you to estimate the cost to the major utility companies and the effect on jobs in the coal-mining industry. The following data are available:*

Strategy	Annual cost to utilities ($ billions)	Cost in jobs (number of jobs lost)
8-million-ton rollback	20.4	14,100
10-million-ton rollback	34.5	21,900
12-million-ton rollback	93.6	13,400[†]

Your assignment is to use the figures above to give projections of the annual cost to utilities and the cost in jobs if the regulations were to be enacted.

You decide to consider the annual cost C to utilities and the job loss J separately. After giving the situation some thought, you decide that you would like to have two equations, one giving C in terms of the rollback tonnage t, and the other giving J in terms of t. Your first inclination is to try linear equations: that is, an equation of the form

$$C = at + b \quad (a \text{ and } b \text{ constants})$$

and a similar one for J, but you discover to your consternation that the data simply won't fit, no matter what the choice of the constants. The reason for this can be seen graphically by plotting C and J versus t (Figure 1).

In neither case do the three points lie on a straight line and, to make matters worse, the job data are not even close to being linear. Thus, you will need a curve to model these data.

After giving the matter further thought, you remember something your mathematics instructor once said: the simplest curve passing through any

---

*Source: Congress of the United States, Congressional Budget Office, *Curbing Acid Rain: Cost, Budget, and Coal Market Effects* (Washington, DC: U.S. Government Printing Office, 1986): xx, xxii, 23, 80.

[†] The reason that job losses drop when the rollback is increased to 12 million tons is that a rollback of this magnitude requires that expensive scrubbers be installed to filter emissions, even if a utility company has switched from coal to other energy sources. Once the scrubbers are installed, it pays the utility companies to switch back to (cheaper) coal as a primary source of energy. A 10-million-ton reduction, on the other hand, results in a massive move away from coal—this being cheaper than installing scrubbers—and hence a dramatic job loss in the coal-mining industry.

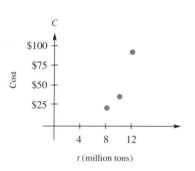 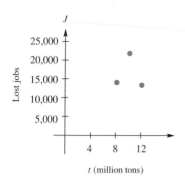

FIGURE 1

three points is a parabola, and given three points not on the same straight line, they lie on some parabola. A general parabola has the equation

$$C = at^2 + bt + c$$

where a, b, and c are constants. The problem now is: what are a, b, and c? You decide to try substituting the values of C and t into the general equation, and you get the following.

$$t = 8, \ C = 20.4 \quad \text{gives} \quad 20.4 = 64a + 8b + c$$
$$t = 10, \ C = 34.5 \quad \text{gives} \quad 34.5 = 100a + 10b + c$$
$$t = 12, \ C = 93.6 \quad \text{gives} \quad 93.6 = 144a + 12b + c$$

Now you notice to your surprise that you have three linear equations in three unknowns! You solve the system, obtaining

$$a = 5.625, \quad b = -94.2, \quad c = 414.$$

Thus, your cost equation becomes

$$C = 5.625t^2 - 94.2t + 414.$$

You now substitute $t = 15$ to get $C = 266.625$. In other words, you are able to predict an annual cost to the utility industry of $266.625 billion. You next turn to the jobs equation and write

$$J = at^2 + bt + c.$$

Substituting the data from the second column of the chart, you obtain

$$14{,}100 = 64a + 8b + c$$
$$21{,}900 = 100a + 10b + c$$
$$13{,}400 = 144a + 12b + c.$$

You solve this and find that

$$a = -2{,}037.5, \quad b = 40{,}575, \quad c = -180{,}100.$$

Thus, your jobs equation is

$$J = -2{,}037.5t^2 + 40{,}575t - 180{,}100.$$

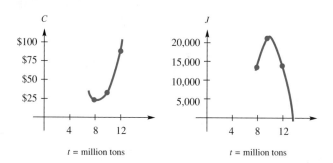

FIGURE 2

Substituting $t = 15$, you get $J = -29{,}912.5$. Uh oh! A negative number! Just what does this signify? Well, your reasoning tells you, if there were to be $-29{,}913$ jobs lost, then this means that there would be 29,913 new jobs *created!** Figure 2 shows the parabolas superimposed on the data points.

You thus submit the following projection: a 15-million-ton rollback will result in an annual cost of \$266.625 billion to the utility industry, but will also result in the creation of 29,913 new jobs in the coal-mining industry.

Exercises

1. Repeat the above computations using the following rounded figures.

Strategy	Annual cost to utilities ($ billions)	Cost in jobs (number of jobs lost)
8-million-ton rollback	20	14,000
10-million-ton rollback	35	22,000
12-million-ton rollback	94	13,000

2. If the 8-million-ton rollback data had not been available, what projections would you have given? (Use the original data.)

3. If the 10-million-ton rollback data had not been available, what projections would you have given? (Use the original data.)

4. Find the equation of the parabola that passes through the points $(1, 2)$, $(2, 9)$, and $(3, 19)$.

5. Is there a parabola that passes through the points $(1, 2)$, $(2, 9)$, and $(3, 16)$?

6. Is there a parabola that passes though the points $(1, 2)$, $(2, 9)$, $(3, 19)$, and $(-1, 2)$?

▼ * The logic behind it is this: since the stringent new emission standards would require that even more scrubbers be installed, the utility industry might as well increase its use of coal (which tends to be far cheaper than other energy sources), thus creating a boom for the coal industry.

7. Use a graphing calculator to estimate the size of the rollback that results in the lowest annual cost to utilities.

8. Use a graphing calculator or computer. You submit your projections on the cost of regulating sulfur emissions, and the EPA tells you, "Thank you very much, but we made a small mistake: a 15-million-ton rollback has in fact already been in effect for the past year and has resulted in a cost of $250 billion to the utilities and the creation of 8,000 jobs in the coal mining industry." (Apparently utilities have been switching back to coal at a greater rate than anticipated.) In view of this, the EPA is now considering a 20-million-ton rollback. You are to come up with projections by tomorrow. [*Hint:* since you now have four data points on each graph, try a general cubic instead: $C = at^3 + bt^2 + ct + d$.]

▶ Review Exercises

In Exercises 1 through 38, use any method to solve the given systems. If you think you need additional practice in a particular method, this is the time to get it.

1. $x + 2y = 4$
$2x - y = 1$

2. $x + y = 2$
$x - 3y = 2$

3. $3x - y = 0$
$2x - y = 0$

4. $2x - 3y = 0$
$3x - 2y = 0$

5. $0.2x - 0.1y = 0.3$
$0.2x + 0.2y = 0.4$

6. $-0.3x + 0.5y = 0.2$
$0.1x - 0.2y = -0.1$

7. $2x/3 - 3y/2 = \frac{1}{3}$
$x/4 + 2y = 0$

8. $-2x/3 + 3y/2 = \frac{1}{6}$
$x/4 - y/2 = -\frac{3}{4}$

9. $x - 3y = 0$
$-x - 3y/2 = 0$

10. $2x - 3y = 0$
$6x - 9y = 0$

11. $2x + 3y = 2$
$-x - 3y/2 = \frac{1}{2}$

12. $2x - 3y = 2$
$2x/5 - 3y/5 = \frac{1}{5}$

13. $x - 2y = 1$
$3x + 2y = \frac{1}{2}$
$x + 6y = -\frac{1}{2}$

14. $2x + y = 1$
$6x - 2y = -1$
$10x - y = \frac{1}{5}$

15. $8x - 6y = 0$
$6x - 3y = \frac{1}{2}$
$2x - 3y = -\frac{1}{2}$

16. $x - 2y = 1$
$3x + 2y = -2$
$5x + y = \frac{1}{5}$

17. $x - 0.1y = 0$
$0.2x + 0.1y = 0$
$2x - y = 0$

18. $-0.3x - 0.5y = -0.1$
$x + y = -4$
$x/17 - y/17 = -1$

19. $x + y = 1$
    $2x + y = 0.3$
    $3x + 2y = \frac{13}{10}$

20. $3x + 0.5y = 0.1$
    $6x + y = 0.2$
    $3x/10 - 0.05y = 0.01$

21. $-x + 2y + z = 2$
    $-x + y + 2z = 2$
    $y - z = 0$

22. $x + 2y = -3$
    $x - z = 0$
    $x + 3y - 2z = -2$

23. $x + 2y + z = 40$
    $\frac{1}{3}x + \frac{1}{3}y + \frac{2}{3}z = \frac{40}{3}$
    $\frac{1}{2}x + \frac{1}{2}z = 10$

24. $\frac{1}{2}x - \frac{1}{2}y + \frac{1}{2}z = 6$
    $\frac{1}{3}x + \frac{1}{3}y + \frac{1}{3}z = 12$
    $\frac{1}{2}x + y + \frac{1}{3}z = 22$

25. $x = -2y + z$
    $y = x + 2z$
    $z = -2x - y + 14$

26. $x = -\frac{1}{2}y$
    $\frac{1}{2}x = -\frac{1}{2}z + 2$
    $z = -3x - y$

27. $x + y + 2z = 2$
    $2x - y + 2z = \frac{3}{2}$
    $\frac{1}{2}x + \frac{1}{2}y + \frac{1}{2}z = \frac{3}{4}$
    $-2x - 2z = -2$

28. $x - y + z = 2$
    $7x + y - z = 6$
    $x - \frac{1}{2}y + \frac{1}{3}z = 1$
    $x + y + z = 6$

29. $x + y + 5z + w = 1$
    $y + 2z + w = 0$
    $x + 3y + 7z + 2w = 1$
    $x + y + 5z + w = 1$

30. $x - y + z + 4w = 3$
    $2x - 2y - 3z + 2w = 0$
    $y + z + w = 2$
    $2x + 4y = 4$

31. $-x + 2y - z = 0$
    $-\frac{1}{2}x - \frac{1}{2}y + z = 0$
    $3y - 3z = 0$

32. $x - \frac{1}{2}y + z = 0$
    $\frac{1}{2}x - \frac{1}{2}z = -1$
    $\frac{3}{2}x - \frac{1}{2}y + \frac{1}{2}z = -1$

33. $x + y - 2z = -1$
    $-2x - 2y + 4z = 2$
    $\frac{3}{4}x + \frac{3}{4}y - \frac{3}{2}z = -\frac{3}{4}$

34. $x + y - z = -\frac{1}{3}$
    $-3x - 3y + 3z = 1$
    $\frac{2}{7}x + \frac{2}{7}y - \frac{2}{7}z = -\frac{2}{21}$

35. $x - y + z = 1$
    $y - z + w = 1$
    $x + z - w = 1$
    $2x + z = 3$

36. $x + y + 4w = 1$
    $y + z - 4w = -1$
    $x + z + 4w = 1$
    $2x + 2y + 2z + 4w = 1$

37. $x + y + w = 100$
    $x + z + w = 100$
    $y + z + w = 100$

38. $x + 2y + w = 100$
    $x + 2z + w = 100$
    $y + z + 2w = 100$

In Exercises 39 through 50, use row operations to reduce the given matrices to row echelon form.

39. $\begin{bmatrix} \frac{2}{3} & \frac{2}{3} & -3 \\ \frac{1}{2} & -\frac{2}{3} & \frac{11}{3} \end{bmatrix}$

40. $\begin{bmatrix} -\frac{2}{3} & -\frac{2}{3} & -\frac{2}{3} \\ \frac{1}{2} & -\frac{2}{3} & \frac{1}{2} \end{bmatrix}$

41. $\begin{bmatrix} -2 & 1 & -3 \\ 1 & 3 & 1 \\ -3 & -2 & -5 \end{bmatrix}$

42. $\begin{bmatrix} 1 & 2 & 3 \\ 4 & 5 & 6 \\ 7 & 8 & 9 \end{bmatrix}$

43. $\begin{bmatrix} 3 & 0 & 0 & 4 & -1 \\ 1 & 2 & 0 & 0 & 0 \\ 0 & 2 & 0 & 0 & 0 \\ 1 & 1 & 1 & 1 & 1 \end{bmatrix}$

44. $\begin{bmatrix} 0 & 0 & 0 & 1 & -1 \\ 2 & 2 & 0 & 0 & 0 \\ 0 & 1 & 0 & 0 & 0 \\ 1 & 1 & 1 & 1 & 1 \end{bmatrix}$

45. $\begin{bmatrix} 0 & 1 & 0 & -4 \\ 1 & 2 & 0 & 0 \\ 0 & 2 & 0 & 0 \\ 1 & 1 & 1 & 0 \\ 1 & -1 & 2 & 2 \end{bmatrix}$

46. $\begin{bmatrix} 1 & 2 & 3 & -1 \\ 1 & 2 & 1 & 0 \\ 0 & -2 & 0 & 1 \\ 1 & 1 & 1 & 0 \\ 1 & -1 & 2 & 2 \end{bmatrix}$

47. $\begin{bmatrix} \frac{1}{2} & 0 & 1 & 1 & 0 & 0 \\ 2 & -2 & 1 & 0 & 1 & 0 \\ 3 & 0 & 0 & 0 & 0 & 1 \end{bmatrix}$

48. $\begin{bmatrix} \frac{1}{2} & 0 & \frac{1}{2} & 1 & 0 & 0 \\ 1 & -2 & -1 & 0 & 1 & 0 \\ 3 & 0 & 0 & 0 & 0 & 1 \end{bmatrix}$

49. $\begin{bmatrix} 0 & 1 & 0 & 0 \\ 1 & 0 & 0 & 0 \\ 0 & 0 & 1 & 0 \\ 0 & 0 & 0 & 1 \end{bmatrix}$

50. $\begin{bmatrix} 1 & 2 & 3 & 4 \\ 0 & 1 & 2 & 3 \\ 0 & 0 & 1 & 2 \\ 0 & 0 & 0 & 1 \end{bmatrix}$

APPLICATIONS

51. *Consumption* The college jogging team goes through jogging shoes like water. The coach usually orders three brands of jogging shoes which the team members obtain at cost: Gauss Jordans, Roebecks, and K Scottish. Gauss Jordans cost the team $40 per pair, Roebecks $50, and K Scottish $50. One year, the team went through a total of 120 pairs at a total cost of $5,700. Given that the team went through as many pairs of Gauss Jordans as Roebecks, how many pairs of each brand of jogging shoes did they use?

52. *Consumption* The university soccer team uses three kinds of soccer balls: Econo Practice balls, Superlite Intermural, and Fancy Deluxe. Econo Practice balls cost $60, Superlite Intermural balls cost $80, and Fancy Deluxe balls cost $120. Last soccer season, the team spent $3,680 on 54 soccer balls, spending three times as much on Econo Practice balls as on Superlite Intermural balls. How many of each kind of soccer ball did the team buy?

53. *Consumption (From the GMAT)* At a certain diner, a hamburger and coleslaw cost $3.95, and a hamburger and french fries cost $4.40. If french fries cost twice as much as coleslaw, how much do french fries cost?

54. *Purchasing (From the GMAT)* Three types of pencils, *J*, *K*, and *L*, cost $0.05, $0.10, and $0.25 each, respectively. If a box of 32 of these pencils costs a total of $3.40 and if there are twice as many *K* pencils as *L* pencils in the box, how many *J* pencils are in the box?

55. *Inheritance (From the GMAT)* A man who died left an estate valued at $111,000. His will stipulated that his estate was to be distributed so that each of his three children received from the estate and his previous gifts, combined, the same total amount. If he had previously given his oldest child $15,000, his middle child $10,000, and his youngest $2,000, how much did the youngest child receive from the estate?

56. *Inheritance (Based on a question from the GMAT)* A woman who died left an estate valued at $22,300. Her will stipulated that her estate was to be distributed so that, added to her previous gifts, her oldest child would receive half of what each of her other two children did. If she had previously given her oldest child $1,500, her middle child $1,000, and her youngest $200, how much did each child receive from the estate?

57. *Nielsen Ratings* The *Nielsen* rating for a cable TV network is based on the number of households that get that channel. Thus, a Nielson rating of 1.0 could indicate a different number of viewers for different cable networks. As of June, 1993, each *Nielson* rating of 1.0 reflected approximately 607,000 viewers of *ABC Network*, 600,000 viewers of *TBS Network*, and 611,000 viewers of *ESPN*.* Assume that on a certain evening,

▼ *Source: "Nielsen's New Ratings for Cable T.V. Draw Fire", *The New York Times,* June 28, 1993, p. D5

a total of 6,372,000 viewers were watching one of these three cable network channels and that this figure included 1.5 million viewers of *TBS* alone. If *ABC* and *ESPN* received the same *Nielsen* rating, what was the *Nielsen* rating for each of the three networks?

58. *Nielsen Ratings* In the previous exercise we saw that each *Nielsen* rating of 1.0 reflected approximately 607,000 viewers of *ABC Network*, 600,000 viewers of *TBS Network*, and 611,000 viewers of *ESPN*. On a certain evening, a total of 3.636 million viewers watched one of these cable network channels, and all three received the same *Nielsen* rating. If 15,000 more viewers were watching the *ABC Network* than the *TBS Network*, what was the *Nielsen* rating each network received?

59. *Alcohol Content* According to guidelines issued by public health officials, moderate drinking is defined as the consumption of no more than one ounce of alcohol per day. Typically, beer consists of 4% alcohol, wine consists of 10% alcohol, and sherry consists of 12% alcohol.* Is it possible for four people to consume a total of ten drinks, selected from 12-ounce servings of beer, 4-ounce servings of wine, and 3-ounce servings of sherry, and consume a total of 4 ounces of alcohol? If so, what possibilities involve only whole numbers of servings?

60. *Nutrition* Your nutritionist, Caralee Woods, has decided that you need 350 mg of Vitamin C, 4,200 I.U. of Vitamin A, and 400 I.U. of Vitamin D per day,† and she recommends the following supplements: Megadose Supra, containing 50 mg of Vitamin C, 1,000 I.U. of Vitamin A, and 100 I.U. of Vitamin D per capsule; Crazy Caps, containing 100 mg of Vitamin C, 200 I.U. of Vitamin A, and 100 I.U. of Vitamin D per capsule; and AC D-Free's, containing 50 mg of Vitamin C and 500 I.U. of Vitamin A per capsule, but no vitamin D. How many of each should you take to obtain exactly the recommended daily dosages?

61. *Degree Requirements* Joyce Stingley is having a terrible time coming up with a course schedule. One reason for this is the very complicated Bulletin of Suburban State University. It reads as follows:

> All candidates for the degree of Bachelor of Arts at SSU must take a total of 124 credits from the Sciences, Fine Arts, Liberal Arts, and Mathematics,‡ including an equal number of Science and Fine Arts credits, and four times as many Mathematics credits as Science credits, but with Liberal Arts credits exceeding Mathematics credits by exactly one-third of the number of Fine Arts credits.

What are all the possible degree programs?

62. *More Degree Requirements* Due to complaints by the students at Suburban State University that their degree requirements were not flexible enough, the University Senate met to revise the Bulletin requirement, and this is what they came up with:

> All candidates for the degree of Bachelor of Arts at SSU must take a total of 124 credits from the Sciences, Fine Arts, Liberal Arts, and Mathematics, including an equal number of Science and Fine Arts credits, and twice as many Mathematics credits as Science credits, but with at least 50 Liberal Arts credits.

What now are all the possible degree programs?

▼ *Source: *The New York Times,* August 5, 1992, p. C12.

† The 1993 U.S. Recommended Daily allowances for these supplements were as follows: Vitamin A: 5,000 I.U.; Vitamin C: 60 mg.; and Vitamin D: 400 I.U. Thus, your nutritionist's recommendation is a little low on Vitamin A, generous on Vitamin C—many nutritionists feel that the Vitamin C RDA should at least be doubled—and dead right on Vitamin D.

‡ Strictly speaking, mathematics is not a science; it is the Queen of the Sciences, although we like to think of it as the Mother of all Sciences.

63. *Econometrics (from the GRE economics test)* Consider the following economic model.

$$C = 10 + 0.8Y_D,$$
$$Y_D = Y - T,$$
$$T = 0.25Y,$$
$$Y = C + \bar{G} + \bar{I},$$

where

C = consumption demand,
Y_D = disposable income,
T = net government tax receipts,
Y = total income,
$\bar{G}$ = exogenous government purchases
$\bar{I}$ = exogenous investment demand.

(The exogenous variables $\bar{G}$ and $\bar{I}$ are to be treated as constants, or inputs to the model, whereas the other variable are the unknowns of the system. The **equilibrium values** are determined by solving the system.) If exogenous demand ($\bar{I}$) increases by 10 units, this will cause equilibrium total income (Y) to rise or fall by what amount?

64. *Econometrics (from the GRE economics test)* Using the same model as in the previous exercise: If exogenous government purchases ($\bar{G}$) increase by 10 units, this will cause equilibrium consumption (C) to rise or fall by what amount?

CHAPTER 3

The World Economy of the Year 2000

by Wassily W. Leontief

The term "world economy" (*Weltwirtschaft*) first appeared in Germany on the eve of World War I, when Kaiser Wilhelm II was preparing to challenge the political and economic domination of the British Empire. It was at about the same time that the German economist Bernhard Harms, with the backing of Germany's newly created steel and heavy-chemical industries, founded the Institute for World Economics in Kiel, the first large institution devoted to economic research on a global scale. These two signs of impending change proved to be reliable. Over the next 60 years Germany would lose two wars, Britain would lose its empire and the notion of a world composed of self-sufficient, autonomous national economies would recede into the realm of conventional abstractions.

A dramatic demonstration of the degree of global interdependence that exists today is provided by the current "oil crisis," whose direct and indirect effects are felt in the farthest corners of five continents. The world economy has become a tangible reality, and at the present time its dominant feature is the gap in income (and thus standard of living) between the poorer, less developed countries of the world and the richer, highly industrialized ones. In this article I shall discuss the prospects for accelerating the rate of growth of the less developed countries of the world under some of the most frequently proposed scenarios for world economic development.

In 1973, to provide a quantitative basis for such an investigation, the United Nations, with special financial support from the Netherlands, commissioned the construction of a general-purpose model of the world economy. To transform the vast collection of microeconomic facts that describe the world economy into an organized system from which macroeconomic projections of future growth could be made the model was to rely on the method of input-output, or interindustry, analysis. The input-output method depicts the structure of an economy in terms of the flows among its producing and consuming sectors, real transfers of goods and services. Such transfers can be desplayed in a statistical input-output table for the economy. This table in turn gives rise to a set of structural equations whose simultaneous solution provides a numerical picture of a possible future state of the economy. In order to generate such a projection from the input-output model it is necessary to make a certain number of assumptions about some of the factors that will determine the pace and the shape of the economy's future growth, that is, some of the variables in the set of equations must be fixed. Hence by trying different assumptions it is possible to project a series of alternative paths for the development of the economy. In this way input-output analysis provides a means of taking the quantitative measure of hopes and plans for the future.

Source: From Wassily W. Leontief, "The World Economy of the Year 2000," published in *Nobel Prize Winners: A Collection of Their Articles*, Scientific American, 1990.

Matrix Algebra and Applications

SECTIONS

1. Matrix Addition and Scalar Multiplication
2. Matrix Multiplication
3. Matrix Inversion
4. Input-Output Models

You're the Expert
The Japanese Economy

APPLICATION ▶ Senator Glenn Russel walks into your cubicle in the Congressional Budget Office. "Look here," he says, "I don't see why the Japanese trade representative is getting so upset with my proposal to cut down on our imports of office supplies from Japan. He claims that it'll hurt their mining operations. But just look at their input-output table. Their office supply sector doesn't use any input from their mining sector. How can our cutting down demand for office supplies hurt mining?" How should you respond?

173

> **INTRODUCTION** ▶ We used matrices in Chapter 2 to organize our work, but we have not yet dealt with them as interesting objects in their own right. There is much that we can do with them besides row operations: we can add, subtract, multiply, and even, on occasion, "divide" matrices. We shall use these operations to study input-output models at the end of this chapter.
>
> Many calculators, electronic spreadsheets, and other computer programs can perform these matrix operations, and this is a big help in doing calculations. Nevertheless, you need to know how these operations are defined to see why they are useful and to understand which operation to use in any particular application.

3.1 MATRIX ADDITION AND SCALAR MULTIPLICATION

Let's start by formally defining a matrix and introducing some basic terms.

> **MATRIX, DIMENSION, ENTRIES**
> An $m \times n$ **matrix** A is a rectangular array of real numbers with m rows and n columns. We shall sometimes refer to m and n as the **dimensions** of the matrix A. The numbers that appear in the matrix are called its **entries**.

We shall usually use uppercase letters as names of matrices.

▼ **EXAMPLE 1**

The following is a 2×3 matrix, since it has 2 rows and 3 columns.

$$A = \begin{bmatrix} 2 & 0 & -1 \\ 33 & -22 & 0 \end{bmatrix}$$

Another way of recognizing it as a 2×3 matrix is by its "measurements": it is two numbers high and three numbers across.

The entries of A are the numbers $2, 0, -1, 33, -22, 0$.

Before we go on... Remember that the number of rows is specified first, and the number of columns second. An easy way to remember this is to think of the acronym "RC" (Row Column) as in "RC Cola™." Then again, some people like to remember it this way: "down first, then across."

Most calculators and computer programs will require you to enter the dimensions of any matrix that you want to use. On the TI-81 and TI-82 calculators, for example, matrices are referred to as [A], [B], and so on. Appendix B contains a section on matrix algebra on the TI-82.* Here is a summary: To define a matrix on the TI-82, press [MATRX] to bring up the matrix menu, select EDIT and press [ENTER], and then enter the dimensions of the matrix, followed by its entries. The matrix will be stored under the name [A] (unless you have selected another name). The names of matrices are accessed through the matrix menu using [MATRX]. You can now display the contents of the matrix [A] by entering [A] [ENTER] on the Home Screen.

▼ EXAMPLE 2

The matrix

$$B = \begin{bmatrix} 2 & 3 \\ 10 & 44 \\ -1 & 3 \\ 8 & 3 \end{bmatrix}$$

is a 4 × 2 matrix, since it has 4 rows and 2 columns.

REFERRING TO THE ENTRIES OF A MATRIX

There is a systematic way of referring to particular entries in a matrix. If i and j are numbers, then the entry in the ith row and j^{th} column of the matrix A is called the ***ij^{th} entry*** of A. We usually write this entry as a_{ij} or A_{ij}. (If the matrix was called B, we would write its ij^{th} entry as b_{ij} or B_{ij}.) Notice that this follows the "RC" convention: the row number is specified first, and the column number second.

REFERRING TO THE ENTRIES OF A MATRIX ON A GRAPHING CALCULATOR

To access any entry of a matrix on a TI-82 calculator, enter the matrix name, followed by the row and column numbers in parentheses. For instance, to access the entry A_{23}, enter

$$[A](2,3)$$

If you then press [ENTER], the entry A_{23} will appear on the screen.

▼ *See Paula Young's *Graphing Calculator Lessons for Finite Mathematics* (published by HarperCollins College Publishers in 1993) for more information on the TI-81 and drill exercises in the use of a graphing calculator.

EXAMPLE 3

If
$$A = \begin{bmatrix} 2 & 0 & -1 \\ 33 & -22 & 0 \end{bmatrix},$$

then
$a_{11} = 2$ (first row, first column)
$a_{12} = 0$ (first row, second column)
$a_{13} = -1$ (first row, third column)
$a_{21} = 33$ (second row, first column)
$a_{22} = -22$ (second row, second column)
$a_{23} = 0$ (second row, third column).

In other words, we have labeled the entries of A as follows.

$$A = \begin{bmatrix} a_{11} & a_{12} & a_{13} \\ a_{21} & a_{22} & a_{23} \end{bmatrix}$$

EXAMPLE 4

Locate b_{21} and b_{12} in the matrix B in Example 2.

SOLUTION b_{21} is the entry in the second row and first column, so $b_{21} = 10$. b_{12} is the entry in the first row and second column, so $b_{12} = 3$.

In general, the $m \times n$ matrix A has its entries labeled as follows.

$$A = \begin{bmatrix} a_{11} & a_{12} & a_{13} & \cdots & a_{1n} \\ a_{21} & a_{22} & a_{23} & \cdots & a_{2n} \\ \cdots & \cdots & \cdots & \cdots & \cdots \\ a_{m1} & a_{m2} & a_{m3} & \cdots & a_{mn} \end{bmatrix}$$

We say that two matrices A and B are **equal** if they have the same dimensions and the corresponding entries are equal.

EXAMPLE 5

For $\begin{bmatrix} 7 & 9 & 0 \\ 0 & -1 & 11 \end{bmatrix}$ to be equal to $\begin{bmatrix} 7 & 9 & x \\ 0 & -1 & y+1 \end{bmatrix}$, we must have corresponding entries equal, so that

$x = 0$
$y + 1 = 11$, or $y = 10$.

Before we go on... Note that the matrix equation $\begin{bmatrix} 7 & 9 & 0 \\ 0 & -1 & 11 \end{bmatrix} = \begin{bmatrix} 7 & 9 & x \\ 0 & -1 & y+1 \end{bmatrix}$ is really six equations in one: $7 = 7, 9 = 9, 0 = x, 0 = 0, -1 = -1, 11 = y + 1$. We used only the two that were interesting.

Note that a 3 × 4 matrix can never equal a 3 × 5 matrix, since they do not have the same dimensions.

> **TRANSPOSITION**
>
> If A is any matrix, then its **transpose** is the matrix obtained by writing its rows as columns. We denote the transpose of the matrix A by A^T.

▼ **EXAMPLE 6**

Let $B = \begin{bmatrix} 2 & 3 \\ 10 & 44 \\ -1 & 3 \\ 8 & 3 \end{bmatrix}$. Then $B^T = \begin{bmatrix} 2 & 10 & -1 & 8 \\ 3 & 44 & 3 & 3 \end{bmatrix}$.

Notice what happens to the dimensions: B is a 4 × 2 matrix, while its transpose is a 2 × 4 matrix. In general, the transpose of an $m \times n$ matrix is an $n \times m$ matrix.

A calculator or computer program that can do matrix operations should be able to compute the transpose. On the TI-82, you can access the symbol "T" by pressing [MATRX], selecting MATH, and then pressing [ENTER]. The command

$$[A]^T \rightarrow [B]$$

then stores the transpose of [A] under the name [B].

> **ROW MATRIX, COLUMN MATRIX**
>
> A matrix with a single row is called a **row matrix** or **row vector**. A matrix with a single column is called a **column matrix** or **column vector**.

▼ **EXAMPLE 7**

The 1 × 5 matrix $C = [3 \quad -4 \quad 0 \quad 1 \quad -11]$ is a row matrix, while the 4 × 1 matrix

$$D = \begin{bmatrix} 2 \\ 10 \\ -1 \\ 8 \end{bmatrix}$$

is a column matrix.

Before we go on... Whereas we use capital letters as the names for most matrices, it is common to use lowercase letters as the names of vectors.

Note that the transpose of a row vector is a column vector, and vice versa. For example,

$$C^T = \begin{bmatrix} 3 \\ -4 \\ 0 \\ 1 \\ -11 \end{bmatrix}$$

and

$$D^T = \begin{bmatrix} 2 & 10 & -1 & 8 \end{bmatrix}.$$

MATRIX ADDITION

To begin the study of matrix algebra, we first discuss **matrix addition.** The rules for addition are simple. First, two matrices cannot be added unless they have the same dimensions. Second, to add two matrices of the same dimensions, we add the corresponding entries.

EXAMPLE 8

$$\begin{bmatrix} 2 & -3 \\ 1 & 0 \\ -1 & 3 \\ 8 & 3 \end{bmatrix} + \begin{bmatrix} 9 & -5 \\ 0 & 13 \\ -1 & 3 \\ 5 & 99 \end{bmatrix} = \begin{bmatrix} 11 & -8 \\ 1 & 13 \\ -2 & -6 \\ 13 & 102 \end{bmatrix}$$

On most calculators, adding matrices is similar to adding numbers: the sum of the matrices [A] and [B] is [A] + [B]. If you attempt to add matrices with different dimensions, you will get an error message.

Matrix subtraction is defined similarly: To subtract one matrix from another of the same dimensions, we simply subtract corresponding entries.

EXAMPLE 9

$$\begin{bmatrix} 2 & -3 \\ 1 & 0 \\ -1 & 3 \\ 8 & 3 \end{bmatrix} - \begin{bmatrix} 9 & -5 \\ 0 & 13 \\ -1 & 3 \\ 5 & 99 \end{bmatrix} = \begin{bmatrix} -7 & 2 \\ 1 & -13 \\ 0 & 0 \\ 3 & -96 \end{bmatrix}$$

EXAMPLE 10   Sales

The APlus auto parts store chain has two outlets, one in Vancouver and one in Quebec. Among other things, they sell wiper blades, windshield cleaning fluid, and floor mats. The monthly sales of these items at the two stores for two months is given in the following tables.

3.1 Matrix Addition and Scalar Multiplication

January Sales

	Vancouver	Quebec
Wiper blades	20	15
Cleaning fluid (bottles)	10	12
Floor mats	8	4

February Sales

	Vancouver	Quebec
Wiper blades	23	12
Cleaning fluid (bottles)	8	12
Floor mats	4	5

Use matrix arithmetic to calculate the change in sales of each product in each store from January to February.

SOLUTION The tables suggest two matrices,

$$J = \begin{bmatrix} 20 & 15 \\ 10 & 12 \\ 8 & 4 \end{bmatrix} \text{ and } F = \begin{bmatrix} 23 & 12 \\ 8 & 12 \\ 4 & 5 \end{bmatrix}.$$

We want to subtract corresponding entries in these two matrices. In other words, we want to compute the difference of the two matrices.

$$F - J = \begin{bmatrix} 23 & 12 \\ 8 & 12 \\ 4 & 5 \end{bmatrix} - \begin{bmatrix} 20 & 15 \\ 10 & 12 \\ 8 & 4 \end{bmatrix} = \begin{bmatrix} 3 & -3 \\ -2 & 0 \\ -4 & 1 \end{bmatrix}$$

Thus, the change in sales of each product is given in the following table.

Changes in Sales

	Vancouver	Quebec
Wiper blades	3	−3
Cleaning fluid (bottles)	−2	0
Floor mats	−4	1

We can define these operations formally as follows.

MATRIX ADDITION AND SUBTRACTION

If A and B are $m \times n$ matrices, then $A + B$ is the $m \times n$ matrix whose ij^{th} entry is given by

$$(A + B)_{ij} = A_{ij} + B_{ij},$$

and $A - B$ is the $m \times n$ matrix whose ij^{th} entry is given by

$$(A - B)_{ij} = A_{ij} - B_{ij}.$$

Q What do those formulas mean?

A In words, the first formula says that the ij^{th} entry of the sum $A + B$ is the sum of the ij^{th} entries of A and B. For practice, you should translate the second formula into words yourself.

Q When can a matrix A be added to *itself*?

A Always, since the expression $A + A$ is the sum of two matrices that certainly have the same dimensions.

Q Can't we write $A + A$ as $2A$?

A We certainly can. Notice that when we compute $A + A$, we end up doubling every entry in A. So we can think of the expression $2A$ as telling us to *multiply every element in A by* 2.

In general, to multiply a matrix by a number, multiply every entry in the matrix by that number.

▼ **EXAMPLE 11**

$$6 \begin{bmatrix} \frac{5}{2} & -3 \\ 1 & 0 \\ -1 & \frac{5}{6} \\ 8 & 3 \end{bmatrix} = \begin{bmatrix} 15 & -18 \\ 6 & 0 \\ -6 & 5 \\ 48 & 18 \end{bmatrix}$$

Before we go on... It is traditional when talking about matrices to call individual numbers **scalars.** For this reason, we call the operation of multiplying a matrix by a number **scalar multiplication.**

 On the TI-82 calculator, 6 times the matrix [A] is 6[A] or 6*[A].

▼ **EXAMPLE 12** Sales

The revenue generated by sales in the Vancouver and Quebec branches of the APlus auto parts store (see Example 10) was as follows.

	January Sales in Canadian Dollars	
	Vancouver	**Quebec**
Wiper blades	140.00	105.00
Cleaning fluid (bottles)	30.00	36.00
Floor mats	96.00	48.00

If the Canadian dollar was worth $0.85U.S. at the time, compute the revenue in U.S. dollars.

SOLUTION We need to multiply each revenue figure by 0.85. We can represent this calculation as scalar multiplication with the following matrix.

$$A = \begin{bmatrix} 140.00 & 105.00 \\ 30.00 & 36.00 \\ 96.00 & 48.00 \end{bmatrix}$$

3.1 Matrix Addition and Scalar Multiplication

The revenue figures in U.S. dollars are given by the scalar multiple

$$0.85A = 0.85 \begin{bmatrix} 140.00 & 105.00 \\ 30.00 & 36.00 \\ 96.00 & 48.00 \end{bmatrix} = \begin{bmatrix} 119.00 & 89.25 \\ 25.50 & 30.60 \\ 81.60 & 40.80 \end{bmatrix}.$$

In other words, \$119U.S. worth of wiper blades were sold in Vancouver, \$89.25U.S. worth of wiper blades were sold in Quebec, and so on.

Formally, scalar multiplication is defined as follows.

> **SCALAR MULTIPLICATION**
>
> If A is an $m \times n$ matrix and c is a real number, then cA is the $m \times n$ matrix whose ij^{th} entry is given by
>
> $$(cA)_{ij} = c \cdot (A_{ij}).$$

Q Haven't you just moved the parentheses?

A There is a little more going on here than that. In words, this rule says: To get the ij^{th} entry of cA, multiply the ij^{th} entry of A by c.

▼ **EXAMPLE 13**

Let

$$A = \begin{bmatrix} 2 & -1 & 0 \\ 3 & 5 & -3 \end{bmatrix}, \quad B = \begin{bmatrix} 1 & 3 & -1 \\ 5 & -6 & 0 \end{bmatrix}, \quad C = \begin{bmatrix} x & y & w \\ z & t+1 & 3 \end{bmatrix}.$$

Evaluate the following: $4A$, cB, $A + 3C$

SOLUTION First, $4A$ is obtained by multiplying each entry of A by 4. Thus,

$$4A = \begin{bmatrix} 8 & -4 & 0 \\ 12 & 20 & -12 \end{bmatrix}.$$

Similarly, cB is obtained by multiplying each entry of B by c.

$$cB = \begin{bmatrix} c & 3c & -c \\ 5c & -6c & 0 \end{bmatrix}$$

We get $A + 3C$ in two steps as follows.

$$A + 3C = \begin{bmatrix} 2 & -1 & 0 \\ 3 & 5 & -3 \end{bmatrix} + 3 \begin{bmatrix} x & y & w \\ z & t+1 & 3 \end{bmatrix}$$

$$= \begin{bmatrix} 2 & -1 & 0 \\ 3 & 5 & -3 \end{bmatrix} + \begin{bmatrix} 3x & 3y & 3w \\ 3z & 3t+3 & 9 \end{bmatrix}$$

$$= \begin{bmatrix} 2+3x & -1+3y & 3w \\ 3+3z & 3t+8 & 6 \end{bmatrix}$$

Addition and scalar multiplication have nice properties, reminiscent of the properties of addition and multiplication of real numbers. Before we state them, we need to introduce some more notation.

If A is any matrix, then $-A$ is the matrix $(-1)A$. In other words, $-A$ is obtained from A by multiplying A by the scalar -1. This amounts to changing the signs of all the entries in A. Thus, for example,

$$-\begin{bmatrix} 4 & -2 & 0 \\ 6 & 10 & -6 \end{bmatrix} = \begin{bmatrix} -4 & 2 & 0 \\ -6 & -10 & 6 \end{bmatrix}.$$

For any two matrices A and B, $A - B$ is the same as $A + (-B)$ (why?).

Also, a **zero matrix** is a matrix whose entries are all zero. Thus, for example, the zero 2×3 matrix is

$$O = \begin{bmatrix} 0 & 0 & 0 \\ 0 & 0 & 0 \end{bmatrix}.$$

Now we state the most important properties of the operations that we have been discussing.

LAWS OF MATRIX ADDITION AND SCALAR MULTIPLICATION

If A, B and C are any $m \times n$ matrices, and if O is the zero $m \times n$ matrix, then the following hold.

$A + (B + C) = (A + B) + C$	Associative law
$A + B = B + A$	Commutative law
$A + O = O + A = A$	Additive identity law
$A + (-A) = O = (-A) + A$	Additive inverse law
$c(A + B) = cA + cB$	Distributive law
$(c + d)A = cA + dA$	Distributive law
$1A = A$	Scalar unit
$0A = O$	Scalar zero

Q Aren't these properties obvious? Don't we know them already?

A They would be obvious if we were talking about addition and multiplication of *numbers*, but here we are talking about addition and multiplication of *matrices*. We are doing what in computer science is known as **operator overloading.** This means that we are using + to mean something new: matrix addition. There is no good reason why matrix addition has to obey *all* the laws of addition of numbers. It happens that it does obey many of them, which is why it is convenient to call it addition in the first place. This means that you can manipulate equations involving matrices in much the same way that you manipulate equations involving numbers. One word of caution: We don't yet know how to multiply matrices, and it isn't what you think. Multiplication of matrices does *not* obey all the same laws as multiplication of numbers.

Transposition also has nice properties.

LAWS OF TRANSPOSITION

If A and B are $m \times n$ matrices, then

$$(A + B)^T = A^T + B^T$$
$$(cA)^T = c(A^T).$$

3.1 EXERCISES

Write the dimensions of the matrices given in Exercises 1 through 10.

1. $A = \begin{bmatrix} 1 & -1 & 0 & 2 & \frac{1}{4} \end{bmatrix}$
2. $B = [44]$
3. $C = \begin{bmatrix} \frac{5}{2} \\ 1 \\ -2 \\ 8 \end{bmatrix}$
4. $D = \begin{bmatrix} 15 & -18 \\ 6 & 0 \\ -6 & 5 \\ 48 & 18 \end{bmatrix}$
5. $E = \begin{bmatrix} e_{11} & e_{12} & e_{13} & \cdots & e_{1q} \\ e_{21} & e_{22} & e_{23} & \cdots & e_{2q} \\ \cdots & \cdots & \cdots & \cdots & \cdots \\ e_{p1} & e_{p2} & e_{p3} & \cdots & e_{pq} \end{bmatrix}$
6. $A = \begin{bmatrix} 2 & -1 & 0 \\ 3 & 5 & -3 \end{bmatrix}$
7. $B = \begin{bmatrix} 1 & 3 \\ 5 & -6 \end{bmatrix}$
8. $C = \begin{bmatrix} x & y & w & e \\ z & t+1 & 3 & 0 \end{bmatrix}$
9. $D = \begin{bmatrix} d_1 & d_2 & \cdots & d_n \end{bmatrix}$
10. $E = [d]$

Referring to Exercises 1–5 above, find the following.

11. a_{11}
12. a_{14}
13. B_{11}
14. c_{21}
15. C_{31}
16. d_{22}
17. e_{13}
18. E_{44}

Referring to Exercises 6–10 above, find the following.

19. a_{11}
20. A_{23}
21. B_{12}
22. C_{23}
23. c_{14}
24. D_{1r}
25. d_{1n}
26. E_{11}

27. Solve for x, y, z, and w.

$$\begin{bmatrix} x+y & x+z \\ y+z & w \end{bmatrix} = \begin{bmatrix} 3 & 4 \\ 5 & 4 \end{bmatrix}$$

28. Solve for x, y, z, and w.

$$\begin{bmatrix} x-y & x-z \\ y-w & w \end{bmatrix} = \begin{bmatrix} 0 & 0 \\ 0 & 6 \end{bmatrix}$$

In Exercises 29–36, let $A = \begin{bmatrix} 0 & -1 \\ 1 & 0 \\ -1 & 2 \\ 5 & 0 \end{bmatrix}$, $B = \begin{bmatrix} \frac{1}{4} & -1 \\ 0 & \frac{1}{3} \\ -1 & 3 \\ 5 & 0 \end{bmatrix}$, $C = \begin{bmatrix} 1 & -1 \\ 1 & 1 \\ -1 & -1 \\ 1 & 1 \end{bmatrix}$.

Evaluate:

29. $A + B$ 30. $A - C$ 31. $A + B - C$ 32. $12B$

33. $2A - C$ 34. $2A + \frac{1}{2}C$ 35. $2A^T$ 36. $B^T + 3C^T$

In Exercises 37–44, let $A = \begin{bmatrix} 1 & -1 & 0 \\ 0 & 2 & -1 \end{bmatrix}$, $B = \begin{bmatrix} 3 & 0 & -1 \\ 5 & -1 & 1 \end{bmatrix}$, $C = \begin{bmatrix} x & 1 & w \\ z & r & 4 \end{bmatrix}$.

Evaluate:

37. $A + B$ 38. $B - C$ 39. $A - B + C$ 40. $\frac{1}{2}B$

41. $2A - B$ 42. $2A - 4C$ 43. $3B^T$ 44. $2A^T - C^T$

In Exercises 45–52, use a graphing calculator or matrix computer software.

Let $A = \begin{bmatrix} 1.5 & -2.35 & 5.6 \\ 44.2 & 0 & 12.2 \end{bmatrix}$, $B = \begin{bmatrix} 1.4 & 7.8 \\ 5.4 & 0 \\ 5.6 & 6.6 \end{bmatrix}$, $C = \begin{bmatrix} 10 & 20 & 30 \\ -10 & -20 & -30 \end{bmatrix}$.

Enter all three matrices in your calculator, and then evaluate the following.

45. $A - C$ 46. $C - A$ 47. $1.1B$

48. $-0.2B$ 49. $A^T + 4.2B$ 50. $(A + 2.3C)^T$

51. $(2.1A - 2.3C)^T$ 52. $(A + C)^T - B$

APPLICATIONS

53. *Bank Executives' Incomes* The following table shows the 1991 salaries of the four highest-paid executives of Toronto Dominion Bank, as well as their annual increases for the subsequent two years.*

1991 Salary	Increase in '92	Increase in '93
$890,000	$576	$34,424
$675,000	$411	$24,589
$275,000	$20,822	$54,370
$275,000	$411	$24,781

Use matrix algebra to find their annual salaries in **(a)** 1992; **(b)** 1993.

---

*Source: Company Reports, *Financial Post,* December 11, 1993, p. 3.

54. *Exchange Rates* The following table of exchange rates shows the value of the Canadian dollar as of Nov. 3, 1993, and also subsequent changes.*

Per C$	Nov. 3, 1993	Increase Nov. 3 to Dec. 3	Increase Dec. 3 to Dec. 10
U.S. dollar	0.7612	−0.0003	−0.0005
Deutschesmark	1.2825	0.0127	−0.0109
Japanese yen	80.44	1.22	0.44
U.K. pound	1.9621	0.0184	0.0072

Use matrix algebra to find the exchange rates on **(a)** Dec. 3, 1993; **(b)** Dec. 10, 1993.

Profits Exercises 55 and 56 are based on the following chart, which shows Carnival Corporation revenues and profits for the years 1989–1994.†

	1989	1990	1991	1992	1993	1994
Carnival Air Lines revenue ($ million)	5	25	40	70	110	150
Revenue from *Carnival Cruise* passengers ($ million)	10	20	20	15	10	5
Net profit/loss	0	0.5	−5	2	0	4

55. Use matrix algebra to calculate *Carnival's* total revenues and expenditures for the given years.

56. Use matrix algebra to determine when *Carnival's* revenues from the airlines exceeded its revenues from cruise passengers by the greatest amount.

57. *Population Distribution* In 1970, the population of the U.S., broken down by regions,‡ was 49.1 million in the Northeast, 56.6 million in the Midwest, 62.8 million in the South, and 34.8 million in the West. In 1980, the population was 49.1 million in the Northeast, 58.9 million in the Midwest, 75.4 million in the South, and 43.2 million in the West. Set up the population figures for each year as a row vector, and then show how to use matrix operations to find the net increase or decrease of population in each region from 1970 to 1980.

58. *Population Distribution* In 1980, the population of the U.S., broken down by regions,‡ was 49.1 million in the Northeast, 58.9 million in the Midwest, 75.4 million in the South, and 43.2 million in the West. Between 1980 and 1990, the population in the Northeast grew by 1.7 million, the population in the Midwest grew by 0.8 million, the population in the South grew by 10 million, and the population in the West grew by 9.6 million. Set up the population figures for 1980, and the growth figures for the decade, as row vectors. Assuming that the population will grow by the same numbers from 1990 to 2000 as it did from 1980 to 1990, show how to use matrix operations to find the population in each region in 2000.

---

* Source: Royal Bank of Canada, *Financial Post,* December 11, 1993, p. 23.
† Figures are estimated from a graph and rounded. Source: Company Reports/*The New York Times,* October 23, 1994, p. F5.
‡ Source: Statistical Abstract of the United States 1991, U.S. Dept. of Commerce.

59. *Inventory* The Left Coast Book Store chain has two stores, one in San Francisco and one in Los Angeles. It stocks three kinds of book: hardcover, softcover, and plastic (for infants). At the beginning of January, its central computer showed the following books in stock.

	Hard	Soft	Plastic
San Francisco	1,000	2,000	5,000
Los Angeles	1,000	5,000	2,000

Suppose now that sales in January were as follows: 700 hardcover books, 1,300 softcover books, and 2,000 plastic books sold in San Francisco; 400 hardcover, 300 softcover, and 500 plastic books sold in Los Angeles. Write these sales figures in the form of a matrix, and then show how matrix algebra can be used to compute the inventory remaining in each store at the end of January.

60. *Inventory* The Left Coast Book Store chain discussed in the previous exercise actually maintained the same sales figures for the first six months of the year. Each month it restocked the stores from its warehouse by shipping 600 hardcover, 1,500 softcover, and 1,500 plastic books to San Francisco; and 500 hardcover, 500 softcover, and 500 plastic books to Los Angeles.
(a) Use matrix operations to determine the total sales over the six months, broken down by store and type of book.
(b) Use matrix operations to determine the inventory in each store at the end of June.

61. *Inventory* Microbucks Computer Company makes two computers, the Pomegranate II and the Pomegranate Classic, at two different factories. The Pom II requires two processor chips, 16 memory chips, and 20 vacuum tubes, while the Pom Classic requires one processor chip, 4 memory chips, and 40 vacuum tubes. Microbucks has in stock at the beginning of the year 500 processor chips, 5,000 memory chips, and 10,000 vacuum tubes at the Pom II factory, and 200 processor chips, 2,000 memory chips, and 20,000 vacuum tubes at the Pom Classic factory. The company manufactures 50 Pom II's and 50 Pom Classics each month.
(a) Find the inventory of parts after two months, using matrix operations.
(b) When will Microbucks run out of one of the parts?

62. *Inventory* Microbucks Computer Company, besides having the stock mentioned in the previous exercise, gets shipments of parts every month in the amounts of 100 processor chips, 1,000 memory chips, and 3,000 vacuum tubes at the Pom II factory, and 50 processor chips, 1,000 memory chips, and 2,000 vacuum tubes at the Pom Classic factory.
(a) What will the inventory of parts be after six months?
(b) How long before Microbucks runs out of one of the parts?

COMMUNICATION AND REASONING EXERCISES

63. What does it mean when we say that $(A + B)_{ij} = A_{ij} + B_{ij}$?

64. What would the 5 × 5 matrix A look like if $A_{ii} = 0$ for every i?

65. What would the matrix A look like if $A_{ij} = 0$ whenever $i \neq j$?

66. Describe a scenario (possibly based on one of the above examples or exercises) in which you might wish to compute $A - 2B$ for certain matrices A and B.

67. Describe a scenario (possibly based on one of the above examples or exercises) in which you might wish to compute $A + B - C$ for certain matrices A, B, and C.

68. Why is matrix addition associative?

3.2 MATRIX MULTIPLICATION

Q Why didn't we include matrix multiplication in the first section? After all, can't we just say, for example, that

$$\begin{bmatrix} 2 & -3 \\ 1 & 0 \\ -1 & 3 \\ 8 & 3 \end{bmatrix} \times \begin{bmatrix} 9 & -5 \\ 0 & 13 \\ -1 & 3 \\ 1 & 0 \end{bmatrix} = \begin{bmatrix} 18 & 15 \\ 0 & 0 \\ 1 & 9 \\ 8 & 0 \end{bmatrix}?$$

A We don't do it that way. Now, there is nothing wrong in principle with defining the product of two $m \times n$ matrices to be the matrix whose entries are the products of the corresponding entries as above. The problem is that this is not particularly useful. The most useful way of multiplying matrices is a little more involved than this.

First, we see how to multiply a row matrix by a column matrix.

Row $\times$ Column

When we calculate the product of a row and a column (with the row written on the left) the answer will be a *single number* (actually a 1×1 matrix). Here are some examples.

▼ **EXAMPLE 1**

$$[2 \ \ 4 \ \ 1] \begin{bmatrix} 2 \\ 10 \\ -1 \end{bmatrix} = [2 \times 2 + 4 \times 10 + 1 \times (-1)]$$
$$= [4 + 40 + (-1)] = [43]$$

In other words, we multiply the first entries together, then the second entries and then the third entries, and then add the answers. We can think of this process in two ways:

Using a diagram,

$$[2 \ \ 4 \ \ 1] \begin{bmatrix} 2 \\ 10 \\ -1 \end{bmatrix}$$

$2 \times 2 = 4$	Product of first entries = 4
$4 \times 10 = 40$	Product of second entries = 40
$1 \times (-1) = \underline{-1}$	Product of third entries = -1
43.	Sum of products = 43

Using algebra,

$$[a \ \ b \ \ c] \begin{bmatrix} x \\ y \\ z \end{bmatrix} = [ax + by + cz].$$

On the TI-82 calculator, the format for multiplying matrices is the same as for multiplying numbers: [A][B] or [A] $\times$ [B] will give the product.
Most electronic spreadsheets have built-in commands to multiply matrices as well.

EXAMPLE 2

Calculate the product

$$[2 \quad 1]\begin{bmatrix} -3 \\ 0 \end{bmatrix}.$$

SOLUTION Following the procedure in Example 1, we multiply the first entries together, multiply the second entries together, and add the results.

$$[2 \quad 1]\begin{bmatrix} -3 \\ 0 \end{bmatrix} = [2 \times (-3) + 1 \times 0] = [-6 + 0] = [-6]$$

Thus, the answer is $[-6]$. Once again, we can illustrate this by using a diagram or using algebra.

Diagram:

$$[2 \quad 1]\begin{bmatrix} -3 \\ 0 \end{bmatrix} \quad \begin{array}{l} 2 \times (-3) = -6 \\ 1 \times 0 = 0 \\ \hline -6 \end{array} \quad \begin{array}{l} \text{Product of first entries} = -6 \\ \text{Product of second entries} = 0 \\ \text{Sum of products} = -6 \end{array}$$

Algebra:

$$[a \quad b]\begin{bmatrix} x \\ y \end{bmatrix} = [ax + by]$$

▶ **NOTES**

1. In the discussion so far, *the row is on the left and the column is on the right* (RC again). Thus, for example, we *don't yet* attempt to make any sense of a product such as*

$$\begin{bmatrix} -3 \\ 0 \end{bmatrix}[2 \quad 1].$$

2. The row size has to match the column size. This means that if we have a 1×4 row on the left, then the column on the right must be 4×1 in order for the product to make sense. Thus, for example, the product

$$[a \quad b]\begin{bmatrix} x \\ y \\ z \end{bmatrix}$$

is not defined. ◀

▼ * We shall be able to make sense of this shortly. In fact, the answer will turn out to be a 2×2 matrix.

▼ **EXAMPLE 3**   Sales and Revenue

The APlus auto parts store mentioned in examples in the previous section had the following sales in its Vancouver store.

	Vancouver
Wiper blades	20
Cleaning fluid (bottles)	10
Floor mats	8

If wiper blades cost $7.00 each, cleaning fluid costs $3.00 per bottle, and floor mats cost $12.00 each, use matrix multiplication to find the total revenue generated by sales of these items.

SOLUTION   We need to multiply each sales figure by the corresponding price and then add the resulting revenue figures. We can do this calculation as a matrix product. We represent the sales by a column vector, as suggested by the table.

$$Q = \begin{bmatrix} 20 \\ 10 \\ 8 \end{bmatrix}$$

We put the selling prices in a row vector.

$$P = \begin{bmatrix} 7.00 & 3.00 & 12.00 \end{bmatrix}$$

We can now compute the total revenue as the product

$$R = PQ = \begin{bmatrix} 7.00 & 3.00 & 12.00 \end{bmatrix} \begin{bmatrix} 20 \\ 10 \\ 8 \end{bmatrix}$$

$$= [140.00 + 30.00 + 96.00] = [266.00].$$

So, these sales generated a total revenue of $266.00.

Before we go on... We could also have written the quantity sold as a row vector (which would be Q^T) and the prices as a column vector (which would be P^T) and then multiplied them in the opposite order ($Q^T P^T$). Try this.

▼ **EXAMPLE 4**

Represent the matrix equation

$$\begin{bmatrix} 2 & -4 & 1 \end{bmatrix} \begin{bmatrix} x \\ y \\ z \end{bmatrix} = [5]$$

as an ordinary equation.

SOLUTION If we perform the multiplication on the left, we get the 1×1 matrix $[2x - 4y + z]$. Thus, the equation may be rewritten as

$$[2x - 4y + z] = [5].$$

Saying that these two 1×1 matrices are equal means that their corresponding entries agree, so we get the equation

$$2x - 4y + z = 5.$$

You can do this example on a calculator or computer that can do symbolic manipulations, but not on one that can do only numeric calculations.

It is also possible to work Example 4 in reverse, as shown below.

▼ **EXAMPLE 5**

Express the linear equation $3x + y - z + 2w = 8$ as a matrix equation in a form similar to Example 4.

SOLUTION

$$\begin{bmatrix} 3 & 1 & -1 & 2 \end{bmatrix} \begin{bmatrix} x \\ y \\ z \\ w \end{bmatrix} = [8]$$

Before we go on... The row matrix $\begin{bmatrix} 3 & 1 & -1 & 2 \end{bmatrix}$ is the row of **coefficients** of the original equation. This should begin to look familiar if you think back to Chapter 2.

Next, we turn to the general case.

MULTIPLYING MATRICES BY MATRICES

In general, we can take the product AB only if the number of columns of A matches the number of rows of B (so that we can multiply the rows of A by the columns of B as above). The product AB is then obtained by taking its ij^{th} entry to be as follows:

$$ij^{th} \text{ entry of } AB = \text{Row } i \text{ of } A \times \text{Column } j \text{ of } B.$$

▼ **EXAMPLE 6**

Calculate $\begin{bmatrix} 2 & 0 & -1 & 3 \end{bmatrix} \begin{bmatrix} 1 & 1 & -8 \\ 1 & -6 & 0 \\ 0 & 5 & 2 \\ -3 & 8 & 1 \end{bmatrix}.$

The number of entries in the row on the left matches the number of entries in each column of the matrix on the right, so we can multiply that row by each of those columns. Doing the multiplications gives three numbers, one for each column, and we arrange these numbers in a row.

$$\underset{\downarrow}{\text{Row 1}} \quad \underset{\downarrow}{\text{Col. 1}} \ \underset{\downarrow}{\text{Col. 2}} \ \underset{\downarrow}{\text{Col. 3}}$$

$$[2 \ \ 0 \ -1 \ \ 3] \begin{bmatrix} 1 & 1 & -8 \\ 1 & -6 & 0 \\ 0 & 5 & 2 \\ -3 & 8 & 1 \end{bmatrix} = [\text{Row 1} \times \text{Col. 1} \quad \text{Row 1} \times \text{Col. 2} \quad \text{Row 1} \times \text{Col. 3}]$$

$$= [-7 \quad 21 \quad -15]$$

Before we go on... Note that we continue to have the row on the left and the column(s) on the right. This will always be the case in matrix multiplication:

Rows on the left $\times$ Columns on the right.

Look at the dimensions of the two matrices above.

$$\text{Match}$$
$$(1 \times 4)(4 \times 3) \rightarrow 1 \times 3$$

The fact that the number of columns in the left-hand matrix matches the number of rows in the right-hand matrix amounts to saying that the middle two numbers must match as above. If we *cancel* the middle matching numbers, then we are left with the dimensions of the product.

Before continuing with more examples, we state the rule for matrix multiplication formally.

MULTIPLICATION OF MATRICES

If A is an $m \times n$ matrix and B is an $n \times k$ matrix, then the product AB is the $m \times k$ matrix whose ij^{th} entry is the product

$$\underset{\downarrow}{\text{Row } i \text{ of } A} \times \underset{\downarrow}{\text{Column } j \text{ of } B}$$

$$(AB)_{ij} = [a_{i1} \ \ a_{i2} \ \ a_{i3} \ \ \ldots \ \ a_{in}] \begin{bmatrix} b_{1j} \\ b_{2j} \\ b_{3j} \\ \ldots \\ b_{nj} \end{bmatrix}$$

$$= a_{i1}b_{1j} + a_{i2}b_{2j} + a_{i3}b_{3j} + \ldots + a_{in}b_{nj}.$$

EXAMPLE 7

Calculate $\begin{bmatrix} 2 & 0 & -1 & 3 \\ 1 & -1 & 2 & -2 \end{bmatrix} \begin{bmatrix} 1 & 1 & -8 \\ 1 & 0 & 0 \\ 0 & 5 & 2 \\ -2 & 8 & -1 \end{bmatrix}$.

SOLUTION Before we start the calculation, we check that the dimensions of the matrices match up.

$$\underset{(2 \times 4)}{\begin{bmatrix} 2 & 0 & -1 & 3 \\ 1 & -1 & 2 & -2 \end{bmatrix}} \overset{\text{Match}}{\underset{(4 \times 3)}{\begin{bmatrix} 1 & 1 & -8 \\ 1 & 0 & 0 \\ 0 & 5 & 2 \\ -2 & 8 & -1 \end{bmatrix}}}$$

The product of the two matrices is defined, and the product will be a 2×3 matrix [we remove the matching 4s: $(2 \times 4)(4 \times 3) \rightarrow 2 \times 3$]. To calculate the product, we follow the prescription above.

$$\begin{array}{c} R_1 \rightarrow \\ R_2 \rightarrow \end{array} \begin{bmatrix} 2 & 0 & -1 & 3 \\ 1 & -1 & 2 & -2 \end{bmatrix} \overset{\begin{array}{ccc} C_1 & C_2 & C_3 \\ \downarrow & \downarrow & \downarrow \end{array}}{\begin{bmatrix} 1 & 1 & -8 \\ 1 & 0 & 0 \\ 0 & 5 & 2 \\ -2 & 8 & -1 \end{bmatrix}} = \begin{bmatrix} R_1 \times C_1 & R_1 \times C_2 & R_1 \times C_3 \\ R_2 \times C_1 & R_2 \times C_2 & R_2 \times C_3 \end{bmatrix}$$

$$= \begin{bmatrix} -4 & 21 & -21 \\ 4 & -5 & -2 \end{bmatrix}$$

Before we go on... Notice in this example that we *cannot* multiply these matrices in the opposite order—the dimensions do not match. We say simply that the product in the opposite order is **not defined.**

EXAMPLE 8 Sales and Revenue

The APlus auto parts store from the previous section had sales in January in their Vancouver and Quebec stores as given in the following table.

	Vancouver	Quebec
Wiper blades	20	15
Cleaning fluid (bottles)	10	12
Floor mats	8	4

Consider two sets of prices: The usual prices for these items are $7.00 each for wiper blades, $3.00 per bottle for cleaning fluid, and $12.00 each for floor mats. The prices for APlusClub members are $6.00 each for wiper blades, $2.50 per bottle for cleaning fluid, and $10.00 each for floor mats. Use matrix multiplication to compute the total revenue at each store, assuming first that all items were sold at the usual prices, and then that they all were sold to APlusClub members.

SOLUTION We can do all of the requested calculations at once with a single matrix multiplication. Consider the following two labeled matrices.

$$Q = \begin{array}{c} \text{wb} \\ \text{cf} \\ \text{fm} \end{array} \begin{bmatrix} V & Q \\ 20 & 15 \\ 10 & 12 \\ 8 & 4 \end{bmatrix}$$

$$P = \begin{array}{c} \text{Usual} \\ \text{Club} \end{array} \begin{bmatrix} \text{wb} & \text{cf} & \text{fm} \\ 7.00 & 3.00 & 12.00 \\ 6.00 & 2.50 & 10.00 \end{bmatrix}$$

The first matrix records the quantities sold, and the second records the selling prices under the two assumptions. To compute the revenue at both stores under the two different assumptions, we compute $R = PQ$.

$$R = PQ = \begin{bmatrix} 7.00 & 3.00 & 12.00 \\ 6.00 & 2.50 & 10.00 \end{bmatrix} \begin{bmatrix} 20 & 15 \\ 10 & 12 \\ 8 & 4 \end{bmatrix}$$

$$= \begin{bmatrix} 266.00 & 189.00 \\ 225.00 & 160.00 \end{bmatrix}$$

We can label this matrix as follows.

$$R = \begin{array}{c} \text{Usual} \\ \text{Club} \end{array} \begin{bmatrix} V & Q \\ 266.00 & 189.00 \\ 225.00 & 160.00 \end{bmatrix}$$

In other words, if the items were sold at the usual price, then Vancouver had a revenue of $266 while Quebec had a revenue of $189, and so on.

Before we go on... We were able to multiply P times Q since the dimensions matched correctly: $(2 \times 3)(3 \times 2) \rightarrow 2 \times 2$. Notice that we could also have multiplied them in the opposite order and gotten a 3×3 matrix. However, this product would have been meaningless. In an application like this, not only do the dimensions have to match, but the *labels* have to match for the result to be meaningful. The labels on the three columns of P were the parts that were sold, and these were also the labels on the three rows of Q. Therefore, we can "cancel labels" at the same time that we cancel the dimensions in the product. However, the labels on the two columns of Q are completely different from the labels on the two rows of P, and there is no useful interpretation of the product QP in this example.

EXAMPLE 9

Let $A = \begin{bmatrix} 1 & -1 \\ 0 & 2 \end{bmatrix}$ and $B = \begin{bmatrix} 3 & 0 \\ 5 & -1 \end{bmatrix}$. Find AB and BA.

SOLUTION Note first that A and B are both 2×2 matrices, so the products AB and BA are both defined and are also 2×2 matrices. We first calculate AB.

$$AB = \begin{matrix} R_1 \to \\ R_2 \to \end{matrix} \begin{bmatrix} 1 & -1 \\ 0 & 2 \end{bmatrix} \begin{matrix} C_1 & C_2 \\ \downarrow & \downarrow \end{matrix} \begin{bmatrix} 3 & 0 \\ 5 & -1 \end{bmatrix} = \begin{bmatrix} R_1 \times C_1 & R_1 \times C_2 \\ R_2 \times C_1 & R_2 \times C_2 \end{bmatrix} = \begin{bmatrix} -2 & 1 \\ 10 & -2 \end{bmatrix}$$

You might now wonder what is the point in calculating BA, since surely $AB = BA$? Well, let's calculate it anyway, if only for the sake of practice.

$$BA = \begin{matrix} R_1 \to \\ R_2 \to \end{matrix} \begin{bmatrix} 3 & 0 \\ 5 & -1 \end{bmatrix} \begin{matrix} C_1 & C_2 \\ \downarrow & \downarrow \end{matrix} \begin{bmatrix} 1 & -1 \\ 0 & 2 \end{bmatrix} = \begin{bmatrix} R_1 \times C_1 & R_1 \times C_2 \\ R_2 \times C_1 & R_2 \times C_2 \end{bmatrix} = \begin{bmatrix} 3 & -3 \\ 5 & -7 \end{bmatrix}$$

Notice that BA has no resemblance to AB! Thus, we have discovered that

Matrix multiplication is not commutative.

In other words, $AB \neq BA$ in general. (There are instances when $AB = BA$ for particular matrices A and B, but this is an exception, not the rule.)

EXAMPLE 10

As promised earlier, we now multiply a *column* by a *row*:

$$\begin{matrix} R_1 \to \\ R_2 \to \end{matrix} \begin{bmatrix} -3 \\ 1 \end{bmatrix} \begin{matrix} C_1 & C_2 \\ \downarrow & \downarrow \end{matrix} \begin{bmatrix} 2 & 1 \end{bmatrix} = \begin{bmatrix} R_1 \times C_1 & R_1 \times C_2 \\ R_2 \times C_1 & R_2 \times C_2 \end{bmatrix} = \begin{bmatrix} -6 & -3 \\ 2 & 1 \end{bmatrix}.$$

Before we go on... We can multiply these matrices in the opposite order, but of course we get a 1×1 matrix if we do so. Again, order matters when multiplying matrices. In general, if AB is defined, then BA need not even be defined. If BA is also defined, it may not have the same dimensions as AB. And even if both products have the same dimensions, we will usually have $AB \neq BA$.

EXAMPLE 11

Evaluate the products AI and IA, where

$$A = \begin{bmatrix} a & b & c \\ d & e & f \\ g & h & i \end{bmatrix}, I = \begin{bmatrix} 1 & 0 & 0 \\ 0 & 1 & 0 \\ 0 & 0 & 1 \end{bmatrix}.$$

SOLUTION First notice that A is arbitrary—it could be any 3×3 matrix.

$$AI = \begin{bmatrix} a & b & c \\ d & e & f \\ g & h & i \end{bmatrix} \begin{bmatrix} 1 & 0 & 0 \\ 0 & 1 & 0 \\ 0 & 0 & 1 \end{bmatrix} = \begin{bmatrix} a & b & c \\ d & e & f \\ g & h & i \end{bmatrix}$$

$$IA = \begin{bmatrix} 1 & 0 & 0 \\ 0 & 1 & 0 \\ 0 & 0 & 1 \end{bmatrix} \begin{bmatrix} a & b & c \\ d & e & f \\ g & h & i \end{bmatrix} = \begin{bmatrix} a & b & c \\ d & e & f \\ g & h & i \end{bmatrix}$$

(Do the algebra to check these.)

Before we go on... In both cases, the answer is the matrix A we started with. In other words,

$$A \cdot I = A$$
$$\text{and} \quad I \cdot A = A,$$

no matter what 3×3 matrix A you start with. Now this should remind you of a familiar fact from arithmetic,

$$a \cdot 1 = a$$
$$\text{and} \quad 1 \cdot a = a.$$

We call the matrix I the 3×3 **identity matrix,** since it appears to play the same role for 3×3 matrices that the identity 1 does for numbers.

There is also a 2×2 identity matrix, as the following example shows.

▼ **EXAMPLE 12**

Let $A = \begin{bmatrix} a & b \\ c & d \end{bmatrix}, I = \begin{bmatrix} 1 & 0 \\ 0 & 1 \end{bmatrix}$. Find AI and IA.

SOLUTION

$$AI = \begin{bmatrix} a & b \\ c & d \end{bmatrix} \begin{bmatrix} 1 & 0 \\ 0 & 1 \end{bmatrix} = \begin{bmatrix} a & b \\ c & d \end{bmatrix} = A$$

$$IA = \begin{bmatrix} 1 & 0 \\ 0 & 1 \end{bmatrix} \begin{bmatrix} a & b \\ c & d \end{bmatrix} = \begin{bmatrix} a & b \\ c & d \end{bmatrix} = A$$

Before we go on... The matrix I in this example is the 2×2 identity matrix. It is interesting to notice that $AI = A$ also if A is a 3×2 matrix (try one). In fact, $AI = A$ for any $m \times 2$ matrix, and $IA = A$ for any $2 \times n$ matrix.

Q What does the 4 × 4 identity matrix look like?

A The answer suggested by the two examples above is

$$I = \begin{bmatrix} 1 & 0 & 0 & 0 \\ 0 & 1 & 0 & 0 \\ 0 & 0 & 1 & 0 \\ 0 & 0 & 0 & 1 \end{bmatrix}.$$

This matrix does indeed have the property that $AI = IA = A$ for every 4 × 4 matrix A. (Try it.)

Q What about the 1 × 1 identity matrix?

A The 1 × 1 identity matrix is simply [1], since $[1][a] = [a][1] = [a]$. Notice that addition and multiplication of 1 × 1 matrices is just like addition and multiplication of ordinary real numbers, except that they happen to have brackets around them.

In general, the $n \times n$ identity matrix I has 1s down the **main diagonal,** and 0s everywhere else. In symbols,

$$I_{ii} = 1,$$
$$\text{and} \quad I_{ij} = 0 \text{ if } i \neq j.$$

Identity matrices are always **square** matrices, meaning that they have the same number of rows as columns. There is no such thing, for example, as the "2 × 4 identity matrix." We can now add to the list of properties we gave for matrix algebra at the end of the last section by writing down properties of matrix multiplication. In stating these properties, we shall assume that all matrix products we write are defined: that is, the matrices have correctly matching dimensions. The first eight rules are the ones we've already seen, while the remainder are new.

LAWS OF MATRIX ADDITION AND SCALAR MULTIPLICATION

If A, B, and C are matrices, if O is the zero matrix, and if I is an identity matrix, then the following hold.

$A + (B + C) = (A + B) + C$	Additive associative law
$A + B = B + A$	Additive commutative law
$A + O = O + A = A$	Additive identity law
$A + (-A) = O = (-A) + A$	Additive inverse law
$c(A + B) = cA + cB$	Distributive law
$(c + d)A = cA + dA$	Distributive law
$1A = A$	Scalar unit
$0A = O$	Scalar zero
$A(BC) = (AB)C$	Multiplicative associative law

3.2 Matrix Multiplication

$AI = IA = A$ *Multiplicative identity law*
$A(B + C) = AB + AC$ *Distributive law*
$(A + B)C = AC + BC$ *Distributive law*
$OA = AO = O$ *Multiplication by zero matrix*

Q What about the commutative law $AB = BA$?

A It is not included because, as we have already seen, it *does not hold* in general. In other words, matrix multiplication is *not* exactly like multiplication of numbers. You have to be a bit careful, as it's easy to use the commutative law by mistake.

We should also say a bit more about transposition. Transposition and multiplication have an interesting relationship. We write down the properties of transposition again, adding one new one.

LAWS OF TRANSPOSITION

$$(A + B)^T = A^T + B^T$$
$$(cA)^T = c(A^T)$$
$$(AB)^T = B^T A^T$$

Notice the change in order in the last law. The order is crucial.

▼ **EXAMPLE 13**

Let $A = \begin{bmatrix} 1 & -1 \\ 0 & 2 \end{bmatrix}$, $B = \begin{bmatrix} 3 & 0 \\ 5 & -1 \end{bmatrix}$. Find $(AB)^T$, $A^T B^T$, and $B^T A^T$.

SOLUTION

$$(AB)^T = \left(\begin{bmatrix} 1 & -1 \\ 0 & 2 \end{bmatrix} \begin{bmatrix} 3 & 0 \\ 5 & -1 \end{bmatrix} \right)^T = \begin{bmatrix} -2 & 1 \\ 10 & -2 \end{bmatrix}^T = \begin{bmatrix} -2 & 10 \\ 1 & -2 \end{bmatrix}$$

$$A^T B^T = \begin{bmatrix} 1 & -1 \\ 0 & 2 \end{bmatrix}^T \begin{bmatrix} 3 & 0 \\ 5 & -1 \end{bmatrix}^T = \begin{bmatrix} 1 & 0 \\ -1 & 2 \end{bmatrix} \begin{bmatrix} 3 & 5 \\ 0 & -1 \end{bmatrix} = \begin{bmatrix} 3 & 5 \\ -3 & -7 \end{bmatrix}$$

$$B^T A^T = \begin{bmatrix} 3 & 0 \\ 5 & -1 \end{bmatrix}^T \begin{bmatrix} 1 & -1 \\ 0 & 2 \end{bmatrix}^T = \begin{bmatrix} 3 & 5 \\ 0 & -1 \end{bmatrix} \begin{bmatrix} 1 & 0 \\ -1 & 2 \end{bmatrix} = \begin{bmatrix} -2 & 10 \\ 1 & -2 \end{bmatrix}.$$

Before we go on... If we fill in the missing steps in the computations done here, we would see that the computations involved in $(AB)^T$ are exactly the same as those in $B^T A^T$ and are quite different from those used to compute $A^T B^T$. If the matrices are not square, but AB is defined, then $A^T B^T$ need not even be defined. Try an example where A is a 2×3 matrix and B is a 3×1 matrix.

In writing down these properties, what we are doing, besides showing you what you can and cannot do with matrices, is giving you a glimpse of the field of mathematics known as **abstract algebra.** Algebraists study operations that resemble the operations on numbers, but differ in some way, like the lack of commutativity seen here. They study many situations in which some familiar properties about numbers continue to hold, but others—such as the commutative law for products—fail.

We end this section with more on the relationship between linear equations and matrix equations, which is one of the important applications of matrix multiplication.

▼ **EXAMPLE 14**

If $A = \begin{bmatrix} 1 & -2 & 3 \\ 2 & 0 & -1 \\ -3 & 1 & 1 \end{bmatrix}$, $X = \begin{bmatrix} x \\ y \\ z \end{bmatrix}$ and $B = \begin{bmatrix} 3 \\ -1 \\ 0 \end{bmatrix}$,

rewrite the matrix equation $AX = B$, or

$$\begin{bmatrix} 1 & -2 & 3 \\ 2 & 0 & -1 \\ -3 & 1 & 1 \end{bmatrix} \begin{bmatrix} x \\ y \\ z \end{bmatrix} = \begin{bmatrix} 3 \\ -1 \\ 0 \end{bmatrix}$$

as a system of linear equations.

SOLUTION As in Example 4, we first evaluate the left-hand side and then set it equal to the right-hand side.

$$\begin{bmatrix} 1 & -2 & 3 \\ 2 & 0 & -1 \\ -3 & 1 & 1 \end{bmatrix} \begin{bmatrix} x \\ y \\ z \end{bmatrix} = \begin{bmatrix} x - 2y + 3z \\ 2x - z \\ -3x + y + z \end{bmatrix}$$

Since this must equal the right-hand side, we equate them.

$$\begin{bmatrix} x - 2y + 3z \\ 2x - z \\ -3x + y + z \end{bmatrix} = \begin{bmatrix} 3 \\ -1 \\ 0 \end{bmatrix}$$

Since these two matrices are equal, their corresponding entries must be equal.

$$x - 2y + 3z = 3$$
$$2x - z = -1$$
$$-3x + y + z = 0$$

In other words, the matrix equation $AX = B$ is equivalent to this system of linear equations. Notice that the coefficients of the left-hand sides of these equations are the entries of the matrix A. We call A the **coefficient matrix** of the system of equations. The entries of X are the unknowns, while the entries of B are the right-hand sides 3, -1, and 0.

EXAMPLE 15

Express the following system of equations as a matrix equation of the form $AX = B$.

$$2x + y = 3$$
$$4x - y = -1$$

SOLUTION The coefficient matrix A has entries equal to the coefficients of the left-hand sides of the equations. Thus,

$$A = \begin{bmatrix} 2 & 1 \\ 4 & -1 \end{bmatrix}.$$

X is the column matrix consisting of the unknowns, while B is the column matrix consisting of the right-hand sides of the equations, so

$$X = \begin{bmatrix} x \\ y \end{bmatrix} \quad \text{and} \quad B = \begin{bmatrix} 3 \\ -1 \end{bmatrix}.$$

The system of equations can be rewritten as the matrix equation $AX = B$ with this A, X, and B. In other words,

$$\begin{bmatrix} 2 & 1 \\ 4 & -1 \end{bmatrix} \begin{bmatrix} x \\ y \end{bmatrix} = \begin{bmatrix} 3 \\ -1 \end{bmatrix}$$

Before we go on... We can check this by multiplying out AX, setting it equal to B, and seeing that we get the original system of equations.

This translation of systems of linear equations into matrix equations is really the first step in the method of solving linear equations discussed in Chapter 2. There we worked with the **augmented matrix** of the system, which is simply A with B adjoined as an extra column. When we write a system in matrix form, as $AX = B$, another thought occurs to us: Could we not solve for the unknown X by *dividing* by A? The next section answers this question.

▶ ### 3.2 EXERCISES

In Exercises 1–20, compute the products.

1. $\begin{bmatrix} 1 & 3 & -1 \end{bmatrix} \begin{bmatrix} 9 \\ 1 \\ -1 \end{bmatrix}$

2. $\begin{bmatrix} 4 & 0 & -1 \end{bmatrix} \begin{bmatrix} -4 \\ 1 \\ 8 \end{bmatrix}$

3. $\begin{bmatrix} -1 & \frac{1}{2} \end{bmatrix} \begin{bmatrix} -\frac{1}{3} \\ 1 \end{bmatrix}$

4. $\begin{bmatrix} -1 & 1 \end{bmatrix} \begin{bmatrix} \frac{3}{4} \\ \frac{1}{4} \end{bmatrix}$

5. $[0 \quad -2 \quad 1] \begin{bmatrix} x \\ y \\ z \end{bmatrix}$

6. $[4 \quad -1 \quad 1] \begin{bmatrix} -x \\ x \\ y \end{bmatrix}$

7. $[-1 \quad 1] \begin{bmatrix} -3 & 1 & 4 & 3 \\ 0 & 1 & -2 & 1 \end{bmatrix}$

8. $[2 \quad -1] \begin{bmatrix} -3 & 1 & 4 & 3 \\ 4 & 0 & 1 & 3 \end{bmatrix}$

9. $[1 \quad -1 \quad 2 \quad 3] \begin{bmatrix} -1 & 2 & 0 \\ 2 & -1 & 0 \\ 0 & 5 & 2 \\ -1 & 8 & 1 \end{bmatrix}$

10. $[0 \quad 1 \quad -1 \quad 2] \begin{bmatrix} 1 & -2 & 1 \\ 0 & 1 & 3 \\ 6 & 0 & 2 \\ -1 & -2 & 11 \end{bmatrix}$

11. $\begin{bmatrix} 1 & 0 & -\frac{1}{2} & 1 \\ -1 & 1 & \frac{1}{4} & -2 \end{bmatrix} \begin{bmatrix} 0 & 1 & -1 \\ 1 & 0 & 1 \\ 4 & 8 & 0 \\ -2 & 8 & -1 \end{bmatrix}$

12. $\begin{bmatrix} \frac{1}{2} & 0 & -1 & 3 \\ 1 & -1 & \frac{1}{4} & -2 \end{bmatrix} \begin{bmatrix} 1 \\ 1 \\ 4 \\ 2 \end{bmatrix}$

13. $\begin{bmatrix} 1 & 0 \\ 1 & -1 \end{bmatrix} \begin{bmatrix} 0 & 1 \\ 0 & 1 \end{bmatrix}$

14. $\begin{bmatrix} 1 & -1 \\ 1 & -1 \end{bmatrix} \begin{bmatrix} 3 & -3 \\ 5 & -7 \end{bmatrix}$

15. $\begin{bmatrix} 1 & -1 \\ 1 & -1 \end{bmatrix} \begin{bmatrix} 2 & 3 \\ 2 & 3 \end{bmatrix}$

16. $\begin{bmatrix} 0 & 1 \\ 1 & 0 \end{bmatrix} \begin{bmatrix} 3 & -3 \\ 2 & -1 \end{bmatrix}$

17. $\begin{bmatrix} 1 & 0 & -1 \\ 2 & -2 & 1 \\ 0 & 0 & 1 \end{bmatrix} \begin{bmatrix} 1 & -1 & 4 \\ 1 & 1 & 0 \\ 0 & 4 & 1 \end{bmatrix}$

18. $\begin{bmatrix} 1 & 2 & 0 \\ 4 & -1 & 1 \\ 1 & 0 & 1 \end{bmatrix} \begin{bmatrix} 1 & 2 & -4 \\ 4 & 1 & 0 \\ 0 & -2 & 1 \end{bmatrix}$

19. $\begin{bmatrix} 1 & 0 & 1 & 0 \\ -1 & 1 & 0 & 1 \\ -2 & 0 & 1 & 4 \\ 0 & -1 & 0 & 1 \end{bmatrix} \begin{bmatrix} 1 \\ -3 \\ 2 \\ 0 \end{bmatrix}$

20. $\begin{bmatrix} 1 & 1 & -7 & 0 \\ -1 & 0 & 2 & 4 \\ -1 & 0 & -2 & 1 \\ 1 & -1 & 1 & 1 \end{bmatrix} \begin{bmatrix} 1 \\ -3 \\ 2 \\ 1 \end{bmatrix}$

21. Find $A^2 = A \cdot A$, $A^3 = A \cdot A \cdot A$, A^4, and A^{100},* given that

$$A = \begin{bmatrix} 0 & 1 & 1 & 1 \\ 0 & 0 & 1 & 1 \\ 0 & 0 & 0 & 1 \\ 0 & 0 & 0 & 0 \end{bmatrix}$$

22. Repeat Exercise 21 with $A = \begin{bmatrix} 0 & 2 & 0 & -1 \\ 0 & 0 & 2 & 0 \\ 0 & 0 & 0 & 2 \\ 0 & 0 & 0 & 0 \end{bmatrix}$.

In Exercises 23–26, let $A = \begin{bmatrix} 0 & -1 & 0 & 1 \\ 10 & 0 & 1 & 0 \end{bmatrix}$, $B = \begin{bmatrix} 0 & -1 \\ 1 & 1 \\ -1 & 3 \\ 5 & 0 \end{bmatrix}$, $C = \begin{bmatrix} 1 & -1 \\ 1 & 1 \\ 1 & 1 \\ 1 & 1 \end{bmatrix}$. Evaluate each expression.

23. AB **24.** AC **25.** $A(B - C)$ **26.** $(B - C)A$

▼ *$A \cdot A \cdot A$ is $A(A \cdot A)$, or the equivalent $(A \cdot A)A$ by the associative law. Similarly, $A \cdot A \cdot A \cdot A = A(A \cdot A \cdot A) = (A \cdot A \cdot A)A = (A \cdot A)(A \cdot A)$; it doesn't matter where we place parentheses.

In Exercises 27–30, let $A = \begin{bmatrix} 1 & -1 \\ 0 & 2 \\ 0 & -2 \end{bmatrix}$, $B = \begin{bmatrix} 3 & 0 & -1 \\ 5 & -1 & 1 \end{bmatrix}$, $C = \begin{bmatrix} x & 1 & w \\ z & r & 4 \end{bmatrix}$. Evaluate each expression.

27. AB **28.** AC **29.** $A(B + C)$ **30.** $(B + C)A$

In Exercises 31–46, use a graphing calculator or computer matrix software.

Let $A = [1.1 \quad -2.1 \quad 4.5]$, $B = \begin{bmatrix} -1.2 & 0 & 0 \\ 0 & -1.2 & 0 \\ 0 & 0 & -1.2 \end{bmatrix}$,

$C = \begin{bmatrix} 0.01 & 1.1 \\ 1.2 & -1.1 \\ 0 & 0 \end{bmatrix}$ $D = \begin{bmatrix} 350 \\ 591 \\ 911 \end{bmatrix}$.

Calculate each of the following.

31. BD **32.** AB **33.** AD

34. DA **35.** BC **36.** $B^2 = B \cdot B$

37. $A(BD)$ **38.** $(AB)D$ **39.** $B^4 = B \cdot B \cdot B \cdot B$

40. $(A + D^T)C$ **41.** $-\dfrac{1}{1.2}(BC)$ **42.** $5BD + 60D$

43. Let $P = \begin{bmatrix} \frac{1}{2} & \frac{1}{2} \\ \frac{1}{4} & \frac{3}{4} \end{bmatrix}$. Calculate $P^2 = P \cdot P$, $P^4 = P^2 \cdot P^2$ and P^8. (Use a graphing calculator or computer software. Round all entries to four decimal places.) Without computing it explicitly, find $P^{1,000}$.

44. Repeat Exercise 43 with $P = \begin{bmatrix} \frac{1}{3} & \frac{2}{3} \\ \frac{1}{2} & \frac{1}{2} \end{bmatrix}$.

45. Repeat Exercise 43 with $P = \begin{bmatrix} 0.25 & 0.25 & 0.50 \\ 0.25 & 0.25 & 0.50 \\ 0.25 & 0.25 & 0.50 \end{bmatrix}$.

46. Repeat Exercise 43 with $P = \begin{bmatrix} 0.1 & 0.2 & 0.7 \\ 0.2 & 0.3 & 0.5 \\ 0 & 0 & 1 \end{bmatrix}$.

In Exercises 47–50, translate the given matrix equations into systems of linear equations.

47. $\begin{bmatrix} 2 & -1 & 4 \\ -4 & \frac{3}{4} & \frac{1}{3} \\ -3 & 0 & 0 \end{bmatrix} \begin{bmatrix} x \\ y \\ z \end{bmatrix} = \begin{bmatrix} 3 \\ -1 \\ 0 \end{bmatrix}$ **48.** $\begin{bmatrix} 1 & -1 & 4 \\ -\frac{1}{3} & -3 & \frac{1}{3} \\ 3 & 0 & 1 \end{bmatrix} \begin{bmatrix} x \\ y \\ z \end{bmatrix} = \begin{bmatrix} -3 \\ -1 \\ 2 \end{bmatrix}$

49. $\begin{bmatrix} 1 & -1 & 0 & 1 \\ 1 & 1 & 2 & 4 \end{bmatrix} \begin{bmatrix} x \\ y \\ z \\ w \end{bmatrix} = \begin{bmatrix} -1 \\ 2 \end{bmatrix}$ **50.** $\begin{bmatrix} 0 & 1 & 6 & 1 \\ 1 & -5 & 0 & 0 \end{bmatrix} \begin{bmatrix} x \\ y \\ z \\ w \end{bmatrix} = \begin{bmatrix} -2 \\ 9 \end{bmatrix}$

In Exercises 51–54, translate the given systems of equations into matrix form.

51. $x - y = 4$
$2x - y = 0$

52. $2x + y = 7$
$-x = 9$

53. $x + y - z = 8$
$2x + y + z = 4$
$\frac{3}{4}x + \frac{1}{2}z = 1$

54. $x + y + 2z = -2$
$4x + 2y - z = -8$
$\frac{1}{2}x - \frac{1}{3}y = 4$

APPLICATIONS

55. *Revenue* Your T-shirt operation is doing a booming trade. Last week you sold 50 tie-dye shirts for $15 apiece, 40 Suburban State University Crew shirts for $10 each, and 30 Lacrosse Ts for $12 each. Use matrix operations to calculate your total revenue for the week.

56. *Revenue* Lucinda Turley, your competitor in Suburban State U's T-shirt market, has apparently been undercutting your prices and outperforming you in sales. Last week she sold 100 tie-dye shirts for $10 each, 50 (low quality) Crew shirts at $5 apiece, and 70 Lacrosse Ts for $8 each. Use matrix operations to calculate her total revenue for the week.

57. *Foreign Currency* You own 500 British pounds, 100,000 Japanese yen, 100 Canadian dollars, 50 U.S. dollars, 200 French francs, 20 Dutch guilders and 20,000 Italian lira. Use matrix algebra and the following table to calculate your total worth in U.S. dollars.*

	British pound	Japanese yen	Canadian dollar	French franc	Dutch guilder	Italian lira
Value in U.S. $	1.4950	0.009159	0.7526	0.1716	0.5256	0.000597

58. *Foreign Currency* Referring to Exercise 57 and to the table that follows, use matrix algebra to calculate the value of your foreign currency holdings in Canadian dollars.*

	British pound	Japanese yen	U.S. dollar	French franc	Dutch guilder	Italian lira
Value in Canadian $	1.9664	0.01217	1.3287	0.2283	0.6984	0.000793

Currency Transactions Exercises 59–62 are based on the following table.*

	British pound	Japanese yen	Canadian dollar	U.S. dollar
Value in British £	1	0.006127	0.5034	0.6689
Value in Japanese ¥	163.22	1	82.17	109.18
Value in Can $	1.9664	0.01217	1	1.3287
Value in U.S. $	1.4950	0.009159	0.7526	1

Let A be the 4×4 matrix suggested by the above table.

59. Let $C = [10 \ \ 0 \ \ 100 \ \ 10]$. What does the matrix AC^T represent?

60. What is the relationship between A_{ij} and A_{ji}?

For Exercises 61 and 62, use a graphing calculator or computer software.

 61. Calculate $\frac{1}{4}A^2$. What does the answer approximate? Give a reason for this answer.

---

*Source for Exercises 57–62: Bank of Montreal Treasury Group (*The Globe and Mail*, Dec. 13, 1993, p. B9.)

62. You and your friend each own 10 British pounds and 1000 Japanese yen. In addition, you own $10 (U.S.) and no Canadian currency while your friend owns $10 (Can.) and no U.S. currency. Use a single matrix operation to compute the total value of your currency holdings and the total value of your friend's holdings in both British pounds and Japanese yen.

63. *Revenue* Recall the Left Coast Book Store from the last section. In January it sold 700 hardcover books, 1,300 softcover books, and 2,000 plastic books in San Francisco; it sold 400 hardcover, 300 softcover, and 500 plastic books in Los Angeles. Now hardcover books sell for $30 each, softcover books sell for $10 each, and plastic books sell for $15 each. Write down a column matrix with the price data, and show how matrix multiplication (using the sales and price data matrices) may be used to compute the total revenue at the two stores.

64. *Costs* Continuing with the Left Coast Book Store chain, each hardcover book costs the store $10, each softcover book costs it $5, and each plastic book costs it $10. Use matrix operations to compute the total *profit* at each store in January.

65. *Costs* Microbucks Computer Co. makes two computers, the Pomegranate II and the Pomegranate Classic. The Pom II requires two processor chips, 16 memory chips, and 20 vacuum tubes, while the Pom Classic requires one processor chip, 4 memory chips, and 40 vacuum tubes. There are two companies that can supply these parts: Motorel can supply them at $100 per processor chip, $50 per memory chip, and $10 per vacuum tube, while Intola can supply them at $150 per processor chip, $40 per memory chip, and $15 per vacuum tube. Write down all of these data in two matrices, one showing the parts required for each model computer, and the other showing the prices for each part from each supplier. Then show how matrix multiplication allows you to compute the total cost for parts for each model when parts are bought from either supplier.

66. *Profits* Continuing with the Pomegranate II and the Pomegranate Classic, it actually costs Motorel only $25 to make each processor chip, $10 for each memory chip, and $5 for each vacuum tube. It costs Intola $50 per processor chip, $10 per memory chip, and $7 per vacuum tube. Use matrix operations to find the total profit each supplier would make on each model.

For Exercises 67 and 68 use a graphing calculator or computer software.

67. *Population Movement* In 1987, the population of the U.S., broken down by regions,* was 49.3 million in the Northeast, 59.1 million in the Midwest, 82.2 million in the South, and 48.2 million in the West. From 1987 to 1988, 0.23% of the population in the Northeast moved to the Midwest, 0.9% moved to the South, and 0.25% moved to the West (the remainder stayed in the Northeast). In that same year, 0.15% of the population in the Midwest moved to the Northeast, 0.89% moved to the South, and 0.36% moved to the West. Also, 0.3% of the population in the South moved to the Northeast, 0.5% moved to the Midwest, and 0.36% moved to the West. Finally, 0.22% of the population in the West moved to the Northeast, 0.47% moved to the Midwest, and 0.48% moved to the South. Set up the 1987 population figures as a row vector, and set up the population movement figures as a matrix. Then use matrix multiplication to compute the population in each region in the year 1988.

68. *Population Movement* Assuming that the percentages given in Exercise 67 also describe the population movements from 1988–1989, use matrix multiplication to predict from the data in Exercise 67 the population in each region in 1989.

COMMUNICATION AND REASONING EXERCISES

69. Comment on the following claim: Every matrix equation represents a system of equations.

70. When is it true that both AB and BA are defined, even though neither A nor B is a square matrix?

71. Find a scenario in which it would be useful to "multiply" two row vectors according to the rule $[a \ b \ c][d \ e \ f] = [ad \ be \ cf]$.

72. Make up an application whose solution reads as follows.

$$\text{Total revenue} = [10 \ \ 100 \ \ 30] \begin{bmatrix} 10 & 0 & 3 \\ 1 & 2 & 0 \\ 0 & 1 & 40 \end{bmatrix}$$

▼ *Source: Statistical Abstract of the United States 1991, U.S. Dept. of Commerce.

3.3 MATRIX INVERSION

Now that we know all about matrix addition, subtraction, and multiplication, you may well be wondering about matrix *division*. Remember first that division is really a form of multiplication. For example, if we want to divide 3 by 7, we can instead multiply 3 by $\frac{1}{7}$, the inverse of 7. In other words, $3 \div 7 = 3 \times \frac{1}{7}$, or 3×7^{-1}. Thus, in order to be able to imitate division of real numbers in the realm of matrices, we need to find a way of calculating the multiplicative **inverse**, A^{-1}, of a matrix A.

Before we can try to find the inverse of a matrix A, we must first be sure that we know what we *mean* by the inverse. Let us first satisfy ourselves that we know what we mean by the inverse a^{-1} of a number a. The **inverse** of a number a is the number, which we sometimes write as a^{-1}, that has the property that $a^{-1} \cdot a = a \cdot a^{-1} = 1$. For example, the inverse of 76 is the number $76^{-1} = \frac{1}{76}$, since $76 \cdot \frac{1}{76} = 1$. Not all numbers have an inverse. For example (and this is the only example), the number 0 has no inverse, since you cannot get 1 by multiplying 0 by anything.

We can now say what we mean by the inverse of a matrix, and to make life easier, we shall restrict attention to **square** matrices, that is, matrices that have the same number of rows as columns.*

> **THE INVERSE OF A MATRIX**
>
> The **inverse** of an $n \times n$ matrix A is that $n \times n$ matrix A^{-1} which, when multiplied by A on either side, yields the $n \times n$ identity matrix I. Thus,
>
> $$A^{-1}A = AA^{-1} = I.$$
>
> If A has an inverse, it is said to be **invertible.** Otherwise, it is said to be **singular.**

Q What entitles us to speak of *the* inverse of a matrix A? Can an invertible matrix have two or more different inverses?

A A matrix A can have no more than one inverse. This is not hard to prove: If B and C were both inverses of A, then

B	$= BI$	Property of the identity
	$= B(AC)$	Since C is an inverse of A
	$= (BA)C$	Associative law
	$= IC$	Since B is an inverse of A
	$= C.$	Property of the identity

In other words, if B and C were both inverses of A, then B and C would have to be equal!

---

* Non-square matrices *cannot* have inverses in the sense that we are talking about here. This is not a trivial fact to prove.

EXAMPLE 1

Verify that the inverse of $\begin{bmatrix} 1 & -1 \\ -1 & -1 \end{bmatrix}$ is $\begin{bmatrix} \frac{1}{2} & -\frac{1}{2} \\ -\frac{1}{2} & -\frac{1}{2} \end{bmatrix}$.

SOLUTION We are claiming that if $A = \begin{bmatrix} 1 & -1 \\ -1 & -1 \end{bmatrix}$, then $A^{-1} = \begin{bmatrix} \frac{1}{2} & -\frac{1}{2} \\ -\frac{1}{2} & -\frac{1}{2} \end{bmatrix}$. We compute

$$AA^{-1} = \begin{bmatrix} 1 & -1 \\ -1 & -1 \end{bmatrix} \begin{bmatrix} \frac{1}{2} & -\frac{1}{2} \\ -\frac{1}{2} & -\frac{1}{2} \end{bmatrix} = \begin{bmatrix} 1 & 0 \\ 0 & 1 \end{bmatrix} = I.$$

Also,

$$A^{-1}A = \begin{bmatrix} \frac{1}{2} & -\frac{1}{2} \\ -\frac{1}{2} & -\frac{1}{2} \end{bmatrix} \begin{bmatrix} 1 & -1 \\ -1 & -1 \end{bmatrix} = \begin{bmatrix} 1 & 0 \\ 0 & 1 \end{bmatrix} = I.$$

Before we go on... It is possible to show that if A and B are square matrices with $AB = I$, then it must also be true that $BA = I$. In other words, once we have checked that $AA^{-1} = I$, we know that we have the inverse matrix. So, the second check above (that $A^{-1}A = I$) was unnecessary.

▶ **NOTE** We saw above that the inverse of $A = \begin{bmatrix} 1 & -1 \\ -1 & -1 \end{bmatrix}$ is $B = \begin{bmatrix} \frac{1}{2} & -\frac{1}{2} \\ -\frac{1}{2} & -\frac{1}{2} \end{bmatrix}$. This also means that the inverse of B is A. Thus, we sometimes refer to such a pair of matrices as an **inverse pair** of matrices. ◀

EXAMPLE 2

Can $A = \begin{bmatrix} 1 & 1 \\ 0 & 0 \end{bmatrix}$ have an inverse?

SOLUTION No. To see why, notice that AB will have its second row all 0 no matter what B is. So AB cannot equal I, no matter what B is.

Before we go on... If you think about it, you can write down many such examples. There is only one number with no multiplicative inverse (0), but there are many matrices with no inverses.

Q How on earth did we find the inverse of A in Example 1?

A We consider first a rather strange way of finding the inverse of an ordinary number, say, 76. (Of course we already *know* that the inverse of 76 is $\frac{1}{76}$, but bear with us anyway.) The inverse of 76 is a number x such that $76x = 1$. This is a linear equation which we can solve using Gauss-Jordan reduction (yes, this is overkill).

$$\begin{bmatrix} 76 & 1 \end{bmatrix} \xrightarrow{\frac{1}{76}R_1} \begin{bmatrix} 1 & \frac{1}{76} \end{bmatrix}$$

The final matrix says $x = \frac{1}{76}$. In the next two examples, we shall see how this generalizes to a calculation of the inverse of a matrix, and then we shall explain exactly why the technique works.

▼ **EXAMPLE 3**

Find the inverse of the matrix $\begin{bmatrix} 1 & -1 \\ -1 & -1 \end{bmatrix}$. (We already know what the inverse is from Example 1, but now we shall see where it came from.)

SOLUTION Generalizing the procedure we used to find the inverse of 76, we put the matrix A on the left and the identity matrix I on the right to get a 2×4 matrix.

$$\begin{bmatrix} 1 & -1 & | & 1 & 0 \\ -1 & -1 & | & 0 & 1 \end{bmatrix}$$
$$\uparrow \uparrow$$
$$A I$$

We now row-reduce the whole matrix:

$$\begin{bmatrix} 1 & -1 & 1 & 0 \\ -1 & -1 & 0 & 1 \end{bmatrix} \xrightarrow{R_2+R_1} \begin{bmatrix} 1 & -1 & 1 & 0 \\ 0 & -2 & 1 & 1 \end{bmatrix} \xrightarrow{2R_1 - R_2}$$

$$\begin{bmatrix} 2 & 0 & 1 & -1 \\ 0 & -2 & 1 & 1 \end{bmatrix} \xrightarrow[-\frac{1}{2}R_2]{\frac{1}{2}R_1} \begin{bmatrix} 1 & 0 & | & \frac{1}{2} & -\frac{1}{2} \\ 0 & 1 & | & -\frac{1}{2} & -\frac{1}{2} \end{bmatrix}.$$
$$\uparrow \uparrow$$
$$I A^{-1}$$

We are left with the identity matrix I on the left and A^{-1} on the right, as was the case when we found the inverse of 76. Thus,

$$A^{-1} = \begin{bmatrix} \frac{1}{2} & -\frac{1}{2} \\ -\frac{1}{2} & -\frac{1}{2} \end{bmatrix}.$$

Before we go on... Note that we have already checked, in Example 1, that this really is the inverse of A.

A calculator or computer program that can compute matrix inverses will appear to do so in one step. (On a TI-81 or TI-82, you can invert the square matrix [A] by the sequence [A] [x⁻¹] [ENTER].) However, the calculator is going through the above procedure or some variation of it to find the inverse. Of course, you can use a calculator that can do row operations to help you go through the row reduction, just as in Chapter 2.

Most electronic spreadsheets can also invert matrices in a single step. In general, technology is indispensable when the matrices are large and the entries look like telephone numbers. (Take a look at the 13×13 matrices in the "You're the Expert" section of this chapter, and you will understand why we had to use an electronic spreadsheet to do all the calculations.)

3.3 Matrix Inversion

We now find the inverse of a 3×3 matrix using the same technique.

▼ **EXAMPLE 4**

Find the inverse of the matrix $A = \begin{bmatrix} 1 & 0 & 1 \\ 2 & -2 & -1 \\ 3 & 0 & 0 \end{bmatrix}$.

SOLUTION We place A on the left and the 3×3 identity matrix on the right, then reduce.

$$\begin{array}{c} A \\ \downarrow \end{array} \quad \begin{array}{c} I \\ \downarrow \end{array}$$

$$\left[\begin{array}{ccc|ccc} 1 & 0 & 1 & 1 & 0 & 0 \\ 2 & -2 & -1 & 0 & 1 & 0 \\ 3 & 0 & 0 & 0 & 0 & 1 \end{array}\right] \begin{array}{l} \\ R_2 - 2R_1 \\ R_3 - 3R_1 \end{array} \longrightarrow \left[\begin{array}{ccc|ccc} 1 & 0 & 1 & 1 & 0 & 0 \\ 0 & -2 & -3 & -2 & 1 & 0 \\ 0 & 0 & -3 & -3 & 0 & 1 \end{array}\right] \begin{array}{l} 3R_1 + R_3 \\ R_2 - R_3 \\ \end{array} \longrightarrow$$

$$\left[\begin{array}{ccc|ccc} 3 & 0 & 0 & 0 & 0 & 1 \\ 0 & -2 & 0 & 1 & 1 & -1 \\ 0 & 0 & -3 & -3 & 0 & 1 \end{array}\right] \begin{array}{l} \frac{1}{3}R_1 \\ -\frac{1}{2}R_2 \\ -\frac{1}{3}R_3 \end{array} \longrightarrow \left[\begin{array}{ccc|ccc} 1 & 0 & 0 & 0 & 0 & \frac{1}{3} \\ 0 & 1 & 0 & -\frac{1}{2} & -\frac{1}{2} & \frac{1}{2} \\ 0 & 0 & 1 & 1 & 0 & -\frac{1}{3} \end{array}\right]$$
$$\qquad\qquad\qquad\qquad\qquad\qquad\qquad\qquad\qquad\qquad\quad \uparrow \qquad\qquad \uparrow$$
$$\qquad\qquad\qquad\qquad\qquad\qquad\qquad\qquad\qquad\qquad\quad I \qquad\qquad A^{-1}$$

Thus,

$$A^{-1} = \begin{bmatrix} 0 & 0 & \frac{1}{3} \\ -\frac{1}{2} & -\frac{1}{2} & \frac{1}{2} \\ 1 & 0 & -\frac{1}{3} \end{bmatrix}.$$

Before we go on... Check this answer by showing that $AA^{-1} = I$.

Q Why does the row reduction procedure to find A^{-1} actually work?

A Let us go back again to Example 3 and the problem of finding the inverse of

$$A = \begin{bmatrix} 1 & -1 \\ -1 & -1 \end{bmatrix}.$$

We can think of this as a problem of finding four unknowns, the unknown entries of A^{-1}:

$$A^{-1} = \begin{bmatrix} x & y \\ z & w \end{bmatrix}.$$

These unknowns must satisfy the equation $AA^{-1} = I$, or

$$\begin{bmatrix} 1 & -1 \\ -1 & -1 \end{bmatrix} \begin{bmatrix} x & y \\ z & w \end{bmatrix} = \begin{bmatrix} 1 & 0 \\ 0 & 1 \end{bmatrix}.$$

If we were to try to find the first column of A^{-1}, consisting of x and z, we would have to solve

$$\begin{bmatrix} 1 & -1 \\ -1 & -1 \end{bmatrix} \begin{bmatrix} x \\ z \end{bmatrix} = \begin{bmatrix} 1 \\ 0 \end{bmatrix}$$

or

$$x - z = 1$$
$$-x - z = 0.$$

To solve this system by the techniques of Chapter 2 we would row-reduce the augmented matrix, which is A with the column B adjoined.

$$\begin{bmatrix} 1 & -1 & | & 1 \\ -1 & -1 & | & 0 \end{bmatrix} \longrightarrow \begin{bmatrix} 1 & 0 & | & x \\ 0 & 1 & | & z \end{bmatrix}$$

To find the second column of A^{-1}, we would similarly row-reduce the augmented matrix obtained by tacking onto A the second column of the identity matrix.

$$\begin{bmatrix} 1 & -1 & | & 0 \\ -1 & -1 & | & 1 \end{bmatrix} \longrightarrow \begin{bmatrix} 1 & 0 & | & y \\ 0 & 1 & | & w \end{bmatrix}$$

Now the row operations used in doing these two reductions would be exactly the same. To save duplication of work, we could do both reductions simultaneously by "doubly augmenting" A, putting both columns of the identity matrix to the right of A.

$$\begin{bmatrix} 1 & -1 & | & 1 & 0 \\ -1 & -1 & | & 0 & 1 \end{bmatrix} \longrightarrow \begin{bmatrix} 1 & 0 & | & x & y \\ 0 & 1 & | & z & w \end{bmatrix}$$

This is exactly the procedure we used in Examples 3 and 4. In other words, this row reduction from $[A \; I]$ to $[I \; A^{-1}]$ is just the work that has to be done to solve the systems of linear equations that define the entries of A^{-1}.

Q Surely there must be a formula for the inverse of a matrix, so that we do not have to row-reduce each time?

A There is such a formula, but not a simple one. In fact, using the formula for anything larger than a 3 × 3 matrix is so time-consuming that row reduction is the method of choice. For 2 × 2 matrices, the formula is very simple, however.

$$\begin{bmatrix} a & b \\ c & d \end{bmatrix}^{-1} = \frac{1}{ad - bc} \begin{bmatrix} d & -b \\ -c & a \end{bmatrix}$$

For instance,

$$\begin{bmatrix} 1 & 2 \\ 3 & 4 \end{bmatrix}^{-1} = \frac{1}{(1)(4) - (2)(3)} \begin{bmatrix} 4 & -2 \\ -3 & 1 \end{bmatrix}$$

$$= -\frac{1}{2} \begin{bmatrix} 4 & -2 \\ -3 & 1 \end{bmatrix} = \begin{bmatrix} -2 & 1 \\ \frac{3}{2} & -\frac{1}{2} \end{bmatrix}.$$

3.3 Matrix Inversion

Q What happens if the quantity $ad - bc$ is zero?

A If $ad - bc = 0$, then the matrix has no inverse. (If you refer to the matrix in Example 2, you will find that $ad - bc = 0$ there.) The quantity $ad - bc$ is called the **determinant** of the matrix $\begin{bmatrix} a & b \\ c & d \end{bmatrix}$.

Q Where does this formula for the inverse of a 2×2 matrix come from?

A It can be obtained using the technique of row reduction. (See the Communication and Reasoning Exercises at the end of the section.)

▼ **EXAMPLE 5**

Find the inverse of the matrix $S = \begin{bmatrix} 1 & 1 & 2 \\ -2 & 0 & 4 \\ 3 & 1 & 2 \end{bmatrix}$.

SOLUTION We proceed as before.

$$\begin{array}{c} S \\ \downarrow \end{array} \quad \begin{array}{c} I \\ \downarrow \end{array}$$

$$\left[\begin{array}{ccc|ccc} 1 & 1 & 2 & 1 & 0 & 0 \\ -2 & 0 & 4 & 0 & 1 & 0 \\ 3 & 1 & -2 & 0 & 0 & 1 \end{array}\right] \begin{array}{c} \\ R_2 + 2R_1 \\ R_3 - 3R_1 \end{array} \longrightarrow \left[\begin{array}{ccc|ccc} 1 & 1 & 2 & 1 & 0 & 0 \\ 0 & 2 & 8 & 2 & 1 & 0 \\ 0 & -2 & -8 & -3 & 0 & 1 \end{array}\right] \begin{array}{c} 2R_1 - R_2 \\ \\ R_3 + R_2 \end{array} \longrightarrow$$

$$\left[\begin{array}{ccc|ccc} 2 & 0 & -4 & 0 & -1 & 0 \\ 0 & 2 & 8 & 2 & 1 & 0 \\ 0 & 0 & 0 & -1 & 1 & 1 \end{array}\right]$$

We stopped here even though the reduction is incomplete for the simple reason that there is *no hope* of getting the identity on the left-hand side. Completing the row reduction will not change the three zeros in the bottom row. So what did we do wrong? Nothing. As in Example 2, we have here a matrix that has no inverse. Any square matrix that, after row reduction, winds up with a row of zeros has no inverse. As we said earlier, we call such a matrix a **singular** matrix.

If you try to invert this matrix using a calculator or computer program, you should get an error.

In practice, deciding whether a given matrix is invertible or singular is easy: simply try to find its inverse. If the process works, then the matrix is invertible, and we get its inverse. If the process fails, then the matrix is not invertible, it's singular, and that's that.

Having used systems of equations and row reduction to find the inverse, we will now show that we can use inverses to solve systems of equations. Recall that, at the end of the previous section, we saw that a system of linear equations could be written in the form

$$AX = B,$$

where A is the coefficient matrix, X is the column matrix of unknowns, and B is the matrix of right-hand sides. Now suppose that there are as many unknowns as equations, so that A is a square matrix. Since the object is to solve for the matrix X of unknowns, we multiply both sides of the equation by the inverse A^{-1} of A (our replacement for dividing by A), getting

$$A^{-1}AX = A^{-1}B.$$

Notice that we put A^{-1} on the left on both sides of the equation. Order matters when multiplying matrices, so we have to be careful to really do the same thing to both sides of the equation. But now, $A^{-1}A = I$, so this is

$$IX = A^{-1}B.$$

Also, $IX = X$ (I being the identity matrix), so we really have

$$X = A^{-1}B,$$

and we have solved for X!

▶ **NOTE** The above argument also shows that if A is an invertible matrix, and $AX = B$ represents a system of equations, then X *has to equal* $A^{-1}B$. That is, there is *one and only one solution* to the system—a unique solution. Thus, if the system $AX = B$ has infinitely many solutions or no solutions, then A cannot be invertible. ◀

▼ **EXAMPLE 6**

Solve the following system using matrix inversion.

$$2x + z = 9$$
$$2x + y - z = 6$$
$$3x + y - z = 9$$

SOLUTION The system has the form $AX = B$, where

$$A = \begin{bmatrix} 2 & 0 & 1 \\ 2 & 1 & -1 \\ 3 & 1 & -1 \end{bmatrix}, \quad X = \begin{bmatrix} x \\ y \\ z \end{bmatrix}, \quad \text{and} \quad B = \begin{bmatrix} 9 \\ 6 \\ 9 \end{bmatrix}.$$

As we saw above, its solution is $X = A^{-1}B$, so we need to calculate this. We first calculate A^{-1} and then multiply the answer by B. The first step, calculation of A^{-1}, we do as before.

$$\begin{array}{c} A \\ \downarrow \end{array} \quad \begin{array}{c} I \\ \downarrow \end{array}$$

$$\left[\begin{array}{ccc|ccc} 2 & 0 & 1 & 1 & 0 & 0 \\ 2 & 1 & -1 & 0 & 1 & 0 \\ 3 & 1 & -1 & 0 & 0 & 1 \end{array}\right] \begin{array}{c} R_2 - R_1 \\ 2R_3 - 3R_1 \end{array} \longrightarrow \left[\begin{array}{ccc|ccc} 2 & 0 & 1 & 1 & 0 & 0 \\ 0 & 1 & -2 & -1 & 1 & 0 \\ 0 & 2 & -5 & -3 & 0 & 2 \end{array}\right] R_3 - 2R_2 \longrightarrow$$

$$\begin{bmatrix} 2 & 0 & 1 & 1 & 0 & 0 \\ 0 & 1 & -2 & -1 & 1 & 0 \\ 0 & 0 & -1 & -1 & -2 & 2 \end{bmatrix} \begin{matrix} R_1 + R_3 \\ R_2 - 2R_3 \\ \end{matrix} \longrightarrow \begin{bmatrix} 2 & 0 & 0 & 0 & -2 & 2 \\ 0 & 1 & 0 & 1 & 5 & -4 \\ 0 & 0 & -1 & -1 & -2 & 2 \end{bmatrix} \begin{matrix} \frac{1}{2}R_1 \\ \\ -R_3 \end{matrix} \longrightarrow$$

$$\begin{bmatrix} 1 & 0 & 0 & | & 0 & -1 & 1 \\ 0 & 1 & 0 & | & 1 & 5 & -4 \\ 0 & 0 & 1 & | & 1 & 2 & -2 \end{bmatrix}$$
$$\uparrow \uparrow$$
$$I A^{-1}$$

Thus,

$$A^{-1} = \begin{bmatrix} 0 & -1 & 1 \\ 1 & 5 & -4 \\ 1 & 2 & -2 \end{bmatrix}$$

and

$$X = A^{-1}B = \begin{bmatrix} 0 & -1 & 1 \\ 1 & 5 & -4 \\ 1 & 2 & -2 \end{bmatrix} \begin{bmatrix} 9 \\ 6 \\ 9 \end{bmatrix} = \begin{bmatrix} 3 \\ 3 \\ 3 \end{bmatrix}.$$

It follows that $x = 3$, $y = 3$, and $z = 3$ is the solution to the system.

Before we go on... Of course, we should check this by substituting back into the original system. This amounts to calculating $A(A^{-1}B)$ and making sure that we get B (why?).

▼ **EXAMPLE 7**

Solve the following four systems of equations.

(a) $2x + z = 1$ (b) $2x + z = 0$
$\ 2x + y - z = 1$ $2x + y - z = 1$
$\ 3x + y - z = 1$ $3x + y - z = 2$
(c) $2x + z = \frac{1}{2}$ (d) $2x + z = 0$
$\ 2x + y - z = 1$ $2x + y - z = 0$
$\ 3x + y - z = 0$ $3x + y - z = 0$

SOLUTION We *could* go ahead and row-reduce all four augmented matrices as we did in Chapter 2, but this would require a lot of work. Notice that the coefficients are the same in all four systems. In other words, we can write the four systems in matrix form as

(a) $AX = B$ (b) $AX = C$
(c) $AX = D$ (d) $AX = E$,

where the matrix A is the same in all four cases:

$$A = \begin{bmatrix} 2 & 0 & 1 \\ 2 & 1 & -1 \\ 3 & 1 & -1 \end{bmatrix}.$$

Now the solutions to these systems are

(a) $X = A^{-1}B$ (b) $X = A^{-1}C$
(c) $X = A^{-1}D$ (d) $X = A^{-1}E$,

so that all we need is the single matrix A^{-1}, which we have already calculated.

$$A^{-1} = \begin{bmatrix} 0 & -1 & 1 \\ 1 & 5 & -4 \\ 1 & 2 & -2 \end{bmatrix}$$

Thus, the four solutions are

(a) $X = A^{-1}B = \begin{bmatrix} 0 & -1 & 1 \\ 1 & 5 & -4 \\ 1 & 2 & -2 \end{bmatrix} \begin{bmatrix} 1 \\ 1 \\ 1 \end{bmatrix} = \begin{bmatrix} 0 \\ 2 \\ 1 \end{bmatrix}$

(b) $X = A^{-1}C = \begin{bmatrix} 0 & -1 & 1 \\ 1 & 5 & -4 \\ 1 & 2 & -2 \end{bmatrix} \begin{bmatrix} 0 \\ 1 \\ 2 \end{bmatrix} = \begin{bmatrix} 1 \\ -3 \\ -2 \end{bmatrix}$

(c) $X = A^{-1}D = \begin{bmatrix} 0 & -1 & 1 \\ 1 & 5 & -4 \\ 1 & 2 & -2 \end{bmatrix} \begin{bmatrix} \frac{1}{2} \\ 1 \\ 0 \end{bmatrix} = \begin{bmatrix} -1 \\ 5\frac{1}{2} \\ 2\frac{1}{2} \end{bmatrix}$

(d) $X = A^{-1}E = \begin{bmatrix} 0 & -1 & 1 \\ 1 & 5 & -4 \\ 1 & 2 & -2 \end{bmatrix} \begin{bmatrix} 0 \\ 0 \\ 0 \end{bmatrix} = \begin{bmatrix} 0 \\ 0 \\ 0 \end{bmatrix}.$

INVERTING A MATRIX

If A is a **square** matrix, A may have an **inverse,** written A^{-1}, with the property

$$AA^{-1} = A^{-1}A = I.$$

If A has an inverse we say that A is **invertible;** otherwise, we say that A is **singular.** When A is invertible, we can solve $AX = B$ by multiplying both sides by A^{-1}, which gives us $X = A^{-1}B$.

3.3 Matrix Inversion

In order to determine whether an $n \times n$ matrix A is invertible or not, and to find A^{-1} if it does exist, follow this procedure:

1. Write the $n \times 2n$ matrix $[A\,|\,I]$ (this is A with the $n \times n$ identity matrix set next to it).
2. Row-reduce this matrix.
3. If the reduced form is $[I\,|\,B]$ (i.e., has the identity matrix in the left part), then A is invertible and $B = A^{-1}$. If you cannot obtain I in the left part, then A is singular.

3.3 EXERCISES

In Exercises 1 through 6, check whether the given pairs of matrices are inverse pairs.

1. $A = \begin{bmatrix} 0 & 1 \\ 1 & 0 \end{bmatrix}, B = \begin{bmatrix} 0 & 1 \\ 1 & 0 \end{bmatrix}$

2. $A = \begin{bmatrix} 2 & 0 \\ 0 & 3 \end{bmatrix}, B = \begin{bmatrix} \frac{1}{2} & 0 \\ 0 & \frac{1}{3} \end{bmatrix}$

3. $A = \begin{bmatrix} 2 & 1 & 1 \\ 0 & 1 & 1 \\ 0 & 0 & 1 \end{bmatrix}, B = \begin{bmatrix} \frac{1}{2} & -\frac{1}{2} & 0 \\ 0 & 1 & -1 \\ 0 & 0 & 1 \end{bmatrix}$

4. $A = \begin{bmatrix} 1 & 1 & 1 \\ 0 & 1 & 1 \\ 0 & 0 & 1 \end{bmatrix}, B = \begin{bmatrix} 1 & -1 & 0 \\ 0 & 1 & -1 \\ 0 & 0 & 1 \end{bmatrix}$

5. $A = \begin{bmatrix} a & 0 & 0 \\ 0 & b & 0 \\ 0 & 0 & 0 \end{bmatrix}, B = \begin{bmatrix} a^{-1} & 0 & 0 \\ 0 & b^{-1} & 0 \\ 0 & 0 & 0 \end{bmatrix}$ $(a, b \neq 0)$

6. $A = \begin{bmatrix} a & 0 & 0 \\ 0 & b & 0 \\ 0 & 0 & c \end{bmatrix}, B = \begin{bmatrix} a^{-1} & 0 & 0 \\ 0 & b^{-1} & 0 \\ 0 & 0 & c^{-1} \end{bmatrix}$ $(a, b, c \neq 0)$

In Exercises 7 through 34, find the inverses of the given matrices if they exist, and check your answers by multiplication.

7. $\begin{bmatrix} 1 & 1 \\ 2 & 1 \end{bmatrix}$

8. $\begin{bmatrix} 0 & 1 \\ 1 & 1 \end{bmatrix}$

9. $\begin{bmatrix} 0 & 1 \\ 1 & 0 \end{bmatrix}$

10. $\begin{bmatrix} 4 & 0 \\ 0 & 2 \end{bmatrix}$

11. $\begin{bmatrix} 1 & 0 \\ 0 & 1 \end{bmatrix}$

12. $\begin{bmatrix} 2 & 1 \\ 1 & 1 \end{bmatrix}$

13. $\begin{bmatrix} 1 & 1 \\ 1 & -1 \end{bmatrix}$

14. $\begin{bmatrix} 4 & 1 \\ 0 & 2 \end{bmatrix}$

15. $\begin{bmatrix} 3 & 0 \\ 0 & \frac{1}{2} \end{bmatrix}$

16. $\begin{bmatrix} 2 & 1 \\ 4 & 2 \end{bmatrix}$

17. $\begin{bmatrix} 1 & 1 \\ 6 & 6 \end{bmatrix}$

18. $\begin{bmatrix} 1 & 3 \\ 2 & 4 \end{bmatrix}$

19. $\begin{bmatrix} \frac{1}{6} & -\frac{1}{6} \\ 0 & \frac{1}{6} \end{bmatrix}$

20. $\begin{bmatrix} 2 & 1 \\ 4 & 2 \end{bmatrix}$

21. $\begin{bmatrix} 1 & 0 \\ \frac{3}{4} & 0 \end{bmatrix}$

22. $\begin{bmatrix} 1 & 1 \\ 1 & 1 \end{bmatrix}$

23. $\begin{bmatrix} 1 & 1 & 1 \\ 0 & 1 & 1 \\ 0 & 0 & 1 \end{bmatrix}$

24. $\begin{bmatrix} 1 & 2 & 3 \\ 0 & 1 & 2 \\ 0 & 0 & 1 \end{bmatrix}$

25. $\begin{bmatrix} 1 & 1 & 1 \\ 1 & 0 & 2 \\ 1 & -1 & 1 \end{bmatrix}$

26. $\begin{bmatrix} 1 & 2 & 3 \\ 0 & 2 & 3 \\ 1 & 0 & 1 \end{bmatrix}$

27. $\begin{bmatrix} 1 & 1 & 1 \\ 1 & -1 & 0 \\ 1 & 2 & 3 \end{bmatrix}$
28. $\begin{bmatrix} 1 & -1 & 3 \\ 0 & 1 & 3 \\ 1 & 1 & 1 \end{bmatrix}$
29. $\begin{bmatrix} 1 & 1 & 1 \\ 1 & 0 & 1 \\ 1 & -1 & 1 \end{bmatrix}$
30. $\begin{bmatrix} 1 & 1 & 1 \\ 0 & 1 & 1 \\ 1 & 0 & 0 \end{bmatrix}$

31. $\begin{bmatrix} 1 & 0 & 1 & 0 \\ -1 & 1 & 0 & 1 \\ -1 & 0 & 0 & 1 \\ 0 & -1 & 0 & 1 \end{bmatrix}$
32. $\begin{bmatrix} 0 & 1 & 1 & 0 \\ -1 & 1 & 1 & 1 \\ -1 & 1 & 0 & 1 \\ 0 & -1 & 0 & 1 \end{bmatrix}$
33. $\begin{bmatrix} 1 & 2 & 3 & 4 \\ 0 & 1 & 2 & 3 \\ 0 & 0 & 1 & 2 \\ 0 & 0 & 0 & 1 \end{bmatrix}$
34. $\begin{bmatrix} 0 & 0 & 0 & 1 \\ 0 & 0 & 1 & 0 \\ 0 & 1 & 0 & 0 \\ 1 & 0 & 0 & 0 \end{bmatrix}$

In Exercises 35–42, we suggest that you use a graphing calculator or computer. Find the inverse of each of the following matrices (when they exist). Round all entries in your answer to two decimal places.

35. $\begin{bmatrix} 1.1 & 1.2 \\ 1.3 & -1 \end{bmatrix}$
36. $\begin{bmatrix} 0.1 & -3.2 \\ 0.1 & -1.5 \end{bmatrix}$

37. $\begin{bmatrix} 3.56 & 1.23 \\ -1.01 & 0 \end{bmatrix}$
38. $\begin{bmatrix} 9.09 & -5.01 \\ 1.01 & 2.20 \end{bmatrix}$

39. $\begin{bmatrix} 1.1 & 3.1 & 2.4 \\ 1.7 & 2.4 & 2.3 \\ 0.6 & -0.7 & -0.1 \end{bmatrix}$
40. $\begin{bmatrix} 2.1 & 2.4 & 3.5 \\ 6.1 & -0.1 & 2.3 \\ -0.3 & -1.2 & 0.1 \end{bmatrix}$

41. $\begin{bmatrix} 0.01 & 0.32 & 0 & 0.04 \\ -0.01 & 0 & 0 & 0.34 \\ 0 & 0.32 & -0.23 & 0.23 \\ 0 & 0.41 & 0 & 0.01 \end{bmatrix}$
42. $\begin{bmatrix} 0.01 & 0.32 & 0 & 0.04 \\ -0.01 & 0 & 0 & 0.34 \\ 0 & 0.32 & -0.23 & 0.23 \\ 0.01 & 0.96 & -0.23 & 0.65 \end{bmatrix}$

In Exercises 43 through 52, use matrix inversion to solve the given systems of linear equations. (You have already solved many of these systems using row reduction in Chapter 2.)

43. $x + y = 4$
    $x - y = 2$

44. $2x + y = 2$
    $2x - 3y = 2$

45. $\frac{x}{3} - \frac{y}{2} = 1$
    $\frac{x}{4} + y = -2$

46. $-\frac{2x}{3} + \frac{y}{2} = -\frac{1}{6}$
    $\frac{x}{4} - y = -\frac{3}{4}$

47. $-x + 2y - z = 0$
    $-x - y + 2z = 0$
    $2x \quad - z = 6$

48. $x + 2y \quad = 4$
    $\quad y - z = 0$
    $x + 3y - 2z = 5$

49. $x + y + 6z = -1$
    $\frac{1}{3}x - \frac{1}{3}y + \frac{2}{3}z = 1$
    $\frac{1}{2}x \quad + z = 0$

50. $x - \frac{1}{2}y \quad = 0$
    $\frac{1}{3}x + \frac{1}{3}y + \frac{1}{3}z = 2$
    $\frac{1}{2}x \quad - \frac{1}{2}z = -1$

51. $x + y + 5z \quad = 1$
    $\quad y + 2z + w = 1$
    $x + 3y + 7z + 2w = 2$
    $x + y + 5z + w = 1$

52. $x + y \quad + 4w = 1$
    $2x - 2y - 3z + 2w = -1$
    $\quad 4y + 6z + w = 4$
    $2x + 4y + 9z \quad = 6$

APPLICATIONS

In Exercises 53–56 (versions of many of these appeared as exercises and examples in Chapter 2), use matrix inverses to obtain the solutions.

53. *Nutrition* A four-ounce serving of *Campbell's* Pork & Beans® contains 5 grams of protein and 21 grams of carbohydrates.* A typical slice of "lite" rye bread contains 4 grams of protein and 12 grams of carbohydrates. The U.S. RDA is 60 grams of protein per day.
    (a) I am planning a meal of "beans-on-toast" and wish to have it supply 20 grams of protein and 80 grams of carbohydrates. How should I prepare my meal?
    (b) If I require A grams of protein and B grams of carbohydrates, give a formula that tells me how many slices of bread and how many servings of Pork & Beans to use.

54. *Nutrition* According to the nutritional information on a package of Honey Nut Cheerios® brand cereal, each 1-ounce serving of Cheerios contains 3 grams of protein and 24 grams of carbohydrates.† Each half cup serving of enriched skim milk contains 4 grams of protein and 6 grams of carbohydrates.
    (a) I am planning a meal of cereal and milk and wish to have it supply 20 grams of protein, and 80 grams of carbohydrates. How should I prepare my meal?
    (b) If I require A grams of protein and B grams of carbohydrates, give a formula that tells me how many servings of milk and Cheerios to use.

55. *Resource Allocation* The Arctic Juice Company makes three juice blends: PineOrange, using 2 quarts of pineapple juice and 2 quarts of orange juice per gallon; PineKiwi, using 3 quarts of pineapple juice and 1 quart of kiwi juice per gallon; and OrangeKiwi, using 3 quarts of orange juice and 1 quart of kiwi juice per gallon. The amount of stock of each kind of juice on hand varies from day to day. How many gallons of each blend can the company make if it has:
    (a) 800 quarts of pineapple juice, 650 quarts of orange juice, 350 quarts of kiwi juice;
    (b) 850 quarts of pineapple juice, 600 quarts of orange juice, 400 quarts of kiwi juice;
    (c) A quarts of pineapple juice, B quarts of orange juice, C quarts of kiwi juice?

56. *Resource Allocation* You manage an ice cream factory that makes three flavors: Creamy Vanilla, Continental Mocha, and Succulent Strawberry. Into each batch of Creamy Vanilla go two eggs, one cup of milk, and two cups of cream. Into each batch of Continental Mocha go one egg, one cup of milk, and two cups of cream, while into each batch of Succulent Strawberry go one egg, two cups of milk, and one cup of cream. Your stocks of eggs, milk, and cream vary from day to day. How many batches of each flavor should you make in order to use up all of your ingredients if you have in stock:
    (a) 350 eggs, 350 cups of milk, and 400 cups of cream;
    (b) 400 eggs, 400 cups of milk, and 400 cups of cream;
    (c) A eggs, B cups of milk, and C cups of cream?

57. *Population Movement* In 1987, the population of the U.S., broken down by regions,‡ was 49.3 million in the Northeast, 59.1 million in the Midwest, 82.2 million in the South, and 48.2 million in the West. From 1987 to 1988, 0.23% of the population in the Northeast moved to the Midwest, 0.9% moved to the South, and 0.25% moved to the West (the remainder stayed in the Northeast). In that same year, 0.15% of the population in the Midwest moved to the Northeast, 0.89% moved to the South, and 0.36% moved to the West. Also, 0.3% of the population in the South moved to the Northeast, 0.5% moved to the Midwest, and 0.36% moved to the West. Finally, 0.22% of the population in the West move to the Northeast, 0.47% moved to the Midwest, and 0.48% moved to the South. Set up the 1987 population figures as a row vector, and set up the population movement figures as a matrix as in Exercise 67 in the last section. Assuming that these percentages also describe the population movements from 1986 to 1987, show how matrix inversion and multiplication allow you to compute the population in each region in the year 1986. (Round all answers to the nearest 0.1 million.)

58. *Population Movement* Assuming that the percentages given in Exercise 57 also describe the population movements from 1985–1986, show how matrix inversion and multiplication allow you to compute from the data in Exercise 57 the population in each region in 1985. (Round all answers to the nearest 0.1 million.)

---

* According to the label information on a 16 oz. can
† Actually, it is 23 g. carbohydrates. We made it 24 g. in order to simplify the calculation.
‡ Source: Statistical Abstract of the United States 1991, U.S. Dept. of Commerce

COMMUNICATION AND REASONING EXERCISES

59. Why is a 2×2 matrix of the form $\begin{bmatrix} a & b \\ a & b \end{bmatrix}$ necessarily singular?

60. If you think of numbers as 1×1 matrices, which numbers are singular 1×1 matrices?

61. Use matrix multiplication to check that the inverse of a general 2×2 matrix is given by

$$\begin{bmatrix} a & b \\ c & d \end{bmatrix}^{-1} = \frac{1}{ad-bc}\begin{bmatrix} d & -b \\ -c & a \end{bmatrix} \quad \text{(assuming } ad - bc \neq 0\text{)}.$$

62. Derive the formula in the previous exercise using row reduction. (Assume that $ad - bc \neq 0$.)

63. If a square matrix A row-reduces to a matrix with a row of zeros, can it be invertible? If so, give an example of such a matrix, and if not, say why not.

64. If a square matrix A row-reduces to the identity matrix, must it be invertible? If so, say why, and if not, give an example of such a (singular) matrix.

65. Your friend has come up with two square matrices A and B, neither of them the zero matrix, with the property that AB is the zero matrix. You immediately tell him that neither A nor B can possibly be invertible. How can you be so sure?

66. A **diagonal** matrix D has the following form.

$$D = \begin{bmatrix} d_1 & 0 & 0 & \cdots & 0 \\ 0 & d_2 & 0 & \cdots & 0 \\ 0 & 0 & d_3 & \cdots & 0 \\ \cdots & \cdots & \cdots & \cdots & \cdots \\ 0 & 0 & 0 & \cdots & d_n \end{bmatrix}$$

When is D singular?

67. If A and B are invertible, check that $B^{-1}A^{-1}$ is the inverse of AB. $[(AB)(B^{-1}A^{-1}) = \ldots]$

68. Solve the matrix equation $A(B + CX) = D$ for X. (You may assume that A and C are square and invertible.)

3.4 INPUT-OUTPUT MODELS

In this section we look at an application of matrix algebra developed by Wassily Leontief in the middle of this century. He won the Nobel prize in economics in 1973 for this work. The application involves analyzing national and regional economies by looking at how various parts of the economy interrelate. We'll work out some of the details by looking at a simple scenario.

First, we think of the economy of a country or a region as being composed of various **sectors,** or groups of one or more industries. Typical sectors are the manufacturing sector, the utilities sector, or the agricultural sector. To introduce the basic concepts, we shall consider two specific sectors: the coal-mining sector (Sector 1) and the utilities sector (Sector 2). Both produce a commodity: the coal-mining sector produces coal, and the utilities sector produces electricity. We measure these products by their dollar value. By **one unit** of a product, we shall mean $1 worth of that product.

Here is the scenario.

1. To produce one unit ($1 worth) of coal, assume that the coal-mining sector uses 50¢ of coal (in its furnaces, say) and 10¢ of electricity.
2. To produce one unit ($1 worth) of electricity, assume that the utilities sector uses 25¢ of coal and 25¢ of electricity.

These are *internal* usage figures. In addition to this, assume that there is an *external* demand (from the rest of the economy) of 7,000 units ($7,000 worth) of coal and 14,000 units ($14,000 worth) of electricity over a specific time period (say, one year).

Q How much should each of the two sectors supply in order to meet both internal and external demand?

A The key to answering this question is to set up equations of the form

$$\text{Total supply} = \text{Total demand}$$

Since we don't know yet how much each sector should supply, the unknowns are

$$x_1 = \text{the total supply (in units) from Sector 1 (coal)},$$

and $x_2 = $ the total supply (in units) from Sector 2 (utilities).

Our equations then take the form

Total supply from Sector 1 = Total demand for Sector 1 products
$$x_1 = 0.50x_1 + 0.25x_2 + 7{,}000$$

Coal required by Sector 1   Coal required by Sector 2   External demand for coal

Total supply from Sector 2 = Total demand for Sector 2 products
$$x_2 = 0.10x_1 + 0.25x_2 + 14{,}000$$

Electricity required by Sector 1   Electricity required by Sector 2   External demand for electricity

This is a system of linear equations in two unknowns.

$$x_1 = 0.50x_1 + 0.25x_2 + 7{,}000$$
$$x_2 = 0.10x_1 + 0.25x_2 + 14{,}000$$

We could now go ahead and rewrite these equations in standard form and then solve for x_1 and x_2, thus answering the question.

Instead of doing this, notice that we can rewrite the system of equations in matrix form as follows:

$$\underbrace{\begin{bmatrix} x_1 \\ x_2 \end{bmatrix}}_{\text{Production}} = \underbrace{\begin{bmatrix} 0.50 & 0.25 \\ 0.10 & 0.25 \end{bmatrix} \begin{bmatrix} x_1 \\ x_2 \end{bmatrix}}_{\text{Internal demand}} + \underbrace{\begin{bmatrix} 7{,}000 \\ 14{,}000 \end{bmatrix}}_{\text{External demand}}.$$

In symbols,

$$X = AX + D.$$

Here,

$$X = \begin{bmatrix} x_1 \\ x_2 \end{bmatrix}$$

is called the **production vector.** Its entries are the amounts produced by the two sectors. The vector

$$D = \begin{bmatrix} 7{,}000 \\ 14{,}000 \end{bmatrix}$$

is called the **external demand** vector, and the matrix

$$A = \begin{bmatrix} 0.50 & 0.25 \\ 0.10 & 0.25 \end{bmatrix}$$

is called the **technology matrix.** The entries of the technology matrix mean the following.

a_{11} = units of Sector 1 needed to produce one unit of Sector 1
a_{12} = units of Sector 1 needed to produce one unit of Sector 2
a_{21} = units of Sector 2 needed to produce one unit of Sector 1
a_{22} = units of Sector 2 needed to produce one unit of Sector 2

You can remember this by the slogan "In the side, out the top."

ENTRIES OF THE TECHNOLOGY MATRIX

The i, j entry of the **technology matrix** A is

a_{ij} = units of Sector i needed to produce one unit of Sector j.

To meet an external demand of D, the economy must produce X, where X satisfies the matrix equation

$$X = AX + D.$$

Now that we have a matrix equation, we can solve it as follows. Subtract AX from both sides.

$$X - AX = D$$

Since $X = IX$, where I is the 2 × 2 identity matrix, we can rewrite this as
$$IX - AX = D.$$
Now factor out X.
$$(I - A)X = D$$
If we multiply both sides by the inverse of $(I - A)$, we get the following.

SOLVING FOR THE PRODUCTION VECTOR

If A is the technology matrix of an economy and D is the external demand vector, then the production vector X that meets the demand can be calculated by
$$X = (I - A)^{-1}D.$$

This is the matrix method of solving for the production vector, and it is the one we shall use.

▼ **EXAMPLE 1** Energy Production

To produce $1 worth of coal, the coal-mining sector uses 50¢ of coal and 10¢ of electricity. To produce $1 worth of electricity, the utilities sector uses 25¢ of coal and 25¢ of electricity. Assume that there is an external demand of $7,000 worth of coal and $14,000 worth of electricity over the period of one year. How much coal and electricity should the two sectors produce in order to meet the demand?

SOLUTION Take the coal sector as Sector 1, and the utilities sector as Sector 2. The technology matrix has entries a_{ij} = units of Sector i to produce one unit of Sector j. Thus,

$$A = \begin{bmatrix} 0.50 & 0.25 \\ 0.10 & 0.25 \end{bmatrix} \quad \text{and} \quad D = \begin{bmatrix} 7{,}000 \\ 14{,}000 \end{bmatrix}.$$

We use the formula
$$X = (I - A)^{-1}D.$$
Now,
$$I - A = \begin{bmatrix} 1 & 0 \\ 0 & 1 \end{bmatrix} - \begin{bmatrix} 0.50 & 0.25 \\ 0.10 & 0.25 \end{bmatrix} = \begin{bmatrix} 0.50 & -0.25 \\ -0.10 & 0.75 \end{bmatrix}.$$

We use the methods of the last section to find the inverse, getting
$$(I - A)^{-1} = \begin{bmatrix} \frac{15}{7} & \frac{5}{7} \\ \frac{2}{7} & \frac{10}{7} \end{bmatrix}.$$

Thus,
$$X = (I - A)^{-1}D$$
$$= \begin{bmatrix} \frac{15}{7} & \frac{5}{7} \\ \frac{2}{7} & \frac{10}{7} \end{bmatrix} \begin{bmatrix} 7{,}000 \\ 14{,}000 \end{bmatrix} = \begin{bmatrix} 25{,}000 \\ 22{,}000 \end{bmatrix}.$$

The matrix X tells us that to meet the demand, the economy must produce $25,000 worth of coal and $22,000 worth of electricity.

Before we go on... How much coal and electricity is used internally? We could calculate the internal usage directly by looking at the original matrix equation, where we saw that AX gives the internal demand. Alternatively, we could calculate the amount left over once external demand is satisfied, as follows.

$$\text{Internal demand} = \text{Production} - \text{External demand}$$
$$= X - D$$
$$= \begin{bmatrix} 25{,}000 \\ 22{,}000 \end{bmatrix} - \begin{bmatrix} 7{,}000 \\ 14{,}000 \end{bmatrix} = \begin{bmatrix} 18{,}000 \\ 8{,}000 \end{bmatrix}.$$

Thus, $18,000 worth of coal and $8,000 worth of electricity were used internally, and were thus unavailable to the outside economy. We should check that AX gives the same answer.

The following examples use real data from the U.S. economy. The use of a graphing calculator or computer software is highly recommended. (Even if you don't have access to the appropriate software, you should still follow these examples.)

EXAMPLE 2

Consider two sectors of the U.S. economy, *crude petroleum and natural gas* (crude), and *petroleum refining and related industries* (refining). According to government figures,* in 1977 the crude sector used $2,302 million worth of its own products, and $259 million worth of the products of the refining sector, in producing $43,523 million worth of goods (crude oil and natural gas). The refining sector in the same year used $60,050 million worth of the products of the crude sector, and $7,754 million worth of its own products, in producing $96,114 million worth of goods (refined oil and the like). What was left over from each of these sectors for use by other parts of the economy or for export?

SOLUTION First, for convenience, we record the given data in the form of a table, called the **input-output table.**

▼ * *Survey of Current Business*, Vol. 59, No. 2, May 1984, U.S. Department of Commerce.

3.4 Input-Output Models

To From	Crude	Refining	Total Output
Crude	2,302	60,050	43,523
Refining	259	7,754	96,114

(All figures are in millions of dollars.) The entries in the left-hand portion are arranged in the same way as those of the technology matrix: the i,j entry represents the number of units of Sector i that went to Sector j. Thus, for instance, the $60,050 million entry in the 1,2 position represents the number of units of Sector 1 (crude) that were used by Sector 2 (refining). ("In the side, out the top.")

We now construct the technology matrix. The technology matrix has entries a_{ij} = units of Sector i to produce *one* unit of Sector j. Thus,

a_{11} = units of crude to produce one unit of crude. We are told that 2,302 million units of crude were used to produce 43,523 million units of crude. Thus, to produce *one* unit of crude, $2{,}302/43{,}523 = 0.05289$ units of crude were used, and so $a_{11} = 0.05289$.

a_{12} = units of crude to produce one unit of refined
    $= 60{,}050/96{,}114 = 0.62478$

a_{21} = units of refined to produce one unit of crude
    $= 259/43{,}523 = 0.00595$

a_{22} = units of refined to produce one unit of refined
    $= 7{,}754/96{,}114 = 0.08068.$

This gives the technology matrix,

$$A = \begin{bmatrix} 0.05289 & 0.62478 \\ 0.00595 & 0.08068 \end{bmatrix}.$$

In short, *we obtained the technology matrix from the input-output table by dividing the Sector 1 column by the Sector 1 total, and the Sector 2 column by the Sector 2 total.*

Now we also know the total output from each sector, so we have already been given the production vector.

$$X = \begin{bmatrix} 43{,}523 \\ 96{,}114 \end{bmatrix}$$

What we are asked for is the external demand vector D, the amount available for the outside economy. To find D, we again use the equation

$$X = AX + D,$$

Production = Internal demand + External demand

where, this time, we are given A and X, and must solve for D. Solving for D gives

$$D = X - AX$$

$$= \begin{bmatrix} 43{,}523 \\ 96{,}114 \end{bmatrix} - \begin{bmatrix} 0.05289 & 0.62478 \\ 0.00595 & 0.08068 \end{bmatrix} \begin{bmatrix} 43{,}523 \\ 96{,}114 \end{bmatrix}$$

$$= \begin{bmatrix} 43{,}523 \\ 96{,}114 \end{bmatrix} - \begin{bmatrix} 62{,}352 \\ 8{,}013 \end{bmatrix} = \begin{bmatrix} -18{,}829 \\ 88{,}101 \end{bmatrix}.$$

The second of these numbers, $88,101 million, is the amount produced by the refining sector that is available to be used by other parts of the economy or to be exported (in fact, since something has to happen to all that refined oil, this is the amount actually used or exported, where use can include stockpiling). The first number, $-18,829$, also represents the amount produced by the crude sector that is available to be used by other parts of the economy or to be exported, but what does it mean that it is negative? It means that there must have been considerable *importing* of crude oil or natural gas, the products of the crude sector, from outside the country.

Before we go on... We could have calculated these numbers more simply from the input-output table. The internal use of units from the crude sector was the sum of the outputs from that sector,

$$2{,}302 + 60{,}050 = 62{,}352.$$

Since only 43,523 units were actually produced by the sector, that left a deficit of $62{,}352 - 43{,}523 = 18{,}829$ units that must have been imported. We could compute the surplus from the refining sector similarly. The calculation in the next example cannot be done as trivially.

INPUT-OUTPUT TABLE

The i, j entry in the left-hand portion of the **input-output table** is the number of units that went from Sector i to Sector j.

We obtain the technology matrix from the input-output table by dividing the Sector 1 column by the total Sector 1 output, and the Sector 2 column by the total Sector 2 output, and so on.

▼ **EXAMPLE 3**

Suppose that external demand for refined petroleum rose to $90,000 million, but that the demand for crude remained the same. How would the production levels of the two sectors considered in Example 2 have to change?

SOLUTION We are being told that now

$$D = \begin{bmatrix} -18{,}829 \\ 90{,}000 \end{bmatrix}$$

and we are being asked to find X, as in Example 1. Remember that we can calculate X from the formula

$$X = (I - A)^{-1}D.$$

Now

$$I - A = \begin{bmatrix} 1 & 0 \\ 0 & 1 \end{bmatrix} - \begin{bmatrix} 0.05289 & 0.62478 \\ 0.00595 & 0.08068 \end{bmatrix} = \begin{bmatrix} 0.94711 & -0.62478 \\ -0.00595 & 0.91932 \end{bmatrix}.$$

We now find the inverse.

$$(I - A)^{-1} = \begin{bmatrix} 1.06037 & 0.72064 \\ 0.006864 & 1.09242 \end{bmatrix}$$

We now calculate X.

$$X = (I - A)^{-1}D = \begin{bmatrix} 1.06037 & 0.72064 \\ 0.006864 & 1.09242 \end{bmatrix} \begin{bmatrix} -18{,}829 \\ 90{,}000 \end{bmatrix} = \begin{bmatrix} 44{,}891 \\ 98{,}189 \end{bmatrix}.$$

Comparing this vector to the production vector used in the previous example, we see that production in the crude sector will have to rise $1,368 million, from $43,523 million to $44,891 million, while production in the refining sector will have to rise $2,075 million, from $96,114 million to $98,189 million.

Before we go on... Using the matrix $(I - A)^{-1}$, there is a slightly different way of answering this question. We are asking for the effect on production of a *change* in final demand of 0 for crude and $90,000 - 88,101 = \$1{,}899$ million for refined products. If we multiply $(I - A)^{-1}$ by the matrix representing this *change,* we obtain

$$\begin{bmatrix} 1.06037 & 0.72064 \\ 0.006864 & 1.09242 \end{bmatrix} \begin{bmatrix} 0 \\ 1{,}899 \end{bmatrix} = \begin{bmatrix} 1{,}368 \\ 2{,}075 \end{bmatrix}.$$

We see the changes required in production: an increase of $1,368 million in the crude sector and an increase of $2,075 million in the refining sector. You should try to see why this works, using matrix algebra.

Notice that the increase in external demand for the products of the refining sector requires the crude sector to also increase production, even though there is no increase in the *external* demand for crude. The reason is that, in order to increase production, the refining sector will need to use more crude oil, so that the *internal* demand for crude oil goes up. The inverse matrix $(I - A)^{-1}$ takes these **indirect effects** into account in a nice way. You can see in the computation we just did that each dollar increase in external demand for refined products will require an increase in production of $0.72064 in the crude sector, as well as an increase in production of $1.09242 in the refining sector. This is how we interpret the entries in $(I - A)^{-1}$, and this is why it is useful to look at this matrix inverse in the first place. We can also say what the effects of an increase of $1 in external demand for crude would be: an increase in production of $1.06037 in the crude sector and an increase of $0.006864 in the refining sector.

Here are some questions to think about: Why are the diagonal entries (slightly) larger than 1? Why is the entry in the lower left so small compared to the others?

In the above examples we used only two sectors of the economy. The data for Examples 2 and 3 were taken from an input-output table published by the U.S. Department of Commerce, in which the whole U.S. economy was broken into 85 sectors. This in turn was a simplified version of a model in which the economy was broken into about 500 sectors. Obviously, computers are necessary to make a realistic input-output analysis possible. Many governments now collect data for and publish input-output tables as part of their national planning. You can find input-output tables for many countries in *National Accounts Statistics: Study of Input-Output Tables, 1970–1980*, United Nations, NY, 1987.

Before looking at a larger example, let's summarize what we've done.

SUMMARY OF INPUT-OUTPUT MODELS

The **technology matrix** A is an $n \times n$ matrix, where n is the number of sectors of the economy being studied. The i,j entry of the technology matrix A is given by

$$a_{ij} = \text{Units of Sector } i \text{ to produce one unit of Sector } j.$$

The **production vector** X is a column vector with n entries. Each entry records the total production of the corresponding sector.
The **external demand vector** D is a column vector with n entries. Each entry records the demand from outside the economy for the products of the corresponding sector.

These matrices are related by the equation

$$D = X - AX,$$

or

$$X = (I - A)^{-1}D.$$

The ijth entry in the matrix $(I - A)^{-1}$ shows us, for each unit increase in demand for Sector j, how much production must rise in Sector i.

3.4 Input-Output Models

> **INPUT-OUTPUT TABLE**
>
> The i,j entry in the left-hand portion of the input-output table is the number of units that went from Sector i to Sector j.
>
> We obtain the technology matrix from the input-output table by dividing the Sector 1 column by the total Sector 1 output, the Sector 2 column by the total Sector 2 output, and so on.

EXAMPLE 4

Consider four sectors of the economy of Kenya*: (1) the traditional economy, (2) agriculture, (3) manufacture of metal products and machinery, and (4) wholesale and retail trade. The input-output table for these four sectors for the year 1976 looks like this (all numbers are 1,000s of K£):

	To	1	2	3	4	Total Output
From	1	8,600	0	0	0	87,160
	2	0	19,847	24	0	531,131
	3	1,463	530	15,315	660	112,780
	4	814	8,529	5,773	2,888	178,911

Suppose external demand for agriculture rose by K£50,000,000 and external demand for metal products and machinery rose by K£10,000,000. How would production in these four sectors have to change to meet this rising demand?

SOLUTION We first need to convert this input-output matrix into a technology matrix. Again, we do this by dividing each column by the total output of that sector. This gives

$$A = \begin{bmatrix} 0.09867 & 0 & 0 & 0 \\ 0 & 0.037367 & 0.00021 & 0 \\ 0.01679 & 0.000998 & 0.13580 & 0.00369 \\ 0.00934 & 0.016058 & 0.05119 & 0.01614 \end{bmatrix}.$$

Now, to see how each sector will react to rising external demand, we must calculate the inverse matrix $(I - A)^{-1}$:

$$I - A = \begin{bmatrix} 0.90133 & 0 & 0 & 0 \\ 0 & 0.962633 & -0.00021 & 0 \\ -0.01679 & -0.000998 & 0.86420 & -0.00369 \\ -0.00934 & -0.016058 & -0.05119 & 0.98385 \end{bmatrix},$$

---

*Source: *Input-Output Tables for Kenya 1976*, Central Bureau of Statistics of the Ministry of Economic Planning and Community Affairs, Kenya.

so

$$(I - A)^{-1} = \begin{bmatrix} 1.10947 & 0 & 0 & 0 \\ 0.000005 & 1.03882 & 0.00252 & 0.000001 \\ 0.021605 & 0.001272 & 1.15740 & 0.004341 \\ 0.011657 & 0.017021 & 0.060223 & 1.01663 \end{bmatrix}.$$

Now we can see the effect of rising demand. Let

$$D = \begin{bmatrix} 0 \\ 50{,}000 \\ 10{,}000 \\ 0 \end{bmatrix}$$

represent the rise in demand (all our figures are in 1,000s of K£). Then the rise in production is

$$(I - A)^{-1}D = \begin{bmatrix} 1.10947 & 0 & 0 & 0 \\ 0.000005 & 1.03882 & 0.00252 & 0.000001 \\ 0.021605 & 0.001272 & 1.15740 & 0.004341 \\ 0.011657 & 0.017021 & 0.060223 & 1.01663 \end{bmatrix} \begin{bmatrix} 0 \\ 50{,}000 \\ 10{,}000 \\ 0 \end{bmatrix}$$

$$= \begin{bmatrix} 0 \\ 51{,}943 \\ 11{,}638 \\ 1{,}453 \end{bmatrix}.$$

Thus, the traditional economy will be unaffected, production in agriculture will rise by K£51,943,000, production in the manufacture of metal products and machinery will rise by K£11,638,000, and activity in wholesale and retail trade will rise by K£1,453,000.

Before we go on... Can you see why the traditional economy was unaffected? Although it takes inputs from other parts of the economy, it is not itself an input to any other part. In other words, there is no demand for the products of the traditional economy coming from any other part of the economy, and so an increase in production in any other sector of the economy will require no increase from the traditional economy. On the other hand, the wholesale and retail trade sector does provide input to the agriculture and manufacturing sectors, so increases in those sectors do require an increase in the trade sector.

One more point: Notice how small the off-diagonal entries in $(I - A)^{-1}$ are. This says that increases in each sector have relatively small effects on the other sectors. We say that these sectors are **loosely coupled.** Regional economies, where many products are destined to be shipped out to the rest of the country, tend to show this phenomenon even more strongly. Notice in Example 2 that the two sectors there are **strongly coupled,** since a rise in demand for refined products requires a comparable rise in the production of crude.

3.4 EXERCISES

In Exercises 1–10, you are given a technology matrix A and an external demand vector D. Find the corresponding production vector X.

1. $A = \begin{bmatrix} 0.5 & 0.4 \\ 0 & 0.5 \end{bmatrix}, D = \begin{bmatrix} 10,000 \\ 20,000 \end{bmatrix}$

2. $A = \begin{bmatrix} 0.4 & 0.4 \\ 0 & 0.5 \end{bmatrix}, D = \begin{bmatrix} 20,000 \\ 10,000 \end{bmatrix}$

3. $A = \begin{bmatrix} 0.1 & 0.4 \\ 0.2 & 0.5 \end{bmatrix}, D = \begin{bmatrix} 25,000 \\ 15,000 \end{bmatrix}$

4. $A = \begin{bmatrix} 0.1 & 0.2 \\ 0.4 & 0.5 \end{bmatrix}, D = \begin{bmatrix} 24,000 \\ 14,000 \end{bmatrix}$

5. $A = \begin{bmatrix} 0.5 & 0.1 & 0 \\ 0 & 0.5 & 0.1 \\ 0 & 0 & 0.5 \end{bmatrix}, D = \begin{bmatrix} 1,000 \\ 1,000 \\ 2,000 \end{bmatrix}$

6. $A = \begin{bmatrix} 0.5 & 0.1 & 0 \\ 0.1 & 0.5 & 0.1 \\ 0 & 0 & 0.5 \end{bmatrix}, D = \begin{bmatrix} 3,000 \\ 3,800 \\ 2,000 \end{bmatrix}$

7. $A = \begin{bmatrix} 0.2 & 0.2 & 0 \\ 0.2 & 0.4 & 0.2 \\ 0 & 0.2 & 0.2 \end{bmatrix}, D = \begin{bmatrix} 16,000 \\ 8,000 \\ 8,000 \end{bmatrix}$

8. $A = \begin{bmatrix} 0.2 & 0.2 & 0.2 \\ 0.2 & 0.4 & 0.2 \\ 0.2 & 0.2 & 0.2 \end{bmatrix}, D = \begin{bmatrix} 7,000 \\ 14,000 \\ 7,000 \end{bmatrix}$

9. $A = \begin{bmatrix} 0.4 & 0.1 & 0 & 0 \\ 0.1 & 0.4 & 0 & 0 \\ 0 & 0 & 0.3 & 0.2 \\ 0 & 0 & 0.2 & 0.3 \end{bmatrix}, D = \begin{bmatrix} 70,000 \\ 63,000 \\ 90,000 \\ 63,000 \end{bmatrix}$

10. $A = \begin{bmatrix} 0.4 & 0.1 & 0 & 0 \\ 0.1 & 0.4 & 0 & 0 \\ 0 & 0.3 & 0.3 & 0.2 \\ 0 & 0 & 0.2 & 0.3 \end{bmatrix}, D = \begin{bmatrix} 70,000 \\ 63,000 \\ 90,000 \\ 63,000 \end{bmatrix}$

11–20. For each of the technology matrices A in Exercises 1–10, determine the changes in production that would result from a unit increase in demand for the products of the first sector, the changes that would result from a unit increase in demand for the products of the second sector, and so on.

APPLICATIONS

21. *United States Input-Output Table** Two of the sectors of the U.S. economy are (1) radio, TV, and communication equipment, and (2) electronic components and accessories. In 1977, the input-output table involving these two sectors was as follows (all figures are in millions of dollars).

	To	1	2	Total Output
From	1	1,653	2	28,333
	2	4,742	2,123	14,790

Determine how both sectors would react to an increase in demand for communication equipment (Sector 1) of $1,000 million, and how both would react to an increase in demand for electronic components (Sector 2) of $1,000 million.

---

*Source for Exercises 21–24: *Survey of Current Business,* Vol. 59, No. 2, May 1984, U.S. Dept. of Commerce.

22. *United States Input-Output Table* Another two sectors of the U.S. economy are (1) finance and insurance, and (2) real estate and rental. In 1977, the input-output table involving these two sectors was as follows (all figures are in millions of dollars).

To From	1	2	Total Output
1	25,270	6,970	128,578
2	2,938	17,816	279,243

Determine how both sectors would react to an increase in demand for finance (Sector 1) of $10,000 million, and how both would react to an increase in demand for real estate (Sector 2) of $10,000 million.

23. *United States Input-Output Table* Four of the sectors of the U.S. economy are (1) livestock and livestock products, (2) other agricultural products, (3) forestry and fishery products, and (4) agricultural, forestry, and fishery services. In 1977, the input-output table involving these four sectors was as follows (all figures are in millions of dollars).

To From	1	2	3	4	Total Output
1	8,905	1,336	0	213	48,603
2	13,769	2,498	0	162	63,176
3	0	0	32	16	5,170
4	1,900	2,520	457	60	8,293

Determine how these four sectors would react to an increase in demand for livestock (Sector 1) of $1,000 million, how they would react to an increase in demand for other agricultural products (Sector 2) of $1,000 million, and so on.

24. *United States Input-Output Table* Another four sectors of the U.S. economy are (1) iron and ferroalloy ores mining, (2) nonferrous metal ores mining, (3) coal mining, and (4) stone and clay mining and quarrying. In 1977 the input-output table involving these four sectors was as follows (all figures are in millions of dollars).

To From	1	2	3	4	Total Output
1	174	0	0	0	2,213
2	10	286	0	0	3,147
3	9	10	2,451	7	16,646
4	11	2	0	169	4,883

Determine how these four sectors would react to an increase in demand for iron ore (Sector 1) of $1,000 million, how they would react to an increase in demand for nonferrous ore (Sector 2) of $1,000 million, and so on.

25. *Australia Input-Output Table** Two of the sectors of the Australian economy are (1) textiles, and (2) clothing and footwear. In 1977, the input-output table involving these two sectors was as follows (all figures are in millions of Australian dollars).

To	1	2	Total Output
From 1	364.9	282.5	1,971.1
2	7.6	318.0	2,226.0

Determine how both sectors would react to an increase in demand for textiles (Sector 1) of $100 million, and how both would react to an increase in demand for clothing (Sector 2) of $100 million.

26. *Australia Input-Output Table* Another two sectors of the Australian economy are (1) community services, and (2) recreation services. In 1978–79, the input-output table involving these two sectors was as follows (all figures are in millions of Australian dollars).

To	1	2	Total Output
From 1	96.6	33.9	14,841.1
2	69.9	249.5	6,188.5

Determine how both sectors would react to an increase in demand for community services (Sector 1) of $1,000 million, and how both would react to an increase in demand for recreation services (Sector 2) of $1,000 million.

27. *Australia Input-Output Table* Four of the sectors of the Australian economy are (1) agriculture, (2) forestry, fishing, hunting, (3) meat and milk products, and (4) other food products. In 1978–79, the input-output table involving these four sectors was as follows (all figures are in millions of Australian dollars).

To	1	2	3	4	Total Output
From 1	678.4	3.7	3,341.5	1,023.5	9,401.3
2	15.5	6.9	17.1	124.5	685.8
3	47.3	4.3	893.1	145.8	6,997.3
4	312.5	22.1	83.2	693.5	4,818.3

Determine how these four sectors would react to an increase in demand for agriculture (Sector 1) of $100 million, how they would react to an increase in demand for forestry (Sector 2) of $100 million, and so on.

▼ *Source for 25–28: *Australian National Accounts and Input-Output Tables, 1978–1979,* Australian Bureau of Statistics.

28. *Australia Input-Output Table* Another four sectors of the Australian economy are (1) petroleum and coal products, (2) non-metallic mineral products, (3) basic metals and products, and (4) fabricated metal products. In 1978–79, the input-output table involving these four sectors was as follows (all figures are in millions of Australian dollars).

	To	1	2	3	4	Total Output
From	1	174.1	30.5	120.3	14.2	3,278.0
	2	0	190.1	55.8	12.6	2,188.8
	3	2.1	40.2	1,418.7	1,242.0	6,541.7
	4	0.1	7.3	40.4	326.0	4,065.8

Determine how these four sectors would react to an increase in demand for petroleum products (Sector 1) of $1,000 million, how they would react to an increase in demand for non-metallic mineral products (Sector 2) of $1,000 million, and so on.

29. *Indonesia Input-Output Table** Two of the sectors of the Indonesian economy are (1) oil refinery, and (2) electricity, gas, and water supply. In 1980, the input-output table involving these two sectors was as follows (all figures are in millions of rupiahs).

	To	1	2	Total Output
From	1	1,462.9	127,914.0	2,927,501.9
	2	16,776.5	73,232.9	523,447.3

Determine how these two sectors would react to an increase in demand for refined oil (Sector 1) of 100,000 million rupiahs, and how they would react to an increase in demand for electricity (Sector 2) of 100,000 million rupiahs.

30. *Indonesia Input-Output Table* Another two sectors of the Indonesian economy are (1) transport and communications, and (2) financing, real estate, and business services. In 1980, the input-output table involving these two sectors was as follows (all figures are in millions of rupiahs).

	To	1	2	Total Output
From	1	359,487.1	47,906.1	4,051,533.4
	2	136,658.5	70,824.2	3,108,093.7

Determine how these two sectors would react to an increase in demand for transportation (Sector 1) of 100,000 million rupiahs, and how they would react to an increase in demand for financing (Sector 2) of 100,000 million rupiahs.

▼ **Source for 29–32: Input-Output Table for Indonesia, 1980, Vol. I.* Indonesian Central Bureau of Statistics.

31. *Indonesia Input-Output Table* Four sectors of the Indonesian economy are (1) paddy (rice), (2) other food crops, (3) other agricultural crops, and (4) livestock and livestock products. In 1980, the input-output table involving these four sectors was as follows (all figures are in millions of rupiahs).

To From	1	2	3	4	Total Output
1	61,432.8	1,035,167.7	0	5,940.4	3,436,221.0
2	0	133,184.3	1,316.7	9,448.0	4,471,902.0
3	0	420.7	589,918.8	10,186.3	3,643,267.5
4	235.0	1,886.6	8,809.7	1,013.3	1,929,474.8

Determine how these four sectors would react to an increase in demand for rice (Sector 1) of 100,000 million rupiahs, how they would react to an increase in demand for other food crops (Sector 2) of 100,000 million rupiahs, and so on.

32. *Indonesia Input-Output Table* Another four sectors of the Indonesian economy are (1) oil refinery, (2) electricity, gas and water supply, (3) construction, and (4) trade. In 1980, the input-output table involving these four sectors was as follows (all figures are in millions of rupiahs).

To From	1	2	3	4	Total Output
1	1,462.9	127,914.0	649,604.5	44,438.9	2,927,501.9
2	16,776.5	73,232.9	6,072.8	35,618.1	523,477.3
3	6,291.3	15,969.5	12,784.4	37,076.8	7,532,682.1
4	6,492.3	11,086.8	1,003,915.1	11,577.3	6,375,656.9

Determine how these four sectors would react to an increase in demand for oil (Sector 1) of 100,000 million rupiahs, how they would react to an increase in demand for electricity (Sector 2) of 100,000 million rupiahs, and so on.

COMMUNICATION AND REASONING EXERCISES

33. What would it mean if the total output figure in the last column of an input-output table was equal to the sum of the other figures in its row?

34. What would it mean if the total output figure in the last column of an input-output table was less than the sum of the other figures in its row?

35. What does it mean if an entry in the matrix $(I - A)^{-1}$ is zero?

36. Why do we expect the diagonal entries in the matrix $(I - A)^{-1}$ to be slightly larger than 1?

37. Why do we expect the off-diagonal entries of $(I - A)^{-1}$ to be less than 1?

38. Why do we expect all the entries of $(I - A)^{-1}$ to be nonnegative?

▶ You're the Expert

THE JAPANESE ECONOMY

Senator Glenn Russel walks into your cubicle in the Congressional Budget Office. "Look here," he says, "I don't see why the Japanese trade representative is getting so upset with my proposal to cut down on our imports of office supplies from Japan. They claim that it'll hurt their mining operations. But just look at their input-output table. Their office supply sector doesn't use any input from their mining sector. How can our cutting down demand for office supplies hurt mining?" Indeed, the senator is right about the input-output table, which you have hanging on your wall. Here is what it looks like (all figures are in 100 million ¥).*

	1	2	3	4	5	6	7	8	9	10	11	12	13
1	20,119	54	126,431	855	0	0	0	10	28	11,923	0	0	125
2	1	46	155,558	11,126	20,238	0	0	1	3	84	0	0	704
3	35,743	2,703	1,027,004	207,368	31,048	20,257	163	116,929	13,125	114,849	6,747	19,682	36,772
4	739	84	5,442	597	2,789	22,084	1,113	2,162	2,173	4,708	0	19	442
5	776	649	52,866	5,030	4,029	7,453	335	6,899	4,418	20,148	0	156	2,979
6	7,217	1,605	140,864	41,987	10,548	58,153	513	36,360	2,120	45,328	3,430	1,994	11,800
7	141	145	6,045	3,101	573	20,457	0	5,450	593	10,189	0	62	1,028
8	5,334	6,088	53,615	23,342	4,160	44,008	185	40,015	7,272	23,861	237	634	8,205
9	0	0	0	0	0	0	0	0	0	0	0	0	2,448
10	164	222	54,291	18,758	3,558	35,065	985	9,982	6,213	39,348	0	159	5,553
11	143	26	4,733	422	320	2,040	4	729	437	1,555	0	26	0
12	1,265	2	23,181	0	0	2,689	0	204	3	10	0	0	103
13	1,904	534	39,354	6,500	1,459	14,415	155	4,888	58	10,693	23	66	0

The sectors are as follows.

1	Agriculture, Forestry & Fishery
2	Mining
3	Manufacturing
4	Construction
5	Electricity, Gas & Water
6	Commerce, Finance, Insurance & Real Estate
7	Real Estate Rent
8	Transportation & Communication
9	Public Services
10	Services
11	Office Supplies
12	Packing
13	Others

The total output from each sector is given in the following table.

▼ *Source: *1980 Input-Output Tables,* Administrative Management Planning, Government of Japan.

1	2	3	4	5	6	7	8	9	10	11	12	13
161,114	206,012	2,396,528	552,574	147,505	891,706	47,783	372,545	132,752	708,326	10,437	28,949	74,178

"If I look at just the mining and office supply sectors," says the senator, "I'm looking at this input-output table."

	To	2	11	Total Output
From	2	46	0	206,012
	11	26	0	10,437

"That gives me

$$A = \begin{bmatrix} 0.000223 & 0 \\ 0.000126 & 0 \end{bmatrix},$$

so

$$(I - A)^{-1} = \begin{bmatrix} 1.00022 & 0 \\ 0.00013 & 1 \end{bmatrix}.$$

That last column tells me that any change in demand for office supplies will have no effect on demand for mining."

Now you have to explain to the senator a point that we've fudged a bit in Section 4. What he said assumes that changing the external demand for office supplies (that is, the demand from outside of these two sectors) will not change the external demand for mining. But in fact, that is unlikely to be true. Changing the demand for office supplies will change the demand for other sectors in the economy directly (the manufacturing sector, for example), which in turn may change the demand for mining. To see these indirect effects properly, you tell the senator that he must look at the whole Japanese economy. The technology matrix A is then

$$\begin{bmatrix}
0.124874 & 0.000262 & 0.052756 & 0.001547 & 0 & 0 & 0 & 0.000027 & 0.000211 & 0.016833 & 0 & 0 & 0.001685 \\
0.000006 & 0.000223 & 0.064910 & 0.020135 & 0.137202 & 0 & 0 & 0.000003 & 0.000023 & 0.000119 & 0 & 0 & 0.009491 \\
0.221849 & 0.013121 & 0.428538 & 0.375276 & 0.210488 & 0.022717 & 0.003411 & 0.313865 & 0.098869 & 0.162141 & 0.64645 & 0.679885 & 0.495726 \\
0.004587 & 0.000408 & 0.002271 & 0.001080 & 0.018908 & 0.024766 & 0.023293 & 0.005803 & 0.016369 & 0.006647 & 0 & 0.000656 & 0.005959 \\
0.004816 & 0.003150 & 0.022059 & 0.009103 & 0.027314 & 0.008358 & 0.007011 & 0.018519 & 0.033280 & 0.028445 & 0 & 0.005389 & 0.040160 \\
0.044794 & 0.007791 & 0.058778 & 0.075984 & 0.071509 & 0.065215 & 0.010736 & 0.097599 & 0.015970 & 0.063993 & 0.328638 & 0.068880 & 0.159077 \\
0.000875 & 0.000704 & 0.002522 & 0.005612 & 0.003885 & 0.022941 & 0 & 0.014629 & 0.004467 & 0.014385 & 0 & 0.002142 & 0.013859 \\
0.033107 & 0.029552 & 0.022372 & 0.042242 & 0.028202 & 0.049353 & 0.003872 & 0.107410 & 0.054779 & 0.033686 & 0.022708 & 0.021901 & 0.110612 \\
0 & 0 & 0 & 0 & 0 & 0 & 0 & 0 & 0 & 0 & 0 & 0 & 0.033002 \\
0.001018 & 0.001078 & 0.022654 & 0.033947 & 0.024121 & 0.039323 & 0.020614 & 0.026794 & 0.046802 & 0.055551 & 0 & 0.005492 & 0.074860 \\
0.000888 & 0.000126 & 0.001975 & 0.000764 & 0.002169 & 0.002288 & 0.000084 & 0.001957 & 0.003292 & 0.002195 & 0 & 0.000898 & 0 \\
0.007852 & 0.000010 & 0.009673 & 0 & 0 & 0.003016 & 0 & 0.000548 & 0.000023 & 0.000014 & 0 & 0 & 0.001389 \\
0.011818 & 0.002592 & 0.016421 & 0.011763 & 0.009891 & 0.016166 & 0.003244 & 0.013121 & 0.000437 & 0.015096 & 0.002204 & 0.002280 & 0
\end{bmatrix}$$

and $(I-A)^{-1}$ is

$$\begin{bmatrix}
1.17757 & 0.00372 & 0.11919 & 0.05231 & 0.03159 & 0.01083 & 0.00335 & 0.04690 & 0.01918 & 0.04660 & 0.08184 & 0.08352 & 0.07381 \\
0.04064 & 1.00465 & 0.13468 & 0.07833 & 0.17706 & 0.01365 & 0.00492 & 0.05583 & 0.02579 & 0.03460 & 0.09303 & 0.09524 & 0.09593 \\
0.56919 & 0.05496 & 1.95538 & 0.81463 & 0.50583 & 0.16014 & 0.04647 & 0.75513 & 0.29265 & 0.42680 & 1.33636 & 1.36653 & 1.14913 \\
0.01084 & 0.00142 & 0.01131 & 1.00937 & 0.02558 & 0.02967 & 0.02445 & 0.01547 & 0.02061 & 0.01332 & 0.01746 & 0.01106 & 0.02114 \\
0.02330 & 0.00573 & 0.05193 & 0.03502 & 1.04549 & 0.01770 & 0.00989 & 0.04494 & 0.04559 & 0.04541 & 0.04059 & 0.04366 & 0.08092 \\
0.11114 & 0.01770 & 0.15685 & 0.16048 & 0.13312 & 1.10143 & 0.02125 & 0.18883 & 0.05798 & 0.12150 & 0.46830 & 0.18927 & 0.29208 \\
0.00701 & 0.00196 & 0.01159 & 0.01440 & 0.01072 & 0.02813 & 1.00138 & 0.02499 & 0.00909 & 0.02110 & 0.01737 & 0.01277 & 0.02930 \\
0.07149 & 0.03679 & 0.07664 & 0.08961 & 0.06645 & 0.07306 & 0.01010 & 1.16247 & 0.07971 & 0.06690 & 0.10037 & 0.08393 & 0.18977 \\
0.00089 & 0.00015 & 0.00129 & 0.00103 & 0.00078 & 0.00076 & 0.00018 & 0.00109 & 1.00030 & 0.00091 & 0.00118 & 0.00104 & 0.03400 \\
0.02503 & 0.00486 & 0.06116 & 0.06898 & 0.04959 & 0.05583 & 0.02572 & 0.06405 & 0.06417 & 1.08045 & 0.05963 & 0.05348 & 0.13220 \\
0.00270 & 0.00038 & 0.00477 & 0.00322 & 0.00389 & 0.00318 & 0.00035 & 0.00451 & 0.00444 & 0.00379 & 1.00424 & 0.00452 & 0.00400 \\
0.01516 & 0.00065 & 0.02042 & 0.00887 & 0.00561 & 0.00503 & 0.00055 & 0.00893 & 0.00324 & 0.00495 & 0.01509 & 1.01454 & 0.01411 \\
0.02690 & 0.00447 & 0.03907 & 0.03130 & 0.02361 & 0.02303 & 0.00532 & 0.03311 & 0.00923 & 0.02750 & 0.03585 & 0.03157 & 1.03040
\end{bmatrix}$$

Now you tell the senator to look at the 11th column to see the effects of a change in demand for office supplies. There is indeed an effect on mining: Every 1¥ increase in demand for office supplies produces a 0.09303¥ increase in demand for mining. This also means that every 1¥ *decrease* in demand for office supplies produces a 0.09303¥ *decrease* in demand for mining. So the Japanese trade representative is right to complain that the senator's plan will hurt their mining companies.

Exercises

1. What does the (2, 11) entry in the matrix A tell you?
2. What does the (2, 11) entry in the matrix $(I - A)^{-1}$ tell you?
3. Why are none of the entries of $(I - A)^{-1}$ negative?
4. Why are the diagonal entries of $(I - A)^{-1}$ close to, and larger than 1?
5. An increase in demand for the products of which sector of the Japanese economy would have the least impact on the mining sector?
6. An increase in demand for the products of which sector of the Japanese economy would have the most impact on the mining sector?
7. Which sector of the Japanese economy has the greatest percentage of its total output available for external consumption?
8. Referring to the conclusion of "You're the Expert," try to account for most of the 0.09303¥ by looking at the technology matrix A. For example, a 1¥ increase in demand for office equipment produces directly a 0.64645¥ increase in demand for manufacturing, which in turn produces a $(0.64645)(0.06491) = 0.04196$¥ increase in demand for mining. What other two-step effects are there? Do they account for all of the 0.09303¥?

Review Exercises

In each of Exercises 1 through 8, write the dimensions and the transpose of the given matrix.

1. $A = \begin{bmatrix} 1 & 2 & 3 \\ 6 & 5 & 4 \end{bmatrix}$
2. $B = \begin{bmatrix} 3 \\ 2 \\ 1 \end{bmatrix}$
3. $C = \begin{bmatrix} 1 & 3 \\ 5 & 7 \\ 9 & 11 \end{bmatrix}$

4. $D = \begin{bmatrix} 10 & 20 & 30 & 40 \end{bmatrix}$
5. $E = \begin{bmatrix} 1 & -1 \\ -1 & 2 \end{bmatrix}$
6. $F = \begin{bmatrix} 1 & 2 & 3 \\ 2 & 4 & 6 \\ 3 & 6 & 9 \end{bmatrix}$

7. $G = \begin{bmatrix} 1 & -1 & 2 & -2 \\ -1 & 1 & -2 & 2 \end{bmatrix}$
8. $H = \begin{bmatrix} 2 & 4 & 6 \\ 6 & 8 & 10 \\ 10 & 12 & 14 \end{bmatrix}$

In Exercises 9–28, let

$$A = \begin{bmatrix} 1 & 2 & 3 \\ 4 & 5 & 6 \end{bmatrix}, B = \begin{bmatrix} 1 & -1 \\ 0 & 1 \end{bmatrix}, C = \begin{bmatrix} -1 & 0 \\ 1 & 1 \\ 0 & 1 \end{bmatrix} \text{ and } D = \begin{bmatrix} -3 & -2 & -1 \\ 1 & 2 & 3 \end{bmatrix}.$$

For each of the following, determine whether the expression is defined, and if it is, evaluate it.

9. $A + B$
10. $A + C$
11. $A - D$
12. $A - C^T$
13. $2A^T + C$
14. $C^T + 3D$
15. AB
16. BA
17. A^TB
18. BA^T
19. BC
20. CB
21. BC^T
22. C^TB
23. A^2
24. B^2
25. AA^T
26. D^TD
27. BB^T
28. B^TB

For each of the matrices in Exercises 29 through 40, find the inverse or determine that the matrix is singular.

29. $\begin{bmatrix} 1 & -1 \\ 0 & 1 \end{bmatrix}$
30. $\begin{bmatrix} 1 & -1 \\ -1 & 1 \end{bmatrix}$
31. $\begin{bmatrix} 1 & 2 \\ 0 & 0 \end{bmatrix}$

32. $\begin{bmatrix} 1 & 2 \\ 3 & 4 \end{bmatrix}$
33. $\begin{bmatrix} 1 & 2 & 3 \\ 0 & 4 & 1 \\ 0 & 0 & 1 \end{bmatrix}$
34. $\begin{bmatrix} 1 & 2 & 3 \\ 1 & 3 & 2 \\ 0 & 1 & 1 \end{bmatrix}$

35. $\begin{bmatrix} 1 & 2 & 3 \\ 1 & 3 & 2 \\ 1 & 1 & 4 \end{bmatrix}$
36. $\begin{bmatrix} 1 & 2 & 3 \\ 4 & 5 & 6 \\ 7 & 8 & 9 \end{bmatrix}$
37. $\begin{bmatrix} 1 & 2 & 3 & 4 \\ 0 & 1 & 2 & 3 \\ 0 & 0 & 1 & 2 \\ 0 & 0 & 0 & 1 \end{bmatrix}$

38. $\begin{bmatrix} 1 & 2 & 3 & 4 \\ 1 & 3 & 4 & 2 \\ 0 & 1 & 2 & 3 \\ 0 & 0 & 1 & 2 \end{bmatrix}$
39. $\begin{bmatrix} 1 & 2 & 3 & 4 \\ 2 & 3 & 3 & 3 \\ 0 & 1 & 2 & 3 \\ 0 & 0 & 1 & 2 \end{bmatrix}$
40. $\begin{bmatrix} 1 & 2 & 3 & 4 \\ 1 & 1 & 2 & 3 \\ 1 & 1 & 1 & 2 \\ 1 & 4 & 6 & 7 \end{bmatrix}$

In Exercises 41–50, write each system of linear equations as a matrix equation, and solve by inverting the coefficient matrix.

41. $x + 2y = 0$
    $3x + 4y = 2$

42. $x + y = 6$
    $x - y = 2$

43. $x + y + z = 3$
    $\phantom{x + {}}y + 2z = 4$
    $\phantom{x + {}}y - z = 1$

44. $x + y + z = 2$
    $x + 2y + z = 6$
    $\phantom{x + {}}y + z = 1$

45. $x + y + z = 2$
    $x + 2y + z = 3$
    $x + y + 2z = 1$

46. $x + y + z = 0$
    $x + 2y + z = -1$
    $x + 2y + 3z = -1$

47. $x + y = 0$
    $\phantom{x + {}}y + z = 1$
    $\phantom{x + y + {}}z + w = 0$
    $x - w = 3$

48. $x - y = 0$
    $\phantom{x + {}}y - z = 1$
    $\phantom{x + y + {}}z - w = 0$
    $x + w = 3$

49. $x + 2y + 3z + 4w = 0$
    $x + 3y + 4z + 2w = 3$
    $\phantom{x + {}}y + 2z + 3w = -1$
    $\phantom{x + y + {}}z + 2w = -1$

50. $x + y + z + w = 2$
    $x - y + z - w = 0$
    $x + y - z - w = -4$
    $x + 2y + 3z + 5w = 10$

APPLICATIONS

51. **Germany Input-Output Table*** Two of the sectors of the former West German economy were (1) iron and steel, and (2) road vehicles. In 1980, the input-output table involving these two sectors was as follows (all figures are in millions of deutschesmarks).

To From	1	2	Total Output
1	63,080	4,956	112,120
2	73	24,475	143,955

    Determine how these two sectors would react to an increase in demand for steel (Sector 1) of DM10,000 million, and how they would react to an increase in demand for vehicles (Sector 2) of DM10,000 million.

▼ *Source for Exercises 51–54: *Input-Output Tabellen 1980,* Statistisches Bundesamt, Wiesbaden, Federal Republic of Germany.

52. *Germany Input-Output Table** Another two sectors of the former West German economy were (1) textiles, and (2) apparel. In 1980, the input-output table involving these two sectors was as follows (all figures are in millions of deutschesmarks).

	To	1	2	Total Output
From	1	9,123	6,596	48,820
	2	6	879	33,326

Determine how these two sectors would react to an increase in demand for textiles (Sector 1) of DM10,000 million, and how they would react to an increase in demand for apparel (Sector 2) of DM10,000 million.

53. *Germany Input-Output Table** Four of the sectors of the former West German economy were (1) glass and glass products, (2) iron and steel, (3) road vehicles, and (4) aircraft and space vehicles. In 1980, the input-output table involving these four sectors was as follows (all figures are in millions of deutschesmarks).

	To	1	2	3	4	Total Output
From	1	1,377	58	886	3	11,172
	2	2	63,080	4,956	77	112,120
	3	72	73	24,475	3	143,955
	4	0	0	0	1,604	13,569

Determine how these four sectors would react to an increase in demand for glass (Sector 1) of DM10,000 million, how they would react to an increase in demand for steel (Sector 2) of DM10,000 million, and so on.

54. *Germany Input-Output Table** Another four sectors of the former West German economy were (1) textiles, (2) apparel, (3) wholesaling services and recycling, and (4) retailing services. In 1980, the input-output table involving these four sectors was as follows (all figures are in millions of deutschesmarks).

	To	1	2	3	4	Total Output
From	1	9,123	6,596	191	43	48,820
	2	6	879	1	24	33,326
	3	796	1,417	2,683	529	117,933
	4	227	88	62	63	101,522

Determine how these four sectors would react to an increase in demand for textiles (Sector 1) of DM10,000 million, how they would react to an increase in demand for apparel (Sector 2) of DM10,000 million, and so on.

▼ **Source for Exercises 51–54: Input-Output Tabellen 1980, Statistisches Bundesamt, Wiesbaden, Federal Republic of Germany.*

CHAPTER 4

Mechanic Strike Snarls USAir Flights

By Kenneth C. Crowe

Tens of thousands of USAir passengers yesterday found the normal stress of air travel turned into an agony of canceled flights as union mechanics struck the nation's sixth largest airline in a contract dispute over wages and work rules.

As picket lines began forming at LaGuardia, Kennedy and Newark airports at 7 a.m. yesterday, USAir announced a contingency plan to operate 60 percent of its normal schedule of 2,600 daily flights throughout the strike. The airline usually flies from 160,000 to 180,000 passengers daily. The company is continuing to maintain its planes using nonunion personnel.

Under the strike contingency, USAir is operating 85 of its normal 143 jet flights out of LaGuardia; three instead of seven flights out of Kennedy; 38 instead of 67 out of Newark, and two instead of four out of Long Island-MacArthur Airport. The USAir Shuttle from LaGuardia to Washington and Boston and USAir Express flights are not affected by the strike. The shuttle is covered by other union contracts, while the 12 different companies operating the Express are under different contracts or are nonunion.

Several airlines, including Delta and Continental announced that they would accept USAir tickets to the destinations they service. USAir spokesman Dave Shipley said some passengers from canceled flights were directed onto USAir Express flights or to other uncanceled USAir flights going to the same airport.

But the contingency plan could be tossed awry by the Association of Flight Attendants' decision yesterday ordering the 9,000 flight attendants to honor the picket lines of the International Association of Machinists immediately or as soon as they completed their most recent flight, which could take up to four days, a union spokesman said.

The company declined to speculate on the impact on the schedule if the bulk of the flight attendants fail to report to work. The USAir unit of the Airline Pilots Association said its members would continue to fly.

Negotiations resumed yesterday afternoon in Washington, D.C., and the union expressed optimism that the dispute could be resolved soon.

USAir, which had a strike-free history of relatively good relations with its workers, appeared on the verge of resolving the contract dispute as the negotiators went into a night-long bargaining session Sunday. The talks broke off shortly before 7 a.m.

The sticking points in the talks on the contract covering the 8,300 machinists, aircraft cleaners and clerks include company demands that the workers pay 10 percent of the cost of their health insurance when the company moved into the black. USAir lost $675 million over the past two years.

Under the pay reduction formula there would be no reduction in the first $20,000 of wages; 10 percent of the next $30,000; 15 percent of the next $50,000; and 30 percent for everything in excess of $100,000. The company says the average machinist, who earns $38,000 a year, hasn't had a raise in two years.

USAir spokesman Buckley said that all employees, union and nonunion, are being asked to sacrifice.

Source: From Kenneth C. Crowe, "Mechanic Strike Snarls USAir Flights," *New York Newsday*, October 6, 1992, p. 5.

Linear Programming

APPLICATION ▶ Fly-by-Night Airlines flies airplanes between 5 cities: Los Angeles, Chicago, Atlanta, New York, and Boston. Because of a strike, the airline has not had sufficient crews to fly its planes, so some of them have been stranded in Chicago and in Atlanta. In fact, it has 15 extra planes in Chicago and 30 extra planes in Atlanta. Los Angeles needs at least 10 planes to get back on schedule, New York needs at least 20, and Boston needs at least 10. It costs $50,000 to fly a plane from Chicago to Los Angeles, $10,000 to fly from Chicago to New York, and $20,000 to fly from Chicago to Boston. It costs $70,000 to fly a plane from Atlanta to Los Angeles, $20,000 to fly from Atlanta to New York, and $50,000 to fly from Atlanta to Boston. How should Fly-by-Night rearrange its planes to get back on schedule at the least cost?

SECTIONS

1 Graphing Linear Inequalities

2 Solving Linear Programming Problems Graphically

3 Solving Standard Linear Programming Problems Using the Simplex Method

4 Solving Nonstandard Linear Programming Problems

You're the Expert

Airline Scheduling

INTRODUCTION ▶ In this chapter, we begin to look at one of the most important types of problems for business and the sciences: the problem of finding the largest or smallest possible value of some quantity (such as profit or cost) under certain constraints (for example, limited resources). We call such problems **optimization** problems because we are trying to find the best, or optimum, value. The optimization problems we look at in this chapter involve linear functions only and are known as **linear programming** (LP) problems. One of the main purposes of calculus, which you may study later, is to solve nonlinear optimization problems.

When a linear programming problem involves only two unknowns, it can often be solved by a graphical method that we discuss in Sections 1 and 2. When there are three or more unknowns, we must use a more algebraic method, as we had to do for systems of linear equations. The method we use is called the **simplex method.** It was invented in 1947 by George Dantzig, and is still the most commonly used method to solve linear programming problems in real applications, from finance to the computation of trajectories for guided missiles. Every student of business is eventually exposed to the simplex method in some form.

When the numbers are fairly small and the unknowns are few, the calculations can be done by hand, as we shall do in this chapter. On the other hand, practical problems often involve large numbers and many unknowns. Problems such as the routing of telephone calls or airplane flights, or the allocation of resources in a manufacturing process, can involve tens of thousands of unknowns. Solving such problems by hand is obviously impractical, and so computers are regularly used. While computer programs most often use the simplex method, mathematicians are always seeking faster methods. One example of a new method is **Karmarkar's algorithm,** invented by N. Karmarkar in 1984. This method caused considerable excitement when first published, but it has not replaced the simplex method as the method of choice in most applications.

Much of what we said about calculators and computers while studying systems of linear equations applies to this chapter as well. In particular, it is very useful to have some technological help in doing the simplex method. In practice, there are computer programs that do most of the work for you, so again you can think of what we teach you here as a peek inside a "black box." What the computers cannot do for you is convert a real situation into a mathematical problem, so the most important thing for you to get out of this chapter is how to recognize and set up a linear programming problem, and how to interpret the results.

4.1 GRAPHING LINEAR INEQUALITIES

By the end of the next section we will be solving linear programming problems with two unknowns. We shall work toward that goal gradually, beginning with a review of the algebra of inequalities. We shall use inequalities to describe the constraints in a problem, such as limitations on resources. First, we recollect the basic notation.

> **INEQUALITIES**
>
> $a \leq b$ means that a **is less than or equal to** b. For example, $3 \leq 99$ and $-2 \leq -2$. $a \geq b$ means that a **is greater than or equal to** b. For example, $\frac{3}{4} \geq \frac{1}{2}$ and $-55 \geq -55$.

(There are, of course, also the inequalities $<$ and $>$, called **strict** inequalities since they do not permit equality. We shall not use them in this chapter.)

Here are some of the basic rules of inequalities. Although we illustrate all of them with the inequality $\leq$, they apply equally well to inequalities with $\geq$.

> **RULES FOR MANIPULATING INEQUALITIES**
>
> **(a)** We can add the same quantity to both sides of an inequality.
>
> If $x \leq y$, then $x + a \leq y + a$ for any real number a.
>
> Example: Since $2 \leq 6$, $2 + (-5) \leq 6 + (-5)$.
>
> **(b)** We can multiply both sides of an inequality by a positive constant.
>
> If $x \leq y$ and a is positive, then $ax \leq ay$.
>
> Example: Since $\frac{1}{2} \leq 90$, $3(\frac{1}{2}) \leq 3(90)$.
>
> **(c)** We can multiply both sides of an inequality by a negative constant if the inequality is *reversed*.
>
> If $x \leq y$ and a is negative, then $ax \geq ay$.
>
> Examples: Since $-4 \leq 2$, $4 \geq -2$.
>
> Since $\frac{1}{2} \leq 90$, $-3(\frac{1}{2}) \geq -3(90)$.

An **inequality in the unknown** x is a statement that one expression involving x is less than or equal to (or greater than or equal to) another. For example,

$$2x + 8 \geq 89$$
$$x^2 - 3x - 1 \leq 0$$
$$2x^3 \leq x^3 - 1.$$

Similarly, we can have inequalities in x and y, such as

$$3x - 2y \geq 8$$
$$x^2 + y^2 \leq 19$$
$$xy \leq \frac{x^2}{y}.$$

To **solve** an inequality in the unknown x is to find all values of x that make the inequality true.* Similarly, to solve an inequality in the unknowns x and y is to find all pairs of values of x and y that make the inequality true. We could now go on to describe inequalities in any number of variables, such as $3x - y + 4z \leq 0$ and so on, but you probably have the idea.

In this chapter, we shall use only *linear* inequalities in two or more unknowns—that is, inequalities of the form

$$ax + by \leq c, \quad a, b, \text{ and } c \text{ real constants}$$
$$(\text{or } ax + by \geq c),$$
$$ax + by + cz \leq d, \quad a, b, c, \text{ and } d \text{ real constants,}$$
$$ax + by + cz + dw \leq e, \quad a, b, c, d, \text{ and } e \text{ real constants,}$$

and so on.

Our first goal is to solve linear inequalities in two variables: in other words, inequalities of the form $ax + by \leq c$.

▼ **EXAMPLE 1**

Solve the inequality $2x + 3y \leq 6$.

SOLUTION We already know how to solve the *equation* $2x + 3y = 6$. As we saw in Chapter 1, the solution of this equation may be pictured as the set of all points (x, y) on the straight-line graph of the equation. This straight line has x-intercept 3 (found by putting $y = 0$ in the equation) and y-intercept 2 (found by putting $x = 0$ in the equation) and is shown in Figure 1.

Notice that if (x, y) is any point on the line, then x and y not only satisfy the *equation* $2x + 3y = 6$, but they also satisfy the *inequality* $2x + 3y \leq 6$, since being equal to 6 qualifies as being less than or equal to 6.

▼ *Compare this with solving an equation in x.

4.1 Graphing Linear Inequalities **243**

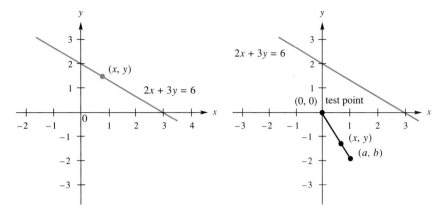

FIGURE 1   **FIGURE 2**

Q Do the points on the line give all possible solutions to the inequality?

A No. For example, try the origin $(0, 0)$ as a "test point." Since $2(0) + 3(0) = 0 \leq 6$, it follows that the point $(0, 0)$ also represents a solution.

Q So how do we get all the other solutions?

A Here is a surprising fact: If a single test point on one side of the line $2x + 3y = 6$ represents a solution of $2x + 3y \leq 6$, then so does every point on that side of the line. In other words, since $(0, 0)$ represents a solution and is not on the line, it follows that every point on the same side of the line as $(0, 0)$ represents a solution as well. There is, of course, nothing special about $2x + 3y \leq 6$. This principle is true for any linear inequality.

Thus, we get a whole **region** full of solutions.

Q Why is this so?

A Well, suppose we have any point (a, b) on the same side of the line as the test point $(0, 0)$. Then we can join it to $(0, 0)$ with a straight line segment not crossing the line $2x + 3y \leq 6$, as in Figure 2. If (a, b) *failed* to satisfy the given inequality, then we would have $2a + 3b > 6$. Now consider the number $2x + 3y$ for points (x, y) along the line segment. As we slide the point (x, y) from the origin down to (a, b), the quantity $2x + 3y$ changes from being < 6 at $(0, 0)$ to being > 6 at (a, b). Thus there must be some point (x, y) along the line where it is *exactly* 6 (how could it change from being < 6 to being > 6 without ever being exactly 6?).* At

▼ * We are using a very important result here from mathematical analysis, called the "intermediate value theorem," which simply states: If the value of a quantity (for example, the temperature) changes continuously along a line, and it is $< c$ at one point and $> c$ at another, then it has to be *exactly c* at some point in between. For example, if it was 0 degrees at midnight last night and is now 8 degrees at 7 A.M., there has to have been a time in between midnight and 7 A.M. when the temperature was *exactly* 6 degrees!

that point, we would have $2x + 3y$ *equals* 6. But wait a minute! The only points where $2x + 3y = 6$ lie on the original line, and the line segment that we drew never touches that line. In other words, our assumption that (a, b) *failed* to satisfy the inequality must have been wrong. Thus, (a, b) must have satisfied the inequality to begin with, as claimed.

Q What about points on the other side of the line?

A The answer is the same: *If a single point not on the line $2x + 3y = 6$ fails to represent a solution of $2x + 3y \leq 6$, then so does every point on that side of the line.* Again, this is true of any linear inequality.

It follows that no points on the other side of the line are included in the solution, since, for example, the test point $(100, 100)$ (which lies well to the right of the line) fails to satisfy the given inequality.

Now we are going to do something that will appear backwards at first. To draw the region of solutions of $2x + 3y \leq 6$ we are going to *shade the part that we do not want*. Think of covering over the unwanted points, leaving those that you do want in full view. The result is Figure 3.

The reason we do this will become clear soon.

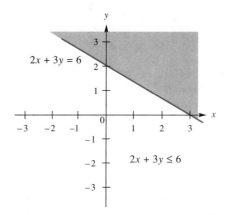

FIGURE 3

A few graphing calculators (and computer programs) can graph inequalities, but most cannot. Even those that do will probably shade the region that you *do* want, rather than the part you do not want, as we have suggested. You might be better off just graphing the line and using test points yourself to decide what the shading should be. The only advantage to using a calculator or computer program comes when you have many lines to graph (as we shall have shortly), when the exact placement of the lines becomes crucial.

Summarizing, we have the following procedure.

> **HOW TO SKETCH THE REGION REPRESENTED BY A LINEAR INEQUALITY IN TWO VARIABLES**
>
> 1. Sketch the straight line obtained by replacing the inequality with an equality.
> 2. Choose a test point not on the line. The origin, (0, 0), is a good choice if the line does not pass through it, and if the line does pass through the origin a point on one of the axes would be a good choice.
> 3. If the test point satisfies the inequality, then the set of solutions is the entire region on the same side of the line as the test point. Otherwise it is the region on the other side of the line. In either case, shade out the side that does *not* contain the solutions, leaving the solution region showing.

▼ **EXAMPLE 2**

Graph the region $3x - 2y \leq 6$.

SOLUTION The boundary line has x-intercept 2 and y-intercept -3. We again use (0, 0) as a test point (since it is not on the line) and see that it is a solution. The graph is shown in Figure 4.

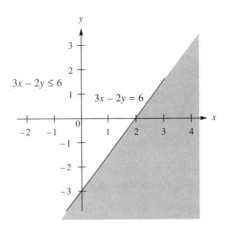

FIGURE 4

EXAMPLE 3

Sketch the regions $x \leq -1$ and $y \geq 0$.

SOLUTION The first region has as boundary the vertical line $x = -1$. Also, the test point $(0, 0)$ is not in the region, so the graph looks like Figure 5.

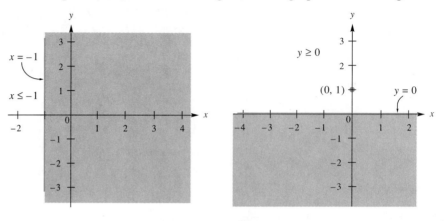

FIGURE 5   **FIGURE 6**

The second region has as boundary the horizontal line $y = 0$ (i.e., the x-axis). We cannot use $(0, 0)$ for the test point, since it lies on the line $y = 0$, so we use $(0, 1)$ instead. Since $1 \geq 0$, this point is in the region, giving us the picture in Figure 6.

EXAMPLE 4

Sketch the regions $x \geq 3y$ and $4x - y \leq 0$.

SOLUTION The first region has as boundary the line $x = 3y$, or, solving for y,

$$y = \frac{1}{3}x.$$

This line passes through the origin with slope $\frac{1}{3}$, so we cannot use the origin as a test point. The next simplest type of point we could test would be a point on one of the axes, such as $(1, 0)$. Substituting these coordinates in the inequality gives $1 \geq 3(0)$, which is true, so $(1, 0)$ is inside the region, as shown in Figure 7.

The second region has boundary line $4x - y = 0$, which we solve for y to get

$$y = 4x.$$

This line has slope 4 and also passes through the origin, so again we can't use $(0, 0)$ as a test point. Let us use $(1, 0)$ once again. Substituting in the

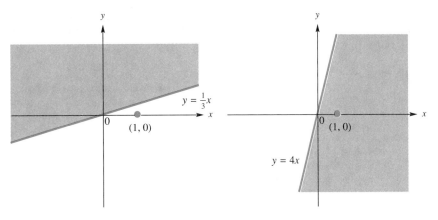

FIGURE 7 **FIGURE 8**

inequality $4x - y \leq 0$ gives $4 - 0 \leq 0$, which is false. Thus, $(1, 0)$ is not in the region, and we get Figure 8.

EXAMPLE 5

Sketch the region of points that satisfy both of the following inequalities.

$$2x - 5y \leq 10$$
$$x + 2y \leq 8$$

SOLUTION Each of the two inequalities corresponds to a region like those above. If a point is to satisfy *both* inequalities, it must lie in both sets of solutions. Put another way, if we cover over the points that are not solutions to $2x - 5y \leq 10$, and then cover over the additional points that are not solutions to $x + 2y \leq 8$, the points that remain uncovered must be solutions to both inequalities, and these are the points that we want. The result is shown in Figure 9, where the unshaded region is the set of solutions.

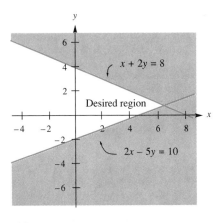

FIGURE 9

Before we go on... As a check on what we've done, we can look at points in various regions in Figure 9. For example, our graph shows that $(0, 0)$ should satisfy both inequalities, and it does.

$$2(0) - 5(0) = 0 \leq 10 \quad ✔$$
$$0 + 2(0) = 0 \leq 8 \quad ✔$$

On the other hand, $(0, 5)$ should fail to satisfy one of the inequalities.

$$2(0) - 5(5) = -25 \leq 10 \quad ✔$$
$$0 + 2(5) = 10 > 8 \quad ✘$$

One more: $(5, -1)$ should fail one of the inequalities.

$$2(5) - 5(-1) = 15 > 10 \quad ✘$$
$$5 + 2(-1) = 3 \leq 8 \quad ✔$$

 Although these graphs are quite easy to do by hand, the more lines we have to graph, the more difficult it becomes to get everything in the right place, and a graphing calculator or computer program can become important. This is especially true when, for instance, three or more lines intersect in points that are very close together and hard to distinguish in hand-drawn graphs.

EXAMPLE 6

Sketch the region of solutions of the following system of inequalities, and list the coordinates of all the corner points.

$$3x - 2y \leq 6$$
$$x + y \geq -5$$
$$y \leq 4$$

SOLUTION Shading the points that we do not want leaves us with the triangle shown in Figure 10. We label the corner points A, B, and C.

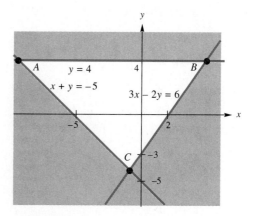

FIGURE 10

4.1 Graphing Linear Inequalities

What remains is to find the coordinates of the points *A*, *B*, and *C*. To do this, we notice that each corner point lies at the intersection of two of the bounding lines. So, to find the coordinates of that point, we need to solve the system of equations given by those two lines. To do this systematically, we make the following table.

Point	Lines Through Point	Coordinates
A	$y = 4$ $x + y = -5$	$(-9, 4)$
B	$y = 4$ $3x - 2y = 6$	$(\frac{14}{3}, 4)$
C	$x + y = -5$ $3x - 2y = 6$	$(-\frac{4}{5}, -\frac{21}{5})$

Here, we have solved the system of equations in the middle column to get the points on the right, using the techniques of Chapter 2. You should do this for practice.

Before we go on... As a partial check on our work, let's make sure that these three points satisfy all of the inequalities.

$$3(-9) - 2(4) = -35 \leq 6 \quad \checkmark$$
$$-9 + 4 = -5 \geq -5 \quad \checkmark$$
$$4 \leq 4 \quad \checkmark$$

$$3(\tfrac{14}{3}) - 2(4) = 6 \leq 6 \quad \checkmark$$
$$\tfrac{14}{3} + 4 = \tfrac{26}{3} \geq -5 \quad \checkmark$$
$$4 \leq 4 \quad \checkmark$$

$$3(-\tfrac{4}{5}) - 2(-\tfrac{21}{5}) = 6 \leq 6 \quad \checkmark$$
$$-\tfrac{4}{5} + (-\tfrac{21}{5}) = -5 \geq -5 \quad \checkmark$$
$$-\tfrac{21}{5} \leq 4 \quad \checkmark$$

As we check these, we are also checking that each of these points makes two of the inequalities into equations but satisfies the third strictly.

Using the trace feature makes it easy to locate corner points graphically. Remember to zoom in for additional accuracy when appropriate. Of course, you can also use a calculator or computer program to help solve the systems of equations, as we talked about in Chapter 2.

EXAMPLE 7 Resource Allocation

The Hi Profile cable TV company broadcasts two types of news programs: Plainfolks News and Highbrow News. Each program requires production and scripting, and the following table lists the number of person-hours needed per program, as well as the total number of person-hours available per week.

	Plainfolks	Highbrow	Total Available
Production	2	1	20
Scripting	10	20	220

Use a graph to show the possible numbers of each type of program Hi Profile can produce per week. This region is called the **feasible region.**

SOLUTION   As we did in Chapter 2, we start by identifying the unknowns: Let x be the number of Plainfolks News programs produced per week, and let y be the number of Highbrow News programs produced per week.

Now, because of our experience with systems of linear equations, we are tempted to say that for production hours, $2x + y = 20$, and for scripting hours, $10x + 20y = 220$. However, who says that Hi Profile has to use up all the available hours? The company might choose to use less than the total available hours if this proves more profitable. Thus, $2x + y$ can be anything *up to a total of* 20. In other words,

$$2x + y \leq 20,$$

and similarly,

$$10x + 20y \leq 220.$$

There are two more restrictions not explicitly mentioned: neither x nor y can be negative (Hi Profile cannot produce a negative *number* of programs*). Thus we have the additional restrictions

$$x \geq 0, y \geq 0.$$

These two inequalities tell us that we are restricted to the first quadrant, since in all the other quadrants either x or y, or both x and y, are negative. So instead of blocking out all the other quadrants, we'll simply restrict our drawing to the first quadrant.

The feasible region shown in Figure 11 is thus a graphical representation of the limitations the company faces.

Before we go on... Every point in the feasible region represents a value for x and a value for y that do not violate any of the company's restrictions. For example, the point (5, 6) lies well inside the region, and this says that

---

▼ *They have, however, been accused of producing *negative* programs!

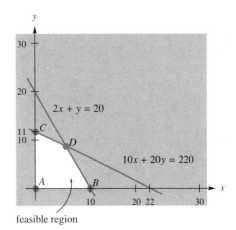

FIGURE 11

Hi Profile can produce 5 Plainfolks programs and 6 Highbrow programs without exceeding its limitations in person-hours. The corner points A, B, C, and D are significant if the company wishes to realize the greatest profit, as we shall see in the next section. We can find the corners as in the following table.

Point	Lines Through Point	Coordinates
A		(0, 0)
B		(10, 0)
C		(0, 11)
D	$2x + y = 20$ $10x + 20y = 220$	(6, 8)

(We have not listed the lines through the first three corners, since they can be read easily from the graph.) Points on the segment DB represent full use of production hours (since the segment lies on the line $2x + y = 20$), and similarly, points on the segment CD represent full use of scripting hours. Note that the point D is the only solution that uses all of the available hours.

▶ 4.1 EXERCISES

Sketch the regions corresponding to the inequalities in Exercises 1–26, and find the coordinates of all corner points (if any).

1. $2x + y \leq 10$ **2.** $4x - y \leq 12$ **3.** $-x - 2y \leq 8$

4. $-x + 2y \geq 4$ **5.** $3x + 2y \geq 5$ **6.** $2x - 3y \leq 7$

7. $x \leq 3y$

8. $y \geq 3x$

9. $\frac{3}{4}x - \frac{1}{4}y \leq 1$

10. $\frac{1}{3}x + \frac{2}{3}y \geq 2$

11. $x \geq -5$

12. $y \leq -4$

13. $4x - y \leq 8$
    $x + 2y \leq 2$

14. $2x + y \leq 4$
    $x - 2y \geq 2$

15. $3x + 2y \geq 6$
    $3x - 2y \leq 6$
    $x \geq 0$

16. $3x + 2y \leq 6$
    $3x - 2y \geq 6$
    $-y \geq 2$

17. $x + y \geq 5$
    $x \leq 10$
    $y \leq 8$
    $x \geq 0, y \geq 0$

18. $2x + 4y \geq 12$
    $x \leq 5$
    $y \leq 3$
    $x \geq 0, y \geq 0$

19. $20x + 10y \leq 100$
    $10x + 20y \leq 100$
    $10x + 10y \leq 60$
    $x \geq 0, y \geq 0$

20. $30x + 20y \leq 600$
    $10x + 40y \leq 400$
    $20x + 30y \leq 450$
    $x \geq 0, y \geq 0$

21. $20x + 10y \geq 100$
    $10x + 20y \geq 100$
    $10x + 10y \geq 80$
    $x \geq 0, y \geq 0$

22. $30x + 20y \geq 600$
    $10x + 40y \geq 400$
    $20x + 30y \geq 600$
    $x \geq 0, y \geq 0$

23. $-3x + 2y \leq 5$
    $3x - 2y \leq 6$
    $x \leq 2y$
    $x \geq 0, y \geq 0.$

24. $-3x + 2y \leq 5$
    $3x - 2y \geq 6$
    $y \leq \frac{1}{2}x$
    $x \geq 0, y \geq 0$

25. $2x - y \geq 0$
    $x - 3y \leq 0$
    $x \geq 0, y \geq 0$

26. $-x + y \geq 0$
    $4x - 3y \geq 0$
    $x \geq 0, y \geq 0$

In Exercises 27–32, we suggest you use a graphing calculator or computer program. Graph the regions corresponding to the inequalities, and find the coordinates of all corner points (if any) to two decimal places.

27. $2.1x - 4.3y \geq 9.7$

28. $-4.3x + 4.6y \geq 7.1$

29. $-0.2x + 0.7y \geq 3.3$
    $1.1x + 3.4y \geq 0$

30. $0.2x + 0.3y \geq 7.2$
    $2.5x - 6.7y \leq 0$

31. $4.1x - 4.3y \leq 4.4$
    $7.5x - 4.4y \leq 5.7$
    $4.3x + 8.5y \leq 10$

32. $2.3x - 2.4y \leq 2.5$
    $4.0x - 5.1y \leq 4.4$
    $6.1x + 6.7y \leq 9.6$

APPLICATIONS

33. *Resource Allocation* You manage an ice cream factory that makes two flavors: Creamy Vanilla and Continental Mocha. Into each quart of Creamy Vanilla go two eggs and three cups of cream. Into each quart of Continental Mocha go one egg and three cups of cream. You have in stock 500 eggs and 900 cups of cream. Draw the feasible region showing the number of quarts of vanilla and number of quarts of chocolate that can be produced. Find the corner points of the region.

34. *Resource Allocation* Podunk Institute of Technology's Math Department offers two courses: Finite Math and Calculus. Each section of Finite Math has 60 students, while each section of Calculus has 50. The department is allowed to offer a total of up to 110 sections. Further, there are no more than 6,000 students who would like to take a math course. Draw the feasible region showing the number of sections of each class that can be offered. Find the corner points of the region.

35. *Nutrition* Ruff, Inc., makes dog food out of chicken and grain. Chicken has 10 grams of protein and 5 grams of fat per ounce, while grain has 2 grams of protein and 2 grams of fat per ounce. A bag of dog food must contain at least 200 grams of protein and at least 150 grams of fat. Draw the feasible region showing the number of ounces of chicken and number of ounces of grain the company can mix into each bag of dog food. Find the corner points of the region.

36. *Purchasing* The Enormous State University's Business School is buying computers. There are two models to choose from, the Pomegranate and the Ami. Each Pomegranate comes with 4 MB of memory and 80 MB of disk space, while each Ami has 3 MB of memory and 100 MB of disk space. For reasons related to its accreditation, the school would like to be able to say that it has a total of at least 480 MB of memory and at least 12,800 MB of disk space. Draw the feasible region showing the number of each kind of computer the school can buy. Find the corner points of the region.

37. *Nutrition* Each serving of *Gerber* Mixed Cereal for Baby contains 60 calories and 11 grams of carbohydrates. Each serving of *Gerber* Mango Tropical Fruit Dessert contains 80 calories and 21 grams of carbohydrates.* You want to provide your child with at least 140 calories and at least 32 grams of carbohydrates. Draw the feasible region showing the number of servings of cereal and number of servings of dessert that you can give your child. Find the corner points of the region.

38. *Nutrition* Each serving of *Gerber* Mixed Cereal for Baby contains 60 calories, 11 grams of carbohydrates, and no Vitamin C. Each serving of *Gerber* Apple Banana Juice contains 60 calories, 15 grams of carbohydrates, and 120% of the U.S. Recommended Daily Allowance (RDA) of Vitamin C for infants.* You want to provide your child with at least 120 calories, at least 26 grams of carbohydrates, and at least 50% of the U.S. RDA of Vitamin C for infants. Draw the feasible region showing the number of servings of cereal and number of servings of juice that you can give your child. Find the corner points of the region.

39. *Investments* Two of the municipal bond funds traded by *Merrill Lynch* are the Muniyield Fund and the *Merrill Lynch* Municipal Bond Fund Class B. Each unit of the Muniyield Fund costs $15 and yields an annual return of 6%, while each unit of the *Merrill Lynch* Municipal Bond Fund Class B costs $12 and yields an annual return of 5%.† You would like to invest a total of up to $42,000 in these funds and would like to earn at least $2,400 in interest in the coming year. Draw the feasible region showing how many units of each fund you can buy. Find the corner points of the region.

40. *Investments* Two of the mutual funds offered by *Prudential Securities* are the Weingarten Fund and the Fidelity Investment High Income Municipal Portfolio. Each unit of the Weingarten Fund costs $17 and yields an estimated annual return (appreciation plus dividends) of 8%, and each unit of the municipal portfolio costs $12 and yields an estimated annual return of 5%.‡ You have up to $58,000 to invest in these funds, and your goal is to appreciate your capital by at least $3,920 in the coming year. Draw the feasible region showing how many units of each fund you can buy. Find the corner points of the region. (Round each coordinate to the nearest whole number.)

COMMUNICATION AND REASONING EXERCISES

41. Find a system of inequalities whose solution set is unbounded (extends infinitely in some direction in the plane).

42. Find a system of inequalities whose solution set is empty.

▼ *Source: nutrition information printed on the package.
† Both quoted as of May 1993 (yields rounded to the nearest percentage point).
‡ The yield for the Weingarten Fund was estimated by a broker. The yield for the municipal fund is quoted as of May 1993 and rounded to the nearest percentage point.

43. Create an interesting scenario leading to the system of inequalities
$$20x + 40y \leq 1,000$$
$$30x + 20y \leq 1,200$$
$$x \geq 0, y \geq 0.$$

44. Create an interesting scenario leading to the system of inequalities
$$20x + 40y \geq 1,000$$
$$30x + 20y \geq 1,200$$
$$x \geq 0, y \geq 0$$

45. How would you use linear inequalities to describe the triangle with corner points (0, 0), (2, 0), and (0, 1)?

46. Explain the advantage of shading the region containing points that do not satisfy the given inequalities. Illustrate with an example.

47. You are setting up a system of inequalities in the unknowns x and y. The inequalities represent constraints faced by Fly-by-Night Airlines, while x represents the number of first-class tickets it should issue for a specific flight and y represents the number of business-class tickets it should issue for that flight. You find that the feasible region is empty. How do you interpret this?

48. In the situation described in the preceding exercise, is it possible instead for the feasible region to be unbounded (to extend infinitely in some direction in the plane)? Explain your answer.

▶ 4.2 SOLVING LINEAR PROGRAMMING PROBLEMS GRAPHICALLY

As we saw in the last example in Section 1, some scenarios lead to a restriction of possibilities described by a system of linear inequalities. In that example, it would be natural to go on and ask which of the various possibilities gives the company the largest profit. This kind of problem is known as a **linear programming problem** (commonly referred to as an LP problem).

> **LINEAR PROGRAMMING PROBLEMS**
>
> A **linear programming problem** in two unknowns x and y is one in which we are to find the maximum or minimum value of a linear expression
>
> $$ax + by$$
>
> (called the **objective function**), subject to a number of linear **constraints** of the form
>
> $$cx + dy \leq e \quad \text{or} \quad cx + dy \geq e.$$
>
> The largest or smallest value of the objective function is called the **optimal value,** and a pair of values of x and y that gives the optimal value constitutes an **optimal solution.**

4.2 Solving Linear Programming Problems Graphically

Here is an example of such a problem and a way of solving it.

▼ **EXAMPLE 1**

Maximize $p = x + y$ subject to the constraints

$$x + 2y \leq 12$$
$$2x + y \leq 12$$
$$x \geq 0, y \geq 0.$$

SOLUTION We begin by drawing the **feasible region** for the problem, which is the set of points representing solutions to the constraints. We do this using the techniques of Section 1, and we get Figure 1.

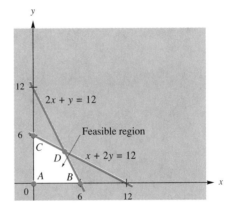

FIGURE 1

Each point in the feasible region gives an x and a y satisfying all of the constraints. The question now is, which of these points gives the largest value of $x + y$? The crucial fact to know is that if the objective function has an optimal value, *one of the corner points will be an optimal solution.* (See Example 5 for a discussion of the fact that there is no optimal value in certain types of problems.) We shall explain in a moment why one of the corner points has to be an optimal solution, but let us use this fact now to solve the problem.

In the following table we find the coordinates of each corner point and also compute the value of the objective function at each corner.

Point	Lines Through Point	Coordinates	$p = x + y$
A		(0, 0)	0
B		(6, 0)	6
C		(0, 6)	6
D	$x + 2y = 12$ $2x + y = 12$	(4, 4)	8

Since one of these points must give the optimal value, we simply pick the one that gives the largest value for p, which is D. Therefore, the optimal value of p is 8, and an optimal solution is $x = 4$ and $y = 4$.

Before we go on... Now we will explain why one of the corner points should be an optimal solution. The question is, which points in the feasible region give the largest possible value of $p = x + y$? Consider first the question, which points give particular values of p? For example, which points make $p = 2$? These would be the points on the line $x + y = 2$, which is the line labeled $p = 2$ in Figure 2.

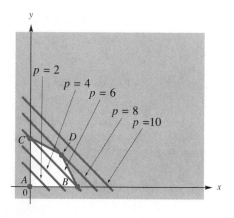

FIGURE 2

Now suppose that we want to know which points make $p = 4$. These would be the points on the line $x + y = 4$, which is the line labeled $p = 4$ in Figure 2. Notice that this line is parallel to but higher than the line $p = 2$. (If p represented profit in an application, we would call these **isoprofit lines,** or **constant-profit lines.**) Imagine moving this line up or down in the picture. As we move it down we see smaller values of p, and as we move it up we see larger values. Several more of these lines are drawn in Figure 2. In particular, look at the line labeled $p = 10$. This line does not meet the feasible region, meaning that no feasible point makes p as large as 10. Starting with the line $p = 2$, as we move the line up, increasing p, there will be a last line meeting the feasible region. In the figure it is clear that this is the line $p = 8$, and this meets the feasible region in only one point, which is the corner point D. Therefore, D gives the largest p value of all feasible points.

Q What if we had been asked to maximize something else, such as $p = x + 3y$?

A Then the optimal solution might be different. Figure 3 shows some of the isoprofit lines for this objective function. This time the last point hit as p increases is C, not D. This tells us that the optimal solution is $x = 0$ and $y = 6$, giving the optimal value $p = 18$.

4.2 Solving Linear Programming Problems Graphically

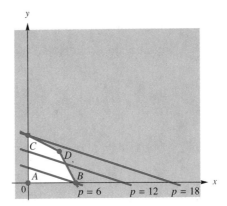

FIGURE 3

Q What if we had been asked to minimize* something, such as $c = x - y$?

A We can draw a similar picture showing the lines along which c is constant. As Figure 4 makes clear, the value of c becomes larger as we move the line to the right and smaller as we move the line to the left.

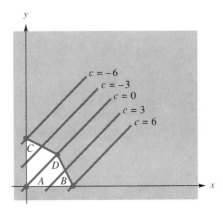

FIGURE 4

The smallest possible value of c will occur at the corner C, so the optimal solution this time is $x = 0$ and $y = 6$, giving the optimal value $c = -6$.

These examples should convince you that the optimal value will always occur at one of the corner points. By the way, it is possible for it to occur at *two* corner points, and at all points along an edge connecting them (do you see why?). We shall see an example later.

▼ * We shall usually use p to denote something to be maximized (such as profit), and c to denote something to be minimized (such as cost).

Here is a summary of the method we have just been using.

> **GRAPHICAL METHOD FOR SOLVING LINEAR PROGRAMMING PROBLEMS IN TWO UNKNOWNS**
>
> 1. Graph the feasible region.
> 2. Compute the coordinates of the corner points.
> 3. Substitute the coordinates of the corner points into the objective function to see which gives the optimal value.

▶ CAUTION   If the feasible region is not bounded, there may be no solution, and the procedure above can be misleading. See Example 5 below for more about this. ◀

We now look at some applications.

▼ **EXAMPLE 2**   Resource Allocation

Acme Babyfoods mixes two strengths of apple juice. One quart of Beginner's juice is made from 30 fluid ounces of water and 2 fluid ounces of apple juice concentrate. One quart of Advanced juice is made from 28 fluid ounces of water and 4 fluid ounces of concentrate. Every day Acme has available 215 gallons (27,520 fluid ounces) of water and 25 gallons (3,200 fluid ounces) of concentrate. Acme makes a profit of 20¢ on each quart of Beginner's juice and 30¢ on each quart of Advanced juice. How many quarts of each should Acme make each day in order to get the largest profit? How would this change if Acme made a profit of 30¢ on Beginner's and 20¢ on Advanced juice?

SOLUTION   Looking at the question that we are asked, we see that our unknown quantities are

$x =$ the number of quarts of Beginner's juice, made each day;

$y =$ the number of quarts of Advanced juice made each day.

(In this context x and y are often called the **decision variables,** since we must decide what their values should be in order to get the largest profit.) We can write the data given in the form of a table (the numbers in the first two columns are per quart of juice).

	Beginner's (x quarts)	Advanced (y quarts)	Amount Available
Water (fl oz)	30	28	27,520
Concentrate (fl oz)	2	4	3,200
Profit (¢)	20	30	

Since nothing in the problem says that Acme must use up all of the water or concentrate, just that it can use no more than what is available, the first two rows of the table give us two inequalities.

$$30x + 28y \leq 27{,}520$$
$$2x + 4y \leq 3{,}200$$

We also have that $x \geq 0$ and $y \geq 0$ because they can't make a negative amount of juice. To finish setting up the problem, we are asked to maximize the profit, which is

$$p = 20x + 30y,$$

expressed in cents. This gives us our linear programming problem.

Maximize $p = 20x + 30y$

subject to the constraints

$$30x + 28y \leq 27{,}520$$
$$2x + 4y \leq 3{,}200$$
$$x \geq 0, y \geq 0.$$

The feasible region is shown in Figure 5.

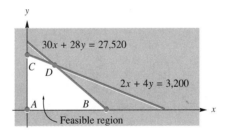

FIGURE 5

The corners and the values of the objective function are given in the following table.

Point	Lines Through Point	Coordinates	$p = 20x + 30y$
A		(0, 0)	0
B		$(917\tfrac{1}{3}, 0)$	$18{,}346\tfrac{2}{3}$
C		(0, 800)	24,000
D	$30x + 28y = 27{,}520$ $2x + 4y = 3{,}200$	(320, 640)	25,600

(Point D was found in an example in Chapter 2, Section 1.) From this table we see that the company should make 320 quarts of Beginner's juice and 640 quarts of Advanced juice for a largest possible profit of 25,600¢, or $256.

If instead the company made a profit of 30¢ on each quart of Beginner's juice and 20¢ on each quart of Advanced juice, then we would have $p = 30x + 20y$. This would give the following table.

Point	Lines Through Point	Coordinates	$p = 30x + 20y$
A		(0, 0)	0
B		$(917\frac{1}{3}, 0)$	27,520
C		(0, 800)	16,000
D	$30x + 28y = 27,520$ $2x + 4y = 3,200$	(320, 640)	22,400

We can see that in this case, Acme should make $917\frac{1}{3}$ quarts of Beginner's juice and no Advanced juice for a largest possible profit of $275.20.

Before we go on... Notice that in the first version of this problem, the company used up all of the water and juice concentrate. In the second version it did not. How could this happen?

▼ **EXAMPLE 3** Maximizing Return on Investments

The Solid Trust Savings & Loan Company has set aside $25 million for loans to home buyers. Its policy is to allocate at least $10 million annually for luxury condominiums. On the other hand, Solid Trust has received a government housing development grant stipulating that at least one-third of the company's loans be set aside for low-income housing. The S&L's return on condominiums is 12%, and its return on low-income housing is 10%. How much should the company allocate for each type of housing to maximize its total return?

SOLUTION We first identify the unknowns: let x be the annual amount (in millions of dollars) allocated to luxury condos, and let y be the annual amount allocated to low-income housing. We now look at the constraints.

The first constraint is mentioned in the first sentence; the company can invest a total of $25 million. Thus,

$$x + y \leq 25.$$

(We used $\leq$ rather than an equality, since the company is not required to invest all of the $25 million; rather, it can invest *up to* $25 million.)

Next, the company has allocated at least $10 million to condos. Rephrasing this in terms of our unknowns, we get

The amount allocated to condos is at least $10 *million.*

The phrase "is at least" means $\geq$. Thus,

$$x \geq 10.$$

The second constraint is that at least one-third of the total financing must be for low-income housing. Rephrasing this,

The amount allocated to low-income housing is at least $\frac{1}{3}$ of the total.

Since the total investment will be $x + y$, we get

$$y \geq \tfrac{1}{3}(x + y).$$

We want to put this in the standard form of a linear inequality, so we subtract $\frac{1}{3}(x + y)$ from both sides of the inequality to get

$$y - \tfrac{1}{3}(x + y) \geq 0,$$

or

$$y - \tfrac{1}{3}x - \tfrac{1}{3}y \geq 0.$$

In other words,

$$-\tfrac{1}{3}x + \tfrac{2}{3}y \geq 0.$$

Multiplying by 3 to clear fractions, we get

$$-x + 2y \geq 0.$$

There are no further constraints, so we have

$$\begin{aligned} x + y &\leq 25 \\ x &\geq 10 \\ -x + 2y &\geq 0 \\ x \geq 0,\ y &\geq 0. \end{aligned}$$

(Do you see why the inequalities $x \geq 0$ and $y \geq 0$ were slipped in here?) Now, what about the return on these investments? According to the data, the annual return is given by

$$P = 0.12x + 0.10y.$$

We want to make this quantity P as large as possible. In other words, we must

Maximize $P = 0.12x + 0.10y$

subject to the constraints

$$\begin{aligned} x + y &\leq 25 \\ x &\geq 10 \\ -x + 2y &\geq 0 \\ x \geq 0,\ y &\geq 0. \end{aligned}$$

The feasible region is shown in Figure 6.

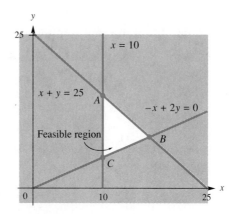

FIGURE 6

We now make a table showing the return on investment at each corner point.

Point	Lines Through Point	Coordinates	$P = 0.12x + 0.10y$
A	$x = 10$ $x + y = 25$	(10, 15)	2.7
B	$x + y = 25$ $-x + 2y = 0$	$(\frac{50}{3}, \frac{25}{3})$	2.833
C	$x = 10$ $-x + 2y = 0$	(10, 5)	1.7

We did not really need to bother with C. If you look at the graph, point C is directly under point A. Thus, it has the same x-coordinate, but a smaller y-coordinate. This means the resulting return for point C will be less than that for point A. Since we are seeking the largest return, we could ignore it (and save ourselves the additional calculation).

From the table, we see that the values of x and y that maximize the return are $x = \frac{50}{3}$ and $y = \frac{25}{3}$, giving a total return of $2.833 million. In other words, the most profitable course of action is to invest $16.667 million in condos and $8.333 million in low-income housing, giving a maximum annual return of $2.833 million.

Before we go on... What breakdown of investments would lead to the *lowest* return?

EXAMPLE 4 Management*

You are the service manager for a supplier of closed-circuit television systems. Your company can provide up to 160 hours per week of technical service for your customers, although the demand for technical service far exceeds this amount. As a result, you have been asked to develop a model to allocate service technicians' time between new customers—those still covered by service contracts—and old customers, whose service contracts have expired. To insure that new customers are satisfied with your company's service, the sales department has instituted a policy that at least 100 hours per week be allocated to servicing new customers. At the same time, your superiors have informed you that the company expects your department to generate at least $1,200 per week in revenues. Technical service time for new customers generates an average of only $5 per hour (because much of the service is still under warranty) and $30 per hour for old customers. Further, service visits take an average of two hours for new customers and one hour for old customers. How many visits should you allocate to each type of customer per week in order to generate the most revenue?

SOLUTION The unknown quantities are as follows.

x = the number of visits to new customers

y = the number of visits to old customers

Turning to the constraints, we first see that there are a total of 160 hours available per week. Although it might be tempting to say that $x + y \leq 160$, we must be careful, since x and y are measured in visits, not hours. So, back to the drawing board: we need some data relating number of visits and hours. This information is buried toward the end of the question: "service visits take an average of two hours for new customers and one hour for old customers." Thus, we do a little cost accounting.

x visits to new customers at 2 hours per visit is a total of $2x$ hours.

y visits to old customers at 1 hour per visit is a total of $1y = y$ hours.

Thus,

$2x + y \leq 160$.

The next restriction is the policy "that at least 100 hours per week be allocated to servicing new customers." Once again, this is expressed in hours, not visits. Rephrasing this in our customary manner we get

The number of hours allocated to servicing new customers is at least 100.

▼ *This example is loosely based on a similar problem in *An Introduction to Management Science*, Sixth Edition, by D. R. Anderson, D. J. Sweeny, and T. A. Williams (St. Paul: West, 1991).

Since we have already seen that the number of hours allocated to servicing new customers is $2x$, this constraint is

$$2x \geq 100,$$

or

$$x \geq 50.$$

The next restriction is that your department must generate at least \$1,200 per week in revenues. So we need to do some more cost accounting, using the amount of revenue generated by each kind of visit (in dollars per hour).

$2x$ hours per week spent at new customers at \$5 per hour gives a revenue of $10x$.

y hours per week spent at old customers at \$30 per hour gives a revenue of $30y$.

Thus,

$$10x + 30y \geq 1{,}200,$$

or

$$x + 3y \geq 120.$$

Finally, the revenue we wish to maximize is given by

$$R = 10x + 30y.$$

(Notice that we already calculated the revenue as $10x + 30y$ in formulating the last constraint.)

Thus, the linear programming problem is the following.

Maximize $R = 10x + 30y$ subject to
$$2x + y \leq 160$$
$$x \geq 50$$
$$x + 3y \geq 120$$
$$x \geq 0, y \geq 0.$$

We proceed as before. Figure 7 shows the feasible region.

We now set up the table for the corner points.

Point	Lines Through Point	Coordinates	$R = 10x + 30y$
A	$x = 50$ $2x + y = 160$	(50, 60)	2,300
B	$x = 50$ $x + 3y = 120$	$(50, \frac{70}{3})$	1,200
C	$x + 3y = 120$ $2x + y = 160$	(72, 16)	1,200

4.2 Solving Linear Programming Problems Graphically

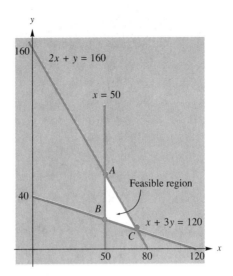

FIGURE 7

Since option A maximizes weekly revenue, you should plan to arrange for 50 service visits per week to new customers and 60 per week to old customers.

Q How would you allocate service technicians' time in order to *minimize* weekly revenue?

A Looking at the table, we see that a curious thing has happened: we get the same minimum revenue at both B and C. Thus we could choose either option to minimize revenue. In fact, we could also choose any point along the line segment BC, and it will also yield a weekly revenue of $1,200. For example, (58.8, 20.4) lies on the line segment BC and yields a weekly revenue of $1,200. This happens because the isorevenue lines are parallel to that edge.

Here is an example for which a graphing calculator or computer might come in handy in drawing the feasible region. Without finding the coordinates, how do you know whether point C is to the left or to the right of the line $x = 50$? If you draw the picture very carefully it will fall to the right, as it should. A graphing calculator or computer will show you right away where the intersection points are relative to the other lines, which is what you need to know to decide which intersections are corner points and which are not.

EXAMPLE 5 Cost Minimization

Steve is mixing baby food for his young son. He is concerned that the meal have at least 1 gram of protein and 5 milligram of iron. He is mixing together cereal, with 0.5 gram of protein and 1 milligram of iron per ounce, and fruit, with 0.2 gram of protein and 2 milligram of iron per ounce. The cereal costs 20¢ per ounce, and the fruit costs 30¢ per ounce.

(a) What mixture will provide the proper nutrition at the least cost?

(b) Is there a most expensive mixture that provides the proper nutrition?

SOLUTION (a) The unknowns are

$x =$ number of ounces of cereal, and

$y =$ number of ounces of fruit in the mixture.

In order for the mixture to have the right amount of protein, we must have

$$0.5x + 0.2y \geq 1.$$

In order for it to have the right amount of iron, we must have

$$x + 2y \geq 5.$$

Also, $x \geq 0$ and $y \geq 0$ for obvious reasons. Finally, the cost is given by

$$C = 20x + 30y.$$

Our linear programming problem is then to

Minimize $C = 20x + 30y$

subject to the constraints

$0.5x + 0.2y \geq 1$

$x + 2y \geq 5$

$x \geq 0, y \geq 0.$

The feasible region is shown in Figure 8.

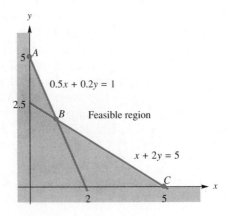

FIGURE 8

Notice that this region is **unbounded:** it extends without bound up and to the right. This just says that huge amounts of cereal and fruit will guarantee enough protein and iron. But in this problem we are really interested in how *little* cereal and fruit we can use, to keep the cost down, so the fact that the feasible region is unbounded does not concern us. We find the corners of the region as usual.

Point	Lines Through Point	Coordinates	$C = 20x + 30y$
A	$x = 0$ $0.5x + 0.2y = 1$	(0, 5)	150
B	$0.5x + 0.2y = 1$ $x + 2y = 5$	(1.25, 1.875)	81.25
C	$x + 2y = 5$ $y = 0$	(5, 0)	100

The cheapest mixture that provides the proper nutrition is 1.25 ounces of cereal and 1.875 ounces of fruit.

(b) For this part, our linear programming problem is the following.

$$\text{Maximize } C = 20x + 30y$$

subject to the constraints

$$0.5x + 0.2y \geq 1$$
$$x + 2y \geq 5$$
$$x \geq 0, y \geq 0.$$

The only difference here is that we are seeking a *maximum* value of C rather than a minimum value. You might be tempted to refer to the above table and answer that corner point A represents the most expensive mixture meeting the nutritional requirements, but you would be wrong! Indeed, we could make the mixture as expensive as we like by using huge quantities of cereal and fruit, which would easily meet the nutritional requirements (even though Steve's son would not be too happy!).

In this case the table above is very misleading. In particular, the point A does *not* give the most expensive mixture possible, and there is no optimal solution.

Before we go on... The possibility that no optimal solution exists can only arise when the feasible region is unbounded or is the empty set. When you find that the feasible region is unbounded, you should always consider the possibility that there is no optimal solution.

> **BOUNDED AND UNBOUNDED FEASIBLE REGION, EXISTENCE OF OPTIMAL SOLUTIONS**
>
> A feasible region is **unbounded** if it extends infinitely in some direction. Otherwise, it is **bounded.**
>
> 1. If the feasible region is bounded (and nonempty), then there is always an optimal solution.
> 2. If the feasible region is unbounded, there is a possibility that there will be no optimal solution. You *must* consider this when using the graphical method.
> 3. The table of values at the corner points is only useful when you know that there is an optimal solution.

▼ **EXAMPLE 6**   Resource Allocation

You are composing a very avant-garde ballade for violins and bassoons. In your ballade, each violinist plays a total of two notes and each bassoonist only one note. In order to make your ballade long enough, you decide that it should contain at least 200 notes. Further, after playing the requisite two notes, each violinist will sing one soprano note, while each bassoonist will sing three soprano notes.* To make the ballade sufficiently interesting, you have decided on a minimum of 300 soprano notes. Finally, in order to give your composition a sense of balance, you wish to have no more than three times as many bassoonists as violinists. Violinists charge $200 per performance and bassoonists $400 per performance. How many of each should your ballade call for in order to minimize personnel costs?

SOLUTION   First, the unknowns are as follows: $x =$ the number of violinists, $y =$ the number of bassoonists. The constraint on the number of notes implies that

$$2x + y \geq 200,$$

since the total number is to be *at least* 200. Similarly, the constraint on the number of soprano notes is

$$x + 3y \geq 300.$$

The next one is a little tricky. As usual, we reword it in terms of the quantities x and y.

> *The number of bassoonists should be no more than three times the number of violinists.*

Thus, $y \leq 3x$.

▼ * The question of whether these musicians are capable of singing decent soprano notes will be left to chance. You reason that a few bad notes will add character to the ballade.

Finally, the total cost per performance will be

$$C = 200x + 400y,$$

and we wish to minimize total cost. Thus, our linear programming problem is as follows.

Minimize $C = 200x + 400y$ subject to
$2x + y \geq 200$
$x + 3y \geq 300$
$3x - y \geq 0$
$x \geq 0, y \geq 0.$

We get the feasible region shown in Figure 9.

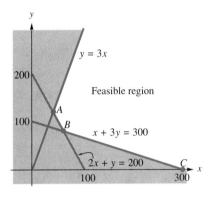

FIGURE 9

Again the feasible region is unbounded, but since we are *minimizing* cost, this is not a problem (the farther we go to the right and up in the feasible region, the larger the cost gets). We first identify the corner points A, B, and C as shown, and proceed to tabulate.

Point	Lines Through Point	Coordinates	$C = 200x + 400y$
A	$2x + y = 200$ $3x - y = 0$	(40, 120)	56,000
B	$2x + y = 200$ $x + 3y = 300$	(60, 80)	44,000
C		(300, 0)	60,000

From the table, we see that the minimum cost is $44,000 per performance, employing 60 violinists and 80 bassoonists. (Quite a wasteful ballade, one might say.)

Here is another example where a graphing calculator or computer would help in determining the corner points. Unless you are very confident of the accuracy of your sketch, how do you know that the line $y = 3x$ falls to the left of the point B, for example? If it were to fall to the right, then B would not be a corner point, and the solution would be different. You could (and should) check that B satisfies the inequality $3x - y \geq 0$, so that the line falls to the left of B as shown. However, if you use a graphing calculator or computer you can be fairly confident of the picture produced without doing further calculations.

▶ 4.2 EXERCISES

Solve the linear programming problems in Exercises 1–20.

1. Maximize $p = x + y$
subject to $x + 2y \leq 9$
$2x + y \leq 9$
$x \geq 0, y \geq 0.$

2. Maximize $p = x + 2y$
subject to $x + 3y \leq 24$
$2x + y \leq 18$
$x \geq 0, y \geq 0.$

3. Minimize $c = x + y$
subject to $x + 2y \geq 6$
$2y + y \geq 6$
$x \geq 0, y \geq 0.$

4. Minimize $c = x + 2y$
subject to $x + 3y \geq 30$
$2x + y \geq 30$
$x \geq 0, y \geq 0.$

5. Maximize $p = 3x + y$
subject to $3x - 7y \leq 0$
$7x - 3y \geq 0$
$x + y \leq 10$
$x \geq 0, y \geq 0.$

6. Maximize $p = x - 2y$
subject to $x + 2y \leq 8$
$x - 6y \leq 0$
$3x - 2y \geq 0$
$x \geq 0, y \geq 0.$

7. Maximize $p = 3x + 2y$
subject to $20x + 10y \leq 100$
$10x + 20y \leq 100$
$10x + 10y \leq 60$
$x \geq 0, y \geq 0.$

8. Maximize $p = x + 2y$
subject to $30x + 20y \leq 600$
$10x + 40y \leq 400$
$20x + 30y \leq 450$
$x \geq 0, y \geq 0.$

9. Minimize $c = 2x + 3y$
subject to $20x + 10y \geq 100$
$10x + 20y \geq 100$
$10x + 10y \geq 80$
$x \geq 0, y \geq 0.$

10. Minimize $c = 4x + y$
subject to $30x + 20y \geq 600$
$10x + 40y \geq 400$
$20x + 30y \geq 600$
$x \geq 0, y \geq 0.$

11. Maximize and minimize $p = x + 2y$
subject to $x + y \geq 2$
$x + y \leq 10$
$x - y \leq 2$
$x - y \geq -2.$

12. Maximize and minimize $p = 2x - y$
subject to $x + y \geq 2$
$x - y \leq 2$
$x - y \geq -2$
$x \leq 10, y \leq 10.$

13. Maximize $p = 2x + 3y$
subject to $x + 2y \geq 10$
$2x + y \geq 10$
$x \geq 0, y \geq 0.$

14. Maximize $p = 3x + 2y$
subject to $x + y \geq 2$
$y \leq 10$
$x \geq 0, y \geq 0.$

15. Minimize $c = 2x + 4y$
subject to $x + y \geq 10$
$x + 2y \geq 14$
$x \geq 0, y \geq 0$.

16. Minimize $c = 3x + y$
subject to $10x + 20y \geq 100$
$30x + 10y \geq 100$
$x \geq 0, y \geq 0$.

17. Minimize $c = 3x - 3y$
subject to $\dfrac{x}{4} \leq y$
$y \leq \dfrac{2x}{3}$
$x + y \geq 5$
$x + 2y \leq 10$
$x \geq 0, y \geq 0$.

18. Minimize $c = -x + 2y$
subject to $y \leq \dfrac{2x}{3}$
$x \leq 3y$
$y \geq 4$
$x \geq 6$
$x + y \leq 16$.

19. Maximize $p = x + y$
subject to $x + 2y \geq 10$
$2x + 2y \leq 10$
$2x + y \geq 10$
$x \geq 0, y \geq 0$.

20. Maximize $p = 2x + 3y$
subject to $x + y \geq 10$
$x + 2y \leq 12$
$2x + y \leq 12$
$x \geq 0, y \geq 0$.

APPLICATIONS

21. *Resource Allocation* You manage an ice cream factory that makes two flavors: Creamy Vanilla and Continental Mocha. Into each quart of Creamy Vanilla go two eggs and three cups of cream. Into each quart of Continental Mocha go one egg and three cups of cream. You have in stock 500 eggs and 900 cups of cream. You make a profit of $3 on each quart of Creamy Vanilla and $2 on each quart of Continental Mocha. How many quarts of each flavor should you make in order to produce the largest profit?

22. *Resource Allocation* Podunk Institute of Technology's Math Department offers two courses: Finite Math and Calculus. Each section of Finite Math has 60 students, while each section of Calculus has 50. The department is allowed to offer a total of up to 110 sections. Further, there are no more than 6,000 students who would like to take a math course. Suppose that the university makes a profit of $100,000 on each section of Finite Math and $50,000 on each section of Calculus (the profit being the difference between what the students are charged and what the professors are paid). How many sections of each course should the department offer in order to make the largest profit?

23. *Nutrition* Ruff, Inc., makes dog food out of chicken and grain. Chicken has 10 grams of protein and 5 grams of fat per ounce, while grain has 2 grams of protein and 2 grams of fat per ounce. A bag of dog food must contain at least 200 grams of protein and at least 150 grams of fat. If chicken costs 10¢ per ounce and grain costs 1¢ per ounce, how many ounces of each should the company use in each bag of dog food in order to minimize cost?

24. *Purchasing* The Enormous State University's Business School is buying computers. There are two models to choose from, the Pomegranate and the Ami. Each Pomegranate comes with 4 MB of memory and 80 MB of disk space, while each Ami has 3 MB of memory and 100 MB of disk space. For reasons related to its accreditation, the school would like to be able to say that it has a total of at least 480 MB of memory and at least 12,800 MB of disk space. If both the Pomegranate and the Ami cost $1000 each, how many of each should the school buy to keep the cost as low as possible?

25. *Nutrition* Each serving of *Gerber* Mixed Cereal for Baby contains 60 calories and 11 grams of carbohydrates.* Each serving of *Gerber* Mango Tropical Fruit

---
▼ * Source: nutrition information printed on the package.

Dessert contains 80 calories and 21 grams of carbohydrates.* If the cereal costs 30¢ per serving and the dessert costs 50¢ per serving, and you want to provide your child with at least 140 calories and at least 32 grams of carbohydrates, how can you do so at the least cost? (Fractions of servings are permitted.)

26. *Nutrition* Each serving of *Gerber* Mixed Cereal for Baby contains 60 calories, 11 grams of carbohydrates, and no Vitamin C. Each serving of *Gerber* Apple Banana Juice contains 60 calories, 15 grams of carbohydrates, and 120% of the U.S. Recommended Daily Allowance (RDA) of Vitamin C for infants.* The cereal costs 10¢ per serving and the juice costs 27¢ per serving. If you want to provide your child with at least 120 calories, at least 26 grams of carbohydrates, and at least 50% of the U.S. RDA of Vitamin C for infants, how can you do so at the least cost? (Fractions of servings are permitted.)

27. *Investments* Two of the municipal bond funds traded by *Merrill Lynch* are the Muniyield Fund and the *Merrill Lynch* Municipal Bond Fund Class B. Each unit of the Muniyield Fund costs $15, yields an annual return of 6%, and has a risk index of 2 per unit. Each unit of the *Merrill Lynch* Municipal Bond Fund Class B costs $12, yields an annual return of 5%, and has a risk index of 1.5 per unit.† You would like to invest a total of up to $42,000 in these funds and would like to earn at least $2,400 in interest in the coming year. How many units of each fund (to the nearest tenth of a unit) should you purchase in order to meet your requirements and minimize the total risk index for your portfolio?

28. *Investments (graphing calculator or computer software required)* Two of the mutual funds offered by *Prudential Securities* are the Weingarten Fund and the Fidelity Investment High Income Municipal Portfolio. Each unit of the Weingarten Fund costs $17 and yields an estimated annual return (appreciation plus dividends) of 8%, and each unit of the municipal portfolio costs $12 and yields an estimated annual return of 5%.‡ You have up to $58,000 to invest in these funds, and your goal is to appreciate your capital by at least $3,920 in the coming year. According to your investment consultant, the Weingarten Fund has a risk index of 10 per unit, while the municipal fund has a risk index of 2 per unit. How many units of each fund (to the nearest tenth of a unit) should you purchase in order to meet your requirements and minimize the total risk index for your portfolio?

29. *Resource Allocation* Your salami manufacturing plant can order up to 1,000 pounds of pork and 2,400 pounds of beef per day for use in manufacturing its two specialties: "Count Dracula Salami" and "Frankenstein Sausage." Production of the Count Dracula variety requires 1 pound of pork and 3 pounds of beef for every salami, while the Frankenstein variety requires 2 pounds of pork and 2 pounds of beef for every sausage. In view of your heavy investment in advertising Count Dracula Salami, you have decided that at least one-third of the total production should be Count Dracula. On the other hand, due to the health-conscious consumer climate out there, your Frankenstein Sausage (sold as having less beef) is earning your company a profit of $3 per sausage, while sales of the Count Dracula variety are down, and it is earning your company only $1 per salami. Given the above restrictions, how many of each kind of sausage should you produce to maximize profits, and what is the maximum possible profit?

30. *Advertising Strategy* According to published statistics in 1993, "Rambo" movies shown on cable TV networks typically attract 3 million viewers, while episodes of "Murder, She Wrote" attract approximately 2.5 million viewers.§ Your marketing services firm has been hired to promote Bald No More Inc.'s hair replacement process by buying at least 30 commercial spots during cable TV showings of

---

▼ * Source: nutrition information printed on the package.

† Risk indices are fictitious, price per unit and yields quoted as of May 1993 (yields rounded to the nearest percentage point).

‡ The yield for the Weingarten Fund was estimated by a broker, and the yield for the municipal fund is quoted as of May 1993 and rounded to the nearest percentage point.

§ Source: "Nielsen's New Ratings for Cable TV Draw Fire," *The New York Times*, June 28, 1993, p. D5.

"Rambo" movies and episodes of "Murder, She Wrote." The cable company running a series of "Rambo" movies has quoted a price of $2,000 per spot, while the cable company showing "Murder, She Wrote" has quoted a price of $1,500 per spot. Bald No More's advertising budget for TV commercials is $70,000, and they would like at least 50% of the total number of spots to appear on "Rambo" shows. How many spots should you purchase on each show in order to reach the most viewers?

31. *Advertising Strategy* According to published statistics in 1993, wrestling features shown on the USA TV network attracted about 2 million viewers, while mid-season baseball features shown on the TBS cable network attracted approximately 1.5 million viewers.* Your marketing services firm has been hired to promote Gauss Jordan sneakers by buying at least 30 commercial spots during cable TV showings of wrestling and baseball features. Assume that USA TV network has quoted a price of $1,000 per spot, while TBS has quoted a price of $1,500 per spot. Gauss Jordan's advertising budget for TV commercials is $40,000, and they would like at least 75% of the total number of spots to appear on wrestling shows. How many spots should you purchase from each network in order to reach the most viewers?

32. *Project Design* The Megabuck Hospital Corp. of Tuscaloosa, Alabama is to build a state-subsidized nursing home in Manhattan, catering to homeless patients as well as high-income patients. State regulations require that every subsidized nursing home must house a minimum of 1,000 homeless patients and no more than 750 high-income patients in order to qualify for state subsidies. The overall capacity of the hospital is to be 2,100 patients. The board of directors, under pressure from a neighborhood group, insists that the number of homeless patients should not exceed twice the number of high-income patients. Due to the state subsidy, the hospital will make an average profit of $10,000 per month for every homeless patient they house, while the profit per high-income patient is estimated at $8,000 per month. How many of each type of patient should they house in order to maximize profit?

33. *Gasoline Stations* The following chart shows 1993 statistics on gasoline sales in New York and Connecticut.[†]

	Annual Sales (billions of gallons)	Number of Gas Stations	State Tax Rate ($ per gallon)	Average Retail Price[a] ($ per gallon)
New York	5.6	7,000	0.20	1.20
Connecticut	2	2,000	0.30	1.20

[a] Price includes state taxes.

(a) Calculate the average annual gasoline sales per gas station in each state, and hence the average annual revenue of a gas station in each state from sales of gas.

(b) Your Connecticut-based firm is planning to open a total of up to 20 gas stations in New York and Connecticut. You would like the firm to generate at least $20.4 million per year in revenues from sales of gasoline while at the same time keeping total state taxes as low as possible. Given these requirements, how many gas stations should you open in each state?

34. *Gasoline Stations* Refer to the table in the previous exercise.

(a) Calculate the average annual gasoline sales per gas station in each state, and hence the average *after-tax* annual revenue of a gas station in each state from sales of gas.

(b) Your Connecticut-based firm is planning to open a total of up to 30 gas stations in New York and Connecticut. You would like the firm to pay no more than $6.2 million in taxes, and would like the firm's after-tax revenues (from sales of gas) to be as large as possible. Given these requirements, how many gas stations should you open in each state?

▼ * Source: "Nielsen's New Ratings for Cable TV Draw Fire," The New York Times, June 28, 1993, p. D5.

† Annual sales, number of gas stations and state taxes, though adjusted for hand computation, are accurate to one significant digit. Average retail price is for unleaded regular gasoline and is accurate to two significant digits. Source: *National Petroleum News*/U.S. Energy Information Administration, State energy offices/*The New York Times*, Oct. 31, 1994, p. B1.

35. *Scheduling* The Scottsville Textile Mill produces several different fabrics on 8 dobbie mills that operate 24 hours per day and are scheduled for 30 days in the coming month. Scottsville Textile Mill will produce only Fabric 1 and Fabric 2 during the coming month. Each dobbie mill can turn out 4.63 yards of either fabric per hour. Assume that there is a monthly demand of 16,000 yards of Fabric 1 and 12,000 yards of Fabric 2. Profits are calculated as 33¢ per yard for each fabric produced on the dobbie mills.*
   (a) Will it be possible to satisfy total demand?
   (b) In the event that total demand is not satisfied, Scottsville Textile Mill will need to purchase the fabrics from another mill to make up the shortfall. Its profits on resold fabrics ordered from another mill amount to 20¢ per yard for Fabric 1 and 16¢ per yard for Fabric 2. How many yards of each fabric should it produce in order to maximize profits?

36. *Scheduling* Repeat the preceding exercise for scheduling production of Fabrics 3 and 4, given the following data: Each dobbie mill can turn out 5.23 yards of either fabric per hour. There is a monthly demand of 16,000 yards of Fabric 3 and 12,000 yards of Fabric 4. Profits are calculated as 70¢ per yard for each fabric produced on the dobbie mills. The profits on resold fabrics ordered from other mills amount to 50¢ per yard for Fabric 3 and 54¢ per yard for Fabric 4.

37. *Planning* My Friends: I, the mighty Brutus, have decided to prepare for retirement by instructing young warriors in the arts of battle and diplomacy. For each hour spent in battle instruction, I have decided to charge 50 ducats, while for each hour in diplomacy instruction I shall charge 40 ducats. Due to my advancing years, I can spend no more than 50 hours per week instructing the youths, though the great Jove knows that they are sorely in need of instruction! Due to my fondness for physical pursuits, I have decided to spend no more than one-third of the total time in diplomatic instruction. On the other hand, the present border crisis with the Gauls is a sore indication of our poor abilities as diplomats. As a result, I have decided to spend at least 10 hours per week instructing in diplomacy. Finally, to complicate things further, there is the matter of Scarlet Brew: I have estimated that each hour of battle instruction will require 10 gallons of Scarlet Brew to quench my students' thirst, and that each hour of diplomacy instruction, being less physically demanding, requires half that amount. Since my harvest of red berries has far exceeded my expectations, I estimate that I'll have to use at least 400 gallons per week in order to avoid storing the fine brew at great expense. Given all these restrictions, how many hours per week should I spend in each type of instruction to maximize my income?

38. *Planning* Repeat the preceding exercise with the following changes: I would like to spend no more than half the total time in diplomatic instruction, and I have at least 600 gallons of Scarlet Brew to use.

39. *Resource Allocation* One day Gillian the magician summoned the wisest of her women. "Devoted followers," she began, "Here I have a quandary: As you well know, I possess great expertise in sleep spells and shock spells, but unfortunately, these are proving to be a drain on my aural energy resources; each sleep spell costs me 500 pico-shirleys of aural energy, while each shock spell requires 750 pico-shirleys. Clearly, I would like to hold my overall expenditure of aural energy to a minimum, and still meet my commitments in protecting the Sisterhood from the ever-present threat of trolls. Specifically, I have estimated that each sleep spell keeps us safe for an average of 2 minutes, while every shock spell protects us for about 3 minutes. We certainly require enough protection to last 24 hours of each day, and possibly more, just to be safe. At the same time, I have noticed that each of my sleep spells can immobilize 3 trolls at once, while one of my typical shock spells (having a narrower range) can immobilize only 2 trolls at once. We are faced, my sisters, with an onslaught of 1200 trolls per day! Finally, as you are no doubt aware, the By-Laws dictate that for a Magician of the Order to remain in Good Standing, the number of shock spells be between one quarter and one-third the number of shock and sleep spells combined. What do I do, Oh Wise Ones?"

▼ *Adapted from *The Calhoun Textile Mill Case* by J. D. Camm, P. M. Dearing, and S. K. Tadisina as presented for case study in *An Introduction to Management Science*, Sixth Edition, by D. R. Anderson, D. J. Sweeny and T. A. Williams (St. Paul: West, 1991). Our exercise uses a subset of the data given in the cited study.

40. *Risk Management* The Grand Vizier of the Kingdom of Um is being blackmailed by numerous individuals, and is having a very difficult time keeping his blackmailers from going public. He has been keeping them at bay with two kinds of payoff: gold bars from the Royal Treasury and political favors. Through bitter experience, he has learned that each payoff in gold gives him peace for an average of about one month, while each political favor seems to earn him about $1\frac{1}{2}$ months of reprieve. In order to maintain his flawless reputation in the Court, he feels he cannot afford any revelations about his tainted past to come to light within the next year. Thus, it is imperative that his blackmailers be kept at bay for 12 months. Further, he would like to keep the number of gold payoffs at no more than one quarter of the combined number of payoffs, since the outward flow of gold bars might arouse suspicion on the part of the Royal Treasurer. On the other hand, he feels that he can do no more than 7 political favors per year without arousing undue suspicion in the Court. The gold payoffs tend to deplete his travel budget. (The treasury has been subsidizing his numerous trips to the Himalayas.) He estimates that each gold bar removed from the treasury will cost him 4 trips. On the other hand, since the administering of political favors tends to cost him valuable travel time, he suspects that each political favor will cost him about 2 trips. Now, he would obviously like to keep his blackmailers silenced and lose as few trips as possible. What is he to do? How many trips will he lose in the next year?

COMMUNICATION AND REASONING EXERCISES

41. Create a linear programming problem in two variables with no optimal solution.

42. Create a linear programming problem in two variables with more than one optimal solution.

43. Create an interesting scenario leading to the following linear programming problem.

$$\text{Maximize } p = 10x + 10y \text{ subject to}$$
$$20x + 40y \leq 1,000$$
$$30x + 20y \leq 1,200$$
$$x \geq 0, y \geq 0.$$

44. Create an interesting scenario leading to the following linear programming problem.

$$\text{Minimize } c = 10x + 10y \text{ subject to}$$
$$20x + 40y \geq 1,000$$
$$30x + 20y \geq 1,200$$
$$x \geq 0, y \geq 0.$$

45. Use an example to explain why there may be no optimal solution to a linear programming problem if the feasible region is unbounded.

46. Use an example to explain why, in the event that an optimal solution does occur despite an unbounded feasible region, that solution corresponds to a corner point of the feasible region.

47. You are setting up an LP problem for Fly-by-Night Airlines with the unknowns x and y. x represents the number of first-class tickets it should issue for a specific flight and y represents the number of business-class tickets it should issue for that flight, and the problem is to maximize profit. You find that there are two different corner points that correspond to the maximum profit. How do you interpret this?

48. In the situation described in the previous exercise, you find that there are no optimal solutions. How do you interpret this?

4.3 SOLVING STANDARD LINEAR PROGRAMMING PROBLEMS USING THE SIMPLEX METHOD

The method discussed in Section 2 works quite nicely for linear programming problems in two unknowns, but what about three or more unknowns? Since we need an axis for each unknown, we would need to draw graphs in three dimensions (where we have x-, y-, and z- coordinates) to deal with problems in three unknowns, and draw in hyperspace to answer questions involving four or more unknowns. Given the state of technology when this book was written, we can't easily do this. So we need another method for solving LP problems that will work for any number of unknowns. One such method, called the **simplex method,** has been the method of choice since it was invented by G. Dantzig in 1947.* In order to illustrate it best, we first use it to solve only so-called "standard maximization problems."

STANDARD MAXIMIZATION PROBLEMS

A standard maximization problem in n unknowns is one in which we are required to maximize (not minimize) an objective function of the form

$$p = ax + by + cz + \ldots (n \text{ terms}),$$

where $a, b, c, \ldots$ are numbers, subject to the constraints

$$x \geq 0, y \geq 0, z \geq 0, \ldots,$$

and further constraints of the form

$$Ax + By + Cz + \ldots \leq N,$$

where $A, B, C, \ldots$ and N are numbers with N *nonnegative.* Note that the inequality here must be a "$\leq$," and not "$=$" or "$\geq$."[†]

---

[*] The first radically different method of solving LP problems was the ellipsoid algorithm published in 1979 by the Soviet mathematician L. G. Khachiyan. In 1984, N. Karmarkar, a researcher at Bell Labs, created a much more efficient method now known as Karmarkar's algorithm. Although these methods (and others since developed) can be shown to be faster than the simplex method in the worst cases, it seems to be true that the simplex method is still the fastest in the applications that arise in practice.

[†] As in the chapter on linear equations, we will seldom use the traditional subscripted variables $x_1, x_2, \ldots$. These are very useful names when you start running out of letters of the alphabet, but we should not find ourselves in that predicament.

4.3 Solving Standard Linear Programming Problems Using The Simplex Method

▼ **EXAMPLE 1**

Here is a standard maximization problem.

Maximize $p = 2x - 3y + 3z$ subject to
$$2x + z \leq 7$$
$$-x + 3y - 6z \leq 6$$
$$x \geq 0, y \geq 0, z \geq 0.$$

▼ **EXAMPLE 2**

The following is *not* a standard maximization problem.

Maximize $p = 2x - 3y + 3z$ subject to
$$2x + z \geq 7$$
$$-x + 3y - 6z \leq 6$$
$$x \geq 0, y \geq 0, z \geq 0.$$

The inequality $2x + z \geq 7$ cannot be written in the required form. If we reverse the inequality by multiplying both sides by -1, we get

$$-2x - z \leq -7,$$

and -7 is not positive.

The idea behind the simplex method is this. In any LP problem, there is a feasible region. If there are only two unknowns, we can draw it; if there are three unknowns, it is a solid region in space; and if there are four or more unknowns, it is a somewhat abstract higher-dimensional region. But it is a faceted region with corners (think of a diamond), and it is at one of these corners that we will find the optimal solution. Geometrically, what the simplex method does is to start at the corner where all the unknowns are 0 (possible because we are talking of *standard* problems) and then walk around the region, from corner to adjacent corner, always increasing the value of the objective function, until the best corner is found. In practice, we will visit only a small number of the corners before finding the right one. Algebraically, as we are about to see, this walking around is accomplished by matrix manipulations presented in the chapter on systems of linear equations.

We describe the method while working through an example.

▼ **EXAMPLE 3**

Maximize $p = 3x + 2y + z$ subject to
$$2x + 2y + z \leq 10$$
$$x + 2y + 3z \leq 15$$
$$x \geq 0, y \geq 0, z \geq 0.$$

SOLUTION

Step 1 *Convert to a system of linear equations.* The inequalities $2x + 2y + z \leq 10$ and $x + 2y + 3z \leq 15$ are not as convenient to work with as equations would be. Now look at the first inequality. It says that the left-hand side, $2x + 2y + z$, must have some positive number (or 0) *added to it* if it is to equal 10. Since we don't yet know what x, y, and z are, we are not yet sure what number to add to the left-hand side. So we invent a new unknown, $s \geq 0$, called a **slack variable,** to "take up the slack," so that

$$2x + 2y + z + s = 10.$$

Turning to the next inequality, $x + 2y + 3z \leq 15$, we now add a slack variable to its left-hand side, to get it up to the value of the right-hand side. We might have to add a different number than we did the last time, so we use a new slack variable, $t \geq 0$, and obtain

$$x + 2y + 3z + t = 15.$$

Now we write the system of equations we have (including the one that defines the objective function) in standard form.

$$\begin{aligned} 2x + 2y + z + s &= 10 \\ x + 2y + 3z + t &= 15 \\ -3x - 2y - z + p &= 0 \end{aligned}$$

Note two things: First, we have written all the variables neatly in columns, as we did in Chapter 2. Second, in rewriting the objective function $p = 3x + 2y + z$, we have left the coefficient of p as $+1$ and brought over all the other quantities to the same side of the equation as p. This will be our standard procedure from now on. *Don't* write $3x + 2y + z - p = 0$ (even though it amounts to the same thing), because the negative coefficients will be important in the following steps.

Step 2 *Set up the initial tableau.* We represent our system of equations by the following table (which is simply the augmented matrix in disguise), called **the initial tableau.**

x	y	z	s	t	p	
2	2	1	1	0	0	10
1	2	3	0	1	0	15
−3	−2	−1	0	0	1	0

The labels along the top keep track of which columns belong to which variables. Now notice a peculiar thing. *If the equations were rewritten with the variables s, t, and p first, then the matrix would be in reduced form.* Why? Because the matrix would be identical to the one above, except that the s, t, and p columns would be on the left, so that the leading entries would be 1s

4.3 Solving Standard Linear Programming Problems Using The Simplex Method

and the columns cleared. This means that we should be able to read off solutions to the system of equations that led to this matrix. According to our usual procedure, we get a whole family of solutions, one for each choice of x, y, and z. Since x, y, and z can be chosen arbitrarily, let's choose them to be zero. Now we solve for the remaining variables by translating the rows back into the original equations. This gives, with $x = y = z = 0$,

$$s = 10 - 2x - 2y - z = 10$$
$$t = 15 - x - 2y - 3z = 15$$
$$p = 0 + 3x + 2y + z = 0.$$

This solution is called the **basic solution** associated with the tableau. The variables s and t are called the **active** variables, and x, y, and z are the **inactive** variables (other terms used are **basic** and **nonbasic** variables). As an aid to recognizing which variables are active and which are inactive, we label each row with the name of the corresponding active variable. The complete initial tableau looks like this.

	x	y	z	s	t	p	
s	2	2	1	1	0	0	10
t	1	2	3	0	1	0	15
p	-3	-2	-1	0	0	1	0

The rightmost column gives the value for each variable listed on the left (including p, which will always be the label for the last row). Thus, we don't need to solve any equations—we can just read off the answers. What about the other variables not listed down the side? As we agreed, they are all zero. Thus, the tableau tells us that the basic solution for this step is

$$x = 0,\ y = 0,\ z = 0,\ s = 10,\ t = 15,\ p = 0.$$

One note of caution: In future steps, reading off the basic solution will have a complication. If the number appearing in, say, the s column were not a 1 but a 2, then the equation would say $2s = 10$, and so $s = \frac{10}{2} = 5$. In general, we have to divide the number in the rightmost column by the number in the column having the same label as appears on the left.

This basic solution represents our starting position $x = y = z = 0$ in the feasible region in xyz space.

Q How do we move to another corner point?

A We choose some pivot in one of the first three columns of the tableau and clear its column. Then we will get a different basic solution, which corresponds to another corner point. Thus, in order to move from corner point to corner point, all we have to do is to choose suitable pivots and clear columns in the usual manner.

The next two steps give the procedure for choosing the pivot.

Step 3 *Select the pivot column* (the column that contains the pivot we are seeking).

SELECTING THE PIVOT COLUMN

Choose the negative number with the largest magnitude in the bottom row. Its column is the pivot column. (If there are two or more candidates, choose any one of them.) If all the numbers in the bottom row are zero or positive, then you are done, and the basic solution is the optimal solution.

Simple enough. The most negative number is -3, so we choose the x-column as the pivot column:

	x	y	z	s	t	p	
s	2	2	1	1	0	0	10
t	1	2	3	0	1	0	15
p	-3	-2	-1	0	0	1	0

↑
Pivot column

Q Why choose the pivot column this way?

A The variable labeling the pivot column is going to be increased from 0 to something positive. In the equation $p = 3x + 2y + z$, the fastest way to increase p is to increase x, since p would increase by 3 units for every unit of increase in x. (If we chose to increase y, then p would only increase by 2 units for every unit of increase in y, and if we increased z instead, p would grow even more slowly.)

Step 4 *Select the pivot in the pivot column.*

SELECTING THE PIVOT

1. The pivot must always be a positive number. (This rules out zeros and negative numbers, such as the -3 in the bottom row.)
2. For each positive entry b in the pivot column, compute the ratio a/b, where a is the number in the rightmost column in that row. We call this a **test ratio**.
3. Of these ratios, choose the smallest one. The corresponding number b is the pivot.

4.3 Solving Standard Linear Programming Problems Using the Simplex Method

In our example, the test ratio in the first row is $\frac{10}{2} = 5$, while the test ratio in the second row is $\frac{15}{1} = 15$. Here, 5 is the smallest, so the 2 in the upper left is our pivot and appears in color.

	x	y	z	s	t	p	
s	2	2	1	1	0	0	10
t	1	2	3	0	1	0	15
p	−3	−2	−1	0	0	1	0

Q Why select the pivot this way?

A The rules given above guarantee that, after pivoting, all variables will be nonnegative in the basic solution. In other words, they guarantee that we will remain in the feasible region. We will explain further after finishing this example.

Step 5 *Use the pivot to clear the column as in Gauss-Jordan reduction* (taking care to follow the exact prescription for formulating the row operations described in the chapter on systems of equations) *and then relabel the pivot row with the label from the pivot column.*

In other words, the s on the left of the pivot will be replaced by x. We call s the **departing** or **exiting variable** and x the **entering variable** for this step.

entering variable ↓

departing variable →

	x	y	z	s	t	p		
s	2	2	1	1	0	0	10	
t	1	2	3	0	1	0	15	$2R_2 - R_1$
p	−3	−2	−1	0	0	1	0	$2R_3 + 3R_1$

giving

	x	y	z	s	t	p	
x	2	2	1	1	0	0	10
t	0	2	5	−1	2	0	20
p	0	2	1	3	0	2	30

This is the second tableau (*tableau* is just the French word for matrix).

Step 6 *Go to Step 3.* But wait! According to Step 3, we are finished, since there are no negative numbers in the bottom row. Thus, we can read off the answer. Remember, though, that the solution for x, the first active variable, is not just $x = 10$, but is $x = \frac{10}{2} = 5$, since the pivot has not been reduced to a 1. Similarly, $t = \frac{20}{2} = 10$ and $p = \frac{30}{2} = 15$. All the other variables are zero, since they are inactive. Thus, the solution is as follows:

> p has a maximum value of 15, and this occurs when $x = 5$, $y = 0$, and $z = 0$. (The slack variables then have the values $s = 0$ and $t = 10$.)

Q Why can we stop when there are no negative numbers in the bottom row, and why must this be an optimal solution?

A The bottom row corresponds to the equation $2y + z + 3s + 2p = 30$, or

$$p = 15 - y - \tfrac{1}{2}z - \tfrac{3}{2}s.$$

Think of this as part of the general solution to our system of equations, with y, z, and s as the parameters. Since these variables must be nonnegative, the largest possible value of p in any feasible solution of the system comes when all three of the parameters are 0. Thus, the current basic solution must be an optimal solution.

Calculators or computers could obviously be a big help in the calculations here, just as in Chapter 2. We'll say more about that after the next couple of examples.

We owe you some further explanation for Step 4. After Step 3, we knew that x would be the entering variable, and we needed to choose the departing variable. In the next basic solution, x was to have some positive value and we wanted this value to be as large as possible (to make p as large as possible) without making any other variables negative. Look again at the equations written in Step 2:

$$s = 10 - 2x - 2y - z$$
$$t = 15 - x - 2y - 3z.$$

We needed to make either s or t into an inactive variable, and hence 0. Also, y and z were to remain inactive. If we had made s inactive, then we would have had $0 = 10 - 2x$, so $x = \frac{10}{2} = 5$. This would have made $t = 15 - 5 = 10$, which would be fine. On the other hand, if we had made t inactive, then we would have had $0 = 15 - x$, so $x = 15$, and this would have made $s = 10 - 2 \cdot 15 = -20$, which would *not* be fine, since slack variables must be nonnegative. In other words, we had a choice of making $x = \frac{10}{2} = 5$ or $x = \frac{15}{1} = 15$, but making x larger than 5 would have made a variable negative. We were thus compelled to choose the smaller ratio, 5, and make s the departing variable. Of course, we do not have to think it through this way every time. We just use the rules stated in Step 4.

4.3 Solving Standard Linear Programming Problems Using the Simplex Method

EXAMPLE 4

Find the maximum value of $p = 12x + 15y + 5z$ subject to the constraints

$$2x + 2y + z \leq 8$$
$$x + 4y - 3z \leq 12$$
$$x \geq 0, y \geq 0, z \geq 0.$$

SOLUTION Following Step 1, we introduce slack variables and rewrite the constraints and objective function in standard form.

$$2x + 2y + z + s = 8$$
$$x + 4y - 3z + t = 12$$
$$-12x - 15y - 5z + p = 0$$

We now follow with Step 2, setting up the initial tableau as follows.

	x	y	z	s	t	p	
s	2	2	1	1	0	0	8
t	1	4	-3	0	1	0	12
p	-12	-15	-5	0	0	1	0

For Step 3, we select as the pivot column the one over the negative number with the largest magnitude in the bottom row, which is the y-column. For Step 4, finding the pivot, we see that the test ratios are $\frac{8}{2}$ and $\frac{12}{4}$, the smallest being $\frac{12}{4} = 3$. So we select the pivot in the t row and set up instructions to clear its column.

	x	y	z	s	t	p		
s	2	2	1	1	0	0	8	$2R_1 - R_2$
t	1	4	-3	0	1	0	12	
p	-12	-15	-5	0	0	1	0	$4R_3 + 15R_2$

The departing variable is t, and the entering variable is y. This gives the second tableau.

	x	y	z	s	t	p	
s	3	0	5	2	-1	0	4
y	1	4	-3	0	1	0	12
p	-33	0	-65	0	15	4	180

We now go back to Step 3. Since we still have negative numbers in the bottom row, we choose the one with the largest magnitude (which is -65) and thus our pivot column is the z-column. Since negative numbers can't be pivots, the only possible choice for the pivot is the 5. (We need not compute the test ratios, since there would be only one to choose from.) We now set up instructions for clearing this column, and proceed to do so, not neglecting to take care of the departing and entering variables:

	x	y	z	s	t	p		
s	3	0	5	2	-1	0	4	
y	1	4	-3	0	1	0	12	$5R_2 + 3R_1$
p	-33	0	-65	0	15	4	180	$R_3 + 13R_1$

giving

	x	y	z	s	t	e	
z	3	0	5	2	-1	0	4
y	14	20	0	6	2	0	72
p	6	0	0	26	2	4	232

Notice how the value of p keeps climbing: it started at 0 in the first tableau, went up to $\frac{180}{4} = 45$ in the second, and is currently at $\frac{232}{4} = 58$. As there are no more negative numbers in the bottom row, we are done, and we can write the solution:

p has a maximum value of $\frac{232}{4} = 58$, and this occurs when

$x = 0$,

$y = \frac{72}{20} = \frac{18}{5}$, and

$z = \frac{4}{5}$.

(The slack variables are all zero.)

Before we go on... As a partial check on our answer we can substitute these values into the objective function and the constraints:

$$58 = 12(0) + 15(\tfrac{18}{5}) + 5(\tfrac{4}{5}) \checkmark$$
$$2(0) + 2(\tfrac{18}{5}) + (\tfrac{4}{5}) = 8 \leq 8 \checkmark$$
$$0 + 4(\tfrac{18}{5}) - 3(\tfrac{4}{5}) = 12 \leq 12. \checkmark$$

We say that this is only a partial check, because this only shows that our solution is feasible and that we have correctly calculated p. This check will *usually* catch any arithmetic mistakes you make, but it is not foolproof.

4.3 Solving Standard Linear Programming Problems Using the Simplex Method

▼ **EXAMPLE 5**

Acme Babyfoods makes two puddings, vanilla and chocolate. Each serving of vanilla pudding requires 2 teaspoons of sugar and 25 fluid ounces of water, and each serving of chocolate pudding requires 3 teaspoons of sugar and 15 fluid ounces of water. Acme has available each day 1,200 teaspoons of sugar and 7,500 fluid ounces of water. Further, Acme makes no more than 200 servings of vanilla pudding, as that is all that it can sell per day. If Acme makes a profit of 10¢ on each serving of vanilla pudding and 7¢ on each serving of chocolate, how many servings of each should it make in order to maximize profit?

SOLUTION We first identify our unknowns.

Let x = the number of servings of vanilla pudding.

Let y = the number of servings of chocolate pudding.

The objective function is the profit $p = 10x + 7y$, and we need to maximize this. For the constraints, we start with the fact that Acme will make no more than 200 servings of vanilla: $x \leq 200$. We can put the remaining data in a table as follows.

	Vanilla	Chocolate	Available
Sugar (tsp)	2	3	1,200
Water (fl oz)	25	15	7,500

Since Acme can use no more sugar and water than is available, we get the two constraints

$$2x + 3y \leq 1{,}200$$

and

$$25x + 15y \leq 7{,}500.$$

Thus, our linear programming problem is as follows.

Maximize $p = 10x + 7y$ subject to

$x \leq 200$

$2x + 3y \leq 1{,}200$

$25x + 15y \leq 7{,}500$

$x \geq 0, y \geq 0.$

Introducing the slack variables and setting up the initial tableau, we obtain

$$\begin{aligned} x \phantom{{}+3y} + s \phantom{{}+t+u+p} &= 200 \\ 2x + 3y \phantom{{}+s} + t \phantom{{}+u+p} &= 1200 \\ 25x + 15y \phantom{{}+s+t} + u \phantom{{}+p} &= 7500 \\ -10x - 7y \phantom{{}+s+t+u} + p &= 0. \end{aligned}$$

Note that we have had to introduce a third slack variable, u. In general, there will be as many slack variables as there are constraints (other than the $x \geq 0$ variety). We now go through the procedure as before. Try to see if you can do all the remaining steps on your own, referring to the steps below to check your work.

	x	y	s	t	u	p	
s	1	0	1	0	0	0	200
t	2	3	0	1	0	0	1,200
u	25	15	0	0	1	0	7,500
p	-10	-7	0	0	0	1	0

$R_2 - 2R_1$
$R_3 - 25R_1$
$R_4 + 10R_1$

	x	y	s	t	u	p	
x	1	0	1	0	0	0	200
t	0	3	-2	1	0	0	800
u	0	15	-25	0	1	0	2,500
p	0	-7	10	0	0	1	2,000

$5R_2 - R_3$

$15R_4 + 7R_3$

	x	y	s	t	u	p	
x	1	0	1	0	0	0	200
t	0	0	15	5	-1	0	1,500
y	0	15	-25	0	1	0	2,500
p	0	0	-25	0	7	15	47,500

$15R_1 - R_2$

$3R_3 + 5R_2$

$3R_4 + 5R_2$

	x	y	s	t	u	p	
x	15	0	0	-5	1	0	1,500
s	0	0	15	5	-1	0	1,500
y	0	45	0	25	-2	0	15,000
p	0	0	0	25	16	45	150,000

Thus, the solution is as follows: the maximum value of p is $150,000/45 = 3,333.333$, and it occurs when $x = 1500/15 = 100$, $y = 15,000/45 = 333.333$. (The slack variables are $s = 1500/15 = 100$ and $t = u = 0$.)

Before we go on... Since this was a problem with only two variables, we could have solved it graphically. It is interesting to think about the relationship between the two methods. Figure 1 shows the feasible region.

4.3 Solving Standard Linear Programming Problems Using the Simplex Method

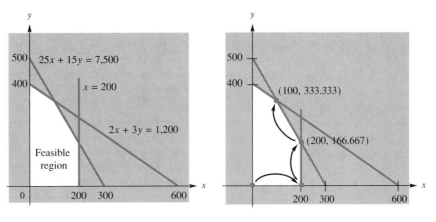

FIGURE 1

FIGURE 2

Each tableau in the simplex method corresponds to a corner of the feasible region, given by the corresponding basic solution. In this example, the sequence of basic solutions is

$$(x, y) = (0, 0), (200, 0), (200, 166.667), (100, 333.333).$$

This is the sequence of corners shown in Figure 2.

In general, we can think of the simplex method as walking from corner to corner of the feasible region until it locates the optimal solution. In problems with many variables and many constraints, it will usually visit only a small fraction of the total number of corners.

It is also interesting in an application like this to think about the meaning of the slack variables. Take s, for example. We can rewrite the original equation $x + s = 200$ as

$$s = 200 - x.$$

In other words, s is the difference between the maximum 200 servings of vanilla that might be made and the number that is actually made. In the optimal solution, $x = 100$, so $s = 200 - 100 = 100$, which says that 100 fewer servings of vanilla were made than the maximum possible. On the other hand, t appears in the equation $2x + 3y + t = 1,200$, which we can write as

$$t = 1,200 - (2x + 3y).$$

This is the difference between the 1,200 teaspoons of sugar available and the amount of sugar that is actually used. In the optimal solution, $t = 0$, which tells us that all of the available sugar is used. If t were not zero, then we would know that some sugar, namely t teaspoons, would go unused. The fact that $u = 0$ tells us that all of the available water is used as well.

THE ROLE OF GRAPHING CALCULATORS AND COMPUTERS

Calculators

A tableau is just a matrix, and the main step in the simplex method is pivoting. Since many calculators can now do row operations on matrices, you can use them to help you with the simplex method, just as you can use them to help with Gauss-Jordan reduction. Just be sure that your calculator can handle the large matrices that arise, since some calculators are very limited in the size of matrix they can hold.

Special-Purpose Computer Software

Any software package capable of doing row operations can be used for the simplex method. In particular, the Macintosh software supplied with this text is specially designed for the simplex method, and it even relabels the rows for you as you pivot. There are also software packages that perform the whole simplex method automatically. Their value for showing the workings of the simplex method is negligible, though, because they do *all* the work for you. On the other hand, these are the programs that are used to solve the really huge LP problems that arise in practice.

Spreadsheet Software

Electronic spreadsheets are proving to be useful tools for doing matrix operations, and they are naturally suited to the simplex method. The text *Topics in Finite Mathematics—An Introduction to the Electronic Spreadsheet* by Samuel W. Spero (published by HarperCollins College Publishers in 1993) will show you how to program your spreadsheet for the simplex method, as well as for other procedures discussed in this text.

▶ NOTE   Since all of the above technology makes the handling of decimals easy, you might find the following convenient when using a graphing calculator or computer.

1. There is no need to get rid of decimals—calculators can handle these with ease.
2. After selecting your pivot, and prior to clearing the pivot column, *divide the pivot row by the value of the pivot, thereby turning the pivot into a* 1. Although this might well result in decimals, it makes the row operations easy to specify. ◀

The following example illustrates the use of a graphing calculator to perform the simplex method.

▼ **EXAMPLE 6**

Maximize $p = 1.3x + 4.5y + 11.3z$ subject to

$$2.4x - 0.21y + 11.7z \leq 3$$
$$3.3x + 5.6y - 2.35z \leq 5$$
$$x - y + 2.3z \leq 10$$
$$x \geq 0, y \geq 0, z \geq 0.$$

4.3 Solving Standard Linear Programming Problems Using the Simplex Method

SOLUTION Pretend that you are using a calculator to solve this problem. One consequence is that you won't be able to label the rows or columns. You can mentally label them as you go, and we'll see that you can do without the row labels anyway. First introduce the slack variables s, t, and u as usual, and set up the initial tableau in the form of a matrix. Your display will show something like the following. (Graphing calculators could not display the whole matrix at once but have scrolling features that permit you to view any part of the matrix.)

2.4	−0.21	11.7	1	0	0	0	3
3.3	5.6	−2.35	0	1	0	0	5
1	−1	2.3	0	0	1	0	10
−1.3	−4.5	−11.3	0	0	0	1	0

The most negative entry in the bottom row is -11.3, so you must select the pivot in its column (Column 3). The smallest test ratio is $3/11.7 = 0.2564$, so the pivot is the entry 11.7 in position 1,3.

Next, as per the note above, turn the pivot into a 1 by dividing Row 1 by the value of the pivot: 11.7.

$$R_1 \longrightarrow \frac{1}{11.7} R_1.$$

This gives

0.20512	−0.0179	1	0.08547	0	0	0	0.25641
3.3	5.6	−2.35	0	1	0	0	5
1	−1	2.3	0	0	1	0	10
−1.3	−4.5	−11.3	0	0	0	1	0

Now clear the pivot column in the usual manner, using the following operations.

$$R_2 \longrightarrow R_2 + 2.35\, R_1$$
$$R_3 \longrightarrow R_3 - 2.3\, R_1$$
$$R_4 \longrightarrow R_4 + 11.3\, R_1$$

This gives

0.205128	−0.017948	1	0.0854701	0	0	0	0.25641
3.78205	5.55782	0	0.200855	1	0	0	5.60256
0.528205	−0.958718	0	−0.196581	0	1	0	9.41206
1.01795	−4.70282	0	0.965812	0	0	1	2.89744

The only negative number in the bottom row is -4.70282, and we see that the pivot is the entry 5.55782 in position 2,2. Turn the pivot into a 1 by the operation

$$R_2 \longrightarrow \frac{1}{5.55782} R_2,$$

giving

0.205128	−0.017948	1	0.0854701	0	0	0	0.25641
0.680492	1	0	0.0361391	0.179927	0	0	1.00805
0.528205	−0.958718	0	−0.196581	0	1	0	9.41026
1.01795	−4.70282	0	0.965812	0	0	1	2.89744

Now pivot on the 2,2 entry by clearing the second column.

$$R_1 \longrightarrow R_1 + 0.017948\, R_2$$
$$R_3 \longrightarrow R_3 + 0.958718\, R_2$$
$$R_4 \longrightarrow R_4 + 4.70282\, R_2$$

(*Note:* the numbers below were found using the exact values as stored in the calculator, rather than the rounded-off values shown on the screen, so if you are following along with hand calculations your answers may differ in the fifth or sixth significant digit.)

0.217342	0	1	0.0861187	0.0032294	0	0	0.274503
0.680492	1	0	0.0361391	0.179927	0	0	1.00805
1.1806	0	0	−0.161934	0.172499	1	0	10.3767
4.21818	0	0	1.13577	0.846163	0	1	7.63812

We see that there are no negative numbers in the bottom row, so we are finished.

Q We don't have labels. So how do we read off the solution?

A Look at the columns containing the pivots. They are the y-column, the z-column, the u-column, and the p-column. We can now read off the solution as follows.

y-column: The pivot is in Row 2, which means that Row 2 would have been labeled with y. So we scan across Row 2 to the answer column to read off the value of y: $y = 1.00805$.

z-column: The pivot is in Row 1, so we scan across Row 1 to the answer column to read off the value of z: $z = 0.274503$.

We ignore the u-column, since that is a slack variable.

p-column: The pivot is in Row 4, so we scan across Row 4 to the answer column to read off the value of p: $p = 7.63812$.

Thus, we have the following solution: the maximum value of p is 7.63812, and this is attained when $x = 0$, $y = 1.00805$, and $z = 0.274503$.

Before we go on... Appendix B contains a graphing calculator program that pivots on any entry of a matrix and also divides a row by a positive number, and so it is particularly suited for the simplex method. This program is written for the TI-82 and more advanced models. (The TI-81 does not accept matrices larger than 6 × 6.) To use it on matrices that are too large to fit on the display, you will need to stop the program after each step in order to examine the entire matrix (called [A]).

SUMMARY OF SIMPLEX METHOD

To solve a standard maximization problem using the simplex method, we take the following steps:

1. Convert to a system of equations by introducing **slack variables** to turn the constraints into equations and rewriting the objective function in standard form.
2. Write down the initial **tableau.**
3. Select the pivot column: Choose the negative number with the largest magnitude in the left-hand side of the bottom row. Its column is the pivot column. (If there are two or more candidates, choose any one of them.) If all the numbers in the left-hand side of the bottom row are zero or positive, then you are finished: the basic solution maximizes the objective function. (See below for the basic solution.)
4. Select the pivot in the pivot column: The pivot must always be a positive number. For each positive entry b in the pivot column, compute the ratio a/b, where a is the number in the answer column in that row. Of these **test ratios**, choose the smallest one. The corresponding number b is the pivot.
5. Use the pivot to clear the column as in Gauss-Jordan reduction (taking care to follow the exact prescription for formulating the row operations described in Chapter 2) and then relabel the pivot row with the label from the pivot column. The variable originally labeling the pivot row is the **departing** or **exiting variable,** and the variable labeling the column is the **entering variable.**
6. Go to Step 3.

To get the **basic solution** corresponding to any tableau in the simplex method, set to zero all variables that do not appear as row labels. The value of a variable that does appear as a row label (an **active variable**) is the number in the rightmost column in that row divided by the number in that row in the column labeled by the same variable.

Here are some last points.

Q What if there is no candidate for the pivot. For example, what do I do with a tableau like the following?

	x	y	z	s	t	p	
s	0	0	5	2	0	0	4
y	−8	20	0	6	5	0	72
p	−20	0	0	26	15	4	232

A Here, the pivot column is the x-column, but there is no suitable entry for a pivot (since zeros and negative numbers can't be pivots). In cases such as this, the feasible region is unbounded and there is also no optimal solution. In other words, p can be made as large as we like without violating the constraints.

Q What should I do if there is a negative number in the rightmost column?

A A negative number will not appear above the bottom row in the rightmost column unless an error has been made. (The bottom-right entry is allowed to be negative.) The most likely errors leading to this situation are these.

(a) The pivot was chosen incorrectly (don't forget to choose the *smallest* test ratio).
(b) The row operation instruction was written backwards or performed backwards (e.g., instead of $R_2 - R_1$, you did $R_1 - R_2$).
(c) An arithmetic error occurred. (We all make those annoying errors from time to time.)

Q What about zeros in the rightmost column?

A Zeros are permissible in the rightmost column. For example, the constraint $x + y \leq 0$ will lead to a zero in the rightmost column.

One last suggestion: If it is possible to do a simplification step (dividing a row by a positive number) *at any stage,* you are perfectly free to do so. As we saw in Chapter 2, this can help prevent the numbers from getting out of hand.

▶ 4.3 EXERCISES

1. Maximize $p = 2x + y$ subject to
$$x + 2y \leq 6$$
$$-x + y \leq 4$$
$$x + y \leq 4$$
$$x \geq 0, y \geq 0.$$

4.3 Solving Standard Linear Programming Problems Using the Simplex Method

2. Maximize $p = x$ subject to
$$x - y \leq 4$$
$$-x + 3y \leq 4$$
$$x \geq 0, y \geq 0.$$

3. Maximize $p = x - y$ subject to
$$5x - 5y \leq 20$$
$$2x - 10y \leq 40$$
$$x \geq 0, y \geq 0.$$

4. Maximize $p = 2x + 3y$ subject to
$$3x + 8y \leq 24$$
$$6x + 4y \leq 30$$
$$x \geq 0, y \geq 0.$$

5. Maximize $z = 3x_1 + 7x_2 + 8x_3$ subject to
$$5x_1 - x_2 + x_3 \leq 1500$$
$$2x_1 + 2x_2 + x_3 \leq 2500$$
$$4x_1 + 2x_2 + x_3 \leq 2000$$
$$x_1 \geq 0, x_2 \geq 0, x_3 \geq 0.$$

6. Maximize $z = 3x_1 + 4x_2 + 6x_3$ subject to
$$5x_1 - x_2 + x_3 \leq 1500$$
$$2x_1 + 2x_2 + x_3 \leq 2500$$
$$4x_1 + 2x_2 + x_3 \leq 2000$$
$$x_1 \geq 0, x_2 \geq 0, x_3 \geq 0.$$

7. Maximize $p = 5x - 4y + 3z$ subject to
$$5x + 5z \leq 100$$
$$5y - 5z \leq 50$$
$$5x - 5y \leq 50$$
$$x \geq 0, y \geq 0, z \geq 0.$$

8. Maximize $p = 6x + y + 3z$ subject to
$$3x + y \leq 15$$
$$2x + 2y + 2z \leq 20$$
$$x \geq 0, y \geq 0, z \geq 0.$$

9. Maximize $p = 7x + 5y + 6z$ subject to
$$x + y - z \leq 3$$
$$x + 2y + z \leq 8$$
$$x + y \leq 5$$
$$x \geq 0, y \geq 0, z \geq 0.$$

10. Maximize $p = 3x + 4y + 2z$ subject to
$$3x + y + z \le 5$$
$$x + 2y + z \le 5$$
$$x + y + z \le 3$$
$$x \ge 0, y \ge 0, z \ge 0.$$

11. Maximize $p = x + y + z + w$ subject to
$$x + y + z \le 3$$
$$y + z + w \le 4$$
$$x + z + w \le 5$$
$$x + y + w \le 6$$
$$x \ge 0, y \ge 0, z \ge 0, w \ge 0.$$

12. Maximize $p = x - y + z + w$ subject to
$$x + y + z \le 3$$
$$y + z + w \le 3$$
$$x + z + w \le 4$$
$$x + y + w \le 4$$
$$x \ge 0, y \ge 0, z \ge 0, w \ge 0.$$

13. Maximize $p = x + y + z + w + v$ subject to
$$x + y \le 1$$
$$y + z \le 2$$
$$z + w \le 3$$
$$w + v \le 4$$
$$x \ge 0, y \ge 0, z \ge 0, w \ge 0, v \ge 0.$$

14. Maximize $p = x + 2y + z + 2w + v$ subject to
$$x + y \le 1$$
$$y + z \le 2$$
$$z + w \le 3$$
$$w + v \le 4$$
$$x \ge 0, y \ge 0, z \ge 0, w \ge 0, v \ge 0.$$

In Exercises 15 through 20 use a calculator, a special-purpose program (such as that available with this text) or an electronic spreadsheet. Round all answers to two decimal places.

15. Maximize $p = 2.5x + 4.2y + 2z$ subject to
$$0.1x + y - 2.2z \le 4.5$$
$$2.1x + y + z \le 8$$
$$x + 2.2y \le 5$$
$$x \ge 0, y \ge 0, z \ge 0.$$

16. Maximize $p = 2.1x + 4.1y + 2z$ subject to
$$3.1x + 1.2y + z \leq 5.5$$
$$x + 2.3y + z \leq 5.5$$
$$2.1x + y + 2.3z \leq 5.2$$
$$x \geq 0, y \geq 0, z \geq 0.$$

17. Maximize $p = x + 2y + 3z + w$ subject to
$$x + 2y + 3z \leq 3$$
$$y + z + 2.2w \leq 4$$
$$x + z + 2.2w \leq 5$$
$$x + y + 2.2w \leq 6$$
$$x \geq 0, y \geq 0, z \geq 0, w \geq 0.$$

18. Maximize $p = 1.1x - 2.1y + z + w$ subject to
$$x + 1.3y + z \leq 3$$
$$1.3y + z + w \leq 3$$
$$x + z + w \leq 4.1$$
$$x + 1.3y + w \leq 4.1$$
$$x \geq 0, y \geq 0, z \geq 0, w \geq 0.$$

19. Maximize $p = x - y + z - w + v$ subject to
$$x + y \leq 1.1$$
$$y + z \leq 2.2$$
$$z + w \leq 3.3$$
$$w + v \leq 4.4$$
$$x \geq 0, y \geq 0, z \geq 0, w \geq 0, v \geq 0.$$

20. Maximize $p = x - 2y + z - 2w + v$ subject to
$$x + y \leq 1.1$$
$$y + z \leq 2.2$$
$$z + w \leq 3.3$$
$$w + v \leq 4.4$$
$$x \geq 0, y \geq 0, z \geq 0, w \geq 0, v \geq 0.$$

APPLICATIONS

21. *Purchasing* You are in charge of purchases at the student-run used book supply program at your college, and you must decide how many introductory calculus, history, and marketing texts should be purchased from students for resale. Because of budget limitations, you cannot purchase more than 650 of these textbooks per semester. There are also shelf space limitations: calculus texts occupy 2 units of shelf space each, history books 1 unit each, and marketing texts 3 units each, and you can spare at most 1,000 units of shelf space for the texts. Assuming the used book program makes a profit of $10 on each calculus text, $4 on each history text, and $8 on each marketing text, how many of each type of text should you purchase to maximize profit? What is the maximum profit the program can make in a semester?

22. *Sales* The Marketing Club at your college has decided to raise funds by selling three types of T-shirts: one with a single-color "ordinary" design, one with a two-color "fancy" design, and one with a three-color "very fancy" design. The club feels that it can sell up to 300 T-shirts. "Ordinary" T-shirts will cost the club $6 each, "fancy" T-shirts $8 each, and "very fancy" T-shirts $10 each, and the club has a total purchasing budget of $3,000. It will sell "ordinary" T-shirts at a profit of $4 each, "fancy" T-shirts at a profit of $5 each, and "very fancy" T-shirts at a profit of $4 each. How many of each kind of T-shirt should the club order to maximize profit? What is the maximum profit the club can make?

23. *Resource Allocation* The Arctic Juice Company makes three juice blends: PineOrange, using 2 portions of pineapple juice and 2 portions of orange juice per gallon; PineKiwi, using 3 portions of pineapple juice and 1 portion of kiwi juice per gallon; and OrangeKiwi, using 3 portions of orange juice and 1 portion of kiwi juice per gallon. Each day it has 800 portions of pineapple juice, 650 portions of orange juice, and 350 portions of kiwi juice available. Its profit on PineOrange is $1 per gallon, its profit on PineKiwi is $2 per gallon, and its profit on OrangeKiwi is $1 per gallon. How many gallons of each blend should it make each day in order to maximize profit? What is the largest possible profit it can make?

24. *Purchasing* Trans Global Tractor Trailers has decided to spend up to $1,500,000 on a fleet of new trucks and is considering three models: the Gigahaul, which has a capacity of 6,000 cubic feet and is priced at $60,000, the Megahaul, with a capacity of 5,000 cubic feet and priced at $50,000, and the Picohaul, with a capacity of 2,000 cubic feet, priced at $40,000. The anticipated annual revenues are $600,000 for each Gigahaul, $550,000 for each Megahaul, and $500,000 for each Picohaul it buys. Trans Global would like a total capacity of up to 130,000 cubic feet and feels that it cannot provide drivers and maintenance for more than 30 trucks. How many of each should it purchase in order to maximize annual revenue? What is the largest possible revenue it can make?

25. *Resource Allocation* The Enormous State University History Department offers three courses, Ancient, Medieval, and Modern History, and the department chairperson is trying to decide how many sections of each to offer this semester. They may offer up to 45 sections total, up to 5000 students would like to take a course, and there are 60 professors to teach them (each professor teaches at most one section). Sections of Ancient History hold 100 students each, sections of Medieval History hold 50 students each, and sections of Modern History hold 200 students each. Modern History sections are taught by a team of 2 professors, while Ancient and Medieval History need only one professor per section. Ancient History nets the university $10,000 per section, Medieval nets $20,000, and Modern History nets $30,000 per section. How many sections of each course should the department offer in order to generate the largest profit? What is the largest profit possible? Will there be any unused sections, any students who did not get into classes, or any professors without anything to teach?

26. *Resource Allocation* You manage an ice cream factory that makes three flavors: Creamy Vanilla, Continental Mocha and Succulent Strawberry. Into each batch of Creamy Vanilla go two eggs, one cup of milk and two cups of cream. Into each batch of Continental Mocha go one egg, one cup of milk and two cups of cream, while into each batch of Succulent Strawberry go one egg, two cups of milk and two cups of cream. You have in stock 200 eggs, 120 cups of milk, and 200 cups of cream. You make a profit of $3 on each batch of Creamy Vanilla, $2 on each batch of Continental Mocha, and $4 on each batch of Succulent Strawberry.
    (a) How many batches of each flavor should you make in order to maximize your profit?
    (b) In your answer to part (a), have you used all your ingredients?
    (c) Due to the poor strawberry harvest this year, you cannot make more than 10 batches of Succulent Strawberry. Does this affect your maximum profit?

27. *Agriculture* Your small farm encompasses 100 acres, and you are planning to grow tomatoes, lettuce, and carrots in the coming planting season. Fertilizer costs per acre are as follows: $5 for tomatoes, $4 for lettuce, and $2 for carrots. Based on past experience, you estimate that each acre of tomatoes will require an average of 4 hours of labor per week, while tending to lettuce and carrots will each require an average of 2 hours per week. You estimate a profit of $2,000 for each acre of tomatoes, $1,500 for each acre of lettuce and $500 for each acre of carrots. You can afford to

spend no more than $400 on fertilizer, and your farm laborers can supply 500 hours per week. How many acres of each crop should you plant in order to maximize total profits? In this event, will you be using all 100 acres of your farm?

28. *Agriculture* Your farm encompasses 500 acres, and you are planning to grow soybeans, corn and wheat in the coming planting season. Fertilizer costs per acre are as follows: $5 for soybeans, $2 for corn and $1 for wheat. You estimate that each acre of soybeans will require an average of 5 hours of labor per week, while tending to corn and wheat will each require an average of 2 hours per week. Based on past yields and current market prices, you estimate a profit of $3,000 for each acre of soybeans, $2,000 for each acre of corn, and $1,000 for each acre of wheat. You can afford to spend no more than $3,000 on fertilizer, and your farm laborers can supply 3,000 hours per week. How many acres of each crop should you plant in order to maximize total profits? In this event, will you be using all the available labor?

29. *Resource Allocation* (Note that the following exercise is almost identical to Exercise 6 in Section 2.4 except for one important detail. Refer to your solution of that problem—if you did it—and then attempt this one.) The Enormous State University Choral Society is planning its annual Song Festival, at which it will serve three kinds of delicacies: granola treats, nutty granola treats, and nuttiest granola treats. The following table shows the ingredients required for a single serving of each delicacy, as well as the total amount of each ingredient available. Amounts are in ounces.

	Granola	Nutty Granola	Nuttiest Granola	Total Available
Toasted Oats	1	1	5	1,500
Almonds	4	8	8	10,000
Raisins	2	4	8	4,000

The Choral Society makes a profit of $6 on each serving of granola, $8 on each serving of nutty granola, and $3 on each serving of nuttiest granola. Assuming that it can sell all that it makes, how many servings of each should it make in order to maximize profits? How much of each ingredient will be left over?

30. *Resource Allocation* Repeat the preceding exercise, but this time assume that the Choral Society makes a $3 profit on each of its delicacies.

31. *Recycling* Safety-Kleen operates the world's largest oil re-refinery at Elgin, Illinois. You have been hired by the company to determine how to allocate its intake of up to 50 million gallons of used oil to its three refinery processes: A, B and C. You are told that electricity costs for process A amount to $150,000 per million gallons treated, while for processes B and C the costs are respectively $100,000 and $50,000 per million gallons treated. Process A can recover 60% of the used oil, process B can recover 55%, and process C can recover only 50%. Assuming a revenue of $4 million per million gallons of recovered oil and an annual electrical budget of $3 million, how much used oil would you allocate to each process in order to maximize total revenues?*

32. *Recycling* Repeat the previous exercise, but this time assume that process C can handle no more than 20 million gallons per year.

33. *Loan Planning*[†] Enormous State University's employee credit union has $5,000,000 available for loans in the coming year. As VP in charge of university finances, you must decide how much capital to allocate to each of four different kinds of loans, as shown in the following table.

Type of Loan	Annual Rate of Return (%)
Automobile loans	8
Furniture loans	10
Signature loans	12
Other secured loans	10

▼ *These figures are accurate: *Safety-Kleen*'s actual 1993 capacity was 50 million gallons, its recycled oil sold for approximately $4 per gallon, its recycling process could recover approximately 55% of the used oil, and its electrical bill was $3 million. Source: "Oil Recycler Greases Rusty City's Economy," *The Chicago Tribune*, May 30 1993, Section 7, p.1.

[†] Adapted from an exercise in *An Introduction to Management Science*, Sixth Edition, by D. R. Anderson, D. J. Sweeny, and T. A. Williams (St. Paul: West, 1991).

State laws and credit union policies impose the following restrictions:
(a) Signature loans may not exceed 10% of the total investment of funds.
(b) Furniture loans plus other secured loans may not exceed automobile loans
(c) Other secured loans may not exceed 200% of automobile loans

How much would you allocate to each type of loan in order to maximize the annual return?

34. *Investments* You have $100,000 and are considering buying stocks in *American Express, CBS,* and *Chemical Bank.* The New York Stock Exchange has quoted the following data.*

Company	Price per Share	Yield (%)
American Express	30	3.5
CBS	240	0.4
Chemical Bank	40	3.4

Your broker has made the following suggestions:
(a) At least 50% of your total investment should be in *CBS;* there is a rumor that the stock will split.
(b) Invest no more than 10% of your total investment in *American Express.*

How many shares of each stock should you purchase in order to maximize your anticipated returns while following your broker's advice? (Assume that it is possible to buy fractions of shares.)
[*Warning:* Some of the numbers in your tableau will get quite large, but the calculation is not difficult to do by hand with the help of an ordinary calculator.]

35. *Portfolio Management* If x dollars is invested in a company that controls, say, 30% of the market with 5 brand-names, then $0.30x$ is a measure of market exposure, while $5x$ is a measure of brand-name exposure. Now suppose you are a broker at a large securities firm, and one of your clients would like to invest up to $100,000 in recording industry stocks. You decide to recommend a combination of stocks in four of the world's largest companies: *Warner Music, Sony, Polygram* and *EMI.* (See the table.†)

	Warner Music	Sony
Country	U.S.A.	Japan
Market share	22%	15%
Number of labels	8	2

	Polygram	EMI
Country	Netherlands	Britain
Market share	12%	11%
Number of labels	7	8

You would like your client's brand-name exposure to be as large as possible but her total market exposure to be limited to 15,000 or less. (This would reflect an average of 15%.) Further, you would like at least 25% of the investment to be in *Polygram,* as you feel that its control of the DGG and Phillips labels is advantageous for its classical music operations. How much should you advise your client to invest in each company?

36. *Portfolio Management* Repeat the previous exercise using the following 1992 figures for market share.†

	Warner Music	Sony	Polygram	EMI
Market share	25%	17%	14%	12%

---

* *The New York Times,* May 17, 1993; prices rounded to the nearest dollar.

† Figures are for 1994. Market share precentages are rounded to the nearest 1%. Source: *The New York Times,* Nov. 1, 1994, p. D1.

37. *Transportation Scheduling* (This exercise is almost identical to one in Chapter 2, Section 4, but is somewhat more realistic; one cannot always expect to exactly fill all orders and keep all plants operating at 100% capacity.) The Tubular Ride Boogie Board Company has manufacturing plants in Tucson, Arizona, and Toronto, Ontario. You have been given the job of coordinating distribution of its latest model, the Gladiator, to its outlets in Honolulu and Venice Beach. The Tucson plant, when operating at full capacity, can manufacture 620 Gladiator boards per week, while the Toronto plant, beset by labor disputes, can only produce 410 Gladiator boards per week. The outlet in Honolulu orders 500 gladiator boards per week, while Venice Beach orders 530 boards per week. Transportation costs are as follows:

Tucson to Honolulu: $10 per board;

Tucson to Venice Beach: $5 per board;

Toronto to Honolulu: $20 per board;

Toronto to Venice Beach: $10 per board.

Your manager has informed you that the company's total transportation budget is $6,550. You realize that it may not be possible to fill all the orders, but you would like the total number of boogie boards shipped to be as large as possible. Given this, how many Gladiator boards should you order shipped from each manufacturing plant to each distribution outlet?

38. *Transportation Scheduling* Repeat the preceding exercise, but use a transportation budget of $5,050.

39. *Transportation Scheduling (Graphing calculator or computer software required)* Your publishing company is about to start its promotional blitz for its new book, *Physics for the Liberal Arts*. You have 10 salespeople stationed in Chicago and 20 in Denver. You would like to fly at most 10 into Los Angeles, and at most 15 into New York. A round-trip plane flight from Chicago to LA costs $428;* from Chicago to NY costs $305; from Denver to LA costs $338; and from Denver to NY costs $493. You want to spend at most $8,895 on plane flights. How many salespeople should you fly from each of Chicago and Denver to each of LA and NY in order to have the most salespeople on the road?

40. *Transportation Scheduling (Graphing calculator or computer software required)* Repeat the previous exercise, but this time spend at most $6,430. [*Warning:* You will have a lot of leeway in choosing pivots in this problem. If you choose the right pivots, you can solve this very quickly, but the wrong pivots will take you the long way around. Think about which variables *should* enter early on.]

COMMUNICATION AND REASONING EXERCISES

41. Find a linear programming problem with three variables and two constraints that requires only one pivot to solve using the simplex method.

42. Find a linear programming problem with three variables and two constraints that requires two pivots to solve using the simplex method.

43. Why is the simplex method useful? (After all, we do have the graphical method for solving LP problems.)

44. Are there any types of linear programming problems that cannot be solved with the methods of this section, but that can be solved using the methods of the previous section? Explain.

45. What is a "basic solution"?

46. In a typical linear programming tableau, there are more unknowns than equations, and we know from the chapter on systems of linear equations that this often implies the existence of infinitely many solutions. How are these solutions related to the basic solutions?

47. Can the value of the objective function decrease in passing from one tableau to the next? Explain.

48. Can the value of the objective function remain unchanged in passing from one tableau to the next? Explain.

▼ * Prices from the OAG Electronic Edition, for the week of Sept. 13, 1993, checked on July 25, 1993.

4.4 SOLVING NONSTANDARD LINEAR PROGRAMMING PROBLEMS

As we saw in Section 2, not all LP problems are standard maximization problems. For example, you might have constraints such as $2x + 3y \geq 4$ or perhaps $2x + 3y = 8$. Or, you might have to minimize, rather than maximize, the objective function. Nonstandard problems are almost as easy to deal with as the standard kind—there is a modification of the simplex method that works very nicely. The best way to illustrate this is by means of examples.

▼ EXAMPLE 1 Nonstandard Constraints

Maximize $p = 12x + 4y + 6z$ subject to

$$x + y + z \leq 100$$
$$10x + 4y + 7z \leq 500$$
$$x + y + z \geq 60$$
$$x \geq 0, y \geq 0, z \geq 0$$

SOLUTION We begin by turning the first two inequalities into equations as usual, since they have the standard form. We get

$$x + y + z + s = 100$$
$$10x + 4y + 7z + t = 500.$$

Although we are tempted to use a slack variable for the third inequality, $x + y + z \geq 60$, we notice that it is nonstandard. In order to turn this into an equation, *adding* something to the left-hand side will not work, since this will make it even bigger. To get it to equal 60, we must *subtract* some nonnegative number. We shall call this number u (since we have already used s and t), and refer to u as a **surplus variable** rather than a slack variable. Thus, we write

$$x + y + z - u = 60.$$

Continuing with the setup, we have

$$x + y + z + s = 100$$
$$10x + 4y + 7z + t = 500$$
$$x + y + z - u = 60$$
$$-12x - 4y - 6z + p = 0.$$

This leads to the initial tableau.

	x	y	z	s	t	u	p	
s	1	1	1	1	0	0	0	100
t	10	4	7	0	1	0	0	500
*u	1	1	1	0	0	-1	0	60
p	-12	-4	-6	0	0	0	1	0

4.4 Solving Nonstandard Linear Programming Problems

Q Why have we put a star next to the third row?

A If we look at the basic solution corresponding to this tableau, we see that it is

$$x = y = z = 0, \ s = 100, \ t = 500, \ u = \frac{60}{-1} = -60.$$

There are several things wrong with this basic solution. First, the values $x = y = z = 0$ do not satisfy the third inequality $x + y + z \geq 60$. Thus the present basic solution is *not feasible*, because it fails to satisfy all the constraints. Second—and this is really the same problem—the surplus variable u is negative, whereas we said that it should be nonnegative. The reason we placed a star next to the row corresponding to u is therefore to alert us to the fact that the present basic solution is not feasible, and that the problem is located in the starred row, where the active variable u is negative.

RULE FOR STARRING ROWS

Whenever an active variable is negative, we star the corresponding row.

In setting up the initial tableau, this just amounts to starring those rows coming from $\geq$ inequalities.

Our use of the simplex method described in the previous section assumed that we were in the feasible region, but now we are not. Thus we must first do something to get ourselves into the feasible region. In practice we can think of this as getting rid of the stars on the rows. Once we get into the feasible region, we go back to the method of the previous section.

Phase 1 *Moving into the feasible region (getting rid of stars on rows)*
The rule for getting rid of a star on a row is quite simple:

To remove the star from a row, find a pivot in that row, and use it to pivot.

Q Fine, but how do we find a pivot there?

A There are several methods we could use, but the following is one of the easier to use in hand calculations.

SELECTING A PIVOT IN A STARRED ROW

Plan A *Ignoring the values in the bottom row,* locate any column in which the pivot, *selected as per test ratios,* happens to be in the starred row.

Plan B If Plan A fails (so that there are no candidate pivots in the starred row) then, *and only then,* pick *any* positive number in that row as the pivot.

If we were to pivot on the 1 in the x-column, for example, the result would be

	x	y	z	s	t	u	p	
x	1	1	1	0	0	-1	0	60

and we would remove the star, since now $x = 60/1 = 60$. The star would be removed if we pivoted on *any* positive number in the row.

However, choosing the wrong pivot can introduce a star in *another* row. For example, pivoting on the 1 in the x-column as above would produce in the second row

	x	y	z	s	t	u	p	
*t	0	-6	-3	0	1	10	0	-100

We would have to put a star on this row since now $t = -100/1 = -100$. We saw why this would happen in the previous section when we discussed the reason for looking at the test ratios. In fact, that discussion applies here to show that using a pivot satisfying the test ratio condition is the only way to keep from introducing stars on other rows.

The reasoning behind Plan A is therefore this: if we can use a pivot satisfying the test ratio condition, we will remove one star without introducing any others. On the other hand, this is not always possible, and then we have to be satisfied with removing one star at the expense of introducing another somewhere else (Plan B).

USING THE SIMPLEX METHOD FOR NONSTANDARD CONSTRAINTS

RULE FOR STARRING ROWS
Whenever an active variable is negative, we star the corresponding row. In setting up the initial tableau, this just amounts to starring those rows coming from $\geq$ inequalities.

PHASE 1

Eliminate Starred Rows
To remove the star from a row, select a pivot in that row, and use it to pivot. To select such a pivot, use the following plan.

Selecting a Pivot in a Starred Row
Plan A *Ignoring the values in the bottom row,* locate any column in which the pivot, *selected as per test ratios,* happens to be in the starred row.

4.4 Solving Nonstandard Linear Programming Problems

Plan B If Plan A fails (so that there are no candidate pivots in the starred row) then, *and only then,* pick *any* positive number in that row as the pivot.

Once there are no starred rows left, proceed to Phase 2.

PHASE 2

In Phase 2, we apply the standard simplex method we studied in the preceding section.

Here is an example that begins with two starred rows.

▼ **EXAMPLE 2**

Maximize $p = -x - y - 4z$ subject to

$$x + 3y + 8z \geq 60$$
$$2x + y + z \leq 200$$
$$x + z \geq 50$$
$$x \geq 0, y \geq 0, z \geq 0.$$

SOLUTION We first introduce slack and surplus variables as in the first example, and then set up the initial tableau.

$$x + 3y + 8z - s = 60$$
$$2x + y + z + t = 200$$
$$x + z - u = 50$$
$$x + y + 4z + p = 0$$

	x	y	z	s	t	u	p	
*s	1	3	8	-1	0	0	0	60
t	2	1	1	0	1	0	0	200
*u	1	0	1	0	0	-1	0	50
p	1	1	4	0	0	0	1	0

Q Because there are no negative numbers in the bottom row, aren't we already finished?

A No, because there are starred rows. The present basic solution is not feasible, since both s and u are negative.

Thus, we need to invoke Plan A (if possible) to find a pivot in one of the starred rows; it doesn't really matter which row. Looking first at the x-column, we see that the pivot is the 1 in the bottom row. Since it is in a starred row, it will do just fine for our pivot.

	x	y	z	s	t	u	p		
*s	1	3	8	-1	0	0	0	60	$R_1 - R_3$
t	2	1	1	0	1	0	0	200	$R_2 - 2R_3$
*u	1	0	1	0	0	-1	0	50	
p	1	1	4	0	0	0	1	0	$R_4 - R_3$

	x	y	z	s	t	u	p	
*s	0	3	7	-1	0	1	0	10
t	0	1	-1	0	1	2	0	100
x	1	0	1	0	0	-1	0	50
p	0	1	3	0	0	1	1	-50

The present basic solution is $x = 50$, $s = -10$, $t = 100$, *everything else* $= 0$, so we lose one of the stars and still retain the star in the first row. We must now find a pivot in the starred row. There are several possible candidates, as shown here.

	x	y	z	s	t	u	p	
*s	0	3	7	-1	0	1	0	10

All the numbers in color happen to satisfy the test ratio requirement to qualify as the pivot in their respective columns. The 3 in the y-column already has part of its column cleared, so we can save ourselves some work by choosing it as the pivot.

	x	y	z	s	t	u	p		
*s	0	3	7	-1	0	1	0	10	
t	0	1	-1	0	1	2	0	100	$3R_2 - R_1$
x	1	0	1	0	0	-1	0	50	
p	0	1	3	0	0	1	1	-50	$3R_4 - R_1$

	x	y	z	s	t	u	p	
y	0	3	7	-1	0	1	0	10
t	0	0	-10	1	3	5	0	290
x	1	0	1	0	0	-1	0	50
p	0	0	2	1	0	2	3	-160

4.4 Solving Nonstandard Linear Programming Problems

Along the way p has become negative. Don't panic. We are not restricting p to be ≥ 0. The variables that can't be negative are all the others: x, y, z, s, t, u. (That's why we draw a line above the p-row in the tableau.)

Now, all of the stars have been eliminated, so we can go on to Phase 2. In choosing the biggest-valued negative number in the bottom row, we ignore the number in the rightmost column and choose from the rest. But there aren't any! Thus, we are done and can simply read off the solution:

$$p = -\frac{160}{3} = -53\frac{1}{3}, \quad x = \frac{50}{1} = 50,$$
$$y = \frac{10}{3} = 3\frac{1}{3}, \quad z = 0.$$

We should check our answers as usual, but we shall leave that to you.

▼ **EXAMPLE 3**

Maximize $p = 2x + 3y$ subject to

$$x + y \geq 35$$
$$x + 2y \leq 60$$
$$2x + y \leq 60$$
$$x \geq 0, y \geq 0.$$

SOLUTION We introduce slack and surplus variables and write down the initial tableau.

$$\begin{aligned} x + y - s &= 35 \\ x + 2y + t &= 60 \\ 2x + y + u &= 60 \\ -2x - 3y + p &= 0 \end{aligned}$$

	x	y	s	t	u	p	
*s	1	1	-1	0	0	0	35
t	1	2	0	1	0	0	60
u	2	1	0	0	1	0	60
p	-2	-3	0	0	0	1	0

We have marked in color the possible pivots (the entries in each column that satisfy the test ratio condition), but none of them is in the starred row. Thus, Plan A fails, and we must fall back on Plan B. We pivot on the 1 in the first row and first column, and hope for the best.

	x	y	s	t	u	p		
*s	1	1	-1	0	0	0	35	
t	1	2	0	1	0	0	60	$R_2 - R_1$
u	2	1	0	0	1	0	60	$R_3 - 2R_1$
p	-2	-3	0	0	0	1	0	$R_4 + 2R_1$

	x	y	s	t	u	p	
x	1	1	−1	0	0	0	35
t	0	1	1	1	0	0	25
*u	0	−1	2	0	1	0	−10
p	0	−1	−2	0	0	1	70

Although we have removed the star from the first row, we have introduced a star in the third row, since u now has the value $-\frac{10}{1} = -10$. This is the "penalty" for using Plan B: it will force one of the variables to change sign.* (Don't forget to check the basic solution at each step, even if it is a quick mental check.)

We have also introduced a negative number in the rightmost column, in the third row, which is easily fixed by multiplying that row by −1 (this will always happen when Plan B has to be used).

	x	y	s	t	u	p	
x	1	1	−1	0	0	0	35
t	0	1	1	1	0	0	25
*u	0	1	−2	0	−1	0	10
p	0	−1	−2	0	0	1	70

Multiplying the row by −1 does not remove the star (why?), so we are still in Phase I. We have marked in color the possible pivots for the next step. Fortunately, one of them lies in the starred row, so we can use Plan A.

	x	y	s	t	u	p		
x	1	1	−1	0	0	0	35	$R_1 - R_3$
t	0	1	1	1	0	0	25	$R_2 - R_3$
*u	0	1	−2	0	−1	0	10	
p	0	−1	−2	0	0	1	70	$R_4 + R_3$

	x	y	s	t	u	p	
x	1	0	1	0	1	0	25
t	0	0	3	1	1	0	15
y	0	1	−2	0	−1	0	10
p	0	0	−4	0	−1	1	80

▼ *If that variable had been negative—so that the corresponding row had been starred—this would have the pleasant effect of causing that variable to become positive, thus removing two stars at once.

4.4 Solving Nonstandard Linear Programming Problems

We have now removed all of the stars, so we are in the feasible region. We proceed with Phase II.

	x	y	s	t	u	p		
x	1	0	1	0	1	0	25	$3R_1 - R_2$
t	0	0	3	1	1	0	15	
y	0	1	-2	0	-1	0	10	$3R_3 + 2R_2$
p	0	0	-4	0	-1	1	80	$3R_4 + 4R_2$

	x	y	s	t	u	p	
x	3	0	0	-1	2	0	60
s	0	0	3	1	1	0	15
y	0	3	0	2	-1	0	60
p	0	0	0	4	1	3	300

Thus, the optimal solution is

$$x = 20,\ y = 20,\ p = 100 \quad (s = 5,\ t = 0,\ u = 0).$$

Before we go on... Since this is an example with only two unknowns, we can picture the sequence of basic solutions on the graph of the feasible region. This is shown in Figure 1.

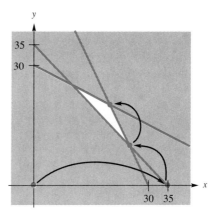

FIGURE 1

You can see that there was no way to jump from (0, 0) in the initial tableau directly into the feasible region, since the first jump must be along an axis (why?).

Now that we know how to deal with nonstandard constraints, we consider **minimization** problems, problems in which we have to minimize, rather than maximize, the objective function. The method of dealing with these is simple.

MINIMIZATION PROBLEMS

We convert a minimization problem into a maximization problem by taking the negative of the objective function.

Here is an example.

▼ **EXAMPLE 4**   Purchasing

You are are in charge of ordering furniture for your company's new headquarters. You need to buy at least 200 tables, 500 chairs, and 300 computer desks. Wall-to-Wall Furniture (WWF) is offering a package of 20 tables, 25 chairs and 18 computer desks for $3,200, while rival Acme Furniture (AF) is offering a package of 10 tables, 50 chairs, and 24 computer desks for $4,000. How many packages should you order from each company in order to minimize your total cost?

SOLUTION   The unknowns here are

x = the number of packages ordered from WWF;

y = the number of packages ordered from AF.

If we put the information about the various kinds of furniture in a table, we get the following.

	WWF	AF	Needed
Tables	20	10	200
Chairs	25	50	500
Computer desks	18	24	300
Cost ($)	3,200	4,000	

From this table, we get the following linear programming problem.

Minimize $c = 3{,}200x + 4{,}000y$ subject to

$$20x + 10y \geq 200$$
$$25x + 50y \geq 500$$
$$18x + 24y \geq 300$$
$$x \geq 0, \; y \geq 0.$$

4.4 Solving Nonstandard Linear Programming Problems

Before we start solving this problem, notice that all the inequalities may be simplified by dividing each by some positive number. The first is divisible by 10, the second by 25, and the third by 6. Dividing through gives the following simpler problem.

$$\text{Minimize } c = 3{,}200x + 4{,}000y \text{ subject to}$$
$$2x + y \geq 20$$
$$x + 2y \geq 20$$
$$3x + 4y \geq 50$$
$$x \geq 0, y \geq 0.$$

How exactly do we convert this to a maximization problem? The technique is as follows: define a new variable p by taking p to be the negative of c, so that $p = -c$. Then, the larger we make p, the smaller c becomes. For example, if we can make p increase from -10 to -5, then c will decrease from 10 to 5. So, if we are looking for the smallest value of c, we might as well look for the largest value of p instead. More concisely,

Minimizing c is the same as maximizing $p = -c$.

Now, since $c = 3{,}200x + 4{,}000y$, we have $p = -3{,}200x - 4{,}000y$, and the problem may be translated into the following maximization problem.

$$\text{Maximize } p = -3{,}200x - 4{,}000y \text{ subject to}$$
$$2x + y \geq 20$$
$$x + 2y \geq 20$$
$$3x + 4y \geq 50$$
$$x \geq 0, y \geq 0.$$

We first introduce surplus variables.

$$\begin{aligned} 2x + y - s &= 20 \\ x + 2y - t &= 20 \\ 3x + 4y - u &= 50 \\ 3{,}200x + 4{,}000y + p &= 0 \end{aligned}$$

The initial tableau is then

	x	y	s	t	u	p	
*s	2	1	−1	0	0	0	20
*t	1	2	0	−1	0	0	20
*u	3	4	0	0	−1	0	50
p	3,200	4,000	0	0	0	1	0

We have marked the possible pivots in color. Since all three rows are starred, either of the color entries will do. Looking at the numbers involved, the second looks a little easier, so we take that.

	x	y	s	t	u	p		
*s	2	1	−1	0	0	0	20	$2R_1 - R_2$
*t	1	2	0	−1	0	0	20	
*u	3	4	0	0	−1	0	50	$R_3 - 2R_2$
p	3,200	4,000	0	0	0	1	0	$R_4 - 2000R_2$

	x	y	s	t	u	p		
*s	3	0	−2	1	0	0	20	
y	1	2	0	−1	0	0	20	$3R_2 - R_1$
*u	1	0	0	2	−1	0	10	$3R_3 - R_1$
p	1,200	0	0	2,000	0	1	−40,000	$R_4 - 400R_1$

	x	y	s	t	u	p		
x	3	0	−2	1	0	0	20	$R_1 + R_3$
y	0	6	2	−4	0	0	40	$R_2 - R_3$
*u	0	0	2	5	−3	0	10	
p	0	0	800	1,600	0	1	−48,000	$R_4 - 400R_3$

	x	y	s	t	u	p	
x	3	0	0	6	−3	0	30
y	0	6	0	−9	3	0	30
s	0	0	2	5	−3	0	10
p	0	0	0	−400	1,200	1	−52,000

This completes Phase 1, and we are now in the feasible region. We are not yet at the optimal solution, however, because of the −400 in the bottom row. We now proceed to find the optimal solution.

	x	y	s	t	u	p		
x	3	0	0	6	−3	0	30	$5R_1 - 6R_3$
y	0	6	0	−9	3	0	30	$5R_2 + 9R_3$
s	0	0	2	5	−3	0	10	
p	0	0	0	−400	1,200	1	−52,000	$R_4 + 80R_3$

4.4 Solving Nonstandard Linear Programming Problems

	x	y	s	t	u	p	
x	15	0	−12	0	3	0	90
y	0	30	18	0	−12	0	240
t	0	0	2	5	−3	0	10
p	0	0	160	0	960	1	−51,200

We have now found the optimal solution (remember that the rightmost entry in last row is allowed to be negative, and this does not indicate that we are at a non-optimal solution). The solution is

$$x = \frac{90}{15} = 6, \ y = \frac{240}{30} = 8, \ p = -51{,}200, \text{ so } c = 51{,}200$$
$$(s = 0, \ t = \frac{10}{5} = 2, \ u = 0)$$

Thus, you should buy 6 packages from Wall-to-Wall Furniture and 8 from Acme Furniture, for a minimum cost of $51,200.

Before we go on... The surplus variables here represent pieces of furniture over and above the minimum requirements. The order you place will result in the correct number of tables ($s = 0$) and computer desks ($u = 0$), but will give you 50 extra chairs ($t = 2$, but t was introduced after we divided the original constraint by 25, so the actual surplus is $2 \cdot 25 = 50$).

The above LP problem is an example of a **standard minimization problem.**

STANDARD MINIMIZATION PROBLEMS

A standard minimization problem in n unknowns is one in which we are required to *minimize* (not maximize) a linear objective function subject to the constraints

$$x \geq 0, \ y \geq 0, \ z \geq 0, \ldots,$$

and further constraints of the form

$$Ax + By + Cz + \ldots \geq N,$$

where $A, B, C, \ldots$ and N are numbers with N *nonnegative.* Note that the inequality here must be a " $\geq$," and not "=" or "$\leq$."

In a standard minimization problem, there is a shortcut that we are allowed to take in Phase 1. Since all of the rows will be starred to begin with, we do not have to worry about creating new starred rows with the first pivot. This frees us to drop the test ratio condition. In fact, it allows us to *reverse* the test ratio condition. If we choose a pivot where the test ratio is *largest,* rather than smallest, we can often remove all the stars at once. (See if you can explain why. The reasoning is similar to the original argument for the test ratio condition.)

To illustrate this shortcut, we redo Phase 1 for the previous example.

▼ **EXAMPLE 5** Shortcut for Phase 1 of Standard Minimization Problems

Redo Phase 1 of the last example by selecting the pivot whose test ratio is largest.

SOLUTION The initial tableau in the last example was the following.

	x	y	s	t	u	p	
*s	2	1	−1	0	0	0	20
*t	1	2	0	−1	0	0	20
*u	3	4	0	0	−1	0	50
p	3,200	4,000	0	0	0	1	0

We have set in color the positive entries whose test ratios are the *largest* in their columns. We select the entry in color in the first column as pivot.

	x	y	s	t	u	p		
*s	2	1	−1	0	0	0	20	$R_1 - 2R_2$
*t	1	2	0	−1	0	0	20	
*u	3	4	0	0	−1	0	50	$R_3 - 3R_2$
p	3,200	4,000	0	0	0	1	0	$R_4 - 3{,}200R_2$

	x	y	s	t	u	p	
s	0	−3	−1	2	0	0	−20
x	1	2	0	−1	0	0	20
u	0	−2	0	3	−1	0	−10
p	0	−2,400	0	3,200	0	1	−64,000

As in Example 3, we get negative numbers in the rightmost column, which we now take care of by multiplying the first and third rows by −1.

	x	y	s	t	u	p	
s	0	3	1	−2	0	0	20
x	1	2	0	−1	0	0	20
u	0	2	0	−3	1	0	10
p	0	−2,400	0	3,200	0	1	−64,000

As advertised, we removed all of the stars at once, so we are now in the feasible region. We could now proceed with Phase 2 to find the optimal solution. (We shall leave that to you for practice.)

Here are some notes that you may find helpful in some of the exercises.

POINTS TO REMEMBER

1. **Cycling.** It is theoretically possible that if you are forced to use Plan B during Phase 1, you will keep getting the same matrix over and over again. This phenomenon is called **cycling.** Although this will rarely happen, if your tableaus do keep cycling, the way to break out of the cycle is to choose a different pivot from the one you chose when the cycle began. (This phenomenon is so rare that many computer versions of the simplex method are not programmed to deal with it.) Cycling can occur only if you are forced to use Plan B.
2. If you are forced to use Plan B, you will wind up with a negative entry in the rightmost column above the last row. When this happens, multiply that row by -1 and check the basic solution. You may have to star that row.
3. Although we haven't given examples of *equality constraints,* such as $2x + 7y - z = 90$, they can be treated by the following trick: *Replace an equality by two inequalities.* In the example just cited, we replace the equality $2x + 7y - z = 90$ by the two inequalities $2x + 7y - z \leq 90$ and $2x + 7y - z \geq 90$. A little thought will convince you that they amount to the same thing as the original equality!
4. If at any stage, it is impossible to choose a pivot as per Plan A or Plan B, then the problem has no solution. In this case, the feasible region is empty.

▶ **4.4 EXERCISES**

1. Maximize $p = x + y$ subject to
$$x + 2y \geq 6$$
$$-x + y \leq 4$$
$$2x + y \leq 8$$
$$x \geq 0, y \geq 0.$$

2. Maximize $p = 3x + 2y$ subject to
$$x + 3y \geq 6$$
$$-x + y \leq 4$$
$$2x + y \leq 8$$
$$x \geq 0, y \geq 0.$$

3. Maximize $p = 12x + 10y$ subject to
$$x + y \leq 25$$
$$x \geq 10$$
$$-x + 2y \geq 0$$
$$x \geq 0, y \geq 0.$$

4. Maximize $p = x + 2y$ subject to
$$x + y \leq 25$$
$$y \geq 10$$
$$2x - y \geq 0$$
$$x \geq 0, y \geq 0.$$

5. Maximize $p = 2x + 5y + 3z$ subject to
$$x + y + z \leq 150$$
$$x + y + z \geq 100$$
$$x \geq 0, y \geq 0, z \geq 0.$$

6. Maximize $p = 3x + 2y + 2z$ subject to
$$x + y + 2z \leq 38$$
$$2x + y + z \geq 24$$
$$x \geq 0, y \geq 0, z \geq 0.$$

7. Maximize $p = 2x + 3y + z + 4w$ subject to
$$x + y + z + w \leq 40$$
$$2x + y - z - w \geq 10$$
$$x + y + z + w \geq 10$$
$$x \geq 0, y \geq 0, z \geq 0, w \geq 0.$$

8. Maximize $p = 2x + 2y + z + 2w$ subject to
$$x + y + z + w \leq 50$$
$$2x + y - z - w \geq 10$$
$$x + y + z + w \geq 20$$
$$x \geq 0, y \geq 0, z \geq 0, w \geq 0.$$

9. Minimize $c = 6x + 6y$ subject to
$$x + 2y \geq 20$$
$$2x + y \geq 20$$
$$x \geq 0, y \geq 0.$$

10. Minimize $c = 3x + 2y$ subject to
$$x + 2y \geq 20$$
$$2x + y \geq 10$$
$$x \geq 0, y \geq 0.$$

11. Minimize $c = 2x + y + 3z$ subject to
$$x + y + z \geq 100$$
$$2x + y \geq 50$$
$$y + z \geq 50$$
$$x \geq 0, y \geq 0, z \geq 0.$$

12. Minimize $c = 2x + 2y + 3z$ subject to
$$x \phantom{{}+y} + z \geq 100$$
$$2x + y \phantom{{}+z} \geq 50$$
$$y + z \geq 50$$
$$x \geq 0, y \geq 0, z \geq 0.$$

13. Minimize $c = 50x + 50y + 11z$ subject to
$$2x \phantom{{}+y} + z \geq 3$$
$$2x + y - z \geq 2$$
$$3x + y - z \leq 3$$
$$x \geq 0, y \geq 0, z \geq 0.$$

14. Minimize $c = 50x + 11y + 50z$ subject to
$$3x \phantom{{}+y} + z \geq 8$$
$$3x + y - z \geq 6$$
$$4x + y - z \leq 8$$
$$x \geq 0, y \geq 0, z \geq 0.$$

15. Minimize $c = x + y + z + w$ subject to
$$5x - y \phantom{{}+z} + w \geq 1000$$
$$z + w \leq 2000$$
$$x + y \phantom{{}+z+w} \leq 500$$
$$x \geq 0, y \geq 0, z \geq 0, w \geq 0.$$

16. Minimize $c = 5x + y + z + w$ subject to
$$5x - y \phantom{{}+z} + w \geq 1000$$
$$z + w \leq 2000$$
$$x + y \phantom{{}+z+w} \leq 500$$
$$x \geq 0, y \geq 0, z \geq 0, w \geq 0.$$

In Exercises 17 through 22 use a graphing calculator, a special-purpose program or an electronic spreadsheet. Round all answers to two decimal places.

17. Maximize $p = 2x + 3y + 1.1z + 4w$ subject to
$$1.2x + y + z + w \leq 40.5$$
$$2.2x + y - z - w \geq 10$$
$$1.2x + y + z + 1.2w \geq 10.5$$
$$x \geq 0, y \geq 0, z \geq 0, w \geq 0.$$

18. Maximize $p = 2.2x + 2y + 1.1z + 2w$ subject to
$$x + 1.5y + 1.5z \phantom{{}+1.5} + w \leq 50.5$$
$$2x + 1.5y \phantom{{}+1.5z} - z - w \geq 10$$
$$x + 1.5y \phantom{{}+1.5z} + z + 1.5w \geq 21$$
$$x \geq 0, y \geq 0, z \geq 0, w \geq 0.$$

19. Minimize $c = 2.2x + y + 3.3z$ subject to
$$x + 1.5y + 1.2z \geq 100$$
$$2x + 1.5y \phantom{{}+1.2z} \geq 50$$
$$1.5y + 1.1z \geq 50$$
$$x \geq 0, y \geq 0, z \geq 0.$$

20. Minimize $c = 50.3x + 10.5y + 50.3z$ subject to
$$3.1x + 1.1z \geq 28$$
$$3.1x + y - 1.1z \geq 23$$
$$4.2x + y - 1.1z \geq 28$$
$$x \geq 0, y \geq 0, z \geq 0.$$

21. Minimize $c = 1.1x + y + 1.5z - w$ subject to
$$5.12x - y + w \leq 1000$$
$$z + w \geq 2000$$
$$1.22x + y \leq 500$$
$$x \geq 0, y \geq 0, z \geq 0, w \geq 0.$$

22. Minimize $c = 5.45x + y + 1.5z + w$ subject to
$$5.12x - y + w \geq 1000$$
$$z + w \geq 2000$$
$$1.12x + y \leq 500$$
$$x \geq 0, y \geq 0, z \geq 0, w \geq 0.$$

APPLICATIONS

23. *Cost Minimization* Succulent Citrus, Inc. produces orange juice and orange concentrate. This year, the company anticipates a demand of at least 10,000 quarts of orange juice and 1,000 quarts of orange concentrate. Each quart of orange juice requires 10 oranges, while each quart of concentrate requires 50 oranges. The company anticipates using at least 200,000 oranges for these products. Each quart of orange juice costs the company 50¢ to produce, while each quart of concentrate costs $2.00 to produce. How many quarts of each product should the company produce to meet the demand and minimize total costs?

24. *Cost Minimization* Fancy Pineapple, Inc. produces pineapple juice and canned pineapple rings. This year, the company anticipates a demand of at least 10,000 pints of pineapple juice and 1,000 cans of pineapple rings. Each pint of pineapple juice requires 2 pineapples, while each can of pineapple rings requires 1 pineapple. The company anticipates using at least 20,000 pineapples for these products. Each pint of pineapple juice costs the company 20¢ to produce, while each can of pineapple rings costs 50¢ to produce. How many pints of pineapple juice and cans of pineapple rings should the company produce to meet the demand and minimize total costs?

25. *Nutrition* Each serving of *Gerber* Mixed Cereal for Baby contains 60 calories and no Vitamin C. Each serving of *Gerber* Mango Tropical Fruit Dessert contains 80 calories and 45% of the U.S. Recommended Daily Allowance (RDA) of Vitamin C for infants. Each serving of *Gerber* Apple Banana Juice contains 60 calories and 120% of the U.S. RDA of Vitamin C for infants.* The cereal costs 10¢ per serving, the dessert costs 53¢ per serving, and the juice costs 27¢ per serving. If you want to provide your child with at least 120 calories and at least 120% of the U.S. RDA of Vitamin C, how can you do so at the least cost?

26. *Nutrition* Each serving of *Gerber* Mixed Cereal for Baby contains 60 calories, no Vitamin C, and 11 grams of carbohydrates. Each serving of *Gerber* Mango Tropical Fruit Dessert contains 80 calories, 45% of the U.S. Recommended Daily Allowance (RDA) of Vitamin C for infants, and 21 grams of carbohydrates. Each serving of *Gerber* Apple Ba-

▼ *Source: nutrition information printed on the box.

nana Juice contains 60 calories, 120% of the U.S. RDA of Vitamin C for infants and 15 grams of carbohydrates.* Assume that the cereal costs 11¢ per serving, the dessert costs 50¢ per serving, and the juice costs 30¢ per serving. If you want to provide your child with at least 180 calories, at least 120% of the U.S. RDA of Vitamin C, and at least 37 grams of carbohydrates, how can you do so at the least cost?

27. *Cost Minimization* The political pollster Canter, Inc., is preparing for a national election. It would like to poll at least 1500 Democrats and 1500 Republicans. Each mailing to the East Coast gets responses from 100 Democrats and 50 Republicans. Each mailing to the Midwest gets responses from 100 Democrats and 100 Republicans. Each mailing to the West Coast gets responses from 50 Democrats and 100 Republicans. Mailings to the East Coast cost $40 each to produce and mail; mailings to the Midwest cost $60 each; and mailings to the West Coast cost $50 each. How many mailings should Canter send to each area of the country in order to get the responses it needs at the least possible cost? What will be the cost?

28. *Cost Minimization* Bingo's Copy Center needs to buy white paper and yellow paper. There are three suppliers that it can buy from. Harvard Paper sells a package of 20 reams of white and 10 reams of yellow for $60; Yale Paper sells a package of 10 reams of white and 10 reams of yellow for $40; and Dartmouth Paper sells a package of 10 reams of white and 20 reams of yellow for $50. If Bingo's needs 350 reams of white and 400 reams of yellow, how many packages should it buy from each supplier so as to minimize the cost? What is the least possible cost?

29. *Cost Minimization* Cheapskate Electronics Store needs to update its inventory of stereos, TVs, and VCRs. There are three suppliers it can buy from: Nadir, Inc., offers a bundle consisting of 5 stereos, 10 TVs, and 15 VCRs for $3,000. Blunt, Inc., offers a bundle consisting of 10 stereos, 10 TVs, and 10 VCRs for $4,000. Sonny, Inc., offers a bundle consisting of 15 stereos, 10 TVs, and 10 VCRs for $5,000. Cheapskate Electronics needs at least 150 stereos, 200 TVs, and 150 VCRs. How can it update its inventory at the least possible cost? What is the least possible cost?

30. *Cost Minimization* Federal Rent-a-Car is putting together a new fleet. It is considering package offers from three car manufacturers. Fred Motor Co. is offering 5 small cars, 5 medium cars, and 10 large cars for $500,000. Admiral Motors is offering 5 small, 10 medium, and 5 large cars for $400,000. Chrysalis is offering 10 small, 5 medium, and 5 large cars for $300,000. Federal would like to buy at least 550 small cars, at least 500 medium cars, and at least 550 large cars. How many packages should it buy from each car maker in order to keep the total cost as small as possible? What will it cost?

31. *Subsidies* In 1993, the Miami Beach City Council was offering to give developers 20% of the cost of building new hotels in Miami Beach, up to a maximum of $50 million. In addition, the city was hoping for at least two hotels with a total capacity of at least 1,400.† Suppose that you were a developer interested in taking advantage of this offer by building a small group of hotels in Miami Beach. You are thinking of three prototypes: a convention-style hotel with 500 rooms costing $100 million, a vacation-style hotel with 200 rooms costing $20 million, and small motels with 50 rooms costing $4 million. The City Council will approve your plans provided you build at least one convention-style hotel and no more than two small motels. How many of each type of hotel should you build in order to satisfy all of the City Council's wishes and stipulations while minimizing your total costs? Will the City's $50 million subsidy be sufficient to cover 20% of your total costs?

32. *Subsidies* Referring to the preceding exercise, you are about to begin the financial arrangements for your new hotels when the Miami Beach City Council informs you that they have changed their mind and now require at least two convention-style hotels and no more than one small motel. (They feel that Miami Beach already has too many small motels.) How many of each type of hotel should you build in order to satisfy all of the City Council's wishes and stipulations while minimizing your total costs? Will the City's $50 million subsidy still be sufficient to cover 20% of your total costs? [In the event that the simplex method tells you to build fractions of hotels, you will

▼ * Source: nutrition information printed on the jar.
† *Chicago Tribune*, June 20, 1993, Section 7, p. 8.

have to round up to the nearest whole number. (Why?)]

33. *Hospital Staffing* The staff director of a new hospital is planning to hire cardiologists, rehabilitation specialists and infectious disease specialists. According to recent data, each cardiology case averages $12,000 revenue, each physical rehabilitation case $19,000 and each infectious disease case $14,000.* The staff director estimates that each specialist will expand the hospital caseload by about 10 patients per week. There are already 3 cardiologists on staff, and the hospital is equipped to admit up to 200 patients per week. Based on past experience, each cardiologist and rehabilitation specialist brings in one government research grant per year, while each infectious disease specialist brings in 3. The board of directors would like to see a total of at least 30 grants per year and would like weekly revenue to be as large as possible. How many of each kind of specialist should be hired?

34. *Hospital Staffing (graphing calculator or computer software required)* The staff director referred to in the preceding exercise was fired, as he had totally misjudged the number of patients each type of specialist would bring to the hospital per week. It turned out that each cardiologist brought in 120 new patients per *year,* each rehabilitation specialist brought in 90 per year, and each infectious disease specialist brought in 50 per year.† It also turned out that the hospital could deal with no more than 1960 new patients per year. Repeat Exercise 33 in the light of this corrected data.

35. *Nutrition (graphing calculator or computer software required)* The Enormous State University Cafeteria, under pressure from the Student Council, decided to begin serving nutritious food for a change. As a goal, the cafeteria mangers decided that each serving of a main course should contain at least 100 grams of protein, at most 350 grams of carbohydrate, at least 3,180 milligrams of calcium, and no more than 1,120 calories. The cafeteria has decided, with typical cynicism, to provide these requirements by serving hamburgers for every meal, mixed with powdered milk and eggs for nutrition. It can get ground beef at 10¢ an ounce, powdered milk at 15¢ per cup, and eggs at 5¢ each. Following is the chart of food values:

	Protein	Carbo-hydrates	Calcium	Calories
Ground beef (1 oz)	10 g	0 g	10 mg	50
Powdered milk (1 cup)	20 g	35 g	1,020 mg	100
Egg (one)	10 g	0 g	30 mg	80

How much of each ingredient should be combined to produce hamburgers at minimum cost? Comment on the answer you get.

36. *Nutrition (graphing calculator or computer software required)* The football team complained that they weren't getting enough calories from the Enormous State University cafeteria's hamburgers (see the previous exercise), and they demanded that the cafeteria change the composition of its hamburgers so as to supply at least 350 grams of carbohydrates per serving. Does this requirement affect the composition of the hamburgers in any way? If so, how?

37. *Finance* (Compare with Example 3 in Section 2.) The Solid Trust Savings & Loan Company has set aside $25 million for loans to finance luxury condos, low-income housing, and urban development. Its policy is to allocate at least $10 million annually for

---

* These (rounded) figured are based on an Illinois survey of 1.3 million hospital admissions (*Chicago Tribune*, March 29, 1993, Section 4, p. 1). Source: Lutheran General Health System, Argus Associates, Inc.

† These (rounded) figures were obtained from the survey referenced in the above footnote by dividing the average hospital revenue per physician by the revenue per case.

luxury condominiums. On the other hand, a lucrative government grant requires that at least one-third of the total financing be for low-income housing. The S&L's return on condos is 12%, its return on low-income housing is 10%, and its return on urban development projects is 5%. How much should it allocate for each type of project to maximize the return?

38. *Scheduling* Since Joe Slim's brother was recently elected to the state Senate, Joe's financial advisement concern, Inside Information Inc. has been doing a booming trade, even though the financial counseling he offers is quite worthless. (None of his seasoned clients pay the slightest attention to his advice). Slim charges different hourly rates to different categories of individuals: $5,000 per hour for private citizens, $50,000 per hour for corporate executives and $10,000 per hour for presidents of universities. Because of his taste for leisure, he feels that he can spend no more than 40 hours per week in consultation. On the other hand, Slim feels that it would be best for his intellect were he to devote at least 10 hours of consultation per week to university presidents. However, Slim always feels somewhat uncomfortable dealing with academics, so he would prefer to spend no more than half his consultation time with university presidents. Further, he likes to think of himself as representing the interests of the common citizen, so he wishes to offer at least two more hours of his time per week to private citizens than to corporate executives and university presidents combined. Given all these restrictions, how many hours per week should he spend with each type of client to maximize his income?

39. *Transportation Scheduling (graphing calculator or computer software required)* Your publishing company is about to start its promotional blitz for its new book, *Physics for the Liberal Arts*. You have 10 salespeople stationed in Chicago and 20 in Denver. You would like to fly at least 10 into Los Angeles and at least 15 into New York. A round-trip plane flight from Chicago to LA costs $428; from Chicago to NY costs $305; from Denver to LA costs $338; and from Denver to NY costs $493.* How many salespeople should you fly from each of Chicago and Denver to each of LA and NY in order to spend the least amount on plane flights?

40. *Transportation Scheduling (graphing calculator or computer software required)* Repeat the preceding exercise, but suppose that you would like at least 15 salespeople in Los Angeles.

41. *Transportation Scheduling* We return to your exploits coordinating distribution for the Tubular Ride Boogie Board Company.† You will recall that the company has manufacturing plants in Tucson, Arizona and Toronto, Ontario, and you have been given the job of coordinating distribution of its latest model, the Gladiator, to its outlets in Honolulu and Venice Beach. The Tucson plant can manufacture up to 620 Gladiator boards per week, while the Toronto plant, beset by labor disputes, can produce no more than 410 Gladiator boards per week. The outlet in Honolulu orders 500 gladiator boards per week, while Venice Beach orders 530 boards per week. Transportation costs are as follows:

Tucson to Honolulu: $10 per board;
Tucson to Venice Beach: $5 per board;
Toronto to Honolulu: $20 per board;
Toronto to Venice Beach: $10 per board.

Your manager has informed you that you must fill all orders and ship the boogie boards at a minimum total transportation cost. How will you do it?

42. *Transportation Scheduling* In the situation described in the preceding exercise, you have just been notified that workers at the Toronto boogie board plant have gone on strike, resulting in a total work stoppage. You are to come up with a revised delivery schedule by tomorrow with the understanding that the Tucson plant can push production to a maximum of 1,000 boards per week. What should you do?

▼ * Prices from the OAG Electronic Edition, for the week of Sept. 13, 1993, checked on July 25, 1993.

† See Exercises in Section 2 of Chapter 2 and also Section 3 of this chapter. This time, we will use the simplex method to solve the version of this problem we first considered in Chapter 2.

COMMUNICATION AND REASONING EXERCISES

43. Find a linear programming problem in three variables that requires one pivot in Phase 1.

44. Find a linear programming problem in three variables that requires two pivots in Phase 1.

45. Find a linear programming problem in two or three variables with no optimal solution, and show what happens when you try to solve it using the simplex method.

46. Find a linear programming problem in two or three variables with more than one optimal solution, and investigate which solution is found by the simplex method.

47. Explain the need for Phase 1 in a nonstandard linear programming problem.

48. Explain the need for Phase 2 in a nonstandard linear programming problem.

49. Explain why, in Phase 2, we choose a pivot corresponding to the smallest test ratio in its column.

50. Explain why, in Phase 1 for a standard minimization problem, we can save time by choosing a pivot corresponding to the largest test ratio.

▶ ## You're the Expert

AIRLINE SCHEDULING

You are the traffic manager for Fly-by-Night Airlines, which flies airplanes between 5 cities: Los Angeles, Chicago, Atlanta, New York, and Boston. As a result of a strike, you have not had sufficient crews to fly your planes, so some of your planes have been stranded in Chicago and in Atlanta. In fact, you have 15 extra planes in Chicago and 30 extra planes in Atlanta. Los Angeles needs at least 10 planes to get back on schedule, New York needs at least 20, and Boston needs at least 10. It costs $50,000 to fly a plane from Chicago to Los Angeles, $10,000 to fly from Chicago to New York, and $20,000 to fly from Chicago to Boston. It costs $70,000 to fly a plane from Atlanta to Los Angeles, $20,000 to fly from Atlanta to New York, and $50,000 to fly from Atlanta to Boston. How should you rearrange your planes to get back on schedule at the least cost?

As always, you remember to start by identifying your unknowns. You need to decide how many planes you will fly from each of Chicago and Atlanta to each of LA, NY, and Boston. This gives you six unknowns.

x = the number of planes to fly from Chicago to LA
y = the number of planes to fly from Chicago to NY
z = the number of planes to fly from Chicago to Boston
u = the number of planes to fly from Atlanta to LA
v = the number of planes to fly from Atlanta to NY
w = the number of planes to fly from Atlanta to Boston

Los Angeles needs at least 10 planes, so you know that

$$x + u \geq 10.$$

Similarly, New York and Boston give you the inequalities

$$y + v \geq 20$$
$$z + w \geq 10.$$

Since Chicago has only 15 extra planes, you have

$$x + y + z \leq 15.$$

Since Atlanta has 30 extra planes,

$$u + v + w \leq 30.$$

Finally, you want to minimize the cost, which is given by

$$C = 50{,}000x + 10{,}000y + 20{,}000z + 70{,}000u + 20{,}000v + 50{,}000w.$$

This gives you your linear programming problem.

$$\text{minimize } C = 50{,}000x + 10{,}000y + 20{,}000z + 70{,}000u + 20{,}000v + 50{,}000w$$

subject to the constraints

$$\begin{aligned}
x \phantom{{}+y+z} + u \phantom{{}+v+w} &\geq 10 \\
y \phantom{{}+z} + v \phantom{{}+w} &\geq 20 \\
z \phantom{{}+u+v} + w &\geq 10 \\
x + y + z \phantom{{}+u+v+w} &\leq 15 \\
u + v + w &\leq 30
\end{aligned}$$

$$x \geq 0, \, y \geq 0, \, z \geq 0, \, u \geq 0, \, v \geq 0, \, w \geq 0.$$

Since you have six unknowns, you know that you will have to use the simplex method, so you start converting into the proper form. You change the minimization problem into the problem of maximizing $P = -C$. Being a clever sort, you notice that it would actually be enough to maximize

$$P = -5x - y - 2z - 7u - 2v - 5w$$

(this will keep the numbers small). You subtract the surplus variables p, q, and r from the left-hand sides of the first three inequalities, and you add the slack variables s and t to the last two. Then it's time to go to work. Remembering all the rules for pivoting, and using a computer to help with the calculations, you get the following sequence of tableaus.

	x	y	z	u	v	w	p	q	r	s	t	P	
*p	1	0	0	1	0	0	−1	0	0	0	0	0	10
*q	0	1	0	0	1	0	0	−1	0	0	0	0	20
*r	0	0	1	0	0	1	0	0	−1	0	0	0	10
s	1	1	1	0	0	0	0	0	0	1	0	0	15
t	0	0	0	1	1	1	0	0	0	0	1	0	30
P	5	1	2	7	2	5	0	0	0	0	0	1	0

	x	y	z	u	v	w	p	q	r	s	t	P	
x	1	0	0	1	0	0	-1	0	0	0	0	0	10
*q	0	1	0	0	1	0	0	-1	0	0	0	0	20
*r	0	0	1	0	0	1	0	0	-1	0	0	0	10
s	0	1	1	-1	0	0	1	0	0	1	0	0	5
t	0	0	0	1	1	1	0	0	0	0	1	0	30
P	0	1	2	2	2	5	5	0	0	0	0	1	-50

	x	y	z	u	v	w	p	q	r	s	t	P	
x	1	0	0	1	0	0	-1	0	0	0	0	0	10
v	0	1	0	0	1	0	0	-1	0	0	0	0	20
*r	0	0	1	0	0	1	0	0	-1	0	0	0	10
s	0	1	1	-1	0	0	1	0	0	1	0	0	5
t	0	-1	0	1	0	1	0	1	0	0	1	0	10
P	0	-1	2	2	0	5	5	2	0	0	0	1	-90

	x	y	z	u	v	w	p	q	r	s	t	P	
x	1	0	0	1	0	0	-1	0	0	0	0	0	10
v	0	1	0	0	1	0	0	-1	0	0	0	0	20
w	0	0	1	0	0	1	0	0	-1	0	0	0	10
s	0	1	1	-1	0	0	1	0	0	1	0	0	5
t	0	-1	-1	1	0	0	0	1	1	0	1	0	0
P	0	-1	-3	2	0	0	5	2	5	0	0	1	-140

	x	y	z	u	v	w	p	q	r	s	t	P	
x	1	0	0	1	0	0	-1	0	0	0	0	0	10
v	0	1	0	0	1	0	0	-1	0	0	0	0	20
w	0	-1	0	1	0	1	-1	0	-1	-1	0	0	5
z	0	1	1	-1	0	0	1	0	0	1	0	0	5
t	0	0	0	0	0	0	1	1	1	1	1	0	5
P	0	2	0	-1	0	0	8	2	5	3	0	1	-125

	x	y	z	u	v	w	p	q	r	s	t	P	
x	1	1	0	0	0	-1	0	0	1	1	0	0	5
v	0	1	0	0	1	0	0	-1	0	0	0	0	20
u	0	-1	0	1	0	1	-1	0	-1	-1	0	0	5
z	0	0	1	0	0	1	0	0	-1	0	0	0	10
t	0	0	0	0	0	0	1	1	1	1	1	0	5
P	0	1	0	0	0	1	7	2	4	2	0	1	-120

Now you see what to do: You should fly 5 planes from Chicago to LA, none from Chicago to NY, and 10 from Chicago to Boston. You should fly 5 planes from Atlanta to LA, 20 from Atlanta to NY, and none from Atlanta to Boston. The total cost will be $1,200,000 to get the airline back on schedule.

Exercises

1. Your boss calls you on the carpet for not flying any planes from Chicago to New York, which is the cheapest run. Come up with a convincing reason for having done that. Your job may be on the line.
2. Your boss *insists* that you fly at least 5 planes from Chicago to New York. What is your best option now?
3. If LA needed 15 planes rather than 10, what would be your best option?
4. If NY needed 25 planes rather than 20, what would be your best option?
5. If Boston needed 15 planes rather than 10, what would be your best option?
6. If there were only 10 planes in Chicago, what would be your best option?
7. If there were 35 planes in Atlanta, what would be your best option?
8. Suppose that the Boston airport was closed because of a snowstorm, so you only needed to send planes to LA and NY. What would be your best option?
9. Suppose that the NY airport was closed because of a snowstorm, so you only needed to send planes to LA and Boston. What would be your best option?
10. A larger cost per flight from Atlanta to NY would change the answer obtained in the text. By how much would the cost have to rise to change the answer?

▸ **Review Exercises**

Sketch the regions corresponding to the inequalities in Exercises 1 through 18.

1. $x + 5y \leq 10$
2. $2x - 3y \leq 12$
3. $-x + 4y \leq 8$
4. $-x - 4y \geq 4$
5. $6x + 2y \geq 4$
6. $4x - 3y \leq 7$
7. $x \leq 2y$
8. $y \geq -x$
9. $x \geq 1$
10. $y \leq 5$
11. $x - 2y \leq 8$
    $x + 2y \leq 8$
12. $x + 2y \leq 4$
    $x - 8y \geq 8$
13. $3x + 2y \geq 6$
    $2x - 3y \leq 6$
    $y \geq 0$
14. $3x + 2y \leq 6$
    $2x - 3y \geq 6$
    $x \geq 0$
15. $x + 2y \leq 20$
    $3x + 2y \leq 30$
    $x \geq 0, y \geq 0$
16. $3x + 4y \leq 120$
    $5x + 2y \leq 100$
    $x \geq 0, y \geq 0$
17. $2x + 3y \geq 60$
    $4x + y \geq 80$
    $x \geq 0, y \geq 0$
18. $3x + 2y \geq 120$
    $x + 2y \geq 60$
    $x \geq 0, y \geq 0$

Solve the linear programming problems in Exercises 19 through 44.

19. Maximize $P = x + y$ subject to the constraints
$$2x + y \leq 20$$
$$x + 2y \leq 20$$
$$x \geq 0, y \geq 0.$$

20. Maximize $P = 2x + y$ subject to the constraints
$$3x + y \leq 30$$
$$x + 2y \leq 20$$
$$x \geq 0, y \geq 0.$$

47. *(adapted from the actuarial exam on operations research)* You are given the following linear programming problem.

$$\text{Maximize } Z = x_1 + 4x_2 + 2x_3 - 10 \text{ subject to}$$
$$4x_1 + x_2 + x_3 \leq 45$$
$$-x_1 + x_2 + 2x_3 \leq 0$$
$$x_1, x_2, x_3 \geq 0.$$

Determine the optimal value of the objective function.

48. *(adapted from on the actuarial exam on operations research)* You are given the following linear programming problem.

$$\text{Maximize } Z = x_1 + 4x_2 + 2x_3 + 10 \text{ subject to}$$
$$x_1 + x_2 + 4x_3 \leq 45$$
$$x_1 + x_2 - 2x_3 \leq 10$$
$$x_1, x_2, x_3 \geq 0.$$

Determine the optimal value of the objective function.

APPLICATIONS

49. *Resource Allocation* You manage an ice cream factory that makes two flavors: Continental Mocha and Succulent Strawberry. Into each quart of Continental Mocha go two cups of cream and two cups of sugar, and into each quart of Succulent Strawberry go three cups of cream and one cup of sugar. You have in stock 800 cups of cream and 400 cups of sugar. You make a profit of $3 on each quart of Continental Mocha and $1 on each quart of Succulent Strawberry. How many quarts of each flavor should you make in order to make the largest profit?

50. *Resource Allocation* Enormous State University's Math Department offers two courses: Finite Math and Topology. Each section of Finite Math has 60 students, and each section of Topology has 10. The department can offer a total of up to 100 sections in a semester, and there are up to 5500 students who would like to take a math course. Each section of Finite Math raises the happiness level of students a total of 20 units,* while each section of Topology raises the happiness level of its students a total of 10 units.[†] How many sections of each course should the department add in order to maximize the happiness of its students?

51. *Cost Minimization* Meow, Inc., makes cat food from fish and cornmeal. Fish has 8 grams of protein and 4 grams of fat per ounce, while cornmeal has 4 grams of protein and 8 grams of fat per ounce. A jumbo can of cat food must contain at least 48 grams of protein and 48 grams of fat. If fish and cornmeal both cost 5¢ per ounce, how many ounces of each should they use in each can of cat food in order to minimize costs?

52. *Cost Minimization* Oz, Inc., makes lion food out of giraffe and gazelle meat. Giraffe meat has 18 grams of protein and 36 grams of fat per pound, while gazelle meat has 36 grams of protein and 18 grams of fat per pound. A batch of lion food must contain at least 36,000 grams of protein and 54,000 grams of fat. Giraffe meat costs $2 per pound, and gazelle meat costs $4 per pound. How many pounds of each should go into each batch of lion food in order to minimize costs?

53. *Resource Allocation* Your umbrella company makes three models: the Sprinkle, the Storm, and the Hurricane. The amounts of cloth, metal, and wood used in making each model are given in this table:

	Sprinkle	Storm	Hurricane	Available
Cloth (sq yd)	1	2	2	600
Metal (lb)	2	1	3	600
Wood (lb)	1	3	6	600
Profit ($)	1	1	2	

The table also shows the amounts of each material available in a given day and the profits to be made from each model. How many of each model should you build in order to maximize your profit?

▼ * Thus averaging 1/3 unit per student.
  [†] Thus averaging 1 unit per student.

54. *Resource Allocation* Fullcourt Press prints three kinds of books: paperback, quality paperback, and hardcover. The amounts of paper, ink, and time on the presses required for each kind of book are given in this table:

	Paper	Quality Paper	Hardcover	Available
Paper (lb)	3	2	1	600
Ink (gal)	2	1	3	600
Time (min)	10	10	10	2200
Profit ($)	1	2	3	

The table also shows the amounts of paper, ink, and time available in a given week, and the profits made on each kind of book. How many of each kind of book should be printed to maximize profit?

55. *Investment* *(graphing calculator or computer software required)* Teachers Insurance and Annuity Association-College Retirement Equities Fund (TIAA-CREF) is one of the largest pension funds in the country. It offers a number of investments, among them TIAA (a guaranteed annuity), which over the last 5 years has averaged a yearly return of 8.99%;* a stock fund, which over the last 5 years has averaged a return of 14.45%; a money market fund, which over its lifetime (of less than 5 years) has averaged 7.01%; a bond fund, which over its lifetime has averaged 11.32%; and a social choice stock fund, which over its lifetime has averaged 14.03%. Suppose that you estimate the risks involved in each fund, on a scale of 0 to 10, as follows: TIAA: 0; stock fund: 8; money market: 5; bond fund: 6; social choice fund: 8. (When determining your total risk, multiply the amount of money you invest in a fund by the risk factor for that fund.) You have $20,000 to invest and would like to earn at least $2,000 from your investments over the next year. How should you allocate your money among these funds so as to minimize your total risk? (Give answers to the nearest $1.)

56. *Investment* *(graphing calculator or computer software required)* How would your answer to the preceding exercise change if you revised your estimate of the risk of TIAA up to 4 and your estimate of the risk of the bond fund down to 5?

57. *Transportation* Federal Rent-a-Car has four locations in the city: Eastside, Westside, Northside, and the Airport. Cars tend to pile up at its Northside and Airport locations, and there tend to be shortages at its Eastside and Westside locations. One day there are 10 extra cars at the Northside location and 15 extra cars at the Airport, while 12 more cars are needed at the Eastside location and 7 more cars at the Westside location to fill reservations. It costs $1 to drive a car from Northside to Eastside, $2 to drive one from Northside to Westside, $2 to drive a car from the Airport to Eastside, and $4 to drive one from the Airport to Westside. How many cars should Federal drive from each of the Northside and Airport locations to the Eastside and Westside locations in order to fill its reservations at the lowest cost?

58. *Transportation* Redo the preceding exercise if the costs to drive a car from the Airport to the Eastside and the Westside locations are both $3.

59. *Investment* JoAnn Smith is considering investing up to $10,000 in three ways: a stock fund run by Integrity Investments, a bond fund sold by Citizen's

60. *Profit Maximization* The "Uclean Meclean" carwash company does two types of carwash: the "Superwax Gleam Job" and the "Econo Rinse." On a certain day, it has 10 cars available to clean (although the company doesn't feel it necessary to clean them all on that day). It happens that on that day, the only employee who shows up for work is Joe Slo, who takes $1\frac{1}{2}$ hours to do a Superwax Gleam Job and 1 hour to do an Econo Rinse. His total workday is 9 hours long. You are the manager of Uclean Meclean, your name is Chuck Fastbuck, two of the cars belong to your favorite customer, Desmorelda Smith, and you tell Joe to be sure to give her cars the Superwax Gleam Job. You make a profit of $25.00 on each Superwax Gleam Job and $10.00 on each Econo Rinse. How many jobs of each kind would you instruct Joe Slo to do in order to generate the most profit?

▼ *Source: The Participant (quarterly newsletter of TIAA-CREF), February 1993.

CHAPTER 5

Toys 'R' Us Announces Plan to Build 115 Stores in 1994

PARAMUS, N.J., Jan.11 (Reuters)—Toys "R" Us Inc., the largest toy retailer in the world, announced plans today to grow even bigger this year by opening another 115 new stores and signing franchise agreements with foreign merchandisers.

The company also said its board had authorized spending as much as $1 billion to buy back its stock during the next several years.

The retailer said it would open 45 new company-owned stores in the United States and 70 in other countries, in addition to the franchise outlets.

The company will open its first outlets in the Persian Gulf region through franchise agreements signed with groups in Saudi Arabia and the United Arab Emirates.

"Through these franchise agreements, we will be using the resources of highly respected and successful local business people throughout parts of the world that would not normally be developed by Toys 'R' Us itself for many years." Charles Lazarus, the chairman of the company, said in a statement announcing the moves.

Toys "R" Us said it had signed a franchise agreement with the Al Futtaim Sons Company in the United Arab Emirates to operate stores in that country, as well as in Qatar, Bahrain, Oman and Kuwait.

It signed another franchise agreement with what it described as a large Saudi Arabian group to develop toy stores in the kingdom.

The retailer, based here, said it would receive "significant royalty and other related service fees" in the two deals, but did not elaborate....

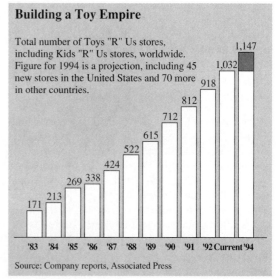

Building a Toy Empire

Total number of Toys "R" Us stores, including Kids "R" Us stores, worldwide. Figure for 1994 is a projection, including 45 new stores in the United States and 70 more in other countries.

Source: Company reports, Associated Press

The New York Times

Source: From "Toys 'R' Us Announces Plan To Build 115 Stores in 1994," *The New York Times*, January 12, 1994, p. D2 and Reuters.

Sets and Counting

SECTIONS

1. Sets
2. The Number of Elements in a Set
3. The Multiplication Principle; Permutations
4. The Addition Principle; Combinations

You're the Expert

Designing a Puzzle

APPLICATION ▶ As Product Design Manager for Cerebral Toys, Inc., you are constantly tracking down ideas for intellectually stimulating yet inexpensive toys. Your design team recently came up with an idea for a puzzle consisting of a number of plastic cubes. Each cube will have two faces colored red, two white, and two blue, and there will be exactly two cubes with each possible configuration of colors. The goal of the puzzle is to seek out the matching pairs, enhancing a child's geometric intuition and three-dimensional manipulation skills. If the kit is to include every possible configuration of colors, how many cubes will the kit contain?

INTRODUCTION

▶ The theory of sets is the foundation for most of mathematics. It also has direct applications such as in searching computer data bases. We shall use set theory extensively in the chapter on probability, and with this in mind much of this chapter revolves around the idea of a **set of outcomes** of a procedure such as rolling a pair of dice or choosing names from a database. Also important in probability is the question of **counting** the number of elements in a set. The theory of counting the number of elements in a set is called **combinatorics.**

To show that counting elements is not a trivial proposition, consider the following example. The betting game lotto (used in many state lotteries) has you pick six numbers from the range 0–54. What is your chance of winning? How many lotto tickets would you need to buy to guarantee that you would win? By the end of this chapter we will be able to answer these questions.

5.1 SETS

We assume that you have some familiarity already with sets. In this section we shall summarize the basic ideas of set theory, with examples emphasizing some particular applications that we will explore further in the rest of the chapter.

SETS AND ELEMENTS

A **set** is a collection of items, referred to as the **elements** of the set.

$x \in A$ means that x is an element of the set A.

$x \notin A$ means that x is not an element of the set A.

$B = A$ means that A and B have the same elements.

$B \subseteq A$ means that B is a **subset** of A—every element of B is also an element of A.

$B \subset A$ means that B is a **proper subset** of A; that is, $B \subseteq A$, but $B \neq A$.

$\emptyset$ is the **empty set,** the set containing no elements. It is a subset of every set.

For example, the set of all characters in Shakespeare's play *The Tempest* is the set

{Alonso, Sebastian, Prospero, Antonio, Ferdinand, Gonzalo, Adrian, Francisco, Caliban, Trinculo, Stephano, Ariel, Miranda}.

Antonio is an element of this set, but Romeo is not. We use braces to enclose the elements of a set. We usually use a capital letter to name a set.

EXAMPLE 1

A company might keep a computer database of its customers. We can think of the database as an electronic embodiment of the abstract set of all of its customers.

EXAMPLE 2

If you flip a coin and observe which side lands up, there are two possible outcomes: heads (H) and tails (T). The **set of outcomes** of flipping a coin once can thus be written

$$S = \{H, T\}.$$

EXAMPLE 3

Suppose that you toss a die and observe which number is topmost. Give the set of outcomes.

SOLUTION Since the faces of a die show a series of dots (from one through six), we *could* write

$$S = \{\boxdot, \boxdot, \boxdot, \boxdot, \boxdot, \boxdot\}$$

but we could much more easily write

$$S = \{1, 2, 3, 4, 5, 6\}.$$

EXAMPLE 4

Now suppose that you toss *two* dice. What is the set of outcomes?

SOLUTION We assume that we can distinguish the dice in some way; let's say one is white and one is colored. A systematic way of laying out the set of outcomes is given in Figure 1.

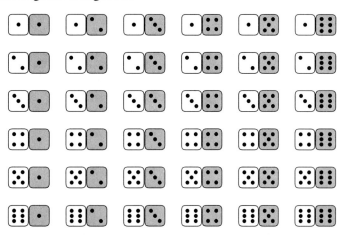

FIGURE 1

In the first row, all the white dice show a 1, in the second row a 2, in the third row a 3, and so on. Similarly, in the first column, all the colored dice show a 1, in the second column a 2, and so on. The diagonal pairs (top left to bottom right) show all the doubles.

We can write the set of outcomes as follows.

$$S = \begin{Bmatrix} (1,1) & (1,2) & (1,3) & (1,4) & (1,5) & (1,6) \\ (2,1) & (2,2) & (2,3) & (2,4) & (2,5) & (2,6) \\ (3,1) & (3,2) & (3,3) & (3,4) & (3,5) & (3,6) \\ (4,1) & (4,2) & (4,3) & (4,4) & (4,5) & (4,6) \\ (5,1) & (5,2) & (5,3) & (5,4) & (5,5) & (5,6) \\ (6,1) & (6,2) & (6,3) & (6,4) & (6,5) & (6,6) \end{Bmatrix}$$

This seems like the best way to organize things. Notice that the set has a total of 36 elements. Also, $(2, 6) \in S$, but $(2, 0) \notin S$.

EXAMPLE 5

Let $B = \{0, 2, 4, 6, 8\}$. B is the set of all nonnegative even integers less than 10. If we don't want to list the individual elements of B, we can instead write

$$B = \{n \mid n \text{ is a nonnegative even integer less than } 10\}.$$

This is read "*B is the set of all n such that n is a nonnegative even integer less than 10.*" Here is the correspondence between the words and the symbols:

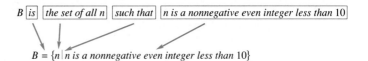

Before we go on... The **nonnegative** integers *include* 0, whereas the **positive** integers *exclude* 0. We use the letter **N** to denote the set of all positive integers, also known as the set of **natural numbers.**

EXAMPLE 6

(a) Which of the following sets are subsets of others in the list?
(b) Which of the following sets are proper subsets of others in the list?

$A = \{5, -9, 1, 3\}$
$B = \{5, 4, -1, 2, 7\}$
$C = \{0, -1, 2, -9, 5, 3, 1\}$
$D = \emptyset$
$E = \{5\}$
$F = \{-9, 1, 3, 5\}$

SOLUTION

(a) Let us look at A first. It is not a subset of B, since -9 and 3 are in A but not in B. It is a subset of C, since C has everything that A has (and more). Thus,

$$A \subseteq C.$$

A is certainly not a subset of D or E, since they have fewer elements than A does. Finally,

$$A \subseteq F$$

since A and F are actually equal (notice that the order in which the elements are listed is not important).

Turning to B, it is not a subset of A, since A has fewer elements, nor is it a subset of C, since 4 is missing from C. There is no hope of it being a subset of either D or E, and we know it is not a subset of F since $F = A$. Thus, B is not a subset of any of the other sets in the list.

Turning to C, it cannot be a subset of any of the other sets, since it has the most elements.

Since D has no elements, it is a subset of *all* the others—the empty set is a subset of every set. Thus,

$$D \subseteq A, \quad D \subseteq B, \quad D \subseteq C, \quad D \subseteq E, \quad \text{and} \quad D \subseteq F.$$

E is a subset of every set listed other than the empty set, so

$$E \subseteq A, \quad E \subseteq B, \quad E \subseteq C, \quad \text{and} \quad E \subseteq F.$$

Finally, since $F = A$, we have

$$F \subseteq A \quad \text{and} \quad F \subseteq C.$$

(b) We now go through our collection of answers to part (a) and check to see which inclusions are proper inclusions. We find the following:

$$A \subset C$$
$$D \subset A, \quad D \subset B, \quad D \subset C, \quad D \subset E, \quad \text{and} \quad D \subset F$$
$$E \subset A, \quad E \subset B, \quad E \subset C, \quad \text{and} \quad E \subset F$$
$$F \subset C.$$

Notice that the two inclusions that are not proper are $A \subseteq F$ and $F \subseteq A$, since F contains no elements that are not in A, and vice-versa.

Before we go on... Although the question asked which sets are subsets of *others* in the list, we should notice that any set is a subset of itself. Therefore, we could also have written the inclusions $A \subseteq A$, $B \subseteq B$, and so on. None of these inclusions are proper. In fact, we sometimes refer to a set as the **improper subset** of itself.

VENN DIAGRAMS

We can visualize sets and relations between sets using **Venn diagrams.** In a Venn diagram, we represent a set as a region, usually a circle (Figure 2).

Everything inside the circle is thought of as being in A. If $x \in A$, then x can be represented by a point inside this circle. A point outside the circle represents an $x \notin A$. If B is a subset of A, then the circle representing B is included in, or possibly equal to, the circle representing A.

FIGURE 2

VENN DIAGRAMS

Each diagram in Figure 3 represents the relation listed below it. For a proper inclusion, the region representing B must be smaller than the region representing A. For simplicity, we usually take the drawing in Figure 3 representing $B \subset A$ to also represent $B \subseteq A$.

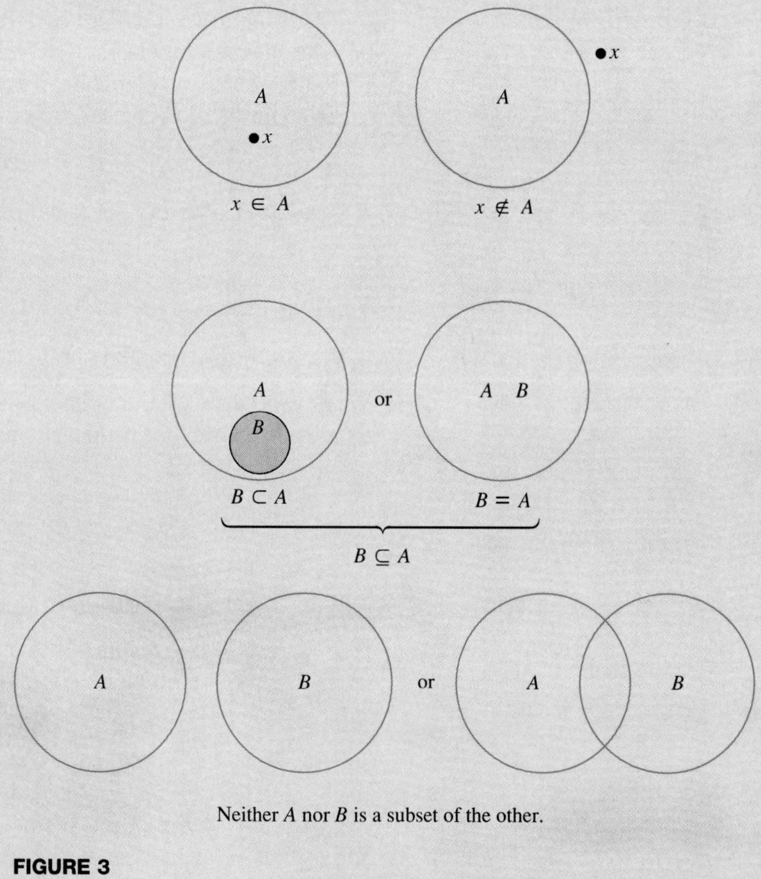

$x \in A$ $x \notin A$

$B \subset A$ or $B = A$

$B \subseteq A$

Neither A nor B is a subset of the other.

FIGURE 3

5.1 Sets

EXAMPLE 7   Debts

A department store has a policy stating that charge account customers who are behind on their payments and have charged more than $1,000 are not permitted to charge any more. In the store's database is the set of customers

$$S = \{\text{Marquez, Siefert, Soyinka, Mahfouz, Paz}\}.$$

A search of the database for customers behind on their payments returns the subset

$$A = \{\text{Marquez, Siefert, Mahfouz}\}.$$

Another search, this time for customers who have charged more than $1,000, returns the subset

$$B = \{\text{Siefert, Mahfouz, Paz}\}.$$

Which customers will not be permitted to charge any more?

SOLUTION   We can picture these relationships using the Venn diagram in Figure 4.

The customers who will not be permitted to charge any more are those who are simultaneously in A and B: Siefert and Mahfouz.

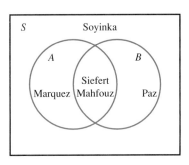

FIGURE 4

SET OPERATIONS

SET OPERATIONS

$A \cup B$ is the **union** of A and B: the set of all elements that are either in A or B (or both).

$$A \cup B = \{x \mid x \in A \text{ or } x \in B\}$$

$A \cap B$ is the **intersection** of A and B: the set of all elements that are common to A and B.

$$A \cap B = \{x \mid x \in A \text{ and } x \in B\}$$

If $A \subseteq S$ and S is the set of all things currently under consideration, then A' is the **complement** of A in S, the set of all elements of S not in A.

$$A' = \{x \in S \mid x \notin A\}$$

LOGICAL EQUIVALENTS

Union   For an element to be in $A \cup B$, it must be in A *or* in B.
Intersection   For an element to be in $A \cap B$, it must be in A *and* in B.
Complement   For an element to be in A', it must *not* be in A.

▶ **NOTE** Mathematicians always use "or" in its *inclusive* sense: one thing or another *or both*. ◀

The union, intersection and complement are illustrated in the Venn diagrams in Figures 5, 6, and 7.

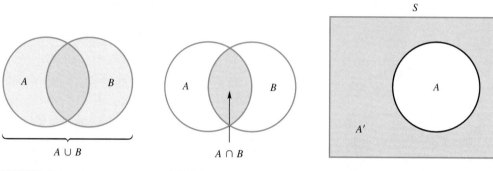

FIGURE 5   **FIGURE 6**   **FIGURE 7**

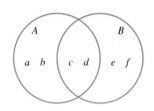

FIGURE 8

▼ **EXAMPLE 8**

Let $A = \{a, b, c, d\}$ and $B = \{c, d, e, f\}$. Find $A \cup B$ and $A \cap B$.

SOLUTION Figure 8 shows a Venn diagram for this example.

By definition, $A \cup B$ contains every element that is in at least one of A and B. Thus,

$$A \cup B = \{a, b, c, d, e, f\}.$$

The intersection, $A \cap B$, consists of all the elements that A and B have in common. Thus,

$$A \cap B = \{c, d\}.$$

▼ **EXAMPLE 9**

Let $A = \{\text{IBM, GE, GM, TWA}\}$ and $B = \{\text{IBM, MGM, GE, GM, TWA}\}$. Find $A \cup B$ and $A \cap B$.

SOLUTION In this example, A is a subset of B because every element in A is also in B. Thus,

$$A \cup B = \{\text{IBM, MGM, GE, GM, TWA}\} = B.$$

On the other hand,

$$A \cap B = \{\text{IBM, GE, GM, TWA}\} = A.$$

We can represent this example with the Venn diagram in Figure 9.

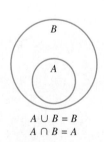

FIGURE 9

EXAMPLE 10

Let $A = \{\text{Apache, Navajo, Seminole, Sioux}\}$ and $B = \{\text{Sotho, Swazi, Xhosa, Zulu}\}$. Find $A \cup B$ and $A \cap B$.

SOLUTION

$$A \cup B = \{\text{Apache, Navajo, Seminole, Sioux, Sotho,}\\ \text{Swazi, Xhosa, Zulu}\}$$

$$A \cap B = \emptyset$$

We can represent this example with the Venn diagram in Figure 10.

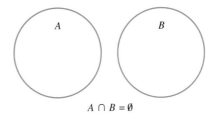

$A \cap B = \emptyset$

FIGURE 10

Before we go on... When $A \cap B = \emptyset$, we say that A and B are **disjoint.** Here is one of the places where it is very useful to consider the empty set to be a valid set. If we did not, then we would have to say that $A \cap B$ was defined only when A and B had something in common, and having to deal with this special case would quickly get tiresome.

EXAMPLE 11 Debts

As in Example 7, suppose that a department store keeps a computer database of its charge customers, that the set of customers is

$$S = \{\text{Marquez, Siefert, Soyinka, Mahfouz, Paz}\},$$

that the subset of its customers who are late on their payments is

$$A = \{\text{Marquez, Siefert, Mahfouz}\},$$

and that the subset of its customers who have charged more than $1,000 is

$$B = \{\text{Siefert, Mahfouz, Paz}\}.$$

If the credit department searched the database for customers who were either late on their payments or who had charged more than $1,000 with the company, what subset would they get? If they searched for customers who were late on their payments *and* had charged more than $1,000, what subset would they get?

SOLUTION Since "or" corresponds to union, the set of customers who were either late on their payments or who had charged more than $1,000 is the union

$$A \cup B = \{\text{Marquez, Siefert, Mahfouz, Paz}\}.$$

Since "and" corresponds to intersection, the set of customers who were late on their payments and who had charged more than $1,000 is the intersection

$$A \cap B = \{\text{Siefert, Mahfouz}\}.$$

Before we go on... Most computer databases can be searched using what are called Boolean or logical operators, such as "and" and "or." Using "and" gives an intersection of subsets, while "or" gives a union.

▼ **EXAMPLE 12**

Let

$$A = \{\text{Shang, Chou, Ch'in, Han}\},$$
$$B = \{\text{Ch'in, Han, T'ang, Sung}\}, \text{ and}$$
$$C = \{\text{Shang, Yuan, Ming, Manchu}\}.$$

(These are Chinese dynasties.) Find $A \cap B$, $A \cap C$, $A \cap (B \cap C)$, and $A \cap (B \cup C)$.

SOLUTION First, $A \cap B$ is the set of dynasties common to A and B. Thus,

$$A \cap B = \{\text{Ch'in, Han}\},$$

as illustrated in Figure 11.

Similarly, $A \cap C$ is the set of dynasties common to A and C, so that

$$A \cap C = \{\text{Shang}\}.$$

To find $A \cap (B \cap C)$, we first find the set inside the parentheses. We find

$$B \cap C = \emptyset$$

because B and C have no elements in common. Thus,

$$A \cap (B \cap C) = A \cap \emptyset = \emptyset,$$

because A and the empty set have no elements in common (the empty set has no elements at all!).

Finally, to find $A \cap (B \cup C)$, we first find the set inside the parentheses.

$$B \cup C = \{\text{Ch'in, Han, T'ang, Sung, Shang, Yuan, Ming, Manchu}\}.$$

We take the intersection of this set with A, getting

$$A \cap (B \cup C) = \{\text{Shang, Ch'in, Han}\}.$$

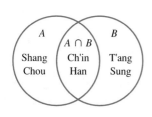

FIGURE 11

▼ **EXAMPLE 13**

Let S be the set of outcomes when a pair of distinguishable dice is cast, as in Example 4. Let E be the set of all outcomes in which the exposed numbers add to 5. Describe E'.

SOLUTION   In Figure 12 we see the set S with the elements of the subset E circled.

$$S = \begin{Bmatrix} (1,1)(1,2)(1,3)\underline{(1,4)}(1,5)(1,6) \\ (2,1)(2,2)\underline{(2,3)}(2,4)(2,5)(2,6) \\ (3,1)\underline{(3,2)}(3,3)(3,4)(3,5)(3,6) \\ \underline{(4,1)}(4,2)(4,3)(4,4)(4,5)(4,6) \\ (5,1)(5,2)(5,3)(5,4)(5,5)(5,6) \\ (6,1)(6,2)(6,3)(6,4)(6,5)(6,6) \end{Bmatrix} \qquad E' = \begin{Bmatrix} (1,1)(1,2)(1,3)(1,5)(1,6) \\ (2,1)(2,2)(2,4)(2,5)(2,6) \\ (3,1)(3,3)(3,4)(3,5)(3,6) \\ (4,2)(4,3)(4,4)(4,5)(4,6) \\ (5,1)(5,2)(5,3)(5,4)(5,5)(5,6) \\ (6,1)(6,2)(6,3)(6,4)(6,5)(6,6) \end{Bmatrix}$$

(elements of E circled)

FIGURE 12   **FIGURE 13**

(Notice how the elements of E all lie along a diagonal line.) The set we're after consists of everything that is *not* circled, as shown in Figure 13.

Before we go on... When we take the complement E' we need to know a **universal set** S containing E. Recall the definition:

$$E' = \{x \in S \mid x \notin E\}.$$

Q Why can't we just take S to be the set of *all things* and take E' to be the set of all things not in E?

A Although this is tempting, talking about entities such as the "set of all things" leads to paradoxes.* Instead, we think of S as being the set of all objects *under consideration*. For instance, when talking about integers, we could take S to be the set of all integers. If we are discussing the outcomes of throwing a die, we can take $S = \{1, 2, 3, 4, 5, 6\}$. In other words, our choice of universal set depends on the context.

▼ * One famous such paradox is called "Russell's Paradox" after the mathematical logician Bertrand Russell. It goes like this: If there were a set of all things, then there would also be a (smaller) set of all sets. Call it S. Now, since S is the set of *all* sets, it must contain itself as a member. In other words, $S \in S$. Let P be the subset of S consisting of all sets that are *not* members of themselves. Now we pose the following question: Is P a member of itself? If it is, then, since it is the set of all sets that are *not* members of themselves, it is not. On the other hand, if it is *not* a member of itself, then it qualifies as an element of P. In other words, it *is* a member of itself! Since neither can be true, something is wrong. What is wrong is the assumption that there is such a thing as the set of all sets.

49. $A \cup (B \cap C) = (A \cup B) \cap (A \cup C)$ ⎱ Distributive laws
50. $A \cap (B \cup C) = (A \cap B) \cup (A \cap C)$ ⎰
51. $S' = \emptyset$
52. $\emptyset' = S$

COMMUNICATION AND REASONING EXERCISES

53. A pair of dice is called **indistinguishable** if we have no way of distinguishing one from the other. Give the set of outcomes when a pair of indistinguishable dice is cast.

54. Explain, illustrating with an example, why $(A \cap B) \cup C \neq A \cap (B \cup C)$.

55. Explain, making reference to operations on sets, why the statement "He plays soccer or rugby and cricket" is ambiguous.

56. Explain the meaning of a universal set, and give two different universal sets that could be used in a discussion about sets of positive integers.

57. Design a database scenario that leads to the following statement: In order to keep the factory operating at maximum capacity, the plant manager should select the suppliers in $A \cap (B \cup C')$.

58. Design a database scenario that leads to the following statement: In order to keep her customers happy, the bookstore owner should stock periodicals in $A' \cup (B \cap C')$.

59. Rewrite the statement "She prefers movies that are not violent, are shorter than two hours, and have neither a tragic ending nor an unexpected ending" in the language of sets.

60. Rewrite the statement "He will cater for any event, as long as there are no more than 1,000 people, it lasts for at least three hours, and it is within a 50-mile radius of Toronto" in the language of sets.

5.2 THE NUMBER OF ELEMENTS IN A SET

Most of this chapter is devoted to studying techniques for calculating the number of elements in a set. If A is a **finite set**—that is, a set with finitely many elements—then the **cardinality** of A is just the number of elements it contains. We will denote this number by $n(A)$. In other words, $n(A)$ is the number of elements in A.

EXAMPLE 1

(a) Let $S = \{a, b, c\}$. Then $n(S) = 3$.

(b) Let S be the set of outcomes when two dice are thrown. Then $n(S) = 36$. (See Example 4 in Section 1.)

(c) $n(\emptyset) = 0$, since the empty set has no elements.

5.2 The Number of Elements in a Set

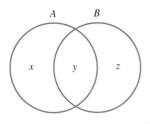

FIGURE 1

In this section we shall look at some simple counting techniques involving the set operations. In the next sections we shall study more sophisticated counting techniques.

Q How can we calculate $n(A \cup B)$ knowing $n(A)$ and $n(B)$?

A Look at the Venn diagram shown in Figure 1.

In this diagram, x represents the number of elements in A but not in B, y is the number of elements in $A \cap B$, and z is the number of elements in B but not in A. Thus, the number of elements in A is $x + y$, or

$$n(A) = x + y.$$

Similarly,

$$n(B) = y + z,$$
$$n(A \cup B) = x + y + z,$$

and

$$n(A \cap B) = y.$$

Thus,

$$\begin{aligned} n(A) + n(B) &= (x + y) + (y + z) \\ &= (x + y + z) + y \\ &= n(A \cup B) + n(A \cap B). \end{aligned}$$

In other words, $n(A) + n(B)$ counts the elements in $A \cap B$ twice. We can rewrite this equation as

$$n(A \cup B) = n(A) + n(B) - n(A \cap B).$$

So to calculate $n(A \cup B)$ we actually need three quantities: $n(A)$, $n(B)$, and $n(A \cap B)$.

▶ **NOTE** If $A \cap B = \emptyset$, that is, A and B are *disjoint*, then $n(A \cap B) = 0$ and so

$$n(A \cup B) = n(A) + n(B). \quad \blacktriangleleft$$

We can summarize as follows.

CARDINALITY OF A UNION

If A and B are finite sets, then

$$n(A \cup B) = n(A) + n(B) - n(A \cap B).$$

In particular, if A and B are disjoint, then

$$n(A \cup B) = n(A) + n(B).$$

(When A and B are disjoint, we say that $A \cup B$ is a **disjoint union**.)

EXAMPLE 2   Surveys

In a recent poll of Canadian voters, 400 favored expanded trade agreements with the U.S., 300 favored expanded trade agreements with Europe, and 50 favored expanded trade agreements with both the U.S. and Europe. How many favored expanded trade with either the U.S. or Europe or both?

SOLUTION   We first translate the question into the language of set theory. Let A be the set of those Canadian voters polled who favored expanded trade agreements with the U.S., and let B be the set of those Canadian voters polled who favored expanded trade agreements with Europe. We are told that

$$n(A) = 400,$$
$$n(B) = 300,$$

and

$$n(A \cap B) = 50.$$

We are asked for $n(A \cup B)$. By the formula,

$$n(A \cup B) = n(A) + n(B) - n(A \cap B)$$
$$= 400 + 300 - 50$$
$$= 650.$$

Thus, 650 of the Canadian voters who were polled favored expanded trade with either the U.S. or Europe.

Before we go on... We can picture this scenario in a Venn diagram (Figure 2). To find the numbers written in the diagram, we first write 50 in $A \cap B$, since that number is given to us. We can next write 350 in the part of A not overlapping B, because there are a total of 400 people in A, 50 of whom have been accounted for, leaving $400 - 50 = 350$. Similarly, there are $300 - 50 = 250$ people in the part of B not overlapping A. This gives us once again $350 + 50 + 250 = 650$ people total.

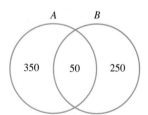

FIGURE 2

Another useful formula gives the cardinality of a complement. If S is our universal set and $A \subseteq S$, then S is the disjoint union of A and its complement. That is,

$$S = A \cup A' \quad \text{and} \quad A \cap A' = \emptyset.$$

Applying our cardinality formula for a disjoint union, we get

$$n(S) = n(A) + n(A').$$

If we solve for $n(A')$ or for $n(A)$, we get the following results.

> **CARDINALITY OF A COMPLEMENT**
>
> If S is a finite universal set, and A is a subset of S, then
>
> $$n(A') = n(S) - n(A)$$
>
> and
>
> $$n(A) = n(S) - n(A').$$

Although the next example can be easily solved without cardinality formulas, it is a useful exercise to express the situation in terms of set theory.

▼ **EXAMPLE 3** Foreign Investment

In 1990, a total of $444 million was invested in South Africa by African countries, European countries, and the U.S. Of this amount, $383 million was invested by European and African countries.* How much was invested by the U.S.?

SOLUTION Let us take S to be the set of all dollars invested in South Africa in 1990. Thus, $n(S) = 444{,}000{,}000$. If A is the set of dollars invested by the U.S., then we know that

$$n(A') = 383{,}000{,}000.$$

Since

$$n(A) = n(S) - n(A'),$$

we get

$$n(A) = 444{,}000{,}000 - 383{,}000{,}000 = 61{,}000{,}000.$$

Thus, a total of $61 million was invested by the U.S.

Q What happens when we take the union of three or more sets? For example, how do we express $n(A \cup B \cup C)$ in terms of the cardinalities of the individual pieces A, B, and C?

A Although there are formulas for the cardinalities of unions of any number of sets, these formulas get more and more complicated as the number of sets grows.† In many applications, we can use Venn diagrams and a little algebra instead.

▼ * Source: *Johannesburg Sunday Times*, Business Section, August 14, 1994, p. 6.
  † In the exercises, you will be asked to derive the simplest of these: the formula for $n(A \cup B \cup C)$.

EXAMPLE 4 Soccer

Of the 18 teams in the South African soccer league, 14 had won at least five games as of August 14, 1994, 14 had lost at least five games, and 12 had drawn at least five games. Eleven of these teams had won at least five games and lost at least five games, 10 teams had both lost and drawn at least five games, and 10 teams had both won and drawn at least five games. Eight teams had won at least five games, drawn at least five games, and lost at least five games.* How many teams had won at least five games and drawn at least five games but lost no more than four games?

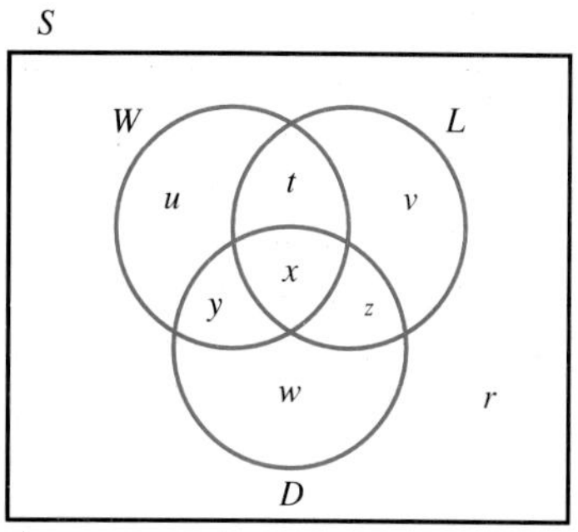

SOLUTION The universal set that suggests itself is the set S consisting of all the teams in the soccer league. Three subsets of S are mentioned: the set W of teams that had won at least five games, the set L of teams that had lost at least five games, and the set D of teams that had drawn at least five games. Let us represent this by a Venn diagram (Figure 3).

In the figure, each small letter represents the number of elements in the smallest subset enclosing the letter. For example, u represents the number of elements in W but in neither L nor D, and z represents the number of elements in both L and D but not in W.

FIGURE 12

The information we are given translates into the following equations.

$$
\begin{aligned}
18 &= x + y + z + u + v + w + t + r & (&= n(S)) \\
14 &= x + y \phantom{{}+z} + u \phantom{{}+v} + t & (&= n(W)) \\
14 &= x \phantom{{}+y} + z \phantom{{}+u} + v + t & (&= n(L)) \\
12 &= x + y + z \phantom{{}+u+v} + w & (&= n(D)) \\
11 &= x \phantom{{}+y+z+u+v+w} + t & (&= n(W \cap L)) \\
10 &= x \phantom{{}+y} + z & (&= n(L \cap D)) \\
10 &= x + y & (&= n(W \cap D)) \\
8 &= x & (&= n(W \cap L \cap D))
\end{aligned}
$$

If we work from the last equation up, we can solve for each of the unknowns by substitution. (You should also fill in these numbers in the Venn diagram in this order. Doing so makes it clearer where the numbers come from.) Substituting the last equation in each of the three equations above it gives $x = 8$, $y = 2$, $z = 2$, $t = 3$. Continuing upwards, we get $w = 0$, $v = 1$, $u = 1$, and finally $r = 1$. Finding all of these unknowns is called **solving** the Venn diagram. However, all we are asked to find is the number of elements in $W \cap D \cap L'$. According to our Venn diagram, this is

$$y = 2.$$

▼ *Source: *Johannesburg Sunday Times*, August 14, 1994, p. 28. The 18 teams are Cape Town Spurs, Pirates, Sundowns, Kaizer Chiefs, Hellenic, QwaQwa Stars, Vaal Professionals, Umtata Bucks, Wits University, Swallows, Aces, Amazulu, Bloemfontein Celtics, Rovers, Rangers, Callies, Pretoria City, Tigers.

Thus, there were 2 teams that won at least 5 games and lost at least 5 games but drew no more than 4 games.

Before we go on... What does the number r in the diagram represent?

▼ **EXAMPLE 5** Reading Lists

In a survey of 400 high school seniors, 60 had not read *Macbeth* but had read either *As You Like It* or *Romeo and Juliet*. 61 had read *As You Like It* but not *Romeo and Juliet*, 15 had read both *Macbeth* and *As You Like It*, 14 had read *As You Like It* and *Romeo and Juliet*, and 9 had read *Macbeth* and *Romeo and Juliet*. Only 5 had read *Macbeth* and *Romeo and Juliet* but not *As You Like It*, and 40 had read only *Macbeth*. Draw a Venn diagram illustrating all of these categories, and solve the diagram.

SOLUTION Let us take the universal set to consist of the 400 students surveyed. We can draw the diagram as shown in Figure 4.

From the information we are given, we get the following equations.

$$400 = x + y + z + u + v + w + t + r$$
$$120 = z + v + w$$
$$61 = v + t$$
$$15 = x + t$$
$$14 = x + z$$
$$9 = x + y$$
$$5 = y$$
$$40 = u$$

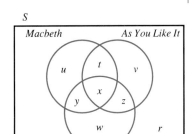

FIGURE 4

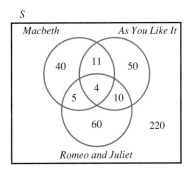

FIGURE 5

Once again, we work from the last equation upwards, solving for the unknowns as we go. We have $u = 40$ and $y = 5$. This gives $x = 4$, $z = 10$, and $t = 11$. Continuing upward, we get $v = 50$, $w = 60$, and $r = 220$. The completely solved Venn diagram is shown in Figure 5.

Before we go on... What does r represent in this example?

Q The examples above showed us how to set up a Venn diagram for a scenario involving three categories. How would we set up a Venn diagram for a situation that involves four categories?

A It is not easy! No matter how you arrange four *circles*, there will be at least one possible overlap not accounted for by the diagram. More complicated shapes may be used, but the diagram quickly becomes difficult to draw for larger numbers of sets. However, the algebraic method we used in the examples above can easily be adapted to work for any number of categories.

5.2 EXERCISES

Let $A = \{$Dirk, Johan, Frans, Sarie$\}$, $B = \{$Frans, Sarie, Tina, Klaas, Henrika$\}$, $C = \{$Hans, Frans$\}$. *Find the numbers indicated in Exercises 1–6.*

1. $n(A) + n(B)$
2. $n(A) + n(C)$
3. $n(A \cup B)$
4. $n(A \cup C)$
5. $n(A \cup (B \cap C))$
6. $n(A \cap (B \cup C))$
7. Verify that $n(A \cup B) = n(A) + n(B) - n(A \cap B)$ with A and B as above.
8. Verify that $n(A \cup C) = n(A) + n(C) - n(A \cap C)$ with A and C as above.
9. If $n(A) = 43$, $n(B) = 20$, $n(A \cap B) = 3$, find $n(A \cup B)$.
10. If $n(A) = 60$, $n(B) = 20$, $n(A \cap B) = 1$, find $n(A \cup B)$.
11. If $n(A \cup B) = 100$ and $n(A) = n(B) = 60$, find $n(A \cap B)$.
12. If $n(A) = 100$, $n(A \cup B) = 150$, $n(A \cap B) = 40$, find $n(B)$.

Let $S = \{$Barnsley, Manchester United, Southend, Sheffield United, Liverpool, Maroka Swallows, Witbank Aces, Royal Tigers, Dundee United, Lyon$\}$ *be a universal set*, $A = \{$Southend, Liverpool, Maroka Swallows, Royal Tigers$\}$, $B = \{$Barnsley, Manchester United, Southend$\}$. *Find the numbers indicated in Exercises 13–18.*

13. $n(A')$
14. $n(B')$
15. $n((A \cap B)')$
16. $n((A \cup B)')$
17. $n(A' \cap B')$
18. $n(A' \cup B')$
19. With S, A, and B as above, verify that $n((A \cap B)') = n(A') + n(B') - n((A \cup B)')$.
20. With S, A, and B as above, verify that $n(A' \cap B') + n(A \cup B) = n(S)$.

In Exercises 21–24, use the given information to complete the solution of the partially solved Venn diagrams.

21.

[Venn diagram with three circles A, B, C in universal set S. Values shown: 4 in A only, 2 in A∩C, 10 in region]

$n(A) = 20$, $n(B) = 20$, $n(C) = 28$, $n(B \cap C) = 8$, $n(S) = 50$

22.

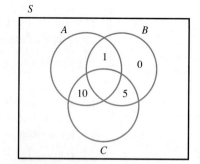

$n(A) = 16$, $n(B) = 11$, $n(C) = 30$, $n(S) = 40$

23.

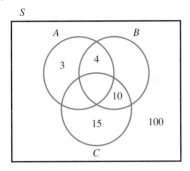

$n(A) = 10$, $n(B) = 19$, $n(S) = 140$

24.

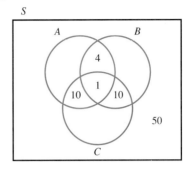

$n(A \cup B) = 30$, $n(B \cup C) = 30$, $n(A \cup C) = 35$

APPLICATIONS

25. *Amusement* On a particularly boring trans-Atlantic flight, one of the authors amused himself by counting the heads of the people in the seats in front of him. He noticed that all 37 of them either had black hair or had a whole row to themselves (or both). Of this total, 33 had black hair and 6 were fortunate enough to have a whole row of seats to themselves. How many black-haired people had whole rows to themselves?

26. *Restaurant Menus* Your favorite restaurant offers a total of 14 desserts, of which 8 have ice cream as a main ingredient and 9 have fruit as a main ingredient. Assuming that all of them have either ice cream or fruit or both as a main ingredient, how many have both?

27. *Psychology* Of a group of 400 subjects in a psychological study, 180 are highly motivated, 220 have attended preschool, and 84 are highly motivated individuals who have attended preschool.
   (a) How many highly motivated individuals did *not* attend preschool?
   (b) How many individuals in the study were neither highly motivated nor had attended preschool?

28. *Medicine* A dermatalogical study finds that of 300 patients with rashes of unknown etiology, 200 cases responded to cortisone cream, 120 responded to an antibiotic cream, and 80 responded to both forms of treatment.
   (a) How many cases responded to the antibiotic, but not the cortisone?
   (b) How many cases did not respond to either form of treatment?

29. *Mutual Funds* Of a group of 30 categories of South African mutual funds quoted on the Johannesburg Stock Exchange on August 15, 1994, mutual funds in 10 of these categories were offered by Old Mutual, Inc., 6 by Sanlam, Inc., and 6 by Syfrets, Inc. Old Mutual and Sanlam offered mutual funds in 4 categories in common, Sanlam and Syfrets offered 2 in common, and Syfrets and Old Mutual offered 2 in common. All three offered funds in two of these categories.* How many categories were offered by Old Mutual and none of the other two companies?

30. *Mutual Funds* Referring to the preceding exercise, how many categories were offered by none of the three companies?

31. *Entertainment* In a survey of 100 Enormous State University students, 21 enjoyed classical music, 22 enjoyed popular rock music, and 27 enjoyed heavy metal music. Further, 5 of the students enjoyed both classical and popular rock. How many of those who enjoyed popular rock did not enjoy classical music?

▼ *As quoted in the *Johannesburg Star*, August 15, 1994, p. 16.

32. *Entertainment* Referring to the preceding exercise, you are also told that 5 students enjoyed all three kinds of music and 53 enjoyed music in none of these categories. How many students enjoyed both classical and popular rock, but disliked heavy metal?

33. *Entertainment* According to a survey of 100 people regarding their movie attendance in the last year, 40 had seen a science fiction movie, 55 had seen an adventure movie, and 35 had seen a horror movie. Moreover, 25 had seen a science fiction movie and an adventure movie, 5 had seen an adventure movie and a horror movie, and 15 had seen a science fiction movie and a horror movie. Only 5 people had seen a movie from all three categories.
   (a) Use the above information to set up an associated Venn diagram and solve it.
   (b) Complete the following sentence: The survey suggests that _____% of science fiction movie fans are also horror movie fans.

34. *Athletics* Of the 4,700 students at Medium Suburban College (MSC), 50 play collegiate soccer, 60 play collegiate lacrosse, and 96 play collegiate football. Only 4 students play both collegiate soccer and lacrosse, 6 play collegiate soccer and football, and 16 play collegiate lacrosse and football. No students play all three sports.
   (a) Use the above information to set up an associated Venn diagram and solve it.
   (b) Complete the following sentence: _____% of the college soccer players also play one of the other two sports at the collegiate level.

COMMUNICATION AND REASONING EXERCISES

35. When is $n(A \cup B) \neq n(A) + n(B)$?

36. Use a Venn diagram to obtain a formula for $n(A \cup B \cup C)$ in terms of $n(A)$, $n(B)$, $n(C)$, $n(A \cap B)$, $n(A \cap C)$, $n(B \cap C)$, and $n(A \cap B \cap C)$.

37. Formulate an interesting application whose answer is $n(A \cup B) = 40$.

38. Formulate an interesting application whose answer is $n(A \cap B) = 20$.

▶ ▬▬ 5.3 THE MULTIPLICATION PRINCIPLE; PERMUTATIONS

Many of the sets of outcomes we shall encounter will be so large that it would be a painful task to list all their elements. For example, if S is the set of possible outcomes in the selection of a sequence of six digits 0–9 (with repetitions allowed), then S has a million elements! In situations like this, it is usually enough to know $n(S)$ and useless to try to list all the elements of S.

The following example illustrates a very useful counting technique.

▼ **EXAMPLE 1**

Suppose we have decided to buy some yogurt, and we notice that it comes in three flavors: Vanilla, Chocolate, and Brazilian Mocha. Suppose further that there are two possible sizes: small and large. How many possible selections can we make?

SOLUTION We attack this question as systematically as we can. The key is to pretend that we are going to make up a serving of yogurt and examine the decisions we must make. We need two steps:

Step 1: Select a flavor.
Step 2: Select a size.

Let us look at the possible outcomes in two ways.

5.3 The Multiplication Principle; Permutations

Analytical Step 1 has three possible outcomes (for the three flavors). For *each* of the three flavors, we can select one of two sizes in Step 2. Thus, the total number of outcomes is 3 × 2 = 6 (3 flavors, with 2 sizes per flavor). In other words, we have *multiplied* the numbers of outcomes of each step, and we get $n(S) = 6$. In fact, here is the entire set of outcomes:

$$S = \{\text{small vanilla, large vanilla, small chocolate, large chocolate, small mocha, large mocha}\}.$$

Tree We can picture this two-step decision process using a **decision tree** X(Figure 1).

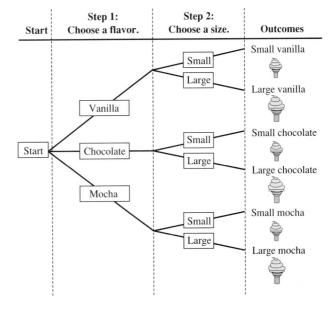

FIGURE 1

We read this diagram as follows. We start on the left at the "Start" box, and we begin with Step 1. Since there are three possible flavors to choose from, there are three possible branches we can take, as shown under the heading "Step 1." Now suppose for argument's sake that we have chosen vanilla. That would take us to the end of the top branch under Step 1. At that point, we commence with Step 2: choosing a size. Since there are two possible sizes to choose from, we take one of two further branches under the Step 2 heading, "small" or "large." If we pick large, that takes us to the "large vanilla" outcome second from the top. Once again, we see that there are six possible outcomes: three branches for Step 1, and two further branches for each outcome of Step 1, giving a total of 6 outcomes.

Before we go on... What if we chose the size first? We would get a different decision tree but the same six outcomes. Try it for yourself.

Q What is the principle behind this example?

A We had a **multi-step process.** (The example was a two-step process, but why limit ourselves to only two steps?) We then did the following calculation.

Step 1 had 3 possible outcomes.

Step 2 had 2 possible outcomes.

Therefore, the entire process had $3 \times 2 = 6$ possible outcomes. This calculation generalizes as follows.

MULTIPLICATION PRINCIPLE

Suppose we have a multi-step process in which

Step 1 has n_1 possible outcomes,

Step 2 has n_2 possible outcomes,

Step 3 has n_3 possible outcomes,

. . .

. . .

Step r has n_r possible outcomes.

Then, the entire process has $n_1 \times n_2 \times n_3 \times \ldots \times n_r$ possible outcomes.

▼ **EXAMPLE 2**

In a last-ditch attempt to organize your huge collection of floppy disks, you decide on a simple procedure for naming them: each disk is to have a three-character code consisting of two letters followed by a digit 0–9. How many possible codes are there?

SOLUTION   We'll apply the multiplication principle to this one. As in the case of the yogurt example, *we pretend that we were constructing such a code.* In order to do so, we must follow a three-step procedure:

Step 1:   Select the first letter — 26 possible choices.
Step 2:   Select the second letter — again, 26 possible choices.
Step 3:   Select the digit — 10 choices.

Thus, by the multiplication principle, there are a total of $26 \times 26 \times 10 = 6{,}760$ possible codes. That will account for quite a large collection of floppies! Since the set of outcomes is so huge, we are not going to show you a tree for this one.

Before we go on... We shall call a procedure for constructing elements of a set a **counting procedure,** since our main use for such a procedure will be in counting the number of elements in the set.

It is important that a counting procedure pass the following test.

> **ACID TEST FOR A COUNTING PROCEDURE**
>
> "Suppose I went through the counting procedure twice, but the second time I made a different choice at one or more of the steps. Would I then get a different outcome?" If the answer is always "yes," then your counting procedure passes the test. If your procedure does not pass the test, it is invalid.

Looking at Example 2, you can see that making a different choice at any of the steps results in a different code. For instance, changing from the letter *J* in Step 2 to the letter *K* results in a different code. Thus, the procedure passes the Acid Test. We shall see in Example 4 what happens if a counting procedure fails the test.

EXAMPLE 3

You are playing Scrabble™ and have the following letters to work with: A, R, E, G. Since you are losing the game, you would like to use all your letters to make a single word, but can't think of any four-letter words using all these letters. In desperation, you decide to list *all* the four-letter sequences possible to see if there are any valid words among them. How large is your list?

SOLUTION We are seeking the cardinality of the set *S* of all four-letter sequences formed with the letters A, R, E, G. To determine $n(S)$, we must come up with a counting procedure. As before, *we pretend that we are constructing a four-letter sequence.*

Step 1: Select the first letter—there are 4 choices for this.

Step 2: Select the second letter. By the time we get to Step 2, there are only 3 letters left to choose from, so there are only 3 possible choices for Step 2.

Step 3: Select the third letter—only 2 choices left.

Step 4: Select the last letter—only 1 letter left to choose from.

Thus, by the multiplication principle, there are a total of $4 \times 3 \times 2 \times 1 = 24$ possible 4-letter sequences. Figure 2 shows the tree for this process. (You can see why we're not going to draw too many of these!)

Before we go on... We once again apply the Acid Test. If at any step we made a different choice, we would have a different letter in that position, and hence a different "word." Another way of seeing this is by looking at the tree and noticing that all the outcomes are different! Thus, our counting procedure passes the Acid Test, and our calculation is valid.

356 Chapter 5 Sets and Counting

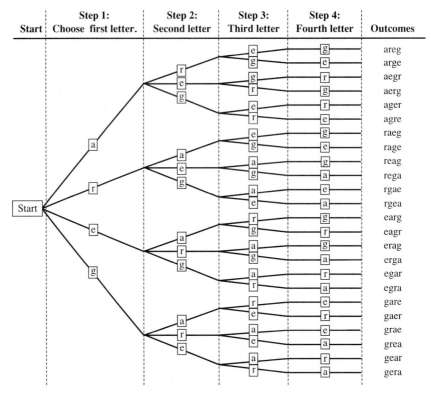

(How many of the outcomes are actual words in English?)

FIGURE 2

Here is our working strategy for counting. We shall use this strategy throughout this section and the next.

> **GUIDE TO COUNTING**
>
> 1. **Formulate a counting procedure.** If you are asked how many possible objects there are, pretend that you are *constructing* such an object and come up with a step-by-step procedure for doing so. List the steps you would take, showing the number of choices at each step.
> 2. **Apply the Acid Test to your counting procedure.** Ask yourself the following question: "Suppose I went through the counting procedure twice, but the second time I made a different choice at one or more of the steps. Would I then get a different outcome?" If the answer is always "yes," your counting procedure passes the test. If your procedure does not pass the test, it is invalid.

5.3 The Multiplication Principle; Permutations

The next example shows the importance of the Acid Test.

▼ EXAMPLE 4

Let us repeat the Scrabble™ question of the last example, but this time with the letters K, E, R, E. Recall that we want to know how many *different* four-letter words can be formed using the given letters.

SOLUTION We first try the counting procedure used in the last example.

Step 1: Select the first letter: 4 choices.
Step 2: Select the second letter: 3 choices.
Step 3: Select the third letter: 2 choices.
Step 4: Select the last letter: 1 choice.

This gives $4 \times 3 \times 2 \times 1 = 24$ choices, as in the last example.

Q This seems okay. What could be wrong here?

A We might suspect that *something* is funny here, because there are two Es. In the last example, all the letters were different. If we use the Acid Test, we see what goes wrong. Suppose, for example, that we selected the first E in Step 1, the second E in Step 2, and then the K and the R. This would result in the word eekr. Now suppose that instead, we selected the *second* E in Step 1, the *first* E in Step 2, and then the K and R. This would result in the same word: eekr. In other words, our procedure has failed the Acid Test, since different choices lead to the same outcome.

Since our original procedure failed the Acid Test, we need a new procedure. Here is a strategy that works nicely for this example. Imagine as before that we are going to construct a word. This time we are going to imagine that we have a sequence of four empty slots: ☐☐☐☐, and instead of selecting letters to fill the slots from left to right, we are going to select *slots* in which to place each of the letters. Remember that we have to use these letters: K, E, R, E. We proceed as follows, leaving the Es until last.

Step 1: Select an empty slot for the K: 4 choices (e.g., ☐☐K☐)

Step 2: Select an empty slot for the R: 3 choices (e.g., R☐K☐)

Step 3: place the Es in the remaining two slots: 1 choice!

Thus, the multiplication principle yields $4 \times 3 \times 1 = 12$ choices.

Before we go on... Use the Acid Test to verify that this procedure is valid. This method is fine when a single letter repeats, but it will not work if there is more than one letter that repeats. In the next section, we'll develop a better strategy for dealing with groups of repeated letters.

You should try constructing a decision tree for this example. If you take the original approach, choosing the letters in order, and you treat the two Es as different, you will see each word appear twice. Alternatively, when you have two Es to choose from, you could count "picking an E" as one possible choice and construct a tree where each word appears once. How many outcomes do you find? Can you apply the multiplication principle to this decision tree? You should also draw the decision tree for the counting procedure we used to find the correct answer.

▼ **EXAMPLE 5**

Ms. Birkitt, the English teacher at Brakpan Girls High School, wanted to stage a production of R. B. Sheridan's play, *The School for Scandal*. The casting was going well until she was left with five unfilled characters and five seniors who were yet to be assigned roles. The characters were Lady Sneerwell, Lady Teazle, Mrs. Candour, Maria, and Snake, while the unassigned seniors were April, May, June, Julie, and Augusta. How many possible assignments were there?

SOLUTION Following the Guide to Counting, let us pretend that we were Ms. Birkitt, and were proceeding to assign seniors to characters. We could then proceed as follows.

Step 1: Choose a senior to play Lady Sneerwell: 5 choices.
Step 2: Choose a senior to play Lady Teazle: 4 choices.
Step 3: Choose a senior to play Mrs. Candour: 3 choices.
Step 4: Choose a senior to play Maria: 2 choices.
Step 5: Choose a senior to play Snake: 1 choice.

Thus, there are $5 \times 4 \times 3 \times 2 \times 1 = 120$ possible assignments. We'll let you use the Acid Test on this one!

The solutions to Examples 3 and 5 were similar. In Example 3, we wound up with $4 \times 3 \times 2 \times 1$, while in Example 5, we wound up with $5 \times 4 \times 3 \times 2 \times 1$.

Q What was it, exactly, that these examples had in common?

A In Example 3, we were counting the number of four-letter sequences (using different letters), or *lists* of four different letters. In Example 5, we were counting the number of *lists* of five different actors.

Q What exactly do you mean by a list?

A A list is a sequence of numbered entries—like "*The New York Times* Bestseller List." Here are two lists that we might have used in Examples 3 and 5.

5.3 The Multiplication Principle; Permutations

List of 4 letters:
1. g
2. e
3. r
4. a

List of 5 actors:
1. May
2. Augusta
3. June
4. Julie
5. April

The list on the left gives the outcome "gera" in Example 3, while the list on the right gives the following cast:

Cast
Lady Sneerwell .... May
Lady Teazle .... Augusta
Mrs. Candour .... June
Maria .... Julie
Snake .... April

We call such a list a **permutation**. For example, the list on the left is a permutation of the letters a, r, e and g. A **permutation of n items** is just a list of those n items.

Q So we have n items to list. What is the total number of possible permutations of those n items?

A The Guide to Counting tells us to pretend that we are making up such a list, so we construct one in several steps. In the first step we choose the first item in the list, and we have n choices. For the second item we have $(n - 1)$ choices, for the third $(n - 2)$, and so on, until we get to the last item, where we have only a single choice left. Thus, the answer is $n \times (n - 1) \times (n - 2) \times \ldots \times 2 \times 1$. This is the product of all the integers from 1 to n and is written as $n!$ (n **factorial**).

> **PERMUTATIONS OF n ITEMS**
> A **permutation of n items** is an ordered list of those items. The number of permutations of n items is given by n factorial, which is
> $$n! = 1 \times 2 \times 3 \times \cdots \times (n - 1) \times n.$$

Q What happens if $n = 0$? In other words, suppose we had *no* items to list.

A This may seem like a foolish question, but it turns out to be useful occasionally (much as the empty set is useful). The answer to the question is that there can be only *one* list containing no items—namely, the *empty list*. In other words, it is logical to take 0! to be 1. Thus, we agree that
$$0! = 1.$$

EXAMPLE 6

In February 1993, Chicago's top ten public companies (rated by market capitalization) were, in alphabetical order: *Abbott Labs, Ameritech, Amoco, Baxter International, Commonwealth Edison, McDonald's, Motorola, Sara Lee, Sears-Roebuck,* and *Waste Management.** How many possibilities are there for the order in which they would fall if listed from largest market capitalization to smallest?

SOLUTION All we are really asked for is the number of possible lists of the 10 companies. In other words, we are looking for the number of permutations of 10 items, so the answer is $10! = 10 \times 9 \times 8 \times 7 \times 6 \times 5 \times 4 \times 3 \times 2 \times 1 = 3{,}628{,}800$. (*Don't* try to write down all the possible lists!)

EXAMPLE 7

Continuing Example 6, how many possibilities are there for the list of the top 6 companies?

SOLUTION Now we are seeking the number of possible lists of six companies chosen from the original 10. The above formula no longer applies, since we are not using all 10, so we use a new counting procedure.

Step 1: Choose the first company: 10 choices.
Step 2: Choose the second company: 9 choices.
Step 3: Choose the third one: 8 choices.
Step 4: Choose the fourth one: 7 choices.
Step 5: Choose the fifth one: 6 choices.
Step 6: Choose the sixth one: 5 choices.

Thus, there are $10 \times 9 \times 8 \times 7 \times 6 \times 5 = 151{,}200$ possible choices. We call this number $P(10, 6)$, the *number of permutations of* 6 *items chosen from* 10, or the *number of permutations of* 10 *items taken* 6 *at a time.*

This last example suggests the following general formula.

PERMUTATIONS OF *n* ITEMS TAKEN *r* AT A TIME

A **permutation of *n* items taken *r* at a time** is an ordered list of *r* items chosen from a set of *n* items. The number of permutations of *n* items taken *r* at a time is given by

$$P(n, r) = n \times (n - 1) \times (n - 2) \times \cdots \times (n - r + 1).$$

(There are *r* terms multiplied together.)

▼ *Source: *The Chicago Tribune,* Feb. 22, 1993, Section 4, p. 4

5.3 The Multiplication Principle; Permutations

EXAMPLE 8

Here are a few quick calculations.

(a) $P(8, 5) = 8 \times 7 \times 6 \times 5 \times 4 = 6{,}720$. Notice that we can also write $P(8, 5)$ as the quotient

$$P(8, 5) = \frac{8!}{(8-5)!} = \frac{8!}{3!}$$
$$= \frac{8 \times 7 \times 6 \times 5 \times 4 \times \cancel{3} \times \cancel{2} \times \cancel{1}}{\cancel{3} \times \cancel{2} \times \cancel{1}}$$
$$= 8 \times 7 \times 6 \times 5 \times 4 = 6{,}720.$$

This is an example of the following alternate formula for $P(n, r)$.

$$P(n, r) = \frac{n!}{(n-r)!}$$

(b) The number of lists of two items chosen from 11 is $P(11, 2) = 11 \times 10 = 110$.

(c) $P(8, 8) = 8! = 40{,}320$. Notice that the alternate formula gives $8!/0!$, but we said that $0! = 1$.

(d) $P(8, 1) = 8 =$ the number of lists of one item chosen from 8.

(e) $P(8, 0) = 1 =$ the number of empty lists chosen from 8.

▶ **NOTE** Although it is convenient to have some formulas available, remember that they will not work in all situations. (Look, for instance, at Examples 4 and 2.) Some situations involve permutations, and some do not. If in doubt, fall back on the Guide to Counting. ◀

5.3 EXERCISES

Evaluate the numbers in Exercises 1–8.

1. 6!
2. 7!
3. 8!/6!
4. 10!/8!
5. $P(6, 4)$
6. $P(8, 3)$
7. $P(6, 4)/4!$
8. $P(8, 3)/3!$

9. Use the Guide to Counting to confirm that the number of possible outcomes when two distinguishable dice are thrown (and the numbers uppermost are observed) is 36.

10. Use the Guide to Counting to confirm that the number of possible outcomes when a coin is tossed four times (and the sequence of heads and tails is observed) is 16.

11. How many five-letter sequences are possible using the letters m, o, n, e, y?

12. How many six-letter sequences are possible using the letters p, r, o, f, i, t?
13. How many six-letter sequences are possible using the letters a, e, a, a, u, k?
14. How many five-letter sequences are possible using the letters f, f, u, a, f?

APPLICATIONS

15. *Multiple-Choice Tests* Professor Easy's final examination has 10 true/false questions followed by 2 multiple-choice questions. In each of the multiple-choice questions you must select the correct answer from a list of five. How many answer sheets are possible?

16. *Multiple-Choice Tests* Professor Tough's final examination has 20 true/false questions followed by 3 multiple-choice questions. In each of the multiple-choice questions you must select the correct answer from a group of six. How many answer sheets are possible?

17. *Tournaments* How many ways are there of filling in the blanks for the following (fictitious) soccer tournament?

18. *Tournaments* How many ways are there of filling in the blanks for a (fictitious) soccer tournament involving the four teams San Diego State, De Paul, Colgate, and Hofstra?

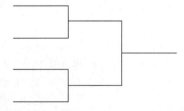

19. *Telephone Numbers* A telephone number consists of 7 digits.
    (a) How many telephone numbers are possible?
    (b) How many of them begin with 463?
    (c) How many telephone numbers are possible if no consecutive pairs of repeated digits are allowed? (For example, 235–9350 is permitted, but not 223–6789.)

20. *Social Security Numbers* A Social Security Number is a sequence of 9 digits.
    (a) How many Social Security Numbers are possible?
    (b) How many of them begin with 023?
    (c) How many Social Security Numbers are possible if no consecutive pairs of repeated digits are allowed? (For example, 235–93–2345 is permitted, but not 126–67–8189.)

21. *Itineraries* Your international diplomacy trip requires stops in Thailand, Singapore, Hong Kong, and Bali. How many possible itineraries are there?

22. *Itineraries* Referring to Exercise 21, how many possible itineraries are there in which the last stop is Thailand?

23. *Menus* The local diner offers a meal combination consisting of an appetizer, a soup, a main course, and dessert. There are five appetizers, two soups, four main courses, and five desserts. Your diet restricts you to choosing between a dessert and an appetizer. (You cannot have both.) Given this restriction, how many three-course meals are possible?

24. *Stock Portfolios* Your broker has suggested that you diversify your investments by splitting your portfolio between mutual funds, municipal bond funds, stocks, and precious metals. She suggests four good mutual funds, three municipal bond funds, eight stocks, and three precious metals (gold, silver, and platinum).
    (a) Assuming your portfolio is to contain one of each type of investment, how many different portfolios are possible?
    (b) Assuming that your portfolio is to contain three mutual funds, two municipal bond funds, one stock, and two precious metals, how many different portfolios are possible?

25. *Product Design* Your company has patented an electronic digital padlock that a user can program

with his or her own four-digit code. (Each digit can be 0 through 9.) It is designed to open either if the correct code is keyed in, or—and this is helpful for forgetful people—if exactly one of the digits is incorrect. How many incorrect codes will open a programmed padlock?

26. *Product Design* Your company has patented an electronic digital padlock that has a telephone-style keypad: each digit from 2 through 9 corresponds to three letters of the alphabet. (See the figure for Exercise 30 below.) How many different four-letter sequences correspond to a single four-digit sequence using digits in the range 2 through 9?

27. *DNA Chains* DNA (deoxyribonucleic acid) is the carrier of genetic information. A DNA chain is a sequence of chemicals called *bases*. There are four possible bases: thymine (T), cytosine (C), adenine (A), and guanine (G).
    (a) How many three-element DNA chains are possible?
    (b) How many *n*-element DNA chains are possible?
    (c) A human DNA chain has 2.1×10^{10} elements. How many human DNA chains are possible?

28. *Credit Card Numbers** Each customer of *Mobil Credit Corporation* is given a 9-digit number for computer identification purposes.
    (a) If each digit can be any number from 0 to 9, are there enough different account numbers for 10 million credit-card holders?
    (b) Would there be enough account numbers if the digits were only 0 or 1?

29. *Codes* A hexadecimal code consists of a sequence of integers ranging from 0 to 15. How many different encodings are possible using sequences of length 4?

30. *Telephone Numbers* When telephone service was first introduced in the U.S., a local telephone number consisted of a sequence of two letters followed by five digits. Three letters were associated with each number from 2 to 9 (just as in the standard telephone layout shown in the figure). How many different telephone numbers were available?

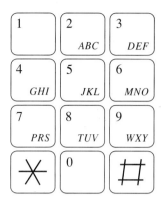

31. *(from the GMAT exam)* If 10 persons meet at a reunion and each person shakes hands exactly once with each of the others, what is the total number of handshakes?
    (a) $10 \cdot 9 \cdot 8 \cdot 7 \cdot 6 \cdot 5 \cdot 4 \cdot 3 \cdot 2 \cdot 1$
    (b) $10 \cdot 10$
    (c) $10 \cdot 9$
    (d) 45
    (e) 36

32. *(based on a question from the GMAT exam)* If 8 couples meet at a reunion and each person dances exactly once with a person of the opposite gender, what is the total number of couples that dance?
    (a) $8 \cdot 7 \cdot 6 \cdot 5 \cdot 4 \cdot 3 \cdot 2 \cdot 1$
    (b) 16
    (c) 4^4
    (d) 2^8
    (e) $4 \cdot 3 \cdot 2 \cdot 1$

33. *Calendars* The World Almanac[†] features a "perpetual calendar": a collection of 14 possible calendars. Why does this suffice to be sure of a calendar for every conceivable year?

34. *Calendars* How many possible calendars are there with February 12 falling on either a Sunday, Monday, or Tuesday?

35. *Car Engines*[‡] In a six-cylinder V6 engine, the even-numbered cylinders are on the left and the odd-numbered cylinders are on the right. A good firing

---

▼ * Taken from an exercise in *Applied Combinatorics* by F. S. Roberts (Englewood Cliffs, N.J.: Prentice-Hall, 1984).

† Source: *The World Almanac and Book of Facts* (New York, NY: Pharos Books, 1992)

‡ Adapted from an exercise in *Basic Techniques of Combinatorial Theory* by D. I. A. Cohen (New York: John Wiley and Sons, 1978)

order is a permutation of the numbers 1 through 6 in which right and left sides are alternated.
(a) How many possible good firing sequences are there?
(b) How many good firing sequences are there that start with a cylinder on the left?

36. *Car Engines* Repeat Exercise 35 for an eight-cylinder V8 engine.

37. *License Plates* Many U.S. license plates display a sequence of three letters followed by 3 digits.
(a) How many such license plates are possible?
(b) In order to avoid confusion of letters with digits, some states do not issue standard plates with the last letter an I, O, or Q. How many license plates are still possible?
(c) Assuming that the letter combinations VET, MDZ, and DPZ are reserved for disabled veterans, medical practitioners and disabled persons respectively, and also taking the restriction in part (b) into account, how many license plates are possible?

38. *License Plates* License plates in Montana have a sequence consisting of (1) a number from 1 to 56; (2) a dot; (3) a positive number of up to five digits.
(a) How many different license plates are possible?
(b) How many different license plates are available for citizens if numbers that end with 0 are reserved for official state vehicles?

39. How many four-letter sequences are possible containing only the letters R and D?

40. How many six-letter sequences are possible containing only the letters A, B, and C?

41. *Mazes*
(a) How many four-letter sequences are possible containing only the letters R and D, with D occurring only once?
(b) Use part (a) to calculate the number of possible routes from Start to Finish in the maze shown in the figure, where each move is either to the right or down.
(c) Comment on what would happen if we allowed left and/or up moves.

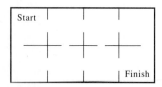

42. *Mazes*
(a) How many six-letter sequences are possible containing only the letters R and D, with D occurring only once?
(b) Use part (a) to calculate the number of possible routes from Start to Finish in the maze shown in the figure, where each move is either to the right or down.
(c) Comment on what would happen if we allowed left and/or up moves.

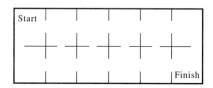

43. *Romeo and Juliet* Here is a list of the main characters in Shakespeare's *Romeo and Juliet*.

Escalus, *prince of Verona*

Paris, *kinsman to the prince*

Romeo, *of Montague Household*

Mercutio, *friend of Romeo*

Benvolio, *friend of Romeo*

Tybalt, *nephew to Lady Capulet*

Friar Lawrence, *a Franciscan*

Lady Montague, *of Montague Household*

Lady Capulet, *of Capulet Household*

Juliet, *of Capulet Household*

Juliet's Nurse

There are a total of ten male and eight female actors available to play these roles. How many possible casts are there? (All roles are to be played by actors of the correct gender.)

44. *Swan Lake* The Enormous State University's Accounting Society has decided to produce a version of the ballet *Swan Lake*, in which all the female roles (including all of the swans) will be danced by males, and vice versa. Here are the main characters:

Prince Siegfried

Prince Siegfried's Mother

Princess Odette, *the White Swan*

The Evil Duke Rotbart

Odile, *the Black Swan*
Cygnette #1, *young swan*
Cygnette #2, *young swan*
Cygnette #3, *young swan*

The ESU Accounting Society has on hand a total of four female dancers and twelve male dancers who are to be considered for the main roles. How many possible casts are there?

45. *Minimalist Art* You are exhibiting your collection of minimalist paintings. Art critics have raved about your paintings, each of which consists of ten vertical colored lines set against a white background. You have used the following rule to produce your paintings: Every second line, starting with the first, is to be either green or red, while the remaining five lines are to be either all yellow, all pink, or all purple. Your collection is complete: every possible combination that satisfies the rules occurs. How many paintings are you exhibiting?

46. *Combination Locks* Dripping wet after your shower, you have completely forgotten the combination of your lock. It is one of those "standard" combination locks, which uses a three-number combination with each number in the range 0 through 39. All you remember is that the second number is either 27 or 37, while the third number ends in a 5. In desperation, you decide to go through all possible combinations using the information you remember. Assuming that it takes about 10 seconds to try each combination, what is the longest possible time you may have to stand dripping in front of your locker?

47. *Traveling Salesperson* Suppose you are a salesperson who must visit the 23 cities Dallas, Tampa, Orlando, Fairbanks, Seattle, Detroit, Chicago, Houston, Arlington, Grand Rapids, Urbana, San Diego, Aspen, Little Rock, Tuscaloosa, Honolulu, New York, Ithaca, Charlottesville, Lynchville, Raleigh, Anchorage, and Los Angeles. Leave all answers to the following in factorial form.
    (a) How many possible itineraries are there that visit each city exactly once?
    (b) Repeat part (a) in the event that the first five stops have already been determined.
    (c) Repeat part (a) in the event that your itinerary must include the sequence Anchorage, Fairbanks, Seattle, Chicago, and Detroit, in that order (though not necessarily as the first five stops).

48. *Traveling Salesperson* Referring to the situation in Exercise 47 (and leaving all answers in factorial form),
    (a) How many possible itineraries are there that start and end at Detroit, visiting each other city exactly once?
    (b) How many possible itineraries are there that start and end at Detroit, visiting Chicago twice and every other city once?
    (c) Repeat part (a) in the event that your itinerary must include the sequence Anchorage, Fairbanks, Seattle, Chicago, and New York, in that order.

49. *Matrices* (Some knowledge of matrices is assumed for this exercise.) Use the Guide to Counting to show that an $m \times n$ matrix must have $m \cdot n$ entries.

50. *Building Blocks* Use the Guide to Counting to show that a rectangular solid with dimensions $m \times n \times r$ can be constructed with $m \cdot n \cdot r$ cubical $1 \times 1 \times 1$ blocks. (See the figure.)

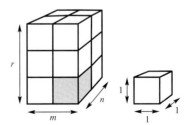

Rectangular solid made up of $1 \times 1 \times 1$ cubes

51. *Computer Codes* A computer "byte" consists of eight "bits," each bit being either a 0 or a 1. If characters are represented using a code that uses a byte for each character, how many different characters can be represented?

52. *Computer Codes* Some written languages, such as Chinese and Japanese, use tens of thousands of different characters. If a language uses roughly 50,000 characters, a computer code for this language would have to use how many bytes per character?

53. *Programming in BASIC* (some programming knowledge assumed for this exercise) How many iterations will be carried out in the following routine?

```
For i = 1 to 10
    For j = 2 to 20
        For k = 1 to 10
            Print i, j, k
        Next k
    Next j
Next i
```

54. *Programming in PASCAL* (some programming knowledge assumed for this exercise) How many iterations will be carried out in the following routine?

```
For i := 1 to 2
    For j := 1 to 2
        For k := 1 to 2
            sum := i+j+k;
```

COMMUNICATION AND REASONING EXERCISES

55. Explain how the multiplication principle leads to the formula for the number of permutations of n items.

56. Explain how the multiplication principle leads to the formula for the number of permutations of n items taken r at a time.

57. If you were hard-pressed to study for an exam on counting and had only enough time to study one topic, would you choose the formulas for the number of permutations, or the multiplication principle? Give reasons for your choice.

58. Which of the following represent a permutation?
(a) An arrangement of books on a shelf
(b) A group of 10 people in a bus
(c) A committee of 5 Senators chosen from 100
(d) A Presidential Cabinet of 5 Secretaries chosen from 20

▶ 5.4 THE ADDITION PRINCIPLE; COMBINATIONS

THE ADDITION PRINCIPLE

Up to this point we have looked at counting procedures using a sequence of steps. It sometimes happens that we are faced with several alternatives, each of which involves a different number of choices. Consider, for example, the following scenario: you are in the cereal aisle of the local supermarket and are undecided as to whether to buy bran flakes or corn flakes. There are five varieties of bran flakes and eleven varieties of corn flakes. Further, each of the items comes in two sizes: large and small. You want to calculate how many possible choices there are. If you try to create a step-by-step counting procedure, you won't succeed. Here is an attempt:

Step 1: Choose the type of cereal: 2 choices.

Step 2: Choose the variety. The number of choices in this step is either 5 or 11 and depends on your first choice, so we can't say for certain what number to use.

Q What do we do?

A Let us try to calculate the answer by common sense first: There are five varieties of bran flakes, each of which comes in two sizes, giving a total of 10 possibilities. Similarly, there are 11 varieties of corn flakes, each of which also comes in two sizes, giving a total of 22 possibilities. Thus, altogether, there are 10 + 22 = 32 possibilities. We have *added* the number of outcomes from the two alternatives. This is the **addition principle.**

> **ADDITION PRINCIPLE**
>
> Suppose we have a counting procedure with several alternatives, in which
>
> > Alternative 1 has n_1 possible outcomes,
> > Alternative 2 has n_2 possible outcomes,
> > Alternative 3 has n_3 possible outcomes,
> > . . .
> > . . .
> > Alternative r has n_r possible outcomes.
>
> Then the entire process has $n_1 + n_2 + n_3 + \cdots + n_r$ possible outcomes.

Q How does the addition principle compare with the multiplication principle?

A The key difference is this: when you have several *steps to follow,* you use the multiplication principle, but when you have several *alternatives to choose from,* you use the addition principle. Here is another way of thinking about it. When following a sequence of steps, you do Step 1 *and* Step 2 *and* Step 3 . . . When following a list of alternatives, you do Alternative 1 *or* Alternative 2 *or* Alternative 3 . . . Thus, "and" translates to "multiply," and "or" translates to "add."

Q Does the addition principle have anything to do with the rule $n(A \cup B) = n(A) + n(B)$ (when A and B are disjoint) from Section 2?

A It has everything to do with that rule. Suppose that we have a counting procedure with two alternatives, and suppose that A represents the set of outcomes corresponding to the first of two alternatives and B the set of outcomes corresponding to the second alternative. Then A and B are disjoint,* and $A \cup B$ is the set of all possible outcomes. Therefore, the rule $n(A \cup B) = n(A) + n(B)$ is the addition principle for two alternatives. The case of three alternatives corresponds to the rule

$$n(A \cup B \cup C) = n(A) + n(B) + n(C)$$

(A, B, C mutually disjoint).

This argument generalizes to any number of alternatives.

▼ * They are disjoint because they correspond to two different alternatives, and the Acid Test tells us that different alternatives should lead to different outcomes if we are using a valid counting procedure.

We shall use counting procedures just as in the previous section, except that now such a procedure may involve choosing between several alternatives.

EXAMPLE 1

There are 5 male students and 6 female students willing to serve on the college's delegation to the 12th Annual Sci-Fi Symposium (to be held in Mars, IL). The delegation is to consist of a president, a vice-president, a treasurer, and a secretary. Since the treasurer and the secretary are to room together, it would be best if they were of the same gender. How many possible delegations are there?

SOLUTION   Following the Guide to Counting, we imagine that we are constructing a delegation. The complicating factor is that the treasurer and the secretary are to be of the same gender. This gives us two alternatives.

Alternative 1: Both are female.
To calculate the number of possibilities for this alternative, let us start with the treasurer and the secretary, since then we will know how many males and females are left for the rest of the positions.

Step 1:   Choose the treasurer: 6 possible choices.

Step 2:   Choose the secretary: 5 possible choices.

Step 3:   Choose the president: 9 people left to choose from.

Step 4:   Choose the vice-president: 8 people left to choose from.

Applying the multiplication principle gives us a total of 2,160 possible choices for this alternative.

Alternative 2: Both are male.

Step 1:   Choose the treasurer: 5 possible choices.

Step 2:   Choose the secretary: 4 possible choices.

Step 3:   Choose the president: 9 people left to choose from.

Step 4:   Choose the vice-president: 8 people left to choose from.

Applying the multiplication principle gives us a total of 1,440 possible choices for this alternative.

Now we can apply the addition principle, getting a total of 2,160 + 1,440 = 3,600 possible delegations.

Before we go on...   We may apply the Acid Test to our counting procedure: If we chose Alternative 2 instead of Alternative 1, we would have a different delegation, since the roommates would be male instead of female. We can check the steps within each alternative using the Acid Test as well.

5.4 The Addition Principle; Combinations

EXAMPLE 2

An exam has the following structure: the student is required to answer either Part A or Part B, but not both. Part A consists of 10 true/false questions followed by 5 multiple-choice questions, each of which requires the student to choose the correct answer from four alternatives. Part B consists of 5 true/false questions followed by 8 multiple-choice questions, each with four alternatives. How many possible combinations of answers are there?

SOLUTION   Suppose we were filling out one of the answer sheets. To start, we must choose between two alternatives.

Alternative 1:   **Choose Part A.**

Step 1:   10 true/false decisions: $2^{10} = 1,024$ possibilities.

Step 2:   5 multiple-choice decisions: $4^5 = 1,024$ possibilities.

This gives $1,024 \times 1,024 = 1,048,576$ possibilities for this alternative.

Alternative 2:   **Choose Part B.**

Step 1:   5 true/false decisions: $2^5 = 32$ possibilities

Step 2:   8 multiple-choice decisions: $4^8 = 65,536$ possibilities

This gives $32 \times 65,536 = 2,097,152$ possibilities for this alternative.

By the addition principle, we get $1,048,576 + 2,097,152 = 3,145,728$ possibilities.

COMBINATIONS

We consider another common counting problem typified by the following.

You are a portfolio manager for a client who wishes to invest in three of the top 10 income funds. Based on a three-year performance record, you have the following list:* Franklin Income, Berwyn Income, Linder Dividend, National Multi-Sector F.I.A., Seligman Income A, Putnam Diversified Inc., Vanguard Wellesley Income, Income Fund of America, Vanguard Preferred Stock, and Prudential Flex-A-Fund Consolidated. How many different portfolios of three out of these 10 funds are possible?

At first glance, you might think that we were asking you to find the number of lists of 3 funds chosen from 10, and you might reply that the answer is $10 \times 9 \times 8 = 720$. But many of these lists contain the same funds. For instance, the two lists

1. Berwyn Income
2. Seligman Income A
3. Vanguard Preferred Stock

*Source: *The New York Times*, Aug. 14, 1993, p. 31.

and

1. Vanguard Preferred Stock
2. Seligman Income A
3. Berwyn Income

are really the same for our purposes, since they represent the same portfolio or set of three funds. (In other words, the Acid Test fails here. Several different lists can give the same set of funds.) Thus, what we are really after is the number of *sets* of three funds chosen from 10, rather than the number of *lists* of three funds chosen from 10.

> **PERMUTATIONS AND COMBINATIONS OF n ITEMS TAKEN r AT A TIME**
>
> A **permutation** of n items taken r at a time is an *ordered* list of r items chosen from n. We write the number of permutations of n items taken r at a time as $P(n, r)$.
>
> A **combination** of n items taken r at a time is an *unordered* set of r items chosen from n. We write the number of combinations of n items taken r at a time as $C(n, r)$.

▶ **NOTE** We are using the terms "list" and "set" deliberately. A *list* is usually understood to be ordered (there is a first item, a second, and so on) while a *set* is, mathematically speaking, an unordered collection. Permutations are lists, while combinations are sets. ◀

Q So we have given a name to our problem: we want to calculate the number $C(10, 3)$ of combinations (or sets) of 3 funds chosen from the 10. How do we calculate this number?

A Let us start by calculating the number $P(10, 3)$ of *lists* of 3 funds by using a somewhat peculiar counting procedure.

Step 1: Choose a *set* of 3 funds chosen from 10.
We don't yet know what this number is, but we have called it $C(10, 3)$.

Step 2: Choose an ordering of the set of 3 funds to form a list.
We *do* know how many ways this can be done: 3!.

Thus, by the multiplication principle, we get

$$P(10, 3) = C(10, 3) \times 3!.$$

We can solve for $C(10, 3)$ to obtain

$$C(10, 3) = \frac{P(10, 3)}{3!}$$

5.4 The Addition Principle; Combinations

$$= \frac{10 \times 9 \times 8}{3 \times 2 \times 1}$$

$$= \frac{720}{6} = 120.$$

Thus, the number of portfolios of three funds chosen from 10 is $C(10, 3) = 120$.

There is nothing special about the numbers 10 and 3 in the above example.

COMBINATIONS OF *n* ITEMS TAKEN *r* AT A TIME

The number of combinations of n items taken r at a time is given by

$$C(n, r) = \frac{P(n, r)}{r!}.$$

We often read $C(n, r)$ as "n choose r." Since $P(n, r) = n!/(n - r)!$, we can also write

$$C(n, r) = \frac{n!}{r!(n - r)!}.$$

▶ **NOTE** Other common notations for $C(n, r)$ are $\binom{n}{r}$ and $_nC_r$ (the latter is one you often see on calculators). When written $\binom{n}{r}$, this number is often called a **binomial coefficient.** ◀

EXAMPLE 3

How many sets of 6 swans can be selected from 10?

SOLUTION A set is unordered, so we use the formula

$$C(10, 6) = \frac{P(10, 6)}{6!}$$

$$= \frac{10 \times 9 \times 8 \times 7 \times 6 \times 5}{6 \times 5 \times 4 \times 3 \times 2 \times 1} = 210.$$

Some people like to calculate $C(n, r)$ by the formula. We prefer to think of it in the following way.

CALCULATING $C(n, r)$

To calculate $C(n, r)$, begin with the fraction n/r and continue multiplying by decreasing numbers on top and bottom until you hit 1 in the denominator.

EXAMPLE 4

Find $C(11, 3)$ and $C(11, 8)$.

SOLUTION

$$C(11, 3) = \frac{11 \times 10 \times 9}{3 \times 2 \times 1} = \frac{11 \times \overset{5}{\cancel{10}} \times \overset{3}{\cancel{9}}}{\cancel{3} \times \cancel{2} \times 1} = 165$$

Notice how much easier it is to calculate this by canceling than by first calculating the top and the bottom and then dividing.*

$$C(11, 8) = \frac{11 \times 10 \times 9 \times \cancel{8} \times \cancel{7} \times \cancel{6} \times \cancel{5} \times \cancel{4}}{\cancel{8} \times \cancel{7} \times \cancel{6} \times \cancel{5} \times \cancel{4} \times 3 \times 2 \times 1}$$

$$= \frac{11 \times \overset{5}{\cancel{10}} \times \overset{3}{\cancel{9}}}{\cancel{3} \times \cancel{2} \times 1} = 165 \text{ again}$$

Before we go on...

Q Why is $C(11, 3)$ the same as $C(11, 8)$?

A $C(11, 3)$ is the number of groups of 3 that can be selected from 11. For instance, we might be selecting a group of 3 Navy Seals from a group of 11 to go on a dangerous mission. This is the same as selecting the group of 8 who are *not* going on the mission. In other words, each group of 3 chosen from 11 corresponds to a group of 8 chosen from 11. Thus, $C(11, 3) = C(11, 8)$. Similarly, $C(42, 40) = C(42, 2)$, and $C(50, 41) = C(50, 9)$. In general,

$$C(n, r) = C(n, n - r).$$

By choosing the one with the smaller denominator, you can make your calculations easier.

Some of these calculations involve large numbers, and this is where a good calculator comes in handy. Your calculator may be equipped with the combination and permutation functions $P(-, -)$ and $C(-, -)$ (they are often labeled $_nP_r$ and $_nC_r$), and this will make your life easier. Otherwise, you may need to enter the factorials by hand.

Q Does $C(5, 0)$ make sense? What is it?

A There are several ways of making sense of $C(5, 0)$, and they all give the same answer: $C(5, 0) = 1$. First, we could ask how many sets with no elements can be chosen from a set of 5. The answer is: only one, the empty set. Second, it should also be true that $C(5, 0) = C(5, 5)$ (see the "Before we go on" discussion in the example above) and $C(5, 5) = 1$ is easy to calculate. Third, we can use the formula directly.

---

*Since $C(n, r)$ is a whole number, you will always be able to cancel everything in the denominator.

5.4 The Addition Principle; Combinations

$$C(5, 0) = \frac{P(5, 0)}{0!} = \frac{1}{1} = 1$$

Here is one of the places where the convention $0! = 1$ is convenient.

▼ **EXAMPLE 5**

You are dealt a five-card poker hand from a standard deck of 52 different cards. How many different hands are possible?

SOLUTION A "hand" of cards is just a set of cards, since the order in which you are dealt them is not important.* Thus, we are asking for the number of sets of 5 cards chosen from a deck of 52. This is

$$C(52, 5) = \frac{52 \times 51 \times 50 \times 49 \times 48}{5 \times 4 \times 3 \times 2 \times 1} = 2{,}598{,}960.$$

Thus, there are 2,598,960 possible poker hands.

▼ **EXAMPLE 6**   Lotto

The betting game lotto (used in many state lotteries) involves choosing a collection of six different numbers in the range 0–54. The order in which you choose them is irrelevant. If lotto tickets cost $1 for two sets of six numbers, and you decide to buy tickets with every possible combination, how much money would you have to spend?

SOLUTION First, we want to know the number of sets of 6 numbers chosen from 55, which is $C(55, 6)$. Now,

$$C(55, 6) = \frac{55 \times 54 \times 53 \times 52 \times 51 \times 50}{6 \times 5 \times 4 \times 3 \times 2 \times 1}$$
$$= 28{,}989{,}675 \text{ numbers.}$$

Thus, you would have to plunk down 28,989,675/2 = $14,494,838 (rounding up to the nearest dollar) to be assured of a win! Since this is considerably more than the typical winnings, which rarely exceed about $5 million, don't bother. On the other hand, buying on the order of 14 million tickets will cause the pot to become larger, and this might result in a frenzy of buying, making the pot larger still (although then you have to worry about splitting the pot among multiple winners).

Before we go on... Since there is only one winning combination in a total of 28,989,675, your chances of winning by buying a single ticket are one in 28,989,675, roughly 1 in 29 million.†

▼ * We shall say more about playing cards below. For this example, all you need to know is that there are 52 different cards in a standard deck.

† In its advertisements, the New York State Lotto used to suggest that you "make a wish upon a star."

Let us emphasize the following once more.

> **THE DIFFERENCE BETWEEN LISTS AND SETS**
>
> 1. By a **list** of 5 items chosen from 22 we mean a *numbered list* of 5 items selected from the 22. Thus, there is a first item, a second one, a third, a fourth, and a fifth. Order is important: if the same items occur on two lists, but in a different order, then the lists are different. The total number of such lists is $P(22, 5) = 22 \times 21 \times 20 \times 19 \times 18$.
> 2. By a **set** of 5 items chosen from 22, we mean an *unordered collection* of 5 items selected from the 22. Two sets are the same if they consist of the same 5 items. The total number of such sets is $C(22, 5) = \dfrac{P(22, 5)}{5!}$.
>
> *Be careful to decide in each situation whether it refers to lists or sets.*

We now look at counting examples that use several of the techniques we've discussed.

▼ **EXAMPLE 7**   Bags of Marbles

A bag contains 3 red marbles, 2 blue ones, and 4 yellow ones.

(a) How many sets of five marbles chosen from this bag are possible?

(b) How many sets of five marbles are there such that 2 are red, 2 are blue, and 1 is yellow?

SOLUTION   For part (a), we are asked to calculate the total number of sets of five chosen from a total of 9, which is $C(9, 5) = C(9, 4) = 126$.

For part (b), we appeal to the Guide to Counting. Suppose you were choosing such a set of marbles. You could then use the following procedure.

Step 1:   Select a set of 2 red marbles from the 3 red ones in the bag: $C(3, 2) = C(3, 1) = 3$ choices.

Step 2:   Select a set of 2 blue marbles from the 2 blue ones in the bag: $C(2, 2) = 1$ choice.

Step 3:   Select 1 yellow marble from the four in the bag: $C(4, 1) = 4$ choices.

Thus, the total number of such sets is $3 \times 1 \times 4 = 12$.

Before we go on... The Acid Test works nicely here. If, for instance, we had chosen a different set of 2 red marbles in part (b), the set we would have wound up with would be different. It might *look* the same, as it would also consist of 2 reds, 2 blues, and 1 yellow, but it would be a different set because it has different red marbles in it. Thus, we are assuming that the marbles are *distinguishable* even when they have the same color. For our purposes in the chapter on probability, it makes no difference in the end whether we assume that the marbles are distinguishable or not, and it makes the calculations easier if we assume that they are.

▼ **EXAMPLE 8** More Bags of Marbles

A bag contains 3 red marbles, 3 blue ones, 3 green ones, and 2 yellow ones.

(a) How many sets of four marbles are possible?
(b) How many sets of four are there such that each one is a different color?
(c) How many sets of four are there in which at least two are red?
(d) How many sets of four are there in which none are red, but at least 1 is green?

SOLUTION For part (a), we have $C(11, 4) = 330$ possible sets of 4 marbles. For part (b), we pretend that we are constructing such a set of marbles.

Step 1: Choose one red one from the 3 red ones: $C(3, 1) = 3$ choices.

Step 2: Choose one blue one from the 3 blue ones: $C(3, 1) = 3$ choices.

Step 3: Choose one green one from the 3 green ones: $C(3, 1) = 3$ choices.

Step 4: Choose one yellow one from the 2 yellow ones: $C(2, 1) = 2$ choices.

This gives a total of $3 \times 3 \times 3 \times 2 = 54$ possible sets.

For part (c), we once again pretend that we are constructing such a set. Since at least 2 must be red, this means that either 2 are red or 3 are red (there being a total of 3 red ones). In other words, we have two *alternatives*.

Alternative 1: Exactly 2 red marbles.

Step 1: Choose 2 red ones: $C(3, 2) = C(3, 1) = 3$ choices.

Step 2: Choose 2 non-red ones. There are 8 of these, so we get $C(8, 2) = 28$ possible choices.

Thus, the total number of choices for this alternative is $3 \times 28 = 84$.

Alternative 2: 3 red marbles.

Step 1: Choose the 3 red ones: $C(3, 3) = 1$ choice.
Step 2: Choose 1 non-red one: $C(8, 1) = 8$ choices.

Thus, the total number of choices for this alternative is $1 \times 8 = 8$.
By the addition principle, we get a total of $84 + 8 = 92$ sets.

Turning to (d), the phrase "at least 1 green" tells us that we again have some alternatives.

Alternative 1: 1 green marble.

Step 1: Choose 1 green marble from the 3: $C(3, 1) = 3$ choices.
Step 2: Choose 3 non-green, non-red marbles: $C(5, 3) = 10$ choices.

Thus, the total number of choices for Alternative 1 is $3 \times 10 = 30$.

Alternative 2: 2 green marbles.

Step 1: Choose 2 green marbles from the 3: $C(3, 2) = 3$ choices.
Step 2: Choose 2 non-green, non-red marbles: $C(5, 2) = 10$ choices.

Thus, the total number of choices for Alternative 2 is $3 \times 10 = 30$.

Alternative 3: 3 green marbles.

Step 1: Choose 3 green marble from the 3: $C(3, 3) = 1$ choice.
Step 2: Choose 1 non-green, non-red marble: $C(5, 1) = 5$ choices.

Thus, the total number of choices for Alternative 3 is $1 \times 5 = 5$.
The addition principle now tells us that the number of such sets is $30 + 30 + 5 = 65$.

Before we go on... Here is an easier way to answer (d). First, the total number of sets having *no* red marbles is $C(8, 4) = 70$. Next, of those, the number containing no green marbles is $C(5, 4) = C(5, 1) = 5$. This leaves $70 - 5 = 65$ groups containing no red marbles but having at least one green marble. (We have really used here the formula for the cardinality of the complement of a set.)

The next example concerns poker hands. For those unfamiliar with playing cards, here is a short description.

> **PLAYING CARDS**
>
> A standard deck consists of 52 playing cards. Each card is in one of 13 **denominations:** Ace, 2, 3, 4, 5, 6, 7, 8, 9, 10, Jack (J), Queen (Q), and King (K), and in one of four **suits:** hearts (♥), diamonds (♦), clubs (♣), and spades (♠). Thus, for instance, the Jack of Spades, J♠, refers to the denomination of Jack in the suit of Spades. The entire deck of cards is thus
>
> | A♥ | 2♥ | 3♥ | 4♥ | 5♥ | 6♥ | 7♥ | 8♥ | 9♥ | 10♥ | J♥ | Q♥ | K♥ |
> | A♦ | 2♦ | 3♦ | 4♦ | 5♦ | 6♦ | 7♦ | 8♦ | 9♦ | 10♦ | J♦ | Q♦ | K♦ |
> | A♣ | 2♣ | 3♣ | 4♣ | 5♣ | 6♣ | 7♣ | 8♣ | 9♣ | 10♣ | J♣ | Q♣ | K♣ |
> | A♠ | 2♠ | 3♠ | 4♠ | 5♠ | 6♠ | 7♠ | 8♠ | 9♠ | 10♠ | J♠ | Q♠ | K♠ |

5.4 The Addition Principle; Combinations

▼ **EXAMPLE 9** Poker Hands

Those of you who play poker know that a poker hand consists of five cards from a standard deck of 52 and that a **full house** consists of three of one denomination ("three of a kind," e.g., three 10s) and two of another ("two of a kind," e.g., two Queens). Here is an example of a full house: 10♣, 10♦, 10♠, Q♥, Q♣.

(a) How many different full houses are there containing three 10s and two Queens?

(b) How many different full houses are there altogether?

(c) What are your chances of being dealt a full house?

SOLUTION

(a) Pretend that you are constructing a full house with three 10s and two Queens.

Step 1: Choose 3 10s: Since there are four 10s to choose from, you get $C(4, 3) = 4$ choices.

Step 2: Choose 2 Queens: $C(4, 2) = 6$ choices.

Thus, there are $4 \times 6 = 24$ possible full houses with 3 10s and 2 Queens.

(b) Pretend that you are constructing a full house. This involves several steps.

Step 1: Choose a denomination for the three of a kind: 13 choices.

Step 2: Choose 3 cards of that denomination: since there are 4 cards of each denomination (one for each suit), we get $C(4, 3) = C(4, 1) = 4$ choices.

Step 3: Choose a different denomination for the two of a kind: there are only 12 denominations left for this, so we have 12 choices.

Step 4: Choose 2 of that denomination: $C(4, 2) = 6$ choices.

Thus, by the multiplication principle, there are a total of $13 \times 4 \times 12 \times 6 = 3{,}744$ possible full houses.

(c) (We are really sneaking in a topic from the next chapter here.) Recall from Example 5 that the total number of possible poker hands is 2,598,960. Of these, only 3,744 are full houses, by part (b). Thus the chances of being dealt a full house are 3,744 in 2,598,960. We say that the **probability** of being dealt a full house is $3{,}744/2{,}598{,}960 \approx 0.001441$.

Before turning to the exercises, we summarize all of our counting techniques.

SUMMARY OF COUNTING TECHNIQUES

Counting Procedure
Always use the Guide to Counting: if asked to calculate the number of possible objects, pretend that you are constructing such an object and come up with a counting procedure to construct one.

Acid Test
Check the validity of your counting procedure by using the Acid Test: a different choice at any stage of your procedure should result in a different outcome.

Multiplication Principle
If you have a sequence of several *steps*, the total number of outcomes (or choices) is obtained by multiplying the number of outcomes (or choices) at each step.

Addition Principle
If you have a list of several *alternatives*, the total number of outcomes is obtained by adding the number of outcomes for each alternative.

Lists
A *list* of r items chosen from n is an ordered list of r items, selected from the n. Thus, there is a first item, a second one, a third, a fourth and a fifth. Order is important: if the same items occur on two lists, but in different orders, then the lists are different. The total number of such lists is $P(n, r) = n!/(n - r)!$. (You can also get this using the multiplication principle.)

Sets
A *set* of r items chosen from n is an unordered collection of r items selected from the n. Two sets are the same if they consist of the same r items. The total number of such sets is

$$C(n, r) = \frac{P(n, r)}{r!} = \frac{n!}{r!(n - r)!}.$$

5.4 EXERCISES

Calculate each of the numbers in Exercises 1–8. (Use a calculator when necessary.)

1. $C(3, 2)$
2. $C(4, 3)$
3. $C(10, 8)$
4. $C(11, 9)$
5. $C(20, 1)$
6. $C(30, 1)$
7. $C(100, 98)$
8. $C(100, 97)$

APPLICATIONS

9. *Floppy Disks* Floppy disks come in two popular sizes (3.5" and 5.25") and two densities (DD, HD). However, the 3.5" ones also come in three colors, while the 5.25" ones come only in black. If you are purchasing a box of disks, how many possibilities do you have to choose from?

10. *Radar Detectors* Radar detectors are either battery powered or the type you plug into a cigarette lighter socket. All radar detectors come in two models: no-frills and fancy. In addition, battery-powered detectors detect either radar, laser or both, while the plug-in types come in models that detect either radar or laser, but not both. How many possible radar detectors can you buy?

11. *Multiple-Choice Tests* A multiple-choice test requires that you answer either Part A, Part B, or Part C. Part A consists of 8 true/false questions, Part B consists of 5 questions with 1 correct answer out of 5, and Part C requires you to match 5 questions with 5 different answers. How many different completed answer sheets are possible?

12. *Multiple-Choice Tests* A multiple-choice test requires that you first answer Part A, and then either Part B or Part C. Part A consists of 4 true/false questions, Part B consists of 4 questions with 1 correct answer out of 5, and Part C requires you to match 6 questions with 6 different answers. How many different completed answer sheets are possible?

13. *(from the GMAT)* Ben and Ann are among 7 contestants from which 4 semifinalists are to be selected. Of the different possible selections, how many contain neither Ben nor Ann?
    (a) 5 (b) 6 (c) 7 (d) 14 (e) 21

14. *(based on a question from the GMAT)* Ben and Ann are among 7 contestants from which 4 semifinalists are to be selected. Of the different possible selections, how many contain Ben but not Ann?
    (a) 5 (b) 8 (c) 9 (d) 10 (e) 20

15. *Variables in BASIC** A variable name in the programming language BASIC can be either a letter or a letter followed by a decimal digit, that is, one of the numbers 0, 1, ..., 9. How many different variables are possible?

16. *File Names in MS-DOS* A file name in *Microsoft DOS*® 5.0 may consist of an 8-letter sequence, or an 8-letter sequence followed by a period and a 3-letter extension. How many different file names of this type are possible? (The actual rules allowed more possibilities than these.)

Marbles In Exercises 17–30, a bag contains 3 red marbles, 2 green ones, 1 lavender one, 2 yellows, and 2 orange ones. Answer the following questions.

17. How many possible groups of four marbles are there?

18. How many possible groups of three marbles are there?

19. How many groups of four marbles include all the red ones?

20. How many groups of three marbles include all the yellow ones?

21. How many groups of four marbles include none of the red ones?

22. How many groups of three marbles include none of the yellow ones?

▼ *From *Applied Combinatorics* by F. S. Roberts (Englewood Cliffs, N.J.: Prentice-Hall, 1984)

23. How many groups of four marbles include one of each color other than lavender?
24. How many groups of five marbles include one of each color?
25. How many groups of five marbles include at least two red ones?
26. How many groups of five marbles include at least one yellow one?
27. How many groups of five marbles include at most one of the yellow ones?
28. How many groups of five marbles include at most one of the red ones?
29. How many groups of five marbles include either the lavender one or a yellow one, but not both?
30. How many groups of five marbles include at least one yellow one but no green ones?

If a die is thrown 30 times, there are 6^{30} different sequences possible. Exercises 31–34 ask how many of these sequences satisfy certain conditions.

31. What fraction of these sequences have exactly 5 ones?
32. What fraction of these sequences have exactly 5 ones and 5 twos?
33. What fraction of these sequences have exactly 15 even numbers?
34. What fraction of these sequences have exactly 10 numbers less than or equal to 2?
35. *Morse Code* In Morse Code, each letter of the alphabet is encoded by a different sequence of dots and dashes. Different letters may have sequences of different lengths. How long should the longest sequence be in order to allow for every possible letter of the alphabet?
36. *Numbers* How many odd numbers between 10 and 99 have distinct digits?
37. *Product Design* The Honest Lock company plans to introduce what it refers to as the "true combination lock." The lock will open if the correct set of three numbers from 0 to 39 are entered in any order.
    (a) How many different combinations of three different numbers are possible?
    (b) If it is allowed that a number appear twice (but not three times), how many more possibilities are created?
    (c) If it is allowed that any two or all three of the numbers may be the same, how many total possibilities are there?
38. *Product Design* A vending machine company is planning production of a new vending machine where the customer will be required to enter a sequence of one or two digits 0–9 followed by one of the letters A, B, C, or D. How many different products could the machine offer for sale?

Poker Hands Recall that a poker hand consists of five cards from a standard deck of 52. (See the chart preceding Example 9.) In Exercises 39–44, find the number of different poker hands of the specified type.

39. Two pairs (two of one denomination, two of another denomination, and one of a third)
40. Three of a kind (three of one denomination, one of another denomination, and one of a third)
41. Two of a kind (two of one denomination and three of different denominations)
42. Four of a kind (all four of one denomination and one of another)
43. Straight (5 cards of consecutive denominations: A, 2, 3, 4, 5 up through 10, J, Q, K, A, not all of the same suit. Note that the Ace counts either as a "1" or as the denomination above King.)
44. Flush (5 cards all of the same suit, but not consecutive denominations)
45. *Committees* The Judicial Committee of the Student Senate is to consist of
    1 Chief Investigator (Party Party),
    2 Assistant Investigators (Study Party),
    2 Rabble Rousers, and
    3 Do-nothing members.

The committee is to be selected from a pool of 20 Senators, half of whom were elected on the Party Party ticket and half on the Study Party ticket, including Freshman Sen. Boondoggle (Study Party). Senator Boondoggle is hoping desperately to serve on the committee, preferably as a Do-nothing member. Unfortunately, Boondoggle's roommate and bitterest enemy, Sen. Porkbarrel (Party Party), is also in the pool of candidates. After giving the matter some thought, Boondoggle finally decides that he will refuse to serve unless
(a) he is a Do-nothing member, and

(b) Porkbarrel is not also serving on the committee. How many possible committees are there that would make Senator Boondoggle happy? [Leave your answer as a product of terms of the form $C(n, r)$.]

46. **Committees** Senator Porkbarrel is furious about having been dropped from the Judicial Committee of the Student Senate and has decided to retaliate by forming his own committee, the Committee to Judge the Judicial Committee. It is to have the following members.

    1 Supreme Investigator (Sen. Porkbarrel)
    2 Semi-supreme Investigators (either party)
    2 Demi-semi-supreme Investigators (Party Party).

There are a total of 8 members of the Party Party (including himself) and 9 members of the Study Party (excluding Sen. Boondoggle—see the previous exercise) he is prepared to consider as committee members. How many committees are possible? [Leave your answer as a product of terms of the form $C(n, r)$.]

In Exercises 47–52, calculate how many different sequences can be formed using the letters of the given words. [Leave your answer as a product of terms of the form $C(n, r)$.]

47. Mississippi
48. Mesopotamia
49. Megalomania
50. Schizophrenia
51. Casablanca
52. Desmorelda

53. **Symmetries of a Six-Pointed Star** A six-pointed star will appear unchanged if it is rotated through any one of the angles 0°, 60°, 120°, 180°, 240°, or 300°. It will also appear unchanged if it is flipped about the axis shown in the figure. A *symmetry* of the six-pointed star consists of either a rotation as above, or a rotation followed by a flip. How many different symmetries are there altogether?

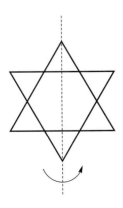

54. **Symmetries of a Five-Pointed Star** A five-pointed star will appear unchanged if it is rotated

through any one of the angles 0°, 72°, 144°, 216°, or 288°. It will also appear unchanged if it is flipped about the axis shown in the figure. A *symmetry* of the five-pointed star consists of either a rotation as above, or a rotation followed by a flip. How many different symmetries are there altogether?

55. **Theory of Linear Programming** (Some familiarity with linear programming is assumed for this exercise.) Suppose you have a linear programming problem with 2 unknowns and 20 constraints. You decide that graphing the feasible region would take a lot of work, but then you recall that corner points are obtained by solving a system of two equations in two unknowns obtained from two of the constraints. Thus, you decide that it might pay instead to locate all the possible corner points by solving all possible combinations of two equations, and then checking whether each solution is a feasible point.
    (a) How many systems of two equations in two unknowns will you be required to solve?
    (b) Generalize this to n constraints.

56. **More Theory of Linear Programming** (Some familiarity with linear programming is assumed for this exercise.) Before the advent of the simplex method for solving linear programming problems, the following method was used. Suppose you have a linear programming problem with 3 unknowns and 20 constraints. You locate "corner points" as follows. Selecting three of the constraints, turn them into equations (by replacing the inequalities with equations), solve

the resulting system of three equations in three unknowns, and then check to see if the solution is feasible.

COMMUNICATION AND REASONING EXERCISES

57. You are tutoring your friend for a test on sets and counting, and she asks the following question: "How do I know what formula to use for a given problem?" What is a good way to respond?

58. Say what is wrong with the following counting procedure for selecting a group of three students from a class of 50 to act as proctors, and suggest a better counting procedure. *Step 1:* Select the first student proctor. *Step 2:* Select a second student proctor. *Step 3:* Select a third student proctor.

59. A textbook has the following exercise. "Three students from a class of 50 are selected to take part in a play. How many choices are possible?" Comment on this exercise.

(a) How many systems of three equations in three unknowns will you be required to solve?
(b) Generalize this to n constraints.

60. Complete the following. If a counting procedure has 5 alternatives, each of which has four steps of 2 choices each, then there are _____ outcomes. On the other hand, if there are 5 steps, each of which has four alternatives of 2 choices each, then there are _____ outcomes.

61. Show by direct calculation that the coefficient of $a^r b^{4-r}$ in $(a+b)^4$ is $C(4, r)$.

62. Explain why the coefficient of $a^r b^{n-r}$ in $(a+b)^n$ is $C(n, r)$ (this is called the **binomial theorem**). [*Hint:* In the product $(a+b)(a+b)\ldots(a+b)$ (n times), in how many different ways can you pick r as and $n-r$ bs to multiply together?]

▶ **You're the Expert**

DESIGNING A PUZZLE

As Product Design Manager for Cerebral Toys, Inc., you are constantly tracking down ideas for intellectually stimulating yet inexpensive toys. You recently received the following memo from Felix Frost, the developmental psychologist on your design team.

To: Felicia
From: Felix
Subject: Crazy Cubes

We've hit on an excellent idea for a new educational puzzle (which we are calling "Crazy Cubes" until Marketing comes up with a better name). Basically, Crazy Cubes will consist of a set of plastic cubes. Each cube will have two faces colored red, two white and two blue, and there will be exactly two cubes with each possible configuration of colors. The goal of the puzzle is to seek out the matching pairs, enhancing a child's geometric intuition and three-dimensional manipulation skills. We are, however, a little stumped on the following question: How many cubes will the kit contain? In other words, how many possible ways can one color the faces of a cube so that two faces are red, two are blue, and two are white?

Looking at the problem, you decide to consult the Guide to Counting, and pretend that you were painting the faces of a cube. You reason that the following three-step procedure ought to suffice:

Step 1: Choose a pair of faces to color red, $C(6, 2) = 15$ choices.

Step 2: Choose a pair of faces to color blue, $C(4, 2) = 6$ choices.

Step 3: Choose a pair of faces to color white, $C(2, 2) = 1$ choice.

This procedure gives a total of $15 \times 6 = 90$ possible cubes. However, before sending your reply to Felix, you run this procedure through the Acid Test (to make sure you haven't made a possibly embarrassing error). You then realize that something is wrong, because there are different choices that result in the same cube. To describe some of these choices, imagine a cube oriented so that four of its faces are facing the four compass directions (Figure 1).

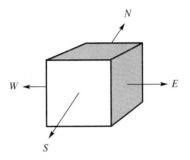

FIGURE 1

Choice 1: Top and bottom faces red, north and south faces white, east and west faces blue

Choice 2: Top and bottom faces red, north and south faces blue, east and west faces white

These cubes are actually the same, as you see by rotating the second cube 90 degrees (Figure 2).

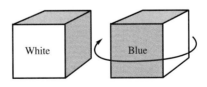

FIGURE 2

You therefore decide that what you need is a more sophisticated counting procedure in order to pass the Acid Test. Here is one that works.

Alternative 1: Faces with the same color opposite each other. Place one of the red faces down. Then the top face is also red. The cube must look like the one drawn in Figure 2. Thus there is only one choice here.

Alternative 2: Red faces opposite each other and the other colors on adjacent pairs of faces. Again there is only one choice, as you can see by putting the red faces on the top and bottom and then rotating.

Alternative 3: White faces opposite each other and the other colors on adjacent pairs of faces. One possibility.

Alternative 4: Blue faces opposite each other and the other colors on adjacent pairs of faces. One possibility.

Alternative 5: Faces with the same color adjacent to each other. Look at the cube so that the edge common to the two red faces is facing you and horizontal (Figure 3). Then the faces on the left and right must be of different colors because they are opposite each other. Assume that the face on the right is white. (If it's blue, then rotate the die with the red edge still facing you until it looks like Figure 4.)

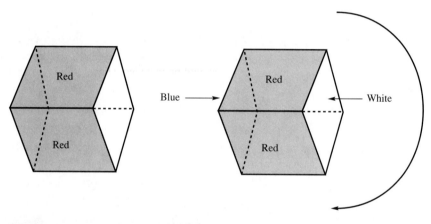

FIGURE 3   **FIGURE 4**

This leaves two choices for the other white face, on the upper or the lower of the two back faces. This alternative gives 2 choices.

It follows that there are $1 + 1 + 1 + 1 + 2 = 6$ choices.

Since the Crazy Cubes kit will feature two of each, the kit will require 12 different cubes.*

▼ *There is a more algebraic way of calculating this number, called Polya enumeration, but it requires a discussion of topics outside the scope of this book.

Exercises

1. In order to enlarge the kit, Felix suggests including two-colored cubes (using two of the colors red, white, and blue) with three faces one color and three another. How many additional cubes will be required?

2. If Felix now suggests adding cubes with two faces one color, one face another color, and three faces the third color, how many additional cubes will be required?

3. Felix changes his mind and suggests the kit use tetrahedral blocks with two colors instead (Figure 5). How many of these would be required?

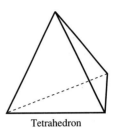

Tetrahedron

FIGURE 5

4. Once Felix finds the answer to the previous exercise, he decides to go back to the cube idea, but this time insists that all possible combinations of up to three colors should be included. (For instance, some cubes will be all one color, others will be two colors.) How many cubes should the kit contain?

▶ Review Exercises

List the elements in each set in Exercises 1–8.

1. The set I of all positive even integers no larger than 10
2. The set N of all negative integers greater than or equal to -3
3. $A = \{n \mid n \in \mathbf{N}, n \text{ odd and } 2 \leq n \leq 10\}$
4. $A = \{n \mid n \in \mathbf{N}, n \text{ even and } 1 < n < 9\}$
5. The set of all outcomes of tossing a coin five times
6. The set of all outcomes of tossing a coin six times
7. The set of all outcomes of tossing two dice such that the two numbers are different
8. The set of all outcomes of tossing two dice such that at least one of the numbers is odd

Let $A = \{1, 2, 3, 4, 5\}$, $B = \{3, 4, 5\}$, and $C = \{1, 2, 5, 6, 7\}$. Find the sets indicated in Exercises 9–18.

9. $A \cup C$
10. $A \cup B$
11. $A \cup (B \cup C)$
12. $(A \cup B) \cup C$
13. $A \cap B$
14. $C \cap A$
15. $A \cap \emptyset$
16. $\emptyset \cap B$
17. $(A \cap B) \cap C$
18. $A \cap (B \cap C)$

Let S be the set of outcomes when two distinguishable dice are thrown, let E be the subset of outcomes in which at most one die shows an even number, and let F be the subset of outcomes in which the sum of the numbers is 7. Describe the sets in Exercises 19–24 in words.

19. E' **20.** F' **21.** $(E \cup F)'$ **22.** $(E \cap F)'$
23. $E' \cup F'$ **24.** $E' \cap F'$

25. How many four-letter sequences are possible using the letters *b, a, l, l*?

26. How many five-letter sequences are possible using the letters *h, e, l, l, o*?

APPLICATIONS

27. *NYSE* The New York Stock Exchange uses 3-letter codes for the companies traded there. How many companies can they list at any one time?

28. *NYSE* Referring to the preceding exercise, how many companies can the NYSE list with codes having three different letters?

29. *License Plates* A license plate in the Gotham City Metropolitan Area consists of a bat logo followed by two letters and two digits. The two letters must be different and the first digit cannot be a zero. How many different license plates are possible?

30. *License Plates* A license plate in the Metropolis Metropolitan Area consists of two letters and two digits followed by either a bat logo or an *S*. The two letters must be different, with the first letter not an *S*, while neither digit is permitted to be a zero. How many different license plates are possible?

31. *Customer IDs* A business would like to use 5-character IDs to keep track of customers. Each ID will consist of some number of letters followed by some digits. If the business anticipates between 500,000 and 700,000 customers, how many letters and how many digits should it use?

32. *Customer IDs* A mail-order company wants to print 6-character customer IDs on its catalogs. Each ID will consist of some number of digits followed by some letters. If it mails out between 10,000,000 and 15,000,000 catalogs, how many digits and how many letters should the company use in its IDs?

33. *English Spelling* The English language contains about 800,000 words (including technical terms). If every word contained the same number of letters, how long would each word have to be?

34. *English Spelling* It is estimated that the average English speaker uses only around 10,000 words. Shakespeare used 33,000 words. If every word contained the same number of letters, how long would each word have to be to accommodate the average English speaker? To accommodate Shakespeare?

35. *Mazes*
  (a) How many four-letter sequences are possible containing two *R*s and two *D*s?
  (b) Use part (a) to calculate the number of possible routes from Start to Finish in the following maze where each move is either to the right or down.

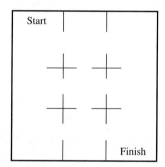

36. *Mazes*
   (a) How many five-letter sequences are possible containing three *R*s and two *D*s?
   (b) Use part (a) to calculate the number of possible routes from Start to Finish in the following maze, where each move is either to the right or down.

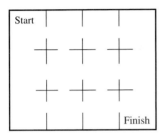

Marbles In Exercises 37–48, a bag contains 4 *red marbles*, 2 *green ones*, 1 *transparent one*, 3 *yellows*, and 2 *orange ones*. Answer the following questions.

37. How many possible groups of five marbles are there?

38. How many possible groups of four marbles are there?

39. How many groups of five marbles include all the red ones?

40. How many groups of four marbles include all the yellow ones?

41. How many groups of five marbles do not include all the red ones?

42. How many groups of six marbles include one of each color?

43. How many groups of six marbles do not include one of each color?

44. How many groups of five marbles include at least two orange ones?

45. How many groups of five marbles include at least two yellow ones?

46. How many groups of five marbles include at most one of the yellow ones but no red ones?

47. How many groups of five marbles include at most one of the red ones but no yellow ones?

48. How many groups of five marbles include either the transparent one or all the red ones, and possibly both?

Poker Hands Recall that a poker hand consists of five cards from a standard deck of 52. In Exercises 49–52, find the number of different poker hands of the specified type. Leave all answers as products of combinations. (See the examples and exercises in Section 4 for the definitions of various poker hands.)

49. A full house with either two Kings and three Queens or two Queens and three Kings

50. Three of a kind with no Aces

51. Two of a kind with no Aces

52. Straight Flush (5 cards of the same suit with consecutive denominations: A, 2, 3, 4, 5 up through 10, J, Q, K, A)

Committees Exercises 53 through 56 deal with the Last Soviet Executive Committee, which is to be selected from a group of ten candidates, and is to be composed of
   1 Supreme Soviet,
   2 Semi-supreme Soviets, and
   3 Demi-semi-supreme Soviets.
Leave all answers as products of combinations.

53. Vladimir Cuchenko Dritznikov, Jr., is one of the ten candidates. How many possible committees have him as a Semi-supreme Soviet?

54. Ivana Blotznikov, another of the ten candidates, is willing to serve on the committee but will not serve as a Semi-supreme Soviet unless Vladimir is also a Semi-supreme Soviet. How many committees are possible under these conditions?

55. Nikolai Sergei Diladigenski, yet another candidate, has persuaded his colleagues that he should be Supreme Soviet. How many committees are possible, assuming that Ivana Blotznikov's wishes are still respected?

56. One day before the committee membership is announced, both Nikolai Sergei Diladigenski and Vladimir Cuchenko Dritznikov Jr. are posted to the diplomatic mission in Tahiti. How many committees are now possible if Ivana Blotznikov's wishes are still respected?

CHAPTER 6

Behind Monty Hall's 3 Doors
A puzzle, a debate and, perhaps, an answer

By John Tierney   Special to the New York Times

BEVERLY HILLS, Calif., July 20— Perhaps it was only an illusion, but for a moment here it seemed that an end might be in sight to the debate raging among mathematicians, readers of Parade magazine and fans of the television game show "Let's Make a Deal."

They began arguing last September after Marilyn vos Savant published a puzzle in Parade. As readers of her "Ask Marilyn" column are reminded each week, Ms. vos Savant is listed in the Guinness Book of World Records Hall of Fame for "Highest I.Q.," but that credential did not impress the public when she answered this question from a reader:

"Suppose you're on a game show, and you're given the choice of three doors: Behind one door is a car; behind the others, goats. You pick a door, say No. 1, and the host, who knows what's behind the other doors, opens another door, say No. 3, which has a goat. He then says to you, 'Do you want to pick door No. 2?' Is it to your advantage to take the switch?"

Since she gave her answer, Ms. vos Savant estimates she has received 10,000 letters, the great majority disagreeing with her. The most vehement criticism has come from mathematicians and scientists, who have alternated between gloating at her ("You are the goat!") and lamenting the nation's innumeracy.

Her answer—that the contestant should switch doors—has been debated in the halls of the Central Intelligence Agency and the barracks of fighter pilots in the Persian Gulf. It has been analyzed by mathematicians at the Massachusetts Institute of Technology and computer programmers at Los Alamos National Laboratory in New Mexico.

But it was not until Thursday afternoon that a truly realistic simulation of the problem was conducted. The experiment took place at the Beverly Hills home of Monty Hall, the host of 4,500 programs of "Let's Make a Deal" from 1963 to 1990.

Ms. vos Savant's vitriolic critics, including the mathematics professors, are dead wrong. But Ms. vos Savant is not entirely correct either, because there is a small flaw in her wording of the problem that was detected not only by Mr. Hall but also by some of the experts. . . .

The experts responded in force to Ms. vos Savant's column. Of the critical letters she received, close to 1,000 carried signatures with Ph.D.'s, and many were on letterheads of mathematics and science departments. . . .

Robert Sachs, a professor of mathematics at George Mason University in Fairfax, Va., expressed the prevailing view that there was no reason to switch doors.

"You blew it!" he wrote. "Let me explain: If one door is shown to be a loser, that information changes the probability of either remaining choice—*neither of which has any reason to be more likely*—to $1/2$. As a professional mathematician, I'm very concerned with the general public's lack of mathematical skills. Please help by confessing your error and, in the future, being more careful." . . .

Source: From John Tierney, "Behind Monty Hall's 3 Doors: A Puzzle, a Debate, and, Perhaps, an Answer," *The New York Times*, July 21, 1991, pp. 1 and 20.

Probability

SECTIONS

1. Sample Spaces and Events
2. Introduction to Probability
3. Principles of Probability
4. Counting Techniques and Probability
5. Conditional Probability and Independence
6. Bayes' Theorem and Applications
7. Bernoulli Trials

You're the Expert
The Monty Hall Problem

APPLICATION ▶ On the game show *Let's Make a Deal,* you are shown three doors, A, B, and C, behind one of which is the Big Prize. After you select one of them—say, door A—to make things more interesting the host (Monty Hall) opens one of the other doors (say, door B) revealing that the Big Prize is not there. He then offers you the opportunity to change your selection to the remaining door, door C. Should you switch or stick with your original guess? Does it make any difference?

INTRODUCTION ▶ Here are some questions you might have wondered about. What is the probability of winning the lottery twice? What are the chances that a college athlete whose drug test is positive for steroid use is actually using steroids? You are playing poker and have been dealt two Jacks. What is the likelihood that one of the next three cards you are dealt will also be a Jack?

Probability theory has become extremely important in many fields, ranging from risk management in business through hypothesis testing in psychology to quantum mechanics in physics. It is the goal of this chapter to familiarize you with the basic concepts of probability theory and to give you a working knowledge that you can apply in a variety of situations.

In the first two sections, the emphasis is placed on translating real-life situations into the language of sample spaces, events, and probability. Once we have mastered the language of probability theory, we spend the rest of the chapter studying some of its theory and applications.

6.1 SAMPLE SPACES AND EVENTS

SAMPLE SPACES

Let's start with a familiar situation: If you toss a coin and observe which side lands up, there are two possible results: heads (H) and tails (T). These are the *only* possible results, ignoring the (remote) possibility that the coin lands on its edge. The act of tossing a coin is an example of an **experiment.** The two possible results H and T are the possible **outcomes** of the experiment, and the set $S = \{H, T\}$ of all possible outcomes is the **sample space** for the experiment.

> **EXPERIMENTS, OUTCOMES, AND SAMPLE SPACES**
> An **experiment** is an occurrence we observe whose result is uncertain. We observe some specific aspect of the occurrence and there will be several possible results, or **outcomes.** The set of all possible outcomes is called the **sample space*** for the experiment.

▼ * In this chapter we shall consider only *finite* sample spaces: that is, ones with only finitely many possible outcomes.

EXAMPLE 1

(a) What is the set of outcomes of the experiment in which a die is thrown and we observe the number on the side facing up?

(b) What is the set of outcomes of the experiment in which two dice (one red, one green) are thrown and we observe the number facing up on each die?

SOLUTION

(a) Since a die can land with any of the numbers from 1 to 6 facing up, the set of outcomes, or sample space, is

$$S = \{1, 2, 3, 4, 5, 6\}.$$

(b) We saw this example in the last chapter, where we arranged the sample space in the following way.

$$S = \begin{Bmatrix} (1,1) & (1,2) & (1,3) & (1,4) & (1,5) & (1,6) \\ (2,1) & (2,2) & (2,3) & (2,4) & (2,5) & (2,6) \\ (3,1) & (3,2) & (3,3) & (3,4) & (3,5) & (3,6) \\ (4,1) & (4,2) & (4,3) & (4,4) & (4,5) & (4,6) \\ (5,1) & (5,2) & (5,3) & (5,4) & (5,5) & (5,6) \\ (6,1) & (6,2) & (6,3) & (6,4) & (6,5) & (6,6) \end{Bmatrix}$$

Here, for instance, the outcome (4, 3) indicates that the red die shows a 4 and the green one a 3.

Before we go on... It is important to specify what is being observed. If we tossed two dice and observed the *sum* of the numbers facing up, then our sample space would be

$$S = \{2, 3, 4, 5, 6, 7, 8, 9, 10, 11, 12\}.$$

Another point: We assumed in (b) that the dice were *distinguishable* (they were different colors). This allowed us to list (2, 4) and (4, 2) as different elements. No two dice are identical, so it is not unreasonable to use the same sample space given in (b) even for dice of the same color. This will be useful for calculating probabilities.

EXAMPLE 2

A factory worker in 1993 may or may not have been covered by a medical insurance plan. If the worker was covered, the coverage could either have been under the employer's group plan or an individual plan. If the worker was covered by an individual plan, the plan could either have been in the worker's name or in that of his or her spouse. Find the sample space for the experiment "select a worker and classify his or her medical coverage."

SOLUTION All we need to do is decide on the various possibilities by using a counting procedure. A tree is useful (Figure 1).

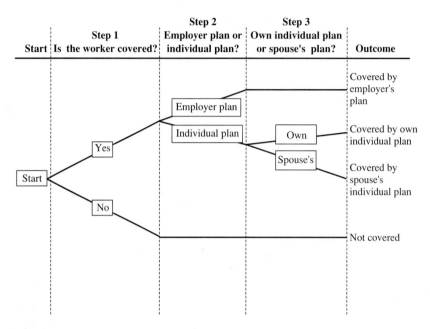

FIGURE 1

This gives us the sample space

$$S = \{\text{covered by employer's plan, covered by own individual plan, covered by spouse's individual plan, not covered}\}.$$

EVENTS

Looking at the last example, suppose that we are interested in the outcomes in which the factory worker was covered by some form of medical insurance. In mathematical language, we are interested in the subset consisting of all outcomes in which the worker was covered.

> **EVENT**
>
> Given a sample space S, an **event** E is a subset of S. The outcomes in E are called the **favorable** outcomes. We say that E **occurs** in a particular experiment if the outcome of that experiment is one of the elements of E: that is, if the outcome of the experiment is favorable.

EXAMPLE 3

Let S be the sample space of Example 2.

(a) Describe the event E that a factory worker was covered by some form of medical insurance.

(b) Describe the event F that a factory worker was not covered by an individual medical plan.

(c) Describe the event G that a factory worker was covered by a government medical plan.

SOLUTION

(a) We had the following sample space.

$$S = \{\text{covered by employer's plan, covered by own individual plan, covered by spouse's individual plan, not covered}\}$$

We are asked for the event that a factory worker was covered by some form of medical insurance. Whenever we encounter a phrase involving "the event that . . . ," we mentally translate this into mathematical language by making the following change of wording.

Replace the phrase "the event that . . ." with the phrase "the subset of the sample space consisting of all outcomes in which . . ."

Thus, we are interested in the subset of the sample space consisting of all outcomes in which the worker was covered by some form of medical insurance. This gives

$$E = \{\text{covered by employer's plan, covered by own individual plan, covered by spouse's individual plan}\}.$$

Put another way, the favorable outcomes are (1) covered by employer's plan, (2) covered by own individual plan, and (3) covered by spouse's individual plan.

(b) We are looking for the event that a factory worker was not covered by an individual medical plan. Translating this into mathematical language, we want the subset of the sample space consisting of all outcomes in which the worker was not covered by an individual medical plan. Thus,

$$F = \{\text{covered by employer's plan, not covered}\}.$$

(c) Translating part (c) into mathematical language, we are interested in the subset of the sample space consisting of all outcomes in which the worker was covered by a government medical plan. Since this is true for *none* of the outcomes in S, there are no outcomes in the event. In other words,

$$G = \emptyset,$$

the empty set. There are no favorable outcomes.

EXAMPLE 5

You roll a red die and a green die and observe the numbers facing up. Describe the following events as subsets of the sample space.

(a) E: Both dice show the same number.

(b) F: The sum of the numbers showing is 6.

(c) G: The sum of the numbers showing is 2.

SOLUTION   We wrote out the sample space for the experiment of throwing two dice in Example 1(b), and here it is again.

$$S = \begin{Bmatrix} (1,1) & (1,2) & (1,3) & (1,4) & (1,5) & (1,6) \\ (2,1) & (2,2) & (2,3) & (2,4) & (2,5) & (2,6) \\ (3,1) & (3,2) & (3,3) & (3,4) & (3,5) & (3,6) \\ (4,1) & (4,2) & (4,3) & (4,4) & (4,5) & (4,6) \\ (5,1) & (5,2) & (5,3) & (5,4) & (5,5) & (5,6) \\ (6,1) & (6,2) & (6,3) & (6,4) & (6,5) & (6,6) \end{Bmatrix}$$

(a) Translating into mathematical language, E is the subset of S consisting of all those outcomes in which both dice show the same number. These outcomes appear down the main diagonal (top left to bottom right) in S:

$$E = \{(1,1), (2,2), (3,3), (4,4), (5,5), (6,6)\}.$$

(b) We are asked for the subset of S consisting of all those outcomes in which the sum of the numbers showing is 6. Here is the sample space once again, with the outcomes in question in color.

$$S = \begin{Bmatrix} (1,1) & (1,2) & (1,3) & (1,4) & (1,5) & (1,6) \\ (2,1) & (2,2) & (2,3) & (2,4) & (2,5) & (2,6) \\ (3,1) & (3,2) & (3,3) & (3,4) & (3,5) & (3,6) \\ (4,1) & (4,2) & (4,3) & (4,4) & (4,5) & (4,6) \\ (5,1) & (5,2) & (5,3) & (5,4) & (5,5) & (5,6) \\ (6,1) & (6,2) & (6,3) & (6,4) & (6,5) & (6,6) \end{Bmatrix}$$

Thus,

$$F = \{(1,5), (2,4), (3,3), (4,2), (5,1)\}.$$

(c) The only outcome in which the numbers showing add to 2 is $(1,1)$. Thus,

$$G = \{(1,1)\}.$$

COMPLEMENT OF AN EVENT

Events may often be described in terms of other events, using set operations. An example is the **negation** of an event E, the event that E does not occur. If in a particular experiment E does not occur, then the outcome of that experiment is not in E, so it is in its *complement*.

As an example, we have part (b) of Example 3. The event F was described as the event that a factory worker was *not* covered by an individual medical plan. This is the negation of the event H that an individual *was* covered by an individual medical plan. H is the subset

$$H = \{\text{covered by own individual plan, covered by spouse's individual plan}\}$$

and $F = H'$.

In general, we have the following.

> **COMPLEMENT OF AN EVENT**
>
> The complement E' of an event E is the event that E does not occur.

EXAMPLE 5

You roll a red die and a green die and observe the two numbers facing up. Describe the event that the sum of the numbers is not 6.

SOLUTION This event is the complement of the event labeled F in Example 4. The event can therefore be obtained from S by deleting all outcomes that are in F.

$$F' = \begin{cases} (1,1) & (1,2) & (1,3) & (1,4) & & (1,6) \\ (2,1) & (2,2) & (2,3) & & (2,5) & (2,6) \\ (3,1) & (3,2) & & (3,4) & (3,5) & (3,6) \\ (4,1) & & (4,3) & (4,4) & (4,5) & (4,6) \\ & (5,2) & (5,3) & (5,4) & (5,5) & (5,6) \\ (6,1) & (6,2) & (6,3) & (6,4) & (6,5) & (6,6) \end{cases}$$

UNIONS AND INTERSECTIONS OF EVENTS

Besides taking the complement of an event, we can also combine events by taking unions and intersections.

Q If E and F are events, how can we describe the event $E \cup F$?

A Consider a simple example: the experiment of throwing a die. Let E be the event that the outcome is a 5, and let F be the event that the outcome is an even number. Thus,

$$E = \{5\},$$
$$F = \{2, 4, 6\}.$$

So,
$$E \cup F = \{5, 2, 4, 6\}.$$

In other words, $E \cup F$ is the event that the outcome is *either* a 5 *or* an even number. In general, we can say the following.

> **UNION OF EVENTS**
>
> If E and F are events, then $E \cup F$ is the event that either E occurs or F occurs (or both).

▶ **NOTE** As in the previous chapter, when we use the word "or" we agree to mean one or the other *or both*. This is called the **inclusive or,** and mathematicians have simply agreed to take this as the meaning of "or" in order to avoid confusion. ◀

EXAMPLE 6

A 1992 study by the Environmental Protection Agency classifies municipal solid waste into several categories, shown here with their percentages: Metals: 8.5%, Glass: 7.0%, Plastics: 8.0%, Yard Waste: 17.6%, Food Waste: 7.3%, Paper: 40.0%, Other: 11.6%.* Consider an experiment in which a particular kind of waste is selected. Let E be the event that the waste is in a category contributing more than 10% to the total, and let F be the event that the waste is either yard waste, food waste, or metals. Describe the event $E \cup F$ both as a set and in words.

SOLUTION Here,
$$S = \{\text{metals, glass, plastics, yard waste, food waste, paper, other}\}.$$

E is the event that the waste is in a category contributing more than 10% to the total. Thus,
$$E = \{\text{yard waste, paper, other}\}.$$

F is the event that the waste is either yard waste, food waste, or metals. Thus,
$$F = \{\text{yard waste, food waste, metals}\}.$$

So,
$$E \cup F = \{\text{metals, yard waste, food waste, paper, other}\}.$$

In words, $E \cup F$ is the event that either the waste is in a category contributing more than 10% to the total, or is either yard waste, food waste, or metals. (Remember that "or" means the inclusive or.)

▼ *The percentages are by weight. Source: Environmental Protection Agency/*World Almanac and Book of Facts, 1992* (New York: Pharos Books, 1992).

It is now natural to ask the following question.

Q If E and F are events, how can we describe the event $E \cap F$?

A Looking at the example above, we get

$$E \cap F = \{\text{yard waste}\}.$$

In words, $E \cap F$ is the event that the waste is in a category contributing more than 10% to the total, *and* it is either yard waste, food waste, or metals. In general, we can say the following.

> **INTERSECTION OF EVENTS**
> If E and F are events, then $E \cap F$ is the event that both E and F occur.

EXAMPLE 7

In the experiment in Example 2, describe the event that a worker was covered by some form of medical insurance but was not covered by an individual medical plan.

SOLUTION Notice that "but" is just a more emphatic form of "and." The event described is the event that both E and F occur, where E and F are the events described in Example 3:

$$E = \{\text{covered by employer's plan, covered by own individual plan, covered by spouse's individual plan}\},$$
$$F = \{\text{covered by employer's plan, not covered}\}.$$

The event we wish to describe is therefore

$$E \cap F = \{\text{covered by employer's plan}\}.$$

The case when $E \cap F$ is empty is interesting, and we give it a name.

> **MUTUALLY EXCLUSIVE EVENTS**
> If E and F are events, then E and F are said to be **disjoint** or **mutually exclusive** if $E \cap F$ is empty.

EXAMPLE 8

A coin is tossed three times, and the sequence of heads and tails is recorded. Decide whether the following pairs of events are mutually exclusive.

(a) E: the first toss shows a head; F: the second toss shows a tail.

(b) *E*: all three tosses land the same way up; *F*: one toss shows heads and the other two show tails.

SOLUTION   The sample space is

$$S = \{HHH, HHT, HTH, HTT, THH, THT, TTH, TTT\}.$$

(a) We have

$$E = \{HHH, HHT, HTH, HTT\}$$

and

$$F = \{HTH, HTT, TTH, TTT\}.$$

Thus,

$$E \cap F = \{HTH, HTT\},$$

and is non-empty. So *E* and *F* are not mutually exclusive.

(b) We have

$$E = \{HHH, TTT\}$$

and

$$F = \{HTT, THT, TTH\},$$

giving

$$E \cap F = \emptyset.$$

So *E* and *F* are mutually exclusive.

Before we go on... The term "mutually exclusive" suggests that the two events in question cannot both occur at the same time. This is the case in part (b): we cannot have all three tosses with the same side up, and at the same time have one showing heads and the others tails. In part (a), we see that it certainly is possible for the first toss to show a head and the second a tail. These events are *not* mutually exclusive since they can both occur at the same time.

SAMPLE SPACES AND EVENTS

1. An **experiment** is an occurrence we observe whose result is uncertain. The set of all possible outcomes of an experiment is its **sample space,** or its **set of outcomes.**
2. Given a sample space *S*, an **event** *E* is a subset of *S*.
3. Whenever we encounter a phrase involving "the event that . . . ," we mentally translate this into mathematical language by making the following change of wording:

Replace the phrase "the event that . . ." with the phrase "the subset of the sample space consisting of all outcomes in which . . ."

4. If E is an event, then its complement E' is the event that E does not occur.
5. If E and F are events, then:

   $E \cup F$ is the event that either E occurs, or F occurs (or both).

   $E \cap F$ is the event that both E and F occur.

6. If E and F are events, then E and F are said to be **disjoint** or **mutually exclusive** if $E \cap F$ is empty.

6.1 EXERCISES

In Exercises 1–20, describe the sample space S of the given experiment, and describe the given event as a subset of S. (Assume that the coins and dice are distinguishable and that what is observed are the faces or numbers uppermost.)

1. Two coins are tossed, and the result is at most one tail.
2. Two coins are tossed, and the result is one or more heads.
3. Three coins are tossed, and the result is at most one head.
4. Three coins are tossed, and the result is more tails than heads.
5. Four coins are tossed, and the result is that a head is never followed by a tail.
6. Four coins are tossed, and the result is that the first shows a head and the last a tail.
7. Two dice are thrown, and the numbers add to 5.
8. Two dice are thrown, and the numbers add to 9.
9. Two dice are thrown, and the numbers add to 1.
10. Two dice are thrown, and one of the numbers is even and the other odd.
11. Two dice are thrown, and both numbers are prime.
12. Two dice are thrown, and neither number is prime.
13. A letter is chosen at random from those in the word *Mozart*, and the letter is a vowel.
14. A letter is chosen at random from those in the word *Mozart*, and the letter is neither an *a* nor an *m*.
15. A sequence of two different letters is randomly chosen from those of the word *sore*, and the first letter is a vowel.
16. A sequence of two different letters is randomly chosen from those of the word *hear*, and the second letter is not a vowel.
17. A sequence of two different letters is randomly chosen from those of the word *sore*, and at most one letter is a vowel.
18. A sequence of two different letters is randomly chosen from those of the word *hear*, and at least one letter is a vowel.
19. A sequence of two different digits is randomly chosen from the digits 0–4, and the first digit is larger than the second.
20. A sequence of two different digits is randomly chosen from the digits 0–4, and the first digit is no smaller than the second.

APPLICATIONS

21. You are considering purchasing either a domestic car, an imported car, a van, an antique car, or an antique truck. Describe the event that you do not buy a car.
22. You are deciding whether to enroll in Psychology 1, Psychology 2, Economics 1, General Economics, or Math for Poets. Describe the event that you decide to avoid economics.

Hospital Finance Exercises 23–28 are based on the following table, which shows the annual revenue generated for hospitals by each specialist in the specified field:*

Type of Specialist	Annualized Revenue Generated
Physical/Rehabilitation	$1,653,704
Pulmonary (lung)	$1,204,796
Hematology/Oncology	$1,619,546
Endocrine (gland)	$ 823,421
Cardiology (heart)	$1,604,623
Infectious Diseases	$ 661,152
Nephrology (kidney)	$1,242,879

23. Your hospital is considering hiring a specialist. Describe the event E that the specialist hired will generate revenues of at least $1 million per year for your hospital.

24. Your hospital is considering firing a specialist. Describe the event F that the specialist fired generates revenues of at most $1 million per year for your hospital.

25. Your hospital is considering hiring a specialist. Let E be the event that your hospital hires a specialist who will generate revenues of at least $1 million per year, and let F be the event that your hospital does not hire either a cardiologist or a nephrologist. Describe the events $E \cup F$ and $E \cap F$ in words, and also by listing the outcomes of each.

26. Your hospital is considering firing a specialist. Let E be the event that your hospital fires a specialist who generates revenues of at most $1 million per year, and let F be the event that your hospital does not fire either a cardiologist or a endocrinologist. Describe the events $E \cup F$ and $E \cap F$ in words, and also by listing the outcomes of each.

27. Your hospital is considering hiring a specialist. Which of the following pairs of events are mutually exclusive?
    (a) E: Your hospital hires a specialist selected from the three top earning specialties;
    F: Your hospital hires a specialist who will generate less than $1,600,000 per year.
    (b) E: Your hospital hires a specialist selected from the three top earning specialties;
    F: Your hospital hires a specialist who will generate less than $1,650,000 per year.

28. Your hospital is considering firing a specialist. Which of the following pairs of events are mutually exclusive?
    (a) E: Your hospital fires a specialist selected from the three lowest earning specialties;
    F: Your hospital fires a specialist who generates more than $1,000,000 per year.
    (b) E: Your hospital fires a specialist selected from the three lowest earning specialties;
    F: Your hospital fires a specialist who generates more than $1,300,000 per year.

Sports Team Finances Exercises 29–34 are based on the following table, which shows revenues and costs for five professional Chicago-based sports teams in millions of dollars in 1992.†

Team	Gate Receipts	Media Revenue	Player Costs	Operating Expenses	Operating Income
Bears	$13.7	$38.1	$28.8	$46.5	$ 9.3
White Sox	28.3	26.2	35.0	61.2	16.7
Bulls	26.7	17.5	19.8	35.6	17.1
Cubs	23.1	28.0	37.4	59.8	5.1
Blackhawks	25.7	2.5	11.0	22.7	6.1

▼ *Sources: Lutheran General Health System, Argus Associates, Inc., *Chicago Tribune*/Stephen Ravenscraft and Celeste Schaefer. These figures were published in the *Chicago Tribune*, March 29, 1993, Section 4, p. 1.

† Source: *Financial World Magazine,* as quoted in the *Chicago Tribune*, May 5, 1993, Section 4, p. 1.

29. Your syndicate is considering making an offer to buy one of the above Chicago-based teams. Describe the event that the team it selects is among the top three in gate receipts and among the lowest three in player costs.

30. Your syndicate is considering making an offer to buy one of the above Chicago-based teams. Describe the event that the team it selects is among the top three in gate receipts and among the top three in media revenue.

31. Your syndicate is considering making an offer to buy one of the above Chicago-based teams. Let E be the event that the team is among the top four in both gate receipts and operating income, and let F be the event that the team has player costs not exceeding $20 million. Describe the events $E \cap F$ and F' in words and also by listing the outcomes of each.

32. Your syndicate is considering making an offer to buy one of the above Chicago-based teams. Let E be the event that the team is among the top four in both gate receipts and media revenue, and let F be the event that the team has operating expenses not exceeding $50 million. Describe the events $E \cap F$ and F' in words and also by listing the outcomes of each.

33. Which of the following pairs of events are mutually exclusive?
    (a) E: Your syndicate offers to buy a team selected from the three top media revenue earners;
    F: Your syndicate offers to buy a team whose operating expenses are less than $50 million per year.
    (b) E: Your syndicate offers to buy a team selected from the two top media revenue earners;
    F: Your syndicate offers to buy a team whose operating expenses are less than $40 million per year.

34. Which of the following pairs of events are mutually exclusive?
    (a) E: Your syndicate offers to buy a team selected from the three top gate receipt earners;
    F: Your syndicate offers to buy a team whose player costs are less than $20 million per year.
    (b) E: Your syndicate offers to buy a team selected from the three top gate receipt earners;
    F: Your syndicate offers to buy a team whose player costs are less than $15 million per year.

35. *Poker Hands* Recall that a poker hand consists of a group of five cards chosen from a standard deck of 52 playing cards. (See the chapter on counting.) Complete the following sentences:
    (a) The sample space is the set of . . .
    (b) The event "a full house" is the set of . . . (Recall that a full house refers to three cards of one denomination and two of another)

36. *Committees* President Bush's cabinet consisted of the Secretaries of State, Treasury, Defense, Interior, Agriculture, Commerce, Labor, Health and Human Services, Housing and Urban Development, Transportation, Energy, Education, Veterans Affairs, and the Attorney General.* Assuming that President Bush had 20 candidates, including James A. Baker, III, to fill these posts, complete the following sentences:
    (a) The sample space is the set of . . .
    (b) The event that James A. Baker, III, is the Secretary of State is the set of . . .

Investments Exercises 37 and 38 are based on the following chart, which shows the top 10 income funds, based on three-year performance, in August 1993:[†]

Top 10 Income Funds	1 Year	3 Years	5 Years
Franklin Income	16.3%	18.2%	14.0%
Berwyn Income	16.9%	18.0%	13.7%
Lindner Dividend	18.9%	17.4%	13.7%
National Multi-Sector FIA	14.2%	17.0%	—
Seligman Income A	15.0%	15.6%	12.4%
Putnam Diversified Inc. A	14.7%	15.0%	—
Vanguard Wellesley Income	14.8%	14.6%	13.8%
Income Fund of America	13.1%	14.5%	13.2%
Vanguard Preferred Stock	12.1%	14.3%	13.5%
Pru. Flex-A-Fund Con. Mgd. A	15.7%	13.0%	—

▼ * Source: *World Almanac and Book of Facts, 1992* (New York: Pharos Books, 1992).
[†] Reproduced from *The New York Times*, Aug. 14, 1993, p. 31. Source: Morningstar Inc.

37. In an experiment to select a Top 10 income fund, describe the following events:
    (a) E: that the 1-year annualized return exceeds the 3-year return;
    (b) F: that the fund's annualized return exceeds 18% for either 1, 3, or 5 years;
    (c) G: that the fund's annualized return does not exceed 18% for either 1, 3, or 5 years.
38. In an experiment to select a portfolio (or collection) of two Top 10 income funds, describe the following events:
    (a) E: that the 3-year annualized return of each fund in the portfolio is at least 18%;
    (b) F: that either the 3-year annualized return of both funds is at least 18% or the 1-year annualized return of both funds is at least 16.5%;
    (c) G: that each of the funds in the portfolio shows either a 1-year or 3-year annualized return of at least 18%.

Animal Psychology Exercises 39–44 concern the following chart, which shows the way in which a dog moves its facial muscles when torn between the drives of fight and flight.* The "fight" drive increases from left to right, while the "flight" drive increases from top to bottom. (Notice that an increase in the "fight" drive causes its upper lip to lift, while an increase in the "flight" drive draws its ears downwards.)

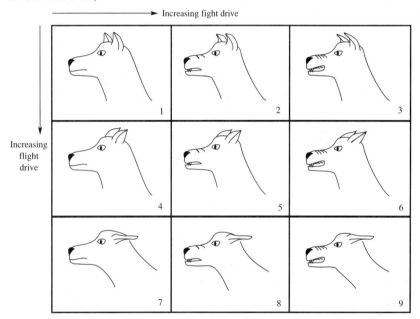

39. Let E be the event that the dog's "flight" drive is the strongest, let F be the event that the dog's "flight" drive is weakest, let G be the event that the dog's "fight" drive is the strongest, and let H be the event that the dog's "fight" drive is weakest. Describe the following events in terms of E, F, G, and H using the symbols ∩, ∪, and ′.
    (a) The dog's "flight" drive is not strongest and its "fight" drive is weakest.
    (b) The dog's "flight" drive is strongest or its "fight" drive is weakest.
    (c) Neither the dog's "flight" drive nor "fight" drive are strongest.
40. Let E be the event that the dog's "flight" drive is the strongest, let F be the event that the dog's "flight" drive is weakest, let G be the event that the dog's "fight" drive is the strongest, and let H be the event that the dog's "fight" drive is weakest. Describe the following events in terms of E, F, G, and H using the symbols ∩, ∪, and ′.

▼ * Source: *On Aggression* by Konrad Lorenz (University Paperback Edition, Cox & Wyman Limited, Fakenham, Norfolk, England: 1967)

(a) The dog's "flight" drive is weakest and its "fight" drive is not weakest.
(b) The dog's "flight" drive is not strongest or its "fight" drive is weakest.
(c) Either the dog's "flight" drive or "fight" drive fail to be strongest.

41. Describe the following events explicitly (as subsets of the sample space).
    (a) The dog's "fight" and "flight" drives are both strongest.
    (b) The dog's "fight" drive is strongest, but its "flight" is neither weakest nor strongest.

42. Describe the following events explicitly (as subsets of the sample space).
    (a) Neither the dog's "fight" drive nor its "flight" drive is strongest.
    (b) The dog's "fight" drive is weakest, but its "flight" is neither weakest nor strongest.

43. Describe the following events in words.
    (a) {1, 4, 7}
    (b) {1, 9}
    (c) {3, 6, 7, 8, 9}

44. Describe the following events in words.
    (a) {7, 8, 9}
    (b) {3, 7}
    (c) {1, 2, 3, 4, 7}

Exercises 45–52 use counting arguments from the last chapter.

45. *Marbles* A bag contains 6 distinguishable marbles. Suzy picks four at random. How many possible outcomes are there? If one of the marbles is red, how many of these outcomes include the red marble?

46. *Chocolates* My couch potato friend enjoys sitting in front of the TV and grabbing handfuls of five chocolates at random from his snack jar. Unbeknownst to him, I have replaced one of the 20 chocolates in his jar with a cashew. (He hates cashews with a passion.) How many possible outcomes are there the first time he grabs five chocolates? How many of these include the cashew?

47. *Horse Races* The 7 contenders in the first horse race at Saratoga on August 19, 1993 were Reappeal, Quickest Blade, Boom Towner, Club Da Nocche, Start a Fight, Senor Cielo, and Game Wager.* You are interested in the first three places (winner, second place, and third place) for the race.
    (a) Find the cardinality $n(S)$ of the sample space S of all possible finishes of the race. (A finish for the race consists of a first-, second-, and third-place winner.)
    (b) Let E be the event that Quickest Blade is in second or third place, and let F be the event that Boom Towner is the winner. Express the event $E \cap F$ in words, and find its cardinality.

48. *Intramurals* The following five teams will be participating in the Urban University's hockey intramural tournament: the Independent Wildcats, the Phi Chi Bulldogs, the Gate Crashers, the Slide Rule Nerds, and the City Slickers. Prizes will be awarded for the winner and runner-up.
    (a) Find the cardinality $n(S)$ of the sample space S of all possible outcomes of the tournament. (An outcome of the tournament consists of a winner and a runner-up.)
    (b) Let E be the event that the City Slickers are runners-up, and let F be the event that the Independent Wildcats are neither the winners nor runners-up. Express the event $E \cup F$ in words, and find its cardinality.

Marbles In Exercises 49–52, Suzy randomly picks three marbles from a bag of 8 marbles (four red ones, two green ones, and two yellow ones). The marbles are all distinguishable. That is, the red ones are distinguishable from one another, and so on.

49. How many outcomes are there in the sample space?

50. How many outcomes are there in the event that Suzy grabs three red marbles?

51. How many outcomes are there in the event that Suzy grabs one marble of each color?

52. How many outcomes are there in the event that Suzy's marbles are not all the same color?

COMMUNICATION AND REASONING EXERCISES

53. In order to determine a sample space for an experiment, what must be specified about that experiment?

54. Give an example of an experiment with two different sample spaces.

▼ *Source: *Newsday,* Aug. 18, 1993, p. 130

55. Complete the following. An event is a _____ of the sample space. If E and F are events, then $(E \cap F)'$ is the event that _____.

56. Two dice are thrown. Could there be two mutually exclusive events that both contain outcomes in which the numbers facing up add to 7?

6.2 INTRODUCTION TO PROBABILITY

EXPERIMENTAL PROBABILITY

Suppose that you have a coin, and you would like to measure the likelihood that heads will come up when it is tossed. You could measure this likelihood experimentally by tossing the coin a large number of times and counting the number of times heads comes up. Suppose, for instance, that in 100 tosses of the coin, heads comes up 58 times. The fraction of times that heads comes up, $58/100 = 0.58$, is the **relative frequency** or **experimental probability** of heads coming up when the coin is tossed. In other words, saying that the experimental probability of heads coming up is 0.58 is the same as saying that heads came up 58% of the time in your series of experiments.

Now let's think about this example in terms of sample spaces and events. First of all, there is an experiment that has been repeated $N = 100$ times: toss the coin and observe the side facing up. The sample space for this experiment is $S = \{H, T\}$. Also, there is an event E in which we are interested: the event that heads comes up, which is $E = \{H\}$. The number of times E has occurred is $fr(E) = 58$ ($fr(E)$ is called the **frequency** of E). The experimental probability of the event E is then

$$P(E) = \frac{fr(E)}{N}$$

$$= \frac{58}{100} = 0.58.$$

▶ **NOTE**

1. The experimental probability $P(E) = 0.58$ gives us an *estimate* of the likelihood that heads will come up when that particular coin is tossed.
2. The larger the number of times the experiment is performed, the more accurate we expect the experimental probability to be.
3. One always speaks of the probability *of an event E*. For every event E there is an experimental probability $P(E)$. ◄

EXPERIMENTAL PROBABILITY

If an experiment is performed N times, and the event E occurs $fr(E)$ times, then the ratio

$$P(E) = \frac{fr(E)}{N}$$

is called the **relative frequency** or **experimental probability** of E. The number $fr(E)$ is called the **frequency** of E. The number of times that the experiment is performed, N, is called the number of **trials** or the **sample size**. If E consists of a single outcome s, then we refer to $P(E)$ as the **experimental probability of the outcome** s, and write $P(s)$.

▼ **EXAMPLE 1** Surveys

In a 1994 survey of 706 adult New York residents, 254 of those interviewed felt that Governor Mario M. Cuomo was doing an excellent or good job as governor of New York.* What is the experimental probability that an adult New York resident felt that Governor Mario M. Cuomo was doing an excellent or good job as governor of New York?

SOLUTION   Here, the experiment consisted of asking an adult New York resident whether Governor Cuomo was doing an excellent or good job as Governor of New York, the sample space is $S = \{yes, no\}$, and the event $E = \{yes\}$. The sample size was $N = 706$, and the frequency of E was $fr(E) = 254$. Thus the experimental probability of E is

$$P(E) = \frac{fr(E)}{N}$$
$$= \frac{254}{706} \approx 0.36.$$

In other words, Governor Cuomo had an approval rating of approximately 36% at the time of the survey.

Before we go on... You might ask how reliable this survey was. In other words, how accurate is this estimate of 36%, or how well does this reflect the feelings of all residents of New York? Statistics provides the tools needed to say to what extent the experimental probability can be trusted.

▶ NOTE   Probabilities are often expressed as percentages, as when we said above that Governor Cuomo had an approval rating of 36%. Similarly, if the weather service announces that there is a 30% chance of rain tomorrow, they are saying that the probability that it will rain tomorrow is 0.3. ◀

▼ *Source: Marist College Institute for Public Opinion/*The New York Times,* August 21, 1994. (The figure of 254 is an estimate based on a graph.)

EXAMPLE 2

The following chart shows the results of 38 throws of a pair of distinguishable dice. E is the event that the sum of the numbers is 5.

Trial	1	2	3	4	5	6	7	8	9	10	11	12	13	14	15	16	17	18	19
Outcome	3, 3	4, 4	3, 2	5, 4	1, 3	5, 1	4, 3	5, 6	6, 3	3, 3	1, 2	5, 5	3, 6	1, 3	6, 2	2, 4	5, 5	4, 3	6, 5
$fr(E)$	0	0	1	1	1	1	1	1	1	1	1	1	1	1	1	1	1	1	1

Trial	20	21	22	23	24	25	26	27	28	29	30	31	32	33	34	35	36	37	38
Outcome	1, 2	6, 6	4, 5	3, 5	2, 6	5, 4	4, 5	1, 3	2, 1	2, 3	1, 5	3, 2	5, 5	1, 5	5, 4	5, 6	3, 6	6, 4	1, 2
$fr(E)$	1	1	1	1	1	1	1	1	1	2	2	3	3	3	3	3	3	3	3

Calculate the experimental probabilities $P(E)$ using the first 10 throws, the first 20 throws, and all 38 throws.

SOLUTION In the table we have also counted the number of favorable outcomes up to each throw (in the line labeled $fr(E)$). For example, in the first 10 throws there was only one favorable outcome, so

$$P(E) = \frac{fr(E)}{N} = \frac{1}{10} = 0.1.$$

In the first 20 throws there was still only one favorable outcome, so

$$P(E) = \frac{1}{20} = 0.05.$$

Finally, in all 38 throws there were 3 favorable outcomes, so

$$P(E) = \frac{3}{38} \approx 0.0789.$$

Before we go on... Notice how wildly the experimental probability changes with N. This is an indication that none of these estimates of the likelihood that the sum will be 5 are very accurate. To increase our confidence in our estimates, we would need to increase the number of trials until the experimental probability "settles down." Again, the theory of statistics can be used to determine how many trials we should use to get a reasonably accurate estimate.

EXAMPLE 3 Medical Insurance

In Example 2 of Section 1, we considered the various forms of medical coverage for a factory worker in 1993. The following chart shows the health insurance status of a sample of 10,000 workers aged 18–64 employed in "large" factories (factories with 1,000 or more workers).*

Uninsured	Covered by Individual Insurance	Covered by Spouse's Employer	Covered by Employer
842	964	1,176	7,018

(a) Find the experimental probabilities of the individual outcomes.

(b) Find the experimental probability that a worker will be covered either by his or her employer or by the spouse's employer.

SOLUTION

(a) We find the experimental probabilities of each of the outcomes by dividing the frequencies by $N = 10,000$.

$$P(\text{uninsured}) = \frac{842}{10,000} = 0.0842$$

$$P(\text{covered by individual insurance}) = \frac{964}{10,000} = 0.0964$$

$$P(\text{covered by spouse's employer}) = \frac{1,176}{10,000} = 0.1176$$

$$P(\text{covered by employer}) = \frac{7,018}{10,000} = 0.7018$$

(b) Here, $E = \{\text{covered by spouse's employer, covered by employer}\}$. Thus,

$$P(E) = \frac{1,176 + 7,018}{10,000}$$

$$= \frac{1,176}{10,000} + \frac{7,018}{10,000}$$

$$= 0.1176 + 0.7018 = 0.8194.$$

Before we go on... In this example, we can observe two important properties of experimental probabilities.

From part (a): The sum of the experimental probabilities of all the outcomes is 1. (Why?)

From part (b): The experimental probability of an event E is the sum of the experimental probabilities of the individual outcomes in E.

---

* An approximation of data published in Employee Benefits Research Institute, as quoted in *The New York Times*, August 23, 1993, p. A9. We have adjusted the figures to correspond to a (fictitious) sample size of 10,000.

> **PROPERTIES OF EXPERIMENTAL PROBABILITY**
>
> Let $S = \{s_1, s_2, \ldots, s_n\}$ be a sample space and let $P(s_i)$ be the experimental probability of the event $\{s_i\}$. Then
>
> (a) $0 \leq P(s_i) \leq 1$, and
> (b) $P(s_1) + P(s_2) + \ldots + P(s_n) = 1$.
>
> In words, the experimental probability of each outcome is a number between 0 and 1 (inclusive), and the experimental probabilities of all the outcomes add up to 1. Also, we can obtain the experimental probability of an event E by adding up the experimental probabilities of the outcomes in E.

THEORETICAL PROBABILITY

Let us return to the example at the start of Section 1: tossing a coin. A "fair" coin is one that is as likely to come up heads as it is to come up tails. In other words, we expect heads to come up 50% of the time if such a coin is tossed many times. Put more precisely, we expect the experimental probability to approach 0.5 as the number of trials gets larger. (See Figure 1.)

We say that the **theoretical probability,** or just the **probability,** of heads coming up in a toss of the coin is one-half, or

$$P(H) = \frac{1}{2}.$$

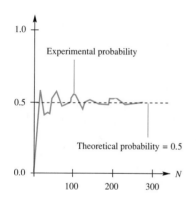

FIGURE 1

We also say that there is a one in two chance of heads coming up for each toss of a coin. Similarly, the probability of tails coming up is one-half, or

$$P(T) = \frac{1}{2}.$$

Notice that we are using the same letter P that we used for experimental probability. It should be clear from the context whether we are talking about experimental probability or theoretical probability.

Here is another example. If you throw a "balanced" die (that is, one that is equally likely to land with any of its six faces up) then you expect to roll a "1" one-sixth of the time. Thus,

$$P(1) = \tfrac{1}{6}.$$

Similarly,

$$P(2) = \tfrac{1}{6}, P(3) = \tfrac{1}{6}, \ldots, P(6) = \tfrac{1}{6}.$$

Also, since half the outcomes are even numbers, you expect to roll an even number one-half of the time, so

$$P(\{2, 4, 6\}) = \tfrac{1}{2}.$$

THEORETICAL PROBABILITY

The **theoretical probability,** or simply **probability,** $P(E)$ of an event E is a probability determined from the nature of the experiment rather than through actual experimentation. The experimental probability approaches the theoretical probability as the number of trials gets larger and larger.

▶ **NOTE** When we use the word "probability" alone we shall always mean *theoretical* probability. When we wish to consider experimental probability, we shall say so explicitly. ◀

Q How do we calculate theoretical probabilities?

A We usually calculate theoretical probabilities as the name suggests: theoretically. That is, we use our knowledge about a given situation rather than experiment to obtain the probability. However, in some situations it is impossible to calculate theoretical probability—for instance, the probability that it will snow in Toronto on November 1—so we must rely on experimental probabilities.

▼ **EXAMPLE 4**

You roll a pair of fair dice. Find the probability that you roll a double: that is, that both dice show the same number.

SOLUTION The sample space for this experiment is

$$S = \begin{Bmatrix} (1,1) & (1,2) & (1,3) & (1,4) & (1,5) & (1,6) \\ (2,1) & (2,2) & (2,3) & (2,4) & (2,5) & (2,6) \\ (3,1) & (3,2) & (3,3) & (3,4) & (3,5) & (3,6) \\ (4,1) & (4,2) & (4,3) & (4,4) & (4,5) & (4,6) \\ (5,1) & (5,2) & (5,3) & (5,4) & (5,5) & (5,6) \\ (6,1) & (6,2) & (6,3) & (6,4) & (6,5) & (6,6) \end{Bmatrix}.$$

(We assume that the dice are distinguishable, or see the *Before we go on* comment in Example 1 of the previous section.) The event that both dice show the same number is

$$E = \{(1,1), (2,2), (3,3), (4,4), (5,5), (6,6)\}.$$

Since the sample space S has 36 outcomes, all equally likely, and E consists of 6 of these outcomes, we expect a double to come up 6 times in every 36 throws. Thus,

$$P(E) = \tfrac{6}{36} = \tfrac{1}{6}.$$

In other words, there is a one-in-six chance of rolling a double.

Q That was easy. All we did was compute *(number of favorable outcomes)/(total number of outcomes)* = $n(E)/n(S)$. Is this all there is to calculating (theoretical) probability?

A Not quite, but we have discovered an important principle.

> **EQUALLY LIKELY OUTCOMES**
>
> In an experiment in which all outcomes are equally likely, the probability of an event E is given by
>
> $$P(E) = \frac{\text{Number of favorable outcomes}}{\text{Total number of outcomes}} = \frac{n(E)}{n(S)}.$$

Remember that this will work *only* when the outcomes are equally likely. If, for example, a die is *weighted*, then the outcomes may not be equally likely, and the formula above will not apply. We shall see such examples below.

▼ **EXAMPLE 5**

You roll a pair of fair dice. Find the probability that the numbers shown add up to 5.

SOLUTION The sample space S is the same one we used in Example 4. The event that the numbers shown add up to 5 is

$$E = \{(1, 4), (2, 3), (3, 2), (4, 1)\}.$$

Therefore,

$$P(E) = \frac{n(E)}{n(S)}$$
$$= \frac{4}{36} = \frac{1}{9} \approx 0.1111.$$

Before we go on... In Example 2, we observed an experimental probability of 0.0789 for this event. If we had rolled the dice, say, several hundred times, the experimental probability would have been much closer to $\frac{1}{9}$.

COMPARING EXPERIMENTAL AND THEORETICAL PROBABILITIES: USING TECHNOLOGY TO GENERATE RANDOM DATA

Most scientific and graphing calculators are equipped with "random number generators." That is, they have the ability to produce a "random" number, usually a decimal between 0 and 1, when you press the appropriate key (usually labeled "RAND" or "RAN#"). Each time you press the key, you get a new random number between 0 and 1. Here is a sequence of 27 random numbers produced by a *Radio Shack*® calculator:

0.555   0.753   0.182   0.336   0.196   0.335   0.921   0.209   0.356
0.536   0.999   0.893   0.104   0.337   0.987   0.575   0.281   0.252
0.519   0.195   0.450   0.639   0.428   0.273   0.340   0.878   0.760

The next example shows how we can use such data to replace actual tossing of coins or rolling of dice.

EXAMPLE 6

Use a random number generator to check the following.

(a) The probability of throwing an even number on a die is $\frac{1}{2}$.

(b) The probability of heads coming up in two consecutive tosses of a coin is $\frac{1}{4}$.

SOLUTION

(a) We know from above that the theoretical probability of rolling an even number is $\frac{1}{2}$. To verify this experimentally, we can use any sequence of random numbers, such as the sequence of 27 decimals given above. Since the numbers are random, their digits are random as well, so we list the sequence of digits occurring in the above sequence:

5, 5, 5, 7, 5, 3, 1, 8, 2, 3, 3, 6, 1, 9, 6, 3, 3, 5, 9, 2, 1, 2,
0, 9, 3, 5, 6, 5, 3, 6, 9, 9, 9, 8, 9, 3, 1, 0, 4, 3, 3, 7, 9, 8,
7, 5, 7, 5, 2, 8, 1, 2, 5, 2, 5, 1, 9, 1, 9, 5, 4, 5, 0, 6, 3, 9,
4, 2, 8, 2, 7, 3, 3, 4, 0, 8, 7, 8, 7, 6, 0.

We throw out those numbers that are not in the range 1–6, getting the following 51 integers, which we can take as the outcomes of 51 throws of a die:

5, 5, 5, 5, 3, 1, 2, 3, 3, 6, 1, 6, 3, 3, 5, 2, 1, 2, 3, 5, 6, 5,
3, 6, 3, 1, 4, 3, 3, 5, 5, 2, 1, 2, 5, 2, 5, 1, 1, 5, 4, 5, 6, 3,
4, 2, 2, 3, 3, 4, 6.

Of these, 18 are even (count them!) and so the experimental probability of an even number coming up was $\frac{18}{51} = 0.35$. This does not agree very well with the theoretical result of 0.50, but we can improve the accuracy by generating more random numbers. Here are another 49 random digits in the range 1–6, generated by the same method:

4, 1, 3, 6, 6, 1, 3, 4, 1, 4, 3, 2, 4, 1, 3, 1, 5, 2, 1, 3, 5, 6,
4, 2, 6, 5, 6, 6, 4, 4, 6, 4, 6, 5, 1, 3, 5, 4, 3, 2, 6, 2, 1, 5,
4, 1, 5, 4, 6.

Of the 100 digits we have now generated, 44 are even, giving an experimental probability of $\frac{44}{100} = 0.44$. This is somewhat closer to the theoretical result. In general, the larger the number of trials, the closer the experimental probability should approximate the theoretical probability.

(b) We can use the above sequence of 100 digits between 1 and 6 to simulate tosses of two coins in a variety of ways. For instance, we can break the sequence into a sequence of pairs, and replace each digit by *H* if it is in the range 1–3 and by *T* if it is in the range 4–6. This gives us the following 50 pairs:

TT, TT, HH, HH, HT, HT, HH, TH, HH, HT, TT, HT,
HH, TH, HT, TH, HH, TH, TH, HT, TT, TH, TH, HH,
HT, TT, HH, TT, HH, TH, TH, HT, HH, HT, HH, HT,
TT, HT, TT, TT, TT, TT, TH, HT, TH, HT, HH, TT,
HT, TT.

The sequence *HH* appears 12 times in the 50 sequences, giving a probability of $\frac{12}{50} = 0.24$. This is very close to the theoretical probability of 0.25.

Before we go on... If you are using a programmable calculator, a graphing calculator, or a computer, you could enter a simple instruction to convert each random decimal into an integer in the desired range (1–6 for a die or 0–1 for a coin) without having to do so by hand. For instance, on the TI-82, the instruction

```
int(10*rand)
```

gives a random digit between 0 and 9, while

```
int(6*rand)+1
```

gives a random digit between 1 and 6 (why?). Pressing ENTER repeatedly will give you as many of these as you want.

You could also program your graphing calculator to have it generate thousands of simulated trials, instead of the measly 100 we have here.*

PROBABILITY DISTRIBUTIONS

We saw that experimental probability had the following properties for a sample space $S = \{s_1, s_2, \ldots, s_n\}$:

(a) $0 \leq P(s_i) \leq 1$

(b) $P(s_1) + P(s_2) + \ldots + P(s_n) = 1$

(c) We can obtain the probability of an event *E* by adding up the probabilities of the outcomes in *E*.

▼ *See, for example, *Graphing Calculator Lessons for Finite Mathematics* by Paula G. Young (New York: HarperCollins College Publishers, 1993).

6.2 Introduction to Probability

Do these properties hold for theoretical probability as well?

A Yes. Since the probability of an outcome is the fraction of times we expect it to occur, probabilities are always numbers between 0 and 1. To get the fraction of times an event will occur, we clearly need to add the fractions of times each outcome in the event will occur. And, since the event $E = S$ *must* occur every time, $P(S) = P(s_1) + \ldots + P(s_n) = 1$.

Here is an interesting hypothetical situation: Pretend that we could purchase, to our exact specifications, a *weighted* die—one in which the six numbers were *not* equally likely to land facing up. For instance, we might wish to purchase a die with

$$P(1) = P(2) = \tfrac{1}{4},$$

and

$$P(3) = P(4) = P(5) = P(6) = \tfrac{1}{8}.$$

All we need to ensure is that the numbers we choose for the probabilities are between 0 and 1, and that the probabilities of all the outcomes add to 1 (check that this is true above). Once we have done that, we can obtain the probability of any event E by adding up the probabilities of the outcomes in E. For example, the probability of rolling a 1 or a 2 with this die will be $\tfrac{1}{4} + \tfrac{1}{4} = \tfrac{1}{2}$.

By specifying the probabilities of all the outcomes in a sample space, we are specifying a **probability distribution.**

PROBABILITY DISTRIBUTION

1. A **probability distribution** is an assignment of a number $P(s_i)$ to each outcome s_i in a sample space $\{s_1, s_2, \ldots, s_n\}$, so that
   (a) $0 \leq P(s_i) \leq 1$, and
   (b) $P(s_1) + P(s_2) + \cdots + P(s_n) = 1$.
   In words, the probability of each outcome must be a number between 0 and 1 (inclusive), and the probabilities of all the outcomes must add up to 1.

2. Given a probability distribution, we can obtain the probability of an event E by adding up the probabilities of the outcomes in E.

▶ **NOTE** Since the probability of an outcome can be zero, we are also allowing the possibility that $P(E) = 0$ for an event E. If $P(E) = 0$, we call E an **impossible event.** The event $\emptyset$ is always impossible, since *something* must happen. ◀

EXAMPLE 7 Weighted Dice

In order to impress your friends with your dice-throwing skills, you have surreptitiously weighted your die in such a way that 6 is three times as likely to come up as any one of the other numbers. Find the probability distribution, and use it to calculate the probability of an even number coming up.

SOLUTION We haven't a clue yet as to what the individual probabilities are—these are our *unknowns*. All we know is that they must add to 1. Let us label our unknowns. There appear to be two of them.

Let x be the probability of rolling a 6.

Let y be the probability of rolling any one of the other numbers.

We are first told that "6 is three times as likely to come up as any other number." If we rephrase this in terms of our unknown probabilities we get:

"The probability of rolling a 6 is three times the probability of rolling any other number."

In other words,

$$x = 3y.$$

We must also use a piece of information not given to us, but one we know must be true: the sum of the probabilities of all the outcomes is 1. So

$$x + y + y + y + y + y = 1,$$

or

$$x + 5y = 1.$$

We now have two linear equations in two unknowns, and we solve for x and y. Substituting the first equation ($x = 3y$) in the second ($x + 5y = 1$) gives

$$8y = 1,$$

or

$$y = \tfrac{1}{8}.$$

To get x, we substitute the value of y back into either equation, and obtain

$$x = \tfrac{3}{8}.$$

Thus, the probability distribution is the one shown in the following table.

Outcome	1	2	3	4	5	6
Probability	$\tfrac{1}{8}$	$\tfrac{1}{8}$	$\tfrac{1}{8}$	$\tfrac{1}{8}$	$\tfrac{1}{8}$	$\tfrac{3}{8}$

We can use the distribution to calculate the probability of an even number coming up by adding the probabilities of the favorable outcomes.

$$P(\{2, 4, 6\}) = \tfrac{1}{8} + \tfrac{1}{8} + \tfrac{3}{8} = \tfrac{5}{8}$$

Thus, there is a $\tfrac{5}{8} = 0.625$ chance that an even number will come up.

Before we go on... We should check that the distribution satisfies the requirements: 6 is indeed three times as likely to come up as any other number. Also, the probabilities we calculated do add up to 1:

$$\tfrac{1}{8} + \tfrac{1}{8} + \tfrac{1}{8} + \tfrac{1}{8} + \tfrac{1}{8} + \tfrac{3}{8} = \tfrac{8}{8} = 1. \checkmark$$

This example concerns a *theoretical* probability distribution. The next one will deal with an experimental probability distribution.

You can use a random number generator to produce a sequence of trials for this weighted die as follows.

Step 1: Produce a sequence of digits in the range 1–8 (not 1–6). (See Example 6.)

Step 2: Turn all the 7s and 8s into 6s.

This gives you a sequence of random digits in the range 1–6 with the digit 6 three times as likely to come up as any of the other digits (why?).

▼ **EXAMPLE 8** Medical Insurance

Here is a more complete version of the table we saw in Example 3 (10,000 of each size of firm were sampled).*

Size of Firm	Uninsured	Covered by Individual Insurance	Covered by Spouse's Employer	Covered by Employer
100–499	1,305	975	1,504	6,216
500–999	903	895	1,489	6,713
1,000 or More	842	964	1,176	7,018

Use the table to give three experimental probability distributions, one for each size firm, showing the probability that a worker was in one or another category of insurance coverage.

SOLUTION We already computed the experimental probability distribution for the largest firm size (1,000 or more) in Example 3. All we did was divide each of the given frequencies by $N = 10,000$, as follows.

$$P(\text{uninsured}) = \frac{842}{10,000} = 0.0842$$

$$P(\text{covered by individual insurance}) = \frac{964}{10,000} = 0.0964$$

▼ * See the footnote there for source information.

$$P(\text{covered by spouse's employer}) = \frac{1{,}176}{10{,}000} = 0.1176$$

$$P(\text{covered by employer}) = \frac{7{,}018}{10{,}000} = 0.7018$$

Thus the experimental probability distribution for a factory with 1,000 or more workers can be summarized by the following table.

		1,000 OR MORE WORKERS		
Outcome	Uninsured	Covered by Individual Insurance	Covered by Spouse's Employer	Covered by Employer
Probability	0.0842	0.0964	0.1176	0.7018

This *is* a probability distribution: the probability of each outcome is between 0 and 1, and the sum of the probabilities is 1. Similarly, the other two experimental probability distributions are the following.

		100–499 WORKERS		
Outcome	Uninsured	Covered by Individual Insurance	Covered by Spouse's Employer	Covered by Employer
Probability	0.1305	0.0975	0.1504	0.6216

		500–999 WORKERS		
Outcome	Uninsured	Covered by Individual Insurance	Covered by Spouse's Employer	Covered by Employer
Probability	0.0903	0.0895	0.1489	0.6713

Before we go on... Before leaving this example, consider the following question.

Q In a factory with 500–999 workers, what is the probability that an employee will be insured by either his or her employer, or by his or her spouse's employer?

A As we saw, we can get the experimental probability of any event by adding the experimental probabilities of its individual outcomes. Looking at the distribution for a factory with 500–999 workers, we have

$$P(\text{Covered by employer}) = 0.6713,$$
$$P(\text{Covered by spouse's employer}) = 0.1489.$$

Thus,

$$P(\{\text{Covered by employer, Covered by spouse's employer}\})$$
$$= 0.6713 + 0.1489$$
$$= 0.8202.$$

6.2 EXERCISES

In Exercises 1–4, calculate the experimental probability $P(E)$ using the given information.

1. $N = 100, fr(E) = 40$
2. $N = 500, fr(E) = 300$
3. 800 adults are polled, and 687 of them support universal health care coverage. E is the event that an adult supports universal health care coverage.
4. 800 adults are polled, and 687 of them support universal health care coverage. E is the event that an adult does not support universal health care coverage.

Exercises 5–10 are based on the following table, which shows the frequency of outcomes when two distinguishable coins are tossed 5,000 times and the uppermost faces are observed.

Outcome	HH	HT	TH	TT
Frequency	1,621	856	1,521	1,002

5. Determine the experimental probability of each of the outcomes.
6. What is the experimental probability that heads comes up at least once?
7. What is the experimental probability that the second coin lands with heads up?
8. What is the experimental probability that the first coin lands with heads up?
9. Would you judge the second coin to be fair? Give a reason for your answer.
10. Would you judge the first coin to be fair? Give a reason for your answer.

In Exercises 11–16, calculate the (theoretical) probability $P(E)$ using the given information, assuming that all outcomes are equally likely.

11. $n(S) = 20, n(E) = 5$
12. $n(S) = 8, n(E) = 4$
13. $n(S) = 10, n(E) = 10$
14. $n(S) = 10, n(E) = 0$
15. $S = \{a, b, c, d\}, E = \{a, b, d\}$
16. $S = \{1, 3, 5, 7, 9\}, E = \{3, 7\}$

In Exercises 17–28, an experiment is given together with an event. Find the probability of each event, assuming that the coins and dice are distinguishable and fair, and that what is observed are the faces or numbers uppermost. (Compare with Exercises 1–12 in Section 1)

17. Two coins are tossed, and the result is at most one tail.
18. Two coins are tossed, and the result is one or more heads.
19. Three coins are tossed, and the result is at most one head.
20. Three coins are tossed, and the result is more tails than heads.
21. Four coins are tossed, and the result is that a head is never followed by a tail.
22. Four coins are tossed, and the result is that the first shows a head and the last a tail.

23. Two dice are thrown, and the numbers add to 5.
24. Two dice are thrown, and the numbers add to 9.
25. Two dice are thrown, and the numbers add to 1.
26. Two dice are thrown, and one of the numbers is even, the other is odd.
27. Two dice are thrown, and both numbers are prime.
28. Two dice are thrown, and neither number is prime.

APPLICATIONS

Student Loans Exercises 29–34 are based on the following table, which shows the number of people who were due to begin paying federal student loans in 1992, and the subsequent default rate.* Round all answers to 4 decimal places.

	Due to Begin Payment in 1992	Defaulted 1992–1993
4-Year Colleges	1,113,079	74,945
2-Year Colleges	254,623	36,918
Trade Schools	621,412	187,739

29. What is the probability that a randomly selected student due to begin payment in 1992 was in a 4-year college?
30. What is the probability that a randomly selected student due to begin payment in 1992 was not in a trade school?
31. Find the probability that a randomly selected student who defaulted in 1992–1993 was not in a trade school.
32. Find the probability that a randomly selected student who defaulted in 1992–1993 was not in a 2-year college.
33. Find the probability that a randomly selected trade school student due to begin payment in 1992 defaulted in 1992–1993.
34. Find the probability that a randomly selected college student due to begin payment in 1992 defaulted in 1992–1993.

Exercises 35 and 36 use counting techniques based on permutations and/or combinations.

35. *Marbles* A bag contains 6 distinguishable marbles. Suzy chooses four at random. How many possible outcomes are there? If one of the marbles is red, how many of these outcomes include the red marble? What is the probability that Suzy grabs the red marble?

36. *Chocolates* My couch potato friend enjoys sitting in front of the TV and grabbing handfuls of five chocolates at random from his snack jar. Unbeknownst to him, I have replaced one of the 20 chocolates in his jar with a cashew. (He hates cashews with a passion.) How many possible outcomes are there the first time he grabs five chocolates? How many of these include the cashew? What is the probability that he will grab the cashew?

37. Complete the following probability distribution table and then calculate the stated probabilities.

Outcome	a	b	c	d	e
Probability	0.1	0.05	0.6	0.05	

(a) $P(\{a, c, e\})$
(b) $P(E \cup F)$, where $E = \{a, c, e\}$ and $F = \{b, c, e\}$
(c) $P(E')$, where E is as in part (b)
(d) $P(E \cap F)$, where E and F are as in part (b).

38. Repeat Exercise 37 using the following table.

Outcome	a	b	c	d	e
Probability	0.10		0.65	0.10	0.05

39. *Opinion Polls* A *New York Times*/CBS News poll of 1,368 people interviewed in March, 1993 showed that 85% of all respondents favored a national law requiring a seven-day waiting period for handgun purchases, 13% opposed it, and 2% were undecided.[†] Considering the poll as an experiment, write down the sample space and the associated experimental probability distribution, and give the experimental probability that a randomly chosen resident has a definite opinion about such a law.

---

*Source: U.S. Department of Education/*Long Island Newsday*, Sept. 2, 1994, p. A17.
[†] Source: *The New York Times*, August 15, 1993, Section 4, p. 4

40. *Opinion Polls* The opinion poll referred to in Exercise 39 also showed that 41% of the respondents would favor an outright ban on the sale of handguns (law enforcement officers excepted), 55% would oppose such a ban, and 4% were undecided. Considering the poll as an experiment, write down the sample space and the associated experimental probability distribution, and give the experimental probability that a randomly chosen resident has a definite opinion about such a law.

41. *Financing Long-Term Medical Care* In 1991, payment of nursing-home expenses was divided as shown in the following pie chart.*

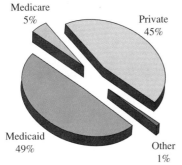

1991 Payments to Nursing Homes
Medicare 5%
Private 45%
Medicaid 49%
Other 1%

Write down the probability distribution that shows the probability that $1 spent on a nursing home originated from each of these categories. What is the probability that $1 spent on a nursing home originated from either Medicare or Medicaid?

42. *Exports* In 1993, U.S. farm exports to Mexico totaled $3,676 million, divided as shown in the following pie chart.†

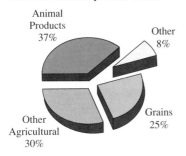

1993 U.S. Farm Exports to Mexico
Animal Products 37%
Other 8%
Grains 25%
Other Agricultural 30%

Write down the probability distribution that shows the probability that $1 worth of exports to Mexico belonged to each of these categories. What is the probability that $1 worth of exports to Mexico belonged to agricultural or animal products?

43. *Stock Prices* The following table shows the end-of-week closing values of the Dow Jones Industrial Average over the 10-week period‡ beginning January 1, 1992:

Week	1	2	3	4	5
Dow	3,210	3,200	3,270	3,220	3,220
Week	6	7	8	9	10
Dow	3,250	3,280	3,270	3,220	3,240

Use these data to formulate an experimental probability distribution with the following three outcomes:

Low: The index is at or below 3,220.
Middle: The index is above 3,220 but not above 3,260.
High: The index is above 3,260.

44. *Stock Prices* The following table shows the end-of-week closing values of the Dow Jones Industrial Average over the 10-week period‡ beginning January 1, 1993:

Week	1	2	3	4	5
Dow	3,250	3,260	3,250	3,310	3,460
Week	6	7	8	9	10
Dow	3,390	3,310	3,380	3,400	3,420

Use these data to formulate an experimental probability distribution with the following three outcomes:

Low: The index is at or below 3,300.
Middle: The index is above 3,300 but not above 3,400.
High: The index is above 3,400.

▼ *Source: Health Care Financing Administration/*The New York Times*, August 28, 1993, p. 32.
† Source: Mexican Consulate/News Reports/*Chicago Tribune*, May 2, 1993, Section 7, p. 1.
‡ Source: *The New York Times*, July 18, 1993, Section 3, p. 7. Averages are rounded to the nearest 10 points.

45. *AIDS Testing* A pharmaceutical company is testing a new AIDS test. They try it on 400 patients known to have AIDS and 200 patients known not to be infected. Of those with AIDS, the new test is positive for 390 and negative for 10. Of those not infected, the test is positive for 10 and negative for 190. What is the experimental probability of a **false negative** result (the probability that a person with the disease will test negative)? What is the experimental probability of a **false positive** result (the probability that an uninfected person will test positive)?

46. *Lie Detectors* A manufacturer of lie detectors is testing its newest design. The researchers ask 300 subjects to deliberately lie and another 500 to tell the truth. Of those who lied, the lie detector caught 200. Of those who told the truth, the lie detector accused 200 of lying. What is the experimental probability of the machine wrongly letting a liar go, and what is the probability that it will falsely accuse someone who is telling the truth?

Gambling In Exercises 47–52 are detailed some of the nefarious dicing practices of the Win Some/Lose Some Casino. In each case, find the probabilities of all the possible outcomes, and also the probability that an odd number or an odd sum, as appropriate, comes up.

47. Some of the dice are cleverly weighted so that each of 2, 3, 4, and 5 is twice as likely to come up as 1 is, and 1 and 6 are equally likely.

48. Other dice are weighted so that each of 2, 3, 4, and 5 is half as likely to come up as 1 is, and 1 and 6 are equally likely.

49. Some pairs of dice are magnetized so that any pair of mismatching numbers is twice as likely to come up as any pair of matching numbers.

50. Other pairs of dice are so strongly magnetized that matching numbers never come up.

51. Some dice are constructed in such a way that deuce (2) is five times as likely to come up as 4 and three times as likely to come up as any of 1, 3, 5, and 6.

52. Other dice are constructed in such a way that deuce (2) is six times as likely to come up as 4 and four times as likely to come up as any of 1, 3, 5, and 6.

Exercises 53–60 require the use of a calculator or computer with a random number generator.

53. Generate 100 random digits in the range 0–9, and calculate the experimental probability that a random digit is in the range 0–4.

54. Use a random number generator to simulate 100 tosses of a fair coin, and compute the experimental probability that the uppermost side is heads.

55. Use a random number generator to simulate the probability distribution in Exercise 41 using 100 samples. (Generate integers in the range 1–100. The outcome is determined by the range. For instance, if the number is in the range 1–49, regard the outcome as Medicaid, if it is in the range 50–54, regard it as Medicare, and so on.)

56. Repeat Exercise 55, but this time simulate the probability distribution in Exercise 42.

57. Generate four sequences of 50 random digits in the range 1–6, and use your sequences to find the experimental probability that the uppermost digits will add to 10 if two dice are cast. You should give four answers:
(a) using 25 trials;
(b) using 50 trials;
(c) using 75 trials;
(d) using 100 trials.

58. Repeat the method of Exercise 57 to calculate the experimental probability that the digits add to 7 when two dice are thrown.

59. Use a random number generator to simulate the weighted die in Exercise 47. You should use at least 100 trials and give an (experimental) probability distribution table based on these data.

60. Repeat Exercise 59, but this time simulate the weighted die in Exercise 48.

COMMUNICATION AND REASONING EXERCISES

61. Complete the following. The _____ probability is an estimate of the _____ probability. This estimate improves as _____ .

62. Complete the following. Experimental probability is defined to be _____ .

63. If the weather service says that there is a 50% chance of rain today, does this suggest experimental probability or theoretical probability? Explain your answer.

64. How would you measure the experimental probability that the weather service accurately predicts the next day's temperature?

65. Interpret your grade point average as an experimental probability by specifying an appropriate experiment including what is observed.

66. How many probability distributions are there on $S = \{a, b, c\}$ with $P(a) = P(b) = 0.75$? Explain your answer.

6.3 PRINCIPLES OF PROBABILITY

In this section we look at some properties of probabilities that help us in calculations. Let's begin by considering the sample space $S = \{HHH, HHT, HTH, HTT, THH, THT, TTH, TTT\}$ of possible outcomes when a coin is tossed three times (and the uppermost face is observed), and let

$E = \{HHT, THH, TTH, HTT\}$ (two consecutive heads or two consecutive tails)

$F = \{HTH, THT\}$ (alternating heads and tails).

Q What is the probability of $E \cup F$?

A We know that we can find the probability $P(E)$ by adding the probabilities of all the outcomes in E, getting $P(E) = \frac{4}{8} = \frac{1}{2}$. Similarly, we can find $P(F)$ by adding the probabilities of all the outcomes in F, getting $P(F) = \frac{2}{8} = \frac{1}{4}$. To find $P(E \cup F)$, we must add the probabilities of all the outcomes in $E \cup F$. Since we have already done this for the outcomes in E and F separately, and since E and F have no outcomes in common, we can simply compute

$$P(E \cup F) = P(E) + P(F) = \frac{1}{2} + \frac{1}{4} = \frac{3}{4}.$$

It is crucial here that E and F are *mutually exclusive* (or *disjoint*) events: that is, $E \cap F = \emptyset$.

This is a special case of the **addition principle.**

ADDITION PRINCIPLE FOR MUTUALLY EXCLUSIVE EVENTS

If E and F are mutually exclusive events, then

$$P(E \cup F) = P(E) + P(F).$$

This holds true also for more events: If $E_1, E_2, \ldots, E_n$ are mutually exclusive events (that is, the intersection of any pair of them is empty) and $E = E_1 \cup E_2 \cup \ldots \cup E_n$, then

$$P(E) = P(E_1) + P(E_2) + \cdots + P(E_n).$$

GENERAL ADDITION PRINCIPLE

If E and F are any events, then

$$P(E \cup F) = P(E) + P(F) - P(E \cap F).$$

We shall justify these principles after the next three examples.

EXAMPLE 1

If a pair of dice is cast, let E be the event that the numbers add to 7, let F be the event that the numbers add to 8, and let G be the event that the numbers add to 9. Find the following:

(a) The probability that the numbers add to 7 or 8.

(b) The probability that the numbers add to 7, 8, or 9.

SOLUTION

(a) We could answer this by counting all the favorable outcomes (those in which the numbers add to 7 or 8) but we shall instead use the addition principle. We are asked for the probability of E or F: that is, $P(E \cup F)$. Since the events E and F are mutually exclusive (the dice cannot add up to *both* 7 and 8), the addition principle for mutually exclusive events applies. We have

$$P(E) = \frac{n(E)}{n(S)} = \frac{6}{36},$$

$$P(F) = \frac{n(F)}{n(S)} = \frac{5}{36}.$$

Thus, by the addition principle,

$$P(E \cup F) = P(E) + P(F)$$

$$= \frac{6}{36} + \frac{5}{36} = \frac{11}{36}.$$

(b) We are asked for the probability of $E \cup F \cup G$. Since these three events are mutually exclusive, and $P(G) = \frac{4}{36}$, the addition principle gives us

$$P(E \cup F \cup G) = P(E) + P(F) + P(G)$$

$$= \frac{6}{36} + \frac{5}{36} + \frac{4}{36}$$

$$= \frac{15}{36} = \frac{5}{12}.$$

EXAMPLE 2 Astrology

The astrology software package Turbo Kismet* works by first generating random number sequences and then interpreting them numerologically. When I ran it yesterday, it informed me that there was a $\frac{1}{3}$ probability that I would meet a tall, dark stranger this month, a $\frac{2}{3}$ probability that I would travel

---

* The name and concept were borrowed from a hilarious (as yet unpublished) novel by the science fiction writer William Orr, who also happens to be a member of the Math Department at Hofstra University.

within the next month, and a $\frac{1}{6}$ probability that I would meet a tall, dark stranger and travel this month. What is the probability that I will either meet a tall, dark stranger or that I will travel this month?

SOLUTION The first two events are

E, the event that I will meet a tall, dark stranger this month, with $P(E) = \frac{1}{3}$,

and

F, the event that I will travel this month, with $P(F) = \frac{2}{3}$.

The third event mentioned—that I will meet a tall, dark stranger and travel this month—is the event $E \cap F$. Thus,

$$P(E \cap F) = \tfrac{1}{6}.$$

We are asked to calculate the event that either I will travel this month, or that I will meet a tall, dark, stranger this month. This is the event $E \cup F$. The general addition principle gives us

$$P(E \cup F) = P(E) + P(F) - P(E \cap F)$$
$$= \tfrac{1}{3} + \tfrac{2}{3} - \tfrac{1}{6}$$
$$= \tfrac{5}{6}.$$

EXAMPLE 3 Salaries

Your company's statistics show that 30% of your employees earn between $20,000 and $39,999, while 20% earn between $30,000 and $59,999. Given that 40% of the employees earn between $20,000 and $59,999,

(a) What percentage earn between $30,000 and $39,999?
(b) What percentage earn between $20,000 and $29,999?

SOLUTION

(a) Think of the sample space of all your employees, with the experiment being to pick one at random. We are given information about two events,

E, the event that an employee earns between $20,000 and $39,999, has $P(E) = 0.3$.

F, the event that an employee earns between $30,000 and $59,999, has $P(F) = 0.2$.

We are asked for the probability that an employee earns between $30,000 and $39,999. That is, we are asked for $P(E \cap F)$. (See Figure 1.)

We are told that the probability of an employee earning between $20,000 and $59,999 is 0.4. That is, $P(E \cup F) = 0.4$. How can we use

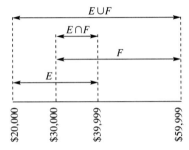

FIGURE 1

our formula for $P(E \cup F)$ if we already *know* $P(E \cup F)$? We take advantage of the fact that the formula

$$P(E \cup F) = P(E) + P(F) - P(E \cap F)$$

can be used to calculate any one of the four quantities appearing in it, as long as we know the other three. Substituting the quantities we know, we get

$$0.4 = 0.3 + 0.2 - P(E \cap F),$$

so

$$P(E \cap F) = 0.3 + 0.2 - 0.4 = 0.1.$$

Thus, 10% of your employees earn between $30,000 and $39,999.

(b) We are asked for the probability of a new event:

G, the event that an employee earns between $20,000 and $29,999.

But $G \cup (E \cap F) = E$ (Figure 2) and G and $E \cap F$ are mutually exclusive.

Thus, applying the addition principle,

$$P(G) + P(E \cap F) = P(E).$$

From part (a), we know that $P(E \cap F) = 0.1$, so

$$P(G) + 0.1 = 0.3,$$

giving

$$P(G) = 0.2.$$

In other words, 20% of your employees are earning between $20,000 and $29,999.

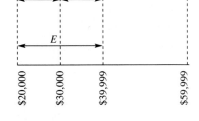

FIGURE 2

Q Why does the addition principle for mutually exclusive events work?

A We already saw an example at the beginning of this section. In general, we will have $E = \{s_1, s_2, \ldots, s_m\}$ and $F = \{t_1, t_2, \ldots, t_n\}$ with E and F having no outcomes in common. This makes $E \cup F = \{s_1, s_2, \ldots, s_m, t_1, t_2, \ldots, t_n\}$. Since we get the probability of an event by adding the probabilities of its outcomes, we have

$$P(E \cup F) = P(s_1) + P(s_2) + \cdots + P(s_m) + P(t_1) \\ + P(t_2) + \cdots + P(t_n) \\ = P(E) + P(F).$$

Q Where does the general addition principle come from?

A First notice that the addition principle is analogous to the formula

$$n(E \cup F) = n(E) + n(F) - n(E \cap F)$$

for the cardinality of a union discussed in Section 2 of the previous chapter. We obtained the formula for $n(E \cup F)$ by reasoning with a Venn diagram, and we can obtain the corresponding formula for $P(E \cup F)$ in the same way. Looking at the Venn diagram in Figure 3, notice that the three events A, $E \cap F$, and B shown there are mutually exclusive, and that their union is $E \cup F$.

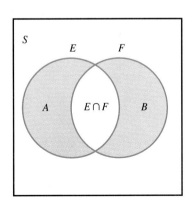

FIGURE 3

Applying the special case of the addition principle for mutually exclusive events gives

$$P(E \cup F) = P(A) + P(E \cap F) + P(B)$$
$$= \underbrace{P(A) + P(E \cap F)}_{P(E)} + \underbrace{P(B) + P(E \cap F)}_{P(F)} - P(E \cap F)$$
$$= P(E) + P(F) - P(E \cap F),$$

which is the general addition principle. A nice way to think about this principle is to notice that $P(E) + P(F)$ counts $P(E \cap F)$ *twice,* so to get $P(E \cup F)$ we need to correct by subtracting $P(E \cap F)$ once.

We can use the addition principle to deduce other useful properties of a probability distribution.

FURTHER PRINCIPLES OF PROBABILITY DISTRIBUTIONS

The following are true for any sample space S and any event E:

(1) $P(S) = 1$ (The probability of *something* happening is 1.)

(2) $P(\emptyset) = 0$ (The probability of *nothing* happening is 0.)

(3) $P(E') = 1 - P(E)$ (The probability of E *not* happening is 1 minus the probability of E.)

Q Can you convince me that all of these are true?

A Let us take them one at a time.

(1) S is the event that something happens. Clearly, something must happen (or our sample space is wrong), so $P(S) = 1$. We used this fact in the previous section to say that the sum of the probabilities of all the outcomes in S must be 1.

(2) Note that $S \cap \emptyset = \emptyset$, so that S and $\emptyset$ are mutually exclusive. Applying the addition principle gives

$$P(S) = P(S \cup \emptyset) = P(S) + P(\emptyset).$$

Subtracting $P(S)$ from both sides gives $0 = P(\emptyset)$.

(3) If E is any event in S, then we can write
$$S = E \cup E',$$
where E and E' are mutually exclusive (why?). Thus, by the addition principle,
$$P(S) = P(E) + P(E').$$
Since $P(S) = 1$, we get
$$1 = P(E) + P(E'),$$
or
$$P(E') = 1 - P(E).$$

EXAMPLE 4

If you roll two fair distinguishable dice, find the probability of each of the following events.

(a) Either the first number is odd or the second is even.

(b) It is *not the case* that either the first number is odd or the second is even.

SOLUTION We *could* find these probabilities by writing out the set of all outcomes and then counting the favorable cases. Instead, we shall use the principles above to get the probabilities.

(a) There are two events under consideration:

E, the event that the first number is odd, with $P(E) = \frac{1}{2}$ (half the outcomes have the first number odd, the other half have it even), and

F, the event that the second number is even, with $P(F) = \frac{1}{2}$.

The event we are interested in is $E \cup F$. We use the addition principle.
$$P(E \cup F) = P(E) + P(F) - P(E \cap F)$$
We already know $P(E)$ and $P(F)$, and we still need $P(E \cap F)$. Now $E \cap F$ is the event that the first number is odd and the second number is even. We can get the probability of this either by directly counting the number of favorable outcomes or by using the following counting procedure:

Step 1: Choose an odd number for the first die; there are 3 choices (1, 3, or 5)

Step 2: Choose an even number for the second die; there are 3 choices (2, 4, or 6).

Thus, there are $3 \times 3 = 9$ favorable outcomes, so $P(E \cap F) = \frac{9}{36} = \frac{1}{4}$. We therefore have

$$P(E \cup F) = \frac{1}{2} + \frac{1}{2} - \frac{1}{4} = \frac{3}{4}.$$

(b) We are asked for the probability of the *complement* of the event $E \cup F$. Since

$$P(E \cup F) = \frac{3}{4}$$

by part (a), it follows that

$$P((E \cup F)') = 1 - P(E \cup F)$$
$$= 1 - \frac{3}{4}$$
$$= \frac{1}{4}.$$

6.3 EXERCISES

In Exercises 1–16, use the given information to find the indicated probability.

1. $E \cap F = \emptyset$, $P(E) = 0.3$, $P(F) = 0.4$. Find $P(E \cup F)$.
2. $E \cap F = \emptyset$, $P(E) = 0.2$, $P(F) = 0.3$. Find $P(E \cup F)$.
3. $E \cap F = \emptyset$, $P(E) = 0.3$, $P(E \cup F) = 0.4$. Find $P(F)$.
4. $E \cap F = \emptyset$, $P(F) = 0.8$, $P(E \cup F) = 0.8$. Find $P(E)$.
5. $P(E) = 0.1$, $P(F) = 0.6$, $P(E \cap F) = 0.05$. Find $P(E \cup F)$.
6. $P(E) = 0.3$, $P(F) = 0.4$, $P(E \cap F) = 0.02$. Find $P(E \cup F)$.
7. $P(E \cup F) = 0.9$, $P(F) = 0.6$, $P(E \cap F) = 0.1$. Find $P(E)$.
8. $P(E \cup F) = 1.0$, $P(E) = 0.6$, $P(E \cap F) = 0.1$. Find $P(F)$.
9. $P(E) = 0.75$. Find $P(E')$.
10. $P(E) = 0.22$. Find $P(E')$.
11. E, F, and G are mutually disjoint. $P(E) = 0.3$, $P(F) = 0.4$, $P(G) = 0.3$. Find $P(E \cup F \cup G)$.
12. E, F, and G are mutually disjoint. $P(E) = 0.2$, $P(F) = 0.6$, $P(G) = 0.1$. Find $P(E \cup F \cup G)$.
13. E and F are mutually disjoint. $P(E) = 0.3$, $P(F) = 0.4$. Find $P((E \cup F)')$.
14. E and F are mutually disjoint. $P(E) = 0.4$, $P(F) = 0.4$. Find $P((E \cup F)')$.
15. $E \cup F = S$ and $E \cap F = \emptyset$. Find $P(E) + P(F)$.
16. $P(E \cup F) = 0.3$ and $P(E \cap F) = 0.1$. Find $P(E) + P(F)$.

In Exercises 17–22, use the general addition principle to find $P(E \cup F)$ in the given situation. (Assume that the coins and dice are distinguishable and fair, and that what is observed are the faces or numbers uppermost.)

17. Two coins are tossed. E is the event that the result is one tail, and F is the event that the result is one head.
18. Two coins are tossed. E is the event that the result is two heads, and F is the event that the result is one head.
19. Three coins are tossed. E is the event that the result is at least one tail, and F is the event that the result is at least one head.

20. Three coins are tossed. *E* is the event that the result is at most one tail, and *F* is the event that the result is at most one head.

21. Two dice are thrown. *E* is the event that the numbers add to 6 or 7, and *F* is the event that the numbers add to 7 or 2.

22. Two dice are thrown. *E* is the event that the numbers add to 2 or 3, and *F* is the event that the numbers add to 3 or 4.

APPLICATIONS

23. *Opinion Polls* A *New York Times*/CBS News poll of 1,368 people interviewed in March 1993 showed that 85% of all respondents favored a national law requiring a seven-day waiting period for handgun purchases, 13% opposed it, and 2% were undecided.* Considering the poll as an experiment, find the probability that a randomly chosen resident was not opposed to such a law.

24. *Opinion Polls* The opinion poll referred to in Exercise 23 also showed that 41% of the respondents would favor an outright ban on the sale of handguns (law enforcement officers excepted), 55% would oppose such a ban, and 4% were undecided. Considering the poll as an experiment, find the probability that a randomly chosen resident was not opposed to such a law.

25. *Greek Life* The TΦΦ Sorority has a tough pledging program—it requires its pledges to master the Greek alphabet forward, backward, and sideways. During the last pledge period, two thirds of the pledges failed to learn it backward and three-quarters of them failed to learn it sideways, while five of the twelve pledges failed to master it either backward or sideways. Since admission into the sisterhood requires both backward and sideways mastery, what fraction of the pledges were disqualified on this basis?

26. *Swords and Sorcery* Lance the Wizard has been informed that tomorrow there will be a 50% chance of encountering the evil Myrmidons and a 20% chance of meeting up with the dreadful Balrog. Moreover, Hugo the elf has predicted that there is a 10% chance of encountering both tomorrow. What is the probability that Lance will be lucky tomorrow and encounter neither the Myrmidons nor the Balrog?

27. *Railroad Transportation* According to a study,[†] 38% of freight in the U.S. was transported (at least part of the way) by rail in 1991. According to the same study, 53% of freight in the U.S. was transported (at least part of the way) by another form of surface transportation (boat or truck). Assuming that 80% of freight was transported by some form of surface transportation, find the probability that a unit of freight traveled by a combination of rail and another form of surface transportation in 1991.

28. *Population* In 1980, the distribution of cities in Montana with 5,000–64,999 inhabitants was as follows:[‡]

   5,000–24,999: 13 cities
   15,000–64,999: 6 cities
   5,000–64,999: 17 cities

   What percentage of these cities had a population of 15,000–24,999?

29. *Judges and Juries* A study[§] revealed that the probability of a judge acquitting a randomly chosen defendant was 0.17, while the probability that a jury would acquit a randomly chosen defendant was 0.33. Further, the probability that both a judge and a jury would have acquitted a randomly chosen defendant was 0.14. Find the probability of the following events.
   (a) A judge would not have acquitted a randomly chosen defendant.
   (b) A jury would not have acquitted a randomly chosen defendant.

---

* Source: *The New York Times*, August 15, 1993, Section 4, p. 4

[†] Source: Association of American Railroads, Eno Transportation Foundation, Railroad Facts, Conrail, reported in *The New York Times*, Aug. 30, 1993.

[‡] Source: U.S. Bureau of the Census.

[§] Source: *Parking Tickets and Missing Women: Statistics and the Law* by Hans Zeisel and Harry Kalven; *Statistics: A Guide to the Unknown*, by J. A. Tanur, et al., eds. (San Francisco: Holden-Day, Inc., 1972), pp. 102–11, as cited in *Finite Mathematics* by D. E. Zitarelli and R. F. Coughlin (Fort Worth: Saunders College Publishing, 1987).

(c) Neither the judge nor a jury would have acquitted a randomly chosen defendant.
(d) A jury would have acquitted a defendant, but a judge would not have acquitted him or her. [*Hint:* $E = (E \cap F') \cup (E \cap F)$.]

30. *Judges and Juries* Repeat Exercise 29 using the following data: the probability of a judge acquitting a randomly chosen defendant was 0.21, while the probability that a jury would acquit a randomly chosen defendant was 0.30. Further, the probability that both a judge and a jury would have acquitted a randomly chosen defendant was 0.15.

31. *Public Health* A study shows that 80% of the population has been vaccinated against the Venusian flu. Of this group, 2% get the flu anyway. If 10% of the total population get this flu, what percent of the population has been exposed to the virus either by the vaccine or by getting the disease?

32. *Public Health* A study shows that 75% of the population has been vaccinated against the Martian ague. Of this group 4% get this disease anyway. If 10% of the total population get this disease, what is the probability that a randomly selected person has not been exposed to this virus (either by the vaccine or by getting the disease)?

COMMUNICATION AND REASONING EXERCISES

33. Complete the following sentence. The probability of the union of two events is the sum of the probabilities of the two events if _____ .

34. If you know $P(E)$ and $P(F)$, what additional information would you need in order to calculate $P(E \cap F)$, and how would you calculate it?

35. Give an example of a sample space S, a probability distribution on S, and two events E and F with the property that E and F are not mutually exclusive, and yet $P(E) + P(F) = P(E \cup F)$.

36. Explain how the addition principle for mutually exclusive events follows from the general addition principle.

37. Explain how the property $P(E') = 1 - P(E)$ follows directly from the properties of a probability distribution.

38. A friend of yours asserted at lunch today that according to the weather forecast for tomorrow, there is a 52% chance of rain and a 60% chance of snow. "But that's impossible!" you blurted out, "the percentages add up to more than 100%." Explain why you were wrong.

6.4 COUNTING TECHNIQUES AND PROBABILITY

In many of the situations we have seen in the previous sections, all outcomes in the sample space were equally likely. This is the case, for example, when you toss a fair coin or roll a fair die. In such a situation, we used the following formula to calculate the probability of an event.

EQUALLY LIKELY OUTCOMES

If all the outcomes in a sample space S are equally likely, and if E is an event, then

$$P(E) = \frac{n(E)}{n(S)} = \frac{\text{Number of favorable outcomes}}{\text{Total number of outcomes}}.$$

Although this formula is simple enough, calculating $n(E)$ and $n(S)$ is not always a simple task, and this (optional) section gives some examples in which we must use the counting techniques we developed in the last chapter.

EXAMPLE 1

A bag contains four red marbles and two green ones. Upon seeing the bag, Suzy (who has compulsive marble-grabbing tendencies) sticks her hand in and grabs three at random. Find the probability that she will get both green ones.

SOLUTION According to the formula, if all the events in S are equally likely, then all we need to know is

(a) the number of elements in the sample space S, and

(b) the number of elements in the event E.

But first of all, what is the sample space? The sample space is the set of all possible outcomes, and each outcome consists of a set of three marbles (in Suzy's hand). So, the set of outcomes is the set of all sets of three marbles chosen from a total of six (four red and two green). We are assuming that the marbles are all distinguishable, so two different sets of three marbles will give two different outcomes, for example. As we commented in Example 1 of Section 1, this is a reasonable assumption. Now there are a large number of possible sets of marbles, so we are not going to try to list them all. After all, all we need to know is the *number* of outcomes in S. We saw how to calculate $n(S)$ in the preceding chapter:

$$n(S) = C(6, 3) = 20.$$

Now what about E? This is the event that she gets both green ones. We must *rephrase this as a subset of S* in order to deal with it.

>E is the collection of sets of three marbles such that one is red and two are green.

Thus, $n(E)$ is the *number* of such sets. We learned how to calculate this in the last chapter also. We pretend that we are constructing a set of three marbles with the desired specifications.

Step 1: Choose a red marble. $C(4, 1) = 4$ possible outcomes.

Step 2: Choose the two green marbles. $C(2, 2) = 1$ possible outcome.

Thus, $n(E) = 4 \times 1 = 4$.

We now get

$$P(E) = \frac{n(E)}{n(S)} = \frac{4}{20} = \frac{1}{5}.$$

Thus, there is a one-in-five chance of Suzy getting both the green ones.

6.4 Counting Techniques and Probability

▼ **EXAMPLE 2** Investment Lottery

The May 5, 1993 issue of the *Chicago Tribune* published the following information on what it termed "high-risk" four- and five-star mutual funds as a guide to investors:

Fund	1-Year Return	5-Year Return	Net Assets (millions of dollars)
20th Century Ultra Inv.	15.24%	21.87%	5,665.8
T. Rowe Internat. Stock	9.24	9.04	2,183.8
AIM Constellation	18.04	21.00	1,831.6
Berger 100	14.72	22.76	846.4
Columbia Special	14.99	18.71	549.6

You decide to select a group of 3 of these funds at random in order to give your portfolio a little "zest." Find the probabilities of the following events:

(a) Your portfolio will include only funds that showed at least a 15% 5-year return.

(b) At least two of the funds in your portfolio show assets of $1,000 million or more.

(c) Two of the funds in your portfolio show assets of $1,000 million or more, and none of them yields a single-digit 1-year return.

SOLUTION First, the sample space is the set of all collections of three funds chosen from the 5. Thus,

$$n(S) = C(5, 3) = 10.$$

(a) The event E of interest is the event that your portfolio will include only funds that showed at least a 15% 5-year return. Thus, E is the set of all groups of three funds with at least a 15% 5-year return. Looking at the chart, we find that there are four such funds (all except T. Rowe). Thus,

$$n(E) = C(4, 3) = 4.$$

We now have

$$P(E) = \frac{n(E)}{n(S)} = \frac{4}{10} = \frac{2}{5}.$$

(b) Let F be the event that at least two of the funds in your portfolio show assets of $1,000 million or more. Thus, F is the set of all groups of three funds of which at least two show assets of $1,000 million or more. To calculate $n(F)$, we pretend that we are assembling such a group of funds. We would then have two alternatives.

Alternative 1: Exactly two of the funds show assets of $1,000 million or more (there are 3 of these funds available).

Step 1: Choose two such funds: $C(3, 2) = 3$ possibilities.

Step 2: Choose one fund that does *not* show assets of at least $1,000 million: $C(2, 1) = 2$ possibilities.

This gives $3 \times 2 = 6$ possible choices for this alternative.

Alternative 2: All three of the funds show assets of $1,000 million or more.

Step 1: Choose three of them: $C(3, 3) = 1$ possibility.
This gives 1 possible choice for this alternative.

By the addition principle of counting, we have a total of $6 + 1 = 7$ possible portfolios. Thus,

$$n(F) = 7,$$

so

$$P(E) = \frac{n(E)}{n(S)} = \frac{7}{10}.$$

(c) Let G be the event that two of the funds in your portfolio show assets of $1,000 million or more, and none of them yields a single-digit 1-year return. Thus, G is the set of all such portfolios. To calculate $n(G)$, we pretend that we are constructing such a portfolio. Since none of them should yield a single-digit 1-year return, we eliminate T. Rowe, leaving us with 4 funds to choose from. We now formulate a counting procedure.

Step 1: Choose two funds showing assets of $1,000 million or more: $C(2, 2) = 1$ choices. (There are only two funds among the four we are working with that have this property, since T. Rowe has been eliminated.)

Step 2: Choose one fund not showing assets of $1,000 million or more: $C(2, 1) = 2$ choices.

This gives a total of

$$n(G) = 1 \times 2 = 2.$$

Thus,

$$P(G) = \frac{n(G)}{n(S)} = \frac{2}{10} = \frac{1}{5}.$$

6.4 Counting Techniques and Probability

▼ **EXAMPLE 3**   Poker Hands

You are dealt five cards from a well-shuffled standard deck of 52. Find the probability that you have a full house. (Recall from the counting chapter that a full house consists of three cards of one denomination and two of another.)

SOLUTION   The first thing to do is to think of S and E as sets.

S is the set of all possible 5-card hands dealt from a deck of 52.

Thus, $\quad n(S) = C(52, 5) = 2{,}598{,}960.$

If the deck is thoroughly shuffled, then each of these 5-card hands is equally likely. E is the set of all possible 5-card hands that constitute a full house. To calculate $n(E)$, we go back to the principles of counting and find

Choose first denomination.   *Choose second denomination.*
     | *Choose 3 of that denomination.* | *Choose 2 of that denomination.*
     ↓   ↓   ↓   ↓

$$n(E) = C(13, 1) \times C(4, 3) \times C(12, 1) \times C(4, 2) = 3{,}744.$$

Thus,

$$P(E) = \frac{n(E)}{n(S)} = \frac{3{,}744}{2{,}598{,}960} \approx 0.0014406.$$

In other words, there is approximately a 0.14% chance that you will be dealt a full house.

▼ **EXAMPLE 4**   Poker

You are playing solitaire poker, and you have dealt yourself the following hand:

$$J\spadesuit,\ J\diamondsuit,\ J\heartsuit,\ 2\clubsuit,\ 10\spadesuit.$$

You decide to exchange the last two cards. The exchange works as follows: The two cards are discarded (not replaced in the deck) and you deal yourself two new cards. Find each of the following:

(a) the probability that you end up with a full house.
(b) the probability that you end up with four Jacks.
(c) the probability that you end up with either a full house or four Jacks.

SOLUTION

(a) In order to get a full house, you must deal yourself two of a kind. We need to find S and E.
    S is the set of all pairs of cards selected from what remains of the original deck of 52. Remembering that you dealt five cards originally, there are $52 - 5 = 47$ cards left in the deck. Thus, $n(S) = C(47, 2) = 1{,}081.$

E is the set of all pairs of cards that constitute two of a kind. Note that you cannot get two Jacks, since there is only one left in the deck. Also, there are only three 2s and three 10s left in the deck.

Choose a denomination other than Jacks, 2s and 10s.   *Choose either 2s or 10s.*
          ↓   ↓
    Choose 2 of that denomination.   OR   *Choose 2 of that denomination.*
          ↓   ↓

$$n(E) = C(10, 1) \times C(4, 2) \quad + \quad C(2, 1) \times C(3, 2)$$
$$= 66$$

Thus,

$$P(E) = \frac{n(E)}{n(S)} = \frac{66}{1{,}081} \approx 0.0611.$$

(b) We have the same sample space as in (a). Let F be the set of all pairs of cards that include the missing Jack of clubs. So,

Choose the Jack of clubs.
          ↓
        Choose 1 card from the remaining 46.
                      ↓

$$n(F) = C(1, 1) \times C(46, 1) = 46.$$

Thus,

$$P(F) = \frac{n(F)}{n(S)} = \frac{46}{1{,}081} \approx 0.0426.$$

(c) We are asked to calculate the probability of the event $E \cup F$. From the general addition principle, we have

$$P(E \cup F) = P(E) + P(F) - P(E \cap F).$$

We already know what $P(E)$ and $P(F)$ are, but what about $P(E \cap F)$? Remembering that $E \cap F$ means "E and F," we see that $E \cap F$ is the event that the pair of cards you are dealt are two of a kind and include the Jack of clubs. But this is impossible, since there is only one Jack left. Thus, $E \cap F = \emptyset$, and so $P(E \cap F) = 0$. This gives us

$$P(E \cup F) = P(E) + P(F) = \frac{66}{1{,}081} + \frac{46}{1{,}081} = \frac{112}{1{,}081} \approx 0.1036.$$

In other words, there is slightly better than a one in ten chance that you will wind up with either a full house or four of a kind.

Before we go on... In a real game of poker there would be fewer cards left in the deck because your opponents would have been dealt cards as well. A further complication is that the exact number of cards left would depend on how many cards your opponents discarded, which in turn depends on exactly what cards they had been dealt originally.

EXAMPLE 5 Committees

The University Senate bylaws at Hofstra University* state the following:

"The University Senate Student Affairs Committee shall consist of one elected faculty Senator, one faculty Senator-at-Large, one elected Student Senator, five student Senators-at-Large (including one from the graduate school), two delegates from the Student Government Association, and the President of the Student Government Association or his/her designate. It shall be chaired by the elected student Senator on the Committee and it shall be advised by the Dean of Students or his/her designate."

You are an undergraduate student and, even though you are not an elected student Senator, you would very much like to serve on the Student Affairs Committee. The student Senators-at-Large as well as the Student Government delegates are chosen by means of a random drawing from a list of candidates. There are already 13 undergraduate candidates for the position of Senator-at-Large and 6 candidates for Student Government delegates, and you have been offered a position on the Student Government Association, should you wish to join it. (This would make you ineligible for a Senator-at-Large position.) What should you do?

SOLUTION You have two options. Option 1 is to include your name on the list of candidates for the Senator-at-Large position. Option 2 is to join the Student Government Association (SGA) and add your name to its list of candidates. Let's look at the two options separately.

Option 1: Add your name to the Senator-at-Large list. This will result in a list of 14 undergraduates for four undergraduate positions. The sample space is the set of all possible outcomes of the random drawing. Each outcome consists of a group of four lucky students chosen from 14. Thus,

$$n(S) = C(14, 4) = 1{,}001.$$

We are interested in the probability that you are among the chosen four. Thus, E is the set of groups of 4 that include you.

$$\underset{\text{Choose yourself.}}{\downarrow} \quad \underset{\text{Choose 3 from the remaining 13.}}{\downarrow}$$
$$n(E) = C(1, 1) \times C(13, 3) = 286$$

So

$$P(E) = \frac{n(E)}{n(S)} = \frac{286}{1{,}001} = \frac{2}{7} \approx 0.2857.$$

Option 2: Join the SGA and add your name to its list. This results in a list of 7 candidates from which two are selected. For this case, the sample space

▼ *As of 1993.

consists of all groups of 2 chosen from 7, so that

$$n(S) = C(7, 2) = 21,$$

and

$$n(E) = \underbrace{C(1, 1)}_{\text{Choose yourself.}} \times \underbrace{C(6, 1)}_{\text{Choose 1 from the remaining 6.}} = 6.$$

Thus,

$$P(E) = \frac{n(E)}{n(S)} = \frac{6}{21} = \frac{2}{7} \approx 0.2857.$$

In other words, the probability of being selected is exactly the same for Option 1 as it is for Option 2! Thus, you can choose either option, and you will have slightly less than a 29% chance of being selected.

▼ **EXAMPLE 6**   True/False Tests

You have neglected to study for your true/false Finance test and you decide to show up completely unprepared. When you read the test, you realize that you haven't a clue as to what the 10 questions are asking, so you decide to guess all the answers by tossing a coin. In order to get an A, you must get at least 8 correct answers, and you will fail if you get fewer than 6 correct answers. What is the probability that you will succeed in getting an A? What is the probability that you will fail?

SOLUTION   Let us be systematic as usual, and find the sample space first. Remembering that S is the set of all possible outcomes, and that an outcome consists of a sequence of ten choices of T/F, we get from the Guide to Counting that

$$n(S) = 2^{10} = 1{,}024.$$

These outcomes are all equally likely, assuming that the coin you toss is fair. Now look at the first event—the event that you get an A. This is the event that at least eight of your answers are correct. Thus, E is the event that either 8, 9, or 10 of the answers are the correct ones. To find $n(E)$, pretend as usual that you are *constructing* a sequence of answers. There are three alternatives:

Alternative 1:   Choose exactly 8 correct answers: $C(10, 8) = 45.$
Alternative 2:   Choose exactly 9 correct answers: $C(10, 9) = 10.$
Alternative 3:   Choose exactly 10 correct answers: $C(10, 10) = 1.$

Thus, there are $45 + 10 + 1 = 56$ possible A answer sheets, so that $n(E) = 56$. Thus,

$$P(E) = \frac{n(E)}{n(S)} = \frac{56}{1{,}024} \approx 0.05469.$$

In other words, you have a 5.5% chance of getting an A.

Turning to the probability that you will fail, note that there are more alternatives for a failing paper (any score from 0 through 5) than there are for a passing one (any score from 6 through 10). Thus, if F is the event that you fail, it is easier to start with F', the event that you pass.

Alternative 1:   Choose exactly 6 correct answers: $C(10, 6) = 210$.
Alternative 2:   Choose exactly 7 correct answers: $C(10, 7) = 120$.
Alternative 3:   Choose exactly 8 correct answers: $C(10, 8) = 45$.
Alternative 4:   Choose exactly 9 correct answers: $C(10, 9) = 10$.
Alternative 5:   Choose exactly 10 correct answers: $C(10, 10) = 1$.

Thus, $n(F') = 210 + 120 + 45 + 10 + 1 = 386$, and

$$P(F') = \frac{n(F')}{n(S)} = \frac{386}{1{,}024}.$$

Since we are interested in the probability that you *don't* pass, we really want

$$P(F) = 1 - P(F') = 1 - \frac{386}{1{,}024} = \frac{638}{1{,}024} \approx 0.6230.$$

Thus, you have approximately a 62% chance of failing the test.

6.4 EXERCISES

Exercises 1–10 concern Suzy the marble-grabber. Recall from Example 1 that whenever she sees a bag of marbles, she grabs a handful at random. In these exercises, she has seen a bag containing 4 red marbles, 3 green ones, 2 white ones and 1 purple one. She grabs five of them. Find the probabilities of the following events, expressing each as a fraction in lowest terms.

1. She has all the red ones.
2. She has none of the red ones.
3. She has at least one white one.
4. She has at least one green one.
5. She has two red ones and one of each of the other colors.
6. She has two green ones and one of each of the other colors.
7. She has at most one green one.
8. She has no more than one white one.
9. She does not have all the red ones.
10. She does not have all the green ones.

APPLICATIONS

Investments Exercises 11–16 are based on the following table, which shows a selection of companies in the Chicago Tribune Top 100 together with their earnings per share and percentage growth in income for the fourth quarter ended December 31, 1992.*

Company	Earning per Share	Growth Rate
Motorola Inc.	$0.68	44%
McDonald's Corp.	0.61	14
Tootsie Roll Industries, Inc.	0.65	31
Abbott Laboratories	0.42	11
Sears, Roebuck & Co.	1.35	16
People's Energy Corp.	0.89	1

11. Assuming that you selected three of these companies' stocks at random, what is the probability that all the stocks in your selection posted quarterly earnings of at least 60¢ per share?

12. Assuming that you selected four of these companies' stocks at random, what is the probability that all the stocks in your selection showed double-digit growth rates?

13. If you selected four of these stocks at random, find the probability that they included *Sears, Roebuck & Co.*, but not *Abbott Labs*.

14. If you selected all but one of these stocks at random, what is the probability that your selection included the company with the largest growth rate and excluded the company showing the smallest growth rate?

15. If your portfolio included 100 shares of *Motorola* and you then purchased 100 shares each of any two companies on the list at random, find the probability that you have 200 shares of Motorola.

16. If your portfolio included 100 shares of Motorola and you then purchased 100 shares each of any three companies on the list at random, find the probability that you have 200 shares of Motorola.

Poker Hands In Exercises 17–22, you are asked to calculate the probability of being dealt various poker hands. (Recall that a poker player is dealt 5 cards at random from a standard pack of 52.) Express each of your answers as a decimal rounded to four decimal places. (On the other hand, if, for example, the answer is 0.000012, then round to 0.00001.)

17. **Two of a Kind:** two cards with the same denomination and three cards with other denominations (different from each other and that of the pair)
Example: K♣, K♥, 2♠, 4♦, J♠

18. **Three of a Kind:** three cards with the same denomination and two cards with other denominations (different from each other and that of the three)
Example: Q♣, Q♥, Q♠, 4♦, J♠

19. **Two Pair:** two cards with one denomination, two with another, and one with a third.
Example: 3♣, 3♥, Q♠, Q♥, 10♠

20. **Straight Flush:** five cards of the same suit with consecutive denominations.
Examples: A♣, 2♣, 3♣, 4♣, 5♣ and 9♦, 10♦, J♦, Q♦, K♦, but *not* 10♦, J♦, Q♦, K♦, A♦ (The last is called a royal flush.)

21. **Flush:** five cards of the same suit, but not a straight flush or royal flush.
Example: A♣, 5♣, 7♣, 8♣, K♣

22. **Straight:** five cards with consecutive denominations, but not all of the same suit.
Examples: 9♦, 10♦, J♣, Q♥, K♦, and 10♥, J♦, Q♦, K♦, A♦.

23. *Lotto* The Sorry State Lottery requires you to select five different numbers from zero through 49. (Order is not important.) You are a Big Winner if the five numbers you select agree with those in the drawing, and you are a Small-Fry Winner if four of your five num-

---

▼ *Source: *Chicago Tribune*, Feb. 22, 1993, Section 4, p. 4.

bers agree with those in the drawing. What is the probability of each outcome? What is the probability that you win something?

24. *Lotto* The Sad State Lottery requires you to select a sequence of three different numbers from zero through 49. (Order is important.) You are a winner if your sequence agrees with that in the drawing, and you are a booby prize winner if your selection of numbers is correct, but in the wrong order. What is the probability of each outcome? What is the probability that you are either a winner or a booby prize winner?

25. *Transfers* Your company is considering offering 400 employees the opportunity to transfer to its new headquarters in Ottawa and, as personnel manager, you decide that it would be fairest if the transfer offers were decided by means of a lottery. Assuming that your company currently employs 100 managers, 100 factory workers and 500 miscellaneous staff, find the following probabilities, leaving the answers as formulas:
(a) All the managers will be offered the opportunity.
(b) You will be offered the opportunity.

26. *Transfers* (See Exercise 25.) After thinking about your proposed method of selecting employees for the opportunity to move to Ottawa, you decide it might be a better idea to select 50 managers, 50 factory workers, and 300 miscellaneous staff, all chosen at random. Find the probability that you will be offered the opportunity. (Leave the answer as a formula.)

27. *Lotteries* In a New York State daily lottery game, three (not necessarily different) digits in the range 0–9 are selected at random. Find the probability that all three are different.

28. *Lotteries* In Exercise 27, find the probability that two of the three digits are the same.

29. *Sports* The following table shows the results of the Big Eight Conference for the 1988 college football season.*

Team	Won	Lost
Nebraska (NU)	7	0
Oklahoma (OU)	6	1
Oklahoma State (OSU)	5	2
Colorado (CU)	4	3
Iowa State (ISU)	3	4
Missouri (MU)	2	5
Kansas (KU)	1	6
Kansas State (KSU)	0	7

This is referred to as a "perfect progression." Making the unreasonable assumption that the "Won" scores were chosen at random in the range 0–7, find the probability of a perfect progression in a Big Eight Conference.[†] Leave your answer as a formula.

30. *Sports* Referring to Exercise 29, find the probability of a perfect progression with Nebraska scoring 7 wins and 0 losses. Leave your answer as a formula.

31. *The Monkey at the Typewriter* Suppose that a monkey is seated at a computer keyboard and randomly strikes the 26 letter keys and the space bar. Find the probability that its first 39 characters (including spaces) will be "to be or not to be that is the question." (Leave the answer as a formula.)

32. *The Cat at the Piano* A standard piano keyboard consists of 88 different keys. Find the probability that a cat, jumping on four keys at random (possibly with repetition), will strike the first four notes of Beethoven's Fifth Symphony. (Leave the answer as a formula.)

33. *Contests (based on a question from the GMAT)* Tyler and Gebriella are among 7 contestants from

---

▼ *Source: "On the Probability of a Perfect Progression," *The American Statistician,* August 1991, vol. 45, no. 3, p. 214.

[†] Even if all the teams are equally likely to win each game, the chances of a perfect progression actually coming up are a little more difficult to estimate, since the number of wins by one team directly affects the number of wins by the others. For instance, it is impossible for all eight teams to show a score of 7 wins and 0 losses at the end of the season—someone must lose! It is, however, not too hard to come up with a counting argument to estimate the total number of win-lose scores actually possible.

which 4 semifinalists are to be selected at random. Find the probability that neither Tyler nor Gebriella is selected.

34. *Contests (based on a question from the GMAT)* Tyler and Gebriella are among 7 contestants from which 4 semifinalists are to be selected at random. Find the probability that Tyler, but not Gebriella, is selected.

35. *Graph Searching* A graph consists of a collection of **nodes** (the heavy dots in the figure) connected by **edges** (line segments from one node to another). A **move on a graph** is a move from one node to another along a single edge. Find the probability of going from Start to Finish in a sequence of two random moves in the graph shown. (*Note:* You may retrace an edge.)

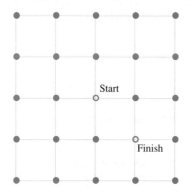

36. *Graph Searching* Referring to Exercise 35, find the probability of going from Start to one of the Finish nodes in a sequence of two random moves in the graph shown in the following figure.

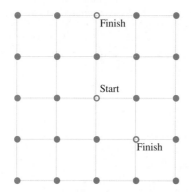

37. *Multiple-Choice Tests* A test has three parts. Part A consists of 8 true/false questions, Part B consists of 5 multiple-choice questions with 5 choices each, and Part C requires you to match 5 questions with 5 different answers. Assuming that you make random guesses in filling out your answer sheet, what is the probability that you will earn 100% on the test? (Express the answer in terms of a formula.)

38. *Multiple-Choice Tests* A test has three parts. Part A consists of 4 true/false questions, Part B consists of 4 multiple-choice questions with 5 choices each, and Part C requires you to match 6 questions with 6 different answers. Assuming that you make random choices in filling out your answer sheet, what is the probability that you will earn 100% on the test? (Express the answer in terms of a formula.)

39. *Tournaments* What is the probability that North Carolina will beat Central Connecticut but lose to Virginia in the following (fictitious) soccer tournament? (Assume that all outcomes are equally likely.)

40. *Tournaments* In a (fictitious) soccer tournament involving the four teams San Diego State, De Paul, Colgate, and Hofstra, find the probability that Hofstra will play Colgate in the finals and win. (Assume that all outcomes are equally likely and that the teams not listed in the first-round slots are placed at random.)

41. *Product Design* Your company has patented an electronic digital padlock that a user can program with his or her own four-digit code. (Each digit can be 0 through 9.) It is designed to open if either the correct code is keyed in, or—and this is helpful for forgetful people—if exactly one of the digits is incorrect. What is the probability that a randomly chosen sequence of 4 digits will open a programmed padlock?

42. *Product Design* Assume that you already know the first digit of the combination for the lock described in Exercise 41. Find the probability that a random guess of the remaining three digits will open the lock.

43. *Quality Control* A manufacturer of lightbulbs tests the quality of its bulbs by choosing at random and testing 5 bulbs from each crate of 20. If any bulb is defective, the crate is rejected. If 3 bulbs in a crate are defective, what is the probability that the manufacturer will reject the crate?

44. *Quality Control* A manufacturer of rocket engines tests the quality of its rockets by choosing at random and testing 4 engines from each crate of 12. If any rocket is defective, the crate is rejected. If 2 rockets in a crate are defective, what is the probability that the manufacturer will reject the crate?

45. *Committees* An investigatory Committee in the Kingdom of Utopia consists of:
   A Chief Investigator (a Royal Party member)
   An Assistant Investigator (a Birthday Party member)
   2 At-Large Investigators (either party)
   5 Ordinary Members (either party).
Royal Party member Larry Sifford is hoping to avoid serving on the committee unless he is the Chief Investigator and Otis Taylor, a Birthday Party member, is the Assistant Investigator. The committee is to be selected at random from a pool of 12 candidates, including Larry Sifford and Otis Taylor, half of whom are Royal Party and half of whom are Birthday Party.
   (a) How many different committees are possible?
   (b) How many committees are possible in which Larry's hopes are fulfilled? (This includes the possibility that he's not on the committee at all.)
   (c) What is the probability that he'll be happy with a randomly selected committee?

46. *Committees* A committee is to consist of a Chair, three Hagglers, and four Do-Nothings. The committee is formed by choosing randomly from a pool of 10 people and assigning them to the various "jobs."
   (a) How many different committees are possible?
   (b) Norm is eager to be the Chair of the committee. What is the probability that he will get his wish?
   (c) Norm's girlfriend Norma is less ambitious and would be happy to hold any position on the committee provided Norm is also selected as a committee member. What is the probability that she will get her wish?
   (d) Norma does not get along with Oona (who is also in the pool of prospective members) and would be most unhappy if Oona were to chair the committee. Find the probability that all her wishes will be fulfilled: she and Norm are on the committee and it is not chaired by Oona.

6.5 CONDITIONAL PROBABILITY AND INDEPENDENCE

CONDITIONAL PROBABILITY

Cyber Video Games, Inc., has been running a television ad for its latest game, "Ultimate Hockey." As Cyber Video's director of marketing, you would like to assess the ad's effectiveness, so you ask your market research team to make a survey of video game players. The results of their survey of 50,000 video game players are summarized in the following chart.

	Saw Ad	Did Not See Ad
Purchased Game	1,200	2,000
Did Not Purchase Game	3,800	43,000

The market research team concludes in their report that the ad campaign is highly effective.

Q But wait! How could the campaign possibly have been effective? Only 1,200 people who saw the ad purchased the game, while 2,000 people purchased the game without seeing the ad! It looks as though potential customers are being *put off* by the ad.

A Let us analyze these figures a little more carefully. First, we can look at the event E that a randomly chosen video game player purchased Ultimate Hockey. In the "Purchased Game" row we see that a total of 3,200 people purchased the game. Thus, the experimental probability of E is

$$P(E) = \frac{fr(E)}{N} = \frac{3{,}200}{50{,}000} = 0.064.$$

To test the effectiveness of the television ad, let's compare this figure with the experimental probability that *a video game player who saw the ad purchased Ultimate Hockey*. This means that we restrict attention to the "Saw Ad" column. This is the following fraction:

$$\frac{\text{Number of people who saw the ad and purchased the game}}{\text{Total number of people who saw the ad}} = \frac{1{,}200}{5{,}000}$$
$$= 0.24.$$

In other words, 24% of those surveyed who saw the ad bought Ultimate Hockey, while overall, only 6.4% of those surveyed bought it. Thus, it appears that the ad campaign *was* highly successful.

Let us introduce some terminology. In this example there were two related events of importance,

E, the event that a video game player purchased Ultimate Hockey, and

F, the event that a video game player saw the ad.

The two probabilities we compared were the experimental probability $P(E)$ and the experimental probability that a video game player purchased Ultimate Hockey *given that* he or she saw the ad. We call the latter probability the (experimental) **probability of E, given F,** and we write it as $P(E \mid F)$. We call $P(E \mid F)$ a **conditional probability**—it is the probability of E under the condition that F occurred.

Q How do we calculate conditional probabilities?

A In the example above we used the ratio

$$P(E \mid F) = \frac{\text{Number of people who saw the ad and bought the game}}{\text{Total number of people who saw the ad}}$$
$$= \frac{\text{Number of favorable outcomes in } F}{\text{Total number of outcomes in } F}.$$

The numerator is the frequency of $E \cap F$, while the denominator is the frequency of F. Thus, we can say the following.

6.5 Conditional Probability and Independence

CONDITIONAL EXPERIMENTAL PROBABILITY
If E and F are events and P is the experimental probability, then
$$P(E \mid F) = \frac{fr(E \cap F)}{fr(F)}.$$

EXAMPLE 1

If you throw a fair die twice and observe the numbers uppermost, find the probability that the sum of the numbers is eight, given that the first number is a 4.

SOLUTION We begin by recalling that the sample space when we throw a fair die twice is the set $S = \{(1, 1), (1, 2), \ldots, (6, 6)\}$ containing the 36 different equally likely outcomes.

Consider two events,

E, the event that the sum of the numbers is 8, and

F, the event that the first number is a 4.

We shall again write $P(E \mid F)$ for the probability that E occurred, given that F occurred. Since we are *given* that F occurred, we may as well *reduce the sample space* to $F = \{(4, 1), (4, 2), \ldots, (4, 6)\}$. Then we are interested in the event that the sum of the numbers is 8 and the first number is a 4, that is, the event $E \cap F = \{(4, 4)\}$. Since all the outcomes in F are still equally likely, we can say that

$$P(E \mid F) = \frac{\text{Number of favorable outcomes (in } F)}{\text{Total number of outcomes (in } F)} = \frac{n(E \cap F)}{n(F)} = \frac{1}{6}.$$

The analysis in Example 1 works whenever the outcomes are all equally likely. By reducing the sample space from S to F we see that $P(E \mid F)$ is the probability of $E \cap F$ when the sample space is taken to be F.

CONDITIONAL PROBABILITY FOR EQUALLY LIKELY OUTCOMES
If all the outcomes in S are equally likely, then
$$P(E \mid F) = \frac{n(E \cap F)}{n(F)}.$$

We can write this formula in another way:

$$P(E \mid F) = \frac{n(E \cap F)}{n(F)} = \frac{n(E \cap F)/n(S)}{n(F)/n(S)} = \frac{P(E \cap F)}{P(F)}.$$

We use this last formula as our general definition of conditional probability.

> **CONDITIONAL PROBABILITY**
>
> If E and F are events, then the **conditional probability of E given F** is defined by
>
> $$P(E|F) = \frac{P(E \cap F)}{P(F)}.$$

▶ **NOTES**

(1) Remember that in the expression $P(E|F)$, E is the event you want to find the probability of, given that you know the event F has occurred.

(2) From the formula, notice that $P(E|F)$ cannot be defined if $P(F) = 0$. Think about whether $P(E|F)$ could make any sense if the event F were impossible. ◀

▼ **EXAMPLE 2** Job Opportunities

Of Suburban College's 4,000 students, 2,500 receive some form of financial aid, 2,200 work part-time, and 1,400 work part-time and also receive financial aid. What is the probability that a randomly chosen student who works part-time also receives financial aid?

SOLUTION Here, the experiment consists of selecting a student at random from among those at Suburban College. We rephrase the question to give us the standard wording, "Find the probability that _____ given that _____."

Find the probability that a student receives financial aid, given that the student works part-time.

This is $P(E|F)$ if we take E to be the event that a student receives financial aid and F to be the event that a student works part-time. Then $E \cap F$ is the event that a student receives financial aid and works part-time. Since all students are equally likely to be chosen in our experiment, we compute

$$P(E|F) = \frac{n(E \cap F)}{n(F)}$$

$$= \frac{1{,}400}{2{,}200} \approx 0.6364.$$

In other words, approximately 63.64% of all students who work part-time also receive financial aid.

Before we go on... Compare $P(E|F)$ to

$$P(E) = \frac{n(E)}{n(S)} = \frac{2{,}500}{4{,}000} = 0.625.$$

A student working part-time is slightly more likely to be receiving financial aid than a student chosen at random.

EXAMPLE 3   Mergers and Acquisitions

You have invested in Mini Co. stocks, as you suspect that Gigantic Conglomerate is about to acquire Mini Co. in a hostile takeover. There is a 90% chance that Gigantic Conglomerate will acquire Mini if Mini shows a profit in next month's financial statement, and there is an 80% chance that Mini will show a profit. What is the probability that Mini will show a profit and be acquired by Gigantic Conglomerate?

SOLUTION   The events in question are as follows.

E = the event that Gigantic Conglomerate will acquire Mini Co.

F = the event that Mini Co. will show a profit.

We are given that $P(F) = 0.8$. We also know that the probability that Gigantic Conglomerate will acquire Mini *given that* Mini shows a profit is 0.9. That is, we are told that $P(E|F) = 0.9$ and are asked to find the probability of E and F, or $E \cap F$. The definition

$$P(E|F) = \frac{P(E \cap F)}{P(F)}$$

can be used to find $P(E \cap F)$ if we solve for $P(E \cap F)$.

$$P(E \cap F) = P(F)P(E|F)$$
$$= (0.8)(0.9) = 0.72$$

Thus, there is a 72% chance that Mini will show a profit and be acquired by Gigantic Conglomerate.

THE MULTIPLICATION PRINCIPLE AND TREES

In the last example, we saw that the formula

$$P(E|F) = \frac{P(E \cap F)}{P(F)}$$

could be used to calculate $P(E \cap F)$ if we rewrite the formula in the following form, known as the **multiplication principle**.

MULTIPLICATION PRINCIPLE
If E and F are events, then
$$P(E \cap F) = P(F)P(E|F).$$

To illustrate this principle, let us return to the scenario with which we began this section: Cyber Video Games, Inc., and its television ad campaign. Cyber's marketing survey was concerned with the following events.

E, the event that a video game player purchased Ultimate Hockey, and

F, the event that a video game player saw the ad.

We can illustrate the various possibilities by means of a two-stage tree (Figure 1).

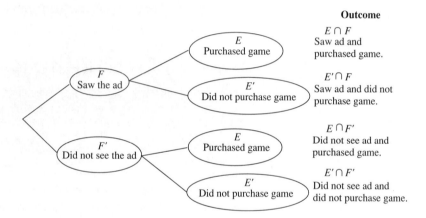

FIGURE 1

Consider the outcome $E \cap F$. To get there from the starting position on the left, you must first travel up to the F node. (In other words, F must occur.) Then you must travel up the branch from the F node to the E node. We are now going to associate a probability with each branch of the tree: the probability of traveling up that branch *given that you have gotten to the node at the beginning of the branch*. For instance, the probability of traveling up the branch from the starting position to the F node is $P(F) = 5{,}000/50{,}000 = 0.1$. (We obtained this from the data given in the survey.) The probability of going up the branch from the F node to the E node is the probability that E occurs, given that F has occurred. In other words, it is the *conditional* probability $P(E|F) = 0.24$. The probability of the outcome $E \cap F$ can then be computed using the multiplication principle:

$$P(E \cap F) = P(F)P(E|F) = (0.1)(0.24) = 0.024.$$

6.5 Conditional Probability and Independence

In other words, *to obtain the probability of the outcome $E \cap F$, we multiply the probabilities on the branches leading to that outcome* (Figure 2).

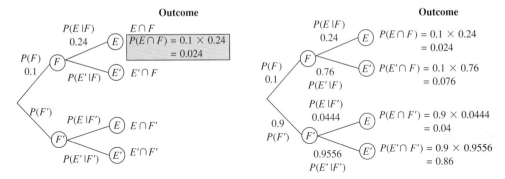

FIGURE 2

FIGURE 3

The same argument holds for the remaining three outcomes, and we can use the table given at the beginning of this section to calculate all the conditional probabilities shown in Figure 3.

▶ **NOTE** Notice that the sum of the probabilities on the branches leaving any node is always 1 (why?). This observation often speeds things up, since once we have labeled one branch we can easily label the other. ◀

EXAMPLE 4

An experiment consists of tossing two coins. The first coin is fair, and the second coin is twice as likely to land with heads facing up as it is with tails facing up. Draw a tree diagram to illustrate all the possible outcomes, and use the multiplication principle to compute the probabilities of all the outcomes.

SOLUTION A quick calculation shows that the probability distribution for the second coin is $P(H) = \frac{2}{3}$ and $P(T) = \frac{1}{3}$. (How did we get that?) Figure 4 shows the tree and the calculations of the probabilities of the outcomes.

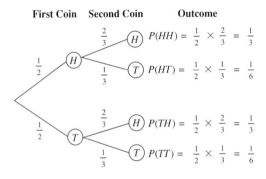

FIGURE 4

INDEPENDENCE

Let us go back once again to Cyber Video Games, Inc., and its ad campaign. We would like to assess the ad's effectiveness. As before, we consider

E, the event that a video game player purchased Ultimate Hockey, and

F, the event that a video game player saw the ad.

As we saw, we could use survey data to calculate

$P(E)$, the probability that a video game player purchased Ultimate Hockey, and

$P(E|F)$, the probability that a video game player *who saw the ad* purchased Ultimate Hockey.

When these probabilities are compared, one of three things can happen.

Case 1, $P(E|F) > P(E)$: This is what the survey data actually showed: a video game player was more likely to purchase Ultimate Hockey if he or she saw the ad. This indicates that the ad is effective—seeing the ad had a positive effect on a player's decision to purchase the game.

Case 2, $P(E|F) < P(E)$: If this had happened, then a video game player would be *less* likely to purchase Ultimate Hockey if he or she were to see the ad. This would indicate that the ad has "backfired": it has, for some reason, put potential customers off. In this case, just as in the first case, the event F has an effect—a negative one—on the event E.

Case 3, $P(E|F) = P(E)$: In this case, seeing the ad would have had absolutely no effect on a potential customer's buying Ultimate Hockey. Put another way, the event F would have had no effect at all on the event E. We would say that the events E and F are **independent.**

In general, we say that two events E and F are independent if $P(E|F) = P(E)$. When this happens, we have

$$P(E) = P(E|F) = \frac{P(E \cap F)}{P(F)},$$

so

$$P(E \cap F) = P(E)P(F).$$

Conversely, if $P(E \cap F) = P(E)P(F)$, then, assuming $P(F) \neq 0$,* $P(E) = P(E \cap F)/P(F) = P(E|F)$.

▼ *We shall discuss the independence of two events only in cases where their probabilities are both nonzero.

6.5 Conditional Probability and Independence

> **INDEPENDENT EVENTS**
>
> The events E and F are **independent** if
> $$P(E|F) = P(E)$$
> or, equivalently,
> $$P(E \cap F) = P(E)P(F).$$
> If two events E and F are not independent, then they are **dependent.**

▶ **NOTES**

(1) The formula $P(E \cap F) = P(E)P(F)$ also says that $P(F|E) = P(F)$. Thus, if F has no effect on E, then likewise E has no effect on F.

(2) Sometimes it is obviously the case that two events, by their nature, are independent. For example, the event that a die you roll comes up 1 is clearly independent of whether or not a coin you toss comes up heads. In some cases, though, we need to check for independence by comparing $P(E \cap F)$ to $P(E)P(F)$. If they are equal, then E and F are independent, but if they are unequal, then E and F are dependent. ◀

EXAMPLE 5

You throw two fair dice, one green and one red, and observe the numbers uppermost. Which of the following pairs of events are independent?

(a) E, the event that the sum is 7, and F, the event that the green die shows a 2 or a 3.

(b) E, the event that the green die is even, and F, the event that both dice have the same parity.*

(c) E, the event that the dice have the same parity, and F, the event that the sum of the numbers is 6.

SOLUTION We check in each case whether $P(E \cap F) = P(E)P(F)$.

(a) Since $n(E) = 6$,
$$P(E) = \frac{n(E)}{n(S)} = \frac{6}{36} = \frac{1}{6}.$$

Since there are 12 outcomes in which the green die shows either 2 or 3,
$$P(F) = \frac{12}{36} = \frac{1}{3}.$$

▼ * Two numbers have the **same parity** if both are even or both are odd. Otherwise, they have **opposite parity.**

Since $E \cap F$ is the event that the sum is 7 and the green die is either 2 or 3, $E \cap F = \{(2, 5), (3, 4)\}$ and $n(E \cap F) = 2$. Thus,

$$P(E \cap F) = \frac{n(E \cap F)}{n(S)} = \frac{2}{36} = \frac{1}{18}.$$

Now

$$P(E)P(F) = \left(\frac{1}{6}\right)\left(\frac{1}{3}\right) = \frac{1}{18} = P(E \cap F),$$

so the events E and F are independent.

Another way of seeing the independence of these two events is to notice that no matter what number the green die shows, the chance of getting a total of 7 is 1 in 6 (why?).

(b) Here, since the green die is even for half the outcomes, $n(E) = 18$, so

$$P(E) = \frac{n(E)}{n(S)} = \frac{18}{36} = \frac{1}{2}.$$

Since the dice have the same parity in half the outcomes,

$$P(F) = \frac{1}{2}.$$

$E \cap F$ is the event that both dice are even, and this is true for 9 of the outcomes, so

$$P(E \cap F) = \frac{n(E \cap F)}{n(S)} = \frac{9}{36} = \frac{1}{4}.$$

Since $P(E)P(F) = P(E \cap F)$ once again, we conclude that the events are independent.

(c) The event E is the same as the event F in part (b), so

$$P(E) = \frac{1}{2}.$$

Since there are 5 outcomes with a sum of 6, $n(F) = 5$, and

$$P(F) = \frac{5}{36}.$$

$E \cap F$ is the event that both dice have the same parity and add to 6. Since, in fact, $E \cap F = F$,

$$P(E \cap F) = \frac{5}{36}.$$

Now $P(E)P(F) \neq P(E \cap F)$, and we conclude that the events are dependent. In fact, if F occurs then E *must* also occur (why?).

6.5 Conditional Probability and Independence

▼ **EXAMPLE 6** Weather Prediction

According to the weather service, there is a 50% chance of rain in New York and a 30% chance of rain in Honolulu. Assuming that New York's weather is independent of Honolulu's, find the probability that it will rain in at least one of these cities.

SOLUTION We take E to be the event that it will rain in New York and F to be the event that it will rain in Honolulu. We are asked to find the probability of $E \cup F$, the event that it will rain in at least one of the two cities. We use the addition principle,

$$P(E \cup F) = P(E) + P(F) - P(E \cap F).$$

We know that $P(E) = 0.5$ and $P(F) = 0.3$. But what about $P(E \cap F)$? Since the events E and F are independent, we can compute

$$P(E \cap F) = P(E)P(F)$$
$$= (0.5)(0.3) = 0.15.$$

Thus,

$$P(E \cup F) = P(E) + P(F) - P(E \cap F)$$
$$= 0.5 + 0.3 - 0.15$$
$$= 0.65.$$

In other words, there is a 65% chance that it will rain either in New York or in Honolulu.

Before we go on... Another way to solve this example is to first calculate $P((E \cup F)')$, the probability that it will *not* rain in either city. This is the same as $P(E' \cap F')$, the probability that it will not rain in New York and it will not rain in Honolulu. Since the weather in New York is independent of the weather in Honolulu, we can calculate

$$P(E' \cap F') = P(E')P(F')$$
$$= [1 - P(E)][1 - P(F)]$$
$$= (0.5)(0.7)$$
$$= 0.35.$$

Therefore, $P(E \cup F) = 1 - P(E' \cap F') = 0.65$ again.

The property $P(E \cap F) = P(E)P(F)$ can be extended to three or more independent events: if, for example, E, F, and G are three mutually independent events (that is, each of them is independent of each of the other two) then, among other things,

$$P(E \cap F \cap G) = P(E)P(F)P(G).$$

17. She gets no more than one of any color, given that she gets 2 red ones.

18. She gets at least one green one, given that she gets no red, yellow, or orange ones.

19. She gets no more than one of any color and not the fluorescent pink one, given that she gets one yellow and one orange one.

20. She gets two red ones and two green ones, given that she gets at least one green one.

APPLICATIONS

Pollution Cleanup Exercises 21–24 are based on the pie chart in the figure, which gives a probability distribution for the possible end results of a tax dollar spent on the Environmental Protection Agency Superfund.* (A total of $7.1 billion was spent as of June 1993). Round all answers to three decimal places.

Possible Uses of Tax Dollars by Superfund

Source: U.S. Environmental Protection Agency, Associated Press. June, 1993.

21. Find the probability that one tax dollar spent on the Superfund will be recovered, given that it is recoverable.

22. Find the probability that one tax dollar spent on the Superfund will be written off, given that it will not be recovered.

23. What percentage of Superfund tax dollars that were neither recovered nor sought in litigation were ultimately written off?

24. What percentage of Superfund tax dollars not sought in litigation were either yet to be pursued or written off?

Student Loans Exercises 25–30 are based on the following table, which shows the number of people who were due to begin paying federal student loans in 1992, and the subsequent default rate.[†] Round all answers to 4 decimal places.

	Due to Begin Payment in 1992	Defaulted 1992–1993
4-Year Colleges	1,113,079	74,945
2-Year Colleges	254,623	36,918
Trade Schools	621,412	187,739

25. Find the probability that a student loan due for repayment in 1992 was given to a student in a 2-year college.

26. Find the probability that a student loan due for repayment in 1992 was given to a student in a trade school.

27. Find the probability that a student loan at a 4-year college due for repayment in 1992 was defaulted in 1992–1993.

28. Find the probability that a student loan at a 2-year college due for repayment in 1992 was defaulted in 1992–1993.

29. Find the probability that a defaulted loan was given to a student in a trade school.

30. Find the probability that a defaulted loan was given to a student in a 4-year college.

Education and Occupation Exercise 31–40 are based on the following tables, which show the breakdown of U.S. occupations by sex and educational attainment for the year 1989.[‡] All numbers are in thousands of civilians aged 25 years and over. Round all answers to 4 decimal places.

▼ *Source: U.S. Department of Education/*Long Island Newsday*, Sept. 2, 1994, p. A17.

[†] Source: U.S. Environmental Protection Agency, Associated Press, June, 1993.

[‡] Source: *Statistical Abstract of the United States, 1991* (111th Ed.), U.S. Department of Commerce, Economics and Statistics Administration, and Bureau of the Census.

Males

	Total[a]	Managerial/ Professional	Technical/Sales/ Administrative	Service	Precision Production	Operators/ Fabricators
Less Than 4 Years High School	8,307	476	643	1,022	2,268	3,074
4 Years of High School Only	19,722	2,393	3,469	1,793	5,788	5,348
1 to 3 Years of College	10,569	2,748	2,928	946	2,200	1,443
4 or More Years of College	15,441	10,160	3,443	459	735	462
Totals	54,039	15,777	10,483	4,220	10,991	10,327

[a] Includes occupations not listed.

Females

	Total[a]	Managerial/ Professional	Technical/Sales/ Administrative	Service	Precision Production	Operators/ Fabricators
Less Than 4 Years High School	5,137	260	1,084	2,016	258	1,398
4 Years of High School Only	18,499	2,426	9,511	3,596	569	2,126
1 to 3 Years of College	9,422	2,690	5,075	1,080	156	353
4 or More Years of College	10,574	7,210	2,756	377	71	105
Totals	43,632	12,586	18,426	7,069	1,054	3,982

[a] Includes occupations not listed.

31. Find the probability that a male was employed in a managerial or professional occupation, given that he had completed at least 4 years of college.

32. Find the probability that a female was employed in a managerial or professional occupation, given that she had completed at least 4 years of college.

33. Find the probability that a male who had not completed 4 or more years of college was employed in a managerial or professional occupation.

34. Find the probability that a female who had not completed 4 or more years of college was employed in a managerial or professional occupation.

35. Find the probability that a male employed in the service industry had completed 4 or more years of college.

36. Find the probability that a female employed in the service industry had completed 4 or more years of college.

37. Find the probability that a male who had attained 4 years of high school only, and was not in the managerial or professional category, was employed as an operator or fabricator.

38. Find the probability that a female who had attained 4 years of high school only, and was not in the managerial or professional category, was employed as an operator or fabricator.

39. Your friend claims that women with less than four years of high school education are more likely to be employed in the service industry than men. Respond to this claim by citing actual probabilities.

40. Your friend claims that women with four or more years of college education are more likely to be employed in the service industry than men. Respond to this claim by citing actual probabilities.

Card Hands Endora, the red headed Queen of Atlantis, summoned her handsome suitors before her, and addressed them as follows: "Here, my bold suitors, is a pack of fifty-two standard cards. You are each to select five at random from the deck. If the outcome includes exactly three hearts and one spade, and includes the Queen of Hearts but excludes the five of spades, you will win my hand in marriage. If, on the other hand, the outcome includes exactly two spades and a heart, you will be banished from my presence for eternity!" In Exercises 41–44, find the probability of each of the stated outcomes. Express each answer as a formula involving combinations without evaluating.

41. A suitor will win her hand in marriage, assuming he is not banished.

42. A suitor will be banished from her presence, assuming he does not win her hand in marriage.

43. A suitor will earn her hand in marriage, assuming that his hand includes 5♦ 5♥ J♠.

44. A suitor whose hand includes exactly two spades will be banished.

In Exercises 45–50, set up a probability tree and use the multiplication principle to calculate the probabilities of all the outcomes.

45. *Sales* Each day, there is a 40% chance that you will sell an automobile. Thirty percent of all the automobiles you sell are two-door models, and the rest are 4-door models.

46. *Product Reliability* You purchase Brand X floppy disks one quarter of the time and Brand Y floppies the rest of the time. Brand X floppy disks have a 1% failure rate, and Brand Y floppy disks have a 3% failure rate.

47. *Car Rentals* Your auto rental company rents out 30 small cars, 24 luxury sedans, and 46 slightly damaged "budget" vehicles. The small cars break down 14% of the time, the luxury sedans break down 8% of the time, and the "budget" cars break down 40% of the time.

48. *Travel* It appears that there is only a one in five chance that you will be able to take your spring vacation in the Greek Islands. If you are lucky enough to go, you will visit either Corfu (20% chance) or Rhodes. On Rhodes, there is a 20% chance of meeting a tall, dark stranger, while on Corfu, there is no such chance.

49. *Employment* In a 1987 survey of married couples with earnings, 95% of all husbands were employed. Of all employed husbands, 71% of their wives were also employed.* (Note that either the husband or wife in a couple with earnings must be employed.)

50. *Salaries* In 1989, 48.0% of all employees in the precision production industry were mechanics and repairers, 46.4% of them were construction tradespersons, and the rest were employed in other production occupations.* Twenty percent of mechanics and repairers, 15% of construction tradespersons, and 10% of other precision production workers had college degrees.

Games You are playing Dungeons and Dragons$^{(TM)}$ and have a pair of tetrahedral dice (four-faced pyramid-shaped dice, each numbered 1, 2, 3, 4). One is red, and the other is green. You have been challenged by the evil magician Donna Beserka, and you wish to raise a protective barrier against her evil powers. In order to be successful, you must throw anything but a double. In Exercises 51–56, test the given pairs of events for independence.

51. You are successful; one die is even and the other odd.

52. You fail; one die is even and the other odd.

53. You fail; both dice are even.

54. You fail; both dice are odd.

---

* Sources: *Statistical Abstract of the United States, 1991,* 111th Ed., U.S. Department of Commerce, Economics and Statistics Administration, and Bureau of the Census.

55. You are successful; the red die is a 4.
56. You are successful; the red die is not a 4.
57. *Weather Prediction*   There is a 50% chance of rain today and a 50% chance of rain tomorrow. Assuming that the event that it rains today is independent of the event that it rains tomorrow, draw a tree diagram showing the probabilities of all outcomes. What is the probability that there will be no rain today or tomorrow?
58. *Weather Prediction*   There is a 20% chance of snow today and tomorrow. Assuming that the event that it snows today is independent of the event that it snows tomorrow, draw a tree diagram showing the probabilities of all outcomes. What is the probability that it will snow by the end of tomorrow?
59. If a coin is tossed eleven times, find the probability of the sequence (H, T, T, H, H, H, T, H, H, T, T).
60. If a die is cast four times, find the probability of the sequence (4, 3, 2, 1).
61. *Drug Tests*   If 90% of the athletes who test positive for steroids in fact use them, and 10% of all athletes use steroids and test positive, what percentage of athletes test positive?
62. *Fitness Tests*   If 80% of candidates for the soccer team pass the fitness test, and only 20% of all athletes are soccer team candidates who pass the test, what percentage of the athletes are candidates for the soccer team?
63. *Medical Tests*   A pharmaceutical company is testing a new AIDS test. It finds that it has a false positive rate of 5% (the probability that someone tests positive for AIDS although he or she does not have the disease), and a false negative rate of 1% (the probability that someone tests negative for AIDS although he or she does have the disease). Suppose that 1% of the population is infected. What is the probability that a randomly selected person both has the disease and tests positive? What is the probability that a randomly selected person does not have the disease but tests positive anyway?
64. *Lie Detectors*   A manufacturer of lie detectors determines that its newest design has a probability of 0.1 of saying that someone has lied if he or she really told the truth, and a probability of 0.2 of saying that someone has told the truth if he or she really lied. On a question that 50% of the population lies about, what is the probability that a person lies and is caught by the machine? What is the probability that a person tells the truth but is accused of lying anyway?
65. *Marketing*   A market survey shows that 40% of the population uses Brand X laundry detergent, 5% of the population gave up doing its laundry last year, and 4% of the population used Brand X and then gave up doing laundry. Are the events of using Brand X and giving up doing laundry independent? Is a user of Brand X detergent more or less likely to give up doing laundry than a user of another brand?
66. *Marketing*   A market survey shows that 60% of the population uses Brand Z computers, 5% of the population quit their jobs last year, and 3% of the population used Brand Z computers and then quit their jobs. Are the events of using Brand Z computers and quitting one's job independent? Is a user of Brand Z computers more or less likely to quit his or her job than a user of another brand?

COMMUNICATION AND REASONING EXERCISES

67. What is the relationship between conditional probability and the concept of a reduced sample space?
68. You wish to ascertain the probability of an event E, but you happen to know that the event F has occurred. Is the probability you are seeking $P(E)$ or $P(E|F)$? Give the reason for your answer.
69. Your television advertising campaign has apparently been very successful: 10,000 people who saw the ad purchased your product, while only 2,000 people purchased the product without seeing the ad. Explain how additional data could show that your ad campaign was, in fact, a failure.
70. Name three pairs of independent events when a pair of distinguishable and fair dice are thrown and the uppermost numbers are observed.
71. If $A \subseteq B$ and $P(B) \neq 0$, why is $P(A|B) = \dfrac{P(A)}{P(B)}$?
72. If $B \subseteq A$ and $P(B) \neq 0$, why is $P(A|B) = 1$?
73. Your best friend thinks that if two mutually exclusive events are independent, then at least one of them is impossible. Establish whether or not she is correct.
74. Another of your friends thinks that two mutually exclusive events with nonzero probabilities can never be independent. Establish whether or not he is correct.

6.6 BAYES' THEOREM AND APPLICATIONS

Should schools test their athletes for drug use? The problem with drug testing is that there are always false positive results, so one can never be certain that an athlete whose test comes up positive is in fact using drugs. This is a very real problem that school administrators have to face. Here is a typical scenario.

▼ **EXAMPLE 1** Drug Tests

Gamma Pharmaceuticals advertises its steroid detection test as being 95% effective, meaning that it will show a positive result for 95% of all steroid users. It also boasts that its test has a low false positive rate: only 6%. This means that the probability of the test showing positive for a non-user is only 0.06.* Estimating that about 30% of its athletes are using steroids, Enormous State University (ESU) begins testing its football players. The quarterback, Hugo V. Huge, tests positive and is promptly dropped from the team. Hugo claims that he is not using steroids. What is the probability that he is telling the truth?

SOLUTION There are two events of interest here: the event that the test comes up positive—call it T—and the event that the person tested uses steroids—call it R (for "'roids").†

Let us put all the information on a tree. For the first branching we can use R and R', since we know that $P(R) = 0.3$ and so $P(R') = 0.7$ (Figure 1). For the second branching we use the outcomes of the drug test: positive (T) or negative (T'). The probabilities on these branches are conditional probabilities, because they depend on whether an athlete uses steroids or not (see Figure 2). These conditional probabilities are given to us, as, for example, the probability of a positive result given that an athlete does use steroids (0.95).

FIGURE 1

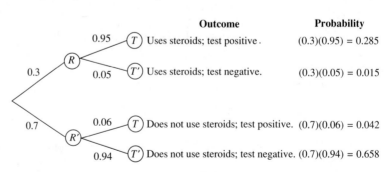

FIGURE 2

▼ * Figures such as these are easy to come by: to get the first figure, the company tests a large group of steroid users and counts the number that test positive, and to get the second, it tests a large group of non-users and again counts the number of positive results.

† We don't want to use S, since we usually use S for the sample space.

(We fill in the probabilities not supplied by remembering that the sum of the probabilities on the branches leaving any node must be 1.)

We are asked for the probability that an athlete uses steroids, given that the test is positive. To calculate this probability, we need to compute

$$P(R|T) = \frac{P(R \cap T)}{P(T)} = \frac{P(\text{uses steroids and test is positive})}{P(\text{test is positive})}.$$

From the tree we see that $P(R \cap T) = 0.285$. To calculate $P(T)$, the probability of the test being positive, notice that there are two outcomes on the tree that reflect a positive test result. The probabilities of these events are 0.285 and 0.042. Since these two events are mutually exclusive (an athlete either uses steroids or does not, but not both), the probability of a test being positive (regardless of whether or not steroids are used) is the sum of these probabilities, 0.327. Thus,

$$P(R|T) = \frac{0.285}{0.327} \approx 0.87.$$

Thus there is an 87% chance that Hugo is using steroids, and hence a 13% chance that he is telling the truth.

Before we go on... Note that the correct answer is 13%, *not* the 6% we might suspect from the test's false positive rating. In fact, we can't answer the question asked without knowing the percentage of athletes actually using steroids. For instance, if *no* athletes at all use steroids, then Hugo must be telling the truth, and so the test result has no significance whatsoever. On the other hand, if *all* athletes use steroids, then Hugo is definitely lying.

BAYES' THEOREM

The calculation we used to answer the question in Example 1 can be recast as a formula known as **Bayes' theorem.** Figure 3 shows a general form of the tree we used in Example 1.

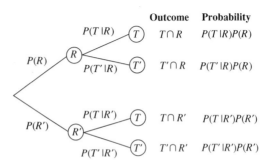

FIGURE 3

We calculated

$$P(R \mid T) = \frac{P(R \cap T)}{P(T)}$$

as follows. We first calculated $P(R \cap T)$ using the multiplication principle,

$$P(R \cap T) = P(T \mid R)P(R).$$

We then calculated $P(T)$ by the addition principle for mutually exclusive events, together with the multiplication principle.

$$P(T) = P(R \cap T) + P(R' \cap T)$$
$$= P(T \mid R)P(R) + P(T \mid R')P(R')$$

Substituting gives

$$P(R \mid T) = \frac{P(T \mid R)P(R)}{P(T \mid R)P(R) + P(T \mid R')P(R')}.$$

This is the short form of Bayes' theorem.

BAYES' THEOREM (SHORT FORM)
If R and T are events, then

$$P(R \mid T) = \frac{P(T \mid R)P(R)}{P(T \mid R)P(R) + P(T \mid R')P(R')}.$$

Although this looks like a complicated formula at first sight, it is not hard to remember if you notice the pattern. We want the left-hand side: $P(R \mid T)$. The numerator on the right has it the other way around, $P(T \mid R)$, multiplied by $P(R)$. This expression also appears in the denominator, added to a similar one with R replaced by R'.

▼ **EXAMPLE 2**   Drug Tests

Suppose that ESU's estimate that 30% of its athletes use steroids was overly pessimistic, and that in fact only 5% use steroids. What now is the probability that Hugo V. Huge is lying about taking steroids?

SOLUTION   Let us find this probability using Bayes' theorem. Using R again as the event that an athlete uses steroids and T as the event that the test is positive, we are given

$$P(T \mid R) = 0.95$$
$$P(T \mid R') = 0.06$$
$$P(R) = 0.05$$
$$P(R') = 0.95.$$

By Bayes' theorem,

$$P(R \mid T) = \frac{P(T \mid R)P(R)}{P(T \mid R)P(R) + P(T \mid R')P(R')}$$

$$= \frac{(0.95)(0.05)}{(0.95)(0.05) + (0.06)(0.95)}$$

$$\approx 0.45.$$

Thus, given that Hugo tested positive, there is only a 45% chance that he actually used steroids, and so a 55% chance that he is telling the truth and did not use steroids. In other words, it is actually more likely that he did *not* use steroids than that he did.

Before we go on... Without knowing the results of the test, we would have said that there was a probability of $P(R) = 0.05$ that Hugo was using steroids. The positive test result raises the probability to $P(R \mid T) = 0.45$, but the test gives too many false positives for us to be any more certain than that that Hugo was actually using steroids.

▼ **EXAMPLE 3** Lie Detectors

The Sherlock Lie Detector Company manufactures the latest in lie detectors, and the Count-Your-Pennies (CYP) store chain is eager to use them to screen its employees for theft. Sherlock's advertising claims that its test misses a lie only once in every 100 instances. On the other hand, an analysis by a consumer group reveals 20% of people telling the truth fail the test anyway.* The local police department estimates that 1 out of every 200 employees has engaged in theft. When the CYP store first screened its employees, the test indicated Mrs. Prudence V. Good was lying when she claimed that she had never stolen from CYP. What is the probability that she had in fact stolen from the store?

SOLUTION Let's use a tree to answer this question. We are asked for the probability that Mrs. Good was lying, and in the preceding sentence we are told that the lie detector test showed her to be lying. So, we are looking for a conditional probability: the probability that she is lying, given that the lie detector test is positive (showed her to be lying). Now we can start to give names to the events.

Let L be the event that a subject is lying, and

let T be the event that the test is positive.

We are looking for $P(L \mid T)$. We know $P(L)$ and the conditional probabilities $P(T' \mid L)$ and $P(T \mid L')$. Figure 4 shows the tree.

▼ *The reason for this is that many people show physical signs of distress when asked accusatory questions. Many people are nervous around police officers even if they have done nothing wrong.

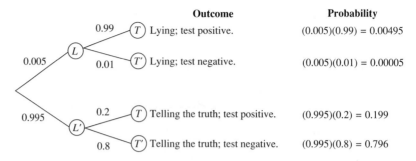

FIGURE 4

From the tree we see that

$$P(L \mid T) = \frac{P(L \cap T)}{P(T)}$$

$$= \frac{0.00495}{0.00495 + 0.199} \approx 0.024.$$

This means that there was only a 2.4% chance that poor Mrs. Good was lying!

Before we go on... Of course the expression on the last line of the calculation above is just Bayes' formula,

$$P(L \mid T) = \frac{P(T \mid L)P(L)}{P(T \mid L)P(L) + P(T \mid L')P(L')}.$$

Q We have seen the "short form" of Bayes' theorem. What is the "long form?"

A There *is* a long form of Bayes' theorem. In order to motivate it, look again at the short formula,

$$P(R \mid T) = \frac{P(T \mid R)P(R)}{P(T \mid R)P(R) + P(T \mid R')P(R')}.$$

The events R and R' form a **partition** of the sample space S. That is, their union is the whole of S, and their intersection is empty. (See Figure 5.)

For the long form of Bayes' theorem, we'll be given a partition of S into three or more events, as shown in Figure 6.

By saying that the events R_1, R_2, and R_3 form a partition of S, we mean that their union is the whole of S and the intersection of any two of them is empty, as in the figure. When dealing with a partition into three events as shown, the formula will give us $P(R_1 \mid T)$ in terms of $P(T \mid R_1)$, $P(T \mid R_2)$, $P(T \mid R_3)$, and $P(R_1)$, $P(R_2)$, $P(R_3)$.

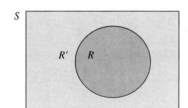

R and R' form a partition of S.

FIGURE 5

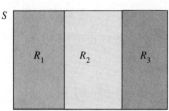

R_1, R_2, and R_3 form a partition of S.

FIGURE 6

6.6 Bayes' Theorem and Applications

BAYES' THEOREM (EXPANDED FORM)

$$P(R_1 \mid T) = \frac{P(T \mid R_1)P(R_1)}{P(T \mid R_1)P(R_1) + P(T \mid R_2)P(R_2) + P(T \mid R_3)P(R_3)}$$

As for why this is true, and what happens when we have a partition into *four or more* events, we will leave these discussions for the exercises. In practice, as was the case with a partition into two events, we can often compute $P(R_1 \mid T)$ by constructing a probability tree.

▼ **EXAMPLE 4**   Grades

Professor X has divided his class into three categories before the final exam: those likely to pass (20% of the class), those likely to fail (60% of the class), and those of whom he is not sure (the rest). After grading the final, he is shocked to discover that 90% of those he classified as "likely to fail" got an A on the final, while only 10% of the students he classified as "likely to pass" got an A. Eighty percent of the students he was unsure about also got A's. What fraction of the students who got an A on the final were in his top category?

SOLUTION   Let R_1 be the event that a student is in the "likely to pass" category, so $P(R_1) = 0.2$,

let R_2 be the event that a student is in the "likely to fail" category, so $P(R_2) = 0.6$, and

let R_3 be the event that a student is in the "not sure" category, so $P(R_3) = 0.2$.

These three categories partition the sample space, which we take to be the set of all students, into three events. The other event in which we are interested is the set of those students who earned an A on the final: call it A.

We are looking for the probability that a student was in the top category ("likely to pass"), given that the student got an A in the final, so we are looking for $P(R_1 \mid A)$. We can now set up the tree (Figure 7).

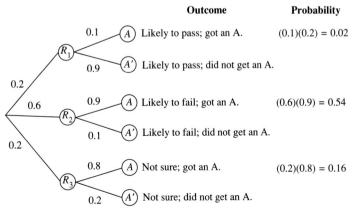

FIGURE 7

So

$$P(R_1 \mid A) = \frac{P(R_1 \cap A)}{P(A)}$$

$$= \frac{0.02}{0.02 + 0.54 + 0.16} \approx 0.028.$$

Thus, we conclude that only 2.8% of the A's were earned by students Prof. X thought likely to get an A.

6.6 EXERCISES

In Exercises 1–8, use Bayes' theorem or a probability tree to calculate the indicated probability. Round all answers to 4 decimal places.

1. $P(A \mid B) = 0.8$, $P(B) = 0.2$, $P(A \mid B') = 0.3$. Find $P(B \mid A)$.

2. $P(A \mid B) = 0.6$, $P(B) = 0.3$, $P(A \mid B') = 0.5$. Find $P(B \mid A)$.

3. $P(X \mid Y) = 0.8$, $P(Y') = 0.3$, $P(X \mid Y') = 0.5$. Find $P(Y \mid X)$.

4. $P(X \mid Y) = 0.6$, $P(X') = 0.4$, $P(X \mid Y') = 0.3$. Find $P(Y \mid X)$.

5. Y_1, Y_2, Y_3 form a partition of S. $P(X \mid Y_1) = 0.4$, $P(X \mid Y_2) = 0.5$, $P(X \mid Y_3) = 0.6$, $P(Y_1) = 0.8$, $P(Y_2) = 0.1$. Find $P(Y_1 \mid X)$.

6. Y_1, Y_2, Y_3 form a partition of S. $P(X \mid Y_1) = 0.2$, $P(X \mid Y_2) = 0.3$, $P(X \mid Y_3) = 0.6$, $P(Y_1) = 0.3$, $P(Y_2) = 0.4$. Find $P(Y_1 \mid X)$.

7. Y_1, Y_2, Y_3 form a partition of S. $P(X \mid Y_1) = 0.4$, $P(X \mid Y_2) = 0.5$, $P(X \mid Y_3) = 0.6$, $P(Y_1) = 0.8$, $P(Y_2) = 0.1$. Find $P(Y_2 \mid X)$.

8. Y_1, Y_2, Y_3 form a partition of S. $P(X \mid Y_1) = 0.2$, $P(X \mid Y_2) = 0.3$, $P(X \mid Y_3) = 0.6$, $P(Y_1) = 0.3$, $P(Y_2) = 0.4$. Find $P(Y_2 \mid X)$.

APPLICATIONS

9. *Athletic Fitness Tests* Any athlete failing the Enormous State University's women's soccer fitness test is automatically dropped from the team. The fitness test is traditionally given at 5 AM on a Sunday morning. Last year, Mona Header failed the test, but she claimed that this was due to the early hour. In fact, a study by the ESU Physical Education Department suggested that 50% of athletes fit enough to play on the team would fail the soccer test for precisely that reason (and would pass the test were it given at a more civilized hour), although no unfit athlete could possibly pass the test. It also estimated that 45% of the athletes who take the test are fit enough to play soccer. What is the probability that Mona was justifiably dropped?

10. *Academic Testing* Professor Frank Nabarro insists that all senior physics majors take his notorious physics aptitude test. The test is so tough that anyone *not* going on to a career in physics has no hope of passing, whereas 60% of the seniors who do go on to a career in physics also fail the test. Further, 75% of all senior physics majors in fact go on to a career in physics. Assuming that you fail the test, what is the probability that you will not go on to a career in physics?

11. *HIV Testing* Some people have estimated that about 1 in every 100 people living in New York City are now infected with the HIV virus. The currently used HIV antibody blood test shows a positive result for about 95 percent of people infected with the virus and also shows a false positive 5 percent of the time.
    (a) What is the probability that a New Yorker who has a positive blood test result is infected with the virus?
    (b) Now calculate the same probability if the epidemic reaches the point where one in every five New Yorkers are infected.

12. *At Risk*   We want to calculate the probability that a history of frequent high-risk activities will lead to HIV infection in the following hypothetical situation: Assume that a survey of HIV-infected individuals shows that 95 percent of them have a history of frequent high-risk encounters, whereas only 5 percent of HIV-free individuals have such a history.
    (a) If one in every thousand people is infected with the HIV virus, what is the probability that a person with a history of frequent high-risk encounters has the virus?
    (b) What happens to this figure if 1% of the population has the virus?

13. *Grade Complaints*   Two of the mathematics professors at Enormous State are Professor A (known for easy grading) and Professor F (known for tough grading). Last semester, roughly three quarters of Professor F's class consisted of former students of Professor A—these students apparently feeling encouraged by their (utterly undeserved) high grades. (Professor F's own former students had fled in droves to Professor A's class in order to try to shore up their grade point averages.) At the end of the semester, as might have been predicted, all of Professor A's former students wound up with a C− or lower. The rest of the students in the class—former students of Professor F who had decided to "stick it out"—fared better, and two thirds of them earned higher than a C−. After discovering what had befallen them, all the students who earned C− or lower got together and decided to send a contingent to the Department Chair in order to complain that their grade point averages had been ruined by this callous and heartless beast! The contingent was to consist of ten representatives, selected at random from among them. How many of the ten would you estimate to have been former students of Professor A?

14. *Weather Prediction*   A local TV station employs Desmorelda, "Mistress of the Zodiac," as their weather forecaster. Now, when it rains, Sagittarius is in the shadow of Jupiter one third of the time, and it rains on four out of every fifty days. On the other hand, Sagittarius falls in Jupiter's shadow on only one in every five rainless days. The powers that be at the station notice a disturbing pattern to Desmorelda's weather predictions. It seems that she always predicts that it will rain when Sagittarius is in the shadow of Jupiter. What percentage of the time is she correct? Should they replace her?

15. *Market Surveys*   A *New York Times* survey* of homeowners showed that 86% of those with swimming pools are married couples, and the other 14% are single. It also showed that 15% of all homeowners had pools.
    (a) Assuming that 90% of all homeowners without pools are married couples, what percentage of homes owned by married couples have pools?
    (b) Would it best pay pool manufacturers to go after single homeowners or married homeowners? Explain.

16. *Crime and Preschool*   Another *New York Times* survey† of needy and disabled youths showed that 51% of those who had no preschool education were arrested or charged with a crime by the time they were 19, while only 31% who had preschool education wound up in this category. The survey did not specify what percentage of the youths in the survey had preschool education, so let us take a guess at that, and estimate that 20% of them had attended preschool.
    (a) What percentage of the youths arrested or charged with a crime had no preschool education?
    (b) What would this figure be if 80% of the youths had attended preschool? Under these circumstances, would youths who had preschool education be more likely to be arrested or charged with a crime than those who did not? Support your answer by quoting statistics.

17. *Family Values*   At one stage in the 1992 Presidential campaign, a poll showed Clinton favored by 48% of the voters surveyed, and Bush favored by 32%. The rest were undecided. Twenty percent of the Clinton supporters rated "family values" as the number-one issue, while 40% of Bush supporters and 30% of the undecided voters felt the same way.‡
    (a) What percentage of those for whom family values was the number-one issue favored Bush?
    (b) Would a further emphasis on family values have helped the Bush campaign?

---

* "All About Swimming Pools," *The New York Times*, Sept. 13, 1992.

† "Governors Develop Plan to Help Preschool Children," *The New York Times*, Aug. 2, 1992.

‡ The figures given are fictitious.

18. *Undecided Voters*   Prior to the 1992 presidential election, Bush had been trailing Clinton by 12 percentage points at one stage, with 50% of the voters still undecided. Eighty percent of the Clinton supporters rated the economy as the number-one issue, while 75% of Bush supporters and 80% of the undecided voters felt the same way.*
    (a) What percentage of voters for whom the economy was the number-one issue were undecided?
    (b) Would it have been wise for the Clinton campaign to continue emphasizing the economy?

19. *Car Rentals*   On September 9, 1993, *The New York Times* reported that *Hertz*, *Avis*, and other car rental companies were screening the driving records of potential customers and rejecting those whose records showed them to be at risk. Assume that before the policy was implemented, 60% of all customers who had accidents in rental cars would have been classified as "at risk," and 80% of customers who had no accidents would have been classified as "not at risk." If 5% of all car rental customers had accidents before implementation of the policy, predict the percentage of customers that will have accidents after the policy is implemented.

20. *Car Rentals*   Repeat Exercise 19 using the following data: Before the policy was implemented, 80% of all customers who had accidents in rental cars would have been classified as "at risk," and 40% of customers who had no accidents would have been classified as "not at risk." Five percent of all car rental customers had accidents before implementation of the policy.

21. *Employment*   In 1989, 6.9% of all employees in the service industry were employed in private households, 5.0% of them were employed in a protective capacity (such as security guards) and the rest were employed in other service capacities.† If 5% of household workers, 25% of protective workers, and 30% of other service workers had college degrees, find (to four decimal places) the probability that a service employee with a college degree was employed in a household.

22. *Employment*   In 1989, 48.0% of all employees in the precision production industry were mechanics and repairers, 46.4% of them were construction tradespersons, and the rest were employed in other production occupations.† If 20% of mechanics and repairers, 15% of construction tradespersons, and 10% of other precision production workers had college degrees, find (to four decimal places) the probability that a college degree holder in the precision production industry was a construction tradesperson.

23. *Employment*   In a 1987 survey of married couples with earnings, 95% of all husbands were employed. Of all employed husbands, 71% of their wives were also employed.† Noting that either the husband or wife in a couple with earnings had to be employed, find the probability that the husband of an employed woman was also employed.

24. *Employment*   Repeat Exercise 23 in the event that 50% of all husbands were employed.

25. *Population Migration*   In 1987, the population of the U.S., broken down by regions, was 59.1 million in the Midwest, 49.3 million in the Northeast, 82.2 million in the South, and 48.2 million in the West. From 1987 to 1988, 0.89% of the population in the Midwest moved to the South, 0.9% of the population in the Northeast moved to the South, 98.84% of the population of the South stayed there, and 0.48% of the population in the West moved to the South.† What percentage of the population of the South moved there from the Northeast that year? (Round the answer to two decimal places.)

26. *Population Migration*   In 1987, the population of the U.S., broken down by regions, was 59.1 million in the Midwest, 49.3 million in the Northeast, 82.2 million in the South, and 48.2 million in the West. From 1987 to 1988, 0.15% of the population in the Midwest moved to the Northeast, 98.62% of the population of the Northeast stayed there, 0.3% of the population of the South moved to the Northeast, and 0.22% of the population in the West moved to the Northeast.† What percentage of the population of the Northeast moved from the South that year? (Round the answer to two decimal places.)

▼ * The figures given are fictitious.

†Sources: *Statistical Abstract of the United States*, 111th Ed., 1991, U.S. Dept. of Commerce/U.S. Bureau of Labor Statistics. Figures rounded to the nearest 0.1%.

27. *Benefits of Exercise* According to a study in the July 18, 1991 issue of *The New England Journal of Medicine*,* it was found that, of 5,990 sedentary middle-aged men, 202 had developed diabetes. It also found that men who were very active (burning about 3,500 calories daily) were half as likely to develop diabetes as men who were sedentary. Assume that one-third of all middle-aged men are very active. What is the probability that a middle-aged man with diabetes is very active?

28. *Benefits of Exercise* Repeat Exercise 27, assuming that only one in ten middle-aged men are very active.

COMMUNICATION AND REASONING EXERCISES

29. Your friend claims that the probability of A given B is the same as the probability of B given A. How would you convince him that he is wrong?

30. Complete the following sentence. To use Bayes' theorem to compute $P(E|F)$, you need to be given _____ .

31. Give an example in which a steroid test gives a false positive only 1 percent of the time, and yet if an athlete tests positive, the chance that he or she has used steroids is under 10%.

32. Give an example in which a steroid test gives a false positive 30% percent of the time, and yet if an athlete tests positive, the chance that he or she has used steroids is over 90%.

33. Derive the expanded form of Bayes' theorem for a partition of the sample space S into three events R_1, R_2, and R_3.

34. Write down an expanded form of Bayes' theorem that applies to a partition of the sample space S into four events R_1, R_2, R_3, and R_4.

6.7 BERNOULLI TRIALS

An important question in probability and statistics is this: Given a large number of scores—for example, student grades on an exam, or the ratings of a professor by students—how should we expect the scores to be distributed? For example, suppose that students are asked to rate a professor on a scale of 0 to 5. Not all the ratings will be the same—some will be high, and others low, with most of them perhaps tending to clump somewhere around the middle (such as 2–3 for an average professor).

Now suppose that you (the students) have gotten together and decided to fill out the evaluation forms randomly. Here are two ways you might proceed.

Method 1

Each student chooses a number from 0 to 5 randomly by throwing a die and subtracting one from the number rolled. Since each outcome is equally likely, the scores will have a "flat" distribution: there will be the same number of 0s, 1s, 2s, 3s, 4s, and 5s. We say in this case that the scores are **uniformly distributed,** which just means that all the scores are equally likely.

Method 2 (A more interesting way)

You each toss a coin five times and count the number of heads you get. You then fill in that number on the scoring sheet. Since it is highly unlikely that you will get five heads in a row (what is the probability?), the professor can expect to get very few 5s. Similarly, he or she can expect to get very few 0s, since this would correspond to 5 tails in a row. Most of the scores would be around 2 or 3, since you expect to get heads about half the time. We refer to

▼ *As cited in a *New York Times* article of the same date.

this type of distribution of scores as a **binomial** distribution. Binomial distributions were first studied by the Swiss mathematician Jakob Bernoulli (1654–1705) who was one of the pioneers of probability theory. A **Bernoulli trial** is an experiment, like the tossing of a coin, with exactly two possible outcomes.

The kind of question we will answer in this section is this: In a class of 100 students, how many of each score can the professor expect to get if the students use coins? That is, how many 0s, how many 1s, etc.? We already have all the mathematical ingredients we need to answer this question: a little counting and a little probability. To keep it simple for the moment, we shall first calculate the probability that a student gives the professor a rating of 2. We'll return to the full question in a while.

▼ **EXAMPLE 1**

If a fair coin is tossed 5 times, what is the probability of heads coming up exactly twice?

SOLUTION   We go straight back to first principles on this one, and look at the sample space and the event we are interested in. Since an outcome is a sequence of 5 choices of H or T, we take

S to be the set of sequences of five choices of H or T, so $n(S) = 2^5$, and

E to be the set of those sequences containing two Hs and three Ts.

To calculate $n(E)$, we pretend, using the Guide to Counting (see the chapter on sets and counting), that we are constructing such a sequence. We start with five empty slots and select two slots for the Hs. There are $C(5, 2)$ possible choices. Once we have chosen the two slots for the Hs, we are done, since the remaining three slots are automatically assigned to the Ts. Thus,

$$n(E) = C(5, 2) = 10.$$

It follows that

$$P(E) = \frac{n(E)}{n(S)} = \frac{C(5, 2)}{2^5} = \frac{10}{32} = \frac{5}{16}.$$

Now we'll make things a little more complicated and see what happens if we use an unfair coin to make the decision.

▼ **EXAMPLE 2**

Suppose that we have a possibly unfair coin, so that the probability of heads is p (we won't tell you what its value is) and the probability of tails is q.* Give a formula for the probability of getting heads exactly twice in five throws.

▼ *Of course, $p + q = 1$.

SOLUTION The sets S and E are the same as in the first example, but there is a problem: we now have a situation in which the outcomes need not be equally likely. For instance, if $p = \frac{2}{3}$, then heads are more likely to come up than tails, and so sequences with more Hs are more likely than those with fewer Hs. Thus we can't use the formula $P(E) = n(E)/n(S)$. Let's back up instead and look at a simpler question.

Q What is the probability that we will get the sequence *HHTTT*?

A The probability that the first toss will come up heads is p.
The probability that the second toss will come up heads is also p.
The probability that the third toss will come up tails is q.
The probability that the fourth toss will come up tails is q.
The probability that the fifth toss will come up tails is q.

The probability that the first toss will be heads *and* the second will be heads *and* the third will be tails *and* the fourth will be tails *and* the fifth will be tails equals the probability of the *intersection* of these five events. Since these are independent events, the probability of the intersection is the product of the probabilities, and thus equals

$$p \times p \times q \times q \times q = p^2 q^3.$$

Now *HHTTT* is only one of several outcomes with two heads and three tails. Others are *HTHTT, TTTHH,* and so on.

Q How many such outcomes are there altogether?

A This is the number of "words" with two Hs and three Ts, and we know that the answer is $C(5, 2)$.

Each of these $C(5, 2)$ (= 10) outcomes has the same probability: $p^2 q^3$ (why?). As we saw in Section 2, the probability of the event E is just the sum of these probabilities. In other words, the probability we are after is

$$p^2 q^3 + p^2 q^3 + \ldots + p^2 q^3 \; (C(5, 2) \text{ times})$$
$$= C(5, 2) p^2 q^3.$$

Before we go on... The structure of this formula is as follows.

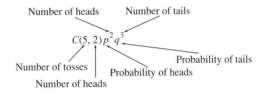

There seems to be a small lack of symmetry here: $C(5, 2)$ refers to the number of heads, but not tails. However, notice that $C(5, 2) = C(5, 3)$, and so this answer is the same as:

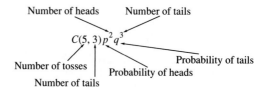

Take your pick! Notice also that if the coin happened to be a fair one, then $p = q = \frac{1}{2}$, and we get the same answer as in Example 1.

We'll now generalize this situation and give a name to what we have been doing. First, a **Bernoulli trial** is an experiment with two possible outcomes, called **success** and **failure**.* Each outcome has a specified probability: p for success and q for failure (so that $p + q = 1$). If we perform a sequence of n independent Bernoulli trials, then some of them result in success (such as heads coming up in the above example) and the rest of them in failure.

Q In a sequence of n independent Bernoulli trials, what is the probability of r successes and $(n - r)$ failures?

A We generalize the formula found in Example 2, using the same sort of analysis.

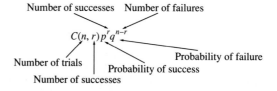

PROBABILITY OF OUTCOMES IN BERNOULLI TRIALS

In a sequence of n independent Bernoulli trials with probabilities p of success and q of failure, the probability of exactly r successes is given by

$$P(\text{exactly } r \text{ successes}) = C(n, r)p^r q^{n-r}.$$

---

* These are customary names for the two possible outcomes and often do not indicate actual success or failure at anything.

EXAMPLE 3   Sports Injuries

Suppose that you have a 1-in-50 chance of being injured every time you participate in a soccer match.

(a) What are the chances that you will be injured exactly twice in a 10-game season?

(b) What are the chances that you will be injured at most twice in a 10-game season?

SOLUTION   Part (a) can be answered by a direct application of the formula, since each soccer game can be treated as a Bernoulli trial with p = probability of an injury $= \frac{1}{50}$, and $q = 1 - \frac{1}{50} = \frac{49}{50}$. Thus, the probability of two injuries is

$$C(10, 2) \left(\frac{1}{50}\right)^2 \left(\frac{49}{50}\right)^8 \approx 45(.0004)(.8508) \approx 0.0153.$$

In part (b), we are asked for the event of either 0, 1, or 2 injuries. We calculate the probabilities separately.

$$\text{Probability of 0 injuries} = C(10, 0) \left(\frac{1}{50}\right)^0 \left(\frac{49}{50}\right)^{10}$$
$$\approx (1)(1)(0.8171) = 0.8171$$

$$\text{Probability of 1 injury} = C(10, 1) \left(\frac{1}{50}\right)^1 \left(\frac{49}{50}\right)^9$$
$$\approx (10)(0.02)(0.8337) \approx 0.1667$$

$$\text{Probability of 2 injuries} = C(10, 2) \left(\frac{1}{50}\right)^2 \left(\frac{49}{50}\right)^8$$
$$\approx 45(0.0004)(0.8508) \approx 0.0153$$

Thus, the probability of zero, one, or two injuries is the probability of the union of these three events and thus equals the sum of the three probabilities: 0.9991. In other words, you have a pretty good chance of surviving 10 games with no more than two injuries!

Before we go on... Our estimate of a 1-in-50 chance is optimistic. Collegiate soccer is usually quite rough, so an estimate of 1 in 8 or so is probably more accurate. How does this change your chances of being injured at most twice? Another issue: Is it realistic to assume that the probability of injury in any one game is independent of the outcome of the previous game?

Chapter 7 Statistics

INTRODUCTION ▶ Statistics is the branch of mathematics concerned with organizing, analyzing, and interpreting numerical data. For example, suppose you were given the current annual incomes of 1,000 lawyers selected at random. You might then wish to answer questions such as the following: If I become a lawyer, what is the income I am most likely to earn? Do lawyers' salaries vary widely? How can I measure this variation?

To answer questions such as these, it is extremely helpful to begin by organizing your data in the form of tables or graphs. This is the topic of the first section of the chapter. Once the data are organized, the next step is to apply mathematical tools for analyzing the data to answer questions such as those we posed above. Numbers such as the **mean** and the **standard deviation** can be computed from the data to reveal interesting things about the distribution of the data. These numbers can then be used make predictions about future events.

The chapter ends with a section on one of the most important distributions in statistics, the **normal distribution.** This distribution describes many sets of data and also plays an important role in the underlying mathematical theory.

7.1 RANDOM VARIABLES AND DISTRIBUTIONS

In many experiments, the outcomes can be assigned numerical values. For instance, if you roll a die, the possible outcomes have the numerical values 1 through 6. If you select a lawyer and ascertain her annual income, then the outcome is again a number. A rule that assigns a numerical value to each outcome of an experiment is called a **random variable.**

> **RANDOM VARIABLE**
>
> A **random variable** X is a rule that assigns a numerical value to each outcome in the sample space of an experiment.

A random variable may have only finitely many values, such as the outcome of a roll of a die. Or its possible values may be infinite but discrete, such as the number of times it takes you to roll a 6 if you keep rolling until you get one. Or the variable may be continuous, as we shall see in the last section of this chapter. For the moment, we shall assume that our random variables take on only finitely many values. We call such variables **finite random variables.**

EXAMPLE 1

Let X be the number of heads that come up when a coin is tossed three times. List the value of X for each possible outcome. What are the possible values of X?

SOLUTION First, we describe X as a random variable.

X is the rule that associates to each outcome the number of heads that come up.

We shall take as the outcomes of this experiment the sequences of heads and tails that may be obtained. Then, for instance, if the outcome is *HTH*, the value of X is 2. We list the values of X for all the outcomes in a table.

Outcome	Value of X
HHH	3
HHT	2
HTH	2
HTT	1
THH	2
THT	1
TTH	1
TTT	0

From the table, we also see that the possible values of X are 0, 1, 2, and 3.

Before we go on... Remember that X is just a rule we decided on. We could have taken X to be a different rule, such as the number of tails, or perhaps the number of heads minus the number of tails. These different rules are examples of different random variables associated with the same experiment.

EXAMPLE 2

You have purchased $10,000 worth of stock in a biotech company whose newest arthritis drug is awaiting approval by the FDA. If the drug is approved this month, the value of the stock will double by the end of the month. If the drug is rejected this month, the stock's value will decline by 80%, and if no decision is reached this month, its value will decline by 10%. Let X be the value of your investment at the end of this month. List the value of X for each possible outcome.

SOLUTION There are three possible outcomes: the drug is approved this month, it is rejected this month, or no decision is reached. Once again, we express the random variable as a rule.

The random variable X is the rule that assigns to each outcome the value of your investment at the end of this month.

We can now tabulate the values of X as follows:

Outcome	Value of X
Approved this month	$ 20,000
Rejected this month	2,000
No decision	9,000

▼ EXAMPLE 3   Bank Deposits

In a random survey of ten commercial banks in the U.S., the banks surveyed reported the following total deposits: $60 billion, $20 billion, $23 billion, $11 billion, $6 billion, $40 billion, $55 billion, $2 billion, $1 billion, $6 billion.* Interpret these as the values of a suitable random variable X.

SOLUTION   There are really two sample spaces under consideration here:

1. The set consisting of the ten banks in the survey. We call this the **sample.**
2. The set of *all* commercial banks in the U.S. We call this the **population.**

In either event, X is the random variable that assigns to each bank the value of its total deposits.

Before we go on... In cases such as this, when we are given a collection of values of a random variable X, we refer to the values as **X-scores.** We also call such data **raw data,** as these are the original values on which we often perform statistical analysis. One important purpose of statistics is to interpret the raw data from the sample to get information about the entire population.

PROBABILITY DISTRIBUTION OF A RANDOM VARIABLE

Given a random variable X, it is natural to look at certain *events*: for instance, the event that $X = 2$. By this, we mean the event consisting of all outcomes whose X-value is 2. Looking once again at the chart in Example 1, X being the number of heads that come up when a coin is tossed three times, we obtain

▼ * These figures are typical for deposits in large U.S. commercial banks. For instance, in 1990 *Citibank* reported total deposits of $112,586 million, while *Bank of America* reported deposits of $77,027 million. [Source: *World Almanac and Book of Facts 1992* (New York: Pharos Books, 1992) and *American Banker*.]

the event that $X = 0$ is $\{TTT\}$,
the event that $X = 1$ is $\{HTT, THT, TTH\}$,
the event that $X = 2$ is $\{HHT, HTH, THH\}$,
the event that $X = 3$ is $\{HHH\}$,
the event that $X = 4$ is $\emptyset$ (since there are no outcomes with four heads).

Each of these events has a certain probability. For instance, the probability that $X = 2$ is $\frac{3}{8}$, since the event in question consists of three of the eight possible (equally likely) outcomes*. We shall abbreviate this by writing

$$P(X = 2) = \frac{3}{8}.$$

Thus, $P(X = 2)$ is the probability of the event $X = 2$, or simply the *probability that $X = 2$*. Similarly,

$$P(X = 4) = 0.$$

The collection of probabilities of each of these events is called the **probability distribution** of the random variable X.

EXAMPLE 4

Let X be the number of heads in three tosses of a coin. Give the probability distribution of X.

SOLUTION X is the random variable of Example 1. The probability distribution of X is given in the following chart.

x	$P(X = x)$
0	$\frac{1}{8}$
1	$\frac{3}{8}$
2	$\frac{3}{8}$
3	$\frac{1}{8}$

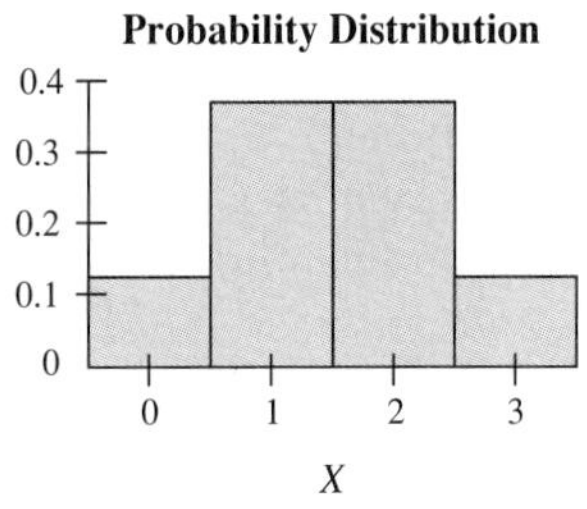

HISTOGRAMS

We can use a bar graph to visualize a probability distribution. Figure 1 shows the bar graph for the probability distribution in Example 4. Such a graph is sometimes called a **histogram.**

FIGURE 1

---

* We could also have calculated this using the formula for Bernoulli trials:
$P(2 \text{ heads}) = C(3, 2)(\frac{1}{2})^2(\frac{1}{2})^1 = 3 \times \frac{1}{8} = \frac{3}{8}$.

Most graphing calculators and spreadsheet programs include software to graph histograms. If you are using a graphing calculator, you can enter the data using an appropriate menu and then instruct the calculator to display the histogram.* If you are using a spreadsheet program, you can obtain a bar graph from a probability distribution by entering the appropriate graphing instruction.†

▼ **EXAMPLE 5**   Income

Suppose that there are only 1,000 lawyers and that the following table gives the number of lawyers in each of several income brackets.

Income Bracket	$20,000–$29,999	$30,000–$39,999	$40,000–$49,999	$50,000–$59,999	$60,000–$69,999	$70,000–$79,999	$80,000–$89,999
Number	20	80	230	400	170	70	30

Think of the experiment of choosing a lawyer at random (all being equally likely). Assign to each lawyer the number X that is the midpoint of his or her bracket. Find the probability distribution of X.

SOLUTION   Since the first bracket contains incomes that are at least $20,000, but less than $30,000 (it would include an income of $29,999.99), its midpoint is $25,000. Similarly, the second bracket has midpoint $35,000, and so on. We can rewrite the table using the midpoints, as follows.

x	25,000	35,000	45,000	55,000	65,000	75,000	85,000
Frequency	20	80	230	400	170	70	30

We have used the term "frequency" rather than "number," although it means the same thing. This table is called a **frequency table.** This is *almost* the probability distribution for X, except that we must replace frequencies by probabilities (we did this in calculating experimental probabilities in the preceding chapter). Start with the lowest income bracket: since 20 of the 1,000 lawyers fall in this group, we compute

$$P(X = 25{,}000) = \frac{20}{1000} = 0.02.$$

▼ *See Appendix B for details on using the TI-82. Another reference, geared to the TI-81, is *Graphing Calculator Lessons for Finite Mathematics* by Paula G. Young (New York: HarperCollins, 1993), which devotes several sections to the graphing of statistical data.
†For further information on this and other statistics topics, see *Topics in Finite Mathematics: An Introduction to the Electronic Spreadsheet* by Samuel W. Spero (New York: Harper-Collins, 1993).

Similarly, we can calculate the remaining probabilities to obtain the following distribution.

x	25,000	35,000	45,000	55,000	65,000	75,000	85,000
$P(X = x)$	0.02	0.08	0.23	0.40	0.17	0.07	0.03

In Figure 2 we see the histogram of the frequency distribution and the histogram of the probability distribution. The only difference between the two graphs is in the scale of the vertical axis (why?).

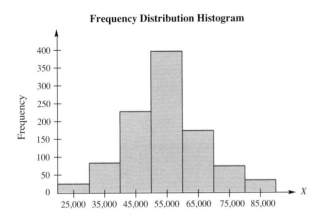
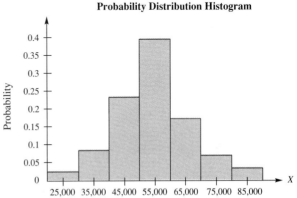

FIGURE 2

Before we go on... We shall often be given a distribution involving categories with *ranges* of values (such as salary brackets), rather than individual values. When this happens, we shall always take X to be the *midpoint* of a category, as we did above. This is a reasonable thing to do, particularly when we have no information about how the scores were distributed within each range.

Now, these 1,000 lawyers may actually have been randomly selected from the set of *all* currently practicing lawyers.* If this is the case, we call the probability distribution we found above the **experimental probability distribution** (as we did in the previous chapter), as opposed to a **theoretical probability distribution** such as the one we found in Example 4. The experimental probability distribution may be used as an *estimate* of the actual distribution of incomes among all practicing lawyers. In fact, it can be shown that, in a very precise sense, the experimental probability distribution obtained from a collection of X-scores is the *best* possible estimate of the actual distribution, if the X-scores are all the information we have.

If you are using a graphing calculator, see Appendix B, where this example is redone using technology.

---

* We would then refer to the 1,000 lawyers selected as the *sample* and to the set of all currently practicing lawyers as the *population*, just as in Example 3.

▶ **NOTE** In the last example, we were given a table showing the number of lawyers in each of several income brackets. This kind of table can be easily constructed from the raw data consisting of the incomes of the 1,000 lawyers: simply count how many fall into each income bracket. Many of the frequency distribution tables in the examples and exercises below were constructed from raw data in this way. We *could* have simply taken the individual salaries as X-values along the lines of Example 3, but then our frequency table would have been huge, with up to 1,000 different X-values. Grouping the data into income brackets has thus saved us a lot of work. There is a trade-off at work here: the larger the income brackets we choose, the more information we lose. (For instance, we could have grouped all 1,000 lawyers in the single income bracket $20,000–$89,999, but then we would not have had a very interesting distribution!) ◀

▼ **EXAMPLE 6** Class Size

The Department of Mathematics at Enormous State University offers classes of various sizes, that is, having various numbers of students enrolled. One semester the distribution of classes was as follows.

Class Size (number of students)	1–9	10–19	20–39	40–59	60–99	100–149
Number	5	5	30	50	10	10

Choose a mathematics class at random and let X be the midpoint of the bracket in which its class size falls. Find the probability distribution of X.

SOLUTION There is an important difference between this example and the preceding one. In Example 5, when we wrote $20,000–$29,999 for the first bracket, we really included any income that was at least $20,000 and less than $30,000, including incomes with fractions of dollars. The midpoint of that bracket, which is also the average of all the possible values in that bracket, was therefore ($20,000 + $30,000)/2 = $25,000. In this example, consider the bracket labeled 10–19. Since the size of a class must be an integer, the midpoint of the bracket should be the average of the integers 10 through 19, which is (10 + 19)/2 = 14.5, *not* (10 + 20)/2 = 15. In Example 5 we had *continuous* data, while in this example we have *discrete* data. With this in mind, we rewrite the table using the midpoints, as follows.

Class Size (number of students)	5	14.5	29.5	49.5	79.5	124.5
Frequency	5	5	30	50	10	10

Since there are a total of 110 classes, we turn this frequency table into a probability distribution by dividing each frequency by 110.

x	5	14.5	29.5	49.5	79.5	124.5
$P(X = x)$	0.0455	0.0455	0.2727	0.4545	0.0909	0.0909

Before we go on... You will have to judge in a given situation whether you are dealing with discrete data or continuous data, and therefore how to compute the midpoints of the brackets. Data that can take on only integer values, for example, are discrete. Data that can take on fractional values are usually continuous. In the exercises we give guidance where it may not be clear how to treat the data.

▼ **EXAMPLE 7** Darts

Joe Student is a terrible darts player. In fact, every time he throws a dart, he has only a 1 in 5 chance of hitting the dartboard! Let X be the number of times he hits the dartboard in 10 attempts. Graph the probability distribution of X and answer the following questions.

(a) What is the most likely number of times he will hit the dartboard?

(b) What is the probability of his hitting the dartboard less than half the time?

SOLUTION We can think of Joe's 10 dart throws as a sequence of 10 Bernoulli trials, with

$$p = \text{probability of success (hitting the board)} = \frac{1}{5}, \text{ and}$$

$$q = \text{probability of failure (missing)} = \frac{4}{5}.$$

For the sample space we may take the set of all sequences of 10 hits or misses, and the random variable X assigns to each sequence the number of times Joe hits the board.

We need to calculate $P(X = x)$ for each value of x. (Here, x can range from 0 to 10.) In the last chapter we saw that the probability of x successes in 10 Bernoulli trials is given by the formula

$$P(X = x) = C(10, x) p^x q^{10-x}.$$

We now use our calculators to generate the following data.

$$P(X = 0) = C(10, 0) \left(\frac{1}{5}\right)^0 \left(\frac{4}{5}\right)^{10} = 0.1074$$

$$P(X = 1) = C(10, 1)\left(\frac{1}{5}\right)^1 \left(\frac{4}{5}\right)^9 = 0.2684$$

$$P(X = 2) = C(10, 2)\left(\frac{1}{5}\right)^2 \left(\frac{4}{5}\right)^8 = 0.3020$$

$$P(X = 3) = C(10, 3)\left(\frac{1}{5}\right)^3 \left(\frac{4}{5}\right)^7 = 0.2013$$

$$P(X = 4) = C(10, 4)\left(\frac{1}{5}\right)^4 \left(\frac{4}{5}\right)^6 = 0.08808$$

$$P(X = 5) = C(10, 5)\left(\frac{1}{5}\right)^5 \left(\frac{4}{5}\right)^5 = 0.02642$$

$$P(X = 6) = C(10, 6)\left(\frac{1}{5}\right)^6 \left(\frac{4}{5}\right)^4 = 0.005505$$

$$P(X = 7) = C(10, 7)\left(\frac{1}{5}\right)^7 \left(\frac{4}{5}\right)^3 = 0.0007864$$

$$P(X = 8) = C(10, 8)\left(\frac{1}{5}\right)^8 \left(\frac{4}{5}\right)^2 = 0.00007373$$

$$P(X = 9) = C(10, 9)\left(\frac{1}{5}\right)^9 \left(\frac{4}{5}\right)^1 = 0.000004096$$

$$P(X = 10) = C(10, 10)\left(\frac{1}{5}\right)^{10} \left(\frac{4}{5}\right)^0 = 0.0000001024$$

The probability distribution is shown in the following chart.

x	0	1	2	3	4	5	6	7	8	9	10
$P(X = x)$	0.1074	0.2684	0.3020	0.2013	0.08808	0.02642	0.005505	0.0007864	0.00007373	0.000004096	0.0000001024

Graphing these data gives the histogram in Figure 3.

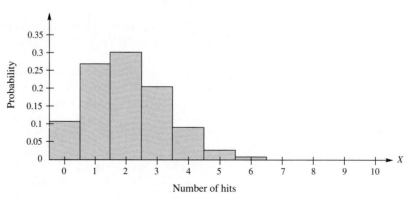

FIGURE 3

The probabilities of the events $X = 7, 8, 9$, and 10 are so small that they aren't visible on the histogram.

Now we can answer part (a). The highest rectangle corresponds to $X = 2$; thus, Joe is most likely to hit the dartboard twice. As for (b), we calculate

$$P(X \leq 4) = P(X = 0) + P(X = 1) + P(X = 2)$$
$$+ P(X = 3) + P(X = 4)$$
$$= 0.1074 + 0.2684 + 0.3020 + 0.2013$$
$$+ 0.08808$$
$$= 0.9672.$$

We used a programmable calculator to compute the probabilities in this distribution. Graphing calculators are not only programmable but have the ability to graph the data.

In the above example, we calculated the probability distribution for a sequence of Bernoulli trials. The resulting distribution is called a **binomial distribution** and can be calculated by the following formula, which we saw in the last chapter.

BINOMIAL DISTRIBUTION

In a sequence of n independent Bernoulli trials, with probability p of success and q of failure, let X be the number of successes. X is called a **binomial variable.** The probability distribution of X is called a **binomial distribution** and is given by

$$P(X = x) = C(n, x) p^x q^{n-x}.$$

Q Where does the name "binomial" come from?

A From the fact that $C(n, x) p^x q^{n-x}$ happens to be $(x + 1)$st term in the expansion of the binomial power $(p + q)^n$. This is not at all a coincidence, but we'll leave it to you to puzzle out the connection.

Q What other kinds of distributions are there?

A There are many. Here are two of them. A **uniform** distribution is one whose histogram is flat. For example, if you roll a fair die, each of 1, 2, 3, 4, 5, and 6 has a probability of $\frac{1}{6}$, so the histogram is flat. A **normal** distribution is one whose histogram is "bell-shaped." We shall be studying normal distributions in Section 4.

7.1 EXERCISES

In each of Exercises 1 through 8,
(a) *say what an appropriate sample space would be;*
(b) *complete the following sentence:*
   X is the rule that . . . ;
(c) *list the value of X for each possible outcome.*

1. X is the number of tails that comes up when a coin is tossed twice.

2. X is the largest number of consecutive times heads comes up in a row when a coin is tossed three times.

3. X is the sum of the numbers when two dice are cast.

4. X is the value of the largest number when two dice are cast.

5. X is the number of red marbles that Tonya has in her hand after she selects four marbles from a bag containing 4 indistinguishable red marbles and 2 indistinguishable green ones.

6. X is the number of green marbles that Suzy has in her hand after she selects four marbles from a bag containing 3 indistinguishable red marbles and 2 indistinguishable green ones.

7. The mathematics final exam scores for the students in your study group are 89%, 85%, 95%, 63%, 92%, 80%.

8. The capacities of the hard drives of your dormitory suite mates' computers are 180MB, 160MB, 180MB, 500MB, 240MB, 160MB.

In Exercises 9 through 14, give the probability distribution for the indicated random variable.

9. A red and a green die are rolled, and X is the number on the red die minus the number on the green die.

10. Two dice are rolled, and X is the absolute value of the difference of the two numbers obtained.

11. Two dice are rolled, and X is the largest of the two numbers obtained.

12. Two dice are rolled, and X is the sum of the numbers, except that a "double" is counted twice. For example, two 3s make $X = 12$. (This is used in the game of backgammon.)

13. Three coins are tossed, and X is the number of heads minus the number of tails.

14. Three coins are tossed, and X is twice the number of heads plus half the number of tails.

APPLICATIONS

15. *Grade Point Averages* The grade point averages of the members of your mathematics class are

   3.2, 3.5, 4.0, 2.9, 2.0, 3.3, 3.5, 2.9, 2.5, 2.0, 2.1, 3.2, 3.6, 2.8, 2.5, 1.9, 2.0, 2.2, 3.9, 4.0.

   Use these raw data to construct a frequency table with the following brackets: 1.01–2.0, 2.01–3.0, 3.01–4.0, and find the probability distribution using the midpoint values as the values of X. (Treat grade point averages as continuous data.)

16. *Test Scores* Your scores (out of 10 points) for the 20 surprise math quizzes last semester were

4.5, 9.5, 10.0, 3.5, 8.0, 9.5, 7.5, 6.5, 7.0, 8.0, 8.0, 8.5, 7.5, 7.0, 8.0, 9.0, 10.0, 8.5, 7.5, 8.0.

Use these raw data to construct a frequency table with the following brackets: 2.1 − 4.0, 4.1 − 6.0, 6.1 − 8.0, 8.1 − 10.0, and find the probability distribution using the midpoint values as the values of X. (Treat your scores as continuous data.)

In Exercises 17 through 22, construct a probability distribution, draw the resulting histogram, and calculate the given probabilities.

17. Five darts are thrown at a dartboard. The probability of hitting the bull's eye is $\frac{1}{10}$. Let X be the number of bull's eyes hit. Find $P(X \geq 3)$.

18. Six darts are thrown at a dartboard. The probability of hitting the bull's eye is $\frac{1}{5}$. Let X be the number of bull's eyes hit. Find $P(X \leq 3)$.

19. Select five cards without replacement from a standard deck of 52, and let X be the number of Queens you draw. Find $P(X \leq 2)$.

20. Select five cards without replacement from a standard deck of 52, and let X be the number of red cards you draw. Find $P(X \leq 2)$.

21. A bag contains 4 red marbles, 2 green ones, and 1 yellow one. Suzy grabs four of them at random. Let X be the number of red marbles she holds. Find $P(X \geq 2)$.

22. A bag contains 3 red marbles, 5 green ones, and 1 yellow one. Suzy grabs four of them at random. Let X be the number of green marbles she holds. Find $P(X \leq 2)$.

Rainfall Exercises 23 through 30 are based on the following chart, which shows the number of wet days during the month of September 1993 in 28 cities around the world.* Before doing any of these exercises, you should first find the probability distribution and draw the histogram with X = the number of wet days. (Assume that the experiment is to select one of these cities at random.)

City	Wet Days	City	Wet Days	City	Wet Days	City	Wet Days
Athens	4	Beijing	7	Bermuda	10	Boston	9
Budapest	7	Buenos Aires	8	Cairo	0	Chicago	9
Delhi	4	Dublin	12	Frankfurt	13	Geneva	10
Hong Kong	12	Houston	8	Jerusalem	0	Johannesburg	2
London	13	Los Angeles	1	Madrid	6	Moscow	13
New York	9	Paris	13	Rio de Janeiro	11	Rome	5
San Francisco	2	Sydney	12	Tokyo	12	Toronto	12

▼ *Source: The New York Times*, September 5, 1993. We have omitted the following (very wet) cities from the list: Mexico City: 23 wet days, Miami: 18 wet days, San Juan: 18 wet days, Stockholm: 14 wet days. Also, Washington D.C. (8 wet days) was omitted to trim the list to an even number.

23. Find $P(X \geq 10)$.

24. Find $P(X \geq 11)$.

25. Find $P(1 \leq X \leq 5)$.

26. Find $P(3 \leq X \leq 6)$.

27. If you had spent September 1993 in a randomly selected city on the list, what is the probability that you would have experienced rain for at least seven days?

28. If you had spent September 1993 in a randomly selected city on the list, what is the probability that you would have experienced less than four days of rain?

29. If you had spent September 1993 in a randomly selected city on the list, what is the most likely number of wet days you would have experienced?

30. If you had spent September 1993 in a randomly selected city on the list, what is the least likely number of wet days you would have experienced?

31. *Car Purchases* In order to persuade her parents to contribute to her new car fund, Carmen has spent the last week surveying the ages of 2,000 cars on campus. Her findings are reflected in the following frequency table.

Age of Car in Years	0	1	2	3	4	5	6	7	8	9	10
Number of Cars	140	350	450	650	200	120	50	10	5	15	10

Carmen's jalopy is 6 years old. She would like to make the following claim to her parents: "x percent of students have newer cars than I." Use a probability distribution to find x.

32. *Car Purchases* Carmen's parents, not convinced of her need for a new car, produced the following statistics showing the ages of cars owned by students on the Dean's List.

Age of Car in Years	0	1	2	3	4	5	6	7	8	9	10
Number of Students	0	2	5	5	10	10	15	20	20	20	40

They then claimed that if she kept her 6-year-old car for another year, her chances of getting on the Dean's List would be increased by x percent. Find x.

33. *Health Care* According to a survey, 11% of U.S. males are expected to spend up to three months in a nursing home, 8% are expected to spend at least three months but less than a year in one, 10% are expected to spend at least one year but less than five years in one, and 4% are expected to spend at least five years but less than ten years in one* (the remainder will spend no time in a nursing

▼ *Source: *New England Journal of Medicine*, as reported in *The New York Times*, August 28, 1993, p. 32.

home). Let X be the random variable that assigns to each U.S. male the midpoint, in months, of the range into which his time spent in a nursing home falls. Use the data to construct a probability distribution table for X.

34. *Health Care* According to a survey, 11% of U.S. females are expected to spend less than three months in a nursing home, 10% are expected to spend at least three months but less than a year in one, 18% are expected to spend at least one year but less than five years in one, and 13% are expected to spend at least five years but less than ten years in one* (the remainder will spend no time in a nursing home). Let X be the random variable that assigns to each U.S. female the midpoint, in months, of the range into which her time in a nursing home falls. Use the data to construct a probability distribution table for X.

35. *Employment* A survey found the following distribution for the sizes of the firms where workers aged 18–64 were employed:†

Size of Firm (Number of Employees)	1	2–9	10–24	25–99	100–499	500–999	1,000–5,000
Percentage of Workers	10%	12%	8%	12%	14%	6%	38%

(Thus, for example, 14% of all workers aged 18–64 were employed in a firm with 100 to 499 employees.) Let X be the midpoint in each category of firm size. Construct a histogram showing the probability distribution for X. (Note that the size of a firm is a discrete value.)

36. *Health Insurance* In a survey of workers in different-sized firms, the following data show the number of workers without medical insurance:†

Size of Firm (Number of Employees)	1	2–9	10–24	25–99	100–499	500–999	1,000–5,000
Number of Uninsured Workers	2,100	434	300	292	202	80	392

Let X be the midpoint in each category of firm size. Construct a histogram showing the probability distribution for X. (Note that the size of a firm is a discrete value.)

▼ * Source: *New England Journal of Medicine*, as reported in *The New York Times*, August 28, 1993, p. 32.
† Source: Employee Benefits Research Institute, cited in *The New York Times*, August 23, 1993, p. A9. Figures given are based on the data cited but have been adjusted for computational reasons. The exact number of workers surveyed is fictitious.

37. *Income Distributions* A 1992 *Chicago Tribune* survey of annual incomes of college presidents in 29 research universities reported the following data:

Income Bracket	$100,000–$199,999	$200,000–$299,999	$300,000–$399,999	$400,000–$499,999
Number	2	13	12	2

   (a) Use this information to construct a probability distribution and histogram with X taken to be the midpoint of a university president's income bracket.
   (b) Shade the area of your histogram corresponding to the probability that a college president earns at least $200,000 per year. What is this probability?

38. *Passing the Bar Exam* A *New York Times* survey of all 72,461 people who took the bar exam in 1992 showed the following data on success rates in the 50 states and the District of Columbia:*

Percentage Who Passed	50–59%	60–69%	70–79%	80–89%	90–99%
Number of States	2	7	19	22	1

   (a) Use this information to construct a probability distribution and histogram with X taken to be the midpoint of a percentage bracket.
   (b) Shade the area of your histogram corresponding to the probability that a randomly selected state enjoyed a pass rate of at least 70%. What is this probability?

In Exercises 39 through 42, construct the binomial distribution for a sequence of n Bernoulli trials with p and q as given.

39. $n = 4, p = \frac{1}{2}, q = \frac{1}{2}$
40. $n = 5, p = \frac{1}{2}, q = \frac{1}{2}$
41. $n = 5, p = \frac{1}{3}, q = \frac{2}{3}$
42. $n = 4, p = \frac{1}{3}, q = \frac{2}{3}$

43. *Testing Your Calculator* Use your calculator to generate a sequence of 100 random digits in the range 0–9, and test the random number generator for uniformness by drawing the distribution histogram.

44. *Testing Your Dice* Repeat Exercise 43, but this time use a die to generate a sequence of 50 random numbers in the range 1–6.

45. *Municipal Bond Funds* Following is a list of the annual return (in %) of the 20 largest municipal bond funds as of August 1993.† Use a histogram to classify

---

▼ *Source: *The New York Times*, August 15, 1993. The two states with the lowest percentage pass rate (50%–59%) were California (56%) and D.C. (53%). Utah had the highest percentage pass rate, with 90% of its entrants passing the bar exam. North Dakota, South Dakota, Nebraska, Montana, Kansas, and South Carolina were all runners-up, with 89% pass rates.
†Source: *The New York Times*, September 12, 1993. All figures are rounded to the nearest 0.1%.

the distribution as approximately normal (bell-shaped), approximately uniform, or neither. Your histogram should be based on the following ranges of data: 1.0–1.4, 1.5–1.9, 2.0–2.4, 2.5–2.9. (Since the data are rounded to the nearest 0.1%, treat them as discrete values.)

2.3, 1.7, 1.6, 2.5, 2.0, 1.9, 2.9, 2.4, 1.8, 2.1, 1.4, 2.1, 2.7, 2.0, 2.2, 2.9, 2.4, 2.3, 2.1, 2.7.

46. *Municipal Bond Funds* Repeat the preceding exercise using the following list of annual returns of the next largest 20 funds:

2.3, 2.5, 2.2, 2.1, 2.2, 2.0, 2.2, 1.1, 2.0, 1.9, 2.8, 2.6, 1.9, 1.9, 2.4, 2.0, 1.9, 2.4, 2.7, 2.6.

COMMUNICATION AND REASONING EXERCISES

47. Give two examples of random variables whose probability distributions are uniform.

48. Give two examples of random variables whose probability distributions are binomial.

49. Find a probability distribution whose histogram is highest in the middle.

50. Find a probability distribution whose histogram is lowest in the middle and highest on either side.

7.2 MEAN, MEDIAN, AND MODE

MEAN, MEDIAN, AND MODE OF A COLLECTION OF DATA

You are probably quite familiar with how to calculate the **average** or **mean** of a collection of scores: you add them up and divide by the number of scores you have. There are two other—perhaps less familiar—numbers you can calculate: the **median** and the **mode**. These are illustrated in the following example.

▼ **EXAMPLE 1** Salaries

You are the manager of a corporate department with a staff of 50 employees whose salaries are given in the following frequency table.

Annual Salary	$15,000	$20,000	$25,000	$30,000	$35,000	$40,000	$45,000
Number of Employees	10	9	3	8	12	7	1

What is the average salary earned by an employee in your department? What is the median salary? What is the mode?

SOLUTION To find the average salary, we first need to find the sum of the salaries earned by your employees.

10 employees at $15,000:	10 × 15,000 =	150,000
9 employees at $20,000:	9 × 20,000 =	180,000
3 employees at $25,000:	3 × 25,000 =	75,000
8 employees at $30,000:	8 × 30,000 =	240,000
12 employees at $35,000:	12 × 35,000 =	420,000
7 employees at $40,000:	7 × 40,000 =	280,000
1 employee at $45,000:	1 × 45,000 =	45,000
	Total =	$1,390,000

Thus, the average annual salary is $\frac{1,390,000}{50} = \$27,800$.

Now, the **median** salary is the *lowest* salary with the following property: at least half of your employees earn the median or less, and at least half of your employees earn the median or more. What we are seeking in this example is the lowest salary such that at least 25 employees earn that salary or less, and at least 25 earn that salary or more. We can find the median salary by checking each possibility in turn.

$15,000: only 10 employees earn $15,000 or less

$20,000: only 19 employees earn $20,000 or less

$25,000: only 22 employees earn $25,000 or less

$30,000: 30 employees earn $30,000 or less, and 28 earn $30,000 or more

It follows that the median is $30,000.

The **mode** is the salary earned by the largest number of employees. From the frequency table, we can see that this is $35,000.

Before we go on... We can see the mode in particular quite clearly in the frequency histogram (Figure 1).

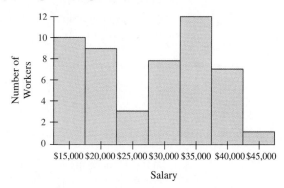

FIGURE 1

The mode is the location of the highest point on the graph. If the histogram had two highest points—say, if 12 people also earned $15,000—then we would say that the data had two modes, or was **bimodal.** Informally, since there is another "peak" in the given data at $15,000, we would often refer to the given distribution of salaries as bimodal. These two peaks might reflect two periods of intense hiring, one recently and one some time ago, with little hiring between those times.

We can also interpret the median and mean directly from the histogram: The median corresponds to the leftmost bar of the histogram with the following properties:

(a) The total area of the median bar and all the bars to the left is at least half the total area of the histogram.
(b) The total area of the median bar and all the bars to the right is at least half the total area of the histogram.

As for the mean, think of the bars in the histogram as having weight proportional to their area. The mean is then the *x*-coordinate of the point on the graph where it would *balance*. In other words, the weight is distributed evenly on either side of the mean.

The calculations of the average and the median from the frequency table are particularly easy to do using either a spreadsheet program or a suitably equipped graphing calculator. Appendix B contains a discussion on using graphing calculators to compute means, medians, and modes. For spreadsheet programs, we shall use Lotus 1-2-3 as an example. We begin by entering the frequency table into the spreadsheet.

	A	B	C	D	E	F	G	H
1	15,000	20,000	25,000	30,000	35,000	40,000	45,000	
2	10	9	3	8	12	7	1	

We then compute in a third row the product *Salary · Number of employees.* To do this, we enter the formula +A1*A2 in cell A3 and then copy this formula into cells B3 through G3. This gives

	A	B	C	D	E	F	G	H
1	15,000	20,000	25,000	30,000	35,000	40,000	45,000	
2	10	9	3	8	12	7	1	
3	150,000	180,000	75,000	240,000	420,000	280,000	45,000	

We can then calculate the total number of employees and the total payroll by entering @SUM(A2..G2) in H2 and @SUM(A3..G3) in H3. This gives

	A	B	C	D	E	F	G	H
1	15,000	20,000	25,000	30,000	35,000	40,000	45,000	
2	10	9	3	8	12	7	1	50
3	150,000	180,000	75,000	240,000	420,000	280,000	45,000	1,390,000

Finally, we can compute the average by dividing H3 by H2, entering +H3/H2 in H4, say. This gives

	A	B	C	D	E	F	G	H
1	15,000	20,000	25,000	30,000	35,000	40,000	45,000	
2	10	9	3	8	12	7	1	50
3	150,000	180,000	75,000	240,000	420,000	280,000	45,000	1,390,000
4								27,800

To compute the median, we begin again with the frequency table. This time we would like to examine, for each salary, the total number of employees making that much *or less*. This is called the **cumulative frequency distribution.** We can calculate the cumulative distribution by first copying A2 into A3, then entering +A3+B2 in B3. Copying this formula into cells C3 through G3 gives us the following.

	A	B	C	D	E	F	G
1	15,000	20,000	25,000	30,000	35,000	40,000	45,000
2	10	9	3	8	12	7	1
3	10	19	22	30	42	49	50

We can then see that the median is $30,000, since this is the lowest salary for which at least 25 employees (half the total) earn that much or less. The cumulative frequency distribution can be graphed as in Figure 2. The median salary is the salary at which the cumulative frequency first touches or crosses the halfway point, 25.

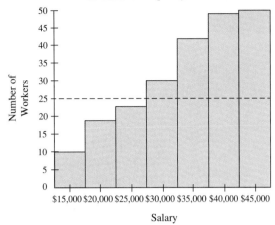

FIGURE 2

Now let us formally define the mean, median, and mode. Before doing that, we introduce a very useful notation. If your collection of scores is $x_1, x_2, \ldots, x_n$, then you calculate their average with the formula

$$\text{Average} = \frac{x_1 + x_2 + \ldots + x_n}{n}.$$

A convenient way of writing the sum that appears in the numerator is to use **summation** or **sigma notation.** We write

$$x_1 + x_2 + \ldots + x_n = \sum_{i=1}^{n} x_i.$$

The notation $\sum_{i=1}^{n} x_i$ is read "the sum, from $i = 1$ to n, of x_i." We think of i as taking on the values $1, 2, \ldots, n$ in turn, making x_i equal $x_1, x_2, \ldots, x_n$ in turn, and we then add up these values.

MEAN, MEDIAN, AND MODE OF A SET OF DATA

Consider a collection of scores $x_1, x_2, \ldots, x_n$. The **mean** or **average** of the collection is

$$\text{Average} = \frac{\left(\sum_{i=1}^{n} x_i\right)}{n}$$

The **median** of the collection is the least score m with the following property: $x_i \leq m$ for at least half of the x_i, and $x_i \geq m$ for at least half the x_i.

The **mode** of the collection is the score that appears most often in the collection.

The mean and the median are both attempts to measure where the "center" of the data is (the mode is also, though to a lesser extent). Which is more appropriate depends on the data. The mean tends to give more weight to scores that are farther away from the center than does the median. For example, if you take the largest score in a collection and make it larger, the mean will increase but the median will remain the same. For this reason the median is often preferred for collections containing a wide range of scores.

EXPECTED VALUE, MEDIAN, AND MODE OF A RANDOM VARIABLE

Now, instead of looking at a collection of data such as X-scores, we will look at the probability distribution of a random variable. We can then calculate numbers analogous to the mean, median, and mode discussed above.

▼ **EXAMPLE 2**

Suppose that you roll a fair die a large number of times. What do you expect the average of the numbers rolled to be? What do you expect the median to be? What do you expect the mode to be?

SOLUTION Suppose we roll the die n times (n is large). We can imagine recording the numbers rolled and then writing down the frequency table. Because the probability of rolling a 1 is $\frac{1}{6}$, we would expect that we would roll a 1 approximately one-sixth of the time, or $n/6$ times. Similarly, each number should appear approximately $n/6$ times. The frequency table should then look like this:

x	1	2	3	4	5	6
Number of times x rolled	$\frac{n}{6}$	$\frac{n}{6}$	$\frac{n}{6}$	$\frac{n}{6}$	$\frac{n}{6}$	$\frac{n}{6}$

To calculate the average, we would next compute the products $1 \cdot (n/6)$, $2 \cdot (n/6)$, and so on, then add these products together and divide by n. This gives us

$$\text{Average} = \frac{1 \cdot \left(\frac{n}{6}\right) + 2 \cdot \left(\frac{n}{6}\right) + 3 \cdot \left(\frac{n}{6}\right) + 4 \cdot \left(\frac{n}{6}\right) + 5 \cdot \left(\frac{n}{6}\right) + 6 \cdot \left(\frac{n}{6}\right)}{n}$$

$$= 1 \cdot \left(\frac{1}{6}\right) + 2 \cdot \left(\frac{1}{6}\right) + 3 \cdot \left(\frac{1}{6}\right) + 4 \cdot \left(\frac{1}{6}\right) + 5 \cdot \left(\frac{1}{6}\right) + 6 \cdot \left(\frac{1}{6}\right)$$

$$= 3.5.$$

This is the expected average value of a large number of rolls, or in short, the **expected value** of a roll of the die. More precisely, we shall say that this is the expected value of the random variable X whose value is the number we get by rolling a die. Notice that n, the number of rolls, does not appear in the expected value. In fact, we could redo the calculation a bit more simply by dividing the frequencies in the table by *n before* adding. Doing this replaces the frequencies with the *probabilities*, $\frac{1}{6}$. That is, it replaces the frequency distribution with the probability distribution.

x	1	2	3	4	5	6
$P(X = x)$	$\frac{1}{6}$	$\frac{1}{6}$	$\frac{1}{6}$	$\frac{1}{6}$	$\frac{1}{6}$	$\frac{1}{6}$

The expected value of X is then the sum of the products $x \cdot P(X = x)$. We shall say more about this after we finish the example.

To find the median of a large number of rolls, we look again at the frequency table above. The value $m = 3$ is the smallest value such that at least half the rolls will be m or less, and at least half will be m or more. The median is therefore $m = 3$.

We can also see the median in the probability distribution. The probability of rolling a 1, 2, or 3 is $\frac{1}{2}$, and the probability of rolling a 4, 5, or 6 is $\frac{1}{2}$. The median m is the lowest number such that $P(X \leq m) \geq \frac{1}{2}$ and $P(X \geq m) \geq \frac{1}{2}$. This is again satisfied by $m = 3$.

Now, the mode should be the number rolled most often. However, since all of the numbers are equally likely to be rolled, none should be rolled more often than the others. Therefore, all six possibilities satisfy the definition of the mode (and the mode is not terribly useful in this example).

> **EXPECTED VALUE, MEDIAN, AND MODE OF A RANDOM VARIABLE**
>
> If X is a finite random variable taking on values $x_1, x_2, \ldots, x_n$, the **expected value** of X, written $E(X)$, is
>
> $$E(X) = x_1 \cdot P(X = x_1) + x_2 \cdot P(X = x_2) + \ldots + x_n \cdot P(X = x_n)$$
> $$= \sum_{i=1}^{n} x_i \cdot P(X = x_i).$$
>
> The **median** of X is the least number m such that
>
> $$P(X \leq m) \geq \tfrac{1}{2} \quad \text{and} \quad P(X \geq m) \geq \tfrac{1}{2}.$$
>
> A **mode** of X is a number m such that $P(X = m)$ is largest. This is the most likely value of X or one of the most likely values if X has several values with the same largest probability.

The expected value of X is what we expect to find for the average of a large number of samples of X. The median is what we expect to find for the median of a large number of samples of X, and a mode is what we expect to find for a mode of a large number of samples of X.

▶ **NOTE** We shall reserve the word *average* for the average of a collection of numbers or *X*-scores. We shall reserve the phrase *expected value* for the value $E(X)$ defined above. The word *mean* is used for both by various people. Some authors use the term **sample mean** for the average. If we refer to the mean *of a random variable* it should be clear that we are referring to the expected value. Again, we shall use *median* both for the median of a collection of numbers and for the median of a random variable, as defined above. Which we mean should be clear from context. The same is true of the word *mode*. ◀

▼ **EXAMPLE 3**

If you roll a pair of dice, what is the expected value of the sum of the two numbers you get? What are the median and the mode?

SOLUTION We take the sample space for this experiment to be the set of all pairs of integers in the range 1–6.

$$S = \begin{Bmatrix} (1,1) & (1,2) & (1,3) & (1,4) & (1,5) & (1,6) \\ (2,1) & (2,2) & (2,3) & (2,4) & (2,5) & (2,6) \\ (3,1) & (3,2) & (3,3) & (3,4) & (3,5) & (3,6) \\ (4,1) & (4,2) & (4,3) & (4,4) & (4,5) & (4,6) \\ (5,1) & (5,2) & (5,3) & (5,4) & (5,5) & (5,6) \\ (6,1) & (6,2) & (6,3) & (6,4) & (6,5) & (6,6) \end{Bmatrix}.$$

Each of these outcomes is equally likely. The possible values of *X* are in the range 2–12. Counting possibilities, we obtain the following probability distribution.

x	2	3	4	5	6	7	8	9	10	11	12
$P(X = x)$	$\frac{1}{36}$	$\frac{2}{36}$	$\frac{3}{36}$	$\frac{4}{36}$	$\frac{5}{36}$	$\frac{6}{36}$	$\frac{5}{36}$	$\frac{4}{36}$	$\frac{3}{36}$	$\frac{2}{36}$	$\frac{1}{36}$

To help you see the pattern in these probabilities, we have not reduced the fractions to lowest terms. (Also, we shall be adding fractions in the next step, so we want to keep the denominators the same.) We now add another row in which we compute the products $x \cdot P(X = x)$ and then add these products together.

x	2	3	4	5	6	7	8	9	10	11	12	
$P(X = x)$	$\frac{1}{36}$	$\frac{2}{36}$	$\frac{3}{36}$	$\frac{4}{36}$	$\frac{5}{36}$	$\frac{6}{36}$	$\frac{5}{36}$	$\frac{4}{36}$	$\frac{3}{36}$	$\frac{2}{36}$	$\frac{1}{36}$	**Total**
$xP(X = x)$	$\frac{2}{36}$	$\frac{6}{36}$	$\frac{12}{36}$	$\frac{20}{36}$	$\frac{30}{36}$	$\frac{42}{36}$	$\frac{40}{36}$	$\frac{36}{36}$	$\frac{30}{36}$	$\frac{22}{36}$	$\frac{12}{36}$	$\frac{252}{36}$

The expected value is $\frac{252}{36} = 7$. In other words, if you repeatedly roll a pair of dice, the average value of the sum of the numbers will be approximately 7.

To calculate the median, we calculate the **cumulative probability distribution,** given by the numbers $P(X \leq x)$.

x	2	3	4	5	6	7	8	9	10	11	12
$P(X = x)$	$\frac{1}{36}$	$\frac{2}{36}$	$\frac{3}{36}$	$\frac{4}{46}$	$\frac{5}{36}$	$\frac{6}{36}$	$\frac{5}{36}$	$\frac{4}{36}$	$\frac{3}{36}$	$\frac{2}{36}$	$\frac{1}{36}$
$P(X \leq x)$	$\frac{1}{36}$	$\frac{3}{36}$	$\frac{6}{36}$	$\frac{10}{36}$	$\frac{15}{36}$	$\frac{21}{36}$	$\frac{26}{36}$	$\frac{30}{36}$	$\frac{33}{36}$	$\frac{35}{36}$	$\frac{36}{36}$

We can now see that the median is 7 because $P(X \leq 7) \geq \frac{1}{2}$ and 7 is the smallest number for which this is true.

Finally, we can see from the probability distribution that the most likely roll is 7, so the mode is 7.

Before we go on... These answers are not surprising in view of the symmetry of the probability distribution (Figure 3).

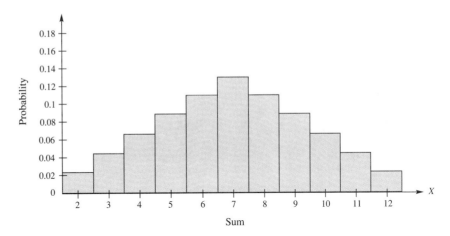

FIGURE 3

Just as we said in Example 1, the mean is the point at which the graph would balance, and by the symmetry this is clearly at $X = 7$. The median is the point at which half the area is to the left and half to the right, and again this is at 7 because of the symmetry. Finally, the mode is the highest point on the graph, which happens to come at 7, right in the center.

A roulette wheel (of the kind used in the U.S.) has the numbers 1 through 36, 0 and 00. A bet on a single number pays 35 to 1. This means that if you place a $1 bet on a single number and win (that is, your number comes up), you get your $1 back plus $35 in winnings. If you lose, you simply lose the $1 that you bet.

EXAMPLE 4 Roulette

What is the expected gain from a $1 bet on a single number? What is the median gain, and what is the mode?

SOLUTION The probability of winning is $\frac{1}{38}$, so the probability of losing is $\frac{37}{38}$. Let X be the gain from a $1 bet. X has two possible values: $X = -1$ if you lose and $X = 35$ if you win. $P(X = -1) = \frac{37}{38}$ and $P(X = 35) = \frac{1}{38}$. This probability distribution and the calculation of the expected value are given in the following table.

X	-1	35	
$P(X = x)$	$\frac{37}{38}$	$\frac{1}{38}$	Total
$xP(X = x)$	$-\frac{37}{38}$	$\frac{35}{38}$	$-\frac{2}{38}$

So we expect to average a small loss of $\frac{2}{38} \approx \$0.053$ on each spin of the wheel.

From the probability distribution we see that the median gain is -1: that is, a loss of $1. The mode is also a loss of $1.

Before we go on... Of course, you cannot actually lose the expected $0.053 on one spin of the wheel. However, if you play many times, this is what you expect to *average*.

A betting game in which the expected value is 0 is called a **fair game.** For example, if you and I flip a coin, and I give you $1 each time it comes up heads, but you give me $1 each time it comes up tails, then the game is fair. Over the long run, we expect to come out even. On the other hand, a game such as roulette in which the expected value is not 0 is **biased** in favor of one of the players. All casino games are slightly biased in favor of the house (except possibly blackjack for a skilled player). Their medians are also often slightly in favor of the house. Thus, any individual gambler has close to a 50% chance of winning, and many gamblers will actually win something (and then return to play some more). However, averaged over the huge numbers of people playing, the house is guaranteed to come out ahead. This is how casinos make (lots of) money.

EXAMPLE 5 Insurance

Floyds of Cape May insures supertankers against major oil spills. The premium is $200,000 (= $0.2 million) per transatlantic crossing. There are three categories of major oil spills: near Atlantic City, coastal but away from Atlantic City, and out to sea. In the event of a coastal spill, the supertanker company will claim $10 million to cover cleanup operations unless the spill was near Atlantic City, in which case the expected claim would be $100

7.2 Mean, Median, and Mode

million due to lawsuits by casino owners. In the case of a spill out to sea, the claim will be the cost of the oil cargo—$1 million on average. Floyds' crack actuarial team estimates that a typical vessel has a 1-in-1,000 chance of having an oil spill per transatlantic crossing. It also estimates that one-tenth of all major oil spills are coastal and one-half of major coastal spills are near Atlantic City. How much profit (or loss) does Floyds expect to make per transatlantic crossing? What is its median profit per crossing? What is its most likely profit per crossing?

SOLUTION First, we consider the sample space. One of the possible outcomes (and surely the one preferred by Floyds) is not mentioned: that an oil tanker experiences no major oil spill, in which case there is no claim. Thus, the sample space is

$$S = \{\text{no oil spill, near Atlantic City, coastal but away from Atlantic City, out to sea}\}$$

We need to find the probabilities of each of these outcomes. Some of our information is in the form of conditional probabilities. For instance, "one-tenth of all major oil spills are coastal and ... one-half of major coastal spills are near Atlantic City." A good way to organize these data is to use a tree (Figure 4).

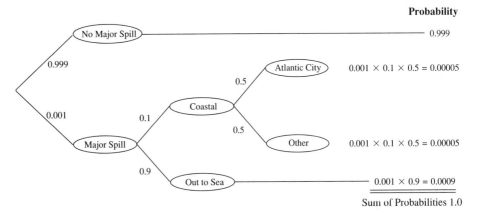

FIGURE 4

As we saw in the last chapter, we obtain the probability of each outcome by multiplying the probabilities along the branches. For example, the probability of a major coastal spill near Atlantic City is $0.001 \times 0.1 \times 0.5 = 0.00005$. Figure 4 also includes a check that the sum of the probabilities of all the outcomes is 1, as it should be.

The random variable X is the profit Floyds will make (in millions). To calculate the values of X, we need to do a little accounting.

No major oil spill: The premium is $0.2 million, and Floyds pays out nothing, so $X = 0.2$.

Major spill near Atlantic City: The premium is $0.2 million, and Floyds pays out $100 million, so $X = 0.2 - 100 = -99.8$.

Coastal spill away from Atlantic City: The premium is $0.2 million, and Floyds pays out $10 million, so $X = 0.2 - 10 = -9.8$.

Major spill out to sea: The premium is $0.2 million, and Floyds pays out $1 million, so $X = 0.2 - 1 = -0.8$.

Now we can set up a table to calculate the expected value of X.

x	−99.8	−9.8	−0.8	0.2	**Total**
$P(X = x)$	0.00005	0.00005	0.0009	0.999	
$xP(X = x)$	−0.00499	−0.00049	−0.00072	0.1998	0.1936

Thus, Floyds can expect to make $193,600 per Atlantic crossing.

From the table, we can see that the median profit is $200,000 and that this is also the mode (the most likely profit).

▼ **EXAMPLE 6**   Guessing on an Exam

An exam has 10 multiple-choice questions, each having 4 choices, and each question is worth 10 points. What is the expected grade of a student who guesses on each question? What is the median grade, and what is the most likely grade?

SOLUTION Let X be the grade received, so X has possible values 0, 10, 20, ... , 100. X has a binomial distribution: the probability $P(X = 10r)$ is the probability of r successes (successful guesses) in $n = 10$ trials with probability $p = \frac{1}{4}$ of success and $q = \frac{3}{4}$ of failure. Thus,

$$P(X = 10r) = C(10, r)\left(\frac{1}{4}\right)^r \left(\frac{3}{4}\right)^{10-r}.$$

Computing these probabilities gives us the following probability distribution.

x	0	10	20	30	40	50	60	70	80	90	100
$P(X = x)$	0.056	0.188	0.282	0.250	0.146	0.058	0.016	0.003	0.0004	0.00003	0.000001

The histogram for this probability distribution is shown in Figure 5.

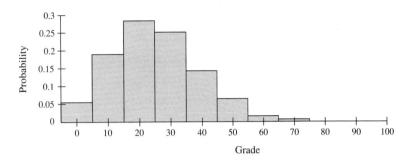

FIGURE 5

We calculate the expected value using the following table.

x	0	10	20	30	40	50	60	70	80	90	100
$P(X = x)$	0.056	0.188	0.282	0.250	0.146	0.058	0.016	0.003	0.0004	0.00003	0.000001
$xP(X = x)$	0	1.88	5.64	7.50	5.84	2.90	0.96	0.21	0.032	0.0027	0.0001
										$E(X)$	25

Thus, a student who guesses on all 10 questions can expect to earn a grade of 25 out of 100. We find the median by calculating the cumulative distribution.

x	0	10	20	30	40	50	60	70	80	90	100
$P(X = x)$	0.056	0.188	0.282	0.250	0.146	0.058	0.016	0.003	0.0004	0.00003	0.000001
$P(X \leq x)$	0.056	0.244	0.526	0.776	0.922	0.980	0.996	0.999	1	1	1

(The sum of the probabilities is not exactly 1 because the probabilities were rounded off. With a programmable calculator or computer it is easier and more accurate not to round off in the middle of a calculation like this.) From the cumulative probabilities, we see that the median grade is 20. Finally, from the probability distribution or from its histogram, we see that the mode (the most likely grade) is 20.

Before we go on... Of course, the expected grade, 25, is not a possible grade on this test. However, if many students use the strategy of guessing, they will average a grade of 25. Or if one student uses this strategy many times, he or she will average a grade of 25.

There is an easy formula for the expected value of a binomial distribution:

> **EXPECTED VALUE OF A BINOMIAL DISTRIBUTION**
>
> If X is the number of successes in a sequence of n independent Bernoulli trials, with probability p of success in each trial, then
>
> $$E(X) = np.$$

For instance, in the example above, there are 10 trials, with $p = \frac{1}{4}$ and $n = 10$. Thus, if Y is the variable that counts the number of questions answered correctly, we get

$$E(Y) = 10 \times \frac{1}{4} = 2.5.$$

Because each question is worth 10 points, we must multiply this answer by 10. Thus,

$$E(X) = 25,$$

as we obtained the hard way in the example.

Q Where does this formula come from?

A We *could* try to compute the sum

$$E(X) = 0 \cdot C(n, 0)p^0 q^n + 1 \cdot C(n, 1)p^1 q^{n-1} + 2 \cdot C(n, 2)p^2 q^{n-2} + \ldots + n \cdot C(n, n)p^n q^0$$

in general, but this is one of the many places in mathematics where a less direct approach is much easier. X is the number of successes in a sequence of n Bernoulli trials, each with probability p of success. Thus, p is the fraction of time we expect a success, so out of n trials we expect np successes. Since X counts successes, we expect the value of X to be np. (With a little more effort, this can be made into a formal proof that the sum above equals np.)

> **MEDIAN AND MODE OF A BINOMIAL DISTRIBUTION**
>
> If X is the number of successes in a sequence of n independent Bernoulli trials, with probability p of success in each trial, then to find its mode we take $(n + 1)p$ and round down (if necessary) to get an integer. If $(n + 1)p$ is already an integer, then both $(n + 1)p - 1$ and $(n + 1)p$ will be modes. If n is large and p is not too close to 0 or 1, the median is approximately equal to the mean, np.

The statements about n large and p not too close to 0 or 1 will be made more precise in Section 4, where we will also see an explanation of that approximation of the median. In general, it is difficult to write a formula for the median, but if n is small it is not too hard to find it directly. The formula for the mode can be found by considering the ratio $P(X = x + 1)/P(X = x)$. As long as this ratio is greater than 1 the probability is increasing, but when it becomes less than 1 the probability decreases. The point where it equals 1 is not hard to find.

▼ **EXAMPLE 7**   Income

A survey is made of the incomes of 1,000 lawyers chosen randomly from all currently practicing lawyers in the U.S. The results are shown in the following table.

Income Bracket	$20,000–$29,999	$30,000–$39,999	$40,000–$49,999	$50,000–$59,999	$60,000–$69,999	$70,000–$79,999	$80,000–$89,999
Number	20	80	230	400	170	70	30

Estimate the mean, median, and mode of the incomes of all currently practicing lawyers in the U.S.

SOLUTION   We first interpret the question in terms of a random variable. Let X be the income of a lawyer selected at random from among all currently practicing lawyers in the U.S. What we are given is a collection of 1,000 X-scores, and we are asked to find the mean, median, and mode of X. It can be shown that in a very precise sense, the best estimates of the mean, median, and mode of X are the average, median, and mode of the collection of X-scores, if the X-scores are all that we know about X. Calculating these in the usual way, we get

$$\text{Average salary} \approx \$54,500$$
$$\text{Median salary} \approx \$55,000$$
$$\text{Mode (most common salary)} \approx \$55,000.$$

▶ **7.2 EXERCISES**

In Exercises 1 through 6, calculate the mean of the given sets of data.

1. 2, 5, 6, 8, −1, −3
2. 3, 1, 6, −3, 0, 5
3. $\frac{1}{2}, \frac{3}{2}, -4, \frac{5}{4}$
4. $-\frac{3}{2}, \frac{3}{8}, -1, \frac{5}{2}$
5. 2.5, −5.4, 4.1, −0.1, 1.1
6. 4.2, −3.2, 0, 1.7, −11.5

In Exercises 7 through 14, calculate the expected value, median, and mode of X for each of the given probability or frequency distributions.

7.

x	10	20	30	40
$P(X = x)$	$\frac{15}{50}$	$\frac{20}{50}$	$\frac{10}{50}$	$\frac{5}{50}$

8.

x	2	4	6	8
$P(X = x)$	$\frac{1}{20}$	$\frac{15}{20}$	$\frac{2}{20}$	$\frac{2}{20}$

9.

x	−5	−1	0	2	5	10
$P(X = x)$	0.2	0.3	0.2	0.1	0.2	0.0

10.

x	−20	−10	0	10	20	30
$P(X = x)$	0.2	0.4	0.2	0.1	0.0	0.1

11.

Income Bracket	$0–$19,999	$20,000–$29,999	$30,000–$39,999	$40,000–$49,999	$50,000–$59,999
Probability	0.2	0.3	0.2	0.2	0.1

12.

Cost of a Car	$0–$1,999	$2,000–$2,999	$3,000–$3,999	$4,000–$4,999	$5,000–$5,999
Probability	0.1	0.3	0.3	0.2	0.1

13.

Population	0–4,999	5,000–9,999	10,000–14,999	15,000–19,999	20,000–24,999
Number of Cities	4	7	2	5	2

14.

Student Population	0–1,999	2,000–3,999	4,000–5,999	6,000–7,999	8,000–9,999
Number of Colleges	2	5	8	5	5

In Exercises 15 through 26, calculate the expected value, median, and mode of the given random variable X.

15. X is the number of tails that come up when a coin is tossed twice.

16. X is the number of tails that come up when a coin is tossed three times.
17. X is the highest number when two dice are cast.
18. X is the lowest number when two dice are cast.
19. X is the number of red marbles that Suzy has in her hand after she selects four marbles from a bag containing 4 red marbles and 2 green ones.
20. X is the number of green marbles that Suzy has in her hand after she selects four marbles from a bag containing 3 red marbles and 2 green ones.
21. A red and a green die are rolled, and X is the number on the red die minus the number on the green die.
22. Two dice are rolled, and $X =$ the absolute value of the difference of the two numbers.
23. Twenty darts are thrown at a dartboard. The probability of hitting the bull's eye is $\frac{1}{10}$. Let X be the number of bull's eyes hit.
24. Thirty darts are thrown at a dartboard. The probability of hitting the bull's eye is $\frac{1}{5}$. Let X be the number of bull's eyes hit.
25. Select five cards without replacement from a standard deck of 52, and let X be the number of Queens you draw.
26. Select five cards without replacement from a standard deck of 52, and let X be the number of red cards you draw.

APPLICATIONS

27. *Investments* *Fidelity Investments* presented the following pie chart in an ad for its Fidelity Asset Manager™ mutual fund, showing the fund's approximate investment breakdown as of June 30, 1993:*

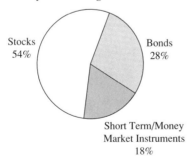

Fidelity Asset Manager™ Mutual Fund

Stocks 54%
Bonds 28%
Short Term/Money Market Instruments 18%

Find the expected annual return (in %) from the fund, assuming that stock investments yield 15%, bonds yield 10%, and short term/money market investments yield 20%. (Take $X =$ annual yield in %.)

---
* The ad appeared in the *The New York Times*, September 16, 1993, p. D15.

28. *Sales Diversification* Dial Corporation's 1993 sales by business line were estimated as shown in the following chart (figures are in millions).*

Dial Corporation 1993 Sales

- Personal Care: $515
- Household and laundry: $685
- Transportation manufacturing: $580
- Consumer services: $1,600
- Consumer products: $1,435

Sales in millions of dollars

Assuming that personal care sales are expected to increase by 10%, household and laundry sales are expected to decrease by 10%, and sales in all the other categories are expected to remain the same, estimate the expected percentage increase or decrease in overall sales. (Take X = percentage increase in sales.)

29. *Farm Population, Female* The following table shows the number of females residing on U.S. farms in 1990, broken down by age.† Numbers are in thousands.

Age	0–14	15–24	25–34	35–44	45–54	55–64	65–74	75–94
Number	459	265	247	319	291	291	212	126

What is the average age of a female residing on a U.S. farm? What is the median age? What is the mode?

30. *Farm Population, Male* The following table shows the number of males residing on U.S. farms in 1990, broken down by age.† Numbers are in thousands.

Age	0–14	15–24	25–34	35–44	45–54	55–64	65–74	75–94
Number	480	324	285	314	302	314	247	118

What is the average age of a male residing on a U.S. farm? What is the median age? What is the mode?

Exercises 31 through 36 are based on exercises in the last section, but these go a little further.

31. *Car Purchases* In order to persuade her parents to contribute to her new car fund, Carmen has spent the last week surveying the ages of 2,000 cars on campus. Her findings are reflected in the following frequency table.

▼ *Source: Company Reports; Donaldson, Lufkin & Jenrette, as cited in *The New York Times*, July 18, 1993.
† Source: Economic Research Service, U.S. Department of Agriculture and Bureau of the Census, U.S. Department of Commerce, 1990.

Age of Car in Years	0	1	2	3	4	5	6	7	8	9	10
Number	140	350	450	650	200	120	50	10	5	15	10

Carmen's jalopy is 6 years old. She would like to support her request by concluding that her car is considerably older than the average campus car. How much older is it? How much older than the median age is it? How much older than the mode?

32. *Car Purchases* Carmen's parents, not convinced of her need for a new car, produced the following statistics showing the ages of cars owned by students on the Dean's List.

Age of Car in Years	0	1	2	3	4	5	6	7	8	9	10
Number	0	2	5	5	10	10	15	20	20	20	40

They then told her that they would help subsidize a new car as soon as her car was two years older than the average age of a Dean's List student's car. How long will she have to wait? How long would she have to wait if they used the median? How long if they used the mode?

33. *Employment* A survey found the following distribution for the sizes of the firms where workers aged 18–64 were employed:*

Size of Firm (Number of Employees)	1	2–9	10–24	25–99	100–499	500–999	1,000–5,000
Percentage of Workers	10%	12%	8%	12%	14%	6%	38%

What was the average firm size for workers aged 18–64? The median firm size? The most common firm size? (Note that the size of a firm is a discrete value.)

34. *Health Insurance* In a survey of 3,800 workers in different-sized firms, the following data show the number of workers without medical insurance:†

Size of Firm (Number of Employees)	1	2–9	10–24	25–99	100–499	500–999	1,000–5,000
Number of Uninsured Workers	2,100	434	300	292	202	80	392

Assuming that these frequencies are representative of the whole labor force, what was the average firm size for an uninsured worker? The median firm size? The most common firm size? (Note that the size of a firm is a discrete value.)

▼ *Source: Employee Benefits Research Institute, cited in *The New York Times*, August 23, 1993.

† Source: Ibid. Figures given are based on the data cited but have been adjusted for computational reasons. The exact number of workers surveyed is fictitious.

35. *Income Distribution* A 1992 *Chicago Tribune* survey of annual incomes of college presidents in 29 research universities reported the following data:

Income Bracket	$100,000–$199,999	$200,000–$299,999	$300,000–$399,999	$400,000–$499,999
Number	2	13	12	2

(a) What was the average salary of a college president in a research university?

(b) How would this figure be affected if the two top salaries were known to be $400,000?

36. *Passing the Bar Exam* A *New York Times* survey of all 72,461 people who took the bar exam in 1992 showed the following data on success rates in the different states (including the District of Columbia):*

Percentage Who Passed	50%–59%	60%–69%	70%–79%	80%–89%	90%–99%
Number of States	2	7	19	22	1

(a) If a group of bar exam candidates was randomly scattered among the states, what percentage would you expect to pass?

(b) How would your answer to part (a) change if all of the states in the 70–79% bracket had pass rates of exactly 70%?

37. *Insurance Schemes* The Acme Insurance Company is launching a drive to generate greater profits, and decides to insure racetrack drivers against wrecking their cars. Acme's research shows that, on average, a racetrack driver races four times a year and has a 1 in 10 chance of wrecking a vehicle, worth an average of $100,000, in every race. The annual premium is $5,000, and Acme automatically drops any driver who is involved in an accident (after paying for a new car), but does not refund the premium. How much profit (or loss) can Acme expect to earn from a typical driver in a year? [*Hint:* Use a tree to compute the probabilities of the various outcomes.]

38. *Insurance* The Blue Sky Flight Insurance company insures passengers against air disasters, charging a prospective passenger $20 for coverage on a single plane ride. In the event of a fatal air disaster, the company pays out $100,000 to the named benefactor. In the event of a non-fatal disaster, it pays out an average of $25,000 for hospital expenses. Given that the probability of a plane crashing on a single trip is 0.00000087,† and that a passenger involved in a plane crash has a 0.9 chance of being killed, determine the profit (or loss) per passenger that the insurance company expects to make on each trip. [*Hint*: Use a tree to compute the probabilities of the various outcomes.]

---

*Source: *The New York Times*, August 15, 1993.

†This was the probability of a passenger plane crashing per departure in 1990. (Source: National Transportation Safety Board.)

39. *Roulette* A roulette wheel has the numbers 1 through 36, 0 and 00. Half of the numbers from 1 through 36 are red, and a bet on red pays even money (that is, if you win you will get back your $1 plus another $1). How much do you expect to win with a $1 bet on red?

40. *Roulette* A roulette wheel has the numbers 1 through 36, 0 and 00. A bet on two numbers pays 17 to 1 (that is, if you win you get back your $1 plus another $17). How much do you expect to win with a $1 bet on two numbers?

41. *Football* The following chart lists the number of passes caught in a season by NFL conference leaders in the years 1960–1990:*

Passes Caught	50–59	60–69	70–79	80–89	90–99	100–109
Frequency	2	4	13	7	3	2

Assuming that these frequencies continue to hold in future seasons, what is the average number of passes you would expect an NFL conference leader to catch in a football season?

42. *Football* The following chart lists the number of passes completed in a season by NFL conference leaders in the years 1960–1990:*

Passes Completed	50–99	100–149	150–199	200–249	250–299	300–349
Frequency	1	3	10	6	9	2

Assuming that these frequencies continue to hold in future seasons, what is the average number of passes you would expect an NFL conference leader to complete in a football season?

43. *Airline Safety* In 1989, the probability that a plane would be involved in a fatal accident was 0.00000165 for a single trip.† Assume that once a plane is involved in a fatal accident, it is scrapped.
   (a) What is the probability that a plane would be involved in a fatal accident in 1 trip, 2 trips or less, 3 trips or less, n trips or less? $\left[\text{Hint: Use a tree. The formula for the sum } a + ar + ar^2 + ar^3 + \ldots + ar^n \text{ is given by } \dfrac{a(1 - r^{n+1})}{1 - r}.\right]$
   (b) How many trips do you expect a plane to make up to and including the time that it is involved in a fatal accident (assuming its age does not affect the likelihood of an accident)? $\left[\text{Hint: The formula for the infinite sum } a + 2ar + 3ar^2 + 4ar^3 + \ldots \text{ is } \dfrac{a}{(1 - r)^2} \text{ if } |r| < 1.\right]$

44. *Airline Safety* Repeat the preceding exercise using the 1990 figure of 0.00000087.†

---
* Source: *World Almanac and Book of Facts 1992* (New York: Pharos Books, 1992)
† Source: National Transportation Safety Board.

COMMUNICATION AND REASONING EXERCISES

45. Give an example of a random variable whose expected value and median are equal.

46. Give an example of a random variable whose expected value is larger than its median.

47. Give an example of a random variable whose expected value is larger than its mode.

48. Teachers sometimes complain when the grade distribution in their classes is bimodal. Why would this present a problem?

7.3 VARIANCE AND STANDARD DEVIATION

VARIANCE AND STANDARD DEVIATION FOR RAW DATA

The following chart lists typical point values of grades in college.*

Grade	A	A–	B+	B	B–	C+	C	C–	D+	D	F
Grade Points	4.0	3.7	3.3	3.0	2.7	2.3	2.0	1.7	1.3	1.0	0.0

Suppose Jodi has earned grades in her first year of A–, B+, B+, B, B, B, B, B–, B–, and C+ in her courses. Then her average is 3.0 (a B). Suppose that Safwan has earned grades of A, A, A–, B+, B+, B–, B–, C+, C, and C in his first year. Then his average is also 3.0. Figure 1 shows histograms for both student's first-year grades.

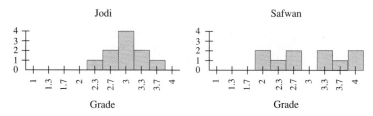

FIGURE 1

Although the two students have the same average, their grades are distributed rather differently. Jodi's grades are concentrated close to her average, while Safwan's are more spread out. We would say that there is more **variation** in Safwan's grades or that they are more widely **dispersed**.

▼ *These are the values at Hofstra University. See your college bulletin for the figures at your school.

7.3 Variance and Standard Deviation

Q Is there a way of *measuring* the dispersion of a set of numbers, such as the sets of Jodi's grades and Safwan's grades?

A If the numbers are $x_1, x_2, \ldots, x_n$ and their average is $\bar{x}$, we are really interested in the distribution of the differences $x_i - \bar{x}$. We could compute the *average* of these differences, but this average will always be 0 (why?). It is really the *sizes* of these differences that interest us, so we might try computing the average of the absolute values of the differences. This idea is reasonable but leads to technical difficulties avoided by a slightly different approach. We shall compute the average of the *squares* of the differences. This average is called the **population variance.** Its square root is called the **population standard deviation.** We shall write σ for the population standard deviation and σ^2 for the population variance. Another notation sometimes used is σ_n for σ and σ_n^2 for σ^2. At the end of this section we will briefly discuss a common variation called the **sample variance.**

POPULATION VARIANCE AND POPULATION STANDARD DEVIATION

Given a set of numbers $x_1, x_2, \ldots, x_n$ with average $\bar{x}$, the **population variance** is

$$\sigma^2 = \frac{(x_1 - \bar{x})^2 + (x_2 - \bar{x})^2 + \ldots + (x_n - \bar{x})^2}{n} = \frac{1}{n}\sum_{i=1}^{n}(x_i - \bar{x})^2$$

and the **population standard deviation** is

$$\sigma = \sqrt{\sigma^2}.$$

▼ **EXAMPLE 1**

Calculate the population variance and population standard deviation for each set of grades given above.

SOLUTION We work with Jodi's grades first. We can organize the calculations in a table.

x_i	$x_i - \bar{x}$	$(x_i - \bar{x})^2$
3.7	0.7	0.49
3.3	0.3	0.09
3.3	0.3	0.09
3.0	0	0
3.0	0	0
3.0	0	0
3.0	0	0
2.7	−0.3	0.09
2.7	−0.3	0.09
2.3	−0.7	0.49
Total 30.0	0	1.34

The second column, $x_i - \bar{x}$, is obtained by subtracting the average, $\bar{x} = 3.0$, from each of the entries in the first column. The entries in the last column are the squares of the entries in the second column.

The population variance, σ^2, is the average of the entries in the right-hand column. We compute this average by dividing the rightmost total by the number of grades, getting

$$\sigma^2 = \frac{1.34}{10} = 0.134$$

The population standard deviation, σ, is the square root of σ^2.

$$\sigma = \sqrt{0.134} \approx 0.37$$

Turning to Safwan's grades, we create the following table.

x_i	$x_i - \bar{x}$	$(x_i - \bar{x})^2$
4.0	1.0	1
4.0	1.0	1
3.7	0.7	0.49
3.3	0.3	0.09
3.3	0.3	0.09
2.7	−0.3	0.09
2.7	−0.3	0.09
2.3	−0.7	0.49
2.0	−1.0	1
2.0	−1.0	1
Total 30.0	0	5.34

This gives

$$\sigma^2 = \frac{5.34}{10} = 0.534,$$

and

$$\sigma = \sqrt{0.534} \approx 0.73.$$

Notice how much larger both the variance and standard deviation of Safwan's grades are than those of Jodi's grades.

Before we go on... In this example the variance and standard deviation measure how *consistent* the two students are. Jodi is more consistent: her grades are clustered around her average. Safwan, on the other hand, is more likely to earn a grade farther away from his average.

Notice that we can divide the total of the x_i-column by n in either table above to compute the average. Further, the sum of the values of $x_i - \bar{x}$ is zero. *This will always be the case* (why?). This provides a useful check of our arithmetic. If the total of the $(x_i - \bar{x})$-column fails to be zero, then we either made an error in calculating the values $x_i - \bar{x}$, or there is a slight error due to rounding.

 Most spreadsheet programs and many calculators can calculate the population variance and standard deviation of a set of numbers automatically. All you do is enter the numbers and instruct the program to display the population variance or standard deviation. (See Appendix B for a discussion on using a graphing calculator.) Alternatively, you can use your spreadsheet program to duplicate the tables above by entering the numbers in a single row and then programming it to calculate the remaining rows.*

Q Why should we take the square root of the variance? Why do we bother calculating the standard deviation?

A Taking the square root corrects for the fact that the variance is the average of the *squares* of the differences $x_i - \bar{x}$. The variance will have units that are the squares of the units of the data. For example, if the units of the data are grade points, as in the example above, the variance will be measured in (grade points)^2. Taking the square root makes the standard deviation have the same units as the original data (in this case, grade points), so the standard deviation gives a value that may be easier to interpret. We shall have more to say about the significance of the standard deviation below, when we discuss the standard deviation of a random variable.

Before returning to random variables, we derive another formula for σ^2 that is sometimes easier to compute.

Take the formula

$$\sigma^2 = \frac{(x_1 - \bar{x})^2 + (x_2 - \bar{x})^2 + \ldots + (x_n - \bar{x})^2}{n}$$

and expand each of the terms $(x_i - \bar{x})^2$ into $x_i^2 - 2x_i\bar{x} + \bar{x}^2$, getting

$$\sigma^2 = \frac{(x_1^2 - 2x_1\bar{x} + \bar{x}^2) + (x_2^2 - 2x_2\bar{x} + \bar{x}^2) + \ldots + (x_n^2 - 2x_n\bar{x} + \bar{x}^2)}{n}.$$

▼ *See *Topics in Finite Mathematics: An Introduction to the Electronic Spreadsheet* by Samuel W. Spero (New York: HarperCollins, 1993).

Grouping like terms together gives

$$\sigma^2 = \frac{x_1^2 + x_2^2 + \ldots + x_n^2 - 2\bar{x}(x_1 + x_2 + \ldots + x_n) + (\bar{x}^2 + \bar{x}^2 + \ldots + \bar{x}^2)}{n}$$

$$= \frac{x_1^2 + x_2^2 + \ldots + x_n^2}{n} - 2\bar{x}\frac{x_1 + x_2 + \ldots + x_n}{n} + \frac{n\bar{x}^2}{n}.$$

Now notice three things. The first term is the average of the x_i^2. The second term is $2\bar{x}$ times the average of the x_i, which means the second term is really $2\bar{x}^2$. Finally, the last term simplifies to $\bar{x}^2$. Therefore, we can write

$$\sigma^2 = \frac{1}{n}\sum_{i=1}^{n} x_i^2 - \bar{x}^2.$$

In words, the population variance is the average of the squares of the numbers minus the square of the average of the numbers (these are not the same thing!).

VARIANCE AND STANDARD DEVIATION OF A RANDOM VARIABLE

Suppose that X is a random variable and that $x_1, x_2, \ldots, x_n$ are (a large number of) X-scores. What do we expect to get for the population variance and standard deviation? We know that for the mean $\bar{x}$ we expect to get $E(X)$, the expected value of X. We can expect the variance then to be the average of the numbers $[x_i - E(X)]^2$. But this is the average of some samples of the values of the random variable $[X - E(X)]^2$, so we expect it to be $E([X - E(X)]^2)$.

VARIANCE AND STANDARD DEVIATION OF A RANDOM VARIABLE

If X is a random variable, its **variance** is defined to be

$$Var(X) = E([X - E(X)]^2).$$

Its **standard deviation** is defined to be

$$\sigma(X) = \sqrt{Var(X)}.$$

Q How do we go about computing $Var(X)$ and $\sigma(X)$?

A There is a straightforward way and an easier way. Suppose, as we still are doing, that X is a finite random variable taking on the values $x_1, x_2, \ldots, x_n$. Then

$$P(X - E(X) = x_i - E(X)) = P(X = x_i),$$

so

$$Var(X) = E([X - E(X)]^2) = \sum_{i=1}^{n}[x_i - E(X)]^2 P(X = x_i).$$

This is the straightforward calculation. There is an easier formula analogous to the alternate formula for population variance we derived earlier. If we take the formula for variance just above, expand $[x_i - E(X)]^2$, and collect terms, we find that

$$Var(X) = E(X^2) - [E(X)]^2.$$

In words, the variance of X is the expected value of X^2 minus the square of the expected value of X (these are not the same thing!).

Of course, once you have calculated $Var(X)$ one way or another, you find $\sigma(X)$ by taking the square root.

EXAMPLE 2

Calculate the variance and standard deviation for the following probability distribution.

x	10	20	30	40	50	60	70
$P(X = x)$	0.1	0.2	0.3	0.2	0.1	0.1	0

SOLUTION We first compute the expected value, $E(X)$, in the usual way.

x	10	20	30	40	50	60	70	
$P(X = x)$	0.1	0.2	0.3	0.2	0.1	0.1	0	$E(X)$
$xP(X = x)$	1	4	9	8	5	6	0	33

Next, we add an extra row for the values of X^2, which we get by squaring the values of X.

x	10	20	30	40	50	60	70	Totals
$P(X = x)$	0.1	0.2	0.3	0.2	0.1	0.1	0	1
$xP(X = x)$	1	4	9	8	5	6	0	$E(X) = 33$
x^2	100	400	900	1,600	2,500	3,600	4,900	

Finally, we calculate $E(X^2)$ by multiplying the entries in the X^2-row by the corresponding probabilities and then adding.

x	10	20	30	40	50	60	70	Totals
$P(X = x)$	0.1	0.2	0.3	0.2	0.1	0.1	0	1
$xP(X = x)$	1	4	9	8	5	6	0	$E(X) = 33$
x^2	100	400	900	1,600	2,500	3,600	4,900	
$x^2P(X = x)$	10	80	270	320	250	360	0	$E(X^2) = 1,290$

We now have all the data we need.

$$\begin{aligned} Var(X) &= E(X^2) - [E(X)]^2 \\ &= 1{,}290 - [33]^2 \\ &= 1{,}290 - 1{,}089 = 201 \\ \sigma(X) &= \sqrt{Var(X)} = \sqrt{201} \approx 14.1774 \end{aligned}$$

 The way we have set up the table lends itself naturally to the use of a spreadsheet program. You can program a spreadsheet to calculate the bottom three rows from the top two.

▼ **EXAMPLE 3**

Calculate the variance and standard deviation of the binomial distribution with $p = \frac{1}{2}$, $q = \frac{1}{2}$, and $n = 10$.

SOLUTION We can calculate $E(X)$ using the formula from the previous section.

$$E(X) = np = 10 \cdot \frac{1}{2} = 5$$

To set up a table to calculate the variance, we need the probabilities $P(X = x)$. For these, remember that the probability of x successes in $n = 10$ trials is

$$\begin{aligned} P(x \text{ successes}) &= C(n, x) p^x q^{n-x} \\ &= C(10, x)\left(\frac{1}{2}\right)^x \left(\frac{1}{2}\right)^{10-x} \\ &= C(10, x)\left(\frac{1}{2}\right)^{10} \\ &= C(10, x) \frac{1}{1{,}024}. \end{aligned}$$

7.3 EXERCISES

In each of Exercises 1 through 8, find the average, population variance, and population standard deviation of the given set of X-scores.

1. 0, 2, 4, 6, 8, 10

2. 1, 3, 5, 7, 9, 11

3. 3.0, 3.0, 4.0, 3.5, 4.5

4. 5.5, 5.0, 6.0, 4.5, 5.0

5. $\frac{1}{2}, \frac{3}{2}, -\frac{5}{4}, \frac{1}{2}, \frac{5}{2}, 0$

6. $-\frac{1}{4}, \frac{3}{4}, \frac{5}{2}, \frac{1}{2}, \frac{5}{2}, 0$

7. 1, 1, 1, 1, 1, 1, 1

8. 2, 3, 2, 3, 2, 3, 2, 3

In each of Exercises 9 through 16, calculate the expected value, the variance, and the standard deviation of X. (You already calculated the expected value in the last exercise set.)

9.

x	10	20	30	40
$P(X = x)$	$\frac{15}{50}$	$\frac{20}{50}$	$\frac{10}{50}$	$\frac{5}{50}$

10.

x	2	4	6	8
$P(X = x)$	$\frac{1}{20}$	$\frac{15}{20}$	$\frac{2}{20}$	$\frac{2}{20}$

11.

x	-5	-1	0	2	5	10
$P(X = x)$	0.2	0.3	0.2	0.1	0.2	0.0

12.

x	-20	-10	0	10	20	30
$P(X = x)$	0.2	0.4	0.2	0.1	0.0	0.1

13.

Income Bracket	$0–$19,999	$20,000–$29,999	$30,000–$39,999	$40,000–$49,999	$50,000–$59,999
Probability	0.2	0.3	0.2	0.2	0.1

14.

Cost of a Car	$0–$1,999	$2,000–$2,999	$3,000–$3,999	$4,000–$4,999	$5,000–$5,999
Probability	0.1	0.3	0.3	0.2	0.1

15.

Population	0–4,999	5,000–9,999	10,000–14,999	15,000–19,999	20,000–24,999
Number of Cities	4	7	2	5	2

16.

Student Population	0–1,999	2,000– 3,999	4,000– 5,999	6,000– 7,999	8,000– 9,999
Number of Colleges	2	5	8	5	5

In each of Exercises 17 through 28, calculate the expected value, the variance, and the standard deviation of the given random variable X. (You already calculated the expected value in the last exercise set.)

17. X is the number of tails that come up when a coin is tossed twice.

18. X is the number of tails that come up when a coin is tossed three times.

19. X is the highest number when two dice are cast.

20. X is the lowest number when two dice are cast.

21. X is the number of red marbles that Suzy has in her hand after she selects four marbles from a bag containing 4 red marbles and 2 green ones.

22. X is the number of green marbles that Suzy has in her hand after she selects four marbles from a bag containing 3 red marbles and 2 green ones.

23. A red and a green die are rolled, and X is the number on the red die minus the number on the green die.

24. Two dice are rolled, and X = the absolute value of the difference of the two numbers.

25. Twenty darts are thrown at a dartboard. The probability of hitting the bull's eye is $\frac{1}{10}$. Let X be the number of bull's eyes hit.

26. Thirty darts are thrown at a dartboard. The probability of hitting the bull's eye is $\frac{1}{5}$. Let X be the number of bull's eyes hit.

27. Select five cards without replacement from a standard deck of 52, and let X be the number of Queens you draw.

28. Select five cards without replacement from a standard deck of 52, and let X be the number of red cards you draw.

In Exercises 29 through 32, use Chebyshev's inequality to calculate the probability that X is in the given range.

29. $E(X) = 10.0$, $\sigma(X) = 0.3$, $9.4 \leq X \leq 10.6$

30. $E(X) = 10.0$, $\sigma(X) = 0.2$, $9.6 \leq X \leq 10.4$

31. $E(X) = 0$, $\sigma(X) = 3.1$, $-12.4 \leq X \leq 12.4$

32. $E(X) = 0$, $\sigma(X) = 0.1$, $-0.6 \leq X \leq 0.6$

In Exercises 33 through 36, use Chebyshev's inequality to find a range of values within which X will fall with a probability of at least 90%.

33. $E(X) = 10.0$, $\sigma(X) = 0.3$

34. $E(X) = 10.0$, $\sigma(X) = 0.2$

35. $E(X) = 0$, $\sigma(X) = 3.1$

36. $E(X) = 0$, $\sigma(X) = 0.1$

APPLICATIONS

37. *Grade Point Averages* Julian's transcript shows a grade point average of 3.00, with a standard deviation of 0.25. What fraction of his total credits were certain to have a grade point average of 2.00 or better?

38. *Grade Point Averages* Julia's transcript shows a grade point average of 3.20, with a standard deviation of 0.20. What fraction of her total credits were certain to have a grade point average of 2.00 or better?

39. *Salaries* If Chebyshev's inequality predicts that at least 75% of all lawyers earn between $80,000 and $280,000, give a range of salaries that covers at least 90% of all lawyers.

40. *Housing Costs* If Chebyshev's inequality predicts that at least 75% of all new homes cost between $100,000 and $280,000, give a range of costs that covers at least 80% of all new homes.

Finance Exercises 41 through 48 are based on the following table, which shows the annual percentage yields and loan rates of 18 of the larger banks in Atlanta, Georgia in 1993.* Assume for these exercises that these banks are representative of all the banks you are interested in

Bank	Money Market Yield	1-yr CD Yield	Home Loan Rate	Bank	Money Market Yield	1-yr CD Yield	Home Loan Rate
Bank South	2.59	3.46	6.75	Bank of Canton	2.94	3.82	8.50
Bank of Covington	2.53	3.44	7.46	Barnett Bank of Atlanta	2.48	3.41	7.75
Chattahoochee Bank	2.53	3.41	6.95	Citizens Trust Bank	2.48	3.40	6.75
Cobb American	2.58	3.44	6.95	Enterprise National	2.28	3.49	7.50
Fidelity National	2.78	3.67	7.75	First State Stockbridge	3.00	3.70	7.50
First Union National	2.48	3.40	7.40	Gwinnett Federal Bank	2.53	3.65	6.50
HomeBanc	2.53	3.40	7.75	Merchant Bank	2.43	3.44	6.98
Metro Bank	2.38	3.30	6.90	Mountain National Bank	2.79	3.46	7.25
NationsBank	2.48	3.39	7.00	Tara State Bank	2.58	3.53	8.50

41. Estimate the expected value and standard deviation of the money market yield to the nearest 0.01%.

42. Estimate the expected value and standard deviation of the 1-year CD yield to the nearest 0.01%.

43. Estimate the expected value and standard deviation, to the nearest 0.01%, of the random variable $X =$ (Home loan rate) $-$ (Money market yield).

44. Estimate the expected value and standard deviation, to the nearest 0.01%, of the random variable $X =$ (Home loan rate) $-$ (1-year CD yield).

45. Use Chebyshev's inequality with $k = 2$ to estimate a range within which at least 75% of money market yields will fall.

46. Use Chebyshev's inequality with $k = 2$ to estimate a range within which at least 75% of 1-year CD yields will fall.

47. Use Chebyshev's inequality to estimate a range within which at least 90% of money market yields will fall.

48. Use Chebyshev's inequality to estimate a range within which at least 90% of 1-year CD yields will fall.

▼ *Source: *The Atlanta Journal/Atlanta Constitution*, August 30, 1993.

49. *Mergers and Acquisitions* The following chart shows the total value, in billions of dollars, of all mergers and acquisitions in the United States from 1981 to 1993.*

Corporate Mergers and Acquisitions in U.S.

Total paid, in billions of dollars

(a) Taking the value to represent the "frequency" of mergers and acquisitions in a year, find the expected value and standard deviation of the year of a merger or acquisition.
(b) Use Chebyshev's inequality with $k = 2$ to complete the following sentence: At least 75% of all mergers and acquisitions, measured by value, took place in the years 19____ to 19____. (Round to the nearest year.)

50. *Cable Television Subscriptions* The following chart shows the number of homes (in millions) that hooked up to cable television during the given year.†

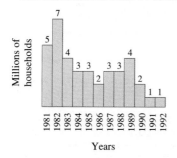

New Subscribers to Cable TV in U.S.

Years

(a) Use the data to find the expected value and standard deviation of the year of a new cable hookup.
(b) Use Chebyshev's inequality with $k = 2$ to complete the following sentence: At least 75% of new cable hookups took place in the years 19____ to 19____. (Round to the nearest year.)

Exercises 51 through 54 are based on exercises in the last exercise set, but these go a little further.

▼ *Source: Securities Data Company, as cited in *The New York Times*, Sept. 19, 1993, Section 3, p.1. All figures are correct to within ±$5 billion. (We read them off a bar chart. The 1993 figure is an estimate based on the figure through September.)

† Source: National Cable Television Association, as cited in the *Chicago Tribune*, Sept.1, 1993, p. 1. Figures are correct to within ±0.5. (We read them off a chart.)

51. *Employment* A survey found the following distribution for the sizes of the firms where workers aged 18–64 were employed:*

Size of Firm (Number of Employees)	1	2–9	10–24	25–99	100–499	500–999	1,000–5,000
Percentage of Workers	10%	12%	8%	12%	14%	6%	38%

In what range of firm sizes were at least 80% of all workers employed?

52. *Health Insurance* In a survey of 3,800 workers in different-sized firms, the following data show the number of workers without medical insurance:*

Size of Firm (Number of Employees)	1	2–9	10–24	25–99	100–499	500–999	1,000–5,000
Number of Uninsured Workers	2,100	434	300	292	202	80	392

In what range of firm sizes were at least 80% of all uninsured workers employed?

53. *Income Distribution* A 1992 *Chicago Tribune* survey of annual incomes of college presidents in 29 research universities reported the following data:

Income Bracket	$100,000–$199,999	$200,000–$299,999	$300,000–$399,999	$400,000–$499,999
Number	2	13	12	2

In what income range were the salaries of at least half of all college presidents?

54. *Passing the Bar Exam* A *New York Times* survey of all 72,461 people who took the bar exam in 1992 showed the following data on success rates in the 50 states plus the District of Columbia:†

Percentage Who Passed	50%–59%	60–69%	70–79%	80–89%	90–99%
Number of States	2	7	19	22	1

Assuming equal numbers of candidates in all the states, find a range of success rates within which at least 80% of candidates fell, and interpret your answer.

---

* Source: Employee Benefits Research Institute, cited in *The New York Times*, August 23, 1993. Figures given are based on the data cited but have been adjusted for computational reasons. The exact number of workers surveyed is fictitious.

† Source: *The New York Times*, August 15, 1993.

55. *Breast Cancer*   It has been estimated that 1 in 8 women will get breast cancer in their lifetimes. * Use Chebyshev's inequality to estimate the probability that, in a neighborhood of 100 women, more than 20 women will get breast cancer.

56. *Breast Cancer*   It has been estimated that 1 in 8 women will get breast cancer in their lifetimes.* Use Chebyshev's inequality to estimate the probability that, in a neighborhood of 500 women, more than 75 women will get breast cancer.

COMMUNICATION AND REASONING EXERCISES

57. In one Finite Math class, the average grade was 75 and the standard deviation of the grades was 5. In another Finite Math class, the average grade was 65 and the standard deviation of the grades was 20. What conclusions can you draw about the distributions of the grades in each class?

58. You are a manager in a precision manufacturing firm and you must evaluate the performance of two employees. You do so by examining the quality of the parts that they produce. One particular item should be exactly 50 mm long. The first employee produces parts that are an average of 50.1 mm long, with a standard deviation of 0.15 mm. The second employee produces parts that are an average of 50.0 mm long, with a standard deviation of 0.4 mm. Which employee do you rate higher, and why?

59. If a finite random variable has an expected value of 10 and a standard deviation of 0, what must its probability distribution be?

60. Find a random variable whose possible values are 1, 2, and 3 with an expected value of 2 and a standard deviation of 1.

7.4 NORMAL DISTRIBUTIONS

CONTINUOUS RANDOM VARIABLES

Figure 1 shows the probability distributions for the number of successes in sequences of 10 and 15 Bernoulli trials each with probability of success

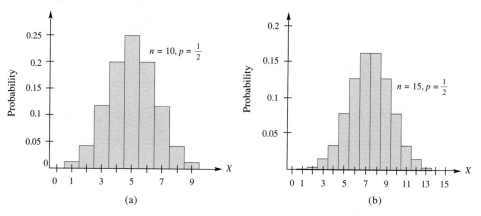

FIGURE 1

▼ *Newsday Discovery section, Dec. 7, 1993.

7.4 Normal Distributions

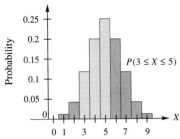

FIGURE 2

$p = \frac{1}{2}$. Since each column is one unit wide, its area is numerically equal to its height. Thus, the area of each rectangle can be interpreted as a probability. For example, in Figure 1(a) the area of the rectangle over $X = 3$ represents $P(X = 3)$. If we wanted to find $P(3 \leq X \leq 5)$, we could add up the areas of the three rectangles over 3, 4, and 5, shown shaded in Figure 2.

Notice that if we added the areas of *all* the rectangles, the total would be 1, since $P(0 \leq X \leq 10) = 1$. We can summarize these observations as follows.

> **PROPERTIES OF THE PROBABILITY DISTRIBUTION HISTOGRAM**
>
> In a probability distribution histogram where each column is one unit wide:
>
> 1. The total area enclosed by the histogram is 1 square unit.
> 2. $P(a \leq X \leq b)$ is the area enclosed by the rectangles lying between $X = a$ and $X = b$.

This discussion is motivation for considering another kind of random variable, one whose probability distribution is specified not by a bar graph as above, but by an arbitrary curve.

FIGURE 3

> **CONTINUOUS RANDOM VARIABLE**
>
> A **continuous random variable** X may take on any real value whatsoever. The probabilities $P(a \leq X \leq b)$ are specified by means of a **probability density curve,** a curve lying above the x-axis with the total area between the curve and the x-axis being 1. The probability $P(a \leq X \leq b)$ is given by the area enclosed by the curve, the x-axis, and the lines $x = a$ and $x = b$ (see Figure 3).

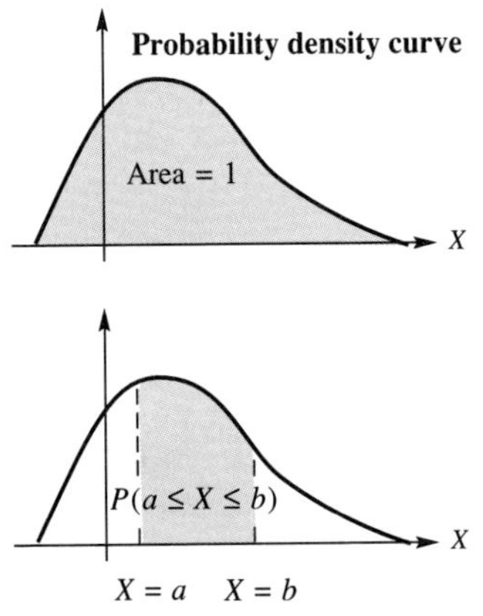

Among all the possible probability density curves, there is an important class of curves collectively called **normal curves,** or **normal distributions.** A variable with such a density curve is called a **normal variable.** These curves are all obtained from the following complicated-looking formula.

> **NORMAL DISTRIBUTION FUNCTION**
>
> $$y = \frac{1}{\sigma\sqrt{2\pi}} e^{-(x-\mu)^2/2\sigma^2}$$

The quantity μ is called the **mean** and can be any real number. The quantity σ is called the **standard deviation** and can be any positive real

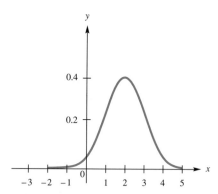

Normal distribution with $\mu = 2$ and $\sigma = 1$

FIGURE 4

number. The number $e = 2.71828 \ldots$ is a very interesting constant that shows up many places in mathematics, much as the constant π does. Finally, the number $1/(\sigma\sqrt{2\pi})$ that appears in front is there to make the total area come out to be 1.

The graph of this function, for $\mu = 2$ and $\sigma = 1$, is shown in Figure 4.

This is the graph known as the "bell curve." In general, the graph of a normal distribution is "bell-shaped" with the peak (the mode) at $X = \mu$. It is symmetric around $X = \mu$, with the consequence that μ is also the mean and the median of X. As suggested by the name, it can be shown that $\sigma(X) = \sigma$.

Figures 5 through 8 show several other normal curves.

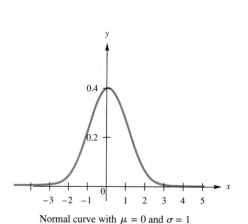

Normal curve with $\mu = 0$ and $\sigma = 1$

FIGURE 5

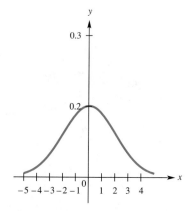

Normal curve with $\mu = 0$ and $\sigma = 2$

FIGURE 6

7.4 Normal Distributions

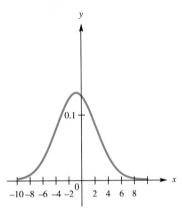

Normal curve with $\mu = -1$ and $\sigma = 3$

FIGURE 7

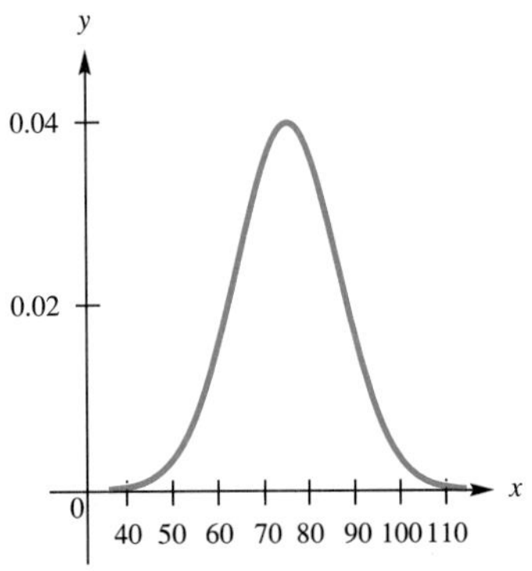

Normal curve with $\mu = 75$ and $\sigma = 10$

FIGURE 8

CALCULATING PROBABILITIES USING THE STANDARD NORMAL DISTRIBUTION

The **standard** normal distribution has $\mu = 0$ and $\sigma = 1$. The corresponding variable, called the **standard normal variable,** we shall always denote by Z. Recall that, to calculate the probability $P(a \leq Z \leq b)$, we need to find the area under the distribution curve between the vertical lines $Z = a$ and $Z = b$. We can use the table in Appendix C to look up these areas. Here is an example.

▼ **EXAMPLE 1**

Let Z be the standard normal variable. Calculate the following probabilities.

(a) $P(0 \leq Z \leq 2.4)$
(b) $P(0 \leq Z \leq 2.43)$
(c) $P(-1.37 \leq Z \leq 2.43)$
(d) $P(1.37 \leq Z \leq 2.43)$

SOLUTION

(a) We are asking for the shaded area in Figure 9.

We can find this area correct to four decimal places by looking at the table in Appendix C, which lists the area under the standard normal curve from $Z = 0$ to $Z = b$ for values of b between 0 and 3.09. To use the table, write 2.4 as 2.40, and read the entry in the row labeled 2.4 and the column labeled .00. (2.4 + .00 = 2.40) Here is the relevant portion of the chart:

Standard Normal Curve

$\mu = 0, \sigma = 1$

FIGURE 9

Z	.00	.01	.02	.03
2.3	.4893	.4896	.4898	.4901
→ 2.4	.4918	.4920	.4922	.4925
2.5	.4938	.4940	.4941	.4943

Thus, $P(0 \leq Z \leq 2.40) = 0.4918$.

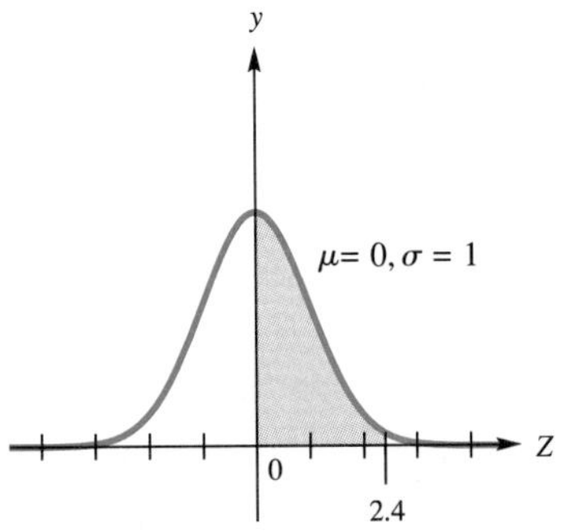

(b) The area we require can be read from the same portion of the chart shown above: write 2.43 as 2.4 + 0.03, and read the entry in the row labeled 2.4 and the column labeled .03:

Z	.00	.01	.02	.03
2.3	.4893	.4896	.4898	.4901
→ 2.4	.4918	.4920	.4922	.4925
2.5	.4938	.4940	.4941	.4943

Thus, $P(0 \leq Z \leq 2.43) = 0.4925$.

(c) Here we cannot use the chart directly, since the range $-1.37 \leq Z \leq 2.43$ does not start at 0. But we can break it up into 2 smaller ranges that start or end at 0.

$$P(-1.37 \leq Z \leq 2.43) = P(-1.37 \leq Z \leq 0) + P(0 \leq Z \leq 2.43)$$

In terms of the graph, we are splitting the desired area into two smaller areas (Figure 10).

We already calculated the area of the right-hand piece in part (b).

$$P(0 \leq Z \leq 2.43) = 0.4925$$

For the left-hand piece, the symmetry of the normal curve tells us that this is the same as the corresponding area on the right, $P(0 \leq Z \leq 1.37)$. This we can find on the table: look at the row labeled 1.3 and the column labeled .07, and read

$$P(-1.37 \leq Z \leq 0) = P(0 \leq Z \leq 1.37) = 0.4147.$$

Thus,

$$P(-1.37 \leq Z \leq 2.43) = P(-1.37 \leq Z \leq 0) + P(0 \leq Z \leq 2.43).$$
$$= 0.4147 + 0.4925$$
$$= 0.9072.$$

(Actually, 0.9071. We lose some accuracy because the figures in the table are rounded off.)

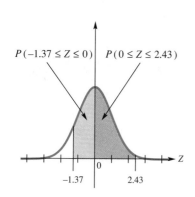

FIGURE 10

(d) The range $1.37 \leq Z \leq 2.43$ does not contain 0, so the technique of (c) cannot be used. Instead, the corresponding area can be computed as the *difference* of two areas.

$$P(1.37 \leq Z \leq 2.43) = P(0 \leq Z \leq 2.43) - P(0 \leq Z \leq 1.37)$$
$$= 0.4925 - 0.4147$$
$$= 0.0778$$

There are several ways in which these calculations can be done by calculator or computer. One that is sometimes available is the **error function,** usually called **erf.** This is related to (but not quite the same as) the area under the normal curve. Precisely,

$$P(0 \leq Z \leq b) = \frac{1}{2}\text{erf}\left(\frac{b}{\sqrt{2}}\right).$$

If your software or calculator can calculate erf, this is easy to use. Alternatively, some computer software and some graphing calculators are equipped to calculate areas under arbitrary curves. Others have to be programmed to do this. The area under the curve $y = f(X)$ between the vertical lines $X = a$ and $X = b$ is called the **definite integral** of $f(X)$ with limits a and b. To calculate $P(a \leq Z \leq b)$, you would want to calculate the area under the curve

$$f(Z) = \frac{1}{\sqrt{2\pi}} e^{-Z^2/2}$$

between a and b. Using a TI-82, for example, you would enter

$$Y_1 = 1/(2\pi)^{\wedge}0.5)e^{\wedge}(-X^{\wedge}2/2)$$

and then enter

$$\text{fnInt}(Y_1, X, 0, 2.40)$$

to solve part (a).

Q What about the area under *nonstandard* normal distributions? For example, if $\mu = 2$ and $\sigma = 3$, then how would we calculate $P(0.5 \leq X \leq 3.2)$?

A We can use the following conversion formula.

STANDARDIZING NORMAL DISTRIBUTIONS
If X has a normal distribution with mean μ and standard deviation σ, and if Z is the standard normal variable, then

$$P(a \leq X \leq b) = P\left(\frac{a-\mu}{\sigma} \leq Z \leq \frac{b-\mu}{\sigma}\right).$$

For example, if $\mu = 2$ and $\sigma = 3$, then

$$P(0.5 \leq X \leq 3.2) = P\left(\frac{0.5-2}{3} \leq Z \leq \frac{3.2-2}{3}\right)$$
$$= P(-0.5 \leq Z \leq 0.4)$$
$$= 0.1915 + 0.1554 = 0.3469.$$

Q Where does the standardizing formula come from?

A To justify it completely requires more mathematics than we shall discuss here. Briefly, if X is normal with mean μ and standard deviation σ, then $X - \mu$ is normal with mean 0 and standard deviation still σ, while $(X - \mu)/\sigma$ is normal with mean 0 and standard deviation 1. In other words, $(X - \mu)/\sigma = Z$. Therefore,

$$P(a \leq X \leq b) = P\left(\frac{a - \mu}{\sigma} \leq \frac{X - \mu}{\sigma} \leq \frac{b - \mu}{\sigma}\right)$$

$$= P\left(\frac{a - \mu}{\sigma} \leq Z \leq \frac{b - \mu}{\sigma}\right).$$

▼ **EXAMPLE 2** Quality Control

Pressure gauges manufactured by Precision Corp. must be checked for accuracy before being placed on the market. To test a pressure gauge, a worker uses it to measure the pressure of a sample of compressed air known to be at a pressure of exactly 50 pounds per square inch. If the gauge reading is off by more than 1% (0.5 pounds), it is rejected. Assuming that the reading of a pressure gauge under these circumstances is a normal random variable with mean 50 and standard deviation 0.5, find the percentage of gauges rejected.

SOLUTION If X is the reading of the gauge, then X has a normal distribution with $\mu = 50$ and $\sigma = 0.5$. We are asking for $P(X < 49.5$ or $X > 50.5) = 1 - P(49.5 \leq X \leq 50.5)$. To calculate this, we calculate

$$P(49.5 \leq X \leq 50.5) = P\left(\frac{49.5 - 50}{0.5} \leq Z \leq \frac{50.5 - 50}{0.5}\right)$$

$$= P(-1 \leq Z \leq 1)$$
$$= 2 \cdot P(0 \leq Z \leq 1)$$
$$= 2 \cdot 0.3413$$
$$= 0.6826.$$

So

$$P(X < 49.5 \text{ or } X > 50.5) = 1 - P(49.5 \leq X \leq 50.5)$$
$$= 1 - 0.6826$$
$$= 0.3174.$$

In other words, 31.74% of the gauges will be rejected.

Before we go on... Again, there is a slight roundoff error. In fact, $P(-1 \leq Z \leq 1) = 0.6827$ to four decimal places, so 31.73% of the gauges will be rejected.

To calculate $P(49.5 \leq X \leq 50.5)$ directly using a TI-82, you would enter

$$Y_1 = (1/0.5(0.5(2\pi)\wedge 0.5))e\wedge(-(X-50)\wedge 2/0.5)$$

and then enter

$$\text{fnInt}(Y_1, X, 49.5, 50.5)$$

We can look at the last example as asking for the probability that X is farther than one standard deviation from its mean, and to find this we had to find the probability that X would lie within one standard deviation of its mean. Notice that if X has a normal distribution with mean μ and standard deviation σ, then

$$P(\mu - k\sigma \leq X \leq \mu + k\sigma) = P(-k \leq Z \leq k)$$

by the standardizing formula. We can easily compute these probabilities for various values of k using the table in Appendix C.

PROBABILITY OF A NORMAL DISTRIBUTION BEING WITHIN k STANDARD DEVIATIONS OF ITS MEAN

$$P(\mu - \sigma \leq X \leq \mu + \sigma) = 0.6827$$
$$P(\mu - 2\sigma \leq X \leq \mu + 2\sigma) = 0.9545$$
$$P(\mu - 3\sigma \leq X \leq \mu + 3\sigma) = 0.9973$$

You should compare these numbers with those given by Chebyshev's inequality. The probabilities above are a good deal larger then the lower bounds given by Chebyshev's inequality. Chebyshev's inequality must work for distributions that are flatter and not concentrated as close to their means as the normal distribution is. Figures 11 through 13 show the regions $-k \leq Z \leq k$ for $k = 1, 2,$ and 3.

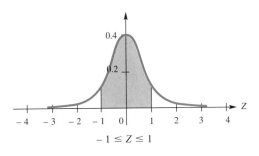

FIGURE 11

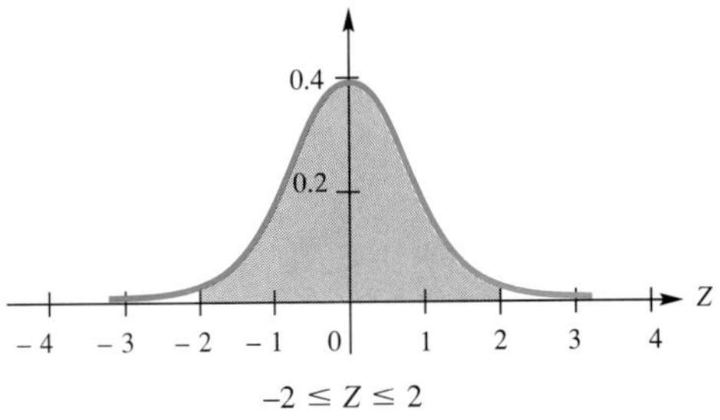

FIGURE 12

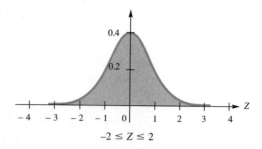

$-2 \leq Z \leq 2$

FIGURE 13

▼ EXAMPLE 3   Loans

The values of mortgage loans made by a certain bank in 1989 were normally distributed with a mean of $120,000 and a standard deviation of $40,000.

(a) What is the probability that a randomly selected mortgage loan was in the range $40,000–$200,000?

(b) You would like to state that 50% of all mortgage loans were in a certain range. What is that range?

SOLUTION

(a) We are asking for the probability that a loan was within 2 standard deviations ($80,000) of the mean. By the calculation done above, this probability is 0.9545.

(b) We look for the k such that

$$P(120{,}000 - k \cdot 40{,}000 \leq X \leq 120{,}000 + k \cdot 40{,}000) = 0.5.$$

Since

$$P(120{,}000 - k \cdot 40{,}000 \leq X \leq 120{,}000 + k \cdot 40{,}000)$$
$$= P(-k \leq Z \leq k)$$

for a variable Z with a standard normal distribution, we look in Appendix C to see for what k we have

$$P(0 \leq Z \leq k) = 0.25.$$

That is, we look *inside* the table to see where 0.25 is and find the corresponding k. We find

$$P(0 \leq Z \leq 0.67) = 0.2486$$

and

$$P(0 \leq Z \leq 0.68) = 0.2517,$$

Therefore, the k we want is about halfway between 0.67 and 0.68: call it 0.675. This tells us that 50% of all mortgage loans were in the range

$$120{,}000 - 0.675 \cdot 40{,}000 = \$93{,}000$$

to
$$120{,}000 + 0.675 \cdot 40{,}000 = \$147{,}000.$$

Before we go on... You should compare these answers with those in Example 5 in the previous section.

You might have noticed that the histograms of binomial distributions (for example, those in Figure 1) have a very rough bell shape. In fact, in many cases it is possible to draw a normal curve that closely approximates a given binomial distribution.

NORMAL APPROXIMATION TO A BINOMIAL DISTRIBUTION

If X is the number of successes in a sequence of n independent Bernoulli trials, with probability p of success in each trial, and if the range of values of X three standard deviations above and below the mean lies entirely within the range 0 to n (the possible values of X), then

$$P(a \leq X \leq b) \approx P(a - 0.5 \leq Y \leq b + 0.5)$$

where Y has a normal distribution with the same mean and standard deviation as X— that is, $\mu = np$ and $\sigma = \sqrt{np(1-p)}$.

▶ **NOTES**

1. The condition that $0 \leq \mu - 3\sigma < \mu + 3\sigma \leq n$ is satisfied if n is sufficiently large and p is not too close to 0 or 1, and ensures that most of the normal curve lies in the range 0 to n.
2. In the formula $P(a \leq X \leq b) \approx P(a - 0.5 \leq Y \leq b + 0.5)$ we assume that a and b are integers. The use of $a - 0.5$ and $b + 0.5$ is called the **continuity correction.** To see that it is necessary, consider what would happen if you wanted to approximate, say, $P(X = 2) = P(2 \leq X \leq 2)$. ◀

Figures 14 through 16 show several binomial distributions with their normal approximations superimposed.

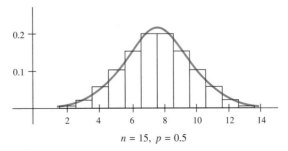

$n = 15, \ p = 0.5$

FIGURE 14

544 Chapter 7 Statistics

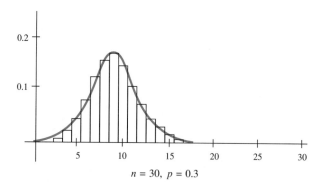

FIGURE 15

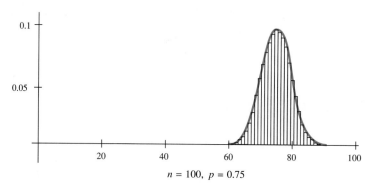

FIGURE 16

EXAMPLE 4

(a) If you flip a fair coin 100 times, what is the probability of getting more than 55 heads or less than 45 heads?

(b) What number of heads (out of 100) would make you suspect that the coin is not fair?

SOLUTION

(a) We are asking for

$$P(X < 45 \text{ or } X > 55) = 1 - P(45 \leq X \leq 55).$$

We *could* compute this by calculating

$$C(100, 45)(0.5)^{45}(0.5)^{55} + C(100, 46)(0.5)^{46}(0.5)^{54} + \cdots,$$

but we can much more easily *approximate* it by looking at a normal distribution with mean $\mu = 50$ and standard deviation $\sigma = \sqrt{100 \cdot 0.5 \cdot 0.5} = 5$. (Notice that three standard deviations above and below the mean is the range

35 to 65, which is well within the range of possible values for X, which is 0 to 100, so the approximation should be a good one.) Let Y have this normal distribution. Then

$$P(45 \leq X \leq 55) \approx P(44.5 \leq Y \leq 55.5)$$
$$= P(-1.1 \leq Z \leq 1.1)$$
$$= 0.7286.$$

Therefore,

$$P(X < 45 \text{ or } X > 55) \approx 1 - 0.7286 = 0.2714.$$

(b) This deep question touches on the question of **statistical significance:** what evidence is strong enough to overturn a reasonable assumption (the assumption that the coin is fair)?

Statisticians have developed various sophisticated ways of answering this question, but we can look at one simple test now. Suppose that we tossed a coin 100 times and threw 66 heads. If the coin were fair, then $P(X > 65) \approx P(Y > 65.5) = P(Z > 3.1) \approx 0.001$. This is small enough to raise a reasonable doubt that the coin is fair. However, we should not be too surprised if we threw 56 heads, since we can calculate that $P(X > 55) \approx 0.1357$, which is not such a small probability. As we said, the actual tests of statistical significance are more sophisticated than this, but we shall not go into them.

▶ **7.4 EXERCISES**

In Exercises 1–14, X has a normal distribution with the given mean and standard deviation. Find the indicated probabilities.

1. $\mu = 0$, $\sigma = 1$. Find $P(0 \leq X \leq 0.5)$.
2. $\mu = 0$, $\sigma = 1$. Find $P(0 \leq X \leq 1.5)$.
3. $\mu = 0$, $\sigma = 1$. Find $P(-0.71 \leq X \leq 0.71)$.
4. $\mu = 0$, $\sigma = 1$. Find $P(-1.71 \leq X \leq 1.71)$.
5. $\mu = 0$, $\sigma = 1$. Find $P(-0.71 \leq X \leq 1.34)$.
6. $\mu = 0$, $\sigma = 1$. Find $P(-1.71 \leq X \leq 0.23)$.
7. $\mu = 0$, $\sigma = 1$. Find $P(0.5 \leq X \leq 1.5)$.
8. $\mu = 0$, $\sigma = 1$. Find $P(0.71 \leq X \leq 1.82)$.
9. $\mu = 50$, $\sigma = 10$. Find $P(35 \leq X \leq 65)$.
10. $\mu = 40$, $\sigma = 20$. Find $P(35 \leq X \leq 45)$.
11. $\mu = 50$, $\sigma = 10$. Find $P(30 \leq X \leq 62)$.
12. $\mu = 40$, $\sigma = 20$. Find $P(30 \leq X \leq 53)$.
13. $\mu = 100$, $\sigma = 15$. Find $P(110 \leq X \leq 130)$.
14. $\mu = 100$, $\sigma = 15$. Find $P(70 \leq X \leq 80)$.

15. Find the probability that a normal variable takes values within 0.5 standard deviations of its mean.

16. Find the probability that a normal variable takes values within 1.5 standard deviations of its mean.

17. Find the probability that a normal variable takes values more than $\frac{2}{3}$ standard deviations away from its mean.

18. Find the probability that a normal variable takes values more than $\frac{5}{3}$ standard deviations away from its mean.

19. If you roll a die 100 times, what is the probability that you will roll a 1 between 10 and 15 times?

20. If you roll a die 100 times, what is the probability that you will roll a 1 between 15 and 20 times?

21. If you roll a die 200 times, what is the probability that you will roll a 1 fewer than 25 times?

22. If you roll a die 200 times, what is the probability that you will roll a 1 more than 40 times?

APPLICATIONS

23. *IQ Scores* IQ scores (as measured by the Stanford-Binet intelligence test) are normally distributed with a mean of 100 and a standard deviation of 16. Find the number of people in the U.S. (assuming a total population of 250,000,000) with an IQ higher than 120.

24. *IQ Scores* IQ scores (as measured by the Stanford-Binet intelligence test) are normally distributed with a mean of 100 and a standard deviation of 16. Find the number of people in the U.S. (assuming a total population of 250,000,000) with an IQ higher than 140.

25. *Product Repairs* The new copier your business bought lists a mean time between failures of 6 months, with a standard deviation of 1 month. One month after a repair, it breaks down again. Is this surprising? (Assume that the times between failures are normally distributed.)

26. *Product Repairs* The new computer your business bought lists a mean time between failures of 1 year, with a standard deviation of 2 months. Nine months after a repair, it breaks down again. Is this surprising? (Assume that the times between failures are normally distributed.)

27. *Baseball* The mean batting average in major league baseball is about 0.250. Supposing that batting averages are normally distributed, that the standard deviation in the averages is 0.01, and that there are 250 batters, what is the expected number of batters with an average of at least 0.400?*

28. *Baseball* The mean batting average in major league baseball is about 0.250. Supposing that batting averages are normally distributed, that the standard deviation in the averages is 0.05, and that there are 250 batters, what is the expected number of batters with an average of at least 0.400?*

29. *Marketing* Your pickle company rates its pickles on a scale of spiciness from 1 to 10. Market research shows that customer preference for spiciness is normally distributed, with a mean of 7.5 and a standard deviation of 1. Assuming that you sell 100,000 jars of pickles, how many jars with a spiciness of 9 or above do you expect to sell?

30. *Marketing* Your hot sauce company rates its sauce on a scale of spiciness of 1 to 20. Market research shows that customer preference for spiciness is normally distributed, with a mean of 12 and a standard deviation of 2.5. Assuming that you sell 300,000 bottles of sauce, how many bottles with a spiciness below 9 do you expect to sell?

31. *Aviation* The probability of a plane crashing on a single trip in 1989 was 0.00000165.[†] Find the probability that in 100,000,000 flights, there will be fewer than 180 crashes.

32. *Aviation* The probability of a plane crashing on a single trip in 1990 was 0.00000087. Find the probability that in 100,000,000 flights, there will be more than 110 crashes.

▼ *The last time that a batter ended the year with an average above 0.400 was in 1941. The batter was Ted Williams of the Boston Red Sox, and his average was 0.406. Over the years, as pitching and batting have improved, the standard deviation in batting averages has declined.
[†] Source for Exercises 31–34: National Transportation Safety Board.

33. *Insurance* Your company issues flight insurance. You charge $2, and in the event of a plane crash you will pay out $1,000,000 to the victim or his or her family. In 1989, the probability of a plane crashing on a single trip was 0.00000165. Assuming that 10 people per flight buy insurance from you, what was your probability of losing money over the course of 100,000,000 flights in 1989? (*Hint*: First determine how many crashes there must be for you to lose money.)

34. *Insurance* Assuming the figures of the preceding exercise, what is your probability of losing money over the course of 10,000,000 flights?

35. *Cancer Rates* It has been estimated that 1 in 8 women will get breast cancer in their lifetimes.* Suppose that in a neighborhood of 100 women, 16 women get breast cancer. Should you suspect some unusual cause for these cancers?

36. *Cancer Rates* As in the previous problem, assume that 1 in 8 women will get breast cancer in their lifetimes. Suppose that, in a neighborhood of 300 women, 60 women get breast cancer. Should you suspect some unusual cause for these cancers?

37. *Casinos* In the casino game of roulette, there are 38 equally likely outcomes: 1 through 36, plus 0 and 00. A player betting $1 on 00, say, will win $36 if 00 comes up (that is, the player gets back the $1 bet plus $35 more), and similarly for all the other numbers. Find the probability that over the course of 10,000 independent bets, the casino will lose money. (*Hint*: First determine how many times the casino must lose the bet for it to lose money.)

38. *Casinos* Repeat the preceding exercise, except now find the probability that over the course of 100,000 independent bets, the casino will lose money.

39. *Polls* In a certain political poll, each person polled has a 90% probability of telling his or her real preference. Suppose that 55% of the population really prefer candidate Goode, and 45% prefer candidate Slick. First find the probability that a person polled will say that he or she prefers Goode, then find the probability that if 1,000 people are polled, more than 52% will say they prefer Goode.

40. *Polls* In a certain political poll, each person polled has a 90% probability of telling his or her real preference. Suppose that 1,000 people are polled, and 51% say that they prefer candidate Goode, while 49% say that they prefer candidate Slick. Find the probability that Goode could do this well if, in fact, only 49% prefer Goode.

41. *IQ Scores* Mensa is a club for people with high IQs. To qualify, you must be in the top 2% of the population. One way of qualifying is by having an IQ of at least 148, as measured by the Cattell intelligence test. Assuming that scores on this test are normally distributed with a mean of 100, what is the standard deviation? (*Hint*: Use the table in Appendix C "backwards.")

42. *SAT Scores* Another way to qualify for Mensa (see the previous exercise) is to score at least 1250 on the SAT, which puts you in the top 2%. Assuming that SAT scores are normally distributed with a mean of 1000, what is the standard deviation? (See the hint for the preceding exercise.)

COMMUNICATION AND REASONING EXERCISES

43. A uniform continuous distribution is one whose probability density curve is a horizontal line. If X takes on values between the numbers a and b with a uniform distribution, find the height of its probability density curve.

44. If X takes on values between the numbers a and b with a uniform distribution (see the preceding exercise), find $E(X)$ and the median of X.

▼ *Newsday* Discovery section, Dec. 7, 1993.

▶ You're the Expert

LIES, DAMNED LIES, AND STATISTICS

There are three kinds of lies: lies, damned lies, and statistics.
— Benjamin Disraeli (1804–1881)

Statistics can be used to support just about anything—including statisticians.
— Anonymous

Since all your friends know you as an expert on probability and statistics, they bring to you puzzling statistics that they read in the paper or see on TV. It's up to you to enlighten them.

Q I read in the paper that 70% of left-handed people come from the northern hemisphere, and only 30% come from the southern hemisphere. Are northerners that much more likely to be left-handed than southerners?

A Not necessarily. It depends on the total population in the northern hemisphere compared to the population in the south. This is really a problem in conditional probability. If we pick a person at random, let L be the event that he or she is left-handed, let R be the event that he or she is right-handed, let N be the event that he or she comes from the northern hemisphere, and let S be the event that he or she comes from the southern hemisphere. We are told that

$$P(N|L) = 0.7$$

and

$$P(S|L) = 0.3.$$

What we are really interested in are $P(L|N)$ and $P(L|S)$. By the definition of conditional probability,

$$P(L|N) = \frac{P(L \cap N)}{P(N)} = \frac{P(N|L)P(L)}{P(N)}.$$

Similarly,

$$P(L|S) = \frac{P(L \cap S)}{P(S)} = \frac{P(S|L)P(L)}{P(S)}.$$

Dividing,

$$\frac{P(L|N)}{P(L|S)} = \frac{P(S)}{P(N)} \cdot \frac{P(N|L)}{P(S|L)}.$$

Now, since we are told that $P(N|L)/P(S|L) = 0.7/0.3$, the equation above says that if $P(S)/P(N) = 0.3/0.7$, then $P(L|N)/P(L|S)$ will equal 1 and $P(L|N)$ will equal $P(L|S)$. In other words, if 70% of people live in the northern hemisphere, it is no surprise that 70% of left-handed people live in the northern hemisphere. If more than 70% of people live in the northern hemisphere, then it will actually be less likely for a northerner to be left-handed than a southerner.

Q Every day, at some random time, I wander down to the train tracks and wait for a train to go by. Then I wander back home. The peculiar thing is that I've seen twice as many trains going east as west. Why are they running so many more trains in one direction than in the other?

A They may not be. Besides being a layabout, you are guilty of poor experimental design. The train schedule confirms that in fact there is exactly one train each hour going in either direction. However, the eastbound train always passes your point on the tracks on the hour, and the westbound train always passes twenty minutes after the hour. If you show up at a random time, you are twice as likely to arrive between twenty after the hour and the hour than you are to arrive between the hour and twenty past, so it is twice as likely that you will see an eastbound train as it is that you will see a westbound train. I suggest you spend a fixed time at the tracks—say, half an hour each day—and then see what happens.

Q While watching a debate on health policy, I heard that in 1980 Sweden had an infant mortality rate of about 7 deaths per 1,000 live births, while the United States had a rate of about 12 deaths per 1,000 live births. Is the U.S. doing that much worse than Sweden on health care?

A Not necessarily. This may be a case of comparing apples and oranges. The issue is this: what is counted as a live birth? In the U.S., physicians make a concerted effort to save premature and low-birthweight babies, even down to one pound in weight. When such a baby dies, as it is highly likely to, it adds to the infant mortality rate. Many countries are not as eager to spend the resources necessary to attempt to save these babies, and they do not count them as live births. This is really again a question of conditional probability. The chance of a baby surviving in Sweden, given that Sweden counts it as a live birth, is higher than the chance of a baby surviving in the U.S., given that the U.S. counts it as a live birth, at least partly because of the different definitions of "live birth."

Q They said on TV that the life expectancy in 1900 in the United States was 50 years. Was a 50-year-old considered ancient then?

A No, not really. This is an example of how misleading the average or expected value can sometimes be. What made the average lifespan so low was the extremely high infant mortality rates of the times. Simply put, many children did not live to become adults. However, those who did live to become adults could expect to live a good deal longer than 50 years. In fact, many people must have lived considerably longer than 50 years, or the average would have been less than 50 years. Put another way, the distribution of lifespans was probably bimodal, with a peak very low (corresponding to the large number of infants who died) and another peak at the high end (where most people who lived past childhood finally died), almost the opposite of a normal distribution. The average would fall between these peaks, even though not many people died at 50.

Exercise
Find examples of valid and invalid uses of statistics in news reports.

Review Exercises

In each of Exercises 1–10, find the probability distribution for the given random variable and draw a histogram for the distribution.

1. A couple has two children, and $X =$ the number of boys.
2. A couple has three children, and $X =$ the number of girls.
3. A four-sided die (with sides numbered 1 through 4) is cast twice, and $X =$ the sum of the two numbers rolled.
4. A four-sided die (with sides numbered 1 through 4) is cast twice, and $X =$ the first number minus the second.
5. A four-sided die (with sides numbered 1 through 4) is cast 5 times, and $X =$ the number of 1s rolled.
6. A four-sided die (with sides numbered 1 through 4) is cast 10 times, and $X =$ the number of 1s rolled.
7. From a bag containing 5 red marbles and 10 green marbles, 5 marbles are chosen. $X =$ the number of red marbles chosen.
8. From a bag containing 8 red marbles and 10 green marbles, 5 marbles are chosen. $X =$ the number of red marbles chosen.
9. From a bin containing 5 defective light bulbs and 45 good ones, 10 are chosen at random. $X =$ the number of defective bulbs chosen.
10. From a bin containing 5 defective light bulbs and 45 good ones, 20 are chosen at random. $X =$ the number of defective bulbs chosen.
11–20. For each of the variables in Exercises 1–10, find the expected value, median, mode, variance, and standard deviation.

In each of Exercises 21–26, the mean and standard deviation of a random variable X are given. Use Chebyshev's inequality to find a range in which X is guaranteed to lie with a probability of 90%.

21. $E(X) = 100$, $\sigma(X) = 16$
22. $E(X) = 100$, $\sigma(X) = 30$
23. $E(X) = 0$, $\sigma(X) = 1$
24. $E(X) = 0$, $\sigma(X) = 1.5$
25. $E(X) = -1$, $\sigma(X) = 0.5$
26. $E(X) = -10$, $\sigma(X) = 0.2$

In each of Exercises 27–32, the mean and standard deviation of a normal variable X are given. Find the indicated probabilities.

27. $\mu = 100$, $\sigma = 16$, $P(80 \leq X \leq 120)$
28. $\mu = 100$, $\sigma = 30$, $P(80 \leq X \leq 120)$
29. $\mu = 0$, $\sigma = 2$, $P(-1 \leq X \leq 3)$
30. $\mu = 0$, $\sigma = 1.5$, $P(-1 \leq X \leq 3)$
31. $\mu = -1$, $\sigma = 0.5$, $P(-1 \leq X \leq 0)$
32. $\mu = -10$, $\sigma = 0.2$, $P(-10.4 \leq X \leq -10)$
33–38. For each normal variable given in Exercises 27–32, find a range in which X is guaranteed to lie with a probability of 90%.
39. If you toss a coin 100 times, approximately what is the probability that you will toss between 45 and 50 heads (inclusive)?
40. If you toss a coin 200 times, approximately what is the probability that you will toss between 90 and 100 heads (inclusive)?

APPLICATIONS

41. *SAT Scores* On polling your class, you find the following SAT scores: 920, 1000, 930, 1090, 770, 1120, 920, 1200, 930, 810, 1100, 1120, 930, 1030, 850, 900, 890, 990, 800, 1140. Group these into brackets of 50 points each (700–749, 750–799, and so on), and write down the resulting frequency table and histogram. Also, compute from both the original data and your frequency table the average, median, mode, and population standard deviation of these scores.

42. *SAT Scores* On polling another class, you find the following SAT scores: 1100, 1130, 1480, 950, 1300, 1250, 1070, 1420, 1460, 1350, 1180, 1100, 1300, 1070, 950, 1150, 1300, 1390, 1160, 1320. Group these into brackets of 50 points each (900–949, 950–999, and so on), and write down the resulting frequency table and histogram. Also, compute from both the original data and your frequency table the average, median, mode, and population standard deviation of these scores.

43. *Home Runs* Here are the National League home run statistics for 1991:*

Home Runs	60–79	80–99	100–119	120–139	140–159	160–179
Number of Teams	2	1	3	2	3	1

Find the average number of home runs and the standard deviation of these data.

44. *Home Runs* The following chart lists the home run statistics for teams in the American League in 1991.* Find the average number of home runs, the median, the mode, and the sample standard deviation of these data.

Home Runs	60–89	90–119	120–149	150–179	180–209
Number of Teams	1	3	6	3	1

45. *College Population by Age* The following table shows the age distribution of U.S. residents (16 years old and over) attending college in 1980.† Find the average, median, mode, and standard deviation of the students' ages.

Age (years)	16–19	20–24	25–29	30–34	35–49
Number in 1980 (thousands)	2,678	4,786	1,928	1,201	1,763

46. *Armed Forces Population by Age* The following table shows the age distribution of persons (16 years old and over) in the U.S. Armed Forces in 1980.† Find the average, median, mode, and standard deviation of their ages.

Age (years)	16–19	20–24	25–29	30–34	35–39	40–44	45–49	50–69
Number in 1980 (thousands)	249	592	291	208	156	78	35	29

47. *IQ Scores* Within Mensa, a club for people with high IQs, the 3 Sigma Club has scores that are at least 3 standard deviations higher than the mean. Assuming a U.S. population of 250,000,000 and normally distributed IQ scores, how many people in the U.S. are qualified for the 3 Sigma Club?

48. *IQ Scores* Continuing with the preceding exercise, if an IQ of 132 puts you in the top 2% of the population, and hence eligible for Mensa, what score must you have to get into the 3 Sigma Club? (Assume that IQ scores are normally distributed with a mean of 100.)

49. *Cancer Rates* The probability of a woman getting breast cancer by the age of 50 has been estimated at 1 in 50.‡ If, in a neighborhood of 500 women 20 develop breast cancer by the age of 50, should you suspect an unusual cause for these cancers?

50. *Cancer Rates* The probability of a woman getting breast cancer by the age of 60 has been estimated at 1 in 24.‡ If, in a neighborhood of 500 women 24 develop breast cancer by the age of 60, should you suspect an unusual cause for these cancers?

51. *Batting Averages*§ If a baseball player's "true" batting average is 0.300, every time he comes to bat he has a probability of 0.300 of getting a hit. If he comes to bat 400 times in a season, what is the probability that he will hit less than 0.280 for the season?

52. *Batting Averages* Continuing the preceding exercise, what is the probability that the baseball player will hit better than 0.330 for the season?

▼ *Source: *The World Almanac and Book of Facts,* 1992.
 † Source: 1980 Census of Population, U.S. Department of Commerce/Bureau of the Census.
 ‡ Source: *Newsday* Discovery section, Dec. 7, 1993.
 § Thanks to Harold Hastings for suggesting Exercises 51 and 52.

CHAPTER 8

Source: Courtesy The Ford Motor Company

The Mathematics of Finance

SECTIONS

1 Simple Interest
2 Compound Interest
3 Annuities and Loans

You're the Expert
Pricing Auto Loans

APPLICATION ▶ You are shopping for a used car and have found two interesting possibilities. The first car costs $4,000, and the second costs $3,800. The first car dealer offers you a two-year loan at 8% with monthly payments and a down payment of $500. The second dealer offers you a three-year loan at 8.5%, also with monthly payments, but with a down payment of $300. You have $1,000 in a savings account paying 4% interest compounded monthly, and you estimate that you will have $200 per month in extra income that you will use to pay for the car, with any money left over from that being deposited in the savings account. Which car will end up costing you less to buy?

INTRODUCTION ▶ Finance is one of the driving forces of commerce in our society. A knowledge of the mathematics of investments and loans is therefore important not only for business majors but for everyone who deals with money. This chapter is an introduction to the basic concepts of the mathematics of finance. In particular, it is a chapter about *interest:* interest paid by an investment, interest paid on a loan, and variations on these.

This chapter focuses on three forms of investment: investments paying simple interest, investments in which interest is compounded, and annuities. We begin by looking at the simplest form of interest payment—appropriately called simple interest—in which interest on an investment is paid periodically directly to the investor, perhaps in the form of a monthly check. If, instead, the interest is reinvested back into the interest-bearing investment, the interest is *compounded* and the value of the account grows as the interest is added. An *annuity* is an investment into which periodic payments are made (increasing annuities) or from which periodic withdrawals are made (decreasing annuities or, looked at from the other side, loans).

We have included a discussion of the use of graphing calculators to help answer questions such as the following: "If I invest $10,000 in a municipal bond fund paying 6.5% interest, and reinvest the interest each year, when will my investment be worth $25,000?" Traditionally, the solution of such problems involves logarithms, but the availability of graphing technology provides an easy alternative. However, even if you do not have access to graphing technology, you will benefit greatly by carefully reading *all* of the examples. If you *are* using a graphing calculator, refer to Appendix B. We also suggest *Graphing Calculator Lessons for Finite Mathematics* by Paula G. Young (New York: HarperCollins, 1993) as a useful general reference.

You are strongly urged to read through the review sections on the algebra of exponentials and radicals in Appendix A before you begin this chapter, because much of the material in this chapter assumes a familiarity with manipulating exponents.

8.1 SIMPLE INTEREST

You have invested $10,000, called the **principal** or **present value,** into a savings account that pays you 5% interest, in the form of a check, each year.

Q How much interest will you earn each year?

A Since the account is paying 5% interest each year, your annual (yearly) interest will be 5% of $10,000, or $(0.05)(10,000) = \$500$.

This calculation leads us to a formula. If we call the present value P and the rate of interest (expressed as a decimal) r, then the annual interest I is given by

$$I = Pr.$$

If the investment is made for a period of t years, then the total interest accumulated is t times this amount, giving us the following.

SIMPLE INTEREST

The **simple interest** on an investment (or loan) of P dollars at an annual interest rate of r for a period of t years is

$$I = Prt.$$

▶ **NOTES**

1. Again, the interest rate r is in decimal form. For instance, if the interest rate is 6%, then $r = 0.06$.
2. The simple interest formula is based on the assumption that the interest is not being left (reinvested) in the account, but is withdrawn as soon as it is earned. When the interest is reinvested, we use the compound interest formula discussed in the next section.
3. The length of time, t, need not be an integral number of years. For example, the same formula would be used to compute the interest on a six-month CD by taking $t = \frac{1}{2}$ (assuming again that the interest is not compounded). ◀

If we add the accumulated interest to the principal, we get the **future value** or **maturity value** of the investment.

$$A = P + Prt$$
$$= P(1 + rt)$$

FUTURE VALUE FOR SIMPLE INTEREST

The **future value** or **maturity value** of a simple-interest investment (or loan) of P dollars at an annual interest rate r for t years is

$$A = P(1 + rt).$$

EXAMPLE 1 Savings Accounts

In March 1994, the Bank of Nova Scotia was paying 4.75% interest on savings accounts.* If payments at this rate are made on a simple-interest basis, find the value of a $1,000 deposit in four years. What is the total interest over the period?

SOLUTION   We use the future value formula

$$A = P(1 + rt).$$

One of the quantities A, P, r, and t is unknown, and the rest are given. To determine which of these is the unknown, we focus on what we are asked to find: the value after 4 years. Thus, the unknown is the future value, A. It is a good habit to list the values of all the quantities, indicating the unknown with a question mark.

$$A = ?$$
$$P = 1{,}000 \text{ (present value, or principal)}$$
$$r = 0.0475 \text{ (interest rate)}$$
$$t = 4 \text{ (number of years)}$$

Substituting these values in the formula gives

$$\begin{aligned} A &= 1{,}000[1 + (0.0475)(4)] \\ &= 1{,}000[1 + 0.19] \\ &= 1{,}000(1.19) = \$1{,}190. \end{aligned}$$

We can calculate the interest using the formula $I = Prt$. Alternatively, we observe that the interest is just the difference between the present and future values:

$$\begin{aligned} I &= A - P \\ &= 1{,}190 - 1{,}000 \\ &= \$190. \end{aligned}$$

Before we go on... We could have used the formula $I = Prt$ to first calculate the interest and then computed $A = P + I$. In general, either approach can be used.

EXAMPLE 2 Bridge Loans

When buying one house and selling another, it is common to take out a *bridge loan*, a short-term loan to bridge the gap between buying the new house and selling the old one. Suppose that a bank charges 12% simple annual interest on such a loan. How much will be owed at the maturation of a 90-day bridge loan of $90,000?

---

*Source: *Toronto Globe and Mail* Report on Business, March 28, 1994, p. B8.

8.1 Simple Interest

SOLUTION We use the future value formula

$$A = P(1 + rt)$$

with

$$A = ?$$
$$P = 90{,}000$$
$$r = 0.12$$
$$t = \frac{90}{365}.$$

Note that we converted the length of the loan, 90 days, into $\frac{90}{365}$ years. (Many banks will count a year as 360 days for this calculation, which has the effect of making 30 days equivalent to 1 month.) Thus,

$$A = 90{,}000\left(1 + 0.12\left(\frac{90}{365}\right)\right) = \$92{,}663.01.$$

Before we go on... The interest owed on the loan is $92,663.01 − 90,000 = $2,663.01.

▼ **EXAMPLE 3** Bonds

A $100 Government of Canada Bond that matures in three years will pay $118 at maturity.* To what annual rate of simple interest does this correspond?

SOLUTION Since the interest is $18, let us use the interest formula

$$I = Prt.$$

This time we are asked to find the annual interest rate r. The individual quantities are

$$I = 18$$
$$P = 100 \text{ (present value)}$$
$$r = ?$$
$$t = 3 \text{ (number of years)}.$$

Substituting these values in the formula gives

$$18 = 100(3r),$$

or

$$18 = 300r,$$

▼ * These figures were approximately accurate as of March 28, 1994. Source: Royal Bank of Canada/*Toronto Globe and Mail* Report on Business, March 28, 1994, p. B9.

and we divide both sides by 300 to obtain

$$r = \frac{18}{300} = 0.06.$$

Thus, three-year Government of Canada Bonds are paying 6% per year simple interest.

Before we go on... We could have used the formula $A = P(1 + rt)$, but it would have taken a couple of additional steps to solve for r.

▼ **EXAMPLE 4**   Bonds

Calculate the future value after x years of a Treasury Bond that now costs $200 and pays 6% simple interest per year, and graph the result.

SOLUTION   We again use the formula $A = P(1 + rt)$, where the quantities are

$$A = ?$$
$$P = 200$$
$$r = 0.06$$
$$t = x.$$

This gives

$$A = 200(1 + 0.06x)$$

or

$$A = 12x + 200.$$

Thus, the future value A depends linearly on time x. Figure 1 shows a graphing calculator plot of this relationship.

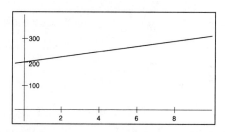

FIGURE 1

Before we go on... We can see from the formula $A = P(1 + rt) = P + Prt$ that A is always a linear function of t (and also a linear function of r and of P).

8.1 Simple Interest

Not all simple-interest investments pay interest on an annual basis. Here is a typical scenario.

▼ **EXAMPLE 5** Retirement Income

Your aunt's retirement account consists of a single $300,000 investment in a municipal bond fund that yields 5.4% per year. As she is now retired, her income consists of the interest, paid to her each month. What is her monthly income?

SOLUTION Since the question is asking for interest, we use the formula

$$I = Prt,$$

with

$$I = ?$$
$$P = 300,000$$
$$r = 0.054$$
$$t = \frac{1}{12}.$$

This will give us the interest for *one month*:

$$I = (300,000)(0.054)\frac{1}{12}$$
$$= 1,350.$$

▼ **EXAMPLE 6** Bridge Loans

You are expecting a tax refund of $800. Since it may take up to 6 weeks to get the refund, the tax preparation firm you use offers, for a fee of $40, to give you a loan of $800 to be paid back with your refund check. Thinking of the fee as interest, what annual rate of interest is the firm charging?

SOLUTION We use the future value formula,

$$A = P(1 + rt).$$

Since you pay the $40 fee up front, the firm is really loaning you only $760. You will repay $800 after 6 weeks. Thus, the quantities are

$$A = 800$$
$$P = 760$$
$$r = ?$$
$$t = \frac{6}{52}.$$

(Note that there are 52 weeks in one year.) Substituting gives

$$800 = 760\left[1 + r\left(\frac{6}{52}\right)\right].$$

To solve for r, we first divide both sides by 760.

$$1 + \frac{6}{52}r = \frac{800}{760} \approx 1.05263$$

Thus,

$$\frac{6}{52}r \approx 1.05263 - 1 = 0.05263,$$

giving

$$r = 0.05263\left(\frac{52}{6}\right) \approx 0.456.$$

In other words, you will be charged 45.6% annual interest! Save your money and wait six weeks for your refund.

8.1 EXERCISES

In Exercises 1–6, compute the simple interest for the specified period. Round all answers to the nearest cent.

1. $2,000 is invested for one year at 6% per year.
2. $1,000 is invested for 10 years at 4% per year.
3. $20,200 is invested for six months at 5% per year.
4. $10,100 is invested for three months at 11% per year.
5. You borrow $10,000 for 10 months at 3% per year.
6. You borrow $6,000 for 5 months at 9% per year.

APPLICATIONS

In Exercises 7–12, compute the specified quantity. Round all answers to the nearest tenth of a year, the nearest cent, or the nearest 0.1%, as the case may be.

7. *Bonds* A bond that has a maturity value of $125 five years from now costs $100. What is the annual rate of simple interest?
8. *Loans* A $400 loan, taken now, with a simple interest rate of 8% per year, will cost a total of $464. When will the loan mature?
9. *Loans* The simple interest on a $1,000 loan at 8% per year amounted to $640. When did the loan mature?
10. *Loans* The simple interest on a $1,000 five-year loan amounted to $350. What was the interest rate charged?
11. *Bonds* Calculate the cost of a bond earning 4.5% simple interest whose value at maturity six years from now will be $1,000.
12. *Loans* Your repayment on a four-year loan, which charged 9.5% simple interest, amounted to $30,500. How much did you originally borrow?

Corporate Income Exercises 13–20 are based on the following chart, which shows Sony Corporation's net income for fiscal years ending in March.

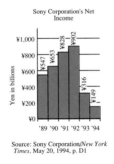

13. Calculate the percentage increase in *Sony's* net income for the period 1989–1990. (Answer to the nearest 0.01%.)
14. Calculate the percentage decrease in *Sony's* net income for the period 1992–1993. (Answer to the nearest 0.01%.)
15. Calculate the annual percentage increase in *Sony's* net income for the period 1989–1992 (to the nearest 0.01%, calculated on a simple-interest basis).
16. Calculate the annual percentage decrease in *Sony's* net income for the period 1989–1994 (to the nearest 0.01%, calculated on a simple-interest basis).
17. During which one-year period did *Sony Corporation* experience the largest percentage increase in net income? What was the percentage increase?
18. During which one-year period did *Sony Corporation* experience the largest percentage decrease in net income? What was the percentage decrease?
19. Did *Sony's* net income undergo simple-interest growth in the years 1989–1992? (Give the reason for your answer.)
20. Did *Sony's* net income undergo simple-interest contraction in the years 1992–1994? (Give the reason for your answer.)

Balance of Trade Exercises 21–26 are based on the following graph, which shows the 12-month moving average of the U.S. Trade Deficit for the period from January 1993 to January 1994.

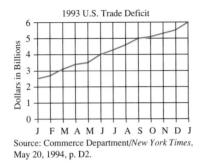

21. At what monthly rate of interest did the average increase for the period shown? (Answer to the nearest 1%.)

22. At what monthly rate did the average increase from June 1993 to September 1993? (Answer to the nearest 1%.)

23. If you used your answer to Exercise 21 as the simple interest rate at which the average was growing since January 1993, what would you predict the value of the average to be in January 1995? (Answer to the nearest $0.5 billion.)

24. If you used your answer to Exercise 22 as the simple interest rate at which the average was growing since January 1993, what would you predict the value of the average to be in January 1995? (Answer to the nearest $0.5 billion.)

25. Use your answer to Exercise 21 to graph the monthly moving average for the period from January 1993 to January 1994.

26. Use your answer to Exercise 22 to graph the monthly moving average for the period from January 1993 to January 1994.

COMMUNICATION AND REASONING EXERCISES

27. One or more of the following graphs represents the future value of an investment earning simple interest. Which one(s)? Give the reason for your choice(s).

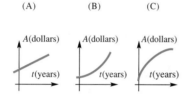

28. *Interpreting the News* You hear the following on your local radio station's business news: "The economy last year grew by 1%. This was the second year in a row in which the economy showed a 1% growth." This means that in dollar terms, the economy grew more last year than the year before. Why?

29. Assume that $100 is invested at a simple interest rate of 5% per year. Graph future value A (in dollars) as a function of time t (in years).

30. Suppose that a $1,000 investment is decreasing in value at a simple interest rate of 15% per year. Graph future value A (in dollars) as a function of time t (in years).

31. Explain why simple interest is not the appropriate way to measure interest on a checking account that pays interest directly into your account.

32. Given that $A = 5t + 400$, for what interest rate is this the equation of future value (in dollars) as a function of time t (in years)?

▶ 8.2 COMPOUND INTEREST

Consider the following scenario: a customer deposits $10,000 at Solid Savings & Loan, which pays 6% interest **compounded monthly.** This means that at the end of each month, the bank determines the (simple) interest earned for one month, *then adds the interest back into the account.* Assuming that the customer lets the interest remain in the account, we shall determine how much will be in the account after 5 years.

Let us begin with the first month. At the end of the first month, the principal plus interest can be calculated by the future value formula from the last section:

$$A = P(1 + rt),$$

where

$$A = ?$$
$$P = 10{,}000$$
$$r = 0.06$$
$$t = \tfrac{1}{12}.$$

Thus, after 1 month:

$$A = 10{,}000\left(1 + \frac{0.06}{12}\right).$$

For reasons that will become clear in a moment, we shall not compute the dollar value but leave the answer in this form. Now, to compute the amount in the account at the end of the second month, we again calculate the principal plus interest, but the principal is the amount that was in the account at the beginning of the month:

$$A = P(1 + rt),$$

where

$$A = ?$$
$$P = 10{,}000\left(1 + \frac{0.06}{12}\right)$$
$$r = 0.06$$
$$t = \tfrac{1}{12}.$$

So after 2 months:

$$A = 10{,}000\left(1 + \frac{0.06}{12}\right)\left(1 + \frac{0.06}{12}\right)$$
$$= 10{,}000\left(1 + \frac{0.06}{12}\right)^2.$$

Notice what is happening: Each month we multiply the balance in the account by $1 + (0.06/12)$. After 5 years we will have done this 60 times (12 months per year for 5 years), so the balance at the end of 5 years will be

$$A = 10{,}000\left(1 + \frac{0.06}{12}\right)^{60} = \$13{,}488.50.$$

This calculation suggests a general formula. We consider an amount P deposited in an account earning interest at an annual rate r compounded m times each year. In the calculation above, $m = 12$ because the interest was compounded each month. We refer to the period of time between each interest payment as the **compounding period.** In the calculation above the compounding period was one month. It is common to write $i = r/m$ for the interest rate per compounding period. In the calculation above, $i = 0.06/12$. If the account earns interest for t years, then it will earn interest for $n = mt$ compounding periods. The amount in the account at the end of this time is given by the following formula.

COMPOUND INTEREST

If an amount P (principal or present value) is deposited in an account earning interest at an annual rate r, compounded m times per year, then the accumulated amount (or future value) after t years is

$$A = P\left(1 + \frac{r}{m}\right)^{mt}$$

or

$$A = P(1 + i)^n,$$

where $i = r/m$ and $n = mt$.

EXAMPLE 1 Mutual Funds

Fidelity Investments advertised that its Fidelity Equity-Income II Fund yielded an average annual return of 23.35%.* Assume that this rate of return will continue for four years and that you invest $10,000 in the fund, reinvesting the interest at the end of each year. How much will your investment be worth at the end of four years?

SOLUTION Since the income is reinvested at the end of each year, this amounts to compounding the interest annually (once per year) for a four-year period. Thus, we use the compound interest formula

$$A = P(1 + i)^n.$$

Since we are compounding once per year, $m = 1$. Thus,

$A = ?$
$P = 10{,}000$ (present value)
$i = r = 0.2335$ (interest per compounding period)
$n = t = 4$ (number of compounding periods).

---

*Rates are accurate as of 6/30/93, as quoted in an advertisement in *The New York Times* on September 26, 1993.

Substituting these values in the formula gives

$$A = 10{,}000(1 + 0.2335)^4$$
$$= 10{,}000(1.2335)^4$$
$$= \$23{,}150.30 \text{ (to the nearest cent)}.$$

Thus, your investment will have more than doubled.

Before we go on... Note that the order of operations in calculating the value of $10{,}000(1.2335)^4$ is the standard one: first calculate the value of the expression in parentheses, then exponentiate, and then multiply. On a standard calculator, we use the following sequence.

[1.2335] [x^y] [4] [×] [10000] [=]

On a graphing calculator or function-based calculator, we enter the following:

$$10000 \times (1.2335\wedge 4)$$

The parentheses are optional, as calculators are programmed to use the standard order of operations.*

You can plot the future value A as a function of the number of years t using the following format.

$$Y_1 = 10000 \times (1.2335\wedge X)$$

If you wish to graph the value of the investment for, say, 5 years, set the X range to $0 \leq X \leq 5$ and the Y range to $0 \leq Y \leq 30000$ (why so large?) and then graph.

▼ **EXAMPLE 2**   Investments Compounded Monthly

You invest \$5,000 in CDs at the Park Avenue Bank, which pay 5% interest compounded monthly. How much money will this give you after 10 years?

SOLUTION   We use the formula

$$A = P\left(1 + \frac{r}{m}\right)^{mt}.$$

The values of the individual quantities are

$$A = ?$$
$$P = 5{,}000$$
$$r = 0.05$$
$$m = 12$$
$$t = 10.$$

▼ *See Section 1 of Appendix A for additional discussion on the order of arithmetic operations and the use of function-based calculators.

Substituting these values in the formula gives

$$A = P\left(1 + \frac{r}{m}\right)^{mt}$$
$$= 5{,}000\left(1 + \frac{0.05}{12}\right)^{(12)(10)}$$
$$= 5{,}000(1.004166667)^{120}$$
$$= \$8{,}235.05.$$

Thus, you will have accumulated a total of $8,235.05 after 10 years.

Before we go on... Compare this answer with what you would expect if the interest were compounded only once a year: $5{,}000(1.05)^{10} = \$8{,}144.47$. When the interest is compounded more often, the interest earns interest beginning earlier, and the net result is more money.

Compare this also with what you would expect if the interest were not compounded at all. Simple interest of 5% for 10 years would yield

$$A = P(1 + rt)$$
$$= 5{,}000(1 + 0.05(10))$$
$$= \$7{,}500.$$

Thus, compounding once per year earns you $644.47 more than simple interest would, while compounding monthly earns you $735.05 more than simple interest would. This is the effect of having the interest earn interest.

 A function-based graphing calculator will allow you to calculate A in one step by entering the formula for A.

$$5000(1+0.05/12)\wedge(12\times10)$$

Note that we have enclosed the whole exponent in parentheses. A common error is to enter instead

$$5000(1+0.05/12)\wedge12\times10 \quad \textbf{WRONG}$$

Why is it wrong? Think about the order of operations. (For additional examples and exercises on the role of parentheses, consult the first section in Appendix A.)

▼ **EXAMPLE 3** Investments

Your friend comes to you and says, "I just don't know what to do with my $100,000 Lotto winnings. I'm thinking of investing the money for five years while I still have my modeling job, and the best I could come up with are the following options: The Megabuck Junk Bond Company, which offers 10% interest compounded annually, and the Trustworthy Savings Association, which offers 9.2% compounded three times per year. Which option would be best for me?"

SOLUTION We must calculate A for each option to see which is better.

Megabuck: $P = \$100{,}000$, $r = 0.1$, $m = 1$, $t = 5$

$$A = P\left(1 + \frac{r}{m}\right)^{mt}$$
$$= 100{,}000(1 + 0.1)^5$$
$$= 100{,}000(1.1)^5$$
$$= 100{,}000(1.61051) = \$161{,}051$$

Trustworthy: $P = \$100{,}000$, $r = 0.092$, $m = 3$, $t = 5$

$$A = P\left(1 + \frac{r}{m}\right)^{mt}$$
$$= 100{,}000\left(1 + \frac{0.092}{3}\right)^{15}$$
$$\approx 100{,}000(1.5731620) = \$157{,}316.20$$

It follows that, although Trustworthy is compounding the interest more frequently, it can't quite match the Megabuck Company's return over the five-year period.

Let us graph both investments as functions of time. Thus, instead of assuming a time period of 5 years, let us leave the number of years as t.

Megabuck: $A = 100{,}000(1.1)^t$

Trustworthy: $A = 100{,}000(1.0307)^{3t}$

(We approximated $1 + \frac{0.092}{3}$ by 1.0307.) The format for the graphing calculator is then

Megabuck: $Y_1 = 100000(1.1)^{\wedge}X$

Trustworthy: $Y_2 = 100000(1.0307)^{\wedge}(3X)$

Graphing these functions in the viewing rectangle $0 \leq X \leq 10$ (for a ten-year period) and $0 \leq Y \leq 300000$ gives us Figure 1. Which curve represents the Megabuck investment?

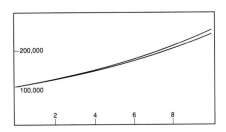

FIGURE 1

Before we go on... The graph also tells us something else: that the graph of A as a function of time x is not linear, but curves upwards. This curve is an example of an **exponential curve**—the graph of an equation of the form $y = p \cdot c^{rx}$, where p, c, and r are nonzero constants (with c positive).

▼ **EXAMPLE 4** Depreciating Investments

Curtis Reynolds made a terrible mistake: acting on the advice of his broker, George Duda, he invested $6,000 in Tarnished Teak Enterprises and discovered that his investment was losing 6% of its value annually. After watching his savings dwindle for two years, and despite repeated assurances from Duda that Teak Enterprises would soon turn around he decided to pull out what was left of the investment and invest it all in Solid Trust at 4% interest, compounded twice a year. Will he have recovered his investment after 6 more years?

SOLUTION This is a two-stage calculation. First, we have the following information:

Tarnished Teak: $P = \$6,000$, $r = -0.06$ (the interest rate is negative because the investment is *declining* at 6% per year), $m = 1$, $t = 2$. Thus,

$$A = P\left(1 + \frac{r}{m}\right)^{mt}$$
$$= 6{,}000(1 - 0.06)^2$$
$$= 6{,}000(.94)^2$$
$$= 6{,}000(.8836) = \$5{,}301.60.$$

So Curtis Reynolds is left with $5,301.60 after two years. He then invests this in Solid Trust, so we calculate as follows.
Solid Trust: $P = 5301.6$, $r = 0.04$, $m = 2$, $t = 6$

$$A = 5301.6\left(1 + \frac{0.04}{2}\right)^{12}$$
$$= 5301.6(1.02)^{12}$$
$$= \$6723.71$$

So we see that Reynolds has not only recouped his losses, but has gained a little extra besides!

It would be interesting to track Reynolds' investments by graphing the two investments.

Teak: $A = 6{,}000(1 - 0.06)^t = 6{,}000(.94)^t$
Solid Trust: $A = 5301.6(1 + 0.02)^{2t} = 5301.6(1.02)^{2t}$

To graph them on your calculator, use the following format:

$Y_1 = 6000(0.94)^\wedge X$
$Y_2 = 5301.6(1.02)^\wedge(2X)$.

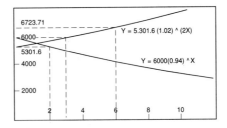

FIGURE 2

The graphs are shown in Figure 2.

You can use the trace feature on the calculator to see when the investment in Solid Trust reaches $6,000, which is when the loss in Teak has been recovered.

▼ **EXAMPLE 5**   Constant Dollars

Inflation in East Avalon is running at 5% per year. My East Avalon municipal bond matures in 10 years' time and will pay me $1,000. What would the equivalent value be now?

SOLUTION   We are given the future value of the bond, and we must calculate the present value P, given the inflation rate. This is also called the value of the bond in **constant dollars.** The effect of inflation is identical to compound interest: prices rise by the same formula. So we use

$$A = P\left(1 + \frac{r}{m}\right)^{mt},$$

where

$$A = 1,000$$
$$P = ?$$
$$r = 0.05$$
$$m = 1$$
$$t = 10.$$

Substituting in the formula gives

$$1,000 = P(1 + 0.05)^{10},$$

so

$$1,000 = P(1.05)^{10}.$$

Solving for P gives

$$P = \frac{1,000}{1.05^{10}} = \$613.91.$$

In other words, the buying power of my $1,000 in 10 years' time will be equivalent to the buying power of $613.91 now.

Before we go on... Converting money into constant dollars allows us to compare prices at different times by compensating for the effects of inflation, as we see in the next example.

▼ **EXAMPLE 6**   Constant Dollars

Inflation in West Avalon is running at 1% per month. TruVision televisions cost $200 today. A newer model, the TruVision II, will cost $220 next year. Which is really the more expensive model?

SOLUTION   In order to compare prices at two different times, we need to convert to constant dollars. Let us convert the $220 from next year into its present value, using the method of Example 5. Since inflation is running at 1% per *month*, the compounding period is one month, and $i = 0.01$ and $n = 12$. We then get

$$P = \frac{A}{(1+i)^n} = \frac{220}{1.01^{12}} = \$195.24.$$

Since this is lower than the $200 price of the original TruVision, the newer model costs less (in constant dollars).

Before we go on... We could also compare the prices by converting the $200 price of the original TruVision to its future value one year from now. In other words, we could calculate the effect of inflation on this price and then compare it to the $220 price of the newer model. Do this for practice.

Let's return to Example 3 for a moment. In that example we compared the results of investing money at 10% compounded annually and at 9.2% compounded three times each year. Here is another way of doing the comparison. Think about the result of investing $1 for one year. In the first case we get 10¢. In the second case we have to use the compound interest formula:

$$A = 1 \cdot \left(1 + \frac{0.092}{3}\right)^3 = 1.09485.$$

So our $1 earns 9.485¢ in the second investment. This is the same amount we would get if we were being paid 9.485% interest compounded only once per year, and so we say that the **effective interest rate** for the second investment is 9.485%. Since 10% is larger than 9.485%, the first investment is better. The effective interest rate gives us a way of directly comparing investments that may be compounded at different intervals. If you look at the calculation we just did, you can see the following formula.

> **EFFECTIVE INTEREST RATE**
>
> $$r_e = \left(1 + \frac{r}{m}\right)^m - 1,$$
>
> where r_e = the effective interest rate, r = the stated interest rate, and m = the number of times per year the interest is compounded.

The stated rate r is also known as the **nominal** interest rate, to distinguish it from the effective rate. The effective rate is also called the **effective yield.**

▼ EXAMPLE 7

Which investment is better: one that pays 20% compounded monthly, or one that pays 20.5% compounded twice per year?

SOLUTION Let us compute the effective interest rates of both investments. First investment: $r = 0.20$, $m = 12$. The effective interest rate is

$$r_e = \left(1 + \frac{r}{m}\right)^m - 1$$
$$= \left(1 + \frac{0.20}{12}\right)^{12} - 1$$
$$= 0.2194, \quad \text{or} \quad 21.94\%.$$

Second investment: $r = 0.205$, $m = 2$. The effective interest rate is

$$r_e = \left(1 + \frac{r}{m}\right)^m - 1$$
$$= \left(1 + \frac{0.205}{2}\right)^2 - 1$$
$$= 0.2155, \quad \text{or} \quad 21.55\%.$$

Thus, the first investment is the better one: each $1 invested will earn 21.94¢ per year, while in the second investment it will earn only 21.55¢ per year.

If you look at newspaper ads for investments or loans, you will see (by law) two rates advertised: the nominal rate and the effective rate. (The effective rate calculation actually used also takes certain fees into account.) You should pay attention to the effective rates, since only these can be compared directly.

USING A GRAPHING CALCULATOR TO ANSWER INVESTMENT QUESTIONS

In the next example, we show how a graphing calculator can be used to answer important questions such as this: "How long will it take me to reach my goal of $5,000 if I invest $3,000 at 6% per year compounded monthly?" Traditional methods of answering questions such as this involve either the use of compound interest tables or the use of logarithms with scientific calculators. However, graphing technology gives us another method to answer such questions.

▼ **EXAMPLE 8**   Investments

How long will it take you to accumulate $5,000 if you invest $3,000 at 6% per year compounded monthly?

SOLUTION   We use the formula $A = P(1 + i)^n$, where

$$A = 5,000$$
$$P = 3,000$$
$$i = r/m = \frac{0.06}{12} = 0.005$$
$$n = ?$$

Substituting gives

$$5,000 = 3,000(1.005)^n,$$

and we are to solve for the unknown number of months n. The first step in solving for n is to graph our investment over time. The standard format for the function we are graphing is the following.

$$Y_1 = 3000(1.005)\wedge X$$

For the X- and Y-ranges, we note that X represents time, so we set $0 \le X \le 120$ to look at the first 10 years (if this is not a long enough time period, we can try a longer one next). For the Y-range, we note that Y represents the value of an investment starting at $3,000 and increasing past $5,000, so we set $0 \le Y \le 5,500$. Figure 3 shows the resulting graph.

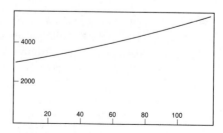

FIGURE 3

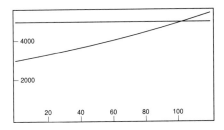
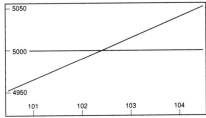

FIGURE 4   **FIGURE 5**

We now wish to know when (X) the investment (Y) has reached $5,000. One way to do this is to superimpose the graph of Y = 5000 and determine the point of intersection by zooming in (Figures 4 and 5).

This yields an *x*-coordinate between 102 and 103. Thus, the investment will have reached $5,000 by the end of the 103rd month, which is to say after 8 years and 7 months.

Before we go on... An alternative way of finding the point of intersection is to use the trace feature. If we choose the *x*-range carefully so that each "pixel" on the screen represents one month (for example, $20 \leq X \leq 114$ on the TI-82, or any range with Xmax = Xmin + 94), then as we trace along the curve we will see the value of the investment at the end of each month given as the *y*-coordinate. When that value rises above $5,000 we will find the answer to the question.

8.2 EXERCISES

In Exercises 1 through 8, calculate, to the nearest cent, the worth of an investment of $10,000 after the stated amount of time.

1. 3% per year, compounded annually, after 10 years
2. 4% per year, compounded annually, after 8 years
3. 2.5% per year, compounded quarterly (4 times per year) after 5 years
4. 1.5% per year, compounded weekly (52 times per year) after 5 years
5. 6.5% per year, compounded daily (assume 365 days per year) after 10 years
6. 11.2% per year, compounded monthly, after 12 years
7. 0.2% per month, compounded monthly, after 10 years
8. 0.45% per month, compounded monthly, after 20 years

In Exercises 9 through 14, calculate the present value of an investment that will be worth $1,000 after the stated amount of time.

9. 10 years, at 5% per year, compounded annually
10. 5 years, at 6% per year, compounded annually

11. 5 years, at 4.2% per year, compounded weekly
12. 10 years, at 5.3% per year, compounded quarterly
13. 4 years, depreciating at 5% per year
14. 5 years, depreciating at 4.5% per year

In Exercises 15 through 20, find the effective annual interest rates of the given annual interest rates. Round your answers to the nearest 0.01%.

15. 5% compounded quarterly
16. 5% compounded monthly
17. 10% compounded monthly
18. 10% compounded daily (assume 365 days per year)
19. 10% compounded hourly
20. 10% compounded every minute

APPLICATIONS

21. *Investments.* You invest $100 in the Lifelong Trust Savings and Loan Company, which pays 6% interest compounded quarterly. By how much will your investment have grown after four years?

22. *Investments* You invest $1,000 in Rapid Growth Funds, which appreciate by 2% per year, with yields reinvested quarterly. By how much will your investment have grown after 5 years?

23. *Depreciating Investments* During the first nine months of 1993, mutual funds in health-related industries depreciated at a rate of 6.8% per year.* Assuming this trend were to continue, how much will a $3,000 investment in this category of funds be worth in 5 years?

24. *Depreciating Stocks* During the first nine months of 1993, the value of *IBM* stocks depreciated at an annual rate of 16.28%.[†] Assuming that this trend were to continue, what will a $10,000 investment in *IBM* stocks be worth in 10 years?

25. *Investments* When I was considering what to do with the $10,000 proceeds from my sale of junk bonds, my broker, Kelly Bell, suggested I invest half of it in gold, whose value was growing by 10% per year, and the other half in CDs, which were yielding 5% per year, compounded every six months. Assuming that these interest rates are sustained, how much will my investment be worth in 10 years?

26. *Investments* When I was considering what to do with the $10,000 proceeds from my sale of *Harper-Collins* stock, my broker, P. Prudence, suggested I invest half of it in municipal bonds, whose value was growing by 6% per year, and the other half in CDs, which were yielding 3% per year, compounded every two months. Assuming that these interest rates are sustained, how much will my investment be worth in 10 years?

27. *Depreciation* During a prolonged recession, property values in Long Island depreciated by 2.5% every six months. If my house cost $200,000 originally, how much was it worth 10 years later?

28. *Depreciation* Stocks in the health industry depreciated by 5.1% in the first nine months of 1993.* Assuming this trend were to continue, how much will a $40,000 investment be worth in nine years? [*Hint:* nine years corresponds to twelve 9-month periods.]

29. *Retirement Planning* I plan to be earning an annual salary of $100,000 when I retire in 15 years' time. I have been offered a job that guarantees an annual salary increase of 4% per year, and the starting salary is negotiable. What salary should I request in order to meet my retirement goals?

▼ *Source: *The New York Times*, October 9, 1993, p. 37.
[†] Source: *The New York Times*, October 9, 1993, p. 40.

30. *Retirement Planning* I plan to be earning an annual salary of $80,000 when I retire in 10 years' time. I have been offered a job that guarantees an annual salary increase of 5% per year, and the starting salary is negotiable. What salary should I request in order to meet my retirement goals?

31. *Stocks* Six years ago, I invested some money in Dracubunny Toy Inc. stocks, acting on the advice of a "friend." As things turned out, the value of the stock decreased by 5% every four months, and I discovered yesterday (to my horror) that my investment was worth $150. How much did I originally invest?

32. *Sales* My recent marketing idea, the *Miracle Algae-Growing Kit,* has been remarkably successful, with monthly sales growing by 6% every six months over the past five years. Assuming that the sales figure at the start of the marketing campaign was 100 kits per month, what is the present rate of sales?

33. *Inflation* Inflation has been running at an effective rate of 5% per year. A car now costs $10,000. How much would it have cost five years ago?

34. *Inflation* Housing prices have been rising at 0.5% each month. A house now costs $200,000. What would it have cost 10 years ago?

35. *Constant Dollars* Inflation is running at 3% per year (effective rate) when you deposit $1000 in an account earning 5% per year compounded annually. *In constant dollars,* how much money will you have two years from now? [*Hint:* First calculate the value of your account in two years' time, and then find its present value based on the inflation rate.]

36. *Constant Dollars* Inflation is running at 1% per month when you deposit $10,000 in an account earning 8% compounded monthly. *In constant dollars,* how much money will you have two years from now? [See the hint for Exercise 35.]

37. *Investments* You are offered two investments. One promises to earn 12% compounded annually. The other will earn 11.9% compounded monthly. Which is the better investment?

38. *Investments* You are offered three investments. The first promises to earn 15% compounded annually, the second will earn 14.5% compounded quarterly, and the third will earn 14% compounded monthly. Which is the best investment?

Inflation Exercises 39 through 46 are based on the following table, which shows the 1992 annual inflation rates in several Latin American countries.* Assume that the rates shown continue indefinitely.

Country	Argentina	Bolivia	Brazil	Colombia	Chile	Ecuador	Mexico	Uruguay
Currency	austral	peso	cruzado	peso	peso	sucre	peso	peso
Inflation Rate (%)	18	11	1,132	26	14	66	13	59

39. If an item in Brazil now costs 100 cruzados, what do you expect it to cost 5 years from now? (Answer to the nearest cruzado.)

40. If an item in Argentina now costs 1,000 australs, what do you expect it to cost 5 years from now? (Answer to the nearest austral.)

41. If an item in Chile will cost 1,000 pesos in 10 years' time, what does it cost now? (Answer to the nearest peso.)

42. If an item in Mexico will cost 20,000 pesos in 10 years' time, what does it cost now? (Answer to the nearest peso.)

▼ * Each rate has been rounded to the nearest percentage point. Source: International Monetary Fund, UN Economic Commission for Latin America and the Caribbean (as cited in the *Chicago Tribune,* June 20, 1993).

43. You wish to invest 1,000 pesos in Columbia at 28% annually, compounded twice a year. Find the value of your investment in 10 years' time, expressing the answer in constant pesos. (Answer to the nearest peso.)

44. You wish to invest 1,000 pesos in Bolivia at 13% annually, compounded twice a year. Find the value of your investment in 10 years' time, expressing the answer in constant pesos. (Answer to the nearest peso.)

45. Which is the better investment: an investment in Colombia yielding 27.5% per year, compounded annually, or an investment in Ecuador, yielding 68.5% per year, compounded every six months? Support your answer with figures showing the future value of an investment of one unit of currency in constant units.

46. Which is the better investment: an investment in Argentina yielding 20% per year, compounded annually, or an investment in Uruguay, yielding 60% per year, compounded every six months? Support your answer with figures showing the future value of an investment of one unit of currency in constant units.

Use a graphing calculator or computer for Exercises 47–68.

Graph the accumulated amount A as a function of the number of years t in each of Exercises 47 through 50.

47. $500 invested at 5% per year, compounded annually; $0 \leq t \leq 40$

48. $600 invested at 7.5% per year, compounded annually; $0 \leq t \leq 40$

49. $500 invested at 5% per year, compounded daily; $0 \leq t \leq 40$

50. $600 invested at 7.5% per year, compounded daily; $0 \leq t \leq 40$

51. *Compound Interest Table* Complete the following table, which shows the value of a $1,000 investment earning 5% per year after the specified time periods. Round all answers to the nearest dollar.

Years	1	2	3	4	5	6	7
Value	$1,050						

52. *Compound Interest Table* Complete the following table, which shows the value of a $1,000 investment earning 6% per year after the specified time periods. Round all answers to the nearest dollar.

Years	1	2	3	4	5	6	7
Value	$1,060						

53. *Competing Investments* I just purchased $5,000 worth of municipal funds that are expected to yield 5.4% every six months. My friend has just purchased $6,000 worth of CDs that are expected to earn 4.8% every six months. Determine when, to the nearest year, the value of my investment will be the same as hers, and what this value will be.

54. *Investments* Determine when, to the nearest year, $3,000 invested at 5% per year, compounded daily, will be worth $10,000.

8.2 Compound Interest

55. *Epidemics* At the start of 1985, the incidence of AIDS was doubling every six months, and 40,000 cases had been reported in the U.S. Assuming this trend were to continue, determine when, to the nearest 0.1 year, the number of cases would have reached 1 million.

56. *Depreciation* My investment in Genetic Splicing, Inc., is now worth $4,354 and is depreciating by 5% every six months. For some reason, I am reluctant to sell the stocks and swallow my losses. Determine when, to the nearest year, my investment will drop below $50.

Government Bonds Exercises 57 through 66 are based on the following table, which lists annual percentage yields on government bonds in several countries.*

Country	U.S.	Japan	Germany	Gr. Britain	Canada	Mexico
Yield (%)	5.16	3.86	5.97	6.81	6.58	13.7

57. Assuming that you invest $10,000 in U.S. government bonds, how long (to the nearest year) must you wait before your investment is worth $15,000 if the interest is compounded annually?

58. Assuming that you invest $10,000 in Japanese government bonds, how long (to the nearest year) must you wait before your investment is worth $15,000 if the interest is compounded annually?

59. If you invest $10,400 in German government bonds and the interest is compounded monthly, how many months will it take for your investment to grow to $20,000?

60. If you invest $10,400 in British government bonds and the interest is compounded monthly, how many months will it take for your investment to grow to $20,000?

61. How long, to the nearest year, will it take an investment in Mexico to double its value if the interest is compounded every 6 months?

62. How long, to the nearest year, will it take an investment in Canada to double its value if the interest is compounded every 6 months?

63. How long will it take a $1,000 investment in Canadian government bonds to be worth the same as an $800 investment in Mexican government bonds? (Assume all interest is compounded annually, and give the answer to the nearest year.)

64. How long will it take a $1,000 investment in Japanese government bonds to be worth the same as an $800 investment in U.S. government bonds? (Assume all interest is compounded annually, and give the answer to the nearest year.)

65. If the interest on U.S. government bonds is compounded daily, how many years will it take for the value of an investment to double? (Assume 365 days in a year.)

▼ * Yields are quoted as of October 18, 1993. (Source: Salamon Brothers, Mexican Government, S. G. Wartburg & Company, J. P. Morgan Global Research, as quoted in *The New York Times*, October 18, 1993.)

66. If the interest on Canadian government bonds is compounded daily, how many years will it take for the value of an investment to double? (Assume 365 days in a year.)

67. *Investments* Berger Funds advertised that its Berger 100 Mutual Fund yielded an average interest rate of 15.9% per year from 1974 to 1993.* Assuming that this rate of return continues, how long, to the nearest tenth of a year, will it take a $5,000 investment in the fund to triple in value?

68. *Investments* Berger Funds also advertised that its Berger 101 Mutual Fund yielded an average interest rate of 14.1% per year from 1974 to 1993*. Assuming that this rate of return continues, how long, to the nearest tenth of a year, will it take a $5,000 investment in the fund to double in value?

COMMUNICATION AND REASONING EXERCISES

69. Why is the graph of the future value of a compound-interest investment as a function of time not a straight line (assuming a nonzero rate of interest)?

70. After how long is the future value of a compound-interest investment the same as the future value of a simple-interest investment at the same annual rate of interest?

71. If two investments have the same effective interest rate, and you graph the future value as a function of time for each of them, are the graphs necessarily the same? Explain your answer.

72. For what kind of compound-interest investment is the effective rate the same as the nominal rate? Explain your answer.

73. For what kind of compound-interest investment is the effective rate greater than the nominal rate? When is it smaller? Explain your answer.

74. If an investment appreciates by 10% per year for five years (compounded annually) and then depreciates by 10% per year (compounded annually) for five more years, will it have the same value that it had originally? Explain your answer.

75. You can choose between two investments in bonds: one maturing in 10 years' time and the other maturing in 15 years' time. If you know the rate of inflation, how do you decide which is the better investment?

76. If you knew the inflation rates for the years 1992 through 1996, how would you convert $100 in 1996 dollars to 1992 dollars?

77. *(graphing calculator or computer software recommended)* On the same set of axes, graph the future value of a $100 investment earning 10% per year as a function of time over a 20-year period, compounded once a year, 10 times a year, 100 times a year, 1,000 times a year, and 10,000 times a year. What do you notice?

78. *(graphing calculator or computer software recommended)* By graphing the future value of a $100 investment that is depreciating by 1% each year, convince yourself that the future value eventually will be less than $1.

▼ *According to an ad placed in *The New York Times* on September 26, 1993.

8.3 ANNUITIES AND LOANS

A typical defined-contribution* pension fund works as follows. Every month while you work, you and your employer deposit a certain amount of money in an account. This money earns interest from the time it is deposited. When you retire, the account continues to earn interest, but you then start withdrawing money at a rate calculated to reduce the account to zero after some number of years. While you are working the account is an example of an **increasing annuity,** an account into which periodic payments are made. After you retire the fund is an example of a **decreasing annuity,** an account from which periodic withdrawals are made.

INCREASING ANNUITIES

Q Suppose that you make a payment of M into an increasing annuity at the end of every month and that the fund earns an annual interest rate of r, compounded monthly. How much money will there be in the account after t years?

A Let us look at a specific example: Suppose that you make a payment of $100 at the end of every month into an account earning 3.6% interest per year, compounded monthly. This means that your investment is earning $3.6/12 = 0.3\%$ per month. As usual, we write $i = 0.036/12 = 0.003$. Let us now calculate the value of the investment at the end of two years (24 months).

Think of the deposits separately. Each earns interest from the time it is deposited, and the total accumulated after two years is the sum of these deposits and the interest they earn. In other words, the accumulated value is the sum of the *future values* of the deposits, taking into account how long each deposit sits in the account. Figure 1 shows a time line with the deposits and the contribution of each to the final value.

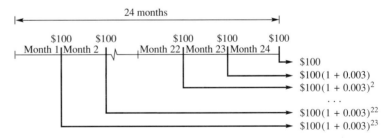

FIGURE 1

---

* Defined-contribution pensions are becoming increasingly common, replacing the defined-benefit pensions that were once the norm. In a defined-benefit pension, you are guaranteed a certain amount as a pension, often a percentage of your final working salary.

The very last deposit (at the end of month 24) has no time to earn interest, so just contributes $100.

The next-to-last deposit (at the end of month 23) earns interest for one month, so, by the compound interest formula, contributes

$$P(1 + i)^n = \$100(1 + 0.003)$$

to the total. The deposit before that one earns interest for two months, so it contributes

$$P(1 + i)^n = \$100(1 + 0.003)^2$$

to the total. Continuing in this way, we come to the very first deposit, which earns interest for 23 months and contributes $\$100(1 + 0.003)^{23}$ to the total.

Adding all of these together gives us

$$A = 100 + 100(1 + 0.003) + 100(1 + 0.003)^2 \\ + \ldots + 100(1 + 0.003)^{23} \\ = 100[1 + (1 + 0.003) + (1 + 0.003)^2 \\ + \ldots (1 + 0.003)^{23}].$$

Fortunately, this sort of sum is well known (to mathematicians, anyway), and is called a **geometric series.** It has been known for a long time that*

$$1 + x + x^2 + \ldots + x^{n-1} = \frac{x^n - 1}{x - 1}.$$

In our case, this gives

$$A = 100 \frac{(1 + 0.003)^{24} - 1}{(1 + 0.003) - 1}$$

or

$$A = 100 \frac{(1 + 0.003)^{24} - 1}{0.003}$$

$$= \$2{,}484.65$$

to the nearest cent. It is now easy to generalize this calculation: we replace 100 with M, the monthly interest rate 0.003 with $i = r/m$, where as usual, m is the number of times interest is compounded (and also the number of times per year that payments are made), and the exponent 24 with n, the number of payments. We get the following.

▼ *The quickest way to convince yourself that this formula is correct is to multiply out $(x - 1)(1 + x + x^2 + \ldots + x^{n-1})$ and see that you get $x^n - 1$. You should also try substituting some numbers. For example, $1 + 3 + 3^2 = 13 = (3^3 - 1)/(3 - 1)$.

8.3 Annuities and Loans

> **INCREASING ANNUITIES**
>
> If M is paid m times per year into an increasing annuity earning interest at the annual rate r compounded m times per year, then the amount A in the annuity after t years is
>
> $$A = M \frac{(1 + i)^n - 1}{i},$$
>
> where, as usual, $i = r/m$ and $n = mt$. (Payments are made at the end of each period.)

▶ **NOTE** There is a notation for this complicated formula commonly used in finance:

$$s_{\overline{n}|i} = \frac{(1 + i)^n - 1}{i}.$$

This is read "s sub n angle i." Calculators designed for use in business often have a single key to calculate this. ◀

EXAMPLE 1 Retirement Accounts

Every month for 10 years, you deposit $100 into a retirement account that pays 5% interest per year compounded monthly. How much money will there be in the account at the end of those 10 years?

SOLUTION This is an increasing annuity with A unknown, $M = \$100$, $r = 0.05$, $m = 12$, and $t = 10$. We compute $i = 0.05/12$ and $n = 12 \cdot 10 = 120$. From the formula,

$$A = M \frac{(1 + i)^n - 1}{i}$$

$$= 100 \frac{\left(1 + \frac{0.05}{12}\right)^{120} - 1}{\frac{0.05}{12}}$$

$$= \$15,528.23.$$

Before we go on... Notice that $\$100 \times 120 = \$12,000$ of this is the money that you deposited, and the other $\$3,528.23$ is the interest that you earned.

With a calculator you can watch the value of the annuity grow. For example, on the TI-82 first enter the initial deposit, $100.

$$100 \;\boxed{\text{ENTER}}$$

Next, add one month's interest and the next deposit.

$$\text{Ans}(1+0.05/12)+100 \;\boxed{\text{ENTER}}$$

If you hit [ENTER] repeatedly, this calculation will be repeated with each new balance, showing you the value of the annuity month by month. A computer spreadsheet can be used in a similar way to compute the monthly balances.

▼ **EXAMPLE 2** Education Fund

Tony and Maria have just had a son, Jose Phillipe. They establish an account to accumulate money for his college education, and they would like to have $100,000 after 17 years. If the account pays 4% interest per year compounded quarterly, and they make deposits every quarter, how large must each deposit be?

SOLUTION This is another increasing annuity. In this case, $A = \$100{,}000$, M is unknown, $t = 17$, $m = 4$, and $r = 0.04$. Therefore, $i = 0.04/4 = 0.01$ and $n = 4 \cdot 17 = 68$.

$$A = M\frac{(1 + i)^n - 1}{i}$$

$$100{,}000 = M\frac{(1 + 0.01)^{68} - 1}{0.01}$$

$$= M\frac{1.01^{68} - 1}{0.01}$$

To solve for M, multiply both sides by the reciprocal of the fraction on the right.

$$M = 100{,}000 \, \frac{0.01}{1.01^{68} - 1}$$

$$= \$1033.89$$

So they must deposit $1,033.89 every quarter in order to meet their goal.

Before we go on... When an increasing annuity is used in this way to accumulate money to pay off an anticipated debt, it is often called a **sinking fund**. The money accumulated will be used in the future to "sink" the debt.

DECREASING ANNUITIES

A **decreasing annuity** is an account, such as a retirement or education account, from which you make periodic withdrawals.

Q You wish to establish a decreasing annuity that will allow you to withdraw W m times a year for t years. What should the present value P of the investment be so that it is drawn down to zero after t years, if it earns an annual interest rate of r paid m times per year?

8.3 Annuities and Loans

A Let us again reason from a specific example: You deposit $\$P$ now in an account earning 3.6% interest per year, compounded monthly. Starting one month from now you will make monthly withdrawals of $100 for two years. What must P be so that the account will be down to $0 in exactly 2 years?

As usual, we write $i = r/m = 0.036/12 = 0.003$. The first withdrawal of $100 will be made one month from now, so its present value is

$$\frac{A}{(1+i)^n} = \frac{100}{1+0.003}$$
$$= 100(1+0.003)^{-1}.$$

In other words, that much of the original $\$P$ goes towards funding the first withdrawal. The second withdrawal, two months from now, has a present value of

$$\frac{A}{(1+i)^n} = \frac{100}{(1+0.003)^2}$$
$$= 100(1+0.003)^{-2}.$$

That much of the original $\$P$ funds the second payment. This continues for two years, at which point you make the last withdrawal, which has a present value of

$$\frac{A}{(1+i)^n} = \frac{100}{(1+0.003)^{24}}$$
$$= 100(1+0.003)^{-24}$$

and exhausts the account. Figure 2 shows a timeline depicting the withdrawals and the present value of each.

Since P must be the sum of these present values, we get

$$P = 100(1+0.003)^{-1} + 100(1+0.003)^{-2} + \ldots$$
$$+ 100(1+0.003)^{-24}$$
$$= 100[(1+0.003)^{-1} + (1+0.003)^{-2} + \ldots$$
$$+ (1+0.003)^{-24}].$$

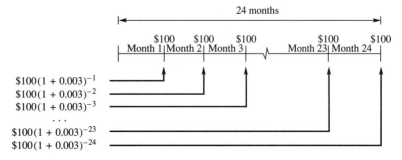

FIGURE 2

We can again find a simpler formula for this sum:

$$x^{-1} + x^{-2} + \ldots + x^{-n} = \frac{1}{x^n}(x^{n-1} + x^{n-2} + \ldots + 1)$$

$$= \frac{1}{x^n} \cdot \frac{x^n - 1}{x - 1}$$

$$= \frac{1 - x^{-n}}{x - 1}.$$

So, in our case,

$$P = 100\,\frac{1 - (1 + 0.003)^{-24}}{(1 + 0.003) - 1},$$

or

$$P = 100\,\frac{1 - (1 + 0.003)^{-24}}{0.003}$$

$$= \$2{,}312.29$$

to the nearest cent. Thus, if you deposit $2,312.29 initially, you can make withdrawals of $100 per month for two years, and your account will be exhausted at the end of that time.

It is now easy to generalize this calculation: replace 100 by W, replace the monthly interest rate 0.003 by $i = r/m$, and the exponent 24 by $n = mt$. We get the following.

> **DECREASING ANNUITIES**
>
> If $P is deposited in an account earning interest at the annual rate r compounded m times per year, and $W is withdrawn m times per year from the account so that the account is drawn down to zero after t years, then
>
> $$P = W\,\frac{1 - (1 + i)^{-n}}{i},$$
>
> where $i = r/m$ and $n = mt$. (Withdrawals take place at the end of each period.)

▶ **NOTE** As with increasing annuities, there is a notation for this formula commonly used in finance. We write

$$a_{\overline{n}|i} = \frac{1 - (1 + i)^{-n}}{i}.$$

Calculators designed for use in business often have a single key to calculate this. ◀

EXAMPLE 3 Trust Funds

You wish to establish a trust fund from which your niece can withdraw $2,000 every six months for 15 years. You wish to invest the trust in tax-free municipal bonds paying 7% interest per year, with the interest paid every six months. How large should the trust be?

SOLUTION If we assume that the trust will be drawn down to 0 after 15 years, we are looking at a decreasing annuity. Thus, we use the decreasing annuity formula

$$P = W\frac{1 - (1 + i)^{-n}}{i}$$

with $P = ?$, $W = 2{,}000$, $r = 0.07$, $m = 2$, and $t = 15$, so $i = 0.07/2 = 0.035$ and $n = 2 \cdot 15 = 30$. Substituting gives

$$P = 2{,}000\,\frac{1 - (1 + 0.035)^{-30}}{0.035}$$

$$= \$36{,}784.09.$$

Thus, the trust should contain at least $36,784.09.

 With a calculator you can watch the value of the annuity shrink. For example, on the TI-82 enter the initial deposit, $36,784.09.

$$36784.09 \;\boxed{\text{ENTER}}$$

Next, add one month's interest and subtract the first withdrawal.

$$\text{Ans}(1+0.07/2)-2000 \;\boxed{\text{ENTER}}$$

If you hit $\boxed{\text{ENTER}}$ repeatedly, this calculation will be repeated with each new balance, showing you the value of the annuity month by month until it shrinks to 0. A computer spreadsheet can be used in a similar way to compute the monthly balances.

EXAMPLE 4 Education

(See Example 2.) Tony and Maria, having accumulated $100,000 for Jose Phillipe's college education, would now like to make quarterly withdrawals over the next four years. How much money can they withdraw each quarter in order to draw down the account to zero at the end of the four years?

SOLUTION Now Tony and Maria's account is acting as a decreasing annuity of $100,000, still earning 4% per year compounded quarterly, to be paid out over 4 years. So, $P = \$100{,}000$, $W = ?$, $r = 0.04$, $m = 4$, and $t = 4$, giving $i = 0.01$ and $n = 16$. We have

$$P = W\frac{1 - (1 + i)^{-n}}{i}$$

$$100{,}000 = W\frac{1 - 1.01^{-16}}{0.01}.$$

Solving for W gives

$$W = 100{,}000 \, \frac{0.01}{1 - 1.01^{-16}}$$

$$= \$6{,}794.46.$$

So if they withdraw $\$6{,}794.46$ each quarter, their account balance will drop to 0 at the end of the four years.

Before we go on... Notice that the total amount they withdraw from the account is

$$\$6{,}794.46 \times 16 = \$108{,}711.36.$$

The extra $\$8{,}711.36$ is interest that the money in the account earned while withdrawals were being made.

▼ **EXAMPLE 5**   Retirement Accounts

Jane Q. Employee has just started her new job with Big Conglomerate, Inc., and is already looking forward to retirement. BCI offers her as a pension plan an annuity that is guaranteed to earn 6% annual interest (compounded monthly). She plans to work for 40 years before retiring, and she would then like to be able to draw an income of $7,000 per month for 20 years. How much do she and BCI together have to deposit per month into the fund to accomplish this?

SOLUTION   Here we have the situation we described at the beginning of this section: an increasing annuity accumulating money to be used later as a decreasing annuity. We need to work backward because we start out knowing what the ultimate payout will be. The first thing we need to do is find out how large a decreasing annuity is needed to support the payout of $7,000 per month. In other words, what should be the present value, P, of an annuity that will earn 6% interest per year and pay out $7,000 per month for 20 years? We use the decreasing annuity formula with $W = 7{,}000$, $i = r/m = 0.06/12 = 0.005$, and $n = mt = 12 \cdot 20 = 240$.

$$P = W \frac{1 - (1 + i)^{-n}}{i}$$

$$= 7{,}000 \, \frac{1 - (1 + 0.005)^{-240}}{0.005} = \$977{,}065.40$$

This is the total that must be accumulated in the increasing annuity during the 40 years Jane plans to work. In other words, this is the *future* value, A, of the increasing annuity. To determine the deposits required for this, we use the increasing annuity formula with $A = 977{,}065.40$, $i = 0.005$ again, and $n = mt = 12 \cdot 40 = 480$:

$$A = M \frac{(1 + i)^n - 1}{i},$$

$$977{,}065.40 = M\frac{1.005^{480} - 1}{0.005},$$

giving

$$M = 977{,}065.40\,\frac{0.005}{1.005^{480} - 1}$$

$$= \$490.62.$$

So, if Jane and BCI between them deposit $490.62 per month in her retirement fund, she can retire in the comfort she desires.

Loans and mortgages can be viewed as decreasing annuities. If you borrow money from a bank, the bank is, in effect, depositing a sum of money with you. It then makes regular withdrawals (your payments) at an agreed-upon interest rate. The process of paying back a loan with equal payments in this way is called **amortizing** (killing) a debt.

▼ **EXAMPLE 6** Home Loans

Marc and Mira are buying a house and have taken out a 30-year, $90,000 mortgage at 8% per year. What will their monthly payments be?

SOLUTION We view this as a $90,000 decreasing annuity, to be paid out over 30 years. Thus, $P = 90{,}000$, $W = ?$, $m = 12$, $t = 30$, and $r = 0.08$, so $i = 0.08/12$ and $n = mt = 360$:

$$P = W\frac{1 - (1 + i)^{-n}}{i}$$

$$90{,}000 = W\,\frac{1 - \left(1 + \dfrac{0.08}{12}\right)^{-360}}{\left(\dfrac{0.08}{12}\right)}$$

giving

$$W = 90{,}000\,\frac{\left(\dfrac{0.08}{12}\right)}{1 - \left(1 + \dfrac{0.08}{12}\right)^{-360}}$$

$$= \$660.39.$$

Before we go on... When we do this calculation, we round off the answer to the nearest cent. This roundoff error accumulates over time, so that by the end of the mortgage it might amount to several dollars. The very last payment is adjusted to correct for this.

As with a decreasing annuity, you can use a graphing calculator to calculate the balance owed each month. First enter the starting balance.

$$90000 \; \boxed{\text{ENTER}}$$

Next, add one month's interest and subtract one month's payment.

$$\text{Ans}(1+0.08/12)-660.39 \; \boxed{\text{ENTER}}$$

This gives you the balance after one month, $89,939.61. If you hit $\boxed{\text{ENTER}}$ repeatedly you will see the balance month by month. This allows you to construct an **amortization table,** such as the following.

Month	Principal	Payment on Principal	Interest Payment
0	$90,000.00		
1	89,939.61	60.39	600.00
2	89,878.82	60.79	599.60
3	89,817.62	61.20	599.19
4	89,756.01	61.61	598.78
5	89,693.99	62.02	598.37
6	89,631.56	62.43	597.96
7	89,568.71	62.85	597.54
8	89,505.44	63.27	597.12
9	89,441.75	63.69	596.70
10	89,377.64	64.11	596.28
11	89,313.10	64.54	595.85
12	89,248.13	64.97	595.42

The column labeled Principal gives the balance remaining on the mortgage initially ($90,000) and after each of the first 12 payments. These are the numbers you calculate in the way described above. The next column gives the amount by which the balance decreases with each payment. For example, the first payment reduces the balance by $60.39, since the balance decreases from $90,000 to $89,939.61. The last column gives the amount of each payment that goes toward interest on the loan, and this is simply the difference between the monthly payment, $660.39, and the payment on principal. You can also calculate the interest payment in another way: it is the interest owed after one month on the balance remaining. For example, on $90,000 at an annual rate of 8%, the amount owed for the first month is

$$90,000\left(\frac{0.08}{12}\right) = \$600,$$

so $600 of the first payment is interest on the loan. You could continue this table for the life of the loan in a straightforward, but tedious, calculation. The next two examples give quicker ways of calculating the interest paid and the balance remaining after a specific length of time.

8.3 Annuities and Loans

▼ EXAMPLE 7 Mortgage Interest

(Continuation of Example 6) Mortgage interest is tax-deductible, so it is important to know how much of a year's mortgage payments represent interest. How much interest will Marc and Mira pay in the first year of their mortgage?

SOLUTION We could answer this question by constructing an amortization table as above and adding up the entries in the column labeled Interest Payment. There is a quicker way, however. First, notice that they will pay a total of $660.39 \times 12 = \$7,924.68$ to the bank in the first year. The question is, how much of this is interest and how much goes to reducing the principal? To find the answer, think about it from the bank's point of view. At the end of the first year it will be holding a 29-year annuity giving payments of $660.39 per month. We can use the decreasing annuity formula to determine the present value of this annuity:

$$P = W \frac{1 - (1 + i)^{-n}}{i}$$

$$= 660.39 \, \frac{1 - \left(1 + \frac{0.08}{12}\right)^{-348}}{\left(\frac{0.08}{12}\right)}$$

$$= \$89,248.43.$$

This is the principal that remains unpaid. So Mark and Mira must have paid $90,000 - 89,248.43 = \$751.57$ on principal in the first year. The rest of their payments, $7,924.68 - 751.57 = \$7,173.11$, was the interest.

Before we go on... This answer does not agree exactly with the numbers in the amortization table above, as a result of roundoff errors in the construction of the table. Recall also that the monthly payment of $660.39 was arrived at after rounding off another calculation. For tax and payoff purposes, the bank must decide exactly how it will deal with these discrepancies. Most banks use an amortization table.

▼ EXAMPLE 8 Mortage Principal

(Continuation of Examples 6 and 7) After 10 years, Mira inherits her aunt's fortune, and she and Mark decide to pay off the mortgage. What principal do they still owe the bank?

SOLUTION We need again to find the principal remaining, this time after 10 years. From the bank's point of view, it now holds a 20-year annuity paying out $660.39 per month. The annuity's present value is

$$P = W\frac{1-(1+i)^{-n}}{i}$$

$$= 660.39\,\frac{1-\left(1+\dfrac{0.08}{12}\right)^{-240}}{\left(\dfrac{0.08}{12}\right)}$$

$$= \$78{,}952.46.$$

So to pay off the mortgage, Marc and Mira will have to send the bank $78,952.46.

Before we go on... It may not look as if much has been accomplished in the first 10 years of paying off the mortgage: just over $10,000 of the original $90,000 debt has been repaid. This is how long-term loans such as mortgages work—most of the borrower's payments cover interest in the early years. We can see this clearly in the amortization table in Example 6. We can also see it by graphing P, the amount of principal left unpaid vs. n, the month. The equation is (with some rounding)

$$P = 660.39\,\frac{1-(1.00667)^{n-360}}{1.00667}$$

$$= 99{,}059 - 9{,}059(1.00667)^n,$$

and its graph (which you can find by plotting points, using a graphing calculator, or using computer software) is shown in Figure 3.

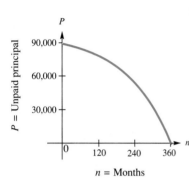

FIGURE 3

USING A GRAPHING CALCULATOR TO FIND LOAN REPAYMENT TIME

Just as in the case of investments earning compound interest, it is not easy to solve for time as the unknown quantity when using the annuity formulas. As before, the traditional methods involve tables or the use of logarithms and scientific calculators, but again we opt for the newer approach using graphing technology.

▼ **EXAMPLE 9** Paying Off a Loan

Discover Card is charging an annual rate of 14.9% interest on outstanding balances.* You have an outstanding balance of $500 and calculate that you can afford to make payments of $20 per month. How long will it take you to settle the debt?

▼ *This is the annual percentage rate quoted on *Discover Card*'s billing statements in April 1994.

SOLUTION We use the decreasing annuity formula

$$P = W\frac{1 - (1 + i)^{-1}}{i},$$

with $P = 500$, $W = 20$, $r = 0.149$, $m = 12$, $i = 0.149/12$, $n = ?$, getting

$$500 = 20\frac{1 - \left(1 + \frac{0.149}{12}\right)^{-n}}{\left(\frac{0.149}{12}\right)}.$$

We can solve for n by graphing

$$P = 20\frac{1 - \left(1 + \frac{0.149}{12}\right)^{-n}}{\left(\frac{0.149}{12}\right)}$$

(what exactly does this function represent?) and using the trace feature to determine the value of n that corresponds to $P = 500$. Using a graphing calculator, we graph

$$Y_1 = 20(1-(1+0.149/12)^{\wedge}(-X))/(0.149/12)$$

with $0 \leq X \leq 60$ and $0 \leq Y \leq 1{,}000$, together with the line $Y = 500$ (Figure 4). We can now zoom in to see where the graphs cross (Figure 5). We can see that the debt will be paid off at the end of 31 months.

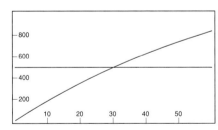
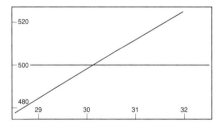

FIGURE 4   **FIGURE 5**

Before we go on... As in Example 8 in the previous section, there is another way of finding this point of intersection. If we choose the *x*-range carefully (for example, $0 \leq X \leq 57$ on the TI-82), then as we trace along the curve we will see the value of the function at the end of each month given as the *y*-coordinate. When that value rises above 500, we will find the answer to our question.

By the way, how much did you actually pay to settle your $500 debt? You paid $20 \times 31 = \$620$. Credit cards are no bargain if you don't pay your balance quickly.

you to eventually repay your debt. (Graph W as a function of n and look at the formula for W to determine W for very large values of n.)

37. *Savings* You are depositing $100 per month in an account that pays 4.5% interest per year (compounded monthly) while your friend Lucinda is depositing $75 per month in an account that earns 6.5% interest per year (compounded monthly). When will her balance pass yours? (Give your answer to the nearest year.)

38. *Car Leasing* You can lease a $15,000 car for $300 per month. For how long should you lease the car so that your monthly payments are lower than purchasing it with an 8% per year loan? (Answer to the nearest year.)

COMMUNICATION AND REASONING EXERCISES

39. Your cousin Simon claims that you have wasted your time studying annuities. He says if you wish to retire on an income of $1,000 per month for 20 years, you need to save $1,000 per month for 20 years. Explain why he is wrong.

40. Your cousin Trevor claims that you will earn more interest by accumulating $10,000 through smaller payments than through larger payments made over a shorter period. Is he correct? Give a reason for your answer.

41. A real estate broker tells you that doubling the period of a mortgage halves the monthly payments. Is she correct? Support your answer by means of an example.

42. Another real estate broker tells you that doubling the size of a mortgage doubles the monthly payments. Is she correct? Support your answer by means of an example.

▶ You're the Expert

PRICING AUTO LOANS

You are shopping for a used car and have found two interesting possibilities. The first car costs $4,000, and the second one costs $3,800. The first car dealer offers you a 2-year loan at 8% with monthly payments, and asks for a down payment of $500. The second dealer offers you a 3-year loan at 8.5%, also with monthly payments, and asks for a down payment of $300. You have $1,000 in a savings account paying 4% interest compounded monthly, and you estimate that you will have $200 per month in extra income that you will use to pay for the car, with any money left over from that being deposited in the savings account. Which car will end up costing you less to buy?

You decide on the following approach. You now have $1,000 in the bank. Three years from now you will have paid off whichever car you decide on, and you will have some amount left in the bank. Whichever car leaves you with more money in the bank three years from now will have cost you less. You therefore need to compute your bank balance three years from now, assuming first that you buy the first car, then assuming that you buy the second car.

You begin with the first car. The $500 down payment immediately reduces your balance to $500. That $500 will sit in the account for the next three years, earning 4% per year compounded monthly. You use the compound interest formula to calculate what $500 will grow to after three years:

$$A_1 = \$500\left(1 + \frac{0.04}{12}\right)^{12(3)}$$
$$= \$563.64.$$

(As you calculate the amounts that add up to your final balance, you label them A_1, A_2, and so on.) You will also be paying off the loan of $3,500 (the cost of the car minus the down payment) at 8% per year for 2 years. You calculate your monthly payment using the decreasing annuity formula:

$$3{,}500 = W \cdot \frac{1 - \left(1 + \frac{0.08}{12}\right)^{-24}}{\left(\frac{0.08}{12}\right)}$$

so

$$W = 3{,}500 \cdot \frac{\left(\frac{0.08}{12}\right)}{1 - \left(1 + \frac{0.08}{12}\right)^{-24}}$$

$$= \$158.30.$$

You will therefore have $200 - 158.30 = \$41.70$ left over each month to deposit in your savings account during the two years it takes you to pay off the car. Using the increasing annuity formula, you determine what these deposits will be worth after two years.

$$A = 41.70 \cdot \frac{\left(1 + \frac{0.04}{12}\right)^{24} - 1}{\left(\frac{0.04}{12}\right)}$$

$$= \$1{,}040.12$$

Since this money will remain in the account for one more year, you can calculate its value at the end of the third year using the compound interest formula.

$$A_2 = 1{,}040.12 \left(1 + \frac{0.04}{12}\right)^{12}$$

$$= \$1{,}082.50$$

Finally, during the third year you will deposit $200 per month into the savings account. Using the increasing annuity formula, you can calculate the amount produced.

$$A_3 = 200 \cdot \frac{\left(1 + \frac{0.04}{12}\right)^{12} - 1}{\left(\frac{0.04}{12}\right)}$$

$$= \$2{,}444.49$$

The total amount you will have in your savings account after three years is therefore

$$A_T = A_1 + A_2 + A_3 = 563.64 + 1{,}082.50 + 2{,}444.49$$
$$= \$4{,}090.63.$$

Now you do a similar calculation for the second car. The $300 down payment reduces your balance to $700, and you calculate the value of $700 after three years (you switch to the letter B so as not to confuse this calculation with the ones you just did).

$$B_1 = 700\left(1 + \frac{0.04}{12}\right)^{36}$$
$$= \$789.09$$

Your monthly payment on a $3,500 loan at 8.5% for 3 years is

$$W = 3{,}500 \, \frac{\left(\frac{0.085}{12}\right)}{1 - \left(1 + \frac{0.085}{12}\right)^{-36}}$$
$$= \$110.49.$$

This leaves $200 - 110.49 = \$89.51$ per month that you will deposit into the account for three years. At the end of the three years this will give you the following amount.

$$B_2 = 89.51 \, \frac{\left(1 + \frac{0.04}{12}\right)^{36} - 1}{\left(\frac{0.04}{12}\right)}$$
$$= \$3{,}417.63$$

This gives you a total balance of

$$B_T = B_1 + B_2 = 789.09 + 3{,}417.63$$
$$= \$4{,}206.72.$$

Looking at the two possible balances, you see that the second car will end up costing you less to buy.

Exercises

1. Suppose that you use your whole $1,000 as down payment on either car. Which one will cost less then?

2. Suppose (in the original discussion) that the interest rate on the second car was 9%. Which car would cost less then?

3. Suppose (in the original discussion) that the interest rate on the second car was 10%. Which car would cost less then?

4. How large must the interest rate on the second car be in order for it to cost more than the first?

5. In the original discussion, on which car will you end up paying more in interest? How does this compare to the original conclusion?

6. Suppose that you have $4,000 in the bank. If you decide to buy the first car, should you buy it with cash or pay the $500 down payment and take out the loan? Justify your answer.

7. Suppose that you have $4,000 in the bank and that your account pays 10% interest compounded monthly. If you decide to buy the first car, should you buy it with cash or pay the $500 down payment and take out the loan? Justify your answer.

8. Referring to the previous two exercises, how high must the interest rate on your savings account be to make it advantageous to take out the loan rather than pay cash?

▶ Review Exercises

APPLICATIONS

1. *Dividends* You invest $5,000 in a stock paying a dividend of 4.75% every year. If you always take the dividend in cash, how much will you earn in 5 years?

2. *Interest* You deposit $7,200 in a savings account paying 3% per year. If you remove the interest as soon as it is credited to your accound, how much interest will you earn over the course of 4 years? Does it matter how often the bank credits your account with interest?

3. *Dividends* You have a chance to invest in a stock paying 4.75% in dividends each year. You intend to take the dividends in cash. If you wish to earn $1,000 in dividends over the next 5 years, how much money must you invest today?

4. *Interest* You have just opened a savings account paying 3% interest per year. You intend to withdraw the interest as soon as it is credited to the account. If you wish to collect $500 in interest over the next 4 years, how much must you deposit now?

5. *Bonds* Calculate the cost of a bond earning 8% simple interest whose value at maturity ten years from now will be $1,000.

6. *Loans* Your repayment on a four-year loan, which charged 8% simple interest, amounted to $30,320. How much did you originally borrow?

7. *Stocks* Stocks of *Cameco Corporation* of Saskatchewan increased more-or-less linearly from $15 at the start of 1992 to $30 at the end of 1994.* What annual rate of simple interest does this represent?

8. *Bonds* The value of your municipal bond fund holdings has declined linearly from $5,000 to $4,000 in four years. What simple-interest rate of depreciation does this represent?

9. *Dividends* You invest $5,000 in a stock paying a dividend of 4.75% every year. If you reinvest the dividend (use it to buy more stock) each time it is paid, how much will you earn in 5 years?

10. *Interest* You deposit $7,200 in a savings account paying 3% per year compounded monthly. If you leave the interest in the account, how much will you earn over the course of 4 years? Does it matter how often the banks compounds the interest?

▼ *Source: Reuter/*Financial Post*, February 12, 1994, p. 13.

11. **Savings** You open a savings account that pays 4% interest compounded monthly. How much must you deposit now if you wish to have $5,000 in 3 years?

12. **Investments** A stock fund is earning 15% yearly. How much must you invest now in order to have $7,000 after 5 years?

13. **Interest** The Passport credit card charges 1.4% interest per month. What is the effective yearly interest rate?

14. **Interest** Your local department store offers a credit card charging "only 5/100% interest per day!" What is the effective yearly interest rate (assuming a 365-day year)?

15. **Effective Rates** Solid Saving & Loan offers an account paying 5.5% compounded monthly. What is the effective interest rate?

16. **Effective Rates** First State Bank offers a loan at 7.5% compounded monthly. What is the effective interest rate?

17. **Effective Rates** Solid Savings & Loan would like to offer an account paying an effective rate of 6%. What should the nominal rate be if interest will be compounded monthly?

18. **Effective Rates** First State Bank would like to offer a loan with an effective interest rate of 9.5%. What should the nominal rate be if interest will be compounded monthly?

19. **Investments** Sherrie Firavich invested her life savings, $10,000, in five-year CDs that paid 6% per year, compounded monthly. After five years she took the proceeds and put it all in the stock market, which proceeded to lose 1% each month. When she took out her money one year later, how much did she have left?

20. **Investments** Earl Karn invested $10,000 in a stock fund with 20% interest reinvested per year. After 5 years, needing ready cash, he removed his money from the fund and put it in a savings account paying 3% compounded monthly. How much money did he have after another 5 years?

21. **Inflation** Inflation in East Camelot is running at 4% per month. If a car costs $10,000 now, how much will an equivalent car cost in two years?

22. **Inflation** Inflation in West Camelot is running at 5% per month. A new house to be put on the market in 4 years will be priced at $100,000. What would the equivalent house cost today?

23. **Inflation** (*graphing calculator recommended*) Inflation in North Camelot is running at 3% per month. How long will it take prices to double?

24. **Inflation** (*graphing calculator recommended*) Prices in South Camelot are *decreasing* 4% per month. How long will it take for prices to be halved?

25. **Savings** If you deposit $150 each month in a savings account paying 6% per year compounded monthly, how much will you have in the account after 3 years? How much of this is interest that you earned?

26. **Investments** You faithfully invest $200 each month in a stock fund earning 15% compounded monthly. How much will your account be worth in 6 years? How much of this amount is attributable to the increase in value of the stocks?

27. **Savings** You would like to save money for a down payment on a house. You need to collect $20,000 over the next 7 years, and you have an account that pays 5% compounded monthly. How much money must you deposit each month to reach your goal?

28. **Savings** You would like to save money for a down payment on a car. You need to collect $4,000 over the next 4 years, and you have an account paying 3% compounded monthly. How much must you deposit each month in order to reach your goal?

29. **Car Loans** Kay LaBarre is buying a $12,000 car. She will make a down payment of $4,000 and finance the balance with a 4-year 7% loan. What will be her monthly payments?

30. **Car Loans** Stacy Thrash would like to buy a $15,000 car. She can afford to make payments of up to $300 per month on a loan. She is offered a 4-year 7% loan. How much of the cost of the car can she afford to finance?

31. **Mortgages** You buy your house with a 30-year $120,000 mortgage at 8%. What will be your monthly payments? What will you pay in interest during the first year of the mortgage? What will you pay in interest over the life of the mortgage?

32. **Loans** You take out a $10,000 loan for 3 years at 8% to buy a car. Assuming monthly payments, how much interest will you pay the first year? How much interest will you pay over the life of the loan?

33. *Credit Cards* You have run up a debt of $4,000 on your credit card. If the card charges 16% per year, how much must you pay per month to retire the debt in two years?

34. *Credit Cards* You have a credit card charging 18% interest per year. Your minimum monthly payment is calculated to retire your debt after three years. If you can only afford to pay $100 per month, what is the largest balance you can carry?

35. *Annuities* A woman nearing retirement wants to buy an annuity that will pay her $2,000 each month for 20 years. If the annuity earns 8% per year, what will it cost her to buy it?

36. *Annuities* A man nearing retirement wants to buy an annuity that will pay him $3,000 per month for 15 years. If the annuity earns 7% per year, what will it cost him to buy it?

37. *Pensions* Leonard Albright has just begun working for a company offering an annuity as a pension plan. This annuity is guaranteed to earn at least 4% per year. He wants to retire in 30 years and then collect $4,000 per month for 20 years. How much must he and his company together deposit in the annuity each month?

38. *Pensions* Sheila Goshorn has just begun working for a company offering an annuity as a pension plan. This annuity is guaranteed to earn at least 5% per year. She wants to retire in 35 years and then collect $5,000 per month for 20 years. How much must she and her company together deposit in the annuity each month?

INTRODUCTION ▶ Exponentials and logarithms are indispensable for an understanding of many processes in economics and nature. Examples include interest, inflation, population growth, spread of epidemics, and radioactive decay. All of these processes are modeled by exponential functions, and logarithms are necessary to answer many of the questions that naturally arise.

You are strongly urged to read through the review sections on the algebra of exponentials and radicals in Appendix A before you begin this chapter, because much of the material in this chapter assumes a familiarity with manipulating exponents.

In this chapter, we will discuss in some detail the use of graphing calculators in the examples and exercises. Even if you do not have access to graphing technology, however, you will benefit greatly by carefully reading *all* of the examples. If you *are* using a graphing calculator, refer to Appendix B and to the first section in Appendix A.

9.1 EXPONENTIAL FUNCTIONS AND APPLICATIONS

We have already seen examples of functions involving exponents, such as $f(x) = x^2$ or $g(x) = (x + 1)^{1/2}$. In each of these examples, the exponent is constant and the base is variable. We are now going to turn the tables and consider functions such as $f(x) = 2^x$, where the base is constant and the exponent is variable. This function is an example of an *exponential function*.

EXPONENTIAL FUNCTION

An **exponential function** is a function of the form

$$f(x) = Ca^x,$$

where C and a are constants and $a > 0$. (We call a the **base** of the exponential function.)

Examples of exponential functions are

$$f(x) = 2^x \quad C = 1, a = 2$$

and

$$\begin{aligned} g(x) &= 3 \cdot 2^{-4x} \\ &= 3(2^{-4})^x \\ &= 3\left(\frac{1}{16}\right)^x. \quad C = 3, a = \frac{1}{16} \end{aligned}$$

9.1 Exponential Functions and Applications

For reference, we repeat below the list of laws of exponents from the algebra review in the appendix. Which of the laws did we use in reformulating $g(x)$ above?

THE LAWS OF EXPONENTS

If a and b are positive and x and y are any real numbers, then the following laws hold.

Law	Example
1. $a^x a^y = a^{x+y}$	$2^3 2^2 = 2^5 = 32$
2. $\dfrac{a^x}{a^y} = a^{x-y}$	$\dfrac{4^3}{4^2} = 4^{3-2} = 4^1 = 4$
3. $a^{-x} = \dfrac{1}{a^x}$	$9^{-0.5} = \dfrac{1}{9^{0.5}} = \dfrac{1}{3}$
4. $a^0 = 1$	$(3.3)^0 = 1$
5. $(a^x)^y = a^{xy}$	$(3^2)^2 = 3^4 = 81$
6. $(ab)^x = a^x b^x$	$(4 \cdot 2)^2 = 4^2 2^2 = 64$
7. $\left(\dfrac{a}{b}\right)^x = \dfrac{a^x}{b^x}$	$\left(\dfrac{4}{3}\right)^2 = \dfrac{4^2}{3^2} = \dfrac{16}{9}$

▼ **EXAMPLE 1**

Let $f(x) = 2^x$, with domain the set $\mathbb{R}$ of real numbers. Then

$$f(3) = 2^3 = 8$$

$$f(-3) = 2^{-3} = \frac{1}{8}$$

$$f(0) = 2^0 = 1$$

$$f(x + h) = 2^{x+h}$$

$$f(20) = 2^{20} = 1{,}048{,}576$$

$$f(-20) = 2^{-20} = \frac{1}{1{,}048{,}567} \approx 0.000000953.$$

Before we go on... We didn't calculate all of these values by hand. You will find that the use of a scientific calculator is indispensable for most of what follows. A graphing calculator would be even better.

Notice something interesting about the values of $f(x)$ we obtained: *they are all positive*. In general,

$$a^x > 0 \text{ for every real number } x.$$

604 Chapter 9 Exponential and Logarithmic Function

▼ **EXAMPLE 2**

On the same set of axes, graph the functions $f(x) = 2^x$ and $g(x) = 2^{-x}$.

SOLUTION Although we can graph these easily on a graphing calculator, we can also graph them easily by hand. Here is a table of values to start with.

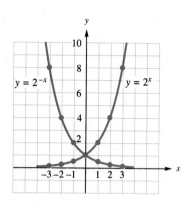

FIGURE 1

x	-3	-2	-1	0	1	2	3
$f(x) = 2^x$	$\frac{1}{8}$	$\frac{1}{4}$	$\frac{1}{2}$	1	2	4	8
$g(x) = 2^{-x}$	8	-4	-2	1	$\frac{1}{2}$	$\frac{1}{4}$	$\frac{1}{8}$

Before we graph the curves, notice the symmetry in the table: values of $f(-x)$ correspond to values of $g(x)$. This is a consequence of the fact that

$$f(-x) = 2^{-x} = g(x)$$

and is reflected by a symmetry between their graphs (see Figure 1).

Notice also how the curve $y = 2^x$ goes "shooting up" for larger values of x. In fact, the y-coordinate doubles for each increase of 1 in x. (For example, if we'd plotted an extra point for $x = 4$, the coordinates would be (4,16), which is much higher than we've allowed for in the picture.) On the other hand, notice how the curve "levels off" for larger and larger negative values of x, since the sequence $2^{-1}, 2^{-2}, 2^{-3}, \ldots$ gets closer and closer to zero. The curve $y = 2^{-x}$ is the mirror image of $y = 2^x$. The curves meet on the y-axis at $y = 1$, since $2^0 = 2^{-0} = 1$.

 If you wish to reproduce these graphs on your graphing calculator, be sure to use the "range" menu to specify $-3 \le x \le 3$ and $0 \le y \le 8$, so that the graphs will use the entire screen. Also, the correct format for entering the equations corresponding to these functions on most graphing calculators is

$$Y_1 = 2\wedge X$$
$$Y_2 = 2\wedge(-X)$$

where the second formula specifies that y_2 is 2 raised to the *quantity* $(-x)$.

▼ **EXAMPLE 3**

On the same set of axes, sketch the functions $f_1(x) = (\frac{1}{2})^x$, $f_2(x) = 1^x$, $f_3(x) = 2^x$, $f_4(x) = 3^x$, and $f_5(x) = 4^x$.

9.1 Exponential Functions and Applications

SOLUTION We can save ourselves some work by noting the following things about these functions.

$y = (\frac{1}{2})^x$ is the same as $y = 2^{-x}$, by the laws of exponents, and we have already drawn that curve.

$y = 1^x$ is the same as $y = 1$, a horizontal line with y-intercept 1.

$y = 2^x$ we have already drawn.

$y = 3^x$ will look like $y = 2^x$, except that the y-values will triple, rather than double, for each increase of 1 in x.

$y = 4^x$ will behave like $y = 2^x$ and $y = 3^x$, except that the y-values will increase by a factor of 4 for each increase of 1 in x.

The graphs are shown in Figure 2. Notice that all the graphs pass through the point $(0, 1)$ because of the identity $a^0 = 1$. If we were to sketch the curve $y = (\frac{3}{2})^x$, it would have the same general shape as $y = 2^x$ but lie between the line $y = 1^x$ and the curve $y = 2^x$, because $\frac{3}{2}$ lies between 1 and 2. Notice also that the graphs *cross* at $(0, 1)$. For example, the curve $y = 3^x$ lies above $y = 2^x$ on the right of the y-axis, but lies below it to the left of the y-axis. (Why?)

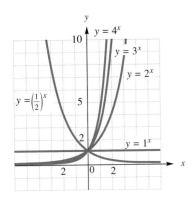

FIGURE 2

▼ **EXAMPLE 4**

Sketch the graphs of $f(x) = 2^x$ and $g(x) = 3 \cdot 2^x$ on the same axes. How are the graphs related?

SOLUTION The graph of $g(x) = 3 \cdot 2^x$ will look like that of $f(x) = 2^x$, except that each point on $y = 3 \cdot 2^x$ will be 3 times as high as the corresponding point on $y = 2^x$. The graphs are shown in Figure 3.

When entering the equation $y = 3 \cdot 2^x$ on your graphing calculator or computer graphing software, you can use the format

$$Y_1 = 3 \times 2\text{\textasciicircum}X$$

rather than the equivalent

$$Y_1 = 3 \times (2\text{\textasciicircum}X)$$

and not bother with parentheses, since calculators are programmed to respect the usual order of operations: exponents first, and then products. Notice that this is *not* the same as

$$Y_1 = (3 \times 2)\text{\textasciicircum}X$$

which is $y = (3 \cdot 2)^x = 6^x$.

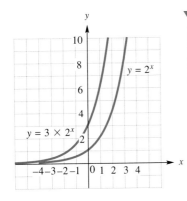

FIGURE 3

EXAMPLE 5

Examples of the values of two functions, f and g, are given in the following table.

x	-2	-1	0	1	2
$f(x)$	-7	-3	1	5	9
$g(x)$	$2/9$	$2/3$	2	6	18

One of these functions is linear and the other is exponential. Which is which?

SOLUTION There are several ways of telling when a function is linear or exponential. We can start with a numerical approach. Remember that a linear function changes by the same amount every time x increases by 1. The values of f behave this way. Every time x increases by 1, the value of $f(x)$ increases by 4. Therefore, f could be a linear function with a *slope* of 4. Since $f(0) = 1$, we see that

$$f(x) = 4x + 1$$

is a linear formula fitting the data.

On the other hand, an exponential function $y = Ca^x$ is *multiplied* by the same amount every time x increases by 1:

$$Ca^{x+1} = Ca^x \cdot a^1 = (Ca^x) \cdot a.$$

The values of g behave this way. Every time x increases by 1, the value of $g(x)$ is multiplied by 3:

$$\frac{2/3}{2/9} = \frac{2}{2/3} = \frac{6}{2} = \frac{18}{6} = 3.$$

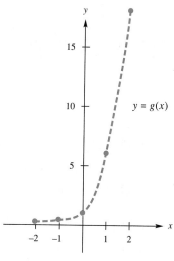

FIGURE 4

Since $g(0) = 2$, we can see that

$$g(x) = 2 \cdot 3^x$$

is an exponential formula fitting the data.

We could also examine the data graphically. Figure 4 shows the graphs of the given points $(x, f(x))$ and $(x, g(x))$. The given values of $f(x)$ clearly lie along a straight line, whereas the values of $g(x)$ lie along a curve. Therefore, f is linear and g is not. To see that g is actually exponential, we need to use the numerical approach above.

Before we go on... The way in which a linear function changes, *adding* the same amount each time x increases by 1, is called **arithmetic growth.** The way in which an exponential function changes, *multiplying* by the same amount each time x increases by 1, is called **geometric growth** or **exponential growth.**

APPLICATIONS

Exponential functions arise in finance and economics mainly through the idea of **compound interest.** We start with a typical scenario: you deposit a **principal** $P = \$10,000$ at Solid Savings & Loan in an account that pays 6% interest **compounded annually.** This means that 6% interest is paid at the end of each year, and *added back into the account.* This money then earns interest the next year along with the original principal.

After one year, you will have the original amount P plus 6% interest:

$$\$10{,}000 + 10{,}000(0.06) = \$10{,}000(1 + 0.06)$$
$$= \$10{,}600.$$

In other words, if we write $r = 0.06$ for the interest rate, the balance after one year is obtained by multiplying the principal by the quantity $1 + r = 1.06$. To obtain the balance after two years, we would multiply the $\$10,600$ that is in the account at the beginning of the year by $1 + r$ again, since the whole amount earns interest in the second year. In other words, the amount of money in the account grows exponentially, being multiplied by $1 + r$ each year. After t years you would have a total of

$$A = P(1 + r)^t$$

in the account. Note that we can think of A as a function of t, and write

$$A(t) = P(1 + r)^t.$$

Then, for instance, $A(10)$ is the accumulated amount after 10 years. How should we interpret, say, $A(-10)$?

▼ **EXAMPLE 6**   **Mutual Funds**

Fidelity Investments advertised that their Fidelity Equity Income II Fund yielded an average annual return of 23.35%.* Assuming that this rate of return will continue indefinitely, express the value of the investment as a function of the number of years t, and use your function to find the value of the investment after 4 years.

SOLUTION   Since the income is reinvested at the end of each year, this amounts to compounding the interest each year, so we use the formula

$$A(t) = P(1 + r)^t.$$

We now substitute the given information: $P = 10,000$ (initial investment) and $r = 0.2335$ (rate of return). This gives

$$A(t) = 10,000(1 + 0.2335)^t,$$

or

$$A(t) = 10,000(1.2335)^t,$$

specifying A as an exponential function of t.

After 4 years, the value is

$$A(4) = 10,000(1.2335)^4 = \$23,150.30.$$

Note that the order of operations in calculating the value of $10,000(1.2335)^4$ is the standard one: first exponentiate, and then multiply. On a standard calculator, we use the following sequence.

$$\boxed{1.2335} \; \boxed{x^y} \; \boxed{4} \; \boxed{=} \; \boxed{\times} \; \boxed{10000} \; \boxed{=}$$

On a graphing calculator or function-based calculator, we enter

$$10000 \times (1.2335\wedge 4)$$

The parentheses are optional, as calculators are programmed to use the usual order of operations.

We can use the graph of the function A to answer interesting questions. First, to graph $A(t) = 10,000(1.2335)^t$, enter the corresponding equation on your graphing calculator using one of the formats shown below.

$$Y_1 = 10000 \times 1.2335\wedge X$$

or

$$Y_1 = 10000(1.2335\wedge X)$$

▼ * Rate was current as of 6/30/93, as quoted in an advertisement in *The New York Times* on September 26, 1993.

You also need to set the viewing window ranges. To graph the value of the investment for a 10-year period, use $0 \leq x \leq 10$. Since the investment starts at \$10,000 and goes up, start with the y-scale $10{,}000 \leq y \leq 100{,}000$. The graph is shown in Figure 5.

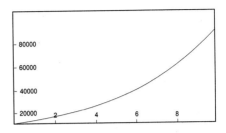

FIGURE 5

Q How long will it take before the investment is worth \$75,000?

A The question is asking for the value of t (x on the graph) such that $A(t) = 75{,}000$. Since $A(t)$ is represented by y, we are asking for the x-coordinate of the point on the graph where $y = 75{,}000$.

To answer this graphically, graph the line

$$Y_2 = 75000$$

along with A, and use the trace feature to approximate the coordinates of the point of intersection (Figure 6).

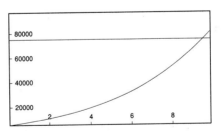

FIGURE 6

You will find that $t \approx 9.6$ years (to the nearest 0.1 year). Thus, it takes a little more than $9\frac{1}{2}$ years to accumulate \$75,000.

In Section 4 we shall see how to answer this question algebraically, using logarithms.

Q Suppose that, instead of compounding the interest once a year, your bank compounds the interest four times a year (once per quarter). What formula should we use to calculate the accumulated amount after t years?

A Look once again at the formula $A = P(1 + r)^t$. We can interpret the exponent as the number of times the interest is added to the account, and r as the interest earned each time. This allows us to answer the question as follows: If the interest is added four times a year for t years, it will be added a total of $4 \times t = 4t$ times. Thus, the exponent for our formula should be $4t$ instead of t. Further, the bank is not going to give you a full year's interest every quarter, but only one-fourth of that. Thus, we should replace r with $r/4$. In other words, the formula we want is

$$A = P\left(1 + \frac{r}{4}\right)^{4t}.$$

This example leads us to the following general formula.

COMPOUND INTEREST

If an amount P (the **present value**) earns interest at an annual interest rate r, compounded m times per year, then the accumulated amount (or **future value**) after t years is

$$A = P\left(1 + \frac{r}{m}\right)^{mt}.$$

▶ NOTES

1. This formula does generalize the previous formula, $A = P(1 + r)^t$, which is the special case $m = 1$.
2. Once again, we can express A as a function of t by writing

$$A(t) = P\left(1 + \frac{r}{m}\right)^{mt}.$$

If we write $C = P$ and $a = (1 + r/m)^m$, then we see that $A(t) = Ca^t$ is an *exponential* function of t. ◀

▼ **EXAMPLE 7** Interest Compounded Monthly

You invest $10,000 in CDs at the Park Avenue Bank, which pays 6% interest compounded monthly. Express the value of your investment as an exponential function of the number of years t that your capital remains invested. How much money will you have after 5 years?

SOLUTION Here, the number of times per year that the interest is compounded is 12. In the compound interest formula

$$A = P\left(1 + \frac{r}{m}\right)^{mt}$$

we substitute $P = 10{,}000$, $r = 0.06$, and $m = 12$. This gives

$$A(t) = P\left(1 + \frac{0.06}{12}\right)^{12t}$$
$$= 10{,}000(1 + 0.005)^{12t}$$
$$= 10{,}000(1.005)^{12t}.$$

After five years, you will have accumulated

$$A(5) = 10{,}000(1.005)^{12(5)}$$
$$= \$13{,}488.50.$$

▼ **EXAMPLE 8** Population Growth

The population of a town was 30,000 in 1990 and was increasing by one-third every year. Express the town's population as a function of the number of years since 1990.

SOLUTION Although it is not money that is growing, the concept is similar. To say that the population is increasing by one-third every year is to say that each year the population is multiplied by $1 + \frac{1}{3}$. Therefore, the population is growing exponentially, as if it were accumulating "interest" at an annual rate of $33\frac{1}{3}\%$ per year. The population is given by the function

$$A(t) = 30{,}000\left(1 + \frac{1}{3}\right)^t$$
$$= 30{,}000\left(\frac{4}{3}\right)^t.$$

Before we go on... Why is the formula *not* $A(t) = 30{,}000(\frac{1}{3})^t$?

▼ **EXAMPLE 9** Suntanning

When DeltaBlock Sunscreen is first applied, it has an SPF rating of 25. According to experimental data, its SPF rating decreases by 20% per hour during exposure to sunlight.

(a) Find and graph DeltaBlock's SPF rating S as a function of x, the number of hours after exposure to the sun.

(b) Your dermatologist advises you not to stay in the sun with anything less than SPF-15 protection. If you use DeltaBlock Sunscreen and wish to follow your dermatologist's advice, how often should you reapply the cream?

SOLUTION

(a) Think of the SPF as a bank balance that is depreciating by 20% every hour. Since r stands for the rate of growth, we must use a negative value in this case: $r = -0.20$. In other words, the SPF rating is being multiplied by $1 - 0.20 = 0.80$ every hour. So

$$S(x) = 25(1 - 0.20)^x$$
$$= 25(0.80)^x.$$

Figure 7 shows a graphing calculator plot of the function S.

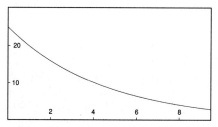

FIGURE 7

(b) We can use the graph to estimate the length of time it takes for the SPF-rating to drop to 15 by graphing the horizontal line $y = 15$ and locating the point of intersection with the curve (using the trace feature if you are using a graphing calculator). As shown in Figure 8, the SPF-rating drops to 15 after approximately 2.29 hours. Thus, you should reapply DeltaBlock after 2.29 hours in order to maintain an SPF of 15 or better.

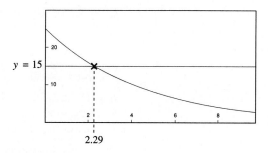

FIGURE 8

9.1 EXERCISES

Sketch the graph of each of the functions given in Exercises 1–12.

1. $f(x) = 2^x$.
2. $f(x) = 3^x$
3. $f(x) = 3^{-x}$
4. $f(x) = 4^{-x}$
5. $g(x) = 2(2^x)$
6. $g(x) = 2(3^x)$
7. $h(x) = -3(2^{-x})$
8. $h(x) = -2(3^{-x})$
9. $r(x) = 1 - 2^x$
10. $r(x) = 2 + 2^{-x}$
11. $s(x) = 2^{x-1}$
12. $s(x) = 2^{1-x}$

13. How are the graphs in Exercises 1 and 11 related?
14. How are the graphs in Exercises 6 and 8 related?
15. How are the graphs in Exercises 1 and 9 related?
16. How are the graphs in Exercises 10 and 12 related?

In Exercises 17–22, sketch the graphs of each of the following pairs of functions on the same set of axes for $-3 \leq x \leq 3$. (Use a graphing calculator or computer program.)

17. $f_1(x) = 1.6^x$, $f_2(x) = 1.8^x$
18. $f_1(x) = 2.2^x$, $f_2(x) = 2.5^x$
19. $f_1(x) = 300(1.1^x)$, $f_2(x) = 300(1.1^{2x})$
20. $f_1(x) = 100(1.01^{2x})$, $f_2(x) = 100(1.01^{3x})$
21. $f_1(x) = 1{,}000(1.045^{-3x})$, $f_2(x) = 1{,}000(1.045^{3x})$
22. $f_1(x) = 1{,}202(1.034^{-3x})$, $f_2(x) = 1{,}202(1.034^{3x})$

In Exercises 23–30:
(a) Express the future value A of a $10,000 investment earning the given interest as a function of time t in years.
(b) Use your function to calculate, to the nearest cent, the future value of the investment after the stated time.

23. 3% per year compounded annually, after 10 years
24. 4% per year compounded annually, after 8 years
25. 2.5% per year compounded quarterly (4 times per year), after 5 years
26. 1.5% per year compounded weekly, after 5 years
27. 6.5% per year compounded daily (365 times per year), after 10 years
28. 11.2% per year compounded monthly, after 12 years
29. 0.2% per month, after 10 years
30. 0.45% per month, after 20 years

APPLICATIONS

31. **Investments** The balance in Susan's bank account is given by $A(t) = 1{,}000(1.056)^t$ after t years. Find the balance in her account after
   (a) 11 years;
   (b) 21 years.

32. **Investments** The balance in Mike's bank account is given by $A(t) = 1{,}000(1.0225)^{2t}$ after t years. Find the balance in his account after
   (a) 11 years;
   (b) 21 years.

33. **Appreciation** The value of my '68 Classic Pontiac in t years' time will be given by
$$V(t) = 6{,}000(1.1)^{2t}.$$
Use a graphing calculator to estimate when the car will be worth $15,000. [*Hint:* Graph the two equations $y_1 = 6{,}000(1.1)^{2x}$ and $y_2 = 15{,}000$ on the same set of axes.]

34. *Depreciation* The value of my '88 Pontiac in t years' time will be given by

$$V(t) = 8{,}000(0.75)^{1.5t}.$$

Use a graphing calculator to estimate when the car will be worth $4,000. [*Hint:* Graph the two equations $y_1 = 8.000(0.75)^{1.5x}$ and $y = 4{,}000$ on the same set of axes.]

35. *Investments* You invest $100 in the Lifelong Trust Savings and Loan Company, which pays 6% interest compounded quarterly. By how much will your investment have grown after four years?

36. *Investments* You invest $1,000 in Rapid Growth Funds, which appreciate by 2% per year, with yields reinvested quarterly. By how much will your investment have grown after 5 years?

37. *Depreciating Investments* During the first nine months of 1993, mutual funds in health-related industries depreciated at a rate of 6.8% per year.* Assuming this trend were to continue, how much will a $3,000 investment in this category of funds be worth in 5 years?

38. *Depreciating Stocks* During the first nine months of 1993, the value of *IBM* stocks depreciated at a rate of 16.28%.† Assuming that this trend were to continue, what will a $10,000 investment in *IBM* stocks be worth in 10 years?

39. *Depreciation* During a prolonged recession, property values in Long Island depreciated by 5% every six months. If a new house cost $200,000, express its future value as a function of its age in years. How much was the house worth 10 years after it was built?

40. *Depreciation* Stocks in the health industry depreciated by 5.1% in the first nine months of 1993.* Assuming this trend were to continue, express the future value of a $40,000 investment as a function of its age in years. How much will it be worth in nine years?

Use a graphing calculator or computer software for Exercises 41 through 48.

Compound Interest *Graph the accumulated amount A as a function of the number of years t in Exercises 41–44.*

41. $550 invested at 1.5% per year, compounded annually; $0 \le t \le 10$

42. $600 invested at 2.2% per year, compounded annually; $0 \le t \le 10$

43. $550 invested at 1.5% per year, compounded daily; $0 \le t \le 10$

44. $600 invested at 2.2% per year, compounded daily; $0 \le t \le 10$

45. *Competing Investments* Bob Carlton just purchased $5,000 worth of municipal funds that are expected to yield 5.4% per year, compounded every six months. Susan Hessney just purchased $6,000 worth of CDs that are expected to earn 4.8% per year, compounded every six months. Determine when, to the nearest year, the value of Bob's investment will be the same as Susan's, and what this value will be.

46. *Investments* Determine when, to the nearest year, $3,000 invested at 5% per year, compounded daily, will be worth $10,000.

47. *Epidemics* At the start of 1985, the incidence of AIDS was doubling every six months, and 40,000 cases had been reported in the U.S. Assuming this trend had continued, determine when, to the nearest 0.1 year, the number of cases would have reached 1 million.

48. *Depreciation* Lee Anne Fisher's investment in Genetic Splicing Inc. is now worth $4,354 and is depreciating by 5% every six months. For some reason, she is reluctant to sell the stocks and swallow her losses. Determine when, to the nearest year, her investment will drop below $50.

---

*Source: *The New York Times*, October 9, 1993, p. 37

†Source: *The New York Times*, October 9, 1993, p. 40.

49. Inflation If the percentage rate of inflation is r per year, then interpret the formula

$$A(r) = 1{,}000(1 + r)^{10}$$

by completing the following sentence: "$A(r)$ is the amount that an item costing \$_____ now will cost in _____ years, given an annual rate of inflation _____." Evaluate $A(0.1)$ and $A(1.15)$ and interpret your answers.

50. Depreciation If the value of a car depreciates every six months at an annual percentage rate of r, interpret the formula

$$A(r) = 10{,}200\left(1 - \frac{r}{2}\right)^{40}$$

by completing the following sentence: "If the car is now worth \$_____ and is depreciating every six months at an annual rate of _____, then $A(r)$ is the amount that it will be worth in _____ years." Evaluate $A(0.1)$ and $A(0.35)$ and interpret your answers.

COMMUNICATION AND REASONING EXERCISES

51. Model the following data using an exponential function $f(x) = Ca^x$.

x	0	1	2
$f(x)$	500	225	101.25

52. Model the following data using an exponential function $f(x) = Ca^x$.

x	-1	0	1
$f(x)$	-97.08737	-100	-103

53. Which of the following three functions will be largest for large values of x?
(a) $f(x) = x^2$
(b) $r(x) = 2^x$
(c) $h(x) = x^{10}$

54. Which of the following three functions will be smallest for large values of x?
(a) $f(x) = x^{-2}$
(b) $r(x) = 2^{-x}$
(c) $h(x) = x^{-10}$

55. What limitations are there to using an exponential function to model growth in real-life situations? Illustrate your answer with an example.

56. Describe two real-life situations in which a linear model would be more appropriate than an exponential one, and two in which an exponential model would be more appropriate than a linear one.

9.2 CONTINUOUS GROWTH AND DECAY AND THE NUMBER e

In the examples in the last section, capital grew (or depreciated) in discrete steps. For instance, in Example 7 interest was added at the end of each month.

In nature, we find examples of growth that occurs *continuously*, as though "interest" is being added more often than every second or fraction of a second. To model this, we need to see what happens to our compound interest formula as we let m (the number of times interest is added per year) become extremely large. Something very interesting happens: instead of getting a bulky formula with very large numbers, we instead get a compact and elegant formula. To begin to see why, let's look at a very simple situation.

Suppose we invest \$1 in the bank for 1 year at 100% interest, compounded m times per year. Then the accumulated capital is

$$A = 1\left(1 + \frac{1}{m}\right)^m = \left(1 + \frac{1}{m}\right)^m.$$

Now, we are interested in what this becomes for large values of m. So let's make a chart that shows how this quantity behaves as m increases.

m	$\left(1 + \dfrac{1}{m}\right)^m$
1	2
10	2.59374246
100	2.70481383
1000	2.71692393
10^4	2.71814593
10^5	2.71826824
10^6	2.71828047
10^7	2.71828169
10^8	2.71828182

Something interesting *does* seem to be happening! The numbers appear to be getting closer and closer to a specific value. In mathematical terminology, we say that the numbers *converge* to a fixed number, 2.7182818 .... This number is one of the most important in mathematics, and is referred to as e. The number e is irrational, just as the more familiar π is, so we cannot write down its exact numerical value. To 20 decimal places, $e = 2.71828182845904523536 \ldots$.

▶ **NOTE** **Evaluating Powers of e on a Calculator**

1. To obtain an approximation to the number e on your calculator, evaluate the quantity e^1. On traditional calculators, you can do this by first entering 1 and then pressing the "e^x" button. This button is sometimes labeled as "inv ln," where "ln" stands for the natural logarithm (which we shall discuss in the next section). On some of the newer calculators, you must enter expressions in formula form: first press the "e^x" button, and then enter 1, followed by "=". In most computer programs, you write exp(x) for e^x, so exp(1) would represent e.

2. To obtain a power of e, such as $e^{-4.5}$ on a traditional calculator, enter the following sequence:

    $\boxed{4.5}\,\boxed{+/-}\,\boxed{e^x}$

    On a formula-based calculator, enter

    $\boxed{e^x}\,\boxed{(}\,\boxed{(-)}\,\boxed{4.5}\,\boxed{)}\,\boxed{=}$

    remembering to enclose the -4.5 in parentheses. In a computer program, enter exp(-4.5) ◀

We now say that if \$1 is invested for 1 year at 100% interest **compounded continuously,** the accumulated capital at the end of that year will amount to \$$e$ = \$2.72 (to the nearest cent). But what about the following more general question?

9.2 Continuous Growth and Decay and the Number e

Q Suppose we invest an amount P for t years at an interest rate of r, compounded continuously. What will the accumulated capital A be at the end of that period?

A In the special case just discussed, we took the compound interest formula and let m get larger and larger. We do the same again, combined with a little of the algebra of exponentials.

$$A = P\left(1 + \frac{r}{m}\right)^{tm}$$

$$= P\left(1 + \frac{1}{(m/r)}\right)^{tm}$$

$$= P\left(1 + \frac{1}{(m/r)}\right)^{(m/r)rt}$$

$$= P\left[\left(1 + \frac{1}{(m/r)}\right)^{(m/r)}\right]^{rt}$$

For continuous compounding of interest, we let m, and hence m/r, get very large. This only affects the term in brackets, which converges to e, and we get the following formula.

COMPOUND INTEREST FORMULA—CONTINUOUS COMPOUNDING

If $\$P$ is invested at an annual interest rate r compounded continuously, then the accumulated amount after t years is

$$A = Pe^{rt}.$$

▶ **NOTE** As in the previous section, we can interpret A as a function of t by writing $A(t) = Pe^{rt}$.

If we write $A(t) = P(e^r)^t$, we see that $A(t)$ is an exponential function of t, where the base is $a = e^r$. ◀

▼ **EXAMPLE 1** Continuous Compounding

Suppose we deposit \$1 in an account yielding 100% interest per year, compounded continuously for x years. Then we have $P = \$1, r = 1.0$, and $t = x$, so

$$A(x) = Pe^{rx}$$
$$= 1 \cdot e^{1.0x}$$
$$= e^x.$$

Thus, the function $A(x) = e^x$ represents something real: the value after x years of a \$1 investment that is growing continuously at 100% per year. We can graph this function by using the following table of values.

x	-3	-2	-1	0	1	2	3
$A(x) = e^x$	0.050	0.135	0.368	1	2.718	7.389	20.08

The graph is shown in Figure 1.

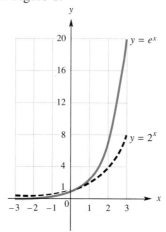

FIGURE 1

Notice the similarity of this graph to the graph of $f(x) = 2^x$ (shown dashed in the figure). Notice also that the investment is more than doubling each year; in fact, it grows by a factor of $e \approx 2.72$ every year.

▼ **EXAMPLE 2**   Continuous Appreciation

You invest \$10,000 at Fastrack Savings & Loan, which pays 6% compounded continuously. Express the balance in your account as a function of the number of years t, and calculate the amount of money you will have after 5 years.

SOLUTION   For the first part, we use the continuous growth formula with $P = 10{,}000$, $r = 0.06$, and t variable, getting

$$A(t) = Pe^{rt}$$
$$= 10{,}000e^{0.06t}.$$

To answer the second part of the question, we calculate

$$A(5) = 10{,}000e^{0.06(5)}$$
$$= 10{,}000e^{0.3}$$
$$\approx \$13{,}498.59.$$

9.2 Continuous Growth and Decay and the Number e

Before we go on... Compare this answer to Example 7 from the last section. Continuous compounding earns more interest than monthly compounding, but not a lot more.

To graph A as a function of t, enter the function on your graphing calculator using the following format.

$$Y_1 = 10000e^{\wedge}(0.06X)$$

Most graphing calculators display "e^" when the $\boxed{e^x}$ key is pressed.

▼ **EXAMPLE 3** Continuous Depreciation

A $100 investment in Constant Growth Funds is continuously declining at a rate of 4% per year. Specify the value of the investment as a function of the number of years, t, and predict the value of the investment in 10 years.

SOLUTION We use the continuous growth formula with $P = 100$, $r = -0.04$, and t variable, getting

$$\begin{aligned} A(t) &= Pe^{rt} \\ &= 100e^{-0.04t}. \end{aligned}$$

After 10 years, the investment will be worth

$$\begin{aligned} A(10) &= 100e^{-0.04(10)} \\ &= 100e^{-0.4} \\ &\approx \$67.03. \end{aligned}$$

Thus, the investment with Constant Growth Funds will have declined to $67.03 in ten years.

▼ **EXAMPLE 4** Investments

(a) You have $100 invested in Quarterly Savings and Loan Company, which pays 5% interest compounded quarterly. By how much will your investment have grown after four years?

(b) Continuity Continental Corp. advertises that they can improve on Quarterly's offer by giving the same interest rate, but compounded continuously. How much more would you have earned had you invested with them?

SOLUTION Part (a) is an ordinary compound interest calculation. We use the compound interest formula $A = P(1 + \frac{r}{m})^{tm}$ with $P = 100$, $r = 0.05$, $m = 4$ and $t = 4$. This gives

$$\begin{aligned} A &= 100(1 + \tfrac{0.05}{4})^{16} \\ &= 100(1.0125)^{16} \\ &\approx 100(1.2199) = 121.99. \end{aligned}$$

Thus, you will earn $21.99 in interest on your $100 investment in four years.

For part (b), we use the continuous compounding formula $A = Pe^{rt}$ with P, t, and r as in part (a), giving

$$A = 100e^{0.20}$$
$$\approx 100(1.2214)$$
$$= 122.14.$$

This shows that, with Continuity Corp., the interest on your investment will amount to $22.14, which is 15¢ better than Quarterly S&L can do. A difference of 15¢ after four years is nothing to get excited about, so you may as well stick with Quarterly.

Before we go on... Would you make the same decision if you had $100,000 to invest?

The **effective annual yield** or **effective interest rate** of an investment is the actual percent by which an investment rises after one year. Another way of thinking of it is as the interest rate that would give the same yield if compounded only once per year. We can calculate the effective yield by calculating the interest earned on an investment of $1 for one year. This gives the following formulas.

EFFECTIVE YIELD FROM COMPOUNDING m TIMES PER YEAR

An investment at an annual interest rate of r, compounded m times per year, has an effective annual yield of

$$r_e = \left(1 + \frac{r}{m}\right)^m - 1.$$

EFFECTIVE YIELD FROM CONTINUOUS COMPOUNDING

An investment at an annual interest rate of r, compounded continuously, has an effective annual yield of

$$r_e = e^r - 1.$$

We can compare investments compounded with different frequencies by comparing their effective yields.

▼ **EXAMPLE 5**   Effective Yield

Which yields more, an investment paying 5% compounded continuously, or an investment paying 5.1% compounded quarterly?

SOLUTION   We compare the effective yields of the two investments. The investment paying 5% compounded continuously yields

$$r_e = e^{0.05} - 1 \approx 0.0513.$$

9.2 Continuous Growth and Decay and the Number e

That is, the effective yield is 5.13%. Said another way, after one year the investment will have appreciated by 5.13%.

The investment paying 5.1% compounded quarterly yields

$$r_e = \left(1 + \frac{0.051}{12}\right)^{12} - 1 \approx 0.0522.$$

Thus the effective yield is 5.22%. This investment is slightly better than the first.

Few banks actually compound interest continuously (although some do!). However, there are other situations where we see exponential growth in which the continuous compounding formula is useful (see also Example 8 in the previous section). In applications beyond finance, it is customary to use the letter k rather than r, and to call k the **fractional rate of growth**. The formula is shown below.

> **EXPONENTIAL GROWTH AT FRACTIONAL GROWTH RATE k**
> If a quantity P at time $t = 0$ grows continuously with a fractional rate of growth k per unit time, then, after a time t, the quantity will measure
> $$A = Pe^{kt}.$$

▼ **EXAMPLE 6** Continuous Population Growth

The population of fleas on my cat Fluffy is increasing continuously at a growth rate of 15% per day, despite all my efforts at eradication! Today, I estimated that Fluffy's fur is host to about 100 fleas, and a week from today I'll be entertaining the Dean of the Faculties (who is fond of cats but is allergic to flea bites). Estimate the flea population on my cat when the Dean comes to visit.

SOLUTION This is an example of exponential growth at a growth rate of $k = 0.15$. We also have $P = 100$ and $t = 7$ (note that time is given in *days*, with the rate of increase as a percentage *per day*). We have

$$\begin{aligned} A &= Pe^{kt} \\ &= 100e^{1.05} \\ &= 100(2.8577) \\ &= 285.77 \approx 286.* \end{aligned}$$

Thus, one can expect the Dean of the Faculties to be unpleasantly surprised when she begins to cuddle Fluffy next week!

▼ *To the nearest flea.

The continuous compounding formula can also be used to describe quantities that are *decreasing*: for example, continuously depreciating assets. We say that such a quantity is showing **exponential decay.**

> **EXPONENTIAL DECAY AT FRACTIONAL DECAY RATE k**
> If a quantity P at time $t = 0$ decays continuously with a **fractional rate of decay** k per unit time, then, after a time t, the amount left is given by
> $$A = Pe^{-kt}.$$

▼ **EXAMPLE 7** Radioactive Decay

Carbon-14, an unstable isotope of carbon, decays continuously to nitrogen at a rate of about 0.0121% per year. If a sample originally contains 50 grams of Carbon-14, how much will be left after 20,000 years?

SOLUTION We use the exponential decay formula with $P = 50$, $k = 0.000121$, and $t = 20,000$.

$$\begin{aligned} A &= Pe^{-kt} \\ &= 50e^{-0.000121(20,000)} \\ &= 50e^{-2.42} \\ &\approx 4.45 \end{aligned}$$

Thus, after 20,000 years, there will still be approximately 4.45 grams of Carbon-14 left in the sample.

Before we go on... Because Carbon-14 decays so slowly, paleontologists use measurements of C-14 to date fossils, in a procedure known as "carbon dating." Carbon dating will be discussed in more detail in the next section.

The next example also involves continuous decay, and it requires either a graphing calculator or graphing software.

▼ **EXAMPLE 8** Exponential Decay

After getting fired from the university two days after an unfortunate experience involving the Dean of the Faculties and my cat Fluffy, I decided to purchase a "Red Flag 1-in-6 Collar" from the pet store. The manufacturer of the collar claims that it works by continuously lowering the flea population at a rate of 1 in every 6 (or 16.67%) per day. By the time I managed to get close enough to Fluffy to put the thing on, his flea population had grown to about 500,000. Graph Fluffy's flea population as a function of time. When will the population be back down to 100 fleas?

9.2 Continuous Growth and Decay and the Number e

SOLUTION We again use the continuous decay formula, with $P = 500{,}000$ and $k = 0.1667$.

$$A = Pe^{-kt}$$
$$= 500{,}000e^{-0.1667t}$$

If you are using a graphing calculator, enter this in the usual format.

$$Y_1 = 500000e\wedge(-0.1667X)$$

The graph of A vs. t, is shown in Figure 2, together with a "zoomed-in" portion of the curve (which you can obtain by using the "zoom" feature on a graphing calculator, or by respecifying the x- and y-ranges for the graph). From the graph (using "trace"), we see that the flea population will be back down to 100 in about 51 days. Maybe I'd better have the cat wear two Red Flag 1-in-6 Collars!

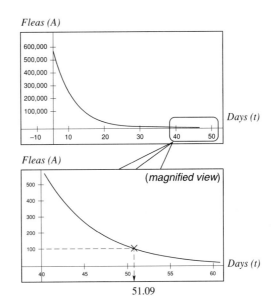

FIGURE 2

Before we go on... We can check our estimate of 51 days by substituting into the decay equation to see what the flea population should be then.

After 51 days: $A = 500{,}000e^{-0.1667 \cdot 51} = 102$ (to the nearest flea).
After 52 days: $A = 500{,}000e^{-0.1667 \cdot 52} = 86$ (to the nearest flea).

We therefore estimate that the flea population will be down to 100 some time on the 52nd day.

We'll be able to answer the question: "Exactly when will the flea population be down to 100 fleas?" without the need for a graphing calculator after we study logarithms in the next section.

9.2 EXERCISES

Compute each of the numbers in Exercises 1–10, rounded to four decimal places.

1. e^3
2. e^4
3. e^{-1}
4. e^{-2}
5. $e^{30 \times 0.001}$
6. $e^{0.014 \times 20}$
7. $100e^{0.0125}$
8. $200e^{-1.124}$
9. $10{,}200e^{-0.025 \times 20}$
10. $20{,}000e^{0.004 \times 40}$

Sketch the graphs of the functions in Exercises 11–16.

11. $f(x) = e^x$
12. $f(x) = e^{-x}$
13. $g(x) = e^{-2x}$
14. $g(x) = e^{2x}$
15. $h(x) = 2e^x$
16. $h(x) = 2e^{-x}$

17. How are the graphs in Exercises 11 and 15 related?
18. How are the graphs in Exercises 12 and 16 related?
19. How are the graphs in Exercises 11 and 13 related?
20. How are the graphs in Exercises 12 and 14 related?

In Exercises 21–26, calculate the value, to the nearest dollar, of an investment of $10,000 after the given time earning the given interest rate compounded continuously.

21. 5%, 10 years
22. 6%, 10 years
23. 2.5%, 50 years
24. 3.5%, 20 years
25. 2.5%, 11.5 years
26. 3.5%, 21.5 years

APPLICATIONS

27. *Investments* The Second Bank of Chicago offers 2.35%, compounded continuously, on its savings accounts. Express the value of a $1,000 deposit as a function of the number of years t, and use your function to find its value after 5 years.

28. *Investments* The Third Bank of Vancouver offers 3.25%, compounded continuously, on its savings accounts. Express the value of a $1,000 deposit as a function of the number of years t, and use your function to find its value after 6 years.

29. *Investments* Rock Solid Bank & Trust is offering a CD that pays 4% compounded continuously. How much interest would a $1,000 deposit earn over 10 years?

30. *Savings* SemiSolid Savings & Loan is offering a savings account that pays $3\frac{1}{2}$% interest compounded continuously. How much interest would a deposit of $2,000 earn over 10 years?

31. *Investments* Rock Solid Bank & Trust is offering a CD that pays 4% compounded continuously. What is the effective yield?

32. *Savings* SemiSolid Savings & Loan is offering a savings account that pays $3\frac{1}{2}$% interest compounded continuously. What is the effective yield?

33. *Loans* Your local loan shark makes an offer you can't refuse: a loan at 20% interest, compounded continuously. What is the effective interest rate?

34. *Credit Cards* Your local credit card company is charging 17% interest compounded continuously. What is the effective interest rate?

35. *Loans* Fifth Federal Bank is offering a loan at 9% compounded monthly, and Ninth National is offering a loan at 8.9% compounded continuously. Which is the better deal?

36. *Credit Cards* PassPort Credit Corp. offers a credit card charging 20% interest compounded monthly, while the competing Eureka card charges 19.9% compounded continuously. Which is the better deal?

37. *Investments* My investment portfolio is as follows:
    (i) $5,000 in Shady Professorial Deals Inc., earning 5.5% per year, compounded continuously;

(ii) $1,000 in Op Art Treasures Funds, depreciating continuously at 30% per year;
(iii) $10,000 in Holiday Magic Cosmetics, Inc., earning a steady 2% per annum, compounded quarterly.

What will the total value of my portfolio be in ten years' time?

38. *Gold Stocks* When I recently inherited my aunt's fortune in gold stocks, worth $300,000, I was told that the stocks had been continuously appreciating at 5.5% per year. My late aunt had bought them 25 years ago. What did she originally pay for the stocks?

39. *Carbon Dating* Carbon-14 decays into nitrogen continuously at a rate of .0121% per year. Express the amount of Carbon-14 in a fossil originally containing 100 grams of Carbon-14 as a function of its age in years. Use your function to estimate the amount left in a 10,000-year-old fossil.

40. *Plutonium Dating* Plutonium-239 decays continuously at a rate of 0.00284% per year. Express the amount of Plutonium-239 left if a sample originally containing 10 grams of the substance is stored for t years. Use your function to estimate the amount that will be left undecayed after 20,000 years.

41. *Bacteria Growth* A dangerous strain of bacteria (Bug XXX) is reproducing in a petrie dish at a rate of 2 new bugs for every bug present per hour. Express this as a percentage growth, and estimate the size of the bacteria culture after 12 hours, assuming there were 1,000 organisms originally present. [This and the next exercise are *not* continuous compounding problems.]

42. *Population Growth* The population of the U.S. in 1990 was about 250,000,000. Assume that for every 1000 citizens, 16 infants are born and 8 citizens die each year. Estimate the population in 2010 to 6 significant figures. [See comment in Exercise 41.]

43. *Varying Inflation* At the start of 1988, prices in Argentina were increasing continuously at a rate of 158% per year. By the start of 1993, the rate was down to 17%.*

(a) Assuming that the inflation rate declined linearly, express the rate of inflation r as a linear function of t, the number of years since 1988.
(b) The average rate $\bar{r}$ of inflation over t years since 1988 is given by the formula
$$\bar{r} = \frac{r(0) + r(t)}{2},$$
where $r(0)$ is the inflation rate in 1988 and $r(t)$ is the inflation rate t years later. Find a formula for $\bar{r}$ in terms of t.
(c) Use your answer to part (b) to find the price $A(t)$ you would expect to pay after t years for an item that cost $P at the start of 1988.
(d) If an item cost $100 on Jan. 1, 1988, what would you expect it to have cost on Jan. 1, 1992? (Use the function you obtained in part (c), and round your answer to the nearest dollar.)

44. *Varying Inflation* Repeat Exercise 43 using the following figures for Mexico: 42% at the start of 1988, 12% at the start of 1992.*

45. *Inflation* Inflation is running at the rate of 3% per year compounded continuously when you deposit $1,000 in an account earning 5% compounded continuously. Determine the purchasing power *in today's dollars* of the amount of money you will have in the account 4 years from now. (In other words, taking into account inflation, what amount of money today would buy the same amount of goods as the money you will have in your account 4 years from now?) Explain why this is the same amount you would get by depositing your money in an account paying 2% interest if there were no inflation. This is called the **real interest rate.**

46. *Inflation* Inflation is running at the rate of 8% per year compounded continuously when you deposit $10,000 in an account earning 5% compounded continuously. Determine the purchasing power *in today's dollars* of the amount of money you will have in the account 4 years from now.

▼ * These figures are based on the actual rates of inflation for the years 1988 and 1992 and have been rounded to the nearest percentage point. Source: International Monetary Fund, U.N. Economic Commission for Latin America and the Caribbean (as cited in *The Chicago Tribune*, June 20, 1993).

In exercises 47–56 use a graphing calculator or computer software.
Plot the graphs of the functions given in Exercises 47–50.

47. $f(x) = 100e^{0.3x}$; $-10 \le x \le 10$
48. $f(x) = 900e^{0.125x}$; $-10 \le x \le 10$
49. $f(x) = 1{,}000e^{-0.005x}$; $-50 \le x \le 50$
50. $f(x) = 9{,}000e^{-0.252x}$; $-20 \le x \le 20$
51. *Carbon Dating* Carbon-14 decays into nitrogen continuously at a rate of 0.0121% per year. If a fossil that originally contained 100 grams of Carbon-14 now contains only 1 gram of Carbon-14, how old (to the nearest year) is the fossil?
52. *Depreciation* Determine when, to the nearest year, a $3,000 investment that is continuously depreciating at a rate of 5% per year will be worth $1,000.
53. *Competing Investments* Mark Badgett just purchased $10,000 worth of municipal funds that are appreciating continuously at a rate of 3% per year. Kathi Callahan just purchased $12,000 worth of CDs that are appreciating continuously at a rate of 2% per year. Determine when, to the nearest year, the value of Mark's investment will be the same as Kathi's, and what this value will be.
54. *Competing Investments* Rich just purchased $100,000 worth of gold coins whose value is appreciating continuously at a rate of 5% per year. Prudence has just purchased $200,000 worth of antiques that are appreciating continuously at a rate of 4.5% per year. Determine when, to the nearest year, the value of Rich's gold coins will pass the value of Prudence's antiques and what this value will be.
55. *Depreciation* Sandra Groff's investment in Leveraged Buyout, Inc. is now worth $7,000 and is depreciating continuously at a rate of 5.5% per year. For some reason, she is reluctant to sell the stocks and swallow her losses. Determine when, to the nearest year, her investment will drop below $1,000.

56. *Bacteria* A strain of bacteria is reproducing continuously at a rate of 0.35% per minute. Determine when, to the nearest minute, a culture containing 2,000 organisms will double in size.
57. *Present Value* Determine the amount of money, to the nearest dollar, you must invest at 6% per year, compounded continuously, so that you will be a millionaire in 10 years' time.
58. *Present Value* Determine the amount of money, to the nearest dollar, you must invest at 7% per year, compounded continuously, so that you will be a millionaire in 10 years' time.

Employment Exercises 59 and 60 are based on the following chart, which shows the number of people employed by Northrop Grumman in Long Island, New York for the period from 1987 to 1995.*

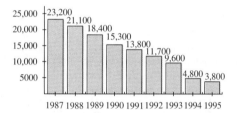

The method of least squares exponential regression[†] *gives the following model.*

$$P = 28{,}131.7e^{-0.220800t},$$

where t is the number of years since 1987.

59. According to the least squares model, when will employment by *Northrop Grumman* Long Island drop to 1,000 employees? (Give the answer to the nearest year.)
60. According to the least squares model, when will employment by *Northrop Grumman* Long Island drop to 500 employees? (Give the answer to the nearest year.)

▼ *The 1987–1994 figures show employment at Grumman Corp. before it was acquired by Northrop Corp. in 1994. The 1994 and 1995 figures are company projections. Source: Grumman Corp., Northrop Grumman Corp./Long Island, *Newsday*, September 23, 1994, p. A5.

[†]This method is described in the chapter on functions of several variables.

COMMUNICATION AND REASONING EXERCISES

61. Explain in words why 5% per year compounded continuously yields more interest than 5% per year compounded monthly.

62. Explain in words why the continuous growth model is appropriate for a population of algae.

63. Your local banker tells you that the reason his bank doesn't compound interest continuously is that it would be too demanding of computer resources, since the computer would need to spend a great deal of time keeping all accounts updated. Comment on his reasoning.

64. Your other local banker tells you that the reason *her* bank doesn't offer continuously compounded interest is that it is equivalent to offering a fractionally higher interest rate compounded daily. Comment on her reasoning.

9.3 LOGARITHMIC FUNCTIONS

LOGARITHMS

Logarithms were invented by John Napier (1550–1617) in the late sixteenth century as a means of aiding calculation. His invention made possible the prodigious hand calculations of astronomer Johannes Kepler (1571–1630), who was the first to describe accurately the orbits and the motions of the planets. Today electronic calculators have done away with that use of logarithms, but many other uses remain.

Consider the equation

$$2^3 = 8.$$

This tells us that the power to which you need to raise 2 in order to get 8 is 3. We shall abbreviate the phrase "*the power to which you need to raise* 2 *in order to get* 8" as "$\log_2 8$." Thus, another way of writing the equation $2^3 = 8$ is

$$\log_2 8 = 3.$$

This is read: "The logarithm of 8 with base 2 is 3."

Now here is the general definition. Let x be any positive number, and let a be a positive number other than 1.

LOGARITHM WITH BASE a

The **logarithm of x with base a, $\log_a x$,** is the power to which you need to raise a in order to get x. In other words, it is *the exponent that turns a into x*. Symbolically,

$$\log_a x = y$$

means

$$a^y = x.$$

▶ **NOTE** If we raise a nonzero number a to any power, the result is always positive. (Recall that the graph of $f(x) = a^x$ lies entirely above the x-axis.) Thus, we cannot speak of the logarithm of a negative number. In other words, the largest possible domain of $f(x) = \log_a x$ is $(0, +\infty)$. ◀

▼ **EXAMPLE 1**

When we say "$\log_3 9 = 2$," we mean: "the power you raise 3 to in order to get 9 is 2." In other words, $3^2 = 9$. The statement $\log_3 9 = 2$ is called the **logarithmic form** of the statement $3^2 = 9$, while the statement $3^2 = 9$ is called the **exponential form** of the statement $\log_3 9 = 2$.

▼ **EXAMPLE 2**

Find $\log_{10} 1{,}000$.

SOLUTION We are asking the question, "What power of 10 gives 1,000?" A moment's thought reveals that $10^3 = 1{,}000$, so $\log_{10} 1{,}000 = 3$. If you prefer a more mechanical way of doing this, proceed as follows.

1. Let $\log_{10} 1{,}000 = x$.
2. In exponent form, this is $10^x = 1{,}000$.
3. But $1{,}000 = 10^3$.
4. Thus, $x = 3$.

▼ **EXAMPLE 3**

Find $\log_4 64$.

SOLUTION Let $\log_4 64 = x$.
Then, in exponent form, $4^x = 64$.
But $64 = 4^3$.
Thus, $x = 3$, so $\log_4 64 = 3$.

Before we go on... Again, you can also ask yourself: What power of 4 gives 64? The answer: $4^3 = 64$, so $\log_4 64 = 3$.

▼ **EXAMPLE 4**

Find $\log_4\left(\dfrac{1}{64}\right)$.

SOLUTION Let $x = \log_4\left(\dfrac{1}{64}\right)$.

Then $\quad 4^x = \dfrac{1}{64}$.

But $\quad \dfrac{1}{64} = \dfrac{1}{4^3} = 4^{-3}$

and so $x = -3$.

EXAMPLE 5

Find $\log_{1/5} 25$.

SOLUTION Let $\log_{1/5} 25 = x$.
In exponent form, this is $\left(\frac{1}{5}\right)^x = 25$.
Using the rules for exponents, this may be written as

$$\frac{1}{5^x} = 25,$$

or $\qquad 5^{-x} = 25.$
But $\qquad 25 = 5^2.$
Thus, $\qquad -x = 2.$
Hence, $\qquad x = -2,$ or $\log_{1/5} 25 = -2.$

EXAMPLE 6

Find $\log_5 1$.

SOLUTION Let $x = \log_5 1$.
Thus, $\qquad 5^x = 1.$
But $\qquad 1 = 5^0$, whence $x = 0$
that is, $\quad \log_5 1 = 0.$

Before we go on... We could repeat this calculation with any base to get $\log_a 1 = 0$ for any $a \neq 1$.

EXAMPLE 7

Find $\log_2 \sqrt{2}$.

SOLUTION Putting $\log_2 \sqrt{2} = x$, we get $2^x = \sqrt{2} = 2^{1/2}$, and so $x = \frac{1}{2}$ (that is, $\log_2 \sqrt{2} = \frac{1}{2}$).

The following are standard abbreviations.

$$\log_{10} x = \log x$$
$$\log_e x = \ln x$$

We call $\log x = \log_{10} x$ the **common logarithm** of x, and $\ln x = \log_e x$ the **natural logarithm** of x.

To obtain $\log_{10} 5$ on a traditional scientific calculator, press $\boxed{5}$ followed by $\boxed{\log}$. (You should get 0.6989 . . .) On a graphing calculator or other function-style calculator, press $\boxed{\log}$ followed by $\boxed{5}$ and then $\boxed{=}$.*

To obtain $\log_e 5$ on a traditional scientific calculator, press $\boxed{5}$ followed by $\boxed{\ln}$. (You should get 1.6094 . . .) On a graphing calculator or other function-style calculator, press $\boxed{\ln}$ followed by $\boxed{5}$ and then $\boxed{=}$.

Q Suppose we want to calculate logs to bases other than 10 and e?

A We use the change-of-base formula.

CHANGE-OF-BASE-FORMULA

$$\log_a b = \frac{\log b}{\log a} = \frac{\ln b}{\ln a}$$

(Take our word for it now. We'll derive this formula later in the section.) We can use this formula to get the logarithm of any number to any base using a calculator. For example, to find $\log_{3.45} 2.261$, we divide log (2.261) by log(3.45), getting 0.6588 (to four significant digits). We get the same answer by dividing ln(2.261) by ln(3.45) (try it).

The following identities for logarithms are as important as the ones for exponents listed in Section 1. We shall discuss why they are true after using them in several examples.

LOGARITHM IDENTITIES

The following identities hold for any positive base $a \neq 1$ and any positive numbers x and y.

Identity
(a) $\log_a (xy) = \log_a x + \log_a y$
(b) $\log_a \left(\frac{x}{y}\right) = \log_a x - \log_a y$
(c) $\log_a (x^r) = r \log_a x$
(d) $\log_a a = 1;\ \log_a 1 = 0$
(e) $\log_a \left(\frac{1}{x}\right) = -\log_a x$
(f) $\log_a x = \frac{\log_b x}{\log_b a}$

Example
$\log_2 16 = \log_2 8 + \log_2 2$
$\log_2 \left(\frac{5}{3}\right) = \log_2 5 - \log_2 3$
$\log_2 (6^5) = 5 \log_2 6$
$\log_2 2 = 1;\ \log_{11} 1 = 0$
$\log_2 \left(\frac{1}{3}\right) = -\log_2 3$
$\log_2 5 = \frac{\log_{10} 5}{\log_{10} 2}$

▼ * We do not guarantee that this will work on *your* calculator—for instance, you might have to use an "Execute" button rather than an "=" button, depending on the brand of calculator—so you should check your instruction manual.

▶ **NOTE** Since these rules hold for any base, they hold in particular for the bases 10 and e, so we can replace "$\log_a$" with either "log" or "ln", as long as we stick to the same base throughout an identity.

Some people like to remember these identities in words. For example, the first identity says that "multiplication on the *inside* corresponds to addition on the *outside*." Another way of remembering them is to notice that "log" converts more difficult operations into simpler operations. ◀

▶ **CAUTION**
1. In all of these identities except (f), the bases of the logarithms must match. For example, rule (a) gives $\log_3 11 + \log_3 4 = \log_3 44$, but does *not* apply to the expression $\log_3 11 + \log_5 4$.
2. People sometimes invent their own identities. Here is one of the most popular ones:

$$\log_a (x + y) = (\log_a x)(\log_a y). \qquad \text{WRONG!}$$

For instance, $\log(99 + 1) \neq \log(99)\log(1)$, since the left-hand side is $\log(100) = 2$, while the right-hand side is $\log(99) \times 0 = 0$.

The following formula is also wrong (we suggest that you give an example to show why).

$$\log_a (x + y) = \log_a x + \log_a y \qquad \text{WRONG!}$$

These just don't work! Just because they're popular doesn't mean they're right.* ◀

Let us now use these identities to help solve some equations having unknowns in the exponent.

▼ **EXAMPLE 8**

Solve $3^{2x} = 4$ for x.

SOLUTION The unknown is in a most inconvenient place up there in the exponent. To bring it down, we shall take the logarithm of both sides and then use rule (c). We could use the logarithm with any base, but we choose to use the natural log.

▼ * Mathematics is not a democracy. Mathematicians try to uncover the truth, and not create it. What is true is true, and what is false is false, and there's absolutely nothing we can do about it! There was an embarrassing incident a few decades ago when a certain state in the Union (that shall remain unnamed) seriously attempted to introduce a law declaring π to be *exactly* 3!

Actually, there is a philosophical controversy here. Platonists believe that mathematics already exists and the mathematician's job is one of exploration and discovery. Another camp believes that mathematicians *invent* mathematics. Both sides agree, however, that once the ground rules have been set, there is an objective criterion of mathematical truth. In other words, what is true is true and what is false is false.

Taking logs,
$$\ln(3^{2x}) = \ln 4.$$

By identity (c),
$$2x \cdot \ln 3 = \ln 4.$$

Dividing,
$$x = \frac{\ln 4}{2 \ln 3} \approx \frac{1.386294}{2.197225} = 0.63093.$$

It is a good idea to leave the actual evaluation of logarithms until the very end, as we did here. This eliminates the need to record long decimal numbers, and it tends to give more accurate answers.

Before we go on... Here is the sequence of calculator operations that will give the answer without having to store intermediate answers.

Traditional calculator: [4][ln][÷][3][ln][=][÷][2][=]

Function-style calculator:
[ln][4][÷][(][2][×][ln][3][)][=]

As usual, we should check the anwer by substituting it back into the original equation.

$$3^{2(0.63093)} \approx 4 \text{ (to six significant digits)} \checkmark$$

▼ EXAMPLE 9

Solve $3^{3t} = 2^{t+1}$ for t.

SOLUTION We shall again take the natural log of both sides, as it worked so nicely last time.

$$\ln(3^{3t}) = \ln(2^{t+1})$$

By identity (c),
$$3t \ln 3 = (t + 1) \ln 2.$$

To solve for t, we must get all the terms with t in them together.
$$3t \ln 3 = t \ln 2 + \ln 2$$

Subtracting $t \ln 2$ from both sides,
$$3t \ln 3 - t \ln 2 = \ln 2.$$

Hence,
$$t(3 \ln 3 - \ln 2) = \ln 2.$$

Thus,
$$t = \frac{\ln 2}{3 \ln 3 - \ln 2} \approx 0.26632.$$

Before we go on... Checking the answer,
$$3^{3(0.26632)} \approx 2.40547$$
$$2^{0.26632+1} \approx 2.40547. \checkmark$$

Q I did the above calculation on my graphing calculator and obtained a different answer: $t = -0.482837$. What did I do wrong?

A This wrong answer, $t = -0.482837$, will result from entering the following information.

$$\ln 2/3 \ln 3 - \ln 2 \quad \text{WRONG!}$$

What is wrong is our representation of the fraction bar: a fraction bar indicates that one *quantity* is divided by another *quantity*. In other words,

$$t = \frac{\ln 2}{3 \ln 3 - \ln 2} = \frac{(\ln 2)}{(3 \ln 3 - \ln 2)}.$$

Thus, the correct calculator format is

$$(\ln 2)/(3 \ln 3 - \ln 2)$$

▼ EXAMPLE 10

Solve $49(1.01)^t = 59$ for t.

SOLUTION Although you are again tempted to take the log of both sides, the problem is the 49 in front of the $(1.01)^t$. If you decide to go ahead and take the logs anyway, you must be careful not to make a common mistake. See if you can determine what is wrong with the following.

$$(1) \ln 49(1.01)^t = \ln 59;$$

therefore,

$$(2) \; t \ln(49(1.01)) = \ln 59.$$

Thus,

$$(3) \; t \ln(49(1.01)) = \ln 59. \quad \text{WRONG!}$$

Hence,

$$(4) \; t = \ln 59 / \ln(49(1.01)) = 1.045.$$

Q How are we sure this is the wrong answer?

A Check it:

$$49(1.01)^{1.045} \approx 49.5 \neq 59.$$

Q What went wrong here?

A The error occurred in Step 2: $\ln 49(1.01)^t$ is *not* equal to $t \cdot \ln(49(1.01))$. It is true that $t \cdot \ln(49(1.01))$ is $\ln(49(1.01))^t$, but this is not what we're evaluating. The point is that $49(1.01)^t$ is *not* the same as $(49(1.01))^t$. The first expression is the *product* of 49 and $(1.01)^t$, while the second is the tth power of the product $49(1.01)$. What we should say is

$$\ln 49(1.01)^t = \ln 49 + \ln(1.01)^t \quad \text{By identity (a)}$$
$$= \ln 49 + t \ln(1.01). \quad \text{By identity (c)}$$

Going back to the beginning, instead of solving for t immediately by taking the log of both sides, we'll first divide by the annoying factor 49.

$$(1.01)^t = \frac{59}{49}$$

Now take the logs.

$$t \cdot \ln 1.01 = \ln\left(\frac{59}{49}\right)$$

Thus,

$$t = \frac{\ln\left(\frac{59}{49}\right)}{\ln 1.01} \approx 18.6644.$$

Before we go on... Check the answer.

$$49(1.01)^{18.6644} \approx 59 \quad \text{(to 5 decimal places)} \quad ✔$$

For practice, solve again for t by first taking the natural log of both sides.

▼ **EXAMPLE 11**

Solve $\frac{1}{2^x} = 2^{x^2}$ for x.

SOLUTION Since 2 is the base of both expressions, let us take logs with base 2 for a change. (This will be more convenient, but you might try solving by taking natural logs of both sides, for practice.)
We have

$$\log_2\left(\frac{1}{2^x}\right) = \log_2(2^{x^2}).$$

By identity (f),

$$-\log_2(2^x) = \log_2(2^{x^2}).$$

Applying identity (c),

$$-x \log_2 2 = x^2 \log_2 2.$$

But identity (d) says that $\log_2 2 = 1$. Thus, we get
$$-x = x^2.$$
This is a quadratic. To solve, we bring all the terms to the same side and factor:
$$x^2 + x = 0,$$
or
$$x(x + 1) = 0.$$
This gives
$$x = 0 \quad \text{or} \quad x = -1.$$
Done!

ALTERNATE SOLUTION (without logarithms)
Rewrite the equation as
$$2^{-x} = 2^{x^2}.$$
Then we must have
$$-x = x^2,$$
giving
$$x^2 + x = 0.$$

Before we go on... Check the answers $x = 0$ and $x = -1$.
$$\frac{1}{2^0} = \frac{1}{1} = 1 \quad \text{and} \quad 2^{0^2} = 2^0 = 1 \quad \checkmark$$
$$\frac{1}{2^{-1}} = 2^1 = 2 \quad \text{and} \quad 2^{(-1)^2} = 2^1 = 2 \quad \checkmark$$

DERIVATION OF THE LOGARITHM IDENTITIES

We now pause to see where (at least some of) the rules for logarithms come from, as we promised. Roughly speaking, they are restatements in logarithmic form of the laws of exponents.

Q Why is $\log_a xy = \log_a x + \log_a y$?

A Let $s = \log_a x$ and $t = \log_a y$. In exponential form, these equations say that
$$a^s = x \quad \text{and} \quad a^t = y.$$
Multiplying these two equations together gives
$$a^s a^t = xy;$$
that is,
$$a^{s+t} = xy.$$

Rewriting this in logarithmic form gives
$$\log_a (xy) = s + t = \log_a x + \log_a y,$$
as claimed.

Here is an intuitive way of thinking about it: Since logs are exponents, this identity expresses the familiar law that the exponent of a product is the sum of the exponents.

Identity (b) is shown in almost the identical way, and we leave it for you for practice.

Q Why is $\log_a (x^r) = r \log_a x$?

A Let $t = \log_a x$. Writing this in exponential form gives
$$a^t = x.$$
Raising this equation to the rth power gives
$$a^{rt} = x^r.$$
Rewriting in logarithmic form gives
$$\log_a (x^r) = rt = r \log_a x,$$
as claimed.

Identity (d) we will leave for you to do as practice.

Q Why is $\log_a \left(\frac{1}{x}\right) = -\log_a x$?

A This follows from identities (b) and (d) (think about it).

Q Why is $\log_a x = \dfrac{\log_b x}{\log_b a}$?

A Let $s = \log_a x$. In exponential form, this says that
$$a^s = x.$$
Take the logarithm with base b of both sides, getting
$$\log_b a^s = \log_b x,$$
then use identity (c):
$$s \log_b a = \log_b x,$$
so
$$s = \frac{\log_b x}{\log_b a}.$$

GRAPHS OF LOGARITHMIC FUNCTIONS

It is useful to have a sense of what the graph of $f(x) = \log_a x$ looks like for various values of a. Recall that the largest possible domain of the function $f(x) = \log_a x$ is $(0, +\infty)$.

9.3 Logarithmic Functions

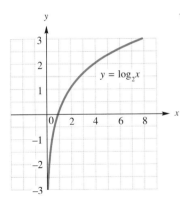

FIGURE 1

▼ EXAMPLE 12

Sketch the graph of $f(x) = \log_2 x$.

SOLUTION To graph this curve, we could use a graphing calculator (as described below) or make a table, as follows. The values of x that we use must be positive, and we have chosen the values shown for convenience. Since $\log_2 x$ is not defined when $x = 0$, we chose several values of x close to zero.

x	$\frac{1}{8}$	$\frac{1}{4}$	$\frac{1}{2}$	1	2	4	8
$f(x) = \log_2 x$	-3	-2	-1	0	1	2	3

Graphing these points gives us Figure 1.

To enter this function in your graphing calculator, use the change-of-base formula

$$\log_2 x = \frac{\log x}{\log 2}.$$

Thus, you will need to use the format

$$Y_1 = (\log X)/(\log 2)$$

(Although the parentheses are not strictly necessary, it is good practice to use them to distinguish the quantities you are dividing, since the fraction bar indicates that one *quantity* is divided by another.) As for the ranges, you could use $0.001 \le x \le 8$ and $-5 \le y \le 5$ to get a facsimile of Figure 1.

▼ EXAMPLE 13

On the same axes, sketch the graphs of

$$f(x) = \log_2 x \quad \text{and} \quad g(x) = 2^x.$$

SOLUTION We have already sketched both curves—the first curve in the last example, and the second curve back in the first section. If we superimpose them on the same set of axes, we notice something interesting (Figure 2).

There is a symmetry between the two curves: one is the mirror image of the other across the diagonal line $y = x$. It is as though the graph of $f(x) = \log_2 x$ was obtained from the graph $g(x) = 2^x$ by switching the x- and y-coordinates. For instance, the point $(2, 1)$ is on the graph of $f(x) = \log_2 x$, while the point $(1, 2)$ is on the graph of $g(x) = 2^x$. It is this property that makes the functions $f(x) = \log_2 x$ and $g(x) = 2^x$ **inverse functions**.

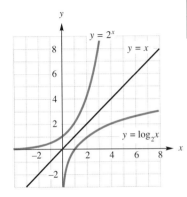

FIGURE 2

Q What does it mean for two functions to be inverse functions?

A Two functions f and g are **inverse functions** if
$$g(f(x)) = x$$
for every x in the domain of f, and
$$f(g(x)) = x$$
for every x in the domain of g. For example, $f(x) = x^2$ with domain $[0, +\infty)$ and $g(x) = \sqrt{x}$ are inverse functions.

Q Are the functions $f(x) = \log_2 x$ and $g(x) = 2^x$ inverse functions?

A Let us check the requirement directly.
$$g(f(x)) = g(\log_2 x)$$
$$= 2^{\log_2 x}$$

Now $\log_2 x$ is the power to which you raise 2 in order to get x. In other words, 2 raised to that power must be x! Thus,
$$g(f(x)) = x.$$

Next,
$$f(g(x)) = f(2^x)$$
$$= \log_2 (2^x)$$
$$= x \log_2 2$$
$$= x,$$

because $\log_2 2 = 1$. Thus, we have shown that the two functions are indeed inverse functions.

There is nothing special about the base 2 in the above discussion. So we can draw the following conclusion.

RELATIONSHIP OF THE FUNCTIONS $f(x) = \log_a x$ **AND** $g(x) = a^x$

If a is any positive number other than 1, then the functions $f(x) = \log_a x$ and $g(x) = a^x$ are inverse functions. This means that
$$a^{\log_a x} = x$$
for all positive x and
$$\log_a (a^x) = x$$
for all real x.

EXAMPLE 14

Using a graphing calculator or graphing software, graph the functions
$$f(x) = \log_a x$$
with $a = \frac{1}{4}, \frac{1}{2}, 2,$ and 4.

SOLUTION Referring to the graphing calculator discussion at the end of Example 12, we can do this using the formula
$$y = \frac{\log x}{\log a},$$
where $a = \frac{1}{4}, \frac{1}{2}, 2,$ and 4. This gives us Figure 3.

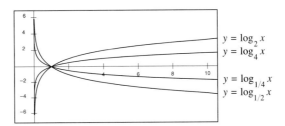

FIGURE 3

Notice that the graphs all pass through the point (1, 0). (Why?) Notice further that the graphs of the logarithmic functions with bases smaller than 1 are upside-down versions of the others.

▶ 9.3 EXERCISES

Rewrite the equations in Exercises 1–12 in logarithmic form.

1. $2^5 = 32$ **2.** $3^2 = 9$ **3.** $3^{-2} = \frac{1}{9}$

4. $2^{-4} = \frac{1}{16}$ **5.** $10^3 = 1{,}000$ **6.** $10^6 = 1{,}000{,}000$

7. $y = e^x$ **8.** $y = e^{-x}$ **9.** $y = x^{-3}$ $(x \ne 1)$

10. $y = x^4$ $(x \ne 1)$ **11.** $1 = e^0$ **12.** $1 = 10^0$

Rewrite the equations in Exercises 13–24 in exponential form.

13. $\log_6 36 = 2$ **14.** $\log_5 125 = 5$

15. $\log_2 \left(\frac{1}{4}\right) = -2$ **16.** $\log_3 \left(\frac{1}{27}\right) = -3$

17. $\log 100{,}000{,}000 = 8$
18. $\log\left(\dfrac{1}{10{,}000}\right) = -4$
19. $\ln\left(\dfrac{1}{e}\right) = -1$
20. $\ln(e^{-3}) = -3$
21. $\log_x y = 3$
22. $\log_3 x = y$
23. $y = \ln(-x) \ (x < 0)$
24. $y = \ln(x^{-1}) \ (x > 0)$

Let $a = \ln 2$, $b = \ln 3$, and $c = \ln 5$. Use the identities for logarithms to express the quantities in Exercises 25–36 in terms of a, b, and c.

25. $\ln 6$
26. $\ln 10$
27. $\ln\left(\dfrac{2}{3}\right)$
28. $\ln\left(\dfrac{3}{5}\right)$
29. $\ln 256$
30. $\ln 81$
31. $\ln\left(\dfrac{3}{10}\right)$
32. $\ln\left(\dfrac{25}{3}\right)$
33. $\ln 0.02$
34. $\ln 0.05$
35. $\ln\left(\dfrac{9}{e}\right)$
36. $\ln 25e$

Complete the equations in Exercises 37–44 by filling in the missing quantity.

37. $\log_a 3 + \log_a 4 = \log_a(\ \ )$
38. $\log_a 3 - \log_a 4 = \log_a(\ \ )$
39. $2\log_a x + \log_a y = \log_a(\ \ )$
40. $2\ln x - 4\ln y = \ln(\ \ )$
41. $2\ln x + 4\ln y - 6\ln z = \ln(\ \ )$
42. $-(2\ln x + \ln y + 6\ln 1) = \ln(\ \ )$
43. $\log_2(x) + 2^x \log_2 y = \log_2(\ \ )$
44. $x^2 \log 2 - \log(2x^2) = \log(\ \ )$

Solve the equations in Exercises 45–56 for the indicated variable.

45. $4 = 2^x$; solve for x.
46. $9 = 3^{-x}$; solve for x.
47. $2^t = 4^{2t}$; solve for t.
48. $9^t = 3^{3t}$; solve for t.
49. $2^t = 2^{-t^2}$; solve for t.
50. $2^{-m} = 2^{m^2} \cdot 4^m$; solve for m.
51. $100 = 50e^{3t}$; solve for t.
52. $20 = 10e^{5t}$; solve for t.
53. $10 = 1{,}000e^{-2x}$; solve for x.
54. $5 = 5{,}000e^{-2.2x}$; solve for x.
55. $200 = 5(2^y)$; solve for y.
56. $1{,}000 = 6(3^{2x})$; solve for x.

Sketch the graphs of the functions in Exercises 57–62.

57. $f(x) = \log_3 x$; $0 < x \le 27$
58. $f(x) = \log_4 x$; $0 < x \le 64$
59. $g(x) = x - \log_2 x$; $0 < x \le 8$
60. $g(x) = \log_3 x - x$; $0 < x \le 27$
61. $h(x) = (\log_2 x)^2$; $0 < x \le 8$
62. $h(x) = (\log_3 x)^3$; $0 < x \le 27$

Use a graphing calculator or graphing software to graph the functions in Exercises 63–66.

63. $f(x) = \log_2(x^2 + 1)$; $-2 \le x \le 2$
64. $f(x) = \log_2((x-1)^2 + 1)$; $-10 \le x \le 10$
65. $f(x) = \ln|x|$; $-10 \le x \le 10 \ (x \ne 0)$
66. $f(x) = \ln|x - 3|$; $-10 \le x \le 10 \ (x \ne 3)$

APPLICATIONS

67. *Richter Scale* The **Richter scale** is used to measure the intensity of earthquakes. The Richter scale rating of an earthquake is given by the formula

$$R = \frac{2}{3}(\log E - 4.4),$$

where E is the energy released by the earthquake (measured in joules*).

(a) The Great San Francisco Earthquake of 1906 registered $R = 8.2$ on the Richter scale. How many joules of energy were released?

(b) In 1989, another San Francisco earthquake registered 7.1 on the Richter scale. What percentage of the energy released in the 1906 earthquake was released in the smaller 1989 earthquake?

(c) Show that if two earthquakes registering R_1 and R_2 on the Richter scale release E_1 and E_2 joules of energy, respectively, then

$$\frac{E_2}{E_1} = 10^{1.5(R_2 - R_1)}.$$

(d) Fill in the missing quantity: If one earthquake registers one point more on the Richter scale than another, then it releases _____ times the amount of energy.

68. *Sound Intensity* The loudness of a sound is measured in **decibels**. The decibel level of a sound is given by the formula

$$D = 10 \log \frac{I}{I_0},$$

where D is the decibel level (dB), I is its intensity in watts per square meter (W/m²) and $I_0 = 10^{-12}$ W/m² is the intensity of a barely audible "threshold" sound. A sound intensity of 90dB or greater causes damage to the average human ear.

(a) Find the decibel levels of each of the following, rounding to the nearest dB:
  Whisper: 115×10^{-12} W/m²
  T.V. (average volume from 10 feet): 320×10^{-7} W/m²
  Loud music: 900×10^{-3} W/m²
  Jet aircraft (from 500 ft.): 100 W/m²

(b) Which of the above sounds would cause damage to the average human ear?

(c) Show that if two sounds of intensity I_1 and I_2 register decibel levels of D_1 and D_2 respectively, then

$$\frac{I_2}{I_1} = 10^{0.1(D_2 - D_1)}.$$

(d) Fill in the missing quantity: If one sound registers one decibel more than another, then it is _____ times as intense.

69. *Sound Intensity* The decibel level of a T.V. set decreases with the distance from the set according to the formula

$$D = 10 \log \left(\frac{320 \times 10^7}{r^2} \right),$$

where D is the decibel level and r is the distance from the T.V. set in feet.

(a) Find the decibel level (to the nearest decibel) at distances of 10, 20, and 50 feet.

(b) Express D in the form $D = A + B \log r$ for suitable constants A and B.

(c) How far must a listener be from a T.V. so that the decibel level drops to 0?

70. *Acidity* The acidity of a solution is measured by its pH. The pH of a solution is given by the formula

$$pH = -\log(H^+),$$

where H^+ measures the concentration of hydrogen ions in moles per liter.† The pH of pure water is 7. A solution is referred to as *acidic* if its pH is below 7 and as *basic* if its pH is above 7.

(a) Calculate the pH of each of the following substances.
  Blood: 3.9×10^{-8} moles/liter
  Milk: 4.0×10^{-7} moles/liter
  Soap solution: 1.0×10^{-11} moles/liter
  Black coffee: 1.2×10^{-7} moles/liter

(b) How many moles of hydrogen ions are contained in a liter of acid rain with a pH of 5.0?

(c) Complete the following sentence: If the pH of a solution increases by 1.0, then the concentration of hydrogen ions _____ .

---

*A joule is a unit of energy. 100 joules of energy would light up a 100-watt light bulb for a second.

†A mole corresponds to about 6.0×10^{23} hydrogen ions. (This number is also known as Avogadro's Number.)

71. *Modeling Demand* You are the owner of a new fried chicken franchise, and you would like to construct a demand equation of the form $q = f(p)$, where q is the number of quarter-chicken servings you sell per hour, and p is the price you charge per serving. You have tried selling the servings at three different prices, and the resulting sales are shown in the following table:

Price per serving (p)	$3.00	$4.00	$5.00
Average hourly sales (q)	4.826	3.889	3.290

(a) Use the sales figures for the prices $3 and $5 to construct a linear demand function of the form $f(p) = mp + b$, where m and b are constants you will need to determine. (Round m and b to three significant digits.)
(b) Use the same sales figures to construct a demand function of the form $f(p) = kp^r$, where k and r are constants you will need to determine. (Round k and r to three significant digits.)
(c) Which of the two demand functions predicts the demand at $4 per serving most accurately?
(d) Use the better of the two demand functions to predict the demand when the price is set at $3.50 per portion.

72. *Modeling Supply* You are the purchasing manager for a large drugstore and have been trying to purchase vitamin supplements at a bulk discount from Back to Nature Supplements, Inc. You have noticed that the number of cases of vitamin C supplement the manufacturer is willing to supply depends on the price you offer, as shown by the following table:

Price per case (p)	$30	$40	$50
Average number of cases supplied (q)	23.40	25.15	26.59

(a) Use the supply figures for the prices $30 and $50 to construct a linear supply function of the form $q = f(p) = mp + b$, where m and b are constants you will need to determine. (Round m and b to three significant digits.)
(b) Use the same sales figures to construct a supply function of the form $f(p) = kp^r$, where k and r are constants you will need to determine. (Round k and r to three significant digits.)
(c) Which of the two supply functions predicts the supply figure for $40 per case most accurately?
(d) Use the better of the two supply functions to predict the number of cases you can order at $60 per case.

COMMUNICATION AND REASONING EXERCISES

73. Explain in words why the logarithm of a product of two numbers is the sum of the logarithms of the individual numbers.

74. Explain in words why the logarithm of a number's reciprocal is the negative of the logarithm of that number.

75. Why is the logarithm of a negative number not defined?

76. Of what use are logarithms, now that they are no longer needed to perform complex calculations?

77. Complete the following: if $y = 4^x$, then $x =$ _____.

78. Complete the following: if $y = \log_6 x$, then $x =$ _____.

79. Complete the following sentence: If y is a linear function of $\log x$, then x is a _____ function of y.

80. If y is an exponential function of x, how are x and $\log y$ related?

▶ 9.4 APPLICATIONS OF LOGARITHMS

You may have noticed in the first section that when we had a compound interest problem in which the number of years was the unknown, we had to solve the problem graphically, not analytically. The reason was that the number of years t appears in the exponent in the expression for compound interest:

$$A = P\left(1 + \frac{r}{m}\right)^{mt},$$

9.4 Applications of Logarithms

and we knew no way of solving for it. As we saw in the previous section, the logarithm function makes solving such equations straightforward. We can now tackle a greater variety of questions about compound interest and also about exponential growth and decay.

▼ **EXAMPLE 1**   Investments

Tax-exempt bonds are yielding an average of 5.2% per year.* Assuming this rate continues, how long will it take a $1,000-dollar investment to be worth $1,500 if the interest is compounded monthly?

SOLUTION Substituting $A = 1{,}500$, $P = 1{,}000$, $r = 0.052$, and $m = 12$ in the compound interest equation gives

$$1{,}500 = 1{,}000\left(1 + \frac{0.052}{12}\right)^{12t}$$

or

$$1{,}500 = 1{,}000(1.004333)^{12t},$$

and we must solve for t. We saw how to solve this kind of equation in the previous section. We first divide both sides by 1,000, getting

$$1.5 = 1.004333^{12t},$$

and then take the natural log of both sides.

$$\ln(1.5) = \ln(1.004333^{12t}) = 12t \ln(1.004333)$$

We can now solve for t.

$$t = \frac{\ln(1.5)}{12 \ln(1.004333)} \approx 7.8 \text{ years}$$

Thus, it will take approximately 7.8 years for a $1,000-dollar investment to be worth $1,500.

▼ **EXAMPLE 2**   Doubling Time

How long does it take to double P dollars invested at 5% interest, compounded continuously?

SOLUTION We want to know when $P becomes $2P. We substitute $A = 2P$, $P = P$, and $r = 0.05$ into the continuous compounding formula $A = Pe^{rt}$ to get

$$2P = Pe^{0.05t}.$$

▼ * In October 1993, according to an index of yields for long-term A-rated general obligation bonds compiled weekly by The Bond Buyer. (Source: *The New York Times*, October 18, 1993).

We need to solve for t. As a first step, we can divide both sides by P, yielding

$$2 = e^{0.05t}.$$

Taking the natural logarithm of both sides,

$$\ln 2 = \ln e^{0.05t},$$

or

$$\ln 2 = 0.05 \, t \ln e.$$

But $\ln e = 1$, by identity (d). Thus, we have

$$\ln 2 = 0.05t.$$

Hence,

$$t = \frac{\ln 2}{0.05} = 13.863 \text{ years}.$$

Before we go on ... What we have just found is known as the **doubling time** t_D. Let's check our answer by substituting it back into the compound interest formula.

$$Pe^{(0.05)(13.863)} = 2P \ \checkmark$$

In the above example, we obtained the doubling time t_D by dividing $\ln 2$ by the interest rate. We can carry through this calculation in general to get the following (the calculation for exponential growth is identical because the basic formula is identical).

DOUBLING TIME FOR CONTINUOUS COMPOUNDING

If $P is invested at an annual interest rate r compounded continuously, the doubling time t_D is given by

$$t_D = \frac{\ln 2}{r}.$$

DOUBLING TIME FOR EXPONENTIAL GROWTH

For a quantity growing exponentially with fractional growth rate k, the doubling time t_D is given by

$$t_D = \frac{\ln 2}{k}.$$

Another way of writing the doubling time formulas is

$$t_D \cdot r = \ln 2$$
$$t_D \cdot k = \ln 2.$$

9.4 Applications of Logarithms

This tells us that once we know one of the two quantities r and t_D (or k and t_D), we can solve for the other.

EXAMPLE 3 Investments

I would like to double my money in 10 years. At what interest rate (compounded continuously) can this be accomplished?

SOLUTION We have the formula

$$t_D \cdot r = \ln 2,$$

where

$$t_D = 10.$$

Thus,

$$r = \frac{\ln 2}{t_D} = \frac{\ln 2}{10} \approx 0.069315.$$

The required rate of interest is therefore 6.9315%.

Before we go on... Check:

$$e^{0.069315 \times 10} = 2. \checkmark$$

EXAMPLE 4 Epidemics

In the early stages of the AIDS epidemic, the number of people infected was doubling every six months. Assuming an exponential growth model, and given that there were an estimated 1 million persons infected at a certain time, estimate the number of infected people 2.5 years later.

SOLUTION Since the doubling time is $t_D = 0.5$ years, this gives us

$$k = \frac{\ln 2}{t_D} = \frac{\ln 2}{0.5} = 2\ln 2 \approx 1.3863.$$

Using the exponential growth model,

$$A = Pe^{tk},$$

we have

$$P = 1{,}000{,}000,\ t = 2.5,\ k = 2\ln 2.$$

This gives

$$A = 1{,}000{,}000\, e^{2.5}(2\ln 2) = 32{,}000{,}000.$$

Thus, the model predicts 32 million infected people 2.5 years later.

Before we go on... Is the exponential growth model reliable here? There are really two questions: (1) Does the model produce the numbers we expect? (2) Are these reasonable numbers?

1. In order to test whether the exponential growth model gives the expected numbers, note that we were told that the number of infected people was doubling every 6 months. Thus, after 6 months, it will have doubled once, giving 2 million cases; after 1 year, it will have doubled twice, giving 4 million cases ($= 2^2$), and so on. After 2.5 years, it will have doubled 5 times, giving $2^5 = 32$ million cases, agreeing with the prediction of the model. Thus, the exponential model produces the expected numbers.

2. On the other hand, the doubling every six months couldn't continue for very long, and this is borne out by observations. If doubling every six months did continue, then in 20 years the number of infected people would be

$$2^{40} \text{ million} \approx 1{,}099{,}511{,}628{,}000{,}000{,}000,$$

a number that is considerably larger than the population of the earth! Thus the exponential model is unreliable for predicting long-term trends.

Epidemiologists use more sophisticated models to measure the spread of epidemics, and these models predict a "leveling-off" phenomenon as the number of cases becomes a significant part of the total population (see the "You're the Expert" section at the end of this chapter). However, the exponential growth model *is* fairly reliable in the early stages of an epidemic.

▼ **EXAMPLE 5** Half-Life

Let's go back to the story of Fluffy's "Red Flag 1-in-6 Flea Collar." (Examples 6 and 8 in Section 2). We saw there that the flea population was decreasing at a fractional rate of $\frac{1}{6}$ per day. Thus, using the exponential decay formula,

$$A = Pe^{(-1/6)t}.$$

How long will it take for the flea population to be cut in half? This time is called the **half-life** t_H of the flea population.

SOLUTION We want to know when the population P becomes $\frac{P}{2}$. Substituting $A = \frac{P}{2}$ gives

$$\frac{P}{2} = Pe^{(-1/6)t} = Pe^{-t/6}.$$

Dividing by P,

$$\frac{1}{2} = e^{-t/6}.$$

Taking the natural log of both sides gives

$$\ln \frac{1}{2} = \ln(e^{-t/6})$$

$$= -\frac{t}{6} \ln e$$

$$= -\frac{t}{6}.$$

In other words, $-\ln 2 = -\dfrac{t}{6}$

or $\qquad \ln 2 = \dfrac{t}{6},$

so $\qquad t = 6 \ln 2 \approx 4.1589$ (days).

Thus, the half-life of the flea population is $t_H \approx 4.2$ days.

Before we go on... Let's check our answer.

$$Pe^{-4.1589/6} = 0.5P \ \checkmark$$

Before we leave this example, let's go back to the step

$$\ln 2 = \frac{1}{6}t.$$

Since $k = \frac{1}{6}$ is the fractional rate of decay, we can see from this equation that

$$\ln 2 = \text{Fractional rate of decay} \times \text{Half-life}$$

or

$$\ln 2 = k \cdot t_H.$$

HALF-LIFE FOR EXPONENTIAL DECAY

For a quantity decaying exponentially with fractional rate of decay k, the half-life t_H is given by

$$t_H = \frac{\ln 2}{k}.$$

Another way of writing this formula is

$$t_H \cdot k = \ln 2.$$

An important form of exponential decay is **radioactive decay.** Some forms of chemical elements, such as uranium, plutonium, cobalt, and carbon, are unstable and decay into more stable elements. The decay takes place on an essentially continuous basis and is modeled with great precision by the exponential decay equation $A = Pe^{-kt}$, where P is the original number of atoms (or the original weight) and A is the number of atoms (or weight) undecayed after time t. Different unstable elements can have vastly different values of k, and hence vastly different half-lives, ranging from millionths of a second to thousands of years.

Carbon dating, one of the methods used by archaeologists to determine the age of a fossil, is based on the following principle. There are two isotopes of carbon: Carbon-12, the stable form, and Carbon-14, which is unstable and gradually decays into nitrogen. Carbon-14 originates in the upper atmosphere when nitrogen is exposed to cosmic radiation and is absorbed in small amounts by all living organisms along with Carbon-12. When an organism dies, absorption of carbon stops, and the Carbon-14 in the organism slowly decays, with a half-life of 5730 years. When a fossil is analyzed, the ratio of Carbon-14 to Carbon-12 in the fossil is measured and compared with the original ratio.* In this way, it is possible to estimate the amount of Carbon-14 that has decayed since the organism was alive, and hence determine the age of the fossil.

▼ **EXAMPLE 6** Carbon Dating

If tests on a fossilized skull reveal that 99.95% of the Carbon-14 has decayed, how old is the skull?

SOLUTION We set this up as a standard decay problem, with $A = Pe^{-kt}$. We know that

A = amount of undecayed C-14 left = 0.05% of P, or 0.0005P.

$P = P$ (unspecified),

$t = ?$ (the unknown),

$k = \ldots$ well, we are told that the half-life is 5730 years. That is, $t_H = 5730$. To get k, we use the formula

$t_H \cdot k = \ln 2$.

So

$$5730k = \ln 2,$$

and

$$k = \frac{\ln 2}{5730} \approx 0.00012097.$$

Now that we have all the constants we need, we substitute them in the decay formula to get

$$0.0005P = Pe^{-0.00012097t}.$$

▼ *The ratio of Carbon-14 to Carbon-12 in any living organism is the same as the ratio in the environment, and this ratio has stayed fairly constant over the millennia, according to geophysical studies of the earth.

As usual, the Ps cancel, giving

$$0.0005 = e^{-0.00012097t}.$$

Taking the natural log,

$$\ln(0.0005) = \ln(e^{-0.00012097t}) = -0.00012097t.$$

Thus,

$$t = -\frac{\ln(0.0005)}{0.00012097} \approx 62{,}833.$$

We conclude that the skull is approximately 63,000 years old.

Before we go on... Let us check our answer.

$$Pe^{-62{,}833 \cdot 0.00012097} = 0.0005P \quad ✔$$

▼ **EXAMPLE 7**   Continuous Population Decline

Given the situation in Example 5, how long will it take for the flea population of 500,000 to get down to 100? (Recall that this was the question we answered graphically in Section 2.)

SOLUTION   Because this is not about half-life, we go back to the original decay equation

$$A = Pe^{-t/6}.$$

Substituting $A = 100$ and $P = 500{,}000$ gives

$$100 = 500{,}000e^{-t/6}.$$

Divide by 500,000.

$$0.0002 = e^{-t/6}$$

Take natural logs.

$$\ln 0.0002 = \ln(e^{-t/6})$$
$$= -\frac{t}{6}$$

Thus,

$$t = -6 \ln 0.0002 \approx 51.103 \text{ days}.$$

Before we go on... This confirms the approximate result we got graphically in the last section. We can check that this is an accurate solution.

$$500{,}000 e^{-51.103/6} \approx 100 \quad ✔$$

9.4 EXERCISES

APPLICATIONS

Government Bonds Exercises 1–10 are based on the following table, which lists annual percentage yields on government bonds in several countries.*

Country	U.S	Japan	Germany	Britain	Canada	Mexico
Yield (%)	5.16	3.86	5.97	6.81	6.58	13.7

1. Assuming that you invest $10,000 in U.S. government bonds, how long (to the nearest year) must you wait before your investment is worth $15,000 if the interest is compounded annually?

2. Assuming that you invest $10,000 in Japanese government bonds, how long (to the nearest year) must you wait before your investment is worth $15,000 if the interest is compounded annually?

3. If you invest $10,400 in German government bonds and the interest is compounded monthly, how many months will it take for your investment to grow to $20,000?

4. If you invest $10,400 in British government bonds and the interest is compounded monthly, how many months will it take for your investment to grow to $20,000?

5. How long, to the nearest year, will it take an investment in Mexico to double its value if the interest is compounded every 6 months?

6. How long, to the nearest year, will it take an investment in Canada to double its value if the interest is compounded every 6 months?

7. How long will it take a $1,000 investment in Canadian government bonds to be worth the same as a $800 investment in Mexican government bonds? (Assume all interest is compounded annually, and give the answer to the nearest year.)

8. How long will it take a $1,000 investment in Japanese government bonds to be worth the same as an $800 investment in U.S. government bonds? (Assume all interest is compounded annually, and give the answer to the nearest year.)

9. If the interest on U.S. government bonds is compounded continuously, how long will it take the value of an investment to double? (Give an answer correct to two decimal places.)

10. If the interest on Canadian government bonds is compounded continuously, how long will it take the value of an investment to double? (Give an answer correct to two decimal places)

11. *Investments* Berger Funds advertised that its Berger 100 Mutual Fund yielded an average interest rate of 15.9% from 1974 to 1993.† Assuming that this rate of return continues, how long, to the nearest tenth of a year, will it take a $5,000 investment in the fund to triple in value?

12. *Investments* Berger Funds also advertised that its Berger 101 Mutual Fund yielded an average interest rate of 14.1% from 1974 to 1993.† Assuming that this rate of return continues, how long, to the nearest tenth of a year, will it take a $5,000 investment in the fund to double in value?

13. *Influenza Epidemics* If each infected student at Enormous State University infects one healthy student with flu every day, how long (to the nearest day) after the first student is infected will it take for the epidemic to spread to 500 students?

▼ *Information was current as of October 18, 1993. (Source: Salamon Brothers, Mexican Government, S.G. Wartburg & Company, J.P. Morgan Global Research, as quoted in *The New York Times,* October 18, 1993.)

† According to an ad placed in *The New York Times* on September 26, 1993.

14. **Cold Epidemics** If each infected student at Enormous State University infects one healthy student with a cold every three days, how long (to the nearest day) after the first student is infected will it take for the epidemic to spread to 500 students?

15. **Investments** How long will it take a $500 investment to be worth $700 if it is continuously compounded at 10% per year? (Give an answer to two decimal places.)

16. **Investments** How long will it take a $500 investment to be worth $700 if it is continuously compounded at 15% per year? (Give an answer to two decimal places.)

17. **Investments** How long, to the nearest year, will it take an investment to triple if it is continuously compounded at 10% per year?

18. **Investments** How long, to the nearest year, will it take me to become a millionaire if I invest $1,000 at 10% interest compounded continuously?

19. **Stocks** Professor Stefan Schwartzenegger invests $1,000 in Tarnished Trade (TT) stocks, which are depreciating continuously at a rate of 20% per year. Luckily for Professor Schwartzenegger, TT declares bankruptcy at the instant his investment has declined to $666. How long after the initial investment did this occur?

20. **Investments**
    (a) The Hofstra choral society's investment of $100,000 in Tarnished Teak Enterprises is depreciating continuously at 16% per year. At this rate, how long, to the nearest year, will it take for the investment to be worth $1?
    (b) After six agonizing months watching its savings dwindle, the choral society pulls out what is left of its investment and buys Sammy Solid Trust Bonds, which earn a steady 6% interest, compounded semi-annually. How long will it take the choral society to recover its losses?

21. **Carbon Dating** The half-life of Carbon-14 is 5730 years. It is found that in a fossilized math professor's skull, 99.875% of the Carbon-14 has decayed since the time it was part of a living math professor (if you call that living). How old is the skull?

22. **Carbon Dating** A fossilized math book has just been dug up, and archeologists find that 99.5% of the Carbon-14 in the paper has decayed. When were the trees cut down to make the paper for the book?

23. **Automobiles** The rate of auto thefts is tripling every 6 months. Find the doubling time.

24. **Televisions** The rate of television thefts is doubling every 4 months. Find the tripling time.

25. **Bacteria** A bacteria culture starts with 1,000 bacteria and doubles in size every 3 hours. Approximately how many bacteria will there be after 2 days?

26. **Bacteria** A bacteria culture starts with 1,000 bacteria. Two hours later there are 1,500 bacteria. Assuming exponential growth, approximately how many critters will there be after two days?

27. **Frogs** Frogs in Nassau County have been breeding like flies! Two years ago, the pledge class of Epsilon Delta was instructed by the brothers to tag all the frogs residing on the ESU campus (Nassau County Branch) as an educational exercise. After several agonizing days, they managed to tag all 50,000 of them (with little Epsilon Delta Fraternity tags). This year's pledge class discovered that the tags had all fallen off, and they wound up tagging a total of 75,000 frogs. Assuming exponential population growth, how many tags should Epsilon Delta order for next year's pledge class?

28. **Flies** Flies in Suffolk County have been breeding like frogs! Three years ago the Health Commission caught 4,000 flies in one hour in a trap. This year it caught 7,000 flies in one hour. Assuming exponential population growth, how many flies should it expect to catch next year?

29. **U.S. Population** The U.S. population was 180,000,000 in 1960 and 250,000,000 in 1990. Assuming exponential population growth, what will be the population in the year 2010?

30. **World Population** World population was estimated at 1.6 billion people in 1900 and 5.3 billion people in 1990. Assuming exponential growth, when were there only two people in the world? Comment on your answer.

31. **Membership** Membership in the Enormous State University Choral Society has been increasing exponentially, doubling every semester. Assuming that there were a total of 20 Choral Society members at the start of this semester, how long do you estimate it would take for its membership to exceed the total U.S. population of 250,000,000?

32. **Membership** Having reached a membership of 10,000,000, the ESU Choral Society began charging

its members a $100 annual membership fee. From that point on, its membership started to decline exponentially, going down by half every year. How long did it take for the membership to reach its original total of 20?

33. *Radioactive Decay* Uranium-235 is used as fuel for some nuclear reactors. It has a half-life of 710 million years. How long would it take 10 grams of Uranium-235 to decay down to 1 gram?

34. *Radioactive Decay* Plutonium-239 is used as fuel for some nuclear reactors, and also as the fissionable material in atomic bombs. It has a half-life of 24,400 years. How long would it take 10 grams of Plutonium-239 to decay to 1 gram?

35. *Radioactive Decay* You are trying to determine the half-life of a new radioactive element you have isolated. You start with 1 gram, and 2 days later determine that it has decayed down to 0.7 gram. What is its half-life?

36. *Radioactive Decay* You have just isolated a new radioactive element. If you can only determine its half-life, you will win the Nobel prize in physics. You purify a sample of 2 grams. One of your colleagues steals half of it, but three days later you find that 0.1 gram of radioactive material is still left. What is the half-life?

Employment Exercises 37–42 are based on the following chart, which shows the number of people employed by Northrop Grumman *in Long Island, New York for the period from 1987 to 1995.**

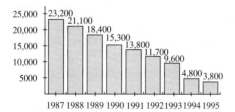

37. Use the 1987 and 1995 figures to construct an exponential decay model for employment at Northrop Grumman/Long Island, and use your model to predict the number of people employed there in 1997. Your model should have the form

$$P = P_0 e^{-kt},$$

where t is the number of years since 1987. (Round your answer to the nearest 100 people.)

38. Repeat Exercise 37 using the 1987 and 1994 figures.

39. The method of least squares exponential regression† gives the model

$$P = 28{,}131.7 e^{-0.220800t},$$

where t is the number of years since 1987. Referring to the model you obtained in Exercise 37, which model predicts a higher employment figure for 1997?

40. Repeat Exercise 39 using your model from Exercise 38.

41. According to the least squares model in Exercise 39, when will employment by *Northrop Grumman*/Long Island drop to 1,000 employees? (Give the answer to the nearest year.)

42. According to the least squares model in Exercise 39, when will employment by *Northrop Grumman*/Long Island drop to 500 employees? (Give the answer to the nearest year.)

43. *Temperature* You just bought a six-pack of non-alcoholic brew, but it is at room temperature (70°F). Since you desire a cold drink, you stick it in your refrigerator, which maintains 35°F. Now the brew will cool down in such a way that the difference between its temperature and the refrigerator's temperature will halve every hour. How long will it take to get down to a drinkable 40°F? [*Hint*: Use the difference D between the temperature of the brew and the refrigerator's temperature as the decaying quantity.]

44. *Temperature* You are heating up grög for your winter party. The drink starts out at 50°F, and you are heating the pot to 400°F. If the difference between the temperature of the grög and the (fixed) temperature of the pot halves every 30 minutes, how long will it take for the grög to reach its ideal temperature of 200°F? [*Hint*: Use the difference D between the temperature of the grög and the temperature of the pot as the decaying quantity.]

---

* The 1987–1994 figures show employment at Grumman Corp. before it was acquired by Northrop Corp. in 1994. The 1994 and 1995 figures are company projections. Source: Grumman Corp., Northrop Grumman Corp. *Newsday*, September 23, 1994, p. A5.

† This method is described in the chapter on functions of several variables.

COMMUNICATION AND REASONING EXERCISES

45. You have decided to use an exponential growth model to describe the student enrollment at your college. What information would you need, and how would you go about constructing the model?

46. Your friend has begun a project to use an exponential growth model to describe the student enrollment at his college. You come to the conclusion that an exponential model is inappropriate for modeling student enrollment. What arguments might you use to support this view?

47. You are midway through your project to describe student enrollment at your college, when a colleague argues that an exponential model is inappropriate. Are there any circumstances that might give you reasons to support the use of such a model?

48. Comment on the following statement: Ultimately, all exponential growth models fail.

▶ You're the Expert

EPIDEMICS

A mysterious epidemic is spreading through the population of the United States. An estimated 150,000,000 people are susceptible to this particular disease. There are 10,000 people already infected, and the number is doubling every two months. As advisor to the Surgeon General, it is your job to predict the course of the epidemic. In particular, the Surgeon General needs to know when the disease will have infected 1,000,000 people, when it will have reached 10,000,000, and when it will have affected 100,000,000 people.

Although the initial spread of an epidemic appears to be exponential, it cannot continue to be so, since the susceptible population is limited. A commonly used model for epidemics is the **logistic curve,** given by the function

$$A(t) = \frac{NP_0}{P_0 + (N - P_0)e^{-kt}},$$

where $A(t) = $ the infected population at time t, $P_0 = $ the population initially infected, and $N = $ the total susceptible population. The number k is a constant that governs the rate of spread of the epidemic. (Its exact meaning will be made clear in a moment.) The graph of this function is shown in Figure 1.

The initial part of this graph shows roughly exponential growth. To see why, multiply the top and bottom of the formula above by e^{kt}, which gives

$$A(t) = \frac{NP_0 e^{kt}}{P_0 e^{kt} + (N - P_0)}.$$

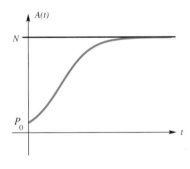

FIGURE 1

Now if P_0 is small compared to N, then for small t the denominator of this expression is approximately N. This gives

$$A(t) \approx P_0 e^{kt}$$

for small t; that is, $A(t)$ grows approximately exponentially in the early part of the epidemic. This also tells us that the number k is the fractional rate of growth that governs the early stages of the epidemic, when $A(t)$ is approximately exponential.

On the other hand, as t gets large, the term e^{-kt} in the original formula gets very small, and $A(t)$ gets close to the number N. You can see this in the rightmost part of the graph as $A(t)$ levels off under the line at height N.* Thus, the line $A = N$ is a horizontal asymptote.

The problems you now face are (1) to determine the constants to use in the logistic curve, and (2) to use the formula to predict the course of the epidemic. From the data you have, you know that $P_0 = 10,000$ and $N = 150,000,000$. The main problem is to find k. But remember that the initial spread of the epidemic is given by $A \approx P_0 e^{kt}$, so k is the fractional rate of growth for this exponential growth formula. You know that the infected population is doubling every two months, so the doubling time is $t_D = 2$ (months). Thus,

$$k = \frac{\ln 2}{t_D} = \frac{\ln 2}{2} = 0.3466.$$

Now we can write the logistic curve governing this particular epidemic:

$$A(t) = \frac{1,500,000,000,000}{10,000 + 149,990,000 e^{-0.3466t}}$$

$$= \frac{150,000,000}{1 + 14,999^{-0.3466t}}.$$

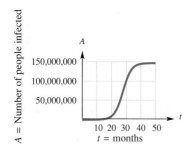

FIGURE 2

The graph of this function is shown in Figure 2.

To check your calculation, calculate $A(2) \approx 20,000$, showing the expected doubling after 2 months.

Now you tackle the question of prediction: When will the disease infect 1,000,000 people? This is asking: When is $A(t) = 1,000,000$? Setting $A(t)$ equal to 1,000,000, you get the equation

$$1,000,000 = \frac{150,000,000}{1 + 14,999 e^{-0.3466t}}.$$

Now you solve for t.

$$1,000,000(1 + 14,999 e^{-0.3466t}) = 150,000,000$$
$$1 + 14,999 e^{-0.3466t} = 150$$
$$14,999 e^{-0.3466t} = 149$$
$$e^{-0.3466t} = \frac{149}{14,999} = 0.009934$$

▼ *If you look at $N - A(t) = \dfrac{N(N - P_0)e^{-kt}}{P_0 + (N - P_0)e^{-kt}} \approx \dfrac{(N - P_0)}{P_0} e^{-kt}$ for large t, you can see that the difference between A and N decays exponentially for large t.

To solve this equation you take the natural log of both sides.

$$-0.3466t = \ln 0.009934 = -4.612$$

$$t = \frac{-4.612}{-0.3466} = 13.3 \text{ months}$$

So, in just over 13 months 1,000,000 people are expected to be infected.

When will 10,000,000 people be infected? Set $A(t) = 10,000,000$, and solve for t as before.

$$10,000,000 = \frac{150,000,000}{1 + 14,999e^{-0.3466t}}$$

$$1 + 14,999e^{-0.3466t} = 15$$

$$e^{-0.3466t} = \frac{14}{14,999} = 0.0009334$$

$$-0.3466t = \ln 0.0009334 = -6.977$$

$$t = \frac{-6.977}{-0.3466} = 20.1 \text{ months}$$

The epidemic will reach 10,000,000 people in just over 20 months. Finally, to determine when 100,000,000 will be infected, you solve

$$100,000,000 = \frac{150,000,000}{1 + 14,999e^{-0.3466t}}$$

and get $t = 29.7$ months. Notice that it took under 7 months to go from 1 million to 10 million infected people, but it takes over 9 months to from 10 million to 100 million. This is the slowing down of the epidemic from the exponential growth in its early stages.

Exercises

1. Track the later stages of the epidemic, by determining when 110 million people will be infected, when 120 million will be, when 130 million will be, and when 140 million will be infected.

2. Referring to Figure 2, when—to the nearest five months—would you estimate the epidemic to be spreading the fastest?

3. Give an estimate of the number of people you expect will be infected during the first year of the epidemic.

4. Explain the role of each of the constants in the logistic function

$$A(t) = \frac{NP_0}{P_0 + (N - P_0)e^{-kt}}.$$

5. Give a logistic model for the following: You have sold 100 "I ♥ Calculus" T-shirts, and sales are going up continuously at a rate of 30% per day. You estimate the total market for "I ♥ Calculus" T-shirts to be 3,000. Now use your model to predict when you will have sold 2,000 T-shirts.

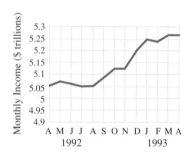

6. In Russia, the average consumer drank two servings of *Coca-Cola*® in 1993. This amount appeared to be increasing exponentially with a doubling time of two years.* Given a long-range market saturation estimate of 100 servings per year, find a logistic model for the consumption of *Coca-Cola* in Russia, and use your model to predict when the average consumption will be 50 servings per year.

7. The graph to the left shows the monthly total U.S. personal income from April 1, 1992 to April 1, 1993, in trillions of dollars.† Obtain a rough logistic model for the income by estimating P_0 and N from the graph, and experimenting with several values of k.

Review Exercises

Rewrite the equations in Exercises 1–6 in logarithmic form.

1. $2^{10} = 1024$
2. $5^{-2} = \frac{1}{25}$
3. $10^{-4} = 0.0001$
4. $y = e^{-2x}$
5. $y = x^e \; (x \neq 1)$
6. $y = 2x^{1/2}$

Rewrite the equations in Exercises 7–12 in exponential form.

7. $\log_3 81 = 4$
8. $\log_{1/3} 81 = -4$
9. $\log 1{,}000 = 3$
10. $\ln e^3 = 3$
11. $\log_x y = -1$
12. $y = -\ln x$

Solve the equations in Exercises 13–18 for x.

13. $4 = 3^x$
14. $10^x = 5^{2x}$
15. $9^x = 3^{-x^2}$
16. $10 = 2^{x^2}$
17. $1000 = 5e^{10x}$
18. $5 = 1000e^{-10x}$

Find the future value of the investments in Exercises 19–24.

19. $2,000 invested at 4% compounded monthly for 4 years
20. $3,000 invested at 5% compounded monthly for 3 years
21. $2,000 invested at 6.75% compounded daily (365 times per year) for 4 years
22. $3,000 invested at 8.25% compounded daily for 5 years
23. $2,000 invested at 3.75% compounded continuously for 4 years
24. $3,000 invested at 4.75% compounded continuously for 3 years

In each of Exercises 25–30, find the time required for the investment to reach the desired goal.

25. $2,000 invested at 4% compounded monthly, goal = $3,000
26. $3,000 invested at 5% compounded monthly, goal = $5,000
27. $2,000 invested at 6.75% compounded daily, goal = $3,000
28. $3,000 invested at 8.25% compounded daily, goal = $5,000
29. $2,000 invested at 3.75% compounded continuously, goal = $3,000
30. $3,000 invested at 4.75% compounded continuously, goal = $5,000

▼ *The doubling time is based on retail sales of Coca-Cola products in Russia. Sales in 1993 were double those in 1991, and were expected to double again by 1995. Source: *The New York Times*, September 26, 1994, p. D2.

† Source: National Association of Purchasing Management, Department of Commerce (*The New York Times*, June 2, 1993, Section 3, p. 1.) Our graph is a facsimile of the published graph (though accuracy may have suffered slightly).

APPLICATIONS

31. *Investments* Diane Blake invested her life savings, $10,000, in five-year CDs that paid 6% per year, compounded monthly. After five years she took her proceeds and put it all in the stock market, which proceeded to lose 1% each month. When she took out her money one year later, how much did she have left?

32. *Investments* Frank Capek invested his life savings, $7,500, in 10-year CDs that paid 7.25% per year, compounded monthly. After 10 years he took his proceeds and put it all in the bond market, which proceeded to lose 1.5% each month. When he took out his money one year later, how much did he have left?

33. *Inflation* Inflation in East Camelot is running at 4% per month. If a car costs $10,000 now, how much will an equivalent car cost in two years?

34. *Inflation* Inflation in West Camelot is running at 3% per month. How long will it take prices to double?

35. *Rabbits* The rabbits on my farm are reproducing like... well, rabbits. I started out with 10, but 4 months later had 30. Assuming exponential growth, when will I be completely overrun with 1,000 bunnies?

36. *Investments* The $1,000 you invested in mutual funds has grown to $1,200 in half a year. Assuming that this rate of return continues, how large will your investment be after 5 years total?

37. *Radioactive Decay* Costenobelium is a highly unstable element. 10µg will decay down to 1µg in one second. What is its half-life?

38. *Radioactive Decay* Wanerium is a highly stable element. 10µg will decay down to 9µg after 10,000 years. What is its half-life?

39. *Investments* Utopia Investments offers an account paying 25% compounded annually, while Erewhon Investments offers an account paying 24% compounded continuously. Which is the better investment?

40. *Investments* Pomegranate Computer's stock is depreciating at a rate of 2% per month, while Mega-Soft's stock dropped 12.5% in the last half year. Which is the worse investment?

41. *Radioactive Decay* Potassium-40 has a half-life of 1.28×10^9 years. If 95% of the Potassium-40 present in a rock when it was formed has decayed, how old is the rock?

42. *Radioactive Decay* A sample of 1 milligram of Einsteinium-246 will decay to 0.9 milligram in 1.11 minutes. How long will it take to decay to 0.1 milligram?

43. *Investments (from the GMAT)* Each month for 6 months the amount of money in a benefit fund is doubled. At the end of the 6 months there is a total of $640 in the fund. How much money was in the fund at the end of 3 months?

44. *Savings (from the GMAT)* On July 1, 1982, Ms. Fox deposited $10,000 in a new account at the annual interest rate of 12 percent compounded monthly. If no additional deposits or withdrawals were made, and if interest was credited on the last day of each month, what was the amount of money in the account on September 1, 1982?

45. *Bacteria (from the GMAT)* The population of a bacteria culture doubles every 2 minutes. Approximately how many minutes will it take for the population to grow from 1,000 to 500,000 bacteria?

46. *Population (based on a question from the GMAT)* The population of a small town doubles every 10 years. Approximately how many years will it take for the population to grow from 10,000 to 25,000 people?

47. *Interest (from the GRE economics exam)* If the interest rate is 10 percent, what is the present discounted value of a dollar due two years from now?

48. *Exponential Decay (from the GRE economics exam)* To estimate the rate at which new instruments will have to be retired, a telephone company uses the "survivor curve":
$$L_x = L_0 e^{-x/t},$$
where
L_x = number of survivors at age x,
L_0 = number of initial installations,
t = average life in years.

All of the following are implied by the curve *except*:
(a) Some of the equipment is retired during the first year of service.
(b) Some equipment survives three average lives.
(c) More than half the equipment survives the average life.
(d) Increasing the average life of equipment by using more durable materials would increase the number surviving at every age.
(e) The number of survivors never reaches zero.

INTRODUCTION ▶ With this chapter we begin to study calculus—one of the most important, most useful, most *used* parts of mathematics. In the world around us, everything is changing, and calculus, at its heart, is the study of *how* things change; how fast and in what direction. Is the Dow Jones average going up, and if so, how fast? If I raise my prices, how many customers will I lose? If I launch this missile here, how high will it go, and where will it come down?

Calculus is concerned first with the *rate of change* of a function. We have already discussed this for linear functions (straight lines), where the *slope* measures the rate of change. But this works only because a straight line maintains the same rate of change along its whole length. Other functions rise faster here than there—or rise in one place and fall in another—so the rate of change varies along the graph. The first and greatest achievement of calculus is that it provides a systematic and straightforward way of *calculating* (hence the name) these rates of change. To describe a changing world, we need a language of change, and that is what calculus gives us.

The history of calculus is an interesting story of personalities, intellectual movements, and controversy. Credit for its invention is given to two mathematicians: Isaac Newton (1642–1727) and Gottfried Leibniz (1646–1716). Newton, an English mathematician and scientist, was the first to invent calculus, probably in the 1660s. We say "probably" because, for various reasons, he did not publish his ideas until much later. This allowed Leibniz to publish his own version of calculus first, in 1684. Fifteen years later, stirred up by nationalist fervor in England and on the continent, controversy erupted over who should get the credit for its invention. The debate got so heated that the Royal Society (of which Newton and Leibniz were both members) set up a commission to investigate the question. The commission decided in favor of Newton, who just happened to be president of the Society at the time. The consensus today is that both mathematicians deserve credit, since they both came to the same conclusions working independently. This is not really surprising: both built on well-known work of other people, and it was almost inevitable that someone would put it all together at about that time.

The controversy did have one unfortunate consequence for English mathematics. Newton's and Leibniz's versions of calculus were not identical. In particular, Leibniz had by far the better notation. This was mainly a consequence of Leibniz's work in philosophy: he believed that all knowledge could be expressed and manipulated in symbolic form and that a good notation would make this easier. He was therefore very interested in finding good notations for his mathematics and worked particularly hard on a notation for calculus. In fact, we still use his notation and no longer use Newton's. The problem for English mathematicians was that they felt compelled to use Newton's version of calculus. Their nationalist fervor also caused them to cut themselves off from the explosion of work on calculus that took place on the continent in the early 1700s, setting back English mathematics for decades.

Another controversy associated with calculus was much more important in the development of mathematics and still influences the way we teach calculus today. Calculus is, a formal way of calculating the rate of change of a function if the formula for that function is known. You can use it without knowing why it works, just as you can do long division or punch the buttons of a calculator without worrying about why you get the right answer. But mathematicians and philosophers of the eighteenth century wanted to know *why* calculus works. (This was part of the intellectual climate of the time. Much of this controversy erupted around the time of the Enlightenment, beginning in the 1730s. We have inherited much of the rationalism of that age.) Nobody could deny that calculus worked. Newton had used it to explain the motions of the planets, and successful applications soon came fast and furiously. Neither Newton nor Leibniz had provided satisfactory justification for their work, and the annoying fact remained that mathematicians could not adequately explain *why* the techniques of calculus worked.* Not until the 1820s did Augustin Louis Cauchy (1789–1857) publish papers showing how calculus could be justified by the use of *limits*. He also gave the careful definition of limits that we use to this day. His work began the great nineteenth-century project of introducing logical rigor into mathematics, establishing a precedent that mathematicians follow to this day.

▼ *Similar problems occur today in the field of physics, which abounds with mathematical techniques that have not been rigorously studied by mathematicians.

10.1 RATE OF CHANGE AND THE DERIVATIVE

There are many situations in which we would like to know how fast a quantity is changing. We saw examples of this when we talked about straight lines and linear functions. In that case, the rate of change is measured by the *slope* of the line. Our first goal is to see what we can use in place of the slope when we have a function that is not linear.

▼ **EXAMPLE 1**   Rate of Change

Suppose we monitor the stock market very closely during one rather active week, and we find that during that week the value of the Dow Jones Average could be accurately described by the function $A(t) = 2{,}500 + 500t - 100t^2$, where t is time in days. The graph of this function is shown in Figure 1 (this is, of course, a parabola).

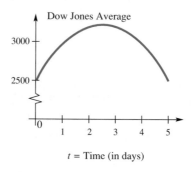

FIGURE 1

Looking at the graph, we can see that the Dow Jones Average ("Dow") rose rather rapidly at the beginning of the week, but by the middle of the week the rise had slowed, until the market faltered and the Dow began to fall more and more rapidly towards the end of the week. Can we calculate exactly how fast the Dow was rising at the beginning of the second day ($t = 1$), and how fast it was falling at the beginning of the fourth day ($t = 3$)?

SOLUTION   One way to answer this question is as follows. Let us first see how fast the Dow rose during the second day.

Start of day 2 ($t = 1$): $A(1) = 2{,}900$
End of day 2 ($t = 2$): $A(2) = 3{,}100$

Change during day 2: $A(2) - A(1) = 3{,}100 - 2{,}900 = 200$

Thus, the Dow increased by 200 points in one day, for a rate of change of 200 points per day. This was its **average rate of change** during the second day.

Now, if we look closely at the graph, we can see that it was actually rising faster at the beginning of the second day than at the end, so its rate of change must have been greater at the beginning of the day than we just calculated. To get a better idea of that rate, let us look at only the first half of the day, from $t = 1$ to $t = 1.5$.

Start of day 2 ($t = 1$): $\quad A(1) = 2{,}900$
Middle of day 2 ($t = 1.5$): $\quad A(1.5) = 3{,}025$

Change during first half of day 2: $A(1.5) - A(1) = 3{,}025 - 2{,}900 = 125$

Thus, the Dow rose 125 points in *half* a day. To translate this into points per day, we divide by the number of days—in this case, 0.5:

$$\text{Average rate of change of } A(t) = \frac{\text{Change in } A(t)}{\text{Time}}$$

$$= \frac{125}{0.5} = 250 \text{ points per day.}$$

As expected, this was greater than the average rate for the whole day.

Still, the market was rising faster at the beginning of the day than at noon, so the rate at which it was rising *right at the beginning of the day* must be even greater than the 250 we just calculated. To get closer to the rate we seek, we must do the same calculation over a smaller part of the day.

To organize our work a little, notice that we are looking at the time interval from time 1 to time $1 + h$, where h is the fraction of the day we are using. We have already calculated the rates for $h = 1$ and $h = 0.5$. The calculation we have been doing is as follows:

$$\text{Average rate of change of } A(t) \text{ from time 1 to time } 1 + h = \frac{A(1 + h) - A(1)}{h}.$$

Here is a table of the average rates of change we get when we choose smaller and smaller values for h:

h	1	0.5	0.1	0.01	0.001	0.0001
$A(1 + h) - A(1)$	200	125	29	2.99	0.2999	0.029999
Ave. rate of change $= \dfrac{A(1 + h) - A(1)}{h}$	200	250	290	299	299.9	299.99

As h decreases, the average rate of change seems to be getting closer and closer to 300 points per day. So it seems reasonable to say that the **instantaneous rate of change** at time $t = 1$ was 300 points per day. This is how fast the Dow was rising at the *instant* day 2 began ($t = 1$).

To see how fast the market was falling at the beginning of the fourth day ($t = 3$), we can do the same calculations with the time intervals from $t = 3$ to $t = 3 + h$, as follows.

$$\text{Average rate of change from time 3 to time } 3 + h = \frac{A(3 + h) - A(3)}{h}.$$

Here is a table of the average rates of change we get for various values of h.

h	1	0.5	0.1	0.01	0.001	0.0001
$A(3+h) - A(3)$	-200	-75	-11	-1.01	-0.1001	-0.010001
Ave. rate of change $= \dfrac{A(3+h) - A(3)}{h}$	-200	-150	-110	-101	-100.1	-100.01

$$\underbrace{A(4) - A(3)}_{1} \quad \underbrace{A(3.5) - A(3)}_{0.5} \quad \underbrace{A(3.1) - A(3)}_{0.1} \quad \underbrace{A(3.01) - A(3)}_{0.01} \quad \underbrace{A(3.001) - A(3)}_{0.001} \quad \underbrace{A(3.0001) - A(3)}_{0.0001}$$

Again, as h decreases, the average rate of change seems to be settling down to a "limiting value," here -100 points per day. Notice that this rate is negative, because the market is *falling* at the beginning of the third day.

Before we go on...

Q Why do we "sneak up" on the value $h = 0$, rather than just setting h equal to 0?

A Look at the formula $(A(1 + h) - A(1))/h$. We cannot simply substitute $h = 0$ because h appears in the denominator and we cannot divide by 0. More fundamentally, in order to see how the Dow is changing, we need to look at it at two different times, so that we can see some change. The single value $A(1) = 2,900$ tells us the Dow at the beginning of the second day, but it tells us absolutely nothing about how fast it is changing.

Notice also that we have measured the rates of change in points per day, that is, *units of A per unit of t*.

With a graphing calculator, you can automate these calculations. A computer spreadsheet is particularly useful for generating tables such as those above. Consult the Appendix for details and examples of the use of a graphing calculator or a spreadsheet program to calculate average rates of change.

We shall see computations similar to the previous example in many different situations. Let us summarize and introduce some terminology.

AVERAGE RATE OF CHANGE; DIFFERENCE QUOTIENT

The **average rate of change** of the function f over the interval $[x, x + h]$ from x to $x + h$ is

$$\text{Average rate of change} = \frac{f(x + h) - f(x)}{h}.$$

We also call this average rate of change the **difference quotient** of f over the interval $[x, x + h]$. Its units of measurement are units of f per unit of x.

The process of letting h get smaller and smaller is called taking the **limit** as h approaches 0. We write "$h \to 0$" as shorthand for "h approaches 0." Taking the limit of the average rates of change gives us the instantaneous rate of change. Here is our notation for this limit.

> **INSTANTANEOUS RATE OF CHANGE; DERIVATIVE**
>
> The **instantaneous rate of change** of f at x is given by taking the limit of the average rates of change (given by the difference quotient) as h approaches 0. We write:
>
> $$\text{Instantaneous rate of change} = \lim_{h \to 0} \frac{f(x+h) - f(x)}{h}.$$
>
> We also call this instantaneous rate of change the **derivative** of f at x and write it as $f'(x)$ (read "f prime of x"). Thus,
>
> $$f'(x) = \lim_{h \to 0} \frac{f(x+h) - f(x)}{h}.$$
>
> The units of $f'(x)$ are units of f per unit of x.

▶ NOTES
1. For now, we shall trust our intuition when it comes to limits. We shall discuss limits in detail in the third section of this chapter.
2. $f'(x)$ is a number we can calculate, or at least approximate, for various values of x, as we have done in the example above. This means that it depends on x, or is a function of x. In other words, f' is another function of x. An old name for this is "the function *derived from f '*", which has been shortened to the *derivative* of f.
3. Finding the derivative of f is called **differentiating** f. ◀

EXAMPLE 2 Demand

The economist Henry Schultz calculated the following demand function for corn:

$$q(p) = \frac{176{,}000}{p^{0.77}}.$$

Here, p is the price (in dollars) per bushel of corn and q is the number of bushels of corn that could be sold at the price p in one year.* How sensitive is demand to price when $p = \$2$ per bushel, and how sensitive is it when $p = \$8$ per bushel?

▼ * This demand function is based on data for the period 1915–1929. Source: Henry Schultz: *The Theory and Measurement of Demand* [as cited in *Introduction to Mathematical Economics* by A. L. Ostrosky, Jr., and J.V. Koch (Waveland Press, Prospect Heights, Illinois, 1979.)]

SOLUTION When we ask how *sensitive* demand is to price, we would like to know how much the demand will change if the price changes. More precisely, we want to know the *rate* at which the demand will change as the price changes.

From the formula above, we have

$$\text{Rate of change of } q = \text{Derivative of } q, \text{ evaluated at } p = 2$$

$$= q'(2) = \lim_{h \to 0} \frac{q(2+h) - q(2)}{h}.$$

The table below shows the average rate of change,

$$\frac{q(2+h) - q(2)}{h} = \frac{\left[\frac{176{,}000}{(2+h)^{0.77}} - \frac{176{,}000}{2^{0.77}} \right]}{h},$$

for various values of h. (On a graphing calculator, enter this function of h as follows, using x instead of h.

$$Y_1 = (176000/(2 + X)\wedge 0.77 - 176000/2\wedge 0.77)/X$$

and obtain the averages by having the calculator evaluate Y_1 at $X = 1, 0.1, 0.01, 0.001,$ and 0.0001.)

h	1	0.1	0.01	0.001	0.0001
Ave. rate of change	−27,678	−38,055	−39,561	−39,718	−39,734

How do we interpret these figures? First, recall that we measure the rate of change in units of q per unit of p; that is, in (demand for) bushels per \$1 increase in price. Thus, the figure for $h = 1$ tells us that when the price of corn is raised from \$2 per bushel to $2 + h = \$3$, demand will drop by an average of 27,678 bushels per \$1 increase in price. Similarly, the figure for $h = 0.1$ tells us that when the price of corn is raised from \$2 per bushel to $2 + h = \$2.10$, demand will drop by an average of 38,055 bushels per \$1 increase in price.

To calculate the instantaneous rate of change $q'(2)$, we need to know what number the average rate approaches as h approaches 0. Although this is not obvious from the table, we can say with some certainty that to three significant digits, $q'(2) \approx -39{,}700$ bushels per dollar. Thus, at a price of \$2 per bushel, the demand for corn will fall at a rate of approximately 39,700 bushels for each dollar increase in price.

We should also be interested in what happens if the price *decreases*. This means that we should also look at negative h:

h	−1	−0.1	−0.01	−0.001	−0.0001
Ave. rate of change	−72,791	−41,579	−39,912	−39,753	−39,737

10.1 Rate of Change and the Derivative

This agrees with what we saw when we looked at positive h: as h approaches 0, the average rate of change approaches about $-39{,}700$. This means that, if the price were to *drop*, the demand would *rise* at a rate of 39,700 bushels per dollar drop in price (draw a graph and think about this for a while).

We can repeat all of these computations for a price of \$8 per bushel. Notice, though, that we are really interested in the limit of the difference quotient, as we were above. We can *approximate* the value of the limit by substituting a small value of h. This is essentially what we did above: What we really paid attention to was the value of the difference quotient for $h = 0.0001$. So, let us compute

$$\frac{q(8 + 0.0001) - q(8)}{0.0001} \approx -3{,}416$$

Therefore, at a price of \$8 per bushel, the demand for corn will fall by approximately 3,416 bushels per \$1 increase in price.

Before we go on... When doing these calculations, it is important to record the calculations to enough significant digits to see the change. For example, in the last calculation the values of $q(8.0001)$ and $q(8)$ differ only in the fifth significant digit. If we write down only three or four significant digits, we would not notice any change at all. It is best to let your calculator or computer use as many digits as it calculates until the very end, when you may round the final answer to as many digits as are reasonable (usually three or four, depending on the accuracy of the data with which you are working).

Q What is the difference in meaning between $q(2)$ and $q'(2)$?

A Briefly, $q(2)$ is the *value of q* when $p = 2$, while $q'(2)$ is the *rate at which q is changing* when $p = 2$. In the above example, $q(2) = 176{,}000/2^{0.77} \approx 103{,}209$ bushels. This means that at a price of \$2 per bushel, the demand for corn (measured in annual sales) is 103,209 bushels. On the other hand, $q'(2) \approx -39{,}700$ bushels per dollar. This means that at a price of \$2 per bushel, the demand *is dropping by* 39,700 *bushels per* \$1 *increase in price.*

Q Do we always need to make a table of difference quotients with decreasing values of h in order to calculate an approximate value for the derivative, or is there a quicker way?

A As we saw in the above example, we can *approximate* the value of the derivative by using a single, small, value of h such as $h = 0.0001$. Graphing calculators do it this way (although they use a far smaller value). The only problem with using $h = 0.0001$ is that (1) we do not get an exact answer, and (2) it is not clear just how accurate our answer is.

Thus we have the following.

> **CALCULATING AN APPROXIMATE VALUE FOR THE DERIVATIVE**
>
> We can calculate an approximate value of $f'(x)$ by using the formula
> $$f'(x) \approx \frac{f(x + 0.0001) - f(x)}{0.0001}.$$

EXAMPLE 3

Use a graphing calculator to approximate $f'(5)$ if $f(x) = x^2 - x^{-0.4}$.

SOLUTION The approximation given above is

$$f'(x) \approx \frac{f(x + 0.0001) - f(x)}{0.0001}$$

$$= \frac{(x + 0.0001)^2 - (x + 0.0001)^{-0.4} - (x^2 - x^{-0.4})}{0.0001}$$

$$= \frac{(x + 0.0001)^2 - (x + 0.0001)^{-0.4} - x^2 + x^{-0.4}}{0.0001}$$

We enter this on the graphing calculator as

$$Y_1 = ((X+0.0001)\wedge 2 - (X+0.0001)\wedge(-0.4) - X\wedge 2 + X\wedge(-0.4))/0.0001$$

Then set $x = 5$ (on the home screen) and evaluate Y_1 by entering*

$$5 \to X$$
$$Y_1$$

to obtain $f'(5) \approx 10.04212$. This answer is accurate to three decimal places; in fact, $f'(5) = 10.04202\ldots$

DELTA NOTATION

We have used the notation $f'(x)$ for the derivative of f at x, but there is another interesting notation. We can write $\Delta f = f(x + h) - f(x)$ for the *change in* f. The Greek letter delta (Δ) is often used to stand for the phrase "the change in." Likewise, we can write $\Delta x = (x + h) - x = h$ for the change in x. Then the difference quotient is

$$\frac{\Delta f}{\Delta x} = \frac{f(x + h) - f(x)}{h},$$

▼ *This applies to TI models of graphing calculators.

which leads to the notation

$$\frac{df}{dx} = \lim_{\Delta x \to 0} \frac{\Delta f}{\Delta x}$$

for the derivative. That is, df/dx is just another notation for $f'(x)$. You should not really think of this as a quotient, but simply as a notation. We read df/dx as "the derivative of f with respect to x." It is necessary to include the phrase "with respect to x" when there are other variables around, and in any case it reminds us which variable is being used as the independent variable. As a last piece of notation, the phrase "the derivative with respect to x" is often abbreviated as d/dx, so that we could write

$$f'(x) = \frac{d}{dx}(f(x)) = \frac{df}{dx}.$$

This notation is most useful if we have a formula for a function that is still unnamed. For example, to refer to the derivative of the function with the formula x^3, we could write

$$\frac{d}{dx}(x^3).$$

We shall say more about this notation in Section 4.

Let us finish this section with another important application.

▼ **EXAMPLE 4** Velocity

If I throw a ball upward at a speed of 100 ft/s, our physicist friends tell us that its height t seconds later will be $s = 100t - 16t^2$. How fast will the ball be rising exactly 2 seconds after I throw it?

SOLUTION The graph of the ball's height as a function of time is shown in Figure 2. Asking "How fast will it be rising?" is really asking for the rate of change of height with respect to time. Why? Think about average velocity for a moment. If we wanted to compute the average velocity of the ball from time 2 to time 3, say, we would first compute the distance the ball rose in that time, or the change in height:

$$\Delta s = s(3) - s(2) = 156 - 136 = 20 \text{ ft}.$$

Since it rose 20 feet in $\Delta t = 1$ second, we use the formula *Speed = Distance/Time* to get an average velocity of

$$\text{Average velocity} = \frac{\Delta s}{\Delta t} = \frac{20}{1} = 20 \text{ ft/s}$$

from time $t = 2$ to $t = 3$. Note that this is just the difference quotient. Now, to get the **instantaneous velocity** at $t = 2$, we take the limit. In other words, we need to calculate the derivative ds/dt at $t = 2$.

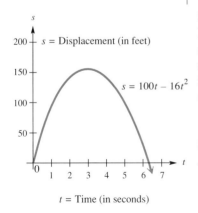

FIGURE 2

Consider each of the functions in Exercises 29–34 as representing the cost to manufacture x items. Find the average costs of manufacturing h more items (i.e., the average rate of change of the total cost) at a production level of x, where x is as indicated and h = 50, 10, and 1. Also, estimate the rate of change of the total cost at the given production level x.

29. $C(x) = 10{,}000 + 5x - \dfrac{x^2}{10{,}000}$; $x = 1{,}000$

30. $C(x) = 20{,}000 + 7x - \dfrac{x^2}{20{,}000}$; $x = 10{,}000$

31. $C(x) = 1{,}000 + 10x - \sqrt{x}$; $x = 100$

32. $C(x) = 10{,}000 + 20x - \sqrt{x}$; $x = 10$

33. $C(x) = 15{,}000 + 100x + \dfrac{1{,}000}{x}$; $x = 100$

34. $C(x) = 20{,}000 + 50x + \dfrac{10{,}000}{x}$; $x = 100$

APPLICATIONS

35. *Demand* Suppose the demand for a new brand of sneakers is given by
$$q = \dfrac{5{,}000{,}000}{p},$$
where p is the price per pair of sneakers, in dollars. Find $q(100)$ and estimate $q'(100)$, and interpret your answers.

36. *Demand* Suppose the demand for an old brand of TV is given by
$$q = \dfrac{100{,}000}{p + 10},$$
where p is the price per TV set, in dollars. Find $q(190)$ and estimate $q'(190)$, and interpret your answers.

37. *Profit* Your monthly profit (in dollars) from selling magazines is given by
$$P = 5n + \sqrt{n},$$
where n is the number of magazines you sell in a month. If you are currently selling $n = 50$ magazines per month, find P and estimate dP/dn. Interpret your answers.

38. *Profit* Your monthly profit (in dollars) from your newspaper route is given by
$$P = 2n - \sqrt{n},$$
where n is the number of subscribers on your route. If you currently have 100 subscribers, find P and dP/dn. Interpret your answers.

39. *Popularity* A study of enrollment at a certain university shows that there is a relationship between enrollment in each professor's classes and the average grade that the professor awards. Measuring grades g on a scale of 0 to 4, the relationship can be expressed by the equation
$$E = 20\sqrt[4]{g},$$
where E is the professor's average enrollment per class. For a professor whose average grade awarded is $g = 2.5$, find E and dE/dg, and interpret your answers.

40. *Popularity* A study of enrollment at another university shows that, again, there is a relationship between enrollment in each professor's classes and the average grade that the professor awards. Measuring grades g on a scale of 0 to 4, the relationship this time can be expressed by the equation
$$E = 15\sqrt[3]{g^2},$$
where E is the professor's average enrollment per class. For a professor whose average grade awarded is $g = 2.5$, find E and dE/dg, and interpret your answers.

41. *Market Average* Joe Downs runs a small investment company from his basement. Every week he

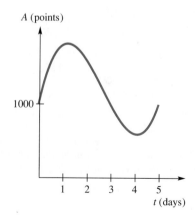

publishes a report on the success of his investments, including the progress of the infamous Joe Downs Average. At the end of one particularly memorable week he reported that the Average for that week had the value $A(t) = 1000 + 1500t - 800t^2 + 100t^3$ points, where t represents the number of days into the week; t ranges from 0 at the beginning of the week to 5 at end of the week. The graph of A is shown above. During the upswing at the beginning of the week, say halfway through the first day, approximately how fast was the value of the Average increasing?

42. *Market Average* Referring to the Joe Downs Average in the previous exercise: During the downswing, say, at the end of the third day, how fast was the value of the Average decreasing?

43. *Learning to Speak* Let $p(t)$ represent the percentage of children who are able to speak at the age of t months.
    (a) It is found that $p(10) = 60$ and $p'(10) = 18.2$.* What does this mean?
    (b) As t increases, what happens to $p(t)$ and $p'(t)$?

44. *Learning to Read* Let $p(t)$ represent the number of children in your class who learned to read at the age of t years.
    (a) Assuming that everyone in your class could read by the age of 7, what does this tell you about $p(7)$ and $p'(7)$?
    (b) Assuming that 25% of the people in your class could read by the age of 5, and that 25.3% of them could read by the age of 5 years and one month, estimate $p'(5)$, remembering to give its units.

45. *Biology—Reproduction* The Verhulst model for population growth specifies the reproductive rate of an organism as a function of the total population according to the following formula:
$$R(p) = \frac{r}{1 + kp}.$$
Here, p is the total population in thousands of organisms, r and k are constants that depend on the particular circumstances and the organism being studied, and $R(p)$ is the reproduction rate in thousands of organisms per hour.† Assume that $k = 0.125$ and $r = 45$ for a particular population, and estimate $R'(4)$. Interpret the result.

46. *Biology—Reproduction* Another model, the predator satiation model for population growth, specifies that the reproductive rate of an organism as a function of the total population varies according to the following formula:
$$R(p) = \frac{rp}{1 + kp}.$$
Here, p is the total population in thousands of organisms, r and k are constants that depend on the particular circumstances and the organism being studied, and $R(p)$ is the reproduction rate in new organisms per hour.† Assume that $k = 0.2$ and $r = 0.08$ for a particular population, and estimate $R'(2)$. Interpret the result.

47. *The Theory of Relativity* In science fiction terminology, a speed of *warp 1* is the speed of light—about 3×10^8 meters per second. (Thus, for instance, a speed of warp 0.8 corresponds to 80% of the speed of light—about 2.4×10^8 meters per second.) According to Einstein's Special Theory of Relativity, a moving object appears to get shorter to a stationary observer as its speed approaches that of light. If a rocket ship whose length is l_0 meters at rest travels at a speed of warp p, its length in meters, as measured by a stationary observer, will be given by
$$l(p) = l_0\sqrt{1 - p^2}, \text{ with domain } [0, 1].$$
Assuming that your rocket is 100 meters long ($l_0 = 100$), estimate $l(0.95)$ and $l'(0.95)$. What do these figures tell you?

48. *Newton's Law of Gravity* The gravitational force exerted on a particle with mass m by another particle with mass M is given by the following function of distance:
$$F(r) = G\frac{Mm}{r^2}, \text{ with domain } (0, +\infty).$$
Here, r is the distance between the two particles in meters, the masses M and m are given in kilograms, $G \approx 0.0000000000667 = 6.67 \times 10^{-11}$, and the resulting force is given in newtons. Given that M and m are both 1,000 kilograms, estimate $F(10)$ and $F'(10)$. What do these figures tell you?

▼ *Based on data presented in the article *The Emergence of Intelligence* by William H. Calvin, *Scientific American*, October, 1994, pp. 101–7.

† Source: *Mathematics in Medicine and the Life Sciences* by F.C. Hoppenstaedt and C.S. Peskin (Springer-Verlag, New York, 1992) pp 20-22.

Exercises 49–56 are based on logarithmic and exponential function.

In the following exercises, estimate the derivative of the given function at the indicated point.

49. $f(x) = e^x$; $x = 0$ **50.** $f(x) = 2e^x$; $x = 1$
51. $f(x) = \ln x$; $x = 1$ **52.** $f(x) = \ln x$; $x = 2$

53. *Sales* Weekly sales of a new brand of sneakers are given by
$$S(t) = 200 - 150e^{-t/10},$$
pairs sold per week, where t is the number of weeks since the introduction of the brand. Estimate $S(5)$ and estimate $S'(5)$, and interpret your answers.

54. *Sales* Weekly sales of an old brand of TV are given by
$$S(t) = 100e^{-t/5},$$
sets per week, where t is the number of weeks after the introduction of a competing brand. Estimate $S(5)$, and $S'(5)$ and interpret your answers.

55. *Logistic Growth in Demand* The demand for a new product can be modeled by a **logistic** curve of the form

$$q(t) = \frac{N}{1 + ke^{-rt}},$$

where $q(t)$ is the total number of units sold t months after the introduction of the new product, and N, k, and r are constants that depend on the product and the market. Assume that the demand for video game units is determined by the above formula, with $N = 10{,}000$, $k = 0.5$, and $r = 0.4$. Estimate $q(2)$ and $q'(2)$, and interpret the results.

56. *Information Highway* The amount of information transmitted each month on the National Science Foundation's Internet network for the years 1988–1994 can be modeled by the equation

$$q(t) = \frac{2e^{0.69t}}{3 + 1.5e^{-0.4t}},$$

where q is the amount of information transmitted each month in billions of data packets, and t is the number of years since the start of 1988.* Estimate the number of data packets transmitted during the first month of 1990, and also the rate at which this number was increasing.

COMMUNICATION AND REASONING EXERCISES

57. Give an algebraic explanation of the fact that if f is a linear function, then the average rate of change over any interval equals the instantaneous rate of change at any point.

58. Give a geometric explanation of the fact that, if f is a linear function, then the average rate of change over any interval equals the instantaneous rate of change at any point.

59. Explain why we cannot put $h = 0$ in the formula

$$f'(x) = \lim_{h \to 0} \frac{f(x+h) - f(x)}{h}$$

for the derivative of f.

60. A manufacturer has manufactured 10,000 surfboards, but the manufacturing level is decreasing at a rate of 4,000 per year. What does this tell you about the number $N(t)$ of surfboards produced as a function of time t (years)?

▶ 10.2 GEOMETRIC INTERPRETATION OF THE DERIVATIVE

As we mentioned at the beginning of the previous section, the rate of change of a linear function is the slope of the corresponding line. For a function whose graph is not a straight line, there is no notion of *the* slope of the graph. Look at the line drawn in Figure 1.

This line is just as steep at the point P as it is at the point Q. Put another way, the line is rising just as fast at P as it is at Q. Now look at the curve shown in Figure 2.

▼ *This is the authors' model, based on figures published in *The New York Times*, Nov. 3, 1993.

10.2 Geometric Interpretation of the Derivative

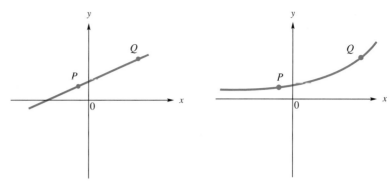

FIGURE 1   **FIGURE 2**

This graph is decidedly steeper at Q than it is at P. This suggests that the "slope" of the curve increases from left to right, and so we cannot assign a single number to measure the steepness of the whole graph. Instead, we shall have to assign a number *to each point of the graph* to measure its steepness *at that point*.

Q Just what do we *mean* by the steepness of a graph at a specified point? In other words, how can we measure it?

A Here is a way to measure steepness. Since we all know how to measure the steepness of a *line* (by its slope), and since we would like to build on what we already know, we will say that the steepness of a graph at a specified point is the slope of the *tangent line to the graph at that point*.

Remember that a tangent line to a circle is a line that touches the circle in just one point. A tangent line gives the circle "a glancing blow," as shown in Figure 3.

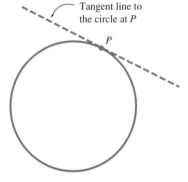

FIGURE 3

A tangent line to an arbitrary curve is similar. Figure 4 shows the lines tangent to the parabola $y = x^2$ at the five points $(-2, 4)$, $(-1, 1)$, $(0, 0)$, $(1, 1)$, and $(2, 4)$.

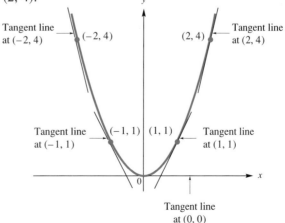

FIGURE 4

The steepness of the parabola at $(-2, 4)$ is the slope of the tangent line we have drawn there. Notice that it has a negative slope of large magnitude. Similarly, the steepness of the parabola at $(-1, 1)$ is the slope of the tangent line at $(-1, 1)$. This line has a negative slope of smaller magnitude. The steepness of the graph at $(0, 0)$ is 0, since the tangent line there is horizontal (it happens to be the x-axis). The steepness of the graph at $(1, 1)$ is positive, since the tangent line there has a positive slope, while the steepness of the graph at $(2, 4)$ is also positive, but larger.

In summary:

> **STEEPNESS OF A GRAPH**
>
> The steepness of a graph at a point P is measured by the slope of the tangent to the curve at P.

We shall often use the abbreviation m_{tan} to denote the slope of the tangent to a graph at a specified point.

The problem we are now faced with is *how to calculate the slope of a tangent line*. The first part of that problem is that we do not know exactly what we *mean* by the tangent line. For the moment we shall not attempt to define it, but continue to use our intuition. We shall return to the *definition* of the tangent line shortly. Now, assuming that we know what we mean by the tangent line, we return to the question of finding its slope. We *could* draw the graph very carefully, draw the tangent line very carefully, and then compute the slope using two points on the line. But can we really draw the graph carefully enough to trust our answers? There is a better way to approach this.

Figure 5 shows what we have in general: the graph of some function, a point P on the graph, and the line tangent to the graph at P. Our task is to find a way of *calculating* the slope of the tangent line.

If we're not told the slope of a line, the only way we can calculate it is by using two points on the line. But, as you can see from the figure, we start out knowing only *one* point—the point on the graph through which the tangent line passes.

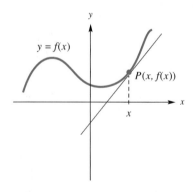

FIGURE 5

Q So how do we go about finding the slope of the tangent line knowing only one point?

A We first find the slope, not of the tangent line, but of an *approximate* tangent line, by selecting a second point Q on the curve close to P. Take a look at Figure 6.

Let us explain what is going on here: The line passing through the points P and Q is an *approximate* tangent line. We get the point Q as follows. Since the original point P has x-coordinate x, we choose a nearby value by adding a small quantity h, getting $x + h$. The point Q is then taken to be the point on the graph with x-coordinate $x + h$ ($h > 0$ in Figure 6, but h may also be negative, which would put Q to the left of P). We obtain the y-coordinate of Q by evaluating the function f there, getting $f(x + h)$. Thus, Q is the point $(x + h, f(x + h))$.

10.2 Geometric Interpretation of the Derivative

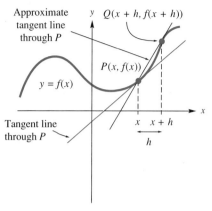

FIGURE 6

The line through P and Q is called a **secant line** of the graph. Its slope can be calculated using the usual slope formula:

$$m_{\text{sec}} = \frac{y_2 - y_1}{x_2 - x_1} = \frac{f(x + h) - f(x)}{(x + h) - x} = \frac{f(x + h) - f(x)}{h}.$$

Do you recognize this? It is the difference quotient we saw in the first section.

Q So we know a formula for the slope of the *secant* line, which is an *approximate* tangent line. How do we get the slope of the *exact* tangent line?

A Here is the key idea: The closer the point Q gets to the point P, the more closely the secant line will approximate the tangent line. In Figure 7 is shown a graph with a tangent line drawn at P and three secant lines drawn through P and points Q_1, Q_2, and Q_3. As Q gets closer to P in the sequence Q_1, Q_2, Q_3, the secant line gets closer to the tangent line, so the slope of the secant line approaches the slope of the tangent line.

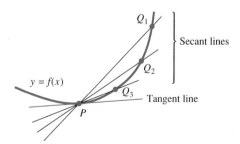

FIGURE 7

This tells us that if we want to find the slope of the tangent line, we need to take the limit of the slopes of the secant lines as h approaches 0. In other words, *the slope of the tangent line at a point is given by the derivative at that point.*

Thus we have the following important formulas:

> **SECANT AND TANGENT LINES**
> The slope of the secant line through $(x, f(x))$ and $(x + h, f(x + h))$ is given by the difference quotient:
> $$m_{\text{sec}} = \frac{f(x + h) - f(x)}{h}$$
> ($=$ Average rate of change of f over the interval $[x, x + h]$).
> The slope of the line tangent to the graph of f at $(x, f(x))$ is given by the derivative:
> $$m_{\text{tan}} = f'(x) = \lim_{h \to 0} \frac{f(x + h) - f(x)}{h}$$
> ($=$ Instantaneous rate of change of f at x).

Q So how do we *define* the tangent line?

A Since we started the discussion above by assuming that there was a tangent line, what we have really shown is that *if* there is a tangent line at a point, its slope ought to be given by the derivative at that point. We turn this around and *define* the tangent line to f through the point $P = (x, f(x))$ to be the line through P with slope $f'(x)$.

It's about time for some examples.

▼ **EXAMPLE 1**

Let $f(x) = x^2$.

(a) Obtain a formula for the slope of the secant line through the points $(x, f(x))$ and $(x + h, f(x + h))$.

(b) Use the answer in part (a) to find the slope of the secant line through the points $(1, 1)$ and $(1 + h, (1 + h)^2)$.

SOLUTION

(a) The slope of any secant line is given by the difference quotient.
$$\begin{aligned} m_{\text{sec}} &= \frac{f(x + h) - f(x)}{h} \\ &= \frac{(x + h)^2 - x^2}{h} \\ &= \frac{x^2 + 2xh + h^2 - x^2}{h} \\ &= \frac{2xh + h^2}{h} \\ &= \frac{h(2x + h)}{h} = 2x + h. \end{aligned}$$

10.2 Geometric Interpretation of the Derivative

This is the slope of the secant line through (x, x^2) and $(x+h, (x+h)^2)$.

(b) For the secant line through $(1, 1)$ and $(1+h, (1+h)^2)$, we have $x = 1$, so

$$m_{sec} = 2x + h = 2 + h.$$

Before we go on... This single formula enables us to find the slope of many secant lines. For instance, choosing $h = 1$, the secant line through $(1, 1)$ and $(2, 4)$ has slope $2 + 1 = 3$. Choosing $h = 2$, the secant line through $(1, 1)$ and $(3, 9)$ has slope $2 + 2 = 4$. The secant line through $(1, 1)$ and $(0, 0)$ has slope $2 - 1 = 1$ (since $h = 0 - 1 = -1$). Figure 8 shows the graph of f, together with these three secant lines.

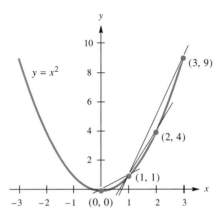

FIGURE 8

In the last instance, we used a negative value for h. There is nothing in our analysis that says h can't be negative. Of course, if h is negative, then $x + h$ is to the *left* of x.

▼ **EXAMPLE 2**

This continues Example 1, with $f(x) = x^2$.

(a) Find a formula for the slope of the tangent line at the point $(x, f(x))$.

(b) Use the answer in part (a) to find the slope of the tangent line at the point $(1, 1)$.

(c) Find an equation of the tangent line at the point $(1, 1)$.

SOLUTION

(a) In Example 1, we calculated the slope of the secant line through $(x, f(x))$ and $(x+h, f(x+h))$ to be

$$m_{sec} = \frac{f(x+h) - f(x)}{h} = 2x + h.$$

To obtain the slope of the *tangent* line, we let h approach 0:

$$m_{\tan} = \lim_{h \to 0} \frac{f(x+h) - f(x)}{h}$$

$$= \lim_{h \to 0} 2x + h.$$

As h gets closer to 0, the sum $2x + h$ gets closer and closer to $2x + 0 = 2x$. Thus,

$$m_{\tan} = \lim_{h \to 0} 2x + h = 2x$$

is the slope of the tangent line at $(x, f(x)) = (x, x^2)$.

(b) For the tangent line at $(1, f(1)) = (1, 1)$, we have $x = 1$, so

$$m_{\tan} = 2x = 2.$$

(c) To find an equation of the tangent line, we may use the point-slope formula, because we know that the tangent line goes through the point (1, 1) with slope 2:

$$y - 1 = 2(x - 1)$$

or

$$y = 2x - 1.$$

This tangent line is shown in Figure 9.

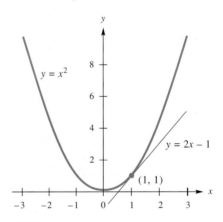

FIGURE 9

Before we go on... We did the following calculation in part (a): If $f(x) = x^2$, then $f'(x) = 2x$. This is our first complete calculation of a derivative function. After we talk about limits in the next section we shall do many more such calculations. In part (b), we calculated $f'(1) = 2$.

If you look back at Figure 4, we can now quantify what we saw there. The tangent line at the point where $x = -2$ has slope $f'(-2) = 2(-2) = -4$.

This is a steeply falling line, as appears in the figure. The tangent line at the point where $x = -1$ has slope $f'(-1) = 2(-1) = -2$. This is also a falling line, but it is less steep than the first. Similarly, the tangent lines at the points where $x = 0$, $x = 1$, and $x = 2$ have slopes $f'(0) = 0$, $f'(1) = 2$, and $f'(2) = 4$, respectively.

ESTIMATING DERIVATIVES GRAPHICALLY USING GRAPHING CALCULATORS OR GRAPHING SOFTWARE

We mentioned in the first section of this chapter that graphing calculators and computers were helpful in approximating derivatives numerically. We can also use graphing calculators or graphing software to approximate derivatives graphically.

▼ EXAMPLE 3

If $f(x) = x^{1.2} - \sqrt{x^{2.2} + x}$, calculate an approximate value for $f'(2)$ graphically.

SOLUTION We shall exploit this idea: If we magnify a small portion of the graph of f near the point of interest, the result is almost indistinguishable from a straight line. In fact, it is almost indistinguishable from the *tangent* line. Therefore, if we calculate the slope of this "line," we will have a close approximation to the slope of the tangent line. Here is what we do.

First, we enter the function in the correct format:

$$Y_1 = X\wedge 1.2 - (X\wedge 2.2 + X) \wedge 0.5$$

Next, we set the viewing window by taking $0 \leq x \leq 5$ (to make sure that the point with $x = 2$ is on the graph) and $-0.6 \leq y \leq 0.6$. (We obtained the y-ranges by experimenting.) Figure 10 shows the graphing calculator output.

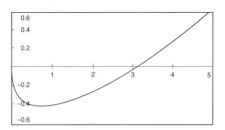

FIGURE 10

Since we are interested in the slope of the tangent line at the point where $x = 2$, we shall zoom into a small portion of the curve centered *exactly* at the point on the graph where x is 2. (This will improve our accuracy.) To do this, we choose a small value for h, say $h = 0.001$, and use

$Xmin = 2 - h = 1.999$, and $Xmax = 2 + h = 2.001$. To specify the y-coordinates of the window, we can use "trace" to give rough values for $f(1.999)$ and $f(2.001)$, and use these values as $Ymin$ and $Ymax$. Thus, let us take $Ymin = -0.271$ and $Ymax = -0.270$.* Figure 11 shows the zoomed-in view.

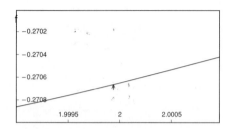

FIGURE 11

As we said, this small portion of the graph is approximately the tangent line. Now we need to measure its slope. To do this, recall that we need the coordinates of two points. *We shall use the two furthest points on the zoomed-in graph:* the points where $x = 1.999$ and $x = 2.001$. Using "trace" to find the y-coordinates, we obtain the points $(1.999, -0.2708359)$ and $(2.001, -0.2704366)$ (alternatively, we could calculate $f(1.999)$ and $f(2.001)$ directly using the calculator). This gives the slope as

$$m = \frac{y_2 - y_1}{x_2 - x_1}$$
$$= \frac{-0.2704366 - (-0.2708359)}{2.001 - 1.999} = 0.19965$$

Since the slope of the tangent is approximately 0.19965, we conclude that $f'(2) \approx 0.19965$.

Before we go on... Why did we use so many decimal places for the y-coordinates? How can we improve the accuracy of our calculation of the slope?

Q Just how accurate is the answer?

A The exact value of $f'(2)$ to five decimal places turns out to be 0.19966. Thus, the answer we obtained is *extremely* accurate—within 0.00001 of the exact answer. As to how we obtained the exact answer, you will find out how to take the derivative of functions as complicated as $f(x)$ in the next chapter.

---

*Alternatively, we can run the "windows" program in Section 1 Appendix C. This program automatically sets $Ymin$ and $Ymax$ for us.

10.2 EXERCISES

In each of the graphs in Exercises 1–8, say at which labeled point the slope of the tangent is **(a)** *greatest and* **(b)** *least (in the sense that* −7 *is less than* 1*).*

1.
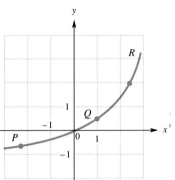

2.
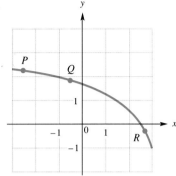

3.
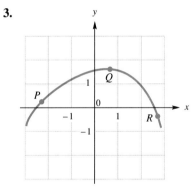

4.
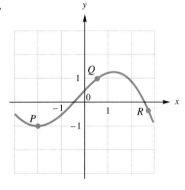

5.
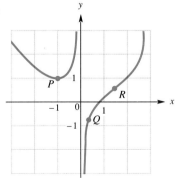

6.
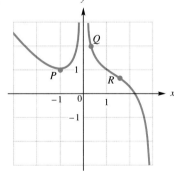

21. Let $f(x)$ have derivative $f'(x)$. Find a formula for the equation of the tangent to the graph of f through the point where $x = a$.

22. Find a formula for the equation of the secant line through the points on the graph of the function f corresponding to $x = a$ and $x = a + h$.

In each of Exercises 23–34:
(a) *find a formula for the slope of the secant line through $(x, f(x))$ and $(x + h, f(x + h))$;*
(b) *find a formula for the slope of the tangent to the graph of f at the point $(x, f(x))$;*
(c) *find the slope of the tangent to the graph of f at the indicated point.*

23. $f(x) = x^2 + 1$; $(2, 5)$
24. $f(x) = x^2 - 3$; $(1, -2)$
25. $f(x) = 2 - x^2$; $(-1, 1)$
26. $f(x) = -1 - x^2$; $(0, -1)$
27. $f(x) = 3x$; $(2, 6)$
28. $f(x) = x - 1$; $(2, 1)$
29. $f(x) = 1 - 2x$; $(2, -3)$
30. $f(x) = \dfrac{x}{3} - 1$; $(-3, -2)$
31. $f(x) = 3x^2 + 1$; $(-1, 4)$
32. $f(x) = 2x^2$; $(-2, 8)$
33. $f(x) = x - x^2$; $(2, -2)$
34. $f(x) = x^2 + x$; $(3, 12)$

In each of Exercises 35–40:
(a) *use any method to find the slope of the tangent to the graph of the given function at the point with the indicated x-coordinate;*
(b) *find an equation of the tangent line in part (a). In each case, sketch the curve together with the appropriate tangent line.*

35. $f(x) = x^3$; $x = -1$
36. $f(x) = x^2$; $x = 0$
37. $f(x) = x + \dfrac{1}{x}$; $x = 2$
38. $f(x) = \dfrac{1}{x^2}$; $x = 1$
39. $f(x) = \sqrt{x}$; $x = 4$
40. $f(x) = 2x + 4$; $x = -1$

Match each of the functions graphed in Exercises 41–46 to the graph of its derivative (shown below Exercise 46).

41.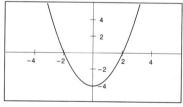

42.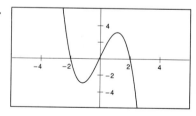

43.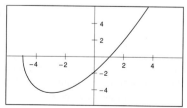

44.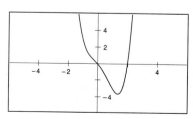

10.2 Geometric Interpretation of the Derivative

45. 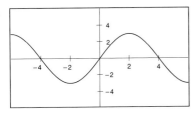   **46.**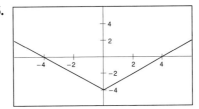

Graphs of derivatives:

(a) 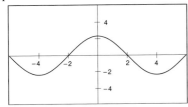   **(b)**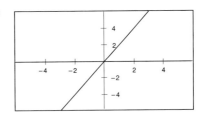

(c) 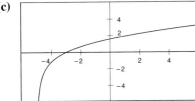   **(d)**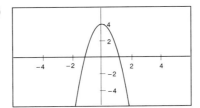

(e) 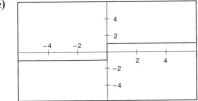   **(f)**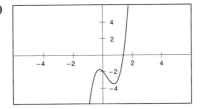

In Exercises 47–50, use a graphing calculator or graphing software to estimate the derivative of

$$f(x) = \frac{1}{x^{1.1} - 4}$$

graphically at the point with the given x-coordinate.

47. $x = 1$ **48.** $x = 2$ **49.** $x = 1.2$ **50.** $x = 4.3$

In Exercises 51–54, use a graphing calculator or graphing software to estimate the derivative of

$$f(x) = \sqrt{1 - x^2}$$

graphically at the point with the given x-coordinate.

51. $x = 0.5$ **52.** $x = 0.75$ **53.** $x = -0.25$ **54.** $x = -0.5$

In Exercises 55–58, use a graphing calculator to graph the given function $f(x)$ over the given range of values of x, and determine the approximate values of x in that range for which $f'(x) = 0$ (if any).

55. $f(x) = x^{3.4} - x^{1.2}$; $0 \le x \le 5$

56. $f(x) = 2x^{4.1} - x^{1.3}$; $0 \le x \le 1$

57. $f(x) = (x-1)(x-2)(x-3)$; $-1 \le x \le 5$

58. $f(x) = (x-1)^2(x-2)^2(x-3)$; $-1 \le x \le 5$

APPLICATIONS

Exercises 59–74 were given in the last section, but this time we ask you to use either a graphing calculator or graphing software to solve them graphically.

59. Demand Suppose that the demand for a new brand of sneakers is given by

$$q = \frac{5,000,000}{p},$$

where p is the price per pair of sneakers in dollars. Graph q for $1 \le p \le 200$, use your graph to find $q(100)$ and estimate $q'(100)$, and interpret your answers.

60. Demand Suppose that the demand for an old brand of TV is given by

$$q = \frac{100,000}{p + 10},$$

where p is the price per TV set, in dollars. Graph q for $0 \le p \le 400$, use your graph to find $q(190)$ and estimate $q'(190)$, and interpret your answers.

61. Profit Your monthly profit (in dollars) from selling magazines is given by

$$P = 5n + \sqrt{n},$$

where n is the number of magazines you sell in a month. Graph P for $0 \le n \le 100$, and use your graph to answer this question: If you are currently selling $n = 50$ magazines per month, find P and estimate dP/dn. Interpret your answers.

62. Profit Your monthly profit (in dollars) from your newspaper route is given by

$$P = 2n - \sqrt{n},$$

where n is the number of subscribers on your route. Graph P for $0 \le n \le 200$, and use your graph to answer this question: If you currently have 100 subscribers, find P and dP/dn. Interpret your answers.

63. Popularity A study of enrollment at a certain university shows that there is a relationship between enrollment in each professor's classes and the average grade that the professor awards. Measuring grades g on a scale of 0 to 4, the relationship can be expressed by the equation

$$E = 20\sqrt[4]{g},$$

where E is the professor's average enrollment per class. Graph E, and use your graph to answer the following question: For a professor whose average grade awarded is $g = 2.5$, find E and dE/dg, and interpret your answers.

64. Popularity A study of enrollment at another university shows that, again, there is a relationship between enrollment in each professor's classes and the average grade that the professor awards. Measuring grades g on a scale of 0 to 4, the relationship this time can be expressed by the equation

$$E = 15\sqrt[3]{g^2},$$

where E is the professor's average enrollment per class. Graph E, and use your graph to answer the following question: For a professor whose average grade awarded is $g = 2.5$, find E and dE/dg, and interpret your answers.

65. Market Average Joe Downs runs a small investment company from his basement. Every week he publishes a report on the success of his investments, including the progress of the infamous Joe Downs Average. At the end of one particularly memorable week he reported that the Average for that week had the value $A(t) = 1000 + 1500t - 800t^2 + 100t^3$ points, where t represents the number of days into the week; t ranges from 0 at the beginning of the week to 5 at end of the week. Graph A, and use your graph to estimate how fast the Average was growing halfway through the first day of the week.

66. Market Average Referring to the Joe Downs Average in the previous exercise: Graph A, and use the graph to estimate how fast the Average was increasing at the end of the third day.

67. Biology—Reproduction The Verhulst model for population growth specifies the reproductive rate of an organism as a function of the total population according to the following formula:

$$R(p) = \frac{r}{1 + kp}.$$

Here, p is the total population in thousands of organisms, r and k are constants that depend on the particular circumstances and the organism being studied, and $R(p)$ is the reproduction rate in thousands of organisms per hour.* Graph this function for $0 \leq p \leq 10$, given that $k = 0.125$ and $r = 45$, and use your graph to find an approximate value of $R'(4)$. Interpret the result.

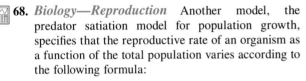68. *Biology—Reproduction* Another model, the predator satiation model for population growth, specifies that the reproductive rate of an organism as a function of the total population varies according to the following formula:

$$R(p) = \frac{rp}{1 + kp}.$$

Here, p is the total population in thousands of organisms, r and k are constants that depend on the particular circumstances and the organism being studied, and $R(p)$ is the reproduction rate in new organisms per hour.* Graph this function for $0 \leq p \leq 10$, given that $k = 0.2$ and $r = 0.08$, and use your graph to find an approximate value of $R'(2)$. Interpret the result.

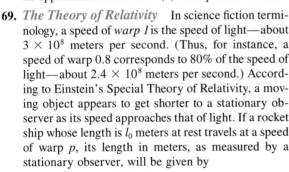69. *The Theory of Relativity* In science fiction terminology, a speed of *warp 1* is the speed of light—about 3×10^8 meters per second. (Thus, for instance, a speed of warp 0.8 corresponds to 80% of the speed of light—about 2.4×10^8 meters per second.) According to Einstein's Special Theory of Relativity, a moving object appears to get shorter to a stationary observer as its speed approaches that of light. If a rocket ship whose length is l_0 meters at rest travels at a speed of warp p, its length in meters, as measured by a stationary observer, will be given by

$$l(p) = l_0 \sqrt{1 - p^2}, \text{ with domain } [0, 1].$$

Graph l as a function of p for a 100-meter rocket ship ($l_0 = 100$), and use your graph to estimate $l(0.95)$ and $l'(0.95)$. What do these figures tell you?

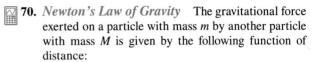70. *Newton's Law of Gravity* The gravitational force exerted on a particle with mass m by another particle with mass M is given by the following function of distance:

$$F(r) = G\frac{Mm}{r^2}, \text{ with domain } (0, +\infty).$$

Here, r is the distance between the two particles in meters, the masses M and m are given in kilograms, $G \approx 0.0000000000667$, or 6.67×10^{-11}, and the resulting force is given in newtons. Graph F as a function of r, given that M and m are both 1,000 kilograms, and use your graph to estimate $F(10)$ and $F'(10)$. What do these figures tell you?

Exercises 71–74 are based on logarithmic and exponential functions and also require the use of a graphing calculator or graphing computer software.

71. *Sales* Weekly sales of a new brand of sneakers are given by

$$S(t) = 200 - 150e^{-t/10},$$

pairs sold per week, *where t is the number of weeks since the introduction of the brand. Graph S as a function of t, and use your graph to estimate $S(5)$ and estimate $S'(5)$. Interpret your answers.

72. *Sales* Weekly sales of an old brand of TV are given by

$$S(t) = 100e^{-t/5},$$

sets per week where t is the number of weeks after the introduction of a competing brand. Graph S as a function of t, and use your graph to estimate $S(5)$ and $S'(5)$. Interpret your answers.

73. *Logistic Growth in Demand* The demand for a new product can be modeled by a **logistic** curve of the form

$$q(t) = \frac{N}{1 + ke^{-rt}},$$

where $q(t)$ is the total number of units sold t months after the introduction of the new product, and N, k, and r are constants that depend on the product and the market. Assume that the demand for video game units

---

* Source: *Mathematics in Medicine and the Life Sciences* by F. C. Hoppensteadt and C. S. Peskin (New York: Springer-Verlag, 1992), pp. 20–22.

is determined by the above formula, with $N = 10{,}000$, $k = 0.5$, and $r = 0.4$. Graph the demand curve using $0 \le t \le 10$ and $5{,}000 \le q \le 10{,}000$, use your graph to estimate $q(2)$ and $q'(2)$, and interpret the results.

 74. *Information Highway* The amount of information transmitted each month on the National Science Foundation's Internet network for the years 1988–1994 can be modeled by the equation

$$q(t) = \frac{2e^{0.69t}}{3 + 1.5e^{-0.4t}},$$

where q is the amount of information transmitted each month in billions of data packets and t is the number of years since the start of 1988.* Graph q as a function of t with $0 \le t \le 4$ and $0 \le q \le 40$, and use your graph to estimate the number of data packets transmitted during the first month of 1990 and also the rate at which this number was increasing.

COMMUNICATION AND REASONING EXERCISES

75. If the derivative of f is zero at a point, what do you know about the graph of f near that point?

76. Sketch the graph of a function whose derivative never exceeds 1.

77. Sketch the graph of a function whose derivative exceeds 1 at every point.

78. Sketch the graph of a function whose derivative is exactly 1 at every point.

79. If the derivative of f is always positive, what do you know about the graph of f?

80. If the derivative of f is increasing, what do you know about the graph of f?

 ## 10.3 LIMITS AND CONTINUITY

The derivative is defined using a limit, and it is now time to say what that means. It is possible to speak of limits by themselves, rather than in the context of the derivative. The story of limits is a long one that we will try to make as concise as possible.

EVALUATING LIMITS NUMERICALLY

Start with a simple example: Look at the function $f(x) = 2 + x$ and ask yourself: "What happens to $f(x)$ as x approaches 3?" The following table shows the value of $f(x)$ for values of x close to, and on either side of 3:

	\multicolumn{4}{c}{x approaching 3 from the left →}		\multicolumn{4}{c}{← x approaching 3 from the right}
x	2.9	2.99	2.999	2.9999	3	3.0001	3.001	3.01	3.1
$f(x) = 2 + x$	4.9	4.99	4.999	4.9999		5.0001	5.001	5.01	5.1

We have left the entry under 3 blank to emphasize that when we are calculating the limit of $f(x)$ as x approaches 3, *we are not interested in its value when x equals 3.*

---

* This is the authors' model, based on figures published in *The New York Times*, Nov. 3, 1993.

10.3 Limits and Continuity

Notice from the table that the closer x gets to 3 from either side, the closer $f(x)$ gets to 5. We write this as

$$\lim_{x \to 3} f(x) = 5,$$

meaning

the limit of $f(x)$, as x approaches 3, equals 5.

Q Why all the fuss? Can't we simply put $x = 3$ and avoid having to use a table?

A This happens to work for *some* functions, but not for *all* functions. The following example illustrates this point.

▼ EXAMPLE 1

Use a table to evaluate $\lim_{x \to 2} \dfrac{x^3 - 8}{x - 2}$.

SOLUTION Notice that we cannot simply substitute $x = 2$, because the function $f(x) = \dfrac{x^3 - 8}{x - 2}$ is not defined at $x = 2$. As above, we can use a table of values, with x approaching 2 from either side.

x approaching 2 from the left → ← x approaching 2 from the right

x	1.9	1.99	1.999	1.9999	2	2.0001	2.001	2.01	2.1
$f(x) = \dfrac{x^3 - 8}{x - 2}$	11.41	11.9401	11.9940	11.9994		12.0006	12.0060	12.0601	12.61

We notice that as x approaches 2 from either side, $f(x)$ approaches 12. This suggests that the limit is 12, and we write

$$\lim_{x \to 2} \frac{x^3 - 8}{x - 2} = 12.$$

Before we go on... Although the table *suggests* that the limit is 12, it by no means establishes that fact conclusively. It is *conceivable* (though not in fact the case here) that putting $x = 1.99999987$ will result in $f(x) = 426$. Using a table can only suggest a value for the limit. We shall talk soon about algebraic techniques for finding limits exactly.

 See "Evaluating a Function Using a Table" in the Evaluating Functions section of Appendix B for a quick and easy method of obtaining tables such as the one above on your graphing calculator.

Before we continue, let us make a more formal definition.

DEFINITION OF A LIMIT

If $f(x)$ approaches the number L as x approaches (but is not equal to) a from both sides, then we say that the **limit** of $f(x)$ as $x \to a$ ("x approaches a") is L. We write

$$\lim_{x \to a} f(x) = L$$

or

$$f(x) \to L \quad \text{as} \quad x \to a.$$

If $f(x)$ *fails* to approach *a single fixed number* as x approaches a from both sides, then we say that $f(x)$ **has no limit** as $x \to a$, or

$$\lim_{x \to a} f(x) \quad \text{does not exist.}$$

▶ **NOTES**

1. It is important that $f(x)$ approach the same number as x approaches a from either side. For instance, if $f(x)$ approaches 5 for $x = 1.9$, 1.99, 1.999, . . ., but approaches 4 for $x = 2.1, 2.01, 2.001,$ . . ., then the limit as $x \to 2$ does not exist.
2. It may happen that $f(x)$ does not approach any fixed number at all as $x \to a$ from either side. In this case, we also say that the limit does not exist.
3. We are deliberately suppressing the exact definition of "approaches," and instead shall trust to your intuition. The following phrasing of the definition of the limit is closer to the one used by mathematicians: *we can make $f(x)$ be as close to L as we like by making x be sufficiently close to a.* ◀

EVALUATING LIMITS GRAPHICALLY

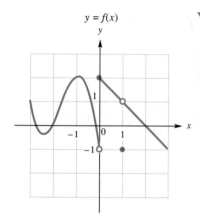

FIGURE 1

▼ **EXAMPLE 2**

The graph of the function f is shown in Figure 1.
From the graph, analyze the following limits.

(a) $\lim_{x \to -2} f(x)$ (b) $\lim_{x \to 0} f(x)$ (c) $\lim_{x \to 1} f(x)$

SOLUTION Since we are given only a graph of f, we must analyze these limits graphically.

(a) Suppose that we had Figure 1 drawn on a graphing calculator.
Graphing calculators are usually equipped with a "trace" feature that allows us to move a cursor along the graph and read the coordinates of points as we go. If we started at a point on the graph to the left of $x = -2$ and moved the cursor along the graph to the right, we could see numerically what the limit of $f(x)$ is. If we just have the graph drawn on paper, we can place a pencil

point on the graph to the left of $x = -2$ and move it along the curve to the right so that the x-coordinate approaches -2. (See Figure 2.)

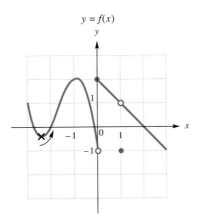

FIGURE 2

As the x-coordinate of the point approaches -2, we can see from the graph that the y-coordinate will approach 0. Similarly, if we place our pencil point to the right of $x = -2$ and move it along the graph to the left, the y-coordinate will again approach 0 (Figure 3).

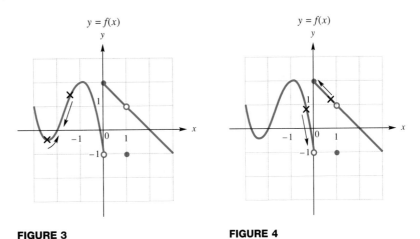

FIGURE 3   **FIGURE 4**

Therefore, as x approaches -2 from either side, $f(x)$ approaches 0, so

$$\lim_{x \to -2} f(x) = 0.$$

(b) Here we move our pencil point toward $x = 0$. If we start from the left of $x = 0$ and approach 0 by moving right, the y-coordinate approaches -1 (Figure 4).

However, if we start from the right of $x = 0$ and approach 0 by moving left, the y-coordinate approaches 2 (see Figure 4 again). So there appear to be two *different* limits: the limit as we approach 0 from the left and the limit as we approach 0 from the right. We write

$$\lim_{x \to 0^-} f(x) = -1,$$

read "the limit as x approaches 0 from the left is -1," and

$$\lim_{x \to 0^+} f(x) = 2,$$

read "the limit as x approaches 0 from the right is 2." These are called the **one-sided limits** of $f(x)$. In order for the **two-sided limit** to exist (the one we are asked to compute), the two one-sided limits must be equal. Since they are not, we conclude that

$$\lim_{x \to 0} f(x) \text{ does not exist.}$$

Limits may or may not exist. In this case there is a "break" in the graph, and we say that the function is **discontinuous** there. We shall return to this concept in more detail later.

(c) Once more, we think about a pencil point (or cursor on a graphing calculator) moving along the graph with x-coordinate approaching $x = 1$ from the left and from the right (Figure 5).

As the x-coordinate of the point approaches 1 from either side, the y-coordinate approaches 1 also. Therefore,

$$\lim_{x \to 1} f(x) = 1.$$

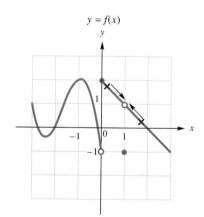

FIGURE 5

Before we go on... Notice that $f(1) = -1$. (Why?) Thus, $\lim_{x \to 1} f(x) \neq f(1)$. In other words, the limit of $f(x)$ as x approaches 1 is not the same as the value of f at $x = 1$. Always bear in mind that when we evaluate a limit as $x \to a$ we do not care about the value of the function at $x = a$. We only care about the value of $f(x)$ as x approaches a. In other words, $f(a)$ may or may not equal $\lim_{x \to a} f(x)$. Part (a) gives us an example where the limit and the value of the function are equal:

$$\lim_{x \to -2} f(x) = 0 = f(-2).$$

We can summarize the graphical method we used in this example as follows.

EVALUATING LIMITS GRAPHICALLY

To decide whether $\lim_{x \to a} f(x)$ exists, and to find its value if it does:

1. Draw the graph of $f(x)$ either by hand or using a graphing calculator.
2. Position your pencil point (or the graphing calculator "trace" cursor) on a point of the graph to the right of $x = a$.
3. Move the point *along the graph* toward $x = a$ from the right, and read the y-coordinate as you go. The value the y-coordinate approaches (if any) is then the limit $\lim_{x \to a^+} f(x)$.
4. Repeat Steps 2 and 3, but this time starting from a point on the graph to the left of $x = a$, and approach $x = a$ along the graph from the left. The value the y-coordinate approaches (if any) is then $\lim_{x \to a^-} f(x)$.
5. If the left and right limits both exist and have the same value L, then

$$\lim_{x \to a} f(x) \text{ exists and equals } L.$$

EVALUATING LIMITS ALGEBRAICALLY

In parts (a) and (c) of Example 2, we saw an example where $\lim_{x \to a} f(x) = f(a)$, and another where $\lim_{x \to a} f(x) \neq f(a)$. In the first case, the graph had no break; in the second, it did have a break. This motivates the following definition.

CONTINUOUS FUNCTION

A function f is **continuous at** a if $\lim_{x \to a} f(x) = f(a)$. The function f is said to be **continuous on its domain** if it is continuous at each point in its domain. If f is not continuous at a particular a, we say that f is **discontinuous** at a or that f has a **discontinuity** at a.

Besides describing a particularly nice kind of function, continuity is useful for evaluating many limits. If we know that a function is continuous at a point a, then we can compute $\lim_{x \to a} f(x)$ by simply substituting $x = a$ into $f(x)$. In order to use this, we need to know some continuous functions. Luckily, there is a large class of functions that are known to be continuous on their domains: roughly speaking, those that are specified by a single formula.

We can be more precise. A **closed-form function** is any function that can be obtained by combining constants, powers of x, exponential functions, radicals, logarithms (and some other functions we shall not encounter in this text) into a *single* mathematical expression by means of the usual arithmetic operations and composition of functions. Examples of closed-form functions are

$$3x^2 - x + 1, \quad \frac{\sqrt{x^2-1}}{6x-1}, \quad e^{-\frac{4x^2-1}{x}}, \quad \sqrt{\log_3(x^2-1)}.$$

They can be as complicated as you like. The following is *not* a closed-form function.

$$f(x) = \begin{cases} -1 & \text{if } x \leq -1 \\ x^2 + x & \text{if } -1 \leq x \leq 1 \\ 2 - x & \text{if } 1 < x \leq 2 \end{cases}$$

The reason for this is that $f(x)$ is not specified by a *single* mathematical expression. What is nice about closed-form functions is the following.

CONTINUITY OF CLOSED-FORM FUNCTIONS

Every closed-form function is continuous on its domain. Thus, *the limit of a closed-form function at a point on its domain can be obtained by substitution.*

For example, if $f(h) = \sqrt{h^2 + 2h + 2}$, then f is continuous on its domain, and so

$$\lim_{h \to 0} f(h) = f(0) = \sqrt{2},$$

since 0 is in the domain of f.

▶ **NOTES**

1. The reason we refer to such functions as continuous on their domain is that their graphs do not break anywhere on their domain (although the graph of a continuous function may break if its domain is broken; see Example 3 below).
2. Mathematics majors spend a great deal of time proving results such as this. We shall ask you to accept it without proof. ◀

In the definition of the derivative, we must take the limit of the difference quotient $(f(x + h) - f(x))/h$ as $h \to 0$. Although the difference quotients we encounter are usually closed-form functions, we cannot evaluate them by substitution because of the following catch: Since h appears in the denominator, $h = 0$ is not in the domain of the difference quotient, and so we cannot evaluate the limit by substitution. However—and this is the the key to finding such limits—some preliminary algebraic simplification may allow us to obtain a closed-form function with $h = 0$ in its domain. We can find the limit by substituting $h = 0$ in the new function. Many limits can be computed by this technique.

▼ **EXAMPLE 3**

Evaluate $\lim_{x \to 2} \dfrac{x^3 - 8}{x - 2}$.

SOLUTION We found this limit in Example 1 using a numerical approach, but we shall now find it algebraically. Notice that $x = 2$ is not in the domain of $f(x) = (x^3 - 8)/(x - 2)$, so we cannot obtain the limit by substitution. We first simplify $f(x)$ to obtain a new function with $x = 2$ in its domain. To do this, notice first that the numerator can be factored as

$$x^3 - 8 = (x - 2)(x^2 + 2x + 4).$$

Thus,

$$\dfrac{x^3 - 8}{x - 2} = \dfrac{(x - 2)(x^2 + 2x + 4)}{x - 2}$$
$$= x^2 + 2x + 4.$$

Once we have canceled the offending $(x - 2)$ in the denominator, we are left with a closed-form function *with 2 in its domain*. Thus,

$$\lim_{x \to 2} \dfrac{x^3 - 8}{x - 2} = \lim_{x \to 2} (x^2 + 2x + 4)$$
$$= 2^2 + 2(2) + 4 = 12.$$

This confirms the answer we found in Example 1.

Before we go on... If the given function fails to simplify, you can always evaluate the limit numerically. It may very well be that the limit does not exist in that case.

▼ **EXAMPLE 4**

Calculate

$$\lim_{h \to 0} \dfrac{(1 + h)^2 - 1}{h}.$$

SOLUTION The given function is not defined when $h = 0$. So, we try to simplify the expression.

$$\lim_{h \to 0} \dfrac{(1 + h)^2 - 1}{h} = \lim_{h \to 0} \dfrac{1 + 2h + h^2 - 1}{h}$$
$$= \lim_{h \to 0} \dfrac{h(2 + h)}{h}$$
$$= \lim_{h \to 0} (2 + h) = 2$$

Before we go on... There is something suspicious about this example and the last one. If 0 was not in the domain before simplifying, but was in the domain after simplifying, we must have changed the function. In fact, when we say that

$$\frac{(1+h)^2 - 1}{h} = 2 + h,$$

we are lying a little bit. What we really mean is that these two expressions are equal *where both are defined*. The functions $f(h) = ((1+h)^2 - 1)/h$ and $g(h) = 2 + h$ are different functions. However, the only difference is that $h = 0$ is in the domain of g and is not in the domain of f. Since $\lim_{h \to 0} f(h)$ explicitly *ignores* any value that f may have at 0, this does not matter. Formally, we can say the following.

> **EQUALITY OF LIMITS**
> If $f(x) = g(x)$ as long as $x \neq a$, then
> $$\lim_{x \to a} f(x) = \lim_{x \to a} g(x).$$

▼ **EXAMPLE 5**

If $f(x) = x^3$, find $f'(x)$.

SOLUTION

$$\begin{aligned}
f'(x) &= \lim_{h \to 0} \frac{f(x+h) - f(x)}{h} \\
&= \lim_{h \to 0} \frac{(x+h)^3 - x^3}{h} \\
&= \lim_{h \to 0} \frac{x^3 + 3x^2 h + 3xh^2 + h^3 - x^3}{h} \\
&= \lim_{h \to 0} \frac{3x^2 h + 3xh^2 + h^3}{h} \\
&= \lim_{h \to 0} \frac{h(3x^2 + 3xh + h^2)}{h} \\
&= \lim_{h \to 0} (3x^2 + 3xh + h^2) \\
&= 3x^2 + 3x(0) + 0^2 \\
&= 3x^2
\end{aligned}$$

Before we go on... When computing derivatives from the definition, as we just did, we always start by writing the definition. This helps get us started in the right direction.

EXAMPLE 6

If $f(x) = \dfrac{1}{x}$, find $f'(x)$.

SOLUTION

$$f'(x) = \lim_{h \to 0} \frac{f(x+h) - f(x)}{h}$$

$$= \lim_{h \to 0} \frac{\left[\dfrac{1}{x+h} - \dfrac{1}{x}\right]}{h}$$

$$= \lim_{h \to 0} \frac{\left[\dfrac{x - (x+h)}{(x+h)x}\right]}{h}$$

$$= \lim_{h \to 0} \frac{x - x - h}{h(x+h)x}$$

$$= \lim_{h \to 0} \frac{-h}{h(x+h)x}$$

$$= \lim_{h \to 0} \frac{-1}{(x+h)x}$$

$$= -\frac{1}{x^2}$$

EXAMPLE 7

If $f(x) = \sqrt{x}$, find $f'(x)$.

SOLUTION

$$f'(x) = \lim_{h \to 0} \frac{f(x+h) - f(x)}{h}$$

$$= \lim_{h \to 0} \frac{\sqrt{x+h} - \sqrt{x}}{h}$$

Now here we encounter a slight problem, since there is no way to "expand" the numerator as we did in Example 5, or combine terms as we did in Example 6. However, there is a technique that you may have seen before called "rationalizing the numerator": multiply top and bottom by the "conjugate," $\sqrt{x+h} + \sqrt{x}$, of the numerator.

$$f'(x) = \lim_{h \to 0} \frac{(\sqrt{x+h} - \sqrt{x})(\sqrt{x+h} + \sqrt{x})}{h(\sqrt{x+h} + \sqrt{x})}$$

The numerator is of the form $(a - b)(a + b) = a^2 - b^2$, so we get

$$f'(x) = \lim_{h \to 0} \frac{(\sqrt{x+h})^2 - (\sqrt{x})^2}{h(\sqrt{x+h} + \sqrt{x})}$$

$$= \lim_{h \to 0} \frac{(x+h) - x}{h(\sqrt{x+h} + \sqrt{x})}$$

$$= \lim_{h \to 0} \frac{h}{h(\sqrt{x+h} + \sqrt{x})}$$

$$= \lim_{h \to 0} \frac{1}{\sqrt{x+h} + \sqrt{x}}$$

$$= \frac{1}{\sqrt{x} + \sqrt{x}}$$

$$= \frac{1}{2\sqrt{x}}.$$

Obviously, you don't want to do too many of these by hand. In the next section we shall start to talk about some shortcuts for finding derivatives, and we will continue this discussion in the next chapter.

▶ 10.3 EXERCISES

Compute the limits in Exercises 1–4 numerically.

1. $\lim\limits_{x \to 0} \dfrac{x^2}{x+1}$

2. $\lim\limits_{x \to 0} \dfrac{x-3}{x-1}$

3. $\lim\limits_{x \to 2} \dfrac{x^2-4}{x-2}$

4. $\lim\limits_{x \to -1} \dfrac{x^2+2x+1}{x+1}$

In each of Exercises 5–20, the graph of f is given. Use the graph to compute the quantities asked for. If a particular quantity fails to exist, say why.

5. (a) $\lim\limits_{x \to 1} f(x)$ (b) $\lim\limits_{x \to -1} f(x)$

6. (a) $\lim\limits_{x \to -1} f(x)$ (b) $\lim\limits_{x \to 1} f(x)$

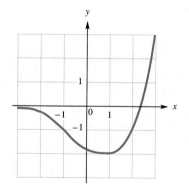
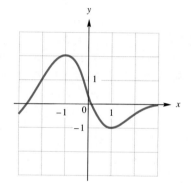

7. (a) $\lim\limits_{x\to 0} f(x)$ (b) $\lim\limits_{x\to 1} f(x)$

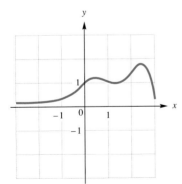

8. (a) $\lim\limits_{x\to -1} f(x)$ (b) $\lim\limits_{x\to 1} f(x)$

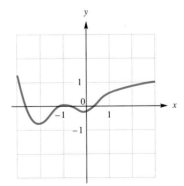

9. (a) $\lim\limits_{x\to 2} f(x)$ (b) $\lim\limits_{x\to 0^+} f(x)$
(c) $\lim\limits_{x\to 0^-} f(x)$ (d) $\lim\limits_{x\to 0} f(x)$ (e) $f(0)$

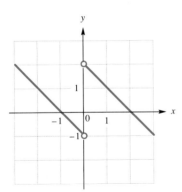

10. (a) $\lim\limits_{x\to 3} f(x)$ (b) $\lim\limits_{x\to 1^+} f(x)$
(c) $\lim\limits_{x\to 1^-} f(x)$ (d) $\lim\limits_{x\to 1} f(x)$ (e) $f(1)$

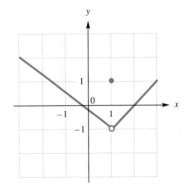

11. (a) $\lim\limits_{x\to -2} f(x)$ (b) $\lim\limits_{x\to -1^+} f(x)$
(c) $\lim\limits_{x\to -1^-} f(x)$ (d) $\lim\limits_{x\to -1} f(x)$ (e) $f(-1)$

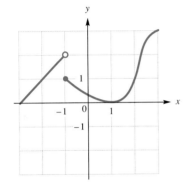

12. (a) $\lim\limits_{x\to -1} f(x)$ (b) $\lim\limits_{x\to 0^+} f(x)$
(c) $\lim\limits_{x\to 0^-} f(x)$ (d) $\lim\limits_{x\to 0} f(x)$ (e) $f(0)$

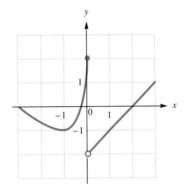

13. (a) $\lim_{x\to -1} f(x)$ (b) $\lim_{x\to 0^+} f(x)$
 (c) $\lim_{x\to 0^-} f(x)$ (d) $\lim_{x\to 0} f(x)$ (e) $f(0)$

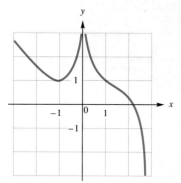

14. (a) $\lim_{x\to 1} f(x)$ (b) $\lim_{x\to 0^+} f(x)$
 (c) $\lim_{x\to 0^-} f(x)$ (d) $\lim_{x\to 0} f(x)$ (e) $f(0)$

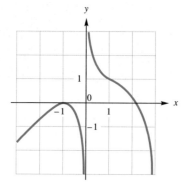

15. (a) $\lim_{x\to -2} f(x)$ (b) $\lim_{x\to 0^+} f(x)$
 (c) $\lim_{x\to 0^-} f(x)$ (d) $\lim_{x\to 0} f(x)$
 (e) $f(0)$ (f) $f(-2)$

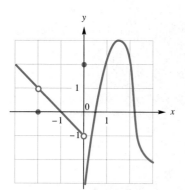

16. (a) $\lim_{x\to 0^-} f(x)$ (b) $\lim_{x\to 2^+} f(x)$
 (c) $\lim_{x\to 0} f(x)$ (d) $\lim_{x\to 2} f(x)$
 (e) $f(0)$ (f) $f(2)$

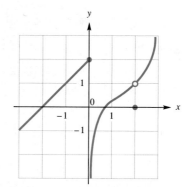

17. (a) $\lim_{x\to 0} f(x)$ (b) $\lim_{x\to 1} f(x)$ (c) $f(0)$

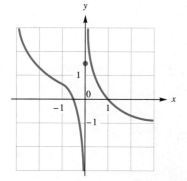

18. (a) $\lim_{x\to 0} f(x)$ (b) $\lim_{x\to 1} f(x)$ (c) $f(0)$

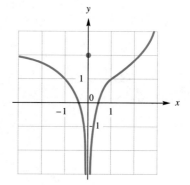

19. (a) $\lim\limits_{x \to 1} f(x)$ (b) $\lim\limits_{x \to 2} f(x)$ **20.** (a) $\lim\limits_{x \to 1} f(x)$ (b) $\lim\limits_{x \to 2} f(x)$

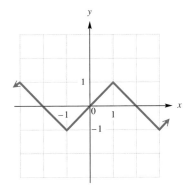
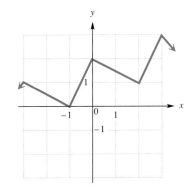

In Exercises 21–36, determine whether each of the functions given in Exercises 5–20 is continuous on its domain. If a particular function is not continuous on its domain, say why.

21. Graph from Exercise 5
22. Graph from Exercise 6
23. Graph from Exercise 7
24. Graph from Exercise 8
25. Graph from Exercise 9
26. Graph from Exercise 10
27. Graph from Exercise 11
28. Graph from Exercise 12
29. Graph from Exercise 13
30. Graph from Exercise 14
31. Graph from Exercise 15
32. Graph from Exercise 16
33. Graph from Exercise 17
34. Graph from Exercise 18
35. Graph from Exercise 19
36. Graph from Exercise 20

Calculate the limits in Exercises 37–44 mentally.

37. $\lim\limits_{x \to 0} (x + 1)$
38. $\lim\limits_{x \to 0} (2x - 4)$
39. $\lim\limits_{x \to 2} \dfrac{2 + x}{x}$
40. $\lim\limits_{x \to -1} \dfrac{4x^2 + 1}{x}$
41. $\lim\limits_{x \to -1} \dfrac{x + 1}{x}$
42. $\lim\limits_{x \to 4} (x + \sqrt{x})$
43. $\lim\limits_{x \to 8} (x - \sqrt[3]{x})$
44. $\lim\limits_{x \to 1} \dfrac{x - 2}{x + 1}$

Calculate each of the limits in Exercises 45–54.

45. $\lim\limits_{h \to 1} (h^2 + 2h + 1)$
46. $\lim\limits_{h \to 0} (h^3 - 4)$
47. $\lim\limits_{h \to 3} 2$
48. $\lim\limits_{h \to 0} -5$
49. $\lim\limits_{h \to 0} \dfrac{h^2}{h + h^2}$
50. $\lim\limits_{h \to 0} \dfrac{h^2 + h}{h^2 + 2h}$
51. $\lim\limits_{x \to 1} \dfrac{x^2 - 2x + 1}{x^2 - x}$
52. $\lim\limits_{x \to -1} \dfrac{x^2 + 3x + 2}{x^2 + x}$
53. $\lim\limits_{x \to 2} \dfrac{x^3 - 8}{x - 2}$
54. $\lim\limits_{x \to -2} \dfrac{x^3 + 8}{x^2 + 3x + 2}$

In each of Exercises 55–78, use the definition to calculate the derivative of the given function.

55. $f(x) = -14$
56. $f(x) = 5$
57. $f(x) = 2x - 3$
58. $f(x) = -3x + 5$
59. $g(x) = -4x - 1$
60. $g(x) = 10x - 100$
61. $g(x) = x^2 - 2x$
62. $g(x) = 3x^2 + 1$
63. $h(x) = -5x^2 + 2x - 1$

64. $h(x) = -3x^2 - x + 5$
65. $f(t) = t^3 + t$
66. $f(t) = 2t^3 - t^2$
67. $g(t) = t^4 - t$
68. $g(t) = 3t^4 + 2t^2$
69. $h(t) = \frac{6}{t}$
70. $h(t) = -\frac{1}{t}$
71. $f(x) = x + \frac{1}{x}$
72. $f(x) = 6 - \frac{6}{x}$
73. $f(x) = \frac{1}{x-2}$
74. $f(x) = \frac{1}{2x+1}$
75. $f(x) = \sqrt{x+1}$
76. $f(x) = \sqrt{x-2}$
77. $g(t) = \frac{1}{\sqrt{t}}$
78. $g(t) = t + \frac{1}{\sqrt{t}}$

COMMUNICATION AND REASONING EXERCISES

79. Describe the three methods of evaluating limits discussed in this section. Give at least one disadvantage of each.

80. Choose one of the limits in Exercises 45–54 and calculate it in three different ways.

81. What is wrong with the following statement? "If $f(a)$ is defined, then $\lim_{x \to a} f(x)$ exists and equals $f(a)$."

82. What is wrong with the following statement? "$\lim_{x \to 4} \frac{\sqrt{x} - 2}{x - 4}$ does not exist, since substituting $x = 4$ yields $0/0$, which is undefined."

83. Give an example of a function f specified by means of algebraic formulas so that f is not continuous at $x = 2$.

84. Give an example of a function f with $\lim_{x \to 1} f(x) = f(2)$.

▶ 10.4 DERIVATIVES OF POWERS AND POLYNOMIALS

So far, we have calculated the derivatives of functions using the definition of the derivative as a limit. These calculations are tedious, so it would be nice to have a quicker way of doing them. In this section, we shall see how to quickly calculate the derivatives of many functions. By the end of the next chapter, we shall be able to find the derivative of almost any function we can write.

First, we review notation we mentioned in Section 1.

Differential Notation is based on an abbreviation for the phrase "the derivative with respect to x." For example, we learned in the preceding section (Example 5) that if $f(x) = x^3$, then $f'(x) = 3x^2$. When we say "$f'(x) = 3x^2$," we mean the following:

"The derivative with respect to x of x^3 equals $3x^2$."

You may wonder why we sneaked in the words "with respect to x." All this means is that the variable of the function is x, and nothing else.* Since

---

▼ *This may seem odd in the case of $f(x) = x^3$, since there are no other variables to worry about. But later, we shall see more complicated expressions than x^3 which involve variables other than x, and it will become necessary to specitfy just what the variable of the function is. This is the same reason that we write "$f(x) = x^3$" rather than just "$f = x^3$."

we shall be using the phrase "the derivative with respect to x" often, we use the following abbreviation:

DERIVATIVE WITH RESPECT TO x

$\dfrac{d}{dx}$ means "the derivative with respect to x."

Thus, for example, the statement

"the derivative with respect to x of x^3 is $3x^2$"

can now be written more compactly as:

$$\frac{d}{dx}(x^3) = 3x^2.$$

Similarly, by Example 7 of the preceding section,

the derivative with respect to x of $\sqrt{x}$ is $\dfrac{1}{2\sqrt{x}}$,

or

$$\frac{d}{dx}\left(\sqrt{x}\right) = \frac{1}{2\sqrt{x}}.$$

Sometimes, we don't want to spell out the function each time we use the phrase "d/dx." For instance, we might have $y = x^{100}$, and wish to talk about "the derivative (with respect to x) of y" rather than "the derivative (with respect to x) of x^{100}." Thus we can write:

$$\frac{d}{dx}(y)$$

or, more compactly,

$$\frac{dy}{dx}.$$

▶ **NOTE** This notation is a little misleading; we must not think of dy/dx as a ratio of two quantities dy and dx, even though it looks like a ratio. Instead, dy/dx is the *limit* of a ratio, as we saw in Section 1:

$$\frac{dy}{dx} = \lim_{\Delta x \to 0} \frac{\Delta y}{\Delta x}. \quad \blacktriangleleft$$

Now we can state the first derivative rule.

POWER RULE

If n is any constant, then the derivative of $f(x) = x^n$ is

$$f'(x) = nx^{n-1}.$$

In differential notation, we write

$$\frac{d}{dx}(x^n) = nx^{n-1}.$$

We shall give reasons for believing this in a moment, but first let us do several quick examples.

EXAMPLE 1

(a) $\dfrac{d}{dx}(x^3) = 3x^2$

(b) $\dfrac{d}{dx}(x^2) = 2x^1 = 2x$

(c) $\dfrac{d}{dx}\left(\dfrac{1}{x}\right) = \dfrac{d}{dx}(x^{-1}) = -x^{-2} = -\dfrac{1}{x^2}$

(d) $\dfrac{d}{dx}\sqrt{x} = \dfrac{d}{dx}(x^{1/2}) = \dfrac{1}{2}x^{-1/2} = \dfrac{1}{2\sqrt{x}}$

(e) If $f(x) = 1/x^2$, then $f'(x) = -2x^{-3} = -2/x^3$.

(f) If $f(x) = x$, then $f'(x) = 1x^0 = 1$.

(g) If $f(x) = 1$, then $f'(x) = 0$ (think of $1 = x^0$).

Before we go on... We did (a), (c) and (d) using the definition of the derivative in Examples 5, 6 and 7 of the previous section and we did (b) in Section 2. You should think about (f) and (g) graphically to see why these are obvious.

Some of the derivatives in the last example are very useful to remember, so we summarize them in a table.

TABLE OF DERIVATIVE FORMULAS

$f(x)$	x^n	x	1	$\dfrac{1}{x}$	$\dfrac{1}{x^2}$	$\sqrt{x}$
$f'(x)$	nx^{n-1}	1	0	$-\dfrac{1}{x^2}$	$-\dfrac{2}{x^3}$	$\dfrac{1}{2\sqrt{x}}$

We suggest that you add to this table as you pick up more information. It is *extremely* helpful to remember the derivatives of common functions such as $1/x$ and $\sqrt{x}$.

EXAMPLE 2

If $f(x) = x^{100}$, find $f'(0), f'(-1)$, and $f'(a)$.

SOLUTION $f'(x) = 100x^{99}$ by the power rule, so we get the required derivatives by substituting:

$$f'(0) = 100 \cdot 0^{99} = 0;$$
$$f'(1) = 100 \cdot 1^{99} = 100;$$
$$f'(a) = 100a^{99}.$$

Now let us see why the power rule is true. We have already checked the rule for x^3 and x^2. What follows is a rather slick proof that works for *any* positive integer. Don't worry if you might not think of doing it this way—neither did we at first! But mathematicians love to find the easiest or cleverest way of doing something. To follow this proof, you must recognize a nice little algebraic fact. First look at these identities.

$$a^1 - b^1 = (a - b)$$
$$a^2 - b^2 = (a - b)(a + b)$$
$$a^3 - b^3 = (a - b)(a^2 + ab + b^2)$$
$$a^4 - b^4 = (a - b)(a^3 + a^2b + ab^2 + b^3)$$
$$\ldots$$

(Use the distributive law to expand the right-hand side in each case.)

These examples generalize to give us the following formula.

DIFFERENCE OF TWO nTH POWERS

If a and b are real numbers, and n is a positive integer, then

$$a^n - b^n = (a - b)(a^{n-1} + a^{n-2}b + a^{n-3}b^2 + \ldots + ab^{n-2} + b^{n-1}).$$

PROOF OF THE POWER RULE FOR POSITIVE INTEGER POWERS

Write $f(x) = x^n$. Then

$$f'(x) = \lim_{h \to 0} \frac{f(x + h) - f(x)}{h}$$
$$= \lim_{h \to 0} \frac{(x + h)^n - x^n}{h}.$$

Think about these graphically to see why they must be true.

▼ EXAMPLE 5

Find the derivative of a general polynomial
$$f(x) = a_0 + a_1 x + a_2 x^2 + \ldots + a_n x^n.$$
$$(a_0, a_1, \ldots, a_n \text{ constants})$$

SOLUTION We take the derivative of each term and then add, getting
$$f'(x) = a_1 + 2a_2 x + \ldots + na_n x^{n-1}.$$

▼ EXAMPLE 6

Assuming the power rule is valid for all real powers, compute
$$\frac{d}{dx}\left(2 - \frac{3}{\sqrt{x}}\right).$$

SOLUTION
$$\frac{d}{dx}\left(2 - \frac{3}{\sqrt{x}}\right) = \frac{d}{dx}(2 - 3x^{-1/2})$$
$$= \frac{d}{dx}(2) - \frac{d}{dx}(3x^{-1/2})$$
$$= 0 - 3\frac{d}{dx}(x^{-1/2})$$
$$= -3\left(-\frac{1}{2}x^{-3/2}\right)$$
$$= \frac{3}{2x^{3/2}} = \frac{3}{2x\sqrt{x}}$$

Before we go on... Try to justify each of the steps above.

With practice, it should become second nature to you to use these rules. You will remember that if you see a sum of terms, you must take the derivative of each term separately, and that if a constant multiplies a function, then that constant will multiply the derivative of that function.

Here is something else to think about: If $y = mx + b$, then $y' = m$. It is not accidental that the derivative gives us the slope of the line. (Why does it happen?)

Finally, let us explain why the rule about sums is true.

PROOF OF THE RULE FOR SUMS

$$\frac{d}{dx}[f(x) + g(x)] = \lim_{h \to 0} \frac{[f(x+h) + g(x+h)] - [f(x) + g(x)]}{h}$$

$$= \lim_{h \to 0} \frac{[f(x+h) - f(x)] + [g(x+h) - g(x)]}{h}$$

$$= \lim_{h \to 0} \left[\frac{f(x+h) - f(x)}{h} + \frac{g(x+h) - g(x)}{h}\right]$$

$$= \lim_{h \to 0} \frac{f(x+h) - f(x)}{h} + \lim_{h \to 0} \frac{g(x+h) - g(x)}{h}$$

$$= \frac{d}{dx}[f(x)] + \frac{d}{dx}[g(x)]$$

The next-to-last step uses a property of limits: if $\lim_{x \to a} F(x) = L$ and $\lim_{x \to a} G(x) = M$, then $\lim_{x \to a}(F(x) + G(x)) = L + M$ (why should this be true?). The last step uses the definition of the derivative again.

We'll leave for you the cases of subtraction (which is essentially the same) and multiplication by a constant (which is easier).

▶ **10.4 EXERCISES**

In Exercises 1–14, calculate the derivative of the given function mentally.

1. $f(x) = x^3$ **2.** $f(x) = x^4$ **3.** $f(x) = 2x^{-2}$
4. $f(x) = 3x^{-1}$ **5.** $f(x) = -x^{1/4}$ **6.** $f(x) = -x^{-1/2}$
7. $f(x) = 2x^4 + 3x^3 - 1$ **8.** $f(x) = -x^3 - 3x^2 - 1$ **9.** $f(x) = -x + \frac{1}{x} + 1$
10. $f(x) = \frac{1}{x} + \frac{1}{x^2}$ **11.** $f(x) = 2\sqrt{x}$ **12.** $f(x) = \frac{2}{x}$
13. $f(x) = 3\sqrt[3]{x}$ **14.** $f(x) = \frac{2}{x^2} + \frac{3}{x^3}$

State the rules you use to obtain the indicated derivative in each of Exercises 15–30. You may use the power rule for any real exponent.

15. $y = 10$; $\frac{dy}{dx}$ **16.** $y = x^3$; $\frac{dy}{dx}$
17. $y = x^2 + x$; $\frac{dy}{dx}$ **18.** $y = x - 5$; $\frac{dy}{dx}$

19. $y = 4x^3 + 2x - 1$; $\dfrac{dy}{dx}$

20. $y = 4x^{-1} - 2x - 10$; $\dfrac{dy}{dx}$

21. $y = x^{104} - 99x^2 + x$; $\dfrac{dy}{dx}$

22. $y = \sqrt{x}(x + x^2)$; $\dfrac{dy}{dx}$

23. $s = \sqrt{t}(t - t^3) + \dfrac{1}{t^3}$; $\dfrac{ds}{dt}$

24. $s = 6t + \dfrac{6}{t}$; $\dfrac{ds}{dt}$

25. $V = \dfrac{4\pi r^3}{3}$; $\dfrac{dV}{dr}$

26. $A = 4\pi r^2 + 2\pi rh$ (h constant); $\dfrac{dA}{dr}$

27. $\dfrac{d}{dt}[t^2 + 4at^5]$ (a constant)

28. $\dfrac{d}{dt}[at^2 + bt + c]$ (a, b, c constant)

29. $\dfrac{d}{dx}\left(\sqrt{x}(1 + x)\right)$

30. $\dfrac{d}{dx}\left(\dfrac{1 + x}{x}\right)$

In Exercises 31–42, find the slope of the tangent to the graph of the given function at the indicated point.

31. $f(x) = x^3$; $(-1, -1)$

32. $g(x) = x^4$; $(-2, 16)$

33. $f(x) = 1 - 2x$; $(2, -3)$

34. $f(x) = \dfrac{x}{3} - 1$; $(-3, -2)$

35. $h(x) = x^{1/4}$; $(16, 2)$

36. $s(x) = x^{2/3}$; $(8, 4)$

37. $f(x) = x^{-1/2}$; $(2, 2^{-1/2})$

38. $f(x) = x^{-1/3}$; $(1, 1)$

39. $g(t) = \dfrac{1}{t^5}$; $(1, 1)$

40. $s(t) = \dfrac{1}{t^3}$; $(-2, -\tfrac{1}{8})$

41. $f(x) = \dfrac{x^2}{4} - \dfrac{x^3}{3}$; $(-1, \tfrac{7}{12})$

42. $f(x) = \dfrac{x^2}{2} + \dfrac{x}{4}$; $(2, \tfrac{5}{2})$

In Exercises 43–48, find the equation of the tangent line to the graph of the given function at the point with the indicated x-coordinate. In each case, sketch the curve together with the appropriate tangent line.

43. $f(x) = x^3$; $x = -1$

44. $f(x) = x^2$; $x = 0$

45. $f(x) = x + \dfrac{1}{x}$; $x = 2$

46. $f(x) = \dfrac{1}{x^2}$; $x = 1$

47. $f(x) = \sqrt{x}$; $x = 4$

48. $f(x) = 2x + 4$; $x = -1$

In Exercises 49–62, find the derivative of the given function.

49. $f(x) = x^2 - 3x + 5$

50. $f(x) = 3x^3 - 2x^2 + x$

51. $f(x) = x + \sqrt{x}$

52. $f(x) = x^{1/2} + 2x^{-1/2}$

53. $g(x) = \dfrac{1}{x^2} + \dfrac{2}{x^3}$ 54. $g(x) = \dfrac{2}{x} - \dfrac{2}{x^3} + \dfrac{1}{x^4}$

55. $h(x) = \dfrac{1}{x} + \dfrac{1}{x^2} + \dfrac{1}{x^3}$ 56. $h(x) = \dfrac{1}{\sqrt{x}}$

57. $r(x) = \sqrt{x} + \dfrac{1}{\sqrt{x}}$ 58. $r(x) = x + \dfrac{7}{\sqrt{x}}$

59. $f(x) = x(x^2 - 1/x)$ 60. $f(x) = x^{-1}(x - 2/x)$

61. $g(x) = \dfrac{x^2 - 2x^3}{x}$ 62. $f(x) = \dfrac{2x + x^2}{x}$

In Exercises 63–70, evaluate the given expression.

63. $\dfrac{d}{dx}\left(x + \dfrac{1}{x^2}\right)$ 64. $\dfrac{d}{dx}\left(2x - \dfrac{1}{x}\right)$

65. $\dfrac{d}{dx}(2x^{1.3} - x^{-1.2})$ 66. $\dfrac{d}{dx}(2x^{4.3} + x^{0.6})$

67. $\dfrac{d}{dt}(at^3 - 4at)$; (*a* constant)   68. $\dfrac{d}{dt}(at^2 + bt + c)$; (*a*, *b*, *c* constant)

69. $\dfrac{d}{dx}(\sqrt{x}(1 + x))$ 70. $\dfrac{d}{dx}\left(\dfrac{1 + x}{x}\right)$

In Exercises 71–76, find the indicated derivative.

71. $y = \dfrac{x^{10.3}}{2} + 99x^{-1}$; find $\dfrac{dy}{dx}$ 72. $y = \dfrac{x^{1.2}}{3} - \dfrac{x^{0.9}}{2}$; find $\dfrac{dy}{dx}$

73. $s = 2.35 + \dfrac{2.1}{t^{1.1}} - \dfrac{t^{0.6}}{2}$; find $\dfrac{ds}{dt}$ 74. $s = \dfrac{2}{t^{1.1}} + t^{-1.2}$; find $\dfrac{ds}{dt}$

75. $V = \dfrac{4}{3}\pi r^3$; find $\dfrac{dV}{dr}$ 76. $A = 4\pi r^2$; find $\dfrac{dA}{dr}$

In Exercises 77–84, find all values of x (if any) where the tangent line to the graph of the given equation is horizontal.

77. $y = 2x^2 + 3x - 1$ 78. $y = -3x^2 - x$

79. $y = 2x + 8$ 80. $y = -x + 1$

81. $y = x + \dfrac{1}{x}$ 82. $y = x + \sqrt{x}$

83. $y = \sqrt{x} - x$ 84. $y = \sqrt{x} + \dfrac{1}{x}$

85. Write out the proof that $\dfrac{d}{dx}[x^4] = 4x^3$.

86. Write out the proof that $\dfrac{d}{dx}[x^5] = 5x^4$.

87. Write out the proof that $\dfrac{d}{dx}[3x^2] = 3\dfrac{d}{dx}[x^2]$.

88. Write out the proof that $\dfrac{d}{dx}\left[\dfrac{1}{2}x^2\right] = \dfrac{1}{2}\dfrac{d}{dx}[x^2]$.

89. Write out the proof that
$$\dfrac{d}{dx}[x^2 + x^3] = \dfrac{d}{dx}[x^2] + \dfrac{d}{dx}[x^3].$$

90. Write out the proof that
$$\dfrac{d}{dx}[2x^2 - 3x^2] = \dfrac{d}{dx}[2x^2] - \dfrac{d}{dx}[3x^3].$$

APPLICATIONS

91. *Cost* Consider the two cost functions $C_1(x) = 10{,}000 + 5x - x^2/10$ and $C_2(x) = 20{,}000 + 10x - x^2/5$. How do rates of change of these cost functions at the same production levels compare?

92. *Cost* The cost of making x teddy bears at the Cuddly Companion Company used to be $C_1(x) = 100 + 40x - 0.001x^2$. Due to rising health insurance costs, it now is $C_2(x) = 1{,}000 + 40x - 0.001x^2$. How does the rate of change of cost at a production level of x teddy bears compare to what it used to be?

93. *Profit* The cost to manufacture x cases of beer per week is $C(x) = 10{,}000 + 30x - 0.01x^2$, while the revenue from selling x cases is $R(x) = 20x$. How must the rate of change of cost and the rate of change of revenue be related when the rate of change of profit is 0? What can you conclude about the cost and revenue when the profit is zero?

94. *Profit* The cost to manufacture x cases of beer per week is $C(x) = 10{,}000 + 30x - 0.01x^2$, while the revenue from selling x cases is $R(x) = 20x$. How must the rate of change of cost and the rate of change of revenue be related when the rate of change of profit is positive? What can you conclude about the cost and revenue when the profit is positive?

95. *Market Average* Joe Downs runs a small investment company from his basement. Every week he publishes a report on the success of his investments, including the progress of the infamous Joe Downs Average. At the end of one particularly memorable week he reported that the Average for that week had the value $A(t) = 1000 + 1500t - 800t^2 + 100t^3$ points, where t represents the number of days into the week; t ranges from 0 at the beginning of the week to 5 at end of the week. The graph of A is shown here. During the upswing at the beginning of the week, say halfway through the first day, how fast was the value of the Average increasing?

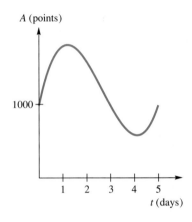

96. *Market Average* Referring to the Joe Downs Average in the previous exercise: During the downswing, say at the end of the third day, how fast was the value of the Average decreasing?

97. *Embryo Development* The oxygen consumption of a bird embryo increases from the time the egg is laid through the time the chick hatches. In the case of a typical galliform bird, the oxygen consumption (in milliliters per hour) can be approximated by

$$c(t) = -0.00271t^3 + 0.137t^2 - 0.892t + 0.149$$
$$(8 \le t \le 30),$$

where t is the time (in days) since the egg was laid.* (An egg will typically hatch at around $t = 28$.) Find $c'(15)$ and $c'(30)$. What do these results tell you about the embryo's oxygen consumption just prior to hatching?

▼ *The model approximates graphical data published in the article *The Brush Turkey* by Roger S. Seymour, *Scientific American*, December, 1991, pp. 108–14.

98. *Embryo Development* The oxygen consumption of a turkey embryo increases from the time the egg is laid through the time the chick hatches. In the case of a brush turkey, the oxygen consumption (in milliliters per hour) can be approximated by

$$c(t) = -0.00118x^3 + 0.119x^2 - 1.83x + 3.972$$
$$(20 \leq t \leq 50),$$

where t is the time (in days) since the egg was laid.* (An egg will typically hatch at around $t = 50$.) Find $c'(30)$ and $c'(50)$. What do these results tell you about the embryo's oxygen consumption just prior to hatching?

99. *Velocity* If a stone is dropped from a height of 100 feet, its height after t seconds is given by $s = 100 - 16t^2$.
  (a) Find its velocity at times $t = 0, 1, 2, 3$, and 4 seconds.
  (b) How long does it take to reach the ground, and how fast is it traveling when it hits the ground?

100. *Velocity* If a stone is thrown down at 120 ft/s from a height of 1,000 feet, its height after t seconds is given by $s = 1,000 - 120t - 16t^2$.
  (a) Find its velocity at times $t = 0, 1, 2, 3$, and 4 seconds.
  (b) How long does it take to reach the ground, and how fast is it traveling when it hits the ground?

101. *Volume* The volume, in cubic centimeters, of a spherical balloon is given by $V = \frac{4}{3}\pi r^3$, where r is its radius in centimeters, and $\pi \approx 3.141592$. Find $V'(10)$, and interpret the result.

102. *Volume* The volume, in cubic centimeters, of an ellipsoid with a circular cross section of radius r centimeters. is given by $V = \frac{4}{3}\pi r^2 s$, where $\pi \approx 3.141592$, and s is as shown in the figure.

If s is fixed at 2 cm, find the rate of change of V with respect to r, and evaluate it at $r = 1$. Interpret the result.

COMMUNICATION AND REASONING EXERCISES

103. *Tangent Lines* What instructions would you give to a fellow student who wanted to accurately graph the tangent line to the curve $y = 3x^2$ at the point $(-1, 3)$?

104. *Tangent Lines* What instructions would you give to a fellow student who wanted to accurately graph a line at right angles to the curve $y = \frac{4}{x}$, at the point where $x = 0.5$?

105. *Tangent Lines* Consider $f(x) = x^2$ and $g(x) = 2x^2$. How do the slopes of the tangent lines of f and g at the same x compare?

106. *Tangent Lines* Consider $f(x) = x^3$ and $g(x) = x^3 + 3$. How do the slopes of the tangent lines of f and g compare?

107. *Tangent Lines* Suppose that $f(x)$ and $g(x)$ are two functions, and $f(x) - g(x)$ has a horizontal tangent at $x = a$. How do the slopes of $f(x)$ and $g(x)$ at $x = a$ compare?

108. *Tangent Lines* Suppose that $f(x)$ and $g(x)$ are two functions, and $f(x) + g(x)$ has a horizontal tangent at $x = a$. How do the slopes of $f(x)$ and $g(x)$ at $x = a$ compare?

109. How would you respond to an acquaintance who says, "I finally understand what the derivative is: it is nx^{n-1}! Why weren't we taught that in the first place instead of the difficult way using limits?"

110. Following is an excerpt from your friend's graded homework:

$$3x^4 + 11x^5 = 12x^3 + 55x^4 \quad \text{✗ WRONG } -8$$

Why was it marked wrong?

---

* The model approximates graphical data published in the article *The Brush Turkey*, by Roger S. Seymour, *Scientific American*, December 1991, pp. 108–14.

▶ **NOTES**

1. In general, the difference quotient $[C(x + h) - C(x)]/h$ gives the **average cost per item** to produce h more items at a current production level of x items.
2. Notice that $C'(x)$ is much easier to calculate than $[C(x + h) - C(x)]/h$ (try it). ◀

As the following examples show, the term "marginal" can apply to quantities other than cost.

▼ **EXAMPLE 2**   Marginal Revenue

Economist Henry Schultz calculated the following demand function for corn:

$$p = \frac{6{,}570{,}000}{q^{1.3}}.$$

Here, p is the price in dollars per bushel, and q is the number of bushels of corn that could be sold at the price p in one year.*

(a) Calculate the annual revenue as a function of the number of bushels q.
(b) Calculate $R(300{,}000)$, and interpret the result.
(c) Calculate the marginal revenue when $q = 300{,}000$ bushels, and interpret the result.

SOLUTION

(a) To calculate the annual revenue R, use $R = pq$ (revenue = price × quantity). Since we want revenue as a function of q only, we substitute for p using the demand equation.

$$R = pq$$

$$R(q) = \frac{6{,}570{,}000}{q^{1.3}} q$$

$$= \frac{6{,}570{,}000}{q^{0.3}}.$$

This gives the annual revenue as a function of the number of bushels of corn sold per year.

▼ *This demand function is based on data for the period 1915–1929. Notice that we have written p as a function of q, rather than the other way around, which would be more natural. Economists often specify demand functions this way, and we will find this convenient for the calculations. Source: Henry Schultz: *The Theory and Measurement of Demand* (as cited in *Introduction to Mathematical Economics* by A. L. Ostrosky, Jr., and J. V. Koch (Waveland Press, Prospect Heights, Illinois, 1979.)

(b) Since

$$R(q) = \frac{6{,}570{,}000}{q^{0.3}},$$

we have

$$R(300{,}000) = \frac{6{,}570{,}000}{(300{,}000)^{0.3}}$$
$$= 149{,}426.97.$$

Thus, if 300,000 bushels of corn were sold in one year, the total revenue for that year would be $149,426.97.

(c) To obtain the *marginal* revenue, we need to take the derivative of the revenue function. Since

$$R(q) = \frac{6{,}570{,}000}{q^{0.3}} = 6{,}570{,}000 q^{-0.3},$$

we have

$$R'(q) = -(0.3)6{,}570{,}000 q^{-1.3}$$
$$= -1{,}971{,}000 q^{-1.3}.$$

This is the marginal revenue function. Its units are units of R per unit of q: that is, dollars per bushel. We are asked to find $R'(300{,}000)$.

$$R'(300{,}000) = -1{,}971{,}000(300{,}000)^{-1.3}$$
$$\approx -0.1494 \text{ dollars per bushel}$$
$$\approx -15¢ \text{ per bushel}$$

Since the derivative of a quantity measures its rate of change, we conclude that, at a sales level of 300,000 bushels per year, the annual revenue is dropping at a rate of 15¢ per additional bushel sold. In other words, each additional bushel a farmer sells will result in a decrease of 15¢ in annual revenue.

Before we go on ...

Q How is it possible for the revenue to decrease with increasing sales?

A In order to sell more corn, a farmer would have to lower his price. That lower price outweighs the increase in sales, and revenue goes down.

Q In order to *increase* revenue, what should a farmer do?

A The fact that the marginal revenue is negative implies that if q *decreases*, then R will *increase* by 15¢ per bushel. In other words, a farmer should raise the price to increase annual revenue, even though this will mean a decrease in the quantity sold.

▼ **EXAMPLE 3** Marginal Profit

In Chapter 1, we found that the demand equation for rubies at Royal Ruby Retailers (RRR) is given by

$$q = -\frac{4p}{3} + 80$$

where p is the retail price it charges per ruby and q is the number of rubies RRR can sell per week at $\$p$ per ruby. Assume that RRR pays $\$15$ per ruby. (These are rather cheap rubies.)

(a) Calculate the weekly revenue and profit as functions of q.
(b) Calculate the marginal profit function, $P'(q)$.
(c) Calculate the profit and marginal profit for $q = 20$, $q = 30$, and $q = 40$, and interpret the results.

SOLUTION

(a) The weekly revenue is given by $R = pq$. Since we need to express R as a function of q only, we must replace p in this equation by a function of q. Since the relationship between p and q is given in the demand equation, and we want p as a function of q, we need to first rewrite the demand equation by solving for p:

$$q = -\frac{4p}{3} + 80$$

gives

$$p = 60 - \frac{3q}{4}.$$

We now substitute this expression for p in $R = pq$ to obtain

$$R(q) = \left(60 - \frac{3q}{4}\right)q = 60q - \frac{3q^2}{4}.$$

This is the revenue as a function of q. For the profit function, recall that

$$P = R - C \text{ (Revenue} - \text{Cost)}.$$

Since RRR pays $\$15$ per ruby, the cost of q rubies is $15q$. Thus,

$$P(q) = \left(60q - \frac{3q^2}{4}\right) - 15q$$

$$= 45q - \frac{3q^2}{4}.$$

(b) The marginal profit function is the derivative,

$$P'(q) = 45 - \frac{3q}{2},$$

and its units are dollars per ruby. Thus, $P'(q)$ gives the rate of change of profit in dollars per ruby sold (per week).

(c) We have

$$P'(20) = 45 - \frac{3(20)}{2} = \$15 \text{ per ruby.}$$

This means that at a demand level of 20 rubies per week, the profit is increasing by \$15 for every additional ruby RRR can sell. It would therefore pay RRR to sell more rubies, which it can do by lowering the price.

$$P'(30) = 45 - \frac{3(30)}{2} = \$0 \text{ per ruby}$$

This means that at a demand level of 30 rubies per week, the profit is neither increasing nor decreasing.

$$P'(40) = 45 - \frac{3(40)}{2} = -\$15 \text{ per ruby}$$

This means that at a demand level of 40 rubies per week, the profit is decreasing by \$15 for every additional ruby they sell. RRR should therefore sell fewer rubies, which they can do by increasing the price.

Before we go on... This analysis shows that RRR should sell more than 20 rubies per week, but less than 40 rubies per week, and the fact that $P'(30) = 0$ tells us that RRR should adjust the price to sell exactly 30 rubies per week. If it sells slightly fewer—say, 29 per week—then the marginal profit will be positive (as you can check by calculating $P'(29)$), indicating that it should sell more than 29. Similarly, $P'(31)$ is negative, indicating that it should sell fewer than 31. In general, *for maximum profit, $P'(q)$ must be 0.* Figure 2 shows the graph of $P(q)$ for this example, and you can see clearly in the graph that $P'(20) > 0$, $P'(40) < 0$, $P'(30) = 0$, and the largest value of $P(q)$ occurs at $q = 30$.

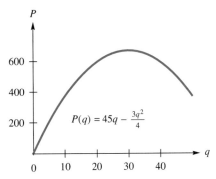

FIGURE 2

▼ **EXAMPLE 4**   Marginal Product

Precision Manufacturers, Inc. is informed by a consultant that its annual profit is given by

$$P = -200{,}000 + 4{,}000q - 0.5q^2,$$

where q is the number of surgical lasers it sells per year. The consultant also informs the company that the number of surgical lasers it can manufacture per year depends on the number n of assembly-line workers it employs according to the equation

$$q = 100n + 0.1n^2.$$

(a) Express P as a function of n, and find $P'(n)$. $P'(n)$ is called the **marginal product** at the employment level of n assembly line workers. What are its units?

(b) Calculate $P(10)$ and $P'(10)$, and interpret the results.

(c) Precision Manufacturers currently employs 10 assembly line workers and is considering laying off assembly line workers. What advice would you give the company's management?

SOLUTION

(a) Since P is given in terms of q, and q is given in terms of n, we can obtain P as a function of n by substituting the expression for q in the expression for P.

$$P = -200{,}000 + 4{,}000q - 0.5q^2$$
$$P(n) = -200{,}000 + 4{,}000(100n + 0.1n^2) - 0.5(100n + 0.1n^2)^2$$
$$= -200{,}000 + 400{,}000n + 400n^2 - 0.5(10{,}000n^2 + 20n^3 + 0.01n^4)$$
$$= -200{,}000 + 400{,}000n + 400n^2 - 5{,}000n^2 - 10n^3 - 0.005n^4$$
$$= -200{,}000 + 400{,}000n - 4{,}600n^2 - 10n^3 - 0.005n^4$$

Then

$$P'(n) = 400{,}000 - 9{,}200n - 30n^2 - 0.02n^3.$$

The units of $P'(n)$ are profit (in dollars) per worker.

(b) The formula

$$P(n) = -200{,}000 + 400{,}000n - 4{,}600n^2 - 10n^3 - 0.005n^4$$

gives

$$P(10) = -200{,}000 + 400{,}000(10) - 4{,}600(10)^2 - 10(10)^3 - 0.005(10)^4$$
$$= \$3{,}329{,}950.$$

Thus, Precision Manufacturers will make an annual profit of $3,329,950 if it employs 10 assembly line workers. The formula

$$P'(n) = 400{,}000 - 9{,}200n - 30n^2 - 0.02n^3$$

gives
$$P'(10) = 400{,}000 - 9{,}200(10) - 30(10)^2 - 0.02(10)^3$$
$$= \$304{,}980 \text{ per worker.}$$

Thus, at an employment level of 10 assembly line workers, annual profit is increasing at a rate of $304,980 per additional worker. In other words, if the company were to employ one more assembly line worker, its annual profit would increase by approximately $304,980.

(c) Since the marginal product is positive, profits will increase if the company increases the number of workers and will decrease if it decreases the number of workers, so your advice would be to hire additional assembly line workers. Downsizing the assembly line work force would reduce the company's annual profits.

Before we go on... The algebra involved in finding P as a function of n was rather messy. In the next chapter we shall see a way to find $P'(n)$ that is not so algebraically intense.

The following question might have occurred to you:

Q How many additional assembly line workers should the company hire to obtain the maximum annual profit?

A Since we have profit as a function of the number of workers n, we can graph the function $P(n)$. Figure 3 shows the graph of $P(n)$ for $0 \leq n \leq 80$, as drawn by a graphing calculator. (For the y-axis scale, we used $-2{,}000{,}000 \leq P \leq 8{,}000{,}000$.)

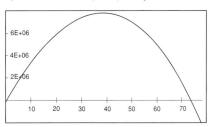

FIGURE 3

The horizontal axis represents the number of assembly line workers, while the vertical axis represents annual profit. At the point on the graph where $x = 10$, the slope is positive (as confirmed by our calculation of $P'(10)$). At approximately $n = 40$, the slope is zero (so the marginal product is zero) and the profit is largest. This tells us that the company should employ approximately 40 assembly line workers for a maximum profit. Thus, the company should hire approximately 30 additional assembly line workers.

Notice that Figure 3 also shows us that the company begins taking a loss at an employment level of about 75 assembly line workers.

10.5 EXERCISES

Consider each of the functions in Exercises 1–6 as representing the cost to manufacture x items. Find the average costs of manufacturing h more items at a production level of x, where x is as indicated and h = 50, 10, and 1. Also, find the marginal cost at the given production level x. (For the average costs, see the note after Example 1.)

1. $C(x) = 10,000 + 5x - \dfrac{x^2}{10,000}$; $x = 1,000$

2. $C(x) = 20,000 + 7x - \dfrac{x^2}{20,000}$; $x = 10,000$

3. $C(x) = 1,000 + 10x - \sqrt{x}$; $x = 100$

4. $C(x) = 10,000 + 20x - \sqrt{x}$; $x = 10$

5. $C(x) = 15,000 + 100x + \dfrac{1,000}{x}$; $x = 100$

6. $C(x) = 20,000 + 50x + \dfrac{10,000}{x}$; $x = 100$

In Exercises 7 and 8, find the marginal cost, marginal revenue, and marginal profit functions, and find all values of x for which the marginal profit is zero. Interpret your answer.

7. $C(x) = 4x$; $R(x) = 8x - \dfrac{x^2}{1,000}$

8. $C(x) = 5x^2$; $R(x) = x^3 + 7x + 10$

APPLICATIONS

9. *Marginal Cost* The cost of producing x teddy bears per day at the Cuddly Companion Company is calculated by their marketing staff to be given by the formula
$$C(x) = 100 + 40x - .001x^2.$$
Find the marginal cost function, and use it to estimate how fast the cost is going up at a production level of 100 Teddy bears. Compare this with the exact cost of producing the 101st teddy bear.

10. *Marginal Cost* Referring to the cost equation in Exercise 9, find the production level for which the marginal cost is zero.

11. *Marginal Net Profit* Suppose that $P(x)$ represents the net profit on the sale of x videocassettes. If $P(1,000) = 3,000$, and $P'(1,000) = -3$, what does this tell you?

12. *Marginal Loss* An automobile retailer calculates that its loss on the sale of Type M cars is given by $L(50) = 5,000$ and $L'(50) = -200$, where $L(x)$ represents the loss on the sale of x Type M cars. What does this tell you?

13. *Marginal Profit* Your monthly profit (in dollars) from selling magazines is given by
$$P = 5n + \sqrt{n},$$
where n is the number of magazines you sell in a month. If you are currently selling $n = 50$ magazines per month, find your profit and your marginal profit. Interpret your answers.

14. *Marginal Profit* Your monthly profit (in dollars) from your newspaper route is given by
$$P = 2n - \sqrt{n},$$
where n is the number of subscribers on your route. If you currently have 100 subscribers, find your profit and your marginal profit. Interpret your answers.

15. *Marginal Product* A car wash firm calculates that its daily profit depends on the number n of workers it employs according to the formula
$$P = 400n - 0.5n^2.$$
Calculate the marginal product at an employment level of 50 workers, and interpret the result.

16. *Marginal Product* Repeat Exercise 15 using the formula
$$P = -100n + 25n^2 - 0.005n^4.$$

17. *Marginal Revenue* Assume that the demand function for tuna in a small coastal town is given by
$$p = \dfrac{50,000}{q^{1.5}},$$

where p is the price (in dollars) per pound of tuna, and q is the number of pounds of tuna that can be sold at the price p in one month.
   (a) Calculate the price that the town's fishery should charge for tuna in order to produce a demand of 500 pounds of tuna per month.
   (b) Calculate the annual revenue R as a function of the number of pounds of tuna q.
   (c) Calculate $R(500)$, and interpret the result.
   (d) Calculate the marginal revenue function and its value at $q = 500$ pounds, and interpret the result.
   (e) If the town fishery's monthly tuna catch amounted to 600 pounds of tuna, and the price is at the level in part (a), would you recommend that the fishery raise or lower the price of tuna in order to increase its revenue?

18. Repeat Exercise 17 assuming a demand equation of
$$p = \frac{80,000}{q^{1.7}}.$$

19. *Marginal Revenue and Marginal Profit* The demand for poultry can be modeled as
$$q = 63.15 - 0.45p + 0.12b$$
where q is the per capita demand for poultry in pounds per year, p is the wholesale price of poultry in cents per pound, and b is the wholesale price of beef in cents per pound.* Assume that the wholesale price of beef is fixed at 45¢ per pound.
   (a) Find the revenue as a function of q, and hence obtain the marginal revenue as a function of q. (Round constants to two decimal places.)
   (b) Find the annual per capita revenue that will result at a demand level of 50 pounds of poultry per year, and estimate the change in revenue that will result if the price is raised to yield a demand level of 49 pounds of poultry per year.
   (c) If a farmer breeds chickens at an average cost of 10¢ per pound, find the annual profit function P in terms of the per capita demand for poultry, evaluate $P'(50)$, and interpret the result.

20. *Demand for Poultry* Referring to the model for demand for poultry in Exercise 19, express the annual per capita revenue R as a function of the price of beef if the wholesale price of poultry is fixed at 40¢ per pound. Calculate $R(45)$ and $R'(45)$, and interpret the results.

21. *Revenue* In Chapter 1, we found that the demand equation for rubies at Royal Ruby Retailers (RRR) is given by
$$q = -\frac{4p}{3} + 80,$$
where p is the retail price RRR charges per ruby and q is the number of rubies RRR can sell per week at $\$p$ per ruby.
   (a) Express the revenue as a function of p, and calculate the marginal revenue as a function of p.
   (b) How fast is the revenue increasing as the price goes up at the following sales prices: (i) $20 per ruby; (ii) $30 per ruby; (iii) $40 per ruby?
   (c) Interpret the results in part (b).

22. *Housing Costs*[†] The cost C of building a house is related to the number k of carpenters used and the number e of electricians used by the formula
$$C = 15,000 + 50k^2 + 60e^2.$$
   (a) Assuming that 10 carpenters are currently being used, find the marginal cost as a function of e.
   (b) If 10 carpenters and 10 electricians are currently being used, use your answer to part (a) to estimate the cost of hiring an additional electrician.
   (c) If 10 carpenters and 10 electricians are currently being used, what is the cost of hiring an additional carpenter?

23. *Emission Control* The cost of controlling emissions at a firm goes up rapidly as the amount of emissions reduced goes up. Here is a possible model:
$$C(q) = 4,000 + 100q^2,$$
where q is the reduction in emissions (in pounds of pollutant per day) and C is the daily cost (in dollars) of this reduction.
   (a) If a firm is currently reducing its emissions by 10 pounds each day, what is the marginal cost of reducing emissions further?

---

*This equation is based on data from poultry sales in the period from 1950 to 1984. (Source: A.H. Studenmund, *Using Econometrics*, Second Ed. (New York: HarperCollins, 1992), pp. 180–81).

[†] Based on an exercise in *Introduction to Mathematical Economics* by A. L. Ostrosky, Jr., and J.V. Koch (Prospect Heights, IL: Waveland Press, 1979).

(b) Government clean-air subsidies to the firm are based on the formula

$$S(q) = 500q,$$

where q is again the reduction in emissions in pounds and S is the subsidy. At what reduction level does the marginal cost surpass the marginal subsidy?

(c) Calculate the net cost function, $N(q) = C(q) - S(q)$, given the cost function and subsidy above, and find the value of q that gives the lowest net cost. What is this lowest net cost? Compare your answer to that for (b), and comment on what you find.

24. *Taxation Schemes* Here is a curious proposal for taxation rates based on income:

$$R(i) = \frac{\sqrt{i}}{1{,}000},$$

where i represents total annual income and $R(i)$ is the income tax rate as a percentage of total annual income. (Thus, for example, an income of $50,000 per year would be taxed at about 22%, while an income of double that amount would be taxed at about 32%.)*

(a) Calculate the after-tax (net) income $N(i)$ an individual can expect to earn as a function of income i.

(b) Calculate an individual's marginal after-tax income at income levels of $100,000 and $500,000.

(c) At what income does an individual's marginal after-tax income become negative? What is the after-tax income at that level, and what does this signify?

(d) What do you suspect is the most anyone can earn after taxes? (See the footnote.)

25. *Fuel Economy* Your Porsche's gas mileage (in miles per gallon) is given as a function $M(x)$ of speed x in mph. It is found that

$$M'(x) = \frac{3600x^{-2} - 1}{(3600x^{-1} + x)^2}.$$

Find $M'(10)$, $M'(60)$, and $M'(70)$. What do the answers tell you about your car?

26. *Marginal Revenue* The estimated marginal revenue for sales of ESU soccer team T-shirts is given by

$$R'(p) = \frac{(8 - 2p)e^{-p^2+8p}}{10{,}000{,}000}$$

where p is the price (in dollars) the soccer players charge for each shirt. Find $R'(3)$, $R'(4)$, and $R'(5)$. What do the answers tell you?

27. *Transportation Costs* Before the Alaskan pipeline was built, there was speculation as to whether it might be more economical to transport the oil by large tankers. The following cost equation was estimated by National Academy of Sciences:

$$C = 0.03 + \frac{10}{T} - \frac{200}{T^2},$$

where C is the cost in dollars of transporting one barrel of oil 1,000 nautical miles, and T is the size of an oil tanker in deadweight tons.†

(a) How much would it cost to transport 100 barrels of oil in a tanker weighing 1,000 tons?

(b) By how much is this cost increasing or decreasing as the weight of the tanker increases from 1,000 tons?

28. *Transportation Costs* Referring to the cost equation in Exercise 27, find the value of T so that $C'(T) = 0$. Interpret the result. By calculating values of $C(T)$ for T close to and on either side of this amount, what more can you say?

29. *Marginal Cost* (from the GRE economics test) In a multiple-plant firm in which the different plants have different and continuous cost schedules, if costs of production for a given output level are to be minimized, which of the following is essential?

(a) Marginal costs must equal marginal revenue.
(b) Average variable costs must be the same in all plants.
(c) Marginal costs must be the same in all plants.
(d) Total costs must be the same in all plants.
(e) Output per man-hour must be the same in all plants.

▼ *This model has the following interesting feature: an income of a million dollars per year would be taxed at 100%, leaving the individual penniless!

† Source: *Use of Satellite Data on the Alaskan Oil Marine Link*, Practical Applications of Space Systems: Cost and Benefits. (National Academy of Sciences, Washington, D.C., 1975, p. B–23)

30. *Study Time (from the GRE economics test)* A student has a fixed number of hours to devote to study and is certain of the relationship between hours of study and the final grade for each course. Grades are given on a numerical scale (e.g., 0 to 100), and each course is counted equally in computing the grade average. In order to maximize his or her grade average, the student should allocate these hours to different courses so that
(a) the grade in each course is the same;
(b) the marginal product of an hour's study (in terms of final grade) in each course is zero;
(c) the marginal product of an hour's study (in terms of final grade) in each course is equal, although not necessarily equal to zero;
(d) the average product of an hour's study (in terms of final grade) in each course is equal;
(e) the number of hours spent in study for each course are equal.

31. *Marginal Product (from the GRE economics test)* Assume that the marginal product of an additional senior professor is 50 percent higher than the marginal product of an additional junior professor and that junior professors are paid one-half the amount that senior professors receive. With a fixed overall budget, a university that wishes to maximize its quantity of output from professors should do which of the following?
(a) Hire equal numbers of senior professors and junior professors.
(b) Hire more senior professors and junior professors.
(c) Hire more senior professors and discharge junior professors.
(d) Discharge senior professors and hire more junior professors.
(e) Discharge all senior professors and half of the junior professors.

32. *Marginal Product (based on a question from the GRE economics test)* Assume that the marginal product of an additional senior professor is twice the marginal product of an additional junior professor and that junior professors are paid two-thirds the amount that senior professors receive. With a fixed overall budget, a university that wishes to maximize its quantity of output from professors should do which of the following?
(a) Hire equal numbers of senior professors and junior professors.
(b) Hire more senior professors and junior professors.
(c) Hire more senior professors and discharge junior professors.
(d) Discharge senior professors and hire more junior professors.
(e) Discharge all senior professors and half of the junior professors.

COMMUNICATION AND REASONING EXERCISES

33. Carefully explain the difference between *cost* and *marginal cost* (a) in terms of their mathematical definition, (b) in terms of graphs, and (c) in terms of interpretation.

34. If your analysis of a manufacturing company yielded positive marginal profit but negative profit at the company's current production levels, what would you advise the company to do?

35. If a company's marginal average cost is zero at the current production level, positive for a slightly higher production level, and negative for a slightly lower production level, what should you advise the company to do?

36. The **acceleration** of cost is defined as the derivative of the marginal cost function: that is, the derivative of the derivative—or *second derivative*—of the cost function. What are the units of acceleration of cost, and how does one interpret this measure?

10.6 MORE ON LIMITS, CONTINUITY, AND DIFFERENTIABILITY

LIMITS

As we saw in Section 3, limits need not always exist. The next two examples show ways in which a limit can fail to exist.

EXAMPLE 1

Let $f(x) = \dfrac{|x|}{x}$. Does $\lim_{x \to 0} f(x)$ exist?

SOLUTION We shall investigate this limit using both the numerical and the graphical approaches.

Numerical Approach We construct a table of values, with x approaching 0 from both sides.

x approaches 0 from the left. → ← x approaches 0 from the right.

x	−0.1	−0.01	−0.001	−0.0001	0	0.0001	0.001	0.01	0.1
$f(x) = \dfrac{|x|}{x}$	−1	−1	−1	−1		1	1	1	1

The table shows that $f(x)$ does not approach the same limit as x approaches 0 from both sides. In fact,

$$\lim_{x \to 0^-} f(x) = -1$$

and

$$\lim_{x \to 0^+} f(x) = 1.$$

Since the one-sided limits have different values, we conclude that $\lim_{x \to 0} f(x)$ does not exist.

Graphical Approach Consider the graph of $f(x)$. The domain of f consists of all real numbers except 0, and, as the table suggests,

$$f(x) = \begin{cases} 1 & \text{if } x > 0 \\ -1 & \text{if } x < 0. \end{cases}$$

(think about what $|x|$ means to see why). The graph of f is shown in Figure 1. As in Section 3, we imagine a pencil point or the "trace" cursor on a graphing calculator approaching $x = 0$ from either side (Figure 2).

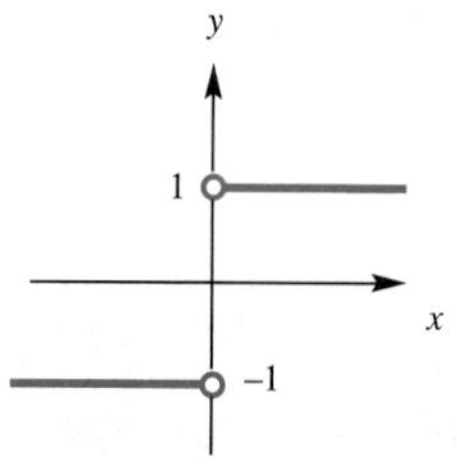

Looking at the y-coordinates, we can see graphically that

$$\lim_{x \to 0^-} f(x) = -1$$

and

$$\lim_{x \to 0^+} f(x) = 1.$$

Since the one-sided limits have different values, we conclude that $\lim_{x \to 0} f(x)$ does not exist.

If your graphing calculator does not have an absolute value function, you can still enter the function $f(x)$ by using the fact that

$$|x| = \sqrt{x^2},$$

FIGURE 1

Graph of $f(x) = \dfrac{|x|}{x}$

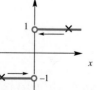

FIGURE 2

so you can enter

$$Y_1 = ((X^2)^\wedge(0.5))/X$$

You can now use "trace" to move the cursor, reading the coordinates as you go.

EXAMPLE 2

Does $\lim_{x \to 0^+} \frac{1}{x}$ exist?

SOLUTION We shall look at this limit using both the numerical and graphical approaches.

Numerical Approach Since we are asked for only the right-hand limit, we need only list values of x approaching 0 from the right.

←x approaches 0 from the right.

x	0	0.0001	0.001	0.01	0.1
$f(x) = \frac{1}{x}$		10,000	1,000	100	10

What seems to be happening as x approaches 0 from the right is that $f(x)$ is increasing **without bound.** That is, if you name any number, no matter how large, $f(x)$ will be even larger if x is sufficiently close to zero. Since $f(x)$ is not approaching a specific real number, the limit does not exist. Since $f(x)$ is becoming arbitrarily large, we also say that the limit **diverges to $+\infty$,** and we write

$$\lim_{x \to 0^+} \frac{1}{x} = +\infty.$$

Graphical Approach We again use the graphing calculator experiment. First, recall that the graph of $f(x) = \frac{1}{x}$ is the standard hyperbola shown in Figure 3.

The figure also shows the pencil point moving so that its x-coordinate approaches 0 from the right. Since the point moves along the graph, it is forced to go higher and higher. In other words, its y-coordinate becomes larger and larger, approaching $+\infty$. Thus, we conclude that

$$\lim_{x \to 0^+} \frac{1}{x} = +\infty.$$

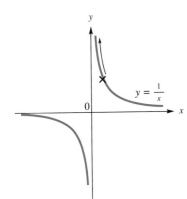

FIGURE 3

Before we go on... You should also check that

$$\lim_{x \to 0^-} \frac{1}{x} = -\infty.$$

We say that as x approaches 0 from the left, $\frac{1}{x}$ diverges to $-\infty$.

CONTINUITY

Recall that f is continuous at a if $\lim_{x \to a} f(x) = f(a)$. We saw in Section 3 that all closed-form functions are continuous on their domains. There are a number of ways in which a function can fail to be continuous at a point in its domain. The next example shows some of them.

▼ **EXAMPLE 3**

Let $f(x)$, $g(x)$, $h(x)$, and $k(x)$ be specified by the graphs in Figure 4. Determine which, if any, are continuous on their domains.

(a)
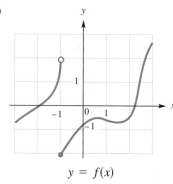
$y = f(x)$

(b)
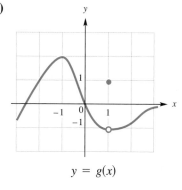
$y = g(x)$

(c)
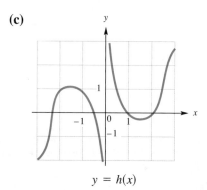
$y = h(x)$

(d)
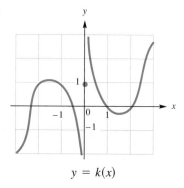
$y = k(x)$

FIGURE 4

SOLUTION Looking at the graph of f, we notice that there is a break at the point where $x = -1$, so we suspect that it is not continuous there. If we use the graphical method to investigate the limit as $x \to -1$, we find that the limit does not exist. Thus, since -1 is a point in the domain of f ($f(-1) = -2$ by the graph) and the limit as $x \to -1$ fails to exist, f is not continuous at -1 and is not continuous on its domain. The particular kind of discontinuity we see here is called a **jump discontinuity**. Since the left and right limits are not equal, the value of f takes a sudden jump when x passes -1.

The graph of g has a "misplaced point" at $x = 1$. We can see graphically that
$$\lim_{x \to 1} g(x) = -1 \neq g(1) = 1.$$
Therefore, g is not continuous at 1, so it is not continuous on its domain. This type of discontinuity is called a **removable discontinuity,** since it could be removed by redefining g at one point: redefining $g(1) = -1$. If we made this change in g, the function would be continuous on its domain.

The graph of h has a break at $x = 0$, but notice that 0 is not in the domain of h: $h(0)$ is not defined. If you choose any a that *is* in its domain, h will be continuous at a, so we conclude that h is continuous on its domain. The reason that its graph is broken is that its domain is broken.

Finally, the function k is almost the same as the function h, except that it is defined at $x = 0$, since $k(0) = 1$. Since the limit of k as $x \to 0$ does not exist, we conclude that k is *not* continuous at 0 and thus not continuous on its domain.

DIFFERENTIABILITY

Since the derivative is defined using a limit, the derivative will not exist if the limit does not. If $f'(a)$ exists, we say that f is **differentiable at a.**

▼ EXAMPLE 4

If $f(x) = |x|$ find $f'(0)$.

SOLUTION   We compute
$$f'(x) = \lim_{h \to 0} \frac{f(x + h) - f(x)}{h}$$
$$f'(0) = \lim_{h \to 0} \frac{f(0 + h) - f(0)}{h}$$
$$= \lim_{h \to 0} \frac{|0 + h| - |0|}{h}$$
$$= \lim_{h \to 0} \frac{|h|}{h}.$$

In Example 1, we saw that this limit does not exist. We say that the derivative of f does not exist at 0, or that f is not differentiable at 0. You can see in the graph of f why this happens (Figure 5).

At $x = 0$ there is a sharp corner. What would be the tangent line there? There is no single reasonable answer. If you look at secant lines through $(0, 0)$, you see lines with slope -1 or $+1$, depending on whether you take the second point to the left or to the right of the origin (try it). We simply say that there is no tangent line, and thus no derivative, at 0.

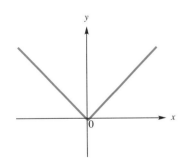

FIGURE 5
Graph of $f(x) = |x|$

(Note that we are only approaching $+\infty$ from the left, as we can hardly approach it from the right!) What seems to be happening is that $f(x)$ is approaching 2. Thus, we write

$$\lim_{x \to +\infty} f(x) = 2.$$

(b) Here, x is approaching $-\infty$, so we make a similar table, this time with x assuming negative values of greater and greater magnitude (read this table from right to left, as you would the graph).

⟵ x approaching $-\infty$.

x	$-100{,}000$	$-10{,}000$	$-1{,}000$	-100	-10
$f(x) = \dfrac{2x^2 - 4x}{x^2 - 1}$	2.0000	2.0004	2.0040	2.0402	2.4242

Once again, $f(x)$ is approaching 2. Thus,

$$\lim_{x \to -\infty} f(x) = 2$$

as well.

Graphical Approach   Figure 8 shows a graphing calculator plot of the function f.

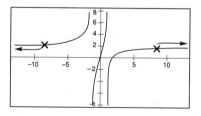

FIGURE 8

We have included the cursor marks for the graphing calculator experiment. To see what happens as $x \to +\infty$, start with your pencil point at the cursor mark on the right, and move to the right toward $+\infty$ while you read off the y-coordinate. The further you move to the right, the closer the y-coordinate gets to 2, showing once again that $f(x) \to 2$ as $x \to +\infty$. Similarly, if you start at the cursor mark on the left and move left toward $-\infty$, the y-coordinate also approaches 2. Thus, $f(x) \to 2$ as $x \to -\infty$.

Algebraic Approach   While calculating the values for the tables in the numerical approach, you might have noticed that the highest power of x in both the numerator and denominator dominated the calculations. For instance, when $x = 100{,}000$, the $2x^2$ in the numerator has the value of $20{,}000{,}000{,}000$, whereas $4x$ has the comparatively insignificant value of $400{,}000$. Similarly, the x^2 in the denominator overwhelms the -1. In other words, for large values of x (or negative values with large magnitude),

$$\frac{2x^2 - 4x}{x^2 - 1} \approx \frac{2x^2}{x^2} = 2.$$

Alternatively, do the following bit of algebra first.

$$\frac{2x^2 - 4x}{x^2 - 1} = \frac{(2x^2 - 4x)/x^2}{(x^2 - 1)/x^2} \quad \text{Divide top and bottom by the highest power of } x.$$

$$= \frac{2 - \dfrac{4}{x}}{1 - \dfrac{1}{x^2}}.$$

Now, as x approaches $+\infty$ or $-\infty$, both $4/x$ and $1/x^2$ approach 0. Therefore,

$$\lim_{x \to \pm\infty} \frac{2x^2 - 4x}{x^2 - 1} = \frac{2}{1} = 2.$$

Before we go on... We say that the graph of f has a **horizontal asymptote** at $y = 2$ because of the limits we have just calculated. This means that the graph approaches the horizontal line $y = 2$ far to the right or left (in this case, both to the right and left). Figure 9 shows the graph of f together with the line $y = 2$.

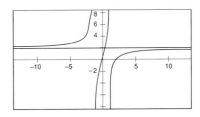

FIGURE 9

The graph also reveals additional interesting information: as $x \to 1^+$, $f(x) \to -\infty$, and as $x \to 1^-$, $f(x) \to +\infty$. Thus,

$$\lim_{x \to 1} f(x) \text{ does not exist.}$$

See if you can determine what is happening as $x \to -1$.

In the above example, $f(x)$ was a **rational function:** a quotient of polynomial functions. In the algebraic approach to that example, we calculated the limit of $f(x)$ at $\pm\infty$ by ignoring all powers of x in both the numerator and denominator except for the largest. It is possible to prove that this procedure is valid for any rational function (using the idea of dividing top and bottom by the highest power of x present).

EVALUATING THE LIMIT OF A RATIONAL FUNCTION AT $\pm\infty$

If $f(x)$ has the form

$$f(x) = \frac{c_n x^n + \ldots + c_2 x^2 + c_1 x + c_0}{d_m x^m + \ldots + d_2 x^2 + d_1 x + d_0}$$

with the c_i and d_i constants ($c_n \neq 0$ and $d_m \neq 0$), then we can calculate the limit of $f(x)$ as $x \to \pm\infty$ by ignoring all powers of x except the highest in both the numerator and denominator. Thus,

$$\lim_{x \to \pm\infty} f(x) = \lim_{x \to \pm\infty} \frac{c_n x^n}{d_m x^m}.$$

▼ **EXAMPLE 8**

Calculate

(a) $\lim\limits_{x \to +\infty} \dfrac{3x^4 - x^3 + 1}{x^3 + 40x^2}$ (b) $\lim\limits_{x \to -\infty} \dfrac{x^3 + 40x^2}{10x^4}$.

SOLUTION

(a) Ignoring all but the highest powers of x, we have

$$\lim_{x \to +\infty} \frac{3x^4 - x^3 + 1}{x^3 + 40x^2} = \lim_{x \to +\infty} \frac{3x^4}{x^3}$$
$$= \lim_{x \to +\infty} 3x.$$

Now we have a far simpler limit to evaluate—we can even say what the limit is without a table: as $x \to +\infty$, then $3x \to +\infty$ as well. Thus,

$$\lim_{x \to +\infty} \frac{3x^4 - x^3 + 1}{x^3 + 40x^2} = +\infty.$$

(b)
$$\lim_{x \to -\infty} \frac{x^3 + 40x^2}{10x^4} = \lim_{x \to -\infty} \frac{x^3}{10x^4}$$
$$= \lim_{x \to -\infty} \frac{1}{10x}.$$

At this stage, a table would be helpful, but once again we can manage without it. If x is, say, $-10{,}000$, then $1/(10x) = -1/100{,}000 = -0.00001$, extremely close to zero. In fact, the larger x gets in magnitude, the smaller $1/(10x)$ must get. Thus,

$$\lim_{x \to -\infty} \frac{x^3 + 40x^2}{10x^4} = 0.$$

10.6 EXERCISES

In each of Exercises 1–12, the graph of f is given. Compute the asked-for limits. If a particular limit fails to exist, say why (for example, it might diverge to $+\infty$).

1. (a) $\lim\limits_{x \to +\infty} f(x)$ (b) $\lim\limits_{x \to -\infty} f(x)$

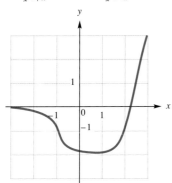

2. (a) $\lim\limits_{x \to +\infty} f(x)$ (b) $\lim\limits_{x \to -\infty} f(x)$

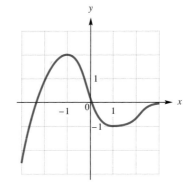

3. (a) $\lim\limits_{x \to +\infty} f(x)$ (b) $\lim\limits_{x \to -\infty} f(x)$

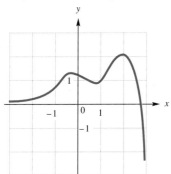

4. (a) $\lim\limits_{x \to \infty} f(x)$ (b) $\lim\limits_{x \to -\infty} f(x)$

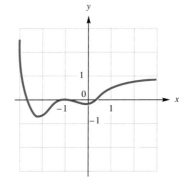

5. (a) $\lim\limits_{x \to 0^-} f(x)$ (b) $\lim\limits_{x \to 0^+} f(x)$ (c) $\lim\limits_{x \to 0} f(x)$

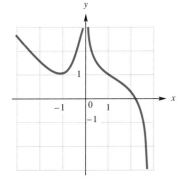

6. (a) $\lim\limits_{x \to 0^-} f(x)$ (b) $\lim\limits_{x \to 0^+} f(x)$ (c) $\lim\limits_{x \to 0} f(x)$

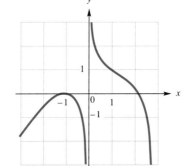

7. (a) $\lim_{x \to 0^-} f(x)$ (b) $\lim_{x \to 0^+} f(x)$ (c) $\lim_{x \to 0} f(x)$ 8. (a) $\lim_{x \to 0^-} f(x)$ (b) $\lim_{x \to 0^+} f(x)$ (c) $\lim_{x \to 0} f(x)$

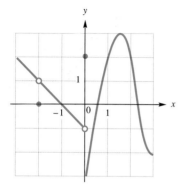

9. (a) $\lim_{x \to 0} f(x)$ (b) $\lim_{x \to -\infty} f(x)$ (c) $\lim_{x \to +\infty} f(x)$ 10. (a) $\lim_{x \to 0} f(x)$ (b) $\lim_{x \to -\infty} f(x)$ (c) $\lim_{x \to +\infty} f(x)$

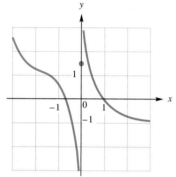
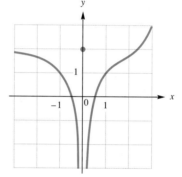

11. (a) $\lim_{x \to -\infty} f(x)$ (b) $\lim_{x \to +\infty} f(x)$ 12. (a) $\lim_{x \to -\infty} f(x)$ (b) $\lim_{x \to +\infty} f(x)$

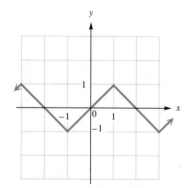
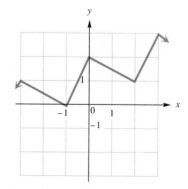

Calculate the limits in Exercises 13–38. If a limit fails to exist, say why. (Remember that you can use the numerical or graphical approaches if an algebraic approach will not work.)

13. $\lim_{x \to 0^+} \dfrac{1}{x^2}$

14. $\lim_{x \to 0^+} \dfrac{1}{x^2 - x}$

15. $\lim_{x \to -1} \dfrac{x^2 + 1}{x + 1}$

16. $\lim_{x \to -1^-} \dfrac{x^2 + 1}{x + 1}$

17. $\lim_{x \to \infty} \dfrac{3x^2 + 10x - 1}{2x^2 - 5x}$

18. $\lim_{x \to \infty} \dfrac{6x^2 + 5x + 100}{3x^2 - 9}$

19. $\lim_{x \to \infty} \dfrac{x^5 - 1{,}000x^4}{2x^5 + 10{,}000}$

20. $\lim_{x \to \infty} \dfrac{x^6 + 3{,}000x^3 + 1{,}000{,}000}{2x^6 + 1{,}000x^3}$

21. $\lim_{x \to \infty} \dfrac{10x^2 + 300x + 1}{5x + 2}$

22. $\lim_{x \to \infty} \dfrac{2x^4 + 20x^3}{1{,}000x^3 + 6}$

23. $\lim_{x \to \infty} \dfrac{10x^2 + 300x + 1}{5x^3 + 2}$

24. $\lim_{x \to \infty} \dfrac{2x^4 + 20x^3}{1{,}000x^6 + 6}$

25. $\lim_{x \to -\infty} \dfrac{3x^2 + 10x - 1}{2x^2 - 5x}$

26. $\lim_{x \to -\infty} \dfrac{6x^2 + 5x + 100}{3x^2 - 9}$

27. $\lim_{x \to -\infty} \dfrac{x^5 - 1{,}000x^4}{2x^5 + 10{,}000}$

28. $\lim_{x \to -\infty} \dfrac{x^6 + 3{,}000x^3 + 1{,}000{,}000}{2x^6 + 1{,}000x^3}$

29. $\lim_{x \to -\infty} \dfrac{10x^2 + 300x + 1}{5x + 2}$

30. $\lim_{x \to -\infty} \dfrac{2x^4 + 20x^3}{1{,}000x^3 + 6}$

31. $\lim_{x \to -\infty} \dfrac{10x^2 + 300x + 1}{5x^3 + 2}$

32. $\lim_{x \to -\infty} \dfrac{2x^4 + 20x^3}{1{,}000x^6 + 6}$

33. $\lim_{x \to 0} |x|$

34. $\lim_{x \to 1} |x - 1|$

35. $\lim_{x \to 1} \dfrac{|x - 1|}{x - 1}$

36. $\lim_{x \to -2} \dfrac{|x + 2|}{x + 2}$

37. $\lim_{x \to 2} e^{x - 2}$

38. $\lim_{x \to +\infty} e^{-x}$

39. $\lim_{x \to +\infty} xe^{-x}$

40. $\lim_{x \to -\infty} xe^{x}$

41. $\lim_{x \to 2^+} \dfrac{e^{-1/(x-2)}}{x - 2}$

42. $\lim_{x \to 2} \dfrac{e^{-1/(x-2)}}{x - 2}$

In each of Exercises 43–50, find all points of discontinuity of the given function. Classify discontinuities as removable, jump, or other.

43. $f(x) = \begin{cases} x + 2 & \text{if } x < 0 \\ 2x - 1 & \text{if } x \geq 0 \end{cases}$

44. $f(x) = \begin{cases} 1 - x & \text{if } x \leq 1 \\ x + 2 & \text{if } x > 1 \end{cases}$

45. $g(x) = \begin{cases} x + 2 & \text{if } x < 0 \\ 2x + 2 & \text{if } x \geq 0 \end{cases}$

46. $g(x) = \begin{cases} 1 - x & \text{if } x \leq 1 \\ x - 1 & \text{if } x > 1 \end{cases}$

47. $h(x) = \begin{cases} x + 2 & \text{if } x < 0 \\ 0 & \text{if } x = 0 \\ 2x + 2 & \text{if } x > 0 \end{cases}$

48. $h(x) = \begin{cases} 1 - x & \text{if } x < 1 \\ 1 & \text{if } x = 1 \\ x + 2 & \text{if } x > 1 \end{cases}$

49. $f(x) = \begin{cases} 1/x & \text{if } x < 0 \\ x & \text{if } x \geq 0 \end{cases}$

50. $f(x) = \begin{cases} x^2 & \text{if } x < 0 \\ x & \text{if } x \geq 0 \end{cases}$

In each of Exercises 51–56, find all points where the given function is not differentiable.

51. $f(x) = \begin{cases} x^2 & \text{if } x < 0 \\ x & \text{if } x \geq 0 \end{cases}$

52. $f(x) = \begin{cases} x^2 & \text{if } x < 0 \\ x^3 & \text{if } x \geq 0 \end{cases}$

53. $g(x) = x^{4/3}$

54. $g(x) = x^{4/9}$

55. $h(x) = |x - 1|$

56. $h(x) = |x + 2|$

27. $g(x) = x^2(2x + 1)$

28. $g(x) = \dfrac{x^2 - 1}{x}$

29. $g(r) = \dfrac{1}{\sqrt{r}} + \dfrac{2}{r}$

30. $g(r) = \dfrac{2}{\sqrt{r}} - \dfrac{\sqrt{r}}{2}$

In each of Exercises 31–40, find the equation of the tangent line at the point on the graph of the given function with the indicated first coordinate.

31. $f(x) = x^2 + 2x - 1$; $x = -2$

32. $g(x) = 3x^3 - x$; $x = -2$

33. $g(t) = \dfrac{1}{5t^4}$; $t = 1$

34. $s(t) = \dfrac{1}{3t^3}$; $x = -2$

35. $h(s) = \dfrac{1}{s} + s$; $s = 2$

36. $h(s) = \sqrt{s} - s$; $s = 9$

37. $r(t) = \dfrac{t^2}{3} - \dfrac{2t^3}{6}$; $t = -1$

38. $r(t) = \dfrac{2t^3}{9} - t$; $t = -1$

39. $h(t) = \dfrac{t^2 - 1}{t}$; $t = 2$

40. $h(t) = \sqrt{t} + 10t$; $t = 4$

In each of Exercises 41–48, find all values of x (if any) where the tangent line to the graph of the given equation is horizontal.

41. $y = -x^2 - 3x - 1$

42. $y = x^2 - x + 4$

43. $y = x^2 + \dfrac{1}{x^2}$

44. $y = \sqrt{x}\,(x - 1)$

45. $y = \sqrt{x} - 1$

46. $y = \sqrt{x} + \dfrac{1}{x}$

47. $y = x - \dfrac{1}{x^2} + 4$

48. $y = 3x^2 - \dfrac{1}{x} + 3$

Calculate the limits in Exercises 49–56 mentally.

49. $\lim\limits_{x \to 0} 3x - 2$

50. $\lim\limits_{x \to 0} 2x + x^2$

51. $\lim\limits_{x \to -1} \dfrac{1 + x}{x}$

52. $\lim\limits_{x \to -1} \dfrac{x + 4}{x}$

53. $\lim\limits_{x \to 9} (x - \sqrt{x})$

54. $\lim\limits_{x \to 1} \dfrac{\sqrt{x} + x^2}{x}$

55. $\lim\limits_{x \to -\infty} (x^2 - 3x + 1)$

56. $\lim\limits_{x \to +\infty} \dfrac{x^2}{2x^2 + 1}$

Investigate each limit in Exercises 57–64 using the numerical approach. If the limit exists, state its value, and if it fails to exist, say why.

57. $\lim\limits_{x \to -1} \dfrac{x^3 + 1}{x + 1}$

58. $\lim\limits_{x \to -6} \dfrac{x^2 + 6x}{x + 6}$

59. $\lim\limits_{x \to 0^-} \dfrac{1}{x^2 - 2x}$

60. $\lim\limits_{x \to 0^+} \dfrac{x}{x^2 - x}$

61. $\lim\limits_{x \to -1^+} \dfrac{x^2 + 1}{x + 1}$

62. $\lim\limits_{x \to -1} \dfrac{x^2 + 1}{x + 1}$

63. $\lim\limits_{x \to +\infty} x^2 e^{-x}$

64. $\lim\limits_{x \to 0^+} \dfrac{x}{e^{-1/x}}$

In Exercises 65–72, the graph of f is given. Compute the indicated quantities graphically. If a particular limit exists, state its value, and if it fails to exist, say why.

65. (a) $\lim_{x \to 1} f(x)$ (b) $\lim_{x \to -1^+} f(x)$
(c) $\lim_{x \to -1} f(x)$ (d) $\lim_{x \to 3} f(x)$
(e) $f(1)$ (f) $f(-1)$

66. (a) $\lim_{x \to -2} f(x)$ (b) $\lim_{x \to 0^+} f(x)$
(c) $\lim_{x \to 0^-} f(x)$ (d) $\lim_{x \to 0} f(x)$
(e) $f(0)$ (f) $f(-2)$

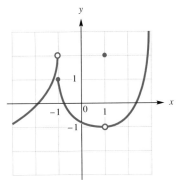

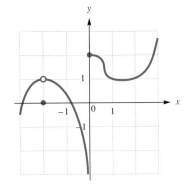

67. (a) $\lim_{x \to 0} f(x)$ (b) $\lim_{x \to +\infty} f(x)$
(c) $\lim_{x \to -\infty} f(x)$

68. (a) $\lim_{x \to 0} f(x)$ (b) $\lim_{x \to +\infty} f(x)$
(c) $\lim_{x \to -\infty} f(x)$

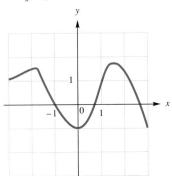

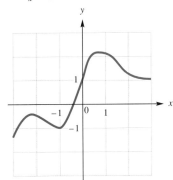

69. (a) $\lim_{x \to -1} f(x)$ (b) $\lim_{x \to -3^+} f(x)$
(c) $\lim_{x \to +\infty} f(x)$ (d) $f(-1)$

70. (a) $\lim_{x \to 1} f(x)$ (b) $\lim_{x \to -\infty} f(x)$
(c) $\lim_{x \to +\infty} f(x)$ (d) $f(1)$

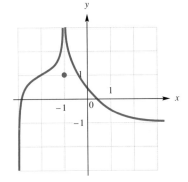

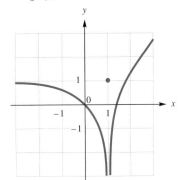

71. (a) $\lim\limits_{x \to 1^-} f(x)$ (b) $\lim\limits_{x \to -\infty} f(x)$
(c) $\lim\limits_{x \to +\infty} f(x)$

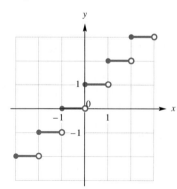

72. (a) $\lim\limits_{x \to 1^+} f(x)$ (b) $\lim\limits_{x \to -\infty} f(x)$
(c) $\lim\limits_{x \to +\infty} f(x)$

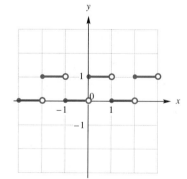

73–80. *Investigate the continuity of the functions whose graphs are given in Exercises 65–72. In each case, say whether f is continuous on its domain. If f fails to be continuous on its domain, say why.*

Calculate each limit in Exercises 81–90. If the limit fails to exist, say why.

81. $\lim\limits_{h \to -1^+} 4$

82. $\lim\limits_{h \to 0^+} -5$

83. $\lim\limits_{h \to 0} \dfrac{h^2 - 2h}{h + h^3}$

84. $\lim\limits_{h \to 0} \dfrac{h^2 + h - 1}{h^2 + h}$

85. $\lim\limits_{x \to -3} \dfrac{2x^2 + 5x - 3}{x + 3}$

86. $\lim\limits_{x \to -1} \dfrac{x^2 + x}{2x^2 + x - 1}$

87. $\lim\limits_{x \to 0} |-x|$

88. $\lim\limits_{x \to 0} \dfrac{1}{|x|}$

89. $\lim\limits_{x \to +\infty} \dfrac{1}{e^x - e^{-x}}$

90. $\lim\limits_{x \to +\infty} \dfrac{e^{-x}}{2}$

APPLICATION EXERCISES

91. *Velocity* If a stone is thrown upward at 100 ft/s, its height after t seconds is given by
$$s = 100t - 16t^2.$$
(a) Find its average velocity over the time intervals $[1, 2]$, $[1, 1.1]$, $[1, 1.01]$, and $[1, 1.001]$.
(b) Find its velocity at time $t = 1$ second.
(c) How long does it take to reach the ground, and how fast is it traveling when it hits the ground?

92. *Velocity* If a stone is thrown up at 120 ft/s from a height of 1,000 feet, its height after t seconds is given by
$$s = 1{,}000 + 120t - 16t^2.$$
(a) Find its average velocity over the time intervals $[2, 3]$, $[2, 2.1]$, $[2, 2.01]$, and $[2, 2.001]$.
(b) Find its velocity at time $t = 2$ seconds.

(c) How long does it take to reach the ground, and how fast is it traveling when it hits the ground?

93. *Marginal Cost* The cost of producing x soccer balls per day at the Taft Sports Company is calculated by its marketing staff to be given by the formula
$$C(x) = 100 + 60x - 0.001x^2.$$
Find the marginal cost function, and use it to estimate how fast the cost is going up at a production level of 50 soccer balls. Compare this with the exact cost of producing the 51st soccer ball.

94. *Marginal Cost* Referring to the cost equation in Exercise 93, find the production level for which the marginal cost is zero.

95. *Rates of Change* A cube is growing in such a way that each of its edges is growing at a rate of 1 centimeter per second. How fast is its volume growing at the instant when the cube has a volume of 1,000 cubic centimeters? [The volume of a cube with edge a is given by $V = a^3$.]

96. **Rates of Change** The volume of a cone with a circular base of radius r and height h is given by $V = \frac{1}{3}\pi r^2 h$. (See the figure.) Find the rate of increase of V with respect to r (assuming that h is fixed). If the quantity r is growing at a uniform rate of 1 cm/s, how fast is the volume growing at the instant when $r = 1$, assuming that $h = 2$ does not change?

97. **Sales Incentives** The Volume Sales Company pays its salespersons a weekly salary of $1,000 plus a sales commission based on the formula

$$C(s) = \frac{\sqrt{s}}{10},$$

where s represents a salesperson's total weekly sales (in dollars), and $C(s)$ is the percentage of total weekly sales paid to the salesperson as a commission.
(a) Calculate the net revenue $R(s)$ the company earns per week from a single salesperson, as a function of sales.
(b) Calculate the company's marginal revenue per salesperson for weekly sales of $100,000 and $500,000.
(c) For what weekly sales figure is the company's marginal net revenue positive? For what sales figures is it negative?
(d) What do you suspect is the most the company can earn in a week from a single salesperson?

98. **Medical Subsidies**
(a) The annual cost C of medical coverage is given by the formula

$$C(q) = 10,000 + 2q^2,$$

where q is the number of employees covered. Find the marginal cost of medical coverage for a 100-employee firm.
(b) The annual government subsidy to the firm is given by the formula

$$S(q) = 200q.$$

At what employee level does the marginal cost surpass the marginal subsidy?
(c) Calculate the net cost function, $N(q)$, given the cost function and subsidy above, graph it, and use your graph to estimate the value of q that gives the lowest net cost. What is this lowest net cost? Compare your answer to that of (b), and comment on what you find.

99. **Frogs** (use of a graphing calculator or computer software required) The population P of frogs in Nassau County is given by the formula

$$P = 10,000e^{0.5t},$$

where t is the time in years since 1995.
(a) Roughly how fast will the frog population be growing in the year 2,000?
(b) In what year will the frog population be growing at a rate of one million per year?
(c) What is the significance of this for the frog leg industry?

100. **Epidemics** (use of a graphing calculator or computer software required) According to the logistic model discussed in the chapter on exponentials, the number of people P infected in an epidemic often follows a curve of the following type:

$$P = \frac{NP_0}{P_0 + (N - P_0)e^{-bt}},$$

where N is the total susceptible population, P_0 is the number of infected individuals at time $t = 0$, and b is a constant that governs the rate of spread. Taking t to be the time in years, and assuming that $b = 0.1$, $N = 1,000,000$, and $P_0 = 1,000$, answer the following questions.
(a) Roughly how fast is the epidemic spreading at the start ($t = 0$), after 10 years, and after 100 years?
(b) When does the spread of the disease slow to 1,000 new cases per year?
(c) What percentage of the population is infected by this time?

INTRODUCTION ▶ In Chapter 3, we studied the concept of the derivative of a function, and we saw some of the things for which it is useful. However, the only functions we could differentiate easily were sums of terms of the form ax^n, where a and n are constants.

In this chapter, we develop techniques that will enable us to differentiate any closed-form function—that is, any function that can be specified by a formula involving powers, radicals, exponents, and logarithms. We also develop techniques for differentiating functions that are only specified *implicitly*—that is to say, functions for which we are not given an explicit formula, but only an equation relating x and y.

The chapter ends with an in-depth retrospective look at the concept of the derivative and approximation of a function by a linear function.

11.1 THE PRODUCT AND QUOTIENT RULES

We know how to find the derivatives of functions that are sums of powers, such as polynomials. However, that leaves many interesting functions whose derivatives we do not know how to find, such as $(x + 1)/(x - 1)$. We have discussed differentiation of sums and differences of functions. We now consider products and quotients.

EXAMPLE 1

The following calculations are wrong.

$$\frac{d}{dx}\left(\frac{x^3}{x}\right) = \frac{3x^2}{1} = 3x^2 \qquad \text{WRONG}$$

$$\frac{d}{dx}(x^3 \cdot x) = 3x^2 \cdot 1 = 3x^2 \qquad \text{WRONG}$$

After all, $x^3/x = x^2$, and we know that $d/dx\,(x^2) = 2x$, not $3x^2$. Our error was the assumption that the derivative of a quotient is the quotient of the derivatives. Similarly, $x^3 \cdot x = x^4$, and its derivative is $4x^3$, not $3x^2$. In other words, the derivative of a product is *not* the product of the derivatives.

11.1 The Product and Quotient Rules

Q If this is not how we find the derivatives of products and quotients, how *do* we find them?

A We use the following rules to differentiate products and quotients.

PRODUCT RULE

$$\frac{d}{dx}[f(x)g(x)] = f'(x)g(x) + f(x)g'(x)$$

QUOTIENT RULE

$$\frac{d}{dx}\left(\frac{f(x)}{g(x)}\right) = \frac{f'(x)g(x) - f(x)g'(x)}{[g(x)]^2}$$

Don't try to remember these rules by the symbols we've used here, but remember them in words. The following slogans are easy to remember, even if the terminology is not precise.

PRODUCT AND QUOTIENT RULES IN WORDS

The derivative of a product is the derivative of the first times the second, plus the first times the derivative of the second.

The derivative of a quotient is the derivative of the top times the bottom, minus the top times the derivative of the bottom, all over the bottom squared.

▶ **CAUTION** One more time: *the derivative of a product is* not *the product of the derivatives, and the derivative of a quotient is* not *the quotient of the derivatives.* To find the derivative of a product, you must use the product rule, and to find the derivative of a quotient, you must use the quotient rule. Forgetting this is a mistake everyone makes from time to time.* ◀

We shall see why the product rule works at the end of the section and why the quotient rule works in the next section. First, let us try some examples.

▼ *Leibniz made this mistake at first, too, so you are in good company.

EXAMPLE 2

Calculate $\frac{d}{dx}(x^3 \cdot x^2)$ two different ways: by multiplying first, and by the product rule.

SOLUTION If we multiply before taking the derivative, we get

$$\frac{d}{dx}(x^3 \cdot x^2) = \frac{d}{dx}(x^5)$$
$$= 5x^4.$$

If we use the product rule, we get

$$\frac{d}{dx}(x^3 \cdot x^2) = \overset{\text{(Derivative of first) (Second)}}{3x^2 \cdot x^2} + \overset{\text{(First) (Derivative of second)}}{x^3 \cdot 2x}$$
$$= 5x^4.$$

Before we go on... This example shows us how the product rule is consistent with the power rule—it leads to the same answer. The first of the two calculations above is obviously easier, so there is no necessity to use the product rule in this example. On the other hand, the product rule will prove indispensable when we study derivatives of more complicated functions, so we cannot do without it.

EXAMPLE 3

Find $\frac{d}{dx}[(x^3 + 2x)(\sqrt{x} + 1)]$.

SOLUTION

$$\frac{d}{dx}[(x^3 + 2x)(\sqrt{x} + 1)]$$
$$= \text{(Derivative of first) (Second)} + \text{(First) (Derivative of second)}$$
$$= (3x^2 + 2)(\sqrt{x} + 1) + (x^3 + 2x)\left(\frac{1}{2\sqrt{x}}\right)$$

Before we go on... Notice that we could also do this example by expanding the whole expression first and *then* taking the derivative. This would enable us to avoid using the product rule. (You should do this for practice.)

11.1 The Product and Quotient Rules

▼ EXAMPLE 4

Find $\dfrac{d}{dx}[(x+1)(x^2+1)(x^3+1)]$.

SOLUTION Here we have a product of three functions, not just two. We can find the derivative by using the product rule twice.

$$\dfrac{d}{dx}[(x+1)(x^2+1)(x^3+1)]$$
$$= \dfrac{d}{dx}(x+1) \cdot [(x^2+1)(x^3+1)] +$$
$$(x+1) \cdot \dfrac{d}{dx}[(x^2+1)(x^3+1)]$$
$$= (1)(x^2+1)(x^3+1) +$$
$$(x+1)[(2x)(x^3+1) + (x^2+1)(3x^2)]$$
$$= (1)(x^2+1)(x^3+1) + (x+1)(2x)(x^3+1) +$$
$$(x+1)(x^2+1)(3x^2)$$

We can see here a more general product rule: the derivative of a product of three functions is found by taking the derivatives of each function in turn and adding the results together. The general formula is

$$(fgh)' = f'gh + fg'h + fgh'.$$

There are similar formulas for products of more than three functions.

▼ EXAMPLE 5

Find $\dfrac{d}{dx}\left(\dfrac{x+1}{x-1}\right)$.

SOLUTION We must use the quotient rule.

(Derivative of top)(Bottom) − (Top)(Derivative of bottom)

$$\dfrac{d}{dx}\left(\dfrac{x+1}{x-1}\right) = \dfrac{(1)(x-1) - (x+1)(1)}{(x-1)^2}$$

↑ Bottom squared

$$= \dfrac{-2}{(x-1)^2}$$

EXAMPLE 6

Find $\dfrac{d}{dx}\left[\dfrac{(x+1)(x+2)}{x-1}\right]$.

SOLUTION We seem to have both a product and a quotient here. Which rule do we use, the product or the quotient rule? Here is a way to decide.

Let us think about how we would calculate (by hand or with a scientific calculator) $(x+1)(x+2)/(x-1)$ for a specific value of x, say $x = 11$. *What would be the last operation we would perform?* Here is how we would probably do the calculation:

1. Calculate $(x+1)(x+2) = (11+1)(11+2) = 156$.
2. Calculate $x - 1 = 11 - 1 = 10$.
3. Divide 156 by 10 to get 15.6.

Thus, the last operation we would perform is division, and so we can regard the whole expression as a quotient—that is, as $(x+1)(x+2)$ *divided by* $x - 1$. Therefore, we will use the quotient rule. The first thing the quotient rule tells us to do is take the derivative of the top. If we examine the top, we notice that it is a product, so we must use the product rule to take its derivative. Here is the calculation.

$$\frac{d}{dx}\frac{(x+1)(x+2)}{x-1} = \frac{\overbrace{[(1)(x+2)+(x+1)(1)]}^{\text{(Derivative of top)}}\overbrace{(x-1)}^{\text{(Bottom)}} - \overbrace{[(x+1)(x+2)]}^{\text{(Top)}}\overbrace{(1)}^{\text{(Derivative of bottom)}}}{(x-1)^2}$$

$$= \frac{(2x+3)(x-1)-(x+1)(x+2)}{(x-1)^2}$$

$$= \frac{x^2 - 2x - 5}{(x-1)^2}$$

What was important here was to determine the *order of operations,* and in particular to determine the last operation to be performed. Pretending to do an actual calculation reminds us of the order of operations, and we shall call this technique the **calculation thought experiment.**

Before we go on... We had to use the product rule to calculate the derivative of the top because the top was $(x+1)(x+2)$, which is a product. Get used to this: differentiation rules often must be used in combination. Once you have determined one rule to use, do not assume that you can forget the others.

Now here is another way we could have done the problem: suppose that our calculation thought experiment took the following form.

1. Calculate $(x+1)/(x-1) = (11+1)/(11-1) = 1.2$.
2. Calculate $x + 2 = 11 + 2 = 13$.
3. Multiply 1.2 by 13 to get 15.6.

Then we would have regarded the expression as a *product*—the product of $(x + 1)/(x - 2)$ and $(x + 2)$—so we could have used the product rule instead. We can't escape the quotient rule, however: we need to use it to take the derivative of the first factor, $(x + 1)/(x - 2)$. You should try this approach as an exercise, and check that you get the same answer.

In the last chapter we proved the power rule only for positive integers. We can now prove it for negative integers as well.

▼ **EXAMPLE 7** **Power Rule for Negative Integers**

Show that if n is a positive integer, then

$$\frac{d}{dx}(x^{-n}) = -nx^{-n-1}.$$

SOLUTION Since we are required here to justify the power rule for negative integers, we can't simply go ahead and *use* it! "Officially," all we can use are the power rule for *positive* integers, and the product and quotient rules (which are also unfinished business at the moment, but we'll attend to them shortly).

What we *can* do is this: write x^{-n} as $1/x^n$ and use the quotient rule. Applying the quotient rule to $1/x^n$ gives

$$\frac{d}{dx}\left(\frac{1}{x^n}\right) = \frac{(0)(x^n) - (1)(nx^{n-1})}{(x^n)^2}$$

$$= \frac{-nx^{n-1}}{x^{2n}}$$

$$= -nx^{n-1-2n} = -nx^{-n-1},$$

and we are done.

Before we go on... Notice that we did use the power rule, but only for a positive integer power, the case that we justified in the previous chapter.

▶ **NOTE** In practice, we should not use the quotient rule for expressions like $3/x^2$. It is much simpler to first rewrite $3/x^2 = 3x^{-2}$, and then use the power rule. ◀

▼ **EXAMPLE 8**

Find $\dfrac{d}{dx}\left(6x^2 + 5\left(\dfrac{x}{x-1}\right)\right)$.

SOLUTION If this seems to be a little complicated at first sight, all we need to do is turn to the calculation thought experiment, which tells us that the expression we are asked to differentiate is a *sum*. Since the derivative of a sum is the sum of the derivatives, we get

$$\frac{d}{dx}\left(6x^2 + 5\left(\frac{x}{x-1}\right)\right) = \frac{d}{dx}(6x^2) + \frac{d}{dx}\left(5\left(\frac{x}{x-1}\right)\right).$$

In other words, we must take the derivatives of $6x^2$ and $5\left(\frac{x}{x-1}\right)$ separately, and then add the answers. The derivative of $6x^2$ is $12x$, and there are two ways of taking the derivative of $5\left(\frac{x}{x-1}\right)$. We could either first multiply the expression $\left(\frac{x}{x-1}\right)$ by 5 to get $\left(\frac{5x}{x-1}\right)$, and then take its derivative using the quotient rule, or we could proceed as follows.

$$\frac{d}{dx}\left(6x^2 + 5\left(\frac{x}{x-1}\right)\right) = \frac{d}{dx}(6x^2) + \frac{d}{dx}\left(5\left(\frac{x}{x-1}\right)\right)$$

$$= 12x + 5\frac{d}{dx}\left(\frac{x}{x-1}\right)$$

We use the quotient rule for the second expression, and we get

$$= 12x + 5\left(\frac{(1)(x-1) - (x)(1)}{(x-1)^2}\right)$$

$$= 12x + 5\left(\frac{-1}{(x-1)^2}\right)$$

$$= 12x - \frac{5}{(x-1)^2}.$$

Now let us see why the product rule works.

PROOF OF PRODUCT RULE

We wish to calculate the derivative of a function $f(x)g(x)$, so we proceed as usual, using the definition of the derivative:

$$\frac{d}{dx}[f(x)g(x)] = \lim_{h \to 0} \frac{f(x+h)g(x+h) - f(x)g(x)}{h}.$$

Q How can we rewrite the numerator so that we can evaluate the limit?

A There are several ways to do this. We choose to use one that is motivated by Figure 1.

The area of the whole rectangle represents $f(x+h)g(x+h)$, while the area of the rectangle labeled ① represents $f(x)g(x)$. The difference between these areas (which is the numerator we are trying to rewrite) is represented by the sum of the areas of the other three rectangles. Rectangle ② has a width of $f(x+h) - f(x)$ and a height of $g(x)$, so its area is $[f(x+h) - f(x)]g(x)$. Similarly, the area of rectangle ③ is

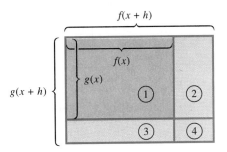

FIGURE 1

$f(x)[g(x+h) - g(x)]$, and the area of rectangle ④ is $[f(x+h) - f(x)][g(x+h) - g(x)]$. In other words,

$$f(x+h)g(x+h) - f(x)g(x)$$
$$= [f(x+h) - f(x)]g(x) + f(x)[g(x+h) - g(x)]$$
$$+ [f(x+h) - f(x)][g(x+h) - g(x)].$$

You should expand the right-hand side and simplify to convince yourself that this is really just an algebraic fact. Returning to the derivative, this gives

$$\frac{d}{dx}[f(x)g(x)]$$
$$= \lim_{h \to 0} \frac{[f(x+h) - f(x)]g(x) + f(x)[g(x+h) - g(x)] + [f(x+h) - f(x)][g(x+h) - g(x)]}{h}$$
$$= \lim_{h \to 0} \left(\frac{f(x+h) - f(x)}{h}\right) g(x) + \lim_{h \to 0} f(x) \left(\frac{g(x+h) - g(x)}{h}\right)$$
$$+ \lim_{h \to 0} \left(\frac{f(x+h) - f(x)}{h}\right)[g(x+h) - g(x)].$$

We already know the following:

$$\lim_{h \to 0} \frac{f(x+h) - f(x)}{h} = f'(x),$$

$$\lim_{h \to 0} \frac{g(x+h) - g(x)}{h} = g'(x),$$

and

$$\lim_{h \to 0} [g(x+h) - g(x)] = 0.$$

The last limit is 0 because if g is differentiable at x, it must be continuous there, so $\lim_{h \to 0} g(x+h) = g(x)$.

Putting this all together,

$$\frac{d}{dx}[f(x)g(x)]$$
$$= f'(x)g(x) + f(x)g'(x) + f'(x) \cdot 0$$
$$= f'(x)g(x) + f(x)g'(x),$$

as we claimed. ◄

It is possible to prove the quotient rule in a very similar way, but this proof would be completely unenlightening. Instead, we shall prove it in the next section using techniques we shall discuss there.

▶ 11.1 EXERCISES

In each of Exercises 1 through 10,
(a) calculate the derivative of the given function mentally without using either the product or quotient rule, and
(b) use the product or quotient rule to find the derivative, and check that you obtain the same answer.

1. $f(x) = 3x$
2. $f(x) = 2x^2$
3. $g(x) = x \cdot x^2$
4. $g(x) = x \cdot x$
5. $h(x) = x(x + 3)$
6. $h(x) = x(1 + 2x)$
7. $r(x) = \dfrac{x^2}{3}$
8. $r(x) = \dfrac{x}{5}$
9. $s(x) = \dfrac{2}{x}$
10. $s(x) = \dfrac{3}{x^2}$

Evaluate $\dfrac{dy}{dx}$ in each of Exercises 11–34.

11. $y = (x + 1)(x^2 - 1)$
12. $y = (x^2 + x)(x - x^2)$
13. $y = (2x^{1/2} + 4x - 5)(x - x^{-1})$
14. $y = (x^{3/4} - 4x - 5)(x^{-1} + x^{-2})$
15. $y = (2x^2 - 4x + 1)^2$
16. $y = (2\sqrt{x} - x^2)^2$
17. $y = (x^2 - \sqrt{x})\left(\sqrt{x} + \dfrac{1}{\sqrt{x}}\right)$
18. $y = (4x^2 - \sqrt{x})\left(\sqrt{x} - \dfrac{2}{\sqrt{x}}\right)$
19. $y = (\sqrt{x} + 1)\left(\sqrt{x} + \dfrac{1}{x^2}\right)$
20. $y = (4x^2 - \sqrt{x})\left(\sqrt{x} - \dfrac{2}{x^2}\right)$
21. $y = \dfrac{2x + 4}{3x - 1}$
22. $y = \dfrac{3x - 9}{2x + 4}$
23. $y = \dfrac{2x^2 + 4x + 1}{3x - 1}$
24. $y = \dfrac{3x^2 - 9x + 11}{2x + 4}$
25. $y = \dfrac{x^2 - 4x + 1}{x^2 + x + 1}$
26. $y = \dfrac{x^2 + 9x - 1}{x^2 + 2x - 1}$
27. $y = \dfrac{\sqrt{x} + 1}{\sqrt{x} - 1}$
28. $y = \dfrac{\sqrt{x} - 1}{\sqrt{x} + 1}$

29. $y = \dfrac{\left(\dfrac{1}{x} + \dfrac{1}{x^2}\right)}{x + x^2}$

30. $y = \dfrac{\left(\dfrac{1}{x} - \dfrac{1}{x^2}\right)}{x^2 + 1}$

31. $y = \dfrac{(x + 3)(x + 1)}{3x - 1}$

32. $y = \dfrac{3x^2 - 9x + 11}{(x - 5)(x - 4)}$

33. $y = \dfrac{(x + 3)(x + 1)(x + 2)}{3x - 1}$

34. $y = \dfrac{3x^2 - 9x + 11}{(x - 5)(x - 4)(x - 1)}$

In Exercises 35 through 42, evaluate the derivatives.

35. $\dfrac{d}{dx}[(x^2 + x)(x^2 - x)]$

36. $\dfrac{d}{dx}[(x^2 + x^3)(x^3 - 2x + 1)]$

37. $\dfrac{d}{dx}[(x^3 + 2x)(x^2 - x)]\big|_{x=2}$

38. $\dfrac{d}{dx}[(x^2 + x)(x^2 - x)]\big|_{x=1}$

39. $\dfrac{d}{dt}\left((t^2 - \sqrt{t})\left(\sqrt{t} + \dfrac{1}{\sqrt{t}}\right)\right)\bigg|_{t=1}$

40. $\dfrac{d}{dt}\left((t^2 + \sqrt{t})\left(\sqrt{t} - \dfrac{1}{\sqrt{t}}\right)\right)\bigg|_{t=1}$

41. $\dfrac{d}{dt}\left(\dfrac{t^2 - \sqrt{t}}{\sqrt{t} + \dfrac{1}{\sqrt{t}}}\right)$

42. $\dfrac{d}{dt}\left(\dfrac{t^2 + \sqrt{t}}{\sqrt{t} - \dfrac{1}{\sqrt{t}}}\right)$

In each of Exercises 43 through 48, find the equation of the line tangent to the graph of the given function at the point with the indicated x-coordinate.

43. $f(x) = (x^2 + 1)(x^3 + x)$, $x = 1$

44. $f(x) = (\sqrt{x} + 1)(x^2 + x)$, $x = 1$

45. $f(x) = \dfrac{x + 1}{x + 2}$, $x = 0$

46. $f(x) = \dfrac{\sqrt{x} + 1}{\sqrt{x} + 2}$, $x = 4$

47. $f(x) = \dfrac{x^2 + 1}{x}$, $x = -1$

48. $f(x) = \dfrac{x}{x^2 + 1}$, $x = 1$

APPLICATIONS

49. *Revenue* The monthly sales of Sunny Electronics' new stereo system are given by $S(x) = 20x - x^2$ hundred units per month, x months after its introduction. The retail price Sunny charges is $p(x) = \$1{,}000 - x^2$, x months after introduction. The revenue Sunny earns must then be $R(x) = 100S(x)p(x)$. Find, five months after the introduction, the rate of change of monthly sales, the rate of change of the price, and the rate of change of revenue. Interpret your answers.

50. *Revenue* The monthly sales of Sunny Electronics' new portable tape player is given by $S(x) = 20x - x^2$ hundred units per month, x months after its introduction. The retail price Sunny charges is $p(x) = \$100 - x^2$, x months after introduction. The revenue Sunny earns must then be $R(x) = 100S(x)p(x)$. Find, six months after the introduction, the rate of change of monthly sales, the rate of change of the price, and the rate of change of revenue. Interpret your answers.

51. *Revenue* Cyndi Keen is currently selling 20 "I ♥ Calculus" T-shirts per day, but sales are dropping at a rate of 3 per day. She is currently charging $7 per T-shirt, but to compensate for dwindling sales, she is increasing the unit price by $1 per day. How fast, and in what direction, is her daily revenue currently changing?

52. *Pricing Policy* Let us turn Exercise 51 around a little: Cyndi Keen is currently selling 20 "I ♥ Calculus" T-shirts per day, but sales are dropping at a rate of 3 per day. She is currently charging $7 per T-shirt, and she wishes to increase her daily revenue by $10 per day. At what rate should she increase the unit price to accomplish this (assuming the price increase does not affect sales)?

53. Bus Travel The Thoroughbred Bus Company finds that its monthly costs for one particular year were given by $C(t) = \$10{,}000 + t^2$ after t months. On the other hand, after t months, the company had $P(t) = 1{,}000 + t^2$ passengers per month. How fast is its cost per passenger changing after 6 months?

54. Bus Travel The Thoroughbred Bus Company finds that its monthly costs for one particular year were given by $C(t) = \$100 + t^2$ after t months. On the other hand, after t months, the company had $P(t) = 1{,}000 + t^2$ passengers per month. How fast is its cost per passenger changing after 6 months?

Some of the following exercises are variations of exercises you have already seen in Chapter 3.

55. Fuel Economy Your Porsche's gas mileage (in miles per gallon) is given as a function $M(x)$ of speed x in mph, where

$$M(x) = \frac{1}{\left(x + \frac{3{,}600}{x}\right)}.$$

Calculate $M'(x)$, and hence $M'(10)$, $M'(60)$, and $M'(70)$. What do the answers tell you about your car?

56. Fuel Economy Your used Chevy's gas mileage (in miles per gallon) is given as a function $M(x)$ of speed x in mph, where

$$M(x) = \frac{10}{\left(x + \frac{3{,}025}{x}\right)}.$$

Calculate $M'(x)$, and hence determine *the sign* of each of the following: $M'(40)$, $M'(55)$, $M'(60)$. Interpret your results.

57. Expansion The number of Toys "R" Us® stores (including Kids "R" Us stores) worldwide increased from 171 at the start of 1984 to 1,032 at the start of 1994.* If the annual revenue at each store was $600,000 in 1984 and was increasing by $50,000 per year, how fast would the company's worldwide revenue have been increasing at the start of 1990? (Use a linear model for the number of stores.)

58. Investments The price of GTE® stock rose from about $22 per share in January 1989 to $35 per share in January 1994.† If you had purchased 100 shares of GTE in January 1989 and steadily purchased additional shares at a rate of 10 shares per month, how fast would the value of your investment have been increasing in January 1994? (Use a linear model for the share price.)

59. Biology—Reproduction The Verhulst model for population growth specifies the reproductive rate of an organism as a function of the total population according to the following formula:

$$R(p) = \frac{r}{1 + kp}.$$

Here, p is the total population in thousands of organisms, r and k are constants that depend on the particular circumstances and the organism being studied, and $R(p)$ is the reproduction rate in thousands of organisms per hour.‡ If $k = 0.125$ and $r = 45$, find $R'(p)$ and hence $R'(4)$. Interpret the result.

60. Biology—Reproduction Another model, the predator satiation model for population growth, specifies that the reproductive rate of an organism as a function of the total population varies according to the following formula:

$$R(p) = \frac{rp}{1 + kp}.$$

Here, p is the total population in thousands of organisms, r and k are constants that depend on the particular circumstances and the organism being studied, and $R(p)$ is the reproduction rate in new organisms per hour.††† Given that $k = 0.2$ and $r = 0.08$, find $R'(p)$ and $R'(2)$. Interpret the result.

▼ *Source: Company Reports/Associated Press (*The New York Times*, Jan. 12, 1994, p. D4.)
†Source: Company Reports, Datastream (*The New York Times*, Jan. 14, 1994, p. D1.)
‡Source: *Mathematics in Medicine and the Life Sciences* by F. C. Hoppensteadt and C. S. Peskin (New York: Springer-Verlag, 1992) pp. 20–22.

61. *Embryo Development* Bird embryos consume oxygen from the time the egg is laid through the time the chick hatches. In the case of a typical galliform bird, the total oxygen consumption (in milliliters) t days after the egg was laid can be approximated by*

$$C(t) = -0.0163t^4 + 1.096t^3 - 10.704t^2 + 3.576t \quad (t \leq 30).$$

(An egg will usually hatch at around $t = 28$.) Suppose that at time $t = 0$ you have a collection of 30 newly hatched eggs and that the number of eggs is decreased linearly to zero at time $t = 30$ days. How fast is the total oxygen consumption of your collection of embryos changing after 25 days? (Answer to the nearest whole number.) Interpret the result.

62. *Embryo Development* Turkey embryos consume oxygen from the time the egg is laid through the time the chick hatches. In the case of a brush turkey, the total oxygen consumption (in milliliters) t days after the egg was laid can be approximated by*

$$C(t) = -0.00708t^4 + 0.952t^3 - 21.96t^2 + 95.328t \quad (t \leq 50).$$

(An egg will typically hatch at around $t = 50$.) Suppose that at time $t = 0$ you have a collection of 100 newly hatched eggs and that the number of eggs is decreased linearly to zero at time $t = 50$ days. How fast is the total oxygen consumption of your collection of embryos changing after 40 days? (Answer to the nearest whole number.) Interpret the result.

COMMUNICATION AND REASONING EXERCISES

63. You have come across the following in a newspaper article: "Revenues of HAL Home Heating Oil Inc. are rising by $4.2 million per year. This is due to an annual increase of 70¢ per gallon in the price HAL charges for heating oil and an increase in sales of 6 million gallons of oil per year." Comment on this analysis.

64. Your friend says that since average cost is obtained by dividing the cost function by the number of units x, it follows that the derivative of average cost is the same as marginal cost, since the derivative of x is 1. Comment on this analysis.

65. Find a demand function $q(p)$ such that at a price per item of $p = \$100$, revenue will rise if the price per item is increased.

66. What must be true about a demand function $q(p)$ so that at a price per item of $p = \$100$, revenue will decrease if the price per item is increased?

67. *Marginal Product* (*from the GRE economics test*) Which of the following statements about average product and marginal product is correct?
(a) If average product is decreasing, marginal product must be less than average product.
(b) If average product is increasing, marginal product must be increasing.
(c) If marginal product is decreasing, average product must be less than marginal product.
(d) If marginal product is increasing, average product must be decreasing.
(e) If marginal product is constant over some range, average product must be constant over that range.

68. *Marginal Cost* (*based on a question from the GRE economics test*) Which of the following statements about average cost and marginal cost is correct?
(a) If average cost is increasing, marginal cost must be increasing.
(b) If average cost is increasing, marginal cost must be decreasing.
(c) If average cost is increasing, marginal cost must be more than average cost.
(d) If marginal cost is increasing, average cost must be increasing.
(e) If marginal cost is increasing, average cost must be larger than marginal cost.

▼ * The model is derived from graphical data published in the article "The Brush Turkey" by Roger S. Seymour, *Scientific American*, December, 1991, pp. 108–14.

11.2 THE CHAIN RULE

We can now find the derivatives of sums, products and quotients of powers of x, but we still cannot take the derivative of an expression such as $\sqrt{3x+1}$. For this we need one more rule, but we need to talk about an idea behind it first. Look at $h(x) = \sqrt{3x+1}$. This function is not a sum, difference, product, or quotient. We can use the calculation thought experiment to find the last operation we would perform in calculating $h(x)$.

1. Calculate $3x + 1$,
2. Take the square root of the answer.

Thus, the last operation is the "square root." We do not yet have a rule for finding the derivative of the square root of a quantity other than x.

There is a way of building $h(x)$ out of two simpler functions: $f(x) = \sqrt{x}$ and $g(x) = 3x + 1$.

$$\begin{aligned} h(x) &= \sqrt{3x+1} \\ &= f(3x+1) \\ &= f(g(x)) \end{aligned}$$

We say that h is the **composite** of f and g. We read $f(g(x))$ aloud as "f of g of x."

In order to compute $h(1)$, say, you would first compute $3 \cdot 1 + 1 = 4$ and then take the square root of 4, giving $h(1) = 2$. In order to compute $f(g(1))$, you would follow exactly the same steps: first compute $g(1) = 4$ then $f(g(1)) = f(4) = 2$. Remember that you must always compute $f(g(x))$ numerically from the inside out: given x, first compute $g(x)$ and then $f(g(x))$.

The reason for writing h as the composite of f and g is that f and g are functions *whose derivatives we know*. Formally, the derivative of the composite function $h(x) = f(g(x))$ is given by the following rule.

FORMAL STATEMENT OF THE CHAIN RULE

If f has derivative f' and g has derivative g', then

$$\frac{d}{dx}[f(g(x))] = f'(g(x)) \cdot g'(x).$$

A more convenient way of writing the chain rule is to introduce a new variable u and let $u = g(x)$. Then we can write $f(g(x)) = f(u)$, $f'(g(x)) = f'(u)$, and $g'(x) = du/dx$.

11.2 The Chain Rule

CHAIN RULE

If u is a function of x, then

$$\frac{d}{dx}[f(u)] = f'(u)\frac{du}{dx},$$

provided both derivatives on the right exist.

In words: *The derivative of f(quantity) is the derivative of f, evaluated at that quantity, times the derivative of the quantity.*

For every function f whose derivative we know, we now get a "generalized" differentiation rule. For example, with $f(x) = x^3$ we get the following.

Original Rule	Generalized Rule
$\frac{d}{dx} x^3 = 3x^2$	$\frac{d}{dx} u^3 = 3u^2 \frac{du}{dx}$

In words:

The derivative of a quantity cubed is 3 times that quantity squared times the derivative of the quantity.

▼ **EXAMPLE 1**

Compute $\frac{d}{dx}(2x^2 + x)^3$.

SOLUTION Using the calculation thought experiment, we find that the last operation we would perform in calculating $(2x^2 + x)^3$ is that of cubing the quantity $(2x^2 + x)$. Thus, we think of $(2x^2 + x)^3$ as "a quantity cubed." We'll calculate its derivative in two ways: using the formula above, and using the verbal form.

Method 1: Using the formula. We think of $(2x^2 + x)^3$ as u^3, where $u = 2x^2 + x$. By the formula,

$$\frac{d}{dx} u^3 = 3u^2 \frac{du}{dx}.$$

Now substitute for u.

$$\frac{d}{dx}(2x^2 + x)^3 = 3(2x^2 + x)^2 \frac{d}{dx}(2x^2 + x)$$
$$= 3(2x^2 + x)^2(4x + 1)$$

Method 2: Using the verbal form.

If we prefer to use the verbal form, we get

The derivative of $(2x^2 + x)$ cubed is three times $(2x^2 + x)$ squared, times the derivative of $(2x^2 + x)$.

In other words,

$$\frac{d}{dx}(2x^2 + x)^3 = 3(2x^2 + x)^2(4x + 1),$$

as we obtained before.

For the next example, recall that the derivative with respect to x of the function $f(x) = \sqrt{x}$ is $1/(2\sqrt{x})$. Now suppose that instead of x under the square root sign, we had some other quantity. The chain rule would then take the following form.

The derivative of the square root of a quantity is one over twice the square root of the quantity, times the derivative of the quantity.

In terms of a formula, if we again let u stand for "the quantity," we can write the following.

Original Rule	Generalized Rule
$\frac{d}{dx}\sqrt{x} = \frac{1}{2\sqrt{x}}$	$\frac{d}{dx}\sqrt{u} = \frac{1}{2\sqrt{u}}\frac{du}{dx}$

EXAMPLE 2

Compute $\frac{d}{dx}\sqrt{3x + 1}$.

SOLUTION The steps in the calculation thought experiment for $\sqrt{3x + 1}$ are

1. Calculate $3x + 1$
2. Take the square root of the answer.

Thus, we are dealing here with *the square root of a quantity*.

Method 1: Using the formula. Think of $\sqrt{3x + 1}$ as $\sqrt{u}$, where $u = 3x + 1$. The formula is

$$\frac{d}{dx}\sqrt{u} = \frac{1}{2\sqrt{u}}\frac{du}{dx}.$$

If we substitute for u, we get

$$\frac{d}{dx}\sqrt{3x + 1} = \frac{1}{2\sqrt{3x + 1}}\frac{d}{dx}(3x + 1)$$

$$= \frac{1}{2\sqrt{3x + 1}}(3)$$

$$= \frac{3}{2\sqrt{3x + 1}}.$$

Method 2: *Using the verbal form.* The verbal form is

The derivative of the square root of a quantity is one over twice the square root of that quantity, times the derivative of the quantity.

In symbols, this is

$$\frac{d}{dx}\sqrt{3x+1} = \frac{1}{2\sqrt{3x+1}}(3),$$

giving the answer in one step!

Before we go on... The formula for $\frac{d}{dx}\sqrt{x}$ is part of the power rule:

$$\frac{d}{dx}\sqrt{x} = \frac{d}{dx}(x^{1/2}) = \frac{1}{2}x^{-1/2} = \frac{1}{2\sqrt{x}}.$$

We can rewrite the calculations we did above as

$$\frac{d}{dx}\sqrt{3x+1} = \frac{d}{dx}[(3x+1)^{1/2}]$$

$$= \frac{1}{2}(3x+1)^{-1/2}(3)$$

$$= \frac{3}{2\sqrt{3x+1}}.$$

These examples show one of the most common uses of the chain rule: to take the derivative of an expression raised to a power. It is worth writing this "generalized power rule" as another rule, recognizing that it is a special case of the chain rule.

GENERALIZED POWER RULE

Power Rule	Generalized Power Rule
$\frac{d}{dx}x^n = nx^{n-1}$	$\frac{d}{dx}u^n = nu^{n-1}\frac{du}{dx}$

In words:
The derivative of a quantity raised to the power n is n times that quantity raised to the power (n − 1), times the derivative of that quantity.

EXAMPLE 3

Find $\dfrac{d}{dx}(x^3 + x)^{100}$.

SOLUTION First, the calculation thought experiment: If we were computing $(x^3 + x)^{100}$, the last operation we would perform is *raising a quantity to the power* 100. Thus we are dealing with *a quantity raised to the power* 100, and so we must use the generalized power rule.

According to the generalized power rule, the derivative of a quantity raised to the power 100 is 100 times that quantity to the power 99, times the derivative of that quantity. In other words,

$$\dfrac{d}{dx}(x^3 + x)^{100} = 100(x^3 + x)^{99}(3x^2 + 1).$$

▶ **CAUTION** The following are examples of common errors.

$$\dfrac{d}{dx}(x^3 + x)^{100} = 100(3x^2 + 1)^{99} \quad \text{WRONG}$$

$$\dfrac{d}{dx}(x^3 + x)^{100} = 100(x^3 + x)^{99} \quad \text{WRONG}$$

Remember that the generalized power rule says that the derivative of a quantity to the power 100 is 100 times *that same quantity* raised to the power 99, *times the derivative of that quantity*. ◀

Q It seems that there are now two formulas for the derivative of an nth power:

$$(1) \;\; \dfrac{d}{dx} x^n = nx^{n-1}$$

and

$$(2) \;\; \dfrac{d}{dx} u^n = nu^{n-1} \dfrac{du}{dx}.$$

Which one do I use?

A Formula (1) is the original power rule, and it only applies to a power of x. Thus, for instance, it does not apply to $(2x + 1)^{10}$, since the quantity that is being raised to a power is not x. Formula (2) applies to a power of any *function of x*, such as $(2x + 1)^{10}$. It can even be used in place of the original power rule. For example, if we take $u = x$ in Formula (2), we obtain

$$\dfrac{d}{dx} x^n = nx^{n-1} \dfrac{dx}{dx}$$

$$= nx^{n-1},$$

since the derivative of x with respect to x is 1. Thus, the generalized power rule is really a generalization of the original power rule, as its name suggests.

▼ **EXAMPLE 4**

Find **(a)** $\dfrac{d}{dx}(2x^5 + x^2 - 20)^{-2/3}$, **(b)** $\dfrac{d}{dx}\left(\dfrac{1}{\sqrt{x+2}}\right)$, and **(c)** $\dfrac{d}{dx}\left(\dfrac{1}{x^2+x}\right)$.

SOLUTION

(a) $\dfrac{d}{dx}(2x^5 + x^2 - 20)^{-2/3} = -\dfrac{2}{3}(2x^5 + x^2 - 20)^{-5/3}(10x^4 + 2x)$

(b) $\dfrac{d}{dx}\left(\dfrac{1}{\sqrt{x+2}}\right) = \dfrac{d}{dx}(x+2)^{-1/2} = -\dfrac{1}{2}(x+2)^{-3/2}(1) = -\dfrac{1}{2(x+2)^{3/2}}$

(c) $\dfrac{d}{dx}\left(\dfrac{1}{x^2+x}\right) = \dfrac{d}{dx}(x^2+x)^{-1} = -(x^2+x)^{-2}(2x+1) = -\dfrac{2x+1}{(x^2+x)^2}$

Before we go on... In the last instance, we could have used the quotient rule instead of the generalized power rule. The reason for this is that we could think of the quantity $1/(x^2 + x)$ in two ways by using the calculation thought experiment:

(1) as 1 divided by something—in other words, as a quotient;
(2) as something raised to the -1 power.

There are two morals to be had from the last example. One is that there are usually several ways to find the right answer. More subtle, though, is that if there are several ways to the answer, those ways may be related. In fact that is the case here, and we can finally give you a reason for believing the quotient rule, if you will believe the chain and product rules.

PROOF OF THE QUOTIENT RULE

By the chain rule, with $u = g(x)$,

$$\dfrac{d}{dx}\left(\dfrac{1}{g(x)}\right) = \dfrac{d}{dx}[g(x)^{-1}]$$

$$= -[g(x)^{-2}]g'(x) = -\dfrac{1}{g(x)^2}g'(x) = -\dfrac{g'(x)}{g(x)^2}.$$

We use this to get the derivative of any quotient.

We have

$$\frac{d}{dx}\left(\frac{f(x)}{g(x)}\right) = \frac{d}{dx}\left(f(x)\frac{1}{g(x)}\right) \qquad \text{Derivative of a product}$$

$$= f'(x)\frac{1}{g(x)} + f(x)\left(-\frac{g'(x)}{g(x)^2}\right) \qquad \text{Product rule}$$

$$= \frac{f'(x)}{g(x)} - \frac{f(x)g'(x)}{g(x)^2}$$

$$= \frac{f'(x)g(x) - f(x)g'(x)}{g(x)^2}$$

which is, of course, the quotient rule. ◀

Thus, if you believe the chain rule, you *must* believe the quotient rule. Of course, we still owe you some reason for believing the chain rule in the first place. Truly rigorous proofs of the chain rule require very careful analysis of limits. Here is a rough outline of a real proof.

IDEA OF PROOF OF THE CHAIN RULE

From the definition of the derivative,

$$\frac{g(x+h) - g(x)}{h} \approx g'(x) \quad \text{for small } h.$$

(Remember that the difference quotient is an approximation to the derivative.) If we multiply both sides by h and add $g(x)$ to both sides, we find

$$g(x+h) \approx g(x) + g'(x)h \quad \text{for small } h.$$

Now the same is true for f.

$$f(y+k) \approx f(y) + f'(y)k \quad \text{for small } k.$$

What we are after is the derivative of $f(g(x))$. Thus, we want to calculate the limit of

$$\frac{f(g(x+h)) - f(g(x))}{h}.$$

If we approximate $g(x+h)$ by $g(x) + g'(x)h$, we get

$$\frac{f(g(x+h)) - f(g(x))}{h} \approx \frac{f(g(x) + g'(x)h) - f(g(x))}{h}.$$

Now we approximate $f(g(x) + g'(x)h)$.

$$f(g(x) + g'(x)h) \approx f(g(x)) + f'(g(x))g'(x)h$$

(Take $f(y+k) \approx f(y) + kf'(y)$ and replace y with $g(x)$ and k with $g'(x)h$.)

Substituting for $f(g(x) + g'(x)h)$, we now get

$$\frac{f(g(x + h)) - f(g(x))}{h} \approx \frac{f(g(x) + g'(x)h) - f(g(x))}{h}$$

$$\approx \frac{f(g(x)) + f'(g(x))g'(x)h - f(g(x))}{h}.$$

The terms $f(g(x))$ cancel to give

$$\frac{f(g(x + h)) - f(g(x))}{h} \approx \frac{f'(g(x))g'(x)h}{h} = f'(g(x))g'(x),$$

which is the chain rule.* ◀

We now look at several more complicated examples.

▼ **EXAMPLE 5**

Find $\dfrac{dy}{dx}$ if $y = (\sqrt{x + 1} + 3x)^{-3}$.

SOLUTION The calculation thought experiment tells us that the last operation we would perform in calculating y is raising the quantity $(\sqrt{x + 1} + 3x)$ to the power -3. Thus, we use the generalized power rule.

$$\frac{dy}{dx} = -3(\sqrt{x + 1} + 3x)^{-4} \frac{d}{dx}(\sqrt{x + 1} + 3x)$$

Notice that we are not yet done, since this equation indicates that we must still find the derivative of $\sqrt{x + 1} + 3x$. We need not do everything in one step. Finding the derivative of a complicated function in several steps helps to keep the problem manageable. Continuing,

$$\frac{dy}{dx} = -3(\sqrt{x + 1} + 3x)^{-4} \frac{d}{dx}(\sqrt{x + 1} + 3x)$$

$$= -3(\sqrt{x + 1} + 3x)^{-4} \left(\frac{d}{dx}\sqrt{x + 1} + \frac{d}{dx}(3x)\right).$$

Now we have two derivatives left to calculate. The second of these we know to be 3, and the first is the derivative of the square root of a quantity. Thus,

$$\frac{dy}{dx} = -3(\sqrt{x + 1} + 3x)^{-4}\left(\frac{1}{2\sqrt{x + 1}} + 3\right).$$

▼ *You might wonder why we never seemed to take the limit as $h \to 0$. In fact, we did this when we replaced quantities by their *approximations*. These approximations become equalities only in the limit as $h \to 0$, so, in effect, we *did* take the limit.

▼ **EXAMPLE 6**

Find $\dfrac{d}{dx}\left((x+10)^3\sqrt{1-x^2}\right)$.

SOLUTION The expression $(x+10)^3\sqrt{1-x^2}$ is a product, so we use the product rule.

$$\frac{d}{dx}\left((x+10)^3\sqrt{1-x^2}\right) = \left(\frac{d}{dx}(x+10)^3\right)\sqrt{1-x^2}$$
$$+ (x+10)^3 \frac{d}{dx}\sqrt{1-x^2}$$

Notice that we left the calculation of the derivatives of the factors for the next step.

$$= 3(x+10)^2\sqrt{1-x^2} + (x+10)^3 \frac{1}{2\sqrt{1-x^2}}(-2x)$$
$$= 3(x+10)^2\sqrt{1-x^2} - \frac{x(x+10)^3}{\sqrt{1-x^2}}$$

The next example is a new treatment of an exercise from the last chapter.

▼ **EXAMPLE 7** Marginal Product

Precision Manufacturers, Inc. is informed by a consultant that its annual profit is given by

$$P = -200{,}000 + 4{,}000q - 0.5q^2,$$

where q is the number of surgical lasers it sells per year. The consultant also informs Precision that the number of surgical lasers it can manufacture per year depends on the number n of assembly-line workers it employs, according to the equation

$$q = 100n + 0.1n^2.$$

Use the chain rule to find the *marginal product* $\dfrac{dP}{dn}$.

SOLUTION Recall that we calculated the marginal product in Chapter 3 by substituting the expression for q in the expression for P to obtain P as a function of n and then finding dP/dn. Alternatively—and this will simplify the calculation—we can use the chain rule. To see how the chain rule applies, notice that P is a function of q, where q in turn is given as a function of n. Thus, by the chain rule,

$$\frac{dP}{dn} = P'(q)\frac{dq}{dn}$$

$$= \frac{dP}{dq} \cdot \frac{dq}{dn}.$$

Now we compute

$$\frac{dP}{dq} = 4{,}000 - q$$

and
$$\frac{dq}{dn} = 100 + 0.2n.$$

Substituting into the equation for $\frac{dP}{dn}$ gives

$$\frac{dP}{dn} = (4{,}000 - q)(100 + 0.2n).$$

Notice that the answer has both q and n as variables. We can express this as a function of n alone by substituting for q.

$$\frac{dP}{dn} = [4{,}000 - (100n + 0.1n^2)](100 + 0.2n)$$

$$= (4{,}000 - 100n - 0.1n^2)(100 + 0.2n)$$

The equation

$$\frac{dP}{dn} = \frac{dP}{dq} \cdot \frac{dq}{dn}$$

is an appealing way of writing the chain rule. In general, we can write the chain rule as follows.

> **CHAIN RULE IN DIFFERENTIAL NOTATION**
>
> If y is a function of u, and u is a function of x, then
>
> $$\frac{dy}{dx} = \frac{dy}{du}\frac{du}{dx}.$$
>
> If y is a function of u, and x is a function of u, then
>
> $$\frac{dy}{dx} = \frac{dy}{du} \bigg/ \frac{dx}{du}.$$

This is one of the reasons we still use Leibniz's differential notation. In this notation the chain rule looks like a simple "cancellation" of du terms.

▼ **EXAMPLE 8** Marginal Revenue

Suppose that a company's weekly revenue R is given as a function of the unit price p, and that p in turn is given as a function of weekly sales q (by means of a demand equation). If

$$\left.\frac{dR}{dp}\right|_{q=1000} = \$40 \text{ per } \$1 \text{ increase in price, and}$$

$$\left.\frac{dp}{dq}\right|_{q=1000} = -\$20 \text{ per additional item sold per week,}$$

find the marginal revenue when sales are 1,000 items per week.

SOLUTION The marginal revenue is $\dfrac{dR}{dq}$. By the chain rule,

$$\frac{dR}{dq} = \frac{dR}{dp}\frac{dp}{dq}.$$

Since we are interested in the marginal revenue at a demand level of 1,000 items per week, we have

$$\left.\frac{dR}{dq}\right|_{q=1000} = (40)(-20) = -\$800 \text{ per additional item demanded.}$$

Thus, if the price is lowered to increase the demand from 1,000 to 1,001 items per week, the weekly revenue will drop by approximately $800.

So far, we have *proved* the (original) power rule only for integer exponents. (Positive integer exponents were dealt with in the previous chapter and negative integer exponents were dealt with in the previous section.) We end this section by proving that the power rule also works for *rational* exponents, those of the form $\frac{p}{q}$, with p and q integers.

▼ **EXAMPLE 9** Power Rule for Rational Exponents

Show that $\dfrac{d}{dx}(x^{p/q}) = \dfrac{p}{q}x^{p/q-1}$.

SOLUTION We first let $y = x^{p/q}$. Thus, the problem is to calculate dy/dx without assuming the power rule for anything but *integer* exponents. Before we do anything, we raise both sides to the power q in order to get integer exponents everywhere.

$$y^q = (x^{p/q})^q = x^p$$

Now we take the equation $y^q = x^p$ and differentiate both sides with respect to x. By the chain rule,

$$\frac{d}{dx}(y^q) = qy^{q-1}\frac{dy}{dx},$$

whereas

$$\frac{d}{dx}(x^p) = px^{p-1}.$$

Equating these derivatives gives

$$qy^{q-1}\frac{dy}{dx} = px^{p-1}.$$

Now remember that we want dy/dx by itself. We can solve for dy/dx by dividing both sides by the quantity qy^{q-1}:

$$\frac{dy}{dx} = \frac{px^{p-1}}{qy^{q-1}}.$$

But $y = x^{p/q}$, and so

$$\frac{d}{dx}(x^{p/q}) = \frac{px^{p-1}}{q(x^{p/q})^{q-1}}$$

$$= \frac{px^{p-1}}{qx^{p-p/q}}$$

$$= \frac{p}{q}x^{p-1-(p-p/q)} = \frac{p}{q}x^{p/q-1}.$$

Done!

Before we go on... This calculation of dy/dx is an example of a technique called **implicit differentiation**. We shall discuss this technique in more detail in Section 4.

With some effort we have now proven the power rule for rational powers, but this still leaves the irrational powers. At the end of the next section we shall see a completely different proof of the power rule that applies to *all* real powers.

▶ ### 11.2 EXERCISES

Mentally calculate the derivatives of the functions in Exercises 1 through 10.

1. $f(x) = (2x + 1)^2$
2. $f(x) = (3x - 1)^2$
3. $f(x) = (x - 1)^{-1}$
4. $f(x) = (2x - 1)^{-2}$
5. $f(x) = (2 - x)^{-2}$
6. $f(x) = (1 - x)^{-1}$
7. $f(x) = \sqrt{2x + 1}$
8. $f(x) = \sqrt{3x - 2}$
9. $f(x) = \dfrac{1}{3x - 1}$
10. $f(x) = \dfrac{1}{(x + 1)^2}$

Calculate the derivatives of the functions in Exercises 11 through 32.

11. $f(x) = (x^2 + 2x)^4$
12. $f(x) = (x^3 - x)^3$
13. $f(x) = (2x^2 - 2)^{-1}$
14. $f(x) = (2x^3 + x)^{-2}$
15. $g(x) = (x^2 - 3x - 1)^{-5}$
16. $g(x) = (2x^2 + x + 1)^{-3}$
17. $h(x) = \dfrac{1}{(x^2 + 1)^3}$
18. $h(x) = \dfrac{1}{(x^2 + x + 1)^2}$
19. $s(t) = (t^2 - \sqrt{t})^4$
20. $s(t) = (2t + \sqrt{t})^{-1}$
21. $f(x) = \sqrt{1 - x^2}$
22. $f(x) = \sqrt{x + x^2}$
23. $f(x) = \sqrt{3\sqrt{x} - \dfrac{1}{\sqrt{x}}}$
24. $f(x) = \sqrt{2\sqrt{x} + \dfrac{1}{2\sqrt{x}}}$
25. $r(x) = (\sqrt{2x + 1} - x^2)^{-1}$
26. $r(x) = (\sqrt{x + 1} + \sqrt{x})^3$
27. $s(x) = \left(\dfrac{2x + 4}{3x - 1}\right)^2$
28. $s(x) = \left(\dfrac{3x - 9}{2x + 4}\right)^3$
29. $h(r) = [(r + 1)(r^2 - 1)]^{-1/2}$
30. $h(r) = [(2r - 1)(r - 1)]^{-1/3}$
31. $f(x) = (x^2 - 3x)^{-2}\sqrt{1 - x^2}$
32. $f(x) = (3x^2 + x)\sqrt{1 - x^2}$

Find the indicated derivative in each of Examples 33 through 40. In each case, the independent variable is an (unspecified) function of t.

33. $y = x^{100} + 99x^{-1}$. Find $\dfrac{dy}{dt}$.
34. $y = \sqrt{x}(1 + x)$. Find $\dfrac{dy}{dt}$.
35. $s = \dfrac{1}{r^3} + \sqrt{r}$. Find $\dfrac{ds}{dt}$.
36. $s = r + r^{-1}$. Find $\dfrac{ds}{dt}$.
37. $V = \dfrac{4}{3}\pi r^3$. Find $\dfrac{dV}{dt}$.
38. $A = 4\pi r^2$. Find $\dfrac{dA}{dt}$.
39. $y = x^3 + \dfrac{1}{x}$, $x = 2$ when $t = 1$, $\left.\dfrac{dx}{dt}\right|_{t=1} = -1$. Find $\left.\dfrac{dy}{dt}\right|_{t=1}$.
40. $y = \sqrt{x} + \dfrac{1}{\sqrt{x}}$, $x = 9$ when $t = 1$, $\left.\dfrac{dx}{dt}\right|_{t=1} = -1$. Find $\left.\dfrac{dy}{dt}\right|_{t=1}$.

APPLICATIONS

41. *Marginal Revenue* We saw in previous chapters that the weekly sales of rubies by Royal Ruby Retailers (RRR) is given by

$$q = -\dfrac{4p}{3} + 80.$$

(a) Express RRR's weekly revenue as a function of p, and hence calculate $\left.\dfrac{dR}{dp}\right|_{q=60}$.

(b) Use the demand equation to calculate $\left.\dfrac{dp}{dq}\right|_{q=60}$.

(c) Use the answers to parts (a) and (b) to find the marginal revenue at a demand level of 60 rubies per week, and interpret the result.

42. *Marginal Revenue* Repeat Exercise 41 for a demand level of $q = 52$ rubies per week.

43. *Marginal Product* Paramount Electronics, Inc. has an annual profit given by

$$P = -100{,}000 + 5{,}000q - 0.25q^2,$$

where q is the number of laptop computers it sells per year. The number of laptop computers it can manufacture per year depends on the number n of electrical engineers Paramount employs, according to the equation

$$q = 30n + 0.01n^2.$$

Use the chain rule to find $\left.\dfrac{dP}{dn}\right|_{n=10}$, and interpret the result.

44. *Marginal Product* Referring to Exercise 43, give a formula for the average profit per computer

$$\bar{P} = \frac{P}{q}$$

as a function of q, and hence determine the **marginal average product**, $\frac{d\bar{P}}{dn}$, at an employee level of 10 engineers. Interpret the result.

45. *Ecology* Manatees are grazing sea mammals sometimes referred to as sea sirens. Increasing numbers of manatees have been killed by boats off the Florida coast. Since 1976, the number M of manatees killed by boats each year is roughly linear, with

$$M(t) \approx 2.27t + 11.5 \qquad (0 \leq t \leq 16).$$

(t is the number of years since 1976*.) Over the same period, the total number B of boats registered in Florida has also been increasing at a roughly linear rate, given by

$$B(t) \approx 19{,}500t + 436{,}000 \qquad (0 \leq t \leq 16).$$

Use the chain rule to give an estimate of $\frac{dM}{dB}$. What does the answer tell you about manatee deaths?

46. *Ecology* The linear models used in the previous exercise are rough, as shown in the following graphs of the actual data.*

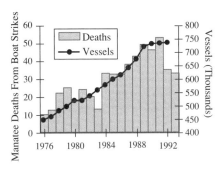

As you can see, the annual number of deaths was declining by 1992, despite an increase in the number of registered boats. This suggests that a linear model is not appropriate for the most recent period. If we consider only the data from 1986 onwards, use a quadratic model for manatee deaths, and use a linear model for the number of vessels, we obtain

$$M(t) \approx -1.35t^2 + 9.64t + 30.92$$
$$\text{and} \quad B(t) \approx 16{,}700t + 641{,}000,$$

where t is the number of years since 1986. Use the chain rule to obtain $\frac{dM}{dB}$. What does the answer tell you about manatee deaths? Give a possible explanation of this result.

47. *Pollution* An offshore oil well is leaking oil and creating a circular oil slick. If the radius of the slick is growing at a rate of 2 mph, find the rate at which the area is increasing when the radius is 3 miles. (Remember that the area of a disc of radius r is $A = \pi r^2$.)

48. *Mold* A mold culture in a dorm refrigerator is circular and growing in size. The radius is growing at a rate of 3 cm/day. How fast is the area growing when the culture is 4 centimeters in radius? (The area of a disc of radius r is $A = \pi r^2$.)

49. *Budget Overruns* The Pentagon is planning to build a new satellite that will be spherical in shape. As is typical in these cases, the specifications keep changing, so that the size of the satellite keeps growing. In fact, the radius of the planned satellite is growing 0.5 feet per week. Now its cost will be $1,000 per cubic foot. At the point when the plans call for a satellite 10 feet in radius, how fast is the cost growing? (Remember that the volume of a solid sphere of radius r is $V = \frac{4}{3}\pi r^3$.)

50. *Soap Bubbles* The soap bubble I am blowing has a surface area that is growing at the rate of 4 cm²/sec. How fast is its radius growing when its surface area is 40 cm²? (The radius of a sphere of surface area S is $r = \sqrt{S/(4\pi)}$.)

51. *Revenue Growth* (graphing calculator or computer software required) The demand for the Cyberpunk II arcade video game is modeled by the logistic curve

$$q(t) = \frac{10{,}000}{1 + 0.5e^{-0.4t}},$$

---

* These models are best-fit linear functions based on graphical data in the article "Manatees," by Thomas J. O'Shea, *Scientific American*, July, 1994, pp. 66–72. Source: Florida Department of Environmental Protection.

where $q(t)$ is the total number of units sold t months after its introduction.
(a) Graph this function for $0 \le t \le 10$ using a range for q of $5{,}000 \le q' \le 10{,}000$, and estimate $q'(4)$.
(b) Assume that the manufacturers of Cyberpunk II sell each unit for $800. What is the company's marginal revenue?
(c) Use the chain rule to estimate the rate at which revenue is growing 4 months after the introduction of the video game.

52. *Information Highway* *(graphing calculator or computer software required)* The amount of information transmitted each month on the National Science Foundation's Internet network for the years from 1988 to 1994 can be modeled by the equation

$$q(t) = \frac{2e^{0.69t}}{3 + 1.5e^{-0.4t}},$$

where q is the amount of information transmitted each month in billions of data packets and t is the number of years since the start of 1988.*
(a) Graph this function with $0 \le t \le 5$ and $0 \le q \le 40$, and estimate $q'(2)$.
(b) Assume that it costs $5 to transmit a million packets of data. What is the marginal cost?
(c) How fast was the cost increasing at the start of 1990?

Money Stock *Exercises 53–56 are based on the following demand function for money (taken from a question on the GRE economics test):*

$$M_d = (2) \times (y)^{0.6} \times (r)^{-0.3} \times (p),$$

where

M_d = demand for nominal money balances;
y = real income;
r = an index of interest rates;
p = an index of prices.

These exercises also use the idea of **percentage rate of growth:** *the percentage rate of growth of f at x is $f'(x)/f(x)$ (this is discussed again in Section 5).*

53. *(from the GRE economics test)* If the interest rate and price level are to remain constant while real income grows at 5 percent per year, the money stock must grow at what percent per year?

54. *(from the GRE economics test)* If real income and the price level are to remain constant while the interest rate grows at 5 percent per year, the money stock must change by what percent per year?

55. *(from the GRE economics test)* If the interest rate is to remain constant while real income grows at 5 percent per year and the price level rises at 5 percent per year, the money stock must grow at what percent per year?

56. *(from the GRE economic test)* If real income grows by 5 percent per year, the interest rate grows by 2 percent per year, and the price level drops by 3 percent per year, the money stock must change by what percent per year?

COMMUNICATION AND REASONING EXERCISES

57. How is the graph of $y = f(x)$ related to the graph of $y = g(x) = f(x + 1)$? How is the slope of the graph of g at $x = a$ related to the slope of the graph of f at $x = a + 1$? Illustrate with $f(x) = x^2 + 2x$.

58. How is the graph of $y = f(x)$ related to the graph of $y = g(x) = f(2x)$? How is the slope of the graph of g at $x = a$ related to the slope of the graph of f at $x = 2a$? Illustrate with $f(x) = x^2 + 2x$.

59. Formulate a simple procedure for deciding whether to first apply the chain rule, the product rule, or the quotient rule when finding the derivative of a function.

60. Give an example of a function f with the property that calculating $f'(x)$ requires use of the following rules in the given order: (1) the chain rule, (2) the product rule, (3) the quotient rule, (4) the chain rule.

61. Give an example of a function f with the property that calculating $f'(x)$ requires use of the chain rule five times in succession.

62. What can you say about composites of linear functions?

---

* This is the authors' model, based on figures published in *The New York Times*, Nov. 3, 1993.

11.3 DERIVATIVES OF LOGARITHMIC AND EXPONENTIAL FUNCTIONS

At this point, we know how to take the derivative of any algebraic expression in x (involving powers, radicals, and so on). We now turn to the derivatives of logarithmic and exponential functions.

DERIVATIVE OF THE LOGARITHM FUNCTION

$$\frac{d}{dx} \log_b x = \frac{1}{x \ln b}$$

An important special case is this:

DERIVATIVE OF THE NATURAL LOGARITHM

$$\frac{d}{dx} \ln x = \frac{1}{x}$$

It is because $\ln x = \log_e x$ has the simplest-looking derivative that it is called the *natural* logarithm.

Q Where do these formulas come from?

A We shall show you at the end of this section. For now, let us look at the graphs of $y = \ln x$ and $y = 1/x$ to see that it is reasonable that the derivative of $\ln x$ should be $1/x$ (Figure 1).

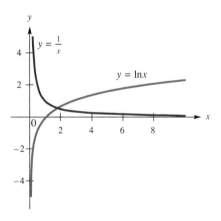

FIGURE 1

When x is close to 0, the graph of $\ln x$ is rising steeply, so the derivative should be a large positive number. And indeed, $1/x$ is positive and large. As x increases, the graph of $\ln x$ becomes less steep, although it continues to rise. Therefore, the derivative should become smaller but remain positive, and this is exactly what happens to $1/x$. So $1/x$ at least behaves the way the derivative of $\ln x$ should.

EXAMPLE 1

Find $\dfrac{d}{dx}[x \ln x]$.

SOLUTION By the calculation thought experiment, we see that $x \ln x$ is a product, so we need to use the product rule.

$$\frac{d}{dx}[x \ln x] = (1)\ln x + x\left(\frac{1}{x}\right) = \ln x + 1$$

(As we cautioned earlier, you must be prepared to use any of the rules in any example.)

If we were to take the derivative of the natural logarithm of a quantity, rather than just x, we would have to use the chain rule. In words, it would read like this:

The derivative of ln*(quantity) is one over that quantity times the derivative of that quantity.*

We can also write this as a formula.

Original Rule	Generalized Rule
$\dfrac{d}{dx}\ln x = \dfrac{1}{x}$	$\dfrac{d}{dx}\ln u = \dfrac{1}{u}\dfrac{du}{dx}$

EXAMPLE 2

Evaluate $\dfrac{d}{dx}\ln(x^2 + 1)$.

SOLUTION If we were to evaluate $\ln(x^2 + 1)$, the last operation we would perform is taking the natural logarithm of something. Thus, the calculation thought experiment tells us that we are dealing with ln *of a quantity*, and so we need the generalized logarithm rule as stated above. Thus,

$$\frac{d}{dx}\ln(x^2 + 1) = \frac{1}{x^2 + 1}\frac{d}{dx}(x^2 + 1)$$

$$= \frac{1}{x^2 + 1}(2x) = \frac{2x}{x^2 + 1}.$$

EXAMPLE 3

Find $\dfrac{d}{dx}\ln\sqrt{x + 1}$.

SOLUTION The calculation thought experiment again tells us that we have

11.3 Derivatives of Logarithmic and Exponential Functions

the natural logarithm of something, so

$$\frac{d}{dx} \ln\sqrt{x+1} = \frac{d}{dx} \ln[(x+1)^{1/2}]$$

$$= \frac{1}{\sqrt{x+1}} \left(\frac{1}{2}(x+1)^{-1/2} \right)$$

$$= \frac{1}{\sqrt{x+1}} \cdot \frac{1}{2\sqrt{x+1}}$$

$$= \frac{1}{2(x+1)}.$$

Before we go on... What happened to the square root? As with many problems involving logarithms, we could have done this one differently and with less bother had we simplified the expression $\ln\sqrt{x+1}$ using the rules of logarithms *before* differentiating. Doing this, we get

$$\ln\sqrt{x+1} = \ln(x+1)^{1/2} = \frac{1}{2}\ln(x+1).$$

Thus,
$$\frac{d}{dx} \ln\sqrt{x+1} = \frac{d}{dx}\left(\frac{1}{2} \ln(x+1) \right)$$

$$= \frac{1}{2}\left(\frac{1}{x+1} \right) = \frac{1}{2(x+1)},$$

the same answer as above.

▼ **EXAMPLE 4**

Evaluate $\frac{d}{dx} \ln[(x+1)(x+2)]$.

SOLUTION This time, we simplify the expression $\ln[(x+1)(x+2)]$ before taking the derivative.

$$\ln[(x+1)(x+2)] = \ln(x+1) + \ln(x+2)$$

Thus,

$$\frac{d}{dx} \ln[(x+1)(x+2)] = \frac{d}{dx} \ln(x+1) + \frac{d}{dx} \ln(x+2)$$

$$= \frac{1}{x+1} + \frac{1}{x+2}$$

$$= \frac{2x+3}{(x+1)(x+2)}.$$

Before we go on... For practice, try doing this example without simplifying first. What other differentiation rule do you have to use?

EXAMPLE 5

Find $\dfrac{d}{dx} \log_2(x^3 + x)$.

SOLUTION Rather than work out a "very generalized logarithm rule" for logs with any base, it is easier to remember that

$$\log_b x = \frac{\ln x}{\ln b},$$

where $\ln b$ is a *constant*. This gives us

$$\frac{d}{dx} \log_2(x^3 + x) = \frac{d}{dx} \left(\frac{\ln(x^3 + x)}{\ln 2} \right)$$

$$= \frac{1}{\ln 2} \frac{d}{dx} \ln(x^3 + x) \qquad \text{because ln 2 is a constant}$$

$$= \frac{3x^2 + 1}{(x^3 + x)\ln 2}.$$

EXAMPLE 6

Find $\dfrac{d}{dx} \ln |x|$.

SOLUTION Before we start, you might ask why we are bothering with the natural log of the absolute value of x to begin with. The reason is this: $\ln x$ is defined only for positive values of x, so its domain is $(0, +\infty)$. The function $\ln |x|$, on the other hand, is defined for *all* values of x other than zero. For example, $\ln |-2| = \ln 2 \approx 0.6931$. Thus the domain of $\ln |x|$ is $(-\infty, 0) \cup (0, +\infty)$. For this reason, it often turns out to be more useful than the ordinary logarithm function.

Now we'll get to work. We might be tempted to use the chain rule here (with $u = |x|$), but there is a slight catch: the function $|x|$ is, as we saw in the last chapter, not differentiable at $x = 0$. We *could* work our way around this, since the logarithm is not defined there anyway, but we'll use the following approach instead.

First,

$$|x| = \begin{cases} x & \text{if } x \geq 0 \\ -x & \text{if } x < 0. \end{cases}$$

11.3 Derivatives of Logarithmic and Exponential Functions

Thus,

$$\ln|x| = \begin{cases} \ln x & \text{if } x > 0 \\ \ln(-x) & \text{if } x < 0. \end{cases}$$

Hence,

$$\frac{d}{dx}\ln|x| = \begin{cases} \dfrac{d}{dx}\ln x = \dfrac{1}{x} & \text{if } x > 0 \\ \dfrac{d}{dx}\ln(-x) = \dfrac{1}{-x}(-1) = \dfrac{1}{x} & \text{if } x < 0. \end{cases}$$

In other words,

$$\frac{d}{dx}\ln|x| = \frac{1}{x}.$$

Before we go on... Figure 2 shows the graphs of $y = \ln|x|$ and $y = 1/x$. Figure 3 shows the graphs of $y = \ln|x|$ and $y = 1/|x|$. You should be able to see from these graphs why the derivative of $\ln|x|$ is $1/x$ and not $1/|x|$.

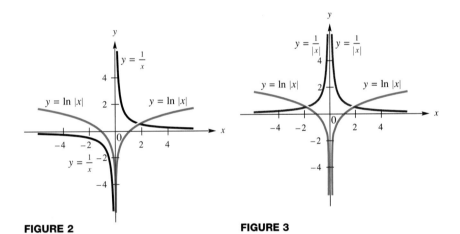

FIGURE 2   **FIGURE 3**

As we said, we could also find the derivative of $\ln|x|$ using the chain rule and the fact that

$$\frac{d}{dx}|x| = \begin{cases} 1 & \text{if } x > 0 \\ -1 & \text{if } x < 0. \end{cases}$$

Try this approach to see how well you understand the chain rule.

This example, in conjunction with the chain rule, gives us the following formulas.

DERIVATIVES OF LOGARITHMS OF ABSOLUTE VALUES

Original Rule	Generalized Rule
$\dfrac{d}{dx} \ln|x| = \dfrac{1}{x}$	$\dfrac{d}{dx} \ln|u| = \dfrac{1}{u}\dfrac{du}{dx}$
$\dfrac{d}{dx} \log_b|x| = \dfrac{1}{x \ln b}$	$\dfrac{d}{dx} \log_b|u| = \dfrac{1}{u \ln b}\dfrac{du}{dx}$

In other words, when taking the derivative of the logarithm of the absolute value of a quantity, we can simply ignore the absolute value sign!

EXAMPLE 7

Find $\dfrac{d}{dx} \ln|x^2 - x + 1|$.

SOLUTION By the generalized logarithm rule,

$$\frac{d}{dx} \ln|x^2 - x + 1| = \frac{2x - 1}{x^2 - x + 1}.$$

We now turn to the derivatives of exponential functions—that is, functions of the form $f(x) = a^x$. We begin by showing how *not* to differentiate them.

▶ **CAUTION** The derivative of a^x is *not* xa^{x-1}. The power rule applies only to *constant* exponents. In this case, the exponent is decidedly *not* constant. Thus, the power rule does not apply, and we shall need a new rule. ◀

DERIVATIVE OF a^x

If a is any positive number, then

$$\frac{d}{dx} a^x = a^x \ln a.$$

In particular:
DERIVATIVE OF e^x

$$\frac{d}{dx} e^x = e^x$$

Thus, e^x has the amazing property that its derivative is itself! (There is another—very simple—function that is its own derivative. What is it?)

EXAMPLE 8

Find $\dfrac{d}{dx}\left(\dfrac{e^x}{x}\right)$.

SOLUTION Using the calculation thought experiment, we see that the expression e^x/x is a quotient, so we use the quotient rule.

$$\frac{d}{dx}\left(\frac{e^x}{x}\right) = \frac{e^x x - e^x(1)}{x^2} = \frac{e^x(x-1)}{x^2}$$

If we were to take the derivative of e raised to a quantity, not just x, we would have to use the chain rule. In words, it would read like this:

The derivative of e raised to a quantity is e raised to that quantity times the derivative of that quantity.

We can also write this as a formula.

DERIVATIVES OF EXPONENTIAL FUNCTIONS

Original Rule	Generalized Rule
$\dfrac{d}{dx} e^x = e^x$	$\dfrac{d}{dx} e^u = e^u \dfrac{du}{dx}$
$\dfrac{d}{dx} a^x = a^x \ln a$	$\dfrac{d}{dx} a^u = a^u \ln a \dfrac{du}{dx}$

EXAMPLE 9

Find $\dfrac{d}{dx} e^{x^2+1}$.

SOLUTION Here the calculation thought experiment says that we have e raised to a quantity. Thus, we will have to use the generalized exponential rule.

$$\frac{d}{dx} e^{x^2+1} = e^{x^2+1}(2x) = 2xe^{x^2+1}$$

EXAMPLE 10

Find $\dfrac{d}{dx} 2^{3x}$.

SOLUTION Using the generalized exponential rule, we get

$$\frac{d}{dx} 2^{3x} = 2^{3x} \ln 2 \, \frac{d}{dx}(3x)$$
$$= 2^{3x}(\ln 2)(3) = (3 \ln 2) 2^{3x}.$$

Before we prove the formulas for the derivatives of logarithmic and exponential functions, here are some applications.

EXAMPLE 11 Epidemics

At the start of 1990, the number of U.S. residents infected with HIV was estimated to be 0.4 million. This number was growing exponentially, doubling every six months. Had this trend continued, how many new cases per month would have been occurring by the start of 1994?

SOLUTION To find the answer, we model this exponential growth using the methods of Chapter 2: after n years, the number of cases is

$$A = 0.4 e^{kn},$$

where k is given by

$$k = \frac{\ln 2}{\text{doubling time}} = \frac{\ln 2}{0.5} \approx 1.3863.$$

Thus,

$$A = 0.4 e^{1.3863n}.$$

We are asking for the number of new cases per month. In other words, we want the rate of change, dA/dn.

$$\frac{dA}{dn} = 0.4(1.3863) e^{1.3863n}$$
$$= 0.55452 e^{1.3863n}$$

At the start of 1994, $n = 4$, so the number of new cases per year is

$$\left.\frac{dA}{dn}\right|_{n=4} = 0.55452 e^{1.3863(4)} \approx 141.96 \text{ million}$$

Thus, the number of new cases *per month* would be $141.96/12 = 11.83$ million!

Before we go on... The reason this figure is astronomically large is that we assumed that exponential growth—the doubling every six months—would continue. A more realistic model for the spread of a disease is the logistic model. (See the "You're the Expert" section in Chapter 2, as well as the next example.)

▼ **EXAMPLE 12**   Sales Growth

The demand for the Cyberpunk II arcade video game is modeled by the logistic curve

$$q(t) = \frac{10{,}000}{1 + 0.5 e^{-0.4t}},$$

where $q(t)$ is the total number of units sold t months after its introduction. How fast were the sales increasing two years after its introduction?

SOLUTION   We are asking for $q'(24)$. We can find the derivative of $q(t)$ using the quotient rule, or we can first write

$$q(t) = 10{,}000(1 + 0.5 e^{-0.4t})^{-1},$$

and then use the generalized power rule.

$$q'(t) = -10{,}000(1 + 0.5 e^{-0.4t})^{-2}(0.5 e^{-0.4t})(-0.4)$$
$$= \frac{2{,}000 e^{-0.4t}}{(1 + 0.5 e^{-0.4t})^2}$$

Thus,

$$q'(24) = \frac{2{,}000 e^{-0.4(24)}}{(1 + 0.5 e^{-0.4(24)})^2} \approx 0.135 \text{ units per month.}$$

Thus, after two years, sales are increasing quite slowly.

You can check this answer graphically. If you plot the sales curve for $0 \le t \le 30$ and $6{,}000 \le q \le 10{,}000$, you will get the curve shown in Figure 4.

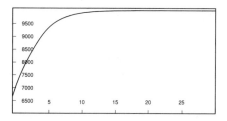

FIGURE 4

Notice that sales level off at about 10,000 units.* We computed $q'(24)$, which is the slope of the curve at the point with t-coordinate 24. If you zoom in to the portion of the curve near $t = 24$, you obtain the graph shown in Figure 5.

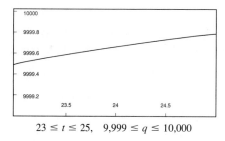

$23 \le t \le 25$, $9{,}999 \le q \le 10{,}000$

FIGURE 5

As you can see, the curve is almost linear. If we pick two points on this segment of the curve, say $(23, 9{,}999.4948)$ and $(25, 9{,}999.7730)$, we can approximate the derivative as

$$\frac{9{,}999.7730 - 9{,}999.4948}{25 - 23} = 0.1391,$$

which is accurate to within 0.004.

It is now time to explain where the formulas for the derivatives of $\ln x$ and e^x come from. We start with $\ln x$.

---

* We can also say this using limits: $\lim_{t \to +\infty} q(t) = 10{,}000$.

11.3 Derivatives of Logarithmic and Exponential Functions

Q Why is $\dfrac{d}{dx} \ln x = \dfrac{1}{x}$?

A To compute $\dfrac{d}{dx} \ln x$, we need to use the definition of the derivative. We shall also use properties of the logarithm to help evaluate the limit.

$$\begin{aligned}
\frac{d}{dx} \ln x &= \lim_{h \to 0} \frac{\ln(x+h) - \ln x}{h} \\
&= \lim_{h \to 0} \frac{1}{h} [\ln(x+h) - \ln x] \\
&= \lim_{h \to 0} \frac{1}{h} \ln\left(\frac{x+h}{x}\right) \\
&= \lim_{h \to 0} \frac{1}{h} \ln\left(1 + \frac{h}{x}\right) \\
&= \lim_{h \to 0} \ln\left(1 + \frac{h}{x}\right)^{1/h} \\
&= \ln\left(\lim_{h \to 0}\left(1 + \frac{h}{x}\right)^{1/h}\right)
\end{aligned}$$

(Switching ln and lim in the last step needs some justification that we'll glide over.*)

$$= \ln\left[\lim_{h \to 0}\left(1 + \frac{1}{(x/h)}\right)^{x/h}\right]^{1/x}$$

As $h \to 0$, the quantity x/h is getting large, and so the limit in brackets is approaching e. (Actually, we should say that $x/h \to +\infty$ as $h \to 0^+$, and $x/h \to -\infty$ as $h \to 0^-$ since x is positive for $\ln x$ to be defined. We should discuss both cases, but we shall leave the second to you.)

$$\begin{aligned}
&= \ln e^{1/x} \\
&= \frac{1}{x} \ln e = \frac{1}{x}
\end{aligned}$$

The rule for the derivative of $\log_b x$ follows from the fact that $\log_b x = \ln x / \ln b$.

▼ * It is actually justified by the fact that the logarithm function is continuous.

Q Why is $\dfrac{d}{dx} e^x = e^x$?

A To find the derivative of e^x we use a shortcut*. Write $g(x) = e^x$. Then

$$\ln g(x) = x.$$

Take the derivative of both sides of this equation to get

$$\frac{g'(x)}{g(x)} = 1,$$

or

$$g'(x) = g(x) = e^x.$$

In other words, the exponential with base e is its own derivative. The rule for exponential functions with other bases follows from the equality $a^x = e^{x \ln a}$ (why?) and the chain rule (try it).

At the end of the last section, we promised to give a proof of the power rule,

$$\frac{d}{dx} x^n = nx^{n-1},$$

that works for all real n. This proof relies on the equality $x^n = e^{n \ln x}$. (You can check this by taking the natural logarithm of both sides.)

$$\begin{aligned}
\frac{d}{dx} x^n &= \frac{d}{dx} e^{n \ln x} \\
&= e^{n \ln x} \frac{d}{dx} [n \ln x] \\
&= e^{n \ln x} \left(\frac{n}{x}\right) \\
&= x^n \left(\frac{n}{x}\right) \\
&= nx^{n-1}
\end{aligned}$$

This proves the power rule for all real powers.

The power rule is a good example of something that often happens in mathematics. Something like the power rule might first be noticed in simple cases, like $(x^2)' = 2x$ and $(x^3)' = 3x^2$. In those simple cases, simple proofs may be found. Other cases can also be proved by simple means, but the straightforward approach starts to get pretty tough (try finding the derivative of $x^{1/4}$ straight from the definition). As mathematicians try to extend a result

▼ * This shortcut is an example of a technique called *logarithmic differentiation*, which is occasionally useful. We will see it again in the next section.

11.3 EXERCISES

Find the derivatives of the functions in Exercises 1–14 mentally.

1. $f(x) = \ln(x - 1)$
2. $f(x) = \ln(x + 3)$
3. $f(x) = \log_2 x$
4. $f(x) = \log_3 x$
5. $g(x) = \ln|x^2 + 3|$
6. $g(x) = \ln|2x - 4|$
7. $h(x) = e^{x+3}$
8. $h(x) = e^{x^2}$
9. $f(x) = e^{-x}$
10. $f(x) = e^{1-x}$
11. $g(x) = 4^x$
12. $g(x) = 5^x$
13. $h(x) = 2^{x^2-1}$
14. $h(x) = 3^{x^2-x}$

Find the derivatives of the functions in Exercises 15–58.

15. $x \ln x$
16. $3 \ln x$
17. $f(x) = (x^2 + 1)\ln x$
18. $f(x) = (4x^2 - x)\ln x$
19. $f(x) = (x^2 + 1)^5 \ln x$
20. $f(x) = \sqrt{x + 1} \ln x$
21. $g(x) = \ln|3x - 1|$
22. $g(x) = \ln|5 - 9x|$
23. $g(x) = \ln|2x^2 + 1|$
24. $g(x) = \ln|x^2 - x|$
25. $g(x) = \ln(x^2 - \sqrt{x})$
26. $g(x) = \ln\left(x + \dfrac{1}{x}\right)$
27. $h(x) = \log_2(x+1)$
28. $h(x) = \log_3(x^2 + x)$
29. $r(t) = (t^2 + 1)\log_3(t + 1/t)$
30. $r(t) = (t^2 - t)\log_3(t + \sqrt{t})$
31. $f(x) = (\ln|x|)^2$
32. $f(x) = \dfrac{1}{\ln|x|}$
33. $r(x) = \ln(x^2) - (\ln(x - 1))^2$
34. $r(x) = (\ln(x^2))^2$
35. $f(x) = xe^x$
36. $f(x) = 2e^x - x^2 e^x$
37. $r(x) = \ln(x + 1) + 3x^3 e^x$
38. $r(x) = \ln|x + e^x|$
39. $f(x) = e^x \ln|x|$
40. $f(x) = e^x \log_2|x|$
41. $f(x) = e^{2x+1}$
42. $f(x) = e^{4x-5}$
43. $h(x) = e^{x^2-x+1}$
44. $h(x) = e^{2x^2-x+\sqrt{x}}$
45. $s(x) = x^2 e^{2x-1}$
46. $s(x) = \dfrac{e^{4x-1}}{x^3 - 1}$
47. $r(x) = (e^{2x-1})^2$
48. $r(x) = (e^{2x^2})^3$
49. $g(x) = \dfrac{e^x + e^{-x}}{e^x - e^{-x}}$
50. $g(x) = \dfrac{1}{e^x + e^{-x}}$

51. $f(x) = \dfrac{1}{x(\ln x)^{1/2}}$

52. $f(x) = \dfrac{e^{-x}}{x(e^x)^{1/2}}$

53. $r(x) = \dfrac{\sqrt{\ln x}}{x}$

54. $r(x) = \dfrac{\sqrt{\ln x}}{x^2 - 1}$

55. $f(x) = \ln|\ln x|$

56. $f(x) = \ln|\ln|\ln x||$

57. $s(x) = \ln\sqrt{\ln x}$

58. $s(x) = \sqrt{\ln(\ln x)}$

Find the equations of the straight lines described in Exercises 59–64.

59. Tangent to the curve $y = e^x \log_2 x$ at the point $(1, 0)$

60. Tangent to the curve $y = e^x + e^{-x}$ at the point $(0, 2)$

61. Tangent to the curve $y = \ln\sqrt{2x + 1}$ at the point on the curve where $x = 0$

62. Tangent to the curve $y = \ln\sqrt{2x^2 + 1}$ at the point where $x = 1$

63. At right angles to the curve $y = e^{x^2}$ at the point where $x = 1$

64. At right angles to the curve $y = \log_2(3x + 1)$ at the point where $x = 1$

APPLICATIONS

65. *Investments* If $10,000 is invested in a savings account yielding 4% per year, compounded continuously, how fast is the balance growing after 3 years?

66. *Investments* If $20,000 is invested in a savings account yielding 3.5% per year, compounded continuously, how fast is the balance growing after 3 years?

67. *Population Growth* The population of Lower Anchovia was 4,000,000 at the start of 1995 and doubling every 10 years. How fast was it growing per year at the start of 1995?

68. *Population Growth* The population of Upper Anchovia was 3,000,000 at the start of 1996 and doubling every 7 years. How fast was it growing per year at the start of 1996?

69. *Radioactive Decay* Plutonium-239 has a half-life of 24,400 years. How fast is a lump of 10 grams decaying after 100 years?

70. *Radioactive Decay* Carbon-14 has a half-life of 5,730 years. How fast is a lump of 20 grams decaying after 100 years?

71. *Investments* If $10,000 is invested in a savings account yielding 4% per year, compounded semiannually, how fast is the balance growing after 3 years?

72. *Investments* If $20,000 is invested in a savings account yielding 3.5% per year, compounded semiannually, how fast is the balance growing after 3 years?

73. *Life Span in Ancient Rome* The percentage $P(t)$ of people surviving to age t years in ancient Rome can be approximated by*

$$P(t) = 92e^{-0.0277t}.$$

Calculate $P'(22)$, and explain what the result indicates.

74. *Communication Among Bees* The audible signals honey bees use to communicate are in the frequency range 0–500 hertz and are generated by their wings. The speed with which bees' wings must move the air to generate these signals depends on the frequency of the signal, and this relationship can be approximated by†

$$V(f) = 95.6e^{0.0049f}.$$

▼ *Based on graphical data in Marvin Minsky's article "Will Robots Inherit the Earth?" *Scientific American*, October, 1994, pp. 109–13.

†Based on graphical data in the article "The Sensory Basis of the Honeybee's Dance Language," by Wolfgang H. Kirchner and William F. Towne, *Scientific American*, June 1994, pp. 74–80.

Here, $V(f)$ is the speed of the air near the bees' wings in millimeters per second, and f is the frequency of the communication signal in hertz. Calculate $V'(200)$, and explain what the result indicates.

75. *Epidemics* The epidemic described in the "You're the Expert" section of the chapter on exponentials followed the curve

$$A = \frac{150,000,000}{1 + 14,999e^{-0.3466t}},$$

where A is the number of people infected and t is the number of months after the start of the disease. How fast is the epidemic growing after 20 months? After 30 months? After 40 months?

76. *Epidemics* Another epidemic follows the curve

$$A = \frac{200,000,000}{1 + 20,000e^{-0.549t}},$$

where t is in years. How fast is the epidemic growing after 10 years? After 20 years? After 30 years?

77. *Information Highway* The amount of information transmitted each month on the National Science Foundation's Internet network for the years from 1988 to 1994 can be modeled by the equation

$$q(t) = \frac{2e^{0.69t}}{3 + 1.5e^{-0.4t}},$$

where q is the amount of information transmitted each month in billions of data packets, and t is the number of years since the start of 1988. How fast was this quantity growing at the start of 1994?

78. *Information Highway* Repeat Exercise 77 using the revised equation

$$q(t) = \frac{2e^{0.72t}}{3 + 1.4e^{-0.4t}}.$$

The following exercises require the use of a graphing calculator or graphing computer software.

79. *Diffusion of New Technology* Numeric control is a technology whereby the operation of machines is controlled by numerical instructions on disks, tapes,

or cards. In a study, W. Mansfield modeled the growth of this technology using the equation

$$p(t) = \frac{0.80}{1 + e^{4.46-0.477t}},$$

where $p(t)$ is the percentage of firms using numeric control in year t.*

(a) Graph this function for $0 \leq t \leq 20$, and estimate $p'(10)$ graphically. Interpret the result.

(b) Use your graph to estimate $\lim_{t \to +\infty} p(t)$, and interpret the result.

(c) Compute $p'(t)$, graph it, and again find $p'(10)$.

(d) Use your graph to estimate $\lim_{t \to +\infty} p'(t)$, and interpret the result.

80. *Diffusion of New Technology* Repeat Exercise 79 using the revised formula

$$p(t) = \frac{0.90e^{-0.1t}}{1 + e^{4.50-0.477t}},$$

which takes into account that in the long term, this new technology will eventually become outmoded and will be replaced by a newer technology.

81. *Growth of HMOs* The enrollment in health maintenance organizations (HMOs) in the years from 1975 to 1992 can be modeled by the equation

$$n(t) = 5 + \frac{40e^{0.002x}}{1 + 25e^{3-0.5x}},$$

where $n(t)$ represents the total number (in millions) of U.S. residents enrolled in HMOs t years after 1975.[†] Calculate $n'(t)$, graph it, and use your graph to determine the value of t ($0 \leq t \leq 20$) when $n'(t)$ was a maximum. Interpret your result.

82. *Demand for Poultry* The demand for poultry can be modeled as

$$q = -60.5 - 0.45p + 0.12b + 12.21 \ln(d),$$

where q is the per capita demand for chicken in pounds per year, p is the wholesale price of chicken in cents per pound, b is the wholesale price of beef in cents per pound, and d is the per capita annual dispos-

---

▼ * Source: "The Diffusion of a Major Manufacturing Innovation," in *Research and Innovation in the Modern Corporation* (W.W. Norton and Company, Inc., New York, 1971, pp. 186–205)

† The authors' model, based on data supplied by the Group Health Association of America (published in *The New York Times*, Oct. 18, 1993).

able income in dollars per year.* Assume that the wholesale prices of chicken and beef are fixed at 25¢ per pound and 50¢ per pound respectively and that the mean disposable income in t years' time will be $25,000 + 1,000t$. Calculate and graph $q'(t)$ for $0 \le t \le 20$, and use your graph to estimate the value of t for which $q'(t) = 0.30$. Interpret the result.

COMMUNICATION AND REASONING EXERCISES

83. A quantity P is growing exponentially with time. Explain the difference between $P(10)$ and $P'(10)$. If P is measured in kilograms and t is measured in days, what are the units of $P'(10)$?

84. The number N of graphing calculators sold on campus is increasing by 3,000 per year at the present time ($t = 0$). Does this mean that $N(0) = 3,000$? Explain your answer.

85. Make the correct selections: If $Q = 100e^{-0.3t}$, then Q is [(a) increasing, (b) decreasing] with increasing t, and Q' is [(a) increasing, (b) decreasing] with increasing t.

86. If $Q = 2,000 - e^{0.3t}$, then
    (a) both Q and Q' are increasing with increasing t;
    (b) both Q and Q' are decreasing with increasing t;
    (c) Q is increasing and Q' is decreasing with increasing t;
    (d) Q is decreasing and Q' is increasing with increasing t.

*The **percentage**, or **fractional**, **rate of change** of a function is defined to be the ratio $f'(x)/f(x)$. (It is customary to express this as a percentage when speaking about percentage rate of change.)*

87. Show that the fractional rate of change of the exponential function e^{kx} is its fractional rate of growth k.

88. Show that the fractional rate of change of $f(x)$ is the rate of change of $\ln(f(x))$.

89. Let $A(t)$ represent a quantity growing exponentially. Show that the percentage rate of growth, $A'(t)/A(t)$, is constant.

90. Let $A(t)$ be the amount of money in an account paying interest compounded some number of times per year. Show that the percentage rate of growth, $A'(t)/A(t)$, is constant. What might this constant represent?

▶ ━━ 11.4 IMPLICIT DIFFERENTIATION

An equation in two variables x and y may or may not determine y as a function of x. For instance, the equation

$$2x^2 + y = 2$$

determines y as a function of x (solve for y to obtain $y = 2 - 2x^2$, a function of x). On the other hand, the equation

$$2x^2 + y^2 = 2$$

does not determine y as a function of x: solving for y yields $y = \pm\sqrt{2 - 2x^2}$. The "$\pm$" sign reminds us that for some values of x there are two corresponding values for y. In other words, y cannot be specified as a single

---

▼ * This equation is based on actual data from poultry sales in the period 1950–1984. Source: A. H. Studenmund, *Using Econometrics*, Second Edition, (New York: HarperCollins, 1992) pp. 180–81.

11.4 Implicit Differentiation

function of x. Even though y is not a function of x, we refer to y as an **implicit function** of x.

The best way to justify the term "implicit function" is to look at the graph of the equation $2x^2 + y^2 = 2$. We can graph it in two ways: by constructing a table along the lines described in Chapter 1, or by using a graphing calculator to superimpose the graphs of

$$y = \sqrt{2 - 2x^2} \quad \text{and} \quad y = -\sqrt{2 - 2x^2}.$$

The graph, an *ellipse*, is shown in Figure 1.

The curve $y = \sqrt{2 - 2x^2}$ constitutes the top half of the ellipse, and $y = -\sqrt{2 - 2x^2}$ constitutes the bottom half. Notice that the graph fails the "vertical line test": vertical lines between $x = -1$ and $x = 1$ pass through two points on the graph, and so the graph of $2x^2 + y^2 = 2$ is not the graph of a function. On the other hand, the graph is made up of the graphs of two very respectable functions, $f(x) = \sqrt{2 - 2x^2}$ and $g(x) = -\sqrt{2 - 2x^2}$. If we choose a point on the top half of the ellipse, then the part of the curve near that point is the graph of the function f. If we choose a point on the lower half, then the part of the curve near that point is the graph of the function g.

Figure 2 shows two selected points on the graph—P on the top half, and Q on the bottom half—together with segments of the graph through those points. Each of these segments is part of the graph of a function, as shown. Thus, given the equation $2x^2 + y^2 = 2$ and a point on its graph, there is a function whose graph is part of the graph of the equation near the point. (There are actually two points on the above graph for which this does not work. Can you see them?) The reason we use the word *implicit* is that we need not—and for many equations cannot—find an explicit formula for a function by solving for y.

We now find the slopes of the tangent lines to the ellipse at P and Q. We could do this by taking the derivative of f or g as appropriate. But there is a far simpler way of finding the derivative of an implicit function of x *without*

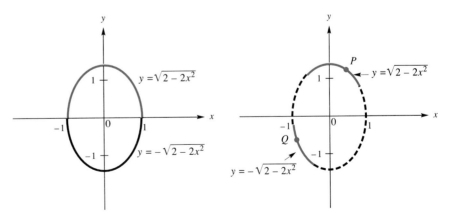

FIGURE 1

FIGURE 2

having to solve the equation for y. (We have already seen an example of this technique when we proved the power rule for rational exponents.)

Q How do we find $\dfrac{dy}{dx}$ if $2x^2 + y^2 = 2$ without first solving for *y*?

A We can do this by using the chain rule and a little cleverness, as follows. We think of *y* as a function of *x* and take the derivative with respect to *x* of both sides of the equation:

$$\frac{d}{dx}\left(2x^2 + y^2\right) = \frac{d}{dx}(2).$$

Recalling the rules for derivatives, we get

$$2\frac{d}{dx}(x^2) + \frac{d}{dx}(y^2) = 0.$$

Now we must be careful. The derivative with respect to *x* of x^2 is $2x$, but the derivative *with respect to x* of y^2 is *not* $2y$. Rather, since *y* is a function of *x*, we must use the chain rule, which tells us that $d/dx\,(y^2) = 2y\,dy/dx$. Thus, we obtain

$$2(2x) + 2y\frac{dy}{dx} = 0.$$

We want to find *dy/dx*, so we treat it as an unknown *and solve for it*: first take the expression not involving *dy/dx* over to the right-hand side.

$$2y\frac{dy}{dx} = -4x$$

Next, divide both sides by $2y$ to obtain

$$\frac{dy}{dx} = -\frac{4x}{2y} = -\frac{2x}{y},$$

and we have found $\dfrac{dy}{dx}$.

Q We calculated the derivative of the implicit function of *x* to be $-2x/y$. But this is not a function of *x*, as there is a "*y*" in the formula. Is this of any use to us?

A First, notice that we can hardly expect to obtain a function of *x* if *y* was not a function of *x* to begin with. The result is still useful, since we can evaluate the derivative at any point on the graph. For instance, if we choose *P* to be the point $(\frac{1}{\sqrt{2}}, 1)$ (which you can check is a point on the graph, since it satisfies $2x^2 + y^2 = 2$), we obtain

$$\frac{dy}{dx} = -\frac{2x}{y} = -\frac{2}{\sqrt{2}} = -\sqrt{2}.$$

Thus, the slope of the tangent to the curve $2x^2 + y^2 = 2$ at the point $P = (\frac{1}{\sqrt{2}}, 1)$ is $-\sqrt{2} \approx -1.414$.

This procedure—finding $\frac{dy}{dx}$ without first solving an equation for y—is called **implicit differentiation.**

EXAMPLE 1

Use implicit differentiation to find dy/dx if $x^2 + 2xy = y$.

SOLUTION Taking the derivative with respect to x of both sides gives

$$\frac{d}{dx}(x^2) + \frac{d}{dx}(2xy) = \frac{d}{dx}(y).$$

Looking at the right-hand side, notice that $\frac{d}{dx}(y) = \frac{dy}{dx}$. Also notice that $2xy$ is the product of $2x$ and y, and so we must apply the product rule to find $\frac{d}{dx}(2xy)$.

$$2x + (2)(y) + (2x)\frac{dy}{dx} = \frac{dy}{dx}$$

That is,

$$2x + 2y + 2x\frac{dy}{dx} = \frac{dy}{dx}.$$

We have marked the term dy/dx in color to remind us that it is the unknown. To solve for it, we bring all the terms containing dy/dx to the left-hand side and all terms not containing it to the right-hand side.

$$2x\frac{dy}{dx} - \frac{dy}{dx} = -2x - 2y$$

Now factor out the common term dy/dx, obtaining

$$\frac{dy}{dx}(2x - 1) = -2x - 2y,$$

and finally, divide by $(2x - 1)$ to obtain

$$\frac{dy}{dx} = \frac{-2x - 2y}{2x - 1} = -\frac{2(x + y)}{2x - 1}.$$

Before we go on... Once again, the derivative is not an explicit function of x. Notice that we could have solved the original equation for y as a function of x and then obtained its derivative. You should do this as an exercise and check that you obtain the same derivative.

EXAMPLE 2

Given that $\ln y = xy$, find the equation of the tangent line to the graph at the point where $y = 1$.

SOLUTION Once again, we differentiate both sides of the equation with respect to x:

$$\frac{d}{dx}(\ln y) = \frac{d}{dx}(xy).$$

Using the chain rule on the left and the product rule on the right, we obtain

$$\frac{1}{y}\frac{dy}{dx} = (1)y + x\frac{dy}{dx}.$$

Bringing the terms with $\frac{dy}{dx}$ to the left gives

$$\frac{1}{y}\frac{dy}{dx} - x\frac{dy}{dx} = y,$$

and so

$$\frac{dy}{dx}\left(\frac{1}{y} - x\right) = y,$$

or

$$\frac{dy}{dx}\left(\frac{1 - xy}{y}\right) = y,$$

so that

$$\frac{dy}{dx} = y\left(\frac{y}{1 - xy}\right) = \frac{y^2}{1 - xy}.$$

The derivative gives us the slope of the tangent line, so we must next evaluate the derivative at the point where $y = 1$. Note that the formula for dy/dx requires values for both the x- and y-coordinates. We get the x-coordinate by substituting $y = 1$ in the original equation.

$$\ln y = xy$$
$$\ln(1) = x \cdot 1.$$

But $\ln(1) = 0$, and so $x = 0$ for this point. Thus,

$$\left.\frac{dy}{dx}\right|_{y=1} = \frac{1^2}{1 - (0)(1)} = 1.$$

Therefore, the tangent line is the line through $(x, y) = (0, 1)$ with slope 1, which is

$$y = x + 1.$$

Before we go on... This is an example where it is simply not possible to solve for *y*. Try it.

EXAMPLE 3

Find $\dfrac{d}{dx}\left[\dfrac{(x+1)^{10}(x^2+1)^{11}}{(x^3+1)^{12}}\right]$ without using the product or quotient rules.

SOLUTION We can do this using a technique called **logarithmic differentiation**. We write

$$y = \frac{(x+1)^{10}(x^2+1)^{11}}{(x^3+1)^{12}}$$

and then take the natural logarithm of both sides.

$$\ln y = \ln\left[\frac{(x+1)^{10}(x^2+1)^{11}}{(x^3+1)^{12}}\right]$$

We can use properties of the logarithm to simplify the right-hand side.

$$\ln y = \ln(x+1)^{10} + \ln(x^2+1)^{11} - \ln(x^3+1)^{12}$$
$$= 10\ln(x+1) + 11\ln(x^2+1) - 12\ln(x^3+1)$$

Now we can find dy/dx using implicit differentiation.

$$\frac{1}{y}\frac{dy}{dx} = \frac{10}{x+1} + \frac{22x}{x^2+1} - \frac{36x^2}{x^3+1}$$

$$\frac{dy}{dx} = y\left(\frac{10}{x+1} + \frac{22x}{x^2+1} - \frac{36x^2}{x^3+1}\right)$$

$$= \frac{(x+1)^{10}(x^2+1)^{11}}{(x^3+1)^{12}}\left(\frac{10}{x+1} + \frac{22x}{x^2+1} - \frac{36x^2}{x^3+1}\right)$$

Before we go on... You should redo this example using the product and quotient rules instead of logarithmic differentiation, and compare the answers. Compare also the amount of work involved in both methods.

Now for an application. Productivity usually depends on both labor and capital. Suppose, for example, you are managing an automobile assembly plant. You can measure its productivity by counting the number of automobiles the plant produces each year. As a measure of labor, you can use the number of employees, and as a measure of capital you can use its operating budget. The **Cobb-Douglas production function** then has the form

$$P = Kx^a y^{1-a},$$

where *P* stands for the number of automobiles produced per year, *x* is the number of employees, and *y* is the operating budget. The numbers *K* and *a* are constants that depend on the particular factory studied, with a between 0 and 1.*

> **EXAMPLE 4**   Cobb-Douglas Production Function

The automobile assembly plant you manage has the Cobb-Douglas production function

$$P = x^{0.3}y^{0.7},$$

where P is the number of automobiles it produces per year, x is the number of employees, and y is the daily operating budget (in dollars). Assume a production level of 1,000 automobiles per year.

(a) Find $\dfrac{dy}{dx}$.

(b) Evaluate this derivative when $x = 80$.

SOLUTION   Since the production level is 1,000 automobiles per year, we have $P = 1{,}000$, and so the equation becomes

$$1{,}000 = x^{0.3}y^{0.7}.$$

(a) We find $\dfrac{dy}{dx}$ by implicit differentiation:

$$0 = \frac{d}{dx}(x^{0.3}y^{0.7})$$

$$= (0.3)x^{-0.7}y^{0.7} + x^{0.3}(0.7)y^{-0.3}\frac{dy}{dx}.$$

Thus,

$$0.7x^{0.3}y^{-0.3}\frac{dy}{dx} = -0.3x^{-0.7}y^{0.7},$$

giving

$$\frac{dy}{dx} = -\frac{0.3x^{-0.7}y^{0.7}}{0.7x^{0.3}y^{-0.3}}$$

$$= -\frac{3y}{7x}.$$

▼ * We shall be studying the Cobb-Douglas production function in detail in the chapter on functions of several variables. In particular, we shall see how to construct a production function to model a real-life situation.

(b) To evaluate this derivative at $x = 80$, we must first find the corresponding value of y. To obtain y, we substitute $x = 80$ in the original equation $1{,}000 = x^{0.3}y^{0.7}$ and solve for y.

$$1{,}000 = (80)^{0.3}y^{0.7} \approx 3.72329 y^{0.7}$$

$$y^{0.7} \approx \frac{1{,}000}{3.72329} \approx 268.580.$$

To obtain y on its own, raise both sides to the power $1/0.7$ to obtain

$$y = (y^{0.7})^{1/0.7} = (268.580)^{1/0.7} \approx 2{,}951.92.$$

Now that we have the corresponding value for y, we evaluate the derivative at $x = 80$ and $y = 2{,}951.92$.

$$\left.\frac{dy}{dx}\right|_{x=80} = -\frac{3y}{7x} \approx -\frac{3(2{,}951.92)}{7(80)} \approx -15.81$$

Before we go on... How do we interpret this result? The first clue is to look at the units of the derivative: recall that the units of dy/dx are units of y per unit of x. Since y is the daily budget, its units are dollars, and since x is the number of employees, its units are employees. Thus,

$$\left.\frac{dy}{dx}\right|_{x=80} = -\$15.81 \text{ per employee.}$$

Next, recall that dy/dx measures the rate of change of y as x changes. Thus (since the answer is negative), the daily budget to maintain production of 1,000 automobiles is decreasing by approximately $15.81 per additional employee at an employment level of 80 employees. Thus, increasing the work force by one worker will result in a saving of about $15.81 per day. Roughly speaking, *a new employee is worth $15.81 per day* at the current levels of employment and production.

11.4 EXERCISES

In Exercises 1–10, find dy/dx using implicit differentiation. In each case, compare your answer with the result obtained by first solving for y as a function of x and then taking the derivative.

1. $2x + 3y = 7$ **2.** $4x - 5y = 9$ **3.** $x^2 - 2y = 6$ **4.** $3y + x^2 = 5$ **5.** $2x + 3y = xy$

6. $x - y = xy$ **7.** $e^x y = 1$ **8.** $e^x y - y = 2$ **9.** $y \ln x + y = 2$ **10.** $\dfrac{\ln x}{y} = 2 - x$

In Exercises 11–30 Find the indicated derivative using implicit differentiation.

11. $x^2 + y^2 = 5$. Find $\dfrac{dy}{dx}$.   **12.** $2x^2 - y^2 = 4$. Find $\dfrac{dy}{dx}$.

13. $x^2y - y^2 = 4$. Find $\dfrac{dy}{dx}$.

14. $xy^2 - y = x$. Find $\dfrac{dy}{dx}$.

15. $3xy - \dfrac{y}{3} = \dfrac{2}{x}$. Find $\dfrac{dy}{dx}$.

16. $\dfrac{xy}{2} - y^2 = 3$. Find $\dfrac{dy}{dx}$.

17. $x^2 - 3y^2 = 8$. Find $\dfrac{dx}{dy}$.

18. $(xy)^2 + y^2 = 8$. Find $\dfrac{dx}{dy}$.

19. $p^2 - pq = 5p^2q^2$. Find $\dfrac{dp}{dq}$.

20. $q^2 - pq = 5p^2q^2$. Find $\dfrac{dp}{dq}$.

21. $xe^y - ye^x = 1$. Find $\dfrac{dy}{dx}$.

22. $x^2e^y - y^2 = e^x$. Find $\dfrac{dy}{dx}$.

23. $e^{st} = s^2$. Find $\dfrac{ds}{dt}$.

24. $e^{s^2t} - st = 1$. Find $\dfrac{ds}{dt}$.

25. $\dfrac{e^x}{y^2} = 1 + e^y$. Find $\dfrac{dy}{dx}$.

26. $\dfrac{x}{e^y} + xy = 9y$. Find $\dfrac{dy}{dx}$.

27. $\ln(y^2 - y) + x = y$. Find $\dfrac{dy}{dx}$.

28. $\ln(xy) - x \ln y = y$. Find $\dfrac{dy}{dx}$.

29. $\ln(xy + y^2) = e^y$. Find $\dfrac{dy}{dx}$.

30. $\ln(1 + e^{xy}) = y$. Find $\dfrac{dy}{dx}$.

In Exercises 31–34, use logarithmic differentiation to find dy/dx.

31. $y = (x^3 + x)\sqrt{x^3 + 2}$

32. $y = \sqrt{\dfrac{x - 1}{x^2 + 2}}$

33. $y = x^x$

34. $y = x^{-x}$

In Exercises 35–46, use implicit differentiation to evaluate dy/dx at the indicated point of the graph (if only the x-coordinate is given, you must also find the y-coordinate).

35. $4x^2 + 2y^2 = 12$, $(1, -2)$

36. $3x^2 - y^2 = 11$, $(-2, 1)$

37. $2x^2 - y^2 = xy$, $(-1, 2)$

38. $2x^2 + xy = 3y^2$, $(-1, -1)$

39. $3x^{0.3}y^{0.7} = 10$, $x = 20$

40. $2x^{0.4}y^{0.6} = 10$, $x = 50$

41. $x^{0.4}y^{0.6} - 0.2x^2 = 100$, $x = 20$

42. $x^{0.4}y^{0.6} - 0.3x^2 = 10$, $x = 10$

43. $e^{xy} - x = 4x$, $x = 3$

44. $e^{-xy} + 2x = 1$, $x = -1$

45. $\ln(x + y) - x = 3x^2$, $x = 0$

46. $\ln(x - y) + 1 = 3x^2$, $x = 0$

APPLICATIONS

47. **Demand** The demand equation for soccer tournament T-shirts is

$$pq - 2{,}000 = q,$$

where q is the number of T-shirts the Enormous State University soccer team can sell for $p apiece.
(a) How many T-shirts can the team sell at $5 apiece?
(b) Find $\left.\dfrac{dq}{dp}\right|_{p=5}$, and interpret the result.

48. **Cost Equations** The cost c (in cents) of producing x gallons of Ectoplasm hair gel is given by the cost equation

$$c^2 - 10cx = 200.$$

(a) Find the cost of producing 1 gallon and 3.5 gallons.
(b) Evaluate dc/dx at $x = 1$ and $x = 3.5$, and interpret the results.

49. *Housing Costs* The cost C of building a house is related to the number k of carpenters used and the number e of electricians used by the formula

$$C = 15{,}000 + 50k^2 + 60e^2.$$

If the cost of the house comes to \$200,000, find $\dfrac{dk}{de}\bigg|_{e=15}$, and interpret your result.

50. *Employment* An employment research company estimates that the value of a recent MBA graduate to an accounting company is

$$V = 3e^2 + 5g^3,$$

where V is the value of the graduate, e is the number of years of prior business experience, and g is the graduate school grade point average. If $V = 200$, find de/dg when $g = 3.0$, and interpret the result.

51. *Grades* A production formula for a student's performance on a difficult English examination is

$$g = 4tx - 0.2t^2 - 10x^2 \quad (x < 30),$$

where g is the grade the student can expect to obtain, t is the number of hours of study for the examination, and x is the student's grade point average.
(a) For how long should a student with a 3.0 grade point average study in order to score 80 on the examination?
(b) Find dt/dx for a student who earns a score of 80, evaluate it when $x = 3.0$, and interpret the result.

52. *Grades* Repeat Exercise 51 using the following production formula for a basket-weaving examination:

$$g = 10tx - 0.2t^2 - 10x^2 \quad (x < 10)$$

Comment on the result.

Exercises 53 and 54 are based on the following demand function for money (taken from a question on the GRE economics test):

$$M_d = (2) \times (y)^{0.6} \times (r)^{-0.3} \times (p),$$

where
$M_d =$ *demand for nominal money balances;*
$y =$ *real income;*
$r =$ *an index of interest rates;*
$p =$ *an index of prices.*

53. *Money Stock* If real income grows while the money stock and the price level remain constant, the interest rate must change at what rate? (First find dr/dy, then dr/dt; your answers will be expressed in terms of r and y.)

54. *Money Stock* If real income grows while the money stock and the interest rate remain constant, the price level must change at what rate?

COMMUNICATION AND REASONING EXERCISES

55. Use logarithmic differentiation to give another proof of the product rule.

56. Use logarithmic differentiation to give another proof of the quotient rule.

57. If y is a specified function of x, then is finding dy/dx directly (that is, explicitly) the same as finding it by implicit differentiation? Explain.

58. Explain why one should not expect dy/dx to be a function of x if y is not a function of x.

59. True or false? If y is a function of x and $dy/dx \neq 0$ at some point, then, regarding x as an implicit function of y,

$$\frac{dx}{dy} = \frac{1}{dy/dx}.$$

Explain your answer.

60. If you are given an equation in x and y such that dy/dx is a function of x only, what can you say about the graph of the equation?

▼ * Based on an exercise in *Introduction to Mathematical Economics* by A. L. Ostrosky, Jr., and J. V. Koch (Prospect Heights, IL: Waveland Press, 1979).

11.5 LINEAR APPROXIMATION AND ERROR ESTIMATION

One of the central themes in this and the previous chapter has been the concept of the derivative as the slope of the line tangent to the graph of a function at a point. When we described how to use a graphing calculator to estimate derivatives, we pointed out that if you zoom in to a portion of a smooth curve near a specified point, it becomes indistinguishable from the tangent line at that point. In other words, the values of the function are close to the values of the linear function whose graph is the tangent line. In this section, we take a careful look at this idea, called the linear approximation of a function.

Let us start with a point $(a, f(a))$ on the graph of a function f. If the curve is smooth at that point—that is, if $f'(a)$ exists—then we have

$$f'(a) = \lim_{h \to 0} \frac{f(a+h) - f(a)}{h}.$$

This means that the smaller h becomes, the closer $(f(a+h) - f(a))/h$ approximates $f'(a)$. Thus, for small values of h (close to zero),

$$f'(a) \approx \frac{f(a+h) - f(a)}{h}.$$

Multiplying both sides by h and solving for $f(a+h)$ gives

$$hf'(a) \approx f(a+h) - f(a),$$

so

$$f(a+h) \approx f(a) + hf'(a).$$

Now, since h is small, $a + h$ is a number close to a. For our purposes it will be more useful to call this number x, so $x = a + h$. This also gives us $h = x - a$. Substituting gives us the approximation

$$f(x) \approx f(a) + (x-a)f'(a).$$

This formula for approximating $f(x)$ is referred to as the **linear approximation of $f(x)$ near $x = a$**. (We saw this formula once before: in our proof of the chain rule.)

LINEAR APPROXIMATION OF f(x) NEAR x = a

If x is close to a, then

$$f(x) \approx f(a) + (x-a)f'(a).$$

The right-hand side,

$$L(x) = f(a) + (x-a)f'(a),$$

which is a linear function of x, is called the **linear approximation of $f(x)$ near $x = a$**.

11.5 Linear Approximation and Error Estimation

Q What is this function $L(x)$?

A Its graph is the line tangent to the graph of f at the point $(a, f(a))$. Indeed, the tangent line through this point has slope $f'(a)$, so the point-slope form of its equation is

$$y - f(a) = f'(a)(x - a)$$

or

$$y = f(a) + (x - a)f'(a).$$

Figure 1 shows the graphs of f and L.

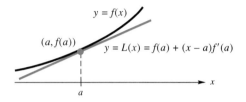

FIGURE 1

EXAMPLE 1

Find the linear approximation of $f(x) = \sqrt{x}$ near $x = 4$.

SOLUTION Since we are interested in $f(x)$ near $x = 4$, we take a to be 4.

$$\begin{aligned}
L(x) &= f(a) + (x - a)f'(a) \\
&= \sqrt{a} + (x - a)\frac{1}{2\sqrt{a}} \\
&= \sqrt{4} + \frac{x - 4}{2\sqrt{4}} \\
&= 2 + \frac{x - 4}{4} = 2 + \frac{x}{4} - \frac{4}{4} = \frac{x}{4} + 1
\end{aligned}$$

Thus, the linear approximation to $\sqrt{x}$ near $x = 4$ is $L(x) = \frac{x}{4} + 1$.

Before we go on... The graphs of $y = \sqrt{x}$ and $y = \frac{x}{4} + 1$ are shown in Figure 2.

We can use $L(x)$ to approximate the square root of any number close to 4 very easily without using a calculator. For example,

$$\sqrt{4.1} \approx L(4.1) = \frac{4.1}{4} + 1 = 2.025$$

(the actual value of $\sqrt{4.1}$ is 2.02485. . . .), and

$$\sqrt{3.9} \approx L(3.9) = \frac{3.9}{4} + 1 = 1.975$$

(the actual value of $\sqrt{3.9}$ is 1.9748. . . .).

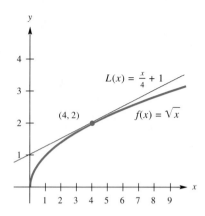

FIGURE 2

▶ **NOTE** Don't forget that in the formula $L(x) = f(a) + (x - a)f'(a)$, $L(x)$ is a function of x, and a is a constant. ◀

▼ **EXAMPLE 2**

Use linear approximation to approximate $\ln(1.134)$.

SOLUTION Here, we are not given a value for a. The key is to use a value close to 1.134 whose natural logarithm we know. Since we know that $\ln(1) = 0$, we take a to be 1.

$$L(x) = f(a) + (x - a)f'(a)$$
$$= \ln a + (x - a)\frac{1}{a}$$
$$= \ln(1) + x - 1 = x - 1$$

Thus, the linear approximation of $\ln x$ near $x = 1$ is $L(x) = x - 1$. In particular,

$$\ln(1.134) \approx L(1.134) = 1.134 - 1 = 0.134.$$

The actual value is $\ln(1.134) = 0.12575\ldots$ Thus, our approximation is off by $0.134 - 0.12575\ldots = 0.00825\ldots$, which is not bad for a quick estimation.

Before we go on... You can use $L(x) = x - 1$ to find approximations to the natural logarithm of *any* number close to 1: for instance,

$$\ln(0.843) \approx 0.843 - 1 = -0.157,$$
$$\ln(0.999) \approx 0.999 - 1 = -0.001.$$

Q $L(x) = f(a) + (x - a)f'(a)$ is an approximation of $f(x)$ near $x = a$. How accurate is this approximation?

A The accuracy depends on how close x is to a. The closer x is to a, the better the approximation. In fact, it can be shown that the error is of the order $(x - a)^2$, meaning that it is no more than a constant times $(x - a)^2$ (the constant depends on the function).

APPROXIMATING RELATIVE CHANGE

Suppose you are a screen printer, and the weekly sales of your T-shirts is given by a demand equation of the form $q = f(p)$. If you are currently charging $\$a$ per shirt, then you sell $f(a)$ T-shirts per week. If you change the price to $\$b$ per shirt, then your weekly sales will change to $f(b)$. Thus, if you change the price from a to b, the change in weekly sales is given by the difference, $f(b) - f(a)$.

Now if the new price b is close to the old price a, our linear approximation formula tells us that

$$f(b) \approx f(a) + (b - a)f'(a).$$

Thus, if you change the price from a to b, the change in weekly sales is

$$f(b) - f(a) \approx (b - a)f'(a),$$

or

$$\text{Change in sales} \approx \text{Change in price times } f'(a).$$

Now we would like to calculate the *percentage* change in f. We do so by dividing by the current sales level $f(a)$.

$$\text{Percentage change in sales}$$
$$= \frac{f(b) - f(a)}{f(a)} \approx (b - a)\frac{f'(a)}{f(a)}$$
$$= \text{Change in price} \times \frac{f'(a)}{f(a)}$$

ESTIMATING THE PERCENTAGE CHANGE IN A FUNCTION f

If x changes from a to b, then the **percentage** or **relative** change in f is approximated by

$$\text{Relative change in } f \approx (b - a)\frac{f'(a)}{f(a)}.$$

The ratio $f'(a)/f(a)$ is called the **percentage rate of change** or **relative rate of change**.

EXAMPLE 3 Sales

The demand equation for your new fraternity T-shirts is given by

$$q = \frac{2{,}000}{p},$$

where q represents the weekly sales of T-shirts at a price of $p. You are currently charging $5 per T-shirt. If you raise the price to $5.50, by what percentage will your sales drop?

SOLUTION Notice first that q is given as a function of p.

$$q = f(p) = \frac{2{,}000}{p}$$

Let us take $a = \$5$ (the old price) and $b = \$5.50$ (the new price). Then $f(a) = f(5) = 400$ shirts per week. The derivative of f is

$$f'(p) = -\frac{2{,}000}{p^2},$$

so $f'(a) = f'(5) = -2{,}000/25 = -80$ shirts per $1 increase in price. The percentage rate of change is

$$\frac{f'(a)}{f(a)} = \frac{-80}{400} = -0.2,$$

or -20% per $1 increase in price. The relative change in f is approximated by

$$(b - a)\frac{f'(a)}{f(a)} = -(5.5 - 5)\frac{80}{400} = -0.1$$

In other words, you can expect your sales to drop by approximately 10% if you raise the price 50¢.

EXAMPLE 4 Measurement Error

Precision Corp. manufactures ball bearings with a radius of 1 millimeter, varying by ±0.01 millimeters. By what percentage can the volume of its ball bearings vary?

SOLUTION We can rephrase the question as follows:

(1) If the radius of a ball bearing changes from 1 millimeter to 1.01 millimeters, by what percentage does the volume change?

(2) If the radius of a ball bearing changes from 1 millimeter to 0.99 millimeter, by what percentage does the volume change?

We can answer both questions using our formula for relative change. Since we are asking for the relative change in volume, we use the formula for the volume of a sphere,

$$V(r) = \frac{4}{3}\pi r^3.$$

Note that this is a function of the radius r. To answer the first question, let us take $a = 1$ millimeter and $b = 1.01$ millimeters. Then $V(1) = 4\pi/3 \approx 4.1888$. To find $V'(a)$, we first take the derivative of V,

$$V'(r) = 4\pi r^2,$$

so $V'(1) = 4\pi \approx 12.5664$. Thus, the percentage rate of change is

$$\frac{V'(1)}{V(1)} = \frac{12.5664}{4.1888} = 3,$$

or 300% per millimeter when the radius is 1 millimeter. The percentage change in V is

$$(b-a)\frac{V'(a)}{V(a)} = (1.01 - 1)\frac{12.5664}{4.1888} = 0.03,$$

or 3%. Similarly, if we *decrease* the radius from 1 millimeter to 0.99 millimeter, the only change in the calculation is that $(b-a)$ becomes $(0.99 - 1) = -0.01$, so that the percentage change in V is -3%. Thus, the volume of the ball bearings will vary by approximately $\pm 3\%$.

We now state some of our results in delta notation. If x is close to a, then we saw that

$$f(x) - f(a) \approx (x - a)f'(a).$$

The quantity $f(x) - f(a)$ represents the change in f corresponding to a change in the independent variable from a to x. In other words,

$$\text{Change in } f \approx \text{Change in } x \times f'(a).$$

Using the delta notation, this becomes

$$\Delta f \approx \Delta x \, f'(x),$$

which is just another way of saying that $f'(x)$ is approximately $\Delta f/\Delta x$. This formula is sometimes written as

$$df = f'(x)\, dx,$$

where df and dx are thought of as very small changes in f and x, respectively, and are referred to as **differentials**. (Actually, $dx = \Delta x$, but df is the linear approximation to Δf. Thus, we can write $\Delta f \approx df = f'(x)\, dx$.)

You can think of df as the change in the linear approximation we called $L(x)$ earlier, while Δf is the actual change in f. Figure 3 shows these quantities on a graph.

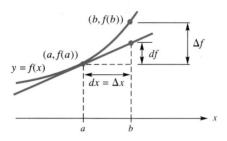

FIGURE 3

EXAMPLE 5

Let $f(x) = \sqrt[3]{x}$. Using the differential notation, approximate $\sqrt[3]{990}$ and $\sqrt[3]{1,010}$.

SOLUTION As in Example 1, we find the linear approximation to f near a point a where $f(a)$ is easy to compute. The obvious choice is $a = 1,000$, since $f(1,000) = 10$. Now the linear approximation, in differential form, is

$$df = \frac{1}{3}x^{-2/3}\,dx = \frac{1}{3\sqrt[3]{x^2}}\,dx.$$

When $x = 1,000$, we get

$$df = \frac{1}{300}\,dx.$$

Now, to approximate $\sqrt[3]{990}$, we take $dx = 990 - 1,000 = -10$, which gives

$$df = \frac{1}{300}(-10) = -0.0333.\ldots$$

This is an approximation to Δf, so

$$\sqrt[3]{990} = f(1,000) + \Delta f \approx f(1,000) + df$$
$$\approx 10 - 0.03333 = 9.96667.$$

Similarly, to approximate $\sqrt[3]{1,010}$ we take $dx = 1,010 - 1,000 = 10$, which gives

$$df = \frac{1}{300}(10) = 0.0333\ldots$$

and

$$\sqrt[3]{1,010} = f(1,000) + \Delta f \approx f(1,000) + df$$
$$\approx 10 + 0.03333 = 10.03333.$$

Before we go on... The actual cube roots are
$$\sqrt[3]{990} = 9.96665$$
and
$$\sqrt[3]{1{,}010} = 10.03322$$
to 5 decimal places (according to a calculator, which is using its own method of approximation . . .).

11.5 EXERCISES

Find the linear approximation near the indicated value for each of the functions in Exercises 1–20.

1. $f(x) = 3x + 5$, $x = 2$
2. $f(x) = -x - 3$, $x = 5$
3. $f(x) = 3x^2 - 4x + 5$, $x = -1$
4. $f(x) = -x^2 + x - 1$, $x = 0$
5. $f(x) = \dfrac{x}{2x + 1}$, $x = 0$
6. $f(x) = \dfrac{2x - 1}{x + 1}$, $x = -2$
7. $f(x) = e^x$, $x = 0$
8. $f(x) = e^{-x}$, $x = 0$
9. $f(x) = \ln(1 + x)$, $x = 0$
10. $f(x) = \ln(1 - x)$, $x = 0$
11. $f(x) = \sqrt{1 + x}$, $x = 0$
12. $f(x) = \sqrt{1 - x}$, $x = 0$
13. $f(x) = x^{1.3}$, $x = 1$
14. $f(x) = x^{2.7}$, $x = 1$
15. $f(x) = \dfrac{e^x + e^{-x}}{2}$, $x = 2$
16. $f(x) = \dfrac{e^x - e^{-x}}{2}$, $x = 2$
17. $f(x) = \dfrac{1}{1 + e^{0.2x}}$, $x = 0$
18. $f(x) = \dfrac{e^{0.5x}}{x}$, $x = 1$
19. $S(r) = 4\pi r^2$, $r = 10$
20. $l(p) = \sqrt{1 - p^2}$, $p = \frac{1}{2}$

Use linear approximation to estimate the numbers in Exercises 21–30.

21. $\sqrt{16.3}$
22. $\sqrt{36.1}$
23. $\sqrt{48.82}$
24. $\sqrt{24.73}$
25. $(8.1)^{2/3}$
26. $(3.9)^{3/2}$
27. $e^{0.3}$
28. $e^{-0.2}$
29. $\ln(0.95)$
30. $\ln(1.05)$

In each of Exercises 31–36, use linear approximation to estimate the percentage change in f(x) that results from a change from $x = a$ to $x = b$.

31. $f(x) = 2x^2 - x$; $a = 3, b = 3.5$
32. $f(x) = 5x^2 + x$; $a = 1, b = 1.2$
33. $f(x) = e^x$; $a = 0, b = -0.3$
34. $f(x) = e^{-x}$; $a = 0, b = 0.3$
35. $f(x) = \dfrac{1}{x}$; $a = 5, b = 6$
36. $f(x) = \dfrac{1}{x^2}$; $a = 5, b = 7$

APPLICATIONS

Some of the themes of the following application exercises may already be familiar to you.

37. *Cost Analysis* The daily cost of manufacturing x camcorders at Consumer Electronics, Inc. is calculated to be

$$C(x) = 1,000 + 150x - 0.01x^2.$$

(a) Find the linear approximation to $C(x)$ near $x = 100$, and use it to estimate $C(105)$.
(b) Estimate the percentage increase in daily costs if production is increased from 100 to 105 camcorders per day.

38. *Cost Analysis* The daily cost of manufacturing x compact discs at the Techno Plus Recording Studio is given by

$$C(x) = 400 + 7x - 0.0001x^3.$$

(a) Find the linear approximation to $C(x)$ near $x = 200$, and use it to estimate $C(210)$.
(b) Estimate the percentage increase in daily costs if production is increased from 200 to 210 CD's per day.

39. *Cost Analysis (based on a GRE economics exam question*)* A firm's average cost function is presented as

$$\text{Average cost} = 350 + \frac{9000}{Q}.$$

Estimate the percentage change in average cost if production increases from 1,000 units to 1,050 units.

40. *Cost Analysis (based on a GRE economics exam question)* A firm's average cost function is presented as

$$\text{Average cost} = 600 + \frac{900}{Q}.$$

Estimate the percentage change in average cost if production increases from 5,000 units to 5,100 units.

41. *Toxic Waste Treatment* The cost of treating waste by removing PCPs is given by

$$C(q) = 5,000 + 120q^2,$$

where q is the reduction in toxicity (in pounds of PCPs removed per day) and $C(q)$ is the daily cost (in dollars) of this reduction. Government subsidies for toxic waste cleanup amount to

$$S(q) = 600q,$$

where q is as above and $S(q)$ is the dollar subsidy.
(a) Calculate the net cost function, $N(q)$, given the cost function and subsidy above.
(b) If a company is currently removing 15 pounds of PCPs per day and decides to increase this amount by 10%, what will be the effect on the daily net cost?

42. *Transportation Costs* Before the Alaskan pipeline was built, there was speculation as to whether it might be more economical to transport the oil by large tankers. The following cost equation was estimated by the National Academy of Sciences:

$$C = 0.03 + \frac{10}{T} - \frac{200}{T^2},$$

where C is the cost in dollars of transporting one barrel of oil 1,000 nautical miles and T is the size of an oil tanker in deadweight tons.[†]
(a) How much would it cost to transport 100 barrels of oil in a tanker weighing 1,000 tons?
(b) If the weight of the tanker is increased by 5%, how will this affect the cost?

43. *Sales Analysis* British statistician Richard Stone published a demand equation for beer in Great Britain that had the form

$$q = Kp^{-1.040},$$

where q is the amount of beer demanded and p is the price of beer. K is a quantity depending on the average consumer's income and the price of other commodi-

---

[*] Source: *GRE Economics* by G. G. Gallagher, G. E. Pollock, W. J. Simeone, and G. Yohe, (Piscataway, N.J.: Research and Education Association, 1989), p. 254.

[†] Source: "Use of Satellite Data on the Alaskan Oil Marine Link," *Practical Applications of Space Systems: Cost and Benefits* (Washington, D.C.: National Academy of Sciences, 1975), p. B–23.

27. $f(x) = \dfrac{x^2 - 2x^{-3} + 1}{x + 1}$

28. $f(x) = \dfrac{2x + x^{-2}}{x - 1}$

29. $r(x) = \dfrac{e^x - e^{-x}}{e^x}$

30. $s(x) = \dfrac{e^{2x} - e^{-2x}}{e^{2x}}$

31. $f(x) = (2x^2 - 2x + 1)^{-4}$

32. $f(x) = (2x^3 + x + 1)^{-3}$

33. $s(t) = (t^2 + e^{3t} + 2\sqrt{t})^4$

34. $s(t) = (2t^2 + e^{-t} + \sqrt{t})^{-1}$

35. $f(x) = \sqrt{e^x - x^2}$

36. $f(x) = \sqrt{e^x + x^2}$

37. $h(x) = \sqrt{1 - \ln x}$

38. $h(x) = \sqrt{e^x + \ln x}$

In each of Exercises 39–50, evaluate the given expression.

39. $\dfrac{d}{dx}\left(\dfrac{x^{1.3}}{1 + x}\right)$

40. $\dfrac{d}{dx}\left(\dfrac{x}{x^{0.1} + 1}\right)$

41. $\dfrac{d}{dx}(e^{0.1x})$

42. $\dfrac{d}{dx}\left(1 + \dfrac{1}{e^{2.1x}}\right)$

43. $\dfrac{d}{dx}\left(\dfrac{1}{1 + 2e^x}\right)$

44. $\dfrac{d}{dx}\left(\dfrac{e^{-x}}{1 + x}\right)$

45. $\dfrac{d}{dt}(b\ln(at^2))$, with a and b constant

46. $\dfrac{d}{dt}(ae^{bt})$, with a and b constant

47. $\dfrac{d}{dx}((\ln x)^2)$

48. $\dfrac{d}{dx}((e^x)^2)$

49. $\dfrac{d}{dx}(e^{x^2 - 3x + 1})$

50. $\dfrac{d}{dx}\left(\dfrac{2e^{x^2+1}}{x}\right)$

In each of Exercises 51–60, find the equation of the tangent line to the graph of the given function at the indicated point.

51. $r(s) = (2s - s^3)(s^2 + 1)$, at $(1, 2)$

52. $r(x) = (x^3 + x)(x - x^2)$, at $(1, 0)$

53. $h(x) = (x^2 - x + 1)e^{-x}$, at $(0, 1)$

54. $h(x) = (2\sqrt{x} - x^2)e^{2x}$, at $(1, e^2)$

55. $s(t) = \dfrac{2t + 5}{t - 1}$, at $(2, 9)$

56. $r(t) = \dfrac{5t - 2}{2t + 4}$, at $(1, \tfrac{1}{2})$

57. $s(t) = \dfrac{e^t \ln|t|}{t^2}$, at $(1, 0)$

58. $r(t) = \dfrac{e^{-t} \ln|t|}{t}$, at $(1, 0)$

59. $f(x) = (x^2 - 2)^{-2}$, at $(2, \tfrac{1}{4})$

60. $f(x) = (x^3 + x)^{-1}$, at $(2, \tfrac{1}{10})$

In each of Exercises 61–68, find $\dfrac{dy}{dx}$.

61. $x^2 - y^2 = x$

62. $2xy + y^2 = y$

63. $e^{xy} + xy = 1$

64. $xe^y - x^2y = 0$

65. $\dfrac{x}{x + y} - x = 1$

66. $\dfrac{x}{xy - 1} = y$

67. $y = x^{x+1}$

68. $y = (x + 1)^x$

In each of Exercises 69–74, find all values of x (if any) where the tangent line to the graph of the given equation is horizontal.

69. $y = x - e^{2x-1}$ **70.** $y = e^{x^2}$ **71.** $y = \dfrac{x}{x+1}$

72. $y = \sqrt{x}(x-1)$ **73.** $f(x) = \sqrt{e^x - x}$ **74.** $f(x) = \sqrt{e^x + x}$

APPLICATIONS

75. *Revenue* You are currently able to sell 100 quarts of ice cream per day at $5 per quart. Increasing the price will cause demand to fall by 15 quarts per dollar increase in price. Should you raise your price?

76. *Revenue* You are currently able to sell 50 quarts of ice cream per day at $5 per quart. Increasing the price will cause demand to fall by 15 quarts per dollar increase in price. Should you raise your price?

77. *Fuel Economy* You are accelerating to enter a highway while, unbeknownst to you, your engine is starting to fail and your gas mileage is decreasing. If you are accelerating at a rate of 5 mph/s at the moment you are driving 40 mph, and your gas mileage is 20 mpg and decreasing at a rate of 2 mpg/s, how fast is your gas use (in gallons per hour) changing?

78. *Fuel Economy* You are decelerating as you leave a highway while, unbeknownst to you, your engine is starting to fail and your gas mileage is decreasing. If you are decelerating at a rate of 5 mph/s at the moment you are driving 40 mph, and your gas mileage is 20 mpg and decreasing at a rate of 2 mpg/s, how fast is your gas use (in gallons per hour) changing?

79. *Marginal Product* In your bicycle factory, your marginal profit at your current production level is $50 per bicycle. Moreover, your production would increase by 10 bicycles per employee you hire. At what rate would your profit change as you hire more employees?

80. *Marginal Product* In your bicycle factory, your marginal profit at your current production level is $60 per bicycle. Moreover, your production would fall by 5 bicycles per employee you hire (due to overcrowding). At what rate would your profit change as you hire more employees?

81. *Rates of Increase* A cube is growing in such a way that each of its edges is growing at a rate of 2 cm/s. How fast is its volume growing at the instant when the cube has a volume of 1,000 cubic centimeters? (The volume of a cube with edge a is given by $V = a^3$.)

82. *Rates of Increase* The volume of a cone with a circular base of radius r and height h is given by $V = \tfrac{1}{3}\pi r^2 h$. (See the figure.)

Find the rate of increase of V with respect to r. If the quantity r is growing at a uniform rate of 3 cm/s, how fast is the volume growing at the instant when $r = 1$, assuming that $h = 2$ does not change?

83. *Investments* If $5,000 is invested in a mutual fund whose value is growing at a rate of 5% per year (compounded once per year), how fast is the investment growing at the end of two years?

84. *Investment* If $5,000 is invested in a mutual fund whose value is declining at a rate of 5% per year (compounded once per year), how fast is the investment declining at the end of two years?

85. *Population Growth* The population P of frogs in Nassau County is given by the formula

$$P = 10{,}000 e^{0.5t},$$

where t is the time in years since 1995.

(a) How fast will the frog population be growing in the year 2,000?

(b) In what year will the frog population be growing at a rate of one million per year?

(c) What is the significance of this for the frog leg industry?

86. *Business Growth* The accompanying graph shows the number of transactions handled by *Western*

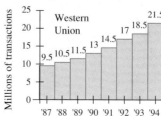

Union for the years 1987-1994.*
These data can be approximated by

$$Q(t) = 9.284e^{0.117t},$$

where t is time in years since 1987, and $Q(t)$ is the number of Western Union transactions (in millions) each year.
(a) According to the model, how fast was Western Union's business (as measured by annual transactions) growing in 1990?
(b) What is the first year that the actual increase (from that year to the next) was exceeded by the rate of increase that year as predicted by the model?

87. Sulfur Emissions Worldwide industrial sulfur emissions since 1860 have followed the pattern shown in the accompanying graph.

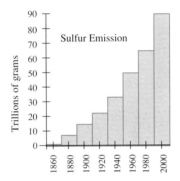

These data are approximately modeled by the function

$$Q(t) = 3.449e^{0.02576t},$$

where t is time in years since 1860, and $Q(t)$ is the amount of sulfur released into the atmosphere by industry each year, in trillions of grams.†
(a) According to the model, how fast was industrial sulfur emission growing in 1960?
(b) When (to the nearest year) did the level of industrial emission of sulfur surpass the earth's natural sulfur emissions level of 28 trillion grams?

88. Epidemics According to the "logistic" model, the number of people P infected in an epidemic often follows a curve of the following type:

$$P = \frac{NP_0}{P_0 + (N - P_0)e^{-bt}},$$

where N is the total susceptible population, P_0 is the number of infected individuals at time $t = 0$, and b is a constant that governs the rate of spread. Taking t to be the time in years, and assuming that $b = 0.1$, $N = 1,000,000$, and $P_0 = 1,000$, answer the following questions.
(a) How fast is the epidemic spreading at the start ($t = 0$), after 10 years, after 100 years?
(b) When does the spread of the disease slow to 1,000 new cases per year?
(c) What percentage of the population is infected by this time?

89. Investments You are considering depositing some money in an account earning 7% compounded continuously. You would like to end up with $20,000 in the account after 10 years.
(a) How much money would you have to invest?
(b) If you were able to increase your investment, how much time could you save in getting to $20,000? Express your answer as a rate, in years per dollar. (Suggestion: use implicit differentiation.)

90. Investment You are considering depositing some money in an account earning 5% compounded continuously. You would like to end up with $30,000 in the account after 20 years.
(a) How much money would you have to invest?
(b) If you shorten the amount of time you want to wait, how much more money would you have to invest? Express your answer as a rate, in dollars per year. (Suggestion: use implicit differentiation.)

91. Demand The price p of a new video game is related to the demand q by the equation

$$100pq + q^2 = 5,000,000.$$

Suppose that the price is set at $40, which will make the demand be 1,000 copies.
(a) Using implicit differentiation and linear approximation, estimate the demand if the price is raised to $42.
(b) Should the price be raised or lowered to increase revenue?

92. Demand According to the demand equation in the previous exercise, if the price is set at $95, then the demand will be 500.
(a) Estimate the demand if the price is raised to $98.
(b) Should the price be raised or lowered to increase revenue?

---
* Source: Company Reports/*The New York Times*, September 24, 1994.
† The exponential model is the authors'. The graphical data was obtained from "Sulfate Aerosol and Climatic Change," Robert J. Charlson and Tom M. L. Wigley, *Scientific American*, February 1994, pp. 48–57.

CHAPTER 12

Source: Courtesy HarperCollins College Publishers.

Applications of the Derivative

SECTIONS

1. Maxima and Minima
2. Applications of Maxima and Minima
3. The Second Derivative: Acceleration and Concavity
4. Curve Sketching
5. Related Rates
6. Elasticity of Demand

You're the Expert
Production Lot Size Management

APPLICATION ▶ Your book publishing company is planning the production of its latest best-seller, which is predicted to sell 100,000 copies per month over the coming year. The book will be printed in several batches of the same number evenly spaced throughout the year. Each printing run has a setup cost of $5,000, a single book costs $1 to produce, and monthly storage costs for books awaiting shipment average 1¢ per book. In order to minimize total cost to your company, how many printing runs should you plan in order to meet the anticipated demand?

Chapter 12 Applications of the Derivative

INTRODUCTION ▶ In this chapter we begin to see the power of calculus as an optimization tool. For instance, if we are given the demand and cost functions for some item we are selling, we wish to price the item so as to get the largest possible profit. We have already seen how to do this in a restricted setting (linear demand and cost functions). However, not all the functions we encounter are linear. The true force of calculus comes into play in finding a maximum or minimum value of a *nonlinear* function.

We begin this chapter with this goal in mind: find the values of a variable that lead to a maximum (or minimum) value of a given function. Once we are familiar with the mechanics, we go on in the second section to apply these techniques to realistic situations.

There is a second, no less important, goal we address in this chapter: using calculus to assist you in understanding the graph of a function. By the time you have completed the material in the first section, you will be well on your way to being able to sketch the important features of a graph. For our section on curve-sketching, we have adopted a six-step "sketch-as-you-go" approach, the aim being to draw the graph of a function as quickly and efficiently as possible. If you use a graphing calculator or computer to draw graphs, the techniques we discuss are still necessary to locate and explain the important features of a graph.

We have also included sections on related rates and elasticity of demand. The first of these examines further the concept of the derivative as a rate of change, while the second discusses optimization of revenue based on the demand equation.

Throughout this chapter, you will notice our emphasis on a systematic approach, not only to the applications, but also to the computational and mechanical problems. This will help demystify a lot of the material and also sharpen your ability to extract the basic patterns underlying the examples.

12.1 MAXIMA AND MINIMA

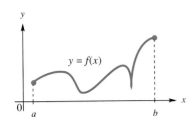

FIGURE 1

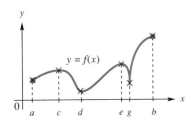

FIGURE 2

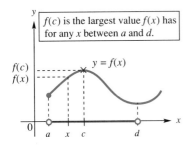

FIGURE 3

Take a look at the graph shown in Figure 1.

This is the graph of a function whose domain is the closed interval $[a, b]$. From the viewpoint of a mathematician, all kinds of exciting things are going on here. There are hills and valleys, and even a small chasm (called a "cusp") toward the right. For many purposes, the important features of this curve are the highs and lows. Suppose, for example, you knew beforehand that the stock price of some company would follow this graph during the course of a week. What would be your best strategy for buying and selling that stock? While you would certainly make a handsome profit if you bought at time a and sold at time b, you would do even better if you followed the old adage to "buy low and sell high," buying at all the lows and selling at all the highs.

Figure 2 shows the graph once again with these important points marked. The points marked on the graph of f give the lows and the highs, illustrating the best stock-trading strategy: buy at a, sell at c, buy at d, sell at e, buy at g, sell at b.

Mathematicians like to give these highs and lows Latin names: the highs (c, e, and b) are referred to as **local maxima**, and the lows (a, d, and g) are referred to as **local minima**. Collectively, these highs and lows are referred to as **local extrema**. (A point of language: the singular forms of minima, maxima, and extrema are minimum, maximum, and extremum.)

Why do we refer to these points as "local" extrema? Take a look at the point corresponding to $x = c$. Compared to nearby portions of the graph, it is the highest point of the graph *in the vicinity*. In other words, if you were an extremely near-sighted mountaineer and were positioned at this point, you would *think* that you were at the highest point of the graph, as you would be totally oblivious of the distant peaks at $x = e$ and $x = b$.

Let us translate this description into mathematical terms. We are talking about the heights of various points on the curve. Now the height of the curve at $x = c$ is measured by $f(c)$, so we are saying that $f(c)$ is larger than any neighboring $f(x)$. More specifically, $f(c)$ is the largest value that $f(x)$ has for all choices of x between a and d. (See Figure 3.)

Formally:

> **LOCAL EXTREMA**
>
> f has a **local** or **relative maximum** at c if there is some interval (r, s) (even a very small one) containing c for which $f(c) \geq f(x)$ for all choices of x between r and s for which $f(x)$ is defined.
>
> f has a **local** or **relative minimum** at c if there is an interval (r, s) (even a very small one) containing c for which $f(c) \leq f(x)$ for all choices of x between r and s for which $f(x)$ is defined.

Figure 4 shows the location of all the local extrema on the graph of f.

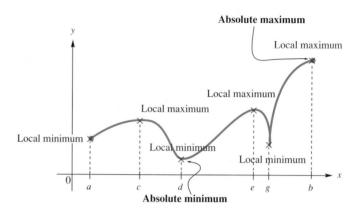

FIGURE 4

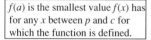

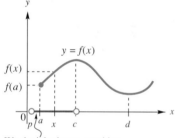

We don't bother to consider an x here, since $f(x)$ is not defined here.

FIGURE 5

You should try to find an interval containing the x-coordinate of each local extreme point as in the definition. For instance, let us show how our definition allows us to say that f has a local minimum at a. In Figure 5 is shown an interval (p, c) containing a. The only values of x in that interval for which $f(x)$ is defined are the numbers in $[a, c)$, and so these are the only ones we need to look at. From the picture, it is clear that $f(a)$ is the smallest value of f in this interval.

In Figure 4 we labeled one of the points an *absolute* minimum and one an absolute maximum. These simply correspond to the smallest and largest values of f overall. Formally:

ABSOLUTE EXTREMA

f has an **absolute maximum** at c if $f(c) \geq f(x)$ for every x in the domain of f.

f has an **absolute minimum** at c if $f(c) \leq f(x)$ for every x in the domain of f.

Q What if $f(x)$ is constant, so the graph of f is a horizontal line?

A According to the definition of a local extremum, *every* point on the line would qualify as both a local minimum and a local maximum (because we use the inequalities $\leq$ and $\geq$ in the definition, rather than strict inequalities). Similarly, by the definition of an absolute extremum, every point would also be an absolute maximum and an absolute minimum.

We have already seen why local maxima and minima are interesting in the context of investment; your best strategy is to buy at the local minima and

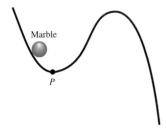

FIGURE 6

sell at the local maxima. (We suggest you spend some time thinking about why this is in fact the best strategy.) Here is another example that illustrates the importance of local extrema. Suppose you have a marble on an incline with the shape shown in Figure 6.

If you place that marble at the point P, it will remain there. Although there are lower points on the hill than P, there is a barrier between the marble and those lower points. The marble remains at the local minimum, which may not be an absolute minimum. Like our near-sighted mountaineer, the marble cannot "see" the lower points. This situation is an example of the scientific principle that systems tend toward the lowest possible energy state. We should be glad that systems can get stuck in local minima—otherwise, we would not stick to the surface of the Earth but all fall into the sun!

This principle also applies to some economic models. Economies are driven by consumers' desire to pay the lowest prices and producers' desire to get the largest profits. But it is possible for economies to fall into local extrema that are not globally best.

Q How do we find these local extreme points?

A We can find them using two approaches:

1. **graphically,** using a graphing calculator or computer graphing software to locate approximate numerical values for the local extrema, or
2. **analytically,** using calculus to obtain exact values for the local extrema. (Sketching the graph or using a graphing calculator is also very helpful in the analytical approach.)

At the end of this section, we shall also see how to combine these two approaches: we shall use analytic methods from calculus to help us improve the accuracy of the graphical method.

GRAPHICAL APPROACH (GRAPHING CALCULATOR OR COMPUTER)

To locate the local extrema graphically, you need some form of graphing technology.

▼ **EXAMPLE 1**

Use a graphing calculator or computer graphing software to locate the local extrema of
$$f(x) = 3x^5 - 25x^3 - 15x^2 + 60x.$$

SOLUTION First, we must graph this function with a wide enough range of x-values to see all of the local extrema. If you experiment a little, you will find

that the ranges
$$-5 \leq x \leq 5$$
and
$$-100 \leq y \leq 100$$
will show you the graph nicely. (See Figure 7.)

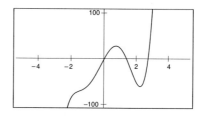

FIGURE 7

There appear to be two local extrema: a local maximum with x somewhere between 0 and 1, and a local minimum a little to the right of $x = 2$. To find the coordinates more precisely, we can use the trace or zoom features. For example, zooming in on the local maximum, we can obtain the graph shown in Figure 8, which uses an x-range of $0.7 \leq x \leq 0.8$ and a y-range of $26.6 \leq y \leq 26.8$.

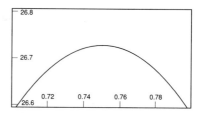

FIGURE 8

We can see from Figure 8 that the local maximum of $f(x)$ is at approximately $x = 0.75$, and $f(0.75) \approx 26.7$ (to three significant digits*). Similarly, zooming in on the local minimum shows that it is at approximately $x = 2.26$, and $f(2.26) \approx -52.7$.

Before we go on... If we want to locate the local extrema more accurately, we will need to zoom in closer. Repeated use of the zoom feature can cause difficulties, because the graph may appear flat near a local extremum, making its location difficult to find. How can we locate the extrema more accurately? Can we find the *exact* location of all local extrema? Using calculus, we shall see how to locate extrema accurately, and often exactly.

▼ *The number of significant digits we use is somewhat arbitrary. Three or four are sufficient for most purposes and are all that are justified in many cases when using real data.

▶ **CAUTION** Many calculators are equipped with built-in procedures to locate absolute maxima and minima. (For instance, the TI-82 has one in the CALC menu and also in the MATH menu.) But these procedures sometimes behave unpredictably and may even miss absolute extrema completely!* In this text, we shall try not to rely on these features but rather on mathematical intuition and graphing technology. ◀

ANALYTICAL APPROACH

In Figure 9 we see once again the graph from Figure 1, but we have now classified the extrema into three types.

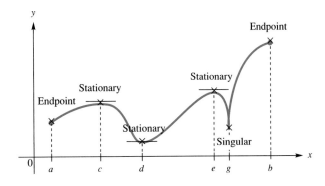

FIGURE 9

Look at the extrema we have labeled as "stationary." Notice that the tangent to the curve at each of these points is horizontal and thus has slope zero. Because the slope of the tangent at a specific value of x is equal to the derivative evaluated there, we conclude that the derivative of f is zero at each of these points. In other words,

$$f'(c) = 0, \quad f'(d) = 0, \quad \text{and} \quad f'(e) = 0.$$

We should be able to find the exact location of each of these extrema by solving the equation $f'(x) = 0$. We call points where $f'(x) = 0$ **stationary points** because the rate of change of f is zero there. We shall call an *extremum* that is a stationary point a **stationary extremum.**

▼ * Here is a glaring example: on a TI-82, the authors plotted the curve $y = x^3 - x^2 - 3x$ with $-5 \leq x \leq 2$ and $-135 \leq y \leq 20$. They then used the CALC menu to locate the minimum with a lower bound of -2.3 and an upper bound of 2, and the calculator located the *local* minimum at approximately 1.39 (whereas the absolute minimum in this range was at -2.3). Yet the same procedure gave the correct absolute minimum (of -5) if the lower bound of -5 was used instead.

> **LOCATING STATIONARY POINTS**
>
> To locate all stationary points (among which are all possible stationary extrema), we solve the equation $f'(x) = 0$ for x and we make sure that x is in the domain of f.

There is a local minimum at $x = g$, but something slightly different happens there: there is no horizontal tangent at that point. In fact, there is no tangent line at all, since the derivative is not defined at $x = g$. (Recall a similar situation with the graph of $f(x) = |x|$ at $x = 0$.) Thus, to locate g, we must look for the values of x (in the domain of f) for which $f'(x)$ does not exist. We call such points **singular points,** and we shall call an extremum that is a singular point a **singular extremum.**

> **LOCATING SINGULAR POINTS**
>
> To locate all singular points (among which are all possible singular extrema), we look for values of x such that $f'(x)$ does not exist, and we make sure that x is in the domain of f.

We call all points of the domain where either $f'(x) = 0$ or $f'(x)$ does not exist **critical points.** The critical points give us *all the possible candidates for stationary and singular local extrema.*

The remaining two extrema are at the endpoints of the domain.* As we see in the picture, they are (almost) always either local maxima or local minima.

> **LOCATING ENDPOINTS**
>
> If the domain of f has any endpoints, these are almost always local extrema.

▶ CAUTION A critical point need not be an extreme point. For example, the graph shown in Figure 10 has two critical points, one of each type, but no local extrema at all!

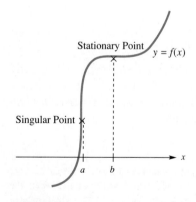

FIGURE 10

---

*Bear in mind that many calculus texts do not count endpoints of the domain as local extrema, although more advanced (analysis) texts do. In view of examples like our stock market investing strategy, we see no good reason not to count endpoints as extrema.

The point at $x = a$ in Figure 10 is a point in the domain at which the tangent is vertical, and so it is a point where the derivative $f'(x)$ is not defined, i.e., a singular point. But $x = a$ is neither a maximum nor a minimum: the graph is higher to the right and lower to the left. Similarly, the point at $x = b$ has a horizontal tangent and thus is a stationary point—yet it is neither a maximum nor a minimum. Thus, solving $f'(x) = 0$ for x and finding points in the domain where $f'(x)$ does not exist only gives us *candidates* for stationary extrema and singular extrema. These candidates need not be actual extrema. ◂

LOCATING CANDIDATES FOR LOCAL EXTREMA

To find all candidates for local extrema, we do the following.

Stationary Points: Solve the equation $f'(x) = 0$ for x, and make sure that x is in the domain of f.

Singular Points: Find all values of x such that $f'(x)$ does not exist and x is in the domain of f.

Endpoints: List all endpoints of the domain (if any).

Q Are there any other types of local extrema?

A The answer is a *qualified* "no." If the function we are looking at happens to be continuous on its domain and differentiable at every point except for a few isolated points, then these are the only kinds of local extrema we need consider. If the function has a discontinuity at some value of x, we need to look at the graph near this value to find out if there are any other local extrema. Because these cases rarely arise in practice, we can consider them on a case-by-case basis.

Q Now we know how to find the candidates for *local* extrema. What about the *absolute* extrema? (Recall that these are the highest and lowest points on the whole graph.)

A Finding these is a little more tricky. We shall see in some of the examples that there need not be any at all. There is, however, a useful theorem which tells us that if the function is continuous and the domain happens to be a closed interval (such as [1, 5]), then the function must have an absolute maximum and an absolute minimum. The absolute maximum is just the highest local maximum, and the absolute minimum is just the lowest local minimum. If the domain is not a closed interval, anything can happen. We see some of the possibilities in the following table (we shall see in the following examples exactly how we determine the extrema).

Function	Graph	Extrema
$f(x) = x^2$, with domain all real numbers		Absolute minimum at $x = 0$; no local or absolute maximum
$f(x) = \frac{1}{x}$, with domain $(0, +\infty)$		No extrema
$f(x) = x^3 - x^2 - 5x$, with domain $(-3, 4)$		A local minimum at $x = \frac{5}{3}$ and a local maximum at $x = -1$, but no absolute extrema

We shall see that no matter what the domain is, a rough sketch of the curve using the methods we shall discuss will often allow us to locate the absolute extrema, if any.

We now turn to several examples of finding maxima and minima analytically. In all of these examples we will follow this procedure: First we find the derivative, then we find the stationary points and singular points. Next, we make a table listing the critical points and endpoints, together with the value of the function at these points, and plot them. From this table and rough sketch we will usually have enough data to be able to say where the extreme points are.

12.1 Maxima and Minima

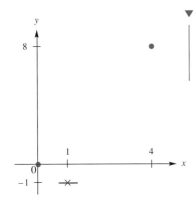

FIGURE 11

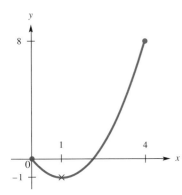

FIGURE 12

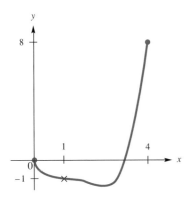

FIGURE 13

EXAMPLE 2

Find the relative and absolute maxima and minima of $f(x) = x^2 - 2x$ on the interval $[0, 4]$.

SOLUTION We begin by locating the stationary points.

Stationary Points To locate these points, we solve the equation $f'(x) = 0$. Because

$$f'(x) = 2x - 2,$$

we solve

$$2x - 2 = 0,$$

getting $x = 1$. The domain of the function is $[0, 4]$, so $x = 1$ is a point of the domain. Thus, $x = 1$ is the only candidate for a stationary local extremum.

Singular Points We look for points where the derivative is not defined. Because the derivative is $2x - 2$, it is defined for every x. Thus, there are no singular points and hence no candidates for singular local extrema.

Endpoints Because the domain is $[0, 4]$, the endpoints are $x = 0$ and $x = 4$.

We record these points in a table, together with the corresponding values of f.

x	0	1	4
$f(x)$	0	-1	8

Plotting these points gives us Figure 11.

The points at $x = 0$ and $x = 4$ are the endpoints, which we show as heavy dots, while the point at $x = 1$ is a stationary point, so that it has a horizontal tangent, and we remind ourselves of this by drawing a horizontal line segment through the point. Connecting these points gives us a graph that must look something like the curve shown in Figure 12.

Notice that the tangent is horizontal at the point $(1, -1)$ but not at the endpoints (because they are not stationary points).

Q All we have are three points! How do we know that the graph doesn't look something like the one in Figure 13, for instance?

A If it did, then there would be an extra stationary point at about $x = 3$. Because stationary points are found by solving $f'(x) = 0$, and we saw that there was only *one* solution, namely, $x = 1$, there can be no other stationary point. In other words, there can't be any maxima or minima not in our table!

Notice that from the correct graph (Figure 12), we see immediately that there is an absolute maximum of 8 at $x = 4$ and an absolute minimum of -1 at $x = 1$. Thus, the extrema are as follows:

local maximum at $(0, 0)$,

absolute minimum at $(1, -1)$,

absolute maximum at $(4, 8)$.

We can also state the result as follows: f has a local maximum of 0 at $x = 0$, an absolute minimum of -1 at $x = 1$, and an absolute maximum of 8 at $x = 4$.

Before we go on... Look once again at the derivative, $f'(x) = 2x - 2$. We know that $f'(1) = 0$ at the minimum. What about $f'(x)$ for values of x on either side of 1? We obtain the following table by choosing a value of x on either side of the stationary point at $x = 1$.

x	0	1	2
$f'(x) = 2x - 2$	-2 (*negative*)	0	2 (*positive*)

Because $f'(0) = -2 < 0$, the graph has negative slope, or f is **decreasing,** for values of x to the left of 1. Because $f'(2) = 2 > 0$, the graph has positive slope, or f is **increasing,** for values of x to the right of 1. We have drawn arrows below the table to show where f is increasing and where it is decreasing, confirming the graph in Figure 12. This table also confirms the fact that there is a local minimum at $x = 1$, since it shows that f decreases approaching $x = 1$ from the left and then increases to the right. Formally, we say that f is decreasing on the interval $[0, 1]$ and increasing on the interval $[1, 4]$.

▼ **EXAMPLE 3**

Locate and classify the extrema of $g(t) = t^3$ on $[-2, 2]$.

SOLUTION By "classifying" the extrema, we mean listing whether each extremum is a local or absolute maximum or minimum as we did in the last example. As before, we consider each type separately.

Stationary Points Solve the equation $g'(t) = 0$. Here,

$$g'(t) = 3t^2,$$

so we solve the equation

$$3t^2 = 0,$$

whose only solution is $t = 0$, which is the only stationary point.

Singular Points Because $g'(t) = 3t^2$ is defined for every t, there are no singular points.

Endpoints Because the domain is given as $[-2, 2]$, the endpoints are $t = -2$ and $t = 2$.

We record these points in a table.

t	−2	0	2
g(t)	−8	0	8

Plotting these three points gives us Figure 14(a), suggesting the curve in Figure 14(b).

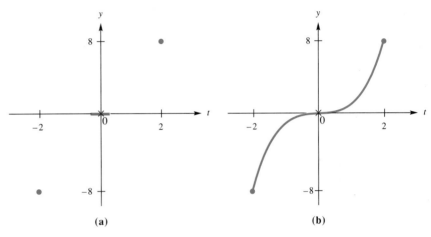

FIGURE 14

Notice something interesting at $t = 0$: we have a stationary point, so we know that the tangent line is horizontal. However, there is a lower point to the left and a higher point to the right, so that this critical point is neither a local maximum nor a local minimum.

Thus, we find the following extrema:

$$\text{absolute minimum of } -8 \text{ at } t = -2, \text{ and}$$
$$\text{absolute maximum of } 8 \text{ at } t = 2.$$

Before we go on... Notice that the shape of the curve is dictated by the requirement that the tangent be horizontal at $t = 0$. It would not be accurate to simply draw a straight line joining the three plotted points.

As in the last example, let us see where g is increasing and where it is decreasing.

t	−1	0	1
$g'(t) = 3t^2$	3 (*positive*)	0	3 (*positive*)
	↗		↗

We find that g is never decreasing, confirming that there can be no local extremum at $t = 0$. Once again, notice how the curve in Figure 14 follows the upward direction of the arrows on either side of the stationary point at $x = 0$.

EXAMPLE 4

Locate and classify the maxima and minima of $f(t) = t^4 - 2t^2$, with domain $[0, +\infty)$.

SOLUTION

Stationary Points We know $f'(t) = 4t^3 - 4t$, so we solve

$$4t^3 - 4t = 0,$$

giving

$$t^3 - t = 0,$$

or

$$t(t-1)(t+1) = 0.$$

Thus, $t = 0, 1,$ or -1. We now have three stationary points. However, the point -1 is not in the domain $[0, \infty)$, so we discard it and keep only the two points 0 and 1.

Once again, there are no singular points.

Endpoints The only endpoint is 0, which is also one of the stationary points.

Our table looks like this:

t	0	1
$f(t)$	0	-1

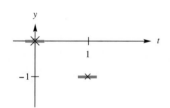

FIGURE 15

Plotting these points, we get Figure 15.

From the graph we can see that f decreases as t goes from 0 to 1. But what happens after that? We saw in the previous example that we cannot *assume* that it goes up again. Therefore, we try a "test point" farther to the right, at 2, say. We add this point to our table.

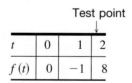

t	0	1	2
$f(t)$	0	-1	8

After plotting these points in Figure 16(a), we can sketch the curve, as shown in Figure 16(b).

Now we can see that f does increase as t goes from 1 to ∞. Thus, f has a local maximum of 0 at 0 and an (absolute) minimum of -1 at 1. There is no absolute maximum, because f continues to increase forever to the right. Notice again that because $(0, 0)$ is a stationary point (in addition to being an endpoint), the tangent to the curve at the origin is horizontal, and so the curve departs from the origin with zero slope as it starts to dip.

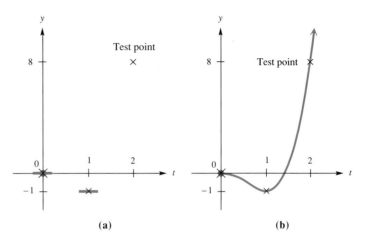

FIGURE 16

Before we go on... Notice two things about the graph in Figure 16.

1. The point (2, 8), is not an endpoint of the graph. The graph continues through that point as t goes to $+\infty$. The point (2, 8) was just a test point and is not a local extremum.
2. There is *no* absolute maximum. The graph climbs without bound as $t \to +\infty$. In mathematical terms, $\lim_{t \to +\infty} f(t) = +\infty$.

Let us once again examine where f is increasing and where it is decreasing, using the derivative.

t	0	$\frac{1}{2}$	1	2
$f'(t) = 4t^3 - 4t$	0	$-\frac{3}{2}$ (*negative*)	0	24 (*positive*)

The function is decreasing on $[0, 1]$ and increasing on $[1, +\infty)$.

Figure 17 shows graphing calculator plots of f and its derivative f' for $0 \le t \le 3$ and $-2 \le y \le 8$.

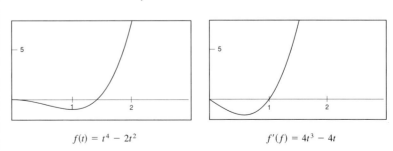

FIGURE 17

Notice several things from these plots:

1. The stationary points at $t = 0$ and $t = 1$ in the graph of f correspond to the points where the graph of f' crosses the t-axis [since this is where $f'(t) = 0$].
2. From the graph of f', we see that $f'(t) \leq 0$ if $0 \leq t \leq 1$, which tells us that the original function f is decreasing for $0 \leq t \leq 1$.
3. Similarly, we see that $f'(t) \geq 0$ if $t \geq 1$, which tells us that the original function f is increasing for $t \geq 1$.

You can have your graphing calculator automatically calculate the derivative of any function numerically and plot it along with the original function. On a TI-82, enter the original function as Y_1 and its derivative as Y_2 using the following format.

$$Y_1 = X^4 - 2X^2$$
$$Y_2 = nDeriv(Y_1, X, X)$$

▼ **EXAMPLE 5**

Locate and classify the extrema of $f(x) = \sqrt[3]{x^2}$ on $[-1, 1]$.

SOLUTION

Stationary Points $f(x) = x^{2/3}$, so $f'(x) = \frac{2}{3}x^{-1/3}$. How do we solve the equation

$$\frac{2}{3}x^{-1/3} = 0?$$

First, *get rid of negative exponents*. Move the $x^{-1/3}$ to the denominator, where its exponent becomes positive, and the equation becomes

$$\frac{2}{3x^{1/3}} = 0.$$

Multiplying by $3x^{1/3}$ yields

$$2 = 0.^*$$

But this is absurd! There is no solution to the equation, which means that there are no stationary points.

▼ * It is useful to remember the following rule:

If $a/b = 0$, then a must be zero.

(We see this by multiplying both sides by b.) The equation given above had $\frac{2}{3x^{1/3}} = 0$, which would imply that 2 would have to be zero, which it isn't. Thus, the equation represents a false statement. In other words, there is no x such that $\frac{2}{3x^{1/3}} = 0$.

Singular Points Look once again at the derivative, written in the form with no negative exponents.

$$f'(x) = \frac{2}{3x^{1/3}}$$

Because the derivative has an $x^{1/3}$ in the denominator, it is not defined when $x = 0$. Since $x = 0$ is in the domain of f, $x = 0$ is a singular point.

Endpoints The endpoints are 1 and -1.

Thus, our table looks like this:

x	-1	0	1
$f(x)$	1	0	1

As usual, we shall plot these points and sketch the curve as best we can. Notice the following, however. The point at $(0, 0)$, being a singular point, is a point at which the derivative is not defined. The derivative is not defined there because we have an x in the denominator of $f'(x)$. However, we can take the *limit* of $f'(x)$ as $x \to 0$ instead of trying to evaluate it at 0.

$$\lim_{x \to 0^+} f'(x) = \lim_{x \to 0^+} \frac{2}{3x^{1/3}} = +\infty$$

$$\lim_{x \to 0^-} f'(x) = \lim_{x \to 0^-} \frac{2}{3x^{1/3}} = -\infty$$

What these limits tell us is that close to $x = 0$ the curve is very steep (because the slope of the tangent is approaching $\pm\infty$). At that point, the tangent must be *vertical*. This will help us draw the curve in Figure 18(b). [Notice the vertical tangent at $(0, 0)$.]

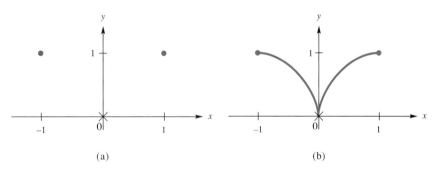

FIGURE 18

Summarizing, we have an

absolute maximum of 1 at $x = -1$,
absolute minimum of 0 at $x = 0$,
absolute maximum of 1 at $x = 1$.

 Figure 19 shows graphing calculator plots of $f(x)$ and f' with $-1 \leq x \leq 1$ and $-2 \leq y \leq 2$.

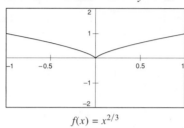
$f(x) = x^{2/3}$

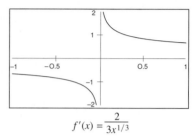
$f'(x) = \dfrac{2}{3x^{1/3}}$

FIGURE 19

Notice several things:

1. At the singular point $x = 0$, the derivative has a vertical asymptote, since it diverges to infinity.
2. For $x < 0$, the derivative is negative, so that f is decreasing, and for $x > 0$, the derivative is positive, so that f is increasing.

▼ EXAMPLE 6

Locate and classify the extrema of $g(x) = x + \dfrac{1}{x}$.

SOLUTION Because no domain was specified, we take the domain to be as large as possible. Since the function cannot be defined when $x = 0$, the largest possible domain consists of all real numbers except 0. In other words, the domain is $(-\infty, 0) \cup (0, +\infty)$.

Stationary Points We have $g'(x) = 1 - \dfrac{1}{x^2}$, so we solve

$$1 - \dfrac{1}{x^2} = 0.$$

Moving the $1/x^2$ over to the other side gives

$$1 = \dfrac{1}{x^2},$$

and multiplying by x^2 gives

$$x^2 = 1, \quad \text{so } x = \pm 1.$$

Singular Points $g'(x)$ is not defined when $x = 0$. However, $x = 0$ is not in the domain. Thus, $x = 0$ is disqualified as a singular point, so there are no singular points.

12.1 Maxima and Minima

Endpoints Since the domain is $(-\infty, 0) \cup (0, +\infty)$, there are no endpoints of the domain that also lie within the domain.

Because $g(0)$ is not defined, there will be a break in the graph where $x = 0$. We shall thus include $x = 0$ in our table as a reminder of this fact. Our table then looks like this:

x	-1	0	1
$g(x)$	-2	✘	2

These points are shown in Figure 20.

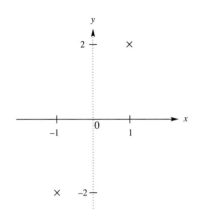

FIGURE 20

We have dotted the vertical line $x = 0$ to indicate where the graph breaks. In other words, the two points we plotted *cannot be joined*. This fact, together with the fact that we know nothing about what happens to the left of -1 and to the right of 1, means that we cannot yet tell whether these points are maxima, minima, or neither. To decide, we shall use test points. This time, we will take them on either side of *both* stationary points because we really have two separate curves. Thus we enlarge our table with some test points.

	test points ↙ ↘				test points ↙ ↘		
x	-2	-1	$-\frac{1}{2}$	0	$\frac{1}{2}$	1	2
$g(x)$	$-\frac{5}{2}$	-2	$-\frac{5}{2}$	✘	$\frac{5}{2}$	2	$\frac{5}{2}$

We plot these points in Figure 21(a), and connect them in the only way possible in Figure 21(b).

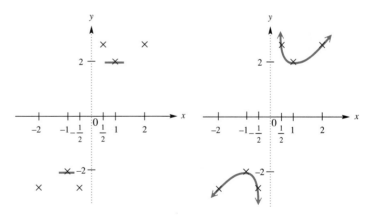

FIGURE 21

Thus we have the following classification:

local maximum of -2 at $x = -1$ and

local minimum of 2 at $x = 1$.

Figure 22 shows a graphing calculator plot of $f'(x) = 1 - \dfrac{1}{x^2}$.

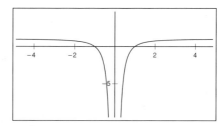

FIGURE 22

How much can you say about the graph of f by looking only at the graph of f'?

COMBINING THE GRAPHING CALCULATOR APPROACH WITH THE ANALYTICAL APPROACH

In the next example, we redo Example 1, this time using calculus together with a graphing calculator to find the local extrema.

EXAMPLE 7

Use a graphing calculator or computer graphing software to locate and classify the local extrema of

$$f(x) = 3x^5 - 25x^3 - 15x^2 + 60x.$$

SOLUTION In Example 1, we graphed this function (Figure 7) and saw by zooming in on the graph that there was one local maximum at approximately $x = 0.75$ and one local minimum at approximately $x = 2.26$. This time, we shall locate the extrema by using the fact that $f'(x) = 0$ at these points. First, we compute

$$f'(x) = 15x^4 - 75x^2 - 30x + 60.$$

Thus, the stationary points occur when

$$15x^4 - 75x^2 - 30x + 60 = 0.$$

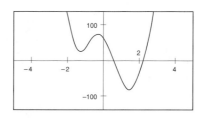

FIGURE 23

Instead of trying to solve this equation analytically, we solve it graphically. Figure 23 shows the graph of f' with the ranges $-5 \leq x \leq 5$ and $-125 \leq y \leq 125$.

Corresponding to the two local extrema we see two places where $f'(x) = 0$. It is somewhat easier to see places where a graph crosses the x-axis than it is to see local extrema. (See the "Before we go on" discussion below.) In this case, we can zoom in on, say, the first such point, getting a picture like Figure 24, with ranges $0.7504 \leq x \leq 0.7506$ and $-0.01 \leq y \leq 0.01$.

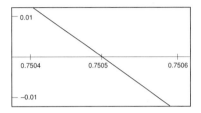

FIGURE 24

Figure 24 shows that the local maximum is at approximately $x = 0.7505$. We obtain the corresponding maximum value of f by computing $f(0.7505) = 26.7276$. Similarly, we can zoom in on the second crossing of the x-axis to find $x \approx 2.2586$. The corresponding minimum value of f is $f(2.2586) = -52.7199$.

Before we go on... If you try to avoid calculus completely by repeatedly zooming in on the local maximum for greater accuracy, you may find (as we did) that after three or four repetitions your graph resembles Figure 25.

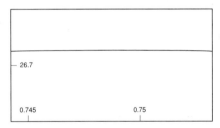

Graph of $f(x) = 3x^5 - 25x^3 - 15x^2 + 60x$ near $x = 75$.

FIGURE 25

As you see, it is difficult to pinpoint the location of the local maximum with the same accuracy we obtained using calculus.

12.1 EXERCISES

Locate and classify all extrema in each of the graphs in Exercises 1–8 and indicate the intervals on which the associated function is increasing or decreasing. (By "classifying" the extrema, we mean listing whether each extremum is a local or absolute maximum or minimum.) Also, locate any stationary points or singular points that are not local extrema.

1.
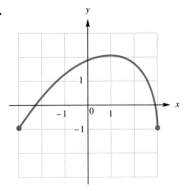

2.
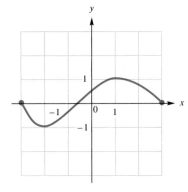

3.
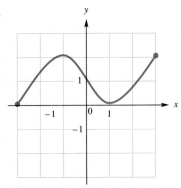

4.
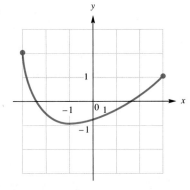

5.
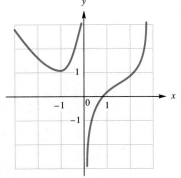

6.
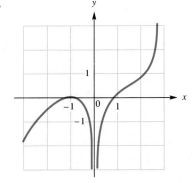

7.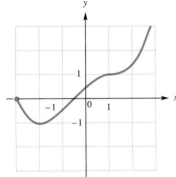

8.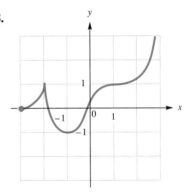

Use a graphing calculator or computer graphing software to find the approximate coordinates (correct to one decimal place) of all extrema for each of the functions in Exercises 9–16.

9. $f(x) = (x-1)(x-2)$ with domain $[0, +\infty)$
10. $f(x) = x(x+3)$ with domain $[-4, 0]$
11. $f(x) = x^x$ with domain $[0.1, +\infty)$
12. $f(x) = \sqrt{x}\ln x$ with domain $[1.1, 5]$
13. $f(x) = x(x-1)^{2/3}$ with domain all real numbers*
14. $f(x) = x + (x-1)^{2/3}$ with domain all real numbers*
15. $f(x) = \dfrac{e^{-x}}{1+e^{-x}}$ with domain all real numbers
16. $f(x) = \dfrac{\sqrt{x}}{1+\sqrt{x}}$ with domain $[0, +\infty)$

Use calculus to find the exact location of all the local and absolute extrema of each of the functions in Exercises 17–40. In each case, give a rough sketch of the curve or use a graphing calculator to help you along.

17. $f(x) = x^2 - 4x + 1$ with domain $[0, 3]$
18. $f(x) = 2x^2 - 2x + 3$ with domain $[0, 3]$
19. $g(x) = x^3 - 12x$ with domain $[-4, 4]$
20. $g(x) = 2x^3 - 6x + 3$ with domain $[-2, 2]$
21. $f(t) = t^3 + t$ with domain $[-2, 2]$
22. $f(t) = -2t^3 - 3t$ with domain $[-1, 1]$
23. $h(t) = 2t^3 + 3t^2$ with domain $[-2, +\infty)$
24. $h(t) = t^3 - 3t^2$ with domain $[-1, +\infty)$
25. $f(x) = x^4 - 4x^3$ with domain $[-1, +\infty)$
26. $f(x) = 3x^4 - 2x^3$ with domain $[-1, +\infty)$
27. $g(t) = \frac{1}{4}t^4 - \frac{2}{3}t^3 + \frac{1}{2}t^2$ with domain $(-\infty, +\infty)$
28. $g(t) = 3t^4 - 16t^3 + 24t^2 + 1$ with domain $(-\infty, +\infty)$
29. $f(t) = (t^2+1)/(t^2-1); \ -2 \le t \le 2, t \ne \pm 1$
30. $f(t) = (t^2-1)/(t^2+1)$ with domain $[-2, 2]$
31. $f(x) = \sqrt{x}(x-1), \quad x \ge 0$
32. $f(x) = \sqrt{x}(x+1), \quad x \ge 0$
33. $g(x) = x^2 - 4\sqrt{x}$
34. $g(x) = \dfrac{1}{x} - \dfrac{1}{x^2}$

▼ *Use the format $(X-1)\wedge(1/3)$ for $x - 1^{2/3}$.

35. $g(x) = x^3/(x^2 + 3)$

36. $g(x) = x^3/(x^2 - 3)$

37. $f(x) = x - \ln x$ with domain $(0, +\infty)$

38. $f(x) = x - \ln(x^2)$ with domain $(0, +\infty)$

39. $g(t) = e^t - t$ with domain $[-1, 1]$

40. $g(t) = e^{-t^2}$ with domain $(-\infty, +\infty)$

In each of Exercises 41–44, a function is specified together with a computer-generated sketch of its graph. Use calculus to locate and classify all extrema.

41. $f(x) = \dfrac{2x^2 - 24}{x + 4}$

42. $f(x) = \dfrac{x - 4}{x^2 + 20}$

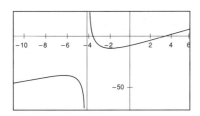
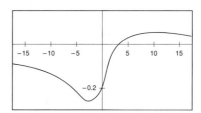

43. $f(x) = xe^{1-x^2}$

44. $f(x) = x \ln x$ with domain $(0, +\infty)$

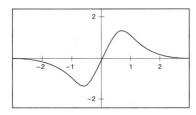
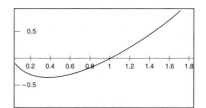

In each of Exercises 45–52, a graphing calculator plot of the derivative of a function is shown. In each case, determine the x-coordinates of all stationary and singular points of the original function, and classify each one as a local maximum, minimum, or neither. (Assume that the function is defined for every x in the viewing window.)

45.

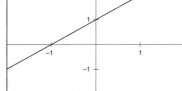

46.

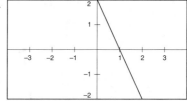

47.

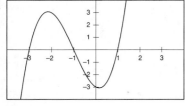

48.

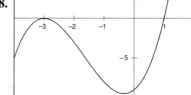

49.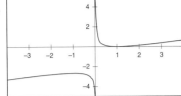

50.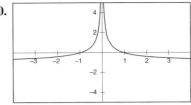

51.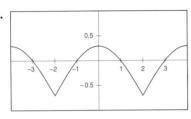

52.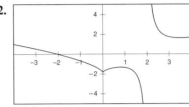

In each of Exercises 53–56, use a graphing calculator or computer to graph both the given function and its derivative, and hence locate all local and absolute extrema, with x-coordinates correct to two decimal places.

53. $y = x^2 + \dfrac{1}{x-2}$ with domain $(-3, 2) \cup (2, 6)$

54. $y = x^2 - 10(x-1)^{2/3}$ with domain $(-4, 4)$

55. $f(x) = (x-5)^2(x+4)(x-2)$ with domain $[-5, 6]$

56. $f(x) = (x+3)^2(x-2)^2$ with domain $[-5, 5]$

COMMUNICATION AND REASONING EXERCISES

57. Draw the graph of a function f with domain all real numbers, such that f is not linear and has no local extrema.

58. Draw the graph of a function g with domain all real numbers, such that g has a local maximum and minimum but no absolute extrema.

59. Draw the graph of a function that has stationary and singular points but no local extrema.

60. Draw the graph of a function that has local, not absolute, maxima and minima, but has no stationary or singular points.

61. If a stationary point is not a local maximum, then must it be a local minimum? Explain your answer.

62. If one endpoint is a local maximum, must the other be a local minimum? Explain your answer.

12.2 APPLICATIONS OF MAXIMA AND MINIMA

There are many times that we would like to find the largest or smallest possible value of some quantity—for instance, the largest possible profit or the lowest cost. We call this the *optimal* (best) value. We can often use calculus to find the optimal value.

In all applications, the first step is to translate a written description into a mathematical problem. The mathematical problem will have the following form. There will be some *unknowns* that we are asked to find, there will be an expression involving those unknowns that must be made as large or as small as possible—the **objective function**—and there may be **constraints**—equations or inequalities relating the variables.*

▼ **EXAMPLE 1** Average Cost

Gymnast Clothing, Inc. manufactures expensive hockey jerseys for sale to college bookstores in runs of up to 500. Its cost function is

$$C(x) = 2000 + 10x + 0.2x^2,$$

where x is the number of hockey jerseys it manufactures. How many jerseys should Gymnast Clothing Inc. produce per run in order to minimize average cost?

SOLUTION Why don't we seek to minimize total cost? The answer would be trivial: to minimize total cost, we would make *no* jerseys at all. Minimizing the average cost is a more practical objective. Here is the procedure we will follow to solve this problem.

1. *Identify the unknown(s) here.* The only unknown is the number x of hockey jerseys Gymnast should manufacture (we know this because the question is "how many jerseys . . .").
2. *Identify the objective function.* The objective function is the quantity that must be made as small (in this case) as possible. In this example, it is the average cost, given by

$$\bar{C}(x) = \frac{C(x)}{x} = \frac{2000 + 10x + 0.2x^2}{x}$$

$$= \frac{2000}{x} + 10 + 0.2x.$$

3. *Identify the constraints (if any).* At most 500 jerseys can be manufactured in a run. Also, $\bar{C}(0)$ is not defined. Thus, our constraint is

$$0 < x \leq 500.$$

Another way of saying this is that the domain of the objective function $\bar{C}(x)$ is $(0, 500]$.

4. *State and solve the resulting optimization problem.* Our optimization problem is to

$$\text{minimize } \bar{C}(x) = \frac{2000}{x} + 10 + 0.2x$$

$$\text{subject to } 0 < x \leq 500.$$

We now proceed to solve this problem as in the previous section.

▼ *If you have studied linear programming, you will notice a similarity here, but unlike the situation in linear programming, neither the objective function nor the constraints need to be linear.

Stationary Points $\bar{C}'(x) = -\dfrac{2000}{x^2} + 0.2$. We set this to zero and solve for x:

$$-\frac{2000}{x^2} + 0.2 = 0,$$

giving

$$0.2 = \frac{2000}{x^2}.$$

Multiplying both sides by x^2,

$$0.2x^2 = 2000,$$

so

$$x^2 = \frac{2000}{0.2} = 10{,}000.$$

Thus, $x = \pm\sqrt{10{,}000} = \pm 100$.

We reject $x = -100$, since it is not in the domain, and so $x = 100$ is the only stationary point.

There are no singular points, since $\bar{C}'(x)$ is defined for all x in the domain $(0, 500]$. The only endpoint is $x = 500$.

Because there is no candidate for an extremum to the left of $x = 100$, we include the testpoint $x = 10$, and we obtain the following table:

x	10	100	500
$\bar{C}(x)$	212	50	114

We see from this table that $\bar{C}(x)$ has an absolute minimum at $x = 100$, and so Gymnast Clothing should manufacture 100 hockey shirts per run in order to minimize average cost. The average cost will be $\bar{C}(100) = \$50$ per jersey.

To obtain the solution graphically, plot the objective function,

$$\bar{C}(x) = \frac{2000}{x} + 10 + 0.2x,$$

with x-range $0.001 \le x \le 500$ (Figure 1).

We notice right away that there is an absolute minimum at about $x = 100$. You can check the accuracy of this answer by plotting $\bar{C}'(x)$ and determining where its graph crosses the x-axis. (See Example 7 of the previous section for a description of this method.)

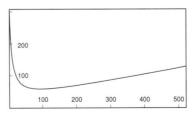

FIGURE 1

We now consider a constrained optimization problem in which the objective function is a function of two variables.

▼ **EXAMPLE 2**

Find x and y minimizing $f = x^2 + y^2$ and satisfying $x + y = 4$.

SOLUTION   Let us follow the procedure used in Example 1.

1. We first identify the unknown(s). In this problem, we are asked to find x and y, so these are the unknowns, and they are already named for us.
2. Next, we find the objective function. This is the quantity that we are required to minimize (or maximize). In this case, it is the function $f = x^2 + y^2$. Note that this is a function of *two* variables, so we can't simply go ahead and set the derivative equal to zero. (Would we take the derivative with respect to x or with respect to y?)
3. Next, we locate the constraint(s). After all, if there were no restrictions on x and y, we could make $x^2 + y^2$ as small as possible by simply choosing x and y to be zero. Looking at the problem, we see the phrase "satisfying $x + y = 4$." This restriction gives us our only constraint: $x + y = 4$.

Our problem is now to

$$\text{minimize } f = x^2 + y^2 \text{ subject to } x + y = 4.$$

Now that we've restated the problem, we can solve it in two simple steps.

First, solve the constraint equation for one of the variables (whichever is convenient), and substitute into the objective function. Doing so will eliminate one variable, giving us the objective as a function of a *single variable*.

Here, our constraint is $x + y = 4$. We solve it for y, getting $y = 4 - x$. Substituting into the objective function gives

$$f = x^2 + (4 - x)^2,$$

a function of the single variable x.

Second, locate the absolute maximum (or minimum) of the objective function as in the previous section.

Stationary Points $f'(x) = 2x + 2(4 - x)(-1) = 2x - 8 + 2x = 4x - 8$. Setting this equal to 0 gives

$$4x - 8 = 0,$$

so that

$$x = 2.$$

Thus, $x = 2$ is the only critical point, as there are no singular points. Also, there are no endpoints.

Plotting this single point will tell us very little, so we choose test points on either side.

x	0	2	4
$f(x)$	16	8	16

Without even drawing the graph,* we can see that f has its minimum value of 8 when $x = 2$.

▶ CAUTION   At this point in the problem, y has been eliminated, so we forget completely about y. We must be careful not to think of $f(x)$ as y as we usually do. The letter y is not playing its customary role here. Instead, y is one of the two unknowns. We are finding the minimum value of f, not y. When you draw the graph of the function f, you should label the vertical axis the f-axis, not the y-axis. ◀

To complete the problem, we must make sure that we answer the question, which was to find both x and y. To get y, we go back to the constraint $y = 4 - x$. Substituting $x = 2$ gives $y = 4 - 2 = 2$ also.

Thus, the minimum value of f is $2^2 + 2^2 = 8$ when x and y are both 2.

EXAMPLE 3

Find x and y maximizing $A = xy$ and satisfying $y = 1 - x^2$ and $0 \leq x \leq 1$.

SOLUTION   Here again, the unknowns are clearly stated, as is the objective function $A = xy$. There are two constraints: a constraint equation $y = 1 - x^2$ and an inequality $0 \leq x \leq 1$. Thus, our problem is to

maximize $A = xy$ *subject to* $y = 1 - x^2$ *and* $0 \leq x \leq 1$.

The constraint equation is already solved for y. We substitute this expression for y into the objective function, getting

$$A = x(1 - x^2) = x - x^3, \quad 0 \leq x \leq 1.$$

Notice that the second constraint does nothing more than specify that the domain of A is $[0, 1]$.

Now we locate the absolute maximum of A in the usual way. The stationary points are the solutions to $A'(x) = 0$, or

$$1 - 3x^2 = 0,$$

that is,

$$x^2 = \frac{1}{3}, \text{ so that } x = \pm\frac{1}{\sqrt{3}}.$$

We reject the negative solution because it is not in the domain of A, leaving us with a single stationary point at $x = 1/\sqrt{3}$. There are also the endpoints 0 and 1 of the domain.

▼ *You should still try to visualize the graph using the table as a reference—a good mental exercise.

Thus, we get the following table:

x	0	$\dfrac{1}{\sqrt{3}}$	1
$A(x)$	0	$\dfrac{2}{3\sqrt{3}}$	0

We can see from this table that we have an absolute maximum of $A = 2/(3\sqrt{3})$ at $x = 1/\sqrt{3}$. To answer the question, we also need y, which we get from the constraint equation: $y = 1 - x^2 = 1 - \frac{1}{3} = \frac{2}{3}$. Therefore, the maximum value of A is $2/(3\sqrt{3})$ and is achieved when $x = 1/\sqrt{3}$ and $y = \frac{2}{3}$.

Before turning to further applications, we summarize the steps we used in these examples.

> **PROCEDURE FOR SOLVING AN OPTIMIZATION PROBLEM**
>
> 1. First, identify the unknown(s) (the quantities asked for in the problem).
> 2. Determine the objective function, the quantity that we are required to minimize (or maximize). The objective function may be a function of one, two, or more variables.
> 3. Determine the constraint(s). These can take the form of equations or inequalities.
> 4. Restate the problem mathematically in the form
>
>   *minimize [maximize] the objective function subject to the constraint(s)*
>
> 5. If the objective function depends on several variables, rewrite it as a function of a single variable. This can be done by solving each constraint equation for one of the variables and substituting into the objective function.
> 6. Now you can locate the absolute maximum (or minimum) of the objective function as in the previous section. (Use the inequality constraints to specify the domain of the objective function.)

▼ **EXAMPLE 4**   Maximizing Area

Sam wants to build a rectangular enclosure as shown in Figure 2 for his pet rabbit, Killer, and he has bought 100 feet of fencing. What are the dimensions of the largest area that he can enclose?

SOLUTION   We must first identify the unknown(s), and for this, we go to the question: "what are the *dimensions* of the largest area he can enclose?" Thus,

 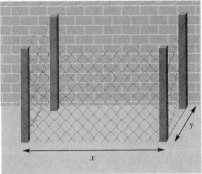

FIGURE 2 **FIGURE 3**

we are asked for the dimensions of the fence. We call these x and y, as shown in Figure 3.

To find the objective function, we look for what it is that we are trying to maximize (or minimize). The phrase "largest area" tells us that our object is to *maximize the area,* which is the product of length and width, so our objective function is

$$A = xy.$$

What about the constraints? If there were no constraints, Sam could simply make the area as large as he wanted by choosing x and y to be enormous. However, he has only 100 feet of electrified fencing to work with. This fact means that the sum of the lengths of the three sides must equal 100, or

$$x + 2y = 100.$$

One further point: because x and y represent lengths of sides of the enclosure, neither can be a negative number. Thus, we can rephrase our problem as follows.

Maximize $A = xy$ subject to $x + 2y = 100$,
$x \geq 0$, and $y \geq 0$.

Now we solve the constraint for one of the variables. We shall solve for x for a change.

$$x = 100 - 2y$$

Substituting this into the objective function gives

$$A = (100 - 2y)y = 100y - 2y^2,$$

and we have eliminated x. What about the inequalities? One says that $x \geq 0$, but we want to eliminate x from this as well. So again, we substitute for x, getting

$$100 - 2y \geq 0.$$

Solving this inequality for y gives $y \leq 50$. The second inequality says that $y \geq 0$. Thus, the constraints give us

$$A = 100y - 2y^2, \quad 0 \leq y \leq 50.$$

So we have A as a function of y this time, and the domain is $[0, 50]$. We now maximize in the usual way. We have

$$A'(y) = 100 - 4y,$$

so we set

$$100 - 4y = 0, \quad \text{giving } y = 25.$$

Adding in the endpoints, we get this table:

y	0	25	50
$A(y)$	0	1,250	0

Thus, the maximum area of 1,250 square feet occurs when $y = 25$ feet. The length of the other fence is obtained by substituting $y = 25$ into the constraint equation $x + 2y = 100$, giving $x = 50$ feet. Thus, the enclosure with the largest area is 50 feet across and 25 feet deep.

▼ **EXAMPLE 5** Revenue

The Cuddly Carriage Co. builds baby strollers. Market Research estimates that if it sets the price at p dollars, then the company can sell $q = 300{,}000 - 10p^2$ strollers per year. What price will bring in the largest annual revenue?

SOLUTION We first identify the unknowns by going to the question. We see that p is our main unknown, but there is also another variable we don't know: the demand q. Thus, we really have two unknowns, p and q. As for the objective function, we look for the quantity that we are trying to maximize (or minimize). We see in the last sentence that our objective is to maximize the annual *revenue,* which is the product of the price per stroller and the number of strollers sold per year. Thus, our objective function is

$$R = pq.$$

We are given the constraint in the form of a demand equation

$$q = 300{,}000 - 10p^2,$$

which is already solved for q. All we have to do, then, is substitute into the objective function.

$$R(p) = (300{,}000 - 10p^2)p = 300{,}000p - 10p^3$$

We should also think about the domain. Price cannot be negative (unless the company is going to *pay* people to take these strollers off its hands!) and neither can demand. Demand will become negative if p gets larger than the point where $q = 0$, that is, $300{,}000 - 10p^2 = 0$, which has a solution $p = 173$. So p must be between 0 and 173, and the domain of $R(p)$ is $[0, 173]$. Now for the calculus:

$$\frac{dR}{dp} = 300{,}000 - 30p^2,$$

so the critical points will be the solutions to $300{,}000 - 30p^2 = 0$. Solving, $30p^2 = 300{,}000$, $p^2 = 10{,}000$, $p = \pm 100$. Now, -100 is outside the domain, so 100 is the only critical point we can use. Together with the endpoints 0 and 173, this gives

p	0	100	173
R	0	20,000,000	170

So, the largest possible annual revenue is $20,000,000, which is achieved by pricing the strollers at $100 each.

The following problem is a classic one that shows the power of calculus.

▼ **EXAMPLE 6**   Maximizing Volume

The Cardboard Box Co. is going to make open-topped boxes out of squares of cardboard 30″ on a side by cutting squares out of the corners and folding up the sides. What is the largest volume box it can make this way?

SOLUTION   Start by drawing a picture (Figure 4).

We are asked to find the largest volume, so this will be our objective, but the real unknowns here are the dimensions of the squares cut out and the dimensions of the resulting box. In the picture we have labeled the sides of the squares cut out as s. This will be the height of the box once the sides are folded up. The bottom edges are labeled w in the picture. It is s and w that we need to find. Our objective function, the volume, is

$$V = s \cdot w \cdot w = sw^2.$$

Our constraint comes from the known dimensions of the original square of cardboard, 30″ on a side. Looking at the picture, you can see that

$$2s + w = 30.$$

Now we solve the constraint for one of the variables. It seems easiest to solve for w.

$$w = 30 - 2s$$

Substituting in the objective function,

$$V = s(30 - 2s)^2.$$

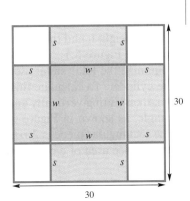

FIGURE 4

EXAMPLE 8 Maximizing Volume

Picky Parcel Service is finicky about the size of the boxes it will accept: The perimeter of the base must be no more than 20 inches, while the perimeter of one side must be no more than 10 inches. What is the largest volume box it will accept?

SOLUTION We select the dimensions of the box as our unknowns, even though these are not mentioned explicitly. Thus, we have three unknowns in this problem: the three dimensions of the box l, w, and h as labeled in Figure 7.

Our objective is to maximize the volume

$$V = lwh.$$

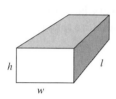

FIGURE 7

Notice that we now need to eliminate *two* variables. To do this, we note that there are two constraints given, on the perimeters of the base and one side, which we will take to be the frontmost side in the picture. Because we want the volume to be as large as possible, we shall take these perimeters to be as large as possible. The stated constraints then translate into the equations

$$2l + 2w = 20$$
$$2h + 2w = 10.$$

The best thing to do now is to write two of the variables in terms of the third. Since w appears in both constraints, it will be easiest to write l and h in terms of w.

$$l = 10 - w$$
$$h = 5 - w$$

Substituting into the objective function, we get

$$V = (10 - w)w(5 - w) = 50w - 15w^2 + w^3.$$

Because none of the dimensions can be negative, the second constraint limits w to be no larger than 5, and so the domain of $V(w)$ is $[0, 5]$.

$$\frac{dV}{dw} = 50 - 30w + 3w^2$$

To solve $3w^2 - 30w + 50 = 0$, we use the quadratic formula, which gives

$$w = \frac{30 \pm \sqrt{900 - 600}}{6} = 5 \pm \frac{5\sqrt{3}}{3}.$$

Of these two solutions, only one is in the correct interval, $w = 5 - 5\sqrt{3}/3 \approx 2.11$.

Together with the endpoints, this gives

w	0	2.11	5
V	0	48.1	0

So the largest acceptable package has a volume of 48.1 cubic inches, with $w = 5 - 5\sqrt{3}/3 \approx 2.11$ inches. The other dimensions will then be $l = 10 - w = 5 + 5\sqrt{3}/3 \approx 7.88$ inches and $h = 5 - w = 5\sqrt{3}/3 \approx 2.88$ inches.

The next example is done with the aid of a graphing calculator.

▼ **EXAMPLE 9**   Labor Resource Allocation

The Gym Sock Company manufactures cotton athletic socks. Production is partially automated through the use of robots. Daily operating costs amount to $50 per laborer and $30 per robot. The number of pairs of socks q the company can manufacture in a day is given by a Cobb-Douglas* production formula

$$q = 50n^{0.6}r^{0.4},$$

where n is the number of laborers and r is the number of robots. Assuming that the company wishes to produce 1,000 pairs of socks per day at a minimum cost, how many laborers and how many robots should it use?

SOLUTION   The unknowns are n, the number of laborers, and r, the number of robots. The objective is to minimize the daily cost,

$$C = 50n + 30r.$$

The constraints are given by the daily quota,

$$1{,}000 = 50n^{0.6}r^{0.4},$$

and the fact that n and r are nonnegative.

We solve the constraint equation for one of the variables—let us solve for n.

$$n^{0.6} = \frac{1{,}000}{50r^{0.4}} = \frac{20}{r^{0.4}}$$

Taking the $1/0.6$ power of both sides gives

$$n = \left(\frac{20}{r^{0.4}}\right)^{1/0.6} = \frac{20^{1/0.6}}{r^{0.4/0.6}} \approx \frac{147.36}{r^{2/3}}.$$

Substituting in the objective equation gives us the cost as a function of r.

$$C = 50\left(\frac{147.36}{r^{2/3}}\right) + 30r$$

$$= 7{,}368r^{-2/3} + 30r$$

▼ *Cobb-Douglas production formulas were discussed in the section on implicit differentiation in the preceding chapter.

Before minimizing, we graph this cost function on a graphing calculator, obtaining the graph shown in Figure 8. (The ranges of the coordinates are shown in the graph.)

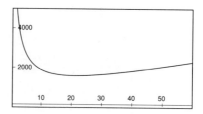

FIGURE 8

From the graph, we see that the cost is minimized when r is approximately 20, and that the minimum occurs at a stationary point. To obtain a more accurate answer, we could either zoom in or set the derivative equal to zero and solve for r. We choose the latter approach. Since

$$C = 7{,}368r^{-2/3} + 30r,$$

$$\frac{dC}{dr} = -4{,}912r^{-5/3} + 30.$$

So we must solve

$$-4{,}912r^{-5/3} + 30 = 0.$$

We can now either solve for r analytically or use our graphing calculator to solve it numerically. Graphing $-4{,}912r^{-5/3} + 30$ gives us Figure 9.

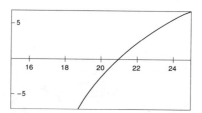

FIGURE 9

The value of r we desire is given by the point where this graph crosses the r-axis. Since r represents numbers of robots, and we are not interested in fractions of a robot, we need only estimate r to the nearest whole number, so we see that $r = 21$.

Now that we know r, we can obtain n by going back to the constraint equation.

$$1{,}000 = 50n^{0.6}r^{0.4} = 50n^{0.6}(21)^{0.4}.$$

Thus,

$$n^{0.6} = \frac{1{,}000}{50(21)^{0.4}} \approx 5.9176.$$

Taking reciprocal powers gives

$$n = 5.9176^{(1/0.6)} \approx 19.360.$$

Thus, to minimize daily operating costs, the company should use 19 laborers and 21 robots.

Before we go on... To find the resulting daily operating cost, we can either substitute the values $n = 19$ and $r = 21$ into the equation $C = 50n + 30r$ for cost, or find the cost from the graph in Figure 8. We leave this as an exercise for you.

12.2 EXERCISES

Solve the optimization problems in Exercises 1–8.

1. Maximize $P = xy$ with $x + y = 10$.

2. Maximize $P = xy$ with $x + 2y = 40$.

3. Minimize $S = x + y$ with $xy = 9$ and both x and $y > 0$.

4. Minimize $S = x + 2y$ with $xy = 2$ and both x and $y > 0$.

5. Minimize $F = x^2 + y^2$ with $x + 2y = 10$.

6. Minimize $F = x^2 + y^2$ with $xy^2 = 16$.

7. Maximize $P = xyz$ with $x + y = 30$ and $y + z = 30$, and $x, y,$ and $z \geq 0$.

8. Maximize $P = xyz$ with $x + z = 12$ and $y + z = 12$ and $x, y,$ and $z \geq 0$.

9. For a rectangle with perimeter 20 to have the largest area, what dimensions should it have?

10. For a rectangle with area 100 to have the smallest perimeter, what dimensions should it have?

APPLICATIONS

11. *Fences* I want to fence in a rectangular vegetable patch. The fencing for the east and west sides costs $4 per foot, while the fencing for the north and south sides costs only $2 per foot. I want to spend $80 on the entire project. What is the largest area that I can enclose?

12. *Fences* Actually, my vegetable garden abuts my house, so that the house itself forms the northern boundary. The fencing for the southern boundary costs $4 per foot, while the fencing for the east and west sides costs $2 per foot. If I want to spend $80 on the project, what is the largest area that I can enclose this time?

13. *Revenue* Hercules Films is deciding on the price of the video release of their film "Son of Frankenstein." They estimate that at a price of p dollars, they can sell a total of $q = 200{,}000 - 10{,}000p$ copies. At what price will they bring in the largest revenue?

14. *Profit* Hercules Films is also deciding on the price of the video release of their film "Bride of the Son of Frankenstein." Again, at a price of p dollars they can sell $q = 200{,}000 - 10{,}000p$ copies, but each copy costs them $4 to make. What should the price be to bring them the largest profit?

15. *Revenue* (Here we revisit Royal Ruby Retailers— see Chapter 1). The demand for rubies at RRR is given by the equation

$$q = -\frac{4}{3}p + 80,$$

where p is the price RRR charges (in dollars) and q is the number of rubies RRR sells per week. At what price should RRR sell its rubies in order to maximize its weekly revenue? (Try not to look at the answer we obtained in Chapter 1 until you have worked through this exercise.)

16. *Revenue* The consumer demand curve for tissues is given by

$$q = (100 - p)^2, \quad 0 \leq p \leq 100$$

where p is the price per case of tissues and q is the demand in weekly sales. At what price should tissues be sold in order to maximize the revenue?

17. *Revenue* Assume that the demand function for tuna in a small coastal town is given by

$$p = \frac{500{,}000}{q^{1.5}},$$

where p is the price (in dollars) per pound of tuna and q is the number of pounds of tuna that can be sold at the price p in one month. Assume that the town's fishery wishes to sell at least 5,000 pounds of tuna per month.

(a) How much should the town's fishery charge for tuna in order to maximize monthly revenue?
(b) How much tuna will it sell per month at that price?
(c) What will its resulting revenue be?

18. *Revenue* Economist Henry Schultz calculated the demand function for corn to be given by

$$p = \frac{6{,}570{,}000}{q^{1.3}},$$

where p is the price (in \$) per bushel of corn, and q is the number of bushels of corn that could be sold at the price p in one year.* Assume that at least 10,000 bushels of corn per year must be sold.

(a) How much should farmers charge per bushel of corn in order to maximize annual revenue?
(b) How much corn can farmers sell per year at that price?
(c) What will the farmers' resulting revenue be?

19. *Revenue* The wholesale price for chicken in the U.S. fell from 25¢ per pound to 14¢ per pound, and at the same time, per capita chicken consumption rose from 22 pounds per year to 27.5 pounds per year.[†] Assuming that the demand for chicken depends linearly on the price, what wholesale price for chicken maximizes revenues for poultry farmers, and what does that revenue amount to?

20. *Revenue* Your underground used book business is doing a booming trade. Your policy is to sell all used versions of *Calculus and You* at the same price (regardless of condition). When you set the price at \$10, sales amounted to 120 volumes during the first week of classes. The following semester, you set the price at \$30 and sold not a single book. Assuming that the demand for books depends linearly on the price, what price gives you the maximum revenue, and what does that revenue amount to?

21. *Profit* As we have seen on several occasions, the demand for rubies at RRR is given by the equation

$$q = -\frac{4}{3}p + 80,$$

where p is the price RRR charges (in dollars) and q is the number of rubies RRR sells per week. Assuming that due to extraordinary market conditions RRR can obtain rubies for \$25 each, how much should it charge per ruby to make the largest possible weekly profit, and what will that profit be?

22. *Profit* The consumer demand curve for tissues is given by

$$q = (100 - p)^2, \quad 0 \leq p \leq 100,$$

where p is the price per case of tissues and q is the demand in weekly sales. If tissues cost \$30 per case, at what price should tissues be sold for the largest possible weekly profit, and what will that profit be?

▼ *Based on data for the period 1915–1929. Source: Henry Schultz, *The Theory and Measurement of Demand* (as cited in *Introduction to Mathematical Economics* by A. L. Ostrosky, Jr., and J.V. Koch (Prospect Heights, Ill.:Waveland Press, 1979.)

[†] Data are provided for the years 1951–1958. Source: U.S. Department of Agriculture, *Agricultural Statistics*.

23. *Profit* A demand equation for your company's virtual reality video headsets is given by
$$p = \frac{1{,}000}{q^{0.3}},$$
where q is the total number of headsets that your company can sell in a week at a price of p dollars. The total manufacturing and shipping cost amounts to $100 per headset.
   (a) What is the largest profit your company can make in a week, and how many headsets will your company sell at this level of profit? (Give answers to the nearest whole number.)
   (b) How much, to the nearest $1, should your company charge per headset for the maximum profit?

24. *Profit* Due to sales by a competing company, your company's sales of virtual reality video headsets have dropped, and your financial consultant revises the demand equation to
$$p = \frac{800}{q^{0.35}},$$
where q is the total number of headsets that your company can sell in a week at a price of p dollars. The total manufacturing and shipping cost still amounts to $100 per headset.
   (a) What is the largest profit your company can make in a week, and how many headsets will your company sell at this level of profit? (Give answers to the nearest whole number.)
   (b) How much, to the nearest $1, should your company charge per headset for the maximum profit?

25. *Box Design* The Chocolate Box Co. is going to make open-topped boxes out of 6″ × 16″ rectangles of cardboard by cutting squares out of the corners and folding up the sides. What is the largest-volume box it can make this way?

26. *Box Design* A packaging company is going to make open-topped boxes with square bases that hold 108 cubic centimeters. What are the dimensions of the box that can be built with the least material?

27. *Asset Appreciation* As the financial consultant to a classic auto dealership, you estimate that the total value of its collection of 1959 Chevrolets and Fords is given by the formula
$$v = 300{,}000 + 1{,}000t^2,$$
where t is the number of years from now. You anticipate an inflation rate running continuously at 5% per year, so that the discounted (present) value of an item that will be worth v in t years' time is given by
$$p = ve^{-0.05t}.$$
When would you advise the dealership to sell the vehicles in order to maximize their discounted value?

28. *Plantation Management* The value of a fir tree in your plantation increases with the age of the tree according to the formula
$$v = \frac{20t}{1 + 0.05t},$$
where t is the age of the tree in years. Given an inflation rate running continuously at 5% per year, the discounted (present) value of a newly planted seedling is given by
$$p = ve^{-0.05t}.$$
At what age (to the nearest year) should you harvest your trees in order to ensure the greatest possible discounted value?

29. *Marketing Strategy* The Feature Software Co. has a dilemma. Its new program, Doors 3.0, is almost ready to go on the market. However, the longer the company works on it, the better they can make it and the more they can charge for it. The company's marketing analysts estimate that if they delay t days they can set the price at $100 + 2t$. On the other hand, the longer they delay, the more market share they will lose to their main competitor (see the next exercise) so that if they delay t days they will be able to sell $400{,}000 - 2{,}500t$ copies of the program. How many days should it delay the release in order to get the largest revenue?

30. *Marketing Strategy* Feature Software's main competitor is Newton Software, and Newton is in a similar predicament. Its product, Walls 5.0, could be sold now for $200, but for each day they delay they could increase the price by $4. On the other hand, they could sell 300,000 copies now, but each day they wait will cut their sales by 1,500. How many days should it delay the release in order to get the largest revenue?

31. *Agriculture* The fruit yield per tree in an orchard containing 50 trees is 100 pounds per tree each year. Due to crowding, the yield decreases by 1 pound per season for every additional tree planted. How many additional trees should be planted for a maximum total annual yield?

32. *Agriculture* Two years ago, your orange orchard contained 50 trees and yielded 75 bags of oranges. Last year, you sold ten of the trees and noticed that the total yield increased to 80 bags. Assuming that the yield of each tree depends linearly on the number of trees in the orchard, what should you do this year in order to maximize your yield?

33. *Average Cost* A cost function for the manufacture of portable CD players is given by

$$C(x) = \$150{,}000 + 20x + \frac{x^2}{10{,}000},$$

where x is the number of CD players manufactured. Interpret each term in this formula. How many CD players should be manufactured in order to minimize average cost? What is the resulting average cost of a CD player? (Give your answer to the nearest dollar.)

34. *Average Cost* Repeat the preceding exercise using the revised cost function

$$C(x) = \$150{,}000 + 20x + \frac{x^2}{100}.$$

35. *Pollution Control* The cost of controlling emissions at a firm goes up rapidly as the amount of emissions reduced goes up. Here is a possible model:

$$C(q) = 4{,}000 + 100q^2,$$

where q is the reduction in emissions (in pounds of pollutant per day) and C is the daily cost (in dollars) of this reduction. Government clean-air subsidies amount to $500 per pound of pollutant removed. How many pounds of pollutant should the company remove each day in order to minimize *net* cost (cost minus subsidy)?

36. *Pollution Control* Repeat the preceding exercise using the following data:

$$C(q) = 2{,}000 + 200q^2,$$

with government subsidies amounting to $100 per pound of pollutant removed per day.

37. *Luggage Dimensions* Fly-by-Night Airlines has a peculiar rule about luggage: the length and width of a bag must add to 45 inches, while the width and height must also add to 45 inches. What are the dimensions of the bag with largest volume that it will accept?

38. *Luggage Dimensions* Fair Weather Airlines has a similar rule. It will accept only bags for which the sum of the length and width is 36 inches, while the sum of length, height, and twice the width is 72 inches. What are the dimensions of the bag with largest volume that it will accept?

39. *Package Dimensions* The U.S. Postal Service (USPS) will accept only packages with a length plus girth no more than 108 inches.* (See figure.)

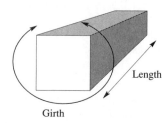

Assuming that the front face of the package (as shown in the figure) is square, what is the largest-volume package that the USPS will accept?

40. *Package Dimensions* The *United Parcel Service* (UPS) will only accept packages with a length no more than 108 inches and length plus girth no more than 130 inches*. (See figure above.) Assuming that the front face of the package (as shown in the figure) is square, what is the largest volume package that UPS will accept?

41. *Average Profit* The Feature Software Company sells its graphing program, Dogwood, with a volume discount: if a customer buys x copies, then they pay† $\$500\sqrt{x}$. It cost the company $10,000 to develop the program and $2 to manufacture each copy. If just one customer buys all the copies of Dogwood, how many copies must the customer buy for Feature Software's average profit per copy to be maximized? How are average profit and marginal profit related at this number of copies?

42. *Average Profit* Repeat the preceding exercise with the charge to the customer being $\$600\sqrt{x}$ and the cost to develop the program being $9,000.

---

* The data were current at the time of this writing.
† This is similar to the site license charge for the program Maple®.

43. *Prison Population* The prison population of the U.S. followed the curve

$$N(t) = 0.028234t^3 - 1.0922t^2 + 13.029t + 146.88 \quad (0 \leq t \leq 39)$$

in the years 1950–1989. Here t is the number of years since 1950 and N is the number of prisoners in thousands.* When, to the nearest year, was the prison population decreasing most rapidly, and when was it increasing most rapidly?

44. *Test Scores* Combined SAT scores in the United States can be approximated by

$$T(t) = -0.01085t^3 + 0.5804t^2 - 10.12t + 962.4 \quad (0 \leq t \leq 22)$$

in the years 1967–1991. Here t is the number of years since 1967 and T is the combined SAT score average for the U.S.[†] Based on this model, when (to the nearest year) was the average SAT score decreasing most rapidly? When was it increasing most rapidly?

45. *Embryo Development* The oxygen consumption of a bird embryo increases from the time the egg is laid through the time the chick hatches. In the case of a typical galliform bird, the oxygen consumption (in milliliters per hour) can be approximated by

$$c(t) = -0.00271t^3 + 0.137t^2 - 0.892t + 0.149 \quad (8 \leq t \leq 30),$$

where t is the time (in days) since the egg was laid.[‡] (An egg will typically hatch at around $t = 28$.) When, to the nearest day, is $c'(t)$ a maximum? What does the answer tell you?

46. *Embryo Development* The oxygen consumption of a turkey embryo increases from the time the egg is laid through the time the chick hatches. In the case of a brush turkey, the oxygen consumption (in milliliters per hour) can be approximated by

$$c(t) = -0.00118t^3 + 0.119t^2 - 1.83t + 3.972 \quad (20 \leq t \leq 50)$$

where t is the time (in days) since the egg was laid.[‡] (An egg will typically hatch at around $t = 50$.) When, to the nearest day, is $c'(t)$ a maximum? What does the answer tell you?

47. *Minimizing Resources* Basic Buckets, Inc., has an order for plastic buckets holding 5,000 cubic centimeters. The buckets are open-topped cylinders, and the company wants to know what dimensions will use the least plastic per bucket. (The volume of an open-topped cylinder with height h and radius r is $\pi r^2 h$, while the surface area is $\pi r^2 + 2\pi rh$.)

48. *Optimizing Capacity* Basic Buckets would like to build a bucket with a surface area of 1,000 square centimeters. What is the volume of the largest bucket it can build? (See the previous exercise.)

The use of either a graphing calculator or graphing computer software is required for Exercises 49–54.

49. *Education* In 1991, the expected income of an individual depended on his or her educational level, according to the following formula.

$$I(n) = 2,928.8n^3 - 115,860n^2 + 1,532,900n - 6,760,800 \quad (12 \leq n \leq 15)$$

Here, n is the number of school years completed, and $I(n)$ is the individual's expected income.[§] Using [12, 15] as the domain, use technology to locate and classify the absolute extrema of $I'(n)$. Interpret the results.

▼ *The model is the authors'. Source for data: *Sourcebook of Criminal Justice Statistics*, 1990, p. 604.

[†] The model is the authors'. Source for data: Educational Testing Service.

[‡] The model approximates graphical data published in the article "The Brush Turkey" by Roger S. Seymour, *Scientific American*, December, 1991, pp. 108–14.

[§] The model is a best-fit cubic based on Table 358, U.S. Department of Education, *Digest of Education Statistics, 1991* (Washington, DC: Government Printing Office, 1991).

50. *Marriage* Based on statistics published by the U.S. Bureau of the Census, the median age of an individual at his or her first marriage can be modeled by the following functions.*

Females: $F(n) = 0.000023453n^3 - 0.0026363n^2 + 0.050582n + 21.766$

Males: $M(n) = 0.000023807n^3 - 0.0025184n^2 + 0.015754n + 0.015754n + 25.966$

n = number of years since 1890; $0 \leq n \leq 102$

(a) Using $[1, 102]$ as the domain, use technology to locate the local extrema of both functions. Round answers to the nearest integer, and interpret the results.

(b) What do the local extrema of $F'(n)$ and $M'(n)$ tell you?

(c) What do the local extrema of $M(n) - F(n)$ tell you?

51. *Asset Appreciation* You manage a small antique store that owns a collection of Louis XVI jewelry boxes. Their value v is increasing according to the formula

$$v = \frac{10000}{1 + 500e^{-0.5t}},$$

where t is the number of years from now. You anticipate an inflation rate of 5% per year, so that the present value of an item that will be worth v in t years' time is given by

$$p = v(1.05)^{-t}.$$

When (to the nearest year) should you sell the jewelry boxes in order to maximize their present value? How much (to the nearest constant dollar) will they be worth at that time?

52. *Harvesting Forests* The following equation models the approximate volume in cubic feet of a typical Douglas fir tree of age t years:†

$$v = \frac{22514}{1 + 22514t^{-2.55}}.$$

The lumber will be sold at $10 per cubic foot, and you do not expect the price of lumber to appreciate in the foreseeable future. On the other hand, you anticipate a general inflation rate of 5% per year, so that the present value of an item that will be worth v in t years' time is given by

$$p = v(1.05)^{-t}.$$

At what age (to the nearest year) should you harvest a Douglas fir tree in order to maximize its present value? How much (to the nearest constant dollar) will a Douglas fir tree be worth at that time?

53. *Resource Allocation* Your automobile assembly plant has a Cobb-Douglas production function given by

$$q = x^{0.4}y^{0.6},$$

where q is the number of automobiles it produces per year, x is the number of employees, and y is the daily operating budget (in dollars). Annual operating costs amount to an average of $20,000 per employee plus the operating budget of $365y$. Assume that you wish to produce 1,000 automobiles per year at a minimum cost. How many employees should you hire?

54. *Resource Allocation* Repeat the preceding exercise using the production formula

$$q = x^{0.5}y^{0.5}.$$

55. *Revenue* (based on a question in the GRE economics test‡) If total revenue (TR) is specified by $TR = a + bQ - cQ^2$, where Q is quantity of output and a, b, and c are positive parameters, then TR is maximized for this firm when it produces Q equal to:
(a) $b/2ac$ (b) $b/4c$ (c) $(a + b)/c$
(d) $b/2c$ (e) $c/2b$

56. *Revenue* (based on a question in the GRE economics test) If total demand (Q) is specified by $Q = -aP + b$, where P is unit price and a and b are positive parameters, then total revenue is maximized for this firm when it charges P equal to:
(a) $b/2a$ (b) $b/4a$ (c) a/b (d) $a/2b$
(e) $-b/2a$.

---

*Source: U.S. Bureau of the Census, "Marital Status and Living Arrangements: March 1992," Current Population Reports, *Population Characteristics*, Series P-20, No. 468, March 1992, p. vii.

† The model is the authors' and is based on data in *Environmental and Natural Resource Economics*, Third Edition, by Tom Tietenberg (New York: HarperCollins, 1992), p. 282.

‡ Source: GRE Economics Test, by G. Gallagher, G. E. Pollock, W. J. Simeone, G. Yohe (Piscataway, N.J.: Research and Education Association, 1989).

COMMUNICATION AND REASONING EXERCISES

57. Explain why the following problem is uninteresting: "A packaging company wishes to make cardboard boxes with open tops by cutting square pieces from the corners of a square sheet of cardboard and folding up the sides. What is the box with the least surface area they can make this way?"

58. Explain why finding the production level that minimizes a cost function is frequently uninteresting.

59. If demand q decreases as price p increases, what does the minimum value of dq/dp measure?

60. Explain why the following problem is uninteresting: "A cost function for the manufacture of portable CD players is given by

$$C(x) = \$150{,}000 + 20x + \frac{x^2}{10{,}000},$$

where x is the number of CD players manufactured. How many CD players should be manufactured in order to minimize total cost?"

61. Explain how you would solve an optimization problem of the following form. "Maximize $P = f(x, y, z)$ subject to $z = g(x, y)$ and $y = h(x)$."

62. Explain how you would solve an optimization problem of the following form. "Maximize $P = f(x, y, z)$ subject to $z = g(x, y)$, $x^2 + y^2 = 1$."

12.3 THE SECOND DERIVATIVE: ACCELERATION AND CONCAVITY

THE SECOND DERIVATIVE AS ACCELERATION

Now that we have seen some of the power of the derivative, we take a look at the derivative of the derivative. For example, if $f(x) = x^3 + 2x^2$, then its derivative is $f'(x) = 3x^2 + 4x$. Since this is also a function of x, we can take the derivative once again. When we do this, we have the derivative of the derivative, or the **second derivative,**

$$f''(x) = 6x + 4.$$

Q What does the second derivative mean?

A The answer to that question is the subject of this entire section.

First, let us go back to the interpretation of the derivative in terms of velocity. Suppose, for example, that the position (i.e., odometer reading) of a car is given by $s(t) = t^3 + 2t^2$ miles, where t is the time in hours. Then, as we saw, the derivative $s'(t)$ gives us the velocity—a measure of *how fast the odometer reading is increasing*—in miles per hour. Thus,

$$v(t) = s'(t) = 3t^2 + 4t.$$

Notice that v also shows up as the reading on the speedometer. Now how fast is *that* reading changing? To measure that, we must take the derivative of v: in other words, the second derivative of s:

$$a(t) = v'(t) = s''(t) = 6t + 4.$$

Here, $a(t)$ is the rate of change of the velocity $v(t)$, and is called the *acceleration*. Thus, just as the velocity measures how fast the odometer is going up (in miles per hour), the acceleration measures how fast the speedometer is changing (in miles per hour per hour).

EXAMPLE 3   Acceleration of Demand

For the first 15 months after its introduction, the total sales of a new video game grows exponentially and can be modeled by the curve

$$S(t) = 20e^{0.4t}.$$

Later, after about 25 months, the total sales follows more closely the curve

$$S(t) = 100{,}000 - 20e^{17-0.4t}.$$

How fast is the total sales accelerating after 10 months, and how fast is it accelerating after 30 months?

SOLUTION   To obtain the acceleration of a quantity, we must take its second derivative. During the first 15 months, the derivative of sales will be

$$S'(t) = 8e^{0.4t}$$

and so the second derivative will be

$$S''(t) = 3.2e^{0.4t}.$$

Thus, after 10 months the total sales will be

$$S(10) = 20e^4 \approx 1{,}092,$$

the rate of change of sales will be

$$S'(10) = 8e^4 \approx 437,$$

and the acceleration of sales will be

$$S''(10) = 3.2e^4 \approx 175.$$

What do these numbers mean? By the end of the 10th month, 1,092 video games will have been sold. The game will continue to sell at the rate of 437 games per month. This rate of sales is increasing by 175 sales per month per month.

To analyze the sales after 30 months, we must use the second formula for sales,

$$S(t) = 100{,}000 - 20e^{17-0.4t}.$$

The derivative is

$$S'(t) = 8e^{17-0.4t}$$

and the second derivative is

$$S''(t) = -3.2e^{17-0.4t}.$$

After 30 months,

$$S(30) = 100{,}000 - 20e^{17-12} \approx 97{,}032,$$
$$S'(30) = 8e^{17-12} \approx 1{,}187,$$

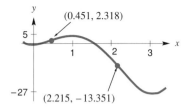

FIGURE 7

To find the y-coordinates of these points, we must, as usual, substitute into the original function $f(x)$: $f(0.451) \approx 2.318$ and $f(2.215) \approx -13.351$. These points are shown in Figure 7.

Before we go on... Recall that setting $f'(x) = 0$ and solving for x only gives us *candidates* for local extrema. Similarly, setting $f''(x) = 0$ and solving for x only gives us candidates for the points of inflection. In this example, we knew by looking at the graph that there had to be two points of inflection: one between 0 and 1 and another between 1 and 3. In general, we need to verify that a candidate for a point of inflection is indeed a point of inflection by checking that f'' changes sign at that point. Actually, testing for a point of inflection is very much like testing for extrema. There is a third-derivative test we could use, but it is unnecessary in most examples. Usually we have plenty of information about the graph by the time we begin testing for points of inflection, or we can obtain enough information with a graphing calculator.

Q Now that we have the second-derivative test for maxima and minima, we can forget about those other, more primitive, approaches (such as using test points). Right?

A Wrong. For example, try the second-derivative test on $f(x) = x^3$ and on $g(x) = x^4$. Both functions have stationary points at $x = 0$, but the second-derivative test tells us nothing about either function, since $f''(0) = 0$ and $g''(0) = 0$. In fact, g has an absolute minimum at $x = 0$, while f has neither. When the second derivative is 0, we need the other approaches.

The second derivative can be written in differential notation as

$$y'' = \frac{d^2y}{dx^2} = \frac{d}{dx}\left(\frac{dy}{dx}\right) = \frac{d^2}{dx^2}(y).$$

We shall use this notation in the next example.

▼ **EXAMPLE 5**

Find all points of inflection of the curve $y = xe^{-x}$ for x in $[0, +\infty)$.

SOLUTION Before taking derivatives, we can use a graphing calculator or a computer to get an idea of where any points of inflection might be. Figure 8 shows a graphing calculator plot of the curve for $0 \le x \le 5$.

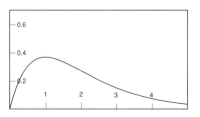

FIGURE 8

It looks as if there is one point of inflection, around $x = 2$. Now let us see what the derivative reveals.

$$\frac{dy}{dx} = e^{-x} - xe^{-x} \quad \text{by the product rule}$$

$$= (1 - x)e^{-x}$$

(Using this derivative we can say that the maximum we can see in the graph is located at $x = 1$.)

$$\frac{d^2y}{dx^2} = -e^{-x} - (1 - x)e^{-x}$$

$$= (x - 2)e^{-x}$$

Setting the second derivative equal to 0 gives

$$(x - 2)e^{-x} = 0.$$

But, if a product of two quantities is zero, at least one of them must be zero. Since e^{-x} is never zero, it must be that the other quantity, $x - 2$, is zero. Thus,

$$x - 2 = 0,$$

so

$$x = 2.$$

This point is the only candidate for a point of inflection. But is this point really a point of inflection? Looking at Figure 8, we can see that the curve is concave down at the maximum at $x = 1$ and concave up at $x = 3$. (We could confirm this by evaluating the second derivative at these points, but let us trust our eyes and our calculator.) Therefore, there is a single point of inflection at $x = 2$.

Before we go on... Another way of telling if there is really a point of inflection is to see if $\frac{d^2y}{dx^2}$ changes sign at $x = 2$, which we can do easily by graphing this second derivative, as was done to get Figure 9.

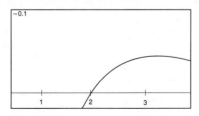

FIGURE 9

This graph (the graph of $(x - 2)e^{-x}$) clearly shows that $\frac{d^2y}{dx^2}$ changes sign at $x = 2$, verifying that there is a point of inflection at $x = 2$ in the graph of y.

12.3 The Second Derivative: Acceleration and Concavity

EXAMPLE 6 Demand for a New Product

The demand for a new product over time often follows a logistic curve of the form

$$Q = \frac{a}{1 + e^{-k(t-c)}},$$

where Q is the total number of items sold up to time t (months, say), and a, c, and k are constants that depend on the particular market. For simplicity, let us look at the example with $a = c = k = 1$:

$$Q = \frac{1}{1 + e^{-(t-1)}}.$$

A graphing calculator plot of this demand equation is shown in Figure 10.

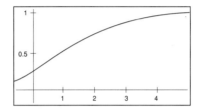

FIGURE 10

From the graph, we see that the sales rise fairly rapidly at first and then level off by about $t = 5$. Notice also that there is a point of inflection at about $t = 1$. Exactly where is this point of inflection, and what does it tell us about sales?

SOLUTION We locate the point of inflection by setting $\dfrac{d^2Q}{dt^2} = 0$ and solving for t.

$$\frac{dQ}{dt} = \frac{e^{-(t-1)}}{(1 + e^{-(t-1)})^2} \quad \text{by the quotient rule}$$

Taking the second derivative,

$$\frac{d^2Q}{dt^2} = \frac{(-e^{-(t-1)})(1 + e^{-(t-1)})^2 - e^{-(t-1)} \cdot 2(1 + e^{-(t-1)})(-e^{-(t-1)})}{(1 + e^{-(t-1)})^4}$$

$$= \frac{e^{-(t-1)}(1 + e^{-(t-1)})(-1 - e^{-(t-1)} + 2e^{-(t-1)})}{(1 + e^{-(t-1)})^4}$$

$$= \frac{e^{-(t-1)}(e^{-(t-1)} - 1)}{(1 + e^{-(t-1)})^3}.$$

Equating the second derivative to zero gives

$$\frac{e^{-(t-1)}(e^{-(t-1)} - 1)}{(1 + e^{-(t-1)})^3} = 0,$$

so that

$$e^{-(t-1)}(e^{-(t-1)} - 1) = 0.$$

Thus, either the first or second factor must be zero. Since the first factor can never be zero (why?) it follows that the second factor must be zero:

$$e^{-(t-1)} - 1 = 0,$$

or

$$e^{-(t-1)} = 1.$$

Because the unknown is in the exponent, we solve by taking the natural logarithm of both sides.

$$\ln(e^{-(t-1)}) = \ln(1)$$

That is,

$$-(t - 1)\ln(e) = 0,$$

or

$$-(t - 1) = 0.$$

Thus, $t = 1$ is the only solution, so the (only) point of inflection is the point on the curve where $t = 1$. By substituting into the equation for Q, we obtain the Q-coordinate.

$$Q = \frac{1}{1 + e^{-(1-1)}} = \frac{1}{1 + 1} = \frac{1}{2}.$$

Thus the point of inflection occurs at $(1, \frac{1}{2})$. What does this tell us about the sales? First, since Q represents total sales, the derivative of Q represents sales per month. Thus, dQ/dt measures *how fast the new product is selling*—a measure of the *demand* for the product. (We have often called this quantity q.)

If you look at the graph in Figure 10, you will notice that the derivative (slope) increases as t goes from 0 to 1 and then begins to decrease. In other words, the rate at which the product is selling increases from $t = 0$ to $t = 1$ and then starts to decrease. Thus, *the rate at which the product is selling reaches a maximum at $t = 1$*. After that, the rate of sales begins to decrease. (It is for this reason that we refer to $t = 1$ as *the point of diminishing returns.*) We can see directly how the rate dQ/dt reaches a maximum at $t = 1$ by plotting it (Figure 11).

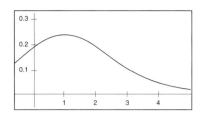

FIGURE 11

Before we go on... Notice something that came up in this discussion: at a point of inflection, the derivative is either a local maximum or a local minimum. In fact, if we were to look for extrema of f', we would do so by setting its derivative, f'', equal to zero—which is exactly what we do to find the points of inflection of f.

If we take the derivative of the second derivative, we get the *third derivative*, which can be written as

$$f'''(x), \quad f^{(3)}(x), \quad \text{or} \quad \frac{d^3y}{dx^3}.$$

Similarly, we can go on differentiating and obtain $f^{(n)}(x)$, or $\frac{d^n y}{dx^n}$, the **n**th **derivative** of f. These derivatives, $f''(x), f'''(x), \ldots, f^{(n)}(x), \ldots$ are referred to as **higher-order derivatives** of f. All of them convey subtle information about the graph of f. (We shall use them when we discuss numerical integration in the next chapter.)

12.3 EXERCISES

Calculate $\frac{d^2y}{dx^2}$ in each of Exercises 1–10.

1. $y = 3x^2 - 6$
2. $y = -x^2 + x$
3. $y = \frac{2}{x}$
4. $y = -\frac{2}{x^2}$
5. $y = 4x^{0.4} - x$
6. $y = 0.2x^{-0.1}$
7. $y = e^{-(x-1)} - x$
8. $y = e^{-x} + e^x$
9. $y = \frac{1}{x} - \ln x$
10. $y = x^{-2} + \ln x$

In Exercises 11–16, the position s of a point (in feet) is given as a function of time t (in seconds). Find **(a)** *its acceleration as a function of t, and* **(b)** *its acceleration at the specified time.*

11. $s = 12 + 3t - 16t^2; \quad t = 2$
12. $s = -12 + t - 16t^2; \quad t = 2$
13. $s = \frac{1}{t} + \frac{1}{t^2}; \quad t = 1$
14. $s = \frac{1}{t} - \frac{1}{t^2}; \quad t = 2$
15. $s = \sqrt{t} + t^2; \quad t = 4$
16. $s = 2\sqrt{t} + t^3; \quad t = 1$

Find the approximate coordinates of all points of inflection (if any) in each of Exercises 17–24.

17.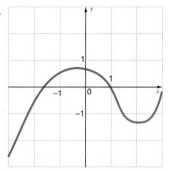

18.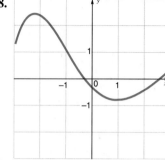

19.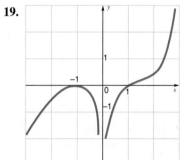

20.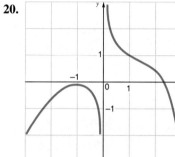

21.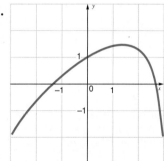

22.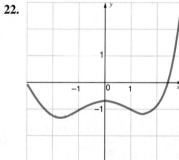

23.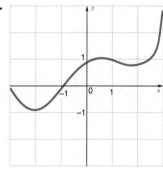

24.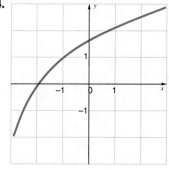

In Exercises 25–36, use the second derivative test to classify critical points where possible. Also, locate all points of inflection.

25. $f(x) = x^2 + 2x + 1$
26. $f(x) = -x^2 - 2x - 1$
27. $f(x) = 2x^3 + 3x^2 - 12x + 1$
28. $f(x) = 4x^3 + 3x^2 + 2$
29. $f(x) = -4x^3 - 3x^2 + 1$
30. $f(x) = -2x^3 - 3x^2 + 12x + 1$
31. $g(x) = (x - 3)\sqrt{x}$
32. $g(x) = (x + 3)\sqrt{x}$
33. $f(x) = x - \ln x$
34. $f(x) = x - \ln(x^2)$
35. $f(x) = x^2 + \ln x^2$
36. $f(x) = 2x^2 \ln x$

In Exercises 37–44, use a graphing calculator to plot the graph of f'' and hence find the approximate coordinates of all the points of inflection (if any) of the given function f. (All coordinates should be correct to two decimal places.)

37. $f(x) = x^3 - 2.1x^2 + 4.3x$
38. $f(x) = 2x^3 + 4.2x^2 - 5.2x$
39. $f(x) = e^{-x^2}$
40. $f(x) = x + e^{-2x^2}$
41. $f(x) = x^4 - 2x^3 + x^2 - 2x + 1$
42. $f(x) = x^4 + x^3 + x^2 + x + 1$
43. $f(x) = x^2 - \ln x$
44. $f(x) = x^2 + \ln x$

APPLICATIONS

45. *Epidemics* The following graph shows the total number, n, of people (in millions) infected in an epidemic as a function of time t (in years).

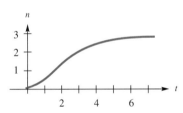

(a) When, to the nearest year, was the rate of infection largest?
(b) When could the Centers for Disease Control announce that the rate of infection was beginning to drop?

46. *Sales* The following graph shows the total number of Pomegranate II computers sold since their release.

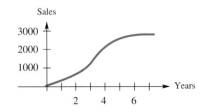

(a) When were the computers selling fastest?
(b) Explain why this graph might look as it does.

47. *Industrial Output* The following graph shows the yearly industrial output of a developing country (mesured in billions of dollars) over a seven-year period.

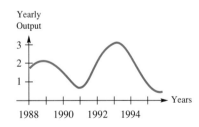

(a) When (to the nearest year) did the rate of change of yearly industrial output reach a maximum?
(b) When (to the nearest year) did the rate of change of yearly industrial output reach a minimum?
(c) When (to the nearest year) did the rate of change of yearly industrial output first start to increase?

48. *Profits* The following graph shows the yearly profits of Gigantic Conglomerate, Inc. (GCI), from 1980 to 1995.

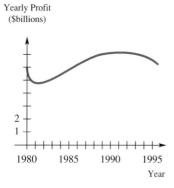

(a) When were the profits rising most rapidly?
(b) When were the profits falling most rapidly?
(c) When could GCI's board of directors legitimately tell stockholders that they had "turned the company around"?

49. *Prison Population* The prison population of the United States followed the curve
$$N(t) = 0.028234t^3 - 1.0922t^2 + 13.029t + 146.88 \quad (0 \le t \le 39)$$
in the years 1950–1989. Here t is the number of years since 1950 and N is the number of prisoners in thousands.* Locate all points of inflection on the graph of N, and interpret the result.

50. *Test Scores* Combined SAT scores in the U.S. can be approximated by
$$T(t) = -0.01085t^3 + 0.5804t^2 - 10.12t + 962.4 \quad (0 \le t \le 22)$$
in the years 1967–1991. Here t is the number of years since 1967 and T is the combined SAT score average for the U.S.† Locate all points of inflection on the graph of T, and interpret the result.

51. *Education and Crime* The following graph shows a striking relationship between the total prison population and the average combined SAT score in the U.S.

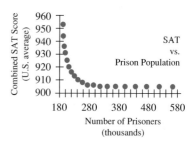

These data can be accurately modeled by
$$S(n) = 904 + \frac{1326}{(n-180)^{1.325}} \quad (192 \le n \le 563).$$
Here, $S(n)$ is the combined U.S. average SAT score at a time when the total U.S. prison population is n thousand.‡
(a) Are there any points of inflection on the graph of S?
(b) What does the concavity of the graph of S tell you about prisons and SAT scores?

52. *Education and Crime* Referring to the model in the previous exercise,
(a) Are there any points of inflection on the graph of S'?
(b) When is S'' a maximum? Interpret your answer in terms of prisoners and SAT scores.

53. *Patents* In 1965, the economist F.M. Scherer modeled the number, n, of patents produced by a firm as a function of the size, s, of the firm (measured in annual sales in millions of dollars). He came up with the following equation based on a study of 448 large firms:§
$$n = -3.79 + 144.42s - 23.86s^2 + 1.457s^3.$$

▼ *The model is the authors'. Source for data: *Sourcebook of Criminal Justice Statistics*, 1990, p. 604.

† The model is the authors'. Source for data: Educational Testing Service.

‡ The model is the authors' based on data for the years 1967–1989. Sources: *Sourcebook of Criminal Justice Statistics*, 1990, p. 604; Educational Testing Service.

§ Source: F. M. Scherer, "Firm Size, Market Structure, Opportunity, and the Output of Patented Inventions," *American Economic Review* 55 (December 1965): 1097–1125.

(a) Find $\dfrac{d^2n}{ds^2}$ and evaluate it at $s = 3$. Is the rate at which patents are produced as the size of a firm goes up increasing or decreasing with size when $s = 3$? Comment on Scherer's words, " we find diminishing returns dominating."

(b) Find $\dfrac{d^2n}{ds^2}\bigg|_{s=7}$ and interpret the answer.

(c) Find the s-coordinate of any points of inflection and interpret the result.

54. Returns on Investments A company finds that the number of new products it develops per year depends on the size of its annual R&D budget, x (in thousands of dollars), according to the formula

$$n(x) = -1 + 8x + 2x^2 - 0.4x^3.$$

(a) Find $n''(1)$ and $n''(3)$, and interpret the results.

(b) Find the size of the budget that gives the largest rate of returns as measured in new products per dollar. (Again called the point of diminishing returns.)

55. Modeling Demand Your marketing group is launching a new 900 telephone service that supplies callers with the correct spelling of any word. You anticipate that the number of calls per day will initially be increasing at a rate of 10 new calls per day and that this rate will drop by 2 calls per day each day. Model this by an equation of the form

$$n = at + bt^2,$$

where n is the total number of phone calls you anticipate, t is time in days and a and b are constants you must determine. [*Hint:* The given information tells you something about $n'(0)$ and $n''(t)$.]

56. Modeling Cost You would like to construct a cost equation for your small tie-dye operation, so you decide on a general cubic equation of the form

$$C = a + bx + cx^2 + dx^3,$$

where C is the daily cost of producing x tie-dye T-shirts. Your daily overheads are $200. The marginal cost at a production level of zero T-shirts is $4.00 per shirt and is decreasing at a rate of 60¢ per T-shirt. The marginal cost reaches a minimum at a production level of 10 T-shirts. What is your cost equation?

57. Modeling Revenue As consultant to a medical research company, you would like to model the anticipated sales of its new antiviral drug Virastat using a logistic equation of the form

$$R = \dfrac{a}{1 + be^{-kt}},$$

where R is the total revenue from sales of Virastat in millions of dollars, t is time (in months), and a, b, and k are constants that you will need to determine. You have the following information to work with. First, the company estimates that it can sell a total of $10 million worth of the drug in the long term. Next, at the present time ($t = 0$) the total sales revenue amounts to $0.5 million and is growing at a rate of $0.02375 million per month.

(a) Use the given information to find a, b, and k and hence the revenue as a function of time. [*Hint:* The given information tells you something about $\lim\limits_{t \to +\infty} R(t)$, $R(0)$ and $R'(0)$.]

(b) Find $R''(0)$, and interpret the result.

Exercises 58 and 59 require the use of either a graphing calculator or graphing computer software.

58. Asset Appreciation You manage a small antique store that owns a collection of Louis XVI jewelry boxes. Their value v is increasing according to the formula

$$v = \dfrac{10000}{1 + 500e^{-0.5t}},$$

where t is the number of years from now. You anticipate an inflation rate of 5% per year, so that the present value of an item that will be worth $$v$ in t years' time is given by

$$p = v(1.05)^{-t}.$$

(a) Graph p as a function of t with $0 \le t \le 40$, $0 \le p \le 6{,}000$ and determine the approximate values of t for all points of inflection (to within ± 2 years).

(b) Now calculate dp/dt, graph it, and hence obtain more accurate estimates of the locations of the points of inflection (to the nearest year).

(c) What is the largest rate of increase of the value of your antiques, and when is this rate attained?

59. *Harvesting Forests* You are considering harvesting a stand of Douglas fir trees. The following equation models the approximate volume in cubic feet of a typical Douglas fir tree of age t years.*

$$V = \frac{22{,}514}{1 + 22{,}514 t^{-2.55}}$$

The lumber will be sold at $10 per cubic foot, and you do not expect the price of lumber to appreciate in the foreseeable future. On the other hand, you anticipate a general inflation rate of 5% per year, so that the present value of an item that will be worth v in t years' time is given by

$$p = v(1.05)^{-t}$$

(a) Graph p as a function of t using the following ranges: $0 \leq t \leq 80$, $0 \leq p \leq 1{,}500$ and determine the approximate values of t for all points of inflection (to within ± 5 years).

(b) Now calculate dp/dt, graph it, and hence obtain more accurate estimates of the locations of the points of inflection (to the nearest year).

(c) What is the largest rate of increase of the value of a fir tree, and when is this rate attained?

COMMUNICATION AND REASONING EXERCISES

60. Complete the following sentence. If the graph of a function is concave up on its entire domain, then its _____ derivative is never _____.

61. Regarding position, s, as a function of time, t, what is the significance of the *third* derivative, $s'''(t)$? Describe an everyday scenario in which it arises.

62. Explain geometrically why the derivative of a function has a local extremum at a point of inflection. Which points of inflection give rise to local maxima in the derivative?

12.4 CURVE SKETCHING

In Chapters 1 and 2 we sketched several curves by plotting points. Plotting points is precisely how most graph-generating software works: the computer or graphing calculator plots hundreds of points and connects them (usually with straight lines), creating the illusion of a smooth curve. There are, however, some drawbacks to this approach. First, it may be difficult to locate the local extrema simply by looking at a curve, but we know already how useful it is to know where they are located. Second, there are many other interesting features you might miss if you were plotting points by hand, such as points of inflection and behavior near points where the function is not defined.

In this section, we use our knowledge about derivatives and limits to analyze a variety of curves. First, we discuss a technique for sketching the graph of a function by hand. Later, we show how these techniques can be used to analyze a curve plotted on a graphing calculator or by computer graphing software. (You do not require graphing technology to benefit from this discussion.)

▼ *The model is the authors' and is based on data in *Environmental and Natural Resource Economics,* Third Edition, by Tom Tietenberg (New York: HarperCollins, 1992), p. 282.

12.4 Curve Sketching

GRAPHING FUNCTIONS BY HAND

Consider the curve $y = 2x - 1 + 2/(2x - 1)$. If we plot the points corresponding to the x-values $-3, -2, -1, 0, 1, 2, 3$, we get the points in the xy-plane shown in Figure 1.

This seems to suggest a curve like the one drawn in Figure 2. But this is not the right graph at all! The actual graph is shown in Figure 3. In other words, we completely missed two local extrema, as well as the fact that the curve breaks into two pieces.

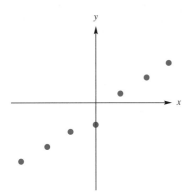

FIGURE 1

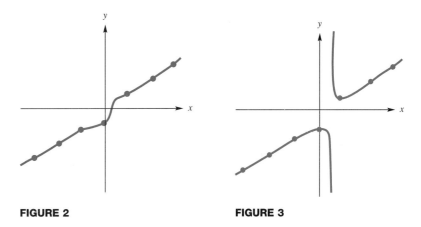

FIGURE 2 **FIGURE 3**

We now describe a simple six-step strategy that enables us to sketch virtually any curve that we encounter and to pinpoint all of its essential features. Most of these steps will not be new: we already know how to locate local extrema and points of inflection. Further, we already know from Chapter 1 what to do near points where the function is not defined. We will simply combine all these techniques and add a few more. Bear in mind the following guiding principle: the graph of a mathematical function is always elegant, so drawing a graph is an artistic exercise as well as a mathematical one.

▼ **EXAMPLE 1**

Sketch the graph of $f(x) = (x - 3)\sqrt{x}$.

SOLUTION Here is our six-step approach:

Step 1: Find the domain of the function if it is not supplied. Here, the domain of the function is not given, so we take its domain to be as large as possible. The domain will be all real numbers unless

(1) there is an expression that can be zero in a denominator, or

(2) there is a square root (or some other even root) of a quantity that can be negative.

Our function has a square root. We know that the function won't be defined where the quantity under the square root sign—namely x—is nega-

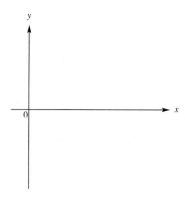

FIGURE 4

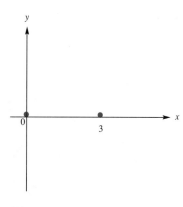

FIGURE 5

tive. Thus, x must be greater than or equal to zero for our function to make sense. So the domain is $[0, +\infty)$.

Because we are concerned only with values of $x \geq 0$, we can place the y-axis on the left of our drawing, as we shall draw nothing to the left of it (see Figure 4). (Let us not get impatient. We are quite aware that there is nothing on the graph yet!)

Step 2: Find the x- and y-intercepts. The intercepts are the points where the curve touches or crosses the two axes. To find these points, we first replace $f(x)$ with y, getting

$$y = (x - 3)\sqrt{x}.$$

To find the y-intercept, we substitute $x = 0$ (points on the y-axis have $x = 0$), which gives us

$$y = (0 - 3)0 = 0.$$

Thus, the y-intercept is 0, so we can mark the point 0 on the y-axis. Similarly, we find the x-intercept by setting $y = 0$ and solving for x.

$$0 = (x - 3)\sqrt{x}$$

From this equation we see that either $x = 3$ or $x = 0$. Thus, there are two x-intercepts, 0 and 3. We plot these points.* We now have the beginnings of a graph (Figure 5).

Step 3: Locate and classify the local extrema, sketching them in. We proceed as we have many times before.

Stationary Points If we take the derivative of the function as it stands, we shall have to use the product rule. Instead, we can rewrite the radical in exponent form and multiply out.

$$y = (x - 3)\sqrt{x} = (x - 3)x^{1/2} = x^{3/2} - 3x^{1/2}$$

Thus,

$$\frac{dy}{dx} = \frac{3}{2}x^{1/2} - \frac{3}{2}x^{-1/2}.$$

We set

$$\frac{3}{2}x^{1/2} - \frac{3}{2}x^{-1/2} = 0.$$

To solve for x, we rewrite the equation with no negative exponents.

$$\frac{3x^{1/2}}{2} - \frac{3}{2x^{1/2}} = 0$$

▼ * In some examples it may be extremely difficult or even impossible to solve for the x-intercepts. On these occasions we either skip this step, or use graphing technology to help us.

12.4 Curve Sketching

We now clear the denominators by multiplying by $2x^{1/2}$, getting

$$3x - 3 = 0, \quad \text{so } x = 1.$$

Singular Points There is a singular point at $x = 0$, because the derivative is not defined there (why not?).

Endpoints $x = 0$ is an endpoint of the domain.

We set up our table, including the intercepts, which make convenient test points.

We can now sketch in the local extrema, getting Figure 6. (Remember to pay special attention to the singular point: the tangent there is vertical.)

Q We seem to have the whole curve already after only three steps! Are we done?

A No. There may still be one or two hidden features, such as points of inflection. Also, we are not yet sure what happens towards the extreme right, as x goes off to infinity.

Step 4: Locate points of inflection, if any. We take $f''(x)$ and set it equal to zero.

$$\frac{dy}{dx} = \frac{3}{2}x^{1/2} - \frac{3}{2}x^{-1/2}$$

$$\frac{d^2y}{dx^2} = \frac{3}{4}x^{-1/2} + \frac{3}{4}x^{-3/2}$$

Setting the second derivative equal to zero gives

$$\frac{3}{4x^{1/2}} + \frac{3}{2x^{3/2}} = 0.$$

But the quantity on the left is positive, so it cannot be zero. Thus the equation has no solutions. The only other candidate for a point of inflection occurs at $x = 0$ where $f''(x)$ is not defined. But this cannot be a point of inflection. (Why?) We conclude that there are no points of inflection. (Our sketch so far suggests this.)

Step 5: Show behavior near points where the function is not defined. The points we have in mind are points where the function is not defined, but where the function is defined just to the left or right. There are no such points in this example, but there is one in the next example.

Step 6: Show behavior as $x \to +\infty$ and $x \to -\infty$. We should see what happens to the y-coordinate for large positive and negative values of x. In

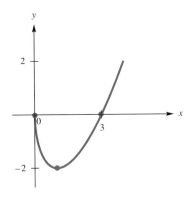

FIGURE 6

other words, *how high does the curve get as we move to the extreme right and left?* We answer this question by taking two limits,

$$\lim_{x \to -\infty} f(x) \quad \text{and} \quad \lim_{x \to +\infty} f(x).$$

In this example, we need only take the limit as $x \to +\infty$ because the domain does not include negative values of x. Now

$$\lim_{x \to +\infty} f(x) = \lim_{x \to +\infty} (x - 3)\sqrt{x}.$$

To calculate this limit, we do a quick mental version of the tabular approach: if x is a very large positive number, both $(x - 3)$ and $\sqrt{x}$ are large positive numbers, and hence, so is their product. Thus,

$$\lim_{x \to +\infty} \sqrt{x}(x - 3) = +\infty.$$

So, as we move farther and farther to the right, the curve gets higher and higher without bound.

Figure 7 shows a picture of an *incorrect* graph.

The graph in Figure 7 seems to be leveling off at about $y = 2$ for large values of x, which would contradict Step 6, which tells us that the graph's height increases without bound. The graph also contradicts Step 4, which says that there are no points of inflection. The curve in Figure 7 seems to have a point of inflection near $x = 4$.

The correct graph was shown in Figure 6.

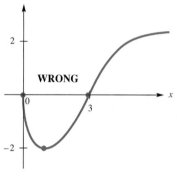

FIGURE 7

To check our work, Figure 8 shows a graphing calculator-generated plot of the curve $y = (x - 3)\sqrt{x}$.

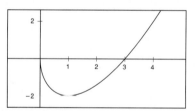

FIGURE 8

Before going on to the next example, we should say that there is nothing wrong with plotting a few additional points if they will help you visualize the curve or confirm your suspicions about its shape. As a rule, however, we'll always try to plot as few as possible.

▼ **EXAMPLE 2**

Sketch the graph of the function $f(x) = \dfrac{1}{x} - \dfrac{1}{x^2}$.

SOLUTION We repeat the six-step process.

12.4 Curve Sketching

Step 1: Find the domain of the function if it is not supplied. Because the domain is not given, we assume the largest possible domain. There is an x in both denominators, so the function is not defined when $x = 0$. Thus, the domain consists of all real numbers except 0, or $(-\infty, 0) \cup (0, +\infty)$.

We can now begin to set up the sketch. Since x is not allowed to be zero, we mark the vertical line $x = 0$ (i.e., the y-axis) as a "forbidden zone"—the curve cannot touch this line. (See Figure 9. In Step 5 we shall see that $x = 0$ is a vertical asymptote.)

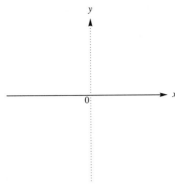

FIGURE 9

Step 2: Find the x- and y-intercepts. To calculate the intercepts, first replace $f(x)$ with y, getting

$$y = \frac{1}{x} - \frac{1}{x^2}.$$

To find the y-intercept, we would have to substitute $x = 0$. But $x = 0$ is not in the domain of f. Thus, there is no y-intercept.

We get the x-intercept(s) by setting $y = 0$ and solving for x.

$$0 = \frac{1}{x} - \frac{1}{x^2}$$

Multiplying by x^2 to clear denominators gives

$$0 = x - 1,$$

so

$$x = 1$$

is the only x-intercept. We mark this point on the x-axis.

Step 3: Locate and classify the local extrema, sketching them in. To find stationary points, we take the derivative and set it equal to zero:

$$f'(x) = -\frac{1}{x^2} + \frac{2}{x^3},$$

so we solve

$$-\frac{1}{x^2} + \frac{2}{x^3} = 0.$$

Multiplying by x^3 to clear denominators gives

$$-x + 2 = 0,$$

so

$$x = 2$$

is our only stationary point. There are no singular points or endpoints. Thus, we need a test point to the right of 2. (We already have one to the left: the x-intercept of 1.)

x	1	2	3
y	0	$\frac{1}{4}$	$\frac{2}{9}$

We can now sketch these points, showing the local maximum at $x = 2$ (Figure 10).

Step 4: Locate points of inflection, if any. To locate the points of inflection, set the second derivative equal to zero and solve for x.

$$f''(x) = \frac{2}{x^3} - \frac{6}{x^4}$$

We thus solve

$$\frac{2}{x^3} - \frac{6}{x^4} = 0.$$

Multiplying by x^4,

$$2x - 6 = 0, \quad \text{so } x = 3.$$

So we have a *possible* point of inflection at $x = 3$ (which happened to be the test point we have already plotted). Now, there may or may not be a point of inflection at $x = 3$. To see that there really is a point of inflection there, we check the concavity to the right of $x = 3$, say, at $x = 4$. We find

$$f''(4) = \frac{2}{64} - \frac{6}{256} > 0,$$

so the curve is concave up to the right of $x = 3$. Because it is concave down to the left, there must be a point of inflection at $x = 3$.

Step 5: Show behavior near points where the function is not defined. Recall that we are interested in points where f is not defined, but is defined just to the left or right. We have such a point: $x = 0$. Thus, we must look at points very close to $x = 0$ on *both* sides and determine what happens to the y-coordinates as x approaches 0. We do so by taking two limits,

$$\lim_{x \to 0^-} f(x) \quad \text{and} \quad \lim_{x \to 0^+} f(x).$$

We can evaluate these limits using the tabular approach. We calculate $f(x)$ for values of x close to 0 (and on either side).

x	$-\frac{1}{100}$	$\frac{1}{100}$
$f(x)$	$-10{,}100$	$-9{,}900$

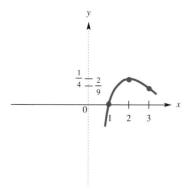

FIGURE 10

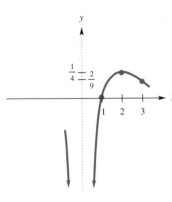

FIGURE 11

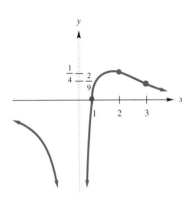

FIGURE 12

This table suggests that

$$\lim_{x \to 0^-} f(x) = -\infty, \text{ and } \lim_{x \to 0^+} f(x) = -\infty.$$

From these limits we see the following.

(1) Immediately to the *left* of $x = 0$, the graph plunges down toward $-\infty$.
(2) Immediately to the *right* of $x = 0$, the graph also plunges down toward $-\infty$.

Thus, the line $x = 0$ is a vertical asymptote. Figure 11 shows our sketch so far.

Step 6: Show behavior as $x \to +\infty$ and $x \to -\infty$. As in the last example, we calculate

$$\lim_{x \to +\infty} f(x) = \lim_{x \to +\infty} \left(\frac{1}{x} - \frac{1}{x^2} \right)$$
$$= 0 - 0 = 0,$$

and

$$\lim_{x \to -\infty} f(x) = \lim_{x \to -\infty} \left(\frac{1}{x} - \frac{1}{x^2} \right)$$
$$= 0 - 0 = 0.$$

Thus, on the extreme left and right of our picture the height of the curve levels off toward zero. Figure 12 shows the completed graph.

Before we go on... Notice that because the curve levels off as $x \to +\infty$, there must be a point of inflection *somewhere* to the right of $x = 2$. In Step 4 we saw that the only place there could be a point of inflection was at $x = 3$. Again, we have verified that there must be a point of inflection at $x = 3$.

Notice another thing: we haven't plotted a single point to the left of the y-axis, and yet we have a pretty good idea of what the curve looks like there!

▼ **EXAMPLE 3**

Sketch the graph of

$$f(x) = \frac{x^2}{(x + 1)(x - 2)}.$$

SOLUTION

Step 1: Find the domain of the function if it is not supplied. Again, the domain is not given to us. Because there is a denominator, we exclude those values of x that make the denominator equal to zero. That is, we exclude both $x = -1$ and $x = 2$. As usual, we begin by sketching in the "forbidden zones": the vertical lines $x = -1$ and $x = 2$ (Figure 13).

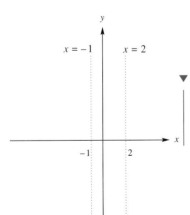

FIGURE 13

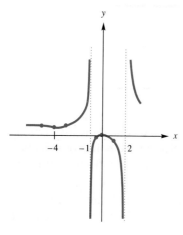

FIGURE 15

We can now sketch the partial graph shown in Figure 15. The lines $x = -1$ and $x = 2$ are vertical asymptotes.

Step 6: Show behavior as $x \to +\infty$ and $x \to -\infty$.
We find

$$\lim_{x \to -\infty} f(x), = \lim_{x \to -\infty} \frac{x^2}{x^2 - x - 2}$$

$$= \lim_{x \to -\infty} \frac{x^2}{x^2} = 1,$$

using the technique of ignoring all but the highest powers of x. Similarly,

$$\lim_{x \to +\infty} f(x) = 1.$$

Thus, on both the extreme left and right, the graph levels off at $x = 1$, so the line $y = 1$ is a horizontal asymptote. The completed graph is shown in Figure 16.

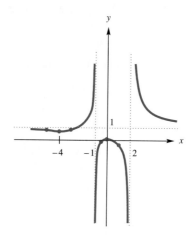

FIGURE 16

From the graph we now discover that there has to be a point of inflection somewhere to the left of the minimum at $x = -4$. Well, if we were forced to find out exactly where it is, we know what we would do. However, instead of trying to find the *exact* location of the point of inflection, let us use technology to locate it numerically.

 To locate the point of inflection, we need to solve $f''(x) = 0$. We can use technology to do this in at least two ways.

Method (a) Graph $f''(x)$ near $x = -5$, and determine where the graph crosses the x-axis.

Method (b) Have your calculator or computer solve $f''(x) = 0$ numerically.

Here is a version of (b) for the TI-82. First enter $f'(x)$ under Y_1 as

$$Y_1 = -(X^2+4X) / (X^2-X-2)^2.$$

Then let Y_2 be its derivative, $f''(x)$.

$$Y_2 = \text{nDeriv}(Y_1, X, X)$$

Now solve $f''(x) = 0$ using the "Solve" command on the Home screen.

$$\text{solve } (Y_2, X, -5) \; \boxed{\text{ENTER}}$$

This tells the calculator to solve the equation $Y_2 = 0$ numerically for x, searching for a solution near $x = -5$. A few seconds later, we find that $x \approx -6.1072$, which is the approximate x-coordinate of the point of inflection. How would you now find the y-coordinate?

SUMMARY: SKETCHING THE GRAPH OF A FUNCTION BY HAND

Step 1: Find the domain of the function if it is not supplied.
If the function is not defined at one or more isolated points, then sketch in vertical dashed lines at those values. These lines break up the curve into disconnected pieces.

Step 2: Find the x- and y-intercepts.
These are the points where the curve touches or crosses the two axes. To find these points, we first replace $f(x)$ with y, getting $y = $ function of x.

y-intercepts: Set $x = 0$ to obtain y.

x-intercepts: Set $y = 0$ and solve for x;

Step 3: Locate and classify the local extrema, sketching them in.
To find the extrema we follow the methods of Section 1 in this chapter.

Step 4: Locate points of inflection, if any.
We set $f''(x) = 0$ and solve for x.

Step 5: Show behavior near points where the function is not defined.
If $x = a$ is such a point, we calculate

$$\lim_{x \to a^-} f(x) \text{ and } \lim_{x \to a^+} f(x).$$

(Remember that $1/small = $ big and $1/big = $ small.)

Step 6: Show behavior as $x \to +\infty$ and $x \to -\infty$.
In other words: *How high does the curve get as we move to the extreme right and left?* To answer, we calculate the limits

$$\lim_{x \to -\infty} f(x) \text{ and } \lim_{x \to +\infty} f(x).$$

ANALYZING THE GRAPH OF A FUNCTION PRODUCED BY A GRAPHING CALCULATOR

If we use a graphing calculator to plot a graph, the actual drawing is done for us, but we still need to analyze its features, such as the exact location of the *x*- and *y*-intercepts, the extrema, and points of inflection. We carry out this analysis in the same way as we did above.

▼ **EXAMPLE 4**

Figure 17 shows a graphing calculator plot of

$$y = 2(x-4)^{2/3} + \frac{x}{2}.$$

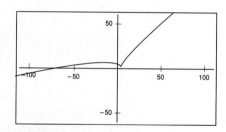

FIGURE 17

(We used the following format:

$$Y_1 = 2((X - 4)\wedge 2) \wedge (1/3) + X/2$$

with the *x*- and *y*-ranges as shown in the figure.)
Analyze this curve. (Round all values to one decimal place.)

SOLUTION First, notice that the domain consists of all real numbers, so we turn to the intercepts and see that there is only a single *x*-intercept. (We can check that there are no intercepts outside the range plotted by enlarging the *x*-range.)

***y*-intercept** Set $x = 0$ in the equation of the curve, obtaining

$$y = 2(-4)^{2/3} + 0 = \sqrt[3]{16} \approx 2.5.$$

***x*-intercept** Set $y = 0$ and solve for *x*.

$$0 = 2(x-4)^{2/3} + \frac{x}{2}$$

If we take the term $x/2$ to the other side of the equation, multiply by 2, and cube both sides, we obtain the cubic equation

$$-x^3 = 64(x-4)^2,$$

or

$$x^3 + 64(x-4)^2 = 0.$$

Unfortunately, there is no easy analytical method for solving cubic equations: the easiest method of solving a general cubic equation is the graphical method—see Chapter 1. Thus, we decide that it is far simpler to find the x-intercept directly from the graph by zooming in for greater accuracy. Figure 18 shows a close-up of the curve.

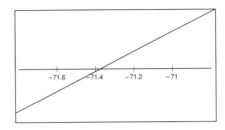

FIGURE 18

We see that the x-intercept occurs at approximately $x = -71.4$ (to the nearest one decimal place).

Extrema We notice in Figure 17 that there appears to be a stationary local maximum at around $x = -20$ and a singular local minimum at around $x = 5$. Analytically,

$$\frac{dy}{dx} = \frac{4}{3}(x-4)^{-1/3} + \frac{1}{2}$$

$$= \frac{4}{3(x-4)^{1/3}} + \frac{1}{2}.$$

Equating this to zero gives

$$\frac{4}{3(x-4)^{1/3}} = -\frac{1}{2}.$$

Cross-multiplying (or clearing the denominators) gives

$$8 = -3(x-4)^{1/3},$$

or

$$(x-4)^{1/3} = -\frac{8}{3}.$$

Cubing both sides,

$$x - 4 = -\left(\frac{8}{3}\right)^3,$$

so

$$x = 4 - \left(\frac{8}{3}\right)^3 \approx -15.0 \text{ (to one decimal place)}.$$

This number is the x-coordinate of the single local maximum. To obtain its y-coordinate we substitute the value of x in the equation of the curve, getting $y \approx 6.7$. Thus, the stationary local maximum occurs at approximately $(-15.0, 6.7)$.

For the singular local minimum, we notice that the derivative is not defined when $x = 4$, so this is where the singular minimum must be. Its y-coordinate (obtained by substituting in the equation of the curve) is 2, so the coordinates of this point are $(4, 2)$.

Points of Inflection Looking at the graph, we suspect that there are no points of inflection, and we can verify this by looking at the second derivative,

$$\frac{d^2y}{dx^2} = -\frac{4}{12}(x-4)^{-4/3} = -\frac{4}{12(x-4)^{4/3}}.$$

The second derivative is always negative (where it is defined), so the graph is always concave down, and there are no points of inflection.

Points Where the Function Is Not Defined There are no such points, since the domain of the function f is the set of all real numbers.

Behavior as $x \to \pm\infty$ Because

$$f(x) = 2(x-4)^{2/3} + \frac{x}{2},$$

a mental version of the tabular method will convince you that, as $x \to +\infty$, $f(x) \to +\infty$ as well (because we are adding larger and larger numbers). Evaluating the limit as $x \to -\infty$ is a little harder because if x is a large negative number, then $f(x)$ is the difference of two large numbers. However, because the graph is concave down everywhere, it must be the case that $\lim_{x \to -\infty} f(x) = -\infty$. (Think about it.)

▶ ## 12.4 EXERCISES

Sketch the graphs of the functions in Exercises 1–26 by hand, analyzing all important features.

1. $f(x) = x^2 - 4x + 1$ with domain $[0, 3]$
2. $f(x) = 2x^2 - 2x + 3$ with domain $[0, 3]$
3. $g(x) = x^3 - 12x$ with domain $[-4, 4]$
4. $g(x) = 2x^3 - 6x$ with domain $[-4, 4]$
5. $f(t) = t^3 + t$ with domain $[-2, 2]$
6. $f(t) = -2t^3 - 3t$ with domain $[-1, 1]$
7. $h(t) = 2t^3 + 3t^2$ with domain $[-2, +\infty)$
8. $h(t) = t^3 - 3t^2$ with domain $[-1, +\infty)$
9. $f(x) = x^4 - 4x^3$ with domain $[-1, +\infty)$
10. $f(x) = 3x^4 - 2x^3$ with domain $[-1, +\infty)$
11. $g(t) = \frac{1}{4}t^4 - \frac{2}{3}t^3 + \frac{1}{2}t^2$ with domain $(-\infty, +\infty)$
12. $g(t) = 3t^4 - 16t^3 + 24t^2 + 1$ with domain $(-\infty, +\infty)$

13. $f(t) = \dfrac{t^2 + 1}{t^2 - 1}$ with $-2 \leq t \leq 2$

14. $f(t) = \dfrac{t^2 - 1}{t^2 + 1}$ with domain $[-2, 2]$

15. $f(x) = (x - 1)\sqrt{x}$

16. $f(x) = (x + 1)\sqrt{x}$

17. $g(x) = x^2 - 4\sqrt{x}$

18. $g(x) = 1/x + 1/x^2$

19. $g(x) = x^3/(x^2 + 3)$

20. $g(x) = x^3/(x^2 - 3)$

21. $f(x) = x - \ln x$ with domain $(0, +\infty)$

22. $f(x) = x - \ln x^2$ with domain $(0, +\infty)$

23. $g(t) = e^t - t$ with domain $[-1, 1]$

24. $g(t) = e^{-t^2}$ with domain $(-\infty, +\infty)$

25. $f(x) = x + \dfrac{1}{x}$

26. $f(x) = x^2 + \dfrac{1}{x^2}$

APPLICATIONS

27. *Asset Appreciation (graphing calculator or computer required)* You are the financial consultant to a classic auto dealership. You estimate that the total value of its collection of 1959 Chevrolets and Fords is given by the formula

$$v = 300{,}000 + 1{,}000t^2,$$

where t is the number of years from now. You anticipate an inflation rate running continuously at 5% per year, so that the discounted (present) value of an item that will be worth $\$v$ in t years' time is given by

$$p = ve^{-0.05t}.$$

Sketch the graph of the discounted value as a function of the time at which the vehicles are sold.
   (a) Calculate the t-coordinates of the extrema analytically, and use a graphing calculator plot of the derivative of the discounted value function to locate the approximate location of points of inflection. What is an appropriate domain for this function?
   (b) Use your graphs [or the information from part (a)] to determine when the value of the collection of classic cars is increasing most rapidly. When is it decreasing most rapidly?
   (c) What is the significance of the asymptote? Is this reasonable?

28. *Plantation Management (graphing calculator or computer required)* The value of a fir tree in your plantation increases with the age of the tree according to the formula

$$v = \dfrac{20t}{1 + 0.05t},$$

where t is the age of the tree in years. Given an inflation rate running continuously at 5% per year, the discounted (present) value of a newly planted seedling is given by

$$p = ve^{-0.05t}.$$

   (a) Sketch the graph of the discounted value of a newly planted seedling as a function of the time at which the tree is harvested. Calculate the t-coordinates of the extrema analytically, and use a graphing calculator plot of the derivative of the discounted value function to locate the approximate location of points of inflection. What is an appropriate domain for this function?
   (b) Use your graphs [or the information from part (a)] to determine when the value of a tree is increasing most rapidly. When is it decreasing most rapidly?
   (c) What is the significance of the asymptote? Is this reasonable?

29. *Average Cost* A cost function for the manufacture of portable CD players is given by

$$C(x) = \$150{,}000 + 20x + \dfrac{x^2}{10{,}000},$$

where x is the number of CD players manufactured. Using an appropriate domain, sketch the graph of the average cost $\bar{C}$ to manufacture x CD players, indicating all absolute extrema and points of inflection (if there are any). What is the significance of the following features?
   (a) absolute minimum
   (b) $\lim\limits_{x \to +\infty} \bar{C}(x)$
   (c) $\lim\limits_{x \to 0^+} \bar{C}$

30. *Average Cost* Repeat Exercise 29 using the revised cost function

$$C(x) = \$150{,}000 + 20x + \dfrac{x^2}{100}.$$

31. *Average Profit* The Feature Software Company sells its graphing program, Dogwood, with a volume discount. If a customer buys x copies, then the customer pays $500\sqrt{x}$. It cost the company \$10,000 to develop the program, and it costs \$2 to manufacture each copy. Assuming that just one customer buys all the copies of Dogwood, sketch the graph of the average profit obtained by selling x copies. On the same set of axes, sketch the graph of the marginal profit function. Locate the point where the two graphs cross each other, and explain its significance.

32. *Average Profit* Repeat Exercise 31 with the charge to the customer being $600\sqrt{x}$ and the cost to develop the program being \$9,000.

33. *Minimizing Resources* Basic Buckets, Inc., has an order for plastic buckets holding 5,000 cubic centimeters. Their buckets are open-topped cylinders. Sketch the graph of the amount of plastic used as a function of the radius. Give a rationale for the behavior of the limits at 0 and infinity. (The volume of an open-topped cylinder with height h and radius r is $\pi r^2 h$, and the surface area is $\pi r^2 + 2\pi rh$.)

34. *Optimizing Capacity* Basic Buckets would like to build a bucket with a surface area of 1,000 square centimeters. Sketch the graph of the volume as a function of the radius. What is an appropriate domain for this function? Why? (See the previous exercise.)

Exercises 35–42 require the use of a graphing calculator or computer software.

35. *Resource Allocation* Your automobile assembly plant has a Cobb-Douglas production function given by

$$q = x^{0.4}y^{0.6},$$

where q is the number of automobiles it produces per year, x is the number of employees, and y is the daily operating budget (in dollars). Annual operating costs amount to an average of \$20,000 per employee plus the operating budget of \$365y. Assume that you wish to produce 1,000 automobiles per year. Sketch the graph of the cost C as a function of the number of employees you hire.

(a) How would you use your graph to estimate the incremental cost per employee at an employment level of 50 employees?

(b) By looking at the graph of C, what can you say about $\lim_{x \to +\infty} C'(x)$? Interpret the answer.

36. *Resource Allocation* Repeat Exercise 35 using the production formula

$$q = x^{0.5}y^{0.5}.$$

37. *Patents* In 1965, the economist F. M. Scherer modeled the number, n, of patents produced by a firm as a function of the size, s, of the firm (measured in annual sales of in millions of dollars). He came up with the following equation based on a study of 448 large firms.*

$$n = -3.79 + 144.42s - 23.86s^2 + 1.457s^3$$

Sketch the graph of n for firms of sizes up to \$10 million in sales. Where on the graph is the smallest rate of increase in numbers of patents produced as a firm's size increases?

38. *Returns on Investments* A company finds that the number of new products it develops per year depends on the size of its annual R&D budget, x (in thousands of dollars), according to the formula

$$n(x) = -1 + 8x + 2x^2 - 0.4x^3.$$

Sketch the graph of n for R&D budgets up to \$400,000. Where is the point of diminishing returns on the graph?

The Normal Curve One of the most useful curves in statistics is the so-called normal distribution curve, which models the distributions of data in a wide range of applications. This curve is given by the function

$$p(x) = \frac{1}{\sigma\sqrt{2\pi}}e^{-(x-\mu)^2/2\sigma^2},$$

where $\pi = 3.14159265\ldots$ and μ and σ are constants called the mean and the standard deviation. Exercises 39 and 40, which require the use of a graphing calculator or computer graphing software, illustrate its use.

▼ *Source: F. M. Scherer, "Firm Size, Market Structure, Opportunity, and the Output of Patented Inventions," *American Economic Review* 55 (December 1965): 1097–1125.

39. *Test Scores* Enormous State University's Calculus I test scores are modeled by the normal distribution with $\mu = 72.6$ and $\sigma = 5.2$. The quantity $p(x)$ represents an approximation of the percentage of students who obtained a score of between $x - 0.5$ and $x + 0.5$ on the test.
(a) Graph the function p on a graphing calculator or computer and determine all its features as outlined in the text. (Find the coordinates of points accurate to one decimal place.)
(b) What percentage of students scored between 89.5 and 90.5 on the test?
(c) What is the most likely test score?
(d) What interesting feature of the curve occurs at the values $x = \mu - \sigma$ and $x = \mu + \sigma$?

40. *Consumer Satisfaction* In a survey, consumers were asked to rate a new toothpaste on a scale of 1–10. The resulting data are modeled by a normal distribution with $\mu = 4.5$ and $\sigma = 1.0$. The quantity $p(x)$ represents an approximation of the percentage of consumers who rated the toothpaste with a score of between $x - 0.5$ and $x + 0.5$ on the test.
(a) Graph the function p on a graphing calculator or computer and determine all its features as outlined in the text. (Find the coordinates of points accurate to one decimal place.)
(b) What percentage of consumers rated the product with scores of between 0.5 and 1.5?
(c) What rating gives the maximum value of p?
(d) What interesting feature of the curve occurs at the values $x = \mu - \sigma$ and $x = \mu + \sigma$?

41. *The Sine Function* Check that your graphing calculator is in "radian mode" and then graph the function $y = \sin x$ for $-10 \le x \le 10$ and $-1.5 \le y \le 1.5$. (You can graph this using the "sin" button: $Y = \boxed{\sin} X$.) Determine the approximate coordinates of two maxima, two minima, and two points of inflection. What can you say about $\lim_{x \to \pm\infty} \sin x$? [Note on the sine function: if you rotate a wheel of radius 1 ft centered at the origin and keep track of the y-coordinate of a point on its rim, then the resulting graph of y versus time, t, will have this shape.]

42. *The Tan Function* Check that your graphing calculator is in "radian mode" and then graph the function $y = \tan x$ for $-10 \le x \le 10$ and $-10 \le y \le 10$. (You can graph this using the "tan" button: $Y = \boxed{\tan} X$.) Determine the approximate coordinates of two vertical asymptotes and two points of inflection. What can you say about $\lim_{x \to \pm\infty} \tan x$?

COMMUNICATION AND REASONING EXERCISES

43. How can you use the graph of $y = f'(x)$ to locate points of inflection on the graph of $y = f(x)$?

44. How can you use the graph of $y = f'(x)$ to locate local extrema on the graph of $y = f(x)$?

45. How can you use the graph of $y = f''(x)$ to calculate points of inflection on the graph of $y = f(x)$?

46. How can you use the graph of $y = f''(x)$ to calculate local extrema on the graph of $y = f'(x)$?

In each of Exercises 47 and 48, a graphing calculator plot of $y = f'(x)$ is shown. Use this graph to give a rough sketch of the graph of $y = f(x)$, assuming the graph passes through the origin. Your sketch should show as many of the features discussed in this section as possible.

47.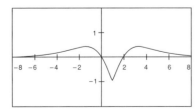

48.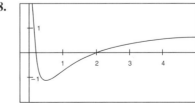

12.5 RELATED RATES

Suppose you are interested in a quantity, Q, that is varying with time. For example, you might be considering the mileage reading on your odometer, the volume of water in a tank, or the height of a space shuttle as it is taking off. Whatever Q stands for, we know that the derivative measures how fast Q is increasing or decreasing. Specifically, if we let t stand for time, then we know the following.

RATE OF CHANGE OF A QUANTITY

$\frac{dQ}{dt}$ is the rate of change of Q.

EXAMPLE 1

If the height (in feet) of a space shuttle is given by the formula

$$h = 3t^2 + 2t,$$

where t is the time in seconds after blast-off, then the rate of ascent of the shuttle at time t seconds is given by

$$\frac{dh}{dt} = 6t + 2 \text{ ft/s}.$$

In particular, after 3 seconds it is ascending at a rate of $6(3) + 2 = 20$ ft/s.

EXAMPLE 2

If the volume (in cm³) of gas in a canister is given by

$$V = e^{2t} + t^2,$$

where t is the time in minutes, then at time t minutes, the rate of change of the volume V is given by

$$\frac{dV}{dt} = 2e^{2t} + 2t \text{ cm}^3/\text{min}.$$

Thus, at the instant $t = 4$ minutes, the volume is increasing at a rate of $2e^8 + 8 \approx 5970$ cm³/min.

In this section, we will be concerned with problems called **related rates** problems. A typical related rates problem is the following.

12.5 Related Rates

The radius of a circle is increasing at a rate of 10 cm/sec. How fast is the area increasing at the instant when the radius has reached 5 cm?

If we look carefully at the problem, we see that there are *two* quantities that are varying with time: the radius, r, and the area, A. Moreover, these two quantities are *related* by the formula

$$A = \pi r^2.$$

The first sentence tells us that r is increasing at a certain rate. When we see a sentence like this referring to speed or rate of change, it is very helpful to rephrase the sentence using the phrase "the rate of change of."

REPHRASING A RELATED RATES PROBLEM

Rephrase all information regarding change using the words "The rate of change of . . . ."

Thus, the first sentence can be rephrased as follows:

The rate of change of r is equal to 10 cm/s.

The sentence can now easily be translated into the equation

$$\frac{dr}{dt} = 10 \text{ cm/s}.$$

The second sentence can be rephrased as follows:

Find the rate of change of A when $r = 5$.

Translating into mathematics, this becomes

$$\text{Find } \frac{dA}{dt} \text{ when } r = 5.$$

Thus, the problem can be read as follows.

$$\frac{dr}{dt} = 10. \text{ Find } \frac{dA}{dt} \text{ when } r = 5.$$

In other words, you are given one rate of change—namely, dr/dt—and are required to find the rate of change dA/dt of the related quantity A. Thus, we call this problem a "related rates" problem.

We shall now describe a systematic way of solving such problems, and we will illustrate our method by means of several examples. First, let's deal with the problem posed above.

EXAMPLE 3

The radius of a circle is increasing at a rate of 10 cm/s. How fast is the area increasing at the instant when the radius has reached 5 centimeters?

SOLUTION Having done the initial analysis above, we now solve this problem using a *tableau* (that is, a fixed arrangement of all the data in the form of a kind of "spreadsheet"). We shall use three headings: The Related Quantities, The Problem, and Solution. We set up the tableau as follows

1. The Related Quantities
Under this heading, we do four things.

  (1) State the quantities that are changing. In this example, they are

   the radius r and the area A.

  (2) Draw a diagram showing these quantities.

▶ **CAUTION** Although the problem asks about $r = 5$, *do not put the 5 in the picture*. If a quantity is changing, it *must* be represented by a *letter* and not a number. Otherwise, you will be tempted to treat the quantity as constant and unchanging. On the other hand, if there was some *fixed* quantity, a quantity that is *not* changing (there is no such quantity here), then feel free to put the fixed number in the diagram. ◀

  (3) Give a formula relating these quantities.

  $$A = \pi r^2$$

  (4) Take the derivative *with respect to time, t,* of both sides. Before we do, notice that d/dt of the left-hand side is just dA/dt. On the right-hand side, we do not simply get $2\pi r$, as this would be the derivative *with respect to r*. Instead, we must use the chain rule.

   The derivative of a quantity squared is twice that quantity times the derivative of that quantity.

Thus, the derivative of the right-hand side is $2\pi r \dfrac{dr}{dt}$.* We get

$$\frac{dA}{dt} = 2\pi r \frac{dr}{dt}.$$

▼ *Recall the generalized power rule: $\dfrac{d}{dx}[u^n] = nu^{n-1}\dfrac{du}{dx}$. Here, instead of x, the variable is t and the role of u is played by r.

Notice that whereas the formula $A = \pi r^2$ gives the relationship between the varying quantities, the derived formula gives the relationship between their *rates of change*.

2. The Problem

Under this heading, we go through the whole problem and restate it in terms of the quantities and their rates of change. Remember to rephrase all statements regarding changing quantities using the phrase "the rate of change of . . . ." We repeat the rephrasings we found in the discussion above.

The first sentence can be rephrased as follows:

The rate of increase of r is equal to 10 cm/s.

In mathematical terms,

$$\frac{dr}{dt} = 10.$$

The second sentence can be rephrased as follows:

Find the rate of increase of A when $r = 5$.

In mathematical terms,

$$\text{Find } \frac{dA}{dt} \text{ when } r = 5.$$

Thus, the whole problem can be stated as follows.

$$\frac{dr}{dt} = 10. \text{ Find } \frac{dA}{dt} \text{ when } r = 5.$$

3. Solution

First copy the relationship between the rates of increase obtained under the first heading.

$$\frac{dA}{dt} = 2\pi r \frac{dr}{dt}$$

Now substitute the data given under heading 2, and solve for the unknown quantity $\frac{dA}{dt}$.

$$\frac{dA}{dt} = 2\pi(5)(10) = 100\pi \approx 314.15 \text{ cm}^2/\text{s}$$

(Note that the units, centimeters and seconds, are specified in the problem.)

Thus, the answer is that the area is increasing at a rate of approximately 314.15 cm²/s.

That's basically all there is to it! (There will sometimes be need for a little more work, as we'll see in the next few examples.)

This is what your tableau in Example 3 should look like when you are done:

FINISHED TABLEAU

1. The Related Quantities

  (1) Changing quantities:

  the radius r and the area A.

  (2) Diagram:

  (3) Formula relating the changing quantities: $A = \pi r^2$
  (4) Derived formula:

$$\frac{dA}{dt} = 2\pi r \frac{dr}{dt}$$

2. The Problem

$$\frac{dr}{dt} = 10. \text{ Find } \frac{dA}{dt} \text{ when } r = 5.$$

3. Solution

$$\frac{dA}{dt} = 2\pi r \frac{dr}{dt}$$

Substituting from heading 2,

$$\frac{dA}{dt} = 2\pi(5)(10) = 100\pi \approx 314.15 \text{ cm}^2/\text{s}.$$

Thus, the area is increasing at a rate of approximately 314.15 cm²/s.

▼ **EXAMPLE 4** Average Cost

The cost to manufacture x portable pagers is

$$C(x) = \$10{,}000 + 3x + \frac{x^2}{10{,}000}.$$

The production level is currently $x = 5{,}000$ and is increasing by 100 units per day. How is the average cost changing?

SOLUTION

1. The Related Quantities

(1) The changing quantities are the production level, x, and the average cost, $\overline{C}$.

(2) In this example, the changing quantities cannot easily be depicted geometrically.

(3) The formula relating the changing quantities must feature both x and the average cost $\overline{C}$. We are given a formula for the *total* cost, so we can obtain one for the *average* cost by dividing by x.

$$\overline{C} = \frac{aC}{x}$$

So

$$\overline{C} = \frac{10{,}000}{x} + 3 + \frac{x}{10{,}000}.$$

(4) Taking derivatives with respect to t of both sides, we obtain the derived formula

$$\frac{d\overline{C}}{dt} = \left(-\frac{10{,}000}{x^2} + \frac{1}{10{,}000}\right)\frac{dx}{dt}.$$

2. The Problem

We see in the second sentence that $x = 5{,}000$ and is increasing by 100 units per day. We rephrase the second fact as follows:

The rate of change of x is 100 units/day.

In mathematical terms,

$$x = 5{,}000 \text{ and } \frac{dx}{dt} = 100.$$

The last sentence asks how the average cost is changing. In other words, we need to

find the rate of increase of $\overline{C}$.

In mathematical terms,

$$\text{Find } \frac{d\overline{C}}{dt}.$$

Thus, the problem reads

$$x = 5{,}000 \text{ and } \frac{dx}{dt} = 100. \text{ Find } \frac{d\overline{C}}{dt}.$$

2. The Problem

$$\frac{dr}{dt} = 1. \text{ Find } \frac{dn}{dt} \text{ when } r = 20.$$

3. Solution
Derived formula:

$$\frac{dn}{dt} = -\frac{2}{3}\left(\frac{n}{r}\right)\left(\frac{dr}{dt}\right)$$

Substituting from heading 2,

$$\frac{dn}{dt} = -\frac{2}{3}\left(\frac{20}{r}\right)(1).$$

(We can compute r by substituting the known value of n in the original formula:

$$20 = n^{0.6}r^{0.4}$$
$$20 = 20^{0.6}r^{0.4}$$
$$r^{0.4} = \frac{20}{20^{0.6}} = 20^{0.4},$$

giving $r = 20$.)
Thus,

$$\frac{dn}{dt} = -\frac{2}{3}\left(\frac{20}{20}\right)(1) = -\frac{2}{3} \text{ laborer per month}$$

Thus, the company is laying off laborers at a rate of $\frac{2}{3}$ per month, or two every three months.

Before we go on... We can interpret this result as saying that at the current level of production and number of laborers, one robot is as productive as $\frac{2}{3}$ of a laborer, or that 3 robots are as productive as 2 laborers.

▶ 12.5 EXERCISES

Translate the statements in Exercises 1–8 into mathematical terms.

1. The population P is currently 10,000 and growing at a rate of 1,000 per year.

2. There are presently 400 cases of Bangkok flu, and the number is growing by 30 new cases every month.

3. The annual revenue of your tie-dye T-shirt operation is currently $7,000 and growing by 10% each year. How fast are annual sales increasing?

4. A ladder is sliding down a wall so that the distance between the top of the ladder and the floor is decreasing at a rate of 3 ft/s. How fast is the base of the ladder receding from the wall?

5. The price of shoes is rising $5 per year. How fast is the demand changing?

6. Stock prices are rising $1,000 per year. How fast is the value of your portfolio increasing?

7. The average global temperature is 60°F and rising by 0.1°F per decade. How fast are annual sales of Bermuda shorts increasing?

8. The country's population is now 260,000,000 and is increasing by 1,000,000 people per year. How fast is the annual demand for diapers increasing?

APPLICATIONS

9. *Doggie Puddles* The area of a circular doggie puddle is growing at a rate of 12 cm²/s.
   (a) How fast is the radius growing at the instant when it equals 10 cm?
   (b) How fast is the radius growing at the instant when the puddle has an area of 49 cm²?

10. *Doggie Puddles* The radius of a circular doggie puddle is growing at a rate of 5 cm/sec.
    (a) How fast is its area growing at the instant when the radius is 10 cm?
    (b) How fast is the area growing at the instant when it equals 36 cm²?

11. *Sliding Ladders* The base of a 50-foot ladder is being pulled away from a wall at a rate of 10 feet per second. How fast is the top of the ladder sliding down the wall at the instant when the base of the ladder is 30 ft from the wall?

12. *Sliding Ladders* The top of a 5-foot ladder is sliding down a wall at a rate of 10 feet per second. How fast is the base of the ladder sliding away from the wall at the instant when the top of the ladder is 3 feet from the ground?

13. *Demand* Assume that the demand function for tuna in a small coastal town is given by

$$p = \frac{50{,}000}{q^{1.5}},$$

where p is the price (in $) per pound of tuna, and q is the number of pounds of tuna that can be sold at the price, p, in one month. The town's fishery finds that the demand for tuna is currently 900 pounds per month and is increasing at a rate of 100 pounds per month. How fast is the price changing?

14. *Demand* In Chapter 1, we found that the demand equation for rubies at Royal Ruby Retailers (RRR) is given by

$$q = -\frac{4p}{3} + 80,$$

where p is the price RRR charges per ruby and q is the number of rubies it can sell per week at p dollars per ruby. RRR finds that the demand for its rubies is currently 20 rubies per week and is dropping at a rate of one per week. How fast is the price changing?

15. *Demand* Demand for your tie-dyed T-shirts is given by the formula

$$p = 5 + \frac{100}{\sqrt{q}},$$

where p is the price (in dollars) you can charge to sell q T-shirts per month. If you currently sell T-shirts for $15 each, and you raise your price by $2 per month, how fast will the demand drop?

16. *Supply* The number of portable CD players you are prepared to supply to the local retail outlet every week is given by the formula

$$p = 0.1q^2 + 3q,$$

where p is the price it offers. The retail outlet is currently offering you $40 per CD player. If the price it offers decreases at a rate of $10 per week, how will this affect the rate of supply?

17. *Revenue* You can now sell 50 cups of lemonade per week at 30¢ per cup, but demand is dropping at a rate of 5 cups per week each week. Assuming that raising the price does not affect demand, how fast do you have to raise your price if you want to keep your weekly revenue constant?

18. *Revenue* You can now sell 40 cars per month at $20,000 per car, and demand is increasing at a rate of 3 cars per month each month. What is the fastest you could drop your price before your monthly revenue starts to drop?

19. *Production* The automobile assembly plant you manage has a Cobb-Douglas production function given by

$$P = 10x^{0.3}y^{0.7},$$

where P is the number of automobiles it produces per year, x is the number of employees, and y is the daily operating budget (in dollars). You maintain a production level of 1,000 automobiles per year. If you currently employ 150 workers and are hiring new workers at a rate of 10 per year, how fast is your daily operating budget changing?

20. *Production* Referring to the Cobb-Douglas production formula in Exercise 19, assume that you maintain a work force of 200 workers and wish to increase production in order to meet a demand that is increasing by 100 automobiles per year. The current demand is 1,000 automobiles per year. How fast should your daily operating budget be increasing?

21. *Balloons* A spherical party balloon is being inflated by pumping in helium at a rate of 3 cubic feet per minute. How fast is the radius growing at the instant when the radius has reached 1 foot? (The volume of a sphere of radius r is $V = \frac{4}{3}\pi r^3$.)

22. *More Balloons* A rather flimsy spherical balloon is designed to pop at the instant its radius has reached 10 cm. Assuming the balloon is filled with helium at a rate of 10 cm³/s, calculate how fast the diameter is growing at the instant it pops. (The volume of a sphere of radius r is $V = \frac{4}{3}\pi r^3$.)

23. *Movement along a Graph* A point on the graph of $y = 1/x$ is moving along the curve in such a way that its x-coordinate is increasing at a rate of 4 units per second. What is happening to the y-coordinate at the instant the y-coordinate is equal to 2?

24. *Motion around a Circle* A point is moving along the circle $x^2 + (y - 1)^2 = 8$ in such a way that its x-coordinate is decreasing at a rate of 1 unit per second. What is happening to the y-coordinate at the instant when the point has reached $(-2, 3)$?

25. *Ships Sailing Apart* The H.M.S. Dreadnought is 40 miles north of Montauk and steaming due north at 20 mph, while the U.S.S. Mona Lisa is 50 miles east of Montauk and steaming due east at an even 30 mph. How fast is their distance apart increasing?

26. *Near Miss* My aunt and I were approaching the same intersection, she from the south and I from the west. She was traveling at a steady speed of 10 mph, while I was approaching the intersection at 60 mph. At a certain instant in time, I was one-tenth of a mile from the intersection, while she was one-twentieth of a mile from it. How fast were we approaching each other at that instant?

27. *Education* In 1991, the expected income of an individual depended on his or her educational level according to the following formula.

$$I(n) = 2{,}928.8n^3 - 115{,}860n^2 + 1{,}532{,}900n \\ - 6{,}760{,}800 \quad (11.5 \le n \le 15.5)$$

Here, n is the number of school years completed, and $I(n)$ is the individual's expected income.* You have completed 13 years of school and are currently a part-time student. Your schedule is such that you will complete the equivalent of one year of college every three years. Assuming that your salary is linked to the above model, how fast is your income going up? (Round your answer to the nearest $1.)

28. *Education* Referring to the model in the previous exercise, assume that someone has completed 14 years of school and that her income is increasing by $10,000 per month. How much schooling per year is this rate of increase equivalent to?

29. *Employment* An employment research company estimates that the value of a recent MBA graduate to an accounting company is estimated as

$$V = 3e^2 + 5g^3,$$

where V is the value of the graduate, e is the number of years of prior business experience, and g is the graduate school grade point average. A company that currently employs graduates with a 3.0 average wishes to maintain a constant employee value of $V = 200$ but finds that the grade point average of its new employees is dropping at a rate of 0.2 per year. How fast must the experience of its new employees be growing in order to compensate for the decline in grade point average?

▼ *The model is a best-fit cubic based on Table 358, U.S. Department of Education, *Digest of Education Statistics, 1991,* Washington, D.C.: Government Printing Office, 1991.

30. *Grades* * A production formula for a student's performance on a difficult English examination is given by

$$g = 4hx - 0.2h^2 - 10x^2,$$

where g is the grade the student can expect to obtain, h is the number of hours of study for the examination, and x is the student's grade point average. The instructor finds that students' grade point averages have remained constant at 3.0 over the years, and that students currently spend an average of 15 hours studying for the examination. However, scores on the examination are dropping at a rate of 10 points per year. At what rate is the average study time decreasing?

31. *Cones* A right circular conical vessel is being filled with green industrial waste at a rate of 100 cubic meters per second. How fast is the level rising after 200π cubic meters have been poured in? (The cone has height 50 m and radius of 30 m at its brim. The volume of a cone of height h and cross-sectional radius r at its brim is given by $V = \frac{1}{3}\pi r^2 h$.)

32. *More Cones* A circular conical vessel is being filled with ink at a rate of 10 cm³/s. How fast is the level rising after 20 cm³ have been poured in? (The cone has height 50 cm and radius of 20 cm at its brim. The volume of a cone of height h and cross-sectional radius r at its brim is given by $V = \frac{1}{3}\pi r^2 h$.)

33. *Cylinders* The volume of paint in a right cylindrical can is given by $V = 4t^2 - t$, where t is time in seconds and V is the volume in cm³. How fast is the level rising when the height is 2 cm? The can has a height of 4 cm and a radius of 2 cm. (*Hint:* To get h as a function of t, first solve the volume $V = \pi r^2 h$ for h.)

34. *Cylinders* A cylindrical bucket is being filled with paint at a rate of 6 cm³ per minute. How fast is the level rising when the bucket starts to overflow? The bucket has radius 30 cm and height 60 cm.

Education and Crime The following graph shows a striking relationship between the total prison population and the average combined SAT score in the United States.

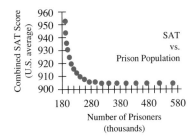

Exercises 35 and 36 are based on the following model for these data:

$$S(n) = 904 + \frac{1326}{(n - 180)^{1.325}} \quad (192 \le n \le 563).$$

Here, $S(n)$ is the combined average SAT score at a time when the total prison population is n thousand.[†]

35. In 1985, the U.S. prison population was 475,000 and increasing at a rate of 35,000 per year. What was the average SAT score, and how fast, and in what direction, was it changing? (Round your answers to two decimal places.)

36. In 1970, the U.S. combined SAT average was 940 and dropping by 10 points per year. What was the U.S. prison population, and how fast, and in what direction, was it changing? (Round your answers to the nearest 100.)

COMMUNICATION AND REASONING EXERCISES

37. If you know how fast one quantity is changing and need to compute how fast a second quantity is changing, what kind of information would you need to know?

38. If three quantities are related by a single equation, how would you go about computing how fast one of them is changing based on a knowledge of the other two?

▼ * Based on an exercise in *Introduction to Mathematical Economics* by A. L. Ostrosky, Jr., and J. V. Koch (Prospect Heights, Ill.: Waveland Press, 1979).

† The model is the authors', based on data for the years 1967–1989. Sources: *Sourcebook of Criminal Justice Statistics, 1990,* p. 604; Educational Testing Service.

39. Why is this section titled "related rates?"

40. In a recent exam, you were given a related rates problem based on an algebraic equation relating two variables x and y. Your friend told you that the correct relationship between dx/dt and dy/dt was given by

$$\left(\frac{dx}{dt}\right) = \left(\frac{dy}{dt}\right)^2.$$

Could he have been correct?

41. Transform the following into a mathematical statement about derivatives: "If my grades are improving at twice the speed of yours, then your grades are improving at half the speed of mine."

42. If two quantities x and y are related by a linear equation, how are their rates of change related?

12.6 ELASTICITY OF DEMAND

Suppose you are manufacturing an extremely popular brand of sneakers, and you are trying to establish how the demand will be affected by an increase in price. Common sense tells you that the demand will decrease as you raise the price. However, it may be the case that the percentage drop in demand is smaller than the percentage increase in price. For example, if you raise the price by 1%, you might suffer only a 0.5% loss in sales. In this case, the loss in sales would be more than offset by the increase in price. Your overall revenue will go up, so a price increase would be in your best interests. In such a situation we say that the demand is **inelastic,** because it is not very sensitive to the increase in price. On the other hand, if your 1% price increase results in a 2% drop in demand, then raising the price will cause a drop in revenues. We would then say that the demand is **elastic** because it is sensitive to a price increase. The wise business executive would do well to *decrease* the price of sneakers in this case.

We can use calculus to predict the response of demand to price changes, provided we have a demand equation for the item we are selling.* To gauge the effect on revenue, we need to look at the percentage drop in demand corresponding to a 1% increase in price. In other words, we want the *percentage drop in demand per percentage increase in price.* This ratio is called the **elasticity of demand**, or **price elasticity of demand**, because it measures the degree to which demand responds to changes in price. The elasticity is usually denoted by the letter E. We will now show you how to get a formula for E starting from a demand equation.

So, let us assume that you are given a demand equation

$$q = f(p),$$

where q stands for the number of items you would sell (per week, per month or what have you) if you set the price per item at p. Now suppose you increase

---

▼ * Coming up with a good demand equation is not always easy. We saw in Chapter 1 that it is possible to find a linear demand equation if we know the sales figures at two different prices. However, such an equation is only a first approximation. To come up with a more accurate demand equation, we might need to gather data corresponding to sales at several different prices and use "curve-fitting" techniques. Another approach would be an analytic one, based on mathematical modeling techniques an economist might use.

the price per item p by a very small amount Δp. Then your percentage increase in price is $\Delta p/p \times 100$ (we multiply by 100 to get a percentage). Now this increase in p will presumably result in a decrease in demand q. Let us denote this corresponding decrease in q by $-\Delta q$ (we use the minus sign because, by convention, Δq stands for the *increase* in demand). Thus, the percentage decrease in demand is $-\Delta q/q \times 100$.

Now recall that E is the ratio

$$E = \frac{\text{Percentage decrease in demand}}{\text{Percentage increase in price}}.$$

Substituting gives

$$E = \frac{\dfrac{-\Delta q}{q} \times 100}{\dfrac{\Delta p}{p} \times 100}.$$

Canceling the 100s and reorganizing, we get

$$E = -\frac{\Delta q}{\Delta p} \cdot \frac{p}{q}.$$

Q This seems fine in theory, but what value should we use for the increase in price, Δp?

Here is an extreme example. Suppose that, delirious with greed, I increase the price of my latest-model sneakers by $1,000,000 per pair. Then I would be very lucky if I sold a single pair! In other words, I should expect the sales to drop to zero. Now this is hardly telling me how the market is going to respond to a modest price increase. In fact, it tells me nothing at all.

A In order to measure the effect most accurately, it seems best to choose the increase in p to be as *small* as possible. The smaller the value of Δp, the more accurately we can gauge the response to a price increase at the current pricing level. In other words, we are interested in the limit as Δp approaches 0.

Now the above equation for E includes the term $\Delta q/\Delta p$, which is precisely the difference quotient associated with the function q of p. Thus, as the value of Δp approaches 0, this quotient approaches the *derivative, dq/dp*, of q with respect to p. It follows that the most *useful* definition of E is the following.

ELASTICITY OF DEMAND

The **elasticity of demand**, E, is the percentage rate of decrease of demand per percentage increase in price. E is given by the formula

$$E = -\frac{dq}{dp} \cdot \frac{p}{q}.$$

▶ **NOTE** We can also think of E another way: as *the percentage rate of increase of demand per percentage drop in price.* ◀

For our first example, we revisit Royal Ruby Retailers.

▼ **EXAMPLE 1** Pricing Policy

The demand for rubies at RRR is given by the equation

$$q = -\frac{4p}{3} + 80,$$

where p is the price RRR charges (in dollars) and q is the number of rubies it sells per week. Find a formula for the elasticity of demand, E, evaluate it at the price level of \$40 per ruby, interpret the answer, and hence determine whether RRR should increase or decrease the price in order to increase revenue.

SOLUTION The first part involves applying the formula

$$E = -\frac{dq}{dp} \cdot \frac{p}{q}.$$

Taking the derivative of q, we see that $\frac{dq}{dp} = -\frac{4}{3}$, so that

$$E = \frac{4p}{3q}.$$

We get E in terms of the price p alone by substituting for q.

$$E = \frac{4p}{3\left(-\frac{4p}{3} + 80\right)}$$

$$= \frac{4p}{240 - 4p} = \frac{p}{60 - p}$$

Now we have the formula for E that we want. When $p = \$40$ per ruby, we get

$$E = \frac{40}{60 - 40} = 2.$$

The percentage drop in demand is twice the percentage increase in price at the current pricing level, so RRR would do well not to increase the price. In fact it should *decrease* the price, because then the percentage *increase* in demand will be twice the percentage decrease in price.

As we said at the start of this section, the demand in the above example is elastic. More precisely:

ELASTIC AND INELASTIC DEMAND AND UNIT ELASTICITY

We say that the demand is
- **elastic** if $E > 1$,
- **inelastic** if $E < 1$, and
- has **unit elasticity** if $E = 1$.

Thus, if the demand is elastic, you can increase revenue by decreasing the price, and if the demand is inelastic you can increase revenue by increasing the price.

Q What about unit elasticity?

A Suppose that you have chosen the unit price exactly right so that the revenue is as large as possible. Then the demand cannot be elastic; otherwise, you would be able to increase the revenue by lowering the unit price. At the same time, the demand cannot be *inelastic*; otherwise, you would be able to increase the revenue by *raising* the unit price. Thus, it must be that the demand is neither elastic nor inelastic. That is, it must have unit elasticity, $E = 1$.

Before we make this more precise, let us apply this reasoning to the previous example.

▼ **EXAMPLE 2** Revenue

Referring to Example 1, find the price RRR should charge in order to maximize revenue.

SOLUTION We are seeking the price level that gives unit elasticity, $E = 1$. We already have an expression for E as a function of price p from the last example,

$$E = \frac{p}{60 - p}.$$

Because we want E to equal 1, we set this expression equal to 1 and obtain

$$1 = \frac{p}{60 - p}.$$

Multiplying by the denominator on the right gives

$$60 - p = p,$$

and so $2p = 60$, or $p = 30$.

Thus we conclude that RRR should sell rubies at $30 apiece in order to maximize revenue. Notice that this is exactly the answer we got in Chapter 1 (when we didn't know any calculus!).

Before going on to the next example, let's look at why the revenue is maximized when $E = 1$. We'll start by treating this as we would any optimization problem. To do this, we must find an expression for the total revenue R in terms of p and then set its derivative with respect to p equal to 0. Recall that

$$\text{Total revenue} = \text{Price per item} \times \text{Total number of items sold}$$

or $\quad R = pq.$

We know that q is a function of p. Thus, R is a function of p although we don't have an explicit formula for this function. We proceed to take the derivative of R with respect to p and set it equal to 0. Because R is a product, we use the product rule.

$$\frac{dR}{dp} = \frac{dq}{dp} p + q \cdot 1 = \frac{dq}{dp} p + q$$

Now, in order to get a maximum revenue, dR/dp must be zero. Thus,

$$\frac{dq}{dp} p + q = 0.$$

Instead of solving for p, we fiddle with the above equation a little. First subtract q from both sides, getting

$$\frac{dq}{dp} p = -q$$

Now divide both sides by $-q$, to get

$$-\frac{dq}{dp} \cdot \frac{p}{q} = 1.$$

Recognize the expression on the left? That's E. Thus, in order to have maximum revenue, we must have $E = 1$, as we claimed.

DETERMINING THE PRICE THAT GIVES MAXIMUM REVENUE

The maximum revenue occurs when $E = 1$. Thus, to determine the price that gives maximum revenue, express elasticity of demand E as a function of p, equate it to 1, and solve for p. (If there is more than one solution for p, choose the solution yielding the largest revenue.)

Q Why does setting $E = 1$ *guarantee* finding the maximum revenue? This seems suspicious in view of the fact that many functions do not have absolute maxima at all.

A Well, we can argue as follows. Setting $p = 0$ always results in zero revenue. On the other hand, in all realistic situations, setting p too high will also result in zero revenue. Thus, there must be *some price p* in between such that the revenue is not zero. One of these intermediate prices will give the *maximum* revenue. Thus, there *is* a value of p that gives a maximum revenue, and setting the derivative of R equal to zero *must* yield that value!* As we saw above, this corresponds to setting $E = 1$.

▼ **EXAMPLE 3** Revenue

Suppose that the demand equation for Bobby Dolls is given by $q = 216 - p^2$, where p is the price per doll in dollars and q represents weekly sales. Find the range of prices for which **(a)** the demand is elastic, **(b)** the demand is inelastic, and **(c)** the weekly revenue is maximized. Also, calculate the maximum weekly revenue.

SOLUTION As in the previous examples, we must first calculate E in terms of p. We have

$$E = -\frac{dq}{dp} \cdot \frac{p}{q}$$

$$= 2p \cdot \frac{p}{216 - p^2}$$

$$= \frac{2p^2}{216 - p^2}.$$

We answer part (c) first. Setting $E = 1$, we get

$$\frac{2p^2}{216 - p^2} = 1,$$

or

$$2p^2 = 216 - p^2,$$

so that

$$3p^2 = 216,$$

or

$$p^2 = 72.$$

Thus, we conclude that the maximum revenue occurs when $p = \sqrt{72} \approx \$8.49$. We can now answer parts (a) and (b) without further calculation: the demand is elastic when $p > \$8.49$ (the price is too high), and the demand is inelastic when $p < \$8.49$ (the price is too low).

Finally, we calculate the maximum weekly revenue, which equals the revenue at the price of $8.49.

$$R = qp = (216 - p^2)p = (216 - 72)\sqrt{72}$$
$$= 144\sqrt{72} \approx \$1,221.88$$

▼ * We are assuming that the demand equation is differentiable (smooth) so that we don't need to worry about singular points or other strange behavior.

12.6 EXERCISES

APPLICATIONS

1. *Demand for Oranges* Given that the weekly sales of Honolulu Red Oranges is given by $q = 1,000 - 20p$, calculate the elasticity of demand for a price of $30 per orange. Interpret your answer and calculate the price that gives a maximum weekly revenue. Also find this maximum revenue.

2. *Demand for Oranges* Repeat Exercise 1 for weekly sales of $1,000 - 10p$.

3. *Tissues* The consumer demand curve for tissues is given by $q = (100 - p)^2$, where p is the price per case of tissues and q is the demand in weekly sales.
   (a) Determine the elasticity of demand E when the price is set at $30, and interpret your answer.
   (b) At what price should tissues be sold in order to maximize the revenue?
   (c) Approximately how many cases of tissues would be demanded at that price?

4. *Bodybuilding* The consumer demand curve for Professor Stefan Schwartzenegger dumbbells is given by $q = (100 - 2p)^2$, where p is the price per dumbbell, and q is the demand in weekly sales. Find the price Professor Schwartzenegger should charge for his dumbbells in order to maximize revenue.

5. *College Tuition* A study of about 1800 colleges and universities in the U.S. resulted in the demand equation $q = 9859.39 - 2.17p$, where q is the enrollment at a college or university, and p is the average annual tuition (plus fees) it charges.*
   (a) The study also found that the average tuition charged by universities and colleges was $2,867. What is the corresponding elasticity of demand? Interpret your answer.
   (b) Based on the study, what would you advise a college to charge its students in order to maximize total revenue, and what would the revenue be?

6. *Demand for Fried Chicken* A fried chicken franchise finds that the demand equation for its new roast chicken product, "Roasted Rooster," is given by

$$p = \frac{40}{q^{1.5}},$$

where p is the price (in dollars) per quarter-chicken serving, and q is the number of quarter-chicken servings that can be sold per hour at this price. Express q as a function of p and find the elasticity of demand when the price is set at $4 per serving. Interpret the result.

7. *Linear Demand Functions* A general linear demand function has the form $f(p) = mp + b$ (m, b constants, $m \neq 0$).
   (a) Obtain a formula for the elasticity of demand at a unit price of p.
   (b) Obtain a formula for the price that maximizes revenue.

8. *Exponential Demand Functions* A general exponential demand function has the form $f(p) = Ae^{-bp}$ (A, b nonzero constants).
   (a) Obtain a formula for the elasticity of demand at a unit price of p.
   (b) Obtain a formula for the price that maximizes revenue.

9. *Hyperbolic Demand Functions* A general hyperbolic demand function has the form $f(p) = \dfrac{k}{p^r}$ (r, k nonzero constants).
   (a) Obtain a formula for the elasticity of demand at unit price p.
   (b) How does E vary with p?
   (c) What does the answer to (b) say about the model?

10. *Quadratic Demand Functions* A general quadratic demand function has the form $f(p) = ap^2 + bp + c$ (a, b, c constants with $a \neq 0$).
    (a) Obtain a formula for the elasticity of demand at a unit price p.
    (b) Obtain a formula for the price or prices that could maximize revenue.

11. *Exponential Demand Functions* The estimated monthly sales of Mona Lisa paint-by-number sets is given by the formula $q = 100e^{-3p^2+p}$, where q is the demand in monthly sales and p is the retail price in ¥.
    (a) Determine the elasticity of demand, E, when the retail price is set at ¥3 and interpret your answer.
    (b) At what price will revenue be a maximum?
    (c) Approximately how many paint-by-number sets will be sold per week at that price?

---

* Based on a study by A. L. Ostrosky, Jr., and J. V. Koch, as cited in their book, *Introduction to Mathematical Economics* (Prospect Heights, Ill.: Waveland Press, 1979), p. 133.

12. *Exponential Demand Functions* Repeat the previous exercise using the demand equation $q = 100e^{p-3p^2 12/p}$.

13. *Modeling Linear Demand (a new look at an old exercise from Chapter 1)* You have been hired as a marketing consultant to Johannesburg Burger Supply, Inc., and you wish to come up with a unit price for its hamburgers in order to maximize the company's weekly revenue. In order to make life as simple as possible, you assume that the demand equation for Johannesburg hamburgers has the linear form $q = mp + b$, where p is the price per hamburger, q is the demand in weekly sales, and m and b are certain constants you'll have to figure out.
    (a) Your market studies reveal the following sales figures: when the price is set at $2.00 per hamburger, the sales amount to 3,000 per week, but when the price is set at $4.00 per hamburger, the sales drop to zero. Use this data to calculate the demand equation.
    (b) Now estimate the unit price in order to maximize weekly revenue and predict what the weekly revenue will be at that price.
    (Compare your answer with the corresponding exercise in Chapter 1, Section 4.)

14. *Modeling Linear Demand* You have been hired as a marketing consultant to Big Book Publishing, Inc., and you have been approached to determine the best selling price for the hit calculus text by Whiner and Istanbul entitled *Fun With Derivatives*. You decide to make life easy and assume that the demand equation for *Fun With Derivatives* has the linear form $q = mp + b$, where p is the price per book, q is the demand in annual sales, and where m and b are certain constants you'll have to figure out.
    (a) Your market studies reveal the following sales figures: when the price is set at $50.00 per book, the sales amount to 10,000 per year; when the price is set at $80.00 per book, the sales drop to 1,000 per year. Use these data to calculate the demand equation.
    (b) Now estimate the unit price in order to maximize annual revenue and predict what Big Book Publishing, Inc.'s annual revenue will be at that price.

15. *Modeling Exponential Demand* As the new owner of a supermarket, you have inherited a large inventory of unsold imported Limburger cheese and would like to set the price so that your revenue from selling it is as large as possible. Previous sales figures of the cheese are shown in the following table:

Price per pound (p)	$3.00	$4.00	$5.00
Monthly sales in pounds (q)	407	287	223

   (a) Use the sales figures for the prices $3 and $5 per pound to construct a demand function of the form $f(p) = Ae^{-bp}$, where A and b are constants you must determine.
   (b) Use your demand function to find the elasticity of demand at each of the prices listed.
   (c) At what price should you sell the cheese in order to maximize monthly revenue?
   (d) If your total inventory of cheese amounts to only 200 pounds, and it will spoil one month from now, how should you price it in order to make the largest revenue? Is this the same answer you got in (c)? If not, give a brief explanation.

16. *Modeling Exponential Demand* Repeat Exercise 15, but this time use the sales figures for $4 and $5 per pound to construct the demand function.

17. *Income Elasticity of Demand (based on a question on the GRE economics test)* If $Q = aP^\alpha Y^\beta$ is the individual's demand function for a commodity, where P is the price of the commodity, Y is the individual's income, and a, α, and β are parameters, explain why β can be interpreted as the **income elasticity of demand**.

18. *College Tuition (from the GRE economics test)* A time-series study of the demand for higher education, using tuition charges as a price variable, yields the following result:
$$\frac{dq}{dp} \cdot \frac{p}{q} = -0.4,$$
where p is tuition and q is the quantity of higher education. Which of the following is suggested by the result?
   (a) As tuition rises, students want to buy a greater quantity of education.
   (b) As a determinant of the demand for higher education, income is more important than price.
   (c) If colleges lowered tuition slightly, their total tuition receipts would increase.
   (d) If colleges raised tuition slightly, their total tuition receipts would increase.
   (e) Colleges cannot increase enrollments by offering larger scholarships.

COMMUNICATION AND REASONING EXERCISES

19. Complete the following sentence. If demand is inelastic, then revenue will decrease if _____.

20. Complete the following sentence. If demand has unit elasticity, the revenue will decrease if _____.

21. Your calculus study group is discussing elasticity of demand, and a member of the group asks the following question. "Since elasticity of demand measures the response of demand to change in unit price, what is the difference between elasticity of demand and the quantity $-dq/dp$?" How would you respond?

22. Another member of your study group claims that unit elasticity of demand need not always correspond to maximum revenue. Is he correct? Explain your answer.

▶ You're the Expert

PRODUCTION LOT SIZE MANAGEMENT

Your publishing company, Knockem Dead Paperbacks, Inc., is planning the production of its latest best-seller, *Henrietta's Heaving Heart* by Celestine A. Lafleur. Sales are projected at 100,000 books per month in the next year. Your job is to coordinate print runs of the book in order to meet the anticipated demand and also minimize total costs to Knockem Dead, Inc.

Each print run has a setup cost of $5,000, each book costs $1 to produce, and monthly storage costs for books awaiting shipment average 1¢ per book. What are you to do?

First you test some scenarios to decide on your strategy. If you decide to print all 1,200,000 books (the estimated demand for the year: 100,000 per month for 12 months) in a single run at the start of the year and sales run as predicted, then the number of books in stock begins at 1,200,000 and decreases to zero by the end of the year, as shown in Figure 1.

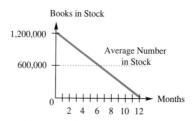

FIGURE 1

On average, you will be storing 600,000 books for 12 months at 1¢ per book, giving a total storage cost of $600,000 \times 12 \times .01 = \$72,000$, while the setup cost for the single print run will be $5,000. When you add to this the total cost of producing 1,200,000 books at $1 per book, your total cost will come out to $1,277,000.

If, on the other hand, you decide to cut down on storage costs by printing the book in two runs of 600,000 each, you will get the picture shown in Figure 2.

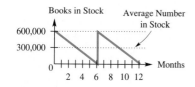

FIGURE 2

As shown in the figure, the storage cost is cut in half, because on average there are only 300,000 books in stock. Thus, the total storage cost is $36,000, while the setup cost has doubled to $10,000 (there are now two runs). The production costs are the same: 1,200,000 books at $1 per book. The total cost is now reduced to $1,246,000, a savings of $31,000 compared to your first scenario.

"Aha!" you say to yourself, "Why not drastically cut costs by setting up a run every month?" You calculate that the setup costs alone would be 12 × $5,000 = $60,000, which is already more than the setup plus storage costs for two runs. Perhaps, then, you should investigate three runs, four runs, and so on, until you reach the lowest cost. This seems a laborious process, especially because you will have to repeat it again when planning for Lafleur's sequel, *Lorenzo's Longing Lips,* due to be released next year. Realizing that this is an optimization problem, you decide to use some calculus to help you come up with a *formula* that you can use for all future plans. So you get to work.

Instead of working with the number 1,200,000, you use the letter N so that you can be as flexible as possible. (What if sales suddenly drop halfway through the year and you have to redo your calculations from scratch?) Thus, you have a total of N books to be produced for the year. You now calculate the total cost of producing them in x production runs (in the first scenario, $x = 1$, while in the second, $x = 2$). Because you are to produce a total of N books in x production runs, you will have to produce N/x books in each of the x runs. N/x is called the **lot size**. As you can see from the diagrams above, the average number of books in storage will be half that amount, $N/(2x)$.

Now you can calculate the total cost for a year. Write P for the setup cost of a single print run ($P =$ $5,000 in your case) and c for the *annual* cost of storing a book (to convert all of the time measurements to years; $c =$ $0.12 here). Finally, write b for the cost of producing a single book ($b =$ $1 here). The cost breakdown is now as follows.

Setup Costs: x print runs @ P dollars per run Px
Storage Costs: $N/(2x)$ books stored @ c dollars per year $cN/(2x)$
Production Costs: N books @ b dollars per book Nb

Total Cost: $Px + \dfrac{cN}{2x} + Nb$

Remember that P, N, c, and b are all constants, while x is the only variable: the number of print runs. Thus, you have the cost function

$$C(x) = Px + \frac{cN}{2x} + Nb,$$

and you are trying to find the value of x that will minimize this total cost. But that's easy! All you need to do is find the local extrema and select the absolute minimum (if any).

First, the domain of the function is $(0, +\infty)$ because there is an x in the denominator, and x can't be negative. Next, you locate the extrema.

Set $C'(x) = 0$ and solve for x:

$$P - \frac{cN}{2x^2} = 0,$$

giving

$$2x^2 = \frac{cN}{P},$$

so

$$x = \sqrt{\frac{cN}{2P}}.$$

You have found only one stationary point. There are no singular points or endpoints. In order to decide whether this represents a local maximum or minimum, you try the second-derivative test.

$$C''(x) = \frac{cN}{x^3}.$$

Because all the numbers (including x) are positive, so is $C''(x)$, so you have a minimum. This also tells you that the whole curve is concave up, and hence that you have an *absolute* minimum.

So now you are practically done! You are absolutely certain that the value of x that gives the lowest total cost is $\sqrt{cN/(2P)}$. You now substitute the numbers to see what this says about *Henrietta's Heaving Heart*.

$$x = \sqrt{\frac{cN}{2P}} = \sqrt{\frac{(0.12)(1{,}200{,}000)}{2(5{,}000)}} \approx 3.79$$

Don't be disappointed that the answer is not a whole number (whole numbers are rarely found in real scenarios). What the answer does indicate is that either 3 or 4 print runs will cost the least money. If you take $x = 3$, you get a total cost of

$$C(3) = (5{,}000)(3) + \frac{(0.12)(1{,}200{,}000)}{(2)(3)} + (1{,}200{,}000)(1)$$
$$= \$1{,}239{,}000,$$

while if you take $x = 4$, you get a total cost of

$$C(4) = (5{,}000)(4) + \frac{(0.12)(1{,}200{,}000)}{(2)(4)} + (1{,}200{,}000)(1)$$
$$= \$1{,}238{,}000.$$

So, four print runs will allow you to minimize your total costs.

Exercises

1. *Lorenzo's Longing Lips* will sell 2,000,000 copies in a year. The remaining costs are the same. How many print runs should you now use?

2. In general, what happens to the number of runs that minimizes cost if both the setup cost and the total number of books are doubled?

3. In general, what happens to the number of runs that minimizes cost if the setup cost increases by a factor of 4?

4. Assuming that the total number of copies and storage costs are as originally stated, find the setup cost that would necessitate a single print run.

5. Assuming that the total number of copies and setup cost are as originally stated, find the storage cost that would necessitate a print run each month.

6. If you look at Figure 2, you will notice that we assumed that all the books in each run were manufactured in a very short time; otherwise the figure might have looked more like Figure 3, which shows the inventory assuming a slower rate of production. How would this affect the answer?

7. Referring to the general situation discussed above, find the cost function (cost as a function of total number of books produced) and average cost function, assuming that the number of runs is chosen to minimize total cost.

8. Let $\bar{C}$ represent the average cost function, calculate $\lim_{N \to +\infty} \bar{C}(N)$, and interpret the result.

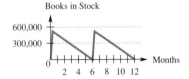

FIGURE 3

Review Exercises

In Exercises 1–16, find all of the local and absolute extrema of the given functions on the given domain (if supplied) or on the largest possible domain (if no domain is supplied).

1. $f(x) = 2x^2 - 2x - 1$ on $[0, 3]$
2. $f(x) = -x^2 - 2x + 3$ on $[0, 3]$
3. $g(x) = 2x^3 - 6x + 1$ on $[-2, \infty)$
4. $g(x) = x^3 - 12x + 1$ on $[-2, \infty)$
5. $g(t) = \frac{1}{4}t^4 + t^3 + t^2$ on $(-\infty, \infty)$
6. $g(t) = 3t^4 + 28t^3 + 72t^2 + 1$ on $(-\infty, \infty)$
7. $f(t) = \frac{t+1}{(t-1)^2}$, $-2 \leq t \leq 2$
8. $f(t) = \frac{t-1}{(t+1)^2}$ on $[-2, 2]$
9. $f(t) = (t-1)^{2/3}$
10. $f(t) = (t^2 + 1)^{2/3}$
11. $g(x) = x - 3x^{1/3}$
12. $g(x) = \frac{1}{x} + \frac{1}{x^2}$
13. $f(r) = \frac{1}{2}r^2 - \ln r$ on $(0, +\infty)$
14. $f(r) = r^2 + \ln r$ on $(0, +\infty)$
15. $g(t) = e^{t^2} + 1$
16. $g(t) = t + e^{-t}$

Exercises 17–32: For these exercises, carefully sketch the graphs of the functions in Exercises 1–16.

17. See Exercise 1.
18. See Exercise 2.
19. See Exercise 3.
20. See Exercise 4.
21. See Exercise 5.
22. See Exercise 6.
23. See Exercise 7.
24. See Exercise 8.
25. See Exercise 9.
26. See Exercise 10.

27. See Exercise 11.
28. See Exercise 12.
29. See Exercise 13.
30. See Exercise 14.
31. See Exercise 15.
32. See Exercise 16.
33. Maximize $P = xy^2$
with $x \geq 0$, $y \geq 0$ and $x^2 + y^2 = 75$.
34. Minimize $P = (1/2)x^2 + y^2$
with $x > 0$, $y > 0$, and $xy^2 = 125$.
35. Minimize $S = 3x + y + z$
with $xy = yz = 9$ and x, y and $z > 0$.
36. Maximize $S = xyz$
with $x^2 + y^2 = x^2 + z^2 = 1$ and $x, y, z > 0$.
37. What are the dimensions of the rectangle with largest area that can be inscribed in the first quadrant of the xy-plane under the curve $y = 1 - x^2$?
38. What are the dimensions of the rectangle with largest area that can be inscribed above the x-axis and under the curve $y = (1 - x^2)^{1/2}$?

APPLICATIONS

39. *Gas Mileage* My Chevy's gas mileage (in miles per gallon) is given as the following function of speed:

$$M(x) = (x/1000 + 1/x)^{-1}$$

(x is the speed in miles per hour, $M(x)$ the gas mileage in miles per gallon). At what speed would you recommend I drive my Chevy in order to maximize fuel economy?

40. *Fast Cars* My Zazna's gas mileage (in miles per gallon) is given as the following function of speed:

$$M(x) = (x/1000 + 4/x)^{-1}$$

(x is the speed in miles per hour, $M(x)$ the gas mileage in miles per gallon). At what speed would you recommend I drive it in order to maximize fuel economy?

41. *Wooden Beams* The strength of a rectangular wooden beam is given by the formula $S = cwt^2$, where c is a constant (depending on the units chosen and the length of the beam), w is its cross-sectional width, and t is its thickness (cross-sectional height). Find the ratio of thickness to width of the strongest beam that can be cut from a circular log. [The equation of a circle of radius r is $x^2 + y^2 = r^2$.]

42. *Wooden Beams* The stiffness of a rectangular wooden beam is given by the formula $S = cwt^3$, where c is a constant (depending on the units chosen and the length of the beam), w is its cross-sectional width, and t is its thickness (cross-sectional height). Find the ratio of thickness to width of the stiffest beam that can be cut from a circular log. [The equation of a circle of radius r is $x^2 + y^2 = r^2$.]

43. *Revenue* The Fancy French Perfume company is deciding on the price of its latest men's cologne, "Deadly." After extensive market research, the company has come up with the demand equation $q = 1,000 - 200p^2 + 20,000p$, where p is the price (in dollars) per 10-oz bottle and q is the number of 10-oz bottles it can sell to a leading department store. What price will bring in the largest revenue?

44. *Profit* The Fancy French Perfume company is also deciding on the price of the after-shave version of its new cologne, "Deadly." Again, at $\$p$ per 10-oz bottle, it can sell $q = 1,000 - 200p^2 + 20,000p$ bottles, but each bottle costs the company $\$10$ to make. What should the price be to bring the largest profit?

45. *Average Cost* The cost to Fullcourt Press of printing q books is

$$C(q) = 10,000 + 20q + \frac{1}{100}q^2.$$

At what production level is the average cost per book the lowest, and what is the least average cost per book?

46. *Average Profit* The demand for Fullcourt Press's latest book is given by

$$q = 100,000 - 2,500p,$$

where q is the number of books it can sell per week at a price of p dollars per book. Using the production cost given in the preceding exercise, find the price that will maximize the average weekly profit.

47. *Hotel Rooms* You have noticed that the occupancy of your 200-room hotel is very sensitive to price increases. If you charge $\$60$ per day for a room, your hotel is usually fully booked, and each $\$10$ increase in the daily fee results in 20 additional vacant rooms. How much should you charge per day in order to maximize your revenue? How many rooms will be vacant at that pricing strategy?

48. *Parking Garage Fees* You manage a parking lot with a 400-car capacity in the theater district and have noticed that you can operate at full capacity if you charge $\$5$ per hour during peak theater hours. Raising the price tends to drive occupancy down, with a loss

of 10 cars for each additional $1 per hour parking fee. How much should you charge to maximize your total revenue, and how many cars will frequent your lot at that price?

49. *Demand* The demand equation for roses at Flower Emporium is

$$9p^2 + 25q^2 = 22{,}500,$$

where q is the number of dozen roses that can be sold per week at p per dozen. If the price is now $30, and Flower Emporium is decreasing the price by $2 per week, how are its sales changing?

50. *Demand* Repeat the preceding exercise, with a demand equation of

$$9p^2 + 16q^2 = 14{,}400.$$

51. *Electronics* Two variable resistances R_1 and R_2 connected in parallel produce a combined resistance R given by the equation

$$\frac{1}{R} = \frac{1}{R_1} + \frac{1}{R_2}.$$

At a certain moment in time, $R_1 = 6$ ohms and is increasing at 2 ohms/s, while $R_2 = 1$ ohm and is decreasing at 1 ohm/s. What is happening to the overall resistance?

52. *Relativity* Einstein's theory of special relativity predicts that the mass m of a particle moving with velocity v is given by

$$m = \frac{m_0}{\sqrt{1 - \frac{v^2}{c^2}}},$$

where m_0 is its rest mass and c is the speed of light ($c \approx 3 \times 10^8$ m./s). If a particle with a rest mass of 100 g is moving at 30% the speed of light and accelerating at 1,000 m/s, how fast is its mass increasing at that instant?

53. *Shadows* A 6-ft-tall man is walking away from a street lamp at 3 ft/s. The street lamp is 15 ft above the ground. How fast is the length of his shadow increasing when he is 10 ft away from the lamp?

54. *Punch Bowls* A hemispherical punch bowl of radius 20 cm is being filled with fruit punch so that its depth is increasing at a rate of 2 cm/s. How fast is its volume increasing at the instant when the depth has reached 10 cm? [The volume of punch at level h is given by $V = \frac{1}{3}\pi h^2(60 - h)$ cm^3.]

55. *Elasticity of Demand* Calculate the elasticity of demand for the demand equation given in Exercise 43 when the price is set at $60 per bottle. Then use the elasticity of demand to show that revenue is maximized when $p = \$66.67$ per bottle.

56. *Elasticity of Demand* Use elasticity of demand to calculate how items should be priced in order to maximize revenue if the demand equation is given by $q = 300 - p$.

57. *Elasticity of Demand* The demand equation for roses at Flower Emporium is

$$9p^2 + 25q^2 = 22{,}500,$$

where q is the number of dozen roses that can be sold per week at p per dozen. Use the elasticity of demand to find the price of roses that maximizes weekly revenue.

58. *Elasticity of Demand* Repeat the preceding exercise, with a demand equation of

$$9p^2 + 16q^2 = 14{,}400.$$

CHAPTER 13

Source: Courtesy Circuit City.

The Integral

APPLICATION ▶ Sunny Electronics Company's chief competitor, Cyberspace Electronics, Inc. has just launched a new home entertainment system that competes directly with Sunny's World Entertainment System. Cyberspace is offering a 2-year limited warranty on its product, so Sunny is thinking of offering a 20-year pro-rated warranty. According to this warranty, if anything goes wrong, Sunny will refund the original $1,000 cost of the system depreciated continuously at an annual rate of 12%. What will it cost Sunny to provide this warranty?

SECTIONS

1. The Indefinite Integral
2. Substitution
3. Applications of the Indefinite Integral
4. Geometric Definition of the Definite Integral
5. Algebraic Definition of the Definite Integral
6. The Fundamental Theorem of Calculus
7. Numerical Integration

You're the Expert
The Cost of Issuing a Warranty

Chapter 13 The Integral

INTRODUCTION ▶ Roughly speaking, calculus is divided into two parts: **differential calculus** (the calculus of derivatives) and **integral calculus,** which is the subject of this chapter and the next. Integral calculus is concerned with problems that are the "reverse" of the problems seen in differential calculus. We begin by studying **antiderivatives,** functions whose derivatives are a given function. The ability to compute antiderivatives allows us to solve many problems in economics, physics, and geometry, including the not-so-obviously related problem of computing areas of complicated regions.

13.1 THE INDEFINITE INTEGRAL

Having studied differentiation in the previous three chapters, we now discuss how to *reverse* the process.

EXAMPLE 1

Given that the derivative of $F(x)$ is $2x$, what is $F(x)$?

SOLUTION After a moment's thought, we recall that the derivative of x^2 is $2x$. Thus, the original function might well have been $F(x) = x^2$. On the other hand, it could have been $F(x) = x^2 + 7$, since the derivative of $x^2 + 7$ is also $2x$. In fact, it could have been $F(x) = x^2 + C$, where C is any constant whatsoever, since the derivative of $x^2 + C$ is $2x$ no matter what the value of C is. Thus, there are *infinitely many answers,* one for each choice of a constant C.

This motivates the following definition.

ANTIDERIVATIVE

By an **antiderivative** of a function $f(x)$, we mean a function $F(x)$ whose derivative is $f(x)$.

EXAMPLE 2

Find an antiderivative of $f(x) = 4x^3 + 1$.

SOLUTION Another way of phrasing the question is this:

$$4x^3 + 1 \text{ is the derivative of what function?}$$

Searching our memories once again, we recall that $4x^3$ by itself is the derivative of x^4, while 1 by itself is the derivative of x. Also, if we add x^4 and x, the derivative of the sum is the sum of the derivatives. So $4x^3 + 1$ is the derivative of $x^4 + x$. In other words, an antiderivative of $4x^3 + 1$ is $F(x) = x^4 + x$.

Before we go on... Notice once again that we could add any constant to this answer and get another answer. For example, $F(x) = x^4 + x - 78.2$ is also an answer. So, any function of the form $F(x) = x^4 + x + C$, where C stands for an arbitrary constant (positive or negative), would be an antiderivative of $4x^3 + 1$.

▼ **EXAMPLE 3**

What possible antiderivatives are there of $f(x) = x$?

SOLUTION The derivative of x^2 is $2x$, which is twice as big as $f(x)$. So we try $\frac{1}{2}x^2$, which has derivative x, as desired. Thus, all the functions represented by $F(x) = \frac{1}{2}x^2 + C$ are antiderivatives of $f(x)$.

Before we go on... In fact, these are *all* the possible antiderivatives of $f(x)$. We'll say more about this in a little while.

We have seen that we get infinitely many antiderivatives of a given function $f(x)$ by adding on an arbitrary constant. We refer to this *collection* of antiderivatives as the **indefinite integral of $f(x)$ with respect to x**. (As with derivatives, the phrase "with respect to x" simply reminds us that the variable is x, and not some other letter.) The term "indefinite" reminds us that there is no single definite answer, but instead a whole *family* of answers, one for each choice of a constant C.

Thus, Example 1 tells us that

the indefinite integral of $2x$ with respect to x is $x^2 + C$.

C will always stand for an arbitrary constant and is usually called the **constant of integration**. Now it would be convenient to use symbols in place of the words "the indefinite integral of $f(x)$ with respect to x" (symbols are both more concise and more precise), and mathematicians have developed such a shorthand: We write the indefinite integral of $f(x)$ with respect to x as $\int f(x)dx$. Here, the symbol $\int$ stands for "the indefinite integral of," and dx stands for "with respect to x."

$$\int \quad f(x) \quad dx$$
$$\uparrow \qquad \uparrow \qquad \uparrow$$
(the indefinite integral of) $f(x)$ (with respect to x)

The function $f(x)$ (of which we are taking the indefinite integral) is called the **integrand.** If this strikes you as a peculiar notation to choose, we shall give some explanation of it in Section 5.

▼ **EXAMPLE 4**

The indefinite integral of $2x$ with respect to x is $x^2 + C$. In symbols,
$$\int 2x \, dx = x^2 + C.$$

Before we go on... In order to check the truth of this statement, remember that it claims that the derivative of $x^2 + C$ is $2x$. In fact,
$$\frac{d}{dx}[x^2 + C] = 2x. \checkmark$$

▼ **EXAMPLE 5**
$$\int [2e^{2x} + 8(x-1)] \, dx = e^{2x} + 4(x-1)^2 + C$$

Before we go on... Again, to check this we compute the derivative of the right-hand side.
$$\frac{d}{dx}[e^{2x} + 4(x-1)^2 + C] = 2e^{2x} + 8(x-1) \checkmark$$

How did we find the antiderivative in the first place? Read on.

Now we would like to make the process of finding indefinite integrals more mechanical. For example, it would be very nice to have a power rule for indefinite integrals similar to the one we already have for derivatives. Let us look at a few powers of x and see if we can notice a pattern.

In Example 3 we saw that the indefinite integral of x is $\frac{1}{2}x^2 + C$, or $x^2/2 + C$. In other words,
$$\int x \, dx = \frac{x^2}{2} + C.$$

What about the indefinite integral of x^2? Let us try a "trial and error" approach. Remember that we are trying to find a function whose derivative is x^2. For a first guess, let us try x^3. Its derivative is $3x^2$, which is three times the answer we want. Thus, we try one-third of that, $\frac{1}{3}x^3$, or $x^3/3$. The derivative of that is exactly x^2, so we have what we wanted,
$$\int x^2 \, dx = \frac{x^3}{3} + C.$$

Now we begin to see a pattern, and we make the following guess.
$$\int x^3 \, dx = \frac{x^4}{4} + C$$

To check that this guess is correct, we take the derivative of the right-hand side, and indeed, we get x^3. Thus, we speculate as follows.

$$\int x^n \, dx = \frac{x^{n+1}}{n+1} + C \quad (n \neq -1)$$

Q Why the restriction on n?

A If we put $n = -1$, the right-hand side of the formula would not make sense because there would be a zero in the denominator.

We'll look more closely at this exception in a moment. First, let us check this formula by taking the derivative of the right-hand side.

$$\frac{d}{dx}\left(\frac{x^{n+1}}{n+1} + C\right) = \frac{(n+1)x^{n+1-1}}{n+1} = x^n$$

Thus, $\frac{x^{n+1}}{n+1} + C$ *is* an antiderivative of x^n, and so the formula works.

Now let us turn to the exceptional case $n = -1$. In this case, we are seeking the indefinite integral of x^{-1}, or $1/x$. Prodding our memories a little, we recall that there is a function whose derivative is $1/x$, namely, $\ln x$. We also know that the derivative of $\ln |x|$ is $1/x$. (This permits x to be negative as well. Put another way, the domain of $\ln |x|$ is the same as the domain of $1/x$, which is not true of $\ln x$.) Thus,

$$\int x^{-1} \, dx = \int \frac{1}{x} \, dx = \ln |x| + C.$$

Summarizing, we have the following.

POWER RULE FOR THE INDEFINITE INTEGRAL

$$\int x^n \, dx = \frac{x^{n+1}}{n+1} + C \quad (n \neq -1)$$

$$\int x^{-1} \, dx = \int \frac{1}{x} \, dx = \ln |x| + C$$

In words, the first formula tells us that *to find the indefinite integral of x^n (if $n \neq -1$) add 1 to the exponent n and then divide by the new exponent.*

Thus, for example,

$$\int \frac{1}{x^{57}} \, dx = \int x^{-57} \, dx = \frac{x^{-56}}{-56} + C = -\frac{x^{-56}}{56} + C$$

$$= -\frac{1}{56x^{56}} + C,$$

since adding one to -57 gives -56 (not -58!).

As another example of this rule, consider the integral $\int 1\,dx$, which is usually written simply as $\int dx$. Since $1 = x^0$, the power rule tells us that

$$\int 1\,dx = \int x^0\,dx = \frac{x^1}{1} + C = x + C.$$

If this seems peculiar, just check that the derivative of $x + C$ is indeed 1.

Here is another antiderivative that is easy to calculate. Since e^x is its own derivative, it is also its own antiderivative. This gives the following rule.

INDEFINITE INTEGRAL OF e^x

$$\int e^x\,dx = e^x + C$$

Q What about more complicated functions, such as $2x^3 + 6x^5 - 1$?

A We need analogs of the theorems on derivatives of sums, differences, and constant multiples.

RULES FOR THE INDEFINITE INTEGRAL

(a) $\int [f(x) \pm g(x)]\,dx = \int f(x)\,dx \pm \int g(x)\,dx$

(b) $\int kf(x)\,dx = k\int f(x)\,dx$ (k constant)

The first rule says what we already saw in Example 2: that the integral of the sum of two functions is the sum of the individual integrals. The same goes for the difference of two functions. In other words, to take the integral of a sum, first take the integrals separately, and then add the answers. The second rule says that to take the integral of a constant times a function, take the integral of the function by itself, and then multiply the answer by that constant. Why are these rules true? For the reason we noticed in Example 2: the derivative of a sum is the sum of the derivatives, and so on.

▼ **EXAMPLE 6**

Find $\int (x^3 + x^5 - 1)\,dx$.

SOLUTION Applying the addition rule (a), we get

$$\int (x^3 + x^5 - 1)\,dx = \int x^3\,dx + \int x^5\,dx - \int 1\,dx$$

$$= \frac{x^4}{4} + \frac{x^6}{6} - x + C.$$

Before we go on... Let us check the answer.

$$\frac{d}{dx}\left(\frac{x^4}{4} + \frac{x^6}{6} - x + C\right) = x^3 + x^5 - 1 \checkmark$$

Q Why is there only a single arbitrary constant C?

A We could have written the answer as $\frac{1}{4}x^4 + D + \frac{1}{6}x^6 + E - x + F$, where D, E, and F are all arbitrary constants. Now suppose that, for example, we set $D = 1$, $E = -2$ and $F = 6$. Then the particular antiderivative we get is $\frac{1}{4}x^4 + \frac{1}{6}x^6 - x + 5$, which has the form $\frac{1}{4}x^4 + \frac{1}{6}x^6 - x + C$. Thus, we could have chosen the single constant C to be 5 and we would have obtained the same answer. In other words, the answer $\frac{1}{4}x^4 + \frac{1}{6}x^6 - x + C$ is just as general as the answer $\frac{1}{4}x^4 + D + \frac{1}{6}x^6 + E - x + F$.

In practice, we don't bother rewriting the integral as several separate integrals but do this in our heads. To put this another way, we just integrate term-by-term, in the same way that we differentiate term-by-term.

▼ **EXAMPLE 7**

Find $\int (4x^3 + 6x^4 - 1)\, dx$.

SOLUTION Applying the addition rule first, we get

$$\int (4x^3 + 6x^4 - 1)\, dx = \int 4x^3\, dx + \int 6x^4\, dx - \int 1\, dx.$$

We now apply the constant multiple rule (b) to the first two terms.

$$= 4\int x^3\, dx + 6\int x^4\, dx - \int 1\, dx$$

$$= 4\frac{x^4}{4} + 6\frac{x^5}{5} - x + C$$

$$= x^4 + \frac{6x^5}{5} - x + C$$

Before we go on... Let us check our answer.

$$\frac{d}{dx}\left(x^4 + \frac{6x^5}{5} - x + C\right) = 4x^3 + 6x^4 - 1 \checkmark$$

Again, we would usually skip the middle steps and write

$$\int (4x^3 + 6x^4 - 1)\, dx = 4\frac{x^4}{4} + 6\frac{x^5}{5} - x + C.$$

It is helpful to think of the constant factors 4 and 6 as "going along for the ride" just as they do when we take derivatives.

EXAMPLE 8

Find $\int \left(\dfrac{x^{3.2}}{3} + \dfrac{1}{x^2} - \dfrac{6}{x} + 7.3e^x \right) dx$.

SOLUTION The integration rules tell us that we can take the indefinite integrals of each term in turn, and add or subtract the answers, as the case may be. First, though, we should rewrite the integrand in exponent form so that we can apply the power rule to the appropriate terms.

$$\int \left(\frac{x^{3.2}}{3} + \frac{1}{x^2} - \frac{6}{x} + 7.3e^x \right) dx = \int \left(\frac{1}{3}x^{3.2} + x^{-2} - 6x^{-1} + 7.3e^x \right) dx$$

$$= \frac{1}{3}\left(\frac{x^{4.2}}{4.2}\right) + \frac{x^{-1}}{(-1)} - 6\ln|x| + 7.3e^x + C$$

$$= \frac{x^{4.2}}{12.6} - \frac{1}{x} - 6\ln|x| + 7.3e^x + C$$

Before we go on... Notice again how the constant factors $\frac{1}{3}$, 6, and 7.3 "went along for the ride."

As usual, let us check our answer.

$$\frac{d}{dx}\left(\frac{x^{4.2}}{12.6} - \frac{1}{x} - 6\ln|x| + 7.3e^x + C \right)$$

$$= \frac{x^{3.2}}{3} + \frac{1}{x^2} - \frac{6}{x} + 7.3e^x \quad \checkmark$$

EXAMPLE 9

Find $\int \left(\dfrac{u^{3.2}}{3} + \dfrac{1}{u^2} - \dfrac{6}{u} + 7.3e^u \right) du$.

SOLUTION This looks very similar to the last example, except for one thing: the letter x has been replaced by the letter u. We are asking for a function *of u* whose derivative *with respect to u* is $u^{3.2}/3 + 1/u^2 - 6/u + 7.3e^u$. There is nothing special about the letter x—we can use u instead if we like. Thus, we get

$$\int \left(\frac{u^{3.2}}{3} + \frac{1}{u^2} - \frac{6}{u} + 7.3e^u \right) du$$

$$= \frac{u^{4.2}}{12.6} - \frac{1}{u} - 6\ln|u| + 7.3e^u + C.$$

Before we go on... In checking our answer, we take the derivative *with respect to u*.

$$\frac{d}{du}\left(\frac{u^{4.2}}{12.6} - \frac{1}{u} - 6\ln|u| + 7.3e^u + C \right)$$

$$= \frac{u^{3.2}}{3} + \frac{1}{u^2} - \frac{6}{u} + 7.3e^u$$

EXAMPLE 10 Cost and Marginal Cost

The marginal cost (in dollars) to produce baseball caps at a production level of x caps is $3.20 - 0.001x$, and it is found that the cost of producing 50 caps is $200. Find the cost function.

SOLUTION We are asked to find the cost function $C(x)$, given that the *marginal* cost function is $3.20 - 0.001x$. Recalling that the marginal cost function is the derivative of the cost function, we have

$$C'(x) = 3.2 - 0.001x,$$

and must find $C(x)$. Now $C(x)$ must be an antiderivative of $C'(x)$, so we write

$$C(x) = \int (3.20 - 0.001x)\, dx$$

$$= 3.20x - 0.001 \frac{x^2}{2} + K \qquad \text{K is the constant of integration.}$$

$$= 3.20x + 0.0005x^2 + K.$$

(Why did we use K and not C for the constant of integration?) Now, unless we know a value for K, we don't really know what the cost function is. However, there is a further piece of information we have ignored: the cost of producing 50 baseball caps is $200. In other words,

$$C(50) = 200.$$

Substituting in our formula for $C(x)$,

$$C(50) = 3.20(50) + 0.0005(50)^2 + K = 200$$

i.e.,

$$161.25 + K = 200$$

so

$$K = 38.75.$$

Now that we know what K is, we can write down the cost function.

$$C(x) = 3.20x + 0.0005x^2 + 38.75$$

Before we go on...

Q What is the significance of the term 38.75?

A If we take $x = 0$, we obtain

$$C(0) = 3.20(0) + 0.0005(0)^2 + 38.75,$$

or

$$C(0) = 38.75.$$

Thus 38.75 is the cost of producing zero items: in other words, the **fixed cost**.

We end this section with a few important ideas.

Let us return for a moment to the beginning of all this, when we said that *all* antiderivatives of $2x$ are of the form $x^2 + C$, for some constant C or the other. If you are as skeptical as we sincerely hope you are, then there should be a nagging doubt at the back of your mind: could there be some mysterious function we've possibly never heard of, *other than* $x^2 + C$, with the property that its derivative is also $2x$? Just because we haven't encountered any such function, it doesn't mean that there isn't one lurking around somewhere. (We might conceivably detect some signal from the Andromeda galaxy telling us about such a function.) Thus, we ought to experience a little discomfort when we claim that *the* indefinite integral of $2x$ is $x^2 + C$.

Well, let us put our minds at ease. We claim that, if $F(x)$ and $G(x)$ are both antiderivatives of $f(x)$, then F and G can differ by at most a constant. In other words, if $F'(x) = G'(x) = f(x)$, then $G(x) = F(x) + C$ for some constant C. To see why, think about what the equation $F'(x) = G'(x)$ says (see Figure 1).

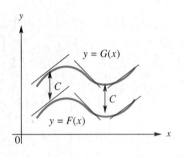

FIGURE 1

If $F'(x) = G'(x)$ for all x, then F and G have the *same slope* at each point. This means that their graphs must be *parallel*, and remain exactly the same vertical distance apart. But that is the same as saying that the functions differ by a constant.*

In summary, if you know one antiderivative $F(x)$ of $f(x)$, then you know them all, since any other antiderivative of f is just $F(x)$ plus some constant. Thus, if we write $F(x) + C$, we can be confident that we have written down all possible antiderivatives of f.

Let us now explore the notation we've developed. Since $\int f(x)\, dx$ means the indefinite integral of $f(x)$, its derivative must be $f(x)$. In other words,

$$\frac{d}{dx}\int f(x)\, dx = f(x).$$

In words, this complicated-looking formula is proclaiming nothing more than the fact that the derivative of a function whose derivative is $f(x)$, is $f(x)$! In other words, *the derivative of the indefinite integral of a function is the function we started with.* For example, if we start with $f(x) = x^3$, take the indefinite integral, which is $\frac{1}{4}x^4 + C$, and then take the derivative, we wind up with what we started with, x^3. This is, after all, how we have been checking our work.

Now, this sounds like we are saying that differentiation and integration are inverse processes: if you start with a function, take its indefinite integral, and then take the derivative of the answer, you will get the function you started with. To be truly inverse, the same thing should happen in reverse, so let us try it. Start with a function $F(x)$, take its derivative, and *then* take the indefinite integral of the answer. For example, if we start with

▼ * This argument can be turned into a more rigorous proof—that is, a proof that does not rely on geometric concepts such as "parallel graphs."

$F(x) = x^2 + 1$, and first take the derivative, we get $2x$. Taking the indefinite integral now gives $x^2 + C$, where we don't know what the constant is anymore. Thus, we lose information by going this route. We don't get the original function $f(x)$ back again, but we get $f(x) + C$. Thus, what we *can* say is:

$$\int \left(\frac{d}{dx} f(x)\right) dx = f(x) + C.$$

This loss of information really takes place when we first take the derivative. For example, the derivative of $x^2 + 1$ is the same as the derivative of $x^2 + 3$. To summarize:

RELATIONSHIP BETWEEN THE INDEFINITE INTEGRAL AND THE DERIVATIVE

$$\frac{d}{dx} \int f(x)\, dx = f(x)$$

$$\int \left(\frac{d}{dx} f(x)\right) dx = f(x) + C$$

13.1 EXERCISES

Evaluate the integrals in Exercises 1–10 mentally.

1. $\int x^5\, dx$
2. $\int x^7\, dx$
3. $\int 6\, dx$
4. $\int (-5)\, dx$
5. $\int x\, dx$
6. $\int (-x)\, dx$
7. $\int (x^2 - x)\, dx$
8. $\int (x + x^3)\, dx$
9. $\int (1 + x)\, dx$
10. $\int (4 - x)\, dx$

Evaluate the integrals in Exercises 11–32.

11. $\int x^{-5}\, dx$
12. $\int x^{-7}\, dx$
13. $\int \left(u^2 - \dfrac{1}{u}\right) du$
14. $\int \left(v^{-2} + \dfrac{2}{v}\right) dv$
15. $\int (3x^4 - 2x^{-2} + x^{-5} + 4)\, dx$
16. $\int (4x^7 - x^{-3} + 1)\, dx$
17. $\int \left(2e^x + \dfrac{5}{x}\right) dx$
18. $\int \left(2e^x + \dfrac{5}{x} + x^{-2}\right) dx$
19. $\int \left(\dfrac{1}{x} + \dfrac{2}{x^2} - \dfrac{1}{x^3}\right) dx$
20. $\int \left(\dfrac{3}{x} - \dfrac{1}{x^5} + \dfrac{1}{x^7}\right) dx$
21. $\int (3x^{0.1} - x^{4.3})\, dx$
22. $\int \left(\dfrac{x^{2.1}}{2} - 2\right) dx$
23. $\int \left(\dfrac{3}{x^{0.1}} - \dfrac{4}{x^{1.1}}\right) dx$
24. $\int \left(\dfrac{1}{x^{1.1}} - \dfrac{1}{x}\right) dx$

25. $\displaystyle\int \frac{x+2}{x^3}\,dx$

26. $\displaystyle\int \frac{x^2-2}{x}\,dx$

27. $\int \sqrt{x}\,dx$

28. $\int \sqrt[3]{x}\,dx$

29. $\displaystyle\int \left(2\sqrt[3]{x} - \frac{1}{2\sqrt{x}}\right)dx$

30. $\displaystyle\int \left(2\sqrt{x} - \frac{1}{2\sqrt{x}}\right)dx$

31. $\displaystyle\int \frac{x - 2\sqrt{x}}{x^2}\,dx$

32. $\displaystyle\int \frac{x^2 - 2\sqrt{x}}{x^3}\,dx$

33. Find $f(x)$ if $f(0) = 1$ and the tangent line at $(x, f(x))$ has slope x.

34. Find $f(x)$ if $f(1) = 1$ and the tangent line at $(x, f(x))$ has slope $\frac{1}{x}$.

35. Find $f(x)$ if $f(0) = 0$ and the tangent line at $(x, f(x))$ has slope $e^x - 1$.

36. Find $f(x)$ if $f(1) = -1$ and the tangent line at $(x, f(x))$ has slope $2e^x + 1$.

APPLICATIONS

37. *Marginal Cost* The marginal cost of producing the xth box of light bulbs is $5 - (x/10{,}000)$, and the fixed cost is \$20,000. Find the total cost function $C(x)$.

38. *Marginal Cost* The marginal cost of producing the xth box of computer disks is $10 + (x^2/100{,}000)$, and the fixed cost is \$100,000. Find the total cost function $C(x)$.

39. *Marginal Cost* The marginal cost of producing the xth roll of film is given by $5 + 2x + \frac{1}{x}$. The total cost to produce one roll is \$1,000. Find the total cost function $C(x)$.

40. *Marginal Cost* The marginal cost of producing the xth box of videotape is given by $10 + x + 1/x^2$. The total cost to produce one hundred boxes is \$10,000. Find the total cost function $C(x)$.

For Exercises 41 and 42, recall that the velocity of a particle moving in a straight line is given by $v = ds/dt$.

41. *Motion in a Straight Line* The velocity of a particle moving in a straight line is given by $v(t) = t^2 + 1$.
(a) Find an expression for the position s after a time t.
(b) Given that $s = 1$ at time $t = 0$, find the constant of integration C, and hence find an expression for s in terms of t without any unknown constants.

42. *Motion in a Straight Line* The velocity of a particle moving in a straight line is given by $v = 3e^t + t$.
(a) Find an expression for the position s after a time t.
(b) Given that $s = 3$ at time $t = 0$, find the constant of integration C, and hence find an expression for s in terms of t without any unknown constants.

COMMUNICATION AND REASONING EXERCISES

43. Give an argument for the rule that the integral of a sum is the sum of the integrals.

44. Is it true that $\displaystyle\int \frac{1}{x^3}\,dx = \ln(x^3) + C$? Give a reason for your answer.

45. Give an example to show that the integral of a product is not the product of the integrals.

46. Give an example to show that the integral of a quotient is not the quotient of the integrals.

47. If x represents the number of items manufactured and $f(x)$ represents the marginal cost per item, what does $\int f(x)\,dx$ represent? In general, how are the units of $f(x)$ and the units of $\int f(x)\,dx$ related?

48. Complete the following: $-1/x$ is an _____ of $1/x^2$, whereas $\ln x^2$ is not. $-1/x + C$ is the _____ of $1/x^2$, because the _____ of $-1/x + C$ is _____.

49. Complete the following sentence. If you take the _____ of the _____ of $f(x)$, you obtain $f(x)$ back. On the other hand, if you take the _____ of the _____ of $f(x)$, you obtain $f(x) + C$.

50. If a Martian told you that the Institute of Alien Mathematics, after a long and difficult search, has announced the discovery of a new antiderivative of $x - 1$ called $M(x)$ (the formula for $M(x)$ is classified information, and so cannot be revealed here) how would you respond?

13.2 SUBSTITUTION

The chain rule for derivatives gives us an extremely useful technique for finding antiderivatives. This technique is called **substitution** or **change of variables.** We'll start with an example to illustrate the mathematics behind the technique of substitution, then discuss how the technique is used in practice.

EXAMPLE 1

Find $\int 2x(x^2 + 1)^{1/2} \, dx$.

SOLUTION The answer is $\frac{2}{3}(x^2 + 1)^{3/2} + C$, and let us see why this is so. We simply need to check the derivative.

$$\frac{d}{dx}\left(\frac{2}{3}(x^2 + 1)^{3/2} + C\right) = \frac{2}{3} \cdot \frac{3}{2}(x^2 + 1)^{1/2} \cdot (2x)$$

$$= 2x(x^2 + 1)^{1/2} \checkmark$$

Before we go on... Notice how we had to use the chain rule to take the derivative above. Let us start again and attempt to compute the integral from scratch. We want to find

$$\int 2x(x^2 + 1)^{1/2} \, dx.$$

Inspired by the chain rule, we let u be the quantity that is raised to the power, $u = x^2 + 1$. Then $du/dx = 2x$. Substituting these in the integral, we get

$$\int \left(\frac{du}{dx} u^{1/2}\right) dx, \text{ or } \int \left(u^{1/2} \frac{du}{dx}\right) dx.$$

Although we are tempted to cancel the dx's, we need to remember that dx is not a real number. The quantity in parentheses reminds us of the chain rule, and in fact

$$\frac{d}{dx}\left(\frac{2}{3} u^{3/2}\right) = u^{1/2} \frac{du}{dx}.$$

This is the same as saying that

$$\int \left(u^{1/2} \frac{du}{dx}\right) dx = \frac{2}{3} u^{3/2} + C.$$

If we now substitute $u = x^2 + 1$ into $\frac{2}{3} u^{3/2} + C$, we get the answer that we gave originally.

Let us reexamine the calculation. We saw that

$$\int \left(u^{1/2} \frac{du}{dx}\right) dx = \frac{2}{3} u^{3/2} + C.$$

On the other hand, we know from the power rule that

$$\int u^{1/2} \, du = \frac{2}{3} u^{3/2} + C$$

also. Therefore,

$$\int \left(u^{1/2} \frac{du}{dx} \right) dx = \int u^{1/2} \, du.$$

There is nothing special about $u^{1/2}$ here. For any function f we can say that

$$\int \left(f(u) \frac{du}{dx} \right) dx = \int f(u) \, du.$$

The point is that $\int f(u) \, du$ may be simpler than the original integral, just as $\int u^{1/2} \, du$ is simpler than $\int 2x(x^2 + 1)^{1/2} \, dx$. Comparing the above pairs of integrals suggests that*

$$\frac{du}{dx} dx = du.$$

We can also write this equation as

$$dx = \frac{1}{du/dx} du.$$

Now let us see how the technique works in practice.

▼ EXAMPLE 2

Calculate $\int x(x^2 + 1)^2 \, dx$.

SOLUTION First, we decide what we are going to take as u. Although there is no rule that always works, the following often works.

Take u to be an expression that is being raised to a power.

In this example, $x^2 + 1$ is being raised to the second power, so let's try setting $u = x^2 + 1$.

The next step is to take the derivative of u with respect to x.

$$\frac{du}{dx} = 2x$$

We can now write the differential equation $du = \frac{du}{dx} dx$, which here is

$$du = 2x \, dx.$$

▼ *In fact, we saw this equation earlier in the section on linear approximation, where we called du and dx **differentials**.

Now we divide both sides by $2x$ to get

$$dx = \frac{1}{2x} du.$$

Let's summarize what we just did.

$u = x^2 + 1$	Decide what to take as u.
$\dfrac{du}{dx} = 2x$	Take the derivative with respect to x.
$du = 2x\, dx$	Write the equation $du = \dfrac{du}{dx} dx$.
$dx = \dfrac{1}{2x} du$	Solve for dx.

Now we are ready for the next step.

Substitute the expression for u in the original integral, and also substitute for dx.

Thus, we get

$$\int x(x^2 + 1)^2\, dx = \int x\, u^2\, \frac{1}{2x}\, du.$$

Next, we do the following.

Simplify the integrand, leaving an integral in u only.

That's easy: the x's cancel.* So we have

$$\int u^2 \frac{1}{2}\, du = \frac{1}{2} \int u^2\, du.$$

Now we have an integral that is easy to calculate:

$$\frac{1}{2} \int u^2\, du = \frac{1}{2}\left(\frac{u^3}{3}\right) + C = \frac{u^3}{6} + C,$$

and we are almost done. We are looking for a function of x, so we substitute $u = x^2 + 1$. Thus, the answer is

$$\frac{(x^2 + 1)^3}{6} + C.$$

▼ * We'll tell you later what to do if they don't.

Before we go on... Here is the solution as we would usually write it.

$$u = x^2 + 1$$
$$\frac{du}{dx} = 2x$$
$$du = 2x\,dx$$
$$dx = \frac{1}{2x}\,du$$

$$\int x(x^2 + 1)^2\,dx = \int x\,u^2\,\frac{1}{2x}\,du \quad \text{Substitute.}$$
$$= \int u^2\,\frac{1}{2}\,du \quad \text{Eliminate } x\text{'s.}$$
$$= \frac{1}{2}\int u^2\,du \quad \text{Simplify integral.}$$
$$= \frac{1}{2}\left(\frac{u^3}{3}\right) + C \quad \text{Evaluate integral.}$$
$$= \frac{u^3}{6} + C$$
$$= \frac{(x^2 + 1)^3}{6} + C \quad \text{Substitute.}$$

We should, as always, check our answer.

$$\frac{d}{dx}\left(\frac{(x^2 + 1)^3}{6} + C\right) = \frac{3(x^2 + 1)^2 \cdot 2x}{6} = x(x^2 + 1)^2 \checkmark$$

▶ **CAUTION**

1. It is important to follow the following steps *in order*:
   First, eliminate *all the x*'s.
   Next, take the antiderivative with respect to *u*.
   Finally, substitute back for *u*.
2. If you can't eliminate *x* or you wind up with a mess, you may have made a bad choice for *u*, so try something else. (See Example 6 in this section for another idea.) ◀

▼ **EXAMPLE 3**

Evaluate $\int 3xe^{x^2}\,dx$.

SOLUTION In the integrand, we have *e* raised to an expression. This is another place where *u*-substitution often works. We take $u = x^2$, the expression to which *e* is being raised.

$$u = x^2$$
$$\frac{du}{dx} = 2x$$
$$du = 2x\,dx$$
$$dx = \frac{1}{2x}\,du$$

$$\int 3xe^{x^2}\,dx = \int 3x\,e^u\,\frac{1}{2x}\,dx$$
$$= \int 3e^u\,\frac{1}{2}\,du$$
$$= \frac{3}{2}\int e^u\,du$$
$$= \frac{3}{2}e^u + C$$
$$= \frac{3}{2}e^{x^2} + C$$

Before we go on... We check our answer.

$$\frac{d}{dx}\left(\frac{3}{2}e^{x^2} + C\right) = \frac{3}{2}\,2x\,e^{x^2} = 3xe^{x^2}\;\checkmark$$

▼ **EXAMPLE 4**

Evaluate $\int (3x + 8)^{-1}\,dx$.

SOLUTION In the integrand, we have the expression $3x + 8$ raised to a power, so we take $u = 3x + 8$.

$$u = 3x + 8$$
$$\frac{du}{dx} = 3$$
$$du = 3\,dx$$
$$dx = \frac{1}{3}\,du$$

$$\int (3x+8)^{-1} dx = \int u^{-1} \frac{1}{3} du$$

$$= \frac{1}{3} \int u^{-1} du$$

$$= \frac{1}{3} \ln|u| + C$$

$$= \frac{1}{3} \ln|3x+8| + C$$

Before we go on... First, we check the answer.

$$\frac{d}{dx}\left(\frac{1}{3}\ln|3x+8| + C\right) = \frac{1}{3} \cdot \frac{1}{3x+8} \cdot 3$$

$$= (3x+8)^{-1} \checkmark$$

Notice something interesting: the answer suggests that

$$\int (ax+b)^{-1} dx = \frac{1}{a} \ln|ax+b| + C$$

for any constants a and b (with $a \neq 0$). In fact, this is easy to see if we simply redo the calculation with $3x + 8$ replaced by $ax + b$. This is a useful result to remember, and we will collect this and similar results at the end of this section.

▼ **EXAMPLE 5**

Evaluate $\int \frac{2x+1}{(3x^2+3x-5)^{1/3}} dx$.

SOLUTION The quantity being raised to a power is in the denominator of the integrand. Thus, we try putting $u = 3x^2 + 3x - 5$. (Putting $u = 2x + 1$ won't work. Try it. If there are two competing candidates for u, it is sometimes best to put u equal to the more complicated of the two.)

$$u = 3x^2 + 3x - 5$$
$$\frac{du}{dx} = 6x + 3$$
$$du = (6x+3) dx$$
$$dx = \frac{1}{6x+3} du$$

$$\int \frac{2x+1}{(3x^2+3x-5)^{1/3}} dx = \int \frac{2x+1}{u^{1/3}(6x+3)} du$$

Now we would normally cancel x's, but the x's don't seem to want to cancel! If we look at what we have for a moment, though, we notice something: $6x + 3 = 3(2x + 1)$, and the quantity $2x + 1$ cancels.

$$= \int \frac{2x + 1}{u^{1/3}3(2x + 1)} \, du$$

$$= \int \frac{1}{3u^{1/3}} \, du$$

$$= \frac{1}{3} \int \frac{1}{u^{1/3}} \, du$$

$$= \frac{1}{3} \int u^{-1/3} \, du \qquad \text{Remember to convert to exponential form.}$$

$$= \frac{1}{3} \cdot \frac{u^{2/3}}{(2/3)} + C$$

$$= \frac{1}{3} \cdot \frac{3}{2} u^{2/3} + C = \frac{1}{2} u^{2/3} + C$$

$$= \frac{1}{2}(3x^2 + 3x - 5)^{2/3} + C$$

Before we go on... We check our answer.

$$\frac{d}{dx}\left(\frac{1}{2}(3x^2 + 3x - 5)^{2/3} + C\right) = \frac{1}{3}(3x^2 + 3x - 5)^{-1/3}(6x + 3)$$

$$= \frac{6x + 3}{3(3x^2 + 3x - 5)^{1/3}}$$

$$= \frac{2x + 1}{(3x^2 + 3x - 5)^{1/3}} \quad ✓$$

▼ **EXAMPLE 6**

Evaluate $\int \frac{2x}{(x - 5)^2} \, dx$.

SOLUTION We try putting $u = x - 5$.

$$u = x - 5$$
$$\frac{du}{dx} = 1$$
$$du = dx$$
$$dx = du$$

$$\int \frac{2x}{(x - 5)^2} \, dx = \int \frac{2x}{u^2} \, du$$

We have arrived at the step where the x's should cancel, and they don't. This is where we need to do something new.

If there are x's left over after cancellation, go back to the equation relating x and u, solve for x, and substitute in the integrand.

In this case, the equation relating x and u is the first equation in the box: $u = x - 5$. We solve for x, getting $x = u + 5$. We now substitute this expression for x into the integral.

$$\int \frac{2x}{u^2} \, du = \int \frac{2(u+5)}{u^2} \, du$$
$$= 2 \int \frac{u+5}{u^2} \, du$$

Now that we have gotten rid of the x's, how do we take the antiderivative? There are two possible ways of doing this. The first method is to break the integrand into two fractions u/u^2 and $5/u^2$ and integrate each term separately. The second method—which really amounts to the same thing—is to move the denominator up to the numerator by changing the sign of the exponent, and then use the distributive law.

$$= 2 \int (u+5) u^{-2} \, du$$
$$= 2 \int (u^{-1} + 5u^{-2}) \, du$$
$$= 2(\ln |u| - 5u^{-1}) + C$$
$$= 2(\ln |x-5| - 5(x-5)^{-1}) + C$$

Before we go on... We check our answer.

$$\frac{d}{dx}(2(\ln|x-5| - 5(x-5)^{-1}) + C) = 2\left(\frac{1}{x-5} + 5(x-5)^{-2}\right)$$
$$= 2\frac{x-5+5}{(x-5)^2}$$
$$= \frac{2x}{(x-5)^2} \checkmark$$

SHORTCUTS

If a and b are constants with $a \neq 0$, then we have the following formulas. (You have already seen one of them in Example 4. All of them can be obtained using the substitution $u = ax + b$. They will appear in the exercises.)

SHORTCUTS: INTEGRALS OF EXPRESSIONS INVOLVING $(ax + b)$

Rule	Example
$\int (ax + b)^n \, dx = \dfrac{(ax + b)^{n+1}}{a(n + 1)} + C$ (if $n \neq -1$)	$\int (3x - 1)^2 \, dx = \dfrac{(3x - 1)^3}{3(3)} + C$
	$= \dfrac{(3x - 1)^3}{9} + C$
$\int (ax + b)^{-1} \, dx = \dfrac{1}{a} \ln\|ax + b\| + C$	$\int (3 - 2x)^{-1} dx = \dfrac{1}{(-2)} \ln\|3 - 2x\| + C$
	$= -\dfrac{1}{2} \ln\|3 - 2x\| + C$
$\int e^{ax+b} \, dx = \dfrac{1}{a} e^{ax+b} + C$	$\int e^{-x+4} \, dx = \dfrac{1}{(-1)} e^{-x+4} + C$
	$= -e^{-x+4} + C$

We end this section with a little advice.

HINTS AND GENERAL GUIDELINES

1. Always remember not to substitute back for u until the end of the calculation, after you have taken the antiderivative.
2. If an x does not cancel in the integrand, do one of the following.
   (a) Try another substitution.
   (b) Go back to the formula for u as a function of x, solve the equation for x as a function of u, and substitute for x.
3. Don't ever bother using the substitution $u = x$. All this does is replace the letter x with the letter u throughout, giving you the same integral you started with!
4. To check your answer, simply take its derivative. You should get the integrand you started with.

13.2 EXERCISES

In Exercises 1–30, evaluate the given integrals.

1. $\int (3x + 1)^5 \, dx$
2. $\int (-x - 1)^7 \, dx$
3. $\int (-2x + 2)^{-2} \, dx$
4. $\int (2x)^{-1} \, dx$
5. $\int x(3x^2 + 3)^3 \, dx$
6. $\int x(-x^2 - 1)^3 \, dx$
7. $\int 2x\sqrt{3x^2 - 1} \, dx$
8. $\int 3x\sqrt{-x^2 + 1} \, dx$
9. $\int xe^{-x^2+1} \, dx$

10. $\int xe^{2x^2-1}dx$

11. $\int (x+1)e^{-(x^2+2x)}\,dx$

12. $\int (2x-1)e^{2x^2-2x}\,dx$

13. $\int \dfrac{-2x-1}{(x^2+x+1)^3}\,dx$

14. $\int \dfrac{x^3-x^2}{3x^4-4x^3}\,dx$

15. $\int \dfrac{x^2+x^5}{\sqrt{2x^3+x^6-5}}\,dx$

16. $\int \dfrac{2(x^3-x^4)}{(5x^4-4x^5)^5}\,dx$

17. $\int 2x\sqrt{x+1}\,dx$

18. $\int \dfrac{x}{\sqrt{x+1}}\,dx$

19. $\int x(x-2)^5\,dx$

20. $\int x(\sqrt[3]{x-2})\,dx$

21. $\int \dfrac{3e^{-1/x}}{x^2}\,dx$

22. $\int \dfrac{2e^{2/x}}{x^2}\,dx$

23. $\int x(x^2+1)^{1.3}\,dx$

24. $\int \dfrac{x}{(3x^2-1)^{0.4}}\,dx$

25. $\int (1+9.3e^{3.1x-2})\,dx$

26. $\int (3.2-4e^{1.2x-3})\,dx$

27. $\int \dfrac{e^{-0.05x}}{1-e^{-0.05x}}\,dx$

28. $\int \dfrac{3e^{1.2x}}{2+e^{1.2x}}\,dx$

29. $\int ((2x-1)e^{2x^2-2x}+xe^{x^2})\,dx$

30. $\int (xe^{-x^2+1}+e^{2x})\,dx$

In Exercises 31–34, derive the equations, where a and b are constants with $a \neq 0$.

31. $\int (ax+b)^n\,dx = \dfrac{(ax+b)^{n+1}}{a(n+1)}+C \quad (n \neq -1)$

32. $\int (ax+b)^{-1}\,dx = \dfrac{1}{a}\ln|ax+b|+C$

33. $\int e^{ax+b}\,dx = \dfrac{1}{a}e^{ax+b}+C$

34. $\int \dfrac{1}{(ax+b)^n}\,dx = -\dfrac{1}{a(n-1)(ax+b)^{n-1}}+C \quad (n \neq 1)$

Use the formulas in Exercises 31 through 34 to calculate the integrals in Exercises 35–44 mentally.

35. $\int e^{-x}\,dx$

36. $\int e^{x-1}\,dx$

37. $\int e^{2x-1}\,dx$

38. $\int e^{-3x}\,dx$

39. $\int (2x+4)^2\,dx$

40. $\int (3x-2)^4\,dx$

41. $\int \dfrac{1}{5x-1}\,dx$

42. $\int (x-1)^{-1}\,dx$

43. $\int (1.5x)^3\,dx$

44. $\int e^{2.1x}\,dx$

45. Find $f(x)$ if $f(0) = 0$ and the tangent line at $(x, f(x))$ has slope $x(x^2+1)^3$.

46. Find $f(x)$ if $f(1) = 0$ and the tangent line at $(x, f(x))$ has slope $\dfrac{x}{x^2+1}$.

47. Find $f(x)$ if $f(0) = \dfrac{1}{2e}$ and the tangent line at $(x, f(x))$ has slope xe^{x^2-1}.

48. Find $f(x)$ if $f(1) = -1 + \dfrac{1}{e}$ and the tangent line at x has slope $(x-1)e^{x^2-2x}$.

APPLICATIONS

49. *Motion in a Straight Line* The velocity of a particle moving in a straight line is given by $v = t(t^2+1)^4 + t$.
    (a) Find an expression for the position s after a time t.
    (b) Given that $s = 1$ at time $t = 0$, find the constant of integration C, and hence find an expression for s in terms of t without any unknown constants.

50. *Motion in a Straight Line* The velocity of a particle moving in a straight line is given by $v = 3te^{t^2}t$.
    (a) Find an expression for the position s after a time t.
    (b) Given that $s = 3$ at time $t = 0$, find the constant of integration C, and hence find an expression for s in terms of t without any unknown constants.

51. *Cost* The marginal cost of producing the xth box of light bulbs is $5 + \sqrt{x+1}$, and the fixed cost is $20,000. Find the total cost function $C(x)$.

52. *Cost* The marginal cost of producing the xth box of computer disks is $(10 + x\sqrt{x^2+1})/100{,}000$, and the fixed cost is $100,000. Find the total cost function $C(x)$.

53. *Cost* The marginal cost of producing the xth roll of film is given by $5 + 1/(x+1)$. The total cost to produce one roll is $1,000. Find the total cost function $C(x)$.

54. *Cost* The marginal cost of producing the xth box of videotape is given by $10 - \dfrac{x}{(x^2+1)^2}$. The total cost to produce one-hundred boxes is $10,000. Find the total cost function $C(x)$.

COMMUNICATION AND REASONING EXERCISES

55. Are there any circumstances under which one should use the substitution $u = x$? Illustrate your answer by means of an example.

56. Give an example of an integral that can be calculated by using the substitution $u = x^2 + 1$. Justify your claim by carrying out the calculation.

57. Give an example of an integral that can be calculated by using the power rule for antiderivatives, and also by using the substitution $u = x^2 + x$. Justify your claim by carrying out the calculations.

58. At what stage of a calculation using a u-substitution should one substitute back for u in terms of x: before or after taking the antiderivative?

59. You are asked to calculate $\displaystyle\int \dfrac{u}{u^2+1}\,du$. What is wrong with the substitution $u = u^2 + 1$?

60. What is wrong with the following "calculation" of $\displaystyle\int \dfrac{1}{x^2-1}\,dx$?

$$\int \dfrac{1}{x^2-1} = \int \dfrac{1}{u} \quad \text{(using the substitution } u = x^2 - 1\text{)}$$
$$= \ln|u| + C$$
$$= \ln|x^2 - 1| + C$$

13.3 APPLICATIONS OF THE INDEFINITE INTEGRAL

APPLICATIONS TO BUSINESS

We have already seen some simple applications of the indefinite integral to business (Example 10 of Section 1 and exercises in Sections 1 and 2). We now look at several other scenarios in which the indefinite integral is useful.

EXAMPLE 1 Volume Discount

A software company offers volume discounts to buyers of its program. If a customer buys a number of copies, then the xth copy will cost the customer $500/\sqrt{x+1}$. What is the total cost to buy x copies?

SOLUTION Let $C(x)$ be the total cost to buy x copies. What we are told is $C'(x)$, the marginal cost.

$$C'(x) = \dfrac{500}{\sqrt{x+1}}$$

Thus,
$$C(x) = \int C'(x)\, dx$$
$$= \int \frac{500}{\sqrt{x+1}}\, dx.$$

We can compute this using a simple substitution.

$$u = x + 1$$
$$du = dx$$

So
$$C(x) = \int \frac{500}{\sqrt{u}}\, du$$
$$= \int 500 u^{-1/2}\, du$$
$$= \frac{500 u^{1/2}}{1/2} + K \quad (K \text{ is the constant of integration.})$$
$$= 1{,}000\,(x+1)^{1/2} + K.$$

We now need to find K. If a customer buys no copies of the program, the customer pays nothing, so $C(0) = 0$, which gives
$$0 = 1{,}000\,(0+1)^{1/2} + K$$
$$= 1{,}000 + K,$$

so
$$K = -1{,}000.$$

Therefore,
$$C(x) = 1{,}000\,(x+1)^{1/2} - 1{,}000.$$

Before we go on... This is in fact a scheme that some software publishers and copy centers do use for volume discounts.

You might object that we should compute the total cost by adding up the selling prices of each copy, so
$$C(x) = \frac{500}{\sqrt{2}} + \frac{500}{\sqrt{3}} + \ldots + \frac{500}{\sqrt{x+1}}.$$

You would be right. However, just as marginal cost is really only an estimate of the cost of an item, so the answer we obtained here by integration is only an estimate of the total revenue. We shall return to the relationship between integration and summation in Section 5.

Q Is there a "cue" to tell us when we need to take the integral? (For instance, how did you know to take the integral to obtain the cost of x items in the above example?)

A Luckily, there is an easy way to tell us whether an integral is needed. First look at the units of measurement of the function we are given and the one we are asked to find. In the above example, we were given cost per item and were asked to calculate cost. The function we were given is measured in dollars per item, while the cost function we want is measured in dollars. We now write an equation relating these units of measurement.

$$\text{Dollars} = \frac{\text{Dollars}}{\text{Item}} \times \text{Items}$$

$$\uparrow \qquad \uparrow \qquad \uparrow$$
$$\text{Cost} = \int \text{Cost per item } dx$$

(The second line is the integral equation we got in the example.) This is the cue we need: *multiplying units of measurement corresponds to taking the integral* (just as dividing units of measurement corresponds to taking the derivative).

▼ **EXAMPLE 2** Total Sales

My publisher tells me that I can expect monthly sales of my epic novel to be given by $1,000(1 - e^{-0.5t})$ copies, where t is the time in months after its publication.

(a) Find a formula for the number of copies $S(t)$ my novel will sell in the first t months.

(b) If there were 20,000 pre-publication orders for my novel, find a formula for the total number of copies $T(t)$ that my novel will have sold t months after publication. How many copies will sell in the first year?

SOLUTION   (a) The formula $1,000(1 - e^{-0.5t})$ gives the number of copies sold per month. In other words, the accumulated sales are going up at a rate of $1,000(1 - e^{-0.5t})$ copies per month, or

$$S'(t) = 1,000(1 - e^{-0.5t}).$$

Thus,

$$S(t) = \int S'(t)\, dt$$

$$= \int 1,000\,(1 - e^{-0.5t})\, dt$$

$$= 1,000 \int (1 - e^{-0.5t})\, dt$$

$$= 1,000 \left(t + \frac{e^{-0.5t}}{0.5} \right) + C$$

$$= 1,000\,(t + 2e^{-0.5t}) + C,$$

and it remains to find the constant C. To do this, note that $S(t)$ represents total sales starting at publication ($t = 0$) and ending at a future time t. Since we are counting only sales after publication,

$$S(0) = 0.$$

Substituting gives

$$S(0) = 1{,}000(0 + 2e^{-0.5(0)}) + C = 0,$$

or

$$2{,}000 + C = 0,$$

so

$$C = -2{,}000.$$

Thus, the total sales are given by

$$S(t) = 1{,}000(t + 2e^{-0.5t}) - 2{,}000 \text{ copies}.$$

(b) Note that the solution to (a) gives the number of copies that will be sold *starting now*. To obtain the total sales T, we must add the 20,000 copies already sold, so

$$T(t) = 1{,}000(t + 2e^{-0.5t}) - 2{,}000 + 20{,}000$$
$$= 1{,}000(t + 2e^{-0.5t}) + 18{,}000.$$

In one year's time, the total sales will therefore amount to

$$T(12) = 1{,}000\,(12 + 2e^{-(0.5)(12)}) + 18{,}000$$
$$\approx 30{,}002 \text{ copies}.$$

Before we go on... How did we recognize that an integral was called for in the first place? We recognized that we were given a rate of change (copies per month). Alternatively, we could have used an analysis of units of measurement: we were given monthly sales, measured in copies per month, and we wanted total sales, measured in copies. So we write

$$\text{Copies} = \frac{\text{Copies}}{\text{Month}} \times \text{Months}$$

$$\text{Total Sales} = \int \text{Monthly sales } dt.$$

Note also that $S(t)$ and $T(t)$ differ by a constant, and so they have the same derivative: $S'(t) = T'(t) = 1{,}000(1 - e^{-0.5t})$.

How reasonable is the model? Figure 1 shows graphing calculator plots of $T'(t)$ and $T(t)$.

The graph of $T'(t)$ on the left shows a reasonable sales pattern for a new product: a fast initial increase, and then a gradual leveling off, in this

13.3 Applications of the Indefinite Integral

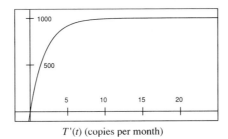
$T'(t)$ (copies per month)

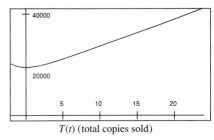
$T(t)$ (total copies sold)

FIGURE 1

case at 1,000 copies per month. The accumulated sales graph on the right thus becomes approximately linear with slope 1,000 for large values of t.

▼ **EXAMPLE 3** Total Demand with Varying Price

Enormous State University (ESU) charges $2,000 per year for tuition and currently has an enrollment of 24,000 undergraduates. It finds that its net annual gain in enrollment can be predicted by the formula $q = 2,400 - 0.25p$, where q is the net annual gain (in students per year) and p is the tuition fee it charges. ESU's financial planning calls for its tuition to be $p = 2,000 + 500t$, t years from now. Express the enrollment as a function of t and predict the enrollment 10 years from now.

SOLUTION The problem asks for enrollment as a function of time, which we denote by $E(t)$. We are given an equation for the net gain in enrollment per year: $q = 2,400 - 0.25p$. Its units, students per year, tell us that this is the derivative, $E'(t)$, so we have

$$E'(t) = 2,400 - 0.25p.$$

The problem now is that $E'(t)$ is given as a function of p rather than t, but we can rectify this by substituting the formula for p, $p = 2,000 + 500t$. This gives

$$E'(t) = 2,400 - 0.25(2,000 + 500t)$$
$$= 1,900 - 125t.$$

Thus,

$$E(t) = \int (1,900 - 125t)\, dt$$
$$= 1,900t - 62.5t^2 + C,$$

and it remains to calculate the constant C. For this, we use the additional piece of information that, at time $t = 0$, the enrollment is 24,000. Thus,

$E(0) = 24,000$, and so

$$24,000 = 1,900(0) - 62.5(0)^2 + C,$$

giving

$$C = 24,000.$$

Thus,

$$E(t) = 1,900t - 62.5t^2 + 24,000.$$

In 10 years' time, the enrollment will be

$$E(10) = 1,900(10) - 62.5(10)^2 + 24,000$$
$$= 36,750 \text{ students.}$$

Before we go on... Note that we did not use directly the information that ESU currently charges $2,000 in annual tuition. This information is implied by the equation $p = 2,000 + 500t$. Notice also that the unit analysis here is similar to that of the previous example: Students = (Students/Year) × Years.

 Figure 2 shows a graphing calculator plot of the graph of $E(t)$ for $t \leq 50$.

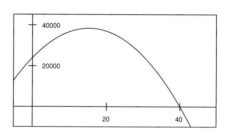

FIGURE 2

The graph predicts that enrollment will peak in about 15 years and that ESU cannot continue to raise fees indefinitely. If it does, it will need to close its doors in 40 years' time! A word of caution: We used a linear demand equation, and we have seen that linear models tend to be accurate only for a small range of values. (We have also not taken into account such factors as inflation and the fees charged by other universities.)

MOTION

Calculus is the language of the physical sciences. In the remainder of this section we discuss in some detail the application of calculus to motion, as an example of the intertwining of mathematics and physics that is an important part of both.

We begin by bringing together some facts scattered through the last several chapters having to do with an object moving in a straight line.

> **POSITION, VELOCITY, AND ACCELERATION**
>
> 1. If $s = s(t)$ is the position of an object at time t, then its velocity is given by the derivative
>
> $$v = \frac{ds}{dt}.$$
>
> In short, *velocity is the derivative of position.*
>
> 2. The acceleration of an object is given by the derivative
>
> $$a = \frac{dv}{dt}.$$
>
> In short, *acceleration is the derivative of velocity.*
>
> 3. On the planet Earth, a freely falling body experiencing no air resistance accelerates at approximately 32 feet per second per second, or 32 ft/s².

Our first goal is to answer the following question. *Suppose that at time $t = 0$, I throw a ball up at a specified velocity v_0 from a specified position s_0. What is its position after t seconds?*

We are going to "work backwards" through the three points above. We shall restate them in reverse order, and also convert the derivative formulas into integral formulas. First, we make the convention that height above the ground is a positive number s, so that if s is increasing, velocity (its derivative) is positive. This means that since the acceleration due to gravity acts downward it is negative.

> **ACCELERATION, VELOCITY, AND POSITION: INTEGRAL FORM**
>
> **I:** $a(t) = -32$ ft/s.²
> **II:** $v(t) = \int a(t)\, dt$
> **III:** $s(t) = \int v(t)\, dt$

Now we work from acceleration to position.
Step 1: Substitute a in II and solve for v:

$$v(t) = \int a(t)\, dt = \int (-32)\, dt = -32t + C,$$

so
$$v(t) = -32t + C.$$

Now we have to deal with the constant C. Looking back at the question we wish to answer, we see that at time $t = 0$, the velocity is v_0 (we call this the **initial velocity**). In other words, when $t = 0$, $v = v_0$. Substituting this into the above equation, we get

$$v_0 = -32(0) + C,$$

so $C = v_0$. Thus,

$$v(t) = v_0 - 32t.$$

Step 2 Substitute the formula for v into III and solve for s:

$$s(t) = \int v(t)\, dt = \int (v_0 - 32t)\, dt = v_0 t - 16t^2 + C,$$

so

$$s(t) = v_0 t - 16t^2 + C.$$

Now we have to deal once again with the constant C. Looking back at the question again, we see that at time $t = 0$, the position is specified as s_0 (we call this the **initial position**). In other words, when $t = 0$, $s = s_0$. Substituting this into the above equation, we get

$$s_0 = v_0(0) - 16(0) + C,$$

so that $C = s_0$. Thus,

$$s(t) = s_0 + v_0 t - 16t^2.$$

This formula tells us exactly where the ball is at time t, and answers our original question.

VELOCITY AND POSITION OF AN OBJECT MOVING VERTICALLY UNDER GRAVITY

The following equations describe the vertical motion of an object under gravity without air resistance.

Velocity Formula   **Position Formula**

$v(t) = v_0 - 32t$ $s(t) = s_0 + v_0 t - 16t^2$

v is velocity in feet per second, s is position (height) in feet, and t is time in seconds.

Remember our convention that up is positive. Thus, v_0 refers to the initial *upward* velocity and s to the height above the ground.

These formulas allow us to answer a number of questions.

13.3 Applications of the Indefinite Integral

▼ **EXAMPLE 4**

A stone is tossed upward at 30 ft/s.

(a) How fast and in what direction is it going after 5 seconds?
(b) Where will it be after 5 seconds?

SOLUTION (a) The question here asks for velocity, so we use the velocity formula

$$v = v_0 - 32t.$$

The first sentence tell us that the initial velocity is 30 ft/s upwards. Thus, $v_0 = 30$. The question asks us what happens at $t = 5$. We substitute these values into the equation and obtain

$$v = 30 - 32(5) = -130 \text{ ft/s}.$$

Thus, at time $t = 5$, the stone is *falling* at a speed of 130 ft/s.

(b) Here, we're looking for the position at $t = 5$, so we use the position formula

$$s = s_0 + v_0 t - 16t^2.$$

To use this formula, we need the values of all the constants on the right-hand side. Now we already know that $t = 5$ and $v_0 = 30$. What about s_0? This is the initial position, the position of the stone at time $t = 0$. Since we are not told this, we are free to set up our own convention and take the initial position as zero. In other words, we take $s_0 = 0$. Now we have all the values we need, and

$$s = 0 + 30(5) - 16(25)$$
$$= -250 \text{ ft}.$$

Thus, the stone is at a location 250 feet *below* its position when tossed. In other words, it has reached its **zenith** (highest point) and then dropped past its initial location, as shown in Figure 3.

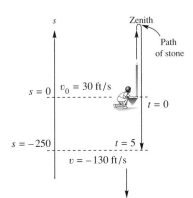

FIGURE 3

Before we go on... You might be curious as to where and when the stone reaches its zenith and starts to fall. The next example shows how we can find this.

▼ **EXAMPLE 5**

A bullet is fired upward at 1,000 ft/s. How high does it get?

SOLUTION We are asked for the position of the bullet at its zenith, so we turn to the position formula

$$s = s_0 + v_0 t - 16t^2.$$

As in the last example, we take $s_0 = 0$, and we are given $v_0 = 1,000$. What about the value of t? We haven't been told what t is, so it appears we are stuck.

When we run into a roadblock, a good strategy is to go back and check whether there is any information that we haven't yet used. The question asks about the highest point reached by the projectile. Although we don't know the value of t at that instant, we *do* know something: the zenith is the point of *maximum* height, so the bullet has velocity zero then (in other words, it has ceased going up and is about to come down). Thus, we look at the *velocity* equation

$$v = v_0 - 32t.$$

Now we know that $v_0 = 1{,}000$, and we want to find the t that will make $v = 0$:

$$0 = 1{,}000 - 32t,$$

so

$$t = 1{,}000/32 = 31.25 \text{ seconds.}$$

Now we can go back to the position equation and substitute for all the quantities on the right-hand side.

$$s = 0 + 1{,}000(31.25) - 16(31.25)^2 = 15{,}625 \text{ ft}$$

The equations we have been using ignore air resistance. The next example shows one way of modeling the effect of air resistance on a falling object: we assume that the acceleration decays exponentially.

▼ **EXAMPLE 6**   Skydiving

A skydiver experiences an acceleration of

$$a(t) = -32e^{-0.2t} \text{ ft/s}^2$$

t seconds after jumping from a plane. If the plane is flying at a height of 15,000 feet when the skydiver jumps, find his height as a function of time. If his parachute fails to deploy, when will he hit the ground, and how fast will he be going at the time?

SOLUTION   As before, we need to work backwards from acceleration to velocity and then to position. We start with

$$v(t) = \int a(t)\, dt = \int (-32e^{0.2t})\, dt$$

$$= -32 \int e^{-0.2t}\, dt$$

$$= -32 \frac{e^{-0.2t}}{-0.2} + C$$

$$= 160 e^{-0.2t} + C.$$

13.3 Applications of the Indefinite Integral

Now, to find C, we think about the initial velocity. As the skydiver jumps out of the plane, his initial velocity will be 0; that is, $v(0) = 0$, so

$$160e^0 + C = 0$$

and

$$160 + C = 0,$$

giving

$$C = -160.$$

Therefore,

$$v(t) = 160e^{-0.2t} - 160.$$

To find the height, we use

$$s(t) = \int v(t)\, dt,$$

so

$$s(t) = \int (160e^{-0.2t} - 160)\, dt$$

$$= 160\frac{e^{-0.2t}}{-0.2} - 160t + C$$

$$= -800e^{-0.2t} - 160t + C.$$

The initial height is 15,000 feet, so $s(0) = 15{,}000$, giving

$$-800e^0 - 160(0) + C = 15{,}000,$$

and hence

$$C = 15{,}800.$$

We can now write

$$s(t) = -800e^{-0.2t} - 160t + 15{,}800.$$

To determine when he will hit the ground, we have to solve $s(t) = 0$, or

$$-800e^{-0.2t} - 160t + 15{,}800 = 0.$$

Unfortunately, this cannot be solved analytically, but we can approximate the answer using a graphing calculator or computer. Figure 4 shows the graph of s.

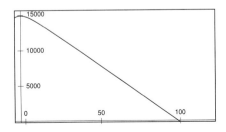

FIGURE 4

The graph shows that the time of impact will be around 100 seconds. Zooming in gives the more accurate estimate of $t = 98.75$ seconds. The velocity at impact is then

$$v(98.75) = 160e^{-0.2(98.75)} - 160 \approx 160 \text{ ft/s}.$$

Before we go on... The graph of velocity is shown in Figure 5.

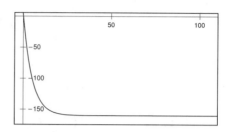

FIGURE 5

Because acceleration decays towards 0, the velocity approaches a limit, in this case -160 ft/s. You can think of the effect of wind resistance as limiting the velocity, keeping it from getting arbitrarily large. The limiting velocity is called the **terminal velocity.**

13.3 EXERCISES

APPLICATIONS

1. *Cost* If your fixed cost to manufacture light bulbs is $10,000 and your marginal cost is $100 + 0.01x$ for the xth box of bulbs, find your total cost for x boxes.

2. *Cost* If your fixed cost to manufacture computer disks is $15,000 and your marginal cost is $200 - 0.05x$ for the xth box of disks, find your total cost for x boxes.

3. *Cost* If your fixed cost to manufacture film is $10,000 and your marginal cost is $100 + 20e^{-0.01x}$ for the xth box of film, find your total cost for x boxes.

4. *Cost* If your fixed cost to manufacture videotape is $15,000 and your marginal cost is $200 + 50e^{-0.05x}$ for the xth box of tape, find your total cost for x boxes.

5. *Volume Discounts* Your office furniture company offers a volume discount on chairs: the first chair costs $200, and you reduce the cost by $2 per chair for each additional chair up to 100 chairs. Find the marginal cost function, and then determine how much you charge for x chairs.

6. *Volume Discounts* Your office furniture company offers a volume discount on tables: the first table costs $1,000, and you reduce the cost by $50 per table for each additional table up to 20 tables. Find the marginal cost function, and then determine how much you charge for x tables.

7. *Volume Discounts* Your software company offers a volume discount on its latest program: you charge $5,000/(x + 10)$ for the xth copy purchased by a customer. How much do you charge for a total of x copies?

8. *Volume Discounts* Your software company plans to offer the following volume discount on its next release: you will charge $50,000/(x + 10)^2$ for the xth

copy bought by a customer. How much will you charge for x copies?

9. **Average Cost** If your fixed cost to manufacture light bulbs is $10,000 and your marginal cost is $200 + 0.02x$ for the xth box of bulbs, find the production level that will minimize your *average* cost per box, and find the minimum average cost.

10. **Average Cost** If your fixed cost to manufacture computer disks is $9,000 and your marginal cost is $50 + 0.2x$ for the xth box of disks, find the production level that will minimize your *average* cost per box, and find the minimum average cost.

11. **Average Revenue** A sporting goods manufacturer offers a volume discount to stores on boogie boards: it charges $20 + 20e^{-0.5x}$ for the xth boogie board. What is the average revenue per boogie board if a store buys x boogie boards?

12. **Average Revenue** A sporting goods manufacturer offers a volume discount to stores on skateboards: it charges $10 + 15e^{-0.01x}$ for the xth skateboard. What is the average revenue per skateboard if a store buys x skateboards?

13. **Total Sales** Daily sales of your T-shirts over the next seven days are predicted by the equation
$$q = 10 - 0.2t^2,$$
where q represents the number of T-shirts sold per day t days from now.
   (a) Give an expression for the total sales $s(t)$ after t days, assuming your total sales now ($t = 0$) are zero.
   (b) How many T-shirts can you expect to sell in the next seven days?

14. **Declining Sales** Five years ago, your rock group's just-released CD, "Galactic Explosion," was selling 10,000 copies per year, but since that time, sales have declined by 2,000 each year, so that annual sales can be modeled by the linear equation
$$q = 10,000 - 2,000t,$$
where q represents annual sales of your CD, and t is the time in years ($t = 0$ corresponds to this date 5 years ago.)
   (a) Give an expression for the total sales $s(t)$ at time t.
   (b) What is the total number of CDs sold to date?

15. **Foreign Investments** The following chart shows the annual flow of private investment to developing countries from more industrialized countries for the years 1986 to 1993.*

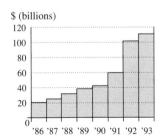

(a) Let $q(t)$ represent the annual quantity of private investments (in billions of dollars), where t is the number of years since the start of 1986. Which of the following models best fits the data shown? (Feel free to use a graphing calculator.)
   (A) $q(t) = 10 + 12.5t$
   (B) $q(t) = 10 + 35.5\sqrt{t}$
   (C) $q(t) = 10 + 1.56t^2$
(b) Use the best-fit model from part (a) to obtain an equation for the total flow $P(t)$ of private investment from January 1, 1986 until t years later.
(c) Use the model to predict how much money will have been sent to developing countries from January 1, 1986 through January 1, 1996.

16. **Foreign Investments** The following chart shows the annual flow of government loans and grants to developing countries from more industrialized countries.*

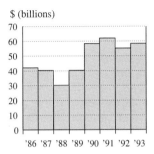

▼ *Source: World Bank (*The New York Times*, Dec. 17, 1993, p. D1.)

(a) Let $q(t)$ represent the annual quantity of loans (in billions of dollars), where t is the number of years since the start of 1986. Which of the following models best fits the data shown? (Feel free to use a graphing calculator.)
- (A) $q(t) = 38 - 0.6(t-3)^2$
- (B) $q(t) = 38 + 0.6(t-3)^2$
- (C) $q(t) = 20 + e^{0.54t}$

(b) Use the best-fit model from part (a) to obtain an equation for the total flow $G(t)$ of government loans and grants from January 1, 1986 until t years later.

(c) Use the model to predict how much money will have been sent to developing countries from January 1, 1986 through January 1, 1996.

17. *Exports* Based on figures released by the U.S. Agriculture Department, the value of U.S. pork exports from 1985 through 1993 can be approximated by the equation

$$q = \frac{460 e^{(t-4)}}{1 + e^{(t-4)}},$$

where q represents annual exports (in millions of dollars) and t the time in years, with $t = 0$ corresponding to January 1, 1985.* The figure shows the actual data with the graph of the equation superimposed.

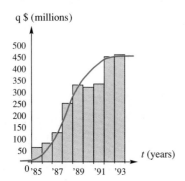

(a) Use the model to obtain an equation for total U.S. pork exports E in billions of dollars as a function of t. (Assume zero total exports as of January 1, 1985; that is, $E(0) = 0$.)

(b) Use your equation to predict the value of total U.S. pork exports to the nearest million dollars from January 1, 1985 to January 1, 2000.

18. *Sales* The weekly demand for your company's Lo-Cal Chocolate Mousse can be modeled by the equation

$$q = \frac{50 e^{2t-1}}{1 + e^{2t-1}},$$

where q is the number of gallons sold per week and t is the time in weeks. At present ($t = 0$) total sales of Lo-Cal mousse amount to 100 gallons.

(a) Express the total sales $s(t)$ of Lo-Cal mousse as a function of time t.

(b) Estimate the total sales of Lo-Cal mousse one year from now.

19. *Production* The production of cellular phones at your electronics plant has a Cobb-Douglas production function given by

$$P = 10 x^{0.3} y^{0.7},$$

where P is the number of cellular phones it produces per year, x is the number of employees, and y is the daily operating budget (in dollars). You have decided to increase the daily operating budget linearly from the present level of $6,000 to $10,000 over the next two years. Your plant employs 100 workers.

(a) Determine the productivity P as a function of t, where t is measured in years from now.

(b) Estimate the total production of cellular phones over the next two years. (Give your answer to the nearest cellular phone.)

20. *Production* Referring to the Cobb-Douglas production formula in the previous exercise, assume that you wish to maintain a daily operating budget of $6,000 but plan to increase the work force linearly from the present 100 employees to 200 over the coming two years.

(a) Determine the productivity P as a function of t, where t is measured in years from now.

(b) Estimate the total production of cellular phones over the next two years. (Give your answer to the nearest cellular phone.)

Using the Logistic Equation to Predict Sales[†] The logistic equation has the form

$$y(x) = \frac{NP_0}{P_0 + (N - P_0)e^{-kx}},$$

where N, P_0 and k are positive constants. (See the accompanying graph.)

---

*This is the authors' model based on figures published by the U.S. Agriculture Department quoting annual sales from 1985 through 1993. (*The New York Times*, Dec. 12, 1993, p. D1.)

[†]See "You're the Expert" at the end of the chapter on logarithmic and exponential functions for a discussion of the use of the logistic equation in modeling epidemics.

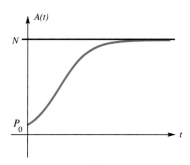

This equation is useful in modeling and predicting the demand for a new product, where

y = demand in annual sales

x = number of years since the introduction of the product to the market

P_0 = initial demand (demand at time $x = 0$)

N = total potential demand

k = approximate initial rate of growth of demand (a good approximation if N is large compared with P_0).

In Exercises 21–26, we use the logistic curve to model the growing demand for several therapeutic drugs.

21. Show that the logistic equation can be rewritten in the form

$$y(x) = \frac{NP_0 e^{kx}}{P_0 e^{kx} + (N - P_0)}.$$

22. Using the result in the preceding exercise and a suitable substitution, show that

$$\int y(x)\, dx = \frac{N}{k} \ln\left(P_0 e^{kx} + (N - P_0)\right) + C.$$

23. *Eli Lilly Corp.*'s human growth hormone Humatrope has a potential market estimated at roughly $N = \$160$ million per year, and its total Humatrope sales can be approximated by the logistic function with $k = 2.7$ and $P_0 = 0.0025$. $y(x)$ represents annual sales in millions of dollars, and x represents the number of years since the drug's approval by the FDA in 1987.* Use the model to give an estimate (to the nearest \$10 million) of the value of total sales of Humatrope over the ten-year period beginning in 1987. (Hint: Use the result from Exercise 22.)

24. *Genentech*'s human growth hormone Protropin has a potential market estimated at roughly $N = \$300$ million per year, and its total Protropin sales can be approximated by the logistic function with $k = 1.7$ and $P_0 = 0.20$. $y(x)$ represents annual sales in millions of dollars, and x represents the number of years since the drug's approval by the FDA in 1986.* Use the model to give an estimate (to the nearest \$10 million) of the value of total sales of Protropin over the ten-year period beginning in 1986. (Hint: Use the result from Exercise 22.)

25. *(Graphing calculator or computer required)* *Genzyme*'s drug against Gaucher's disease, Ceredase, cost the company \$30 million to develop and has a potential market of \$900 million per year. At the time of FDA approval (1991) annual sales were estimated at \$50 million. Assuming an initial rate of growth specified by $k = 0.5$ (that is, approximately a 50% initial growth rate), use a graphing calculator to estimate how long (to the nearest year) it will take *Genzyme* to earn 10 times development costs from sales of Ceredase.[†] (Hint: Use the result from Exercise 22.)

26. *(Graphing calculator or computer required)* Repeat the previous exercise using an initial growth rate specified by $k = 0.3$ (that is, approximately a 30% initial growth rate). (Hint: Use the result from Exercise 22.)

---

* The model is a very crude one, based on 1991 sales data, total sales data through May, 1992, and very rough estimates of the potential market and selling price. Source: Senate Judiciary Committee; Subcommittee on Antitrust and Monopoly/*The New York Times*, May 14, 1992, p. D1.

† Initial sales figure based on sales for 9 months of 1991 immediately after FDA approval. Source: ibid.

Vertical Motion In Exercises 27–38, neglect the effects of air resistance.

27. If a stone is dropped from a rest position above the ground, how fast, and in what direction, will it be traveling after 10 seconds?

28. If a stone is thrown upward at 10 feet per second, how fast, and in what direction, will it be traveling after 10 seconds?

29. Show that if a projectile is thrown upward with a velocity of $v_0/32$ ft/s, then it will reach its highest point after $v_0/32$ seconds.

30. Use the result of the preceding exercise to show that if a projectile is thrown upward with a velocity of v_0 ft/s, then its highest point will be $v_0^2/64$ feet above the starting point.

In Exercises 31–38, use the results of the previous two exercises.

31. I threw a ball up in the air to a height of 20 feet. How fast was the ball traveling when it left my hand?

32. I threw a ball up in the air to a height of 40 feet. How fast was the ball traveling when it left my hand?

33. I threw a ball up in the air to a height of 20 feet. Where was the ball after 4 seconds?

34. I threw a ball up in the air to a height of 40 feet. Where was the ball after 5 seconds?

35. A piece of chalk is tossed vertically upward by Professor Schwartzenegger and hits the ceiling 100 feet above with a *BANG*.
    (a) What is the minimum speed the piece of chalk must have been traveling to enable it to hit the ceiling?
    (b) Assuming that Professor Schwartzenegger in fact tossed the piece of chalk up at 100 ft/s, how fast would it have been moving when it struck the ceiling?
    (c) Assuming that Professor Schwartzenegger tossed the chalk up at 100 ft/s, and that it recoils from the ceiling with the same speed it had at the instant it hit, how long will it take the chalk to make the return journey and hit the ground?

36. A projectile is fired vertically upwards from ground level at 16,000 ft/s.
    (a) How high does it go?
    (b) How long does it take to reach its zenith (highest point)?
    (c) How fast is it traveling when it hits the ground?

37. *Strength* Professor Strong can throw a 10-pound, dumbbell twice as high as Professor Weak can. How much faster can Professor Strong throw it?

38. *Weakness* Professor Weak can throw a computer disk three times as high as Professor Strong can. How much faster can Professor Weak throw it?

39. *Varying Acceleration* A particle in a nuclear accelerator undergoes an acceleration given by
$$a = 3t^2 - 4t + 5 \text{ ft/s}^2.$$
Assuming it started from rest, how far has it traveled in 120 seconds?

40. *Varying Acceleration* The Starship Galactica undergoes an acceleration given by
$$a = e^{2(t-2)} + 5 \text{ ft/s}^2.$$
Assuming it started from rest, how far has it traveled in 60 seconds?

41. *Fast Cars* Professor Hare's Gran Turismo can do 0 to 60 mph in 3 seconds. Assuming constant acceleration, how fast is it accelerating? [60 mph = 88 ft/s]

42. *Slow Cars* Professor Turtle's Slomobile can only do 0 to 60 mph in 23 seconds. Assuming constant acceleration, how fast is it accelerating? [60 mph = 88 ft/s]

COMMUNICATION AND REASONING EXERCISES

43. Why might it be a bad idea to offer a volume discount in which the marginal cost is $100e^{-0.01x}$?

44. Would it be a bad idea to offer a volume discount in which the marginal cost is $10,000/(x + 10)^2$?

45. Complete the following sentence. A unit analysis tells us that if $P(s)$ and $Q(s)$ are functions with the property that units of $P(s) \times$ units of $s =$ units of $Q(s)$, then _____ .

46. Why are the units of measurement of $\int f(x)\, dx$ equal to the units of measurement of $f(x)$ times the units of measurement of x?

13.4 GEOMETRIC DEFINITION OF THE DEFINITE INTEGRAL

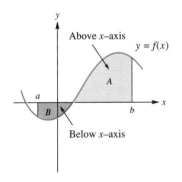

FIGURE 1

This section and the next are in many ways the intellectual climax of the course so far. We are about to see that the antiderivative provides a solution to what appears to be a completely unrelated problem. We call Newton and Leibniz the inventors of calculus largely because they made this connection.

Up to this point we have been talking about the *indefinite* integral. This suggests that there may be such a thing as the *definite* integral. Indeed there is. To introduce it, we are now going to talk—of all things—about *area*. Consider the graph of a function f, as in Figure 1.

We have chosen two values a and b of x and have shaded the region enclosed by the vertical lines $x = a$ and $x = b$, the x-axis, and the curve $y = f(x)$. We would like to find the area of the shaded region, but for several reasons, we shall *subtract* any area that lies below the x-axis (the grey area in the figure). We say that the **definite integral of f from a to b** is the total "net" area: $A - B$. For example, if $A = 5$ square units, and $B = 2$ square units, then the definite integral is equal to $5 - 2 = 3$. We write the definite integral as $\int_a^b f(x)\,dx$. This is read as "the integral from a to b of $f(x)$, with respect to x," or "the integral from $x = a$ to $x = b$ of $f(x)$, with respect to x." We summarize this with the following definition.

GEOMETRIC DEFINITION OF THE DEFINITE INTEGRAL

If f is a function whose domain contains the closed interval $[a, b]$, then the **definite integral of $f(x)$ from $x = a$ to $x = b$** is defined as

(Area between the vertical lines $x = a$ and $x = b$ that is *above* the x-axis and below the graph of $f(x)$) −

(Area between the vertical lines $x = a$ and $x = b$ that is *below* the x-axis and above the graph of $f(x)$)

assuming that these areas exist and are finite. We denote the definite integral of $f(x)$ from a to b by $\int_a^b f(x)\,dx$, where we read the symbols as follows.

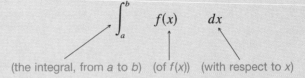

The numbers a and b are called the **limits of integration.** The number a is called the **lower** limit and b the **upper** limit.

▶ NOTES

1. The reason we call this the *geometric* definition of the definite integral is that we have defined it using the geometric notion of area. In the next section we shall give an algebraic definition.

2. The stipulation that the areas in question exist and are finite puts restrictions on the functions f that we can integrate. Although we shall not go into the fine details of the theory, examples of functions whose integrals we can find are the functions that are continuous on $[a, b]$ and the functions that are *bounded* and have only finitely many discontinuities in $[a, b]$.* ◀

Q The notation looks suspiciously like the *indefinite* integral with some new decorations. What on earth has this to do with antiderivatives?

A Therein lies the surprise at the heart of calculus, but we'll keep you in suspense a moment longer. In the meantime, remember this:

$\int f(x) \, dx$ means the *indefinite* integral of $f(x)$ with respect to x—in other words, the general form of the antiderivative of $f(x)$.

$\int_b^a f(x) \, dx$ stands for the *definite* integral of $f(x)$ from a to b—in other words, the area as discussed above.

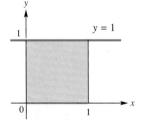

FIGURE 2

▼ **EXAMPLE 1**

Find $\int_0^1 1 \, dx$.

SOLUTION This is the area of the region shaded in Figure 2.
Since this is a square with sides equal to 1, its area is 1, so

$$\int_0^1 1 \, dx = 1.$$

Before we go on... What, then, is $\int_0^2 1 \, dx$? $\int_0^1 2 \, dx$? $\int_0^1 (-1) \, dx$? (Answers: 2, 2, and -1.)

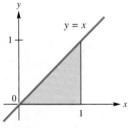

FIGURE 3

▼ **EXAMPLE 2**

Calculate $\int_0^1 x \, dx$.

SOLUTION This is the area of the region shaded in Figure 3.
This is a triangle with height 1 and base 1, so it has area $\frac{1}{2}$. Thus,

$$\int_0^1 x \, dx = \frac{1}{2}.$$

Before we go on... What, then, is $\int_0^2 x \, dx$? $\int_0^1 2x \, dx$? $\int_0^1 (-x) \, dx$? (Answers: 2, 1, and $-\frac{1}{2}$.)

▼ *$f(x)$ is bounded on $[a, b]$ if the part of its graph corresponding to $a \leq x \leq b$ lies entirely between two horizontal lines. For example, the function given by

$$f(x) = \begin{cases} \dfrac{1}{x} & \text{if } x > 0 \\ 0 & \text{if } x = 0 \end{cases}$$

is not bounded on $[0, 1]$, since its graph goes up to infinity near $x = 0$. Thus we shall not (yet) try to compute $\int_0^1 f(x) \, dx$. (See the section on improper integrals in the next chapter for more on this example.)

13.4 Geometric Definition of the Definite Integral

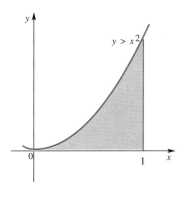

FIGURE 4

Q What about $\int_0^1 x^2 \, dx$?

A This is the area of the region under part of a parabola, as shown in Figure 4. Unfortunately, this is a very difficult area to calculate geometrically. In fact, the culmination of one part of Greek geometry was a calculation by Archimedes of (what amounted to) this area.

So as not to keep you in suspense any longer, we shall tell you how to calculate areas such as $\int_0^1 x^2 \, dx$, and then in Section 6 we'll see just why the calculation works. In order to express the answer in the simplest possible way, we first introduce some notation.

If F is any function, by $[F(x)]_a^b$ we shall mean the difference $F(b) - F(a)$.

Here are some quick examples.

$$[x^2 + 1]_{-1}^1 = [(1^2 + 1) - ((-1)^2 + 1)]$$
$$= [1 + 1 - (1 + 1)] = 0$$

$$[x + e^x]_0^1 = [(1 + e^1) - (0 + e^0)]$$
$$= [1 + e - 0 - 1] = e$$

$$[3(x^3 - x)]_{-2}^2 = 3[(x^3 - x)]_{-2}^2$$
$$= 3[(2^3 - 2) - ((-2)^3 - (-2))]$$
$$= 3[8 - 2 + 8 - 2] = 36$$

Notice the care you must take to get all of the signs right.

Here is one of the most important results in calculus.

THE FUNDAMENTAL THEOREM OF CALCULUS (FTC)

If $F(x)$ is any antiderivative of $f(x)$, and $f(x)$ is continuous on the interval $[a, b]$, then

$$\int_a^b f(x) \, dx = [F(x)]_a^b = F(b) - F(a).$$

Actually, this is only half of the FTC. In Section 6 we shall see the whole thing and also discuss why it is true. For the moment, just appreciate what it says: in order to calculate areas, all we need to be able to do is calculate antiderivatives. But this is not hard, as we've seen. This result has been called the biggest accident in all of mathematics. Who would have thought that the problems of finding areas and finding rates of change should be at all related?

▼ **EXAMPLE 3**

Calculate $\int_0^1 x^2 \, dx$.

SOLUTION To calculate this integral, we follow the FTC, which says: first find an antiderivative, then evaluate at the limits of integration, and then subtract.

$$\int_0^1 x^2 \, dx = \left[\frac{x^3}{3}\right]_0^1 = \frac{1}{3}[x^3]_0^1 = \frac{1}{3}[1^3 - 0^3] = \frac{1}{3}.$$

Thus, the area in Figure 4 is exactly $\frac{1}{3}$ square units.

Before we go on...

Q What happened to the constant of integration C?

A You can leave it out, since the FTC says that you can use *any* antiderivative you like, and so we chose the simplest-looking antiderivative, which is $x^3/3$. Here is what happens if we use another antiderivative, $(x^3/3) + 2$.

$$\int_0^1 x^3 \, dx = \left[\frac{x^3}{x} + 2\right]_0^1 = \left(\frac{1^3}{3} + 2\right) - \left(\frac{0^3}{3} + 2\right)$$
$$= \frac{1}{3} + 2 - 2 = \frac{1}{3}.$$

Thus, we get the same answer as before. Notice how the constant 2 canceled. This is what would happen to C if we left it in, so we leave it out.

AVOIDING ERRORS IN SIGNS

When evaluating the definite integral, you can avoid common errors in signs as follows. When you have found the antiderivative and obtained, say, $\left[x - \frac{x^3}{3}\right]_{-1}^1$, here is what you can do.

1. Make the template

$$[(\quad) - (\quad)],$$

leaving the insides of the parentheses blank.

2. Fill in the value of the antiderivative at the upper limit of integration in the first blank and its value at the lower limit of integration in the second blank, getting, for example,

$$\left[\left(1 - \frac{1^3}{3}\right) - \left((-1) - \frac{(-1)^3}{3}\right)\right].$$

Now we have all the needed parentheses in place, and we will not forget to distribute minus signs correctly.

▼ **EXAMPLE 4**

Calculate the total area enclosed by the x-axis, the vertical lines $x = -2$ and $x = 2$, and the curve $y = x^2 - 1$.

13.4 Geometric Definition of the Definite Integral

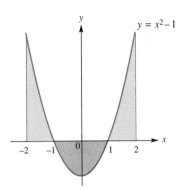

FIGURE 5

SOLUTION We first make a quick sketch of the curve, which we recognize as a parabola with x-intercepts -1 and 1. The graph is shown in Figure 5.

If we were to simply calculate the definite integral $\int_{-2}^{2} (x^2 - 1) \, dx$, we would wind up with the *net* area, the sum of the two pieces on either side *minus* the piece in the middle below the x-axis. This is not what the question is asking for: we are trying to calculate the *total* area, and this means *adding* the areas of the three pieces. To do this, we must calculate the area of each piece separately.

$$\text{Right-hand piece:} \quad \int_{1}^{2} (x^2 - 1) \, dx = \left[\frac{x^3}{3} - x \right]_{1}^{2}$$

$$= \left[\left(\frac{2^3}{3} - 2 \right) - \left(\frac{1^3}{3} - 1 \right) \right]$$

$$= \frac{8}{3} - 2 - \frac{1}{3} + 1$$

$$= \frac{4}{3}$$

$$\text{Middle piece:} \quad \int_{-1}^{1} (x^2 - 1) \, dx = \left[\frac{x^3}{3} - x \right]_{-1}^{1}$$

$$= \left[\left(\frac{1^3}{3} - 1 \right) - \left(\frac{(-1)^3}{3} - (-1) \right) \right]$$

$$= \frac{1}{3} - 1 + \frac{1}{3} - 1$$

$$= -\frac{4}{3}$$

This integral is negative because the area lies below the x-axis. The actual area of the middle piece must be $(+)\frac{4}{3}$ square units.

Left-hand piece: By the symmetry of the curve, this has the same area as the right-hand piece, which is $\frac{4}{3}$.

Thus, the total area is $\frac{4}{3} + \frac{4}{3} + \frac{4}{3} = 4$.

Before we go on... If we had simply calculated $\int_{-2}^{2} (x^2 - 1) \, dx$, we would have obtained

$$\int_{-2}^{2} (x^2 - 1) \, dx = \left[\frac{x^3}{3} - x \right]_{-2}^{2} = \left[\left(\frac{2^3}{3} - 2 \right) - \left(\frac{(-2)^3}{3} - (-2) \right) \right]$$

$$= \frac{8}{3} - 2 + \frac{8}{3} - 2 = \frac{4}{3}.$$

Notice two things. First, this is the wrong answer. Second, it is the sum of the areas of the two pieces above the x-axis, minus the area below the x-axis. That is, it is $\frac{4}{3} + \frac{4}{3} - \frac{4}{3}$.

EXAMPLE 5

Calculate $\int_1^2 (2x - 1)e^{2x^2-2x} \, dx$.

SOLUTION This is one of those integrals that requires a u-substitution. We have two ways we can proceed. We could first make the substitution, then take the antiderivative, then substitute for u, and finally evaluate at $x = 2$ and $x = 1$ and subtract.

We prefer the following procedure. When we make the u-substitution, we calculate the u-values corresponding to $x = 2$ and $x = 1$ and change the limits of integration. This will save us having to put everything back in terms of x before substituting the limits. Here we go.

$$u = 2x^2 - 2x$$
$$\frac{du}{dx} = 4x - 2$$
$$dx = \frac{1}{4x - 2} du$$

When $x = 2$, $u = 4$ (substituting $x = 2$ in the formula $u = 2x^2 - 2x$)

When $x = 1$, $u = 0$ (substituting $x = 1$ in the formula $u = 2x^2 - 2x$)

Now, in the substitution step, we change the limits $x = 1$ and $x = 2$ to the corresponding limits $u = 0$ and $u = 4$.

$$\underset{\underset{x\text{-value}}{\nearrow}}{\int_1^{\overset{x\text{-value}}{\searrow}2}} (2x - 1)e^{2x^2-2x} \, dx = \int_0^{\overset{u\text{-value}}{\searrow}4} (2x - 1)e^u \frac{1}{4x - 2} du$$

$$= \int_0^4 (2x - 1)e^u \frac{1}{2(2x - 1)} du$$

$$= \int_0^4 e^u \frac{1}{2} du$$

$$= \frac{1}{2} \int_0^4 e^u \, du$$

$$= \frac{1}{2} [e^u]_0^4$$

$$= \frac{1}{2}(e^4 - e^0) = \frac{1}{2}(e^4 - 1)$$

13.4 Geometric Definition of the Definite Integral

Before we go on... The alternative process outlined in the first paragraph is this:

$$\int (2x - 1) e^{2x^2 - 2x} \, dx = \int e^u \frac{1}{2} du \quad \text{(after substitution)}$$

$$= \frac{1}{2} e^u + C$$

$$= \frac{1}{2} e^{2x^2 - 2x} + C,$$

and so

$$\int_1^2 (2x - 1) e^{2x^2 - 2x} \, dx = \left[\frac{1}{2} e^{2x^2 - 2x} \right]_1^2 = \frac{1}{2} (e^4 - 1).$$

▼ **EXAMPLE 6**

Calculate $\int_{-1}^{1} x(3x^2 + 3)^3 \, dx$.

SOLUTION We use substitution.

$$u = 3x^2 + 3$$

$$\frac{du}{dx} = 6x$$

$$dx = \frac{1}{6x} du$$

When $x = 1$, $u = 6$.
When $x = -1$, $u = 6$.

Substituting,

$$\int_{-1}^{1} x(3x^2 + 3)^3 \, dx = \int_6^6 xu^3 \frac{1}{6x} \, du$$

$$= \frac{1}{6} \int_6^6 u^3 \, du$$

$$= \frac{1}{6} \left[\frac{u^4}{4} \right]_6^6 = \frac{1}{24} (6^4 - 6^4) = 0.$$

Before we go on... Looking back at the calculation, notice that we could have saved ourselves a lot of work if we had paused to think after getting

$$\frac{1}{6} \int_6^6 u^3 \, du.$$

This is the definite integral of u^3 from $u = 6$ to $u = 6$ and is the area of a region with 0 width. No wonder the answer is 0!

EXAMPLE 7 Displacement

A car traveling down a road has velocity $v(t) = 30 + 2t$ mph at time t hours. How far does it travel between times $t = 1$ and $t = 5$?

SOLUTION The most straightforward way to do this would be to find the position $s(t)$ and then calculate the difference $s(5) - s(1)$ to find the difference in the car's positions at the first hour and the fifth. (Notice that the car never turns around to double back on itself, so $s(5) - s(1)$ *is* the total distance traveled.) There is a small problem with carrying out this calculation, however: we do not know where the car starts, so we can determine s only up to a constant of integration. This is only a small problem, since the constant will cancel out in the difference. On the other hand, since $s(t)$ is an antiderivative of $v(t)$, the difference $s(5) - s(1)$ is the value of $\int_1^5 v(t)\,dt$, so we can instead calculate this definite integral.

$$\int_1^5 v(t)\,dt = \int_1^5 (30 + 2t)\,dt = [30t + t^2]_1^5$$
$$= (150 + 25) - (30 + 1)$$
$$= 144 \text{ miles}$$

The way in which we solved the last example gives another useful interpretation of the integral. Remember that $f'(x)$ gives the rate of change of $f(x)$. If we integrate the rate of change over the closed interval $[a, b]$, that is, compute the definite integral $\int_a^b f'(x)\,dx$, we get the **total change** of $f(x)$ between $x = a$ and $x = b$, which is $f(b) - f(a)$. In the case of motion, the total change of position from one time to another is also known as the **displacement**. So, if $v(t)$ is velocity, $\int_a^b v(t)\,dt$ gives the displacement from time a to time b.

THE DEFINITE INTEGRAL AS TOTAL CHANGE

Given the rate of change $f'(x)$ of a quantity $f(x)$, the total change of $f(x)$ between $x = a$ and $x = b$ is given by:

$$\text{Total change in } f(x) = f(b) - f(a) = \int_a^b f'(x)\,dx.$$

EXAMPLE 8 Total Revenue

According to data published in *The New York Times*, the annual revenue of *United Airlines* increased more-or-less linearly from January 1988 through December 1992 and could be modeled by the equation

$$r(t) = 8.50 + 0.95t,$$

where $r(t)$ is *United Airlines'* annual revenue in billions of dollars, and t is time in years since January 1988.*

(a) What was the total revenue earned by *United Airlines* over the reported period?

(b) Obtain an equation giving the total revenue earned by *United Airlines* since January 1, 1988 as a function of t, where t is the time in years since January 1, 1988.

(c) Assuming the trend continues, use the model to predict the total revenue earned by *United Airlines* from January 1988 through December 1994.

SOLUTION **(a)** Let $R(t)$ be the total revenue earned by *United Airlines*, up to time t. The given function $r(t)$ is revenue per year, and is thus the derivative of $R(t)$. Hence, the total change in $R(t)$ from $t = 0$ to $t = 5$ is

$$R(5) - R(0) = \int_0^5 r(t)\, dt$$

$$= \int_0^5 (8.50 + 0.95t)\, dt$$

$$= [8.50t + 0.475t^2]_0^5$$

$$= (8.50(5) + 0.475(5)^2) - (8.50(0) + 0.475(0)^2)$$

$$= \$54.375 \text{ billion.}$$

(b) We are asked for $R(t) - R(0)$. We have already done most of the work in part (a).

$$R(t) - R(0) = [8.50t + 0.475t^2]_0^t$$

$$= (8.50(t) + 0.475(t)^2) - (8.50(0) + 0.475(0)^2)$$

$$= 8.50t + 0.475t^2$$

(c) We are asked for $R(7) - R(0)$. We can use the answer to part (b) with $t = 7$.

$$R(7) - R(0) = 8.50(7) + 0.475(7)^2$$

$$= \$82.775 \text{ billion}$$

Before we go on... Figure 6 shows a facsimile of the revenue chart that actually appeared in the cited *New York Times* article, with the graph of our linear model superimposed.

Notice something interesting: each rectangle in the shaded region has width one unit and height equal to the revenue earned by *United Airlines* during the specified year. Thus, the area of, say, the rectangle corresponding to 1991 is $1 \times 11.7 = 11.7$, the total revenue for that year. In other words,

▼ * The model is the authors', based on data published in *The New York Times*, Dec. 24, 1993, p. D1. (Source: Company Reports.)

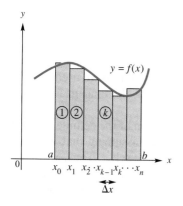

FIGURE 2

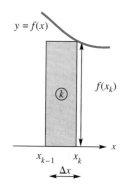

FIGURE 3

Before looking at the heights of the rectangles, we first give names to all the values of x that mark the boundaries between successive rectangles (see Figure 2). The first rectangle has its left edge at $x_0 = a$ and its right edge Δx units to the right, at x_1. Thus, $x_1 = a + \Delta x$. The second rectangle has its left edge at x_1 and its right edge another Δx units to the right, at x_2, with $x_2 = a + 2\Delta x$. In general, the kth rectangle has its left edge at x_{k-1} and its right edge Δx units to the right, at x_k, with $x_k = a + k\Delta x$. The last rectangle (the nth one) has its left edge at x_{n-1} and its right edge at $x_n = b$.

We now turn to the heights of the rectangles. If you look closely at Figures 1 and 2, you will notice that the right edge of each rectangle touches the graph of $y = f(x)$. Thus, the first rectangle has a height of $f(x_1)$, since its right edge has x-coordinate x_1 and the height of the graph at $x = x_1$ is given by $f(x_1)$. Similarly, the kth rectangle has height $f(x_k)$, and the last (nth) rectangle has height $f(x_n)$. Figure 3 shows an enlargement of the kth rectangle.

Its area is height $\times$ width $= f(x_k)\,\Delta x$. Now we add together the areas of all the rectangles, and obtain

$$\text{Approximate value of } \int_a^b f(x)\,dx = f(x_1)\,\Delta x + f(x_2)\,\Delta x$$
$$+ \cdots + f(x_k)\,\Delta x$$
$$+ \cdots + f(x_n)\,\Delta x \ .$$

Since sums such as these are often used in mathematics, a shorthand has been developed for them. We write

$$\sum_{k=1}^{n} f(x_k)\,\Delta x = f(x_1)\,\Delta x + f(x_2)\,\Delta x + \cdots + f(x_k)\,\Delta x$$
$$+ \cdots + f(x_n)\,\Delta x.$$

The symbol "Σ" is the Greek letter sigma, and stands for **summation.** The letter k here is called the index of summation, and we can think of it as counting off the rectangles. We read the notation as "the sum from $k = 1$ to n of the quantities $f(x_k)\,\Delta x$." Think of it as a set of instructions:

Set $k = 1$, and obtain $f(x_1)\,\Delta x$.
Set $k = 2$, and obtain $f(x_2)\,\Delta x$.
. . .
Set $k = n$, and obtain $f(x_n)\,\Delta x_n$.
Now sum all the above quantities.

Q Since this is an approximation to the definite integral $\int_a^b f(x)\,dx$, how do we get the *exact* area?

A By our usual technique of passing to a limit. We can obtain more and more accurate approximations to the area by taking smaller values of Δx (or equivalently, taking larger and larger values for n). This is suggested by Figure 4, which shows the approximations for $n = 6$ and $n = 12$. Thus, to obtain the *exact* area or definite integral, we pass to the limit as $\Delta x \to 0$. In other words, we make the following definition.

13.5 Algebraic Definition of the Definite Integral

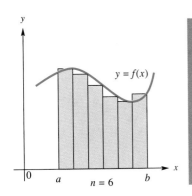

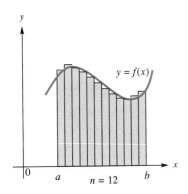

FIGURE 4

> **ALGEBRAIC DEFINITION OF THE DEFINITE INTEGRAL**
>
> If $f(x)$ is any continuous function, then the **definite integral of f from a to b** is defined to be
>
> $$\int_a^b f(x)\,dx = \lim_{\Delta x \to 0} \sum_{k=1}^n f(x_k)\,\Delta x,$$
>
> where $\Delta x = \dfrac{b-a}{n}$, and $x_k = a + k\Delta x$. The limit is obtained by letting $\Delta x \to 0$, or equivalently, by letting $n \to +\infty$.

(Actually, this is one of several definitions, all equivalent for continuous f but different for pathological functions.) By the way, you can see now where the notation for the integral came from: the "$\int$" is an elongated "S," the Roman equivalent of the Greek Σ, $f(x_k)$ becomes $f(x)$, and Δx becomes dx in the limit.

The sum

$$\sum_{k=1}^n f(x_k)\,\Delta x$$

by itself is called the ***n*th right-hand sum** of f (because we have taken the height of each rectangle to be the value of the function at the x-coordinate of its *right* edge). We get the ***n*th left-hand sum** by replacing $f(x_k)$ by $f(x_{k-1})$ in the above formula. The limit is the same as for the right-hand sum. We can get other kinds of "in-between" sums by using $f(x_k^*)$ instead of $f(x_k)$, where x_k^* is any value of x between x_{k-1} and x_k. All these choices lead to the same result in the limit. All these different kinds of sums are referred to as **Riemann sums,** after the nineteenth-century mathematician Bernhard Riemann.

▶ **NOTES**
1. The fact that the limit exists, and is the same for the left- and right-hand sums, is not easy to prove. It depends on the fact that we assumed that $f(x)$ was continuous on the interval $[a, b]$.
2. Our assumption that the function is positive can be dropped. The formula for the nth sum automatically subtracts area below the x-axis, since if the graph lies below the x-axis at $x = x_k$, $f(x_k)$ will be negative.
3. The subdivision of the interval $[a, b]$ into n equal parts is usually referred to as a **partition** of $[a, b]$. ◀

▼ **EXAMPLE 1**

Find the left- and right-hand sums of $f(x) = x^2$ on the interval $[0, 1]$ with $n = 5$. Then calculate the average of the two sums, and compare the answer with the actual integral. How accurate is it?

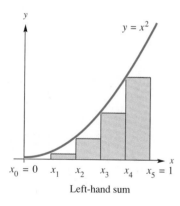
Left-hand sum

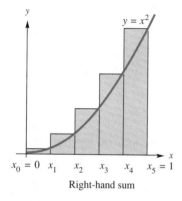
Right-hand sum

FIGURE 5

SOLUTION Before we do the calculations, look at Figure 5, which shows the rectangles for these sums.

Looking at the figure, we expect the left-hand sum be smaller than the actual area and the right-hand sum to be bigger. We now turn to the calculation.

Step 1: First calculate $\Delta x = \dfrac{b-a}{n}$.

Here, $a = 0$, $b = 1$ and $n = 5$, so $\Delta x = (1 - 0)/5 = 0.2$.

Step 2: Calculate the values of x_k and $f(x_k)$ for $0 \leq k \leq n$.

Since $n = 5$, we must calculate 6 values of $f(x_k)$. These are shown in the following table.

k	0	1	2	3	4	5
$x_k = a + k\Delta x$	0	0.2	0.4	0.6	0.8	1.0
$f(x_k)$	0	0.04	0.16	0.36	0.64	1.0

Step 3: For the *left-hand* sum, add the values of $f(x_k)$ starting with the leftmost ($k = 0$) and ending with $k = n - 1 = 4$, and then multiply the sum by Δx. For the *right-hand sum*, add the values of $f(x_k)$, starting with the second ($k = 1$) and ending at $k = n = 5$, and then multiply the sum by Δx.

Left-hand sum: $(0 + 0.04 + 0.16 + 0.36 + 0.64)(0.2) = 0.24$

Right-hand sum: $(0.04 + 0.16 + 0.36 + 0.64 + 1.0)(0.2) = 0.44$

Both sums are approximations of the definite integral $\int_0^1 x^2 \, dx$. The Fundamental Theorem of Calculus tells us that the exact answer is

$$\int_0^1 x^2 \, dx = \left[\frac{x^3}{3}\right]_0^1 = \frac{1}{3} = 0.3333\ldots$$

Although neither the left-hand sum nor the right-hand sum gives a good approximation, we are asked to also look at the average of the two sums.

$$\text{average sum} = \frac{\text{Left-hand sum} + \text{Right-hand sum}}{2}$$

$$= \frac{0.24 + 0.44}{2} = 0.34,$$

which is a far better approximation. For reasons that will become clearer in Section 7, this average is referred to as the **trapezoidal sum**.

Now we can complete the answer to the question: we compare the trapezoidal sum with the actual sum by taking the magnitude of their difference,

$$|0.34 - 0.3333\ldots| = 0.00666\ldots$$

This is the **error term.** Now round this error to a single significant digit.

$$0.00666\ldots \approx 0.007$$

We say our answer is accurate to within ± 0.007.

Before we go on... If we use Riemann sums to approximate an integral, we will usually not be able to determine the exact error (otherwise we would know the exact answer, so there would be little point in computing an approximation!). However, there are ways of computing the maximum possible error. We shall return to this topic in Section 7.

GRAPHING CALCULATORS

Calculating left- and right-hand sums can be a tedious process, especially for large values of n. (If you chose $n = 100$, for example, you would be kept busy filling in values in a table with 101 columns!) Luckily, we have programmable graphing calculators and computers to do the chore for us. In the appendix you will find a graphing calculator program called SUMS that computes both the left and right Riemann sums. (It works on the TI-82 and higher numbered models. You should read through the section on programming in your calculator instruction booklet before attempting to enter this program.)

The next example uses the TI-82 program.

EXAMPLE 2

Use a graphing calculator to calculate the left- and right-hand sums for the integral

$$\int_{-1}^{1} \sqrt{1 - x^2}\, dx,$$

with $n = 100$, 200, and 500.

SOLUTION   In the "Y =" window, we enter the function:

$$Y_1 = (1 - X^{\wedge}2)^{\wedge}(0.5)$$

We then hit [PRGM], select the program SUMS (the name we have given to the program) and enter 100 for N, -1 for the left endpoint, and 1 for the right endpoint. We obtain the following answers.

Left-hand sum: 1.56913 . . .
Right-hand sum: 1.56913 . . .

With $N = 200$, we obtain

Left-hand sum: 1.57020 . . .
Right-hand sum: 1.57020 . . .

With $N = 500$, we obtain

Left-hand sum: 1.57064 . . .
Right-hand sum: 1.57064 . . .

Exercises 17–22 require the use of a graphing calculator with the program described in the text, or a similar computer program. In each case, calculate the left-hand, right-hand and trapezoidal sums for the given integrals with **(a)** *n* = 100 *and* **(b)** *n* = 200. *Round all answers to four decimal places. [Your graphing calculator should have* $\boxed{\sin}$ *and* $\boxed{\pi}$ *buttons.]*

17. $\int_0^1 4\sqrt{1-x^2}\, dx$

18. $\int_0^1 \dfrac{4}{1+x^2}\, dx$

19. $\int_2^3 \dfrac{2x^{1.2}}{1+3.5x^{4.7}}\, dx$

20. $\int_3^4 3xe^{1.3x}\, dx$

21. $\int_0^\pi \sin(x)\, dx$

22. $\int_0^\pi 2(\sin(x))^2\, dx$

APPLICATIONS

23. *Sales* Estimated sales of *IBM*'s mainframe computers and related peripherals showed the following pattern of decline over the years 1990 through 1994.*

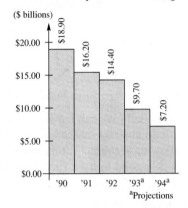

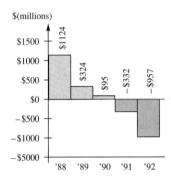

(a) Using the figures for 1990 and 1994, modexl IBM's annual mainframe sales using a linear function $s(t) = mt + b$, where $s(t)$ is the annual sales in billions of dollars and t is time in years since the start of 1990. (Take $t = 0.5$ for the 1990 figure and $t = 4.5$ for the 1994 figure.)

(b) Approximate $\int_0^5 s(t)\, dt$ by a trapezoidal sum with $n = 5$. How accurately does your model estimate IBM's total mainframe sales over the given period? How do you account for this discrepancy?

24. *Total Profit* The profitability of *United Airlines* for the years 1988 through 1992 showed the following decline.†

(a) Using the figures for 1988 and 1992, model United's annual profits using a linear function $p(t) = mt + b$, where $p(t)$ is the annual profit in millions of dollars and t is time in years since the start of 1988. (Take $t = 0.5$ for the 1988 figure and $t = 4.5$ for the 1992 figure.)

(b) Approximate $\int_0^5 p(t)\, dt$ by a trapezoidal sum with $n = 5$. How accurately does your model estimate United's aggregate profit or loss over the given period? Explain any discrepancy you find.

25. *Cost* The marginal cost function for the manufacture of portable CD players is given by

$$C'(x) = 20 - \dfrac{x}{5,000},$$

where x is the number of CD players manufactured. Use the trapezoidal sum with $n = 1$ to estimate the cost of producing CD players 11 through 100. Compare your answer with the answer given by the Fundamental Theorem of Calculus.

26. *Cost* Repeat the previous exercise using the marginal cost function

$$C'(x) = 25 - \dfrac{x}{50}.$$

▼ *Source: Salomon Brothers (*The New York Times*, October 26, 1993, p. D1.)
† Source: Company Reports (*The New York Times*, December 24, 1993, p. D1.)

27. Surveying My uncle intends to build a kidney-shaped swimming pool in his small yard, and the town zoning board will approve the project only if the total area of the pool does not exceed 500 square feet. The accompanying figure shows a diagram of the planned swimming pool, with measurements of its width at the indicated points. Will my uncle's plans be approved? (Use left- and right-hand sum estimates.)

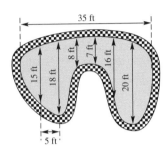

28. Pollution An aerial photograph of an ocean oil spill shows the pattern in the accompanying diagram. Assuming that the oil slick has a uniform depth of 0.01 meters, how many cubic meters of oil would you estimate to be in the spill? (Volume = Area × Thickness)

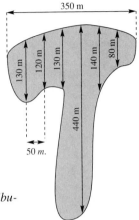

The Normal Curve The normal distribution curve, which models the distributions of data in a wide range of applications, is given by the function

$$p(x) = \frac{1}{\sqrt{2\pi}\,\sigma} e^{-(x-\mu)^2/2\sigma^2},$$

where $\pi = 3.14159265\ldots$ and σ and μ are constants called the standard deviation and the mean. Exercises 29 and 30 illustrate its use and require a graphing calculator programmed to calculate trapezoidal sums.

29. Test Scores Enormous State University's Calculus I test scores are modeled by the normal distribution with $\mu = 72.6$ and $\sigma = 5.2$. The percentage of students who obtained scores between a and b on the test is given by

$$\int_a^b p(x)\,dx.$$

(a) Use a trapezoidal Riemann sum with $n = 40$ to estimate the percentage of students who obtained between 60 and 100 on the test.
(b) What percentage of students scores less than 30?

30. Consumer Satisfaction In a survey, consumers were asked to rate a new toothpaste on a scale of 1–10. The resulting data are modeled by a normal distribution with $\mu = 4.5$ and $\sigma = 1.0$. The percentage of consumers who rated the toothpaste with a score between a and b on the test is given by

$$\int_a^b p(x)\,dx.$$

(a) Use a trapezoidal Riemann sum with $n = 10$ to estimate the percentage of customers who rated the toothpaste 5 or above. (Use the range 4.5 to 10.5.)
(b) What percentage of customers rated the toothpaste 0 or 1? (Use the range -0.5 to 1.5.)

COMMUNICATION AND REASONING EXERCISES

31. Let $f(x) = mx + c$, where m and c are constants. Why does the trapezoidal approximation to $\int_a^b f(x)\,dx$ gives the exact answer with $n = 1$? Convince yourself of this by sketching the left-hand and right-hand sums for $\int_0^1 (1-x)\,dx$.

32. Let $f(x)$ be a decreasing function on the interval $[a, b]$. (That is, $f'(x) \leq 0$ on the interval.) Demonstrate by means of a sketch that the left-hand sum is always greater than or equal to the right-hand sum. What can you say about *increasing* functions?

33. Let $f(x)$ be an increasing function. Draw a picture and demonstrate that the difference between the right- and left-hand sums for $\int_a^b f(x)\,dx$ is $(f(b) - f(a))\Delta x$. Conclude that the difference goes to 0 as $n \to +\infty$

34. Another approximation of the integral is the **midpoint** approximation, in which we compute the sum

$$\sum_{i=1}^{n} f(\bar{x}_k)\,\Delta x,$$

where $\bar{x}_k = (x_{k-1} + x_k)/2$ is the point midway between the left and right endpoints of the interval $[x_{k-1}, x_k]$. Why is it true that the midpoint approximation is exact if f is linear (compare Exercise 31)?

13.6 THE FUNDAMENTAL THEOREM OF CALCULUS

Although we have been using the Fundamental Theorem of Calculus rather heavily, we have not explained where it comes from. Now that we are familiar with the definite integral as an area or limit of a sum, the proof of the Fundamental Theorem will seem less mysterious. Here is the complete statement of the theorem.

> **THEOREM (THE FUNDAMENTAL THEOREM OF CALCULUS, OR FTC)**
>
> Let f be any continuous function defined on the interval $[a, b]$. Then
>
> **1.** $f(x)$ has an antiderivative, namely, $\int_a^x f(t)\,dt$, and
> **2.** if $F(x)$ is any antiderivative of $f(x)$, then
>
> $$\int_a^b f(x)\,dx = F(b) - F(a).$$

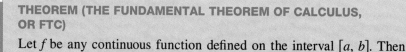

FIGURE 1

We shall outline the proof of these two statement over the next several pages. By the way, what we are about to do was first done by the English mathematician Isaac Barrow (1630–1677), who was Newton's teacher. It was Newton, however, who recognized its importance.

For convenience, we shall take f as a function of t rather than x (see Figure 1). Our figures will assume that f is always positive, but this is not necessary.

Concentrating on the second statement of the FTC, we consider the question: How do we calculate the definite integral $\int_a^b f(t)\,dt$? We start by considering a harder question. We shall try to calculate, for every x between a and b, the area

$$A(x) = \int_a^x f(t)\,dt.$$

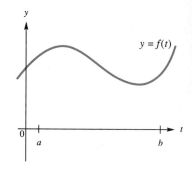

FIGURE 2

This is the area under the curve from $t = a$ to $t = x$, as shown in Figure 2. Observe that varying x means moving the vertical line on the right of the shaded area. This changes the value of $A(x)$. In other words, $A(x)$ *depends on x*. In fact, it is a *rule* that gives us a number—namely, the area—for any specified value of x. Thus, it is a perfectly respectable function

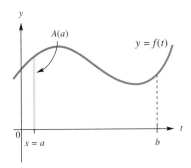
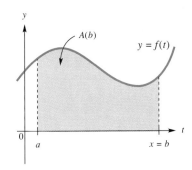

FIGURE 3

of x, even if we do not have a formula for its values. The domain of $A(x)$ is $[a, b]$. Figure 3 shows two specific values of $A(x)$.

From the figure, we see that $A(a) = 0$, and $A(b) = \int_a^b f(t)\, dt$, which is what we are trying to calculate.

Q Now what do we do with this new function $A(x)$?

A Since we are doing calculus, why not take its derivative?

Q But how do you take the derivative of such a strange function?

A The same way you take the derivative of *any* new function: by using the definition

$$A'(x) = \lim_{h \to 0} \frac{A(x + h) - A(x)}{h}.$$

This may not be as easy as it looks. We do not have an algebraic formula for $A(x)$, so how do we deal with the difference quotient? All we know about $A(x + h)$ and $A(x)$ is that they represent areas, so we sketch them in Figure 4.

The figure shows us that $A(x + h) - A(x)$ is equal to the bigger area minus the smaller area, giving the area of the "sliver" on the right.

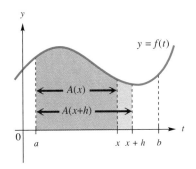
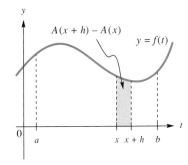

FIGURE 4

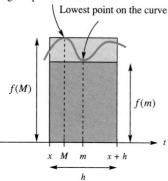

FIGURE 5

Thus, we write

$$A'(x) = \lim_{h \to 0} \frac{A(x+h) - A(x)}{h} = \lim_{h \to 0} \frac{\text{Area of sliver}}{h}.$$

Now we still need to say something about the area of the sliver. To do this, we are going to look at a magnified picture of the sliver, and we'll also allow for the fact that the curve might have been more wobbly than we originally drew it (see Figure 5).

Now what on earth have we done to it? First, notice that our magnified view shows the wobbles we promised. Second, we have chosen two new t-values: M and m. M is the value of t between x and $x + h$ that gives the maximum value of f, while m is the t-value that gives the minimum value of f. These give the highest and lowest points of the curve restricted to the interval $[x, x + h]$. We have also shaded in two rectangles. The taller one, of height $f(M)$ and width h, is clearly an overestimation of the area of the sliver, while the shorter one, of height $f(m)$ and width h, is an underestimation of the area of the sliver. In other words, the area of the sliver is somewhere *between* the areas of the two rectangles. We can represent this by the inequality

$$\text{Area of short rectangle} \leq \text{Area of sliver} \leq \text{Area of tall rectangle},$$

or

$$h\, f(m) \leq \text{Area of sliver} \leq h\, f(M).$$

Now we divide the whole inequality by h to get

$$\frac{h\, f(m)}{h} \leq \frac{\text{Area of sliver}}{h} \leq \frac{h\, f(M)}{h}$$

(we are assuming that $h > 0$ as in the picture, but something similar can be done if $h < 0$) or, canceling the h's,

$$f(m) \leq \frac{\text{Area of sliver}}{h} \leq f(M).$$

Now remember that we are trying to find the limit of the middle term as h approaches zero. First, let us look at what is happening to the two end terms as $h \to 0$. Since both the points m and M are forced to lie between x and $x + h$, it follows that every time we choose a smaller value for h, both m and M must be selected once again, closer and closer to x (since $x + h$ is approaching x). In other words, both m and M approach x as $h \to 0$. Thus (because of the continuity of f), both $f(m)$ and $f(M)$ are approaching $f(x)$ as $h \to 0$. We have the following situation as $h \to 0$.

$$\underset{\underset{\text{approaching } f(x)}{\uparrow}}{f(m)} \leq \frac{\text{Area of sliver}}{h} \leq \underset{\underset{\text{approaching } f(x)}{\uparrow}}{f(M)}$$

Since the two end terms have limit $f(x)$ as $h \to 0$, and since the middle quantity is sandwiched between them, it too must be approaching $f(x)$ as $h \to 0$.

So we conclude

$$A'(x) = \lim_{h \to 0} \frac{A(x+h) - A(x)}{h} = \lim_{h \to 0} \frac{\text{Area of sliver}}{h} = f(x).$$

Thus, we have, with our bare hands, calculated the derivative of $A(x)$ and found that $A'(x) = f(x)$.

Q Good job! But what is the point of all of this?

A We have found that *the derivative of $A(x)$ is $f(x)$*. Rephrasing this,

the area function $A(x)$ is an antiderivative of $f(x)$.

So we are finally beginning to see a connection between area and antiderivatives. This is, in fact, the first statement of the FTC.

Now let us return to the question: How do we calculate the definite integral $\int_a^b f(t)\, dt$? We are asking for the value of $A(b)$, which we now know to be an antiderivative of $f(x)$, evaluated at $x = b$. Let us think for a moment about *any* antiderivative $F(x)$ of $f(x)$. Since both $F(x)$ and $A(x)$ are antiderivatives of $f(x)$, they differ by a constant, so we can write

$$F(x) = A(x) + C.$$

Now compare $F(b) - F(a)$ with $A(b) - A(a)$:

$$F(b) - F(a) = (A(b) + C) - (A(a) + C)$$
$$= A(b) - A(a).$$

But recall that $A(a) = 0$. Thus,

$$F(b) - F(a) = A(b) = \int_a^b f(t)\, dt.$$

This is the second statement of the FTC.

▶ **NOTES**
1. The first statement of the Fundamental Theorem of Calculus is more interesting than you might think at first. It says that if you have any continuous function whatsoever, then there is an antiderivative for it lurking around somewhere.
2. Notice that another way of saying that $\int_a^x f(t)\, dt$ is an antiderivative of $f(x)$ is to say that its derivative is $f(x)$. In other words, we have the following formula.

DERIVATIVE OF THE INTEGRAL

$$\frac{d}{dx} \int_a^x f(t)\, dt = f(x)$$

3. The second part of the theorem justifies the method of calculating areas we used in the last section, and it took mathematicians a long time to realize this! Prior to the discovery of the Fundamental Theorem, mathematicians calculated areas under curves by using Riemann sums. They sliced the area into thin rectangles to get an approximate area and then tried to take the limit as the rectangles got narrower and narrower. As a result, the calculation of even simple integrals such as $\int_b^a e^x \, dx$ became a formidable task, worthy of a published article in a mathematics journal! ◄

▼ EXAMPLE 1

Find the area function $A(x) = \int_1^x f(t) \, dt$ if $f(x) = 1/x^2$. Graph both $f(x)$ and $A(x)$.

SOLUTION We compute $A(x)$ using the Fundamental Theorem of Calculus: we first find an antiderivative of f.

$$A(x) = \int_1^x \frac{1}{t^2} \, dt = \int_1^x t^{-2} \, dt = -t^{-1} \Big|_1^x = -x^{-1} - (-1) = 1 - \frac{1}{x}$$

The graphs of $f(x)$ and $A(x)$ are shown in Figure 6.

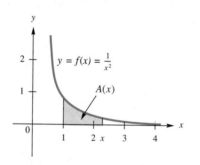
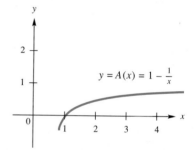

FIGURE 6

Before we go on... Notice that $A(x)$ is an antiderivative of $f(x)$. It is the particular antiderivative such that $A(1) = 0$, since $\int_1^1 1/t^2 \, dt$, the area from 1 to 1, must be 0. Notice also that $A(x)$ increases but never gets larger than 1. We shall see the significance of this in the next chapter when we discuss improper integrals.

▼ EXAMPLE 2

Find $\dfrac{d}{dx} \int_0^x e^{t^2} \, dt$.

SOLUTION We have $\dfrac{d}{dx} \int_0^x e^{t^2} \, dt = e^{x^2}$ by the first statement in the Fundamental Theorem.

Before we go on... The remarkable thing about this example is that it is impossible to write an explicit formula for $\int_0^x e^{t^2}\,dt$ in terms of "elementary" functions (which include all those we have talked about and a few more). Nonetheless, we found an explicit formula for the *derivative* of this function.

▶ **13.6 EXERCISES**

1. Use geometry (not antiderivatives) to compute $A(x) = \int_0^x 1\,dt$, and verify that $A'(x) = 1$.

2. Use geometry (not antiderivatives) to compute $A(x) = \int_0^x t\,dt$, and verify that $A'(x) = x$.

3. Use geometry to compute $A(x) = \int_2^x (t - 2)\,dt$, and verify that $A'(x) = x - 2$.

4. Use geometry to compute $A(x) = \int_2^x 1\,dt$, and verify that $A'(x) = 1$.

5. Use geometry to compute $A(x) = \int_a^x c\,dt$, and verify that $A'(x) = c$. (c is a constant.)

6. Use geometry to compute $A(x) = \int_a^x (t - 2)\,dt$, and verify that $A'(x) = x - 2$.

For each of the functions in Exercises 7–16, use antiderivatives to find the area function $A(x) = \int_a^x f(t)\,dt$. Sketch both $f(x)$ and $A(x)$.

7. $f(x) = x$, $a = 0$

8. $f(x) = 2x$, $a = 1$

9. $f(x) = x^2$, $a = 0$

10. $f(x) = x^2 + 1$, $a = 0$

11. $f(x) = e^x$, $a = 0$

12. $f(x) = e^{-x}$, $a = 0$

13. $f(x) = 1/x$, $a = 1$

14. $f(x) = 1/(x + 1)$, $a = 0$

15. $f(x) = \begin{cases} x & \text{if } 0 \le x \le 1 \\ 2 & \text{if } x > 1 \end{cases}$, $a = 0$

16. $f(x) = \begin{cases} 2x & \text{if } 0 \le x \le 2 \\ 1 & \text{if } x > 2 \end{cases}$, $a = 0$

APPLICATIONS

17. *The Natural Logarithm Returns* Suppose that you had never heard of the natural logarithm function and were therefore stuck when you tried to find an antiderivative of $\frac{1}{x}$. Here is how you might proceed.
    (a) What does the Fundamental Theorem of Calculus give as an antiderivative of $\frac{1}{x}$?
    (b) Choosing $a = 1$, give this "new" function the name $M(x)$, and derive the following properties: $M(1) = 0$; $M'(x) = \frac{1}{x}$.
    (c) Use the chain rule for derivatives to show that if a is any positive constant, then
    $$\frac{d}{dx}(M(ax)) = \frac{1}{x}.$$
    (d) Let $F(x) = M(ax) - M(x)$. Deduce that $F'(x) = 0$, and hence that $F(x) = \text{constant}$.
    (e) By setting $x = 1$, find the value of this constant, and hence deduce that
    $$M(ax) = M(a) + M(x).$$

18. *The Error Function* The **error function**, $\text{erf}(x)$, is defined by
    $$\text{erf}(x) = \frac{2}{\sqrt{\pi}} \int_0^x e^{-t^2}\,dt.$$
    This function is very important in statistics.
    (a) Find $\text{erf}'(x)$.
    (b) Use a trapezoidal sum with $n = 5$ to approximate $\text{erf}(1)$ and $\text{erf}(2)$.
    (c) Find an antiderivative of e^{-x^2} in terms of $\text{erf}(x)$.
    (d) Use the answers to parts (b) and (c) to approximate
    $$\int_1^2 e^{-x^2}\,dx.$$

In the Exercises 19–22, use a graphing calculator or computer to obtain the graphs of the given function with the given ranges. Use "trace" or some other method to evaluate the function as requested. (Your answers should be accurate to two decimal places. The instructions after the exercises refer to the TI-82 and similar models.)

19. $f(x) = \int_0^x e^{t^2}\, dt$ ($-3 \le x \le 3$, $-1 \le y \le 2$). Calculate $f(0)$, $f(0.5)$, and $f(1)$.
[On the TI-82, set $Y_1 =$ fnInt (e^ (−T^ 2), T, 0, X) in the "Y=" menu and hit GRAPH. Have patience! The calculator must compute each y-coordinate by doing the integral numerically.]

20. $f(x) = \int_0^x t^2 e^t\, dt$ ($-3 \le x \le 3$, $-1 \le y \le 2$). Calculate $f(-1)$, $f(0)$, and $f(1)$.
[$Y_1 =$ fnInt(T^ 2e^T, T, 0, X)]

21. $f(x)$ is the antiderivative of $\sqrt{1 - x^2}$ with the property that $f(0) = 0$ ($-1 \le x \le 1$, $-1 \le y \le 1$). Calculate $f(-0.5)$ and $f(0.5)$.

22. $f(x)$ is the antiderivative of $\dfrac{1}{\sqrt{1 - x^2}}$ with the property that $f(0) = 0$ ($-0.5 \le x \le 0.5$, $-2 \le y \le 2$). Calculate $f(-0.5)$ and $f(0.5)$.

COMMUNICATION AND REASONING EXERCISES

23. What does the Fundamental Theorem of Calculus permit one to do?

24. Your friend has just told you that the function $f(x) = e^{-x^2}$ can't be integrated and hence has no antiderivative. Is your friend correct? Explain your answer.

25. Use the FTC to find an antiderivative F of
$$f(x) = \begin{cases} 0 & \text{if } x < 0 \\ 1 & \text{if } x \ge 0. \end{cases}$$
Is it true that $F'(x) = f(x)$ for *all* x?

26. According to the Fundamental Theorem of Calculus as stated in this text, which functions are guaranteed to have antiderivatives? What does the answer to the previous exercise tell you about the theorem as stated in this text?tx

13.7 NUMERICAL INTEGRATION

The Fundamental Theorem of Calculus gives us an exact formula for computing $\int_a^b f(x)\, dx$, *provided* we can find an antiderivative for f. This method of evaluating definite integrals is called the **analytic** method. However, as we have seen, there are times when this is difficult or impossible. In these cases, it is usually good enough to find an approximate, or **numerical** solution, and there are some very simple ways to do this. We have already been using the left-hand and right-hand Riemann sums, as well as their average, which we referred to, mysteriously, as the "trapezoidal" sum.

In this section, we shall take this a step further. First, we shall find out where the term "trapezoidal" comes from. Second, we shall develop a more efficient method of approximating the integral, called Simpson's rule. Finally, we shall discuss the accuracy of both the trapezoid rule and Simpson's rule, so that we can choose the number of partitions with more confidence when approximating the integral by a sum.

THE TRAPEZOID RULE AND SIMPSON'S RULE

The **trapezoid rule** starts by taking a simple partition of $[a, b]$ into n intervals of the same width, just as we did with the left- and right-hand sums. Since the total width must be $b - a$, each small interval must have width $\Delta x = (b - a)/n$. As we did with the Riemann sums, we take

$$x_0 = a$$
$$x_1 = a + \Delta x = a + \frac{b - a}{n}$$
$$x_2 = a + 2\Delta x = a + 2\frac{b - a}{n}$$
$$\ldots$$
$$x_k = a + k\Delta x = a + k\frac{b - a}{n}$$
$$\ldots$$
$$x_n = a + n\Delta x = b.$$

The trapezoid rule gives us the following approximation to the definite integral.

TRAPEZOID RULE

If $x_k = a + k\Delta x = a + k\frac{b - a}{n}$, then

$$\int_a^b f(x)\, dx \approx \frac{b - a}{2n} [f(x_0) + 2f(x_1) + 2f(x_2) + \ldots + 2f(x_{n-1}) + f(x_n)].$$

Q This doesn't look like a Riemann sum at all. Why is it an approximation to the integral?

A Recall that the left-hand and right-hand sums were given by

$$\text{Left-hand sum} = \Delta x[f(x_0) + f(x_1) + f(x_2) + \ldots + f(x_{n-1})]$$

and

$$\text{Right-hand sum} = \Delta x[f(x_1) + f(x_2) + \ldots + f(x_{n-1}) + f(x_n)].$$

The average—or what we referred to as the "trapezoidal" sum, is given by

$$\text{Average} = \frac{\text{Left-hand sum} + \text{Right-hand sum}}{2}$$

$$= \frac{1}{2}\Delta x[f(x_0) + f(x_1) + f(x_2) + \ldots + f(x_{n-1}) + f(x_1)$$
$$+ f(x_2) + \ldots + f(x_{n-1}) + f(x_n)]$$

$$= \frac{\Delta x}{2}[f(x_0) + 2f(x_1) + 2f(x_2) + \ldots + 2f(x_{n-1}) + f(x_n)]$$

$$= \frac{b-a}{2n}[f(x_0) + 2f(x_1) + 2f(x_2) + \ldots + 2f(x_{n-1}) + f(x_n)],$$

which is the trapezoid rule. In other words, the trapezoid rule just gives the trapezoidal sum, or the average of the left- and right-hand sums. Thus, although it is not really a Riemann sum, it is the average of two Riemann sums.

Q Where does the term "trapezoid" come from?

A Look at Figure 1.

Instead of approximating the area in each strip with the area of a rectangle, as we do for Riemann sums, the idea is to use a different shape. Between x_{k-1} and x_k we take the shaded region shown in Figure 2.

This is a **trapezoid,** a four-sided region with two opposite sides parallel. In this case it is the two vertical sides that are parallel. The area of a trapezoid is the average length of the parallel sides, times the distance between them, which in this case gives

$$\frac{f(x_{k-1}) + f(x_k)}{2} \cdot \frac{b-a}{n} = \frac{b-a}{2n}[f(x_{k-1}) + f(x_k)].$$

If we now add up the area of these trapezoids, each $f(x_k)$ will appear twice, except for $f(a)$ and $f(b)$, which appear once each. The sum is then the formula given as the trapezoid rule.

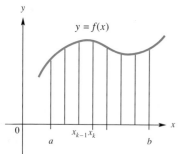

FIGURE 1

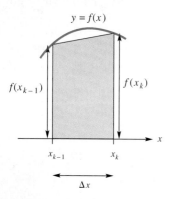

FIGURE 2

▼ **EXAMPLE 1**

Use the trapezoid rule to approximate $\int_0^1 x^2\, dx$, **(a)** using 5 intervals and **(b)** using 10 intervals.

SOLUTION Instead of calculating the left- and right-hand sums separately and then taking the average, we can calculate the trapezoid rule in one step using the same table we used for the Riemann sums.

(a) $n = 5$, so $\Delta x = \dfrac{b-a}{n} = \dfrac{1-0}{5} = 0.2$.

k	0	1	2	3	4	5
$x_k = a + k\Delta x$	0	0.2	0.4	0.6	0.8	1.0
$f(x_k)$	0	0.04	0.16	0.36	0.64	1.0

By the trapezoid rule,

$$\int_0^1 x^2\, dx \approx \dfrac{b-a}{2n}[f(x_0) + 2f(x_1) + 2f(x_2) + \ldots + 2f(x_{n-1}) + f(x_n)]$$
$$= 0.1[0 + 2(0.04) + 2(0.16) + 2(0.36) + 2(0.64) + 1.0]$$
$$= 0.34.$$

(b) $n = 10$, so $\Delta x = \dfrac{b-a}{n} = \dfrac{1-0}{10} = 0.1$.

x	0	1	2	3	4	5	6	7	8	9	10
$x_k = a + k\Delta x$	0	0.1	0.2	0.3	0.4	0.5	0.6	0.7	0.8	0.9	1.0
$f(x_k)$	0	0.01	0.04	0.09	0.16	0.25	0.36	0.49	0.64	0.81	1.0

By the trapezoid rule,

$$\int_0^1 x^2\, dx \approx \dfrac{b-a}{2n}[f(x_0) + 2f(x_1) + 2f(x_2) + \ldots + 2f(x_{n-1}) + f(x_n)]$$
$$= 0.05[0 + 2(0.10) + 2(0.04) + 2(0.09) + 2(0.16) + 2(0.25)$$
$$\quad + 2(0.36) + 2(0.49) + 2(0.64) + 2(0.81) + 1.0]$$
$$= 0.335.$$

Before we go on... The exact answer, of course, is

$$\int_0^1 x^2\, dx = \left[\dfrac{x^3}{3}\right]_0^1 = \dfrac{1}{3} = 0.3333\ldots$$

▼ **EXAMPLE 2**

Use the trapezoid rule with 6 intervals to approximate

$$\int_0^6 e^{-x^2}\, dx.$$

SOLUTION This example is interesting because it has been shown that the antiderivative of e^{-x^2} cannot be written in terms of "elementary" functions (which include those we have talked about here, and a few more).* However, we can easily approximate this integral. Since we are using $n = 6$, we have $\Delta x = (b - a)/n = (6 - 0)/6 = 1$, and so

$$\int_0^6 e^{-x^2} \, dx \approx 0.5[e^0 + 2e^{-1} + 2e^{-4} + 2e^{-9} + 2e^{-16} + 2e^{-25} + e^{-36}]$$
$$\approx 0.886.$$

Before we go on... You would probably be surprised to learn that the exact value of the slightly larger integral, $\int_0^\infty e^{-x^2} \, dx$, is in fact $\sqrt{\pi}/4$. Since $\sqrt{\pi}/4 = 0.8862\ldots$, this agrees with our answer to three decimal places. (The area $\int_6^\infty e^{-x^2} \, dx$ is exceptionally small.)

Simpson's rule is another approximation of the integral. Again, we start by partitioning $[a, b]$ into intervals all of the same width, but this time we must use an even number of intervals, so n will be even.

SIMPSON'S RULE

If n is even, and $x_k = a + k \Delta x = a + k\dfrac{b - a}{n}$, then

$$\int_a^b f(x) \, dx \approx \frac{b - a}{3n}[f(a) + 4f(x_1) + 2f(x_2) + 4f(x_3)$$
$$+ \ldots + 2f(x_{n-1}) + 4f(x_{n-1}) + f(b)].$$

FIGURE 3

Q Why?

A As with the trapezoid rule, we want to approximate the areas in each strip by something more complicated than a rectangle. This time we take the strips in pairs (which is why we need an even number of them) and draw a *parabola* through the three points $(x_{k-1}, f(x_{k-1}))$, $(x_k, f(x_k))$, and $(x_{k+1}, f(x_{k+1}))$, as shown in Figure 3.

It is then not too difficult to find the equation of this parabola (it has the form $y = Ax^2 + Bx + C$), and from that to find the area underneath by integrating. The remarkably simple answer is

$$\text{Area under parabola} = \frac{b - a}{3n}[f(x_{k-1}) + 4f(x_k) + f(x_{k+1})]$$

When we add the area under the parabola over the first two strips to the area under the parabola over the third and fourth strips, and so on, we get Simpson's rule.

---

*In one of the exercises of the previous section, we saw that its antiderivative is a constant multiple of the **error function,** $\text{erf}(x)$, and we calculated a few of its values.

EXAMPLE 3

Use 4 intervals in Simpson's rule to approximate $\int_0^1 x^2 \, dx$.

SOLUTION Since $n = 4$, we have $(b - a)/n = 1/4$, and Simpson's rule tells us that

$$\int_0^1 x^2 \, dx \approx \frac{1}{12}\left[0 + 4\frac{1}{16} + 2\frac{4}{16} + 4\frac{9}{16} + 1\right]$$

$$= \left(\frac{1}{12}\right)\left(\frac{64}{16}\right)$$

$$= \frac{1}{3}.$$

Before we go on... This is the *exact* answer. What is going on? Remember that Simpson's rule is based on approximating the graph by quadratic functions. If the function is already quadratic, as it is here, the approximation is exact.

EXAMPLE 4

Use 6 intervals in Simpson's rule to approximate $\int_0^6 e^{-x^2} \, dx$.

SOLUTION As in Example 2, $(b - a)/n = 1$, so Simpson's rule gives us

$$\int_0^6 e^{-x^2} \, dx \approx \frac{1}{3}[e^0 + 4e^{-1} + 2e^{-4} + 4e^{-9} + 2e^{-16} + 4e^{-25} + e^{-36}]$$

$$\approx 0.8362.$$

USING A GRAPHING CALCULATOR OR SPREADSHEET

We have already seen a graphing calculator program to compute the left- and right-hand Riemann approximations of $\int_a^b f(x) \, dx$. Further, we have seen that the trapezoid rule amounts to the average of the left- and right-hand sums, so instead of writing a new program for the trapezoid rule, all we need do is to add a few lines of code to the program SUMS to compute the average, and we obtain a program that shows all three sums. This is the program TRAP in the appendix. The program following it, SIMP, calculates the sum for Simpson's rule.

▶ **NOTE** To run these programs, don't forget to first enter the function you are integrating under Y_1 in the "Y=" window. ◀

The appendix also contains instructions for using a computer spreadsheet to do these calculations.

EXAMPLE 5

Use the trapezoid rule and Simpson's rule to approximate $\int_{-1}^{1} 2\sqrt{1-x^2}\, dx$ with $n = 200$. How accurate are these approximations?

SOLUTION This integral gives twice the area of a semicircle with radius 1, so it should be equal to π. Using the two rules, we get

Trapezoid rule: $\int_{-1}^{1} 2\sqrt{1-x^2}\, dx \approx 3.14041\ldots$
Simpson's rule: $\int_{-1}^{1} 2\sqrt{1-x^2}\, dx \approx 3.14113\ldots$

The actual value of π is 3.141592654 . . ., so to estimate the errors, we round the differences to one significant figure:

Trapezoid rule error: $|3.14041 - 3.14159| = 0.00118 \approx 0.001$
Simpson's rule error: $|3.14113 - 3.14159| = 0.00046 \approx 0.0005$.

Thus Simpson's rule is more accurate—to within $\pm\, 0.0005$—since it has the smallest error.

Before we go on... Your graphing calculator should also have its own (more sophisticated) algorithm for evaluating a definite integral numerically: On a TI-82, enter

fnInt(Y_1, X, -1, 1)

and it gives the more accurate answer of

3.141593074.

Its error is $|3.141593074 - 3.141592654| = 0.0000004204 \approx 0.0000004$.

ACCURACY

If you look at Example 1, you see that the trapezoid rule gets closer to the right answer for larger n. If you compare Examples 2 and 4, you see that we have two different approximations for the same integral. This raises some interesting and important questions: How large does n have to be to get "close enough" to the right answer? Which of the answers in Examples 2 and 4 is closest to the right answer? To answer these questions, we need to know something about the **error** in these two rules: that is, how far they are from the right answer.

Error = |Approximation − Exact answer|

Q Doesn't this raise a "Catch 22" situation? In order to know the error, we need to know the exact answer. But if we knew the exact answer, then we would hardly need to find a numerical approximation in the first place!

A We remedy this dilemma as follows: since we can't always calculate exactly what the error is, we look instead for a **bound** on the error. For instance, instead of trying to say "the error is exactly 0.001," we say instead, "the error is no larger than 0.001."

The following formulas give bounds on the errors for the rules we have been using.

> **THE ERRORS IN THE TRAPEZOID RULE AND SIMPSON'S RULE**
>
> If $f''(x)$ is continuous in $[a, b]$, then the error in the trapezoid rule is no larger than
>
> $$\frac{(b-a)^3}{12n^2} |f''(M)|,$$
>
> where $|f''(M)|$ is the largest value of $|f''(x)|$ in $[a, b]$.
>
> If $f^{(4)}(x)$ is continuous in $[a, b]$, then the error in Simpson's rule is no larger than
>
> $$\frac{(b-a)^5}{180n^4} |f^{(4)}(M)|,$$
>
> where $|f^{(4)}(M)|$ is the largest value of $|f^{(4)}(x)|$ in $[a, b]$.

We will not talk about where these facts come from, as their derivations are beyond the scope of this book. However, they are not hard to use.

▼ **EXAMPLE 6**

How accurate is the calculation in Example 2?

SOLUTION In that example we used 6 intervals in the trapezoid rule to estimate

$$\int_0^5 e^{-x^2} dx.$$

In order to use the error estimate, we need to know the largest value of $f''(x)$ on the interval $[0, 6]$, for $f(x) = e^{-x^2}$. Calculating,

$$f'(x) = -2xe^{-x^2}, \text{ and}$$
$$f''(x) = 2(2x^2 - 1)e^{-x^2}.$$

Since we want to find the extreme values of f'', we calculate its derivative,

$$f'''(x) = 4x(3 - 2x^2)e^{-x^2}.$$

Now $f'''(x) = 0$ only when $x = 0$ or $3 - 2x^2 = 0$, so $x = 0$ or $\pm\sqrt{\frac{3}{2}} \approx \pm 1.225$. Checking values in the interval $[0, 6]$, we get the following.

x	0	1.225	6
$f''(x)$	-2	0.89	3×10^{-14}

The largest value of $|f''(x)|$ is therefore 2. This tells us that the error is no larger than

$$\frac{(6-0^3)}{12 \cdot 6^2} \cdot 2 = 1.$$

This tells us that all we really know from the calculation we did is that

$$-0.114 \le \int_0^5 e^{-x^2}\, dx \le 1.886.$$

This does not give us much confidence in our calculation.

Before we go on... In fact, chances are that the calculation is more accurate than that, but the mathematics we've just done gives us no reason to believe this.

▼ **EXAMPLE 7**

How accurate is the calculation in Example 4?

SOLUTION   That example used 6 intervals in Simpson's rule to approximate

$$\int_0^5 e^{-x^2}\, dx.$$

In order to estimate the error, we need to find the largest value of $|f^{(4)}(x)|$, for x in $[0, 6]$, with $f(x) = e^{-x^2}$ again. Calculating derivatives beyond those we know from above,

$$f^{(4)}(x) = 4(4x^4 - 12x^2 + 3)e^{-x^2}, \text{ and}$$
$$f^{(5)}(x) = -8x(4x^4 - 15x^2 + 15)e^{-x^2}.$$

To find the extreme values of $f^{(4)}(x)$, we need to know where $f^{(5)}(x) = 0$, but the only place this happens is when $x = 0$. (This can easily be checked graphically. Alternatively, notice that, if $f^{(5)}(x) = 0$, either $x = 0$, or $4x^4 - 15x^2 + 15 = 0$. But this equation has no solutions, since the similar quadratic $4u^2 - 15u + 15 = 0$ has none.) Checking values, we get the following.

x	0	6
$f^{(4)}(x)$	12	4×10^{-12}

The largest value of $|f^{(4)}(x)|$ is 12, so the error in Simpson's rule is no larger than

$$\frac{(6-0)^5}{180 \cdot 6^4} \cdot 12 = 0.4.$$

This means that we can say that

$$0.4362 \leq \int_0^5 e^{-x^2} dx \leq 1.2362.$$

Again, we would not be surprised to find out that our calculation was more accurate than that, but we have no real reason to believe so.

EXAMPLE 8

Going back to Example 1, how large must n be to approximate the answer to 5 decimal places?

SOLUTION We are now asking that the error be no larger than 0.000001. To get the formula for the error, we need to find the largest value of $|f''(x)|$ on [0, 1], but this is easy since $f''(x) = 2$ is constant: its largest value is 2. The error if we use n intervals is then

$$\frac{(1-0)^3}{12n^2} \cdot 2 = \frac{1}{6n^2}.$$

Our question is: How large must n be so that $1/(6n^2)$ is smaller than 0.000001? If we set $1/(6n^2) = 0.000001$ and solve for n, we get $n = 408.2$, meaning that we need $n \geq 409$ to be guaranteed 5 decimal places of accuracy.

Before we go on... This large a sum is much too tedious to compute by hand but easy enough to do with a programmable calculator or computer.

EXAMPLE 9

How large should n be in Example 4 to approximate the answer to 5 decimal places?

SOLUTION Our error estimate is the one we found in Example 6: the error is no larger than $6^5 \cdot 12/(180n^4) = 518.4/n^4$. If we set this equal to 0.000001, we find that $n = 150.9$, so we need n to be at least 151 to be guaranteed 5 digits of accuracy.

USING A GRAPHING CALCULATOR TO FIND THE ERROR ESTIMATES

We have included two further graphing calculator programs in the appendix—called TRAPERR and SIMPERR—that do the calculation of the error estimates for you. The next example demonstrates their use on the TI-82.

EXAMPLE 10

Use a graphing calculator to approximate $\int_{0.5}^{5} x \ln x \, dx$ with both the trapezoid rule and Simpson's rule using 250 partitions. How reliable are these answers?

SOLUTION We must first take care of the "Y=" window, so we need all derivatives up to the fourth.

$$f(x) = x \ln x \qquad f'(x) = \ln x + 1$$

$$f''(x) = \frac{1}{x} \qquad f'''(x) = -\frac{1}{x^2}$$

$$f^{(4)}(x) = \frac{2}{x^3}$$

We enter these in the "Y=" window as follows.

$$Y_1 = X*\ln(X)$$
$$Y_2 = abs(1/X)$$
$$Y_4 = abs(2/(X^3))$$

We then run TRAP and SIMP, with $N = 250$, $A = 0.5$, $B = 5$, and obtain

Trapezoid rule: $\displaystyle\int_{0.5}^{5} x \ln x \, dx \approx 14.01717947$

Simpson's rule: $\displaystyle\int_{0.5}^{1} x \ln x \, dx \approx 14.01711731.$

To obtain the error estimates, we run **TRAPERR** and **SIMPERR** with the above values and find

Trapezoid error estimate: $0.000243 \approx 0.0002$

Simpson error estimate: $0.0000000419904 \approx 0.00000004$

Clearly, the approximation given by Simpson's rule is far more accurate, so we conclude

$$\int_{0.5}^{5} x \ln x \, dx = 14.01711731 \pm 0.00000004.$$

In other words, this estimate is accurate to 7 decimal places!

Before we go on... Let us compare this with the TI-82 graphing calculator's built-in integration algorithm. Enter

$$\text{fnInt}(X\ln X, X, \emptyset.5, 5)$$

and press ENTER. The TI-82 gives the answer 14.0171173, which has one less significant digit than we obtained with Simpson's rule. In other words, Simpson's rule with $n = 250$ gives slightly more accurate information.

▶ **NOTE** You may suspect that the trapezoid rule is always less accurate, but that is not the case. It is possible to construct functions for which the trapezoid rule gives a far more accurate approximation than Simpson's rule (in fact, there is an example in the text above). Further, calculating the error for the trapezoid rule is easier, since taking the second derivative involves less work than taking the fourth derivative. (Think about $f(x) = 1/(1 + x^2)$, for example.) ◀

In this section we have just scratched the surface of a large subject. For just about any problem in which it might be difficult to find an exact answer analytically, people have worked out approximations. These problems include finding solutions to equations, computing integrals, and many more. This field of **numerical methods** has been around for as long as people have wanted to do computations. It became particularly important when high-speed computers and calculators were developed, so that the routine calculations could be done much more quickly.* Today, numerical methods underlie the large-scale computer calculations used for economic forecasting, weather forecasting, and, much more successfully, research in physics. A key to using any approximation is knowing something about how bad an approximation it is, so all approximations have error estimates similar to those we gave for the trapezoid rule and Simpson's rule.

▶ ### 13.7 EXERCISES

In each of Exercises 1–6, use the trapezoid rule with 10 intervals to approximate the integral. Use the error estimate to tell how accurate your answer is. If possible, compute the integral exactly using the Fundamental Theorem of Calculus, and compare the exact answer to your approximation.

1. $\int_0^2 2x \, dx$
2. $\int_{-1}^1 (x - 1) \, dx$
3. $\int_0^3 x^3 \, dx$
4. $\int_1^3 (x^3 + 1) \, dx$
5. $\int_1^5 \ln x \, dx$
6. $\int_2^6 \ln x \, dx$

7–12. Repeat Exercises 1–6, but use Simpson's rule with 10 intervals.

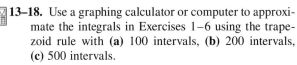

13–18. Use a graphing calculator or computer to approximate the integrals in Exercises 1–6 using the trapezoid rule with **(a)** 100 intervals, **(b)** 200 intervals, **(c)** 500 intervals.

19–24. Use a graphing calculator or computer to approximate the integrals in Exercises 1–6 using Simpson's rule with **(a)** 100 intervals, **(b)** 200 intervals, **(c)** 500 intervals.

25. Consider $\int_0^2 x^4 \, dx$. How large should n be in order to compute this integral to 3 decimal places using the trapezoid rule?

26. Consider $\int_0^2 x^5 \, dx$. How large should n be in order to compute this integral to 3 decimal places using the trapezoid rule?

▼ *The Manhattan Project to develop the atomic bomb, during World War II, used a room full of calculators. This is what they called the women hired to do the lengthy calculations on the mechanical contraptions that were available at the time. The first electronic computers were developed during this war to speed up these and other wartime calculations.

27. Repeat Exercise 25 using Simpson's rule.
28. Repeat Exercise 26 using Simpson's rule.
29. Consider $\int_0^{10} xe^{-x}\, dx$. How large should n be in order to compute this integral to 3 decimal places using the trapezoid rule?
30. Consider $\int_0^{10} x^2 e^{-x}\, dx$. How large should n be in order to compute this integral to 3 decimal places using the trapezoid rule?
31. Repeat Exercise 29 using Simpson's rule.
32. Repeat Exercise 30 using Simpson's rule.

Computing the Length of a Graph *The length of the graph of the function f from $x = a$ to $x = b$ (see the figure) can be shown to be*

$$\text{Length} = \int_a^b \sqrt{1 + [f'(x)]^2}\, dx$$

Use this formula to approximate the lengths of the graphs of the functions in Exercises 33–36. In each case, use Simpson's rule with **(a)** *6 partitions (if you are doing it by hand)* **(b)** *200 partitions (if you are using technology).*

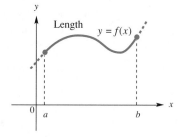

33. $f(x) = x^2$, x in $[0, 1]$
34. $f(x) = x^2/2$, x in $[0, 2]$
35. $f(x) = e^x$, x in $[0, 1]$
36. $f(x) = e^{-x}$, x in $[5, 10]$

COMMUNICATION AND REASONING EXERCISES

37. Looking at the error estimate for the trapezoid rule, by how much will the error shrink if you increase n by a factor of 10? What does this say about the increase in the number of digits of accuracy of the estimate given by the rule?

38. Looking at the error estimate for Simpson's rule, by how much will the error shrink if you increase n by a factor of 10? What does this say about the increase in the number of digits of accuracy of the estimate given by the rule?

▶ You're the Expert

THE COST OF ISSUING A WARRANTY*

You are a consultant to Sunny Electronics Company. Sunny's chief competitor, Cyberspace Electronics, Inc., has just launched a new home entertainment system that competes directly with Sunny's World Entertainment System (WES). Cyberspace is offering a 2-year limited warranty on its product, so Sunny is thinking of offering a 20-year pro-rated warranty. According to this warranty, if anything goes wrong, Sunny will refund the original $1,000 cost of the system depreciated continuously at an annual rate of 12%. Sunny has asked you to estimate what it will cost the company to provide this warranty.

Using the formula for continuous depreciation, you calculate that it will cost Sunny

$$C(t) = 1{,}000 e^{-0.12t}$$

▼ *This situation was inspired by the article "Determination of Warranty Reserves" by W.W. Menke, *Management Sciences,* Vol. 15, No. 10, June, 1969. (The Institute of Management Sciences)

to replace a system that is t years old. You realize that some information is missing—you would like to know how the failure rate of the World Entertainment System (WES) increases with time. For instance, what percentage of products will fail after 5 years? 10 years? After searching the literature for a while, you find a reasonable model that predicts the percentage of items that will fail after time t:

$$p(t) = 1 - e^{-t/N},$$

where t is the time in years and N is the average lifetime of the product. For the WES, the average lifetime happens to be 15 years. So you use $p(t) = 1 - e^{-t/15}$, whose graph is shown in Figure 1.

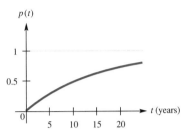

FIGURE 1

This is the total percentage that fail in t years. Thus, for example, the percentage of systems that will fail by the end of the first year is $p(1) = 1 - e^{-1/15} \approx 0.0645$, or 6.45%, while the percentage that will fail by the end of 5 years is $p(5) = 1 - e^{-5/15} \approx 0.283$, or 28.3%.

You start by trying to estimate the cost to the company for small time periods, say once a quarter, and you realize that it would be more helpful to know the *failure rate*—the percentage of entertainment systems that will fail per year. Recalling the ideas of calculus, you remember that the failure rate is given by the derivative,

$$p'(t) = \frac{1}{15} e^{-t/15}.$$

Thus, for example, the percentage of sets that fail in the fifth year is approximately $p'(5) \approx 0.048$, or 4.8%.

For the sake of definiteness, you suppose that the company has sold 100 sets, and you calculate the costs one quarter at a time. During the first quarter ($t = 0.25$ years), the percentage of the 100 sets that will fail is approximately

$$\text{Annual failure rate} \times \frac{1}{4} \approx p'(0.25) \cdot (0.25).$$

Thus, the cost to the company for the 100 sets is

$$\begin{aligned}&\text{Number of failures} \times \text{Cost per failure} \\&\approx 100 p'(0.25) \cdot (0.25) \times 1000 e^{-0.12(0.25)} \\&= 100{,}000 p'(0.25) e^{-0.12(0.25)}(0.25)\end{aligned}$$

For the second quarter, the cost is approximately

$$100{,}000 p'(0.50) e^{-0.12(0.50)}(0.25),$$

since now $t = 0.5$ years. Adding all these costs through the end of the eightieth quarter gives

$$\begin{aligned}&100{,}000 p'(0.25) e^{-0.12(0.25)}(0.25) \\&+ 100{,}000 p'(0.50) e^{-0.12(0.50)}(0.25) \\&+ 100{,}000 p'(0.75) e^{-0.12(0.75)}(0.25) \\&+ \ldots + 100{,}000 p'(80) e^{-0.12(80)}(0.25).\end{aligned}$$

Just as you are about to calculate this awful-looking sum with 80 terms, you notice that it looks suspiciously like a Riemann sum. In fact, you can write it as

$$\sum_{k=1}^{80} 100{,}000\, p'(t_k) e^{-0.12 t_k}\, \Delta t,$$

where Δt is the time period (one quarter $= 0.25$ years) and $t_k = k\, \Delta t$. Further, you realize that you will get more accurate estimates by taking smaller and smaller time periods: in other words, by taking the limit of the sum as $\Delta t \to 0$. The limit is the definite integral

$$\lim_{\Delta t \to 0} \sum_{k=1}^{n} 100{,}000\, p'(t_k) e^{-0.12 t_k}\, \Delta t,$$

$$= \int_0^{20} 100{,}000\, p'(t) e^{-0.12 t}\, dt.$$

(The limits of integration must be in units of t, that is, in years.) Substituting the formula for $p'(t)$ gives

$$\text{Total cost} = \int_0^{20} 100{,}000 \cdot \frac{1}{15} e^{-t/15} e^{-0.12 t}\, dt$$

$$= \frac{100{,}000}{15} \int_0^{20} e^{-t/15} e^{-0.12 t}\, dt$$

$$= \frac{100{,}000}{15} \int_0^{20} e^{-14 t/75}\, dt$$

$$= \frac{100{,}000}{15} \left(-\frac{75}{14}\right) [e^{-14 t/75}]_0^{20}$$

$$= \$34{,}860.25$$

This is the cost of providing warranty coverage for 100 sets. Thus, for one set the cost will be $348.60, more than a third of the cost of a new set.

Exercises

1. Find the cost per system if Sunny decides to limit the warranty to 10 years.

2. How many years should the warranty last to cost Sunny no more than $100 per set?

3. If Sunny kept the 20-year warranty plan but increased the discount rate from 12% to 30%, how would this affect the warranty cost per set?

4. The sales manager at Sunny tells you that the company is prepared to spend no more than $100 per set on warranty costs but would like to keep the 20-year pro-rated warranty. What could you recommend?

5. A junior executive in the company suggests that, under the original plan, extending the warranty to 30 years will make very little difference. How accurate is her assessment?

6. Suppose that Sunny used linear depreciation rather than exponential depreciation. In particular, suppose that Sunny will pay $C(t) = 1{,}000 - 50t$ if a system

fails after t years (up to 20 years). How much will it cost Sunny to provide this warranty? (You will have to use numerical integration. The integral can also be evaluated using integration by parts, a technique that will be discussed in the next chapter.)

7. Suppose that Sunny used linear depreciation over 10 years rather than 20 (so that $C(10) = 0$). How much would it cost to provide this warranty?

Review Exercises

Evaluate the integrals in Exercises 1–34.

1. $\int 10x^{10} \, dx$

2. $\int 2x^7 \, dx$

3. $\int \dfrac{3}{x^4} \, dx$

4. $\int \dfrac{4}{x^5} \, dx$

5. $\int \left(\sqrt{x} + \dfrac{1}{x}\right) dx$

6. $\int \left(\sqrt[3]{x} - \dfrac{2}{x}\right) dx$

7. $\int \left(2e^x + 3x - \dfrac{4}{x}\right) dx$

8. $\int \left(5e^x - 2x^2 + \dfrac{3}{x}\right) dx$

9. $\int (x + 2)^{10} \, dx$

10. $\int (2x-3)^5 \, dx$

11. $\int x\sqrt[3]{1 + x^2} \, dx$

12. $\int x^2\sqrt{x^3 + 2} \, dx$

13. $\int \dfrac{x}{x^2 + 1} \, dx$

14. $\int \dfrac{x^2}{x^3 + 2} \, dx$

15. $\int 5e^{-2x} \, dx$

16. $\int 4e^{2x+1} \, dx$

17. $\int (xe^{x^2} - 3x) \, dx$

18. $\int \left(xe^{x^2+1} - \dfrac{1}{x + 2}\right) dx$

19. $\int \dfrac{4.7x^{0.2}}{x^{1.2} - 4} \, dx$

20. $\int \dfrac{2.3x^{0.1} - 4}{x^{0.9}} \, dx$

21. $\int \dfrac{e^{0.3t}}{1 + 2e^{0.3t}} \, dt$

22. $\int \dfrac{e^{0.3t} - 1}{e^{0.3t}} \, dt$

23. $\int_0^1 (x^3 + 2x) \, dx$

24. $\int_{-1}^1 (x^3 + x^2) \, dx$

25. $\int_1^2 \dfrac{3}{x^2} \, dx$

26. $\int_{-2}^{-1} \dfrac{2}{x} \, dx$

27. $\int_0^1 (e^{-x} + x) \, dx$

28. $\int_0^{10} xe^{-x^2} \, dx$

29. $\int_0^2 x^2\sqrt{x^3 + 1} \, dx$

30. $\int_1^3 x^3\sqrt{x^2 - 1} \, dx$

31. $\int_0^1 (4 + x^{0.2}(2 - x^{1.2})^4) \, dx$

32. $\int_0^1 \left(3 - \dfrac{x^{0.2}}{1 + x^{1.2}}\right) dx$

33. $\int_{-1}^1 xe^{-0.3x^2} \, dx$

34. $\int_0^4 \dfrac{e^{0.3t}}{1 + e^{0.3t}} \, dt$

Find the areas of the regions described in Exercises 35–46.

35. Bounded by $y = 1 - x^2$, the x-axis, and the lines $x = -1$ and $x = 1$

36. Bounded by $y = 1 - x^4$, the x-axis, and the lines $x = -1$ and $x = 1$

37. Bounded by $y = 1/x$, the x-axis, and the lines $x = 1$ and $x = 10$

38. Bounded by $y = 2/x$, the x-axis, and the lines $x = 2$ and $x = 10$

1012 Chapter 13 The Integral

39. Bounded by $y = xe^{-x^2}$, the x-axis, and the lines $x = 0$ and $x = 5$
40. Bounded by $y = 1 - e^{3x-4}$, the x-axis, and the lines $x = 0$ and $x = 5$
41. Bounded by $y = 1 - 2x^2$ and the x-axis
42. Bounded by $y = 1 - 2x^4$ and the x-axis
43. Bounded by $y = x^4 - x^2$ and the x-axis
44. Bounded by $y = x^4 - 5x^2 + 4$ and the x-axis
45. Bounded by $y = x^2 - x^3$, the x-axis, and the line $x = -1$
46. Bounded by $y = e^x - e^{-x}$, the x-axis, and the lines $x = -1$ and $x = 1$

Evaluate the left- and right-hand Riemann approximations and the trapezoid approximation for the integrals in Exercises 47–50 using the stated number of partitions. (Round your answers to two decimal places.)

47. $\displaystyle\int_{-1}^{1} e^{-x^2}\, dx, \quad n = 4$
48. $\displaystyle\int_{0}^{5} e^{-x^2}\, dx, \quad n = 5$
49. $\displaystyle\int_{0.5}^{2} \ln x \, dx, \quad n = 3$
50. $\displaystyle\int_{0}^{1} \sqrt{1 + \sqrt{x}}\, dx, \quad n = 2$

Use a computer or graphing calculator to evaluate the left- and right-hand Riemann approximations for the integrals in Exercises 51–54 using the stated number of partitions. (Round your answers to four decimal places.)

51. $\displaystyle\int_{-1}^{1} e^{x^2}\, dx, \quad n = 100, 200, 500$
52. $\displaystyle\int_{0}^{1} e^{x^2}\, dx, \quad n = 100, 200, 500$
53. $\displaystyle\int_{0.5}^{2} \dfrac{dx}{4 + \ln x}, \quad n = 100, 200, 500$
54. $\displaystyle\int_{0}^{1} \sqrt{1 + x^2}\, dx, \quad n = 100, 200, 500$

55–62. Using a graphing calculator or computer, approximate the integrals in Exercises 47–54 using Simpson's rule with $n = 50, 100,$ and 500.

63–70. For each of the integrals given in 47–54, determine how large n must be to approximate the integral to within ± 0.001 using the trapezoid rule.

71–78. For each of the integrals given in 47–54, determine how large n must be to approximate the integral to within ± 0.001 using Simpson's rule.

APPLICATIONS

79. *Magazine Sales* The accompanying chart shows annual newsstand sales of all magazines in the U.S. (in millions of magazines).*
    (a) Model the annual sales with a linear function $s(t)$ using the data from 1982 ($t = 0$) and 1992 ($t = 10$).
    (b) Compare the actual sales over the given time period with the total sales predicted by your model.
    (c) Use your model to estimate the number of magazines sold in the U.S. from the start of 1980 ($t = -2$) to the start of 1990.

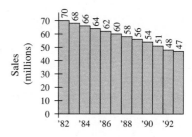

▼ *Source: Vos, Gruppo & Capell Inc./*The New York Times*, Dec. 8, 1993, p. D6.

80. *Oversupply* The following chart shows the number of unsold newsstand magazines in the U.S.*

(a) Model the annual surplus with a linear function $s(t)$ using the data from 1982 ($t = 0$) and 1990 ($t = 8$).
(b) Compare the total number of unsold magazines for the given time period with the figure predicted by your model.
(c) Use your model to estimate the total number of unsold magazines in the U.S. from the start of 1980 ($t = -2$) to the start of 1990.

81. *Acceleration* On the surface of the moon, the downward acceleration due to gravity is 5.3 ft/s², and there is no air resistance. If a ball is thrown upward from the ground at a velocity of 100 ft/s, how high will it get?

82. *Acceleration* If a ball is dropped from a height of 1,000 ft above the surface of the moon (see the preceding exercise), how long will it take to drop to the ground? How fast will it be falling when it hits the ground?

83. *Acceleration* A rocket accelerates at a rate of $320 - 320e^{-t/60}$ ft/s² at t seconds after liftoff. How far will it rise in the first minute?

84. *Acceleration* A race car accelerates from a stop at a rate of $e^{-t/120}$ ft/s² after t seconds. How far will it travel in the second minute?

85. *Cost* The marginal cost of Better Baby Buggies Inc.'s Turbo model is $50 + 40/(x + 40)$ for the xth buggy made in a week, and the company's fixed costs amount to $50,000 each week. Find the total cost to make x Turbo buggies in one week.

86. *Cost* The marginal cost of Better Baby Buggies Inc.'s Gran Turismo model is $75 + 500/(x + 10)^2$ for the xth buggy made in a week, and the company's fixed costs amount to $20,000 each week. Find the total cost to make x Gran Turismo buggies in one week.

87. *Revenue* Due to volume discounts, Better Baby Buggies receives $75 + 25e^{-x}$ for the xth Turbo buggy it sells. Find the total revenue the company receives by selling x Turbo buggies.

88. *Revenue* Due to volume discounts and other effects, Better Baby Buggies receives $100 + 50xe^{-x^2}$ for the xth Gran Turismo buggy it sells. Find the total revenue the company receives by selling x Gran Turismo buggies.

89. *Demand for Solar Cells* Based on data from 1978 through 1993, the demand for solar energy can be modeled by

$$q(p) = \frac{268.227}{p^{1.1990}},$$

where p is the cost of solar cells for each watt of capacity (in constant 1973 dollars) and q is the annual sales of solar cells in millions of watts (or megawatts) of capacity shipped.† At the time of this writing, the *Enron Corporation* is planning to build a 100-megawatt solar power plant in the southern Nevada desert.
(a) Use the demand equation to estimate what the solar cells may cost Enron (to the nearest $1 per watt of capacity).
(b) Assume that the (rounded) cost to Enron is the lowest that is currently (January 1995) technologically feasible, but that this cost will continue to decrease exponentially, halving every 4.342 years.‡ Obtain the price p per watt as a function of time t.
(c) Use the demand equation together with the result from part (b) to estimate the total sales of solar cells (in megawatts of capacity) from January 1995 through January 1998.

90. *Demand for Solar Cells* Repeat the preceding exercise using the rougher demand equation

$$q(p) = \frac{300}{p}.$$

---

* Source: Vos, Gruppo & Capell Inc./*The New York Times*, Dec. 8, 1993, p. D6.
† The model is a power regression based on data gleaned from published graphs. Source: PV News/Worldwatch Institute/*The New York Times*, November 15, 1994, p. D2.
‡ Based on an exponential regression of the above data.

CHAPTER 14

Tapping the Rich May Prove Tricky

By SYLVIA NASAR

As Bill Clinton's new economic team considers how best to turn his campaign promises on taxes into legislative proposals, it may feel the ground shifting.

His campaign tossed around a host of proposals on government financing, from spending cuts to getting more tax revenue from corporations and individuals, all aimed at getting the resources to accomplish goals ranging from middle-class tax relief to public works to halving the deficit. But the consensus among experts was that many of the proposals were unrealistic, and that the proposal most likely to yield significant new money was tax increases on the wealthiest taxpayers.

Getting the really rich—the people who reaped an outsize share of the economic gains of the 1980s—to pay more was a major plank of the Democratic campaign. For Clinton tax purposes, a couple with income of $200,000 and a single taxpayer with $150,000 count as rich.

LIKELY TO FALL SHORT

The problem is that while higher tax rates on high incomes are likely to provide a good deal of new money, they are not likely, barring an unexpectedly buoyant economy, to generate the $92 billion over four years that the Clinton camp has claimed.

Many experts had been skeptical of that claim from the start. Congressional Budget Office estimates put the added revenue at a maximum of $80 billion over four years and estimates from Treasury officials and the Republican side of the Joint Economic Committee of Congress are even lower.

But even as long-term deficit projections are looking gloomier, so also is the outlook for collecting as much from the rich as had been projected even in the lower estimates.

Perhaps the biggest consideration is one raised in a new study of how very rich taxpayers react to higher taxes. The study published by the National Bureau of Economic Research strongly suggests that extremely rich people—the top slice of the top 1 percent of taxpayers—have considerable flexibility to expose less of their income to taxation. . . .

"If there's a significant change in rates, say from 31 percent to 41 percent, people will change their behavior to take advantage of ways to defer income," Professor Poterba said. "It would set into motion a return to pre-1986 tax shelters."

Another reason for thinking that the rich will yield less revenue than Clinton tax planners had hoped is that the rich may not possess as many riches as they used to. The latest Internal Revenue Service summary of tax returns, for 1989, released this year, shows that many high fliers—real estate empire builders, retailers and newly redundant executives—had their wings clipped during the last four years of economic drift.

The number of taxpayers reporting pretax income of $1 million or more dropped from 62,000 in 1988 to 58,000 in 1989, and their share of total income shrank from 5.5 percent to 4.7 percent. . . .

Source: From Sylvia Nasar, "Tapping the Rich May Prove Tricky," *The New York Times*, December 12, 1992, p. 39.

Further Integration Techniques and Applications of the Integral

SECTIONS

1. Integration by Parts
2. Integration Using Tables
3. Area Between Two Curves and Applications
4. Averages and Moving Averages
5. Improper Integrals and Applications
6. Differential Equations and Applications

You're the Expert

Estimating Tax Revenues

APPLICATION ▶ You have just been hired by the incoming administration to coordinate national tax policy, and the so-called experts on your staff can't seem to agree on which of three tax proposals will result in the highest revenue for the government. The data you have are the two income tax proposals (graphs of tax vs. income) and the distribution of incomes in the country. How do you use this information to decide which tax policy will result in the most revenue?

INTRODUCTION ▸ We have seen that integration is a powerful tool for the measurement of area and the computation of total sales, revenues, and profits. In this chapter, we shall first look at some further techniques for the computation of integrals and then at some more applications of the integral. We shall then see how to extend the definition of the definite integral to include integrals over infinite intervals, and show how such integrals can be used for long-term forecasting. Finally, we shall introduce the beautiful theory of differential equations and their numerous applications.

14.1 INTEGRATION BY PARTS

In this section we shall discuss **integration by parts,** an integration technique that comes from the product rule for derivatives. The tabular method we present here has been around for some time and makes integration by parts quite simple, particularly in problems where it has to be iterated. The particular version we present was developed and taught to us by Professor Dan Rosen.

To start with, we introduce a little notation in order to simplify things while we introduce integration by parts (we shall use this notation only in the next few pages). If f is any function, then we shall denote its derivative by $D(f)$ and an antiderivative by $I(f)$. Thus, for example, if $f(x) = 2x^2$, then

$$D(f) = 4x,$$

while

$$I(f) = \frac{2x^3}{3}.$$

[If we wished, we could instead take $I(f) = 2x^3/3 + 46$.]

INTEGRATION BY PARTS FORMULA

$$\int (f \cdot g)\, dx = f \cdot I(g) - \int [D(f) \cdot I(g)]\, dx$$

In words, this says:

The integral of a product of two functions is the first times the integral of the second minus the **integral of** (the derivative of the first times the integral of the second).

If we don't like having to find the integral of $f \cdot g$, integration by parts allows us to consider instead the (possibly) easier integral of $D(f) \cdot I(g)$.

Q Where does the integration by parts formula come from?

A We start by applying the product rule to the function $f \cdot I(g)$:

$$D(f \cdot I(g)) = D(f) \cdot I(g) + f \cdot D(I(g))$$
$$= D(f) \cdot I(g) + f \cdot g$$

since $D(I(g))$ is the derivative of the antiderivative of g. Integrating both sides gives

$$f \cdot I(g) = \int [D(f) \cdot I(g)] \, dx + \int (f \cdot g) \, dx.$$

If we now subtract $\int [D(f) \cdot I(g)] \, dx$ from both sides, we get the integration by parts formula.

It is convenient to arrange the calculation of the integration by parts formula in tabular form:

	D	I
+	f	g
−	$D(f)$ →	$I(g)$

The column marked D is the differentiation column, and the column marked I is the integration column. We place f in the D column and g in the I column to indicate that we will be differentiating f and integrating g (as in the formula). The slanted arrow reminds us that we are to multiply the two functions it links, f and $I(g)$, and the color plus sign on the left reminds us that $(+)f \cdot I(g)$ appears in the answer. The horizontal arrow linking $D(f)$ and $I(g)$ reminds us that we are to take the product of $D(f)$ and $I(g)$, and integrate. The color minus sign on the left reminds us that it is $-\int D(f) \cdot I(g) \, dx$ that appears in the answer. Combining the contributions of the two lines gives us $f \cdot I(g) - \int D(f) \cdot I(g)$, which is the integration by parts formula.

▼ **EXAMPLE 1**

Calculate $\int xe^x \, dx$.

SOLUTION First notice that none of the techniques of integration that we've talked about up to now will help us. In particular, we cannot simply find antiderivatives of x and e^x and multiply them together: you should check that $\frac{x^2}{2} e^x$ is *not* an antiderivative for xe^x. So, let's try integration by parts. We are required to find the integral of the *product* of x and e^x. Now, here we must make a decision: while using integration by parts, which of the two functions gets differentiated, the x or the e^x? Since the derivative of x is just 1, differentiating makes it simpler, so let us try putting x in the D column. This means that e^x has to go in the I column. We then get the following table:

D	I
$+\ x$	e^x
$-\ 1$ →	e^x

We can now read the answer as

$$\int xe^x\,dx = xe^x - \int 1 \cdot e^x\,dx = xe^x - e^x + C.$$

Before we go on... Had we decided to put e^x in the D column instead, we would have had the following table:

D	I
$+\ e^x$	x
$-\ e^x$ →	$\frac{1}{2}x^2$

From this table we get

$$\int xe^x\,dx = \frac{1}{2}x^2 e^x - \int \frac{1}{2}x^2 e^x\,dx.$$

To evaluate this requires computing a worse integral than the one we started with! How do we know beforehand which way to go? We don't. We have to be willing to do a little trial and error. We try it one way, and if it doesn't make things simpler, we try it another way. *Remember, though, that the function we put in the I column must be one that we can integrate.*

▼ **EXAMPLE 2**

Calculate $\int x^2 e^{-x}\,dx$.

SOLUTION   Again, we have a product—the integrand is the product of x^2 and e^x. Since differentiating x^2 makes it simpler, we put it in the D column and get

D	I
$+\ x^2$	e^{-x}
$-\ 2x$ →	$-e^{-x}$

(Recall that the integral of e^{-x} is $-e^{-x}$.) Thus,

$$\int x^2 e^{-x}\, dx = -x^2 e^{-x} - \int 2x(-e^{-x})\, dx$$

The last integral is simpler than the one we started with, but still involves a product. It is a good candidate for another integration by parts. The table we would use would start with $2x$ in the D column and $-e^{-x}$ in the I column, but notice that this is exactly what we see in the last row of the table above. Therefore, we *continue the process,* elongating the table:

D	I
$+\ x^2$	e^{-x}
$-\ 2x$	$-e^{-x}$
$+\ 2$	e^{-x}

Notice how the signs on the left alternate, because we really need to compute $-\int 2x(-e^{-x})dx$. Now we would still have to compute an integral (the integral of the product of the functions in the bottom row) to complete the computation. But why stop here? Let us continue the process one more step:

D	I
$+\ x^2$	e^{-x}
$-\ 2x$	$-e^{-x}$
$+\ 2$	e^{-x}
$-\ 0$	$-e^{-x}$

In the bottom line we see that all that is left to integrate is $0 \cdot (-e^{-x}) = 0$. Since an indefinite integral of 0 is 0, we can simply read the answer as

$$\int x^2 e^{-x}\, dx = x^2(-e^{-x}) - 2x(e^{-x}) + 2(-e^{-x}) + C$$
$$= -x^2 e^{-x} - 2x e^{-x} - 2 e^{-x} + C$$
$$= -e^{-x}(x^2 + 2x + 2) + C.$$

Before we go on... Since that took several steps, let's check our work.

$$\frac{d}{dx}[-e^{-x}(x^2 + 2x + 2) + C] = e^{-x}(x^2 + 2x + 2) - e^{-x}(2x + 2)$$
$$= x^2 e^{-x}\ \checkmark$$

1034 Chapter 14 Further Integration Techniques and Applications of the Integral

> ### FINDING THE AREA BETWEEN THE GRAPHS OF f(x) AND g(x)
>
> 1. Find all points of intersection by solving $f(x) = g(x)$ for x. This either determines the interval over which you will integrate, or breaks up a given interval into regions between the intersection points.
> 2. Find the area of each region between intersection points by integrating the difference of the larger and the smaller function. (If you accidentally take the smaller minus the larger, the integral will calculate the negative of the area, so just take the absolute value.)
> 3. Add together the areas you found in Step 2 to get the total area.

We now turn to some interesting applications to business and finance.

CONSUMERS' SURPLUS

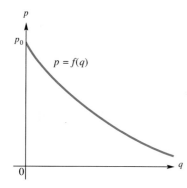

FIGURE 8

Consider a general demand curve presented, as is traditional in economics, as $p = f(q)$, where p is unit price and q is demand measured, say, in annual sales (Figure 8).

We can interpret f as follows:

$f(q)$ is the price at which the demand will be q units per year.

The price p_0 shown on the graph is the highest price that customers are willing to pay.

Now suppose that a manufacturer sets the unit price at $\bar{p} < p_0$. Then the corresponding demand in units per year is $\bar{q}$, where $f(\bar{q}) = \bar{p}$ (Figure 9). There are some consumers who are willing to pay a higher price than $\bar{p}$, and these consumers save money by paying the lower price. This raises the following question.

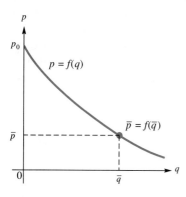

FIGURE 9

Q If the price is set at $\bar{p}$, how much will be saved per year by consumers who were willing to pay a higher price?

A We can calculate this from the graph of the demand function as follows. First partition the interval $[0, \bar{q}]$ into n subintervals of equal length, as we did when we calculated Riemann sums. Figure 10 shows a typical subinterval.

The interval $[q_{k-1}, q_k]$ represents units q_{k-1} through q_k, and the price consumers would have been willing to pay for these units is approximately $f(q_k)$. But they only paid $\bar{p}$ for each of these units, so they saved a total of

Savings in price × Number of units $= (f(q_k) - \bar{p})\Delta q$.

This is the area of the shaded rectangle in Figure 10. Now, these savings apply only to the units between q_{k-1} and q_k. To obtain the total savings, we

14.3 Area Between Two Curves and Applications

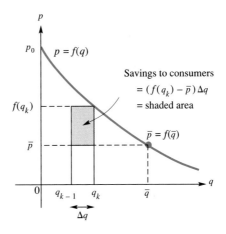

FIGURE 10

must add together the portions corresponding to the entire interval* $[0, \bar{q}]$, and so

$$\text{Total savings} \approx (f(q_1) - \bar{p})\Delta q + (f(q_2) - \bar{p})\Delta q \\ + \ldots + (f(q_n) - \bar{p})\Delta q.$$

But this is the right-hand Riemann sum for the function $f(q) - \bar{p}$ and so, passing to the limit, we obtain the exact savings as

$$\text{Total savings} = \int_0^{\bar{q}} (f(q) - \bar{p})dq.$$

This total savings is called the **consumers' surplus** and is represented by the area enclosed by the graphs of $p = f(q)$ and $p = \bar{p}$ in Figure 11.

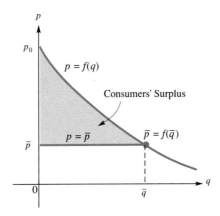

FIGURE 11

▼ * We stop at $\bar{q}$ because if the price is set at $\bar{p}$, only $\bar{q}$ units are sold, and we are only interested in the savings on units actually sold.

To summarize:

> **CONSUMERS' SURPLUS**
>
> The **consumers' surplus** is the total amount saved by consumers who were willing to pay more than the selling price of $\bar{p}$ per unit, and is given by
> $$CS = \int_0^{\bar{q}} (f(q) - \bar{p})\, dq,$$
> where $f(\bar{q}) = \bar{p}$. Geometrically, it is the area of the region enclosed by the graphs of $p = f(q)$ and $p = \bar{p}$ for $0 \le q \le \bar{p}$.

EXAMPLE 5 Consumers' Surplus

Your used CD store has an exponential demand equation of the form
$$p = 15e^{-0.01q},$$
where q represents daily sales of CDs and p is the price you charge per used CD. Calculate the daily consumers' surplus if you sell your used CDs at $5 each.

SOLUTION We are given $f(q) = 15e^{-0.01q}$ and $\bar{p} = 5$. We still need $\bar{q}$. By definition,
$$f(\bar{q}) = \bar{p};$$
that is,
$$15e^{-0.01\bar{q}} = 5,$$
so we must solve for $\bar{q}$:
$$e^{-0.01\bar{q}} = \frac{1}{3},$$
so
$$-0.01\bar{q} = \ln\left(\frac{1}{3}\right) = -\ln 3.$$
Thus,
$$\bar{q} = \frac{\ln 3}{0.01} \approx 109.8612.$$

14.3 Area Between Two Curves and Applications

We now have

$$CS = \int_0^{\bar{q}} (f(q) - \bar{p}) \, dq$$

$$= \int_0^{109.8612} (15e^{-0.01q} - 5) \, dq$$

$$= \left[\frac{15}{-0.01} e^{-0.01q} - 5q \right]_0^{109.8612}$$

$$= (-500 - 549.306) - (-1{,}500 - 0)$$

$$= \$450.69 \text{ per day.}$$

PRODUCERS' SURPLUS

We can also consider extra income earned by producers. Consider a supply equation of the form $p = f(q)$, where now $f(q)$ is the price at which a supplier is willing to supply q items (per time period). Because a producer is generally willing to supply more units at a larger price per unit, a supply curve usually has a positive slope, as shown in Figure 12.

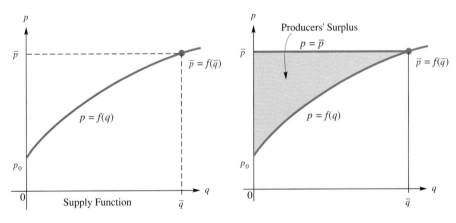

FIGURE 12

FIGURE 13

The price p_0 is the lowest price that a producer is willing to charge. Now suppose an item is priced at $\bar{p}$. Then a producer who is prepared to charge less than that will earn "extra" money as a result of the higher selling price $\bar{p}$. The total extra earnings by such producers is called the **producers' surplus**. Arguing as before, we can say that this is given by the area shown in Figure 13.

> **PRODUCERS' SURPLUS**
>
> The **producers' surplus** is the extra amount earned by producers who were willing to charge less than the selling price of $\bar{p}$ per unit, and is given by
>
> $$PS = \int_0^{\bar{q}} (\bar{p} - f(q))\, dq,$$
>
> where $f(\bar{q}) = \bar{p}$. Geometrically, it is the area of the region enclosed by the graphs of $p = f(q)$ and $p = \bar{p}$ for $0 \leq q \leq \bar{q}$.

▼ **EXAMPLE 6**   Producers' Surplus

My dorm-room tie-dye T-shirt enterprise has grown to the extent that I am now able to produce them in bulk, and several sororities have begun placing orders. I have informed one sorority that I am prepared to supply $20\sqrt{p-4}$ T-shirts at a price of p dollars per shirt. What is the total surplus to my enterprise if I sell them to the sorority at $8 each?

SOLUTION   We need to calculate the producers' surplus when $\bar{p} = 8$. The supply equation is

$$q = 20\sqrt{p-4},$$

but in order to use the formula for producers' surplus we need to express p as a function of q. Therefore, we solve the supply equation for q. First, square both sides to remove the radical sign:

$$q^2 = 400(p-4)$$

so

$$p - 4 = \frac{q^2}{400},$$

giving

$$p = f(q) = \frac{q^2}{400} + 4.$$

We now need the value of $\bar{q}$ corresponding to $\bar{p} = 8$. The easiest way to find it is to substitute $p = 8$ in the original equation, which gives

$$\bar{q} = 20\sqrt{8-4} = 20\sqrt{4} = 40.$$

Thus,

$$PS = \int_0^{\bar{q}} (\bar{p} - f(q))\, dq$$

$$= \int_0^{40} \left(8 - \left(\frac{q^2}{400} + 4\right)\right) dq$$

$$= \int_0^{40} \left(4 - \frac{q^2}{400}\right) dq$$

$$= \left[4q - \frac{q^3}{1{,}200}\right]_0^{40} = \$142.22.$$

Before we go on...

Q How do we interpret this?

A First notice that, since $\bar{q} = 40$, I will be selling the sorority 40 T-shirts at \$8 per shirt. On the other hand, had the sorority been more savvy, it could have said the following: "For the nth T-shirt that you make, charge the lowest price at which you are willing to supply n T-shirts." That request would not violate my supply policy, but would result in a lower price for the total of 40 T-shirts (since the supply curve dictates that smaller amounts correspond to lower prices per unit). The \$142.22 represents the extra amount I will get by charging \$8 for all 40 T-shirts.

▼ **EXAMPLE 7**   *Equilibrium Price*

Referring to Example 6, a representative informs me that the sorority is prepared to order only $\sqrt{200(16 - p)}$ T-shirts at p dollars each. At the same time, I wish to maintain my policy of supplying $20\sqrt{p - 4}$ T-shirts at p dollars each. I would like to produce as many T-shirts for them as possible in order to avoid being left with unsold T-shirts. What price should I charge per T-shirt, and what then are the consumers' and producers' surpluses?

SOLUTION   The price that guarantees neither a shortage nor a surplus of T-shirts is the equilibrium price, the price where supply equals demand. We have

$$\text{Supply:} \quad q = 20\sqrt{p - 4}$$
$$\text{Demand:} \quad q = \sqrt{200(16 - p)}.$$

Equating these gives

$$20\sqrt{p - 4} = \sqrt{200(16 - p)}$$

so

$$400(p - 4) = 200(16 - p),$$

giving

$$400p - 1{,}600 = 3{,}200 - 200p$$
$$200p = 1{,}600,$$

or

$$p = \$8 \text{ per T-shirt.}$$

We therefore take $\bar{p} = 8$. The corresponding value for q is obtained by substituting $p = 8$ into either the demand or supply equation.

$$\bar{q} = 20\sqrt{8-4} = 40$$

Thus, $\bar{p} = 8$ and $\bar{q} = 40$. We must now calculate the consumers' surplus and the producers' surplus. We have already calculated the producers' surplus for $\bar{p} = 8$ in Example 6, getting

$$PS = \$142.22.$$

For the consumers' surplus, we must first express p as a function of q for the demand equation. Thus, we solve the demand equation for p as we did for the supply equation in Example 6, and we obtain

$$\text{Demand: } f(q) = 16 - \frac{q^2}{200}.$$

Therefore,

$$CS = \int_0^{\bar{q}} (f(q) - \bar{p})\, dq$$
$$= \int_0^{40} \left[\left(16 - \frac{q^2}{200}\right) - 8\right] dq$$
$$= \int_0^{40} \left(8 - \frac{q^2}{200}\right) dq$$
$$= \left[8q - \frac{q^3}{600}\right]_0^{40} = \$213.33.$$

Before we go on... Figure 14 shows both the consumers' and producers' surpluses on the same graph.

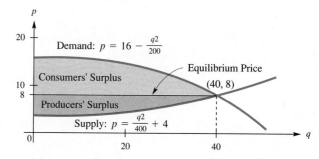

FIGURE 14

Q How do we interpret this result?

A We saw that I earned an extra $142.22 (producers' surplus) due to the fact that the sorority was not as savvy as I had expected. On the other hand, *I* could also have been more savvy and said: "I will charge you as much for the *n*th T-shirt as you are prepared to pay per shirt for *n* T-shirts," and I would have earned $213.33 more (the consumers' surplus). As it turns out, neither I nor the sorority were savvy enough, and it seems that I lost more as a result than they did!

▶ 14.3 EXERCISES

Find the areas of the indicated regions in Exercises 1–24 (a graphing calculator may be useful for Exercises 21–24).

1. Between $y = x^2$ and $y = -1$ for x in $[-1, 1]$.
2. Between $y = x^3$ and $y = -1$ for x in $[-1, 1]$.
3. Between $y = -x$ and $y = x$ for x in $[0, 2]$.
4. Between $y = -x$ and $y = x/2$ for x in $[0, 2]$.
5. Between $y = x$ and $y = x^2$ for x in $[-1, 1]$.
6. Between $y = x$ and $y = x^3$ for x in $[-1, 1]$.
7. Between $y = e^x$ and $y = x$ for x in $[0, 1]$.
8. Between $y = e^{-x}$ and $y = -x$ for x in $[0, 1]$.
9. Between $y = (x - 1)^2$ and $y = -(x - 1)^2$ for x in $[0, 1]$.
10. Between $y = x^2(x^3 + 1)^{10}$ and $y = -x(x^2 + 1)^{10}$ for x in $[0, 1]$.
11. Enclosed by $y = x$ and $y = x^4$.
12. Enclosed by $y = x$ and $y = -x^4$.
13. Enclosed by $y = x^3$ and $y = x^4$.
14. Enclosed by $y = x$ and $y = x^3$.
15. Enclosed by $y = x^2$ and $y = x^4$.
16. Enclosed by $y = x^4 - x^2$ and $y = x^2 - x^4$.
17. Enclosed by $y = e^x$, $y = 2$, and the *y*-axis.
18. Enclosed by $y = e^{-x}$, $y = 3$, and the *y*-axis.
19. Enclosed by $y = \ln x$, $y = 2 - \ln x$, and $x = 4$.
20. Enclosed by $y = \ln x$, $y = 1 - \ln x$, and $x = 4$.
21. Enclosed by $y = e^x$ and $y = 2x + 1$.
22. Enclosed by $y = e^x$ and $y = x + 2$.
23. Enclosed by $y = \ln x$ and $y = \frac{x}{2} - \frac{1}{2}$.
24. Enclosed by $y = \ln x$ and $y = x - 2$.

Calculate the consumers' surplus for each of the demand equations in Exercises 25–36 at the indicated unit price $\bar{p}$.

25. $p = 10 - 2q$, $\bar{p} = 5$
26. $p = 100 - q$, $\bar{p} = 20$
27. $p = 100 - 3\sqrt{q}$, $\bar{p} = 76$
28. $p = 10 - 2q^{1/3}$, $\bar{p} = 6$
29. $p = 500e^{-2q}$, $\bar{p} = 100$
30. $p = 100 - e^{0.1q}$, $\bar{p} = 50$
31. $q = 100 - 2p$, $\bar{p} = 20$
32. $q = 50 - 3p$, $\bar{p} = 10$
33. $q = 100 - 0.25p^2$, $\bar{p} = 10$
34. $q = 20 - 0.05p^2$, $\bar{p} = 5$
35. $q = 500e^{-0.5p} - 50$, $\bar{p} = 1$
36. $q = 100 - e^{0.1p}$, $\bar{p} = 20$

Calculate the producers' surplus for each of the supply equations in Exercises 37–48 at the indicated unit price $\bar{p}$.

37. $p = 10 + 2q$, $\bar{p} = 20$
38. $p = 100 + q$, $\bar{p} = 200$
39. $p = 10 + 2q^{1/3}$, $\bar{p} = 12$
40. $p = 100 + 3\sqrt{q}$, $\bar{p} = 124$
41. $p = 500e^{0.5q}$, $\bar{p} = 1{,}000$
42. $p = 100 + e^{0.01q}$, $\bar{p} = 120$
43. $q = 2p - 50$, $\bar{p} = 40$
44. $q = 4p - 1{,}000$, $\bar{p} = 1{,}000$
45. $q = 0.25p^2 - 10$, $\bar{p} = 10$
46. $q = 0.05p^2 - 20$, $\bar{p} = 50$
47. $q = 500e^{0.05p} - 50$, $\bar{p} = 10$
48. $q = 10(e^{0.1p} - 1)$, $\bar{p} = 5$

APPLICATIONS

49. *College Tuition* A study of about 1800 colleges and universities in the U.S. resulted in the demand equation $q = 9859.39 - 2.17p$, where q is the enrollment at a college or university and p is the average annual tuition (plus fees) it charges.* Officials at Enormous State University have developed a policy whereby the number of students it will accept per year at a tuition level of p dollars is given by $q = 100 + 0.5p$. Find the equilibrium tuition price $\bar{p}$ and the consumers' and producers' surplus at this tuition level. How much less would the university be earning per year if it decided to charge the nth student it admits an annual tuition of $2(n - 100)$? (Express all answers to the nearest dollar.)

50. *Fast Food* A fast-food outlet finds that the demand equation for its new side-dish, "Sweetdough Tidbit" is given by
$$p = \frac{128}{(q + 1)^2},$$
where p is the price (in cents) per serving, and q is the number of servings that can be sold per hour at this price. At the same time, the franchise is prepared to sell $q = 0.5p - 1$ servings per hour at a price of p cents. Find the equilibrium price $\bar{p}$ and the consumers' and producers' surplus at this price level. How much less would the franchise be earning per hour if it sold the nth serving per hour of Sweetdough Tidbits for $2(n + 1)$ cents?

51. *Sales* The following graph shows the demand and supply curves for your "$E = mc^2$" T-shirts. Assuming that you sell your T-shirts at the equilibrium price, who has the larger surplus—you or the consumer?

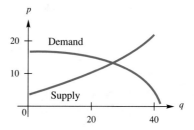

---

* Based on a study by A.L. Ostrosky, Jr., and J. V. Koch, as cited in their book, *Introduction to Mathematical Economics* (Prospect Heights, IL: Waveland Press, 1979), p. 133.

52. *Sales* Repeat the preceding exercise using the following curves:

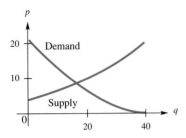

53. *Linear Demand* Given a linear demand equation of the form $q = -mp + b$ ($m > 0$), find a formula for the consumers' surplus at a price level of $\bar{p}$ per unit.

54. *Linear Supply* Given a linear supply equation of the form $q = mp + b$ ($m > 0$), find a formula for the producers' surplus at a price level of $\bar{p}$ per unit.

55. *Cost* Snapple Beverage Corp.'s annual revenue R and profit P in millions of dollars for the period 1989 through 1993 can be approximated by
$$R = 16.15e^{0.87t}$$
and
$$P = 3.93e^t,$$
where t is the number of years since 1989.*
(a) Use these models to estimate *Snapple*'s accumulated costs for this period.
(b) How does this relate to the area between two curves?
(c) What does a comparison of the exponents in the formulas for R and P tell you about *Snapple*?

56. *Cost* Microsoft Corp.'s annual revenue R and profit P in billions of dollars for the period 1986 through 1994 can be approximated by
$$R = 26.27e^{0.37t}$$
and
$$P = 0.6256e^{0.37t},$$
where t is the number of years since 1986.†
(a) Use these models to estimate its accumulated costs for this period.
(b) How does this relate to the area between two curves?
(c) What does a comparison of the exponents in the formulas for R and P tell you about *Microsoft*?

57. *Subsidizing Emission Control* The marginal cost to the utilities industry of reducing sulfur emissions at several levels of reduction is shown in the following table.‡

Reduction (millions of tons)	8	10	12
Marginal Cost (dollars per ton)	270	360	780

(a) Show by substitution that this data can be modeled by
$$C'(q) = 1,000,000(41.25q^2 - 697.5q + 3,210),$$
where q is the reduction in millions of tons of sulfur and $C'(q)$ is the marginal cost of reducing emissions (in dollars per million tons of reduction).§
(b) If the government subsidizes sulfur emissions at a rate of \$400 per ton, find the total net cost to the utilities industry of removing 12 million tons of sulfur.
(c) The answer to part (b) is represented by the area between two graphs. What are the equations of these graphs?

▼ * Obtained using exponential regression on graphical data published in *The New York Times*, July 10, 1994, Section 13, p. 1. (Raw data were estimated from a graph. Source: Snapple Beverage Corp./Datastream/Nielsen North America.)

† These equations were obtained using exponential regression on graphical data published in *The New York Times*, July 18, 1994, p. D1. (Raw data were estimated from a graph. Source: Computer Intelligence Infocorp./Microsoft company records.)

‡ See "You're the Expert" in the chapter "Introduction to the Derivative." These figures were produced in a computerized study of reducing sulfur emissions from the 1980 level by the given amounts. Source: Congress of the United States, Congressional Budget Office, *Curbing Acid Rain: Cost, Budget and Coal Market Effects* (Washington, D.C.: Government Printing Office, 1986): xx, xxii, 23, 80.

§ You can obtain this equation directly from the data (as we did) by assuming that $C'(q) = aq^2 + bq + c$, substituting the values of $C'(q)$ for the given values of q, and then solving for a, b, and c.

58. *Variable Emissions Subsidies* Referring to the marginal cost equation in Exercise 57, assume that the government subsidy for sulfur emissions is given by the formula

$$S'(q) = 500{,}000{,}000q,$$

where q is the level of reduction in millions of tons and $S'(q)$ is the marginal subsidy in dollars per million tons.

(a) Find the total net cost to the utilities industry of removing 12 million tons of sulfur. Interpret your answer.
(b) The answer to part (a) is represented by the net area between two graphs. What are the equations of these graphs?

COMMUNICATION AND REASONING EXERCISES

59. *Foreign Trade* The following graph shows Canada's monthly exports and imports for the year ending February 1, 1994.* What does the area between the export and import curves represent?

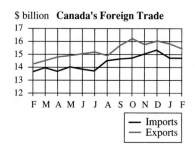

60. *Foreign Trade* The following graph shows a fictitious country's monthly exports and imports for the year ending February 1, 2003. What does the total area enclosed by the export and import curves represent, and what does the definite integral of the difference, Exports − Imports, represent?

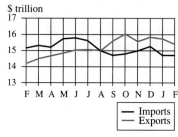

61. What is wrong with the following claim: "I own 100 units of *Abbott Laboratories, Inc.* stocks that originally cost me $22 per share. My net income from this investment over the period March 5–April 27, 1993 is represented by the area between the stock price curve and the purchase price curve as shown on the following graph."[†]

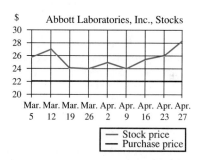

62. Is it always true that a company's total profit over a one-year period is represented by the area between the daily revenue and daily cost curves? Illustrate your answer by means of an example.

▼ *Source: Statistics Canada/*Globe and Mail*, April 20, 1994, p. B3.
[†] Source: News Reports/*Chicago Tribune*, April 28, 1993, Section 3, p. 6.

14.4 AVERAGES AND MOVING AVERAGES

AVERAGES

You probably already know how to find the average of a collection of numbers. If you want to find the average of, say, 20 numbers, you simply add them up and divide by 20. More generally, if you want to find the **average**, or **mean** of the n numbers $y_1, y_2, y_3, \ldots y_n$, you add them up and divide by n.

> **AVERAGE OR MEAN OF A SET OF VALUES**
>
> $$\bar{y} = \frac{y_1 + y_2 + \ldots + y_n}{n}.$$

Let us revisit an example from the last chapter.

▼ **EXAMPLE 1** Pork Exports

According to figures released by the U.S. Agriculture Department, the yearly value of U.S. pork exports were as shown in Figure 1.*

The average value of pork exports for the 9-year period shown is

$$\bar{y} = \frac{y_1 + y_2 + \ldots + y_n}{n}$$

$$= \frac{60 + 80 + 125 + 250 + 330 + 320 + 325 + 450 + 460}{9}$$

$$= \frac{2{,}400}{9} \approx \$266.67 \text{ million per year.}$$

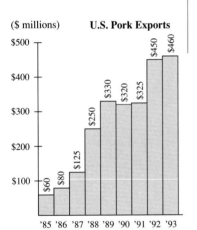

FIGURE 1

Before we go on... The height of each rectangle in the bar graph is equal to the value of exports for that year; the first rectangle has height 60, the second 80, and so on. Since the rectangles have width 1 unit (1 year), the *area* of the first rectangle is 60, the area of the second is 80, and so on. Thus, the sum of the numbers $60 + 80 + \ldots$ is also the sum of the areas of these rectangles. The average is obtained by dividing this number by 9, which happens to be the total width of the bar graph. In other words,

$$\text{Average height of bar graph} = \frac{\text{Area of bar graph}}{\text{Width of bar graph}}.$$

▼ *Source: U.S. Agriculture Department. (*The New York Times*, Dec. 12, 1993, p. D1.) (The figures shown on the graph are estimates based on the published graph.)

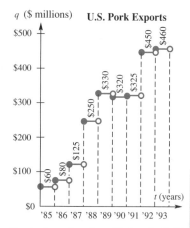

Graph of step function with domain [0, 9]

FIGURE 2

We can also think of the area of the bar graph in Example 1 as the area under a step function (Figure 2). If we call this step function f and take its domain as $[a, b]$, we therefore have

$$\text{Average value of } f = \frac{\text{Area under graph}}{b - a}.$$

Now the area under the graph is calculated by the definite integral, so we make the following definition.

> **AVERAGE VALUE OF A FUNCTION**
>
> The **average**, or **mean**, of a function $f(x)$ on the interval $[a, b]$ is given by
>
> $$\bar{f} = \frac{1}{b - a} \int_a^b f(x)\, dx.$$

This formula is useful in situations where we need to average a continuous function instead of a collection of numbers. Such a situation would arise, for example, if we use a mathematical model to approximate a large collection of data.

▼ **EXAMPLE 2**

Find the average of $f(x) = x$ on the interval $[0, 4]$.

SOLUTION

$$\bar{f} = \frac{1}{b - a} \int_a^b f(x)\, dx$$

$$= \frac{1}{4 - 0} \int_0^4 x\, dx$$

$$= \frac{1}{4} \left[\frac{x^2}{2} \right]_0^4 = \frac{1}{8}(16 - 0) = 2$$

Before we go on... Look at the graph of this function together with the average value 2 (Figure 3).

The area under the graph of f is the area of the triangle shown in the figure, which is $\left(\frac{1}{2}\right)(4)(4) = 8$. We can construct a rectangle with the same area and the same base $[0, 4]$ by taking the height of the rectangle to be 2, the average value of f. Thus, we can think of the average as the height of a rectangle whose area is the same as the area under the graph of f. This equality of areas follows from the equation

$$(b - a)\bar{f} = \int_a^b f(x)\, dx.$$

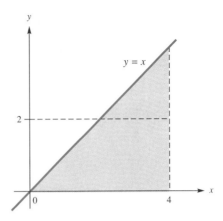

FIGURE 3

▼ **EXAMPLE 3**   More on Pork Exports

In the last chapter we used the following function to model the pork export figures shown in Example 1:

$$q(t) = \frac{460e^{(t-4)}}{1 + e^{(t-4)}},$$

where q represents annual sales (in millions of dollars) and t is the time in years, with $t = 0$ corresponding to Jan. 1, 1985. Figure 4 shows the actual sales data with the graph of the equation superimposed. Calculate $\bar{q}$ for the period Jan. 1, 1985 to Jan. 1, 1994.

SOLUTION   We use the formula

$$\bar{q} = \frac{1}{b-a} \int_a^b q(t)\, dt$$

$$= \frac{1}{9-0} \int_0^9 \frac{460e^{(t-4)}}{1 + e^{(t-4)}}\, dt$$

$$= \frac{1}{9} [460 \ln(1 + e^{(t-4)})]_0^9$$

$$= \frac{460}{9}[(\ln(1 + e^5)) - (\ln(1 + e^{-4}))]$$

$$\approx \$254.97 \text{ million per year.}$$

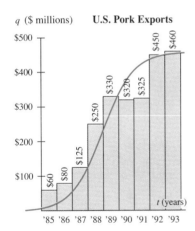

FIGURE 4

Before we go on...   We saw in Example 1 that the average of the actual sales was $266.67 million. Although the accuracy of our model leaves something to be desired, we can use our model to estimate the average value of pork exports for periods into the future for which actual data are not yet available. Also, a more carefully constructed model might yield more accurate estimates.

EXAMPLE 4 Average Balance

The People's Credit Union pays 3% interest, compounded continuously, and at the end of the year pays a bonus of 1% of the average amount in each account during the year. If you deposit $10,000 at the beginning of the year, how much interest and how large a bonus will you get?

SOLUTION Using the continuous compound interest formula, the amount of money you have in the account at time t is

$$A(t) = 10{,}000 e^{0.03t},$$

where t is measured in years. For a period of one year, t runs from 0 to 1. At the end of the year the account will have

$$A(1) = \$10{,}304.55,$$

so you will have earned $304.55 interest. To compute the bonus, we first need to find the average amount in the account. But that means we need to find the average of $A(t)$ over the interval $[0, 1]$. Thus,

$$\bar{A} = \frac{1}{b-a} \int_a^b A(t)\, dt$$

$$= \frac{1}{1-0} \int_0^1 10{,}000 e^{0.03t}\, dt$$

$$= \frac{10{,}000}{0.03} \left[e^{0.03t} \right]_0^1$$

$$\approx 333{,}333.33 \left[e^{0.03} - 1 \right] \approx \$10{,}151.51.$$

The bonus is 1% of this, or $101.52.

Before we go on... The 1% bonus was one-third of the total interest. Think about why this happened. What fraction of the total interest would the bonus be if the interest rate was 4%, 5%, or 10%?

MOVING AVERAGES

Suppose that you follow the performance of a company's stock by recording the daily closing price. These numbers may jump around quite a bit. In order to see any trends, you would like a way to "smooth out" this data. The **moving average** is one common way to do that.

EXAMPLE 5   Stock Prices

The following table shows Colossal Conglomerate Corp.'s closing stock prices for a 20-day period:

Day	1	2	3	4	5	6	7	8	9	10
Price	20	22	21	24	24	23	25	26	20	24
Day	11	12	13	14	15	16	17	18	19	20
Price	26	26	25	27	28	27	29	27	25	24

Plot these prices and also the five-day moving average.

SOLUTION

The five-day moving average is the average of each day's price together with the preceding four days. We can compute this starting on the fifth day. The numbers we get are these.

Day	1	2	3	4	5	6	7	8	9	10
Moving Average					22.2	22.8	23.4	24.4	23.6	23.6
Day	11	12	13	14	15	16	17	18	19	20
Moving Average	24.2	24.4	24.2	25.6	26.4	26.6	27.2	27.6	27.2	26.4

The closing stock prices and moving averages are plotted in Figure 5.

As you can see, the moving average is less volatile. Since it averages the stock's performance over five days, a single day's fluctuation is smoothed out. Look at day 9 in particular. The moving average also tends to lag behind the actual performance, since it takes into account past history. Look at the downturns at days 6 and 18 in particular.

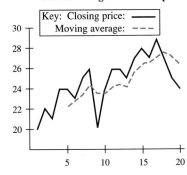

FIGURE 5

Before we go on... The period of 5 days is arbitrary. Using a longer period of time would smooth the data more, but increase the lag. For data used as economic indicators, such as housing prices or retail sales, it is common to compute the 1-year moving average to smooth out seasonal variations.

 You can program your graphing calculator to compute the moving averages of a list of data and to plot the scatter graph shown in the figure. We have included instructions and a program for the TI-82 in Appendix B. Computer spreadsheets are also quite useful, and Appendix C includes an example of this as well.

If we use a mathematical model to describe a large collection of data, we may want to compute the moving average of a continuous function. This we can do using the definite integral.

EXAMPLE 6

Find the 2-unit moving average of $f(x) = x^2$.

SOLUTION We are looking for the function $\bar{f}(x) =$ average of f on $[x - 2, x]$. We compute this using the integral formula we gave at the beginning of this section.

$$\bar{f}(x) = \frac{1}{x - (x - 2)} \int_{x-2}^{x} t^2 \, dt$$

$$= \frac{1}{2} \left[\frac{t^3}{3} \right]_{x-2}^{x} = \frac{1}{6} [x^3 - (x - 2)^3]$$

$$= \frac{1}{6}(x^3 - (x^3 - 6x^2 + 12x - 8)) = x^2 - 2x + \frac{4}{3}$$

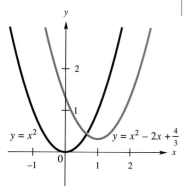

FIGURE 6

The graphs are shown in Figure 6. Notice again that the moving average does not reach the extreme minimum that the original function does, and lags behind it in the sense that its minimum occurs later than that of $f(x)$.

If you look at the first and second steps in the calculation we did in the above example, you will see that the n-unit moving average of a function can be written as follows.

n-UNIT MOVING AVERAGE OF A FUNCTION

The n-unit moving average of a function is given by

$$\bar{f}(x) = \frac{1}{n} \int_{x-n}^{x} f(t) \, dt.$$

USING A GRAPHING CALCULATOR OR COMPUTER

The following example shows how one can use a graphing calculator to calculate and plot the moving average of a function. (See the appendix for a program to compute and plot moving averages of lists of data.)

EXAMPLE 7

Use a graphing calculator to plot the 3-unit moving average of

$$f(x) = \frac{x}{1 + |x|}, \quad -5 \leq x \leq 5.$$

14.4 Averages and Moving Averages

SOLUTION This function is a little tricky to integrate analytically, since $|x|$ is defined differently for positive and negative values of x, so we let our graphing calculator approximate the integral for us. (The format that follows is for the TI-82 calculator. Other graphing calculators should be similar.)
Enter

$$Y_1 = X/(1 + \text{abs}(X))$$
$$Y_2 = (1/3)\,\text{fnInt}(T/(1 + \text{abs}(T)), T, X - 3, X)$$

The Y_2 entry is a numerical approximation of the 3-unit moving average of $f(x)$,

$$\bar{f}(x) = \frac{1}{3}\int_{x-3}^{x} \frac{t}{1 + |t|}\,dt.$$

We set the viewing window ranges to $-5 \leq x \leq 5$ and $-1 \leq y \leq 1$, and plot these curves. (You will need to wait a while for the plot of the moving average—the calculator has to do a numerical integration to obtain each point on the graph.) The result is shown in Figure 7.

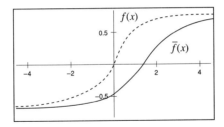

FIGURE 7

Before we go on... We can also use the calculator to evaluate the moving average at any value of x. For instance, to calculate $\bar{f}(1.2)$ on a TI-82 calculator, we enter

$$1.2 \rightarrow X \quad \boxed{\text{ENTER}}$$
$$Y_2 \qquad\quad \boxed{\text{ENTER}}$$

This has the effect of setting $x = 1.2$ and then evaluating the function Y_2. The answer we obtain is

$$\bar{f}(1.2) \approx -0.11960.$$

14.4 EXERCISES

Find the average value of the indicated quantity in Exercises 1 through 4.

1. Total Assets of Utilities Funds

Source: Morningstar Inc./
New York Times, Jan. 2, 1994

2. Number of Personal Computers in the U.S. with CD–ROM Drives

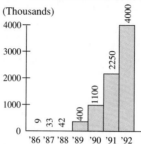

Source: Info Tech/New York Times,
Dec. 28, 1993

3. Value of Announced Mergers and Corporate Transactions

Source: Securities Data Company/
New York Times, Jan. 3, 1994

4. Annual Rate of Increase in Medical Costs in New York City Metropolitan Area

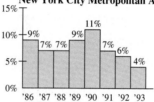

Source: Federal Bureau of
Labor Statistics/New York Times,
Dec. 25, 1993

Find the averages of the functions in Exercises 5–10 over the given intervals. Plot each function and its average on the same graph, as in Example 2.

5. $f(x) = x^3$ over $[0, 2]$ **6.** $f(x) = x^3$ over $[-1, 1]$ **7.** $f(x) = x^3 - x$ over $[0, 2]$

8. $f(x) = x^3 - x$ over $[0, 1]$ **9.** $f(x) = e^{-x}$ over $[0, 2]$ **10.** $f(x) = e^x$ over $[-1, 1]$

Plot the sequences in Exercises 11 and 12 and their 5-unit moving averages, as in Example 5.

11. 1, 2, 3, 4, 3, 2, 3, 4, 5, 6, 5, 6, 7, 8, 9, 8, 7, 8, 9, 10

12. 1, 2, 3, 4, 5, 6, 7, 6, 5, 4, 3, 2, 3, 4, 5, 6, 7, 8, 9, 10

Calculate the 5-unit moving average of each of the functions in Exercises 13–20. Plot each function and its moving average on the same graph, as in Example 6. (You may use a graphing calculator or a computer for these plots, but you should compute the moving averages analytically.)

13. $f(x) = x^3$ **14.** $f(x) = x^3 - x$ **15.** $f(x) = x^{2/3}$ **16.** $f(x) = x^{2/3} + x$

17. $f(x) = e^{0.5x}$ **18.** $f(x) = e^{-0.02x}$ **19.** $f(x) = \sqrt{x}$ **20.** $f(x) = x^{1/3}$

APPLICATIONS

21. *Investments* If you invest $10,000 at 8% interest compounded continuously, what is the average amount in your account over one year?

22. *Investments* If you invest $10,000 at 12% interest compounded continuously, what is the average amount in your account over one year?

23. *Average Balance* Suppose that you have an account (paying no interest) into which you deposit $3,000 at the beginning of each month. You withdraw money so that the amount in the account decreases linearly to 0 by the end of the month. Find the average amount in the account over a period of several months. (Assume that the account starts at $0 at $t = 0$ months.)

24. *Average Balance* Suppose that you have an account (paying no interest) into which you deposit $4,000 at the beginning of each month. You withdraw $3,000 during the course of each month, in such a way that the amount decreases linearly. Find the average amount in the account in the first two months. (Assume that the account starts at $0 at $t = 0$ months.)

In Exercises 25 through 28, use a graphing calculator or graphing computer software to plot the given functions together with their three-unit moving averages.

25. $f(x) = \dfrac{10x}{1 + 5|x|}$ **26.** $f(x) = \dfrac{1}{1 + e^x}$ **27.** $f(x) = \ln(1 + x^2)$ **28.** $f(x) = e^{1-x^2}$

29. *Expansion of Fast-Food Outlets* The following chart shows the approximate number of McDonald's® restaurants in the United States at year-end, in thousands.*

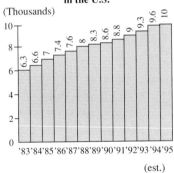

Number of McDonald's Restaurants in the U.S. (Thousands): '83: 6.3, '84: 6.6, '85: 7, '86: 7.4, '87: 7.6, '88: 8, '89: 8.3, '90: 8.6, '91: 8.8, '92: 9, '93: 9.3, '94: 9.6, '95: 10 (est.)

(a) Use the 1985 and 1995 figures to model these data with a linear function.
(b) Find and plot the 4-year moving average of your model.
(c) What can you say about the slope of the moving average?
(d) (For programmed calculator or computer) Use the graphing calculator or spreadsheet program described in Appendices B and C to plot the actual data and 4-year moving averages.

30. *Fast-Food Customers* The following graph shows the declining number of Americans (year-end figures in thousands) per fast-food outlet in the U.S.†

---

* Source: Technomics/U.S. Department of Commerce/*The New York Times*, Jan. 9, 1994, p. F5.
† Source: Technomics/*The New York Times*, Jan. 9, 1994, p. F5.

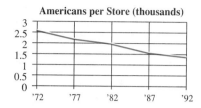

Americans per Store (thousands)

(a) Use the 1972 and 1992 figures to model these data with a linear function. (Give constants correct to two decimal places.)
(b) Find and plot the 5-year moving average of your model.
(c) What can you say about the slope of the moving average?
(d) (For programmed calculator or computer) Use the graphing calculator or spreadsheet program described in Appendices B and C to plot the actual data and 5-year moving averages.

31. *Moving Average of a Linear Function* Find a formula for the a-unit moving average of a general linear function $f(x) = mx + b$.

32. *Moving Average of an Exponential Function* Find a formula for the a-unit moving average of a general exponential function $f(x) = Ae^{kx}$.

33. *Market Share* The following graph shows the market share of *McDonald's*® restaurants in the fast-food business in the U.S.*

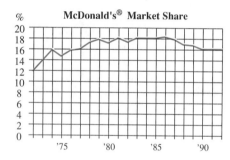

McDonald's® Market Share

(a) With $t = 0$ representing 1972, obtain a quadratic model
$$p(t) = at^2 + bt + c$$
that approximates the percentage market share enjoyed by *McDonald's*. (Use the data for $t = 0$, $t = 10$ and $t = 20$, rounded to the nearest percentage.)
(b) Use your model to estimate *McDonald's* average market share over the period shown in the graph.
(c) Use your model to predict *McDonald's* average market share for the three years ending in the year 2,000.

▼ *Source: Technomics/U.S. Department of Commerce/*The New York Times,* Jan. 9, 1994, p. F5.

34. *Growth in Prozac Sales* The following graph shows the annual growth in new prescriptions for Prozac® since its introduction in 1988.*

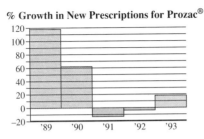

(a) With $t = 0$ representing 1989, obtain a quadratic model
$$p(t) = at^2 + bt + c$$
that approximates the percentage growth in Prozac prescriptions. (Use the data for $t = 0$, $t = 2$ and $t = 4$, rounded to the nearest 10%.)
(b) Use your model to estimate the average annual rate of growth in Prozac prescriptions over the period shown in the graph.
(c) Use your model to predict the average rate of growth of Prozac prescriptions for the three years ending in 1996.

35. *Fair Weather*† The Cancun Royal Hotel's advertising brochure features the following chart showing the year-round temperature.

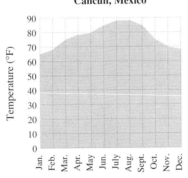

(a) Estimate and plot the two- and three-month moving averages. (Use a graphing calculator program, if available.)
(b) What can you say about the 24-month moving average?
(c) Comment on the limitations of a quadratic model for this data.

▼ *Sources: Mehta and Isaly, Smith Barney Shearson; Lehman Brothers/*The New York Times,* Jan. 9, 1994, p. F7.

† Inspired by an exercise in the Harvard Consortium Calculus project (p. 76). Temperatures are fictitious.

36. *Foul Weather* Repeat Exercise 35 using the following data from the Tough Traveler Lodge in Frigidville.

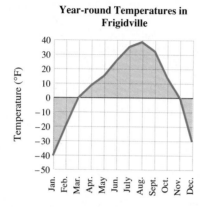

Year-round Temperatures in Frigidville

37. *Average Rate of Increase* In Exercise 4 you computed the average annual rate of increase in medical costs in New York City, but you did it in a somewhat naive way. Here is a better way:
(a) Taking into account compounding, find the total percent increase in medical costs from 1986 through 1993 (using the data in Exercise 4).
(b) Find the annual rate of increase that (taking into account compounding) would have produced the same total percent increase over that period. How does this answer compare with your answer to Exercise 4?

38. *Average Rate of Inflation* Suppose that the government's monthly inflation figures for a year are as follows:

Jan.	Feb.	Mar.	Apr.	May	Jun.	Jul.	Aug.	Sept.	Oct.	Nov.	Dec.
0.5%	1%	0.7%	1.3%	0.8%	0.6%	0.5%	0.2%	0.3%	0.4%	0.1%	0.2%

(a) Taking into account compounding, find the inflation rate for the year.
(b) Find the monthly rate that would have produced the same inflation for the year. How does this compare with the average of the monthly rates?

Modeling with the Cosine Function Exercises 39–42 require the use of a graphing calculator or computer. (Note: Make sure that your calculator is set to radians mode before you begin.) The accompanying figure shows a sketch of the graph of the function

$$f(x) = C + A\cos\left(\frac{2\pi}{k}(x - d)\right) \quad (C, A, k, d \text{ constants}).$$

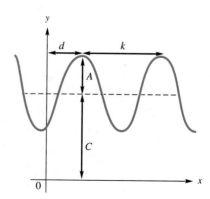

39. Use a graphing calculator or computer to plot the graph of $y = 5 + 4\cos(x - 0.1)$ and its 2-unit moving average.

40. Use a graphing calculator or computer to plot the graph of $y = 2 + 0.5\cos(2(x - 1))$ and its 3-unit moving average.

41. *Electrical Current* The typical voltage V supplied by an electrical outlet in the U.S. is given by
$$V(t) = 165\cos(120\pi t),$$
where t is time in seconds.
(a) Use a graphing calculator to find the average voltage over the interval $[0, \frac{1}{6}]$. How many times does the voltage reach a maximum in one second? (This is referred to as the number of **cycles per second.**)
(b) Use a graphing calculator to plot the function $S(t) = (V(t))^2$ over the interval $[0, \frac{1}{6}]$.
The **root mean square** voltage is given by the formula
$$V_{rms} = \sqrt{\bar{S}},$$
where $\bar{S}$ is the average value of $S(t)$ over one cycle. Calculate V_{rms}.

42. *Tides* The depth of water at my favorite surfing spot varies from 5 ft to 15 ft, depending on the tide. Last Sunday, high tide occurred at 5:00 AM, and the next high tide occurred at 6:30 PM. Use the cosine model to describe the depth of water as a function of time t in hours since midnight on Sunday morning. What was the average depth of the water between 10:00 AM and 2:00 PM?

COMMUNICATION AND REASONING EXERCISES

43. What property does a (non-constant) function have if its average value over an interval is zero? Sketch a graph of such a function.

44. Can the average value of a function f on an interval be larger than its value at any point in that interval? Explain.

45. Explain why it is sometimes more useful to consider the moving average of a stock price rather than the stock price itself.

46. Your monthly salary has been steadily increasing for the past year, and your average monthly salary was x dollars. Would you have earned more money if you were paid x dollars per month? Explain your answer.

47. Criticize the following claim: "the average value of a function on an interval is midway between its highest and lowest value."

48. Your manager tells you that 12-month moving averages gives at least as much information as shorter-term moving averages, and very often more. How would you argue that she is wrong?

49. Which of the following most closely approximates the original function:
(a) its 10-unit moving average (b) its 1-unit moving average (c) its 0.8-unit moving average. Explain your answer.

50. Is an increasing function larger or smaller than its one-unit moving average? Explain.

14.5 IMPROPER INTEGRALS AND APPLICATIONS

All the definite integrals we have seen have the form $\int_a^b f(x)\,dx$, where a and b are finite and $f(x)$ is piecewise-continuous on the closed interval $[a, b]$. There are occasions when we would like to relax these requirements, and when we do so we obtain what are called **improper integrals.** There are various types of improper integrals.

INTEGRALS IN WHICH A LIMIT OF INTEGRATION IS INFINITE

These are integrals of the form

$$\int_a^{+\infty} f(x)\, dx, \quad \int_{-\infty}^{b} f(x)\, dx, \quad \text{or} \quad \int_{-\infty}^{+\infty} f(x)\, dx.$$

Let us concentrate for a moment on the first form, $\int_a^{+\infty} f(x)\, dx$. If, instead of $+\infty$, we took a very large positive number M, then the integral would be just the very ordinary-looking definite integral $\int_a^M f(x)\, dx$. We might be tempted to say that $\int_a^{+\infty} f(x)\, dx$ means $\int_a^M f(x)\, dx$, where M is some very large number. But how large should M be? By now the answer is probably suggesting itself to you: we take the limit, as M approaches $+\infty$, of the integral $\int_a^M f(x)\, dx$.

IMPROPER INTEGRAL WITH AN INFINITE LIMIT OF INTEGRATION

We define

$$\int_a^{+\infty} f(x)\, dx = \lim_{M \to +\infty} \int_a^M f(x)\, dx,$$

provided the limit exists. If the limit exists, we say that $\int_a^{+\infty} f(x)\, dx$ **converges**. Otherwise, we say that $\int_a^{+\infty} f(x)\, dx$ **diverges**. Similarly, we define

$$\int_{-\infty}^{b} f(x)\, dx = \lim_{M \to -\infty} \int_M^b f(x)\, dx,$$

provided the limit exists. Finally, we define

$$\int_{-\infty}^{+\infty} f(x)\, dx = \int_{-\infty}^{a} f(x)\, dx + \int_a^{+\infty} f(x)\, dx$$

for some convenient a, provided *both* integrals on the right converge.

Q We know that the integral can be used to calculate the area under the curve. Is this also true for improper integrals?

A Yes. The best way to illustrate this is by looking at several examples.

▼ **EXAMPLE 1**

Calculate $\int_1^{+\infty} \frac{1}{x^2}\, dx$ and $\int_{-\infty}^{-1} \frac{1}{x^2}\, dx$.

SOLUTION We do the first one, and leave the second—an almost identical calculation—to you. The definition tells us that we must first evaluate $\int_1^M (1/x^2)\, dx$ and then take the limit as $M \to +\infty$. Now

$$\int_1^M \frac{1}{x^2}\, dx = \int_1^M x^{-2}\, dx = -[x^{-1}]_1^M = 1 - \frac{1}{M}.$$

Thus,
$$\int_1^{+\infty} \frac{1}{x^2}\,dx = \lim_{M \to +\infty} \int_1^M \frac{1}{x^2}\,dx$$
$$= \lim_{M \to +\infty} \left(1 - \frac{1}{M}\right) = 1.$$

In other words, the integral converges to 1.

Before we go on... How do we interpret this integral geometrically? The integral $\int_1^M (1/x^2)\,dx$ calculates the area shaded in Figure 1(a).

Geometrically, letting M approach $+\infty$ corresponds to moving the right-hand boundary in Figure 1(a) farther and farther to the right, resulting in Figure 1(b). In other words, the improper integral $\int_1^{+\infty} (1/x^2)\,dx$ calculates the area of an *infinitely long* region.

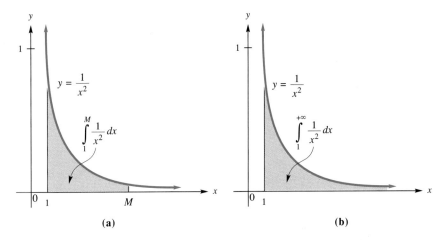

(a)   (b)

FIGURE 1

Q You said $\int_1^{+\infty} (1/x^2)\,dx = 1$. The infinitely long region shown in Figure 1(b) has an area of only 1 square unit?!

A Exactly! Think of it this way: suppose you had enough paint to cover exactly 1 square unit of area. If you paint the region in Figure 1(a) and then keep painting, extending the right edge farther and farther to the right, you would never run out of paint. This is one of the places where mathematics seems to contradict common sense. But common sense is notoriously unreliable when dealing with infinities.

EXAMPLE 2

Compute $\int_1^{+\infty} \frac{1}{x}\,dx$.

SOLUTION Proceeding as before, we first calculate
$$\int_1^M \frac{1}{x}\,dx = [\ln|x|]_1^M = \ln M - \ln 1 = \ln M.$$

Thus,

$$\int_1^{+\infty} \frac{1}{x}\, dx = \lim_{M \to +\infty} (\ln M) = +\infty.$$

The integral diverges to $+\infty$.

Before we go on... Figure 2 shows the area in question, compared with $\int_1^{+\infty} (1/x^2)\, dx$.

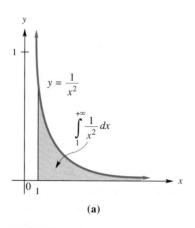 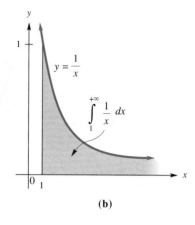

(a)   (b)

FIGURE 2

Although the infinitely long area in Figure 2(a) is exactly 1 square unit, the infinitely long area in Figure 2(b) is infinitely large! This means that, no matter how much paint you were given (even a million gallons) you would eventually run out of paint if you tried to paint the region on the right.

▼ **EXAMPLE 3**   Future Sales of Freon

It is estimated that from the year 2000, sales of freon will decrease continuously at a rate of about 15% per year. Further, sales of freon at the start of the year 2000 are predicted to be 35 million pounds per year.* What will be the total future sales of freon starting in the year 2000?

SOLUTION   We use the continuous decay formula to predict annual sales of freon beginning in the year 2000.

$$\begin{aligned} s(t) &= Ae^{-kt} \\ &= 35e^{-0.15t} \quad (t \text{ is the number of years since 2000}) \end{aligned}$$

▼ *These figures are approximations based on published data. (Source: The Automobile Consulting Group/*The New York Times,* December 26, 1993, p. F23.)

We are asked for *total* sales beginning at $t = 0$. Recalling that total sales can be computed as the definite integral of annual sales, we obtain

$$\begin{aligned}
\text{Total sales beginning in 2000} &= \int_0^{+\infty} 35e^{-0.15t}\, dt = \lim_{M \to +\infty} \int_0^M 35e^{-0.15t}\, dt \\
&= \lim_{M \to +\infty} 35 \int_0^M e^{-0.15t}\, dt \\
&= -35 \lim_{M \to +\infty} \left[\frac{e^{-0.15t}}{0.15} \right]_0^M \\
&= \frac{35}{0.15} \lim_{M \to +\infty} (e^{-0.15M} - e^0) \\
&= -\frac{35}{0.15} (-1) \approx 233.33 \text{ million pounds,}
\end{aligned}$$

(since $\lim_{M \to +\infty} e^{-0.15M} = 0$).

 Although it is not easy to program a graphing calculator to calculate a limit, we can approximate the above result numerically by making a table of values for $\int_0^M 35e^{-0.15t}\, dt$ for larger and larger values of M. The following table shows the result of applying Simpson's rule using the graphing calculator program discussed in the appendix. N is the number of subdivisions. (Why are we using more subdivisions for increasing values of M?)

M	20	40	100
N	40	80	200
$\int_0^M 35e^{-0.15t}\, dt$	221.7	232.8	233.3

Alternatively, we could have used the built-in integration feature ("fnInt" on the TI-82 and TI-85) and let the calculator worry about the value of N.

INTEGRALS IN WHICH THE INTEGRAND BECOMES INFINITE AT AN ENDPOINT

These are integrals of the form

$$\int_a^b f(x)\, dx,$$

where $f(x) \to \pm\infty$ as either $x \to a$ or $x \to b$. We sometimes say that $f(x)$ **becomes infinite** at either $x = a$ or $x = b$.

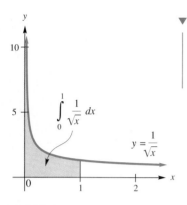

FIGURE 3

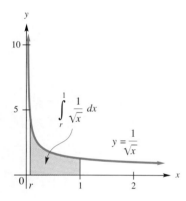

FIGURE 4

EXAMPLE 4

Calculate the integral

$$\int_0^1 \frac{1}{\sqrt{x}}\, dx.$$

SOLUTION Notice that $f(x)$ becomes infinite at $x = 0$, and so the integral is improper. Looking at Figure 3, we see that the corresponding region is infinitely long in the vertical direction rather than the horizontal direction.

Since the problem occurs at $x = 0$ (where the integrand is not defined) we approximate the integral by starting just to the right of $x = 0$ at, say $x = r$, and then letting r approach zero from the right. Figure 4 shows the region corresponding to such an approximation.

We then take

$$\int_0^1 \frac{1}{\sqrt{x}}\, dx = \lim_{r \to 0^+} \int_r^1 \frac{1}{\sqrt{x}}\, dx$$

$$= \lim_{r \to 0^+} \int_r^1 x^{-1/2}\, dx$$

$$= \lim_{r \to 0^+} [2x^{1/2}]_r^1 = \lim_{r \to 0^+} (2 - 2r^{1/2}) = 2.$$

Thus, as in Example 1 we have an infinitely long region with finite area.

In general, we make the following definition.

IMPROPER INTEGRAL WHERE THE INTEGRAND BECOMES INFINITE

If $f(x)$ is defined for all x with $a < x \leq b$ but becomes infinite at $x = a$, we define

$$\int_a^b f(x)\, dx = \lim_{r \to a^+} \int_r^b f(x)\, dx,$$

provided the limit exists. Similarly, if $f(x)$ is defined for all x with $a \leq x < b$ but becomes infinite at $x = b$, we define

$$\int_a^b f(x)\, dx = \lim_{r \to b^-} \int_a^r f(x)\, dx,$$

provided the limit exists. In either case, if the limit exists, we say that $\int_a^b f(x)\, dx$ **converges**. Otherwise, we say that $\int_a^b f(x)\, dx$ **diverges**.

▶ **CAUTION**

If the integrand becomes infinite at both endpoints or becomes infinite at some point strictly between a and b, then the above definition does not apply. We shall see how to deal with such cases in the examples. ◀

EXAMPLE 5

Investigate the convergence of the integral $\int_{-1}^{3} \frac{x}{x^2 - 9}\, dx.$

SOLUTION First, we check to see whether the integrand becomes infinite for any value of x in the range of integration. To do this, we set the denominator equal to zero and solve for x.

$$x^2 - 9 = 0$$

gives

$$(x - 3)(x + 3) = 0,$$

so

$$x = -3 \quad \text{or} \quad x = 3.$$

The first of these solutions is outside the range of integration, so we ignore it. The second, $x = 3$, happens to be one of the endpoints of the range of integration, so we conclude that we do have an improper integral, and use the definition above.

$$\int_{-1}^{3} \frac{x}{x^2 - 9}\, dx = \lim_{r \to 3^-} \int_{-1}^{r} \frac{x}{x^2 - 9}\, dx$$

The integral on the right can be calculated using the substitution $u = x^2 - 9$.

$$u = x^2 - 9$$
$$\frac{du}{dx} = 2x$$
$$dx = \frac{1}{2x}\, du$$
When $x = r$, $u = r^2 - 9$.
When $x = -1$, $u = -8$.

Thus,

$$\int_{-1}^{r} \frac{x}{x^2 - 9}\, dx = \frac{1}{2} \int_{-8}^{r^2-9} \frac{1}{u}\, du$$
$$= \frac{1}{2}\Big[\ln |u|\Big]_{-8}^{r^2-9}$$
$$= \frac{1}{2}[\ln |r^2 - 9| - \ln 8].$$

Taking the limit,

$$\int_{-1}^{3} \frac{x}{x^2 - 9} dx = \lim_{r \to 3^-} \int_{-1}^{r} \frac{x}{x^2 - 9} dx$$

$$= \lim_{r \to 3^-} \left(\frac{1}{2} [\ln |r^2 - 9| - 8] \right) = -\infty.$$

Thus, the integral diverges to $-\infty$.

Before we go on... Figure 5 shows the area in question. The figure makes it clear why the integral diverged to $-\infty$ instead of $+\infty$.

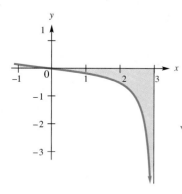

FIGURE 5

In the next example the integrand becomes infinite at a value strictly between the limits of integration.

▼ **EXAMPLE 6**

Investigate the convergence of the integral $\int_{-3}^{2} \frac{1}{x^2} dx$.

SOLUTION We first check to see whether the integrand becomes infinite at any value of x in the range of integration, and we notice that $x = 0$ is the only such value of x. Since 0 lies between -3 and 2, we conclude that we have an improper integral. But $x = 0$ is not an endpoint of the range of integration, so the above definition does not apply.

Q If the integrand becomes infinite at some point c *inside* the interval $[a, b]$ of integration, what do we do?

A Break up the interval into two intervals $[a, c]$ and $[c, b]$. Then deal with each of these intervals separately.

In our case, the integrand is not defined at $x = 0$, so we rewrite the integral as

$$\int_{-3}^{2} \frac{1}{x^2} dx = \int_{-3}^{0} \frac{1}{x^2} dx + \int_{0}^{2} \frac{1}{x^2} dx.$$

Each integral on the right is an improper integral that we know how to handle. The first is

$$\int_{-3}^{0} \frac{1}{x^2} dx = \lim_{r \to 0^-} \int_{-3}^{r} \frac{1}{x^2} dx$$

$$= \lim_{r \to 0^-} \left(\frac{1}{3} - \frac{1}{r} \right) = +\infty,$$

so this integral diverges to $+\infty$. Since this piece of the integral diverges, there is no hope of the original integral converging, so we can stop here and conclude that the given integral diverges.

Before we go on... If we had been sloppy and not bothered to check whether we had an improper integral to begin with, we would have obtained the incorrect answer,

$$\int_{-3}^{2} \frac{1}{x^2} \, dx = \left[-\frac{1}{x} \right]_{-3}^{2} = \frac{1}{6}. \quad \text{WRONG}$$

The moral is that even the most innocent-looking definite integral may be an improper integral in disguise! We must remember to check to see if the integrand becomes infinite at some point between a and b.

▶ **NOTE** Suppose that an improper integral I can be expressed as a sum of two or more improper integrals,

$$I = J + K + \ldots$$

Then (1) if one or more of the integrals $J, K, \ldots$ diverges, so does I, and (2) if all of the integrals $J, K, \ldots$ converge, then so does I. Moreover, I converges to the sum of the individual integrals. ◀

▶ **NOTE ON GRAPHING CALCULATORS AND COMPUTERS**

Q What would happen if we tried to calculate the integral $\int_{-3}^{2} \frac{1}{x^2} \, dx$ using a graphing calculator or computer?

A The best way to answer this question is to try it on a graphing calculator. For example, you can use the TI-82 to evaluate this integral numerically by entering

fnInt (1/X^2, X, −3, 2)

and pressing ENTER. We tried it, waited a while, and then the calculator said that there was an error. (Apparently its algorithm detected that something was amiss.*)

Not willing to give up so easily, we next tried the trapezoid rule with $N = 200$. This time, the calculator gave a "division by zero" error. We quickly realized that the reason for this was that $x = 0$ was one of the points in the subdivision, and the error resulted from the attempt to evaluate $f(0)$. To remedy that, we used $N = 201$ instead (so that 0 would not be a subdivision point) and obtained the answer 437.81. We then tried again with $N = 501$, and obtained the answer 1,092.51. In other words, the larger the choice for N, the larger the value of the integral, suggesting once again that the integral diverges to $+\infty$.

▼ * Most graphing calculator and computer integration algorithms work by using smaller and smaller subdivisions until the answers converge to a fixed real number. If, as is the case here, the integral diverges, then the answer fails to converge, the algorithm will fail, and an error will be returned.

However, do not feel confident that you can use numerical integration to determine the convergence or divergence of an improper integral. For instance, if you evaluate $\int_{-1}^{1} \frac{1}{x}\, dx$ with the trapezoid rule, you will obtain the following answers:

N	201	301	413	665
$\int_{-1}^{1} \frac{1}{x}\, dx$ by trap. rule	5×10^{-12}	1×10^{-11}	4×10^{-11}	-2×10^{-11}

These values are quite small, but getting no closer to zero (this is probably round-off error). It is not true that the value of this integral is zero. If you use the method of Example 6 to investigate this integral, you will find that it diverges: $\int_{-1}^{0} \frac{1}{x}\, dx$ diverges to $-\infty$ and $\int_{0}^{1} \frac{1}{x}\, dx$ diverges to $+\infty$. Why are the trapezoid sums so close to zero, then? ◀

We end this section with an example where the integral is improper in more than one way.

▼ **EXAMPLE 7**

Investigate the convergence of $\int_{0}^{+\infty} \frac{1}{x^{0.6}}\, dx$.

SOLUTION The given integral is improper in two ways: one of the endpoints is infinite, and the integrand becomes infinite at the other endpoint. As in Example 6, we must first break the integral into two improper integrals,

$$\int_{0}^{1} \frac{1}{x^{0.6}}\, dx \text{ and } \int_{1}^{+\infty} \frac{1}{x^{0.6}}\, dx.$$

(We could have broken it at any positive value, but we chose 1 for convenience.) In the first integral the integrand becomes infinite at one of the endpoints, and in the second one of the endpoints is infinite. We now investigate these integrals separately.

$$\int_{0}^{1} x^{-0.6}\, dx = \lim_{r \to 0^+} \int_{r}^{1} x^{-0.6}\, dx$$
$$= \lim_{r \to 0^+} \left[\frac{x^{0.4}}{0.4}\right]_{r}^{1}$$
$$= \lim_{r \to 0^+} \left(\frac{1}{0.4} - \frac{r}{0.4}\right) = \frac{1}{0.4} = \frac{5}{2}$$

$$\int_{1}^{+\infty} x^{-0.6}\, dx = \lim_{M \to +\infty} \int_{1}^{M} x^{-0.6}\, dx$$
$$= \lim_{M \to +\infty} \left[\frac{x^{0.4}}{0.4}\right]_{1}^{M} = +\infty$$

Since the second integral diverges, we conclude that the original integral diverges as well.

14.5 EXERCISES

Decide whether each of the integrals in Exercises 1 through 26 converges. If the integral converges, compute its value.

1. $\int_{1}^{+\infty} x \, dx$

2. $\int_{0}^{+\infty} e^{-x} \, dx$

3. $\int_{-2}^{+\infty} e^{-0.5x} \, dx$

4. $\int_{1}^{+\infty} \frac{1}{x^{1.5}} \, dx$

5. $\int_{-\infty}^{2} e^{x} \, dx$

6. $\int_{-\infty}^{-1} \frac{1}{x^{1/3}} \, dx$

7. $\int_{-\infty}^{-2} \frac{1}{x^2} \, dx$

8. $\int_{-\infty}^{0} e^{-x} \, dx$

9. $\int_{0}^{+\infty} x^2 e^{-6x} \, dx$

10. $\int_{0}^{+\infty} (2x - 4)e^{-x} \, dx$

11. $\int_{0}^{5} \frac{2}{x^{1/3}} \, dx$

12. $\int_{0}^{2} \frac{1}{x^2} \, dx$

13. $\int_{-1}^{2} \frac{3}{(x+1)^2} \, dx$

14. $\int_{-1}^{2} \frac{3}{(x+1)^{1/2}} \, dx$

15. $\int_{-1}^{2} \frac{3x}{x^2 - 1} \, dx$

16. $\int_{-1}^{2} \frac{3}{x^{1/3}} \, dx$

17. $\int_{-2}^{2} \frac{1}{(x+1)^{1/5}} \, dx$

18. $\int_{-2}^{2} \frac{2x}{\sqrt{4-x^2}} \, dx$

19. $\int_{-1}^{1} \frac{2x}{x^2 - 1} \, dx$

20. $\int_{-1}^{2} \frac{2x}{x^2 - 1} \, dx$

21. $\int_{-\infty}^{+\infty} xe^{-x^2} \, dx$

22. $\int_{-\infty}^{\infty} xe^{1-x^2} \, dx$

23. $\int_{0}^{+\infty} \frac{1}{x \ln x} \, dx$

24. $\int_{0}^{+\infty} \ln x \, dx$

25. $\int_{0}^{+\infty} \frac{2x}{x^2 - 1} \, dx$

26. $\int_{-\infty}^{0} \frac{2x}{x^2 - 1} \, dx$

APPLICATIONS

27. *Sales* My financial adviser has predicted that annual sales of BATMAN® T-shirts will continue to decline continuously at a rate of 10% each year. At the moment, I have 3,200 of the shirts in stock and am selling them at a rate of 200 per year. Will I ever sell them all?

28. *Revenue* Alarmed about the sales prospects for my BATMAN T-shirts (see Exercise 27) I will try to make up lost revenues by increasing the price by $1 each year. I now charge $10 per shirt. What is the total amount of revenue I can expect to earn from sales of my T-shirts, assuming the sales levels described in the previous exercise?

29. *Advertising* Spending on cigarette advertising in the United States declined from close to $600 million per year at the start of 1991 to half that amount three years later.* Use a continuous decay model to forecast the total revenue that will be spent on cigarette advertising beginning in January 1991.

30. *Sales* Sales of the text *I Love Calculus* have been declining steadily by 5% per year. Assuming that *I Love Calculus* currently sells 5,000 copies per year and that sales will continue this pattern of decline, calculate total future sales of the text.

31. *Variable Sales* The value of your Chateau Petit Mont Blanc 1963 vintage burgundy is continuously increasing at 40% per year, and you have a supply of 1,000 liter bottles worth $85 each at today's prices. In order to ensure a steady income, you have decided to sell your wine at a diminishing rate—starting at 500 bottles per year and then decreasing this figure exponentially at a fractional rate of 100% per year. How much income (to the nearest dollar) can you expect to generate by this scheme?

32. *Panic Sales* Unfortunately, your large supply of Chateau Petit Mont Blanc is continuously turning to vinegar at a fractional rate of 60% per year! You have thus decided to sell off your Petit Mont Blanc at $50

▼ *Sources: Advertising Age/Competitive Media Reporting (*The New York Times,* March 3, 1994, p. D1.)

Exercises 27–30 require the use of a graphing calculator or computer.

 27. *Market Saturation* You have just introduced a new model of TV. You predict that the market will saturate at 2,000,000 TVs and that your total sales will be governed by the equation

$$\frac{dS}{dt} = \frac{1}{4}S(2 - S),$$

where S is the total sales in millions of TVs and t is measured in months. If you give away 1,000 TV sets when you first introduce the TV, what will S be? Sketch the graph of S as a function of t. About how long will it take to saturate the market? (See Exercise 25.)

 28. *Epidemics* A certain epidemic of influenza is predicted to follow the function defined by

$$\frac{dA}{dt} = \frac{1}{10}A(20 - A),$$

where A is the number of people infected (in millions) and t is the number of months after the epidemic starts. If 20,000 cases are reported initially, find $A(t)$ and sketch its graph. When is A growing fastest? How many people will eventually be affected? (See Exercise 25.)

 29. *Growth of Tumors* The growth of tumors in animals can be modeled by the Gompertz equation

$$\frac{dy}{dt} = -ay \ln\left(\frac{y}{b}\right),$$

where y is the size of a tumor, t is time, and a and b are constants that depend on the type of tumor and the units of measurement.
(a) Solve for y as a function of t.
(b) If $a = 1$, $b = 10$, and $y(0) = 5$ cm^3 (with t measured in days), find the specific solution and graph it.

 30. *Growth of Tumors* Continuing the preceding exercise, suppose that $a = 1$, $b = 10$, and $y(0) = 15$ m^3. Find the specific solution and graph it. Comparing its graph to that obtained in the previous exercise, what can you say about tumor growth in these instances?

COMMUNICATION AND REASONING EXERCISES

31. What is the difference between a particular solution and the general solution of a differential equation? How do we get a particular solution from the general solution?

32. Why is there always an arbitrary constant in the general solution of a differential equation? Why are there not two or more arbitrary constants in a first-order differential equation?

33. Show by example that a **second-order** differential equation, one involving the second derivative y'', usually has two arbitrary constants in its general solution.

34. Find a differential equation that is not separable.

35. Find a differential equation whose general solution is $y = 4e^{-x} + 3x + C$.

36. Explain how, knowing the elasticity of demand as a function of either price or demand, you may find the demand equation.

▶  You're the Expert

ESTIMATING TAX REVENUES

You have just been hired by the incoming administration of a certain country as chief consultant for national tax policy and have been getting conflicting advice from the so-called finance experts on your staff. Several of them have come up with plausible suggestions for new tax structures, and your job is to choose the plan that results in the most revenue for the government.

Before you can evaluate their plans, you realize that it is essential to know the income distribution: that is, how many people earn how much money. One might think that the most useful way of specifying income distribution would be to use a function that gives the exact number $f(x)$ of people earning a given salary x. This would necessarily be a **discrete** function—it only makes sense if x happens to be a whole number of cents. There is, after all, no one earning a salary of exactly $12,000.142567! Further, this function would behave rather erratically, since there are, for example, probably many more people making a salary of exactly $30,000 than exactly $30,000.01. It is far more convenient to start with the function defined by

$$N(x) = \text{total number of people earning between 0 and } x \text{ dollars.}$$

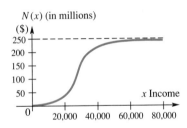

FIGURE 1

The graph of $N(x)$ might look like the one shown in Figure 1.

Notice that the curve climbs most steeply at around $x = \$30,000$, which suggests that most people are earning around that amount. If we take the *derivative* of $N(x)$, we get an **income distribution function.** Its graph might look like the one shown in Figure 2.

The derivative $N'(x)$ peaks at $30,000. Since the derivative measures the rate of change, its value at x is the additional number of taxpayers per $1 increase in salary. Thus, its value at, say, $x = 20,000$ tells us the approximate number of people earning between $20,000 and $20,001, so the fact that $N'(20,000) = 5,000$ tells us that approximately 5,000 people are earning a salary of $20,000. In other words, N' shows the distribution of incomes among the population, hence the name "distribution function."*

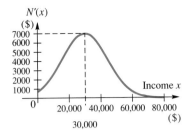

FIGURE 2

You thus send a memo to your experts requesting the income distribution function for the nation. After much collection of data, they tell you that the income distribution function is

$$N'(x) = 7,000\, e^{-(x-30,000)^2/400,000,000}.$$

This is in fact the function whose graph is shown in Figure 2, and it is an example of a **normal** distribution. Notice that the curve is symmetric around the median income of $30,000 and that about 7,000 people are earning between $30,000 and $30,001 annually.[†]

Given this income distribution, your financial experts have come up with two possible tax policies (Figures 3 and 4).

In the first alternative, all taxpayers pay half of their income in taxes, except that no one pays more than $40,000 in taxes. In the second alternative, there are four **tax brackets,** described by the following table.

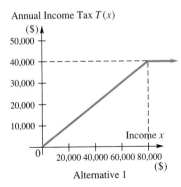

FIGURE 3

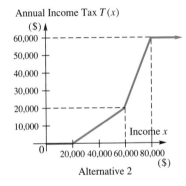

FIGURE 4

▼ * We shall look at a very similar idea later in the chapter on probability and calculus.

† You might find it odd that you weren't given the original function N, but it will turn out that you don't need it. How would you compute it? Is it possible to write down a formula for N?

Income	Marginal tax rate
$0–20,000	0%
$20,000–60,000	50%
$60,000–80,000	200%
Above $80,000	0%

Now you must determine which alternative will generate the largest annual tax revenue.

Each of Figures 3 and 4 is the graph of a function, T. Before calculating the formulas for these functions, you try to work with the general situation. You have an income distribution function N' and a tax function T, both functions of annual income. You need to find a formula for total tax revenues. You begin by trying subdivision. First, you decide to work only with incomes in some finite bracket $[0, M]$ by using a cutoff—say, $M = \$10$ million. (You shall eventually let M approach $+\infty$.) Next, you subdivide the interval $[0, M]$ into a large number of intervals of width Δx. If $[x_{k-1}, x_k]$ is a typical such interval, you need to calculate the approximate tax revenue from people whose total incomes lie between x_{k-1} and x_k. You will then sum over k to get the total revenue.

You need to know how many people are making incomes between x_{k-1} and x_k. Because $N(x_k)$ people are making incomes *up to* x_k and $N(x_{k-1})$ people are making incomes up to x_{k-1}, the number of people making incomes between x_{k-1} and x_k is $N(x_k) - N(x_{k-1})$. Because x_k is very close to x_{k-1}, the incomes of these people are all approximately equal to x_k dollars, so each of these taxpayers is paying an annual tax of about $T(x_k)$. This gives a tax revenue of

$$[N(x_k) - N(x_{k-1})]T(x_k).$$

Now you do a clever thing. You write $x_k - x_{k-1} = \Delta x$ and replace the quantity $N(x_k) - N(x_{k-1})$ by

$$\frac{N(x_k) - N(x_{k-1})}{\Delta x} \Delta x.$$

This gives you a tax revenue of about

$$\frac{N(x_k) - N(x_{k-1})}{\Delta x} T(x_k) \, \Delta x.$$

from wage-earners in the bracket $[x_{k-1}, x_k]$. Summing over k gives an approximate total revenue of

$$\sum_{k=1}^{n} \frac{N(x_k) - N(x_{k-1})}{\Delta x} T(x_k) \, \Delta x.$$

The larger n is, the more accurate your estimate will be, so you take the limit of the sum as $n \to \infty$. When you do this, two things happen. First, the quantity

$$\frac{N(x_k) - N(x_{k-1})}{\Delta x}$$

approaches the derivative, $N'(x_k)$. Second, the sum, which you recognize as a Riemann sum, approaches the integral

$$\int_0^M N'(x)T(x)\,dx.$$

You now take the limit as $M \to +\infty$ and obtain

$$\text{Total tax revenue} = \int_0^{+\infty} N'(x)T(x)\,dx.$$

This indefinite integral is fine in theory, but the actual calculation will have to be done numerically, so you stick with the upper limit of $10 million for now, and you will have to check that it is reasonable at the end (notice that, by the graph of N', it appears that extremely few, if any, people earn that much). Now you already have a formula for $N'(x)$, but you still need to write formulas for the tax functions $T(x)$ for both alternatives.

Alternative 1 The graph in Figure 3 rises linearly from 0 to 40,000 as x ranges from 0 to 80,000, and then stays constant at 40,000. Thus, the slope of the first part is $\frac{40,000}{80,000} = \frac{1}{2}$. The taxation function is therefore

$$T(x) = \begin{cases} \frac{x}{2} & \text{if } 0 \leq x \leq 80,000 \\ 40,000 & \text{if } x \geq 80,000. \end{cases}$$

To perform the integration, you need to break the integral into two pieces, the first from 0 to 80,000 and the second from 80,000 to 10,000,000. In other words,

$$R_1 = \int_0^{80,000} (7{,}000e^{-(x-30{,}000)^2/400{,}000{,}000})\frac{x}{2}\,dx +$$

$$\int_{80,000}^{10,000,000} (7{,}000e^{-(x-30{,}000)^2/400{,}000{,}000})40{,}000\,dx.$$

You decide not to attempt this by hand*! You use numerical integration software to obtain a grand total of $R_1 = \$3{,}732{,}760{,}000{,}000$, or \$3.73276 trillion.[†]

Alternative 2 The graph in Figure 4 rises linearly from 0 to 20,000 as x ranges from 20,000 to 60,000, rises from 20,000 to 60,000 as x ranges from

---

▼ * The first integral requires a substitution and integration by parts. The second cannot be done in elementary terms at all.
† Rounded to six significant digits.

60,000 to 80,000, and then stays constant at 60,000. Thus, the slope of the first incline is $\frac{1}{2}$ and the slope of the second incline is 2 (this is why the *marginal* tax rates are 50% and 200% respectively). The taxation function is therefore

$$T(x) = \begin{cases} 0 & \text{if } 0 \leq x \leq 20{,}000 \\ \dfrac{x - 20{,}000}{2} & \text{if } 20{,}000 \leq x \leq 60{,}000 \\ 20{,}000 + 2(x - 60{,}000) & \text{if } 60{,}000 \leq x \leq 80{,}000 \\ 60{,}000 & \text{if } x \geq 80{,}000 \end{cases}$$

Values of x between 0 and 20,000 do not contribute to the integral, and so

$$R_2 = \int_{20{,}000}^{60{,}000} (7{,}000 e^{-(x-30{,}000)^2/400{,}000{,}000}) \left(\frac{x - 20{,}000}{2} \right) dx$$

$$+ \int_{60{,}000}^{80{,}000} (7{,}000 e^{-(x-30{,}000)^2/400{,}000{,}000})(20{,}000 + 2(x - 60{,}000)) \, dx$$

$$+ \int_{80{,}000}^{10{,}000{,}000} (7{,}000 e^{-(x-30{,}000)^2/400{,}000{,}000}) 60{,}000 \, dx.$$

This gives $R_2 = \$1.52016$ trillion—considerably less than Alternative 1. Thus, even though this alternative taxes the wealthy more heavily, it yields less total revenue.

Now what about the cutoff at \$10,000,000 annual income? If you try either integral again with an upper limit of \$100 million, you will see no change in either one to 6 significant digits. There simply are not enough taxpayers earning an income above \$10,000,000 to make a difference. You conclude that your answers are sufficiently accurate and that the first alternative provides the most tax revenue.

Exercises

(Exercises 1–6 require the use of a graphing calculator or computer.)

In Exercises 1–6, calculate the total tax revenue for a country with the given income distribution and tax policies.

1. $N'(x) = 3{,}000 e^{-(x-10{,}000)^2/10{,}000}$, 25% tax on all income

2. $N'(x) = 3{,}000 e^{-(x-10{,}000)^2/10{,}000}$, 45% tax on all income

3. $N'(x) = 5{,}000 e^{-(x-30{,}000)^2/100{,}000}$, no tax on an income below \$30,000, \$10,000 tax on any income of \$30,000 or above.

4. $N'(x) = 5{,}000 e^{-(x-30{,}000)^2/100{,}000}$, no tax on an income below \$50,000, \$20,000 tax on any income of \$50,000 or above.

5. $N'(x) = 7{,}000e^{-(x-30{,}000)^2/400{,}000{,}000}$, $T(x)$ with the following graph:

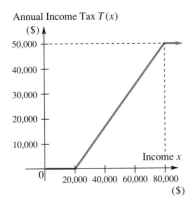

6. $N'(x) = 7{,}000e^{-(x-30{,}000)^2/400{,}000{,}000}$, $T(x)$ with the following graph:

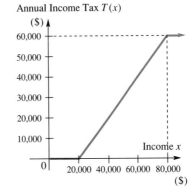

Review Exercises

Evaluate the integrals in Exercises 1–30.

1. $\displaystyle\int (x^2 + 2)e^x \, dx$

2. $\displaystyle\int (x^2 + x)e^{2x} \, dx$

3. $\displaystyle\int x^2 \ln 2x \, dx$

4. $\displaystyle\int \sqrt{x} \ln \sqrt{x} \, dx$

5. $\displaystyle\int_0^1 x^2 e^x \, dx$

6. $\displaystyle\int_{-2}^2 (x^3 + 1)e^{-x} \, dx$

7. $\displaystyle\int_1^e x^2 \ln x \, dx$

8. $\displaystyle\int_1^{2e} (x + 1)\ln x \, dx$

9. $\displaystyle\int \frac{dx}{9 + 4x^2}$

10. $\displaystyle\int \frac{dx}{9 - 4x^2}$

11. $\displaystyle\int \frac{dx}{\sqrt{9 + 4x^2}}$

12. $\displaystyle\int \frac{dx}{\sqrt{9 - 4x^2}}$

13. $\displaystyle\int \sqrt{x^2 + 2x + 2} \, dx$

14. $\displaystyle\int \sqrt{x^2 + 2x - 2} \, dx$

15. $\displaystyle\int \frac{e^{2x}}{1 + e^{4x}} \, dx$

16. $\displaystyle\int \frac{e^{2x}}{e^{4x} - 1} \, dx$

17. $\displaystyle\int_0^1 \frac{e^{2x}}{\sqrt{9 + 4e^{4x}}} \, dx$

18. $\displaystyle\int_{-1}^0 \frac{e^{2x}}{\sqrt{9 - 4e^{4x}}} \, dx$

19. $\displaystyle\int_1^e \frac{\sqrt{(\ln x)^2 + 1}}{x} \, dx$

20. $\displaystyle\int_1^e \frac{\ln x \sqrt{(\ln x)^2 + 1}}{x} \, dx$

21. $\displaystyle\int_1^\infty \frac{1}{x^5} \, dx$

22. $\displaystyle\int_1^\infty \frac{1}{x^{0.6}} \, dx$

23. $\displaystyle\int_{-\infty}^1 e^{x/2} \, dx$

24. $\displaystyle\int_{-\infty}^0 xe^{x/2} \, dx$

25. $\displaystyle\int_{-\infty}^\infty \frac{1}{x^2} \, dx$

26. $\displaystyle\int_{-\infty}^\infty e^{-x} \, dx$

27. $\displaystyle\int_0^1 \frac{1}{(x-1)^2} \, dx$

28. $\displaystyle\int_0^1 \frac{1}{x^2 - 1} \, dx$

29. $\displaystyle\int_0^1 \frac{1}{\sqrt{1 - x}} \, dx$

30. $\displaystyle\int_0^1 \frac{1/\sqrt{x}}{1 - \sqrt{x}} \, dx$

Find the areas indicated in Exercises 31–40.

31. Between $y = x^3$ and $y = 1 - x^3$ for x in $[0, 1]$.
32. Between $y = x^3$ and $y = 1 - x^3$ for x in $[-1, 1]$.
33. Between $y = e^x$ and $y = 2e^{-x}$ for x in $[0, 2]$.
34. Between $y = e^x$ and $y = e^{-x}$ for x in $[0, 2]$.
35. Between $y = 1 - x^2$ and $y = x^2$.
36. Between $y = 1 - x^4$ and $y = x^4$.
37. Between $y = x^3 - x$ and $y = x - x^3$.
38. Between $y = x^4 - x^2 + 1$ and $y = x^4$.
39. Between $y = \dfrac{1}{x^2 + 1}$ and $y = \dfrac{1}{x^2 - 1}$ for x in $[0, \frac{1}{2}]$.
40. Between $y = \dfrac{1}{x^2 + 4}$ and $y = \dfrac{1}{2x^2 + 3}$ for x in $[0, 1]$.

41–44. Find the averages of the functions in Exercises 5–8.
45–48. Find the averages of the functions in Exercises 17–20.

In Exercises 49–52, find the 2-unit moving averages of the given functions. If you are using a graphing calculator or computer software, plot the original function and its moving average on the same graph.

49. $f(x) = x^{2/3}$
50. $f(x) = x^{4/3}$
51. $f(x) = \ln x$
52. $f(x) = x \ln x$

53. $\dfrac{dy}{dx} = x^2 y^2$
54. $\dfrac{dy}{dx} = x^3 y^2$
55. $\dfrac{dy}{dx} = xy + x + y + 1$
56. $\dfrac{dy}{dx} = xy + 2x$
57. $\dfrac{dy}{dx} = \dfrac{1}{y}$
58. $\dfrac{dy}{dx} = \dfrac{1}{y^2}$
59. $xy \dfrac{dy}{dx} = 1$, $y(1) = 1$
60. $x^2 y \dfrac{dy}{dx} = 1$, $y(1) = 1$
61. $y(x^2 + 1) \dfrac{dy}{dx} = xy^2$, $y(0) = 2$
62. $xy^2 \dfrac{dy}{dx} = y(x^2 + 1)$, $y(1) = 10$

APPLICATIONS

63. *Dangerous Weapons* The following graph shows the number of deaths per year in the U.S. resulting from injuries related to motor vehicles and firearms.*
    (a) Estimate the definite integral of the function represented by the top graph from '73 to '92. What does the answer tell you?
    (b) Estimate the area between the graphs from '73 to '92. What does the answer tell you?

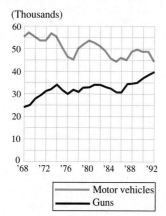

Deaths Per Year in U.S.

64. *Dangerous Weapons* Referring to the graph in the previous exercise, estimate the average difference between the two graphs from '68 to '88. What does the answer tell you?

65. *Cost* Your factory buys nuts, currently priced at $2 per box, but the price is rising 3% per month continuously due to high inflation. You currently buy 1,000 boxes per month, but your purchases are increasing by 100 boxes per month. How much will you spend on nuts in the next 12 months?

66. *Cost* Your factory buys bolts, currently priced at $1.50 per box, but the price is rising 2.5% per month continuously due to high inflation. You currently buy 5,000 boxes per month, but your purchases are declining by 200 boxes per month. How much will you spend on bolts in the next 12 months?

67. *Supply and Demand* The supply equation for fertilizer is
$$p = (q - 20)^{0.5} \quad (q \geq 20),$$
where p is the price (in dollars) per 20-lb bag and q is the weekly supply. The demand equation is
$$p = (70 - q)^{0.5} \quad (q \leq 70).$$
Find the equilibrium price and the consumers' and producers' surpluses at that price.

▼ *Source: National Center for Health Statistics/*The New York Times*, January 26, 1994, p. A12.

68. *Supply and Demand* The supply equation for compost is
$$p = (q - 10)^{0.7} \quad (q \geq 10),$$
where p is the price (in dollars) per 20-lb bag and q is the weekly supply. The demand equation is
$$p = (60 - q)^{0.7} \quad (q \leq 60).$$
Find the equilibrium price and the consumers' and producers' surpluses at that price.

69. *Motion* A cannonball launched into the air from ground level had an upward velocity of $v(t) = 320 - 32t$ ft/s after t seconds.
   (a) Find the area under the graph of v for $0 \leq t \leq 10$. What does this area represent?
   (b) Find the time T at which the ball returns to the ground, and then find the area under the graph for $0 \leq t \leq T$ without further calculation.

70. *Motion* Two cars pass the same point on the New York Thruway at time $t = 0$ hours. Car A maintains a speed of $v_A = 55$ mph, while car B has velocity $v_B(t) = 30 + 20t$ for $0 \leq t \leq 2$ and then $v_B(t) = 70 - 20t$ for $2 \leq t \leq 4$.
   (a) Find the area between the graphs of v_A and v_B for $0 \leq t \leq 1$. What does this area represent?
   (b) When will the two cars meet for the second time? Express your answer both as a time and in terms of the areas between the graphs of v_A and v_B.

71. *Pedestrian Safety* The following chart shows the number of pedestrians killed by vehicles in New York City each year over the given period.*
   (a) Find the average number of pedestrians killed in New York per year over the given period.
   (b) Give a rough sketch of the 2-year moving average without doing any actual computation, and compare your sketch with that of the actual 2-year moving average.

Pedestrians Killed by Vehicles in New York City

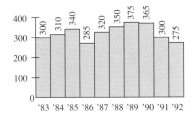

72. *Growth of Medical Costs* Repeat the previous exercise using the following chart, which shows the annual rate of increase in medical costs in the New York City Metropolitan Center.†

Rate of Increase in Medical Costs in the New York City Metropolitan Center

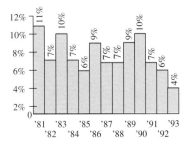

73. *Investments* If your investment of $10,000 in Tarnished Teak Enterprises is continuously depreciating at 6% per year, what is the average amount in your account over one year?

74. *Bacterial Growth* If a culture of bacteria is doubling every hour, what is the average population over the first two hours, assuming that the culture contains a million organisms at the start?

75. *Investments* You are offered an investment whose instantaneous rate of growth (in dollars per year) is always 10% of the square root of the amount in the account. If you invest $10,000 initially, find the amount present as a function of time.

76. *Loans* A junior executive at a bank came up with the following idea: The bank should offer an account accumulating interest at a rate (in dollars per year) equal to 1% of the square of the amount in the account at any time. Looking at what would happen if someone were to deposit $100 in such an account, explain why the executive was fired on the spot.

---

*Source: Department of Transportation/*The New York Times*, January 21, 1994, p. B1.
†Source: Federal Bureau of Labor Statistics/*The New York Times*, December 25, 1993, p. L 36.

CHAPTER 15

"Royal Flush"
After some lean years, luxury homes are starting to sell again on Long Island

A Long Island house that looks like a million bucks may cost a good deal less now. And buyers are starting to notice.

The market for luxury houses, battered by the recession, is experiencing a resurgence. Most of the buyers are Long Island residents who have been able to sell their middle-market houses and take advantage of the not-so-sky-high price of living in luxury.

Last year was the Daniel Gale Agency's best year ever for selling luxury houses—those priced at $400,000 or more—said Bonnie Devendorf, associate broker and manager of the firm's Locust Valley office. She said the firm, which covers the North Shore from Stony Brook to Port Washington, is on its way this year to beat the record, despite a $9 million home sale in 1993.

"Year over year, the market above $400,000 is certainly up from last year," said Jim Linane, regional manager of Chase Manhattan Personal Financial Services, which specializes in mortgages of more than $300,000. He said such areas as Port Washington, Manhasset and Jericho are experiencing a lot of activity. "They're fairly convenient to the railroad and are good commuter towns," he said. "The South Shore Suffolk area is a little bit slow this year. But anything's that's a decent commute into the city seems to be pretty hot."

An easy commute into New York was a major consideration for Steve and Carol Sterneck, who recently moved from Manhattan into an East Hills home. "It was a combination of factors that led to the decision to move out of the city and into a home," Steve Sterneck said. . . .

Source: From Jacqueline Henry, "Royal Flush," *New York Newsday*, 1994, p. D8.

Functions of Several Variables

SECTIONS

1. Functions of Two or More Variables
2. Three-Dimensional Space and the Graph of a Function of Two Variables
3. Partial Derivatives
4. Maxima and Minima
5. Constrained Maxima and Minima and Applications
6. Least-Squares Fit
7. Double Integrals

You're the Expert
Constructing a Best-Fit Demand Curve

APPLICATION ▶ Your market research of real estate investments reveals the following sales figures for new homes of different prices over the past year.

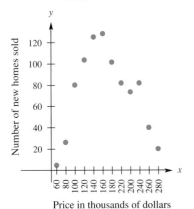

Sales of New Homes

As the chief economist for Sturdy Homes, Inc., you would like to come up with a demand equation for home buying trends.

Chapter 15 Functions of Several Variables

INTRODUCTION ▸ We have studied functions of a single variable quite extensively now. However, not every function that arises is a function of a single variable. For example, suppose you run a company that produces two types of floppy disks, low-density and high-density, and you wish to come up with a cost equation. If your factory produces x low-density disks and y high-density disks, your cost equation might look something like this:

$$C(x, y) = 0.2x + 0.4y + 0.1\sqrt{xy}.$$

This is a function of two variables, since it *depends on* both x and y. As we shall see, we can extend the techniques of calculus to such functions, allowing us to examine marginal costs, for example. One of the main applications we shall look at is optimization: finding the maximum or minimum of a function of two or more variables.

15.1 FUNCTIONS OF TWO OR MORE VARIABLES

Recall that a function of one variable is a rule for manufacturing a new number $f(x)$ from a single number x. In the same vein, a **function, f, of x and y** is a rule for manufacturing a new number $f(x, y)$ from a *pair* of numbers (x, y). For instance, the cost function $C(x, y) = 0.2x + 0.4y + 0.1\sqrt{xy}$ mentioned in the introduction is such a function. The function $f(x, y) = x^2 + y^2$ is another example. Figure 1 illustrates this concept—in goes a pair of numbers and out comes a single number.

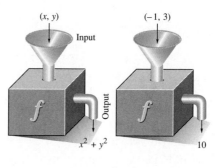

$f(x, y) = x^2 + y^2$

FIGURE 1

For example, with $f(x, y) = x^2 + y^2$, $f(-1, 3) = (-1)^2 + 3^2 = 10$. In other words, we evaluated $f(-1, 3)$ by substituting the quantity (-1) for x and quantity 3 for y.

EXAMPLE 1

Let $f(x, y) = x^2 - 2xy + y - 1$. Find $f(-1, 1), f(0, 0), f(a, 1), f(a, b)$, and $f(a + h, b)$.

SOLUTION To calculate $f(-1, 1)$ we substitute the quantity (-1) for x and 1 for y. Thus,

$$f(-1, 1) = (-1)^2 - 2(-1)(1) + (1) - 1 = 3.$$

Similarly,

$$f(0, 0) = 0^2 - 2(0)(0) + (0) - 1 = -1,$$
$$f(a, 1) = a^2 - 2a(1) + (1) - 1 = a^2 - 2a,$$
$$f(a, b) = a^2 - 2ab + b - 1,$$
$$f(a + h, b) = (a + h)^2 - 2(a + h)b + b - 1.$$

You can have your graphing calculator compute $f(x, y)$ numerically as follows (these instructions are for the TI-82, and other calculators are similar). In the "Y=" screen, enter

$$Y_1 = X^\wedge 2 - 2XY + Y - 1$$

Then, to evaluate, say, $f(-4, 3)$, enter

$$-4 \rightarrow X$$
$$3 \rightarrow Y$$
$$Y_1$$

and the calculator will evaluate the function and give the answer, $f(-4, 3) = 42$.

This procedure can be laborious if we want to calculate $f(x, y)$ for several different values of x and y, so we have supplied a program to remedy this. (See Appendix B.)

EXAMPLE 2   Cost Functions

Suppose that you own a company making two models of stereo speakers: the Ultra Mini and the Big Stack. Your total monthly cost (in dollars) to make x Ultra Minis and y Big Stacks is given by

$$C(x, y) = \$10{,}000 + 20x + 40y.$$

What is the significance of each term in this formula?

SOLUTION The terms have meanings similar to those we saw for cost functions of a single variable. The constant term 10,000 is the **fixed cost,** the amount it costs even if you make no speakers, because

$$C(0, 0) = 10{,}000.$$

The term $20x$ indicates that each Ultra Mini adds $20 to the total cost. We say that $20 is the **marginal cost** of each Ultra Mini. The term $40y$ indicates that each Big Stack adds $40 to the total cost. The marginal cost of each Big Stack is $40.

Before we go on... This is an example of a **linear** function of two variables. The coefficients of x and y play roles similar to that of the slope of a line. In particular, they give the rates of change of the function as each variable increases while the other stays constant (think about it).

▼ **EXAMPLE 3** Cost Functions

Another possibility for the cost function in the previous example is

$$C(x, y) = \$10{,}000 + 20x + 40y + 30\sqrt{x + y}.$$

We shall see later that it is still possible to compute the marginal cost of each model of speaker, and the extra term has the effect that the marginal costs decrease. This might be the result of economies produced by buying common materials in bulk. Notice that the extra term is really a function of the total number of speakers produced, $x + y$.

Before we go on... What values of x and y may we substitute into $C(x, y)$? Certainly we must have $x \geq 0$ and $y \geq 0$, because it makes no sense to speak of manufacturing a negative number of speakers. There is certainly also some upper bound to the number of speakers that can be made in a month. This might take one of several forms. It might be that $x \leq 100$ and $y \leq 75$. The inequalities $0 \leq x \leq 100$ and $0 \leq y \leq 75$ describe the region in the plane shaded in Figure 2.

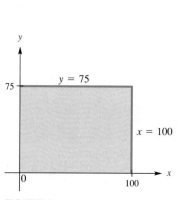
FIGURE 2

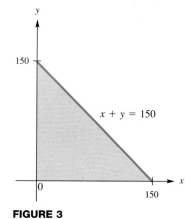
FIGURE 3

Another possibility would be that the *total* number of speakers is bounded—say, $x + y \leq 150$. This, together with $x \geq 0$ and $y \geq 0$, describes the region shaded in Figure 3.

In either case, the region shown represents the pairs (x, y) for which $C(x, y)$ is defined. Just as in the case of a function of one variable, we call this region the **domain** of the function. As before, when the domain is not given explicitly we agree to take the largest possible domain.

15.1 Functions of Two or More Variables

▼ **EXAMPLE 4** Revenue

Revenue is given by the equation

$$R = pq,$$

where p is the price per item and q is the number of items sold. We can now think of this as a function of two variables and write it as $R(p, q)$.

Before we go on... In examples earlier in this book we reduced R to a function of one variable by writing, for example, q as a function of p. This gives

$$R(p) = R(p, q(p)).$$

We used similar methods in, for example, optimization problems to reduce what were really functions of two or more variables to functions of one variable alone. In this chapter we shall see how to apply calculus directly to functions of several variables. ▪

The next example reintroduces the Cobb-Douglas production formula, this time viewed as a function of two variables.

▼ **EXAMPLE 5** Cobb–Douglas Production Function

Productivity usually depends on both labor and capital. For example, we can measure the productivity of an automobile manufacturing plant by counting the number of automobiles it produces per year. As a measure of labor, we can use the number of employees, and we can use the annual operating budget total as our measure of capital. The Cobb-Douglas production function then has the form

$$P(x, y) = Kx^a y^{1-a},$$

where P stands for the number of automobiles produced per year, x is the number of employees, and y is the annual operating budget. The numbers K and a are constants that depend on the particular factory we are looking at, with a between 0 and 1. For instance, we might have $P(x, y) = 10x^{0.3}y^{0.7}$ for our particular automobile factory. Values for these constants can be determined using actual production data. (We shall be doing this in the exercises.)

Before we go on... Note that if we have either $x = 0$ or $y = 0$, then $P(x, y) = 0$. This is consistent with common sense—we can hardly expect anything to be produced without labor or capital. ▪

Functions of several variables can arise in geometry as well. For example, the area of a rectangle with height h and width w is

$$A(h, w) = hw.$$

Here is another useful example.

▼ **EXAMPLE 6** Distance

Express the distance of the point (x, y) to the point $(3, 2)$ as a function of the two variables x and y. Use your function to find the distance from $(-1, 1)$ to $(3, 2)$.

SOLUTION If we plot (x, y) and $(3, 2)$, we might get the picture shown in Figure 4. We need to calculate the distance d.

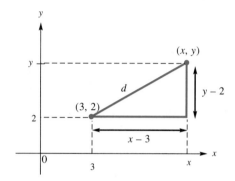

FIGURE 4

By the Pythagorean theorem applied to the right triangle shown, we get

$$d^2 = (x - 3)^2 + (y - 2)^2.$$

Taking square roots (d is supposed to be a distance, so we take the positive square root), we get

$$d = \sqrt{(x - 3)^2 + (y - 2)^2}.$$

Thus, the distance is given by the function

$$d(x, y) = \sqrt{(x - 3)^2 + (y - 2)^2}.$$

Now, to find the distance from $(-1, 1)$ to $(3, 2)$, all we need to do is compute $d(-1, 1)$.

$$d(-1, 1) = \sqrt{(-1 - 3)^2 + (1 - 2)^2} = \sqrt{17}$$

There is nothing special about having chosen the particular point $(3, 2)$ in the above example. If we replace it with the more general point (a, b), we get the following useful formula for the distance between two points.

DISTANCE BETWEEN TWO POINTS IN THE PLANE

The distance between the points (a, b) and (x, y) is

$$d = \sqrt{(x - a)^2 + (y - b)^2}.$$

If (a, b) happens to be the origin, so that $a = b = 0$, we get the following special case.

> **DISTANCE TO THE ORIGIN IN THE PLANE**
> The distance from the point (x, y) to the origin is
> $$d = \sqrt{x^2 + y^2}.$$

This gives the following equation for the circle centered at the origin with radius r.

$$\sqrt{x^2 + y^2} = r$$

Squaring both sides gives the following, which we shall need in later sections.

> **EQUATION OF THE CIRCLE OF RADIUS r CENTERED AT THE ORIGIN**
> $$x^2 + y^2 = r^2$$

We looked briefly at Newton's Law of Gravity in the chapter on functions and their graphs. Recall that according to Newton's Law, the gravitational force exerted on a particle with mass m by another particle with mass M is given by the following function of distance.

$$F(r) = G\frac{Mm}{r^2}$$

Here, r is the distance between the two particles in meters, the masses M and m are given in kilograms, $G \approx 6.67 \times 10^{-11}$, and the resulting force is measured in newtons.*

▼ **EXAMPLE 7** Newton's Law of Gravity Revisited

Find the gravitational force exerted on a particle with mass m situated at the point (x, y) by another particle with mass M situated at the point (a, b). Express the answer as a function of the coordinates of the particle with mass m.

SOLUTION We already have the formula we need—almost. The formula for gravitational force is expressed as a function of the distance, r, between the

---

▼ *A newton is the force that will cause a 1-kilogram mass to accelerate at 1 meter/sec².

two particles. Since we are given the coordinates of the two particles, we can express r in terms of these coordinates by using the formula for distance.

$$r = \sqrt{(x-a)^2 + (y-b)^2}$$

Substituting for r, we get

$$F(x, y) = G\frac{Mm}{(x-a)^2 + (y-b)^2}.$$

Before we go on... Notice that $F(a, b)$ is not defined, since substituting $x = a$ and $y = b$ would make the denominator equal 0. Thus, the largest possible domain of F excludes the point (a, b). Since (a, b) is the only value of (x, y) for which F is not defined, we can deduce that *the domain of F consists of all points (x, y) except for (a, b)*. In other words, the domain of F is the whole xy-plane with the single point (a, b) missing.*

Q What is the difference between the function $F(r)$ and the new function we just obtained, $F(x, y)$?

A The old function required us to calculate the distance between the two particles before we could find the force, as it was a function of distance r. The new function gives the force as soon as we know the coordinates of the particles.

Q Why have we expressed F as a function of x and y only, and not also as a function of a and b?

A It is a matter of interpretation. When we write F as a function of x and y we are thinking of a and b as *constants*. For instance, (a, b) could be taken to be the coordinates of the sun—which we often assume to be fixed in space—while (x, y) could be thought of as the coordinates of the earth—which is moving around the sun. In that case it is most natural to think of x and y as variable and a and b as constant.

In case you have the impression that the only interesting functions are those of one or two variables, consider the following example of a function of *three* variables.

▼ **EXAMPLE 8**   Faculty Salaries

David Katz came up with the following function for the faculty salary of a professor with 10 years of teaching experience in a large university.

$$S(x, y, z) = 13{,}005 + 230x + 18y + 102z$$

Here, S is the salary in 1969–1970 in dollars per year, x is the number of books the professor has published, y is the number of articles published, and

▼ *Mathematicians often refer to this as a "punctured plane."

z is the number of "excellent" articles published.* What salary would you expect a professor with 10 years' experience to have earned in 1969–1970 if she has published two books, 20 articles, and 3 "excellent" articles?

SOLUTION All we need to do is calculate

$$S(2, 20, 3) = 13{,}005 + 230(2) + 18(20) + 102(3)$$
$$= \$14{,}131.$$

Before we go on... This is an example of a **linear** function of three variables. Katz came up with his model by surveying a large number of faculty members and then finding the linear function "best" fitting the data. Such models are called **linear regression** models. We shall see, in the section on **least-squares fit,** how to find the coefficients in such a model given the data obtained in the survey.

What does this model say about the value of a single book or a single article? If a book takes 15 times as long to write as an article, how would you recommend a professor spend her writing time?

The general form of a linear function of several variables is this:

LINEAR FUNCTION OF SEVERAL VARIABLES

A **linear function of the variables** $x, y, z, \ldots$ is a function of the form

$$f(x, y) = a + bx + cy + dz + \ldots \quad (a, b, c, d, \ldots \text{ constants}).$$

15.1 EXERCISES

For each of the functions in Exercises 1–4, evaluate **(a)** $f(0, 0)$; **(b)** $f(1, 0)$; **(c)** $f(0, -1)$; **(d)** $f(a, 2)$; **(e)** $f(y, x)$; **(f)** $f(x + h, y + k)$.

1. $f(x, y) = x^2 + y^2 - x + 1$

2. $f(x, y) = x^2 - y - xy + 1$

3. $f(x, y) = \sqrt{(x - 1)^2 + (y - 2)^2}$

4. $f(x, y) = \sqrt{(x + 1)^2 + (y - 1)^2}$

For each of the functions in Exercises 5–8, evaluate **(a)** $g(0, 0, 0)$; **(b)** $g(1, 0, 0)$; **(c)** $g(0, 1, 0)$; **(d)** $g(z, x, y)$; **(e)** $g(x + h, y + k, z + l)$, *provided such a value exists.*

5. $g(x, y, z) = e^{x+y+z}$

6. $g(x, y, z) = \ln(x + y + z)$

7. $g(x, y, z) = \dfrac{xyz}{x^2 + y^2 + z^2}$

8. $g(x, y, z) = \dfrac{e^{xyz}}{x + y + z}$

---

* David A. Katz, "Faculty Salaries, Promotions and Productivity at a Large University," *American Economic Review,* June 1973, pp. 469–477. Prof. Katz's equation actually included other variables, such as the number of dissertations supervised, so our equation assumes that all of these are zero.

In each of Exercises 9–12, find the distance between the given pairs of points.

9. $(1, -1)$ and $(2, -2)$

10. $(1, 0)$ and $(6, 1)$

11. $(a, 0)$ and $(0, b)$

12. (a, a) and (b, b)

13. Find k so that $(1, k)$ is equidistant from $(0, 0)$ and $(2, 1)$.

14. Find k so that (k, k) is equidistant from $(-1, 0)$ and $(0, 2)$.

15. Describe the set of points (x, y) such that $(x - 2)^2 + (y + 1)^2 = 9$.

16. Describe the set of points (x, y) such that $(x + 3)^2 + (y - 1)^2 = 4$.

17. Describe the set of points (x, y) such that $x^2 + 6x + y^2 + 4y + 7 = 0$.

18. Describe the set of points (x, y) such that $x^2 - 4x + y^2 + 2y - 3 = 0$.

APPLICATIONS

Exercises 19 through 22 involve the Cobb-Douglas production function (see Example 5). Recall that it has the form

$$P(x, y) = Kx^a y^{1-a}$$

where P stands for the number of items produced per year, x is the number of employees, and y is the annual operating budget. (The numbers K and a are constants that may depend on the particular situation we are looking at, with $0 \leq a \leq 1$.)

19. *Productivity* How many items will be produced per year by a company with 100 employees and an annual operating budget of $500,000, if $K = 1,000$, $a = 0.5$?

20. *Productivity* How many items will be produced per year by a company with 50 employees and an annual operating budget of $1,000,000, if $K = 1,000$, $a = 0.5$?

21. *Production Modeling with Cobb-Douglas* Two years ago, my piano manufacturing plant employed 1,000 workers, had an operating budget of $1 million, and turned out 100 pianos. Last year, I slashed the operating budget to $10,000, and production dropped to 10 pianos.
(a) Use the data for each of the two years and the Cobb-Douglas formula to obtain two equations in K and a.
(b) Take logs of both sides in each equation and obtain two linear equations in a and $\log K$.
(c) Solve these equations to obtain values for a and K.
(d) Use these values in the Cobb-Douglas formula to predict production if I increase the operating budget back to $1 million but lay off half the work force.

22. *Production Modeling with Cobb-Douglas* Repeat Exercise 21 using the following data: Two years ago: 1,000 employees, $1 million operating budget, 100 pianos. Last year: 1,000 employees, $100,000 operating budget, 10 pianos.

23. *Modeling Spending with a Linear Function* The following table shows total U.S. personal income and consumer spending in trillions of dollars for three months in 1992.*

	April	July	Nov.
Income	5.05	5.05	5.1
Spending	4	4.05	4.15

Model monthly spending as a function of monthly income and time using a linear function of the form

$$s(i, t) = ai + bt + c \quad (a, b, c \text{ constants}),$$

where s represents monthly consumer spending, i represents monthly income (in trillions of dollars) and t represents time in months since April 1992. (Round the constants to two decimal places.)

24. *Modeling International Investments with a Linear Function* The following table shows the annual flow of private investment and government loans

---

▼ *Source: National Association of Purchasing Management, Department of Commerce (The New York Times, June 2, 1993, Section 3, p. 1.) We have rounded all figures to the nearest $0.05 trillion.*

(in billions of dollars) to developing countries for the indicated years.*

	1986	1990	1992
Private	20	40	100
Government	40	60	60

Model annual private investment as a function of annual government investment and time using a linear function of the form

$$p(g, t) = ag + bt + c \quad (a, b, c \text{ constants}),$$

where p represents annual private investment, g represents annual government investment, and t represents the time in years since 1986.

25. *Demand for Beer* Economist Richard Stone obtained a demand function of the following form for beer in pre-World-War-II Great Britain.

$$Q(y, p, r) = Ky^{-0.023}p^{-1.040}r^{0.939}$$

Here, Q is the value of total annual sales of beer, y is the total real income in Great Britain, p is the average retail price of beer, and r is the average retail price of all other commodities. K is a positive constant that depends on the units of beer and currency.†

(a) Does the demand for beer increase or decrease with increasing values of r?

(b) If $K = 200$, find $Q(2 \times 10^8, 0.5, 500)$ to the nearest whole number, and interpret your answer.

26. *The Logistic Function* One form of the logistic equation‡ is

$$f(r, a, t) = \frac{K}{1 + e^{-r(t-a)}} + L,$$

where $K > 0$ and $L \geq 0$ are constants.

(a) Does the value of f increase or decrease with increasing t?

(b) Does the value of f increase or decrease with increasing a?

(c) Assume that $K = 1$, $L = a = 0$. Use a graphing calculator (or some other method) to determine the effect of increasing r on the graph of f versus t.

27. *Utility* Suppose your newspaper is trying to decide between two competing desktop publishing software packages, "Macro Publish" and "Turbo Publish." You estimate that if you purchase x copies of Macro Publish and y copies of Turbo Publish, your company's daily productivity will be given by the function

$$U(x, y) = 6x^{0.8}y^{0.2} + x.$$

$U(x, y)$ is measured in pages per day (U is called a **utility function**). If $x = y = 10$, calculate the effect of increasing x by one unit, and interpret the result.

28. *Housing Costs*§ The cost C (in dollars) of building a house is related to the number k of carpenters used and the number e of electricians used by

$$C(k, e) = 15{,}000 + 50k^2 + 60e^2.$$

If $k = e = 10$, compare the effects of increasing k by one unit and increasing e by one unit. Interpret the result.

29. The volume of an ellipsoid with cross-sectional radii a, b, and c is given by the formula

$$V(a, b, c) = \tfrac{4}{3}\pi abc.$$

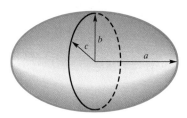

(a) Find at least two sets of values for a, b, and c so that $V(a, b, c) = 1$.

(b) Find the value of a such that $V(a, a, a) = 1$, and sketch the resulting ellipsoid.

---

* Source: World Bank (*The New York Times*, Dec. 17, 1993, p. D1.)

† Source: Richard Stone, "The Analysis of Market Demand," *Journal of the Royal Statistical Society* 108 (1945); 286–382.

‡ See "You're the Expert" in the chapter on logarithmic and exponential functions for a discussion of the logistic equation.

§ Based on an Exercise in *Introduction to Mathematical Economics* by A.L. Ostrosky Jr. and J.V. Koch (Prospect Heights, IL: Waveland Press, 1979.)

30. The volume of a right elliptical cone with height h and radii a and b of its base is given by the formula

$$V(a, b, h) = \tfrac{1}{3}\pi abh.$$

(a) Find at least two sets of values for a, b, and h so that $V(a, b, h) = 1$.

(b) Find the value of a such that $V(a, a, a) = 1$, and sketch the resulting cone.

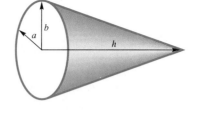

Complete the tables in Exercises 31–34.

The use of a graphing calculator or computer is suggested for Exercises 33 through 38.

31.

x	y	$f(x, y) = x^2\sqrt{1 + xy}$
-1	-1	
1	12	
0.3	0.5	
41	42	

32.

x	y	$f(x, y) = x^2 e^y$
0	2	
-1	5	
1.4	2.5	
11	9	

33.

x	y	$f(x, y) = x \ln(x^2 + y^2)$
3	1	
1.4	-1	
e	0	
0	e	

34.

x	y	$f(x, y) = \dfrac{x}{x^2 - y^2}$
-1	2	
0	0.2	
0.4	2.5	
10	0	

35. *Level Curves* The height of each point in a hilly region is given as a function of its coordinates by the formula

$$f(x, y) = y^2 - x^2.$$

(a) Use a graphing calculator to plot the curves on which the height is 0, 1, and 2 on the same set of axes. These are called **level curves of f**.

(b) *Without* using a graphing calculator, sketch the level curve $f(x, y) = 3$.

(c) *Without* using a graphing calculator, sketch the curves $f(y, x) = 1$ and $f(y, x) = 2$.

36. *Isotherms* The temperature (in degrees Fahrenheit) at each point in a region is given as a function of the coordinates by the formula

$$T(x, y) = 60.5(x - y^2).$$

(a) Use a graphing calculator to sketch the curves on which the temperature is 0°, 30°, and 90°. These curves are called **isotherms**.

(b) *Without* using a graphing calculator, sketch the isotherms corresponding to 20°, 50° and 100°.

(c) What do the isotherms corresponding to negative temperatures look like?

37. *Pollution* The burden of man-made aerosol sulfate in the earth's atmosphere, in grams per square meter, is given by

$$B(x, n) = \frac{xn}{A},$$

where x is the total weight of aerosol sulfate emitted into the atmosphere per year and n is the number of years it remains in the atmosphere. A is the surface area of the earth, approximately 5.1×10^{14} square meters.*

(a) Calculate the burden, given the current (1995) estimated values of $x = 1.5 \times 10^{14}$ grams per year, and $n = 5$ days (0.014 years).

(b) If x and n are as above, what does the function $W(x, n) = xn$ measure?

▼ *Source: Robert J. Charlson and Tom M. L. Wigley, "Sulfate Aerosol and Climatic Change," *Scientific American,* February, 1994, pp. 48–57.

38. *Pollution* The amount of aerosol sulfate (in grams) was approximately 45×10^{12} grams in 1940, and has been increasing exponentially ever since, with a doubling time of approximately 20 years.* Use the model from the previous exercise to give a formula for the atmospheric burden of aerosol sulfate as a function of the time t in years since 1940 and the number of years n it remains in the atmosphere.

39. *Alien Intelligence* Frank Drake, an astronomer at the University of California at Santa Cruz, devised the following equation to estimate the number of planet-based civilizations in our Milky Way galaxy willing and able to communicate with Earth.[†]

$$N(R, f_p, n_e, f_l, f_i, f_c, L) = R f_p n_e f_l f_i f_c L$$

R = the number of new stars formed in our galaxy each year
f_p = the fraction of those stars that have planetary systems
n_e = the average number of planets in each such system that can support life
f_l = the fraction of such planets on which life actually evolves
f_i = the fraction of life-sustaining planets on which intelligent life evolves
f_c = the fraction of intelligent-life-bearing planets on which the intelligent beings develop the means and the will to communicate over interstellar distances
L = the average lifetime of such technological civilizations (in years)

(a) What would be the effect on N if any one of the variables were doubled?
(b) How would you modify the formula if you were interested only in the number of intelligent-life-bearing planets in the galaxy?
(c) How could one convert this function into a linear function?
(d) (For discussion) Try to come up with an estimate of N.

40. *More Alien Intelligence* The formula given in the previous exercise restricts attention to planet-based civilizations in our galaxy. Give a formula that includes intelligent planet-based aliens from the galaxy Andromeda. (Assume that all the variables used in the formula for the Milky Way have the same values for Andromeda.)

41. *Minivan Sales* *Chrysler*'s percentage share of the U.S. minivan market in the period 1993–1994 could be approximated by the linear function

$$c(x, y, z) = 72.3 - 0.8x - 0.2y - 0.7z,$$

where x is the percentage of the market held by foreign manufacturers, y is *General Motors*' percentage share, and z is *Ford*'s percentage share.[‡]

(a) Results for the third quarter of 1994 showed *Chrysaler*'s share as 38.8%, *GM*'s share as 20.1%, and *Ford*'s share as 32.9%. According to the model, what was the share held by foreign manufacturers?
(b) Which of the three competitors would you regard as representing the greatest potential harm to *Chrysler*'s minivan sales?

42. *Minivan Sales* Referring to the model in the previous exercise, use the fact that the variables x, y, z, and c together account for 100% of all minivan sales in the U.S. to obtain c as a function of y and z only. What does your model say about *Chrysler*'s domestic competitors?

COMMUNICATION AND REASONING EXERCISES

43. Illustrate by means of an example how a real-valued function of the two variables x and y gives different real-valued functions of one variable when we restrict y to be different constants.

44. Give an example of a function of the two variables x and y with the property that interchanging x and y has no effect.

---

[*] Source: Robert J. Charlson and Tom M. L. Wigley, "Sulfate Aerosol and Climatic Change," *Scientific American,* February, 1994, pp. 48–57.

[†] Source: "First Contact" (Plume Books/Penguin Group)/*The New York Times,* October 6, 1992, p. C1.

[‡] The model is the authors'. Source for raw data: Ford Motor Company/*New York Times,* November 9, 1994, p. D5.

45. Give an example of a function f of the two variables x and y with the property that $f(x, y) = -f(y, x)$.

46. Suppose that $C(x, y)$ represents the cost of x CDs and y cassettes. If $C(x, y + 1) < C(x + 1, y)$ for every $x \geq 0$ and $y \geq 0$, what does this tell you about the cost of CDs and cassettes?

47. Brand Z's annual sales are affected by the sales of related products X and Y, as follows. Each \$1-million increase in sales of brand X causes a \$2.1-million decline in sales of brand Z, whereas each \$1-million increase in sales of brand Y results in an increase of \$0.4 million in sales of brand Z. Currently, brands X, Y, and Z are each selling \$6 million per year. Model the sales of brand Z using a linear function.

48. Let $f(x, y, z) = 43.2 - 2.3x + 11.3y - 4.5z$. Complete the following sentence. An increase of 1 in the value of y causes the value of f to _____ by _____, whereas increasing the value of x by 1 and _____ the value of z by _____ causes a decrease of 11.3 in the value of f.

15.2 THREE-DIMENSIONAL SPACE AND THE GRAPH OF A FUNCTION OF TWO VARIABLES

Just as functions of a single variable have graphs, so do functions of two or more variables. Let us return for a moment to the graph of a function of a single variable, such as $f(x) = x^2 + 1$. The procedure we used was the following.

1. Replace $f(x)$ with a new variable y, getting $y = x^2 + 1$.
2. Graph the resulting equation. (This is the standard parabola $y = x^2$ raised one unit above the y-axis.)

Now let us try this procedure with a function of two variables, say, $f(x, y) = x^2 + y^2$.

1. Replace $f(x, y)$ with a new variable z (we are already using x and y), getting $z = x^2 + y^2$.
2. Graph the resulting equation. How?

In order to graph this equation, we shall need *three* axes: the x-, y-, and z-axes. In other words, our graph will live in **three-dimensional space,** or **3-space.** *

Just as we had two mutually perpendicular axes in two-dimensional space (the "xy-plane"), so we have three mutually perpendicular axes in three-dimensional space (Figure 1).

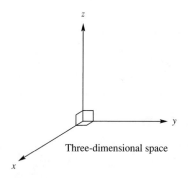

FIGURE 1

▼ * If we were dealing instead with a function of *three* variables, then we would need to go to *four-dimensional* space. Here we run into visualization problems (to say the least!) so we won't discuss the graphs of functions of three or more variables in this text.

15.2 Three-Dimensional Space and the Graph of a Function of Two Variables

In the three-dimensional picture, the x-axis is meant to be coming straight out of the page at you. (Another way to visualize it is to imagine that you are looking at the bottom left-hand corner of your classroom. The corner itself is the origin, while the blackboard is on the "yz-plane.") Notice that, in both 2-space and 3-space, the axis labeled with the last letter always goes up. The z-direction is the "up" direction in 3-space, rather than the y-direction.

There are three important planes associated with these axes: the xy-plane, the yz-plane, and the xz-plane. These planes are shown in Figure 2.

Note that any two of these planes intersect in one of the axes (for example, the xy- and xz-planes intersect in the x-axis) while all three meet at the origin. Also notice that the xy-plane consists of all points with z-coordinate zero, the xz-plane consists of all points with $y = 0$, and the yz-plane consists of all points with $x = 0$.

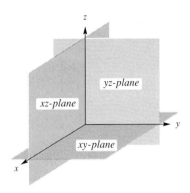

FIGURE 2

Now we describe a way of assigning coordinates to points in 3-space, much as we did in 2-space. In 3-space, each point has *three* coordinates, as you might expect: the x-coordinate, the y-coordinate, and the z-coordinate. To see how this works, look at the following examples.

▼ **EXAMPLE 1**

Locate the point $(1, 2, 3)$ in 3-space.

SOLUTION Our procedure is similar to the one we used in 2-space: start at the origin, proceed 1 unit in the x direction (towards you along the x-axis), then proceed 2 units in the y direction (to the right), and finally, proceed three units in the z direction (straight up). We wind up at the point P shown in Figure 3.

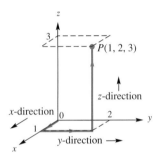

FIGURE 3

Before we go on... There are several other ways of thinking about the location of P. First, the fact that the x-coordinate of P is 1 means that P is 1 unit toward us from the back plane (the yz-plane). The fact that its y-coordinate is 2 means that it is two units to the right of the xz-plane. The fact that its z-coordinate is 3 means that it is three units above the "ground" (the xy-plane).

Here is another, extremely useful, way of thinking about the location of P. First, look at the x- and y-coordinates, obtaining the point $Q(1, 2)$ in the xy-plane. The point we want is then three units vertically above the point Q, since the z-coordinate of a point is just its height. This strategy is shown in Figure 4.

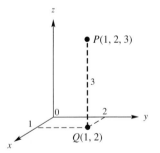

FIGURE 4

Here is yet another approach. Imagine a rectangular box situated with one of its corners at the origin, as shown in Figure 5. The corner opposite the origin is our point P.

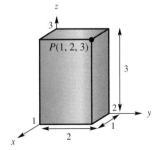

FIGURE 5

▼ **EXAMPLE 2**

Locate the points $P(0, -1, 2)$, $Q(-2, 3, 1)$, and $R(-1, -2, 0)$.

SOLUTION As in 2-space, negative coordinates indicate movement in directions opposite to positive coordinates. For the point P, we go 0 units in the x

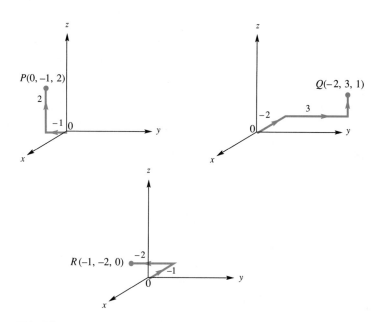

FIGURE 6

direction, −1 units in the y direction (that is, 1 unit to the left) and 2 units up. We locate the other two points in a similar way, as shown in Figure 6.

Always remember this:
The z-coordinate of a point is its height above the xy-plane.

Our next task is to describe the graph of a function $f(x, y)$ of two variables. Recall that the graph is the set of all the points (x, y, z) with $z = f(x, y)$. In other words, for *every* point (x, y) in the domain of f, the z-coordinate is given by evaluating the function at (x, y). Thus, there will be a point on the graph above *every* point in the domain of f, so that the graph will in fact be a *curved surface* of some sort.

▼ **EXAMPLE 3**

Describe the graph of $f(x, y) = x^2 + y^2$.

SOLUTION Your first thought might be to make a table of values. You could choose some values for x and y, and then, for each such pair, calculate $z = x^2 + y^2$. For example, you might get the following table.

(x, y)	$(0, 0)$	$(1, 0)$	$(0, 1)$	$(1, 1)$	$(-1, 0)$	$(0, -1)$	$(-1, -1)$
$z = x^2+y^2$	0	1	1	2	1	1	2

This gives the following seven points on the graph of f: $(0, 0, 0)$, $(1, 0, 1)$, $(0, 1, 1)$, $(1, 1, 2)$, $(-1, 0, 1)$, $(0, -1, 1)$, $(-1, -1, 2)$. These points are shown in Figure 7.

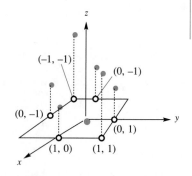

FIGURE 7

The points on the xy-plane we chose for our table are marked with hollow circles, while the corresponding points on the graph are marked with solid dots. The problem is that this small number of points hardly tells us what the surface looks like, and even if we plotted more points it is not clear that we would get anything more than a mass of circles on the page.

What can we do then? There are several alternatives. One place to start is to let a graphing calculator or computer draw the graph. The graph in Figure 8 was generated by a computer.

The computer software produces this drawing by calculating some values of $f(x, y)$, and then joining adjacent points with lines to create the appearance of a surface with a grid drawn on it. Most calculators or computers allow you to rotate the drawing to look at it from different sides and get a better idea what the surface looks like. This particular surface is called a **paraboloid.**

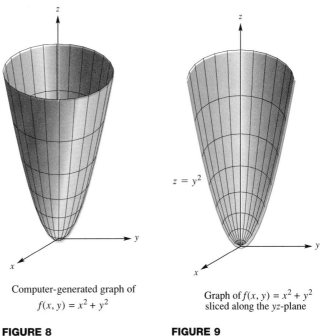

Computer-generated graph of
$f(x, y) = x^2 + y^2$

FIGURE 8

Graph of $f(x, y) = x^2 + y^2$
sliced along the yz-plane

FIGURE 9

Q The name "paraboloid" suggests "parabola," and the surface is reminiscent of a parabola. Why is this?

A In answering this question we shall see a useful analytic technique for understanding a graph. The technique is to slice through the surface with various planes. This will take a little while to explain.

If we slice vertically through this surface along the yz-plane, we get the picture in Figure 9.

The front edge, where we cut, looks like a parabola, and it is one. To see why, notice that the yz-plane is the set of points where $x = 0$. To get the intersection of $x = 0$ and $z = x^2 + y^2$, we substitute $x = 0$ in the second equation, getting

$$z = y^2.$$

This is the equation of a parabola in the yz-plane.

Similarly, we can slice through the surface with the xz-plane by setting $y = 0$. This gives the parabola $z = x^2$ in the xz-plane (Figure 10).

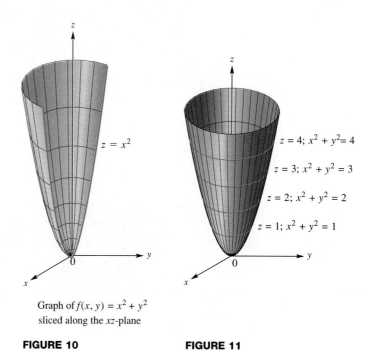

Graph of $f(x, y) = x^2 + y^2$ sliced along the xz-plane

FIGURE 10

FIGURE 11

We can also look at horizontal slices through the surface: that is, slices by planes parallel to the xy-plane. These are given by setting $z = c$ for various numbers c. For example, if we set $z = 1$, we will see only the points with height 1. Substituting in the equation $z = x^2 + y^2$ gives the equation

$$1 = x^2 + y^2,$$

which is the equation of a circle of radius 1. If we set $z = 4$, we get the equation of a circle of radius 2.

$$4 = x^2 + y^2$$

In general, if we slice through the surface at height $z = c$, we get a circle (of radius $\sqrt{c}$). Figure 11 shows several of these circles.

Looking at these circular slices, it becomes clear that this surface is the one we get by taking the parabola $z = x^2$ and spinning it around the z-axis. This is an example of what is known as a **surface of revolution.**

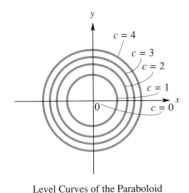

Level Curves of the Paraboloid
$z = x^2 + y^2$

FIGURE 12

Before we go on... Notice that each horizontal slice through the surface was obtained by putting $z = $ *constant*. This gave us an equation in x and y, describing a curve. These curves are called the **level curves** of the surface $z = f(x, y)$. In this example, the equations are of the form $x^2 + y^2 = $ *constant*, and so the level curves are circles. Figure 12 shows the level curves for $c = 0, 1, 2, 3$, and 4.

The level curves give a contour map of the surface. Each curve shows you all of the points on the surface at a particular height c. You can use this contour map to visualize the shape of the surface. In your mind's eye, move the contour at $c = 1$ to a height of 1 unit above the xy-plane, the contour at $c = 2$ to a height of 2 units above the xy-plane, and so on. You will end up with something like Figure 11.

Graphing many level curves by hand is laborious, and this is one place that graphing calculators come in handy.

EXAMPLE 4

Describe the graph of $g(x, y) = \frac{1}{2}x + \frac{1}{3}y - 1$.

SOLUTION Notice first that g is a linear function of x and y. Figure 13 shows a portion of the graph.

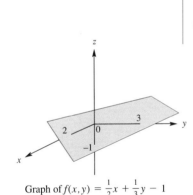

Graph of $f(x, y) = \frac{1}{2}x + \frac{1}{3}y - 1$

FIGURE 13

In fact, the graph is an infinite flat surface. In general, the graph of any linear function of two variables will be a plane. To better picture exactly what plane this is, we can find the three **intercepts,** the places where the plane crosses the coordinate axes, as shown in the figure.

x-intercept This is where the plane crosses the x-axis. Since points on this axis have both their y- and z-coordinates 0, we set $y = 0$ and $z = 0$, which gives

$$0 = \tfrac{1}{2}x - 1.$$

Hence, $x = 2$ is the place where the plane crosses the x-axis.

y-intercept Set $x = 0$ and $z = 0$, getting

$$0 = \tfrac{1}{3}y - 1.$$

This gives $y = 3$ as the place where the plane crosses the y-axis.

z-intercept Set $x = 0$ and $y = 0$ to get $z = -1$ as the place where the plane crosses the z-axis.

Three points are enough to define a plane, so we can say that the plane is the one passing through the three points $(2, 0, 0)$, $(0, 3, 0)$, and $(0, 0, -1)$.

Before we go on... What do the level curves of a linear function of two variables look like?

EXAMPLE 5

Describe the graph of $f(x, y) = \sqrt{4 - x^2 - y^2}$.

SOLUTION Before doing any drawing or analyzing, we need to think about the domain of f. We cannot take the square root of a negative number, so we must have

$$4 - x^2 - y^2 \geq 0.$$

If we write this inequality as

$$x^2 + y^2 \leq 4,$$

we see that it describes the region inside the circle of radius 2 centered at the origin. Thus, the domain is the disc shown in Figure 14.

Therefore, the graph of f will be a surface lying entirely above this disc.

Figure 15 shows a computer-generated drawing of this graph.

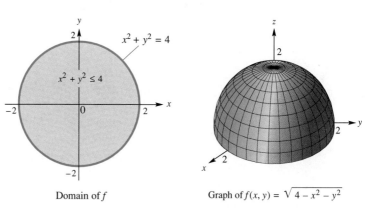

Domain of f

FIGURE 14

Graph of $f(x, y) = \sqrt{4 - x^2 - y^2}$

FIGURE 15

(Actually, we have deliberately generated the graph in an unconventional way to emphasize its "circularity." If we were to draw the graph in the conventional way, we would get something like Figure 16. Why does the bottom look ragged in this drawing?)

The surface in Figure 15 resembles a hemisphere. Let us see if slices confirm this.

Slice by the yz-plane Set $x = 0$ to get the following equations.

$$z = \sqrt{4 - y^2}$$
$$z^2 = 4 - y^2$$
$$y^2 + z^2 = 4$$

The last one is the equation of a circle of radius 2 centered at the origin in the yz-plane. Actually, we see only the top half of this circle, since the equation $z = \sqrt{4 - y^2}$ tells us that we shall take only positive z. (See Figure 17.)

15.2 Three-Dimensional Space and the Graph of a Function of Two Variables

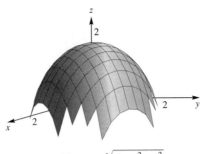

Graph of $f(x, y) = \sqrt{4 - x^2 - y^2}$

FIGURE 16

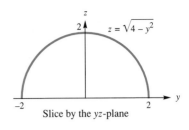

Slice by the yz-plane

FIGURE 17

Slice by the xz-plane This is similar. If we set $y = 0$, we get

$$x^2 + z^2 = 4.$$

This is the circle of radius 2 centered at the origin in the xz-plane. Again, we really see only the top half of the circle.

Let us now look at some level curves.

Slice by $z = 0$ Setting $z = 0$ gives the following equations.

$$0 = \sqrt{4 - x^2 - y^2}$$
$$0 = 4 - x^2 - y^2$$
$$x^2 + y^2 = 4$$

This is the circle of radius 2 in the xy-plane. This time we see the whole circle. In fact, this is the boundary of the domain of f.

Slice by $z = 1$ Now we get these equations.

$$1 = \sqrt{4 - x^2 - y^2}$$
$$1 = 4 - x^2 - y^2$$
$$x^2 + y^2 = 3$$

This is the equation of the circle of radius $\sqrt{3} \approx 1.7$.

Slice by $z = 2$ This gives $x^2 + y^2 = 0$, which says that both x and y must be 0. Thus, at a height of 2 there is only the single point $(0, 0, 2)$ on the surface.

If we slice by a plane $z = c$ for any height c between 0 and 2, we will get a circle.

However, if $c > 2$, then we will end up with an equation with $x^2 + y^2$ equal to a negative number, which is impossible. This means that there are no points on the surface with height greater than 2. Figure 18 shows the level curves $z = c$ with $c = 0, 1, 2$.

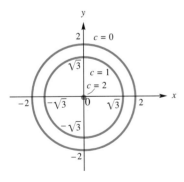

Level curves (slices by $z = c$)

FIGURE 18

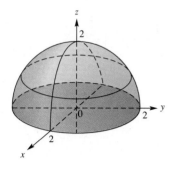

FIGURE 19

Putting these slices together in 3-space gives us Figure 19, confirming that the surface is indeed a hemisphere of radius 2 centered at the origin.

Put another way, this is the surface of revolution obtained by taking the semicircle $z = \sqrt{4 - x^2}$ and spinning it around the z-axis.

Before we go on... If we start with the equation $z = \sqrt{4 - x^2 - y^2}$, squaring both sides and rearranging terms gives

$$x^2 + y^2 + z^2 = 4.$$

By reasoning similar to that in the previous section, this is the equation of a sphere of radius 2 centered at the origin (see the exercises). The original equation tells us, however, that z must be taken to be positive, so the surface is just the upper hemisphere.

▼ **EXAMPLE 6**

Describe the graph of $f(x, y) = xy$.

SOLUTION The graph is shown in Figure 20.

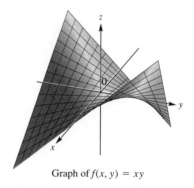

Graph of $f(x, y) = xy$

FIGURE 20

This is an example of a "saddle point" at the origin (we shall return to this in a later section). Slices in various directions show interesting features. Let us begin with the level curves.

Slice by $z = c$ This gives $xy = c$, which is a hyperbola. The level curves for various values of c are shown in Figure 21.

The case $c = 0$ is interesting: $xy = 0$ has as its graph the union of the x-axis and the y-axis (why?).

Slice by $x = c$ This gives $z = cy$, the line with slope c through the origin. It is surprising that even though the surface clearly curves, there are straight lines that can be drawn on it. In fact, all of the grid lines drawn in Figure 20 are straight.

Looking at Figure 20, we see that it might be interesting to look at the slices by vertical planes at 45° angles to the xz- and yz-planes.

Slice by $y = x$ This gives $z = x^2$, the parabola. This is the upward curve in the saddle.

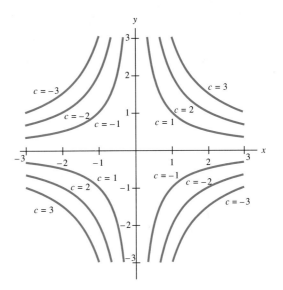

FIGURE 21

Slice by $y = -x$ This gives $z = -x^2$, a downward curving parabola. This is the downward curve in the saddle.

Before we go on... The point of this example is that slices in different directions can show very different aspects of a surface. Try planes parallel to the coordinate planes, but also be willing to look at planes suggested by the surface itself.

15.2 EXERCISES

1. Sketch the cube with vertices $(0, 0, 0)$, $(1, 0, 0)$, $(0, 1, 0)$, $(0, 0, 1)$, $(1, 1, 0)$, $(1, 0, 1)$, $(0, 1, 1)$, and $(1, 1, 1)$.

2. Sketch the cube with vertices $(-1, -1, -1)$, $(1, -1, -1)$, $(-1, 1, -1)$, $(-1, -1, 1)$, $(1, 1, -1)$, $(1, -1, 1)$, $(-1, 1, 1)$, and $(1, 1, 1)$.

3. Sketch the pyramid with vertices $(1, 1, 0)$, $(1, -1, 0)$, $(-1, 1, 0)$, $(-1, -1, 0)$, and $(0, 0, 2)$.

4. Sketch the solid with vertices $(1, 1, 0)$, $(1, -1, 0)$, $(-1, 1, 0)$, $(-1, -1, 0)$, $(0, 0, -1)$, and $(0, 0, 1)$.

Sketch the planes in Exercises 5−10.

5. $z = -2$
6. $z = 4$
7. $y = 2$
8. $y = -3$
9. $x = -3$
10. $x = 2$

Match the graphs with the equations in Exercises 11–18.

11. $f(x, y) = 1 - 3x + 2y$

12. $f(x, y) = 1 - \sqrt{x^2 + y^2}$

13. $f(x, y) = 1 - (x^2 + y^2)$

14. $f(x, y) = y^2 - x^2$

15. $f(x, y) = -\sqrt{1 - (x^2 + y^2)}$

16. $f(x, y) = 1 + (x^2 + y^2)$

17. $f(x, y) = \dfrac{1}{x^2 + y^2}$

18. $f(x, y) = 3x - 2y + 1$

A
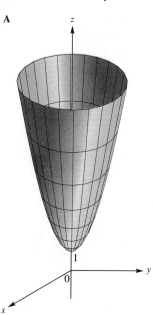

B
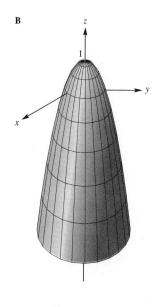

C
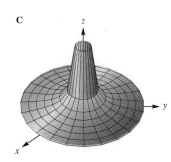

D
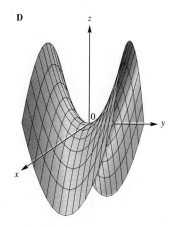

E
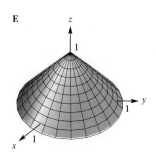

F
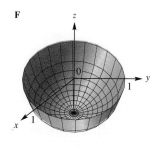

G
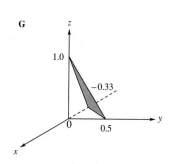

H
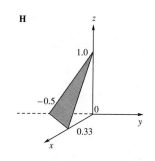

Sketch the graphs of the functions in Exercises 19–40.

19. $f(x, y) = 1 - x - y$
20. $f(x, y) = x + y - 2$
21. $g(x, y) = 2x + y - 2$
22. $g(x, y) = 3 - x + 2y$
23. $h(x, y) = x + 2$
24. $h(x, y) = 3 - y$
25. $r(x, y) = x + y$
26. $r(x, y) = x - y$
27. $s(x, y) = 2x^2 + 2y^2$. Show cross sections at $z = 1$ and $z = 2$.
28. $s(x, y) = -(x^2 + y^2)$. Show cross sections at $z = -1$ and $z = -2$.
29. $t(x, y) = x^2 + 2y^2$. Show cross sections at $x = 0$ and $z = 1$.
30. $t(x, y) = \frac{1}{2}x^2 + y^2$. Show cross sections at $x = 0$ and $z = 1$.
31. $f(x, y) = 2 + \sqrt{x^2 + y^2}$. Show cross sections at $z = 3$ and $y = 0$.
32. $f(x, y) = 2 - \sqrt{x^2 + y^2}$. Show cross sections at $z = 0$ and $y = 0$.
33. $f(x, y) = -2\sqrt{x^2 + y^2}$. Show cross sections at $z = -4$ and $y = 1$.
34. $f(x, y) = 2 + 2\sqrt{x^2 + y^2}$. Show cross sections at $z = 4$ and $y = 1$.
35. $f(x, y) = y^2$
36. $g(x, y) = x^2$
37. $h(x, y) = \dfrac{1}{y}$
38. $k(x, y) = e^y$
39. $f(x, y) = e^{-(x^2+y^2)}$
40. $g(x, y) = \dfrac{1}{\sqrt{x^2 + y^2}}$

APPLICATIONS

A graphing calculator is suggested for Exercises 41 through 50.

41. *Cobb-Douglas Production Function* Graph the level curves at $z = 0, 1, 2,$ and 3 of $P(x, y) = Kx^a y^{1-a}$ if $K = 1$ and $a = 0.5$. Here, x is the number of workers, y is the operating budget, and $P(x, y)$ is the productivity. Interpret the level curve at $z = 3$.

42. *Cobb-Douglas Production Function* Graph the level curves at $z = 0, 1, 2,$ and 3 of $P(x, y) = Kx^a y^{1-a}$ if $K = 1$ and $a = 0.25$. Here, x is the number of workers, y is the operating budget, and $P(x, y)$ is the productivity. Interpret the level curve at $z = 0$.

43. *Utility* Suppose that your newspaper is trying to decide between two competing desktop publishing software packages, "Macro Publish" and "Turbo Publish." You estimate that if you purchase x copies of Macro Publish and y copies of Turbo Publish, your company's daily productivity will be given by
$$U(x, y) = 6x^{0.8}y^{0.2} + x,$$
where $U(x, y)$ is measured in pages per day. (U is called a *utility function*.) Graph the level curves at $z = 0, 10, 20,$ and 30. What does the level curve at $z = 0$ tell you?

44. *Utility* Suppose that your small publishing company is trying to decide between two competing desktop publishing software packages, "Macro Publish" and "Turbo Publish." You estimate that if you purchase x copies of Macro Publish and y copies of Turbo Publish, your company's daily productivity will be given by
$$U(x, y) = 5x^{0.2}y^{0.8} + x,$$
where $U(x, y)$ is measured in pages per day. Graph the level curves at $z = 0, 10, 20,$ and 30. Give a formula for the level curve at $z = 30$ specifying y as a function of x. What does this curve tell you?

45. *Housing Costs** The cost C of building a house is related to the number k of carpenters used and the number e of electricians used by

$$C(k, e) = 15{,}000 + 50k^2 + 60e^2.$$

Describe the level curves $C = 30{,}000$ and $C = 40{,}000$. What do these level curves represent?

46. *Housing Costs** The cost C of building a house (in a different area from that in the previous exercise) is related to the number k of carpenters used and the number e of electricians used by

$$C(k, e) = 15{,}000 + 70k^2 + 40e^2.$$

Describe the slices by the planes $k = 2$ and $e = 2$. What do these slices represent?

47. *Area* The area of a rectangle of height h and width w is $A(h, w) = hw$. Sketch a few level curves of A. Looking at your graph, if the perimeter $h + w$ of the rectangle is constant, what h and w give the largest area? (We suggest you draw in the line $h + w = c$ for several values of c.)

48. *Area* The area of an ellipse with semimajor axis a and semiminor axis b is $A(a, b) = \pi ab$. Sketch the graph of A. If $a^2 + b^2$ is constant, what a and b give the largest area?

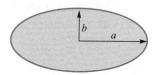

COMMUNICATION AND REASONING EXERCISES

49. Show that the distance between the points (x, y, z) and (a, b, c) is given by the following **three-dimensional distance formula**.

$$d = \sqrt{(x - a)^2 + (y - b)^2 + (z - c)^2}.$$

The following diagram should be of assistance.

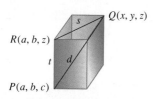

50. Use the result of the previous exercise to show that the sphere of radius r centered at the origin is given by the following equation.

$$x^2 + y^2 + z^2 = r^2$$

51. Why is three-dimensional space used to represent the graph of a function of two variables?

52. Why is it that we can sketch the graphs of functions of two variables on the two-dimensional flat surfaces of these pages?

▶ ══ **15.3** PARTIAL DERIVATIVES

Recall that if f is a function of x, then df/dx measures how fast f changes as x increases. If f is a function of two or more variables we can ask how fast f changes as each variable increases (and the others remain fixed). These rates of change are called the "partial derivatives of f," and measure how each variable contributes to the change in f. Here is a more precise definition.

▼ ∗ Based on an exercise in *Introduction to Mathematical Economics* by A. L. Ostrosky, Jr., and J. V. Koch (Prospect Heights, IL: Waveland Press, 1979.)

PARTIAL DERIVATIVES

The **partial derivative of f with respect to x** is the derivative of f with respect to x, treating all other variables as constant;
the **partial derivative of f with respect to y** is the derivative of f with respect to y, treating all other variables as constant;
and so on for other variables.

The partial derivatives are written as $\partial f/\partial x$, $\partial f/\partial y$, and so on. The symbol "∂" is used (instead of "d") to remind us that there is more than one variable and that we are holding the other variables fixed.

EXAMPLE 1

Let $f(x, y) = x^2y + y^2x - xy + y$. Find $\partial f/\partial x$ and $\partial f/\partial y$.

SOLUTION To find $\partial f/\partial x$, we take the derivative with respect to x, thinking of y as a constant.

$$\frac{\partial f}{\partial x} = 2xy + y^2 - y$$

Notice that $\partial/\partial x[x^2y]$ is the derivative of x^2 times a *constant*, since we are treating y as constant. Therefore, it is $2x$ times that constant—that is, $2xy$. Similarly, $\partial/\partial x[y] = 0$, since the derivative of a constant is 0.

We obtain $\partial f/\partial y$ as the derivative of f with respect to y, regarding x as constant.

$$\frac{\partial f}{\partial y} = x^2 + 2xy - x + 1$$

EXAMPLE 2 Productivity

The Cobb-Douglas production function for the Handy Gadget Company is

$$P = 100x^{0.3}y^{0.7},$$

where P is the number of gadgets it turns out per month, x is the number of employees at the company, and y is the monthly operating budget. Handy Gadgets employs 50 people. How fast is the number of gadgets produced going up as the operating budget increases from a level of $10,000 per month?

SOLUTION We are looking for the rate of increase of production P with respect to the operating budget y, with x held fixed at 50. Thus, we take the partial derivative $\partial P/\partial y$.

$$\frac{\partial P}{\partial y} = 70x^{0.3}y^{-0.3}$$

We must now evaluate this derivative at $x = 50$ and $y = 10{,}000$. We obtain

$$\left.\frac{\partial P}{\partial y}\right|_{(50,\, 10{,}000)} = 70(50)^{0.3}(10{,}000)^{-0.3} \approx 14.28.$$

How do we interpret this? First, $\partial P/\partial y$ is measured in gadgets/dollar (per month), so the answer tells us that productivity goes up by about 14 gadgets per month for every additional dollar added to the monthly operating budget. Not bad!

Before we go on... It might be illuminating to calculate the actual productivity at these figures. This we do by evaluating the original function P at the given values of x and y.

$$P(50, 10{,}000) = 100(50)^{0.3}(10{,}000)^{0.7} \approx 204{,}028 \text{ gadgets per month}$$

Thus, the increase of 14 in the number of gadgets is not so wonderful after all. A \$1 increase in the budget represents a 0.01% increase in operating costs, but this results in an increase of about only 0.0067% in productivity.

GEOMETRIC INTERPRETATION OF PARTIAL DERIVATIVES

Q We know that if f is a function of one variable x, the derivative df/dx gives the slopes of the tangent lines to its graph. What about the *partial* derivatives of a function of several variables?

A Suppose that f is a function of x and y. By definition, $\partial f/\partial x$ is the derivative of the function of x you get by holding y fixed. If you evaluate this derivative at the point (a, b), you are holding y fixed at the value b, taking the ordinary derivative of the resulting function of x, and evaluating this

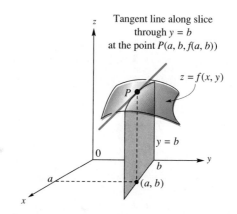

$\left.\dfrac{\partial f}{\partial x}\right|_{(a,\,b)}$ is the slope of the tangent line at the point $P(a, b, f(a, b))$ along the slice through $y = b$.

FIGURE 1

at $x = a$. Now, holding y fixed at b amounts to slicing through the graph of f along the plane $y = b$, resulting in a curve. Thus, the partial derivative is the slope of the tangent line to this curve at the point where $x = a$ and $y = b$, along the plane $y = b$ (Figure 1).

Note that this fits with our interpretation of $\partial f/\partial x$ as the rate of increase of f with increasing x when y is held fixed at b.

The other partial derivative, $\partial f/\partial y|_{(a,b)}$ is, similarly, the slope of the tangent line at the same point $P(a, b, f(a, b))$, but along the slice by the plane $x = a$. You should to draw the corresponding picture for this on your own.

▼ **EXAMPLE 3**

Calculate $\partial f/\partial x$, $\partial f/\partial y$, and $\partial f/\partial z$ if $f(x, y, z) = xy^2z^3 - xy$.

SOLUTION Although we now have three variables, the calculation remains the same: $\partial f/\partial x$ is the derivative of f with respect to x, holding *both* other variables, y and z, constant.

$$\frac{\partial f}{\partial x} = y^2z^3 - y$$

Similarly, $\partial f/\partial y$ is the derivative of f with respect to y, with both x and z held constant.

$$\frac{\partial f}{\partial y} = 2xyz^3 - x$$

Finally, to find $\partial f/\partial z$, we hold both x and y constant and take the derivative with respect to z.

$$\frac{\partial f}{\partial z} = 3xy^2z^2$$

Before we go on... The procedure is the same for any number of variables: to get the partial derivative with respect to any one variable, we treat all the others as constants.

▼ **EXAMPLE 4** Gravity in 3-Space

According to Newton's Law of Gravity, the gravitational force exerted on a particle with mass m situated at the point (x, y, z) by another particle with mass M situated at the point (a, b, c) is given by

$$F(x, y, z) = G\frac{Mm}{(x - a)^2 + (y - b)^2 + (z - c)^2}.$$

(All units of distance are in meters, the masses M and m are in kilograms, $G \approx 6.67 \times 10^{-11}$, and the resulting force is measured in newtons.) Suppose that a 1,000-kg mass is situated at the origin and that your 100-kg space module is at the point (10, 100, 1,000) and traveling in the *x*-direction at one meter per second. How fast is the gravitational force your module experiences decreasing?

SOLUTION   First, substitute for the constants *M*, *m*, *a*, *b*, and *c*.

$$F(x, y, z) = G\frac{100{,}000}{x^2 + y^2 + z^2}$$

We'll substitute for *G* later. Also, since *y* and *z* are not changing, we *could* substitute for them now, but we'll leave them as variables for the time being and substitute later. Now the question we have to answer is really this: How fast is *F* decreasing per unit of increase in *x*? In other words, we need to find the partial derivative, $\partial F/\partial x$, and evaluate it at (10, 100, 1,000). To do this, we hold *y* and *z* constant and differentiate with respect to *x*.

$$\frac{\partial F}{\partial x} = -G\frac{200{,}000x}{(x^2 + y^2 + z^2)^2}$$

Evaluating at (10, 100, 1,000) now gives

$$\left.\frac{\partial F}{\partial x}\right|_{(10, 100, 1{,}000)} = -G\frac{2{,}000{,}000}{(1{,}010{,}100)^2} \approx -1.31 \times 10^{-16}.$$

Thus, the gravitational force is *decreasing* at 1.31×10^{-16} newtons per second.

Before we go on... Here are some variations on this problem. If we were traveling at 2 meters per second instead of one meter per second, the rate of decrease would be twice that amount. More generally, if we were traveling at *v* meters per second, the rate would be *v* times that amount.* If we were traveling in the *y*-direction instead of the *x*-direction, we would have taken the partial derivative with respect to *y* instead of *x*, and similarly for the *z*-direction.

▼ **EXAMPLE 5**   Cost Functions

Returning to an example from Section 1, suppose that you own a company making two models of stereo speakers, the Ultra Mini and the Big Stack. Your total monthly cost (in dollars) to make *x* Ultra Minis and *y* Big Stacks is given by

$$C(x, y) = \$10{,}000 + 20x + 40y.$$

What is the significance of $\partial C/\partial x$ and $\partial C/\partial y$?

▼ *Can you see how this is really a related rates problem in disguise?

SOLUTION First, we compute these partial derivatives.

$$\frac{\partial C}{\partial x} = 20$$

$$\frac{\partial C}{\partial y} = 40$$

These are the amounts that each additional Ultra Mini or Big Stack will add to the total cost, respectively. These are, in other words, the **marginal costs** of each model of speaker.

Before we go on... How much does the cost rise if you increase x by Δx and y by Δy? In this example, the change in cost is given by

$$\Delta C = 20\, \Delta x + 40\, \Delta y = \frac{\partial C}{\partial x} \Delta x + \frac{\partial C}{\partial y} \Delta y.$$

This leads to the **chain rule for several variables.** Part of this rule says that if x and y are both functions of t, then C is a function of t through them, and the rate of change of C with respect to t can be calculated as

$$\frac{dC}{dt} = \frac{\partial C}{\partial x} \cdot \frac{dx}{dt} + \frac{\partial C}{\partial y} \cdot \frac{dy}{dt}.$$

We shall not have a chance to use this interesting result further in this book.

▼ **EXAMPLE 6** Cost Functions

Another possibility for the cost function in the previous example is

$$C(x, y) = \$10{,}000 + 20x + 40y + 30\sqrt{x + y}.$$

Now what are the marginal costs of the two models of speakers?

SOLUTION We compute the partial derivatives.

$$\frac{\partial C}{\partial x} = 20 + \frac{15}{\sqrt{x + y}}$$

$$\frac{\partial C}{\partial y} = 40 + \frac{15}{\sqrt{x + y}}$$

Now the marginal costs depend on the total production level $x + y$. As production increases, the marginal costs decline. For example, if $x + y = 9$, then it will cost an additional $25 for each Ultra Mini and $45 for each Big Stack. On the other hand, if $x + y = 25$, it will cost only $23 for each additional Ultra Mini and only $43 for each Big Stack.

Just as for functions of a single variable, we can calculate second derivatives. Suppose, for example, that we have a function of x and y, say, $f(x, y) =$

$x^2 - x^2y^2$. We know that

$$\frac{\partial f}{\partial x} = 2x - 2xy^2.$$

The symbol $\partial/\partial x$ means "the partial derivative with respect to x," just as d/dx stood for "the derivative with respect to x" in the chapter on derivatives. If we now want to take the partial derivative with respect to x once again, there is nothing to stop us.

$$\frac{\partial}{\partial x}\left(\frac{\partial f}{\partial x}\right) = 2 - 2y^2$$

This is called the **second partial derivative $\partial^2 f/\partial x^2$.**

We get the following derivatives similarly.

$$\frac{\partial f}{\partial y} = -2x^2 y$$

$$\frac{\partial^2 f}{\partial y^2} = -2x^2$$

Now what if we instead take the partial derivative with respect to y of $\partial f/\partial x$?

$$\frac{\partial^2 f}{\partial y\,\partial x} = \frac{\partial}{\partial y}\left(\frac{df}{dx}\right) = \frac{\partial}{\partial y}(2x - 2xy^2) = -4xy$$

Here, $\partial^2 f/\partial y\,\partial x$ means "first take the partial derivative with respect to x, and then with respect to y," and is called a **mixed partial derivative.** If we differentiate in the opposite order, we get

$$\frac{\partial^2 f}{\partial x\,\partial y} = \frac{\partial}{\partial x}\left(\frac{\partial f}{\partial y}\right) = \frac{\partial}{\partial x}(-2x^2 y) = -4xy,$$

the same as $\partial^2 f/\partial y\,\partial x$. This is no coincidence—the mixed partial derivatives $\partial^2 f/\partial x\,\partial y$ and $\partial^2 f/\partial y\,\partial x$ will always agree as long as the first partial derivatives are both differentiable functions of x and y and the mixed partial derivatives are continuous. Since all the functions we shall use are of this type, we can take the derivatives in any order we like when calculating mixed derivatives.

Before we go to the exercises, here is another notation for partial derivatives.

$$f_x \text{ means } \frac{\partial f}{\partial x}.$$

$$f_y \text{ means } \frac{\partial f}{\partial x}.$$

$$f_{xy} \text{ means } (f_x)_y = \frac{\partial^2 f}{\partial y\,\partial x}. \quad \text{Note the order in which the derivatives are taken.}$$

$$f_{yx} \text{ means } (f_y)_x = \frac{\partial^2 f}{\partial x\,\partial y}.$$

15.3 EXERCISES

In each of Exercises 1–16, find $\partial f/\partial x$, $\partial f/\partial y$, $\partial^2 f/\partial x^2$, $\partial^2 f/\partial y^2$, $\partial^2 f/\partial x\,\partial y$, and $\partial^2 f/\partial y\,\partial x$, and then evaluate them at the point $(1, -1)$ if possible.

1. $f(x, y) = 3x^2 - y^3 + x - 1$
2. $f(x, y) = x^{1/2} - 2y^4 + y + 6$
3. $f(x, y) = 3x^2 y$
4. $f(x, y) = x^4 y^2 - x$
5. $f(x, y) = x^2 y^3 - x^3 y^2 - xy$
6. $f(x, y) = x^{-1} y^2 + xy^2 + xy$
7. $f(x, y) = (2xy + 1)^3$
8. $f(x, y) = 1/(xy + 1)^2$
9. $f(x, y) = e^{x+y}$
10. $f(x, y) = e^{2x+y}$
11. $f(x, y) = 5x^{0.6} y^{0.4}$
12. $f(x, y) = -2x^{0.1} y^{0.9}$
13. $f(x, y) = 4.1 x^{1.2} e^{-0.2y}$
14. $f(x, y) = x + \dfrac{e^{0.1y}}{x^2}$
15. $f(x, y) = e^{0.2xy}$
16. $f(x, y) = xe^{xy}$

In each of Exercises 17–28, find $\partial f/\partial x$, $\partial f/\partial y$ and $\partial f/\partial z$, and then evaluate them at the point $(0, -1, 1)$ if possible.

17. $f(x, y, z) = xyz$
18. $f(x, y, z) = xy + xz - yz$
19. $f(x, y, z) = -\dfrac{4}{x + y + z^2}$
20. $f(x, y, z) = \dfrac{6}{x^2 + y^2 + z^2}$
21. $f(x, y, z) = xe^{yz} + ye^{xz}$
22. $f(x, y, z) = xye^z + xe^{yz} + e^{xyz}$
23. $f(x, y, z) = x^{0.1} y^{0.4} z^{0.5}$
24. $f(x, y, z) = 2x^{0.2} y^{0.8} + z^2$
25. $f(x, y, z) = e^{xyz}$
26. $f(x, y, z) = \ln(x + y + z)$
27. $f(x, y, z) = \dfrac{2{,}000z}{1 + y^{0.3}}$
28. $f(x, y, z) = \dfrac{e^{0.2x}}{1 + e^{-0.1y}}$

APPLICATIONS

29. *Minivan Sales* Chrysler's percentage share of the U.S. minivan market in the period 1993–1994 could be approximated by the linear function
$$c(x, y, z) = 72.3 - 0.8x - 0.2y - 0.7z,$$
where x is the percentage share of the market held by foreign manufacturers, y is *General Motors'* percentage share, and z is *Ford's* percentage share.* Calculate the partial derivatives $\partial c/\partial x$, $\partial c/\partial y$, and $\partial c/\partial z$, and interpret the results.

30. *Minivan Sales* In the previous exercise, if we take into account the fact that the variables x, y, z, and c together account for 100% of all minivan sales in the U.S., we obtain
$$c = -38.5 + 3.0y + 0.5z,$$
where y is *General Motors'* percentage share of the market, and z is *Ford's* percentage share. Calculate the partial derivatives $\partial c/\partial y$ and $\partial c/\partial z$ and interpret the results, explaining why the coefficients of y and z are positive.

---

*The model is the authors'. Source for raw data: Ford Motor Company/*The New York Times*, November 9, 1994, p. D5.

31. **Marginal Cost** Your weekly cost to manufacture x cars and y trucks is given by
$$C(x, y) = \$200{,}000 + 6{,}000x + 4{,}000y - 100{,}000e^{-0.01(x+y)}.$$
What is the marginal cost of cars? Of trucks? How do these marginal costs behave as total production rises?

32. **Marginal Cost** Your weekly cost to manufacture x bicycles and y tricycles is given by
$$C(x, y) = \$20{,}000 + 60x + 20y + 50\sqrt{xy}.$$
What is the marginal cost of bicycles? Of tricycles? How do these marginal costs behave as x and y increase?

33. **Average Cost** If you average your costs over your total production, you get the **average cost,** written $\bar{C}$.
$$\bar{C}(x, y) = \frac{C(x, y)}{x + y}$$
Find the average cost for the cost function in Exercise 31. Then find the marginal average cost of a car and the marginal average cost of a truck at a production level of 50 cars and 50 trucks. Round your answer to two decimal places. Interpret your answers.

34. **Average Cost** Find the average cost for the cost function in Exercise 32 (see the previous exercise). Then find the marginal average cost of a bicycle and the marginal average cost of a tricycle at a production level of 5 bicycles and 5 tricycles. Round your answer to two decimal places. Interpret your answers.

35. **Marginal Revenue** As manager of an auto dealership, you offer a car rental company the following deal. You will charge $15,000 per car and $10,000 per truck, but you will then give the company a discount of $5,000 times the square root of the total number of vehicles it buys from you. Looking at your marginal revenue, is this a good deal for the rental company?

36. **Marginal Revenue** As marketing director for a bicycle manufacturer, you come up with the following scheme. You will offer to sell a dealer x bicycles and y tricycles for
$$R(x, y) = \$3{,}500 - 3{,}500e^{-0.02x - 0.01y}.$$
Find your marginal revenue for bicycles and for tricycles. Are you likely to be fired for your suggestion?

37. **Research Productivity** Here we apply a variant of the Cobb-Douglas function to the modeling of research productivity. A mathematical model of research productivity at a particular physics laboratory is given by
$$P = 0.04x^{0.4}y^{0.2}z^{0.4},$$
where P is the annual number of ground-breaking research papers produced by the staff, x is the number of physicists on the research team, y is the laboratory's annual research budget, and z is the annual National Science Foundation subsidy to the laboratory. Find the rate of increase of research papers per government-subsidy-dollar at a subsidy level of $1,000,000 per year and a staff level of 10 physicists if the annual budget is $100,000.

38. **Research Productivity** A major drug company estimates that the annual number P of patents for new drugs developed by its research team is best modeled by the formula
$$P = 0.3x^{0.3}y^{0.4}z^{0.3},$$
where x is the number of research biochemists on the payroll, y is the annual research budget, and z is size of the bonus awarded to discoverers of new drugs. Assuming that the company has 12 biochemists on the staff, an annual research budget of $500,000 and pays $40,000 bonuses to developers of new drugs, calculate the rate of growth in the annual number of patents per new research staff member.

39. **Utility** Your newspaper is trying to decide between two competing desktop publishing software packages, "Macro Publish" and "Turbo Publish." You estimate that if you purchase x copies of Macro Publish and y copies of Turbo Publish, your company's daily productivity will be given by
$$U(x, y) = 6x^{0.8}y^{0.2} + x.$$
$U(x, y)$ is measured in pages per day. (Recall that U is called a *utility function.*)
(a) Calculate $\partial U/\partial x|_{(10, 5)}$ and $\partial U/\partial y|_{(10, 5)}$ to two decimal places, and interpret the results.
(b) What does the ratio $\partial U/\partial x|_{(10, 5)}/\partial U/\partial y|_{(10, 5)}$ tell you about the usefulness of these products?

40. **Grades*** A production formula for a student's performance on a difficult English examination is

---

* Based on an exercise in *Introduction to Mathematical Economics* by A. L. Ostrosky, Jr., and J. V. Koch (Prospect Heights, IL: Waveland Press, 1979).

given by
$$g(t, x) = 4tx - 0.2t^2 - x^2.$$
Here g is the grade the student can expect to obtain, t is the number of hours of study for the examination, and x is the student's grade point average.
(a) Calculate $\partial g/\partial t|_{(10, 3)}$ and $\partial g/\partial x|_{(10, 3)}$ and interpret the results.
(b) What does the ratio $\partial g/\partial t|_{(10, 3)}/\partial g/\partial x|_{(10, 3)}$ tell you about the relative merits of study and grade point average?

41. *Electrostatic Repulsion* If positive electric charges of Q and q coulombs are situated at positions (a, b, c) and (x, y, z) respectively, then the force of repulsion they experience is given by
$$F = K\frac{Qq}{(x - a)^2 + (y - b)^2 + (z - c)^2},$$
where $K \approx 9 \times 10^9$, F is given in newtons, and all positions are measured in meters. Assume that a charge of 10 coulombs is situated at the origin, and that a second charge of 5 coulombs is situated at $(2, 3, 3)$ and moving in the y-direction at one meter per second. How fast is the electrostatic force it experiences decreasing?

42. *Electrostatic Repulsion* Repeat the preceding exercise, assuming that a charge of 10 coulombs is situated at the origin and that a second charge of 5 coulombs is situated at $(2, 3, 3)$ and moving in the negative z-direction at one meter per second.

43. *Investments* Recall that the compound interest formula for annual compounding is
$$A(P, r, t) = P(1 + r)^t,$$
where A is the future value of an investment of P dollars after t years at an interest rate of r.
(a) Calculate $\partial A/\partial P$, $\partial A/\partial r$, and $\partial A/\partial t$, all evaluated at $(100, 0.10, 10)$. (Round answers to two decimal places.) Interpret your answers.
(b) What does the function $\partial A/\partial P|_{(100, 0.10, t)}$ of t tell you about your investment?

44. *Investments* Repeat the preceding exercise using the formula for continuous compounding,
$$A(P, r, t) = Pe^{rt}.$$

45. *Modeling with the Cobb-Douglas Production Formula* Assume you are given a production formula of the form
$$P(x, y) = Kx^ay^b \quad (a + b = 1).$$
(a) Obtain formulas for $\partial P/\partial x$ and $\partial P/\partial y$, and show that $\partial P/\partial x = \partial P/\partial y$ precisely when $x/y = a/b$.
(b) Let x be the number of workers a firm employs and let y be its monthly operating budget in thousands of dollars. Assume that the firm currently employs 100 workers and has a monthly operating budget of \$200,000. If each additional worker contributes as much to productivity as each additional \$1,000 per month, find values of a and b that would model the firm's productivity.

46. *Housing Costs** The cost C of building a house is related to the number k of carpenters used and the number e of electricians used by
$$C(k, e) = 15{,}000 + 50k^2 + 60e^2.$$
If 3 electricians are currently employed in building your new house, and the marginal cost per additional electrician is the same as the marginal cost per additional carpenter, how many carpenters are being used? (Round your answer to the nearest carpenter.)

47. *Nutrient Diffusion* Suppose that one cubic centimeter of nutrient is placed at the center of a circular petri dish filled with water. We might wonder how it is distributed after a time of t seconds. According to the classical theory of diffusion, the concentration of nutrient (in parts of nutrient per part of water) after a time t is given by
$$u(r, t) = \frac{1}{4\pi Dt}e^{-r^2/(4Dt)}.$$
Here D is the *diffusivity*, which we will take to be 1, and r is the distance from the center in centimeters. How fast is the concentration increasing at a distance of 1 cm from the center three seconds after the nutrient is introduced?

48. *Nutrient Diffusion* Referring to the previous exercise, how fast is the concentration increasing at a distance of 4 cm from the center four seconds after the nutrient is introduced?

▼ * Based on an exercise in *Introduction to Mathematical Economics* by A. L. Ostrosky, Jr., and J. V. Koch (Prospect Heights, IL: Waveland Press, 1979).

COMMUNICATION AND REASONING EXERCISES

49. Given that $f(a, b) = r$, $f_x(a, b) = s$, and $f_y(a, b) = t$, complete the following sentence: _____ is increasing at a rate of _____ units per unit of x, _____ is increasing at a rate of _____ units per unit of y, and the value of _____ is _____ when $x =$ _____ and $y =$ _____.

50. A firm's productivity depends on two variables, x and y. Currently, $x = a$ and $y = b$, and the firm's productivity is 4,000 units. Productivity is increasing at a rate of 400 units per unit *decrease* in x, and is decreasing at a rate of 300 units per unit increase in y. What does all of this information tell you about the firm's productivity function $g(x, y)$?

51. Give an example of a function $f(x, y)$ with $f(1, 1) = 10$, $f_x(1, 1) = -2$, and $f_y(1, 1) = 3$.

52. Give an example of a function $f(x, y, z)$, all of whose partial derivatives are nonzero constants.

53. The graph of $z = b + mx + ny$ (b, m, and n constants) is a plane.
    (a) Explain the geometric significance of the numbers b, m, and n.
    (b) Show that the equation of the plane passing through (h, k, l) with slope m in the x-direction (in the sense of $\partial/\partial x$ and slope n in the y-direction is
    $$z - l = m(x - h) + n(y - k).$$

54. The **tangent plane** to the graph of $f(x, y)$ at $P(a, b, f(a, b))$ is the plane containing the lines tangent to the slice through the graph by $y = b$ (as in Figure 1 in the text) and the slice through the graph by $x = a$. Use the result of the preceding exercise to show that the equation of the tangent plane is
$$z = f(a, b) + f_x(a, b)(x - a) + f_y(a, b)(y - b).$$

15.4 MAXIMA AND MINIMA

In the chapter on applications of the derivative we saw how to locate local extrema of a function of a single variable. In this section, we extend our methods to functions of two variables. Similar techniques work for functions of three or more variables.

Figure 1 shows a portion of the graph of the function $f(x, y) = 2(x^2 + y^2) - (x^4 + y^4) + 3$.

The graph resembles a "fling carpet" and there are several interesting points, marked a, b, c, and d on the graph. The point a has coordinates $(0, 0, f(0, 0))$, is directly above the origin $P(0, 0)$, and is the lowest point on the portion of the graph shown. Thus, we say that f has a **local minimum** at $(0, 0)$, since $f(0, 0)$ is smaller than $f(x, y)$ for any (x, y) near the point P. Similarly, the point b is higher than any point in the vicinity, $f(1, 1) \geq f(x, y)$ for any (x, y) near Q. Thus, we say that f has a **local maximum** at $(1, 1)$. The points c and d represent a new phenomenon and are called **saddle points**. They are neither local maxima nor local minima, but seem to be a little of both. To see more clearly what features a saddle point has, look at Figure 2, which shows a portion of the graph near the point c.

It is easy to see from this picture where the term "saddle point" comes from. Now, if we slice through the graph along $y = 1$, we get a curve on which c is the lowest point. Thus, c looks like a local minimum along this slice. On the other hand, if we slice through the graph along $x = 0$, we get another curve at right angles to the first one, and on which c is the *highest* point, so that c looks like a local maximum along this slice. This kind of

15.4 Maxima and Minima

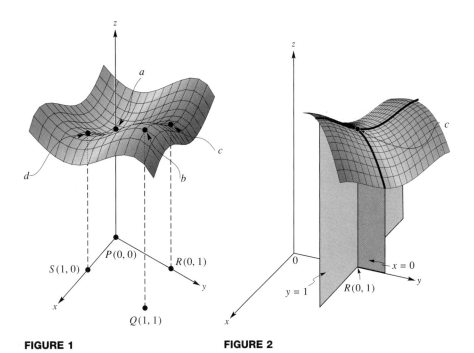

FIGURE 1

FIGURE 2

behavior characterizes a saddle point: f has a **saddle point** at (r, s) if f has a local minimum at (r, s) along some slices through that point and a local maximum along other slices through that point. If you look at the other saddle point, d, in Figure 1, you see the same behavior.

We also notice the following Figure 1.

1. The points P, Q, R, and S are all in the **interior** of the domain of f. That is, none of them lies on the boundary of the domain. Said another way, we can move some distance in any direction from any of these points without leaving the domain of f.
2. The tangent lines along the slices through these points parallel to the x- and y-axes are *horizontal*. Thus, the partial derivatives, $\partial f/\partial x$ and $\partial f/\partial y$, will be zero when evaluated at any of the points P, Q, R, and S. This gives us a way of locating candidates for local extrema and saddle points.

> **LOCATING CANDIDATES FOR LOCAL EXTREMA AND SADDLE POINTS IN THE INTERIOR OF THE DOMAIN OF f**
>
> Set $\partial f/\partial x = 0$ and $\partial f/\partial y = 0$ simultaneously, and solve for x and y. Then check that the resulting points (x, y) are in the interior of the domain of f.

Points at which all the partial derivatives are zero are called **critical points**. Thus, it is the critical points that are the only candidates for local extrema and saddle points in the interior of the domain of f.* Let's apply this principle to the function whose graph was shown in Figure 1.

▼ **EXAMPLE 1**

Locate all the critical points of the function

$$f(x, y) = 2(x^2 + y^2) - (x^4 + y^4) + 3.$$

SOLUTION According to the principle above, we must set $\partial f/\partial x = 0$ and $\partial f/\partial y = 0$ and solve for x and y.

$$\frac{\partial f}{\partial x} = 4x - 4x^3 = 4x(1 - x^2)$$

$$\frac{\partial f}{\partial y} = 4y - 4y^3 = 4y(1 - y^2)$$

Setting these equal to zero gives us the following simultaneous equations.

$$4x(1 - x^2) = 0$$
$$4y(1 - y^2) = 0$$

The first equation has solutions $x = 0$, 1, or -1, while the second has solutions $y = 0$, 1, or -1. Since these are simultaneous equations, we need values of x and y that satisfy *both* equations. We can do this by choosing any one of the values for x that satisfies the first and any one of the values of y that satisfies the second. This gives us a total of *nine* critical points:

$$(0, 0), (0, 1), (0, -1), (1, 0), (1, 1), (1, -1), (-1, 0),$$
$$(-1, 1), \text{ and } (-1, -1).$$

Four of these are the points P, Q, R, and S on the graph in Figure 1. The other five are also points at which f has extreme or saddle points, but Figure 1 shows only a small portion of the actual graph, and the remaining five points are out of range. (We'll see more of the graph of f a little later on.)

To get the points on the graph corresponding to these x- and y-values, we take the z-coordinate to be $f(x, y) = 2(x^2 + y^2) - (x^4 + y^4) + 3$, getting the nine points

$$(0, 0, 3), (0, 1, 4), (0, -1, 4), (1, 0, 4), (1, 1, 5),$$
$$(1, -1, 5), (-1, 0, 4), (-1, 1, 5), \text{ and } (-1, -1, 5).$$

▼ * We'll be looking at extrema on the *boundary* of the domain of a function in the next section. What we are calling critical points correspond to the *stationary* points of a function of one variable. We shall not consider the analogs of the singular points.

15.4 Maxima and Minima

Before we go on...

Q How we can tell whether each of these candidates actually gives a saddle point or a local extremum?

A There is an analog of the second-derivative test that is useful in determining which candidates are extrema and which are saddle points. We shall discuss it after the next example.

EXAMPLE 2

Locate all possible extrema and saddle points on the graph of

$$f(x, y) = e^{-(x^2+y^2)}.$$

SOLUTION The partial derivatives of f are

$$\frac{\partial f}{\partial x} = -2xe^{-(x^2+y^2)}$$

$$\frac{\partial f}{\partial y} = -2ye^{-(x^2+y^2)}.$$

Setting these equal to zero gives the following equations.

$$-2xe^{-(x^2+y^2)} = 0$$
$$-2ye^{-(x^2+y^2)} = 0$$

The first equation implies that $x = 0$,* and the second implies that $y = 0$. Thus, the only critical point is $(0, 0)$, and the corresponding point on the graph is $(0, 0, 1)$, since $f(0, 0) = 1$.

Before we go on...

Q Since we haven't drawn the graph of f, how do we know whether f has an extremum or saddle point there?

A If you look at the function $f(x, y) = e^{-(x^2+y^2)}$, you will notice that the exponent is always zero or negative. If you raise e to a negative number, the result is less than 1, so 1 must be the maximum value the function can take. In other words, $(0, 0, 1)$ is an **absolute maximum** of f. Figure 3 shows the graph of this function.

Notice that this surface is radially symmetric about the z-axis (why?) and levels off toward $z = 0$ as x or y becomes large.

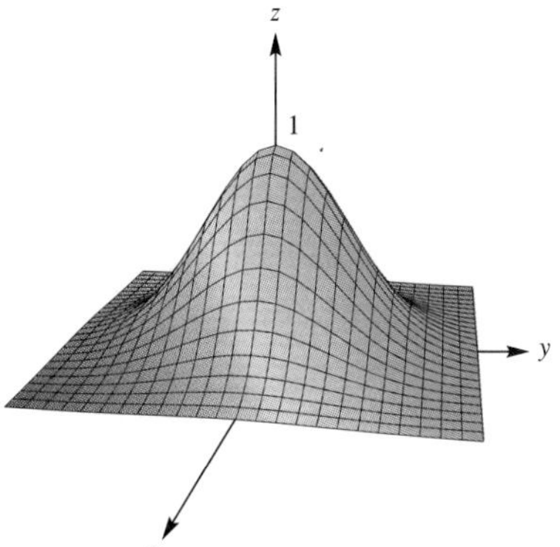

The following test gives us a way of deciding whether a critical point gives a local maximum, minimum, or saddle point.

Graph of $f(x, y) = e^{-(x^2 + y^2)}$

FIGURE 3

▼ * Recall that if a product of two numbers is zero, one or the other must be zero. In our case, the number $e^{-(x^2+y^2)}$ can't be zero (since e^u is never zero), giving the result claimed.

> **SECOND-DERIVATIVE TEST FOR FUNCTIONS OF TWO VARIABLES**
>
> Suppose $f(x, y)$ is a function of two variables and that (a, b) is a critical point in the interior of the domain of f (so that $f_x(a, b) = 0$ and $f_y(a, b) = 0$). Let H be the quantity
>
> $$f_{xx}(a, b) f_{yy}(a, b) - [f_{xy}(a, b)]^2.$$
>
> Then
>
> f has a local minimum at (a, b) if $H > 0$ and $f_{xx}(a, b) > 0$,
> f has a local maximum at (a, b) if $H > 0$ and $f_{xx}(a, b) < 0$, and
> f has a saddle point at (a, b) if $H < 0$.
>
> If $H = 0$, the test tells us nothing, so we need to look at the graph to see what is going on.

▶ **NOTE** There is a second-derivative test for functions of three or more variables, but it is considerably more complicated. We shall stick with functions of two variables for the most part in this book. The justification of the second-derivative test is beyond the scope of this book. ◀

Let us use the second-derivative test to continue Example 1.

▼ **EXAMPLE 3**

Use the second-derivative test to classify all the critical points of the function $f(x, y) = 2(x^2 + y^2) - (x^4 + y^4) + 3$.

SOLUTION We found the following nine critical points in Example 1:

$(0, 0)$, $(0, 1)$, $(0, -1)$, $(1, 0)$, $(1, 1)$, $(1, -1)$, $(-1, 0)$, $(-1, 1)$, and $(-1, -1)$.

We now need to apply the second-derivative test to each point in turn. First, we calculate all the second derivatives we shall need:

$$f_x = 4x - 4x^3, \quad \text{so} \quad f_{xx} = 4 - 12x^2, \quad \text{and} \quad f_{xy} = 0,$$

and

$$f_y = 4y - 4y^3, \quad \text{so} \quad f_{yy} = 4 - 12y^2.$$

The point $(0, 0)$: $f_{xx}(0, 0) = 4$, $f_{yy}(0, 0) = 4$, and $f_{xy}(0, 0) = 0$. Thus, $H = 4(4) - 0^2 = 16$. Since $H > 0$ and $f_{xx}(0, 0) > 0$, we have a local minimum at $(0, 0, 3)$.

The point $(0, 1)$: $f_{xx}(0, 1) = 4$, $f_{yy}(0, 1) = -8$, and $f_{xy}(0, 1) = 0$. Thus, $H = 4(-8) - 0^2 = -32$. Since $H < 0$, we have a saddle point at $(0, 1, 4)$.

The point $(0, -1)$: $f_{xx}(0, -1) = 4$, $f_{yy}(0, -1) = -8$ and $f_{xy}(0, -1) = 0$. These are the same values we got for the last point, so we again have a saddle point at $(0, -1, 4)$.

The point (1, 0): $f_{xx}(1, 0) = -8, f_{yy}(1, 0) = 4$, and $f_{xy}(1, 0) = 0$. Thus, $H = (-8)4 - 0^2 = -32$. Since $H < 0$, we have a saddle point at (1, 0, 4).

The point (1, 1): $f_{xx}(1, 1) = -8, f_{yy}(1, 1) = -8$, and $f_{xy}(1, 1) = 0$. Thus, $H = (-8)^2 - 0^2 = 64$. Since $H > 0$ and $f_{xx}(1, 1) < 0$, we have a local maximum at (1, 1, 5).

The point (1, −1): These are the same values we got for the last point, so we again have a local maximum at (1, −1, 5).

The point (−1, 0): These are the same values we got for the point (1, 0), so we have a saddle point at (−1, 0, 4).

The point (−1, 1): These are the same values we got for the point (1, 1), so we have a local maximum at (−1, 1, 5).

The point (−1, −1): These are again the same values we got for the point (1, 1), so we have a local maximum at (−1, −1, 5).

Before we go on... Figure 4 shows a larger portion of the graph of $f(x, y)$ including all nine critical points. We leave it to you to spot their locations on the graph.

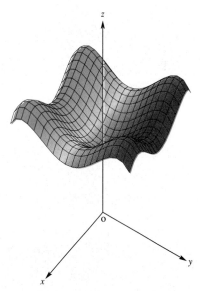

Graph of $f(x, y) = 2(x^2 + y^2) - (x^4 + y^4) + 3$

FIGURE 4

▼ **EXAMPLE 4**

Locate and classify all critical points of

$$f(x, y) = x^2 + 2y^2 + 2xy + 4x.$$

SOLUTION We first calculate the first-order partial derivatives.

$$f_x = 2x + 2y + 4$$
$$f_y = 2x + 4y$$

Setting these equal to zero gives a system of two linear equations in two unknowns.

$$x + y = -2$$
$$x + 2y = 0$$

This system has solution $(-4, 2)$, so this is our only critical point. The z-coordinate of this point is $f(-4, 2) = -8$. The second partial derivatives are

$$f_{xx} = 2, \quad f_{xy} = 2, \quad f_{yy} = 4.$$

Notice that all these derivatives are constants—they have the same value at every point (a, b). Thus, $H = 2 \cdot 4 - 2^2 = 4$. Since $H > 0$ and $f_{xx} > 0$, the second-derivative test tells us that we have a local minimum at $(-4, 2, -8)$.

▼ **EXAMPLE 5**

Locate and classify all critical points of $f(x, y) = x^2 y - x^2 - y^2$.

SOLUTION

$$f_x = 2xy - 2x = 2x(y - 1)$$
$$f_y = x^2 - 2y$$

Setting these equal to zero gives

$$x = 0 \text{ or } y = 1$$
$$x^2 = 2y.$$

We get a solution by choosing either $x = 0$ or $y = 1$ and substituting into $x^2 = 2y$, giving two possibilities: $0 = 2y$ and $x^2 = 2$. The first, $0 = 2y$, gives $(0, 0)$. The second, $x^2 = 2$, gives $x = \pm\sqrt{2}$ and hence the points $(\sqrt{2}, 1)$ and $(-\sqrt{2}, 1)$.

Thus, we have three critical points. To apply the second-derivative test, we calculate the second derivatives.

$$f_{xx} = 2y - 2$$
$$f_{xy} = 2x$$
$$f_{yy} = -2$$

We now look at each critical point in turn.

The point $(0, 0)$: $f_{xx}(0, 0) = -2$, $f_{xy}(0, 0) = 0$, $f_{yy}(0, 0) = -2$, so $H = 4$. Since $H > 0$ and $f_{xx}(0, 0) < 0$, the second-derivative test tells us that f has a local maximum at $(0, 0, 0)$.

The point $(\sqrt{2}, 1)$: $f_{xx}(\sqrt{2}, 1) = 0$, $f_{xy}(\sqrt{2}, 1) = 2\sqrt{2}$, and $f_{yy}(\sqrt{2}, 1) = -2$, so $H = -8$. Since $H < 0$, we know that f has a saddle point at $(\sqrt{2}, 1, -1)$.

The point $(-\sqrt{2}, 1)$: $f_{xx}(-\sqrt{2}, 1) = 0$, $f_{xy}(-\sqrt{2}, 1) = -2\sqrt{2}$, and $f_{yy}(\sqrt{2}, 1) = -2$, so once again $H = -8$, and the point $(-\sqrt{2}, 1, -1)$ is a saddle point.

Before we go on... Figure 5 shows the graph of f. See if you can spot the three critical points.

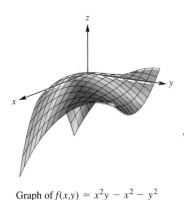

Graph of $f(x,y) = x^2y - x^2 - y^2$

FIGURE 5

EXAMPLE 6

Locate and classify all the critical points of $f(x, y) = 2x^4 - 6x^2y + y^4$.

SOLUTION

$$f_x = 8x^3 - 12xy = 4x(2x^2 - 3y)$$
$$f_y = 4y^3 - 6x^2 = 2(2y^3 - 3x^2)$$
$$f_{xx} = 24x^2 - 12y$$
$$f_{xy} = -12x$$
$$f_{yy} = 12y^2$$

To obtain the critical points, we set f_x and f_y equal to zero and solve. Setting $f_x = 0$ gives

$$x = 0 \text{ or } 3y = 2x^2.$$

Setting $f_y = 0$ gives

$$2y^3 = 3x^2.$$

To solve these equations simultaneously, we choose either $x = 0$ or $3y = 2x^2$ and combine it with $2y^3 = 3x^2$.

The combination $x = 0$ and $2y^3 = 3x^2$ gives the solution $(0, 0)$.

The combination $3y = 2x^2$ and $2y^3 = 3x^2$ is more tricky, but we can solve each of these equations for x^2 and equate the two answers:

$$x^2 = \tfrac{3}{2}y \text{ and } x^2 = \tfrac{2}{3}y^3$$
$$\tfrac{3}{2}y = \tfrac{2}{3}y^3$$

Thus, $y(\tfrac{3}{2} - \tfrac{2}{3}y^2) = 0$, giving $y = 0$ or $y = \pm\tfrac{3}{2}$. If $y = 0$, then $x = 0$, and we have the point $(0, 0)$ once again. If $y = \tfrac{3}{2}$, then $x^2 = (\tfrac{3}{2})^2$, so $x = \pm\tfrac{3}{2}$. If $y = -\tfrac{3}{2}$, then $x^2 = -(\tfrac{3}{2})^2$, giving no real solution for x. Thus the only solutions are the following.

$$(0, 0), (\tfrac{3}{2}, \tfrac{3}{2}) \text{ and } (-\tfrac{3}{2}, \tfrac{3}{2})$$

The point $(0, 0)$: All the second derivatives are zero when evaluated at $(0, 0)$, so the second-derivative test fails. This means that we must look at the function and try to decide whether there is a local extremum at $(0, 0)$. Now $f(0, 0) = 0$. If we slice the graph along $y = 0$, we get $f(x, 0) = 2x^4$, which we know possesses a minimum at $x = 0$. Thus, our critical point is the lowest

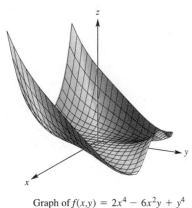

Graph of $f(x,y) = 2x^4 - 6x^2y + y^4$

FIGURE 6

point on the graph for this slice. If we slice along $x = 0$, we get $f(0, y) = y^4$, so there is again a minimum in this slice. Thus, it's beginning to look as though $(0, 0)$ is a local minimum. Now might be a good time to look at the graph, as drawn by computer (see Figure 6).

The graph appears to get lower in the direction of $x = y$, so let us try taking a slice along this plane. If we set $x = y$, we get $f(x, x) = 2x^4 - 6x^3 + x^4 = 3x^4 - 6x^3 = 3x^3(x - 2)$. Now this is negative for all values of x between 0 and 2, showing that the surface dips below the xy-plane along this slice. In other words, $(0, 0)$ can't be a local minimum, since the surface is lower than $z = 0$ at points nearby. We conclude that $(0, 0, 0)$ is neither a local maximum nor a local minimum. Moreover, it isn't a saddle point either. (Why?)

The point $(\frac{3}{2}, \frac{3}{2})$: $f_{xx}(\frac{3}{2}, \frac{3}{2}) = 36$, $f_{xy}(\frac{3}{2}, \frac{3}{2}) = -18$, $f_{yy}(\frac{3}{2}, \frac{3}{2}) = 27$. Thus $H = 648 > 0$, and we have a local minimum at $(\frac{3}{2}, \frac{3}{2}, -\frac{81}{16})$.

The point $(-\frac{3}{2}, \frac{3}{2})$: $f_{xx}(-\frac{3}{2}, \frac{3}{2}) = 36$, $f_{xy}(-\frac{3}{2}, \frac{3}{2}) = -18$, $f_{yy}(-\frac{3}{2}, \frac{3}{2}) = 27$. Thus again $H > 0$, and we have another local minimum at $(-\frac{3}{2}, \frac{3}{2}, -\frac{81}{16})$.

15.4 EXERCISES

In Exercises 1 through 6, classify each labeled point on the graph as either:
 (a) *a local maximum;* (b) *a local minimum;* (c) *a saddle point;*
 (d) *a critical point, but neither a local extremum nor a saddle point;*
 (e) *none of the above.*

1.

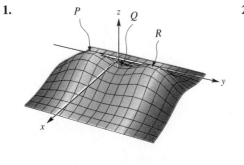

2.

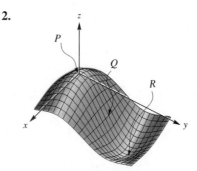

3.

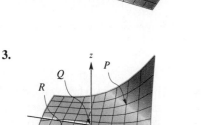

4.

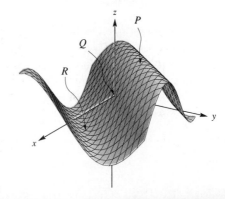

5.

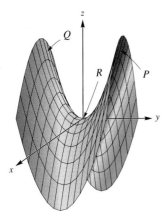

6.

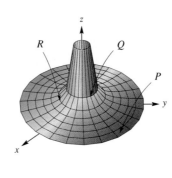

7. Sketch the graph of a function that has one local extremum and no saddle points.
8. Sketch the graph of a function that has one saddle point and one local extremum.

Locate and classify all critical points of each function in Exercises 9–26.

9. $f(x, y) = x^2 + y^2 + 1$
10. $f(x, y) = 4 - (x^2 + y^2)$
11. $g(x, y) = 1 - x^2 - x - y^2 + y$
12. $g(x, y) = x^2 + x + y^2 - y - 1$
13. $h(x, y) = x^2y - 2x^2 - 4y^2$
14. $h(x, y) = x^2 + y^2 - y^2x - 4$
15. $s(x, y) = e^{x^2+y^2}$
16. $s(x, y) = e^{-(x^2+y^2)}$
17. $t(x, y) = x^4 + 8xy^2 + 2y^4$
18. $t(x, y) = x^3 - 3xy + y^3$
19. $f(x, y) = x^2 + y - e^y$
20. $f(x, y) = xe^y$
21. $f(x, y) = e^{-(x^2+y^2+2x)}$
22. $f(x, y) = e^{-(x^2+y^2-2x)}$
23. $f(x, y) = xy + \dfrac{2}{x} + \dfrac{2}{y}$
24. $f(x, y) = xy + \dfrac{4}{x} + \dfrac{2}{y}$
25. $g(x, y) = x^2 + y^2 + \dfrac{2}{xy}$
26. $g(x, y) = x^3 + y^3 + \dfrac{3}{xy}$

APPLICATIONS

27. *Average Cost* Your bicycle factory makes two models, 5-speeds and 10-speeds. Each week, the total cost to make x 5-speeds and y 10-speeds is given by
$$C(x, y) = \$10{,}000 + 50x + 70y + 0.0125xy.$$
At what production levels is your average cost least? (Remember that the average cost is given by $\overline{C}(x, y) = C(x, y)/(x + y)$.)

28. *Average Cost* Your bicycle factory makes two models, 5-speeds and 10-speeds. Each week, the total cost to make x 5-speeds and y 10-speeds is given by
$$C(x, y) = \$140{,}000 + 20x + 90y + 0.07xy.$$
At what production levels is your average cost least?

29. *Average Cost* Let $C(x, y)$ be any cost function. Show that when the average cost is minimized, the marginal costs C_x and C_y both equal the average cost. Explain why this is reasonable.

30. *Average Profit* Let $P(x, y)$ be any profit function. Show that when the average profit is maximized, the marginal profits P_x and P_y both equal the average profit. Explain why this is reasonable.

31. *Revenue* Your company manufactures two models of stereo speakers, the Ultra Mini and the Big Stack. Demand for each depends partly on the price of the other. If one is expensive, more people will buy the other. If p_1 is the price per pair of the Ultra Mini, and

p_2 is the price of the Big Stack, demand for the Ultra Mini is given by

$$q_1(p_1, p_2) = 100{,}000 - 100p_1 + 10p_2,$$

where q_1 represents the number of pairs of Ultra Minis that will be sold in a year. The demand for the Big Stack is given by

$$q_2(p_1, p_2) = 150{,}000 + 10p_1 - 100p_2.$$

Find the prices for the Ultra Mini and the Big Stack that will maximize your total revenue.

32. *Revenue* Repeat the previous exercise, with the following demand functions.

$$q_1(p_1, p_2) = 100{,}000 - 100p_1 + p_2$$
$$q_2(p_1, p_2) = 150{,}000 + p_1 - 100p_2$$

COMMUNICATION AND REASONING EXERCISES

33. Let $H = f_{xx}(a, b) f_{yy}(a, b) - f_{xy}(a, b)^2$. What condition on H guarantees that f has a local extremum at the point (a, b)?

34. Let H be as in the previous exercise. Give an example to show that it is possible to have $H = 0$ and a local minimum at (a, b).

35. Suppose that when the graph of $f(x, y)$ is sliced by a vertical plane through (a, b) parallel to either the xz-plane or the yz-plane, the resulting curve has a local maximum at (a, b). Does this mean that f has a local maximum at (a, b)? Explain your answer.

36. Suppose that f has a local maximum at (a, b). Does it follow that when the graph of f is sliced by a vertical plane parallel to either the xz-plane or the yz-plane, the resulting curve has a local maximum at (a, b)? Explain your answer.

37. The tangent plane to a graph was introduced in the exercises in the previous section. Explain why the tangent plane is parallel to the xy-plane at a local maximum or minimum of $f(x, y)$.

38. Explain why the tangent plane is parallel to the xy-plane at a saddle point of $f(x, y)$.

15.5 CONSTRAINED MAXIMA AND MINIMA AND APPLICATIONS

So far we have looked only at the local extrema of f that lie in the interior of the domain of f. There may also be local extrema on the boundary of the domain (just as, for a function of one variable, the endpoints of the domain may be local extrema). This situation arises, for example, in optimization problems with constraints, similar to those we saw in the chapter on applications of the derivative. Here is a typical example.

$$\text{Maximize } S = xy + 2xz + 2yz$$
$$\text{subject to } xyz = 4$$
$$\text{and } x \geq 0, y \geq 0, z \geq 0.$$

There are two kinds of constraints in this example: equations and inequalities. The inequalities specify a restriction on the domain of S. Our strategy for solving such problems is essentially the same as in the chapter on applications of the derivative. First, we use any equality constraints to eliminate variables. In the examples in this section we will be able to reduce to a function of only two variables. The inequality constraints then help define the domain of this function. Next, we locate any critical points in the interior of the domain. Finally, we look at the boundary of the domain. When there is no boundary to worry about, another method, called the *method of Lagrange multipliers*, comes in handy.

We first look at functions of two variables with restricted domains, so we can see how to handle the boundaries.

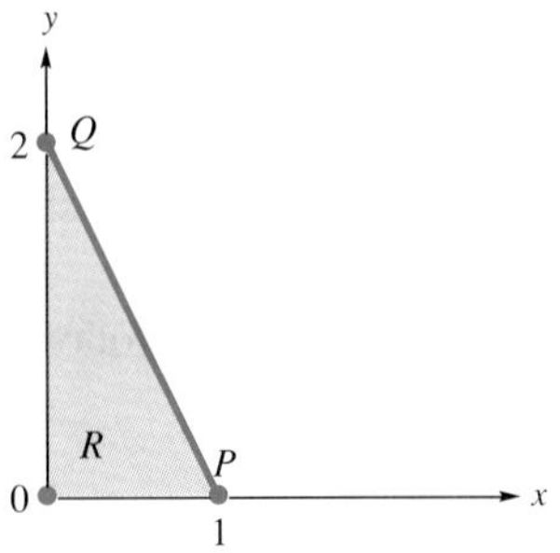

15.5 Constrained Maxima and Minima and Applications

EXAMPLE 1

Find the maximum and minimum value of $f(x, y) = xy - x - 2y$ on the triangular region R with vertices $(0, 0)$, $(1, 0)$, and $(0, 2)$.

SOLUTION The *domain* of f is the region R, which is shown in Figure 1.

Locate critical points in the interior of the domain. We have

$$f_x = y - 1$$
$$f_y = x - 2$$
$$f_{xx} = 0$$
$$f_{xy} = 1$$
$$f_{yy} = 0.$$

FIGURE 1

The only critical point is thus $(2, 1)$. Since this lies outside the domain (the region R) we ignore it. Thus, there are no critical points in the interior of the domain of f.

Locate local extrema on the boundary of the domain. The boundary of the domain consists of three line segments, OP, OQ, and PQ. We deal with these one at a time.

Segment OP This line segment has equation $y = 0, 0 \leq x \leq 1$. Along this segment we see $f(x, 0) = -x$. We find the local extrema of this function of one variable by the methods we used in the chapter on applications of the derivative. There are no critical points, and there are two endpoints, $x = 0$ and $x = 1$. Since $y = 0$, this gives us the following two candidates for local extrema: $(0, 0, 0)$ and $(1, 0, -1)$.

Segment OQ This line segment has equation $x = 0, 0 \leq y \leq 2$. Along this segment we see $f(0, y) = -2y$. We now locate the local extrema of this function of one variable. Once again, there are only the endpoints, $y = 0$ and $y = 2$. Since $x = 0$, this gives us the two candidates $(0, 0, 0)$ and $(0, 2, -4)$.

Segment PQ This line segment has equation $y = -2x + 2$ with $0 \leq x \leq 1$. Along this segment we see

$$f(x, -2x + 2) = x(-2x + 2) - x - 2(-2x + 2)$$
$$= -2x^2 + 5x - 4.$$

This function of x (whose graph is an upside-down parabola) has a stationary maximum when its derivative, $-4x + 5$ is 0, or $x = \frac{5}{4}$. Since this is bigger than 1, it lies outside the domain $0 \leq x \leq 1$. Thus, we reject it. There are no other critical points, and the endpoints are $x = 0$ and $x = 1$. When $x = 0$, $y = -2(0) + 2 = 2$, giving the point $(0, 2, -4)$. When $x = 1$, $y = -2(1) + 2 = 0$, giving the point $(1, 0, -1)$.

Thus, the candidates for maxima and minima are $(0, 0, 0)$, $(1, 0, -1)$, and $(0, 2, -4)$ (which happen to lie over the corner points of the domain R). Since their z-coordinates give the value of f, we see that f has an absolute maximum of 0 at the point $(0, 0)$ and an absolute minimum of -4 at the point $(0, 2)$.

$$x = 100 - 5y, \quad 5 \le y \le 20$$

This gives

$$P = 10(100 - 5y) + 60y + 40\sqrt{100 - 4y}$$
$$= 1{,}000 + 10y + 40\sqrt{100 - 4y}$$
$$P' = 10 - \frac{80}{\sqrt{100 - 4y}}.$$

Setting the derivative equal to 0 and solving, we get $y = 9$. Substituting this in $x = 100 - 5y$, we get $x = 55$. Now we check the second derivative.

$$P'' = -\frac{160}{(100 - 4y)^{3/2}} < 0$$

This tells us that we have found a local maximum. In fact, it must be the maximum value on this line segment. The point on the graph is $(55, 9, 1410)$.

$$x = 90 - 3y, \quad 0 \le y \le 5$$

This gives

$$P = 10(90 - 3y) + 60y + 40\sqrt{90 - 2y}$$
$$= 900 + 30y + 40\sqrt{90 - 2y}$$
$$P' = 30 - \frac{40}{\sqrt{90 - 2y}}.$$

Setting this equal to 0 and solving for y gives $y = 44\frac{1}{9}$, which is outside the range $0 \le y \le 5$. Therefore, there are no critical points on this line segment. Its endpoints are $y = 0$ and $y = 5$, giving $(90, 0, 904.2)$ and $(75, 5, 1407.8)$.

Looking at the heights of all of the points we have found, the highest is $(55, 9, 1410)$, so this gives you the largest profit. In other words, you should make 55 Ultra Minis and 9 Big Stacks each week, giving you the largest possible profit of $1410 per week.

Before we go on... If you have studied linear programming, this should remind you of the problems you solved by that technique. However, since the objective function is not linear, the techniques of linear programming fail to solve this problem. In particular, notice that the solution we found is *not* at a corner of the feasible region, but in the middle of one edge. It is also quite possible that the solution could be in the interior of the region.

THE METHOD OF LAGRANGE MULTIPLIERS

Suppose we have an optimization problem in which it is difficult or impossible to solve a constraint equation for one of the variables. Then we can use the method of **Lagrange multipliers** to avoid this difficulty. We shall restrict attention to the case of a single constraint equation, although the method generalizes to any number of constraint equations.

15.5 Constrained Maxima and Minima and Applications

> **LOCATING LOCAL EXTREMA USING THE METHOD OF LAGRANGE MULTIPLIERS**
>
> To locate the candidates for local extrema of a function $f(x, y, \ldots)$ subject to the constraint $g(x, y, \ldots) = 0$, solve the following system of equations for $x, y, \ldots$, and λ.
>
> $$f_x = \lambda g_x$$
> $$f_y = \lambda g_y$$
> $$\ldots$$
> $$g = 0.$$
>
> The unknown λ is called a **Lagrange multiplier.** The points $(x, y, \ldots)$ that occur in solutions are then the candidates for the local extrema of f subject to $g = 0$.

▼ **EXAMPLE 5**

Use the method of Lagrange multipliers to find the maximum value of $f(x, y) = 2xy$ subject to $x^2 + 4y^2 = 32$.

SOLUTION We start by rewriting the problem in standard form.

Maximize $f(x, y) = 2xy$ *subject to* $x^2 + 4y^2 - 32 = 0$.

Here, $g(x, y) = x^2 + 4y^2 - 32$, and the system of equations we need to solve is

$$f_x = \lambda g_x, \quad \text{or} \quad 2y = 2\lambda x,$$
$$f_y = \lambda g_y, \quad \text{or} \quad 2x = 8\lambda y,$$
$$g = 0, \quad \text{or} \quad x^2 + 4y^2 - 32 = 0.$$

A convenient way to solve such a system is to solve one of the equations for λ and then substitute in the remaining equations. Thus, we start by solving the first equation to obtain

$$\lambda = \frac{y}{x}.$$

(A word of caution: since we divided by x, we made the implicit assumption that $x \neq 0$, so before continuing, we should check what happens if $x = 0$. But if $x = 0$, then the first equation, $2y = 2\lambda x$, tells us that $y = 0$ as well, and this contradicts the third equation: $x^2 + 4y^2 - 32 = 0$. Thus, we can rule out the possibility that $x = 0$.)

Substituting in the remaining equations gives

$$x = \frac{4y^2}{x}, \quad \text{or} \quad x^2 = 4y^2,$$

and $\quad x^2 + 4y^2 - 32 = 0.$

Notice how we have reduced the number of unknowns and also the number of equations by one. We can now substitute $x^2 = 4y^2$ in the last equation, obtaining

$$4y^2 + 4y^2 - 32 = 0,$$

or

$$8y^2 = 32,$$

giving

$$y = \pm 2.$$

We now substitute back to obtain

$$x^2 = 4y^2 = 16,$$

or

$$x = \pm 4.$$

We don't need the value of λ, so we won't solve for it. Thus, the candidates for local extrema are given by $x = \pm 4$ and $y = \pm 2$, giving the four points $(-4, -2)$, $(-4, 2)$, $(4, -2)$, and $(4, 2)$.

Recall that we are seeking the values of x and y that give the maximum value for $f(x, y) = 2xy$. Since we now have only four points to choose from, we compare the values of f at these four points and conclude that the maximum value of f occurs when $(x, y) = (-4, -2)$ or $(4, 2)$.

Before we go on...

Q Something is suspicious here. We didn't check to see whether these candidates were local extrema to begin with, let alone absolute extrema! How do we justify this omission?

A One of the difficulties with using the method of Lagrange multipliers is that it does not provide us with a test analogous to the second-derivative test for functions of several variables. However, if you grant that the function in question does have an absolute maximum, then we require no test, since one of the candidates must give this maximum.

Q But how do we know that the given function has an absolute maximum?

A The best way to see this is by giving a geometric interpretation. The constraint $x^2 + 4y^2 = 32$ tells us that the point (x, y) must lie on the ellipse shown in Figure 5. The function $f(x, y) = 2xy$ gives the area of the rectangle shaded in Figure 5.

Since there is a largest possible rectangle of this type, the function f must have an absolute maximum for at least one pair of coordinates (x, y).

15.5 Constrained Maxima and Minima and Applications

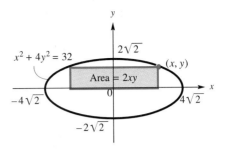

FIGURE 5

Q When can I use the method of Lagrange multipliers? When *should* I use it?

A We have only discussed the method when there is a single equality constraint. There is a generalization, which we shall not discuss, that works when there are more equality constraints (as a hint, we need to introduce one multiplier for each constraint). So, if you have a problem with more than one equality constraint, or with any inequality constraints, you must use the method we discussed earlier in this section. On the other hand, if you have one equality constraint, and it would be difficult to solve it for one of the variables, then you should use Lagrange multipliers. We have noted in the exercises where you might use Lagrange multipliers.

Q Why does the method of Lagrange multipliers work?

A An adequate answer is beyond the scope of this book.

▶ **15.5 EXERCISES**

Find the maximum and minimum values, and the points at which they occur, for each function in Exercises 1–16.

1. $f(x, y) = x^2 + y^2$, $0 \le x \le 2$, $0 \le y \le 2$
2. $g(x, y) = \sqrt{x^2 + y^2}$, $1 \le x \le 2$, $1 \le y \le 2$
3. $h(x, y) = (x - 1)^2 + y^2$, $x^2 + y^2 \le 4$
4. $k(x, y) = x^2 + (y - 1)^2$, $x^2 + y^2 \le 9$
5. $f(x, y) = e^{x^2+y^2}$, $4x^2 + y^2 \le 4$
6. $g(x, y) = e^{-(x^2+y^2)}$, $x^2 + 4y^2 \le 4$
7. $h(x, y) = e^{4x^2+y^2}$, $x^2 + y^2 \le 1$
8. $k(x, y) = e^{-(x^2+4y^2)}$, $x^2 + y^2 \le 4$
9. $f(x, y) = x + y + 1/(xy)$, $x \ge \frac{1}{2}$, $y \ge \frac{1}{2}$, $x + y \le 3$
10. $g(x, y) = x + y + 8/(xy)$, $x \ge 1$, $y \ge 1$, $x + y \le 6$
11. $h(x, y) = xy + 8/x + 8/y$, $x \ge 1$, $y \ge 1$, $xy \le 9$

12. $k(x, y) = xy + 1/x + 8/y$, $x \geq \frac{1}{4}$, $y \geq \frac{1}{4}$, $xy \leq 9$

13. $f(x, y) = x^2 + 2x + y^2$, on the region in the figure.

14. $g(x, y) = x^2 + y^2$, on the region in the figure.

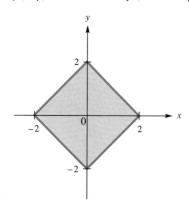

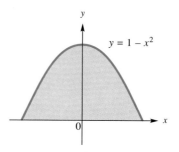

15. $h(x, y) = x^3 + y^3$, on the region in the figure.

16. $k(x, y) = x^3 + 2y^3$, on the region in the figure.

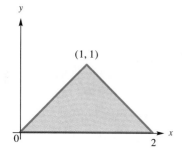

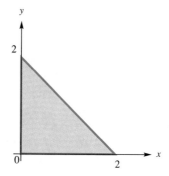

17. At what points on the sphere $x^2 + y^2 + z^2 = 1$ is the product xyz a maximum? (The method of Lagrange multipliers can be used here.)

18. At what point on the surface $z = (x^2 + x + y^2 + 4)^{1/2}$ is the quantity $x^2 + y^2 + z^2$ a minimum? (The method of Lagrange multipliers can be used here.)

APPLICATIONS

19. *Cost* Your bicycle factory makes two models, 5-speeds and 10-speeds. Each week, your total cost to make x 5-speeds and y 10-speeds is given by

$$C(x, y) = \$10{,}000 + 50x + 70y - 0.5xy.$$

You want to make between 100 and 150 5-speeds, and between 80 and 120 10-speeds. What combination will cost you the least? What combination will cost you the most?

20. *Cost* Your bicycle factory makes two models, 5-speeds and 10-speeds. Each week, your total cost to make x 5-speeds and y 10-speeds is given by

$$C(x, y) = \$10{,}000 + 50x + 70y - 0.46xy.$$

You want to make between 100 and 150 5-speeds, and between 80 and 120 10-speeds. What combination will cost you the least? What combination will cost you the most?

21. *Profit* Your software company sells two programs, Walls and Doors. Your profit from selling x copies of Walls and y copies of Doors is given by

$$P(x, y) = 20x + 40y - 0.1(x^2 + y^2).$$

If you can sell a maximum of 200 copies of the two programs together, what combination will bring you the largest profit?

22. *Profit* Your software company sells two programs, Walls and Doors. Your profit from selling x copies of Walls and y copies of Doors is given by
$$P(x, y) = 20x + 40y - 0.1(x^2 + y^2).$$
If you can sell a maximum of 400 copies of the two programs together, what combination will bring you the largest profit?

23. *Temperature* The temperature at the point (x, y) on the square with vertices $(0, 0)$, $(0, 1)$, $(1, 0)$, and $(1, 1)$ is given by $T(x, y) = x^2 + 2y^2$. Find the hottest and coldest points on the square.

24. *Temperature* The temperature at the point (x, y) on the square with vertices $(0, 0)$, $(0, 1)$, $(1, 0)$, and $(1, 1)$ is given by $T(x, y) = x^2 + 2y^2 - x$. Find the hottest and coldest points on the square.

25. *Temperature* The temperature at the point (x, y) on the disc $\{(x, y) \mid x^2 + y^2 \leq 1\}$ is given by $T(x, y) = x^2 + 2y^2 - x$. Find the hottest and coldest points on the disc.

26. *Temperature* The temperature at the point (x, y) on the disc $\{(x, y) \mid x^2 + y^2 \leq 1\}$ is given by $T(x, y) = 2x^2 + y^2$. Find the hottest and coldest points on the disc.

The method of Lagrange multipliers can be used for Exercises 27–38.

27. *Geometry* Find the point on the plane $-2x + 2y + z - 5 = 0$ closest to the point $(-1, 1, 3)$.

28. *Geometry* Find the point on the plane $2x - 2y - z + 1 = 0$ closest to the point $(1, 1, 0)$.

29. *Geometry* What point on the surface $z = x^2 + y - 1$ is closest to the origin?

30. *Geometry* What point on the surface $z = x + y^2 - 3$ is closest to the origin?

31. *Construction Cost* A closed rectangular box is made with two kinds of materials. The top and bottom are made with heavy-duty cardboard costing 20¢ per square foot, while the sides are made with lightweight cardboard costing 10¢ per square foot. Given that the box is to have a capacity of 2 cubic feet, what should its dimensions be if the cost is to be minimized?

32. *Construction Cost* Repeat the previous exercise if the heavy-duty cardboard costs 30¢ per square foot

and the lightweight cardboard costs 5¢ per square foot.

33. *Construction Cost* Referring to Exercise 31, my company wishes to manufacture boxes with a capacity of 2 cubic feet as cheaply as possible, but unfortunately, the company that manufactures the cardboard is unable to give me price quotes for the heavy-duty and lightweight cardboard as yet. Find formulas for the dimensions of the box in terms of the price per square foot of heavy-duty and lightweight cardboard.

34. *Construction Cost* Repeat the previous exercise, assuming that only the bottoms of the boxes are to made using heavy-duty cardboard.

35. *Package Dimensions* The *U.S. Postal Service* (USPS) will accept only packages with a length plus girth no more than 108 inches. (See figure.)
    What is the largest-volume package that the USPS will accept?

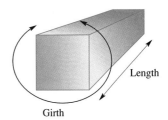

36. *Package Dimensions* The *United Parcel Service* (UPS) will accept only packages with a length no more than 108 inches and length plus girth no more than 130 inches (See figure above.) What is the largest-volume package that UPS will accept?

37. *Geometry* Find the dimensions of the rectangular box with least volume that can be inscribed above the xy-plane and under the paraboloid $z = 1 - (x^2 + y^2)$.

38. *Geometry* Find the dimensions of the rectangular box with least volume that can be inscribed above the xy-plane and under the paraboloid $z = 2 - (2x^2 + y^2)$.

39. *Resource Allocation* You manage an ice cream factory that makes two flavors: Creamy Vanilla and Continental Mocha. Into each quart of Creamy Vanilla go two eggs and three cups of cream. Into each quart of Continental Mocha go one egg and three cups of cream. You have in stock 500 eggs and 900 cups of cream. Your profit on x quarts of vanilla and y quarts

of mocha are $P(x, y) = 3x + 2y - 0.01(x^2 + y^2)$. How many quarts of each flavor should you produce in order to make the largest profit?

40. *Resource Allocation* Repeat the preceding exercise using the profit function $P(x, y) = 3x + 2y - 0.005(x^2 + y^2)$.

41. *Resource Allocation* Urban Institute of Technology's Math Department offers two courses: Finite Math and Calculus. Each section of Finite Math has 60 students, while each section of Calculus has 50. The department is allowed to offer a total of up to 110 sections. Further, there are no more than 6,000 students who would like to take a math course. The university's profit on x sections of Finite Math and y sections of Calculus is

$$P(x, y) = \$5,000,000(1 - e^{-0.02x - 0.01y})$$

(the profit being the difference between what the students are charged and what the professors are paid). How many sections of each course should the department offer in order to make the largest profit?

42. *Resource Allocation* Repeat the preceding exercise using the profit function

$$P(x, y) = \$5,000,000(1 - e^{-0.01x - 0.02y}).$$

43. *Nutrition* Gerber Mixed Cereal for Baby costs 10¢ per serving. Gerber Mango Tropical Fruit Dessert costs 53¢ per serving. If you want the product of the number of servings of each to be at least 10 per day, how can you do so at the least cost?

44. *Nutrition* Repeat the preceding exercise if instead you want the product of the number of servings of cereal and the square of the number of servings of dessert to be at least 10.

45. *Purchasing* The ESU Business School is buying computers. It has two models to choose from, the Pomegranate and the Ami. Each Pomegranate comes with 4 MB of memory and 80 MB of disk space, while each Ami has 3 MB of memory and 100 MB of disk space. For reasons related to its accreditation, the school would like to be able to say that it has a total of at least 480 MB of memory and at least 12,800 MB of disk space. Because of complicated volume pricing, the cost to the school of x Pomegranates and y Amis is

$$C(x, y) = x^2 + y^2 - 200x - 200y + 220,000.$$

How many of each kind of computer should the school buy in order to minimize the average cost per computer?

46. *Purchasing* Repeat the preceding exercise assuming that the cost is

$$C(x, y) = x^2 + y^2 - 300x - 200y + 332,500.$$

COMMUNICATION AND REASONING EXERCISES

47. If the partial derivatives of a function of several variables are never zero, is it possible for the function to have local extrema on some domain? Explain your answer.

48. Suppose we know that $f(x, y)$ has an absolute maximum somewhere in the domain D, and that (a, b) is the only point in D such that $f_x(a, b) = f_y(a, b) = 0$. Must it be the case that f has an absolute maximum at (a, b)? Explain.

49. Under what circumstances would it be necessary to use the method of Lagrange multipliers?

50. Under what circumstances would the method of Lagrange multipliers not apply?

51. A **linear programming problem in two variables** is a problem of the following form.

Maximize (or minimize) $f(x, y)$

subject to constraints $C(x, y) \geq 0$ or $C(x, y) \leq 0$.

Here, the objective function f and the constraints C are linear functions. There may be several such linear constraints in one problem. Explain why the solution cannot occur in the interior of the domain of f.

52. Continuing the preceding exercise, explain why the solution will actually be at a corner of the domain of f (where two or more of the line segments making up the boundary meet). This result—or rather a slight generalization of it—is known as the Fundamental Theorem of Linear Programming.

15.6 LEAST-SQUARES FIT

Throughout this book we have used functions to model relationships between variables, for example the relationship between price and demand. Often these functions were linear. In this section, we discuss how to come up with such a model.

In the simplest case, we have two data points and we only need to find the equation of the line passing through them. However, it often happens that we have many data points that don't quite lie on one line. The problem then is to find the line coming closest to passing through all of the points.

Suppose, for example, that your market research of real estate investments reveals the following sales figures for new homes of different prices over the past year (this is some of the data referred to at the beginning of the chapter).

Price (Thousands of $)	Sales of New Homes This Year
$150–$169	126
$170–$189	103
$190–$209	82
$210–$229	75
$230–$249	82
$250–$269	40
$270–$289	20

If we simplify the situation by replacing each of the price ranges by a single price in the middle of the range, we get the following table:

Price (Thousands of $)	Sales of New Homes This Year
$160	126
$180	103
$200	82
$220	75
$240	82
$260	40
$280	20

We would like to use these data to construct a demand function for the real estate market. (Recall that a demand function gives demand, q—measured here by annual sales—as a function of unit price, p.) Figure 8.6.1 shows a plot of q versus p.

The data definitely suggest a straight line more-or-less, and hence an approximately linear relationship between p and q. Figure 8.6.2 shows several possible "straight-line fits."

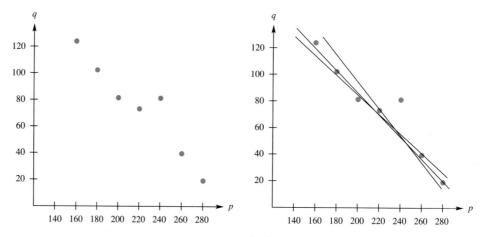

FIGURE 1

FIGURE 2

The question now is, "What line best fits the data?"

To answer this question, we need to be clear as to just what we mean by "best fits." Call the linear function whose graph is the desired straight line $l(p)$, and let $q(p)$ be the actual sales at price p. Since we want $l(p)$ to be as good an approximation to $q(p)$ as possible, what would seem best is that $l(p)$ be as close as possible to the actual sales $q(p)$ for each price p in our table. The distances between the **predicted values** $l(p)$ and the **observed values** $q(p)$ appear as the vertical distances shown in Figure 8.6.3, and are calculated as $|q(p) - l(p)|$ for each value of p for which we have a data point.

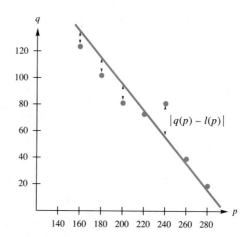

FIGURE 3

Q Since we want the vertical distances to be as small as possible, why can't we set them all to zero and solve?

A If this were possible, then there would be a straight line that passes through all the data points. A look at the graph shows that this is not the case.

Q Then why not find the line that minimizes *all* the vertical distances $|q(p) - l(p)|$?

A This is not possible either. The line that minimizes the first two distances is the line that passes through the first two data points, since it makes the distances 0. But this line certainly does not minimize the distance to the third point. In other words, there is a trade-off: making some distances smaller makes others larger.

Q So what do we do?

A Since we cannot minimize *all* of the distances, we minimize some reasonable combination of them.

Now, one reasonable combination of the distances would be their *sum*, but that turns out the be difficult to work with (because of the absolute values). Instead, we use the sum of the *squares* of the distances. In other words, if we denote the various values of p by $p_1, p_2, \ldots, p_n$, then we shall minimize the quantity

$$S = (l(p_1) - q(p_1))^2 + (l(p_2) - q(p_2))^2 + \ldots + (l(p_n) - q(p_n))^2.$$

The line that minimizes S is called the **least-squares line** or **best-fit line** associated with the given data.

Q Is there a good reason for looking at the sum of the squares of the distances?

A There are several. As we mentioned above, it is technically easier to use the sum of the squares than to use the sum of the distances themselves. On a more theoretical level, we can view the observed values $(q(p_1), q(p_2), \ldots, q(p_n))$ and predicted values $(l(p_1), l(p_2), \ldots, l(p_n))$ as being two points in n-dimensional space, and we can view our goal as the minimization of the distance between these two points. The distance between these points is given by a generalization of the formula for the distance between two points in 2-space or 3-space:

$$d = \sqrt{(l(p_1) - q(p_1))^2 + (l(p_2) - q(p_2))^2 + \cdots + (l(p_n) - q(p_n))^2}.$$

The sum of squares, S, is therefore the square of the distance between two points in n-space, and minimizing S minimizes the distance between these points.

To see how to minimize S we begin with a simpler example.

▼ **EXAMPLE 1**

Find the least-squares line associated with the following data:

p	1	2	3	4
q	1.5	1.6	2.1	3.0

SOLUTION Since we are looking for a linear approximation of q as a function of p, we take l to have the form

$$l(p) = mp + b.$$

Our job is to determine the constants m and b. We must now minimize the quantity

$$\begin{aligned} S &= (l(1) - q(1))^2 + (l(2) - q(2))^2 + (l(3) - q(3))^2 \\ &\quad + (l(4) - q(4))^2 \\ &= (m + b - 1.5)^2 + (2m + b - 1.6)^2 \\ &\quad + (3m + b - 2.1)^2 + (4m + b - 3.0)^2. \end{aligned}$$

The variables here are m and b, so we have a function of two variables to minimize, and we proceed as in Section 4. We begin by taking partial derivatives.

$$\begin{aligned} S_m &= 2(m + b - 1.5) + 4(2m + b - 1.6) \\ &\quad + 6(3m + b - 2.1) + 8(4m + b - 3.0) \\ S_b &= 2(m + b - 1.5) + 2(2m + b - 1.6) \\ &\quad + 2(3m + b - 2.1) + 2(4m + b - 3.0) \end{aligned}$$

We must now set these equal to zero and solve for the two unknowns m and b. In order to make this easier, we first gather terms.

$$\begin{aligned} S_m &= 60m + 20b - 46 = 0 \\ S_b &= 20m + 8b - 16.4 = 0 \end{aligned}$$

This is a system of two linear equations in the two unknowns m and b, and has solution $m = 0.5$, $b = 0.8$ (you can check by solving for yourself!). Thus, our least-squares line is

$$l(p) = 0.5p + 0.8.$$

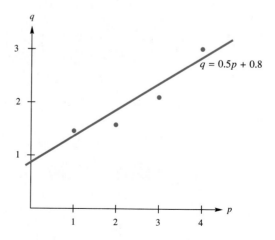

FIGURE 4

Before we go on... We didn't check that these values of m and b gave an absolute minimum of S. But we can argue as follows: There must be a minimum value of S somewhere, it must show up as a critical point, and since we have located the only possible critical point, this must be it!

Figure 8.6.4 shows the data points and the least-squares line.

Notice that the line doesn't pass through even one of the original points, and yet it is the straight line that best approximates them.

Before going on to another example, let us return to the calculation we just did and see if we can come up with a *formula* for m and b. To do this, we consider the general situation with a table of given data.

p	p_1	p_2	$\ldots$	p_n
q	q_1	q_2	$\ldots$	q_n

The linear function we are after has the form $l(p) = mp + b$, and we want to minimize the quantity

$$S = (l(p_1) - q(p_1))^2 + (l(p_2) - q(p_2))^2$$
$$+ \ldots + (l(p_n) - q(p_n))^2$$
$$= (mp_1 + b - q_1)^2 + (mp_2 + b - q_2)^2$$
$$+ \ldots + (mp_n + b - q_n)^2.$$

We set the partial derivatives equal to 0.

$$S_m = 2p_1(mp_1 + b - q_1) + 2p_2(mp_2 + b - q_2)$$
$$+ \ldots + 2p_n(mp_n + b - q_n) = 0$$
$$S_b = 2(mp_1 + b - q_1) + 2(mp_2 + b - q_2)$$
$$+ \ldots + 2(mp_n + b - q_n) = 0$$

Gathering terms and dividing by 2, we get

$$(p_1^2 + p_2^2 + \ldots + p_n^2)m + (p_1 + p_2 + \ldots + p_n)b$$
$$= (p_1q_1 + p_2q_2 + \ldots + p_nq_n)$$
$$(p_1 + p_2 + \ldots + p_n)m + nb = (q_1 + q_2 + \ldots + q_n).$$

We can rewrite these equations more neatly using Σ-notation.

$$\left(\sum p_i^2\right)m + \left(\sum p_i\right)b = \sum p_iq_i$$
$$\left(\sum p_i\right)m + nb = \sum q_i$$

We conclude the following.

LEAST-SQUARES LINE

The **least-squares line** line through $(p_1, q_1), (p_2, q_2), \ldots, (p_n, q_n)$ has the form

$$q = mp + b,$$

where the constants m and b are the solutions to the system of equations

$$\left(\sum p_i^2\right)m + \left(\sum p_i\right)b = \sum p_i q_i$$

$$\left(\sum p_i\right)m + nb = \sum q_i.$$

Let us now return to the data on demand for real estate with which we began this section.

▼ **EXAMPLE 2**   Demand

Find a linear demand equation that best fits the following data, and use it to predict annual sales of homes priced at $140,000.

Price (Thousands of $)	Sales of New Homes This Year
$160	126
$180	103
$200	82
$220	75
$240	82
$260	40
$280	20

SOLUTION   We first calculate the following quantities needed in the formula.

$$\sum p_i^2 = \text{sum of squares of } p\text{-values} = 350{,}000$$

$$\sum p_i = \text{sum of } p\text{-values} = 1{,}540$$

$$\sum q_i = \text{sum of } q\text{-values} = 528$$

$$\sum p_i q_i = \text{sum of products} = 107{,}280$$

Thus, the system of equations we must solve is

$$350{,}000m + 1{,}540b = 107{,}280$$
$$1{,}540m + 7b = 528.$$

Solving this system by hand or with the aid of a calculator or computer, we obtain the following solution (rounded to four significant digits).

$$m \approx -0.7929, \ b \approx 249.9$$

Thus, our least-squares line is

$$q = -0.7929p + 249.9.$$

We can now use this equation to predict the annual sales of homes priced at $140,000, as we were asked to do. Remembering that p is the price in thousands of dollars, we set $p = 140$ and solve for q, getting $q \approx 139$. Thus, our model predicts that approximately 139 homes will have been sold in the range $140,000–$159,000.

Before we go on... We must remember that these figures were for sales in a *range* of prices. For instance, it would be extremely unlikely that 139 homes would have been sold at exactly $140,000. On the other hand, it does predict that, were we to place 139 homes on the market at $140,000, we could expect to sell them all.*

Figure 8.6.5 shows the original data, together with the least-squares line.

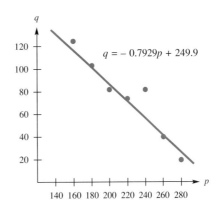

FIGURE 5

The formula for the least-squares line is simple enough that even many non-graphing calculators have built in the ability to find this line. Typi-

---

▼ *This is really a debatable point; we clearly wouldn't sell *any* if they were not perceived by prospective buyers are being worth $140,000 or if there were already a glut of $140,000 homes on the market. Thus, real-life planning is often more complicated than our simple model permits. Still, it remains a powerful predictive tool.

cally, you enter the data points one by one. As you do this, the calculator computes the sums Σp_i, Σq_i, Σp_1^2, and $\Sigma p_i q_i$. One press of a button then calculates m and b. Spreadsheets are also handy for calculating these coefficients and generally can calculate them automatically (under the name **linear regression**).

▶ **NOTE** In the above example, we used a linear demand model based on data in the range $160,000–$280,000 to predict demand at $140,000. This process is called **extrapolation**, since we have chosen a price *outside* the specified range. However, we have already seen in Chapter 1 the errors that arise when we try to extrapolate a linear model—the further we extrapolate the model, the less reliable it becomes. (This does not apply only to linear models, but to any mathematical model.) For instance, if in the previous example we had used the model to predict the demand for $350,000 homes, our linear model would predict a negative demand. Thus, extrapolation of any mathematical model beyond the range for which data are available must be done with caution. We are on firmer ground with **interpolation**, whereby we predict a value *inside* our range of data. ◀

▼ **EXAMPLE 3** Population Growth

The U.S. population, according to the U.S. Bureau of the Census, was as follows in the beginning of this century (all figures in thousands of people).

Year	1900	1910	1920	1930	1940	1950
Population	76,212	92,228	106,022	123,203	132,165	151,326

Fit a function of the form $A = Pe^{kt}$ to this data, where A is the population and t is the number of years after 1900. Use the model to predict the population in 1990.

SOLUTION The equation we are asked to fit is not linear, but there is a clever way to apply the techniques of this section anyway. If we take the natural logarithm of both sides of the equation, we get

$$\ln A = \ln P + kt$$

This equation is linear *in t and* $\ln A$. We can fit a least-squares line to the data we obtain by replacing the population figures with their natural logarithms.

t	0	10	20	30	40	50
$\ln A$	11.24127	11.43202	11.57140	11.72159	11.79181	11.92719

We then compute (rounding to four significant digits)

$$\sum t_i^2 = 5{,}500$$

$$\sum t_i = 150$$

$$\sum t_i \ln A_i = 1{,}765$$

$$\sum \ln A_i = 69.69.$$

This gives us the following system of equations.

$$5{,}500m + 150b = 1{,}765$$
$$150m + 6b = 69.69$$

The solution to this system is $m = 0.013$ and $b = 11.3$, so our least-squares line is

$$\ln A = 11.3 + 0.013t.$$

This means that $k = 0.013$, and $\ln P = 11.3$, so $P = 80{,}800$. Our exponential curve is then

$$A = 80{,}800 e^{0.013t} \quad \text{(population in thousands } t \text{ years after 1900)}.$$

Figure 8.6.6 shows the actual population figures and the curve we just found.

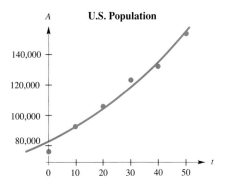

FIGURE 6

Our equation predicts the following population in 1990.

$$A = 80{,}800 e^{0.013(90)} \approx 260{,}000 \text{ thousand}$$

The actual population, according to the 1990 census, was 250,000 thousand people.

Before we go on... Although we extrapolated well beyond the given range of data, the answer was fairly accurate. Is there a moral to this? We doubt it.

Net Income of the Walt Disney Company

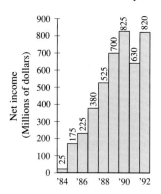

Find a least-squares linear model for this data. (Find profit *p* as a function of the year *t*, with $t = 0$ corresponding to 1980.) Use your model to estimate *Disney*'s profit (to the nearest million dollars) in 1993.

12. *Stock Prices* Repeat the previous exercise, but this time model *Walt Disney*'s stock price. (See the chart.)*

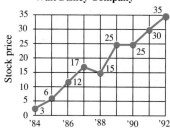

13. *Gross National Product* The GNP (in billions of dollars, to the nearest billion dollars) of the U.S. for the years 1960–1990 is given in the following table.†

1960	1970	1975	1980	1989	1990
515	1,016	1,598	2,732	5,201	5,465

Fit a least-squares line to this data, and graph both it and the data. Predict the GNP in the year 2000. (Answer to the nearest billion dollars).

14. *Gross National Product* Fit an exponential function $A = Pe^{kt}$ to the data in the preceding exercise. Take *t* to be the year since 1960, and round all constants in your answer to four significant digits. Which of these two exercises gives a better fit? (Compare their graphs with the data.)

▼ *Stock prices are rough estimates. Source: Ibid.
† Source: Bureau of Economic Analysis, U.S. Department of Commerce.

15. *Population* The population of the U.S., in thousands, is given by decade in the following table.*

1790	1800	1810	1820	1830	1840	1850
3,929	5,308	7,240	9,638	12,861	17,063	23,192

1860	1870	1880	1890	1900	1910	1920
31,443	38,558	50,189	62,980	76,212	92,228	106,022

1930	1940	1950	1960	1970	1980	1990
123,203	132,165	151,326	179,323	203,302	226,542	248,710

Fit an exponential equation $A = Pe^{kt}$ to this data, taking t to be the time since 1790, and graph both it and the data. Predict the population in the year 2000.

16. *Population* Repeat the preceding exercise, fitting a least-squares line. Which of these two exercises provides the better fit?

17. *Stock Market* The yearly highs of the Dow Jones Industrial Average (to the nearest point) for the decade of the 1980s are given in the following table.

1980	1981	1982	1983	1984	1985
1000	1024	1071	1287	1287	1553

1986	1987	1988	1989	1990
1956	2722	2184	2791	3000

Graph these data, judge whether a line or an exponential function provides the best fit, and then find a function of that kind to fit the data. What would you predict for the high in the year 2000?

18. *Stock Market* The yearly lows of the Dow Jones Industrial Average (to the nearest point) for the decade of the 1980s are given in the following table.

1980	1981	1982	1983	1984	1985
759	824	776	1027	1087	1185

1986	1987	1988	1989	1990
1502	1739	1879	2145	2365

▼ *Source: Bureau of the Census, U.S. Department of Commerce.

Graph these data, judge whether a line or an exponential function provides the best fit, and then find a function of that kind to fit the data. What would you predict for the low in the year 2000?

19. *Inflation*   The consumer price index (CPI) for various years from 1970 to 1990 is given in the following table* (1982–4 = 100).

1970	1975	1980	1985	1987	1988	1989	1990
38.8	53.8	82.4	107.6	113.6	118.3	124.0	130.7

Fit an exponential curve to these data. Predict the CPI for the year 2000.

20. *Inflation*   The buying power of a dollar for various years from 1970 to 1990 is given in the following table (calculated by taking 100/CPI, 1982–4 = 1.00).

1970	1975	1980	1985	1987	1988	1989	1990
2.58	1.86	1.21	0.929	0.880	0.845	0.806	0.765

Fit an exponential curve to these data. Predict the buying power of a dollar in the year 2000. How is your answer here related to the answer to the previous exercise?

Best-Fit Power Function   Exercises 21 and 22 are based on modeling using a "power" function of the form

$$y(x) = ax^b,$$

where a and b are constants to be determined. If we take the natural logarithm of both sides of this equation, we get

$$\ln y = \ln(ax^b) = \ln a + b \ln x,$$

showing a linear relationship between $\ln y$ and $\ln x$. Thus, given a sequence of data points (x_i, y_i), we can therefore fit a power curve $y = ax^b$ to it by finding the best-fit straight line for the points $(\ln x_i, \ln y_i)$. The slope is then b, and the y-intercept is $\ln a$.

21. *Best-Fit Demand Curve*   You have the following data showing the weekly sales at various prices for your fluorescent blue shower curtains.

Price, p	$10	$15	$20	$25	$30
Weekly Sales, q	30	15	10	5	3

Find the best-fit power function, and graph it together with the given data points.

22. *Best-Fit Demand Curve*   Repeat the preceding exercise using the following data for sales of your fluorescent pink shower curtains:

---

* Source: Bureau of Labor Statistics, U.S. Department of Labor.

Price, p	$10	$15	$20	$25	$30
Weekly Sales, q	500	400	300	250	220

23. *Cobb-Douglas Production Function* Recall that the Cobb-Douglas production function has the form

$$P(x, y) = Kx^a y^{1-a},$$

where P stands for the number of items produced per unit time, x is the number of employees, and y is the operating budget for that time. The numbers K and a are constants that depend on the particular situation we are looking at, with a between 0 and 1. Show how, by taking logs of both sides, you can obtain a linear relationship between $\ln(P/y)$ and $\ln(x/y)$. Explain how this relationship allows you to estimate K and a given a number of data points (x_i, y_i, P_i).

24. *Cobb-Douglas Production Function* Suppose that you monitor production levels in your factory, and over a period of seven days you get the following data for daily production.

Workers, x	100	110	90	100	95	105	110
Budget, y	10,000	9,000	9,000	12,000	11,000	9,500	10,000
Production, P	400	410	350	430	400	405	425

Fit a Cobb-Douglas production function of the form $P = Kx^a y^{1-a}$ to this data, and predict your daily production if you use 120 workers and a budget of $12,000. (See Exercise 23.)

COMMUNICATION AND REASONING EXERCISES

25. If the points $(x_1, y_1), (x_2, y_2), \ldots, (x_n, y_n)$ lie on a straight line, what can you say about the least-squares line associated with these points?

26. What can you say about the least-squares line associated with two data points?

27. If all but one of the points $(x_1, y_1), (x_2, y_2), \ldots, (x_n, y_n)$ lie on a straight line, does the least-squares line associated with these points pass through all but one of these points?

28. Must the least-squares line pass through at least one of the data points? Illustrate your answer with an example.

29. Why must care be taken when using mathematical models to extrapolate?

30. Model your mathematics test scores so far using a least-squares line, and use it to predict your score on the next test.

15.7 DOUBLE INTEGRALS

When discussing functions of one variable, we computed the area under a graph by integration. The analog for the graph of a function of two variables would be the *volume* under the graph, as in Figure 1.

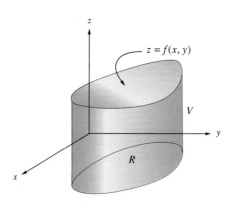

FIGURE 1

By analogy with the definite integral of a function of one variable, we make the following definition.

> **GEOMETRIC DEFINITION OF THE DOUBLE INTEGRAL**
> The **double integral of $f(x, y)$ over the region R in the xy-plane** is defined as
>
> (volume *above* the region R and under the graph of f) −
> (volume *below* the region R and above the graph of f).
>
> We denote the double integral of $f(x, y)$ over the region R by
>
> $$\iint_R f(x, y)\, dx\, dy.$$

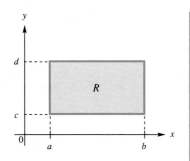

FIGURE 2

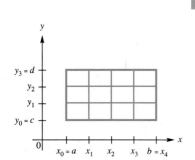

FIGURE 3

As we saw in the case of the definite integral of a function of one variable, we also desire an *algebraic*, or *numerical*, definition for two reasons: to make the mathematical definition more precise—so as not to rely on the notion of "volume"—and for direct computation of the integral using technology.

We start with the simplest case, when the region R is a rectangle $a \leq x \leq b$ and $c \leq y \leq d$. (See Figure 2.)

In order to compute the volume over R, we mimic what we did to find the area under the graph of a function of one variable. We break up the interval $[a, b]$ into m intervals all of width $\Delta x = (b - a)/m$, and we break up $[c, d]$ into n intervals all of width $\Delta y = (d - c)/n$. Figure 3 shows an example with $m = 4$ and $n = 3$.

This gives us mn rectangles defined by $x_{i-1} \leq x \leq x_i$ and $y_{j-1} \leq y \leq y_j$. Over one of these rectangles, f is approximately equal to its value at one corner, say, $f(x_i, y_j)$. The volume under f over this small rectangle is then approximately the volume of the rectangular brick shown in Figure 4.

This brick has height $f(x_i, y_j)$, and its base is Δx by Δy. Its volume is $f(x_i, y_j) \, \Delta x \, \Delta y$. Adding together the volumes of all of the bricks over the small rectangles in R, we get

$$\iint_R f(x, y) \, dx \, dy \approx \sum_{j=1}^{n} \sum_{i=1}^{m} f(x_i, y_j) \, \Delta x \, \Delta y.$$

This double sum is called a **double Riemann sum.** Algebraically, we define the double integral to be the limit of the Riemann sums as m and n go to infinity.

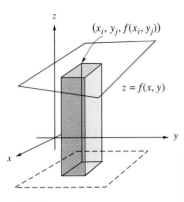

FIGURE 4

ALGEBRAIC DEFINITION OF THE DOUBLE INTEGRAL

$$\iint_R f(x, y) \, dx \, dy \approx \lim_{m \to \infty} \lim_{n \to \infty} \sum_{j=1}^{n} \sum_{i=1}^{m} f(x_i, y_j) \, \Delta x \, \Delta y$$

▶ **NOTE** This definition is adequate (the limit exists) when f is continuous. More elaborate definitions are needed for badly behaved functions.

This definition also gives us a clue about how to compute a double integral. The innermost sum is $\sum_{i=1}^{m} f(x_i, y_j) \, \Delta x$, which is a Riemann sum for $\int_a^b f(x, y_j) \, dx$. The innermost limit is therefore

$$\lim_{m \to \infty} \sum_{i=1}^{m} f(x_i, y_j) \, \Delta x = \int_a^b f(x, y_j) \, dx.$$

The outermost limit is then also a Riemann sum, and we get the following way of calculating double integrals. ◀

COMPUTING THE DOUBLE INTEGRAL OVER A RECTANGLE

If R is the rectangle $a \leq x \leq b$ and $c \leq y \leq d$, then

$$\iint_R f(x, y) \, dx \, dy = \int_c^d \left(\int_a^b f(x, y) \, dx \right) dy = \int_a^b \left(\int_c^d f(x, y) \, dy \right) dx.$$

▶ **NOTE** The second formula comes from switching the order of summation in the double sum. ◀

▼ **EXAMPLE 1**

Let $f(x, y) = xy$, and let R be the rectangle $0 \leq x \leq 1$ and $0 \leq y \leq 1$. Compute

$$\iint_R xy \, dx \, dy.$$

SOLUTION

$$\iint_R xy \, dx \, dy = \int_0^1 \int_0^1 xy \, dx \, dy$$

(We usually drop the parentheses like this.) We compute this **iterated integral** from the inside out. First, we compute

$$\int_0^1 xy \, dx.$$

To do this computation, we do as we did when finding partial derivatives: we treat y as a constant. This gives

$$\int_0^1 xy \, dx = \left[\frac{x^2}{2} \cdot y\right]_{x=0}^1 = \frac{1}{2}y - 0 = \frac{1}{2}y.$$

We can now calculate the outer integral.

$$\int_0^1 \int_0^1 xy \, dx \, dy = \int_0^1 \frac{1}{2}y \, dy = \left[\frac{1}{4}y^2\right]_0^1 = \frac{1}{4}$$

Before we go on... We could also reverse the order of integration.

$$\int_0^1 \int_0^1 xy \, dy \, dx = \int_0^1 \left[x \frac{y^2}{2}\right]_{y=0}^1 dx = \int_0^1 \frac{1}{2}x \, dx = \left[\frac{1}{4}x^2\right]_0^1 = \frac{1}{4}$$

▼ **EXAMPLE 2**

If R is the rectangle $0 \leq x \leq 2$, $1 \leq y \leq 3$, compute

$$\iint_R (e^x + y) \, dx \, dy.$$

SOLUTION

$$\iint_R (e^x + y) \, dx \, dy = \int_1^3 \int_0^2 (e^x + y) \, dx \, dy$$

$$= \int_1^3 \left[e^x + xy\right]_{x=0}^2 dy$$

$$= \int_1^3 [(e^2 + 2y) - (1 + 0)] \, dy$$

$$= \int_1^2 (2y + e^2 - 1) \, dy$$

$$= \left[y^2 + e^2 y - y\right]_1^2$$

$$= 2 + e^2$$

Before we go on... Again, we could have reversed the order of integration. You should do this for practice.

15.7 Double Integrals

Often, we need to integrate over regions R that are not rectangular. There are two cases that come up. First, we may have a region like the one shown in Figure 5.

In this region, the bottom and top sides are defined by functions $y = c(x)$ and $y = d(x)$ respectively, so that the whole region can be described by the inequalities $a \leq x \leq b$ and $c(x) \leq y \leq d(x)$. To evaluate a double integral over such a region, we have the following.

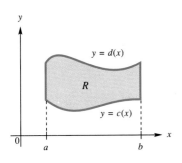

FIGURE 5

COMPUTING THE DOUBLE INTEGRAL OVER A NONRECTANGULAR REGION

If R is the region $a \leq x \leq b$ and $c(x) \leq y \leq d(x)$, then we integrate over R according to the following equation.

$$\iint_R f(x, y) \, dx \, dy = \int_a^b \int_{c(x)}^{d(x)} f(x, y) \, dy \, dx$$

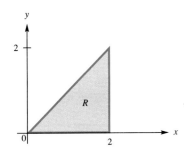

FIGURE 6

▼ **EXAMPLE 3**

If R is the triangle shown in Figure 6, compute

$$\iint_R x \, dx \, dy.$$

SOLUTION R is the region described by $0 \leq x \leq 2, 0 \leq y \leq x$.

$$\iint_R x \, dx \, dy = \int_0^2 \int_0^x x \, dy \, dx$$

$$= \int_0^2 \left[xy \right]_{y=0}^x dx$$

$$= \int_0^2 x^2 \, dx$$

$$= \left[\frac{x^3}{3} \right]_0^2$$

$$= \frac{8}{3}$$

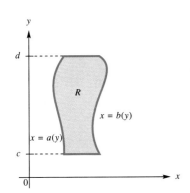

FIGURE 7

Another type of region is shown in Figure 7. This is the region described by $c \leq y \leq d$ and $a(y) \leq x \leq b(y)$. To evaluate a double integral over such a region, we have the following.

> **INTEGRATION OVER A NONRECTANGULAR REGION (CONT.)**
>
> If R is the region $c \leq y \leq d$ and $a(y) \leq x \leq b(y)$, then we integrate over R according to the following equation.
>
> $$\iint_R f(x, y) \, dx \, dy = \int_c^d \int_{a(y)}^{b(y)} f(x, y) \, dx \, dy$$

EXAMPLE 4

Redo Example 3, integrating in the opposite order.

SOLUTION We can do this if we can describe the region in Figure 6 in the way shown in Figure 7. In fact, it is the region $0 \leq y \leq 2$ and $y \leq x \leq 2$. A good way of seeing this description is to draw a horizontal line through the region, as in Figure 8.

The possible heights for such a line are $0 \leq y \leq 2$. The line extends from $x = y$ on the left to $x = 2$ on the right, so $y \leq x \leq 2$. We can now compute the integral.

$$\iint_R x \, dx \, dy = \int_0^2 \int_y^2 x \, dx \, dy$$

$$= \int_0^2 \left[\frac{x^2}{2}\right]_y^2 dy$$

$$= \int_0^2 \left(2 - \frac{1}{2}y^2\right) dy$$

$$= \left[2y - \frac{1}{6}y^3\right]_0^2$$

$$= \left(4 - \frac{8}{6}\right) - 0$$

$$= \frac{8}{3}$$

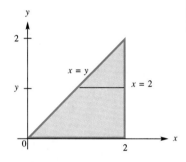

FIGURE 8

Before we go on... Many regions can be described in two different ways. Sometimes one description will be much easier to work with than the other, so it pays to try both.

There are many applications of double integrals besides finding volumes. We can also use them to find *averages*. Remember that the average of $f(x)$ on $[a, b]$ is given by $\int_a^b f(x) \, dx$ divided by $b - a$, the length of the interval.

AVERAGE OF A FUNCTION OF TWO VARIABLES

The **average of $f(x, y)$ on the region R** is

$$\bar{f} = \frac{1}{A} \iint_R f(x, y) \, dx \, dy.$$

Here, A is the area of R.

EXAMPLE 5

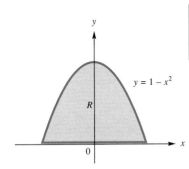

FIGURE 9

Find the average value of $f(x, y) = x + y$ over the region R shown in Figure 9.

SOLUTION It looks easiest to describe R as the region $-1 \leq x \leq 1$ and $0 \leq y \leq 1 - x^2$. We first compute the area of R, which is the area under the graph of $y = 1 - x^2$.

$$A = \int_{-1}^{1} (1 - x^2) \, dx = \frac{4}{3}$$

Now we compute the double integral.

$$\iint_R (x + y) \, dx \, dy = \int_{-1}^{1} \int_0^{1-x^2} (x + y) \, dy \, dx$$

$$= \int_{-1}^{1} \left[xy + \frac{y^2}{2} \right]_{y=0}^{1-x^2} dx$$

$$= \int_{-1}^{1} \left(x(1 - x^2) + \frac{1}{2}(1 - x^2)^2 \right) dx$$

$$= \int_{-1}^{1} \left(\frac{1}{2} + x - x^2 - x^3 + \frac{1}{2}x^4 \right) dx$$

$$= \left[\frac{x}{2} + \frac{x^2}{2} - \frac{x^3}{3} - \frac{x^4}{4} + \frac{x^5}{10} \right]_{-1}^{1}$$

$$= \frac{8}{15}$$

We can now compute the average.

$$\bar{f} = \frac{1}{(4/3)} \cdot \frac{8}{15} = \frac{2}{5}$$

Before we go on... If we wanted to describe the region with the order of integration reversed, it would be $0 \leq y \leq 1$ and $-\sqrt{1 - y} \leq x \leq \sqrt{1 - y}$. This leads to a more difficult integral to evaluate.

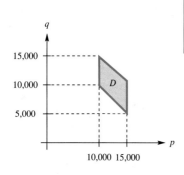

FIGURE 10

▼ **EXAMPLE 6** Average Revenue

Your marketing department estimates that if you price your new line of cars at p dollars per car, you will be able to sell between $q = 20{,}000 - p$ and $q = 25{,}000 - p$ cars in the first year. If you price the cars somewhere between $10{,}000$ and $15{,}000$, what is the average of all the possible revenues you could have in a year?

SOLUTION Revenue is given by $R = pq$ as usual, and we are told that $10{,}000 \le p \le 15{,}000$ and $20{,}000 - p \le q \le 25{,}000 - p$. This domain D of prices and demands is shown in Figure 10.

To average the revenue R over the domain D we need to compute the area A of D. Using either calculus or geometry, we get $A = 25{,}000{,}000$. We then need to integrate R over D.

$$\iint_D pq\, dp\, dq = \int_{10{,}000}^{15{,}000} \int_{20{,}000-p}^{25{,}000-p} pq\, dq\, dp$$

$$= \int_{10{,}000}^{15{,}000} \left[\frac{1}{2}pq^2\right]_{q=20{,}000-p}^{25{,}000-p} dp$$

$$= \frac{1}{2}\int_{10{,}000}^{15{,}000} [p(25{,}000 - p)^2 - p(20{,}000 - p)^2]\, dp$$

$$= \frac{1}{2}\int_{10{,}000}^{15{,}000} (225{,}000{,}000\,p - 10{,}000\,p^2)\, dp$$

$$= 3{,}072{,}916{,}666{,}666{,}667$$

We get the following average.

$$\bar{R} = \frac{3{,}072{,}916{,}666{,}666{,}667}{25{,}000{,}000}$$

$$\approx \$122{,}900{,}000 \text{ per year}$$

Before we go on... To check that this is a reasonable answer, notice that the revenues at the corners of the domain are $100{,}000{,}000$ per year, $150{,}000{,}000$ per year (at two corners), and $75{,}000{,}000$ per year. Some of these are smaller than the average and some larger, as we would expect. You should also check that the maximum possible revenue is $156{,}250{,}000$ per year (what is the minimum possible revenue?)

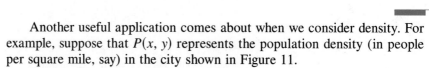

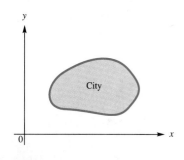

FIGURE 11

Another useful application comes about when we consider density. For example, suppose that $P(x, y)$ represents the population density (in people per square mile, say) in the city shown in Figure 11.

If we break the city up into small rectangles (for example, city blocks), then the population in the small rectangle $x_{i-1} \le x \le x_i$ and $y_{j-1} \le y \le y_j$ will be approximately $P(x_i, y_j)\, \Delta x\, \Delta y$. Adding up all of these population estimates, we get

$$\text{Total population} \approx \sum_{i=1}^{m} \sum_{j=1}^{n} P(x_i, y_j) \, \Delta x \, \Delta y.$$

Since this is a double Riemann sum, we get the following calculation of the population of the city when we take the limit as m and n go to infinity.

$$\text{Total population} = \iint_{\text{City}} P(x, y) \, dx \, dy$$

▼ **EXAMPLE 7**   Population

Squaresville is a city in the shape of a square 5 miles on a side. The population density at a distance of x miles east and y miles north of the southwest corner is $P(x, y) = x^2 + y^2$ thousand people per square mile. Find the total population of Squaresville.

SOLUTION   Squaresville is pictured in Figure 12, in which we put the origin in the southwest corner of the city.

To compute the total population, we integrate the population density over the city S.

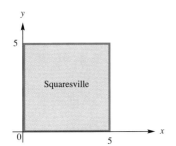

FIGURE 12

$$\begin{aligned}
\text{Population} &= \iint_S P(x, y) \, dx \, dy \\
&= \int_0^5 \int_0^5 (x^2 + y^2) \, dx \, dy \\
&= \int_0^5 \left[\frac{x^3}{3} + xy^2 \right]_{x=0}^{5} dy \\
&= \int_0^5 \left(\frac{125}{3} + 5y^2 \right) dy \\
&= \left[\frac{125}{3} y + \frac{5}{3} y^3 \right]_0^5 \\
&= \frac{1,250}{3} \approx 417 \text{ thousand people}
\end{aligned}$$

Before we go on... Note that the average population density is the total population divided by the area of the city, which is about 17 thousand people per square mile. Compare this calculation with the calculations of averages in the previous two examples.

15.7 EXERCISES

Computed the integrals in Exercises 1–16.

1. $\int_0^1 \int_0^1 (x - 2y)\, dx\, dy$

2. $\int_{-1}^1 \int_0^2 (2x + 3y)\, dx\, dy$

3. $\int_0^1 \int_0^2 (ye^x - x - y)\, dx\, dy$

4. $\int_1^2 \int_2^3 \left(\frac{1}{x} + \frac{1}{y}\right) dx\, dy$

5. $\int_0^3 \int_0^2 e^{x+y}\, dx\, dy$

6. $\int_0^1 \int_0^1 e^{x-y}\, dx\, dy$

7. $\int_0^1 \int_0^{2-y} x\, dx\, dy$

8. $\int_0^1 \int_0^{2-y} y\, dx\, dy$

9. $\int_{-1}^1 \int_{y-1}^{y+1} e^{x+y}\, dx\, dy$

10. $\int_0^1 \int_y^{y+2} \frac{1}{\sqrt{x+y}}\, dx\, dy$

11. $\int_0^1 \int_{-x^2}^{x^2} x\, dy\, dx$

12. $\int_1^4 \int_{-\sqrt{x}}^{\sqrt{x}} \frac{1}{x}\, dy\, dx$

13. $\int_0^1 \int_0^x e^{x^2}\, dy\, dx$

14. $\int_0^1 \int_0^{x^2} e^{x^3+1}\, dy\, dx$

15. $\int_1^3 \int_{1-x}^{8-x} \sqrt[3]{x+y}\, dy\, dx$

16. $\int_1^2 \int_{1-2x}^{x^2} \frac{x+1}{(2x+y)^2}\, dy\, dx$

In each of Exercises 17–24, integrate the given function over the indicated domain. (Remember that you often have a choice as to the order of integration.)

17. $f(x, y) = 2$

18. $f(x, y) = x$

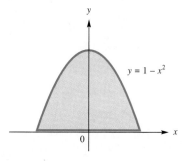

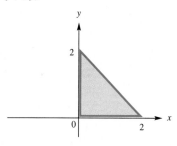

19. $f(x, y) = 1 + y$

20. $f(x, y) = e^{x+y}$

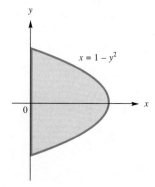

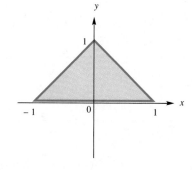

21. $f(x, y) = xy^2$ **22.** $f(x, y) = xy^2$

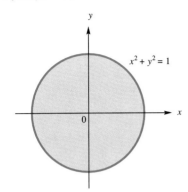
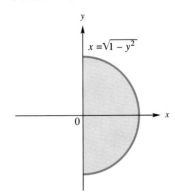

23. $f(x, y) = x^2 + y^2$ **24.** $f(x, y) = x^2$

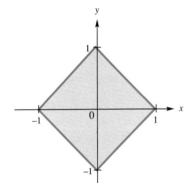
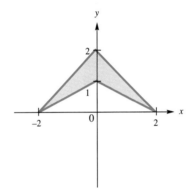

25–32. Find the average value of each function in Exercises 17–24.

In each of Exercises 33–40, sketch the region over which you are integrating, then write down the integral with the order of integration reversed (changing the limits of integration as necessary).

33. $\displaystyle\int_0^1 \int_0^{1-y} f(x, y)\, dx\, dy$ **34.** $\displaystyle\int_{-1}^1 \int_0^{1+y} f(x, y)\, dx\, dy$ **35.** $\displaystyle\int_{-1}^1 \int_0^{\sqrt{1+y}} f(x, y)\, dx\, dy$

36. $\displaystyle\int_{-1}^1 \int_0^{\sqrt{1-y^2}} f(x, y)\, dx\, dy$ **37.** $\displaystyle\int_0^2 \int_1^{4/x^2} f(x, y)\, dy\, dx$ **38.** $\displaystyle\int_1^{10} \int_0^{\ln x} f(x, y)\, dy\, dx$

39. $\displaystyle\int_0^2 \int_{2x}^4 f(x, y)\, dy\, dx$ **40.** $\displaystyle\int_{-2}^2 \int_{-\frac{1}{2}\sqrt{4-x^2}}^{\frac{1}{2}\sqrt{4-x^2}} f(x, y)\, dy\, dx$

41. Find the volume under the graph $z = 1 - x^2$ over the region $0 \le x \le 1$ and $0 \le y \le 2$.

42. Find the volume under the graph of $z = 1 - x^2$ over the triangle $0 \le x \le 1$, $0 \le y \le 1-x$.

43. Find the volume of the tetrahedron shown in the figure (its corners are $(0, 0, 0)$, $(1, 0, 0)$, $(0, 1, 0)$, and $(0, 0, 1)$).

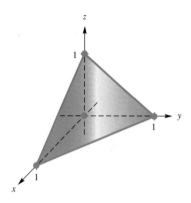

44. Find the volume of the tetrahedron with corners at $(0, 0, 0)$, $(a, 0, 0)$, $(0, b, 0)$, and $(0, 0, c)$.

APPLICATIONS

45. *Cobb-Douglas Production Function* The Cobb-Douglas production function for the Handy Gadget Company is given by

$$P = 10,000x^{0.3}y^{0.7},$$

where P is the number of gadgets it turns out per month, x is the number of employees at the company, and y is the monthly operating budget in thousands of dollars. Because the company hires part-time workers, it uses anywhere between 45 and 55 workers each month, and its operating budget varies from $8,000 to $12,000 per month. What is the average number of gadgets it can turn out per month? (Answer to the nearest gadget.)

46. *Cobb-Douglas Production Function* Repeat the preceding exercise using the production function

$$P = 10,000x^{0.7}y^{0.3}.$$

47. *Revenue* Your latest CD-ROM of clip-art is expected to sell between $q = 8,000 - p^2$ and $q = 10,000 - p^2$ copies if priced at p dollars. You plan to set the price between $40 and $50. What are the maximum and minimum possible revenues you can make? What is the average of all the possible revenues that you can make?

48. *Revenue* Your latest CD-ROM drive is expected to sell between $q = 180,000 - p^2$ and $q = 200,000 - p^2$ units if priced at p dollars. You plan to set the price between $300 and $400. What are the maximum and minimum possible revenues you can make? What is the average of all the possible revenues that you can make?

49. *Revenue* Your self-published novel has demand curves between $p = 15,000/q$ and $p = 20,000/q$. You expect to sell between 500 and 1,000 copies. What are the maximum and minimum possible revenues you can make? What is the average of all the possible revenues that you can make?

50. *Revenue* Your self-published book of poetry has demand curves between $p = 80,000/q^2$ and $p = 100,000/q^2$. You expect to sell between 50 and 100 copies. What are the maximum and minimum possible revenues you can make? What is the average of all the possible revenues that you can make?

51. *Population Density* The town of West Podunk is shaped like a rectangle 20 miles wide and 30 miles tall (see the figure). It has a population density of $P(x, y) = e^{-0.1(x+y)}$ hundred people x miles east and y

miles north of the southwest corner of town. What is the total population of the town?

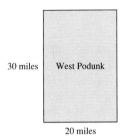

52. *Population Density* The town of East Podunk is shaped like a triangle with a base of 20 miles and a height of 30 miles (see the figure). It has a population density of $P(x, y) = e^{-0.1(x+y)}$ hundred people x miles east and y miles north of the southwest corner of town. What is the total population of the town?

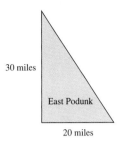

53. *Temperature* The temperature at the point (x, y) on the square with vertices (0, 0), (0, 1), (1, 0), and (1, 1) is given by $T(x, y) = x^2 + 2y^2$. Find the average temperature on the square.

54. *Temperature* The temperature at the point (x, y) on the square with vertices (0, 0), (0, 1), (1, 0), and (1, 1) is given by $T(x, y) = x^2 + 2y^2 - x$. Find the average temperature on the square.

COMMUNICATION AND REASONING EXERCISES

55. Explain how double integrals can be used to compute **(a)** the area between two curves in the xy-plane; **(b)** the volume of solids in 3-space.

56. Complete the following sentence: The first step in calculating an integral of the form $\int_a^b \int_{c/x}^{d(x)} f(x, y) \, dy \, dx$ is to evaluate the integral _____, obtained by holding _____ constant and integrating with respect to _____.

57. Show that if a, b, c, and d are constant, then
$$\int_a^b \int_c^d f(x)g(y) \, dx \, dy = \int_c^d f(x) \, dx \int_a^b g(y) \, dy.$$
Test this result on the integral $\int_0^1 \int_1^2 ye^x \, dx \, dy$.

58. If the units of $f(x, y)$ are bootlags per square meter, and x and y are given in meters, what are the units of $\int_a^b \int_{c/x}^{d(x)} f(x, y) \, dx \, dy$?

▸ You're the Expert

CONSTRUCTING A BEST-FIT DEMAND CURVE

You are a consultant for a new real estate development company that has purchased a large tract of land in Monmouth County, New Jersey, and intends to build a number of homes there. The company would like to target a single price bracket, and it has hired you to determine what its most profitable course of action might be. It typically sells new homes at 30% over cost, and it plans to build as many homes as the market will bear. Further, the company's market research team has come up with the following sales figures for new homes in Monmouth County over the past year.

Price (Thousands of $)	Sales of New Homes This Year
$50–$69	9
$70–$89	25
$90–$109	80
$110–$129	112
$130–$149	125
$150–$169	126
$170–$189	103
$190–$209	82
$210–$229	75
$230–$249	82
$250–$269	40
$270–$289	20

What would you advise the company to do?

You decide that your best strategy is to first find a demand equation based on these data. You first recast the data by replacing each price range with a single price in the middle of the range, coming up with the following table (in which p is in thousands of dollars) and graph (Figure 1).

p	60	80	100	120	140	160	180	200	220	240	260	280
q	9	25	80	112	125	126	103	82	75	82	40	20

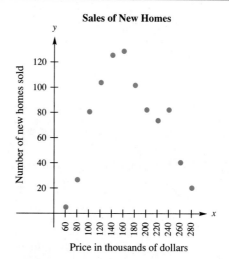

FIGURE 1

You decide at once that fitting a least-squares line is not the way to go. The plotted data reminds you of a parabola, not a straight line. So you decide instead to try a demand equation of the form $q = f(p) = ap^2 + bp + c$.

You now need to find values for a, b, and c that give the parabola that best fits the data.

As a first step, you decide to look at the general situation. You have a collection of data points $(p_1, q_1), (p_2, q_2), \ldots, (p_n, q_n)$, and desire the least-squares fit quadratic function $f(p) = ap^2 + bp + c$. Thus, you need to minimize the quantity

$$\begin{aligned} S &= (f(p_1) - q(p_1))^2 + (f(p_2) - q(p_2))^2 \\ &\quad + \ldots + (f(p_n) - q(p_n))^2 \\ &= (ap_1^2 + bp_1 + c - q_1)^2 + (ap_2^2 + bp_2 + c - q_2)^2 \\ &\quad + \ldots + (ap_n^2 + bp_n + c - q_n)^2. \end{aligned}$$

Noting that the variables are a, b, and c, you take partial derivatives.

$$\begin{aligned} S_a &= 2p_1^2(ap_1^2 + bp_1 + c - q_1) + 2p_2^2(ap_2^2 + bp_2 + c - q_2) \\ &\quad + \ldots + 2p_n^2(ap_n^2 + bp_n + c - q_n) \\ S_b &= 2p_1(ap_1^2 + bp_1 + c - q_1) + 2p_2(ap_2^2 + bp_2 + c - q_2) \\ &\quad + \ldots + 2p_n(ap_n^2 + bp_n + c - q_n) \\ S_c &= 2(ap_1^2 + bp_1 + c - q_1) + 2(ap_2^2 + bp_2 + c - q_2) \\ &\quad + \ldots + 2(ap_n^2 + bp_n + c - q_n) \end{aligned}$$

To locate the absolute minimum, you must set these equal to zero and solve for a, b, and c. You notice right away that these equations are linear in a, b, and c, so you group terms, divide by 2, and get

$$\begin{aligned} (p_1^4 + p_2^4 + \ldots &+ p_n^4)a + (p_1^3 + p_2^3 + \ldots + p_n^3)b \\ + (p_1^2 + p_2^2 + \ldots &+ p_n^2)c \\ &= (p_1^2 q_1 + p_2^2 q_2 + \ldots + p_n^2 q_n) \\ (p_1^3 + p_2^3 + \ldots &+ p_n^3)a + (p_1^2 + p_2^2 + \ldots + p_n^2)b \\ + (p_1 + p_2 + \ldots &+ p_n)c \\ &= (p_1 q_2 + p_2 q_2 + \ldots + p_n q_n) \\ (p_1^2 + p_2^2 + \ldots &+ p_n^2)a + (p_1 + p_2 + \ldots + p_n)b + nc \\ &= (q_1 + q_2 + \ldots + q_n). \end{aligned}$$

Using Σ-notation, these equations read more simply:

$$\left(\sum p_i^4\right)a + \left(\sum p_i^3\right)b + \left(\sum p_i^2\right)c = \sum p_i^2 q_i$$

$$\left(\sum p_i^3\right)a + \left(\sum p_i^2\right)b + \left(\sum p_i\right)c = \sum p_i q_i$$

$$\left(\sum p_i^2\right)a + \left(\sum p_i\right)b + nc = \sum q_i.$$

Now all you need to do is evaluate the coefficients and solve. Going back to the data, we have the following calculations.

$$\sum p_i^4 \approx 20{,}427{,}200{,}000 \qquad \sum p_i^3 = 88{,}128{,}000$$

$$\sum p_i^2 = 404{,}000 \qquad \sum p_i = 2{,}040$$

$$\sum p_i^2 q_i = 27{,}523{,}200 \qquad \sum p_i q_i = 148{,}760$$

$$\sum q_i = 879$$

Thus, your system of equations is

$$20{,}427{,}200{,}000a + 88{,}128{,}000b + 404{,}000c = 27{,}523{,}200$$
$$88{,}128{,}000a + 404{,}000b + 2{,}040c = 148{,}760$$
$$404{,}000a + 2{,}040b + 12c = 879.$$

Using your calculator, you obtain

$$a \approx -0.008626, \quad b \approx 2.921, \quad c \approx -132.9.$$

Thus, your demand equation is

$$q = -0.008626p^2 + 2.921p - 132.9.$$

Before going any further, you have your graphing calculator draw the graph of q superimposed on the data with which you started (Figure 2).

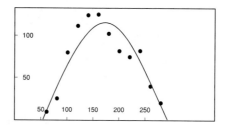

FIGURE 2

Not a bad fit! Now you get to work maximizing the profit. In order to do this, you need to find a formula giving the profit P as a function of the unit price p. Now, you know that the company can sell q homes at a unit price of p, where q is given by the formula $q = -0.008634p^2 + 2.924p - 133.1$. What you must compute is the *profit* it will obtain from the sale of these homes. Since it sells them at 30% over cost, it makes a profit of $30 on every $130 it charges. Since its total revenue will be

$$R = pq = p(-0.008626p^2 + 2.921p - 132.9),$$

its profit is $\frac{30}{130}$ of the revenue.

$$P = \frac{30}{130}(-0.008626p^3 + 2.921p^2 - 132.9p)$$

To maximize the profit, we set the derivative of profit equal to zero, and solve.

$$\frac{30}{130}(-0.025878p^2 + 5.842p - 132.9) = 0$$

Using the quadratic formula, we get two stationary points, $p = 200.1$ and $p = 25.67$. You then check that the lower value gives a local minimum, and it is the higher price that gives a local maximum. Thus, you report that according to your expert opinion, the company should target the $170,000–189,000 market range. Further, you inform the company that it can sell $q(200.1) \approx 106$ homes (assuming there are no competitors in Monmouth County!)

Exercises

1. Find the best *cubic* approximation $q = ap^3 + bp^2 + cp + d$ to the data in this section. Sketch its graph together with the data points. If you used this approximation, would you change your advice to the development company?

2. ***Investments in South Africa*** It is 1994, you are the CEO of a large chain of fast-food stores, and you are considering expanding to South Africa. You have decided that the most prudent policy would be to wait until the total number of U.S. companies with direct investment in South Africa is at least 500. The following chart shows the number of U.S. companies with direct investment in South Africa for the years 1986 through 1994.*
   (a) If you use a best-fit quadratic approximation of the number of companies as a function of the number of years since 1986, when should you invest in South Africa?
   (b) Comment on the long-term implications of your model.

▼ *Source: Investor Responsibility Research Center/*New York Times*, January 31, 1994, p. D1.

When we write an expression involving two or more of these operations, such as

$$2(3 - 5) + 4 \cdot 5,$$

or

$$\frac{2 \cdot 3^2 - 5}{4 - (-1)},$$

we agree to use the following rules to decide on the order in which we do the operations.

STANDARD ORDER OF OPERATIONS

1. **Parentheses and Fraction Bars** Calculate the values of all expressions inside parentheses or brackets first, working from the innermost parentheses out. When dealing with a fraction bar, calculate the numerator and denominator separately and then do the division.
   Examples:

   $$6(2 + [3 - 5] - 4) = 6(2 + (-2) - 4) = 6(-4) = -24$$

   $$\frac{(4 - 2)}{3(-2 + 1)} = \frac{2}{3(-1)} = \frac{2}{-3} = -\frac{2}{3}$$

2. **Exponents** Next, raise all numbers to the indicated powers.
   Examples:

   $$(2 + 4)^2 = 6^2 = 36$$
   $$(2 \cdot 3)^2 = 6^2 = 36$$
   $$2 \cdot 3^2 = 2 \times 9 = 18$$

   Note the distinction here.

   $$2\left(\frac{3}{4 - 5}\right)^2 = 2\left(\frac{3}{-1}\right)^2 = 2(-3)^2 = 2 \times 9 = 18$$

3. **Multiplication and Division** Next, do all the multiplications and divisions from left to right.
   Examples:

   $$6(2[3 - 5] \div 4 \cdot (-1)) \div 2 = 6(2(-2) \div 4 \cdot (-1)) \div 2$$
   $$= 6(-4 \div 4 \cdot (-1)) \div 2$$
   $$= 6((-1) \cdot (-1)) \div 2$$
   $$= 6 \cdot 1 \div 2 = 3$$

$$4\left(\frac{2(4-2)}{3(-2\cdot 5)}\right) = 4\left(\frac{2(2)}{3(-10)}\right) = 4\left(\frac{4}{-30}\right) = \frac{16}{-30} = -\frac{8}{15}$$

$$4\left(\frac{2(4^2\cdot 2)}{3(-2+5)}\right) = 4\left(\frac{2(16\cdot 2)}{3(3)}\right) = 4\left(\frac{64}{9}\right) = \frac{256}{9}$$

4. **Addition and Subtraction** Last, do the remaining additions and subtractions from left to right.
Examples:

$$2(3-5)^2 + 6 - 1 = 2(-2)^2 + 6 - 1$$
$$= 2(4) + 6 - 1 = 8 + 6 - 1 = 13$$

$$\left(\frac{1}{2}\right)^2 - (-1)^2 + 4 = \frac{1}{4} - 1 + 4 = -\frac{3}{4} + 4 = \frac{13}{4}$$

CALCULATORS, GRAPHING CALCULATORS, AND COMPUTERS

Any good calculator will respect the standard order of operations. However, we must be careful with division and exponentiation and often must use parentheses. The following table gives some examples of simple mathematical expressions and their calculator equivalents in the functional format used in most graphing calculators and computer programs. (These expressions would be entered in a scientific calculator in essentially the same way.)

Here is one more fact about calculators (and calculations in general): A calculation can never give you an answer more accurate than the numbers you start with. As a general rule of thumb, if you have numbers measuring something in the real world (time, length, or gross domestic product, for example) and these numbers are accurate only to a certain number of digits, then any calculations you do with them will be accurate only to that many digits (at best). For example, if someone tells you that a rectangle has sides of length 2.2 ft and 4.3 ft, you can say that the area is (approximately) 9.5 sq ft, rounding to two significant digits. If you report that the area is 9.46 sq ft, as your calculator will tell you, the third digit is probably suspect. We shall usually use data that are accurate to three or four digits, and round our answers to that size.

Mathematical Expression	Calculator Equivalent	Comments
$\dfrac{2}{3-5}$	2/(3−5)	Note the use of parentheses instead of the fraction bar. If we omit the parentheses, we get the expression shown next.
$\dfrac{2}{3} - 5$	2/3−5	The calculator automatically follows the usual order of operations.
$\dfrac{2}{3 \times 5}$	2/(3*5)	Putting the denominator in parentheses ensures that the multiplication is carried out first. The symbol "*" is usually used for multiplication in graphing calculators and computers.
$\dfrac{2}{3} \times 5$	(2/3)*5	Putting the fraction in parentheses ensures that it is calculated first. Some calculators will interpret 2/3*5 as $\dfrac{2}{3 \times 5}$ but 2/3(5) as $\dfrac{2}{3} \times 5$.
$\dfrac{2-3}{4+5}$	(2−3)/(4+5)	Note once again the use of parentheses in place of the fraction bar.
2^3	2^3	The caret "^" is commonly used to denote exponentiation.
2^{3-4}	2^(3−4)	Be careful to use parentheses to tell the calculator where the exponent ends. Enclose the *entire exponent* in parentheses.
$2^3 - 4$	2^3−4	Without parentheses, the calculator will follow the usual order of operations: exponentiation first, then subtraction.
3×2^{-4}	3*2^(−4)	The shorter minus sign stands for the negation sign, which on some calculators is a separate key.
$2^{-4 \times 3} \times 5$	2^(−4*3)*5	Note once again how parentheses enclose the entire exponent.
$\dfrac{2^{3-2} \times 5}{2-7}$	2^(3 − 2)*5/(2 − 7) or (2^(3 − 2)*5)/(2 − 7)	Notice again the use of parentheses to hold the denominator together. We could also have enclosed the numerator in parentheses, although this is optional (why?).
$\dfrac{2^{3-6}+1}{2-4^3}$	(2^(3 − 6)+1)/(2 − 4^3)	Here, it is necessary to enclose each of the numerator and denominator in parentheses.
$2^{3-6} + \dfrac{1}{2} - 4^3$	2^(3 − 6)+1/2 − 4^3	This is the effect of leaving out the parentheses around the numerator and denominator in the previous expression.

▶ A-1 EXERCISES

Calculate each of the expressions in Exercises 1–14, expressing your answer as a whole number or a fraction in lowest terms:

1. $2(4 + (-1))(2 \cdot -4)$
2. $3 + ([4 - 2] \cdot 9)$
3. $\dfrac{3 + ([3 + (-5)])}{3 - 2 \times 2}$
4. $\dfrac{12 - (1 - 4)}{2(5 - 1) \cdot 2 - 1}$
5. $2 \cdot (-1)^2 \div 2$
6. $2 + 4 \cdot 3^2$
7. $2 \cdot 4^2 + 1$
8. $1 - 3 \cdot (-2)^2 \times 2$
9. $\dfrac{3 - 2(-3)^2}{-6(4 - 1)^2}$
10. $\dfrac{1 - 2(1 - 4)^2}{2(5 - 1)^2 \cdot 2}$
11. $3\left(\dfrac{-2 \cdot 3^2}{-(4 - 1)^2}\right)$
12. $-\left(\dfrac{8(1 - 4)^2}{-9(5 - 1)^2}\right)$
13. $3\left(1 - \left(-\dfrac{1}{2}\right)^2\right)^2 + 1$
14. $3\left(\dfrac{1}{9} - \left(\dfrac{2}{3}\right)^2\right)^2 + 1$

Convert each of the expressions in Exercises 15–30 into its calculator equivalent as in the table in the text.

15. $3 \times (2 - 5)$
16. $4 + \dfrac{5}{9}$
17. $\dfrac{3}{2 - 5}$
18. $\dfrac{4 - 1}{3}$
19. $\dfrac{3 - 1}{8 + 6}$
20. $3 + \dfrac{3}{2 - 9}$
21. $3 - \dfrac{4 + 7}{8}$
22. $\dfrac{4 \times 2}{\left(\dfrac{2}{3}\right)}$
23. $\dfrac{\left(\dfrac{2}{3}\right)}{5}$
24. $\dfrac{2}{\left(\dfrac{3}{5}\right)}$
25. $3^{4-5} \times 6$
26. $\dfrac{2}{3 + 5^{7-9}}$
27. $3\left(1 + \dfrac{4}{100}\right)^{-3}$
28. $3\left(\dfrac{1 + 4}{100}\right)^{-3}$
29. $3\left(1 - \left(-\dfrac{1}{2}\right)^2\right)^2 + 1$
30. $3\left(\dfrac{1}{9} - \left(\dfrac{2}{3}\right)^2\right)^2 + 1$

▶ A.2 INTEGER EXPONENTS

POSITIVE EXPONENTS

If a is any real number and n is any *positive integer*, then by a^n we mean the quantity $a \cdot a \cdots a$ (n times); thus, $a^1 = a$, $a^2 = a \cdot a$,

$a^5 = a \cdot a \cdot a \cdot a \cdot a$. Here are some examples with actual numbers:

$$3^2 = 9, \quad 2^3 = 8, \quad 0^{34} = 0, \quad (-1)^5 = -1.$$

In the expression a^n, the number n is called the *exponent*, and the number a is called the *base*. The following rules show how to combine such expressions.

EXPONENT IDENTITIES

(a) $a^m a^n = a^{m+n}$ *Example:* $2^3 2^2 = 2^5 = 32$

(b) $\dfrac{a^m}{a^n} = a^{m-n}$ if $m > n$ and $a \neq 0$

                           Example: $\dfrac{4^3}{4^2} = 4^{3-2} = 4^1 = 4$

(c) $(a^n)^m = a^{nm}$ *Example:* $(3^2)^2 = 3^4 = 81$

(d) $(ab)^n = a^n b^n$ *Example:* $(4 \cdot 2)^2 = 4^2 2^2 = 64$

(e) $\left(\dfrac{a}{b}\right)^n = \dfrac{a^n}{b^n}$ if $b \neq 0$ *Example:* $\left(\dfrac{4}{3}\right)^2 = \dfrac{4^2}{3^2} = \dfrac{16}{9}$

▶ **CAUTION**

(a) In identities (a) and (b), the bases of the expressions must be the same. For example, rule (a) gives $3^2 3^4 = 3^6$ but does *not* apply to $3^2 4^2$.

(b) People sometimes invent their own identities, such as $a^m + a^n = a^{m+n}$, which is wrong! (If you don't believe this, try it with $a = m = n = 1$.) If you wind up with something like $2^3 + 2^4$, you are stuck with it—there are no identities around to simplify it further. ◀

▼ **EXAMPLE 1**   Positive Exponents

a. $10^2 10^3 = 10^{2+3}$ By (a)

          $= 10^5 = 100{,}000$

b. $\dfrac{4^6}{4^3} = 4^{6-3}$ By (a)

      $= 4^3 = 64$

c. $\dfrac{(x^2)^3}{x^3} = \dfrac{x^6}{x^3}$ By (c)

$= x^{6-3}$ By (b)

$= x^3$

d. $\dfrac{(x^4 y)^3}{y} = \dfrac{(x^4)^3 y^3}{y}$ By (d)

$= \dfrac{x^{12} y^3}{y}$ By (c)

$= x^{12} y^{3-1}$ By (b)

$= x^{12} y^2$

Q Just where do these identities come from?

A Let us look at them one at a time.

(a) $a^m a^n = a^{m+n}$

We can see why this works by looking at an example:* Let's take $a = 3$, $m = 2$, and $n = 3$. Then the left-hand side is

$$3^2 3^3 = (3 \cdot 3)(3 \cdot 3 \cdot 3) = 3 \cdot 3 \cdot 3 \cdot 3 \cdot 3,$$

while the right-hand side is

$$3^{2+3} = 3^5 = 3 \cdot 3 \cdot 3 \cdot 3 \cdot 3,$$

the same thing.

(b) $\dfrac{a^m}{a^n} = a^{m-n}$

We check this by example again, this time with $a = 3$, $m = 5$, and $n = 2$. Then the left-hand side is

$$\dfrac{3^5}{3^2} = \dfrac{3 \cdot 3 \cdot 3 \cdot \cancel{3} \cdot \cancel{3}}{\cancel{3} \cdot \cancel{3}} = 3 \cdot 3 \cdot 3,$$

while the right-hand side is

$$3^{5-2} = 3^3 = 3 \cdot 3 \cdot 3,$$

the same thing.

(c) $(a^n)^m = a^{nm}$

Let's take $a = 3$, $m = 2$, and $n = 3$. Then the left-hand side is

$$(3^2)^3 = (3 \cdot 3)^3 = (3 \cdot 3)(3 \cdot 3)(3 \cdot 3),$$

while the right-hand side is

$$3^6 = 3 \cdot 3 \cdot 3 \cdot 3 \cdot 3 \cdot 3,$$

▼ *Although we are not giving a formal proof, examples are where the ideas for formal proofs usually come from, and it is the ideas we're interested in here.

while the right-hand side is
$$3^6 = 3 \cdot 3 \cdot 3 \cdot 3 \cdot 3 \cdot 3,$$
the same thing.

(d) $(ab)^n = a^n b^n$ and $\left(\dfrac{a}{b}\right)^n = \dfrac{a^n}{b^n}$

We'll check the first by means of an example, and leave the second for you. Let's take $a = 3$, $b = 5$, and $n = 4$. The left-hand side is then
$$(3 \cdot 5)^4 = (3 \cdot 5)(3 \cdot 5)(3 \cdot 5)(3 \cdot 5),$$
while the right-hand side is
$$3^4 5^4 = (3 \cdot 3 \cdot 3 \cdot 3)(5 \cdot 5 \cdot 5 \cdot 5),$$
which is the same thing.

Thus, these identities are really just observations about how multiplication works.

NEGATIVE AND ZERO EXPONENTS

It turns out to be very useful to allow ourselves to use exponents that are not positive integers. These are dealt with by the following definition.

NEGATIVE EXPONENTS

If a is any real number other than zero and n is any positive integer, then we define
$$a^{-n} = \frac{1}{a^n} = \frac{1}{a \cdot a \cdots a} \quad (n \text{ times}).$$

ZERO EXPONENTS

If a is any real number other than zero, then we define
$$a^0 = 1.$$

Thus, for instance,
$$2^{-3} = \frac{1}{2^3} = \frac{1}{8} \quad \text{and} \quad 3^0 = 1.$$

EXAMPLE 2 Negative and Zero Exponents

(a) $3^0 = 1$

(b) $1{,}000{,}000^0 = 1$

(c) $3^{-3} = \dfrac{1}{3^3} = \dfrac{1}{27}$

(d) $x^{-1} = \dfrac{1}{x^1} = \dfrac{1}{x}$

(e) $(-3)^{-2} = \dfrac{1}{(-3)^2} = \dfrac{1}{9}$

(f) $1^{-27} = \dfrac{1}{1^{27}} = 1$

(g) $y^7 y^{-2} = y^7 \dfrac{1}{y^2} = y^5$

Q Where do these strange definitions come from?

A From the desire to keep the exponent identities true even if we allow 0 or negative exponents. For example, if the first exponent identity, $a^m a^n = a^{m+n}$, is to remain true, we must be able to say

$$a^0 a^n = a^{0+n} = a^n.$$

Dividing both sides by a^n gives $a^0 = 1$. So this *forces* us to say that $a^0 = 1$. On the other hand, the same identity tells us that

$$a^{-n} a^n = a^{-n+n} = a^0 = 1.$$

Dividing both sides by a^n gives $a^{-n} = 1/a^n$. Again, this forces us to say that this is what we mean by a^{-n}.

It turns out that once we make these agreements, all of the exponent identities remain true as stated, without any restrictions on the exponents m and n, as long as we make sure that $a \neq 0$ where having $a = 0$ would not make sense, namely, where 0^0 or 0 raised to a negative power would appear.*

---

* Trying to raise 0 to a negative power would involve division by 0, so this is not defined. But what about raising 0 to the 0 power? If 0^0 had a definite value, then this value should be extremely close to the value of, say, $0^{0.000001}$. But 0 raised to any power is 0. Thus, 0^0 should be 0. On the other hand it should also be extremely close to $(0.000001)^0$, which is 1, since any nonzero number raised to 0 is 1. How can it be close to both 0 and 1?

EXAMPLE 3

Simplify $\dfrac{x^4 y^{-3}}{x^5 y^2}$, and express the answer using no negative exponents.

SOLUTION

$$\dfrac{x^4 y^{-3}}{x^5 y^2} = x^{4-5} y^{-3-2} = x^{-1} y^{-5} = \dfrac{1}{xy^5}$$

EXAMPLE 4

Simplify $\left(\dfrac{x^{-1}}{x^2 y}\right)^5$, and express the answer using no negative exponents.

SOLUTION

$$\left(\dfrac{x^{-1}}{x^2 y}\right)^5 = \dfrac{(x^{-1})^5}{(x^2 y)^5} = \dfrac{x^{-5}}{x^{10} y^5} = \dfrac{1}{x^{15} y^5}$$

▶ A.2 EXERCISES

Evaluate the expressions in Exercises 1–16.

1. 3^3
2. $(-2)^3$
3. $-(2 \cdot 3)^2$
4. $(4 \cdot 2)^2$
5. $\left(\dfrac{-2}{3}\right)^2$
6. $\left(\dfrac{3}{2}\right)^3$
7. $(-2)^{-3}$
8. -2^{-3}
9. $\left(\dfrac{1}{4}\right)^{-2}$
10. $\left(\dfrac{-2}{3}\right)^{-2}$
11. $2 \cdot 3^0$
12. $3 \cdot (-2)^0$
13. $2^3 2^2$
14. $3^2 3$
15. $2^2 2^{-1} 2^4 2^{-4}$
16. $5^2 5^{-3} 5^2 5^{-2}$

Simplify each expression in Exercises 17–30, expressing your answer with no negative exponents.

17. $x^3 x^2$
18. $x^4 x^{-1}$
19. $-x^2 x^{-3} y$
20. $-xy^{-1} x^{-1}$
21. $\dfrac{x^3}{x^4}$
22. $\dfrac{y^5}{y^3}$
23. $\dfrac{x^2 y^2}{x^{-1} y}$
24. $\dfrac{x^{-1} y}{x^2 y^2}$
25. $\dfrac{(xy^{-1} z^3)^2}{x^2 y z^2}$
26. $\dfrac{x^2 y z^2}{(xyz^{-1})^{-1}}$
27. $\left(\dfrac{xy^{-2} z}{x^{-1} z}\right)^3$
28. $\left(\dfrac{x^2 y^{-1} z^0}{xyz}\right)^2$
29. $\left(\dfrac{x^{-1} y^{-2} z^2}{xy}\right)^{-2}$
30. $\left(\dfrac{xy^{-2}}{x^2 y^{-1} z}\right)^{-3}$

▶ A.3 RADICALS AND RATIONAL EXPONENTS

If a is any nonnegative real number, then its **square root** is the nonnegative number whose square is a. For example, the square root of 16 is 4, since $4^2 = 16$. We write the square root of n as $\sqrt{n}$. It is important to remember that $\sqrt{n}$ is never negative. Thus, for instance, $\sqrt{9}$ is 3, and not -3, even though $(-3)^2 = 9$. If we want to speak of the "negative square root" of 9, we write it as $-\sqrt{9} = -3$. If we wanted to write both square roots at once, we would write $\pm\sqrt{9} = \pm 3$.

The **cube root** of a real number a is the number whose cube is a. The cube root of a is written $\sqrt[3]{a}$ so that, for example, $\sqrt[3]{8} = 2$ (since $2^3 = 8$). Note that we can take the cube root of any number, whether positive, negative, or zero. For instance, the cube root of -8 is $\sqrt[3]{-8} = -2$, since $(-2)^3 = -8$. Unlike square roots, the cube root of a number may be negative. In fact, the cube root of a always has the same sign as a.

Higher roots are defined similarly. The **fourth root** of the *nonnegative* number a is defined as the nonnegative number whose fourth power is a, written $\sqrt[4]{a}$. The **fifth root** of any number a is the number whose fifth power is a, and so on.

▶ NOTE

We cannot take an even-numbered root of a negative number, but we can take an odd-numbered root of any number. Even roots are always positive, while odd roots have the same sign as the number we start with. ◀

▼ **EXAMPLE 1**

(a) $\sqrt{4} = 2$, since $2^2 = 4$.
(b) $\sqrt{16} = 4$, since $4^2 = 16$.
(c) $\sqrt{1} = 1$, since $1^2 = 1$.
(d) $\sqrt{2}$ is not a whole number, but is approximately equal to 1.414213562.
(e) $\sqrt{1+1} = \sqrt{2} \approx 1.414213562$. Note that we first added the quantities under the square root sign, and then took the square root. In general, $\sqrt{a+b}$ means the square root of the *quantity* $(a + b)$. The radical sign acts like a pair of parentheses or a fraction bar, telling us to evaluate what is inside before taking the root.
(f) $\sqrt[3]{27} = 3$, since $3^3 = 27$.
(g) $\sqrt[3]{-64} = -4$, since $(-4)^3 = -64$.
(h) $\sqrt[4]{16} = 2$, since $2^4 = 16$, but $\sqrt[4]{-16}$ is not defined.
(i) $\sqrt[5]{-1} = -1$, since $(-1)^5 = -1$. Similarly, $\sqrt[n]{-1} = -1$ if n is any odd number.

▶ CAUTION

It is important to remember that $\sqrt{a+b}$ is *not* equal to $\sqrt{a} + \sqrt{b}$ (consider $a = b = 1$, for example). Equating these expressions is a common error, so be careful! ◀

What *is* true is the following.

> **RADICALS OF PRODUCTS AND QUOTIENTS**
>
> If a and b are any real numbers (nonnegative in the case of even-numbered roots), then
>
> (1) $\sqrt[n]{ab} = \sqrt[n]{a}\sqrt[n]{b}$;
>
> (2) $\sqrt[n]{\dfrac{a}{b}} = \dfrac{\sqrt[n]{a}}{\sqrt[n]{b}}$ if $b \neq 0$.
>
> *Examples:* $\sqrt{9 \cdot 4} = \sqrt{9}\sqrt{4} = 3 \times 2 = 6$.
> (We could also have calculated this as follows: $\sqrt{9 \cdot 4} = \sqrt{36} = 6$)
>
> $\sqrt{\dfrac{9}{4}} = \dfrac{\sqrt{9}}{\sqrt{4}} = \dfrac{3}{2}$.
>
> Rule 1 is similar to the rule $(a \cdot b)^2 = a^2 \cdot b^2$ for the square of a product, and Rule 2 is similar to the rule $(a/b)^2 = a^2/b^2$ for the square of a quotient.

EXAMPLE 2

(a) $\sqrt{9 + 16} = \sqrt{25} = 5$ (It is *not* equal to $\sqrt{9} + \sqrt{16}$, which is $3 + 4 = 7$.)

(b) $\sqrt{4 \cdot 25} = \sqrt{100} = 10$. Also, $\sqrt{4 \cdot 25} = \sqrt{4}\sqrt{25} = 2 \times 5 = 10$.

(c) $2\sqrt{3 - 1} = 2\sqrt{2} \approx 2.828427$

(d) $\sqrt{4(3 + 13)} = \sqrt{4(16)} = \sqrt{4}\sqrt{16} = 2 \cdot 4 = 8$.

(e) If x is nonnegative, then $\sqrt{x^2} = x$.

(f) If x is negative, then $\sqrt{x^2} = -x$. For instance, $\sqrt{(-3)^2} = \sqrt{9} = 3 = -(-3)$.

(g) In general, $\sqrt{x^2} = |x|$, the **absolute value of x,** which is the nonnegative number with the same value as x. For instance, $|-3| = 3, |3| = 3$, and $|0| = 0$.

(h) In general, $\sqrt[n]{x^n} = x$ if n is odd, and $\sqrt[n]{x^n} = |x|$ if n is even.

(i) Whether or not x is negative, $\sqrt{x^4} = x^2$. (Why?)

(j) If x is nonnegative, then
$$\sqrt{x^3} = \sqrt{x^2 \cdot x} = \sqrt{x^2}\sqrt{x} = x\sqrt{x},$$
and
$$\sqrt{x^5} = \sqrt{x^4 \cdot x} = \sqrt{x^4}\sqrt{x} = x^2\sqrt{x}.$$

(k) $\sqrt{\dfrac{x^2+y^2}{z^2}} = \dfrac{\sqrt{x^2+y^2}}{\sqrt{z^2}} = \dfrac{\sqrt{x^2+y^2}}{z}$, if z is positive. (We can't simplify the numerator any further.)

(l) $\sqrt[3]{-216} = \sqrt[3]{(-27)8} = \sqrt[3]{-27}\sqrt[3]{8} = (-3)2 = -6$

EXAMPLE 3   Solving Equations

(a) Solve the equation $x^2 - 4 = 0$.

Adding 4 to both sides gives $x^2 = 4$. Since this says that the square of x is 4, the number x is either 2 or -2 since squaring either one gives 4. Thus, the solution is $x = \pm 2$. (We'll see later that the solution can also be obtained by factoring.)

(b) Solve for x: $x^2 - \tfrac{1}{2} = 0$.

Adding $\tfrac{1}{2}$ to both sides gives $x^2 = \tfrac{1}{2}$. Thus,
$$x = \pm\sqrt{\dfrac{1}{2}} = \pm\dfrac{1}{\sqrt{2}}.$$

(c) Solve the equation $x^3 + 64 = 0$.

Subtracting 64 from both sides of the equation gives $x^3 = -64$. Thus, $x = \sqrt[3]{-64} = -4$.

RATIONAL EXPONENTS

We already know what we mean by expressions such as x^4 and a^{-6}. The next step is to make sense of non-integral exponents as in $a^{1/2}$, 3^π, and similar beasts. First we look at rational exponents—that is, exponents of the form p/q with p and q integers.

Q What should we mean by $a^{1/2}$?

A The overriding concern here is that the exponent identities should remain true. In this case the identity to look at is the one that says that $(a^m)^n = a^{mn}$. This identity tells us that
$$(a^{1/2})^2 = a^1 = a.$$

That is, $a^{1/2}$, when squared, gives us a. But that must mean that $a^{1/2}$ is the *square root* of a—that is,
$$a^{1/2} = \sqrt{a}.$$

Q If q is a positive integer, what should we mean by $a^{1/q}$?

A Do as we did above, replacing 2 with q:
$$(a^{1/q})^q = a^1 = a,$$
so
$$a^{1/q} = \sqrt[q]{a}, \text{ the } q\text{th root of } a.$$

Notice that if a is negative, this makes sense only for q odd. To avoid this problem we usually stick to positive a.

Q If p and q are integers (q being positive), what should we mean by $a^{p/q}$?

A By the exponent identities
$$a^{p/q} = (a^p)^{1/q} = (a^{1/q})^p,$$
which gives us the following.

CONVERSION BETWEEN RATIONAL EXPONENTS AND RADICALS

If a is any nonnegative number, then
$$a^{p/q} = \sqrt[q]{a^p} = (\sqrt[q]{a})^p.$$

In particular,
$$a^{1/q} = \sqrt[q]{a}, \text{ the } q\text{th root of } a.$$

From left to right, we say that we are rewriting the expression in **radical form.** From right to left, we are rewriting the expression in **exponential form.** Again, if a is negative all of this makes sense only if q is odd. Also, the exponent is called a *rational exponent* since it is a rational number, p/q.

▼ **EXAMPLE 4**

$$4^{3/2} = (\sqrt{4})^3 = 2^3 = 8$$

$$8^{2/3} = (\sqrt[3]{8})^2 = 2^2 = 4$$

$$9^{-3/2} = \frac{1}{9^{3/2}} = \frac{1}{(\sqrt{9})^3} = \frac{1}{3^3} = \frac{1}{27}$$

All of the exponent identities continue to work when we allow rational exponents. In other words, we are free to use all the exponent identities even if the exponents are not integers.

EXAMPLE 5

Simplify $2^2 2^{7/2}$.

SOLUTION

$$\begin{aligned}
2^2 2^{7/2} &= 2^{2+\frac{7}{2}} \\
&= 2^{\frac{11}{2}} \\
&= 2^{5\frac{1}{2}} \\
&= 2^{5+\frac{1}{2}} \\
&= 2^5 \cdot 2^{\frac{1}{2}} \\
&= 2^5 \sqrt{2}
\end{aligned}$$

EXAMPLE 6

Simplify $\dfrac{(x^3)^{5/3}}{x^3}$.

SOLUTION $\dfrac{(x^3)^{5/3}}{x^3} = \dfrac{x^5}{x^3} = x^2$

EXAMPLE 7

Simplify $\dfrac{\sqrt{3}}{\sqrt[3]{3}}$.

SOLUTION $\dfrac{\sqrt{3}}{\sqrt[3]{3}} = \dfrac{3^{1/2}}{3^{1/3}} = 3^{\frac{1}{2}-\frac{1}{3}} = 3^{\frac{1}{6}} = \sqrt[6]{3}$

EXAMPLE 8

Simplify $\sqrt[4]{a^6}$.

SOLUTION $\sqrt[4]{a^6} = a^{6/4} = a^{3/2} = a \cdot a^{1/2} = a\sqrt{a}$

▶ **A.3 EXERCISES**

Evaluate the expressions in Exercises 1–16, rounding your answer to four significant digits where necessary.

1. $\sqrt{4}$
2. $\sqrt{5}$
3. $\sqrt{\dfrac{1}{4}}$
4. $\sqrt{\dfrac{1}{9}}$
5. $\sqrt{\dfrac{16}{9}}$
6. $\sqrt{\dfrac{9}{4}}$
7. $\dfrac{\sqrt{4}}{5}$
8. $\dfrac{6}{\sqrt{25}}$

9. $\sqrt{4+9}$ 10. $\sqrt[4]{16-9}$ 11. $\sqrt[3]{27 \div 8}$ 12. $\sqrt[3]{8 \times 64}$

13. $\sqrt{(-2)^2}$ 14. $\sqrt{(-1)^2}$ 15. $\sqrt{\frac{1}{4}(1+5)}$ 16. $\sqrt{\frac{1}{9}(3+33)}$

Simplify the expressions in Exercises 17–24, given that x, y, z, a, b, and c are positive real numbers.

17. $\sqrt{a^2 b^2}$ 18. $\sqrt{\frac{a^2}{b^2}}$ 19. $\sqrt{(x+9)^2}$ 20. $(\sqrt{x+9})^2$

21. $\sqrt[3]{x^3(a^3+b^3)}$ 22. $\sqrt[4]{\frac{x^4}{a^4 b^4}}$ 23. $\sqrt{\frac{4xy^3}{x^2 y}}$ 24. $\sqrt{\frac{4(x^2+y^2)}{c^2}}$

Rewrite the expressions in Exercises 25–32 in exponential form.

25. $\sqrt{3}$ 26. $\sqrt{8}$ 27. $\sqrt{x^3}$ 28. $\sqrt[3]{x^2}$

29. $\sqrt[3]{xy^2}$ 30. $\sqrt{x^2 y}$ 31. $\frac{x^2}{\sqrt{x}}$ 32. $\frac{x}{\sqrt{x}}$

Rewrite the expressions in Exercises 33–38 in radical form.

33. $2^{2/3}$ 34. $3^{4/5}$ 35. $x^{4/3}$ 36. $y^{7/4}$

37. $(x^{1/2} y^{1/3})^{1/5}$ 38. $x^{-1/3} y^{3/2}$

Simplify the expressions in Exercises 39–48.

39. $4^{-1/2} 4^{7/2}$ 40. $2^{1/a}/2^{2/a}$ 41. $3^{2/3} 3^{-1/6} 3^{4/6} 3^{-4/6}$ 42. $2^{1/3} 2^{-1} 2^{2/3} 2^{-1/3}$

43. $\frac{x^{3/2}}{x^{5/2}}$ 44. $\frac{y^{5/4}}{y^{3/4}}$ 45. $\frac{x^{1/2} y^2}{x^{-1/2} y}$ 46. $\frac{x^{-1/2} y}{x^2 y^{3/2}}$

47. $\left(\frac{x}{y}\right)^{1/3} \left(\frac{y}{x}\right)^{2/3}$ 48. $\left(\frac{x}{y}\right)^{-1/3} \left(\frac{y}{x}\right)^{1/3}$

Solve each equation in Exercises 49–56 for x, rounding your answer to four significant digits where necessary:

49. $x^2 - 16 = 0$ 50. $x^2 - 1 = 0$ 51. $x^2 - \frac{4}{9} = 0$

52. $x^2 - \frac{1}{10} = 0$ 53. $x^2 - (1+2x)^2 = 0$ 54. $x^2 - (2-3x)^2 = 0$

55. $x^5 + 32 = 0$ 56. $x^4 - 81 = 0$

A.4 THE DISTRIBUTIVE LAW: MULTIPLYING ALGEBRAIC EXPRESSIONS

DISTRIBUTIVE LAW

The **distributive law** for real numbers states that

$$a(b \pm c) = ab \pm ac;$$
$$(a \pm b)c = ac \pm bc.$$

for any real numbers a, b, and c.

A.4 The Distributive Law: Multiplying Algebraic Expressions

EXAMPLE 1

(a) $2(x - \frac{1}{4})$ is *not* equal to $2x - \frac{1}{4}$, but is equal to $2x - 2(\frac{1}{4}) = 2x - \frac{1}{2}$.
(b) $x(x + 1) = x^2 + x$
(c) $2x(3x - 4) = 6x^2 - 8x$
(d) $(x - 4)x^2 = x^3 - 4x^2$

We can also use the distributive law to rewrite more complicated expressions.

EXAMPLE 2

(a) $(x + 2)(x + 3) = (x + 2)x + (x + 2)3 = (x^2 + 2x) + (3x + 6)$
$= x^2 + 5x + 6$
(b) $(x + 2)(x - 3) = (x + 2)x - (x + 2)3 = (x^2 + 2x) - (3x + 6)$
$= x^2 - x - 6$

There is quicker way of expanding expressions such as this, called the "FOIL" method (First, Outer, Inner, Last). Consider, for instance, the expression $(x + 1)(x - 2)$. The FOIL method says: take the product of the *first* terms: $x \cdot x = x^2$, the product of the *outer* terms: $x \cdot (-2) = -2x$, the product of the *inner* terms: $1 \cdot x = x$, the product of the *last* terms: $1 \cdot (-2) = -2$, and then add them all up, getting $x^2 - 2x + x - 2 = x^2 - x - 2$.

EXAMPLE 3

(a) $(x - 2)(2x + 5) = 2x^2 + 5x - 4x - 10 = 2x^2 + x - 10$
$\quad\quad\quad\quad\quad\quad\quad\quad\uparrow\quad\uparrow\quad\uparrow\quad\uparrow$
$\quad\quad\quad\quad\quad\quad\quad$ First Outer Inner Last
(b) $(x^2 + 1)(x - 4) = x^3 - 4x^2 + x - 4$
(c) $(a - b)(a + b) = a^2 + ab - ab - b^2 = a^2 - b^2$
(d) $(a + b)^2 = (a + b)(a + b) = a^2 + ab + ab + b^2$
$= a^2 + 2ab + b^2$
(e) $(a - b)^2 = (a - b)(a - b) = a^2 - ab - ab + b^2$
$= a^2 - 2ab + b^2$

The last three are particularly important and are worth memorizing.

SPECIAL FORMULAS

$(a - b)(a + b) = a^2 - b^2$ **Difference of two squares**
$(a + b)^2 = a^2 + 2ab + b^2$ **Square of a sum**
$(a - b)^2 = a^2 - 2ab + b^2$ **Square of a difference**

▼ **EXAMPLE 4**

(a) $(2 - x)(2 + x) = 4 - x^2$
(b) $(1 + a)(1 - a) = 1 - a^2$
(c) $(x + 3)^2 = x^2 + 6x + 9$
(d) $(4 - x)^2 = 16 - 8x + x^2$

Here are some longer examples.

▼ **EXAMPLE 5**

(a) $(x + 1)(x^2 + 3x - 4) = (x + 1)x^2 + (x + 1)3x - (x + 1)4$
$= (x^3 + x^2) + (3x^2 + 3x) - (4x + 4)$
$= x^3 + 4x^2 - x - 4$

(b) $\left(x^2 - \dfrac{1}{x} + 1\right)(2x + 5) = \left(x^2 - \dfrac{1}{x} + 1\right)2x + \left(x^2 - \dfrac{1}{x} + 1\right)5$
$= (2x^3 - 2 + 2x) + \left(5x^2 - \dfrac{5}{x} + 5\right)$
$= 2x^3 + 5x^2 + 2x + 3 - \dfrac{5}{x}$

(c) $(x - y)(x - y)(x - y) = (x^2 - 2xy + y^2)(x - y)$
$= (x^2 - 2xy + y^2)x - (x^2 - 2xy + y^2)y$
$= (x^3 - 2x^2y + xy^2) - (x^2y - 2xy^2 + y^3)$
$= x^3 - 3x^2y + 3xy^2 - y^3$

▶ **A.4 EXERCISES**

Expand each expression in Exercises 1–22.

1. $x(4x + 6)$
2. $(4y - 2)y$
3. $(2x - y)y$
4. $x(3x + y)$
5. $(x + 1)(x - 3)$
6. $(y + 3)(y + 4)$
7. $(2y + 3)(y + 5)$
8. $(2x - 2)(3x - 4)$
9. $(2x - 3)^2$
10. $(3x + 1)^2$
11. $\left(x + \dfrac{1}{x}\right)^2$
12. $\left(y - \dfrac{1}{y}\right)^2$
13. $(2x - 3)(2x + 3)$
14. $(4 + 2x)(4 - 2x)$
15. $\left(y - \dfrac{1}{y}\right)\left(y + \dfrac{1}{y}\right)$
16. $(x - x^2)(x + x^2)$
17. $(x^2 + x - 1)(2x + 4)$
18. $(3x + 1)(2x^2 - x + 1)$
19. $(x^2 - 2x + 1)^2$
20. $(x + y - xy)^2$
21. $(y^3 + 2y^2 + y)(y^2 + 2y - 1)$
22. $(x^3 - 2x^2 + 4)(3x^2 - x + 2)$

A.5 FACTORING ALGEBRAIC EXPRESSIONS

We can think of factoring as applying the distributive law in reverse. For example,

$$2x^2 + x = x(2x + 1),$$

which can be checked by using the distributive law. The first technique of factoring is to locate a **common factor**—that is, a term that occurs as a factor in each of the expressions being added or subtracted. For example, x is a common factor in $2x^2 + x$, since it is a factor of both $2x^2$ and x. On the other hand, x^2 is not a common factor, since it is not a factor of the second term, x.

Once we have located a common factor, we can "factor it out" by applying the distributive law.

EXAMPLE 1

(a) $2x^3 - x^2 + x$ has x as a common factor, so

$$2x^3 - x^2 + x = x(2x^2 - x + 1).$$

(b) $2x^2 + 4x$ has $2x$ as a common factor, so

$$2x^2 + 4x = 2x(x + 2).$$

(c) $2x^2y + xy^2 - x^2y^2$ has xy as a common factor, so

$$2x^2y + xy^2 - x^2y^2 = xy(2x + y - xy).$$

(d) $(x^2 + 1)(x + 2) - (x^2 + 1)(x + 3)$ has $x^2 + 1$ as a common factor, so

$$\begin{aligned}(x^2 + 1)(x + 2) - (x^2 + 1)(x + 3) &= (x^2 + 1)[(x + 2) - (x + 3)] \\ &= (x^2 + 1)(x + 2 - x - 3) \\ &= (x^2 + 1)(-1) \\ &= -(x^2 + 1) = -x^2 - 1.\end{aligned}$$

(e) $12x(x^2 - 1)^5(x^3 + 1)^6 + 18x^2(x^2 - 1)^6(x^3 + 1)^5$ has $6x(x^2 - 1)^5(x^3 + 1)^5$ as a common factor, so

$$\begin{aligned}12x(x^2 - 1)^5(x^3 + 1)^6 &+ 18x^2(x^2 - 1)^6(x^3 + 1)^5 \\ &= 6x(x^2 - 1)^5(x^3 + 1)^5[2(x^3 + 1) + 3x(x^2 - 1)] \\ &= 6x(x^2 - 1)^5(x^3 + 1)^5(2x^3 + 2 + 3x^3 - 3x) \\ &= 6x(x^2 - 1)^5(x^3 + 1)^5(5x^3 - 3x + 2).\end{aligned}$$

We would also like to be able to reverse calculations such as $(x + 2)(2x - 5) = 2x^2 - x - 10$. That is, starting with the expression

$2x^2 - x - 10$, we would like to **factor** it and get back the original expression $(x + 2)(2x - 5)$. An expression of the form $ax^2 + bx + c$, where a, b, and c are real numbers, is called a **quadratic** expression in x. Thus, given a quadratic expression $ax^2 + bx + c$, we would like to write it in the form $(dx + e)(fx + g)$ for some real numbers d, e, f, and g. There are some quadratics, such as $x^2 + x + 1$, that cannot be factored in this form at all. Here, we shall consider only quadratics that do factor, and in such a way that the numbers $d, e, f,$ and g are integers (whole numbers). (Other cases are fully discussed in Section 8.) The usual technique of factoring such quadratics is a "trial-and-error" approach, which we illustrate by means of examples.

▼ **EXAMPLE 2**

Factor $x^2 - 6x + 5$.

SOLUTION Concentrate on the first and last terms:

x^2 has factors x and x (since $x \cdot x = x^2$);

5 has factors 5 and 1.

Group them together and make an attempt.

$$(x + 5)(x + 1) = x^2 + 6x + 5$$

This is fine, except for the sign of the middle term. But notice that we can also get the 5 by multiplying (-5) and (-1). In other words, 5 also has factors (-5) and (-1). Using these instead gives

$$(x - 5)(x - 1) = x^2 - 6x + 5,$$

so we have found the correct factorization.

▼ **EXAMPLE 3**

Factor $x^2 - 4x - 12$.

SOLUTION The first term, x^2, has factors x and x.
The last term, -12, factors in many ways—for example, as $(-3) \cdot 4$, $12 \cdot (-1)$ and $(-6) \cdot 2$, among others. Trying the first gives

$$(x - 3)(x + 4) = x^2 + x - 12. \qquad \text{No good.}$$

The second gives

$$(x + 12)(x - 1) = x^2 + 11x - 12. \qquad \text{No good.}$$

The third gives

$$(x - 6)(x + 2) = x^2 - 4x - 12. \qquad \text{Correct!}$$

A.5 Factoring Algebraic Expressions

Notice that in our third trial, we had two factors of the last term whose sum, -4, was the coefficient of x in the middle term. This is what we look for when factoring a quadratic with an initial term of x^2: *Find the numbers whose product is the constant term and whose sum is the coefficient of x.* In the last example, $(-6) \cdot 2 = -12$ and $-6 + 2 = -4$. When the coefficient of x^2 is *not* 1, we have to work a little harder.

▼ EXAMPLE 4

Factor the quadratic $4x^2 - 5x - 6$.

SOLUTION Possible factorizations of $4x^2$ are $2x \cdot 2x$ or $4x \cdot x$. Possible factorizations of -6 are $(-3) \cdot 2$, $3 \cdot (-2)$, $(-6) \cdot 1$, or $6 \cdot (-1)$.

We now systematically try out all the possibilities until we come up with the correct one. For instance, the choices $2x \cdot 2x$ and $(-3) \cdot 2$ give

$$(2x - 3)(2x + 2) = 4x^2 - 2x - 6. \quad \text{No good.}$$

The choices $4x \cdot x$ and $(-3) \cdot 2$ give

$$(4x - 3)(x + 2) = 4x^2 + 5x - 6.$$

This is *almost* correct, except for the sign of the middle term. We can fix this by switching signs.

$$(4x + 3)(x - 2) = 4x^2 - 5x - 6$$

▼ EXAMPLE 5

Factor $4x^2 - 25$.

SOLUTION We recognize this as the difference of two squares.

$$\begin{aligned} 4x^2 - 25 &= (2x)^2 - 5^2 \\ &= (2x - 5)(2x + 5) \end{aligned}$$

Not all quadratic expressions factor in this way. In Section 8 we look at a "test" that will tell us whether or not a given quadratic factors.

▼ EXAMPLE 6

Factor $x^4 - 5x^2 + 6$.

SOLUTION This is not a quadratic, you say? Correct, it's a quartic (a fourth-degree expression). However, it looks rather like a quadratic. In fact, it is quadratic in x^2, meaning that it is

$$(x^2)^2 - 5(x^2) + 6 = y^2 - 5y + 6,$$

On the other hand,

$$x^2 + 1 = \sqrt{},$$
$$y + = 9, \quad \text{and}$$

The quick brown fox jumps over the lazy dog

are not equations.

An **equation** is a statement that two mathematical expressions are equal. In other words, it consists of two mathematical expressions separated by an equal sign.

Of course, this begs the question: What is a mathematical expression? Formally, a **mathematical expression** is a string of symbols including letters, numbers and some other characters ($+, -, \times, \div, \sqrt{}$, and so on) that obeys a lengthy list of "rules of syntax," such as the rule that says there should be expressions both before and after a plus sign. We shall not attempt to write down these rules—it would not be very illuminating. Remember this, though: The rules amount to saying that if you replaced all the letters in an expression with numbers, you would have a calculation that you can actually carry out.

Here is an important point to remember: The letters that occur in an equation signify numbers. Some stand for well-known numbers, such as π, c (the speed of light: 3×10^8m/s) or e (the base of natural logarithms: 2.71828 . . .). Some stand for **variables** or **unknowns.** Variables are quantities (such as length, height, or number of items) that can have many possible values, while unknowns are quantities whose values you may be asked to determine. The distinction between variables and unknowns is fuzzy, and mathematicians often use these terms interchangeably.

▼ **EXAMPLE 1**

$x + y = 7$ can be thought of as an equation in *two unknowns, x* and *y.* For example, x could stand for the number of days per week you attend math class and y for the number of days per week you don't attend math class. The equation $x + y = 7$ then amounts to the statement that there are a total of seven days in the week. If you knew the number x, you could find the remaining unknown, y.

We could also think of this as an equation in *two variables,* as the numbers x and y could vary depending on the week you're talking about.

Before we go on... It's interesting to notice that x and y do not vary randomly—again, if you know x then you know y. We can say that the value of y *depends on* the value of x. It's also common to say that y is a *function of x.*

Here is another important term: a **solution** to an equation in one or more unknowns is an assignment of numerical values to each of the unknowns, so that when these values are substituted for the unknowns, the equation becomes a *true statement about numbers.*

factor what we get. Here goes:

$$x^4 - 5x^2 + 4 = 10$$
$$x^4 - 5x^2 - 6 = 0$$
$$(x^2 - 6)(x^2 + 1) = 0.$$

(Here we used a sometimes useful trick that we mentioned in Section 5: we treated x^2 like x and x^4 like x^2, so factoring $x^4 - 5x^2 - 6$ is essentially the same as factoring $x^2 - 5x - 6$.) Now we are allowed to say that one of the factors must be 0:

$$x^2 - 6 = 0 \text{ has solutions } x = \pm\sqrt{6} = \pm 2.449\ldots$$

and

$$x^2 + 1 = 0 \text{ has no real solutions.}$$

Therefore we get exactly two solutions, $x = \pm\sqrt{6} = \pm 2.449\ldots$.

Before we go on... Substituting into the original equation:

$$((\pm\sqrt{6})^2 - 1)((\pm\sqrt{6})^2 - 4) = (6 - 1)(6 - 4)$$
$$= 5 \cdot 2 = 10. \checkmark$$

In solving equations involving rational expressions, the following rule is very useful.

SOLVING AN EQUATION OF THE FORM $P/Q = 0$

$$\text{If } \frac{P}{Q} = 0, \text{ then } P = 0.$$

How else could a fraction equal 0? If that is not convincing, multiply both sides by Q (which cannot be 0 if the quotient is defined).

▼ **EXAMPLE 5**

Solve

$$\frac{(x + 1)(x + 2)^2 - (x + 1)^2(x + 2)}{(x + 2)^4} = 0.$$

SOLUTION We can immediately set the numerator equal to 0:

$$(x + 1)(x + 2)^2 - (x + 1)^2(x + 2) = 0.$$

Factor:

$$(x + 1)(x + 2)[(x + 2) - (x + 1)] = 0$$

or

$$(x + 1)(x + 2)(1) = 0.$$

This gives us $x + 1 = 0$ or $x + 2 = 0$, so $x = -1$ or $x = -2$. But these are *not* both solutions to the original equation. Think about why for a moment before reading on.

The problem is that $x = -2$ does not make sense in the original equation: it makes the denominator 0. So it is not a solution, and $x = -1$ is the only solution.

Before we go on... Of course, we should check that $x = -1$ really is a solution:

$$\frac{(-1 + 1)(-1 + 2)^2 - (-1 + 1)^2(-1 + 2)}{(-1 + 2)^4} = 0 \ \checkmark$$

One more comment: If we had simplified the left-hand side of the original equation before trying to solve it, we would have factored the top as we just did, but then we could have cancelled an $x + 2$ from the top and bottom, getting

$$\frac{(x + 1)}{(x + 2)^3} = 0.$$

Then, setting the top equal to 0 gives only $x + 1 = 0$, so $x = -1$ is the only answer we see. This involves about the same amount of work but has the advantage of not giving us that fake solution.

EXAMPLE 6

Solve $1 - \dfrac{1}{x^2} = 0$.

SOLUTION Write 1 as $\frac{1}{1}$, so that we now have a difference of two rational expressions,

$$\frac{1}{1} - \frac{1}{x^2} = 0.$$

To combine these we can put both over a common denominator of x^2, which gives

$$\frac{x^2 - 1}{x^2} = 0.$$

Now we can set the numerator, $x^2 - 1$, equal to zero. Thus,

$$x^2 - 1 = 0,$$

so

$$(x - 1)(x + 1) = 0,$$

giving $x = \pm 1$.

Before we go on... Check:
$$1 - 1/(\pm 1)^2 = 1 - 1 = 0. \checkmark$$

This equation could also have been solved by writing

$$1 = \frac{1}{x^2}$$

and then multiplying both sides by x^2.

EXAMPLE 7

Solve

$$\frac{2x - 1}{x} + \frac{3}{x - 2} = 0.$$

SOLUTION We *could* first perform the subtraction on the left and then set the top equal to 0, but here is another approach. Subtracting the second expression from both sides gives

$$\frac{2x - 1}{x} = \frac{-3}{x - 2}.$$

Cross-multiplying (multiplying both sides by both denominators—that is, by $x(x - 2)$), now gives

$$(2x - 1)(x - 2) = -3x,$$

so

$$2x^2 - 5x + 2 = -3x.$$

Adding $3x$ to both sides gives the quadratic equation

$$2x^2 - 2x + 1 = 0.$$

The discriminant is $(-2)^2 - 4 \cdot 2 \cdot 1 = -4 < 0$, so we conclude that there is no real solution.

Before we go on... Notice that when we said $(2x - 1)(x - 2) = -3x$, we were *not* allowed to conclude that $2x - 1 = -3x$ or $x - 2 = -3x$.

EXAMPLE 8

Solve

$$\frac{\left(2x\sqrt{x + 1} - \dfrac{x^2}{\sqrt{x + 1}}\right)}{x + 1} = 0.$$

SOLUTION Setting the top equal to 0 gives

$$2x\sqrt{x+1} - \frac{x^2}{\sqrt{x+1}} = 0.$$

This still involves fractions. To get rid of the fractions, we could put everything over a common denominator ($\sqrt{x+1}$) and then set the top equal to 0, or we could multiply the whole equation by that common denominator in the first place to clear fractions. If we do the second, we get

$$2x(x+1) - x^2 = 0,$$

or

$$2x^2 + 2x - x^2 = 0,$$

or

$$x^2 + 2x = 0.$$

Factoring,

$$x(x+2) = 0,$$

so either $x = 0$ or $x + 2 = 0$, giving us $x = 0$ or $x = -2$. Again, one of these is not really a solution.

The problem is that $x = -2$ cannot be substituted into $\sqrt{x+1}$, since we would then have to take the square root of -1, and we are not allowing ourselves to do that. Therefore, $x = 0$ is the only solution.

Before we go on... As usual, we should check that $x = 0$ really works:

$$\frac{\left(2(0)\sqrt{0+1} - \frac{(0)^2}{\sqrt{0+1}}\right)}{0+1} = 0. \checkmark$$

▶ A.9 EXERCISES

Solve the equations in Exercises 1–26.

1. $x^4 - 3x^3 = 0$
2. $x^6 - 9x^4 = 0$
3. $x^4 - 4x^2 = -4$
4. $x^4 - x^2 = 6$
5. $(x+1)(x+2) + (x+1)(x+3) = 0$
6. $(x+1)(x+2)^2 + (x+1)^2(x+2) = 0$
7. $(x^2+1)^5(x+3)^4 + (x^2+1)^6(x+3)^3 = 0$
8. $10x(x^2+1)^4(x^3+1)^5 - 10x^2(x^2+1)^5(x^3+1)^4 = 0$
9. $(x^3+1)\sqrt{x+1} - (x^3+1)^2\sqrt{x+1} = 0$
10. $(x^2+1)\sqrt{x+1} - \sqrt{(x+1)^3} = 0$
11. $\sqrt{(x+1)^3} + \sqrt{(x+1)^5} = 0$
12. $(x^2+1)\sqrt[3]{(x+1)^4} - \sqrt[3]{(x+1)^7} = 0$

13. $(x + 1)^2(2x + 3) - (x + 1)(2x + 3)^2 = 0$

14. $(x^2 - 1)^2(x + 2)^3 - (x^2 - 1)^3(x + 2)^2 = 0$

15. $\dfrac{(x + 1)^2(x + 2)^3 - (x + 1)^3(x + 2)^2}{(x + 2)^6} = 0$

16. $\dfrac{6x(x^2 + 1)^2(x^2 + 2)^4 - 8x(x^2 + 1)^3(x^2 + 2)^3}{(x^2 + 2)^8} = 0$

17. $\dfrac{2(x^2 - 1)\sqrt{x^2 + 1} - \dfrac{x^4}{\sqrt{x^2 + 1}}}{x^2 + 1} = 0$

18. $\dfrac{4x\sqrt{x^3 - 1} - \dfrac{3x^4}{\sqrt{x^3 - 1}}}{x^3 - 1} = 0$

19. $x - \dfrac{1}{x} = 0$

20. $1 - \dfrac{4}{x^2} = 0$

21. $\dfrac{1}{x} - \dfrac{9}{x^3} = 0$

22. $\dfrac{1}{x} - \dfrac{1}{x + 1} = 0$

23. $\dfrac{x - 4}{x + 1} - \dfrac{x}{x - 1} = 0$

24. $\dfrac{2x - 3}{x - 1} - \dfrac{2x + 3}{x + 1} = 0$

25. $\dfrac{x + 4}{x + 1} + \dfrac{x + 4}{3x} = 0$

26. $\dfrac{2x - 3}{x} - \dfrac{2x - 3}{x + 1} = 0$

APPENDIX

Using a Graphing Calculator

INTRODUCTION ▸ Graphing calculators can display graphs, perform matrix operations, perform computations on tables of statistical data, and be programmed for more specialized tasks. While earlier programmable calculators had some of these features, the main distinguishing feature of the graphing calculator is its graphics display, which is in effect a small computer monitor.

The variety of graphing calculators on the market is large and increasing rapidly. In this appendix, we shall use the Texas Instruments TI-82 an our example. If you have another model or brand you should be able to do everything we do here, but you may have to consult the manual for details on exactly which buttons to press when.

B.1 GRAPHING FUNCTIONS

The graphing programs used by calculators work by plotting hundreds of points on the graph of a given function and then joining them, usually with straight line segments, creating the effect of a smooth curve.

Let us experiment by using the TI-82 to graph the two equations

$$x + 3y = 4,$$
$$y = -x^2 - 2x + 3.$$

A-49

Before we begin, we must first rewrite each equation in the form $y = \text{function of } x$. The second is already in this form, and we solve the first for y to obtain

$$y = \frac{4 - x}{3}$$

To enter these functions of x, press $\boxed{\text{Y=}}$ to get the display "Y=" and enter the equations as they are written.

Keystrokes

Y₁=(4−X)/3 $\boxed{(}\ \boxed{4}\ \boxed{-}\ \boxed{X}\ \boxed{)}\ \boxed{\div}\ \boxed{3}\ \boxed{\text{ENTER}}$

Y₂=-X²−2X+3 $\boxed{(\text{-})}\ \boxed{X}\ \boxed{X^2}\ \boxed{-}\ \boxed{2}\ \boxed{X}\ \boxed{+}\ \boxed{3}\ \boxed{\text{ENTER}}$

The parentheses are absolutely essential in the first equation. Also notice that this calculator distinguishes between "minus," $\boxed{-}$, and "negative," $\boxed{(\text{-})}$.

Before you go on to graph these functions, you should first set the **viewing window coordinates,** which determine the portion of the xy-plane you will view. The viewing window is determined by four numbers: Xmin, Xmax, Ymin, Ymax. Figure 1 shows the graphs of both equations with two viewing windows.

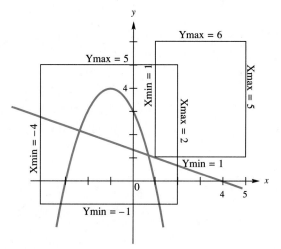

FIGURE 1

Notice that the viewing window specified by

Xmin=1, Xmax=5, Ymin=1, Ymax=6

contains no portion of either graph, so that if these were the settings you used, you would see a blank screen when you press $\boxed{\text{GRAPH}}$. Since this is not what we want, let us use the other window shown in Figure 1. To specify this window, press $\boxed{\text{WINDOW}}$ and enter the values shown.

```
Xmin=-4    [ENTER]  Remember to use negative, [(-)], rather than
                    minus.
Xmax=2     [ENTER]
Xscl=1     [ENTER]  It will place marks at 1-unit intervals on the
                    x-axis.
Ymin=-1    [ENTER]
Ymax=5     [ENTER]
Yscl=1     [ENTER]  It will place marks at 1-unit intervals on the
                    y-axis.
```

You are now ready to draw the graphs, so press [GRAPH]. Figure 2 shows what the display will look like.

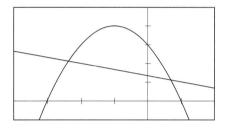

FIGURE 2

We shall now do two things with this graph.

FINDING THE COORDINATES OF THE VERTEX OF THE PARABOLA

(See the applications on maximizing revenue in Chapter 1.)

You can approximate the coordinates of the vertex by pressing [TRACE] and then using the left- and right-direction keys to move the cursor to a position as close to the vertex as possible. (See Figure 3, which shows the vertex at (1, 4).) The up- and down-arrow keys allow you to jump from one graph to the other.

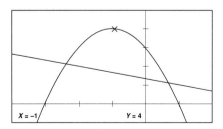

FIGURE 3

We were lucky to find the exact coordinates of the vertex in this example—usually we will not be able to do so. We should always "zoom in" for greater accuracy. To zoom in, use the trace feature to position the cursor at the point you want to examine more closely, press [ZOOM], and select "2: Zoom In" by pressing [2][ENTER]. The calculator then changes the viewing window to a small one centered at the cursor and redraws the graph as shown in the close-up view on the right in Figure 4.

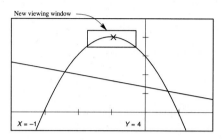
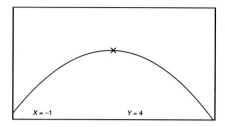

FIGURE 4

After zooming in, you should again press [TRACE] to reposition the cursor on the curve and adjust its position for greater accuracy. You can zoom in again for even more accuracy.

LOCATING POINTS WHERE TWO GRAPHS CROSS

(See the material on locating break-even points in Chapter 1.)

To locate a point where two graphs cross, we can use the same technique we just used in locating the vertex: trace and zoom. We first use trace to position the curvor as close to an intersection point as possible, then use zoom followed by trace for more accuracy, repeating as many times as necessary.

SETTING THE y-COORDINATES OF THE VIEWING WINDOW AUTOMATICALLY

The following TI-82 program calculates the y-coordinates of the highest and lowest points on a graph and then sets the window coordinates so that the whole graph just fits in the window. (Read the section on programming in the manual before entering this program.)

Program to Set Window Coordinates (TI-82)

```
PROGRAM: WINDOW
:Input"ENTER XMIN",M
:Input"ENTER XMAX",N
:M→Xmin
:N→Xmax
:(N-M)/95→D

:M→X
:Y₁→L
```

"Xmin" and "Xmax" are found under "Window" in VARS.

Increments of X (Graph window is 95 pixels wide.)

L is temporary Xmin.

```
:L→H                                H is temporary Xmax.
:For(X,M,N,D)
:If Y₁>H
:Then
:Y₁→H                               Increase H if a point on the graph is
                                    higher.
:End
:If Y₁<L
:Then
:Y₁→L                               Decrease L if a point of the graph is
                                    lower.
:End
:End                                End of loop
:L→Ymin
:H→Ymax
:Stop
```

Before running the program, the function you wish to graph should be entered as Y_1. The program will ask you for Xmin and Xmax, and it will then set the window coordinates automatically. If you press GRAPH after running the program, the calculator will draw the graph of Y_1 using the window determined by the program.

B.2 EVALUATING FUNCTIONS

The calculator can be used to evaluate functions in several different ways. Let us explore the function

$$f(x) = x^{1/3} + (x-1)^{2/3}.$$

Although the TI-82 is quite happy taking the $\frac{1}{3}$ power of negative numbers, it does not evaluate the $\frac{2}{3}$ power of negative numbers. You can trick it into doing so by using the identity $(x-1)^{2/3} = ((x-1)^2)^{1/3}$.

$$Y_1 = X\wedge(1/3) + ((X-1)^2)\wedge(1/3).$$

Figure 1 shows the graph of f using the viewing window defined by

$$\text{Xmin} = -3, \text{Xmax} = 2, \text{Ymin} = 0, \text{Ymax} = 2.$$

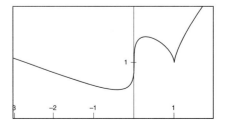

FIGURE 1

EVALUATING A FUNCTION ALGEBRAICALLY

Having entered the function in the "Y=" screen, you can repeatedly access it without having to retype the function. For instance, to evaluate $f(0.1532)$, enter (in the home screen)

0.1532 →X [ENTER]   This sets $X = 0.1532$. The arrow is obtained by pressing [STO→]

Y₁   [ENTER]   This tells it to evaluate y as defined by Y₁ on the "Y="screen. To obtain it on the home screen, press

[2nd] [Y-VARS] [ENTER] [ENTER].

EVALUATING A FUNCTION GRAPHICALLY

Alternatively, you can use the graph to evaluate the function at a value of x and view the corresponding point on the graph at the same time. First graph the function using a suitable viewing window (as above). Although you could now use the trace feature to approximate, say, $f(0.1532)$ by searching for a point whose x-coordinate is as close to 0.1532 as possible, you can accomplish this more easily and accurately by using the "calc" feature, as follows.

With the graph displayed, press [2nd] [CALC], select "1:value" and press [ENTER]. You will then be asked for the value of X, so enter 0.1532 and press [ENTER]. The cursor will now be placed at the point on the graph where $x = 0.1532$ and the corresponding y-coordinate is $f(0.1532)$.

EVALUATING A FUNCTION USING A TABLE

Instead of using the graph, you can use a table to display the values of a function as follows. First, make sure that your function is entered in the "Y=" screen as above. Next, obtain the TABLE SETUP screen by pressing [2nd] [TblSet]. Under "Indpnt" select "Ask." (This gives you the option of entering values for the independent variable x.) Press [ENTER] and quit to return to the home screen. Now press [2nd] [TABLE], and you will obtain a table on which you can enter values of x, and immediately be shown the corresponding values for $f(x)$ in the Y₁ column.

Note that you can use this method to evaluate several functions at once by taking advantage of Y₂, Y₃, ...

B.3 MATRIX ALGEBRA

Graphing calculators are usually equipped to perform all the standard matrix algebra operations. Here, we give examples of how to use the TI-82 to add, subtract, multiply, transpose, and invert matrices. We shall use the following matrices (the TI-82 uses the names [A], [B],... for matrices):

$$[A] = \begin{bmatrix} -1 & 1 & 3 \\ 4 & 5 & 2 \end{bmatrix}, \quad [B] = \begin{bmatrix} 1 & 0 \\ 4 & -2 \\ 3 & 5 \end{bmatrix}, \quad [C] = \begin{bmatrix} 1 \\ -3 \\ 2 \end{bmatrix},$$

$$\text{and } [D] = \begin{bmatrix} 0.1 & 0.2 & 0.7 \\ 0.5 & 0 & 0.5 \\ 0.2 & 0.7 & 0.1 \end{bmatrix}.$$

Start with the 2×3 matrix [A].

ENTERING A MATRIX

Press [MATRIX] to bring up the matrix window. Next, select EDIT, and then select the name of the matrix you want to define. (In this case, [A] is automatically selected, since it is first on the list.) Press [ENTER] to obtain the matrix edit window, which allows you to specify the matrix [A]. First enter the dimensions 2×3 of [A] at the top of the window, and then enter the entries of the matrix. (Remember to press [ENTER] after each entry. You can use the arrow keys to move about the matrix at will.) When you are done, press [2nd] [QUIT] to return to the home screen.

To view the matrix [A] on the home screen, press [MATRIX], select [A], and press [ENTER]. This displays the name "[A]" on the home screen. To view it, press [ENTER] again. (If the matrix is too large for the screen, use the arrow keys to view hidden parts of it.)

The matrices [B], [C], and [D] can now be entered in the same way.

PERFORMING MATRIX OPERATIONS

Here is a list of some operations and how each is entered on the TI-82. (Remember that to use a matrix name on the home screen, you must press [MATRIX], select the name you want, and press [ENTER].)

Operation	TI-82	Comments
AB	[A] [B] or [A] * [B]	
Let $E = AB$	[A] [B] → [E]	→ is [STO▸]
$A \pm B^t$	$[A] \pm [B]^T$	^{T} is found under MATH in the MATRX screen.
D^{-1}	$[D]^{-1}$	Use the [x⁻¹] key.
D^5	[D]^5	

You can experiment with high powers of the matrix [D] by entering

[D]^2 [ENTER]

and then repeatedly pressing [x^2] [ENTER], which has the effect of squaring the result each time, until the matrix stabilizes (stops changing). This matrix is the long-term transition matrix D^∞ discussed in the chapter on Markov processes.

B.4 PIVOTING PROGRAMS

Below are listed two programs—one for the TI-82 and a longer one for the less sophisticated TI-81—that you can use to row-reduce a matrix. To enter these programs, you should consult the section on programming in your user manual. (There is a useful table of instructions and functions, together with the keystroke sequences needed to obtain them, at the back of the manual.)

Graphing Calculator Program to Perform Pivot Operations (TI-82)

PROGRAM: PIVOT

```
:Lbl 1
:
:Pause [A]              Display matrix stored as [A] and allow scrolling
:Disp "1 TO PIVOT"      A little menu for the user
:Disp "2 TO DIVIDE"
:Input ("3 TO STOP ", V)
:
:If V=3
:Then
:Stop                   Self-explanatory
:End
:
:If V=1                 Here, we pivot.
:Then
:Input ("ROW ", R)      Get pivot row.
:Input ("COLUMN ", C)   Get pivot column.
:[A] (R, C) →P          P is the pivot—entry (R,C).
:dim [A] →L₁            Gets dimension of the matrix as a list.
:L₁ (1) →N              N is the number of rows in the matrix.
:For (I, 1, N, 1)       For I = 1 to N in steps of 1
```

:If I≠R	*R* is the pivot row—don't change it.
:Then	
:If [A](I, C)≠0	Only proceed if there is something to clear in the pivot column.
:Then	
:[A] (I, C) →X	We'll need to remember the entry.
:*row(P, [A], I) →[A]	The row operation $R_I \to P \times R_I$
:*row+(-X, [A], R, I) →[A]	The row operation $R_I \to R_I - X \times R_R$.
:End	End of the "Then" three lines up
:End	End of the "Then" seven lines up
:End	End of the loop
:End	End of this case ($V = 1$)
:	
:If V=2	Here, we divide.
:Then	
:Input ("DIVIDE ROW ", R)	
:Input ("BY ", X)	We'll divide *R* by *X*.
:*row(1/X, [A], R) →[A]	Does exactly that
:End	
:	
:Goto 1	Ask for next instruction.

Graphing Calculator Program to Perform Pivot Operations (TI-81)

PROGRAM: PIVOT	
:Disp "NUMBER OF ROWS "	Asks you for the number of rows in the matrix [A]
:Input N	
:Lbl 1	
:	
:Disp "1 TO PIVOT"	A little menu for the user
:Disp "2 TO DIVIDE"	
:Disp "3 TO STOP"	
:Disp [A]	Display matrix stored as [A]
:Input V	
:	
:If V=3	
:Stop	Self-explanatory

```
:
:If V=2
:Goto 2                         That's where we do the divide stuff.
:
:                               If it gets here, V must be 1, so we pivot.
:Disp "ROW"                     Get pivot row.
:Input R
:Disp "COLUMN"                  Get pivot column.
:Input C
:[A] (R, C) →P                  P is the pivot: entry (R, C).
:1→I                            Start of a loop: for I = 1 to N
:Lbl 5                          (I is the row you're on.)
:If I=R
:Goto 7                         R is the pivot row—don't change it.
:If [A] (I, C)=0                Only proceed if there is something to
:Goto 7                             clear in the pivot column.
:[A] (I, C) →X                  We'll need to remember the entry.
:*row (P, [A], I) →[A]          The row operation $R_I \to P \times R_I$
:*row+(-X, [A], R, I) →[A]      The row operation $R_I \to R_I - X \times R_R$.
:Lbl 7
:I+1→I                          Increment I by 1.
:If I≤N
:Goto 5                         End of loop
:
:Goto 1                         Ask for next instruction.
:
:Lbl 2                          Here, we divide.
:Disp "DIVIDE ROW?"             We'll divide R by X.
:Input R
:Disp "BY?"
:Input X
:*row (1/X, [A], R) →[A]        Does it
:
:Goto 1                         Ask for next instruction.
```

To Run the Program

Before running either program listed above, make sure that your matrix is entered under the name [A]. (To do this, follow the procedure in Section B.3.) When you run the program and the matrix is displayed on the TI-82 version, you can use the arrow keys to view hidden parts of the matrix, and then press [ENTER] to continue the program. In either program, press [1] [ENTER] to pivot, [2] [ENTER] to do a simplification step (or to turn the pivots into 1's when you are done) and [3] [ENTER] to terminate.

▶ **NOTE** This program does not allow you to recover the original matrix, so it is a good idea to store your matrix [A] under another name in case you want to examine it later. The instruction

$$[A] \to [B] \quad \boxed{\text{ENTER}}$$

will store your matrix as [B], so you can recover your original matrix by entering

$$[B] \to [A] \quad \boxed{\text{ENTER}}$$

later. ◀

B.5 STATISTICAL DATA

The following scenario is based on an example in the text. You are the manager of a corporate department with a staff of 50 workers whose salaries are given by the following frequency table.

Annual Salary	$15,000	$20,000	$25,000	$30,000	$35,000	$40,000	$45,000
Number of Workers	2	5	8	5	15	10	5

ENTERING STATISTICAL DATA

To enter lists of data on the TI-82, press [STAT], select EDIT, and then press [ENTER] to obtain the "list" screen, which shows three columns, labeled L_1, L_2, L_3. (Three more columns are off the screen.) Enter the salaries in the column under L_1 (pressing [ENTER] after each entry), and enter the numbers of workers in the column under L_2.

When you are done, press [2nd] [QUIT] to return to the home screen. You can now use your calculator to do a variety of calculations involving the data stored in lists L_1 and L_2.

DOING STATISTICAL CALCULATIONS

To find the mean, median, variance, and standard deviation of the entries in either L_1 or L_2, do the following:

1. Press [STAT], select CALC, choose SetUp and press [ENTER] to obtain the SET UP CALCS screen. (This permits you to select the list you want to work with.)
2. Under 1-Var Stats Xlist, select L_1 or L_2, depending on the list you want to work with, select 1 as the frequency, and press [ENTER].
3. Press [2nd] [QUIT] to return to the home screen.
4. Press [STAT], select CALC, choose 1-Var Stats, and press [ENTER].
5. "1-Var Stats" should now appear on the home screen. Press [ENTER] to run this program.

To find the expected value of a worker's salary, as well as the variance, standard deviation, median, and mode, you will need to *weight* the salaries using the numbers of workers as the weights.

1. Press [STAT], select CALC, choose SetUp, and press [ENTER] to obtain the SET UP CALCS screen.
2. Under 1-Var Stats, select L_1 as Xlist (the salaries), and select L_2 as the frequency (pressing [ENTER] each time). This does the weighting.
3. Press [2nd] [QUIT] to return to the home screen.
4. Press [STAT], select CALC, choose 1-Var Stats, and press [ENTER].
5. "1-Var Stats" should now appear on the home screen. Press [ENTER] to run this program.

GRAPHING STATISTICAL DATA

To prepare your calculator for statistical plotting, press [2nd] [Y-VARS], select On/Off, and then select FnOff. (Alternatively, make sure that there are no functions active in the "Y=" screen.) Then quit to go back to the home screen.

▶ CAUTION To reset your calculator when you are done, you will need to repeat the above procedure and select FnOn in order to resume normal plotting of curves. (Alternatively, you can reactivate functions in the "Y=" screen.) ◀

If you want to see a histogram of the data entered above, do the following.

1. Press [2nd] [STAT PLOT], select Plot1, and press [ENTER] to obtain the Plot1 screen.
2. Select On, press [ENTER], and then select the icon representing the histogram (or the kind of graph you want) and press [ENTER].

3. Select L₁ as the Xlist and L₂ as the frequency (pressing [ENTER] each time).
4. When done, select [2nd] [QUIT] to return to the home screen.
5. Now press [WINDOW] to set the viewing coordinates. In this example, use

Xmin=12000	
Xmax=48000	
Xscl=5000	The width of the histogram bars
Ymin=0	
Ymax=16	Look at the frequency figures on the table
Yscl=1	Since the frequencies are whole numbers

6. Now press [GRAPH] for the histogram.

▶ **NOTE** When done with statistical plotting, press [2nd] [STAT PLOT], select Plot1, and select Off to turn off statistical plotting. Then don't forget to press [2nd] [Y-VARS], select On/Off, and then select FnOn to resume normal graphing. ◀

B.6 CALCULATING DIFFERENCE QUOTIENTS

Recall that the **difference quotient** for the function f has the form

$$\frac{f(x+h) - f(x)}{h}.$$

In a typical situation, we might be given a value for x and be required to calculate the difference quotient for several values of h. As an example, we shall use the graphing calculator to find the difference quotient for $f(x) = \sqrt{x^2 - 1}$ at $x = 4$ and $h = \pm 0.1, \pm 0.01, \pm 0.001, \pm 0.0001$.

First, we evaluate the difference quotient for f at $x = 4$.

$$\frac{f(4+h) - f(4)}{h} = \frac{\sqrt{(4+h)^2 - 1} - \sqrt{4^2 - 1}}{h}$$

$$= \frac{\sqrt{(4+h)^2 - 1} - \sqrt{15}}{h}$$

We wish to evaluate this function of h at the given values of h. To do this, we can employ the methods of the previous section: Go to the "Y=" screen and enter the above function with X playing the role of h as the independent variable.

```
Y₁=(((4+X)^2-1)^0.5-15^0.5)/X
```

Evaluating for One Value of h at a Time

On the Home screen, enter

$0.1 \to X$ [ENTER]

Y_1 [ENTER]

The value of the difference quotient will then be displayed as 1.031957242. Now repeat the procedure for all the other values of h listed above. If you wish to enter all the values of h on a table instead, proceed as follows.

Evaluating Using a Table

Follow the instructions in the last section on using a table to evaluate a function. If you then enter the values of h under the X column, the values of the difference quotient will appear on the Y_1 column.

X	Y_1
0.1	1.032
−0.1	1.0327
0.01	1.0327
−0.01	1.0328
0.001	1.0328
−0.001	1.0328
0.0001	1.0328

(The TI-82 permits only seven entries at a time in the table. To evaluate at other values, replace some of the values above with new ones.)

B.7 PROGRAMS FOR LEFT- AND RIGHT-HAND SUMS, TRAPEZOID RULE, AND SIMPSON'S RULE

Below are listed two programs that you can use to calculate approximations to the definite integral. The first of these gives the left- and right-hand Riemann sums and trapezoidal sum of a given function, while the second gives the Simpson's rule approximation. To enter these programs, you should consult the section on programming in your user manual. (There is also a useful table of instructions and functions, together with the keystroke sequences needed to obtain them, at the back of the manual.)

B.7 Programs for Left- and Right-Hand Sums, Trapezoid Rule, and Simpson's Rule

Graphing Calculator Program to Compute Left- and Right-Hand Riemann Sums and Trapezoid Rule (TI-82)

Program	Comment
`PROGRAM: SUMS`	
`:Input ("N? ",N)`	Prompts for the number of rectangles
`:Input ("LEFT ENDPOINT? ",A)`	Prompts for the left endpoint a
`:Input ("RIGHT ENDPOINT? ",B)`	Prompts for the right endpoint b
`:(B-A)/N→D`	D is $\Delta x = (b-a)/n$.
`:0→R`	R will eventually be the right-hand sum.
`:0→L`	L will eventually be the left-hand sum.
`:A→X`	X is the current x-coordinate.
`:For (I,1,N)`	Start of a loop—recall the sigma notation
`:L+Y₁→L`	Increment L by $f(x_{i-1})$
`:A+I*D→X`	Corresponds to our formula $x_i = a + i\Delta x$
`:R+Y₁→R`	Increment R by $f(x_i)$
`:End`	End of loop
`:L*D→L`	Left sum
`:R*D→R`	Right sum
`:(L+R)/2→T`	Trapezoidal sum
`:Disp "LEFT SUM IS ",L`	
`:Disp "RIGHT SUM IS ",R`	
`:Disp "TRAP SUM IS ",T`	
`:Stop`	

Graphing Calculator Program for Simpson's Rule (TI-82)

Program	Comment
`PROGRAM: SIMP`	
`:Input ("EVEN NUMBER N≥2? ",N)`	Make sure that N is even and at least 2.
`:Input ("LEFT ENDPOINT? ",A)`	
`:Input ("RIGHT ENDPOINT? ",B)`	
`:(B-A)/N→D`	
`:0→S`	
`:A→X`	
`:For (I,1,N-1,2)`	The loop increments I in steps of 2.
`:S+Y₁→S`	The next four steps compute the sum $f(x_{i-1}) + 4f(x_i) + f(x_{i+1})$.
`:A+I*D→X`	
`:S+4*Y₁→S`	
`:X+D→X`	

```
:S+Y₁→S
:End
:S*D/3→S                          Multiply by Δx/3.
:Disp "SIMPSON SUM IS ",S
:Stop
```

Using the Programs

To use the above programs, first enter the function $f(x)$ whose sums you wish to compute as Y_1 in the "Y=" window. Then run the program. It will first ask for N, which is the number of partitions you wish to use. Enter the number (which must be at least 1) and then press [ENTER]. (If you are running SIMP, make sure that the number you enter is even.) Next, it will ask for the values of the endpoints A and B in that order, so you enter them in the same way. Then wait a while (the larger N is, the longer the wait), and the answers will appear.

PROGRAMS TO ESTIMATE THE ERROR IN THE TRAPEZOID RULE AND SIMPSON'S RULE

The next two programs can be used to give upper bounds of the error in using either the trapeziod rule or Simpson's rule with a specified number of partitions.

Graphing Calculator Program to Compute the Error in the Trapezoid Rule (TI-82)

```
PROGRAM: TRAPERR
:Input ("EVEN N ",N)              Number of partitions
:Input ("LEFT ENDPOINT? ",A)
:Input ("RIGHT ENDPOINT? ",B)
:fMax(Y₂,X,A,B)→X                 Calculates maximum value of
:Y₂→R                             |f''(x)| on [a, b]
:A→X
:Y₂→S                             The TI-82 sometimes ignores
:B→X                              endpoints, so we look at their
:Y₂→T                             values as well and find the
                                  maximum of all the candidates.
:max({R,S,T})→M
:M*(B−A)^3/(12*N^2)→E             Formula for error in trapezoid
                                  rule
:Disp "ERROR BOUND IS ",E
:Stop
```

Graphing Calculator Program to Compute the Error in Simpson's Rule (TI-82)

```
PROGRAM: SIMPERR
:Input ("EVEN N ",N)              Number of partitions
```

B.8 Calculating and Plotting Averages and Moving Averages   A-65

```
:Input ("LEFT ENDPOINT? ",A)
:Input ("RIGHT ENDPOINT? ",B)
:fMax(Y₄,X,A,B) →X
:Y₄ →R
:A →X
:Y₄ →S
:B →X
:Y₄ →T
:max({R,S,T}) →M
:M*(B−A)^5/(180*N^4) →E

:Disp "ERROR BOUND IS ",E
:Stop
```

Calculates maximum value of $|f^{(4)}(x)|$ on $[a, b]$

The TI-82 sometimes ignores endpoints, so we look at their values as well and find the maximum of all the candidates.

Formula for error in Simpson's rule

Using the Programs

To use TRAPERR, you must first enter the function $|f''(x)|$ as Y_2 in the "Y=" window. (We have used Y_2 since you may already have the original function $f(x)$ as Y_1.) To use SIMPERR, you must first enter the function $|f^{(4)}(x)|$ as Y_4 in the "Y=" window. This allows you to run TRAP, SIMP, TRAPERR, and SIMPERR without having to change anything in the "Y=" window.

▶ B.8 CALCULATING AND PLOTTING AVERAGES AND MOVING AVERAGES

PLOTTING A LIST OF DATA

The following table, from an example in the chapter on applications of the integral, shows Colossal Conglomerate Corp.'s closing stock prices for a 20-day period:

Day	1	2	3	4	5	6	7	8	9	10
Price	20	22	21	24	24	23	25	26	20	24
Day	11	12	13	14	15	16	17	18	19	20
Price	26	26	25	27	28	27	29	27	25	24

To plot these data on a scatter graph, press [STAT], select EDIT, and press [ENTER] to obtain the list screen. If there is data on any of the lists shown, first clear it by selecting that list (L_1, L_2, ...) and pressing [CLEAR].

Next, enter the days 1, 2, 3, ... in L_1 and the prices 20, 22, 21, 24, ... in L_2. (Remember to press [ENTER] after each entry.)

To obtain a line graph of the stock prices, press [STAT PLOT], and select `Plot1` and press [ENTER] to obtain the `Plot1` menu. Next, select `On`, press [ENTER], and then select the icon representing the type of graph you want and again press [ENTER]. Next, press [ZOOM] [9] to set the x- and y-ranges to fit the data. The graph should then appear in the window.

▶ NOTE   If there are any active functions in the "Y=" screen, these too will be plotted if they are in range. To prevent this, either remove them or deactivate them (that is, prevent them from being plotted) by selecting the "=" signs in "Y_n= . . . " and pressing [ENTER]. You can reactivate them later by repeating this procedure. ◀

Graphing Calculator Program to Plot Moving Averages (TI-82)

```
PROGRAM: MOVINGAV
:Input ("NUMBER OF DATA POINTS ",N)
:Input ("PERIOD ",M)                    For an m-unit moving average
:For (I,1,M−1,1)
:L₂(I) →L₃(I)                           Set first m − 1 entries on L₃ to
                                          those of L₂;
:End                                    L₃ will be the moving averages
                                          list.

:For (I,M,N,1)
:∅→S                                    S will be the sum of data
                                          values.
:For(J,∅,M−1,1)
:S+L₂(I−J) →S
:End
:S/M→L₃(I)                              Put m-unit average on list L₃
:End
:Plot1(xyLine,L₁,L₂)
:Plot2(xyLine,L₁,L₃)
:FnOff                                  from "On/Off" in Y-VARS menu
:PlotsOn 1,2                            from STATPLOT menu
:ZoomStat                               "Zoom" menu—sets ranges
:Stop
```

USING THE PROGRAM

To run the program, first enter the data on L_1 and L_2 as above. The program will display line graphs of both the values listed in L_1 and the moving averages in L_3. To access the moving averages directly, go to L_3.

Once you are done with statistical graphing and wish to reset the calculator to its normal mode (so that your statistical graphs will not be superimposed on every graph you draw), you should run the following little program. We have named it "AARESET" to ensure that it appears first on the list of programs, so that you can run it by simply pressing [PRGM], [ENTER], [ENTER].

Program to Reset the TI-82 to Normal Graphing Mode

```
PROGRAM: AARESET
:Plotsoff          From STATPLOT menu)
:FnOn              From "On/Off" in Y-VARS menu
:Stop
```

Be sure to run AARESET whenever you are done with your line graph plotting.

Alternatively, you can reset the calculator manually using the following sequence of keystrokes:

[2nd] [STAT PLOT], 4, [ENTER], [2nd], [Y-VARS], 5, 1, [ENTER].

B.9 EVALUATING A FUNCTION OF SEVERAL VARIABLES

The following little program will permit you to calculate $f(x, y)$ repeatedly with a minimum of keyboard work.

Graphing Calculator Program to Calculate $f(x, y)$ (TI-82)

```
PROGRAM:FXY
:1→A
:While A=1
:Input ("ENTER X ",X)
:Input ("ENTER Y ",Y)
:Disp "ANSWER",Y₈
:Input ("ENTER 1 TO CONTINUE ",A)
:End
:Stop
```

To run this program, first enter the function of two variables as Y_8 in the "Y=" screen. Then run the program. When prompted with "ENTER 1 TO CONTINUE," you can quit the program by entering any number other than 1. Entering 1 will cause it to ask for a new pair (x, y).

▶ NOTE It is easy to adapt this program to calculate values of functions of three or more variables. We leave this to you. ◀

B.10 CALCULATING PROBABILITIES ASSOCIATED WITH CONTINUOUS DISTRIBUTIONS

The following little program calculates the probability $P(A \leq X \leq B)$ associated with a normal distribution for a given mean and standard deviation. It can easily be adapted for use with either an exponential or beta distribution, and we leave that task to the interested reader.

Graphing Calculator Program to Calculate Probabilities Based on Normal Distribution (TI-82)

```
PROGRAM:NORMAL
:Input "ENTER MEDIAN MU ",M
:Input "ENTER ST DEV ",T
:Input "LEFT LIMIT ",A
:Input "RIGHT LIMIT ",B
:fnInt((1/T(2π)
^.5))e^(-(X-M)^2
/(2T²)),X,A,B) →P



:Disp "PROBABILITY IS ",P
:Stop
```

Calculates $P(A \leq X \leq B)$
Value of median μ
Value of standard deviation σ

This is a single instruction line. The line breaks correspond to what should appear on your screen. "fnInt" can be obtained by pressing [MATH] [9]. The short minus sign is "negative" [(-)] and the longer one is "minus" [-].

To Run the Program

When you run this little program, it asks for the mean, standard deviation, and the two endpoints *A* and *B*. It then goes ahead and calculates $P(A \leq X \leq B)$.

APPENDIX C

Using a Computer Spreadsheet

INTRODUCTION ▶ Computer spreadsheets can be used to do many of the numerical calculations discussed in this book. In this appendix we show how to use a spreadsheet to calculate difference quotients, numerical approximations to integrals, and averages and moving averages.

For our examples we shall use the program Lotus 1-2-3, but other computer spreadsheets should be very similar. Consult the manual for your particular program for details on its commands.

C.1 CALCULATING DIFFERENCE QUOTIENTS

Recall that the **difference quotient** for the function f has the form

$$\frac{f(x + h) - f(x)}{h}.$$

In a typical situation, we might be given a value for x and be required to calculate the difference quotient for several values of h. As an example, we shall use the spreadsheet to find the difference quotient for $f(x) = \sqrt{x^2 - 1}$ at $x = 4$ and $h = \pm 0.1, \pm 0.01, \pm 0.001, \pm 0.0001$.

First, we evaluate the difference quotient for f at $x = 4$.

$$\frac{f(4 + h) - f(4)}{h} = \frac{\sqrt{(4 + h)^2 - 1} - \sqrt{4^2 - 1}}{h}$$

$$= \frac{\sqrt{(4 + h)^2 - 1} - \sqrt{15}}{h}$$

There are several ways of conveniently evaluating this at the eight values of h we want. Here is a nice way of doing it in Lotus 1-2-3 (other spreadsheets should be very similar). Start by entering the following in row 1:

	A	B	C
1		4	@SQRT(B1^2−1)

The 4 in position B1 is the value $x = 4$ that we are interested in, and the formula in C1 is $f(x)$. Now enter the following.

	A	B	C
1		4	@SQRT(B1^2−1)
2	−1	+$B$1 + A2	
3	−0.1		
4	−0.01		
5	−0.001		
6	−0.0001		
7	0.0001		
8	0.001		
9	0.01		
10	0.1		
11	1		

The values in the A column are the values of h that we want to use. The formula in B2 calculates $x + h$. If you now copy this formula and paste it into the cells B3 though B11, it will calculate in those cells the values $x + h$ for all the values of h you entered. (*Note:* Entering $B$1 instead of B1 makes Lotus 1-2-3 always take the value of x from B1. However, as you paste the formula into the other cells, the A2 in the formula will be changed to A3, A4, and so on, as appropriate, to pick up the correct value of h.) Now copy the formula in C1 into the cells C2 through C11, and the C column will contain the values $f(x + h)$ for various h. Finally, use the D column to calculate the difference quotients:

A	B	C	D
	4	@SQRT(B1^2−1)	
−1	+$B$1 + A2	@SQRT(B2^2−1)	+(C2−$C$1)/A2

C.2 Calculating Left- and Right-Hand Riemann Sums, the Trapezoid Rule, and Simpson's Rule

The formula in D2 will calculate the difference quotient for the first value $h = -1$. If you now copy this formula into D3 through D11, you will calculate the remaining difference quotients. After calculation, the spreadsheet should look like this:

	A	B	C	D
1		4	3.8729833	
2	−1	3	2.8284271	1.0445562
3	−0.1	3.9	3.7696154	1.0336798
4	−0.01	3.99	3.8626545	1.0328819
5	−0.001	3.999	3.8719505	1.0328042
6	−0.0001	3.9999	3.8728801	1.0327964
7	0.0001	4.0001	3.8730866	1.0327947
8	0.001	4.001	3.8740161	1.032787
9	0.01	4.01	3.8833104	1.0327097
10	0.1	4.1	3.9761791	1.0319572
11	1	5	4.8989795	1.0259961

You can see from this that the derivative $f'(4)$ is approximately 1.03279.

C.2 CALCULATING LEFT- AND RIGHT-HAND RIEMANN SUMS, THE TRAPEZOID RULE, AND SIMPSON'S RULE

In this example we shall calculate the left- and right-hand Riemann sum approximations to

$$\int_0^2 e^{-x^2}\, dx$$

for $n = 10$. We begin by computing the numbers x_k. Enter a in the top left cell of the spreadsheet, and in the cell below it calculate $x_1 = a + \Delta x = a + 2/10$.

	A	B	C	D
1	0			
2	+A1 + 2/10			

Appendix C Using a Computer Spreadsheet

If we now copy the entry in cell A2 into cells A3 through A11, we get all the values of x_k.

	A	B	C	D
1	0			
2	0.2			
3	0.4			
4	0.6			
5	0.8			
6	1			
7	1.2			
8	1.4			
9	1.6			
10	1.8			
11	2			

In B1 we now enter the formula for $f(x) = e^{-x^2}$.

	A	B	C	D
1	0	@EXP(−A1^2)		
2	0.2			
3	0.4			
4	0.6			
5	0.8			
6	1			
7	1.2			
8	1.4			
9	1.6			
10	1.8			
11	2			

C.2 Calculating Left- and Right-Hand Riemann Sums, the Trapezoid Rule, and Simpson's Rule

If we copy this formula into cells B2 through B11, column B will have the values $f(x_k)$.

	A	B	C	D
1	0	1		
2	0.2	0.9607894		
3	0.4	0.8521437		
4	0.6	0.6976763		
5	0.8	0.5272924		
6	1	0.3678794		
7	1.2	0.2369277		
8	1.4	0.1408584		
9	1.6	0.0773047		
10	1.8	0.0391638		
11	2	0.0183156		

We can now compute the left- and right-hand Riemann sums by entering their formulas in any convenient cells—say, C1 and C2.

	A	B	C	D
1	0	1	@SUM(B1..B10)*2/10	
2	0.2	0.9607894	@SUM(B2..B11)*2/10	
3	0.4	0.8521437		
4	0.6	0.6976763		
5	0.8	0.5272924		
6	1	0.3678794		
7	1.2	0.2369277		
8	1.4	0.1408584		
9	1.6	0.0773047		
10	1.8	0.0391638		
11	2	0.0183156		

After calculating the left- and right-hand Riemann sums, we can calculate the trapezoidal sum as their average.

	A	B	C	D
1	0	1	0.9800072	
2	0.2	0.9607894	0.7836703	
3	0.4	0.8521437	+(C1+C2)/2	
4	0.6	0.6976763		
5	0.8	0.5272924		
6	1	0.3678794		
7	1.2	0.2369277		
8	1.4	0.1408584		
9	1.6	0.0773047		
10	1.8	0.0391638		
11	2	0.0183156		

The final spreadsheet then looks like this, with the left-hand sum in C1, the right-hand sum in C2, and the trapezoidal sum in C3.

	A	B	C	D
1	0	1	0.9800072	
2	0.2	0.9607894	0.7836703	
3	0.4	0.8521437	0.8818388	
4	0.6	0.6976763		
5	0.8	0.5272924		
6	1	0.3678794		
7	1.2	0.2369277		
8	1.4	0.1408584		
9	1.6	0.0773047		
10	1.8	0.0391638		
11	2	0.0183156		

C.2 Calculating Left- and Right-Hand Riemann Sums, the Trapezoid Rule, and Simpson's Rule

SIMPSON'S RULE

Now we shall calculate the Simpson's rule approximation to

$$\int_0^2 e^{-x^2}\, dx$$

for $n = 10$. Since every other $f(x_k)$ is treated differently in Simpson's approximation, we enter the x_k's in two separate columns.

	A	B	C	D	E
1	0	+A1+2/10			
2	+A1+2*2/10				

We copy the formula in A2 into A3 through A6, and the formula in B1 into B2 through B5.

	A	B	C	D	E
1	0	0.2			
2	0.4	0.6			
3	0.8	1			
4	1.2	1.4			
5	1.6	1.8			
6	2				

We enter the formula for $f(x_k)$ in C1 and then copy it into C2 through C6 and D1 through D5.

	A	B	C	D	E
1	0	0.2	@EXP(−A1^2)		
2	0.4	0.6			
3	0.8	1			
4	1.2	1.4			
5	1.6	1.8			
6	2				

	A	B	C	D	E
1	0	0.2	1	0.9607894	
2	0.4	0.6	0.8521437	0.6976763	
3	0.8	1	0.5272924	0.3678794	
4	1.2	1.4	0.2369277	0.1408584	
5	1.6	1.8	0.0773047	0.0391638	
6	2		0.0183156		

Finally, in E1 we enter the formula for Simpson's rule, in the form

(@SUM(C1..C5)+@SUM(C2..C6)+4*@SUM(D1..D5))*2/(3*10)

(why does this calculate Simpson's rule?). The final spreadsheet is this, with the result in E1:

	A	B	C	D	E
1	0	0.2	1	0.9607894	0.8820748
2	0.4	0.6	0.8521437	0.6976763	
3	0.8	1	0.5272924	0.3678794	
4	1.2	1.4	0.2369277	0.1408584	
5	1.6	1.8	0.0773047	0.0391638	
6	2		0.0183156		

C.3 CALCULATING AVERAGES AND MOVING AVERAGES

The following table, from Example 5 in the Chapter on Applications of the Integral, shows Colossal Conglomerate Corp.'s closing stock prices for a 20-day period:

Day	1	2	3	4	5	6	7	8	9	10
Price	20	22	21	24	24	23	25	26	20	24
Day	11	12	13	14	15	16	17	18	19	20
Price	26	26	25	27	28	27	29	27	25	24

We shall demonstrate how to plot these data and the 5-day moving average using a computer spreadsheet.

C.3 Calculating Averages and Moving Averages

We begin by entering the data in the spreadsheet, either as a row or a column. For this example, we shall enter the data in a column. (We show only the first 10 entries.)

	A	B
1	20	
2	22	
3	21	
4	24	
5	24	
6	23	
7	25	
8	26	
9	20	
10	24	
...	...	

We can now compute the moving average by first entering in cell B5 the formula @AVG(A1..A5).

	A	B
1	20	
2	22	
3	21	
4	24	
5	24	@AVG(A1..A5)
6	23	
7	25	
8	26	
9	20	
10	24	
...	...	

Appendix D Table: Area Under a Normal Curve

Z	0.00	0.01	0.02	0.03	0.04	0.05	0.06	0.07	0.08	0.09
0.8	0.2881	0.2910	0.2939	0.2967	0.2995	0.3023	0.3051	0.3078	0.3106	0.3133
0.9	0.3159	0.3186	0.3212	0.3238	0.3264	0.3289	0.3315	0.3340	0.3365	0.3389
1.0	0.3413	0.3438	0.3461	0.3485	0.3508	0.3531	0.3554	0.3577	0.3599	0.3621
1.1	0.3643	0.3665	0.3686	0.3708	0.3729	0.3749	0.3770	0.3790	0.3810	0.3830
1.2	0.3849	0.3869	0.3888	0.3907	0.3925	0.3944	0.3962	0.3980	0.3997	0.4015
1.3	0.4032	0.4049	0.4066	0.4082	0.4099	0.4115	0.4131	0.4147	0.4162	0.4177
1.4	0.4192	0.4207	0.4222	0.4236	0.4251	0.4265	0.4279	0.4292	0.4306	0.4319
1.5	0.4332	0.4345	0.4357	0.4370	0.4382	0.4394	0.4406	0.4418	0.4429	0.4441
1.6	0.4452	0.4463	0.4474	0.4484	0.4495	0.4505	0.4515	0.4525	0.4535	0.4545
1.7	0.4554	0.4564	0.4573	0.4582	0.4591	0.4599	0.4608	0.4616	0.4625	0.4633
1.8	0.4641	0.4649	0.4656	0.4664	0.4671	0.4678	0.4686	0.4693	0.4699	0.4706
1.9	0.4713	0.4719	0.4726	0.4732	0.4738	0.4744	0.4750	0.4756	0.4761	0.4767
2.0	0.4772	0.4778	0.4783	0.4788	0.4793	0.4798	0.4803	0.4808	0.4812	0.4817
2.1	0.4821	0.4826	0.4830	0.4834	0.4838	0.4842	0.4846	0.4850	0.4854	0.4857
2.2	0.4861	0.4864	0.4868	0.4871	0.4875	0.4878	0.4881	0.4884	0.4887	0.4890
2.3	0.4893	0.4896	0.4898	0.4901	0.4904	0.4906	0.4909	0.4911	0.4913	0.4916
2.4	0.4918	0.4920	0.4922	0.4925	0.4927	0.4929	0.4931	0.4932	0.4934	0.4936
2.5	0.4938	0.4940	0.4941	0.4943	0.4945	0.4946	0.4948	0.4949	0.4951	0.4952
2.6	0.4953	0.4955	0.4956	0.4957	0.4959	0.4960	0.4961	0.4962	0.4963	0.4964
2.7	0.4965	0.4966	0.4967	0.4968	0.4969	0.4970	0.4971	0.4972	0.4973	0.4974
2.8	0.4974	0.4975	0.4976	0.4977	0.4977	0.4978	0.4979	0.4979	0.4980	0.4981
2.9	0.4981	0.4982	0.4982	0.4983	0.4984	0.4984	0.4985	0.4985	0.4986	0.4986
3.0	0.4987	0.4987	0.4987	0.4988	0.4988	0.4989	0.4989	0.4989	0.4990	0.4990

Answers to Odd-Numbered Exercises

Chapter 1

Section 1.1 (page 14)

1. $P(-2, 4)$; $Q(1, 3)$; $R(-3, -4)$; $S(3, -3)$; $T(0, -2)$; $U(-\frac{9}{2}, 0)$; $V(\frac{5}{2}, \frac{3}{2})$; $W(\frac{5}{2}, 0)$

3. 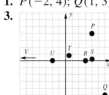   **5.**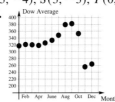

7.

Month	M	A	M	J	J	A	S	O	N	D	J	F	M
Deficit ($Billions)	−5.5	−7	−7.5	−6.5	−7.5	−8.5	−8.5	−7	−7.5	−6.5	−7.5	−8.0	−10.0

9. (a) 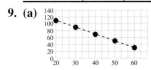   **(b)** A $10 increase in price has the effect of reducing sales by 20,000.

11.

Time	0	1	2	3	4
Height	500	480	440	350	250

13. (a) 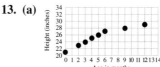   **(b)** 0.3 inches

15. 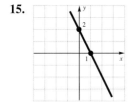   **17.** 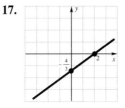   **19.** 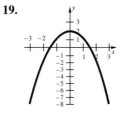   **21.**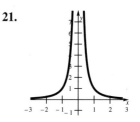

A-85

23. 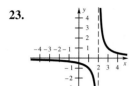 **25.** 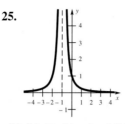 **27.** 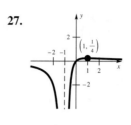 **29.**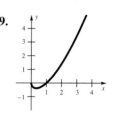

31. (a) 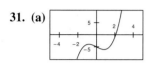 **31 (b)** Nothing is visible, since this viewing window is a square with sides of length 2 centered at the origin, and does not contain any points of the graph.

33. Replacing y with $(y - c)$ moves the curve up c units.   **35.** Replacing x with $(x - c)$ moves the curve to the right c units.

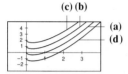   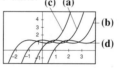

37.    **39.** 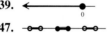   **41.**    **43.**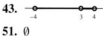

45. ○————○   **47.** ○—○ ●————● ○—○   **49.** ————●————   **51.** ∅
  −4   4   −1 0 1 2   3 4   2

53. $[-1, 2]$ **55.** $(0, +\infty)$ **57.** $(-4, -2) \cup (-2, +\infty)$

Section 1.2 (page 32)

1. (a) $f(0) = 3$ (b) $f(1) = 6$ (c) $f(-1) = 2$ (d) $f(-3) = 6$ (e) $f(a) = a^2 + 2a + 3$
(f) $f(x + h) = (x + h)^2 + 2(x + h) + 3$ (g) $(f(x + h) - f(x))/h = 2x + h + 2$
3. (a) $g(1) = 2$ (b) $g(-1) = 0$ (c) $g(4) = 65/4$ (d) $g(x) = x^2 + \frac{1}{x}$ (e) $g(s + h) = (s + h)^2 + 1/(s + h)$
(f) $g(s + h) - g(s) = (s + h)^2 + 1/(s + h) - (s^2 + \frac{1}{s})$.
5. (a) $\phi(0) = \sqrt{3}$ (b) $\phi(-2) = \sqrt{7}$; (c) $\phi(x + h) = \sqrt{(x + h)^2 + 3}$ (d) $\phi(x) + h = \sqrt{x^2 + 3} + h$
7. $-2x - h - 2$ **9.** $\dfrac{-2}{(x + h + 1)(x + 1)}$ **11.** $1 - \dfrac{1}{x(x + h)}$
13. $-\dfrac{2x + h}{x^2(x + h)^2}$ **15.** (a) yes; $f(4) = \dfrac{63}{16}$ (b) not defined (c) not defined   **17.** (a) not defined
(b) not defined   (c) yes, $f(-10) = 0$ **19.** (a) yes, $f(0) = 1$ (b) not defined (c) yes, $f(-3) = 2$
21. $(-\infty, +\infty)$ **23.** $[0, +\infty)$ **25.** $[1, +\infty)$ **27.** $(-\infty, 0) \cup (0, +\infty)$ **29.** $(-\infty, 0) \cup (0, +\infty)$
31. $(-\infty, 2) \cup (2, +\infty)$
33. (a) $f(1) = 20$ (b) $f(2) = 30$ (c) $f(3) = 30$ (d) $f(5) = 22$ (e) $f(3) - f(2) = 0$
35. (a) $f(1) = 1.3$ (b) $f(-2) = 0$ (c) $f(0) = 2$ (d) $f(3) = 0$ (e) $f(3) - f(2) = -0.7$
37. (a) $f(-3) = -1$ (b) $f(0) = 1.25$ (c) $f(1) = 0$ (d) $f(2) = 1$ (e) $\dfrac{f(3) - f(2)}{3 - 2} = 0$
39. (a) $f(-3) = -0.5$ (b) $f(-2) = 0$ (c) $f(0) = 1$ (d) $f(2) = 0$ (e) $\dfrac{f(2) - f(0)}{2 - 0} = -\dfrac{1}{2}$
41. (a) (I) (b) (IV) (c) (V) (d) (VI) (e) (III) (f) (II)   **43.** Equation: $y = x^3$; graph: a

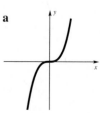

45. Equation: $y = x^4$; graph: 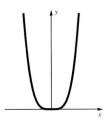 **47.** Equation: $y = \frac{1}{x^2}$; graph: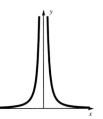

49. 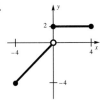 **51.** 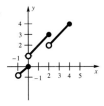 **53.** 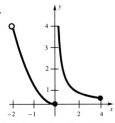 **55.**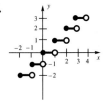

57.   **59.** Domain: $(-\infty, 1) \cup (1, 2) \cup (2, 3) \cup (3, 4) \cup (4, +\infty)$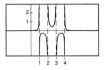

61. 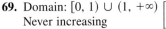 **63.** Domain $[-5, 1) \cup (1, 2) \cup (2, 3) \cup (3, 5]$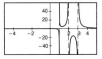

65. Domain: $[0, 1) \cup (1, 2) \cup (2, 5]$ **67.** Lowest point approximately $(0.333, -0.385)$

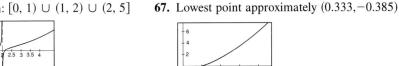

69. Domain: $[0, 1) \cup (1, +\infty)$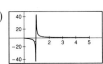
Never increasing

71. (a) 358,600 (b) 361,200 (c) $6.00 **73.** (a) $[0, 8]$ (b) $[0, +\infty)$ is not an appropriate domain, since it would predict investments in South Africa into the indefinite future with no basis. (It would also lead to preposterous results for large values of t.) **75.** (a) $12,000 (b) $N(q) = 2{,}000 + 100q^2 - 500q$; $N(20) = \$32{,}000$
77. (a) $C(x) = 8x + 80/x$, with domain $(0, +\infty)$ (b) Approximate value of x for lowest cost is 3.2 feet.

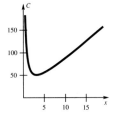

79. (a) $V(h) = \dfrac{1}{27}\pi h^3$, with domain $[0, +\infty)$ **(b)** $h(V) = (27V/\pi)^{1/3}$, with domain $[0, +\infty)$

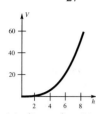

81. (a) $(0, +\infty)$ **(b)** $R(4000) = 30$ per hour **(c)**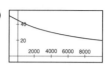

83. $G(Y_p) = 0.55Y_p$, with domain $(0, +\infty)$; $G(2) = 1.1$ trillion dollars. Thus inflation causes the actual GNP to fall below the projected GNP. **85. (a)** 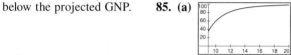 **(b)** 82% **(c)** 13.7 months

87. (a) 31.22; the rocket ship appears to be 31.22 meters in length. **(b)** $p = 0.8660$ warp, or 86.60% the speed of light

Section 1.3 (page 56)

1. 2   **3.** 2   **5.** -2 **7.** -1 **9.** $1/2$ **11.** 0   **13.** $\sqrt{2}$;   **15.** infinite   **17.** $(d-b)/(c-a)$
19. parallel   **21.** neither   **23.** perpendicular   **25.** perpendicular
27. (a) (IV) **(b)** (VII) **(c)** (IX) **(d)** (II) **(e)** (I) **(f)** (V) **(g)** (VI) **(h)** (III) **(i)** (VIII)
29. $f(x) = 3x$ **31.** $f(x) = \tfrac{1}{4}x - 1$ **33.** $f(x) = -5x + 6$ **35.** $f(x) = -3x + \tfrac{9}{4}$
37. $f(x) = -x + 12$ **39.** $f(x) = 2x + 4$ **41.** $f(x) = x + 2$;   **43.** $x = 3$ (not a function)
45. 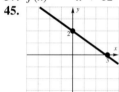 **47.** 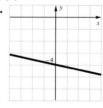 **49.** 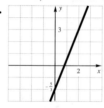 **51.**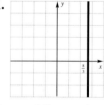

53. 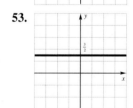 **55.** 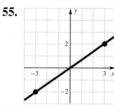 **57.** $x \approx 1.3$ 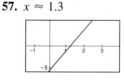 **59.** $x \approx 1.0$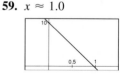

61. $x \approx -0.4$  **63.** $x \approx 1.8$ **65.** $x \approx -10.2$

Section 1.4 (page 72)

1. $C(x) = 1500x + 1200$ **3.** $q = -40p + 2000$ **5. (a)** $R(t) = 8.50 + 0.95t$ **(b)** \$19.9 billion   **(c)** The model becomes unreasonable for large positive and negative values of t. For instance, it gives negative revenue for the year 1979 ($t = -9$) and predicts revenues that rise without bound in the future.   **7.** $f = 9c/5 + 32$; 86°F; 71.6°F; 14°F; 6.8°F   **9.** $C(x) = 88x + 20$ **(a)** \$196   **(b)** \$88   **(c)** \$88   **(d)** \$88 per tuxedo
11. $I(N) = 5N/100 + 50{,}000$; $N = \$1{,}000{,}000$; marginal income is $m = 5¢$ per dollar of net profit   **13.** Fixed cost = \$8,000, marginal cost = \$25 per bicycle   **15. (a)** $v(n) = 60{,}000 - 3{,}000n$ **(b)** 19.67 years   **(c)** 20 years   **(d)** after 20 years; the model predicts negative value   **17.** $t(s) = 16s$ (t = recovery time in hours and s = number of sets). When $t(15) = 240$ hr, or 10 days! This indicates that our linear model is reliable only for small numbers of sets. We need a nonlinear model to predict recovery time for arbitrary numbers of sets.   **19.** $P(x) = 100x - 5{,}132$, with domain $[0, 405]$. For profit, $x \geq 52$.
21. (a) $P(x) = 396 - 0.05x$ (millions of dollars), with domain $[0, 1{,}100{,}000]$ **(b)** 47,920 homes damaged, or 4.36% of all the homes they insured   **23.** 5,000 units   **25.** $FC/(SP - VC)$
27. (a) $P(x) = 30x - 10{,}000$, with domain $[0, +\infty)$ **(b)** Solve $P(1{,}000 + y) = 40y$ giving $y = 2{,}000$ new customers.   **29.** $b = -0.2365n + 55.2$; $b = 0$ when $n = 233.4$; about midway through the year 2053.
31. $p(n) = 3000n + 2500$; 29,500 gal   **33.** $L(n) = 12.2 - 0.28n$, with domain $[0, 43.57]$

35. $P(x) = 579.7x - 20{,}000$, with domain $[0, +\infty)$; $x = 34.50$ g per day for breakeven
37. (a) $q(p) = -0.45p + 70.8$, with domain $[0, +\infty)$ **(b)** 62 pounds per year   **(c)** \$1.57 per pound; $[0, 157]$
(d) $R(p) = -0.45p^2 + 70.8p$, with domain $[0, 157]$ **39. (a)** $n(t) = 245 - 3.583t$ (t = years since 1920)
(b) 1940   **(c)** The model becomes unreliable for values of t beyond 60, predicting a negative number of cases in 1990.   **41. (a)** $m = -1500$, $b = 6000$; $q = -1500p + 6000$ **(b)** 6,000 per week
(c) $R = p(-1500p + 6000) = -1500p^2 + 6000p$ **(d)** \$2 per hamburger; $R = \$6{,}000$
43. (a) $C(x) = 400x + 30{,}000$, with domain $[0, 1{,}000]$ **(b)** $\overline{C}(x) = 400 + \dfrac{30{,}000}{x}$, with domain $(0, 1{,}000]$ **(c)** 600 items per month   **45.** 60 units

Section 1.5 (page 87)

1. vertex: $(-\frac{3}{2}, -\frac{1}{4})$; y-intercept: 2; x-intercepts: $-2, -1$

3. vertex: $(-\frac{1}{2}, -\frac{5}{4})$; y-intercept: -1; x-intercepts: $-1/2 \pm \sqrt{5}/2$

5. vertex: $(-2\sqrt{2}, -3)$; y-intercept: -1; x-intercepts: $2(-\sqrt{2} \pm \sqrt{3})$

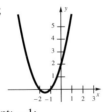

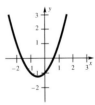

7. vertex: $(-1, 0)$; y-intercept: 1; x-intercept: -1

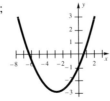

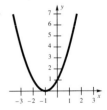

9. vertex: (0,0); y-intercept: 0; x-intercept: 0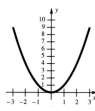

11. vertex (0,1); y-intercept: 1; no x-intercepts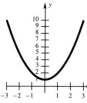

13. $R = -4p^2 + 100p$; maximum revenue when $p = \$12.50$.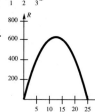

15. $R = -2p^2 + 400p$; maximum revenue when $p = \$100$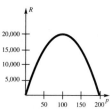

17. maximum revenue when $p = \$140$, $R = \$9,800$
19. maximum revenue with 70 houses, $R = \$9,800,000$
(b) $R = -0.5818p^2 + 36.245p$ **(c)** 31¢ per pound
20 mpg **25.** \$10 per pound **27. (a)** 4 seconds **(b)** True; the time the ball is airborne is given by $t = v_0/16$. Thus, doubling v_0 results in doubling t.
21. (a) $q = -0.5818p + 36.245$
23. maximum efficiency at 60 mph, efficiency =

Section 1.6 (page 94)

1. -3.45 ± 0.05, 1.45 ± 0.05 **3.** -2.55 ± 0.01 **5.** -1.886 ± 0.001, 0.503 ± 0.001, 1.622 ± 0.001
7. 1.17 ± 0.02 **9.** -0.47 ± 0.01, 0.54 ± 0.01, 3.94 ± 0.01 **11.** 1.34 ± 0.05 **13.** No solutions
15. 1.69 ± 0.05 **17.** They make a profit with 32 or more employees.   **19.** 6.95%

Chapter 1 Review Exercises (page 98)

1. $3x - y = 0$ **3.** $3x - y - 4 = 0$ **5.** $5x - 4y - 9 = 0$ **7.** $x + 3y + 5 = 0$ **9.** $x + 2y + 3 = 0$

11. 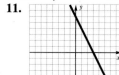   **13.** 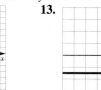   **15.**   **17.**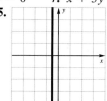

19. 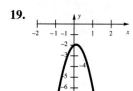   **21.** 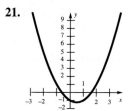   **23.** 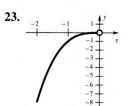   **25.**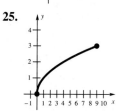

27. 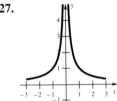 **29.** 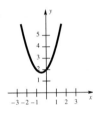 **31.** 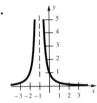 **33.**

35. 1.319 ± 0.005 **37.** $0, 2.924 \pm 0.005$ **39.** $0.68, 16.00 \pm 0.005$ **41.** (a) $\frac{1}{2}$ (b) $\frac{1}{3}$ (c) 1 (d) $\frac{1}{x}$
(e) $\frac{1}{x^2 + x + 2}$ (f) $x + \frac{1}{x^2 + 2}$ **43.** (a) 0 (b) 0 (c) $\sqrt{x + h - 1}$ (d) $\sqrt{x - 1} + h$

45. (a) 2 (b) $\frac{a^4 + 1}{a^2}$ (c) $\frac{(x + h)^2 + 1}{x + h}$ $- h$ (d) $\frac{x + 1}{\sqrt{x}} + h$

47. $2x + 1 + h$ **49.** $\frac{-4}{[2(x + h) - 1][2x - 1]}$ **51.** $p = \frac{9}{2}n + 96$. At age 90, $p = \$388.50$
53. $s = 55t + 45{,}000$ **55.** $v = 32t$; acceleration $= 32$ ft/s/s **57.** $300{,}000$ **59.** 12th year
61. $R = \$1{,}200$ when books were sold at \$10. Demand equation is $q = -6p + 180$; $R = -6p^2 + 180p$; $p = \$15$ maximizes revenue; total revenue at that price is \$1,350.
63. 50¢ **65.** (a) 21 (b) \$2 **67.** (a) $C(x) = 10{,}000 + 10x$ (b) $\bar{C}(x) = 10 + \frac{10{,}000}{x}$
(c) 4,000 fixtures **69.** (a) 43.59m (b) 0.9999, or 99.99% the speed of light (c) Its apparent length would be zero. **71.** (a) $I(x) = 100 + 2.75\sqrt{x}$ (b) $I(1) = 102.75, I(100) = 127.50$ (c) Approximately 1,322 sets per month. They will be lucky to have anyone work for them!

73. (a) $S(x) = 1.75x - 19$, with domain $[28, 48]$ (b) $S(x) = \begin{cases} 1.75x - 19 & \text{if } 28 \le x \le 48 \\ x + 17 & \text{if } 48 \le x \le 54 \\ 71 & \text{if } 54 \le x \le 58 \end{cases}$

75. $q = (500 - 100p)/p$; this gives the demand corresponding to the price p

Chapter 2

Section 2.1 (page 120)

1. $(2, 2)$ **3.** $(-1, 1)$ **5.** $(3, 1)$ **7.** $(6, 6)$ **9.** $(\frac{5}{3}, -\frac{4}{3})$ **11.** $(0, -2)$ **13.** $(x, \frac{1}{3}(1 - 2x))$
15. no solution **17.** $(20, 10)$ **19.** $(10, 10)$ **21.** 242 for and 193 against **23.** 200 gallons of vanilla and 100 gallons of mocha **25.** 2 servings of Mixed Cereal and 1 serving of Mango Tropical Fruit. **27.** \$3,333.33 in Fidelity and \$6,666.67 in Vanguard **29.** 5 soccer games and 7 football games **31.** 7 Brand X and 5 Brand Y
33. (a) 181 acidified lakes in the Adirondacks and 483 in Florida. (b) Approximately 14% of the lakes in the Adirondacks are polluted, while approximately 23% of Florida's lakes are polluted. Thus, Florida has the worse record. **35.** 33 pairs of dirty socks and 11 T-shirts **37.** 15 trips to the Bahamas and 30 VCRs
39. full-time: \$1,200, part-time: \$600 **41.** \$40 **43.** 55 widgets **45.** (a) $3\frac{1}{3}$ servings of beans, and $\frac{5}{6}$ slice of bread (b) yes: no beans and 5 slices of bread **47.** (a) $q = -900p + 9100$ (b) $q = 700p + 500$
(c) $p = \$5.38$ per can (to the nearest cent)

Section 2.2 (page 137)

1. $\begin{bmatrix} 1 & 0 & \frac{13}{4} \\ 0 & 1 & -\frac{5}{2} \end{bmatrix}$ **3.** $\begin{bmatrix} 1 & 0 & \frac{10}{7} \\ 0 & 1 & -\frac{1}{7} \\ 0 & 0 & 0 \end{bmatrix}$ **5.** $\begin{bmatrix} 1 & 0 & 0 \\ 0 & 1 & 0 \\ 0 & 0 & 1 \end{bmatrix}$ **7.** $(3, 1)$ **9.** $(5, 5)$ **11.** $(\frac{5}{3}, -\frac{4}{3})$

13. $(0, -2)$ **15.** $(\frac{1}{2}(1 - 3y), y)$ **17.** no solution **19.** $(20, 10)$ **21.** $(10, 10)$ **23.** $(\frac{1}{4}, \frac{3}{4})$
25. no solution **27.** $(\frac{10}{3}, \frac{1}{3})$ **29.** $(1.7467, -0.1620)$ **31.** $(8.1412, 1.3508)$ **33.** $(0.04153 + 0.0546y, y)$

Section 2.3 (page 147)

1. $(4, 4, 4)$ **3.** $(-1, -3, \frac{1}{2})$ **5.** $x = z, y = z, z$ arbitrary, or $(z, z, z), z$ arbitrary **7.** $(-17, 20, -2)$
9. $x = 1, y = z - 2, z$ arbitrary, or $(1, z - 2, z), z$ arbitrary **11.** $x = 4 + y, y$ arbitrary, $z = -1$, or $(4 + y, y, -1), y$ arbitrary **13.** $x = 4 - \frac{y}{3} + \frac{z}{3}, y$ arbitrary, z arbitrary, or $(4 - \frac{y}{3} + \frac{z}{3}, y, z), y$ arbitrary, z arbitrary **15.** $(-\frac{3}{2}, 0, \frac{1}{2}, 0)$ **17.** $(0, 0, 0)$ **19.** no solution **21.** $(-1, 1, 1)$ **23.** $(2z, 3z, z), z$ arbitrary **25.** $(-3z, 1 - 2z, z, 0), z$ arbitrary **27.** $(\frac{1}{5}(7 - 17x_3 + 8x_4), \frac{1}{5}(1 - 6x_3 - 6x_4), x_3, x_4)$, x_3, x_4 arbitrary **29.** $(6, 5, 4)$ **31.** $(x, y, z, p) = (1, 2, -1, 0)$ **33.** $(x, y, c, s, t) = (100 - (\frac{2}{3})s + (\frac{1}{3})t, 100 + (\frac{1}{3})s - (\frac{2}{3})t, 200 - (\frac{1}{3})s - (\frac{1}{3})t, s, t)$ **35.** $(x, y, z) = (\frac{4}{21}, \frac{5}{21}, \frac{12}{21})$ **37.** $(10, -20, 20)$ **39.** $(1, 2, 3, 4, 5)$
41. $(-2, -2 + z - u, z, u, 0), z, u$ arbitrary **43.** $(1.0, 1.4, 0.2)$ **45.** $(-5.5, -0.9, -7.4, -6.6)$
47. $(16, \frac{12}{7}, -\frac{162}{7}, -\frac{88}{7})$ **49.** $(-\frac{8}{15}, \frac{7}{15}, \frac{7}{15}, \frac{7}{15}, \frac{7}{15})$

Section 2.4 (page 156)

1. 100 batches of vanilla, 50 batches of mocha, 100 batches of strawberry **3.** 5 of each **5.** 7 of each
7. 3 sections of Finite Math, 2 sections of Business Calculus, and 1 section of Computer Methods **9.** Volvo A.B. up $2 million, Volvo Car Corp. down $1 million, Volvo Truck Corp. up $1 million **11.** 1980: 27,628 plans junked, 1985: 86,139 junked, and 1987: 73,443 junked. **13.** 22 tons from Cheesey Cream, 56 tons from Super Smooth & Sons, and 22 tons from Bagel's Best Friend **15.** 10 evil sorcerers, 50 warriors, and 500 orcs **17.** 9.5 million 32-year-olds, 9 million 50-year-olds, and 8 million 60-year-olds **19.** 200 Democrats, 20 Republicans, 13 of other parties **21.** (a) Brooklyn to Manhattan: 500 books; Brooklyn to Long Island: 500 books; Queens to Manhattan: 1,000 books; Queens to Long Island: 1,000 books (b) Brooklyn to Manhattan: 1,000 books; Brooklyn to Long Island: none; Queens to Manhattan: 500 books; Queens to Long Island: 1,500 books, giving a total cost of $8,000
23. Yes; $20 million in Company X; $5 million in Company Y, $10 million in Company Z, and $30 million in Company W **25.** (a) No; The general solution is: Eastward Blvd: $S + 200$; Northwest Lane: $S + 50$; Southwest Lane: S, where S is arbitrary. Thus, it would suffice to know the traffic along Southwest Lane. (b) yes, as it leads to the solution Eastward Blvd: 260; Northwest Lane: 110; Southwest Lane: 60 **27.** (a) If x = traffic on April St., y = traffic on Broadway, z = traffic on Division, and w = traffic on Embankment, then $x = -w + 400$, $y = w - 100$, $z = -w + 300$, and $100 \leq w \leq 300$. (b) Measuring the traffic flow on any of April, Broadway, Division, or Embankment will determine the answer. **29.** $10 billion **31.** It donated $600 to each of the MPBF and the SCN, and $1,200 to the Jets. **33.** x = water, y = grey matter, z = tumor **35.** x = water, y = bone, z = tumor, u = air **37.** tumor **39.** TWA: 30; Northwest: 90; American: 40

Chapter 2 Review Exercises (page 167)

1. $(\frac{6}{5}, \frac{7}{5})$ **3.** $(0, 0)$ **5.** $(\frac{5}{3}, \frac{1}{3})$ **7.** $(\frac{16}{41}, -\frac{2}{41})$ **9.** $(0, 0)$ **11.** no solution **13.** no solution
15. $(\frac{1}{4}, \frac{1}{3})$ **17.** $(0, 0)$ **19.** $(-\frac{7}{10}, \frac{17}{10})$ **21.** $(-2 + 3z, z, z), z$ arbitrary **23.** $(10, 10, 10)$ **25.** no solution
27. $(\frac{1}{2}, \frac{1}{2}, \frac{1}{2})$ **29.** $(1, 0, 0, 0)$ **31.** $(z, z, z), z$ arbitrary **33.** $(-1 - y + 2z, y, z), y, z$ arbitrary
35. $(2 - w, w, -1 + 2w, w), w$ arbitrary **37.** $(50 - w/2, 50 - w/2, 50 - w/2, w), w$ arbitrary

39. $\begin{bmatrix} 1 & 0 & \frac{4}{7} \\ 0 & 1 & -\frac{71}{14} \end{bmatrix}$ **41.** $\begin{bmatrix} 1 & 0 & 0 \\ 0 & 1 & 0 \\ 0 & 0 & 1 \end{bmatrix}$ **43.** $\begin{bmatrix} 1 & 0 & 0 & 0 \\ 0 & 1 & 0 & 0 \\ 0 & 0 & 1 & 0 & \frac{5}{4} \\ 0 & 0 & 0 & 1 & -\frac{1}{4} \end{bmatrix}$ **45.** $\begin{bmatrix} 1 & 0 & 0 & 0 \\ 0 & 1 & 0 & 0 \\ 0 & 0 & 1 & 0 \\ 0 & 0 & 0 & 1 \\ 0 & 0 & 0 & 0 \end{bmatrix}$

47. $\begin{bmatrix} 1 & 0 & 0 & 0 & 0 & \frac{1}{3} \\ 0 & 1 & 0 & \frac{1}{2} & -\frac{1}{2} & \frac{1}{4} \\ 0 & 0 & 1 & 1 & 0 & -\frac{1}{6} \end{bmatrix}$ **49.** $\begin{bmatrix} 1 & 0 & 0 & 0 \\ 0 & 1 & 0 & 0 \\ 0 & 0 & 1 & 0 \\ 0 & 0 & 0 & 1 \end{bmatrix}$

51. 30 Gauss Jordans, 30 Roebecks, and 60 K Scottish   **53.** French fries cost $0.90.   **55.** $44,000
57. ABC: 4.0, TBS: 2.5, ESPN: 4.0   **59.** yes; four possible combinations: 1 beer, 7 servings of wine and 2 servings of sherry; 2 beers, 4 servings of wine and 4 servings of sherry; 3 beers, 1 serving of wine and 6 servings of sherry.
61. A student is forced to take exactly the following combination: Liberal Arts: 52 credits, Sciences: 12 credits, Fine Arts: 12 credits, Mathematics: 48 credits.   **63.** Y will increase by 25 units.

Chapter 3

Section 3.1 (page 183)

1. 1×5 **3.** 4×1 **5.** $p \times q$ **7.** 2×2 **9.** $1 \times n$ **11.** 1   **13.** 44   **15.** -2 **17.** e_{13}
19. 2   **21.** 3   **23.** e **25.** d_n **27.** $x = 1, y = 2, z = 3, w = 4$

29. $\begin{bmatrix} \frac{1}{4} & -2 \\ 1 & \frac{1}{3} \\ -2 & 5 \\ 10 & 0 \end{bmatrix}$ **31.** $\begin{bmatrix} -\frac{3}{4} & -1 \\ 0 & -\frac{2}{3} \\ -1 & 6 \\ 9 & -1 \end{bmatrix}$ **33.** $\begin{bmatrix} -1 & -1 \\ 1 & -1 \\ -1 & 5 \\ 9 & -1 \end{bmatrix}$ **35.** $\begin{bmatrix} 0 & 2 & -2 & 10 \\ -2 & 0 & 4 & 0 \end{bmatrix}$ **37.** $\begin{bmatrix} 4 & -1 & -1 \\ 5 & 1 & 0 \end{bmatrix}$

39. $\begin{bmatrix} -2+x & 0 & 1+w \\ -5+z & 3+r & 2 \end{bmatrix}$ **41.** $\begin{bmatrix} -1 & -2 & 1 \\ -5 & 5 & -3 \end{bmatrix}$ **43.** $\begin{bmatrix} 9 & 15 \\ 0 & -3 \\ -3 & 3 \end{bmatrix}$ **45.** $\begin{bmatrix} -8.5 & -22.35 & -24.4 \\ 54.2 & 20 & 42.2 \end{bmatrix}$

47. $\begin{bmatrix} 1.54 & 8.58 \\ 5.94 & 0 \\ 6.16 & 7.26 \end{bmatrix}$ **49.** $\begin{bmatrix} 7.38 & 76.96 \\ 20.33 & 0 \\ 29.12 & 39.92 \end{bmatrix}$ **51.** $\begin{bmatrix} -19.85 & 115.82 \\ -50.935 & 46 \\ -57.24 & 94.62 \end{bmatrix}$

53. **(a)** Annual salaries in 1992 = Annual salaries in 1991 + Increase in 1992

$$= \begin{bmatrix} 890,000 \\ 675,000 \\ 275,000 \\ 275,000 \end{bmatrix} + \begin{bmatrix} 576 \\ 411 \\ 20,822 \\ 411 \end{bmatrix} = \begin{bmatrix} 890,576 \\ 675,411 \\ 295,822 \\ 275,411 \end{bmatrix}$$

(b) Annual salaries in 1993 = Annual salaries in 1992 + Increase in 1993

$$= \begin{bmatrix} 890,576 \\ 675,411 \\ 295,822 \\ 275,411 \end{bmatrix} + \begin{bmatrix} 34,424 \\ 24,589 \\ 54,370 \\ 24,781 \end{bmatrix} = \begin{bmatrix} 925,000 \\ 700,000 \\ 350,192 \\ 300,192 \end{bmatrix}$$

55. Total revenue = $[5 \quad 25 \quad 40 \quad 70 \quad 110 \quad 150] + [10 \quad 20 \quad 20 \quad 15 \quad 10 \quad 5]$
$= [15 \quad 45 \quad 60 \quad 85 \quad 120 \quad 155]$
Expenditures = Revenue − Profit
$= [15 \quad 45 \quad 60 \quad 85 \quad 120 \quad 155] - [0 \quad 0.5 \quad -5 \quad 2 \quad 0 \quad 4]$
$= [15 \quad 44.5 \quad 65 \quad 83 \quad 120 \quad 151]$

57. 1970 distribution = $A = [49.1 \quad 56.6 \quad 62.8 \quad 34.8]$
1980 distribution = $B = [49.1 \quad 58.9 \quad 75.4 \quad 43.2]$
Net change 1970 to 1980 = $B - A = [0 \quad 2.3 \quad 12.6 \quad 8.4]$ (all net increases)

59. Sales = $\begin{bmatrix} 700 & 1,300 & 2,000 \\ 400 & 300 & 500 \end{bmatrix}$ Inventory − Sales = $\begin{bmatrix} 300 & 700 & 3,000 \\ 600 & 4,700 & 1,500 \end{bmatrix}$

61. (a) $\begin{array}{c}\text{Pom II}\\ \text{Pom Classic}\end{array}\overset{\text{Proc. Mem. Tubes}}{\begin{bmatrix} 2 & 16 & 20 \\ 1 & 4 & 40 \end{bmatrix}}=\text{Use}\quad \text{Inventory}=\begin{bmatrix} 500 & 5{,}000 & 10{,}000 \\ 200 & 2{,}000 & 20{,}000 \end{bmatrix}$

$\text{Inventory}-100\cdot\text{Use}=\begin{bmatrix} 300 & 3{,}400 & 8{,}000 \\ 100 & 1{,}600 & 16{,}000 \end{bmatrix}$

(b) $\text{Inventory}-50x\cdot\text{Use}=\begin{bmatrix} 500-100x & 5{,}000-800x & 10{,}000-1{,}000x \\ 200-50x & 2{,}000-200x & 20{,}000-2{,}000x \end{bmatrix}$

Solving for x to see when the inventory is brought down to 0, this will happen first to Pom Classic processor chips after 4 months.

Section 3.2 (page 199)

1. $[13]$ **3.** $\begin{bmatrix}5\\6\end{bmatrix}$ **5.** $[-2y+z]$ **7.** $[3\ \ 0\ \ -6\ \ -2]$ **9.** $[-6\ \ 37\ \ 7]$

11. $\begin{bmatrix} -4 & 5 & -2 \\ 6 & -15 & 4 \end{bmatrix}$ **13.** $\begin{bmatrix} 0 & 1 \\ 0 & 0 \end{bmatrix}$ **15.** $\begin{bmatrix} 0 & 0 \\ 0 & 0 \end{bmatrix}$ **17.** $\begin{bmatrix} 1 & -5 & 3 \\ 0 & 0 & 9 \\ 0 & 4 & 1 \end{bmatrix}$ **19.** $\begin{bmatrix} 3 \\ -4 \\ 0 \\ 3 \end{bmatrix}$

21. $A^2=\begin{bmatrix} 0 & 0 & 1 & 2 \\ 0 & 0 & 0 & 1 \\ 0 & 0 & 0 & 0 \\ 0 & 0 & 0 & 0 \end{bmatrix}\quad A^3=\begin{bmatrix} 0 & 0 & 0 & 1 \\ 0 & 0 & 0 & 0 \\ 0 & 0 & 0 & 0 \\ 0 & 0 & 0 & 0 \end{bmatrix}\quad A^4=O\quad A^{100}=O$

23. $\begin{bmatrix} 4 & -1 \\ -1 & -7 \end{bmatrix}$ **25.** $\begin{bmatrix} 4 & -1 \\ -12 & 2 \end{bmatrix}$ **27.** $\begin{bmatrix} -2 & 1 & -2 \\ 10 & -2 & 2 \\ -10 & 2 & -2 \end{bmatrix}$ **29.** $\begin{bmatrix} -2+x-z & 2-r & -6+w \\ 10+2z & -2+2r & 10 \\ -10-2z & 2-2r & -10 \end{bmatrix}$

31. $\begin{bmatrix} -420 \\ -709.2 \\ -1{,}093.2 \end{bmatrix}$ **33.** $[3{,}243.4]$ **35.** $\begin{bmatrix} -0.012 & -1.32 \\ -1.44 & 1.32 \\ 0 & 0 \end{bmatrix}$

37. $[-3892.08]$ **39.** $\begin{bmatrix} 2.0736 & 0 & 0 \\ 0 & 2.0736 & 0 \\ 0 & 0 & 2.0736 \end{bmatrix}$ **41.** $\begin{bmatrix} 0.01 & 1.1 \\ 1.2 & -1.1 \\ 0 & 0 \end{bmatrix}$

43. $P^2=\begin{bmatrix} 0.375 & 0.625 \\ 0.3125 & 0.6875 \end{bmatrix}\quad P^4\approx\begin{bmatrix} 0.3359 & 0.6641 \\ 0.3320 & 0.6680 \end{bmatrix}\quad P^8\approx\begin{bmatrix} 0.3333 & 0.6667 \\ 0.3333 & 0.6667 \end{bmatrix}$

$P^{1{,}000}\approx\begin{bmatrix} 0.3333 & 0.6667 \\ 0.3333 & 0.6667 \end{bmatrix}$

45. $P^2=\begin{bmatrix} 0.25 & 0.25 & 0.50 \\ 0.25 & 0.25 & 0.50 \\ 0.25 & 0.25 & 0.50 \end{bmatrix}\quad P^4=\begin{bmatrix} 0.25 & 0.25 & 0.50 \\ 0.25 & 0.25 & 0.50 \\ 0.25 & 0.25 & 0.50 \end{bmatrix}\quad P^8=\begin{bmatrix} 0.25 & 0.25 & 0.50 \\ 0.25 & 0.25 & 0.50 \\ 0.25 & 0.25 & 0.50 \end{bmatrix}$

$P^{1{,}000}=\begin{bmatrix} 0.25 & 0.25 & 0.50 \\ 0.25 & 0.25 & 0.50 \\ 0.25 & 0.25 & 0.50 \end{bmatrix}$

47. $2x-y+4z=3;\ -4x+\dfrac{3y}{4}+\dfrac{z}{3}=-1;\ -3x=0$ **49.** $x-y+w=-1;\ x+y+2z+4w=2$

51. $\begin{bmatrix} 1 & -1 \\ 2 & -1 \end{bmatrix} \begin{bmatrix} x \\ y \end{bmatrix} = \begin{bmatrix} 4 \\ 0 \end{bmatrix}$ **53.** $\begin{bmatrix} 1 & 1 & -1 \\ 2 & 1 & 1 \\ \frac{3}{4} & 0 & \frac{1}{2} \end{bmatrix} \begin{bmatrix} x \\ y \\ z \end{bmatrix} = \begin{bmatrix} 8 \\ 4 \\ 1 \end{bmatrix}$

55. Revenue = Price × Quantity = $[15 \quad 10 \quad 12] \begin{bmatrix} 50 \\ 40 \\ 30 \end{bmatrix} = [1,510]$. So revenue = $1,510.

57. Worth in U.S.$ = $[500 \quad 100{,}000 \quad 100 \quad 50 \quad 200 \quad 20 \quad 20{,}000] \begin{bmatrix} 1.4950 \\ 0.009159 \\ 0.7526 \\ 1 \\ 0.1716 \\ 0.5256 \\ 0.000597 \end{bmatrix} = [\$1{,}845.43]$

59. AC^T represents the value, in each of the four listed currencies, of a wallet containing £10, ¥0, Can.$100 and U.S.$10.

61. $\frac{1}{4}A^2$ is approximately equal to A. The reason for this is that the ijth entry of A^2 gives the value, in units of currency i, of a purse containing the equivalent of four units of currency j, so this is 4 times the value of A_{ij}.

63. Price: $\begin{matrix} \text{Hard} \\ \text{Soft} \\ \text{Plastic} \end{matrix} \begin{bmatrix} 30 \\ 10 \\ 15 \end{bmatrix}$. $\begin{bmatrix} 700 & 1{,}300 & 2{,}000 \\ 400 & 300 & 500 \end{bmatrix} \begin{bmatrix} 30 \\ 10 \\ 15 \end{bmatrix} = \begin{bmatrix} \$64{,}000 \\ \$22{,}500 \end{bmatrix}$

65. $\begin{bmatrix} 2 & 16 & 20 \\ 1 & 4 & 40 \end{bmatrix} \begin{bmatrix} 100 & 150 \\ 50 & 40 \\ 10 & 15 \end{bmatrix} = \begin{bmatrix} \$1{,}200 & \$1{,}240 \\ \$700 & \$910 \end{bmatrix}$

67. 1987 distribution = $A = \begin{matrix} \text{NE} & \text{MW} & \text{S} & \text{W} \\ [49.3 & 59.1 & 82.2 & 48.2] \end{matrix}$

Population movements = P = $\begin{matrix} & \text{To} & \text{To} & \text{To} & \text{To} \\ & \text{NE} & \text{MW} & \text{S} & \text{W} \\ \text{From NE} \\ \text{From MW} \\ \text{From S} \\ \text{From W} \end{matrix} \begin{bmatrix} 0.9862 & 0.0023 & 0.009 & 0.0025 \\ 0.0015 & 0.986 & 0.0089 & 0.0036 \\ 0.003 & 0.005 & 0.9884 & 0.0036 \\ 0.0022 & 0.0047 & 0.0048 & 0.9883 \end{bmatrix}$

Distribution in 1988 = $AP = [49.1 \quad 59.0 \quad 82.4 \quad 48.3]$

Section 3.3 (page 213)

1. yes   **3.** yes   **5.** no   **7.** $\begin{bmatrix} -1 & 1 \\ 2 & -1 \end{bmatrix}$ **9.** $\begin{bmatrix} 0 & 1 \\ 1 & 0 \end{bmatrix}$ **11.** $\begin{bmatrix} 1 & 0 \\ 0 & 1 \end{bmatrix}$ **13.** $\begin{bmatrix} \frac{1}{2} & \frac{1}{2} \\ \frac{1}{2} & -\frac{1}{2} \end{bmatrix}$ **15.** $\begin{bmatrix} \frac{1}{3} & 0 \\ 0 & 2 \end{bmatrix}$

17. singular   **19.** $\begin{bmatrix} 6 & 6 \\ 0 & 6 \end{bmatrix}$ **21.** singular   **23.** $\begin{bmatrix} 1 & -1 & 0 \\ 0 & 1 & -1 \\ 0 & 0 & 1 \end{bmatrix}$ **25.** $\begin{bmatrix} 1 & -1 & 1 \\ \frac{1}{2} & 0 & -\frac{1}{2} \\ -\frac{1}{2} & 1 & -\frac{1}{2} \end{bmatrix}$

27. $\begin{bmatrix} 1 & \frac{1}{3} & -\frac{1}{3} \\ 1 & -\frac{2}{3} & -\frac{1}{3} \\ -1 & \frac{1}{3} & \frac{2}{3} \end{bmatrix}$ **29.** singular   **31.** $\begin{bmatrix} 0 & 1 & -2 & 1 \\ 0 & 1 & -1 & 0 \\ 1 & -1 & 2 & -1 \\ 0 & 1 & -1 & 1 \end{bmatrix}$ **33.** $\begin{bmatrix} 1 & -2 & 1 & 0 \\ 0 & 1 & -2 & 1 \\ 0 & 0 & 1 & -2 \\ 0 & 0 & 0 & 1 \end{bmatrix}$

35. $\begin{bmatrix} 0.38 & 0.45 \\ 0.49 & -0.41 \end{bmatrix}$ **37.** $\begin{bmatrix} 0.00 & -1.00 \\ 0.81 & 2.87 \end{bmatrix}$ **39.** singular **41.** $\begin{bmatrix} 91.35 & -8.65 & 0 & -71.30 \\ -0.07 & -0.07 & 0 & 2.49 \\ 2.60 & 2.60 & -4.35 & 1.37 \\ 2.69 & 2.69 & 0 & -2.10 \end{bmatrix}$

43. $x = 3, y = 1$ **45.** $x = 0, y = -2$ **47.** $(6, 6, 6)$ **49.** $(-1, -3, \frac{1}{2})$ **51.** $(-\frac{3}{2}, 0, \frac{1}{2}, 0)$

53. (a) $3\frac{1}{3}$ servings of beans, and $\frac{5}{6}$ slice of bread

(b) $\begin{bmatrix} -\frac{1}{2} & \frac{1}{6} \\ \frac{7}{8} & -\frac{5}{24} \end{bmatrix} \begin{bmatrix} A \\ B \end{bmatrix}$

55. (a) 100 gallons of PineOrange, 200 gallons of PineKiwi, and 150 gallons of OrangeKiwi each day.

(b) $62\frac{1}{2}$ gallons of PineOrange, $241\frac{2}{3}$ gallons of PineKiwi, and $158\frac{1}{3}$ gallons of OrangeKiwi each day.

(c) $\begin{bmatrix} \frac{1}{4} & \frac{1}{4} & -\frac{3}{4} \\ \frac{1}{6} & -\frac{1}{6} & \frac{1}{2} \\ -\frac{1}{6} & \frac{1}{6} & \frac{1}{2} \end{bmatrix} \begin{bmatrix} A \\ B \\ C \end{bmatrix}$

57. 1987 distribution $= A = \begin{bmatrix} 49.3 & 59.1 & 82.2 & 48.2 \end{bmatrix}$

Population movements $= P = \begin{matrix} \\ \\ \text{From NE} \\ \text{From MW} \\ \text{From S} \\ \text{From W} \end{matrix} \begin{matrix} \text{To} & \text{To} & \text{To} & \text{To} \\ \text{NE} & \text{MW} & \text{S} & \text{W} \end{matrix} \\ \begin{bmatrix} 0.9862 & 0.0023 & 0.009 & 0.0025 \\ 0.0015 & 0.986 & 0.0089 & 0.0036 \\ 0.003 & 0.005 & 0.9884 & 0.0036 \\ 0.0022 & 0.0047 & 0.0048 & 0.9883 \end{bmatrix}$

Distribution in 1986 $= AP^{-1} = \begin{bmatrix} 49.5 & 59.2 & 81.9 & 48.1 \end{bmatrix}$

Section 3.4 (page 227)

1. $X = [52{,}000 \quad 40{,}000]^T$ **3.** $X = [50{,}000 \quad 50{,}000]^T$ **5.** $X = [2{,}560 \quad 2{,}800 \quad 4{,}000]^T$
7. $X = [27{,}000 \quad 28{,}000 \quad 17{,}000]^T$ **9.** $X = [138{,}000 \quad 128{,}000 \quad 168{,}000 \quad 138{,}000]^T$

In Exercises 11–19, read off the answers from the columns of the given matrices.

11. $\begin{bmatrix} 2 & 1.6 \\ 0 & 2 \end{bmatrix}$ **13.** $\begin{bmatrix} 1.351 & 1.081 \\ 0.5405 & 2.432 \end{bmatrix}$ **15.** $\begin{bmatrix} 2 & 0.4 & 0.08 \\ 0 & 2 & 0.4 \\ 0 & 0 & 2 \end{bmatrix}$ **17.** $\begin{bmatrix} 1.375 & 0.5 & 0.125 \\ 0.5 & 2 & 0.5 \\ 0.125 & 0.5 & 1.375 \end{bmatrix}$

19. $\begin{bmatrix} 1.7143 & 0.2857 & 0 & 0 \\ 0.2857 & 1.7143 & 0 & 0 \\ 0 & 0 & 1.5556 & 0.4444 \\ 0 & 0 & 0.4444 & 1.5556 \end{bmatrix}$

In Exercises 21–31, read down the columns of the matrices for the answers.

21. $\begin{bmatrix} 1{,}062 & 0.1677 \\ 207.5 & 1{,}168 \end{bmatrix}$ **23.** $\begin{bmatrix} 1{,}236 & 28.6 & 2.89 & 32.5 \\ 366 & 1{,}050 & 2.68 & 30.1 \\ 0.123 & 0.084 & 1{,}006 & 1.96 \\ 63.4 & 43.3 & 89.8 & 1{,}010 \end{bmatrix}$ **25.** $\begin{bmatrix} 122.8 & 18.2 \\ 0.55 & 116.7 \end{bmatrix}$

27. $\begin{bmatrix} 109.2 & 1.93 & 60.2 & 29.3 \\ 0.29 & 101.1 & 0.49 & 3.1 \\ 0.78 & 0.87 & 115.1 & 4.3 \\ 4.3 & 3.9 & 4.0 & 118.1 \end{bmatrix}$ **29.** $\begin{bmatrix} 100{,}213 & 28{,}472.4 \\ 667.7 & 116{,}456 \end{bmatrix}$

31. $\begin{bmatrix} 101,820 & 24,293.3 & 11.7 & 432.7 \\ 0.035 & 103,070 & 45.9 & 505.2 \\ 0.044 & 11.9 & 119,322 & 630.3 \\ 7.0 & 45.2 & 288.7 & 100,054 \end{bmatrix}$

Chapter 3 Review Exercises (page 235)

1. 2×3; $A^T = \begin{bmatrix} 1 & 6 \\ 2 & 5 \\ 3 & 4 \end{bmatrix}$ **3.** 3×2; $C^T = \begin{bmatrix} 1 & 5 & 9 \\ 3 & 7 & 11 \end{bmatrix}$ **5.** 2×2; $E^T = \begin{bmatrix} 1 & -1 \\ -1 & 2 \end{bmatrix}$

7. 2×4; $G^T = \begin{bmatrix} 1 & -1 \\ -1 & 1 \\ 2 & -2 \\ -2 & 2 \end{bmatrix}$ **9.** undefined   **11.** $\begin{bmatrix} 4 & 4 & 4 \\ 3 & 3 & 3 \end{bmatrix}$ **13.** $\begin{bmatrix} 1 & 8 \\ 5 & 11 \\ 6 & 13 \end{bmatrix}$ **15.** undefined

17. $\begin{bmatrix} 1 & 3 \\ 2 & 3 \\ 3 & 3 \end{bmatrix}$ **19.** undefined   **21.** $\begin{bmatrix} -1 & 0 & -1 \\ 0 & 1 & 1 \end{bmatrix}$ **23.** undefined

25. $\begin{bmatrix} 14 & 32 \\ 32 & 77 \end{bmatrix}$ **27.** $\begin{bmatrix} 2 & -1 \\ -1 & 1 \end{bmatrix}$ **29.** $\begin{bmatrix} 1 & 1 \\ 0 & 1 \end{bmatrix}$ **31.** singular   **33.** $\begin{bmatrix} 1 & -\frac{1}{2} & -\frac{5}{2} \\ 0 & \frac{1}{4} & -\frac{1}{4} \\ 0 & 0 & 1 \end{bmatrix}$

35. singular   **37.** $\begin{bmatrix} 1 & -2 & 1 & 0 \\ 0 & 1 & -2 & 1 \\ 0 & 0 & 1 & -2 \\ 0 & 0 & 0 & 1 \end{bmatrix}$ **39.** singular   **41.** $\begin{bmatrix} 1 & 2 \\ 3 & 4 \end{bmatrix}\begin{bmatrix} x \\ y \end{bmatrix} = \begin{bmatrix} 0 \\ 2 \end{bmatrix}, \begin{bmatrix} x \\ y \end{bmatrix} = \begin{bmatrix} 2 \\ -1 \end{bmatrix}$

43. $\begin{bmatrix} 1 & 1 & 1 \\ 0 & 1 & 2 \\ 0 & 1 & -1 \end{bmatrix}\begin{bmatrix} x \\ y \\ z \end{bmatrix} = \begin{bmatrix} 3 \\ 4 \\ 1 \end{bmatrix}, \begin{bmatrix} x \\ y \\ z \end{bmatrix} = \begin{bmatrix} 0 \\ 2 \\ 1 \end{bmatrix}$ **45.** $\begin{bmatrix} 1 & 1 & 1 \\ 1 & 2 & 1 \\ 1 & 1 & 2 \end{bmatrix}\begin{bmatrix} x \\ y \\ z \end{bmatrix} = \begin{bmatrix} 2 \\ 3 \\ 1 \end{bmatrix}, \begin{bmatrix} x \\ y \\ z \end{bmatrix} = \begin{bmatrix} 2 \\ 1 \\ -1 \end{bmatrix}$

47. $\begin{bmatrix} 1 & 1 & 0 & 0 \\ 0 & 1 & 1 & 0 \\ 0 & 0 & 1 & 1 \\ 1 & 0 & 0 & -1 \end{bmatrix}\begin{bmatrix} x \\ y \\ z \\ w \end{bmatrix} = \begin{bmatrix} 0 \\ 1 \\ 0 \\ 3 \end{bmatrix}, \begin{bmatrix} x \\ y \\ z \\ w \end{bmatrix} = \begin{bmatrix} 1 \\ -1 \\ 2 \\ -2 \end{bmatrix}$ **49.** $\begin{bmatrix} 1 & 2 & 3 & 4 \\ 1 & 3 & 4 & 2 \\ 0 & 1 & 2 & 3 \\ 0 & 0 & 1 & 2 \end{bmatrix}\begin{bmatrix} x \\ y \\ z \\ w \end{bmatrix} = \begin{bmatrix} 0 \\ 3 \\ -1 \\ -1 \end{bmatrix}, \begin{bmatrix} x \\ y \\ z \\ w \end{bmatrix} = \begin{bmatrix} 1 \\ 0 \\ 1 \\ -1 \end{bmatrix}$

In Exercises 51–53, read down each column for the answers.

51. $\begin{bmatrix} 22,864 & 948 \\ 18 & 12,049 \end{bmatrix}$ **53.** $\begin{bmatrix} 11,407 & 14 & 85 & 3 \\ 12 & 22,864 & 948 & 147 \\ 89 & 18 & 12,050 & 3 \\ 0 & 0 & 0 & 11,341 \end{bmatrix}$

49. Make 200 quarts of Continental Mocha and no Succulent Strawberry for a profit of $600.   **51.** Use 4 ounces each of fish and cornmeal, for a total cost of 40¢ per can.   **53.** Make 240 Sprinkles, 120 Storms, and no Hurricanes, for a total profit of $360.   **55.** Invest $16,300 in TIAA and $3,700 in the stock fund.   **57.** Drive 3 cars from Northside to Eastside, 7 cars from Northside to Westside and 9 cars from the Airport to Eastside.
59. She should invest $2,500 in the stock fund and $7,500 in the annuity.

Chapter 5

Section 5.1 (page 342)

1. $F = \{\text{spring, summer, fall, winter}\}$ **3.** $I = \{1, 2, 3, 4, 5, 6\}$ **5.** $A = \{0, 1, 2, 3\}$ **7.** $B = \{0, 2, 4, 6, 8\}$
9. $S = \{(H, H), (H, T), (T, H), (T, T)\}$ **11.** $S = \{H1, H2, H3, H4, H5, H6, T1, T2, T3, T4, T5, T6\}$
13. $S = \{(1, 5), (2, 4), (3, 3), (4, 2), (5, 1)\}$ **15.** $S = \emptyset$ **17.** A **19.** A **21.** {June, Janet, Jill, Justin, Jeffrey, Jello, Sally, Solly, Molly, Jolly}   **23.** {Jello}   **25.** $\emptyset$ **27.** {Jello}   **29.** {Janet, Justin, Jello, Sally, Solly, Molly, Jolly}   **31.** $A \cap B = \{\text{Acme, Crafts}\}$ **33.** $B \cup C = \{\text{Acme, Brothers, Crafts, DeTour, Effigy, Global, Hilbert}\}$ **35.** $A' \cup C = \{\text{DeTour, Hilbert}\}$ **37.** $A \cap B' \cap C' = \emptyset$ **39.** {(1, 1), (1, 3), (1, 5), (3, 1), (3, 3), (3, 5), (5, 1), (5, 3), (5, 5)}   **41.** $\emptyset$ **43.** {(1, 1), (1, 3), (1, 5), (3, 1), (3, 3), (3, 5), (5, 1), (5, 3), (5, 5), (2, 2), (2, 4), (2, 6), (4, 2), (4, 4), (4, 6), (6, 2), (6, 4), (6, 6)}

Section 5.2 (page 350)

1. 9   **3.** 7   **5.** 4   **7.** $n(A \cup B) = 7$; $n(A) + n(B) - n(A \cap B) = 4 + 5 - 2 = 7$ **9.** 60   **11.** 20
13. 6   **15.** 9   **17.** 4   **19.** $n((A \cap B)') = 9$; $n(A') + n(B') - n((A \cup B)') = 6 + 7 - 4 = 9$
21. 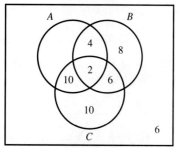   **23.**

25. 2   **27. (a)** 96   **(b)** 84   **29.** 6   **31.** 17

33. (a) 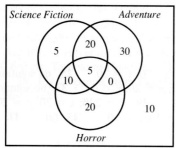   **(b)** 37.5%

Section 5.3 (page 361)

1. 720 **3.** 56 **5.** 360 **7.** 15
9. Step 1: Choose the outcome from the first die: 6 choices.
Step 2: Choose the outcome for the second die: 6 choices.
Thus, the total number of choices is $6 \times 6 = 36$.
11. $5! = 120$ **13.** $6 \times 5 \times 4 \times 1 = 120$ **15.** $2^{10} \times 5^2 = 1{,}024 \times 25 = 25{,}600$ **17.** $2 \times 2 \times 2 = 8$
19. (a) $10^7 = 10{,}000{,}000$ **(b)** $10^4 = 10{,}000$ **(c)** $10 \times 9^6 = 5{,}314{,}410$ **21.** $4! = 24$
23. $2 \times 5 \times 2 \times 4 = 80$ **25.** $4 \times 9 = 36$ **27. (a)** $4^3 = 64$ **(b)** 4^n **(c)** $4^{2.1 \times 10^{10}}$
29. $16^4 = 65{,}536$ **31.** (d)
33. Step 1: Choose a day of the week for January 1: 7 choices.
Step 2: Decide whether or not it is a leap year: 2 choices.
Total: $7 \times 2 = 14$ possible calendars
35. (a) $6 \times 3 \times 2 \times 2 \times 1 \times 1 = 72$ **(b)** $3 \times 3 \times 2 \times 2 \times 1 \times 1 = 36$ **37. (a)** $26^3 \times 10^3 = 17{,}576{,}000$
(b) $26^2 \times 23 \times 10^3 = 15{,}548{,}000$ **(c)** $15{,}548{,}000 - 3 \times 10^3 = 15{,}545{,}000$
39. $2 \times 2 \times 2 \times 2 = 16$ **41. (a)** $4 \times 1 = 4$ **(b)** 4 **(c)** There would be an infinite number of routes.
43. $(10 \times 9 \times 8 \times 7 \times 6 \times 5 \times 4) \times (8 \times 7 \times 6 \times 5) = 1{,}016{,}064{,}000$ possible casts
45. $2 \times 2 \times 2 \times 2 \times 2 \times 3 = 96$ paintings **47. (a)** $23!$ **(b)** $18!$ **(c)** $19 \times 18!$
49. Step 1: Choose a row: m choices.
Step 2: Choose a column: n choices.
Hence, there are $m \cdot n$ possible outcomes. **51.** $2^8 = 256$ **53.** 1,900

Section 5.4 (page 379)

1. 3 **3.** 45 **5.** 20 **7.** 4,950 **9.** $(2 \times 3) + 2 = 8$ **11.** $2^8 + 5^5 + 5! = 3{,}501$ **13.** (a)
15. 286 **17.** $C(10, 4) = 210$ **19.** $C(3, 3) \times C(7, 1) = 7$ **21.** $C(7, 4) = 35$
23. $3 \times 2 \times 2 \times 2 = 24$ **25.** $C(3, 2)C(7, 3) + C(3, 3)C(7, 2) = 126$ **27.** $C(2, 1)C(8, 4) + C(8, 5) = 196$
29. $C(1, 1)C(7, 4) + C(2, 1)C(7, 4) = 105$ **31.** $\dfrac{C(30, 5) \times 5^{25}}{6^{30}} = 0.192$ **33.** $\dfrac{C(30, 15) \times 3^{15} \times 3^{15}}{6^{30}} = 0.144$
35. 4 **37. (a)** 9,880 **(b)** 1,560 **(c)** $9{,}880 + 1{,}560 + 40 = 11{,}480$ **39.** $13 \times 6 \times 12 \times 6 \times 11 = 61{,}776$
41. $13 \times 6 \times 12 \times 11 \times 10 = 102{,}960$ **43.** $10 \times 4 \times 4 \times 4 \times 4 \times 4 - 10 \times 4 = 10{,}200$
45. Steps: (1) Select Boondoggle as a Do-nothing member. (2) Select the Chief Investigator from the Party Party.
(3) Select the Assistant Investigators from the Study Party. (4) Select the Rabble Rousers. (5) Select the other 2
Do-nothings. We thus get $C(1, 1)C(9, 1)C(9, 2)C(15, 2)C(13, 2)$.
*In the answers to Exercises 47–51, each step involves choosing slots for groups of letters. The order of the letters
selected is the order in which they appear in the given words (skipping repeats).*
47. $C(11, 1)C(10, 4)C(6, 4)C(2, 2)$ **49.** $C(11, 2)C(9, 1)C(8, 1)C(7, 3)C(4, 1)C(3, 1)C(2, 1)C(1, 1)$
51. $C(10, 2)C(8, 4)C(4, 1)C(3, 1)C(2, 1)C(1, 1)$ **53.** 12 **55. (a)** $C(20, 2) = 190$ **(b)** $C(n, 2)$

Chapter 5 Review Exercises (page 385)

1. $I = \{2, 4, 6, 8, 10\}$ **3.** $A = \{3, 5, 7, 9\}$
5. $S = \{HHHHH, HHHHT, HHHTH, HHHTT, HHTHH, HHTHT, HHTTH, HHTTT,$
   $HTHHH, HTHHT, HTHTH, HTHTT, HTTHH, HTTHT, HTTTH, HTTTT,$
   $THHHH, THHHT, THHTH, THHTT, THTHH, THTHT, THTTH, THTTT,$
   $TTHHH, TTHHT, TTHTH, TTHTT, TTTHH, TTTHT, TTTTH, TTTTT\}$
7. $S = \{(1, 2), (1, 3), (1, 4), (1, 5), (1, 6), (2, 1), (2, 3), (2, 4), (2, 5), (2, 6),$
   $(3, 1), (3, 2), (3, 4), (3, 5), (3, 6), (4, 1), (4, 2), (4, 3), (4, 5), (4, 6),$
   $(5, 1), (5, 2), (5, 3), (5, 4), (5, 6), (6, 1), (6, 2), (6, 3), (6, 4), (6, 5)\}$
9. $A \cup C = \{1, 2, 3, 4, 5, 6, 7\}$ **11.** $A \cup (B \cup C) = \{1, 2, 3, 4, 5, 6, 7\}$ **13.** $A \cap B = \{3, 4, 5\}$

15. $A \cap \emptyset = \emptyset$ **17.** $(A \cap B) \cap C = \{5\}$ **19.** the set of outcomes in which both dice show an even number   **21.** the set of outcomes in which both dice show an even number   **23.** the set of outcomes in which both dice show an even number or do not add to 7
25. 12   **27.** $26 \times 26 \times 26 = 17{,}576$ **29.** $26 \times 25 \times 9 \times 10 = 58{,}500$ **31.** 2 letters, 3 digits
33. 5 letters   **35. (a)** 6   **(b)** 6   **37.** $C(12, 5) = 792$ **39.** $C(4, 4)C(8, 1) = 8$
41. $C(12, 5) - C(4, 4)C(8, 1) = 784$ **43.** $C(12, 6) - C(4, 1)C(2, 1)C(1, 1)C(3, 1)C(2, 1) = 924 - 48 = 876$
45. $C(3, 2)C(9, 3) + C(3, 3)C(9, 2) = 288$ **47.** $C(4, 0)C(5, 5) + C(4, 1)C(5, 4) = 21$ **49.** $2C(4, 2)C(4, 3)$
51. $C(12, 1)C(4, 2)C(44, 1)C(40, 1)C(36, 1)/3!$ **53.** $C(1, 1)C(9, 1)C(8, 1)C(7, 3)$ **55.** $C(7, 3) + C(8, 2)C(7, 3)$

Chapter 6

Section 6.1 (page 399)

1. $S = \{HH, HT, TH, TT\}$; $E = \{HH, HT, TH\}$ **3.** $S = \{HHH, HHT, HTH, HTT, THH, THT, TTH, TTT\}$;
$E = \{HTT, THT, TTH, TTT\}$ **5.** $S = \{HHHH, HHHT, HHTH, HHTT, HTHH, HTHT, HTTH, HTTT, THHH,$
$THHT, THTH, THTT, TTHH, TTHT, TTTH, TTTT\}$; $E = \{HHHH, THHH, TTHH, TTTH, TTTT\}$
7.
$$S = \begin{matrix} (1,1) & (1,2) & (1,3) & (1,4) & (1,5) & (1,6) \\ (2,1) & (2,2) & (2,3) & (2,4) & (2,5) & (2,6) \\ (3,1) & (3,2) & (3,3) & (3,4) & (3,5) & (3,6) \\ (4,1) & (4,2) & (4,3) & (4,4) & (4,5) & (4,6) \\ (5,1) & (5,2) & (5,3) & (5,4) & (5,5) & (5,6) \\ (6,1) & (6,2) & (6,3) & (6,4) & (6,5) & (6,6) \end{matrix} \; ; E = \{(1,4), (2,3), (3,2), (4,1)\}$$
9. S same as in Exercise 7; $E = \emptyset$
11. S as in Exercise 7; $E = \{(2, 2), (2, 3), (2, 5), (3, 2), (3, 3), (3, 5), (5, 2), (5, 3), (5, 5)\}$
13. $S = \{m, o, z, a, r, t\}$; $E = \{o, a\}$
15. $S = \{(s, o), (s, r), (s, e), (o, s), (o, r), (o, e), (r, s), (r, o), (r, e), (e, s), (e, o), (e, r)\}$; $E = \{(o, s), (o, r), (o, e), (e, s), (e, o), (e, r)\}$
17. $S = \{(s, o), (s, r), (s, e), (o, s), (o, r), (o, e), (r, s), (r, o), (r, e), (e, s), (e, o), (e, r)\}$; $E = \{(s, o), (s, r), (s, e), (o, s), (o, r), (r, s), (r, o), (r, e), (e, s), (e, r)\}$
19. $S = \{01, 02, 03, 04, 10, 12, 13, 14, 20, 21, 23, 24, 30, 31, 32, 34, 40, 41, 42, 43\}$; $E = \{10, 20, 21, 30, 31, 32, 40, 41, 42, 43\}$
21. $E = \{\text{van, antique truck}\}$
23. $E = \{\text{specialists in Physical/Rehabilitation, Pulmonary, Hematology/Oncology, Cardiology, Nephrology}\}$
25. $E \cup F$ is the event that your hospital hires a specialist who will generate revenues of at least $1 million per year, or hires any specialist other than a cardiologist or a nephrologist.
$E \cup F = \{\text{specialists in Physical/Rehabilitation, Pulmonary, Hematology/Oncology, Cardiology, Nephrology, Endocrine, Infectious Disease}\}$
$E \cap F$ is the event that your hospital hires a specialist who will generate revenues of at least $1 million per year, but does not hire either a cardiologist or a nephrologist.
$E \cap F = \{\text{specialists in Physical/Rehabilitation, Pulmonary, Hematology/Oncology}\}$
27. (a) Mutually exclusive   **(b)** Not mutually exclusive   **29.** $E = \{\text{Bulls, Blackhawks}\}$ **31.** $E \cap F$ is the event that the team is among the top four in both gate receipts and operating income, and has player costs not exceeding $20 million; $E \cap F = \{\text{Bulls, Blackhawks}\}$. F' is the event that the team has player costs exceeding $20 million; $F' = \{\text{Bears, White Sox, Cubs}\}$.   **33. (a)** Not mutually exclusive   **(b)** Mutually exclusive   **35. (a)** The sample space is the set of all groups of five cards chosen from a standard deck of 52.   **(b)** The event "a full house" is the set of all groups of 5 cards, chosen from a standard deck of 52, in which three are of one denomination and two are of another denomination.   **37. (a)** $E = \{\text{Lindner Dividend, Vanguard Wellesley Income, Pru. Flex-A-Fund Con. Mgd. A}\}$.

(b) $F = $ {Franklin Income, Lindner Dividend}. (c) $G = F' = $ {Berwyn Income, National Multi-Sector FIA, Seligman Income A, Putnam Diversified Inc. A, Vanguard Wellesley Income, Income Fund of America, Vanguard Preferred Stock, Pru. Flex-A-Fund Con. Mgd. A}   **39.** (a) $E' \cap H$ (b) $E \cup H$ (c) $(E \cup G)' = E' \cap G'$
41. (a) {9} (b) {6}   **43.** (a) The dog's "fight" drive is weakest. (b) The dog's "fight" and "flight" drives are either both strongest or both weakest. (c) Either the dog's "fight" drive is strongest, or its "flight" drive is strongest.
45. $C(6, 4) = 15$; $C(1, 1)C(5, 3) = 10$ **47.** (a) $n(S) = P(7, 3) = 210$ (b) $E \cap F$ is the event that Boom Towner wins and Quickest Blade is in second or third place. In other words, it is the set of all lists of three horses in which Boom Towner is first and Quickest Blade is second or third. $n(E \cap F) = 10$.   **49.** $C(8, 3) = 56$
51. $C(4, 1)C(2, 1)C(2, 1) = 16$

Section 6.2 (page 417)

1. $P(E) = 0.4$ **3.** $P(E) = 0.85875$

5.

Outcome	HH	HT	TH	TT
Exp. Probability	0.3242	0.1712	0.3042	0.2004

7. 0.6284   **9.** The second coin seems biased in favor of heads, since heads comes up approximately 63% of the time.   **11.** $P(E) = \frac{1}{4}$ **13.** $P(E) = 1$ **15.** $P(E) = \frac{3}{4}$ **17.** $P(E) = \frac{3}{4}$ **19.** $P(E) = \frac{1}{2}$
21. $P(E) = \frac{5}{16}$ **23.** $P(E) = \frac{1}{9}$ **25.** $P(E) = 0$ **27.** $P(E) = \frac{1}{4}$ **29.** 0.5596   **31.** 0.3734
33. 0.3021   **35.** $C(6, 4) = 15$; $C(1, 1)C(5, 3) = 10$; $P(E) = \frac{10}{15} = \frac{2}{3}$

37.

Outcome	a	b	c	d	e
Probability	0.1	0.05	0.6	0.05	0.2

(a) 0.9 (b) 0.95 (c) 0.1 (d) 0.8

39. $S = $ {favor the law, oppose the law, undecided}

Outcome	favor	oppose	undecided
Probability	0.85	0.13	0.02

P(definite opinion) $= 0.98$

41.

Outcome	Medicaid	Medicare	Private	Other
Probability	0.49	0.05	0.45	0.01

P(Medicare or Medicaid) $= 0.54$

43.

Outcome	Low	Middle	High
Probability	0.5	0.2	0.3

45. $P(\text{false negative}) = 10/400 = 0.025$, $P(\text{false positive}) = 10/200 = 0.05$
47. $P(1) = P(6) = \frac{1}{10}$; $P(2) = P(3) = P(4) = P(5) = \frac{1}{5}$; $P(\text{odd}) = \frac{1}{2}$
49. $P(1, 1) = P(2, 2) = \ldots = P(6, 6) = \frac{1}{66}$; $P(1, 2) = \ldots = P(6, 5) = \frac{1}{33}$, $P(\text{odd sum}) = \frac{6}{11}$
51. $P(2) = \frac{15}{38}$; $P(4) = \frac{3}{38}$, $P(1) = P(3) = P(5) = P(6) = \frac{5}{38}$; $P(\text{odd}) = \frac{15}{38}$

Section 6.3 (page 427)

1. 0.7 **3.** 0.1 **5.** 0.65 **7.** 0.4 **9.** 0.25 **11.** 1.0 **13.** 0.3 **15.** 1.0 **17.** 0.5 **19.** 1.0
21. $\frac{1}{3}$ **23.** $1 - 0.13 = 0.87$ **25.** all of them **27.** 0.11
29. (a) $1 - 0.17 = 0.83$ **(b)** $1 - 0.33 = 0.67$ **(c)** 0.64 **(d)** 0.19 **31.** $0.8 + 0.1 - 0.016 = 0.884$

Section 6.4 (page 437)

1. $C(4, 4)C(6, 1)/C(10, 5) = \frac{6}{252} = \frac{1}{42}$ **3.** $(C(2, 1)C(8, 4) + C(2, 2)C(8, 3))/C(10, 5) = \frac{196}{252} = \frac{7}{9}$
5. $C(4, 2)C(3, 1)C(2, 1)C(1, 1)/C(10, 5) = \frac{36}{252} = \frac{1}{7}$ **7.** $(C(7, 5) + C(3, 1)C(7, 4))/C(10, 5) = \frac{126}{252} = \frac{1}{2}$
9. $1 - C(4, 4)C(6, 1)/C(10, 5) = \frac{41}{42}$ **11.** $\frac{1}{2}$ **13.** $\frac{4}{15}$ **15.** $\frac{1}{3}$
17. $C(13, 1)C(4, 2)C(12, 3)C(4, 1)C(4, 1)C(4, 1)/C(52, 5) = 1{,}098{,}240/2{,}598{,}960 \approx 0.4226$
19. $C(13, 2)C(4, 2)C(4, 2)C(44, 1)/C(52, 5) = 123{,}552/2{,}598{,}960 \approx 0.0475$
21. $(C(4, 1)C(13, 5) - C(4, 1)C(10, 1))/C(52, 5) = 5{,}108/2{,}598{,}960 \approx 0.0020$
23. Probability of being a big winner $= 1/C(50, 5) = 1/2{,}118{,}760 \approx 0.000000472$
Probability of being a small-fry winner $= C(5, 4)C(45, 1)/C(50, 5) = 225/2{,}118{,}760 \approx 0.000106194$
Probability of being either a big winner or a small-fry winner $\approx 0.000000471 + 0.000106194 = 0.000106665$
25. (a) $C(600, 300)/C(700, 400)$ **(b)** $C(699, 399)/C(700, 400) = \frac{4}{7}$ **27.** $P(10, 3)/10^3 = \frac{18}{25}$ **29.** $8!/8^8$
31. $1/27^{39}$ **33.** $\frac{1}{7}$ **35.** $\frac{1}{8}$ **37.** $1/(2^8 \times 5^5 \times 5!)$ **39.** $\frac{1}{8}$ **41.** $37/10{,}000$ **43.** 0.6009
45. (a) $C(6, 1)C(6, 1)C(10, 2)C(8, 5) = 90{,}720$
(b) $C(1, 1)C(1, 1)C(10, 2)C(8, 5) + C(5, 1)C(6, 1)C(9, 2)C(7, 5) = 25{,}200$ **(c)** $25{,}200/90{,}720 = \frac{5}{18}$

Section 6.5 (page 453)

1. $\frac{1}{10}$ **3.** $\frac{1}{5}$ **5.** $\frac{2}{9}$ **7.** $\frac{1}{2}$ **9.** 0 **11.** $\frac{1}{84}$ **13.** $\frac{5}{21}$ **15.** $\frac{24}{175}$ **17.** 0 **19.** $\frac{2}{5}$
21. 0.199 **23.** 0.050 **25.** 0.1280 **27.** 0.0673 **29.** 0.6266 **31.** 0.6580 **33.** 0.1455
35. 0.1088 **37.** 0.3086
39. The claim is correct. The probability that a woman with less than four years of high school education will be employed in the service industry is 0.3924, while the corresponding figure for men is 0.1230.
41. $C(12, 2)C(12, 1)C(26, 1)/[C(52, 5) - C(13, 2)C(13, 1)C(26, 2)]$ **43.** $C(11, 1)/C(49, 2)$
45.

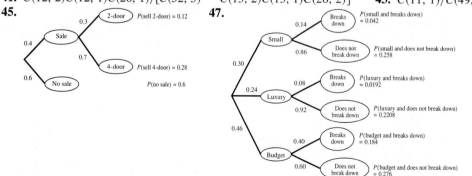

49.

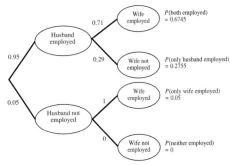

51. dependent **53.** dependent **55.** independent
57. 0.25

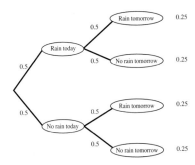

59. $(\frac{1}{2})^{11} = 1/2048$ **61.** 11.11%
63. $P(\text{AIDS and positive}) = 0.0099$, $P(\text{not AIDS and positive}) = 0.0495$
65. not independent; $P(\text{giving up} \mid \text{used Brand X}) = 0.1$ is larger than $P(\text{giving up})$

Section 6.6 (page 464)

1. 0.4   **3.** 0.7887   **5.** 0.7442   **7.** 0.1163   **9.** 0.71   **11.** (a) 0.16 (b) 0.83   **13.** 9
15. (a) 14.43%; (b) 19.81% of single homeowners have pools. Thus, they should go after the single homeowners.
17. (a) 45% (b) The first thought is "no," since only 45% of the people who considered family values as the number-one issue favored Bush. Most of them (the other 55%) did not favor Bush. On the other hand, you can calculate that 34% of the "family values" voters favored Clinton, so that an increased emphasis on family values would wind up decreasing Clinton's support by a wider margin and swelling the undecided category.
19. 2.564%   **21.** 0.0123   **23.** 0.93   **25.** 0.54%   **27.** 0.20

Section 6.7 (page 472)

1. $C(8, 2)(\frac{2}{3})^2(\frac{1}{3})^6$ **3.** $C(8, 0)(\frac{2}{3})^0(\frac{1}{3})^8 = (\frac{1}{3})^8$ **5.** $C(8, 8)(\frac{2}{3})^8(\frac{1}{3})^0 = (\frac{2}{3})^8$
7. $C(8, 3)(\frac{2}{3})^3(\frac{1}{3})^5 + C(8, 2)(\frac{2}{3})^2(\frac{1}{3})^6 + C(8, 1)(\frac{2}{3})^1(\frac{1}{3})^7 + C(8, 0)(\frac{2}{3})^0(\frac{1}{3})^8$
9. $C(8, 0)(\frac{2}{3})^0(\frac{1}{3})^8 + C(8, 1)(\frac{2}{3})^1(\frac{1}{3})^7 + C(8, 2)(\frac{2}{3})^2(\frac{1}{3})^6 + C(8, 3)(\frac{2}{3})^3(\frac{1}{3})^5$
11. $C(6, 2)(\frac{1}{3})^2(\frac{2}{3})^4 = \frac{80}{243}$ **13.** $C(6, 4)(\frac{1}{3})^4(\frac{2}{3})^2 + C(6, 5)(\frac{1}{3})^5(\frac{2}{3})^1 + C(6, 6)(\frac{1}{3})^6(\frac{2}{3})^0 \approx 0.1001$
15. $1 - C(6, 6)(\frac{1}{3})^6(\frac{2}{3})^0 = \frac{728}{729} \approx 0.9986$

17. 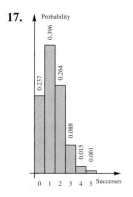 **19.** 0.875   **21.** 0.96   **23.** 0.79   **25.** 0.28   **27.** 0.83   **29.** at least 69 bulbs

Chapter 6 Review Exercises (page 475)

1. $S = \{HHH, HHT, HTH, HTT, THH, THT, TTH, TTT\}$,
   $E = \{HHT, HTH, HTT, THH, THT, TTH, TTT\}$, $P(E) = \frac{7}{8}$
3. $S = \{HHHH, HHHT, HHTH, HHTT, HTHH, HTHT, HTTH, HTTT, THHH, THHT, THTH, THTT,$
   $TTHH, TTHT, TTTH, TTTT\}$,
   $E = \{HTTT, THTT, TTHT, TTTH, TTTT\}$, $P(E) = \frac{5}{16}$
5. $S = \{(1, 1), (2, 1), \ldots, (6, 6)\}$,
   $E = \{(1, 6), (2, 5), (3, 4), (4, 3), (5, 2), (6, 1)\}$, $P(E) = \frac{1}{6}$
7. $S = \{(1, 1), (2, 1), \ldots, (6, 6)\}$,
   $E = \{(1, 1), (1, 3), (1, 5), (2, 2), (2, 4), (2, 6), (3, 1), (3, 3), (3, 5), (4, 2), (4, 4), (4, 6), (5, 1), (5, 3), (5, 5), (6, 2),$
   $(6, 4), (6, 6)\}$, $P(E) = \frac{1}{2}$
9. $P(2) = P(3) = P(4) = P(5) = \frac{1}{8}$; $P(1) = P(6) = \frac{1}{4}$
11. $P(1) = \frac{3}{25}$; $P(6) = \frac{6}{25}$; $P(2) = P(3) = P(4) = P(5) = \frac{4}{25}$ **13.** 0.4
15. 56 possible outcomes; 6 of these include the Krugerrands; probability of grabbing both Krugerrands $= \frac{3}{28}$
17. $2C(1, 1)C(30, 1)/C(31, 2) = \frac{4}{31}$ **19.** $C(25, 3)/C(31, 3)$ **21.** $C(1, 1)C(3, 1)C(27, 1)C(26, 1)/C(31, 2)C(29, 2)$
23. $C(3, 1)C(28, 1)C(2, 1)C(27, 1)/C(31, 2)C(29, 2)$ **25.** $C(8, 5)/C(52, 5)$ **27.** $C(20, 5)/C(52, 5)$
29. $C(6, 1)C(5, 1)C(4, 3)C(4, 2)/C(52, 5)$
31. $\frac{1}{15}$ **33.** $\frac{1}{6}$ **35.** $\frac{1}{4}$ **37.** dependent   **39.** independent   **41.** independent
43. (a) 0.0975   (b) $0.05(1 + 0.95 + (0.95)^2) \approx 0.1426$ (c) $0.05(1 + 0.95 + (0.95)^2 + (0.95)^3) \approx 0.1855$
45. (a) $\frac{2}{31}$ (b) $\frac{1}{300}$ **47.** $4(\frac{1}{11})^1(\frac{10}{11})^3$ **49.** $(\frac{10}{11})^4 + 4(\frac{1}{11})^1(\frac{10}{11})^3 + C(4, 2)(\frac{1}{11})^2(\frac{10}{11})^2$ **51.** 0.034

Chapter 7

Section 7.1 (page 490)

1. (a) $S = \{HH, HT, TH, TT\}$ (b) X is the rule that assigns to each outcome the number of tails.   (c) $X(HH) = 0$, $X(HT) = 1$, $X(TH) = 1$, $X(TT) = 2$ **3.** (a) $S = \{(1, 1), (1, 2), \ldots, (1, 6), (2, 1), (2, 2), \ldots, (6, 6)\}$ (b) X is the rule that assigns to each outcome the sum of the two numbers.   (c) $X(1, 1) = 2$, $X(1, 2) = 3, \ldots, X(6, 6) = 12$
5. (a) $S = \{(4, 0), (3, 1), (2, 2)\}$ (listed in order (red, green))   (b) X is the rule that assigns to each outcome the number of red marbles.   (c) $X(4, 0) = 4$, $X(3, 1) = 3$, $X(2, 2) = 2$ **7.** (a) $S =$ the set of students in the study group   (b) X is the rule that assigns to each student his or her final exam score.   (c) The values of X, in the order given, are 89%, 85%, 95%, 63%, 92%, 80%.

9.

x	−5	−4	−3	−2	−1	0	1	2	3	4	5
$P(X = x)$	$\frac{1}{36}$	$\frac{2}{36}$	$\frac{3}{36}$	$\frac{4}{36}$	$\frac{5}{36}$	$\frac{6}{36}$	$\frac{5}{36}$	$\frac{4}{36}$	$\frac{3}{36}$	$\frac{2}{36}$	$\frac{1}{36}$

11.

x	1	2	3	4	5	6
$P(X = x)$	$\frac{1}{36}$	$\frac{3}{36}$	$\frac{5}{36}$	$\frac{7}{36}$	$\frac{9}{36}$	$\frac{11}{36}$

13.

x	−3	−1	1	3
$P(X = x)$	$\frac{1}{8}$	$\frac{3}{8}$	$\frac{3}{8}$	$\frac{1}{8}$

15.

1.01–2.0	2.01–3.0	3.01–4.0
4	7	9
1.5	2.5	3.5
$\frac{4}{20}$	$\frac{7}{20}$	$\frac{9}{20}$

17.

x	0	1	2	3	4	5
$P(X = x)$	0.59049	0.32805	0.0729	0.0081	0.00045	0.00001

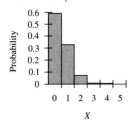

$P(X \geq 3) = 0.00856$

19.

x	0	1	2	3	4
$P(X = x)$	$\frac{1{,}712{,}304}{2{,}598{,}960}$	$\frac{778{,}320}{2{,}598{,}960}$	$\frac{103{,}776}{2{,}598{,}960}$	$\frac{4{,}512}{2{,}598{,}960}$	$\frac{48}{2{,}598{,}960}$

$P(X \leq 2) = \dfrac{2{,}594{,}400}{2{,}598{,}960} \approx 0.998$

21.

x	1	2	3	4
$P(X = x)$	$\frac{4}{35}$	$\frac{18}{35}$	$\frac{12}{35}$	$\frac{1}{35}$

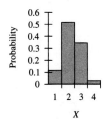

$P(X \geq 2) = \frac{31}{35} \approx 0.886$

23. $P(X \geq 10) = \frac{12}{28}$ **25.** $P(1 \leq X \leq 5) = \frac{6}{28}$ **27.** $\frac{19}{28}$ **29.** 12 wet days

31. $1{,}910/2{,}000 = 95.5\%$

33.

x	0	1.5	7.5	36	90
$P(X = x)$	0.67	0.11	0.08	0.1	0.04

35.

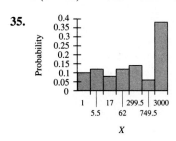

37. (a)

x	150,000	250,000	350,000	450,000
$P(X = x)$	0.06897	0.4483	0.4138	0.06897

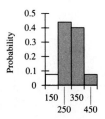

(b) See shaded part of histogram above.
$P(X \geq 200{,}000) = \frac{27}{29} \approx 0.931$

39.

x	0	1	2	3	4
$P(X = x)$	0.0625	0.25	0.375	0.25	0.0625

41.

x	0	1	2	3	4	5
$P(X = x)$	0.131687	0.329218	0.329218	0.164609	0.0411523	0.0041152

45.

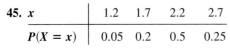

Approximately normal

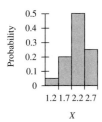

Section 7.2 (page 509)

1. 2.833 **3.** −0.1875 **5.** 0.44 **7.** expected value = 21, median = 20, mode = 20 **9.** expected value = −0.1, median = −1, mode = −1 **11.** expected value = $31,000, median = $25,000, mode = $25,000 **13.** expected value = 10,999.5, median = 7,499.5, mode = 7,499.5 **15.** expected value = 1, median = 1, mode = 1 **17.** expected value = 4.472, median = 5, mode = 6 **19.** expected value = 2.667, median = 3, mode = 3 **21.** expected value = 0, median = 0, mode = 0 **23.** expected value = 2, median = 2, mode = 2 **25.** expected value = 0.385, median = 0, mode = 0 **27.** 14.5% **29.** average = 39.1, median = 40, mode = 7.5 **31.** 3.4 years older than the mean, 3 years older than the median, 3 years older than the mode **33.** average = 1,237, median = 299.5, mode = 3,000 **35.** (a) average = $298,000 (b) average = $295,000 **37.** a loss of $29,390 **39.** expect to lose 5.3¢ **41.** 78 passes **43.** (a) 1 trip: 0.00000165; 2 trips: 0.0000033; 3 trips: 0.00000495; n trips: $1 - (0.99999835)^n$ (b) 606,000 trips

Section 7.3 (page 529)

1. average = 5, variance = 11.7, standard deviation = 3.42 **3.** average = 3.6, variance = 0.34, standard deviation = 0.583 **5.** average = 0.625, variance = 1.37, standard deviation = 1.17 **7.** average = 1, variance = 0, standard deviation = 0 **9.** expected value = 21, variance = 89, standard deviation = 9.43 **11.** expected value = −0.1, variance = 10.69, standard deviation = 3.27 **13.** expected value = $31,000, variance = 199,000,000, standard deviation = $14,100 **15.** expected value = 10,999.5, variance = 42,750,000, standard deviation = 6,538 **17.** expected value = 1, variance = 0.5, standard deviation = 0.707 **19.** expected value = 4.472, variance = 1.97, standard deviation = 1.40 **21.** expected value = 2.667, variance = 0.356, standard deviation = 0.596 **23.** expected value = 0, variance = 5.83, standard deviation = 2.42 **25.** expected value = 2, variance = 1.8, standard deviation = 1.34 **27.** expected value = 0.385, variance = 0.327, standard deviation = 0.572 **29.** $P(9.4 \leq X \leq 10.6) \geq \frac{3}{4}$ **31.** $P(-12.4 \leq X \leq 12.4) \geq \frac{15}{16}$ **33.** $9.05 \leq X \leq 10.95$ **35.** $-9.8 \leq X \leq 9.8$ **37.** 93.75% **39.** $21,886 to $338,114 **41.** expected value = 2.58%, standard deviation = 0.18% **43.** expected value = 4.76%, standard deviation = 0.50% **45.** 2.22% to 2.94% **47.** 2.01% to 3.15% **49.** (a) expected value = 1987.7, standard deviation = 3.1 years (b) 1981 to 1994 **51.** 0 to 4,350 **53.** $196,000 to $400,000 **55.** $P(X < 5 \text{ or } X > 20) \leq 0.194$, so $P(X > 20) \leq 0.194$

Section 7.4 (page 545)

1. 0.1915 **3.** 0.5222 **5.** 0.6710 **7.** 0.2417 **9.** 0.8664 **11.** 0.8621 **13.** 0.2286 **15.** 0.3830 **17.** 0.5052 **19.** 0.3502 **21.** 0.0475 **23.** 26,400,000 **25.** This is surprising, because the time between failures was more than 5 standard deviations away from the mean, which happens with an extremely small probability.

27. 0   **29.** about 6,680   **31.** 0.8708   **33.** 0.0029   **35.** Not necessarily. If X is the number of breast cancer cases in the neighborhood, then $P(X \geq 16) \approx 0.18$, which is not negligible.   **37.** 0.1825   **39.** Probability that a person will say Goode = 0.54. Probability that Goode polls more than 52% ≈ 0.892.   **41.** standard deviation = 23.4

Chapter 7 Review Exercises (page 550)

1.

x	0	1	2
$P(X = x)$	$\frac{1}{4}$	$\frac{1}{2}$	$\frac{1}{4}$

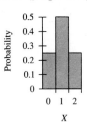

3.

x	2	3	4	5	6	7	8
$P(X = x)$	$\frac{1}{16}$	$\frac{2}{16}$	$\frac{3}{16}$	$\frac{4}{16}$	$\frac{3}{16}$	$\frac{2}{16}$	$\frac{1}{16}$

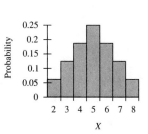

5.

x	0	1	2	3	4	5
$P(X = x)$	0.2373	0.3955	0.2637	0.08789	0.01465	0.00098

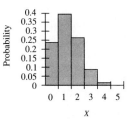

7.

x	0	1	2	3	4	5
$P(X = x)$	0.08392	0.3497	0.3996	0.1499	0.01665	0.0003330

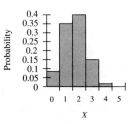

9.

x	0	1	2	3	4	5
$P(X = x)$	0.3106	0.4313	0.2098	0.04418	0.003965	0.0001189

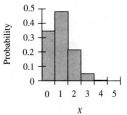

11. expected value = 1, median = 1, mode = 1, variance = 0.5, standard deviation = 0.7071
13. expected value = 5, median = 5, mode = 5, variance = 2.5, standard deviation = 1.581
15. expected value = 1.25, median = 1, mode = 1, variance = 0.9375, standard deviation = 0.9682
17. expected value = 1.667, median = 2, mode = 2, variance = 0.794, standard deviation = 0.891

19. expected value = 1, median = 1, mode = 1, variance = 0.7347, standard deviation = 0.8571
21. $49.4 \leq X \leq 150.6$ **23.** $-3.16 \leq X \leq 3.16$ **25.** $-2.58 \leq X \leq 0.58$
27. $P(80 \leq X \leq 120) = 0.789$ **29.** $P(-1 \leq X \leq 3) = 0.625$ **31.** $P(-1 \leq X \leq 0) = 0.477$
33. $73.7 \leq X \leq 126.3$ **35.** $-3.29 \leq X \leq 3.29$ **37.** $-1.82 \leq X \leq -0.18$ **39.** 0.4041

41.

x	750–799	800–849	850–899	900–949	950–999
fr	1	2	2	6	1
x	1000–1049	1050–1099	1100–1149	1150–1199	1200–1249
fr	2	1	4	0	1

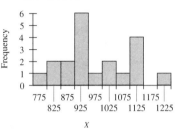

Original data: average = 972, median = 930, mode = 930, standard deviation = 121
Frequency table: average = 977.5, median = 925, mode = 925, standard deviation = 120
43. average = 119.5, standard deviation = 31
45. average = 26, median = 22.5, mode = 22.5, standard deviation = 7.9   **47.** 325,000 people   **49.** Yes, the probability of 20 or more cases is about 0.001.   **51.** 0.177

▶ Chapter 8

Section 8.1 (page 560)

1. $120   **3.** $505   **5.** $250   **7.** 5%   **9.** 8 years   **11.** $787.40   **13.** 19.38%   **15.** 21.63%
17. 1990–1991; 26.8% increase   **19.** No. The net income rose by 106 billion yen from 1989 to 1990 but by 175 billion yen from 1990 to 1991. Since the interest changed, this was not simple-interest growth.   **21.** 12%
23. $9.7 billion
25.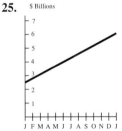

Section 8.2 (page 573)

1. $13,439.16   **3.** $11,327.08   **5.** $19,154.30   **7.** $12,709.44   **9.** $613.91   **11.** $810.65
13. $1,227.74   **15.** 5.09%   **17.** 10.47%   **19.** 10.52%   **21.** $26.90   **23.** $2,109.60
25. $21,161.79   **27.** $120,537.54   **29.** $55,526.45 per year   **31.** $377.63   **33.** $7,835.26
35. $1,039.21   **37.** The better investment is the one earning 11.9% compounded monthly.   **39.** 28,382,689 cruzados   **41.** 270 pesos   **43.** 1,363 pesos   **45.** The Ecuadorian investment is better: it yields 1.0857 units of currency (in constant units) per unit invested as opposed to 1.01190 units for Columbia.

47. 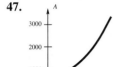 **49.**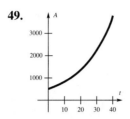

51.

Years	1	2	3	4	5	6	7
Value	$1,050	$1,103	$1,158	$1,216	$1,276	$1,340	$1,407

53. 16 years; approximate value $26,900   **55.** 2.3 years   **57.** 8 years   **59.** 132 months   **61.** 5 years
63. 3 years   **65.** 14 years   **67.** 7.4 years

Section 8.3 (page 592)

1. $15,528.23   **3.** $171,793.82   **5.** $147.05   **7.** $491.12   **9.** $90,155.46   **11.** $69,610.99
13. $554.60   **15.** $1,366.41   **17.** $248.85   **19.** $1,984.65   **21.** $973.54   **23.** $7,451.49
25. You should take the loan from Solid Savings & Loan: it will have payments of $248.85 per month. The payments on the other loan would be more than $300 per month.
27.

Year	Interest	Payments on Principal
1	$3,935.44	$1,798.52
2	3,785.71	1,948.25
3	3,624.00	2,109.96
4	3,448.88	2,285.08
5	3,259.23	2,474.73
6	3,053.82	2,680.14
7	2,831.36	2,902.60
8	2,590.46	3,143.50
9	2,329.55	3,404.41
10	2,046.98	3,686.98
11	1,740.96	3,993.00
12	1,409.55	4,324.41
13	1,050.62	4,683.34
14	661.91	5,072.05
15	240.93	5,493.03

29. first five years: $402.62/month; last 25 years: $601.73   **31.** Original monthly payments were $824.79. The new monthly payments will be $613.46. You will save $36,488.88 in interest.   **33.** 13 years   **35.** 4.5 years
37. 24 years

Chapter 8 Review Exercises (page 597)

1. $1,187.50   **3.** $4,210.53   **5.** $555.56   **7.** 33.33%   **9.** $1,305.80   **11.** $4,435.49   **13.** 18.16%
15. 5.64%   **17.** 5.84%   **19.** $11,956.00   **21.** $25,633.04   **23.** 23.4 months   **25.** $5,900.42; $500.42 of that is interest   **27.** $199.34   **29.** $191.57   **31.** $880.52; $9,564.14; $196,987.20   **33.** $195.85
35. $239,108.58   **37.** $951.07

Chapter 9

Section 9.1 (page 613)

1. 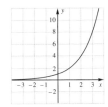 **3.** 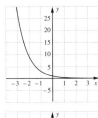 **5.** 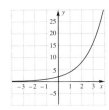 **7.**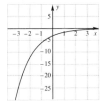

9. 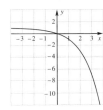 **11.**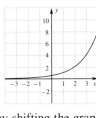

13. The graph of s is obtained by shifting the graph of f one unit to the right, inverting the graph of f in the y-direction, and then shifting it up one unit.   **15.** The graph of r is obtained by

In answers 17 through 21, f_1 is solid and f_2 is dashed.

17. 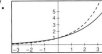 **19.** 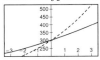 **21.**

23. (a) $A(t) = 10,000\,(1.03)^t$ (b) $13,439.16 **25.** (a) $A(t) = 10,000\,(1.00625)^{4t}$ (b) $11,327.08
27. (a) $A(t) = 10,000\left(1 + \frac{0.065}{365}\right)^{365t}$ (b) $19,154.30   **29.** (a) $A(t) = 10,000\,(1.002)^{12t}$ (b) $12,709.44
31. (a) $1,820.97 (b) $3,140.09   **33.** 4.8 years   **35.** $26.90   **37.** $2,109.60
39. $A(t) = 200,000(0.95)^{2t}$; $71,697.18
41. 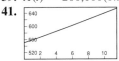 **43.**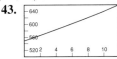

45. 31 years   **47.** 2.3 years   **49.** $A(r)$ is the amount that an item costing $1,000 now will cost in 10 years, given an annual rate of inflation r. We find $A(0.1) = 2,593.74$. Thus, at a rate of inflation of 10% per year, an item costing $1,000 now will cost $2,593.74 in 10 years. We find $A(1.15) = 2,110,496.32$. Thus, at a rate of inflation of 115% per year, an item costing $1,000 now will cost $2,110,496.32 in 10 years.

Section 9.2 (page 624)

1. 20.0855   **3.** 0.3679   **5.** 1.0305   **7.** 101.2578   **9.** 6,186.6127
11. 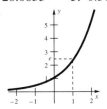   **13.** 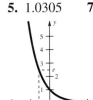   **15.**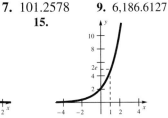

17. The graph of h is obtained from the graph of f by expanding it by a factor of 2 in the y-direction.
19. The graph of g is obtained from the graph of f by compressing it by a factor of 2 in the x-direction, and then reflecting it in the y-axis. **21.** $16,487 **23.** $34,903 **25.** $13,331 **27.** $A(t) = 1,000e^{0.0235t}$; $A(5) = $1,124.68$ **29.** $491.82 **31.** 4.08% **33.** 22.14% **35.** Ninth National has the lower effective rate. **37.** $20,923.99 **39.** $A(t) = 100e^{-0.000121t}$; $A(10,000) \approx 29.82$ grams. **41.** 200% per hour growth rate; 531,441,000 bugs **43.** (a) $r(t) = 1.58 - 0.282t$; (b) $\bar{r} = 1.58 - \dfrac{0.282t}{2}$ (c) $A(t) = Pe^{t(1.58 - \frac{0.282t}{2})}$ (d) $5,820 **45.** $1083.29

47. 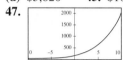 **49.**

51. 38,059 years old **53.** 18 years; value = $17,280 **55.** 35 years **57.** $548,812 **59.** 2002

Section 9.3 (page 639)

1. $\log_2 32 = 5$ **3.** $\log_3(\frac{1}{9}) = -2$ **5.** $\log 1{,}000 = 3$ **7.** $\ln y = x$ **9.** $\log_x y = -3$ **11.** $\ln 1 = 0$
13. $6^2 = 36$ **15.** $2^{-2} = \frac{1}{4}$ **17.** $10^8 = 100{,}000{,}000$ **19.** $e^{-1} = \frac{1}{e}$ **21.** $x^3 = y$ **23.** $e^y = -x$
25. $a + b$ **27.** $a - b$ **29.** $8a$ **31.** $b - (a + c)$ **33.** $-(a + 2c)$ **35.** $2b - 1$ **37.** 12 **39.** x^2y
41. $\dfrac{x^2 y^4}{z^6}$ **43.** xy^{2x} **45.** 2 **47.** 0 **49.** $t = 0, -1$ **51.** 0.23105 **53.** 2.30259 **55.** 5.32193

57. 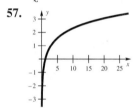 **59.** 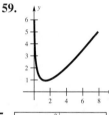 **61.**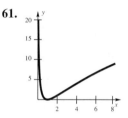

63. 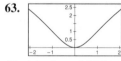 **65.**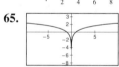

67. (a) about 5.012×10^{16} joules of energy (b) about 2.24% (c) proof (d) 31.62
69. (a) 75 dB, 69 dB, 61 dB (b) $D = 95.05 - 20 \log r$ (c) 56,559 feet **71.** (a) $f(p) = 7.13 - 0.768p$
(b) $f(p) = 11.0p^{-0.750}$ (c) $f(p) = 11p^{-0.75}$ (d) $f(3.50) = 4.30$ servings per hour

Section 9.4 (page 650)

1. 8 years **3.** 132 months **5.** 5 years **7.** 3 years **9.** 13.43 years **11.** 7.4 years **13.** 9 days
15. 3.36 years **17.** 11 years **19.** 2.03 years **21.** 55,259 years old **23.** 3.8 months **25.** 65,536,000 bacteria **27.** 91,856 frogs **29.** 311,000,000 **31.** 24 semesters **33.** 2,360 million years **35.** 3.89 days **37.** $P = 23{,}200e^{-0.2261439t}$; 2,400 people in 1997 **39.** The least squares model gives a higher employment figure for 1997. **41.** 2002 **43.** 2.8 hours

Chapter 9 Review Exercises (page 656)

1. $\log_2 1024 = 10$ **3.** $\log 0.0001 = -4$ **5.** $\log_x y = e$ **7.** $3^4 = 81$ **9.** $10^3 = 1000$ **11.** $x^{-1} = y$
13. $\log_3 4 = \log 4/\log 3 = 1.262$ **15.** 0 or -2 **17.** 0.5298 **19.** $2,346.40 **21.** $2,619.86
23. $2,323.67 **25.** 10.2 years **27.** 6 years **29.** 10.8 years **31.** $11,956.00 **33.** $25,633.04

35. 16.8 months **37.** 0.301 seconds **39.** Erewhon (effective yield = 27.12%) **41.** 5.5 billion years
43. $80 **45.** 18 minutes **47.** $0.83

Chapter 10

Section 10.1 (page 671)

In (1)–(9), d.q. stands for the difference quotient.

1. d.q. = $2h$;

h	d.q.
1	2
0.1	0.2
0.01	0.02
0.001	0.002
0.0001	0.0002
−0.0001	−0.0002
−0.001	−0.002
−0.01	−0.02
−0.1	−0.2
−1	−2

3. d.q. = $\dfrac{\frac{1}{2+h} - \frac{1}{2}}{h} = -\dfrac{1}{2(2+h)}$

h	d.q.
1	−0.1667
0.1	−0.2381
0.01	−0.2488
0.001	−0.2499
0.0001	−0.24999
−0.0001	−0.25001
−0.001	−0.2501
−0.01	−0.2513
−0.1	−0.2632
−1	−0.5

5. d.q. = $\dfrac{(1+h)^3 - 1}{h} = 3 + 3h + h^2$

h	d.q.
1	7
0.1	3.31
0.01	3.0301
0.001	3.0030
0.0001	3.0003
−0.0001	2.9997
−0.001	2.9970
−0.01	2.9701
−0.1	2.71
−1	1

7. d.q. = $8 + h$

h	d.q.
1	9
0.1	8.1
0.01	8.01
0.001	8.001
0.0001	8.0001
−0.0001	7.9999
−0.001	7.999
−0.01	7.99
−0.1	7.9
−1	7

9. −2 **11.** 3 **13.** 0.03125 **15.** −0.17678 **17.** −5 **19.** −1.5

21.

h	1	0.1	0.01
v_{ave}	39	39.9	39.99

$v_{inst} = 40$ mph

23.

h	1	0.1	0.01
v_{ave}	140	66.2	60.602

$v_{inst} = 60$ mph

25.

h	1	0.1	0.01
v_{ave}	59.45	59.41	59.40

$v_{inst} = 59.4$ mph

27.

h	1	0.1	0.01
v_{ave}	9.443	9.938	9.994

$v_{inst} = 10$ mph

29.

h	50	10	1
C_{ave}	4.795	4.799	4.7999

$C'(1,000) = 4.8$

31.

h	50	10	1
C_{ave}	9.955	9.951	9.950

$C'(100) = 9.95$

33.

h	50	10	1
C_{ave}	99.93	99.91	99.90

$C'(100) = 99.9$

35. $q(100) = 50{,}000$, $q'(100) = -500$. A total of 50,000 pairs of sneakers can be sold at a price of \$100, but the demand is decreasing at a rate of 500 pairs of sneakers per \$1 increase in the price.

37. $P = \$257.07$ and $dP/dn = 5.07$. Your current profit is \$257.07 per month, and is increasing at a rate of \$5.07 per additional magazine in sales. **39.** $E = 25.1$, $dE/dg = 2.51$. The professor's average class size is 25.1 students, and this is rising at a rate of 2.51 students per 1-point increase in grades awarded. **41.** $A'(0.5) = 775$ points per day **43.** (a) 60% of children can speak at the age of 10 months. Further, at the age of 10 months, this percentage is increasing by 18.2% per month. (b) As t increases, $p(t)$ approaches 100 (assuming all children eventually learn to speak) and $p'(t)$ approaches 0, since the percentage stops increasing. **45.** (a) $R'(4) = -2.5$ thousand organisms per hour, per 1,000 new organisms. This means that the reproduction rate of organisms in a culture containing 4,000 organisms is declining at a rate of 2,500 organisms per hour for every 1,000 new organisms.
47. $l(.95) = 31.22$ meters and $l'(.95) = -304.24$ meters/warp. Thus, at a speed of warp 0.95, the rocket ship has an observed length of 31.22 meters and its length is decreasing at a rate of 304.24 meters per unit warp, or 3.0424 meters per one percent increase in the speed (measured in warp). **49.** 1.000 **51.** 1.000 **53.** $S(5) = 109$, $S'(5) = 9.10$. After 5 weeks, sales are 109 pairs of sneakers per week, and sales are increasing at a rate of 9.1 pairs of sneakers per week each week. **55.** $q(2) = 8{,}165$, $q'(2) = 599.2$. Thus, two months after the introduction, 8,165 video game units have been sold, and the demand is growing at a rate of 599.2 units per month.

Section 10.2 (page 683)

1. (a) R (b) P **3.** (a) P (b) R **5.** (a) Q (b) P **7.** (a) P (b) Q **9.** (a) Q (b) R (c) P **11.** (a) R (b) Q (c) P **13.** (a) R (b) Q (c) P **15.** (a) $(1, 0)$ (b) none (c) $(-2, 1)$ **17.** (a) $(-2, 2)$, $(0, 1)$, $(2, 0)$ (b) none (c) none
19. The tangent to the graph of the function f at the general point where $x = a$ is the line passing through $(a, f(a))$ with slope $f'(a) = \lim_{h \to 0} \dfrac{f(a+h) - f(a)}{h}$. **21.** $y = f(a) + (x - a)f'(a)$. **23.** (a) $2x + h$ (b) $2x$ (c) 4
25. (a) $-2x - h$ (b) $-2x$ (c) 2 **27.** (a) 3 (b) 3 (c) 3 **29.** (a) -2 (b) -2 (c) -2
31. (a) $6x + 3h$ (b) $6x$ (c) -6 **33.** (a) $1 - 2x - h$ (b) $1 - 2x$ (c) -3
35. (a) 3 (b) $y = 3x + 2$ **37.** (a) $\dfrac{3}{4}$ (b) $3x - 4y = -4$

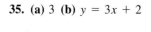
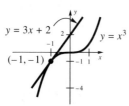
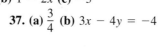
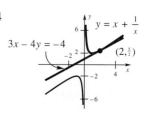

39. (a) $\dfrac{1}{4}$ (b) $x - 4y = -4$ **41.** (b)   **43.** (c)   **45.** (a)   **47.** -0.12 **49.** -0.15

51. -0.58 **53.** 0.26   **55.** 0.625   **57.** 1.425, 2.575   **59.** $q(100) = 50{,}000$, $q'(100) = -500$. A total of 50,000 pairs of sneakers can be sold at a price of \$100, but the demand is decreasing at a rate of 500 pairs of sneakers per \$1 increase in the price.   **61.** $P = \$257.07$ and $dP/dn = 5.07$. Your current profit is \$257.07 per month, and this is increasing at a rate of \$5.07 per additional magazine in sales.   **63.** $E = 25.1$, $dE/dg = 2.51$. The professor's average class size is 25.1 students, and this is rising at a rate of 2.51 students per 1-point increase in grades awarded.   **65.** $A'(0.5) = 775$ points per day   **67.** (a) $R'(4) = -2.5$ thousand organisms per hour, per 1,000 new organisms. This means that the reproduction rate of organisms in a culture containing 4,000 organisms is declining at a rate of 2,500 organisms per hour for every 1,000 new organisms. **69.** $l(.95) = 31.22$ meters and $l'(.95) = -304.24$ meters/warp. Thus, at a speed of warp 0.95, the rocket ship has an observed length of 31.22 meters and its length is decreasing at a rate of 304.24 meters per unit warp, or 3.0424 meters per one-percent increase in the speed (measured in warp). **71.** $S(5) = 109$, $S'(5) = 9.10$. After 5 weeks, sales are 109 pairs of sneakers per week, and sales are increasing at a rate of 9.1 pairs of sneakers per week each week. **73.** $q(2) = 8,165$, $q'(2) = 599.2$. Thus, two months after the introduction, 8,165 video game units have been sold, and the demand is growing at a rate of 599.2 units per month.

Section 10.3 (page 700)

1. 0 **3.** 4 **5. (a)** -2 **(b)** -1 **7. (a)** 1 **(b)** 1 **9. (a)** 0 **(b)** 2 **(c)** -1 **(d)** does not exist, since the left and right limits disagree **(e)** 2 **11. (a)** 1 **(b)** 1 **(c)** 2 **(d)** does not exist, since the left and right limits disagree **(e)** 1 **13. (a)** 1 **(b)** does not exist **(c)** does not exist **(d)** does not exist **(e)** not defined **15. (a)** 1 **(b)** does not exist **(c)** -1 **(d)** does not exist **(e)** 2 **(f)** 0 **17. (a)** does not exist **(b)** 0 **(c)** 1.5 **19. (a)** 1 **(b)** 0 **21.** continuous on its domain **23.** continuous on its domain **25.** not continuous on its domain; discontinuous at $x = 0$ **27.** not continuous on its domain; discontinuous at $x = -1$ **29.** continuous on its domain **31.** not continuous on its domain; discontinuous at $x = 0$ and $x = -2$ **33.** not continuous on its domain; discontinuous at $x = 0$ **35.** continuous on its domain **37.** 1 **39.** 2 **41.** 0 **43.** 6 **45.** 4 **47.** 2 **49.** 0 **51.** 0 **53.** 12 **55.** 0 **57.** 2 **59.** -4 **61.** $2x - 2$ **63.** $-10x + 2$ **65.** $3t^2 + 1$ **67.** $4t^3 - 1$ **69.** $-\dfrac{6}{t^2}$ **71.** $1 - \dfrac{1}{x^2}$ **73.** $-\dfrac{1}{(x-2)^2}$ **75.** $\dfrac{1}{2\sqrt{x+1}}$ **77.** $-\dfrac{1}{2t\sqrt{t}}$

Section 10.4 (page 711)

1. $f'(x) = 3x^2$ **3.** $f'(x) = -4x^{-3}$ **5.** $f'(x) = -\dfrac{1}{4}x^{-3/4}$ **7.** $f'(x) = 8x^3 + 9x^2$
9. $f'(x) = -1 - \dfrac{1}{x^2}$ **11.** $f'(x) = \dfrac{1}{\sqrt{x}}$ **13.** $f'(x) = x^{-2/3}$
15. $\dfrac{dy}{dx} = 10 \cdot 0 = 0$ by constant multiples and the power rule **17.** $\dfrac{d}{dx}(x^2 + x) = \dfrac{d}{dx}(x^2) + \dfrac{d}{dx}(x)$ (sum) $= 2x + 1$ (power rule) **19.** $\dfrac{d}{dx}(4x^3 + 2x - 1) = \dfrac{d}{dx}(4x^3) + \dfrac{d}{dx}(2x) - \dfrac{d}{dx}1$ (sum and difference) $= 4\dfrac{d}{dx}(x^3) + 2\dfrac{d}{dx}x - \dfrac{d}{dx}1$ (constant multiples) $= 12x^2 + 2$ (power rule) **21.** $\dfrac{d}{dx}(x^{104} - 99x^2 + x) = \dfrac{d}{dx}(x^{104}) - \dfrac{d}{dx}(99x^2) + \dfrac{d}{dx}x$ (sum and differences) $= 104x^{103} - 99\dfrac{d}{dx}(x^2) + 1$ (constant multiples and power rule) $= 104x^{103} - 198x + 1$ (power rule) **23.** $s = t^{3/2} - t^{7/2} + t^{-3}$. Thus, $\dfrac{ds}{dt} = \dfrac{d}{dt}(t^{3/2} - t^{7/2} + t^{-3}) = \dfrac{d}{dt}(t^{3/2}) - \dfrac{d}{dt}(t^{7/2}) + \dfrac{d}{dt}(t^{-3})$ (sums and differences) $= \dfrac{3}{2}t^{1/2} - \dfrac{7}{2}t^{5/2} - 3t^{-4}$ (power rule). **25.** $\dfrac{d}{dr}\left(\dfrac{4\pi r^3}{3}\right) = \dfrac{4\pi}{3}\dfrac{d}{dr}(r^3)$ (constant multiple) $= \dfrac{4\pi}{3}(3r^2)$ (power rule) $= 4\pi r^2$ **27.** $2t + 20at^4$ (sum, constant multiples, power rule) **29.** $\tfrac{1}{2}x^{-1/2} + \tfrac{3}{2}x^{1/2}$ (multiply out, then use sum and power rules) **31.** 3 **33.** -2
35. $\tfrac{1}{32}$ **37.** $-\dfrac{1}{4\sqrt{2}}$ **39.** -5 **41.** $-\tfrac{3}{2}$ **43.** $y = 3x + 2$

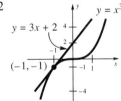

45. $3x - 4y = -4$ 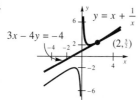 **47.** $x - 4y = -4$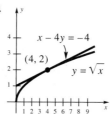

49. $f'(x) = 2x - 3$ **51.** $f'(x) = 1 + \dfrac{1}{2\sqrt{x}}$ **53.** $g'(x) = -\dfrac{2}{x^3} - \dfrac{6}{x^4}$ **55.** $h'(x) = -\dfrac{1}{x^2} - \dfrac{2}{x^3} - \dfrac{3}{x^4}$

57. $r'(x) = \dfrac{1}{2\sqrt{x}} - \dfrac{1}{2x\sqrt{x}}$ **59.** $f'(x) = 3x^2$ **61.** $g'(x) = 1 - 4x$ **63.** $1 - \dfrac{2}{x^3}$ **65.** $2.6x^{0.3} + 1.2x^{-2.2}$

67. $3at^2 - 4a$ **69.** $\dfrac{1}{2\sqrt{x}} + \dfrac{3\sqrt{x}}{2}$ **71.** $5.15x^{9.3} - 99x^{-2}$ **73.** $-\dfrac{2.31}{t^{2.1}} - \dfrac{0.3}{t^{0.4}}$ **75.** $4\pi r^2$ **77.** $-\tfrac{3}{4}$

79. no such values **81.** $x = 1, -1$ **83.** $\dfrac{1}{4}$ **85-89.** Proofs **91.** The rate of change of C_2 is twice the rate of change of C_1. **93.** The rates of change of cost and revenue must be equal. This means that the next case costs as much to make as it will bring in in revenue. **95.** $A'(0.5) = 775$ points per day **97.** $c'(15) \approx 1.389$, $c'(30) \approx 0.011$. Thus, the hourly oxygen consumption is increasing at a slower rate when the chick hatches. (Notice that the oxygen consumption is still increasing, however.) **99. (a)** $0, -32, -64, -96, -128$ ft/s **(b)** 2.5 seconds; downward at 80 ft/s **101.** $V'(10) = 4\pi(100) \approx 1{,}256.63$ cubic centimeters per centimeter increase in r. When $r = 10$, the volume increases by approximately 1,256.63 cubic centimeters for every 1-centimeter increase in the radius.

Section 10.5 (page 724)

1.

h	50	10	1
C_{ave}	4.795	4.799	4.7999

$C'(1{,}000) = 4.8$

3.

h	50	10	1
C_{ave}	9.955	9.951	9.950

$C'(100) = 9.95$

5.

h	50	10	1
C_{ave}	99.93	99.91	99.90

$C'(100) = 100 - 1{,}000/100^2 = 99.9$

7. $C'(x) = 4$; $R'(x) = 8 - x/500$; $P'(x) = 4 - x/500$; $P'(x) = 0$ when $x = 2{,}000$. Thus, at a production level of 2,000, the profit is stationary (neither increasing nor decreasing) with respect to the production level. This may indicate a maximum profit at a production level of 2,000. **9.** $C'(x) = 40 - .002x$. The cost is going up at a rate of \$39.80 per teddy bear. The cost of producing the 101st teddy is $C(101) - C(100) = \$39.799$. **11.** The profit on the sale of 1,000 videocassettes is \$3,000, and is decreasing by approximately \$3 per additional videocassette sold. **13.** $P = \$257.07$ and $dP/dn = 5.07$. Your current profit is \$257.07 per month, and this would increase by approximately \$5.07 for each increase by one magazine in sales. **15.** $P'(50) = \$350$. This means that, at an employment level of 50 workers, the firm's daily profit will increase by approximately \$350 per additional worker it hires. **17. (a)** \$4.47 per pound **(b)** $R(q) = \dfrac{50000}{q^{0.5}}$ **(c)** $R(500) = \$2{,}236.07$. This is the monthly revenue that will result from setting the price at \$4.47 per pound. **(d)** $R'(q) = -\dfrac{25000}{q^{1.5}}$; $R'(500) = -\$2.24$ per additional pound of tuna demanded. Thus, at a demand level of 500 pounds of tuna, the revenue is decreasing by approximately \$2.24 per additional pound. **(e)** The fishery should raise the price in order to decrease demand, since the revenue drops when the price is lowered, even though the sales increase. It would be more economical to store the excess tuna. **19. (a)** $R(q) = 152.33q - 2.22q^2$; $R'(q) = 152.33 - 4.44q$. **(b)** $R'(50) = -69.67$¢ per pound of poultry. Thus if the price is raised, causing a decrease in demand of one pound of poultry, the revenue will increase by approximately 69.67¢. **(c)** $P(q) = 142.33q - 2.22q^2$; $P'(q) = 142.33 - 4.44q$; $P'(50) = -79.67$¢ per pound. This means that, if a farmer lowers the price to increase the per capita demand by one pound, the annual profit will decrease by approximately 79.67¢.

21. (a) $R(p) = -\frac{4p^2}{3} + 80p$; $R'(p) = -\frac{8p}{3} + 80$ **(b)** (i) $R'(20) = \$26.67$ per \$1 increase in price; (ii) $R'(30) = \$0$ per \$1 increase in price; (ii) $R'(40) = -\$26.67$ per \$1 increase in price **(c)** If it charges \$20 per ruby, it can increase its revenue by approximately \$26.67 by raising the price \$1. If it charges \$30 per ruby, the revenue is not moving with the price. Thus, if they raise the price by \$1, the revenue will not change significantly. If it charges \$40 per ruby, the revenue would drop by approximately \$26.67 if it were to raise the price another \$1.
23. (a) $C'(q) = 200q$; $C'(10) = \$2,000$ per one-pound reduction in emissions. **(b)** $S'(q) = 500$. Thus, $S'(q) = C'(q)$ when $500 = 200q$, or $q = \frac{5}{2}$ pounds per day reduction. **(c)** $N(q) = C(q) - S(q) = 100q^2 - 500q + 4,000$. This is a parabola with lowest point (vertex) given by $q = -b/2a = 500/200 = \frac{5}{2}$. The net cost at this production level is $N(\frac{5}{2}) = \$3,375$ per day. The value of q is the same as that for part (b). The net cost to the firm is minimized at the reduction level for which the cost of controlling emissions begins to increase faster than the subsidy. This is why we get the answer by setting these two rates of increase equal to each other. **25.** $M'(10) \approx 0.0002557$ mpg/mph. This means that, at a speed of 10 mph, the fuel economy is increasing at a rate of 0.0002557 miles per gallon for each 1-mph increase in speed. $M'(60) = 0$ mpg/mph. This means that, at a speed of 60 mph, the fuel economy is neither increasing nor decreasing with increasing speed. $M'(70) \approx -0.00001799$. This means that, at 70 mph, the fuel economy is decreasing by approximately 0.00001799 miles per gallon per 1-mph increase in speed. Thus, 60 mph is the most fuel-efficient speed for the car. **27. (a)** \$3.98 per 1,000 nautical miles **(b)** The cost is dropping at a rate of 0.1¢ per extra ton. **29.** (c) **31.** (d)

Section 10.6 (page 737)

1. (a) diverges to $+\infty$ **(b)** 0 **3. (a)** diverges to $-\infty$ **(b)** 0 **5. (a)** diverges to $+\infty$ **(b)** diverges to $+\infty$ **(c)** diverges to $+\infty$ **7. (a)** -1 **(b)** diverges to $-\infty$ **(c)** does not exist, since the left and right limits disagree **9. (a)** does not exist, since the left and right limits disagree **(b)** diverges to $+\infty$ **(c)** -1 **11. (a)** does not exist, since the graph oscillates between -1 and 1 **(b)** does not exist for the reason given in (a) **13.** diverges to $+\infty$ **15.** does not exist, since the left and right limits disagree **17.** $\frac{3}{2}$ **19.** $\frac{1}{2}$ **21.** does not exist **23.** 0 **25.** $\frac{3}{2}$ **27.** $\frac{1}{2}$ **29.** does not exist **31.** 0 **33.** 0 **35.** does not exist **37.** 1 **39.** 0 **41.** 0 **43.** jump discontinuity at $x = 0$ **45.** continuous everywhere **47.** removable discontinuity at $x = 0$ **49.** discontinuity at $x = 0$ **51.** not differentiable at $x = 0$ **53.** differentiable everywhere **55.** not differentiable at $x = 1$ **57.** $\lim_{t \to +\infty} n(t) = 0$. Thus, if the trend were to continue indefinitely, the annual number of DWI arrests in New Jersey will decrease to zero in the long term. **59.** $\lim_{t \to +\infty} p(t) = 100$, $\lim_{t \to +\infty} p'(t) = 0$. This tells us that the percentage of children who learn to speak approaches 100% as their age increases, with the number of additional children learning to speak approaching zero. **61.** $\lim_{t \to +\infty} q(t) = 10,000$, meaning that a total of 10,000 units are sold over the life of the game

Chapter 10 Review Exercises (page 744)

1. (a) P **(b)** Q **(c)** R **3. (a)** Q, R **(b)** P **5. (a)** $(0, -1)$ **(b)** $(2, 1)$ **(c)** $(-2, 1)$ **7. (a)** $(0, 1)$ **(b)** $(-2, 0)$ **(c)** none **9.** $6x + 4x^{-2}$ **11.** x **13.** $1.2x^{0.2} + 2.3x^{1.3}$ **15.** $0.6s^{-0.8} - 44$ **17.** $3at^2 - 2bt$ **19.** $\frac{1}{\sqrt{x}} + \frac{1}{3\sqrt[3]{x^2}}$ **21.** $3 - \frac{3}{x^4}$ **23.** $\frac{1}{5} + \frac{5}{x^2}$ **25.** $\frac{3}{2\sqrt{x}}$ **27.** $6x^2 + 2x$ **29.** $-\frac{1}{2r\sqrt{r}} - \frac{2}{r^2}$ **31.** $y = -2x - 5$ **33.** $y = \frac{-4t}{5} + 1$ **35.** $y = \frac{3s}{4} + 1$ **37.** $y = -\frac{5t}{3} - 1$ **39.** $y = \frac{5}{4}t - 1$ **41.** $x = -\frac{3}{2}$ **43.** $x = \pm 1$ **45.** none **47.** $x = -\sqrt[3]{2}$ **49.** -2 **51.** 0 **53.** 6 **55.** diverges to $+\infty$ **57.** 3 **59.** diverges to $+\infty$

61. diverges to $+\infty$ **63.** 0 **65. (a)** -1 **(b)** 1 **(c)** does not exist; left and right limits disagree **(d)** diverges to $+\infty$ **(e)** 2 **(f)** 1 **67. (a)** -1 **(b)** diverges to $-\infty$ **(c)** 1 **69. (a)** diverges to $+\infty$ **(b)** diverges to $-\infty$ **(c)** -1 **(d)** 1 **71. (a)** 1 **(b)** diverges to $-\infty$ **(c)** diverges to $+\infty$ **73.** discontinuous at -1 and 1 **75.** continuous **77.** discontinuous at -1 **79.** discontinuous at every integer **81.** 4 **83.** -2 **85.** -7 **87.** 0 **89.** 0 **91. (a)** 52, 66.4, 67.84, 67.984 ft/s **(b)** 68 ft/s **(c)** hits the ground after 6.25 seconds, and is traveling 100 ft/s at the time **93.** $C'(x) = 60 - 0.002x$; $C'(50) = \$59.90$; $C(51) - C(50) = \$59.899$ **95.** $V'(1,000) = 300$ cm³/s **97. (a)** $R(s) = s - 1,000 - \dfrac{s^{3/2}}{1,000}$ **(b)** $R'(100,000) = 0.526$; $R'(500,000) = -0.0607$ **(c)** $R'(s) > 0$ for $s < \$444,444.44$; $R'(s) < 0$ for $s > \$444,444.44$ **(d)** $R(444,444.44) = \$147,148.15$ **99. (a)** 60,912 frogs per year **(b)** 2006

Chapter 11

Section 11.1 (page 760)

1. 3 **3.** $3x^2$ **5.** $2x + 3$ **7.** $\dfrac{2x}{3}$ **9.** $-\dfrac{2}{x^2}$ **11.** $(x^2 - 1) + 2x(x + 1) = (x + 1)(3x - 1)$ **13.** $(x^{-1/2} + 4)(x - x^{-1}) + (2x^{1/2} + 4x - 5)(1 + x^{-2})$ **15.** $8(2x^2 - 4x + 1)(x - 1)$ **17.** $\left(2x - \dfrac{1}{2\sqrt{x}}\right)\left(\sqrt{x} + \dfrac{1}{\sqrt{x}}\right) + (x^2 - \sqrt{x})\left(\dfrac{1}{2\sqrt{x}} - \dfrac{1}{2x^{3/2}}\right)$ **19.** $\dfrac{1}{2\sqrt{x}}\left(\sqrt{x} + \dfrac{1}{x^2}\right) + (\sqrt{x} + 1)\left(\dfrac{1}{2\sqrt{x}} - \dfrac{2}{x^3}\right)$ **21.** $-\dfrac{14}{(3x - 1)^2}$ **23.** $\dfrac{6x^2 - 4x - 7}{(3x - 1)^2}$ **25.** $\dfrac{5x^2 - 5}{(x^2 + x + 1)^2}$ **27.** $\dfrac{-1}{\sqrt{x}(\sqrt{x} - 1)^2}$ **29.** $-\dfrac{3}{x^4}$ **31.** $\dfrac{3x^2 - 2x - 13}{(3x - 1)^2}$ **33.** $\dfrac{[(x + 1)(x + 2) + (x + 3)(x + 2) + (x + 3)(x + 1)](3x - 1) - 3(x + 3)(x + 1)(x + 2)}{(3x - 1)^2}$ **35.** $4x^3 - 2x$ **37.** 64 **39.** 3 **41.** $\dfrac{\left(2t - \dfrac{1}{2\sqrt{t}}\right)\left(\sqrt{t} + \dfrac{1}{\sqrt{t}}\right) - (t^2 - \sqrt{t})\left(\dfrac{1}{2\sqrt{t}} - \dfrac{1}{2t\sqrt{t}}\right)}{\left(\sqrt{t} + \dfrac{1}{\sqrt{t}}\right)^2}$ **43.** $y = 12x - 8$ **45.** $y = \dfrac{x}{4} + \dfrac{1}{2}$ **47.** $y = -2$ **49.** $S'(5) = 10$ (sales are increasing at a rate of 1,000 units per month); $p'(5) = -10$ (the price is dropping at a rate of \$10 per month); $R'(5) = 900,000$ (revenue is increasing at a rate of \$900,000 per month) **51.** decreasing at a rate of \$1 per day **53.** decreasing at a rate of \$0.10 per month **55.** $M'(x) = \dfrac{3600x^{-2} - 1}{(3600x^{-1} + x)^2}$. $M'(10) \approx 0.0002557$ mpg/mph. This means that, at a speed of 10 mph, the fuel economy is increasing at a rate of 0.0002557 miles per gallon for each 1-mph increase in speed. $M'(60) = 0$ mpg/mph. This means that at a speed of 60 mph, the fuel economy is neither increasing nor decreasing with increasing speed. $M'(70) \approx -0.00001799$. This means that, at 70 mph, the fuel economy is decreasing at a rate of 0.00001799 miles per gallon for each 1-mph increase in speed. Thus, 60 mph is the most fuel-efficient speed for the car. **57.** \$111,870,000 per year **59.** $R'(p) = -\dfrac{5.625}{(1 + 0.125p)^2}$; $R'(4) = -2.5$ thousand organisms per hour, per 1,000 new organisms. This means that the reproduction rate of organisms in a culture containing 4,000 organisms is declining at a rate of 2,500 organisms per hour for every 1,000 new organisms.

61. Oxygen consumption is decreasing at a rate of 1634 milliliters per day. This is due to the fact that the number of eggs is decreasing, since $C'(30)$ is positive.   **67.** (a)

Section 11.2 (page 775)

1. $4(2x + 1)$ **3.** $-(x - 1)^{-2}$ **5.** $2(2 - x)^{-3}$ **7.** $\dfrac{1}{\sqrt{2x + 1}}$ **9.** $-\dfrac{3}{(3x - 1)^2}$

11. $4(x^2 + 2x)^3(2x + 2)$

13. $-4x(2x^2 - 2)^{-2}$ **15.** $-5(2x - 3)(x^2 - 3x - 1)^{-6}$ **17.** $-\dfrac{6x}{(x^2 + 1)^4}$ **19.** $4\left(2t - \dfrac{1}{2\sqrt{t}}\right)(t^2 - \sqrt{t})^3$

21. $-\dfrac{x}{\sqrt{1 - x^2}}$ **23.** $\dfrac{\left(\dfrac{3}{2\sqrt{x}} + \dfrac{1}{2x\sqrt{x}}\right)}{2\sqrt{3\sqrt{x} - \dfrac{1}{\sqrt{x}}}}$ **25.** $-\dfrac{\left(\dfrac{1}{\sqrt{2x + 1}} - 2x\right)}{(\sqrt{2x + 1} - x^2)^2}$ **27.** $\dfrac{-56(x + 2)}{(3x - 1)^3}$

29. $-\dfrac{1}{2}[(r + 1)(r^2 - 1)]^{-3/2}(3r - 1)(r + 1)$ **31.** $-2(x^2 - 3x)^{-3}(2x - 3)\sqrt{1 - x^2} - \dfrac{x(x^2 - 3x)^{-2}}{\sqrt{1 - x^2}}$

33. $(100x^{99} - 99x^{-2})\dfrac{dx}{dt}$ **35.** $\left(-\dfrac{3}{r^4} + \dfrac{1}{2\sqrt{r}}\right)\dfrac{dr}{dt}$ **37.** $4\pi r^2 r'(t)$ **39.** $-47/4$

41. (a) $R(p) = -\dfrac{4p^2}{3} + 80p;\ \dfrac{dR}{dp}\bigg|_{q=60} = \$40$ per $1 increase in price   (b) $-\$0.75$ per ruby   (c) $-\$30$ per ruby. Thus, at a demand level of 60 rubies per week, the weekly revenue is decreasing at a rate of $30 per additional ruby demanded.   **43.** At an employment level of 10 engineers, it will increase its profit at a rate of $146,454.90 per additional engineer hired.   **45.** $\dfrac{dM}{dB} \approx 0.000116$. This means that approximately 1.16 manatees are killed each year for every 10,000 registered boats.   **47.** 12π sq mi/hr

49. $\$200,000\pi$/week $= \$628,318.53$/week   **51.** (a) $q'(4) = 333.14$ units per month
(b) $R'(q) = 800$ (c) $R'(t) \approx \$266,512$ per month.   **53.** 3% per year   **55.** 8% per year

Section 11.3 (page 791)

1. $\dfrac{1}{x - 1}$ **3.** $\dfrac{1}{x \ln 2}$ **5.** $\dfrac{2x}{x^2 + 3}$ **7.** e^{x+3} **9.** $-e^{-x}$ **11.** $4^x \ln 4$ **13.** $(2^{x^2-1})2x \ln 2$

15. $1 + \ln x$ **17.** $2x \ln x + \dfrac{x^2 + 1}{x}$ **19.** $10x(x^2 + 1)^4 \ln x + \dfrac{(x^2 + 1)^5}{x}$ **21.** $\dfrac{3}{3x - 1}$ **23.** $\dfrac{4x}{2x^2 + 1}$

25. $\dfrac{\left(2x - \dfrac{1}{2\sqrt{x}}\right)}{x^2 - \sqrt{x}}$ **27.** $\dfrac{1}{(x + 1)\ln 2}$ **29.** $2t \log_3\left(t + \dfrac{1}{t}\right) + \dfrac{(t^2 + 1)\left(1 - \dfrac{1}{t^2}\right)}{\left(t + \dfrac{1}{t}\right)\ln 3}$ **31.** $\dfrac{2 \ln|x|}{x}$

33. $\dfrac{2}{x} - \dfrac{2 \ln(x - 1)}{x - 1}$ **35.** $e^x(1 + x)$ **37.** $\dfrac{1}{x + 1} + 3e^x(x^3 + 3x^2)$ **39.** $e^x \ln|x| + \dfrac{e^x}{x}$ **41.** $2e^{2x+1}$

43. $(2x - 1)e^{x^2-x+1}$ **45.** $2xe^{2x-1} + 2x^2e^{2x-1}$ **47.** $4(e^{2x-1})^2$ **49.** $-\dfrac{4}{(e^x - e^{-x})^2}$

51. $\dfrac{-(2(\ln x)^{1/2} + (\ln x)^{-1/2})}{2x^2 \ln x}$ **53.** $\dfrac{1 - 2 \ln x}{2x^2(\ln x)^{1/2}}$ **55.** $\dfrac{1}{x \ln x}$ **57.** $\dfrac{1}{2x \ln x}$

59. $y = (e/\ln 2)(x - 1) \approx 3.92(x - 1)$ **61.** $y = x$ **63.** $y = -\dfrac{1}{2e}(x - 1) + e$ **65.** $451.00 per year
67. $400{,}000 \ln 2 \approx 277{,}000$ people/yr   **69.** 0.000283 g/yr   **71.** $446.02 per year
73. $P'(22) \approx -1.3855$. This indicates that in ancient Rome, the percentage of people surviving was decreasing at a rate of 1.3855% per year at age 22.   **75.** 3,110,000 cases/month; 11,200,000 cases/month; 722,000 cases/month   **77.** Growing at a rate of 28.3309 billion packets per month each year.   **79. (a)** $p'(10) \approx 0.0931$, so that the percentage of firms using numeric control is increasing at a rate of 9.31% per year after 10 years.   **(b)** 0.80. Thus, in the long term, 80% of all firms will be using numeric control.   **(c)** same as (a)   **(d)** 0. Thus, in the long run, the percentage of firms using numeric control will remain constant at 80%.   **81.** $n'(t)$ is a maximum when $t \approx 12.47$. $(n'(12.47) \approx 5.1673)$. This means that the rate of growth of enrollment in HMOs reached a maximum of 5.167 million new enrollments per year at about June of 1987.

Section 11.4 (page 801)

1. $-\dfrac{2}{3}$ **3.** x **5.** $\dfrac{y - 2}{3 - x}$ **7.** $-y$ **9.** $-\dfrac{y}{x(1 + \ln x)}$ **11.** $-\dfrac{x}{y}$ **13.** $-\dfrac{2xy}{x^2 - 2y}$ **15.** $-\dfrac{6 + 9x^2 y}{9x^3 - x^2}$
17. $\dfrac{3y}{x}$ **19.** $\dfrac{p + 10p^2 q}{2p - q - 10pq^2}$ **21.** $\dfrac{ye^x - e^y}{xe^y - e^x}$ **23.** $\dfrac{se^{st}}{2s - te^{st}}$ **25.** $\dfrac{ye^x}{2e^x + y^3 e^y}$ **27.** $\dfrac{y - y^2}{-1 + 3y - y^2}$
29. $-\dfrac{y}{x + 2y - xye^y - y^2 e^y}$ **31.** $(x^3 + x)\sqrt{x^3 + 2}\left(\dfrac{3x^2 + 1}{x^3 + x} + \dfrac{1}{2}\dfrac{3x^2}{x^3 + 2}\right)$ **33.** $x^x(1 + \ln x)$
35. 1   **37.** -2 **39.** -0.03314 **41.** 31.7295   **43.** -0.189783 **45.** 0   **47. (a)** 500 T-shirts
(b) $\left.\dfrac{dq}{dp}\right|_{p=5} = -125$ T-shirts per dollar. Thus, when the price is set at $5, the demand is dropping at a rate of 125 T-shirts per $1 increase in price.
49. $\left.\dfrac{dk}{de}\right|_{e=15} = -0.307$ carpenters per electrician. This means that, for a $200,000 house whose construction employs 15 electricians, adding one more electrician would cost as much as approximately 0.307 additional carpenters. In other words, one electrician is worth approximately 0.307 carpenters.   **51. (a)** 22.93 hours. (The other root is rejected, since it is larger than 30.)
(b) $\left.\dfrac{dt}{dx}\right|_{x=3.0} = -11.2132$ hours per grade point. This means that, for a 3.0 student who scores 80 on the examination, 1 grade point is worth approximately 11.2132 hours.   **53.** $\dfrac{dr}{dy} = 2\dfrac{r}{y}$, so $\dfrac{dr}{dt} = 2\dfrac{r}{y}\dfrac{dy}{dt}$ by the chain rule   **55.** Let $y = f(x)g(x)$. Then $\ln y = \ln f(x) + \ln g(x)$, and $\dfrac{1}{y}\dfrac{dy}{dx} = \dfrac{f'(x)}{f(x)} + \dfrac{g'(x)}{g(x)}$, so
$\dfrac{dy}{dx} = y\left(\dfrac{f'(x)}{f(x)} + \dfrac{g'(x)}{g(x)}\right) = f(x)g(x)\left(\dfrac{f'(x)}{f(x)} + \dfrac{g'(x)}{g(x)}\right) = f'(x)g(x) + f(x)g'(x)$.

Section 11.5 (page 811)

1. $3x + 5$ **3.** $2 - 10x$ **5.** x **7.** $1 + x$ **9.** x **11.** $\dfrac{x}{2} + 1$ **13.** $1.3x - 0.3$
15. $\dfrac{x(e^2 - e^{-2}) + 3e^{-2} - e^2}{2}$ **17.** $0.5 - 0.05x$ **19.** $80\pi(x - 5)$ **21.** 4.0375   **23.** 6.98714
25. 4.0333   **27.** 1.3   **29.** -0.05 **31.** 36.67%   **33.** -30% **35.** -20%
37. (a) $148x + 1{,}100$; $C(105) \approx \$16{,}640$.   **(b)** 4.65%   **39.** The average cost drops by approximately 0.13%.
41. (a) $N(q) = 120q^2 - 600q + 5{,}000$ **(b)** The daily net cost will increase by approximately 19.57%.
43. (a) $K \approx 21{,}930$; $q = 21{,}930 p^{-1.040}$ **(b)** An approximate 20.8% drop in sales will result.   **(c)** Profits will increase by approximately 59.2%.   **45.** $\pm 4\%$

Chapter 11 Review Exercises (page 816)

1. $3(x-1)^2$ **3.** $-2(2x+4)^{-2}$ **5.** $\dfrac{1}{2\sqrt{x+1}}$ **7.** $-\dfrac{2}{(2x+1)^2}$ **9.** $\dfrac{2x}{x^2+1}$ **11.** $\dfrac{2x+1}{(x^2+x)\ln 2}$
13. $-e^{-x}$ **15.** $e^x(1+x)$ **17.** $9x^2-2x$ **19.** $4x\sqrt{x}+\dfrac{2x^2-1}{2\sqrt{x}}$ **21.** $\dfrac{-4}{(2x-1)^2}$
23. $e^x(x^2+2x-2)$ **25.** $4x\ln x + \dfrac{2x^2-1}{x}$ **27.** $\dfrac{(2x+6x^{-4})(x+1)-x^2+2x^{-3}-1}{(x+1)^2}$ **29.** $2e^{-2x}$
31. $-4(2x^2-2x+1)^{-5}(4x-2)$ **33.** $4(t^2+e^{3t}+2\sqrt{t})^3\left(2t+3e^{3t}+\dfrac{1}{\sqrt{t}}\right)$ **35.** $\dfrac{e^x-2x}{2\sqrt{e^x-x^2}}$
37. $-\dfrac{1}{2x\sqrt{1-\ln x}}$ **39.** $\dfrac{1.3x^{0.3}(1+x)-x^{1.3}}{(1+x)^2}$ **41.** $0.1e^{0.1x}$ **43.** $-\dfrac{2e^x}{(1+2e^x)^2}$ **45.** $\dfrac{2b}{t}$
47. $\dfrac{2\ln x}{x}$ **49.** $(2x-3)e^{x^2-3x+1}$ **51.** $y=2$
53. $y=-2x+1$ **55.** $y=-7t+23$ **57.** $y=et-e$ **59.** $y=-x+9/4$ **61.** $\dfrac{2x-1}{2y}$
63. $-\dfrac{y}{x}$ **65.** $\dfrac{y-(x+y)^2}{x}$ **67.** $x^{x+1}\left(1+\ln x+\dfrac{1}{x}\right)$ **69.** $x=\dfrac{1-\ln 2}{2}$ **71.** no such values
73. $x=0$ **75.** Yes: revenue is rising at a rate of $25 per dollar increase in price. **77.** 9/20 gph/s **79.** $500 per employee **81.** $V'(1,000)=600$ cm³/s **83.** increasing at a rate of $924.18 per year **85. (a)** 60,912 frogs per year **(b)** 2006 **87. (a)** growing at a rate of 1.168 trillion grams each year **(b)** 1941 **89. (a)** $9931.71 **(b)** save 0.00144 years per dollar **91. (a)** Demand drops to about 967. **(b)** $dR/dp=q+p\,(dq/dp)$ is positive, so price should be raised.

Chapter 12

Section 12.1 (page 842)

1. Absolute minimum of -1 at $x=3$ and $x=-3$, absolute maximum of 2 at $x=1$; increasing on $[-3,1]$, decreasing on $[1,3]$ **3.** Absolute minimum of 0 at $x=-3$ and $x=1$, absolute maximum of 2 at $x=-1$ and $x=3$; increasing on $[-3,-1]$ and $[1,3]$, decreasing on $[-1,1]$ **5.** Local minimum of 1 at $x=-1$; increasing on $[-1,0)$ and $(0,+\infty)$, decreasing on $(-\infty,-1]$ **7.** Local maximum of 0 at $x=-3$, absolute minimum of -1 at $x=-2$, stationary point at $(1,1)$; increasing on $[-2,+\infty)$, decreasing on $[-3,-2]$ **9.** Local maximum at $(0.0, 2.0)$, absolute minimum at $(1.5, -0.3)$ **11.** Local maximum at $(0.1, 0.8)$, absolute minimum at $(0.4, 0.7)$ **13.** Local maximum at $(0.6, 0.3)$, local minimum at $(1.0, 0.0)$ **15.** No maxima or minima **17.** Absolute maximum of 1 at 0; absolute minimum of -3 at 2; local maximum of -2 at 3 **19.** Absolute minimum of -16 at -4; absolute maximum of 16 at -2; absolute minimum of -16 at 2; absolute maximum of 16 at 4 **21.** Absolute minimum of -10 at -2; absolute maximum of 10 at 2 **23.** Absolute minimum of -4 at -2; local maximum of 1 at -1; local minimum of 0 at 0 **25.** Local maximum of 5 at -1; absolute minimum of -27 at 3 **27.** Absolute minimum of 0 at 0 **29.** Local minimum of $\frac{5}{3}$ at -2; local maximum of -1 at 0; local minimum of $\frac{5}{3}$ at 2 **31.** Local maximum of 0 at $x=0$; absolute minimum of $-2/(3\sqrt{3})$ at $x=1/3$. **33.** Local maximum of 0 at $x=0$; absolute minimum of -3 at $x=1$ **35.** No local extrema **37.** Absolute minimum of 1 at 1 **39.** Local maximum of $1+1/e$ at -1; absolute minimum of 1 at 0; absolute maximum of $e-1$ at 1. **41.** Local maximum at $(-6,-24)$, local minimum at $(-2,-8)$
43. Absolute maximum at $\left(\dfrac{1}{\sqrt{2}},\sqrt{\dfrac{e}{2}}\right)$, absolute minimum at $\left(-\dfrac{1}{\sqrt{2}},-\sqrt{\dfrac{e}{2}}\right)$

45. Stationary minimum at $x = -1$ **47.** Stationary minima at $x = -3$ and $x = 1$, stationary maximum at $x = -1$ **49.** Singular minimum at $x = 0$, stationary non-extreme point at $x = 1$ **51.** Stationary maxima at $x = -3$ and $x = 1$, stationary minima at $x = -1$ and $x = 3$ **53.** Local minima at $(0.15, -0.52)$ and $(2.45, 8.22)$, local maximum at $(1.40, 0.29)$ **55.** Absolute maximum at $(-5, 700)$, local maxima at $(3.10, 28)$ and $(6, 40)$; absolute minimum at $(-2.10, -393)$; local minimum at $(5, 0)$

Section 12.2 (page 854)

1. $x = y = 5$; $P = 25$ **3.** $x = y = 3$; $S = 6$ **5.** $x = 2, y = 4$; $F = 20$ **7.** $x = 20, y = 10, z = 20$; $P = 4,000$ **9.** 5×5 **11.** 5×10 **13.** $p = \$10$ **15.** $p = \$30$ **17. (a)** \$1.41 per pound   **(b)** 5,000 pounds   **(c)** \$7,071.07 per month   **19.** 34.5¢ per pound, for an annual (per capita) revenue of \$5.95.   **21.** \$42.50 per ruby, for a weekly profit of \$408.33   **23. (a)** 656 headsets, for a profit of \$28,120   **(b)** \$143 per headset   **25.** 1600/27 cubic inches   **27.** in 30 years' time   **29.** 55 days   **31.** Plant 25 additional trees.   **33.** 38,730 CD players, giving an average cost of \$28 per CD player.   **35.** It should remove 2.5 pounds of pollutant per day.   **37.** $l = 30$ inches, $w = 15$ inches, $h = 30$ inches   **39.** $l = 36$ inches, $w = h = 18$ inches, $V = 11,664$ cubic inches   **41.** 1,600 copies. At this value of x, average profit equals marginal profit; beyond this the marginal profit is smaller than the average.   **43.** decreasing most rapidly in 1963; increasing most rapidly in 1989   **45.** Maximum when $t = 17$ days. This means that the embryo's oxygen consumption is increasing most rapidly 17 days after the egg is laid.   **47.** $h = r = 11.7$ cm   **49.** Absolute minimum of 5,137 at $n = 13.2$, absolute maximum of 34,040 at $n = 15$. Thus, the salary value per extra year of school is increasing most slowly (\$5,137 per year) at a level of 13.2 years of schooling, and most rapidly (\$34,040 per year) at a level of 15 years of schooling.   **51.** You should sell them in 17 years' time, when they will be worth approximately \$3,960   **53.** 71 employees   **55.** (d)

Section 12.3 (page 877)

1. 6   **3.** $\dfrac{4}{x^3}$ **5.** $-0.96x^{-1.6}$ **7.** $e^{-(x-1)}$ **9.** $\dfrac{2}{x^3} + \dfrac{1}{x^2}$ **11.** $a = -32$ ft/s²; at $t = 2$, $a = -32$ ft/s²   **13.** $a = 2/t^3 + 6/t^4$ ft/s²; at $t = 1$, $a = 8$ ft/s²   **15.** $a = -1/(4t^{3/2}) + 2$ ft/s²; at $t = 4$, $a = 63/32$ ft/s².   **17.** $(1, 0)$ **19.** $(1, 0)$ **21.** none   **23.** $(-1, 0), (1, 1)$ **25.** minimum at $(-1, 0)$, no points of inflection   **27.** maximum at $(-2, 21)$; minimum at $(1, -6)$; point of inflection at $\left(-\frac{1}{2}, \frac{15}{2}\right)$ **29.** minimum at $\left(-\frac{1}{2}, \frac{3}{4}\right)$; maximum at $(0, 1)$; point of inflection at $\left(-\frac{1}{4}, \frac{7}{8}\right)$.   **31.** maximum at $(0, 0)$; minimum at $(1, -2)$; no inflection points   **33.** minimum at $(1, 1)$; no points of inflection   **35.** no local extrema; point of inflection at $(1, 1)$ and $(-1, 1)$.   **37.** $(0.70, 2.32)$ **39.** $(-0.71, 0.61), (0.71, 0.61)$ **41.** $(0.79, -0.55), (0.21, 0.61)$ **43.** no points of inflection   **45. (a)** 2 years into the epidemic   **(b)** 2 years into the epidemic   **47. (a)** 1992   **(b)** 1994   **(c)** 1990   **49.** Point of inflection at $(12.89, 193.8)$. The prison population was declining most rapidly at $t = 12.89$ (that is, in 1963), at which time the prison population was 193,800.   **51. (a)** There are no points of inflection in the graph of S.   **(b)** Since the graph is concave up, the derivative of S is increasing, and so the rate of *decrease* of SAT scores with increasing numbers of prisoners is diminishing. In other words, the apparent effect of more prisoners is diminishing.   **53. (a)** $\left.\dfrac{d^2n}{ds^2}\right|_{s=3} = -21.494$. Thus, for a firm with annual sales of \$3 million, the rate at which new patents are produced decreases with increasing firm size. This means that the returns (as measured in the number of new patents per increase of \$1 million in sales) are diminishing as the firm size increases.   **(b)** $\left.\dfrac{d^2n}{ds^2}\right|_{s=7} = 13.474$. Thus, for a firm with annual sales of \$7 million, the rate at which new patents are produced increases with increasing firm size by 13.474 new patents per \$1 million².   **(c)** There is a point of inflection when $s \approx 5.4587$, so that in a firm with annual sales of \$5,458,700 per year, the number of new patents produced per additional \$1 million in sales is a minimum.   **55.** $n = 10t - t^2$ **57. (a)** $a = 10, b = 19, k = 0.5$; $R = \dfrac{10}{1 + 19e^{-0.5t}}$ **(b)** $R''(0) = \$0.00106875$ million per month². This means that, when $t = 0$, the revenue is accelerating by $0.00106875 million per month each month.   **59. (a)** 10 years and 50 years   **(b)** 9 years and 42 years   **(c)** $23 per year, after 9 years

Section 12.4 (page 896)

1.

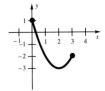

x-intercept at $2 - \sqrt{3}$, y-intercept at 1
Absolute minimum at $(2, -3)$
Absolute maximum at $(0, 1)$
Local maximum at $(3, -2)$
No points of inflection

3.

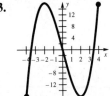

x-intercepts at 0 and $\pm 2\sqrt{3}$, y-intercept at 0
Absolute minima at $(-4, -16)$ and $(2, -16)$
Absolute maxima at $(-2, 16)$ and $(4, 16)$
Point of inflection at $(0, 0)$

5.

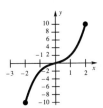

Intercept of 0 on both axes
Absolute minimum at $(-2, -10)$
Absolute maximum at $(2, 10)$
Point of inflection at $(0, 0)$

7.

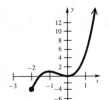

t-intercepts $-\frac{3}{2}$ and 0, y-intercept 0
Absolute minimum at $(-2, -4)$
Local minimum at $(0, 0)$
Local maximum at $(-1, 1)$
Point of inflection at $\left(-\frac{1}{2}, \frac{1}{2}\right)$

9.

x-intercepts at 0 and 4, y-intercept at 0
Local maximum at $(-1, 5)$
Absolute minimum at $(3, -27)$
Points of inflection at $(0, 0)$ and $(2, -16)$

11.

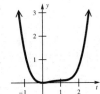

t-intercept 0, y-intercept, 0
Absolute minimum at $(0, 0)$
Points of inflection at $\left(\frac{1}{3}, \frac{11}{324}\right)$ and $\left(1, \frac{1}{12}\right)$

13.

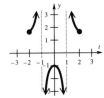

No t-intercept, y-intercept at -1
Local minima at $\left(-2, \frac{5}{3}\right)$ and $\left(2, \frac{5}{3}\right)$
Local maximum at $(0, -1)$
No points of inflection
Vertical asymptotes: $x = \pm 1$

15.

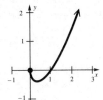

x-intercepts at 0 and 1, y-intercept at 0
Local maximum at $(0, 0)$
Absolute minimum at $\left(\frac{1}{3}, -\frac{2}{3\sqrt{3}}\right)$
No points of inflection

17.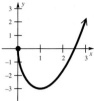
x-intercepts at 0 and $4^{2/3}$, *y*-intercept at 0
Local maximum at $(0, 0)$
Absolute minimum at $(1, -3)$
No points of inflection

19.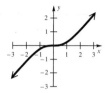
Intercept 0 on both axes
No maxima or minima
Point of inflection at $(0, 0)$

21.
No intercepts
Absolute minimum at $(1, 1)$
No points of inflection
Vertical asymptote at $x = 0$

23.
No *t*-intercepts, *y*-intercept at 1
Absolute minimum at $(0, 1)$
Absolute maximum at $(1, e - 1)$
Local maximum at $(-1, 1 + e^{-1})$
No points of inflection

25.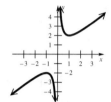
No intercepts
Local minimum at $(1, 2)$
Local maximum at $(-1, -2)$
No points of inflection
Vertical asymptote: $y = 0$

27.
(a) Absolute maximum of 300,000 at $t = 0$. Local minimum of 242,612 at $t = 10$. Local maximum of 267,756 at $t = 30$. Points of inflection at $(17.6393, 252,955)$ and $(62.3607, 185,322)$. An appropriate domain is $[0, +\infty)$.
(b) Increasing most rapidly in 17.64 years, decreasing most rapidly in 0 years. **(c)** The model predicts that the collection of cars will eventually becomes worthless in terms of discounted value. This is saying that the value of the collection of classic cars, although increasing without bound, will eventually become worthless as a result of inflation. This is not reasonable if collectors continue to prize classic cars, so we may conclude that the given model is not a reliable one for long-term prediction.

29.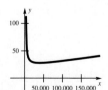
Absolute minimum at $(\sqrt{1,500,000,000}, 27.74597) \approx (38,730, 27.75)$ **(a)** This shows that the average cost per CD player is a minimum of $27.75 per player at a production level of 38,730 CD players. **(b)** $\lim_{x \to +\infty} \overline{C}(x) = +\infty$, showing that, when the production level is pushed higher and higher, the average cost per item rises without bound.

31.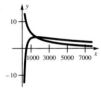

The graphs cross each other at (1,600, 4.25), which is the absolute maximum of the average profit function. When the average profit is maximized, the marginal profit and the average profit are equal.

33.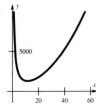

The limits at 0 and infinity are both infinite. Since the limit at zero is infinite, the amount of plastic needed to make very narrow buckets (small radius) would be very large. The reason for this is that the buckets would have to be extremely tall in order to hold the requisite volume. On the other hand, since the limit at infinity is also infinite, this says that a large amount of material would be needed to make very wide buckets (even though they would be very short).

35.

(a) Estimate the derivative at the point where $x = 50$. This is the slope of the tangent: $-\$15,858$ per year per employee. (b) $\lim\limits_{x \to +\infty} C'(x) = +\infty$. This says that as the number of employees becomes large, the additional cost per employee becomes large without bound.

37.

Smallest rate of increase at $s = 5.46$

39.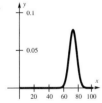

27. $\dfrac{2x^{3/2}}{3} + C$ **29.** $\dfrac{3x^{4/3}}{2} - x^{1/2} + C$ **31.** $\ln |x| + \dfrac{4}{\sqrt{x}} + C$ **33.** $f(x) = \dfrac{x^2}{2} + 1$
35. $f(x) = e^x - x - 1$ **37.** $C(x) = 5x - x^2/20{,}000 + 20{,}000$ **39.** $C(x) = 5x + x^2 + \ln x + 994$
41. (a) $s = \tfrac{1}{3}t^3 + t + C$; **(b)** $C = 1$; $s = \tfrac{1}{3}t^3 + t + 1$

Section 13.2 (page 949)

1. $\tfrac{1}{18}(3x+1)^6 + C$ **3.** $\tfrac{1}{2}(-2x+2)^{-1} + C$ **5.** $\tfrac{1}{24}(3x^2+3)^4 + C$ **7.** $\tfrac{2}{9}(3x^2-1)^{3/2} + C$
9. $-\tfrac{1}{2}e^{-x^2+1} + C$ **11.** $-\tfrac{1}{2}e^{-(x^2+2x)} + C$ **13.** $\tfrac{1}{2}(x^2 + x + 1)^{-2} + C$
15. $\tfrac{1}{3}(2x^3 + x^6 - 5)^{1/2} + C$ **17.** $4(\tfrac{1}{5}(x+1)^{5/2} - \tfrac{1}{3}(x+1)^{3/2}) + C$
19. $\tfrac{1}{7}(x-2)^7 + \tfrac{1}{3}(x-2)^6 + C$ **21.** $3e^{-1/x} + C$ **23.** $\dfrac{(x^2+1)^{2.3}}{4.6} + C$ **25.** $x + 3e^{3.1x-2} + C$ **27.** $20 \ln |1 - e^{-0.05x}| + C$
29. $\tfrac{1}{2}(e^{2x^2-2x} + e^{x^2}) + C$ **35.** $-e^{-x} + C$ **37.** $\tfrac{1}{2}e^{2x-1} + C$
39. $\dfrac{(2x+4)^3}{6} + C$ **41.** $\dfrac{\ln|5x-1|}{5} + C$ **43.** $\dfrac{(1.5x)^4}{6} + C$ **45.** $f(x) = \tfrac{1}{8}(x^2+1)^4 - \tfrac{1}{8}$
47. $f(x) = \dfrac{e^{x^2-1}}{2}$ **49. (a)** $s = \tfrac{1}{10}(t^2+1)^5 + \tfrac{1}{2}t^2 + C$ **(b)** $C = \tfrac{9}{10}$; $s = \tfrac{1}{10}(t^2+1)^5 + \tfrac{1}{2}t^2 + \tfrac{9}{10}$
51. $C(x) = 5x + \tfrac{2}{3}(x+1)^{3/2} + 19{,}999.33$ **53.** $C(x) = 5x + \ln(x+1) + 994.31$

Section 13.3 (page 962)

1. $C(x) = 10{,}000 + 100x + 0.005x^2$ **3.** $C(x) = 12{,}000 + 100x - 2{,}000e^{-0.01x}$
5. $C'(x) = 200 - 2x$, $C(x) = 200x - x^2$ **7.** $C(x) = 5{,}000(\ln(x+10) - \ln 10) = 5{,}000 \ln\left(\dfrac{x}{10} + 1\right)$
9. $x = 1{,}000$, $\overline{C}(1{,}000) = 220$ **11.** $\overline{C}(x) = 20 - 400\dfrac{e^{-0.05x} - 1}{x}$ **13. (a)** $s(t) = 10t - \dfrac{0.2t^3}{3}$
(b) $s(7) \approx 47$ T-shirts **15. (a)** (C) **(b)** $P(t) = 10t + 0.52t^3$ **(c)** $P(10) = \$620$ billion
17. (a) $E(t) = 460 \ln(1 + e^{(t-4)}) - 8.349$; **(b)** $E(15) = \$5{,}052$ million
19. (a) $P(t) = 39.8107(6{,}000 + 2{,}000t)^{0.7}$ **(b)** $42{,}878$ cellular phones
23. \$940 million **25.** 3 years **27.** 320 ft/s downwards. **31.** $(1280)^{1/2} \approx 35.78$ ft/s **33.** 113 feet below me **35. (a)** 80 feet per second **(b)** 60 ft/s
(c) 1.25 seconds **37.** $\sqrt{2} \approx 1.414$ times as fast **39.** 50,724,000 feet **41.** $88/3 \approx 29.33$ ft/s²

Section 13.4 (page 977)

1. $\dfrac{14}{3}$ **3.** 0 **5.** $\dfrac{40}{3}$ **7.** -0.9045 **9.** $2(e-1)$ **11.** $\tfrac{1}{2}(e^1 - e^{-3})$ **13.** $50(e^{-1} - e^{-2}) \approx 11.627$
15. $\tfrac{2}{3}$ **17.** $e^{2.1} - e^{-0.1}$ **19.** 0 **21.** $\tfrac{5}{2}(e^3 - e^2)$ **23.** $\dfrac{3^{5/2}}{10} - \dfrac{3^{3/2}}{6} + \dfrac{1}{15}$ **25.** $[\ln 26 - \ln 7]/3$
27. $\tfrac{1}{2}$ **29.** $16/3$ **31.** $56/3$ **33.** $\tfrac{1}{2}$ **35.** 296 miles **37.** \$783 **39.** \$2.4 billion
41. (a) $c(t) = 9.2 - 0.7636t$ **(b)** \$55.0022 billion **43.** 60.7 milliliters **45.** \$5,052 million **47.** 42,878 cellular phones **49.** 9 gallons **51. (a)** $s(t) = 330e^{0.0872t}$ **(b)** \$9,044 billion (compared to \$8,225 billion)
(c) \$7,011 billion **53.** 926 T-shirts

Section 13.5 (page 987)

1. Left sum: 4; right sum: 8; trap. sum: 6; error: 0 **3.** Left sum: 6; right sum: 6; trap. sum: 6; error: ±1
5. Left sum: -0.4; right sum: 0.4; trap. sum: 0; error: 0 **7.** Left sum: 2.3129; right sum: 0.3130; trap. sum:

1.3130; error: ±1   **9.** Left sum: 0.7456; right sum: 0.6456; trap. sum: 0.6956; error: ±0.002 $\frac{1}{1}$. 3.0370
13. 0.8863   **15.** 0.2648   **17. (a)** Left sum: 3.1604; right sum: 3.1204; trap. sum: 3.1404 **(b)** Left sum:
3.1512; right sum: 3.1312; trap. sum: 3.1412   **19. (a)** Left sum: 0.0258; right sum: 0.0254; trap. sum:
0.0256   **(b)** Left sum: 0.0257; right sum: 0.0255; trap. sum: 0.0256
21. (a) Left sum: 1.9998; right sum: 1.9998; trap. sum: 1.9998   **(b)** Left sum: 2.0000; right sum: 2.0000; trap.
sum: 2.0000   **23. (a)** $s(t) = -2.925t + 18.9625$ **(b)** The model gives total sales of $58.25 billion. The actual
total sales were $66.4 billion, much higher than the sales predicted by the model. One reason for this is that the 1992
sales were considerably higher than predicted by the linear model. (If you join the midpoint of the top of the first
column to that of the last, you will find that the 1992 sales "stick out" above this line.)
25. $1,799.01 (same answer as given by the FTC)
27. Yes. Both estimates give an area of 420 square feet.   **29. (a)** 99.2%   **(b)** 1.5×10^{-14}%

Section 13.6 (page 995)

1. $A(x) = x$ **3.** $A(x) = \dfrac{(x-2)^2}{2}$ **5.** $A(x) = c(x-a)$ **7.** $A(x) = \dfrac{x^2}{2}$ **9.** $A(x) = \dfrac{x^3}{3}$
11. $A(x) = e^x - 1$ **13.** $A(x) = \ln x$ **15.** $A(x) = x^2/2$ if $0 \le x \le 1$; $\frac{1}{2} + 2(x-1)$ if $x > 1$
19. $f(0) = 0, f(0.5) = 0.46, f(1) = 0.75$ **21.** $f(-0.5) = -0.49, f(0.5) = 0.49$

Section 13.7 (page 1007)

1. estimate = 4, error = 0   **3.** estimate = 20.4525, error ≤ 0.405, exact answer = 20.25
5. estimate = 4.037, error ≤ 0.053   **7.** 4, exact   **9.** 20.25, exact   **11.** 4.0470 ± 0.0034
13. all estimates = 4   **15. (a)** 20.252025   **(b)** 20.250506   **(c)** 20.250081
17. (a) 4.0470829   **(b)** 4.0471629   **(c)** 4.0471853   **19.** all estimates = 4   **21.** all
estimates = 20.25   **23. (a)** 4.0471895340   **(b)** 4.0471895604   **(c)** 4.0471895621   **25.** $n \ge 566$
27. $n \ge 15$ **29.** $n \ge 1{,}291$ **31.** $n \ge 69$ **33. (a)** 1.478940   **(b)** 1.478942857
35. (a) 2.003503   **(b)** 2.003497111

Chapter 13 Review Exercises (page 1011)

1. $10x^{11}/11 + C$ **3.** $-1/x^3 + C$ **5.** $(2/3)x^{3/2} + \ln |x| + C$ **7.** $2e^x + 3x^2/2 - 4\ln |x| + C$
9. $(x+2)^{11}/11 + C$ **11.** $(3/8)(x^2+1)^{4/3} + C$ **13.** $(1/2)\ln(x^2+1) + C$ **15.** $-(5/2)e^{-2x} + C$
17. $e^{x^2}/2 - (3/2)x^2 + C$ **19.** $\dfrac{4.7}{1.2} \ln |x^{1.2} - 4| + C$ **21.** $\dfrac{1}{0.6} \ln |1 + 2e^{0.3t}| + C$
23. 5/4   **25.** 3/2   **27.** $3/2 - 1/e$ **29.** 52/9   **31.** $4 + \dfrac{2^5 - 1}{6}$ **33.** 0   **35.** 4/3   **37.** ln 10
39. $\dfrac{1 - e^{-25}}{2}$ **41.** $\dfrac{2\sqrt{2}}{3}$ **43.** 4/15   **45.** 2/3   **47.** Both sums are 1.46; trap. sum: 1.46
49. Left sum: −0.14, right sum: 0.55, trap. sum = 0.20
51. $n = 100$: left sum = 2.9257, right sum = 2.9257; $n = 200$: left sum = 2.9254, right sum = 2.9254
$n = 500$: left sum = 2.9253, right sum = 2.9253   **53.** $n = 100$: left sum = 0.3649, right sum = 0.3636;
$n = 200$: left sum = 0.3646, right sum = 0.3639; $n = 500$: left sum = 0.3644, right sum = 0.3641
55. $n = 50$: 1.49364830; $n = 100$: 1.49364827; $n = 500$: 1.49364827   **57.** $n = 50$: 0.23286788; $n = 100$:
0.23286795; $n = 500$: 0.23286795   **59.** $n = 50$: 2.925305; $n = 100$: 2.925304; $n = 500$: 2.925303
61. $n = 50$: 0.36423405; $n = 100$: 0.36423404; $n = 500$: 0.36423404   **63.** $n \ge 37$ **65.** $n \ge 34$
67. $n \ge 105$ **69.** $n \ge 13$ **71.** $n \ge 8$ **73.** $n \ge 8$ **75.** $n \ge 14$ **77.** $n \ge 6$
79. (a) $S(t) = -2.2t + 70$, t = years since 1982   **(b)** actual = 704 million, estimated = 680 million
(c) 634 million   **81.** 943 ft   **83.** 6,560 ft   **85.** $49,850 + 50x + 40 \ln (x + 40)$
87. $75x + 25(1 - e^{-x})$ **89. (a)** $2 per watt of capacity   **(b)** $p = 2e^{-0.1597t}$ **(c)** 473.5 megawatts

Chapter 14

Section 14.1 (page 1023)

1. $2e^x(x - 1) + C$ 3. $-e^{-x}(2 + 3x) + C$ 5. $e^{2x}(2x^2 - 2x - 1)/4 + C$
7. $-e^{-2x+4}(2x^2 + 2x + 3)/4 + C$ 9. $-e^{-x}(x^2 + x + 1) + C$ 11. $(x^4 \ln x)/4 - x^4/16 + C$
13. $(t^3/3 + t)\ln 2t - t^3/9 - t + C$ 15. $\frac{3}{4}t^{4/3}(\ln t - \frac{3}{4}) + C$ 17. e 19. $38{,}229/286$
21. $(\frac{7}{2})\ln 2 - \frac{3}{4}$ 23. $\frac{1}{4}$ 25. $1 - 11e^{-10}$ 27. $4 \ln 2 - \frac{7}{4}$ 29. $28{,}800{,}000(1 - 2e^{-1})$ ft
31. $500 + 10x - 1/(x + 1) - [\ln(x + 1)]/(x + 1)$ 33. $\$33{,}598$ 35. $\$1{,}478$ million

Section 14.2 (page 1028)

1. $1 + x - \ln|1 + x| + C$ 3. $\frac{1}{30}(6x - 2)(1 + 2x)^{3/2} + C$ 5. $\frac{x}{2}\sqrt{x^2 + 4} + 2\ln|x + \sqrt{x^2 + 4}| + C$
7. $\tan^{-1}x + C$ 9. $\ln|x + \sqrt{3 + x^2}| + C$ 11. $\frac{1}{4}\ln\left|\frac{2x + 1}{2x - 1}\right| + C$ 13. $\frac{1}{3}(2x^2 - 1)^{3/2} + C$
15. $\frac{1}{2\sqrt{3}}\ln\left|\frac{\sqrt{3}x - 1}{\sqrt{3}x + 1}\right| + C$ 17. $\tan^{-1}(x + 3) + C$
19. $\frac{2x - 1}{4}\sqrt{x^2 - x + 3} + \frac{11}{8}\ln\left|x - \frac{1}{2} + \sqrt{x^2 - x + 3}\right| + C$ 21. $-\frac{1}{6x} + \frac{1}{3}\ln\left|\frac{2 + 4x}{x}\right| + C$
23. $\frac{1}{6}\ln\left|\frac{e^x + 3}{e^x - 3}\right| + C$ 25. $\tan^{-1}(\ln x) + C$ 27. $C(x) = 2[15x^2 - 12x + 8](1 + x)^{3/2} + 9{,}984$
29. $C(x) = 30\ln|x + \sqrt{x^2 + 9}| + 10{,}000 - 30\ln 3$ 31. $5{,}000 \ln|x + \sqrt{x^2 + 1}|$
33. $\overline{C}(x) = \frac{10}{x}\left[\ln\left|\frac{x + \sqrt{4 + x^2}}{2}\right|\right] + \frac{10{,}000}{x}$ 35. $119{,}004.17$ ft

Section 14.3 (page 1041)

1. $\frac{8}{3}$ 3. 4 5. 1 7. $e - \frac{3}{2}$ 9. $\frac{2}{3}$ 11. $\frac{3}{10}$ 13. $\frac{1}{20}$ 15. $\frac{4}{15}$ 17. $2\ln 2 - 1$
19. $8 \ln 4 + 2e - 16$ 21. 0.3222 23. 0.3222 25. $\$6.25$ 27. $\$512$ 29. $\$119.53$ 31. $\$900$
33. $\$416.67$ 35. $\$326.27$ 37. $\$25$ 39. $\$0.50$ 41. $\$386.29$ 43. $\$225$ 45. $\$25.50$
47. $\$12{,}684.60$ 49. $\bar{p} = \$3{,}655$, CS $= \$856{,}140$, PS $= \$3{,}715{,}650$. The university would earn $\$3{,}715{,}650$ less.
51. You save more. 53. CS $= \frac{1}{2m}(b - m\bar{p})^2$. 55. (a) $\$373.35$ million (b) This is the area of the region between the graphs of $R(t)$ and $P(t)$ for $0 \le t \le 4$. (c) Since the exponent for P is larger, this tells us that the ratio of profit to revenue was increasing; that is, costs accounted for a decreasing proportion of the revenues.
57. (b) $\$7{,}260$ million (c) The area between $y = 1{,}000{,}000(41.25q^2 - 697.5q + 3{,}210)$ and $y = 400{,}000{,}000$ for $0 \le q \le 12$.

Section 14.4 (page 1052)

1. $\$14{,}403$ million 3. $\$215.714$ billion
5. Average $= 2$ 7. Average $= 1$

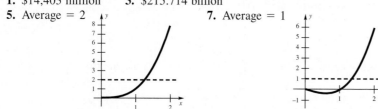

9. Average $= (1 - e^{-2})/2$

11. Moving average: 2.6, 2.8, 3.0, 3.2, 3.4, 4.0, 4.6, 5.2, 5.8, 6.4, 7.0, 7.6, 7.8, 8.0, 8.2, 8.4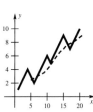

13. Moving average: $\bar{f}(x) = x^3 - (15/2)x^2 + 25x - 125/4$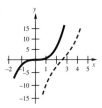

15. Moving average: $\bar{f}(x) = (3/25)[x^{5/3} - (x-5)^{5/3}]$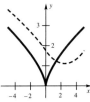

17. $\bar{f}(x) = \frac{2}{5}(e^{0.5x} - e^{0.5(x-5)})$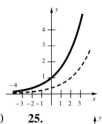

19. $\bar{f}(x) = \frac{2}{15}(x^{3/2} - (x-5)^{3/2})$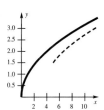

21. $10,410.88 **23.** $1500 **25.** 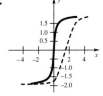 **27.**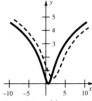

29. (a) $n(t) = 6.4 + 0.3t$, where t is the number of years since 1983 **(b)** $\bar{n}(t) = 5.8 + 0.3t$ **(c)** The slope of the moving average is the same as the slope of the original function. **(d)**

31. $\bar{f}(x) = mx + b - \dfrac{ma}{2}$

33. (a) $p(t) = -0.04t^2 + t + 12$ **(b)** 16.67% **(c)** 10.38% **35. (a)**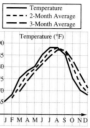

(b) The 24-month moving average is constant and equal to the year-long average of approximately 77°. **(c)** A quadratic model could not be used to predict temperatures beyond the given 12-month period, since temperature patterns are periodic, whereas parabolas are not.

37. (a) 78.10 **(b)** 7.48% Slightly less than the answer to Ex. 4. (This is expected because of compounding over eight years.) **39.** **41. (a)** Average voltage over $[0, \tfrac{1}{6}]$ is zero; 60 cycles per second. **(b)**   **(c)** 116.673 volts.

Section 14.5 (page 1067)

1. diverges **3.** converges to $2e$ **5.** converges to e^2 **7.** converges to $\tfrac{1}{2}$ **9.** converges to $\tfrac{1}{108}$
11. converges to $3 \cdot 5^{2/3}$ **13.** diverges **15.** diverges **17.** converges to $\tfrac{5}{4}(3^{4/5} - 1)$ **19.** diverges
21. converges to 0 **23.** diverges **25.** diverges **27.** No; You will not sell more than 2,000 of them.
29. $2,596.85 million **31.** $70,833 **33.** $25,160 billion **35.** $\int_0^{+\infty} q(t)\, dt$ diverges, indicating that there is no bound to the expected future exports of pork. $\int_{-\infty}^0 q(t)\, dt$ converges to approximately 8.3490, indicating that total exports of pork prior to 1985 amounted to approximately $8.3490 million.
37. (a) 2.467 meteors **(b)** The integral diverges. We can interpret this as saying that the number of impacts by meteors smaller than 1 megaton is very large. (This makes sense because, for example, this number includes meteors no larger than a grain of dust.) **39. (a)** $\Gamma(1) = 1; \Gamma(2) = 1$
41. 1 **43.** 0.1586

Section 14.6 (page 1077)

In the answers to Exercises 1–9, A and C denote arbitrary constants.

1. $y = \dfrac{x^3}{3} + \dfrac{2x^{3/2}}{3} + C$ **3.** $\dfrac{y^2}{2} = \dfrac{x^2}{2} + C$ **5.** $y = Ae^{x^2/2}$ **7.** $y = -\dfrac{2}{(x+1)^2} + C$ **9.** $y = \sqrt{(\ln|x|)^2 + D}$

11. $y = \dfrac{x^4}{4} - x^2 + 1$ **13.** $y = (x^3 + 8)^{1/3}$ **15.** $y = 2x$ **17.** $y = e^{x^2/2} - 1$ **19.** $y = -\dfrac{2}{\ln(x^2 + 1) + 2}$

21. With $s(t) = $ monthly sales after t months, $\dfrac{ds}{dt} = -0.05s$; $s = 1{,}000$ when $t = 0$. Solution: $s = 1{,}000 e^{-0.05t}$.

23. With $S(t) = $ total sales after t months, $\dfrac{dS}{dt} = -0.1(100{,}000 - S)$; $S(0) = 0$. Solution: $S = 100{,}000(1 - e^{-0.1t})$.

27. $S = \dfrac{2/1999}{e^{-0.5t} + 1/1999}$

Graph:

29. (a) $y = be^{Ae^{-at}}$, A = constant   **(b)** $y = 10e^{-0.69315e^{-t}}$

Graph: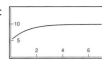

It will take about 27 months to saturate the market.

Chapter 14 Review Exercises (p. 1083)

1. $(x^2 - 2x + 4)e^x + C$ **3.** $(1/3)x^3 \ln 2x - x^3/9 + C$ **5.** $e - 2$ **7.** $(2e^3 + 1)/9$

9. $\dfrac{1}{6}\tan^{-1}\left(\dfrac{2x}{3}\right) + C$ **11.** $\dfrac{1}{2}\ln|2x + \sqrt{9 + 4x^2}| + C$

13. $\dfrac{x+1}{2}\sqrt{x^2 + 2x + 2} + \dfrac{1}{2}\ln|x + 1 + \sqrt{x^2 + 2x + 2}| + C$ **15.** $\tfrac{1}{2}\tan^{-1}(e^{2x}) + C$ **17.** 0.41817

19. 1.14779   **21.** $\tfrac{1}{4}$ **23.** $2\sqrt{e}$ **25.** diverges   **27.** diverges   **29.** 2   **31.** $\dfrac{1}{2}\left[\dfrac{3}{2^{1/3}} - 1\right]$

33. $3 - 4\sqrt{2} + e^2 + \dfrac{2}{e^2}$ **35.** $\dfrac{2\sqrt{2}}{3}$ **37.** 1   **39.** 1.01295   **41.** $e - 2$ **43.** 2.66229

45. 0.418171   **47.** 0.66799

49. $\bar{f}(x) = \dfrac{3}{10}[x^{5/3} - (x-2)^{5/3}]$

51. $\bar{f}(x) = \tfrac{1}{2}[x \ln x - (x-2)\ln(x-2) - 2]$

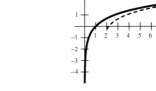

53. $y = -\dfrac{3}{x^3} + C$ **55.** $y = Ae^{(x+1)^2/2} - 1$

57. $\dfrac{y^2}{2} = x + C$ **59.** $y = \sqrt{2\ln|x| + 1}$ **61.** $y = 2\sqrt{x^2 + 1}$ **63. (a)** Approximately 920. There were approximately 920,000 deaths resulting from motor vehicles in the period 1973–1992.   **(b)** Approximately 280. There were approximately 280,000 more deaths resulting from motor vehicles than guns in the period 1973–1992.
65. $47,259.56   **67.** Equilibrium price = $5, CS = $82.11, PS = $41.67   **69. (a)** 1,600. This is the total distance the cannonball has traveled.   **(b)** $T = 20$ secs. Area = 3,200   **71. (a)** Average = 322 pedestrians per year.
(b) 2-year moving average:    **73.** $9,705.91   **75.** $y = (100 + 0.05t)^2$

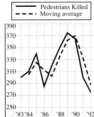

Chapter 15

Section 15.1 (page 1095)

1. $f(0, 0) = 1; f(1, 0) = 1; f(0, -1) = 2; f(a, 2) = a^2 - a + 5; f(y, x) = y^2 + x^2 - y + 1;$
$f(x + h, y + k) = (x + h)^2 + (y + k)^2 - (x + h) + 1$ 3. $f(0, 0) = \sqrt{5}; f(1, 0) = 2; f(0, -1) = \sqrt{10};$
$f(a, 2) = |a - 1|; f(y, x) = \sqrt{(y - 1)^2 + (x - 2)^2}; f(x + h, y + k) = \sqrt{(x + h - 1)^2 + (y + k - 2)^2}$
5. $g(0, 0, 0) = 1; g(1, 0, 0) = g(0, 1, 0) = e; g(z, x, y) = e^{x+y+z}; g(x + h, y + k, z + l) = e^{x+h+y+k+z+l}$
7. $g(0, 0, 0)$ does not exist; $g(1, 0, 0) = g(0, 1, 0) = 0; g(z, x, y) = xyz/(x^2 + y^2 + z^2);$
$g(x + h, y + k, z + l) = (x + h)(y + k)(z + l)/[(x + h)^2 + (y + k)^2 + (z + l)^2]$ 9. $\sqrt{2}$ 11. $\sqrt{a^2 + b^2}$
13. $\frac{1}{2}$ 15. circle with center $(2, -1)$ and radius 3   17. circle with center $(-3, -2)$ and radius $\sqrt{6}$
19. 7,071,068 (to the nearest item)
21. (a) $100 = K(1{,}000)^a(1{,}000{,}000)^{1-a}; 10 = K(1{,}000)^a(10{,}000)^{1-a}$ (b) $\log K - 3a = -4; \log K - a = -3$
    (c) $a = 0.5, K \approx 0.003162$ (d) $P = 71$ pianos (to the nearest piano)
23. $s(i, t) = 0.33i + 0.02t + 2.32$
25. (a) The demand function increases with increasing values of r.
    (b) $Q(2 \times 10^8, 0.5, 500) = 90{,}680$. This means that if the total real income in Great Britain is 2×10^8 units of currency, and if the average retail price of beer is 0.5 units of currency per unit of beer, and if the average retail price of all other commodities is 500 units of currency, then 90,680 units of beer will be sold per year.
27. $U(11, 10) - U(10, 10) \approx 5.75$. This means that, if your company now has 10 copies of Macro Publish and 10 copies of Turbo Publish, then the purchase of one additional copy of Macro Publish will result in a productivity increase of approximately 5.75 pages per day.
29. (a) $(a, b, c) = (3, \frac{1}{4}, \frac{1}{\pi}); (a, b, c) = (\frac{1}{\pi}, 3, \frac{1}{4})$.
    (b) $a = (3/4\pi)^{1/3}$. The resulting ellipsiod is a sphere with radius a.

31.

x	y	$f(x, y) = x^2\sqrt{1 + xy}$
-1	-1	$\sqrt{2}$
1	12	$\sqrt{13}$
0.3	0.5	0.096514
41	42	$1{,}681\sqrt{1{,}723}$

33.

x	y	$f(x, y) = x \ln(x^2 + y^2)$
3	1	$3 \ln 10$
1.4	-1	1.5193
e	0	$2e$
0	e	0

35. (a) 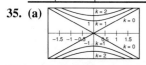   (b) 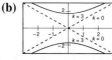   (c)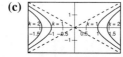

37. (a) 4.118×10^{-3} gram per square meter   (b) the total weight of sulfates in the earth's atmosphere
39. (a) The value of N would be doubled.   (b) $N(R, f_p, n_e, f_l, f_i, L) = R f_p n_e f_l f_i L$, where here L is the average lifetime of an intelligent civilization.   (c) The function is not linear, since it involves a product of variables.
(d) By taking the logarithm of both sides, since this would yield the linear function
$\ln(N) = \ln(R) + \ln(f_p) + \ln(n_e) + \ln(f_l) + \ln(f_i) + \ln(L)$.
41. (a) The model predicts 8.06%. (The actual figure was 8.2%, showing the accuracy of the model.)   (b) Foreign manufacturers, since each 1% gain of the market by foreign manufacturers decreases Chrysler's share by 0.8%—the largest of the three.

Section 15.2 (page 1109)

1. 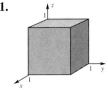 **3.** 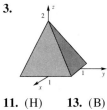 **5.** 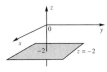 **7.**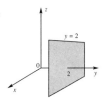

9. 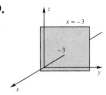 **11.** (H) **13.** (B) **15.** (F) **17.** (C)

19. 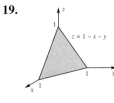 **21.** 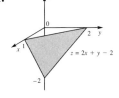 **23.** 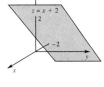 **25.**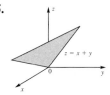

27. 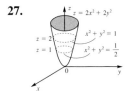 **29.** 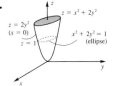 **31.** 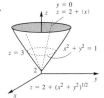 **33.**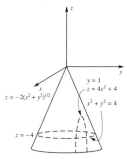

35. 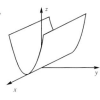 **37.** 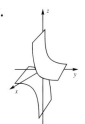 **39.**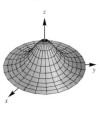

41.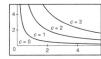

The level curve at $z = 3$ has the form $3 = x^{0.5}y^{0.5}$, or $y = 9/x$, and shows the relationship between the number of workers and the operating budget at a production level of 3 units.

43.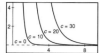

The level curve at $z = 0$ consists of the nonnegative y-axis ($x = 0$) and tells us that zero utility corresponds to zero copies of Macro Publish, regardless of the number of copies of Turbo Publish. (Zero copies of Turbo Publish does not necessarily result in zero utility, according to the formula.)

45. Both slices are quarter-ellipses. (We only see the portion in the first quadrant because $e \geq 0$ and $k \geq 0$.) The level curve $C = 30,000$ represents the relationship between the number of electricians and the number of carpenters used in building a home that costs \$30,000. A similar relationship corresponds to for the level curve $C = 40,000$.

47. The following figure shows several level curves together with several lines of the form $h + w = c$.

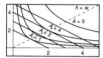

From the figure, thinking of the curves as contours on a map, we see that the largest value of A anywhere along any of the lines $h + w = c$ occurs midway along the line, when $h = w$. Thus, the largest-area rectangle with a fixed perimeter occurs when $h = w$ (that is, when the rectangle is a square).

Section 15.3 (page 1119)

1. $f_x(x, y) = 6x + 1; f_x(1, -1) = 7$
$f_y(x, y) = -3y^2; f_y(1, -1) = -3$
$f_{xx}(x, y) = 6; f_{xx}(1, -1) = 6$
$f_{yy}(x, y) = -6y; f_{yy}(1, -1) = 6$
$f_{xy}(x, y) = f_{yx}(x, y) = 0; f_{xy}(1, -1) = f_{yx}(1, -1) = 0$
3. $f_x(x, y) = 6xy; f_x(1, -1) = -6$
$f_y(x, y) = 3x^2; f_y(1, -1) = 3$
$f_{xx}(x, y) = 6y; f_{xx}(1, -1) = -6$
$f_{yy}(x, y) = 0; f_{yy}(1, -1) = 0$
$f_{xy}(x, y) = f_{yx}(x, y) = 6x; f_{xy}(1, -1) = f_{yx}(1, -1) = 6$
5. $f_x(x, y) = 2xy^3 - 3x^2y^2 - y; f_x(1, -1) = -4$
$f_y(x, y) = 3x^2y^2 - 2x^3y - x; f_y(1, -1) = 4$
$f_{xx}(x, y) = 2y^3 - 6xy^2; f_{xx}(1, -1) = -8$
$f_{yy}(x, y) = 6x^2y - 2x^3; f_{yy}(1, -1) = -8$
$f_{xy}(x, y) = f_{yx}(x, y) = 6xy^2 - 6x^2y - 1; f_{xy}(1, -1) = f_{yx}(1, -1) = 11$
7. $f_x(x, y) = 6y(2xy + 1)^2; f_x(1, -1) = -6$
$f_y(x, y) = 6x(2xy + 1)^2; f_y(1, -1) = 6$
$f_{xx}(x, y) = 24y^2(2xy + 1); f_{xx}(1, -1) = -24$
$f_{yy}(x, y) = 24x^2(2xy + 1); f_{yy}(1, -1) = -24$
$f_{xy}(x, y) = f_{yx}(x, y) = 6(2xy + 1)^2 + 24xy(2xy + 1) = 6(6xy + 1)(2xy + 1);$
$f_{xy}(1, -1) = f_{yx}(1, -1) = 30$
9. $f_x(x, y) = e^{x+y}; f_x(1, -1) = 1$
$f_y(x, y) = e^{x+y}; f_y(1, -1) = 1$
$f_{xx}(x, y) = e^{x+y}; f_{xx}(1, -1) = 1$
$f_{yy}(x, y) = e^{x+y}; f_{yy}(1, -1) = 1$
$f_{xy}(x, y) = f_{yx}(x, y) = e^{x+y}; f_{xy}(1, -1) = f_{yx}(1, -1) = 1$

11. $f_x(x, y) = 3x^{-0.4}y^{0.4}$; $f_x(1, -1)$ not defined
$f_y(x, y) = 2x^{0.6}y^{-0.6}$; $f_y(1, -1)$ not defined
$f_{xx}(x, y) = -1.2x^{-1.4}y^{0.4}$; $f_{xx}(1, -1)$ not defined
$f_{yy}(x, y) = -1.2x^{0.6}y^{-1.6}$; $f_{yy}(1, -1)$ not defined
$f_{xy}(x, y) = f_{yx}(x, y) = 1.2x^{-0.4}y^{-0.6}$; $f_{xy}(1, -1)$ and $f_{yx}(1, -1)$ not defined
13. $f_x(x, y) = 4.92x^{0.2}e^{-0.2y}$; $f_x(1, -1) = 4.92e^{0.2}$
$f_y(x, y) = -0.82x^{1.2}e^{-0.2y}$; $f_y(1, -1) = -0.82e^{0.2}$
$f_{xx}(x, y) = -0.984x^{-0.8}e^{-0.2y}$; $f_{xx}(1, -1) = -0.984e^{0.2}$
$f_{yy}(x, y) = 0.164x^{1.2}e^{-0.2y}$; $f_{yy}(1, -1) = 0.164e^{0.2}$
$f_{xy}(x, y) = f_{yx}(x, y) = -0.984x^{0.2}e^{-0.2y}$; $f_{xy}(1, -1) = f_{yx}(1, -1) = -0.984e^{0.2}$
15. $f_x(x, y) = 0.2ye^{0.2xy}$; $f_x(1, -1) = -0.2e^{-0.2}$
$f_y(x, y) = 0.2xe^{0.2xy}$; $f_y(1, -1) = 0.2e^{-0.2}$
$f_{xx}(x, y) = 0.04y^2e^{0.2xy}$; $f_{xx}(1, -1) = 0.04e^{-0.2}$
$f_{yy}(x, y) = 0.04x^2e^{0.2xy}$; $f_{yy}(1, -1) = 0.04e^{-0.2}$
$f_{xy}(x, y) = f_{yx}(x, y) = 0.2(1 + 0.2xy)e^{0.2xy}$; $f_{xy}(1, -1) = f_{yx}(1, -1) = 0.16e^{-0.2}$
17. $f_x(x, y, z) = yz$; $f_x(0, -1, 1) = -1$
$f_y(x, y, z) = xz$; $f_y(0, -1, 1) = 0$
$f_z(x, y, z) = xy$; $f_z(0, -1, 1) = 0$
19. $f_x(x, y, z) = 4/(x + y + z^2)^2$; $f_x(0, -1, 1)$ not defined
$f_y(x, y, z) = 4/(x + y + z^2)^2$; $f_y(0, -1, 1)$ not defined
$f_z(x, y, z) = 8z/(x + y + z^2)^2$; $f_z(0, -1, 1)$ not defined
21. $f_x(x, y, z) = e^{yz} + yze^{xz}$; $f_x(0, -1, 1) = e^{-1} - 1$
$f_y(x, y, z) = xze^{yz} + e^{xz}$; $f_y(0, -1, 1) = 1$
$f_z(x, y, z) = xy(e^{yz} + e^{xz})$; $f_z(0, -1, 1) = 0$
23. $f_x(x, y, z) = 0.1\dfrac{y^{0.4}z^{0.5}}{x^{0.9}}$; $f_y(x, y, z) = 0.4\dfrac{x^{0.1}z^{0.5}}{y^{0.6}}$; $f_z(x, y, z) = 0.5\dfrac{x^{0.1}y^{0.4}}{z^{0.5}}$; $f_x(0, -1, 1)$ is not defined; $f_y(0, -1, 1)$ is not defined, $f_z(0, -1, 1)$ is not defined.
25. $f_x(x, y, z) = yze^{xyz}$; $f_y(x, y, z) = xze^{xyz}$; $f_z(x, y, z) = xye^{xyz}$;
$f_x(0, -1, 1) = -1, f_y(0, -1, 1) = f_z(0, -1, 1) = 0$
27. $f_x(x, y, z) = 0$; $f_y(x, y, z) = -\dfrac{600z}{y^{0.7}(1 + y^{0.3})^2}$; $f_z(x, y, z) = \dfrac{2{,}000}{1 + y^{0.3}}$;
$f_x(0, -1, 1) = 0; f_y(0, -1, 1)$ is not defined; $f_z(0, -1, 1)$ is not defined.
29. $\partial c/\partial x = -0.8$, showing that Chrysler's percentage of the market decreases by 0.8% for every 1% rise in foreign manufacturers' share. $\partial c/\partial y = -0.2$, showing that Chrysler's percentage of the market decreases by 0.2% for every 1% rise in Ford's share. $\partial c/\partial z = -0.7$, showing that Chrysler's percentage of the market decreases by 0.7% for every 1% rise in G.M.'s share.
31. The marginal cost of cars is $6{,}000 + 1{,}000e^{-0.01(x+y)}$ per car. The marginal cost of trucks is $4{,}000 + 1{,}000e^{-0.01(x+y)}$ per truck. Both marginal costs decrease as production rises.
33. $\bar{C}(x, y) = \dfrac{200{,}000 + 6{,}000x + 4{,}000y - 100{,}000e^{-0.01(x+y)}}{x + y}$
$\bar{C}_x(50, 50) = -\$2.64$ per car. This means that at a production level of 50 cars and 50 trucks per week, the average cost per car is decreasing by $2.64 for each additional car manufactured. $\bar{C}_y(50, 50) = -\$22.64$ per truck. This means that at a production level of 50 cars and 50 trucks per week, the average cost per truck is decreasing by $22.64 for each additional truck manufactured.
35. No; your marginal revenue from the sale of cars is $15{,}000 - 2{,}500/\sqrt{x + y}$ per car and $10{,}000 - 2{,}500/\sqrt{x + y}$ per truck from the sale of trucks. These increase with increasing x and y. In other words, you will earn more revenue per vehicle with increasing sales, and so the rental company will pay more for each additional vehicle it buys.

37. $P_z(10, 100{,}000, 1{,}000{,}000) \approx 0.0001010$ papers/$

39. (a) $U_x(10, 5) = 5.18$, $U_y(10, 5) = 2.09$. This means that if 10 copies of Macro Publish and 5 copies of Turbo Publish are purchased, the company's daily productivity is increasing at a rate of 5.18 pages per day for each additional copy of Macro purchased and by 2.09 pages per day for each additional copy of Turbo purchased.
(b) $U_x(10, 5)/U_y(10, 5) \approx 2.48$ is the ratio of the usefulness of one additional copy of Macro to one of Turbo. Thus, with 10 copies of Macro and 5 copies of Turbo, the company can expect approximately 2.48 times the productivity per additional copy of Macro than Turbo.

41. 6×10^9 n/s

43. (a) $A_P(100, 0.1, 10) = 2.59$; $A_r(100, 0.1, 10) = 2{,}357.95$; $A_t(100, 0.1, 10) = 24.72$. Thus, for a $100 investment at 10% interest invested for 10 years, the accumulated amount is increasing at a rate of $2.59 per $1 of principal, at a rate of $2,357.95 per increase of 1 in r (note that this would correspond to an increase in the interest rate of 100%), and at a rate of $24.72 per year. (b) $A_P(100, 0.1, t)$ tells you the rate at which the accumulated amount in an account bearing 10% interest with a principal of $100 is growing per $1 increase in the principal, t years after the investment.

45. (a) $P_x = Ka(y/x)^b$ and $P_y = Kb(x/y)^a$. They are equal precisely when $a/b = (x/y)^b(x/y)^a$. Substituting $b = 1 - a$ now gives $a/b = x/y$. (b) The given information implies that $P_x(100, 200) = P_y(100, 200)$. By part (a), this occurs precisely when $a/b = x/y = \frac{100}{200} = \frac{1}{2}$. But $b = 1 - a$, so $a/(1 - a) = \frac{1}{2}$, giving $a = \frac{1}{3}$ and $b = \frac{2}{3}$.

47. Decreasing at 0.007457 cc/s

Section 15.4 (page 1130)

1. local maximum at P and R, saddle point at Q **3.** critical point (neither saddle point nor local extremum) at Q, non-critical point at P and R **5.** non-critical point at P and Q, saddle point at R

7.

9. minimum of 1 at $(0, 0)$ **11.** maximum of $\frac{3}{2}$ at $(-\frac{1}{2}, \frac{1}{2})$
13. maximum of 0 at $(0, 0)$, saddle points at $(\pm 4, 2, -16)$
15. minimum of 1 at $(0, 0)$
17. minimum of -16 at $(-2, \pm 2)$; $(0, 0)$ a critical point which is not a local extremum
19. saddle point at $(0, 0, -1)$ **21.** maximum of e at $(-1, 0)$ **23.** minimum of $3(2^{2/3})$ at $(2^{1/3}, 2^{1/3})$
25. minimum of 4 at both $(1, 1)$ and $(-1, -1)$ **27.** 400 5-speeds and 2,000 10-speeds

29. $\overline{C}_x = \partial/\partial x[C/(x + y)] = [(x + y)C_x - C]/(x + y)^2$. This is zero when $(x + y)C_x = C$, or $C_x = \dfrac{C}{x + y} = \overline{C}$.
Similarly, $\overline{C}_y = 0$ when $C_y = \overline{C}$. This is reasonable because if the average cost is decreasing with increasing x, then the average cost is greater than the marginal cost C_x. Similarly, if the average cost is increasing with increasing x, the average cost is less than the marginal cost C_x. Thus, if the average cost is stationary with increasing x, then the average cost equals the marginal cost C_x. (The situation is similar for the case of increasing y.)

31. They should charge $580.81 for the Ultra Mini and $808.08 for the Big Stack.

Section 15.5 (page 1141)

1. maximum value of 8 at $(2, 2)$, minimum value of 0 at $(0, 0)$ **3.** maximum value of 9 at $(-2, 0)$, minimum value of 0 at $(1, 0)$ **5.** maximum value of e^4 at $(0, \pm 2)$, minimum value of 1 at $(0, 0)$ **7.** maximum value of e^4 at $(\pm 1, 0)$, minimum value of 1 at $(0, 0)$ **9.** maximum value of 5 at $(\frac{1}{2}, \frac{1}{2})$, minimum value of 3 at $(1, 1)$
11. maximum value of $\frac{161}{9}$ at $(1, 9)$ and $(9, 1)$, minimum value of 17 at $(1, 1)$ **13.** maximum value of 8 at $(2, 0)$, minimum value of -1 at $(-1, 0)$ **15.** maximum value of 8 at $(2, 0)$, minimum value of 0 at $(0, 0)$
17. $(1/\sqrt{3}, 1/\sqrt{3}, 1/\sqrt{3})$, $(-1/\sqrt{3}, -1/\sqrt{3}, 1/\sqrt{3})$, $(1/\sqrt{3}, -1/\sqrt{3}, -1/\sqrt{3})$, $(-1/\sqrt{3}, 1/\sqrt{3}-1/\sqrt{3})$

19. For minimum cost of $16,600, make 100 5-speeds and 80 10-speeds. For maximum cost of $17,400, make 100 5-speeds and 120 10-speeds.   **21.** For a maximum profit of $4,500, sell 50 copies of Walls and 150 copies of Doors. For a minimum profit of $0, sell nothing.   **23.** hottest point: $(1, 1)$; coldest point: $(0,0)$ **25.** hottest points: $(-1/2, \pm\sqrt{3}/2)$; coldest point: $(1/2, 0)$ **27.** $(-\frac{5}{9}, \frac{5}{9}, \frac{25}{9})$ **29.** $(0, \frac{1}{2}, -\frac{1}{2})$ **31.** $1 \times 1 \times 2$ **33.** $(2l/h)^{1/3} \times (2l/h)^{1/3} \times 2^{1/3}(h/l)^{2/3}$, where l = cost of lightweight cardboard, and h = cost of heavy-duty cardboard per square foot   **35.** 11,664 cubic inches (18 inches $\times$ 18 inches $\times$ 36 inches)   **37.** $1 \times 1 \times 1/2$ **39.** Produce 150 quarts of vanilla and 100 quarts of mocha, for a profit of $325.   **41.** Offer 100 sections of Finite Math and no sections of Calculus.   **43.** Use 7.280 servings of Mixed Cereal and 1.374 servings of Tropical Fruit Dessert.   **45.** Buy 100 of each.

Section 15.6 (page 1154)

1. $y = 3x/2 - 2/3$ 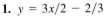   **3.** $y = 0.4118x + 0.9706$ 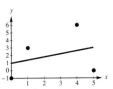   **5.** 3.6%

7. 21.3 billion barrels   **9.** Life expectancy was increasing by 0.291 years each year.   **11.** $P = 101.08t - 330.33$; $P(13) = \$984$ million.   **13.** $6755 billion   **15.** $P = 5{,}835.53e^{0.020769t}$ (t = time since 1790); $P(2{,}000) = 457{,}353{,}000$.   **17.** Linear function; $y = 217.69t + 718.36$ (t = time in years since 1980). High in 2000: 5,072.   **19.** $A = 40.65e^{0.0605t}$ (t = time in years since 1970). Index in the year $2000 = A(30) = 249.63$.   **21.** $q = 3803p^{-2.058}$

23. Taking the natural log of both sides gives $\ln P = \ln K + a \ln x + (1 - a) \ln y$. This leads to the linear relationship $\ln (P/y) = \ln K + a \ln (x/y)$. Thus, given a number of data points (x_i, y_i, P_i), we can use them to calculate the best-fit linear relationship between $\ln (P/y)$ and $\ln (x/y)$. The slope is then a and the intercept is $\ln K$.

Section 15.7 (page 1168)

1. $-\frac{1}{2}$ **3.** $e^2/2 - 7/2$ **5.** $(e^3 - 1)(e^2 - 1)$ **7.** $\frac{7}{6}$ **9.** $\frac{1}{2}[e^3 - e - e^{-1} + e^{-3}]$ **11.** $\frac{1}{2}$ **13.** $\frac{1}{2}(e - 1)$ **15.** $\frac{45}{2}$ **17.** $\frac{8}{3}$ **19.** $\frac{4}{3}$ **21.** 0   **23.** $\frac{2}{3}$ **25.** 2   **27.** 1   **29.** 0   **31.** $\frac{1}{3}$ **33.** $\int_0^1 \int_0^{1-x} f(x, y) \, dy \, dx$ 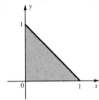   **35.** $\int_0^1 \int_{x^2-1}^1 f(x, y) \, dy \, dx$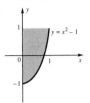

37. $\int_1^4 \int_1^{2/\sqrt{y}} f(x, y) \, dx \, dy$ **39.** $\int_1^4 \int_0^{\ln_2 y} f(x, y) \, dx \, dy$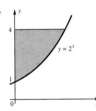

41. $\frac{4}{3}$ **43.** $\frac{1}{6}$ **45.** 161,781 gadgets   **47.** Maximum revenue is $375,500. Minimum revenue is $256,000. Average revenue is $312,750.   **49.** Maximum revenue is $20,000. Minimum revenue is $15,000. Average revenue is $17,500.   **51.** 8,216   **53.** 1 degree

Chapter 15 Review Exercises (page 1176)

1. $g(0, 0, 0) = 0, g(1, 0, 0) = 1, g(0, 1, 0) = 0, g(x, x, x) = x^3 + x^2, g(x, y + k, z) = x(y + k)(x + y + k - z) + x^2$ **3.** $g(0, 0, 0) = 0, g(1, 0, 0) = 1, g(0, 1, 0) = 0, g(x, x, x) = xe^{x^2+x} \; g(x, y + k, z) = xe^{x(y+k)+z}$ **5.** $2\sqrt{2}$
7. $|b - c|$ **9.** 2,511,886   **11.** $f_x = 2x + yz, f_y = xz, f_z = xy, f_{xy} = z, f_{xz} = y, f_{zz} = 0$
13. $f_x = (-x^2 + y^2 + z^2)/(x^2 + y^2 + z^2)^2, f_y = -2xy/(x^2 + y^2 + z^2)^2, f_z = -2xz/(x^2 + y^2 + z^2)^2,$
$f_x(0, 1, 0) = 1$ **15.** 0   **17.** decreasing by 1.02266×10^{-12} newtons per second   **19.** absolute minimum at $(1, \frac{3}{2})$ **21.** saddle point at $(1, 0)$ **23.** saddle point at $(0, 0)$ **25.** critical point at $(0, 0)$ (neither a local extremum nor a saddle point)   **27.** absolute maximum at each point on the circle $x^2 + y^2 = 1$ **29.** $(0, 2, \sqrt{2})$
31. $(1, 0, 1)$ **33.** coldest point: $(1, 0)$; hottest point: $(3, 0)$ **35.** Producing 1,000 pencils and 1,200 pens costs the least. Producing 900 or 1,100 pencils and 1,500 pens costs the most.   **37.** $y = -x/2 + (5/6)$
39. $y = -0.1471x + 1.6176$ **41.** $P = 0.16037e^{0.0013240t}$. The fit is a poor one. $P(2,000) \approx 2.27$ billion, $P(2050) \approx 2.42$ billion.   **43.** $\frac{4}{15}(5^{5/2} - 3^{5/2} - 32 + 2^{5/2})$ **45.** $\frac{1}{2} \ln 5$ **47.** 2
49. $\frac{2}{15}(5^{5/2} - 3^{5/2} - 32 + 2^{5/2})$ **51.** $\frac{1}{4} \ln 5$ **53.** $\frac{2}{\pi}$ **55.** $\frac{40}{3}$ **57.** $230.50

Appendix A

Section A.1 (page A-5)

1. -48 **3.** -1 **5.** 1   **7.** 33   **9.** $\frac{5}{18}$ **11.** 6   **13.** $\frac{43}{16}$ **15.** 3*(2−5)   **17.** 3/(2−5)
19. (3−1)/(8+6)   **21.** 3−(4+7)/8   **23.** (2/3)/5   **25.** 3^(4−5)*6   **27.** 3*(1+4/100)^(−3)
29. 3(1−(−1/2)^2)^2+1

Section A.2 (page A-10)

1. 27   **3.** -36 **5.** $\frac{4}{9}$ **7.** $-\frac{1}{8}$ **9.** 16   **11.** 2   **13.** 32   **15.** 2   **17.** x^5 **19.** $-\frac{y}{x}$ **21.** $\frac{1}{x}$
23. $x^3 y$ **25.** $\frac{z^4}{y^3}$ **27.** $\frac{x^6}{y^6}$ **29.** $\frac{x^4 y^6}{z^4}$

Section A.3 (page A-15)

1. 2   **3.** $\frac{1}{2}$ **5.** $\frac{4}{3}$ **7.** $\frac{2}{5}$ **9.** 3.606   **11.** $\frac{3}{2}$ **13.** 2   **15.** 2   **17.** ab **19.** $x + 9$ **21.** $x\sqrt[3]{a^3 + b^3}$
23. $\frac{2y}{\sqrt{x}}$ **25.** $3^{1/2}$ **27.** $x^{3/2}$ **29.** $(xy^2)^{1/3}$ **31.** $\frac{x^2}{x^{1/2}}$ **33.** $\sqrt[3]{2^2}$ **35.** $\sqrt[5]{x^4}$ **37.** $\sqrt[5]{\sqrt{x}\sqrt[3]{y}}$ **39.** 64
41. $\sqrt{3}$ **43.** $\frac{1}{x}$ **45.** xy **47.** $\left(\frac{y}{x}\right)^{1/3}$ **49.** ± 4 **51.** $\pm \frac{2}{3}$ **53.** $-1, -\frac{1}{3}$ **55.** -2

Section A.4 (page A-18)

1. $4x^2 + 6x$ **3.** $2xy - y^2$ **5.** $x^2 - 2x - 3$ **7.** $2y^2 + 13y + 15$ **9.** $4x^2 - 12x + 9$ **11.** $x^2 + 2 + \frac{1}{x^2}$
13. $4x^2 - 9$ **15.** $y^2 - \frac{1}{y^2}$ **17.** $2x^3 + 6x^2 + 2x - 4$ **19.** $x^4 - 4x^3 + 6x^2 - 4x + 1$ **21.** $y^5 + 4y^4 + 4y^3 - y$

Section A.5 (page A-22)

1. $x(2 + 3x)$ **3.** $2x^2(3x - 1)$ **5.** $(x - 1)(x - 7)$ **7.** $(x - 3)(x + 4)$ **9.** $(2x + 1)(x - 2)$
11. $(2x + 3)(3x + 2)$ **13.** $(3x - 2)(4x + 3)$ **15.** $(x + 2y)^2$ **17.** $(x^2 - 1)(x^2 - 4)$
19. $x = 0, -\frac{2}{3}$ **21.** $x = 0, \frac{1}{3}$ **23.** $x = 1, 7$ **25.** $x = 3, -4$ **27.** $x = -\frac{1}{2}, 2$ **29.** $x = -\frac{3}{2}, -\frac{2}{3}$
31. $x = \frac{2}{3}, -\frac{3}{4}$ **33.** $x = -2y$ **35.** $x = \pm 1, \pm 2$ **37.** $(x + 1)(2x + 5)$
39. $(x^2 + 1)^5(x + 3)^3(x^2 + x + 4)$ **41.** $-x^3(x^3 + 1)\sqrt{x + 1}$ **43.** $(x + 2)\sqrt{(x + 1)^3}$

Section A.6 (page A-25)

1. $\dfrac{2x^2 - 7x - 4}{x^2 - 1}$ **3.** $\dfrac{3x^2 - 2x + 5}{x^2 - 1}$ **5.** $\dfrac{x^2 - x + 1}{x + 1}$ **7.** $\dfrac{x^2 - 1}{x}$ **9.** $\dfrac{2x - 3}{x^2 y}$ **11.** $\dfrac{(x + 1)^2}{(x + 2)^4}$
13. $\dfrac{-1}{\sqrt{(x^2 + 1)^3}}$ **15.** $\dfrac{-(2x + y)}{x^2(x + y)^2}$

Section A.7 (page A-29)

1. -1 **3.** $-\frac{3}{2}$ **5.** -1 **7.** 1 **9.** $-\dfrac{b}{2}$ **11.** $-\dfrac{b}{a}$ **13.** ± 1 **15.** 0 **17.** ± 2 **19.** 1
21. $-1, -6$ **23.** no solutions **25.** no solutions

Section A.8 (page A-40)

1. $x = 5$ **3.** $x = -\dfrac{4}{3}$ **5.** $x = \dfrac{(c - b)}{a}$ **7.** $x = -4, \frac{1}{2}$ **9.** no solutions **11.** $x = \pm\sqrt{\dfrac{5}{2}}$
13. $x = -1$ **15.** $x = -1, 3$ **17.** $x = \dfrac{1 + \sqrt{5}}{2}, \dfrac{1 - \sqrt{5}}{2}$ **19.** $x = 1$ **21.** $x = \pm 1, \pm 3$
23. $x = \pm\sqrt{\dfrac{-1 + \sqrt{5}}{2}}$ **25.** $x = -1, -2, -3$ **27.** $x = -3$ **29.** $x = 1$ **31.** $y = -2$
33. $x = 1, \pm\sqrt{5}$ **35.** $x = \pm 1, \pm\dfrac{1}{\sqrt{2}}$ **37.** $x = -2, -1, 2, 3$

Section A.9 (page A-46)

1. $x = 0, 3$ **3.** $x = \pm\sqrt{2}$ **5.** $x = -1, -\frac{5}{2}$ **7.** $x = -3$ **9.** $x = 0, -1$ **11.** $x = -1$ ($x = -2$ is not a solution.) **13.** $x = -2, -\frac{3}{2}, -1$ **15.** $x = -1$ **17.** $x = \pm\sqrt[4]{2}$ **19.** $x = \pm 1$ **21.** $x = \pm 3$
23. $x = \frac{2}{3}$ **25.** $x = -4, -\frac{1}{4}$

Subject Index

A

Abel, Niels Henrik (1802–1829), A-39
Abstract algebra, 198
Academic testing, 464
Acceleration, 865, 866, 957, 1013
 and second derivative, 865, 869
 due to gravity, 866
 in air, 867
 of cost, 727
 of demand, 868
 varying, 966, 828
Accuracy. *See also* Error
 of approximation to derivative, 682
Acid test, 355
Acidity, 641
Active variable, 279
Addition principle for cardinality, 366–369
 for mutually exclusive events, 421
 for the union of two events, 421
Advertising, 272, 273, 1067
Agriculture, 296–297, 861
AIDS, 645. *See also* HIV testing
Airline safety, 515–516, 546
Airline scheduling, 239, 322–325
Airplane manufacture, 122
Alcohol content, 170
Algebra of rational functions, A-23
Algebra review, A-1–A-47
Algebraic definition of
 definite integral, 981–990
 derivative, 665
 double integral, 1161

Alien intelligence, 1099
Amortization, 727 table
Amusement, 351
Animal psychology, 402–403
Annuity, 574, 599
 decreasing, 579, 582, 584
 increasing, 579, 581
Antiderivative. *See* Integral
Applications of maxima and minima, 845–865
 of the definite integral, 1029–1069
 of the derivative, 821–931
 of the indefinite integral, 951–966
Appreciation, 613, 618, 861
Approximating percentage change, 807
Arc length, 1008
Archery, 16
Area
 between two graphs, 1029–1044
 minimum, 1134
 of an ellipse, 1112
 surface. *See* Surface area
Art, 365
Asset appreciation, 157, 861, 864, 881, 897
Astrology, 422–423, 477
Asymptotes
 horizontal, 37
 vertical, 37
At risk, 465, 481
Automation, 90
Automobile leasing, 594
Automobile manufacturing, 75

Average(s), 1045–1056. *See also* Mean
 average value of a function, 1046
 balance, 1048, 1053
 cost. *See* Cost
 market, 672, 673, 714. *See also* Dow Jones Industrial Average
 moving, 1048–1051, 1054, A-65, A-67, A-78
 of a set of numbers, 1045
 rate of change. *See* Rate of change
 rate of inflation, 1056
 revenue. *See* Revenue
 value of a function of two variables, 1165
 velocity. *See* Velocity
 with a computer spreadsheet, A-76–A-78
 with a graphing calculator A-65–A-67
Aviation, 151–152, 163

B

Bacteria, 625, 626, 651, 657
Bacterial growth, 1085
Balance of trade, 561–562
Balloons, 912
Bank deposits, 482
Bank executives' incomes, 184
Bar exam, 494, 514, 533
Base of logarithm, 627
 change of, 630
Baseball, 546, 551
Basic solution, 279, 282, 291

Basic variable, 179
Bayes' theorem, 459–460
   expanded form, 462
   short form, 460
Beef, price of, 3, 95–97
Benefits of exercise, 467
Bernoulli trial, 470. *See also*
   Binomial distribution
   probabilities of outcomes of a
     sequence, 470
Best fit. *See* Least-squares fit
Biased game, 504
Bimodal distribution, 497
Binomial coefficient, 371
Binomial distribution, 489
   approximation by normal
     distribution, 543
   expected value, 508
   median, 508
   mode, 508
   standard deviation, 524
   variance, 524
Biology, 74
   communication among bees,
     792–793
   embryo development, 714, 715,
     763, 863, 978
   reproduction, 39–40, 673, 689,
     762
Blending
   juices, 112–1114, 150–151
   love potions, 116–117
Bodybuilding, 920
Bonds, 547–548, 597
   government, 717–718, 740
Box design, 861
Break-even, 61, 64, 75, 76, 78, 85,
   88, 100
Breast cancer, 473, 477, 534, 547,
   551
Budget overruns, 777
Budgeting, 122
Building blocks, 365
Bus travel, 762
Business growth, 818–819

C

Cable television, 532
Calculation thought experiment,
   756, 757
Calculus, Fundamental Theorem of,
   969, 990–996
Calendars, 363
Cannonball, height of, 16
Car engines, 363, 365
Car purchases, 492, 512–513
Car rentals, 456, 466
Carbon-14 dating, 625, 626,
   648–649, 651
Cardinality, 344
   addition principle, 366–369
   combinations, 369–378
   of a complement, 347
   guide to counting, 356
   multiplication principle, 354
   permutations, 359–361
   and probability, 429–437
   of a union, 345
   using a counting procedure, 354
   using a decision tree, 353
Cartesian coordinates. *See*
   Coordinates
Cartesian plane. *See* xy-plane
Casting, 358, 364
Cat at the piano, the, 439
Cauchy, Augustin Louis
   (1789–1857), 661
Cellular telephones, 75
Celsius, 72
Chain rule, 764–778
   for several variables, 1117
   in words, 765
Chebyshev's inequality, 525
Chlorofluorocarbons (CFCs), 121
Chocolates, 403, 418
Class scheduling, 120
Class size, 486, 487
Closed-form function, 696
Cobb-Douglas production function,
   799–801, 912, 964, 979,
   1091, 1096, 1111, 1113,
   1121, 1159, 1170

Codes, 363, 365
College
   enrollment, 955
   tuition, 920, 921, 955, 1126
Combination locks, 365
Combinations, 369–378
   vs. Permutations, 370, 374
Commission, 99, 101, 753
Committees, 368, 380–381, 387,
   401, 435–436, 441
Common factor, A-19
Common logarithm, 629
Communication among bees,
   792–793
Complement, 337. *See also* Set
   cardinality of, 347
   of an event, 395
   probability of, 425
Compound interest, 564, 607
Compounding period, 564
Computer programming, 366, 379
Computer spreadsheet
   difference quotients, A-69–A-71
   Riemann sums, A-71–A-74
   Simpson's rule, A-75
   Trapezoid rule, A-74
   using, A-69–A-78
Computerized axial tomographic
   (CAT) scanners, 161–162
Concavity
   and second derivative, 870
   concave up/down, 869
   point(s) of inflection, 870–873
Conditional probability, 441–444
   for equally likely outcomes, 443
   experimental, 443
   general definition, 444
Constant dollars, 569–570, 575
Constant-profit lines, 256
Constrained maxima and minima,
   1132–1144
   Lagrange multipliers, 1138–1141
Constraints, 254, 846
Construction cost, 1143
Consumer price index, 1158
Consumer satisfaction, 899, 989

Consumers' surplus, 1034–1037, 1089
Consumption, 169
Contests, 439–440
Continuity correction, 543
Continuity. *See* Function(s)
Continuous
  appreciation, 618
  compounding, 617. *See also* Exponential growth
  depreciation, 619. *See also* Exponential decay
  function. *See* Function
  growth and decay. *See* Exponential growth and Exponential decay
Coordinates
  in the plane, 5–7
  in three-dimensional space, 1100–1102
  x- and y- coordinates, 6
  viewing windows, 10
  z- coordinate, 1101
Corporate income, 701
Cost
  analysis, 812
  average, 78, 100, 763, 846, 862, 897, 926, 963, 1028, 1120, 1131
  construction, 1143
  equation. *See* Cost, function
  fixed, 59
  function, 24, 39, 58–61, 72–74, 78, 99–100, 714, 802, 881, 937, 940, 951, 962, 977, 980, 1017, 1024, 1028, 1084, 1090, 1116–1117, 1142
  housing. *See* Housing costs
  marginal, 59, 716, 724, 726, 741, 748, 763, 937, 940, 951, 988, 1013, 1120
  maximum/minimum, 1131, 1177
  medical, 1072, 1090
  minimization, 265–267, 318, 319, 320, 334
  modeling. *See* Modeling, cost
  of a warranty, 1008–1011
  of reducing sulfur emissions, 741
  production, 923
  total, 962, 977, 980, 1017, 1024, 1028, 1043, 1089
  transportation. *See* Transportation costs variable, 59
Counting procedure, 354
  acid test for, 355
CPA exam, 75, 104
Credit cards, 363, 624
Crime, 880, 913
  disorganized, 76
  organized, 76
  and preschool, 465
Critical point(s), 828, 1124
Curve sketching, 882–899
Customer IDs, 386

D

Dangerous weapons, 1084
Dantzig, G. (1914–), 276
Darts, 487–489
Database as a set, 333, 339, 342, 343
Dating, 544
Debts, 337, 339
Decay, continuous. *See* Continuous decay
Decibel, 641
Definite integral. *See* Integral
Degree requirements, 170
Delta notation, 668
Demand, 117–119, 122, 123. *See also* Modeling demand and Sales
  acceleration of, 868
  elastic, 917
  elasticity of. *See* Elasticity of demand
  equation/function, 15, 38, 67–72, 77, 78, 84, 99, 100, 101, 642, 665, 672, 688, 763, 793, 802, 808, 812, 813, 819, 881, 911, 920, 921, 927, 1017, 1075, 1085, 1096, 1171–1175
  exponential demand functions, 920–921
  for a new product, 875
  for beer, 1098
  for homes, 1087, 1171–1175
  for fried chicken, 920
  for oranges, 920
  for solar cells, 1013
  growth in, 674, 689
  hyperbolic demand functions, 920
  inelastic, 917
  linear demand functions, 920, 921, 1043
  modeling. *See* Modeling, demand
  total, 955
  unit elasticity of, 917
Dental plans, 76, 94
Departing variable, 22, 281
Depletion of natural gas reserves, 1155
Depreciation, 574, 577, 659
Derivative, 659
  applications of, 821–931
  approximate value of, 668
  as rate of change, 662–674
  chain rule, 764–778; in words, 765
  computing from the definition, 698–700
  delta notation for, 668
  definition of, 665
  existence of, 731
  generalized power rule, 767
  geometric interpretation of, 675–690
  graphically using graphing calculator, 681–682. *See also* Graphing calculator
  implicit differentiation, 775
  notation, 704–705
  numerical approximation with graphing calculator, 666, 668
  of a constant times x, 710
  of a constant, 710
  of exponential functions, 779–794; in words, 785
  of logarithmic functions, 779–794; in words, 780
  of powers and polynomials, 709–715
  of products and quotients, 752–764

of sums, differences and constant multiples, 708
of the integral, 993
partials, 1112–1122. *See also* Partial derivatives
power rule, 706–708
power rule for negative integers, 757
power rule for rational exponents, 775
product rule, 752–764; in words, 753
relationship with indefinite integral, 939
second derivative, 865–882. *See also* Second derivative
units of, 665
with respect to x, 705

Descartes, Rene (1595–1650), 7
Designing a puzzle, 331, 382–385
Dice, 341, 380, 477
Diet, 121
Difference quotient, 669
  with a computer spreadsheet, A-69–A-71
  with a graphing calculator, A-61–A-63
Differentiable function, 731
Differential equation(s), 1069–1078
  elementary, 1070
  general solution of, 1070
  separable, 1072
Differential notation, 704–705, 772
Differentiation, techniques of, 751–823
Diffusion of new technology, 793
Discontinuity, 695, 730
  removable, 731
Discontinuous function. *See* Discontinuity
Disease, 524, 527
Displacement. *See* Motion
Distance
  between two points in the plane, 1092
  between two points in three-dimensional space, 1112
  to the origin in the plane, 1093

Distribution, 412–413
  bimodal, 497
  binomial, 489–508, 524
  cumulative, 498, 499
  expected value of binomial, 508
  experimental, 485
  median of binomial, 508
  mode of binomial, 508
  normal, 489, 535
  of a random variable, 483
  standard normal, 537
  theoretical, 485
  uniform, 489
  visualizing using a histogram, 483
Dividends, 597, 598
DNA, 363
Doggie puddles, 911
Domain, 23–25. *See also* Function, domain of
  continuity on, 695, 696
Donations, 161
Double integrals, 1159–1171
  algebraic definition of, 1161
  computing over a nonrectangular region, 1163, 1164
  computing over a rectangular region, 1161
  geometric definition of, 1150
Doubling time, 643–644
Drug tests, 457–461

E

$E = mc^2$, 980
e, 615–616
  evaluating on a calculator, 616
Earthquakes, 641
Ecology, 76, 777
Econometrics, 160, 161, 171
Economy
  of Australia, 229–230
  of Germany, 236
  of Indonesia, 230–231
  of Japan, 179, 231–234
  of Kenya, 225–226
  of U.S., 220–222, 227–228
Education, 863, 880, 912, 913
Education and occupation, 454–456
Education fund, 582, 585–586

Effective interest rate, 570, 571, 600. *See also* Effective yield
Effective yield, 571, 620
Einstein, Albert (1879–195), 41. *See also* Relativity, theory of
Einstein's energy equation ($E = mc^2$), 980
Elastic demand, 917
Elasticity of demand, 916–922, 927
Electric current, 160, 1057
Electrostatic repulsion, 1121
Element, 332. *See also* Set
Elementary differential equation, 1070
Elementary row operations, 126
Ellipse, 795
Embryo development. *See* Biology
Emission control, 725
Employment, 456, 466, 493, 513, 533, 626, 652, 803, 912
Empty set, 332
  probability of, 425
Endpoints, 828
Energy, 76, 219–220
English spelling, 386
Entering variable, 281
Entertainment, 351, 352
Environmental Protection Agency, 741
Epidemics, 77, 577, 645, 650, 651, 653–656, 749, 786, 793, 819, 879
Equality of limits, 698
Equation(s), A-25–A-29
  cost. *See* Cost
  function, demand. *See* Demand function
  linear, 42
  of a line: point-slope form, 47–48, 55
  of a line: slope-intercept form, 51, 55
  polynomial, A-30
  solution, A-28
  solving, by factoring, A-31; cubic equations, A-36–A-40;

polynomial equations, A-29–A-40; for exponent, 631–635; higher order polynomial equations, A-39; miscellaneous, A-40–A-47; numerically with graphing calculators, 89–94; quadratic equations, 81, A-30–A-36
test for existence of real solutions, A-34
Equilibrium price, 117–119, 1039, 1089
erf. *See* Error, function
Error
  estimation, 804–813, 1005–1007
  function (erf), 539, 995
  in Simpson's rule, 1003
  in the Trapezoid rule, 1003
Estimating mortgage rates, 813–819
Estimating relative change, 807
Event, 392
  complement of, 395
  disjoint, 397
  impossible, 413
  independent, 449
  intersection of two, 397
  mutually exclusive, 397
  union of two, 396
Exchange rates, 185, 202
Exiting variable, 281
Expansion, 762
Expected value, 501
  of a binomial distribution, 508
Experiment, 390
Exponential
  curve, 568
  decay, 615–627, 657
  form, 627
  functions, 602–615; derivatives of, 779–794
  growth, 615–627; doubling time for, 643–645
  half-life, 646–649
Exponents
  identities/laws of, 105, A-6
  integer, A-6–A-10
  negative, A-8
  positive, A-6
  rational, A-10–A-16
  relationship with radicals, A-14
  zero, A-8
Exports, 419, 964, 978, 1045, 1047
Extremum/extrema. *See* Maxima and minima

F

Factoring, A-19–A-23
Faculty salaries, 1094
Fahrenheit, 72
Fair game, 504
Family values, 465
Farm population, 512
Fast cars, 926, 966. *See also* Slow cars
Fast food, 1042
  customers, 1053
  outlets, 1053
Feasible region, 250, 2155
  unbounded, 267
Federal deficit, U.S., 15
Feeding schedules, 156
Fences, 859
File names, 379
Finance, 531
Financing long-term medical care, 419
Fitness tests, 457, 464
Floppy disks, 379
Football, 515
Foreign investments, 465. *See also* Investments
Foreign trade, 1044
Fractional decay rate, 622
Fractional growth rate, 621
Frequency, 404–405
  cumulative, 498–499
  relative, 404–405
  table, 484
Frogs, 749
Fuel economy, 726, 762, 818, 979. *See also* Fuel efficiency and Gas mileage
Fuel efficiency, 88. *See also* Fuel economy and Gas mileage

Function(s), 10, 18–41
  average value of, 1046
  continuous (continuity), 695–696, 730–731
  cost, 24, 39, 58. *See also* Cost, function
  definition of, 18
  demand. *See* Demand, function
  differentiable, 731
  discontinuous, 695
  domain of, 23–25, 883, 884
  exponential vs. linear, 606, 615
  exponential, 104–117
  gamma, 1068
  graph of, 26–41. *See also* Graph
  greatest integer, 36
  implicit, 795
  inverse, 637–638
  linear function of several variables, 1095
  linear, 42, 58
  logarithmic, 627–642
  logistic, 653–656, 674, 689
  normal distribution, 535
  numerical definition of, 23
  of several variables, 1087, 1083, A-67
  piecewise, defined, 29
  profit, 61. *See also* Profit
  quadratic, 79–89, 164–167
  rational, 735
  real-valued, 18
  revenue, 38, 61. *See also* Revenue
  rounding, 37
  sine, 899
  tan, 899
  vertical line test for, 30
  zero(s) of, 89–94
Fundamental Theorem of Calculus, 969, 990–996
  geometric proof of, 990–993
  statement of, 990
Future value, 555, 610

G

Galilei, Galileo (1564–1642), 866
Galois, Evariste (1811–1832), A-39
Gambling, 420
Gamma function, 1068

Gas mileage, 926
Gasoline stations, 273
Gauss, Carl Friedrich (1777–1855), 135
Gauss-Jordan reduction. *See* Row reduction
General solution. *See* Differential equation(s)
Generalized power rule, 767
Geometric definition/interpretation of
   definite integral, 967–980
   derivative, 675–690
   double integral, 1160
   partial derivative, 1114
Geometric series, 581
GMAT, 99, 100, 121, 122, 169, 363, 379, 439–440, 657
Gold coins, 476
Gold stocks, 625
Government bonds, 650
Grades, 463–464, 476, 803, 913, 1120
Gradient. *See* Slope
Graph searching, 440
Graph(s)
   analyzing with the aid of a graphing calculator, 894–896
   of equations, 8–14
   of exponential functions, 604–606, 637
   of functions, 26–41, 882–899
   of logarithmic functions, 636–639
Graphing functions by hand, 882–893
Graphing calculator, A-49–A-62
   accuracy of, 109–110
   analyzing graphs with the aid of a graphing calculator, 894–896
   and simplex method, 288–291
   and statistics, 497, 519, A-59–A-61
   answering investment questions with, 572–573
   approximate value of derivative, 668
   approximating pi (π) with, 1002
   average cost, 847

average rate of change, 666
averages and moving averages, A-65–A-67
break-even, 64, 86
calculating difference quotients, A-61–A-62
calculating normal probabilities, 539, 541, A-61–A-62
calculating Riemann sums, 985–987
college enrollment, 956
comparing experimental and theoretical probability, 410–412
compound interest, 608–609
depreciation, 611–612
derivative, 668, 681–682, 732, 733
entering a matrix, 175, A-55
error in Simpson's rule and Trapezoid rule, 1005–1011
estimating derivatives graphically, 681–682
evaluating a function, A-53–A-59
evaluating e, 616
existence of derivative, 732, 733
exponential decay, 622–623
exponential functions, 604, 605, 619, 732
finding loan repayment time, 590–591
finding points where two graphs cross, A-52
finding the equilibrium point, 119
formatting expressions for, A-4
functions of several variables, 1089, A-67
graphing a function, A-49–A-53
graphing future value, 565
graphing histograms, 484, A-60–A-61
graphing inequalities, 245
improper integrals, 1061, 1065–1066
labor resource allocation, 857
left sums, sums, A-63–A-65
limits at infinity, 734, 735
linear functions, 10
logarithmic functions, 637, 639
matrix operations, 177, 187, A-54–A-56

maxima and minima, 825–827, 835–836, 838, 840–841
maximum profit, 723
maximum revenue, 72
moving average, 1050–1051, A-65–A-67
normal distribution, 989
order of operations, 566
piecewise-defined functions, 29
point(s) of inflection, 873–874
predicting sales with logistic equation, 965
probability, A-67–A-68
quadratic functions, 83
referring to the entries of a matrix, 175
Riemann sums, A-63–A-65
right sums, A-63–A-65
row-reducing a matrix, 131, 132, 788, A-56–A-57
sales growth, 788
sales of freon, 1061
scaling of axes, 49
setting the viewing window automatically, A-52
Simpson's rule, 9901, A-63–A-65
skydiving, 960
solving equations with, 89–94
tracking a decreasing annuity, 585
tracking a mortgage, 588
tracking an increasing annuity, 581–582
Trapezoid rule, 901, A-63–A-65
using a graphing calculator, A-49–A-68
viewing window, 10, A-60
xMin, xMax, yMin, yMax, 10, 13, A-60
zeros of functions, 89–94
Graphing. *See* Graph
Gravity, Newton's Law of. *See* Newton's Law of Gravity
GRE economics exam, 40, 78, 122, 160, 161, 171, 473, 657, 726, 727, 763, 778, 803, 812, 864, 921
Greek life, 428
Growth
   continuous. *See* Continuous growth

of HMOs, 793
of medical costs, 1072, 1090
of tumors, 1078
Guide to counting, 356

H

Half-life, 646–649
Harvesting forests, 864, 881
Health care, 492–493
Health insurance, 593, 513, 533
Health spending, 979
Hemisphere, 1106
Histogram, 483
HIV testing, 420, 464, 473
HMOs, growth of, 793
Horse races, 403
Hospital finance, 400
Hospital staffing, 320
Hotel rooms, 926
Housing costs, 531, 725, 803, 997, 1112, 1121
Hyperbola, 11, 32

I

Implicit differentiation, 775, 794–803
Implicit function, 795
Improper integrals, 1057–1069
Inactive variable, 279
Income, 73, 103, 105, 484–485, 494, 509, 514, 528, 533, 656
   distribution, 1079
   elasticity of demand, 921
   tax, 22
Indefinite integral. *See* Integral
Independent events, 449
Independent variable, 21
Industrial output, 879
Inelastic demand, 917
Inequality, 241–242. *See also* Linear inequality
   in an unknown, 242
   rules for manipulating, 241
   solution of, 242
Infant height, 16

Infant mortality, 549, 1155
Infant weight, 16
Inflation rate(s), 40, 625, 657
   varying, 625
Inflation, 569–570, 575–576, 598
Information highway, 674, 690, 777, 793
Inheritance, 169
Input-output models, 216–226
   external demands vector, 218
   indirect effects, 223
   input-output table, 220, 222
   loosely coupled, 226
   production vector, 218
   strongly coupled, 226
   technology matrix, 218
Insurance, 75, 158, 504, 514, 546
Integral, 539, 929–1017. *See also* Integration
   and area between two graphs, 1029–1044
   antiderivative, 930
   applications of the definite integral, 1029–1069
   applications of the indefinite integral, 951–966
   definite, 967; algebraic definition of, 981–990; as total change, 974; geometric definition of, 967–980
   double. *See* Double integrals
   form of acceleration, velocity, and position, 957
   improper, 1057–1069
   indefinite, 930–940
   limits of, 967
   numerical integration, 996–1008
   of e^x, 934
   power rule for, 917
   relationship with derivative, 939
   rules for, 934
   shortcuts for integrals involving $(ax + b)$, 949
   substitution, 941–951
   Simpson's rule, 1000–1008
   table of, A-81–A-83
   Trapezoid rule, 997–1008
   units of, 953, 954

Integration. *See also* Integral
   by parts, 1016–1025; formula for, 1016; tabular method, 1017
   of a polynomial times a function, 1020
   using tables, 1025–1029
Intelligence, alien, 1099
Intercept
   as fixed cost, 59
   as starting position, 66
   in demand function, 69
   x- and y-intercepts, of a line, 50, 51; of a parabola, 81; of the graph of a function, 884–893; of a plane, 1105
Interest, 597, 598
   compound, 564, 576, 607
   compounding period, 564
   continuous compounding of, 607
   effective rate, 570, 571
   effective yield, 571
   nominal rate, 571
   simple, 555
Interior design, 476
Intermediate value theorem, 243
Interpreting the news, 562
Intersection, 337. *See also* Set
   of two events, 397
Interval(s), 12–14
   closed, 12
   half-open, 13
   open, 12
   unions of, 13
Intramurals, 403
Inventory control, 158, 159, 186
Inverse functions, 637–638
Investments, 94, 121, 154–156, 157, 159, 253, 272, 298, 329, 347, 401–402, 431–432, 438, 511, 553, 598, 613, 614, 624, 643, 650, 651, 657, 762, 792, 793, 880, 1053, 1085, 1121
   compounded, 565–568, 572–576, 578
   depreciating, 568–569
   foreign/international, 963, 1068, 1096

in South Africa, 38, 1075
  maximizing return on, 260–262, 298
IQ scores, 546, 547, 551
Isoprofit lines, 256
Isotherms, 1098
Itineraries, 362

J
Job training, 473
Jordan, Wilhelm (1842–1899), 135
Judges and juries, 428–429

K
Karmarkar's algorithm, 276
Karmarkar, N., 276
Kepler, Johannes (1571–1630), 627
Khachiyan, L.G., 276
Kinetic energy, 980
Kirchoff's current law, 160
Kirchoff, Gustav Robert (1824–1887), 160

L
Labor resource allocation, 857
Ladders, 907, 911
Lagrange multipliers, 1138–1141
Laplace transform, 1169
  largest possible, 24–25
  of a function, 23–25, 695, 696
Law, 121, 428–429
Learning
  to read, 673
  to speak, 673
Least-squares fit
  line, 1145–1159
  parabola, 1087, 1171–1175
  power function, 1158
Left and right sum. See Riemann sum
Leibniz, Gottfried (1646–1716), 660
Length of a curve, 1008
License plates, 364, 386
Lie detectors, 420, 457, 461–462

Life expectancy, 549
Life span, 792, 1155
Limit(s), 690–704, 727–740
  at infinity, 733–736
  definition of, 692
  divergent, 729
  equality of, 698
  evaluating algebraically, 695–700
  evaluating graphically, 692–695, 734, 735
  evaluating numerically, 690–692
  evaluating with a graphing calculator, 693
  existence of, 692, 694, 727
  infinite, 729
  left and right-limits, 693–694
  of rational functions, 736
Limiting velocity, 867
Line
  constant profit, 256
  equation of, 42–55
  horizontal, 46, 54–55
  isoprofit, 256
  slope of, 43–45
  vertical, 46, 54–55
Linear
  approximation, 804–813
  function, 42–55, 58–72. See also Function(s), linear
  function of several variables, 1095
  model(s). See Function(s), linear
  regression, 1145–1159
Linear equation, 42–55, 58–72. See also Function(s), linear
  coefficient row, 124
  in three or more unknowns, 138–147
  system of. See System of linear equations
Linear inequality, graphing, 241–251
Linear programming, 1144
Linear programming problem
  constant profit lines, 256
  constraints, 254
  existence of optimal solutions, 268
  feasible region, 255

graphical method for solving, 258
in two unknowns, 254
isoprofit lines, 256
objective function, 254
optimal solution, 254
optimal value, 254
solving graphically, 254–270
solving using the simplex method, 276–292, 300–315. See also Simplex method
standard maximization problem, 276
standard minimization problem, 3143
with unbounded feasible region, 267
Loan planning, 297
Loans, 418, 454, 526, 542, 593, 597, 598, 624, 1085
  amortization of, 587
  automobile, 593–596, 598
  bridge, 556, 559–560
  credit card, 590–591, 593–594, 599
  home, 587–588. See also Mortgage
Local extrema. See Maxima and minima
Local maxima and minima. See Maxima and minima
Logarithm, 627–642
  as exponent, 627
  base of, 627
  change of base, 630
  common, 629
  function(s), (graph of), 637–639
  identities, 630, 635–636
  natural, 629–779
Logarithmic differentiation, 799
Logarithmic form, 627
Logarithmic function, 627–642
  derivatives of, 779–794
Logistic function, 653, 969, 1077, 1097. See also Function(s)
  derivation of, 1077
  use in modeling epidemics, 653–656
  use in predicting sales, 965
Lorenz, Konrad (1903–1989), 402

Lot size management, 922–925
Lotteries, 439
Lotto, 373, 438–439
Luggage dimensions, 862

M

Magazine sales, 1012, 1013
Management, 122, 263–265
Manatees, 777
Manufacturing, 158
Marbles, 374–376, 379–380, 387, 403, 418
Marginal
   analysis, 716–727
   cost. *See* Cost
   loss, 724
   product, 722, 724, 727, 763, 776, 777, 818
   profit. *See* Profit
   revenue. *See* Revenue
Market average. *See* Average(s)
Market saturation, 1077, 1078
Market share, 1054
Market surveys, 465
Marketing, 457, 546
Marketing strategy, 861
Marriage, 864
Matrix, 127
   addition, 178–179
   augmented, 127, 199
   clearing a column, 127–128
   coefficient, 198
   column, 177
   determinant of, 209
   diagonal, 216
   dimensions, 174
   elementary row operations on, 126
   entries, 174, 175
   equality, 176
   identity, 195
   inverse, 203
   inverse of a 2 X 2, 208
   invertible, 204
   laws of addition and scalar multiplication, 182, 196–197
   laws of transposition, 183, 197
   multiplication, 187–192
   operations using a graphing calculator, A-54–A-56
   reduced row echelon form of, 134
   row, 177
   row-reduced, 130
   row reduction of, 130
   row reduction using a graphing calculator, A-56–A-59
   scalar multiplication, 180, 181
   singular, 204
   subtraction, 178, 179
   of a system of equations, 127
   technology, 218
   transposition, 177
   using to solve a system of linear equations, 123–147
   zero, 182
Maxima and minima, 823–865, 1122–1144
   absolute extremum, 824; maximum, 824, 1125; minimum, 824
   applications of, 845–865, 1134–1144
   constrained, 846, 1132–1133
   constraint (equation), 846
   endpoints, 828
   Lagrange multipliers, 1138–1141
   local (relative) extremum, 823, 1122; maximum, 823, 1122; minimum, 823, 1122 locating extrema analytically, 827–828, 1123; combining graphical and analytical approach, 840–841; graphically, 825–827; in curve sketching, 893
   objective function, 846
   optimization problem, 846; procedure for solving, 850
   relative. *See* Maxima and minima, local
   second-derivative test for, 871, 1126
   singular point, 828
   stationary point, 827
   test point, 834
Maximum. *See* Maxima and minima
Mazes, 364, 386, 387
Mean, 499, 501. *See also* Expected value, Average
   of a collection of data, 495–500
   and histograms, 497
   of a random variable, 501–502
   sample, 502
   of a set of numbers, 1045
Measurement(s), error, 808
Median, 499, 501
   of a collection of data, 495–500
   and histograms, 497
   of a random variable, 501
Medical costs, 1072, 1085
Medical insurance, 391–392, 393, 406–407, 415–417
Medical subsidies, 749
Medicine, 351, 457
Melanoma, 77
Menus, 351, 362
Mergers and acquisitions, 445, 532
Meteor impacts, 1068
Minimizing resources, 854, 863
Minivan sales, 1099, 1119
Mode, 499, 501
   of a collection of data, 495–500
   and histograms, 497
   of a random variable, 501
Modeling
   cost, 881
   demand, 67, 77, 78, 95–97, 642, 881, 921
   employment decline, 652
   enrollment, 653
   epidemics, 653–656
   income, 656
   international investments, 1096–2097
   revenue, 881
   sales, 655–656
   supply, 642
   with the cosine function, 1056
   with the Cobb-Douglas production function, 1096, 1121, 1159
Money stock, 778, 903
Money supply, 1173
Monkey at the typewriter, the, 439

Monty Hall problem, the, 389, 474–475
Morse code, 380
Mortality rates, 158
Mortgage, 567–590, 593, 598
　adjustable rate, 593
　payments, 813–819
　refinancing, 593
Motion, 88, 99, 941, 951, 956–962, 974, 977, 1024, 1085
Moving average, 1048–1051
　with a graphing calculator, A-65–A-67
Multiple-choice tests, 362, 369, 379, 440, 506
Multiplication principle
　for cardinality, 354
　for probability, 445–446
Municipal bonds, 494–495
Municipal waste, 396
Muscle recovery time, 74
Mutual funds, 608–650
Mutual funds, 704–705
Mutually exclusive events, 397
　addition principle for, 421

N

Napier, John (1550–1617), 627
Natural logarithm, 629. *See also* Logarithm, Logarithmic function(s)
Natural numbers, 334
Net income, Sony Corporation, 38
Newton, Isaac (1642–1727), 660
Newton's Law of Cooling, 1073
Newton's Law of Gravity, 41, 101, 673, 689, 1093–1094, 1115
Nominal interest rate, 571
Nonnegative numbers, 334
Normal density function, 989, 1069
Normal distribution (curve). *See* Normal density function
Normal distribution, 489, 535
　approximation to binomial distribution, 543
　calculating from standard distribution, 539
　probability of being within k standard deviations of the mean, 541
　standard, 537
Numbers, 380
Numerical integration, 997–1008. *See also* Integral
Numerical methods, 1007
Nutrient diffusion, 1121
Nutrition, 114–116, 120, 121, 122, 170, 215, 253, 265–267, 271, 272, 318, 320, 1144

O

Objective function, 254, 846
Oil
　exploration costs, 978
　reserves, 74, 76
Operations on the real numbers, A-2–A-5
Operator overloading, 182
Opinion polls, 418–409, 428
Optimal solution, 254
Optimal value, 254
Optimization problem, 845–846
　procedure for solving, 850
Optimizing capacity, 863
Or, 338
Organized crime, 76
Orr, William, 422
Outcomes, 390
　equally likely, 410
　favorable, 392
Oversupply, 1017

P

Package dimensions, 862, 1143
Panic sales, 1067
Parabloid, 1103
Parabola, 31, 32, 81–89
　vertex of, A-51–A-55
Parameter, 144
Parity, 449
Parking garage fees, 926
Partial derivative(s), 1112–1122
　geometric interpretation of, 1114
　higher order, 1118
　mixed, 1118
Partition, 462
Patents, 880, 898
Pedestrian safety, 1085
Pension plans, 157, 592, 593, 599
　defined-benefit, 579
　defined-contribution, 579
Percentage change, 807
Percentage rate of change, 794, 807
Perfect progression, 439
Permutations, 359–361
　of n items, 359
　of n items taken r at a time, 360
　vs. Combinations, 370, 374
Pest control, 158
pH, 641
Pi (π), 1002
Pivoting. *See* Matrix, clearing a column
Plane, 1105
Planning, 274
Plantation management, 861, 897
Playing cards, 376, 456. *See also* Poker hands
Plutonium-239 dating, 625
Point(s) of inflection, 870–871, 893
Point-slope formula, 47–48, 55
Poker hands, 373, 377, 380, 387, 401, 433–434, 438, 476. *See also* Playing cards
Polls, 547
Pollution, 76, 121, 164–167, 454, 777, 989, 1154
　emission control, 725, 862, 1043, 1044
　sulfur emissions, 659, 741–742, 819
Polya enumeration, 384
Popularity, 672, 688
Population migration, 185, 203, 215, 466

Population standard deviation. *See* Standard deviation
Population variance. *See* Variance
Population, 428, 1167
　decline, 649
　density, 1171
　farm, 512
　growth, 76, 611, 621, 625, 651, 753, 792, 818, 1152, 1177
Port exports, 965, 978, 1045, 1046, 1047, 1068
Position, 957, 958
Positive numbers, 334
Poultry, price of, 3, 77, 88, 95–97
Power rule
　for derivatives, 706, 707–708
　for integrals, 933
　for negative exponents, 757
　for rational exponents, 775
　generalized, 767
Present value, 610, 626
Presidential pardons, 477
Pricing policy, 761, 916
Principal, 607
Prison population, 863, 880
Probability. *See also* Random variable
　addition principle, 421
　comparison of experimental and theoretical, 410–412
　of complement, 425
　conditional, 441–444. *See also* Conditional probability
　and counting techniques, 429–437
　density curves, 535
　density function, A-67–A-68
　distribution, 412–413; of a random variable, 483
　of E, given F. *See* Conditional probability
　of empty set, 425
　with equally likely outcomes, 410, 429–437
　experimental, 404–405, 408, 485
　multiplication principle, 445–446
　of outcomes of a sequence of Bernoulli trials, 470
　theoretical, 408–409, 485

　and tree diagrams, 446–447
　of union, 421
　of whole sample space, 425
Producers' surplus, 1038–1041, 1089
Product design, 362–362, 380, 440–441
Product reliability, 456, 546
Product rule, 752, 764
　in words, 753
Production, 912, 964, 979
　costs, 923
　lot size management, 922–925
Profit, 61, 64, 74, 75, 185, 203, 335, 672, 688, 714, 880, 1077, 1155
　average, 898, 926, 1131, 1177
　maximizing, 859, 860, 861, 1136, 1142, 1143
Project design, 273
Psychology, 351
Public health, 429, 473
Publishing, 122
Punch bowls, 927
Purchasing, 121, 136, 169, 253, 271, 295, 296, 310–313, 1144

Q
Quadrants, 7
Quadratic formula, 81, A-33
Quadratic functions, 79–89. *See also* Function(s), quadratic
Quadratic model, 164–167
Quality control, 441, 473, 540, 813
Quotient rule, 752–764 in words, 753 proof of, 769–770

R
Radar detectors, 379
Radicals, A-10–A-16
　relationship with rational exponents, A-14
Radioactive decay, 94, 622, 647, 652, 657, 792
Railroad transportation, 428

Rainfall, 491–492
Random variable, 480
　binomial, 489. *See also* Binomial distribution
　continuous, 535
　expected value, median, and mode of, 501
　experimental probability distribution, 485
　finite, 480
　normal, 489
　probability distribution of, 483
　standard deviation of, 520
　theoretical probability distribution, 485
　uniform, 489
　variance, 520
Rate of change, 65, 66, 72, 748, 749, 900
　as derivative, 662–674, 900
　average, 662–674, 956
　instantaneous, 662–674
　percentage, 794, 807
Rate of increase, 324. *See also* Rate of change
Rational expressions, A-23–A-29
　algebra of, A-23
Raw data, 482
Reading lists, 349
Recycling, 76, 297
Reduced row echelon form of a matrix, 134
Reducing sulfur emissions, 741–744
Related rates, 900–914
　rephrasing a related rates problem, 901
Relative change, 807
Relative extrema, 823. *See also* Maxima and minima
Relative maxima and minima, 823. *See also* Maxima and minima
Relativity, Einstein's theory of, 41, 673, 689, 927, 980
Removable discontinuity, 731. *See also* Discontinuity
Reproduction. *See* Biology
Research productivity, 1120

Resource allocation, 120, 136, 137, 215, 250–252, 258–260, 268–272, 274, 285–287, 296, 328, 329, 869, 1143, 1144
Retirement income, 559
Retirement planning, 574–575, 581–582, 586–587, 593
Returns on investments, 880, 881, 898
Revenue, 38, 61, 73, 77, 78, 84, 87, 99, 100, 202, 204, 761, 763, 777, 818, 859, 881, 911, 917, 919, 926, 974, 977, 978, 1013, 1024, 1067, 1091, 1131, 1132
   average, 963, 1166, 1170
   from taxes, 1015, 1078, 1083
   lost, 1024
   marginal, 718, 724, 725, 774, 776
   maximizing, 71, 78, 852, 859, 860, 864, 918, 921
   modeling. *See* Modeling, revenue
Richter scale, 641
Riemann sum, 983
   left-hand, right-hand, 983, A-63–A-65, A-71–A-74
Riemann, Georg Friedrich Bernhard (1826–1866), 983
Risk management, 275, 589
Roulette, 452, 504, 515
Row operations. *See* Elementary row operations
Row reduction, 130
Row-reduced matrix, 134
Russell's paradox, 347
Russell, Bertrand (1872–1970), 347

S

Saddle point(s), 1123
Salaries, 423–424, 456, 466, 495, 498, 531. *See also* Income
   faculty salaries, 994
   scales in Japan, 101
Sales, 178–181, 296, 456, 516, 655, 656, 674, 689, 787, 879, 953, 963, 979, 988, 1017, 1042, 1043, 1055, 1060, 1067, 1077, 1099. *See also* Demand, function

analysis, 812, 813
incentives, 749. *See also* Commission
   of minivans, 1099
   predicting with logistic function, 965
   and revenue, 189, 192–193
Sample mean, 502
Sample size, 405
Sample space, 390
   event as subset of, 392 finite, 390, probability of whole, 425
Sample variance, 528
SAT scores, 547, 550, 872, 913
Savings accounts, 593–594, 598
Savings, 624, 657
Scalar, 180. *See also* Matrix, scalar multiplication
Scheduling, 274, 321
Scrabble™, 355, 357
Secant line, 677–678
Second derivative, 865–882
   and acceleration, 865–869
   and concavity, 869–877
   and points of inflection, 871
   geometric interpretation of, 869–877
   test for maxima and minima, 871, 1126
Separable differential equation, 1072
Set, 332
   complement, 337
   database as a, 333, 339, 342, 343
   disjoint union, 345
   element of 332
   empty, 332
   finite, 344
   improper subset of, 335
   intersection, 337
   logical equivalents of operations on, 337
   number of elements in. *See* Cardinality
   operations on, 337
   of outcomes, 333, 390
   partition of, 462
   proper subset of, 332
   subset of, 332
   union, 337

universal, 341,
   Venn diagrams of, 336
Setup costs, 923
Shadows, 927
Shakespeare, William (1564–1616), 332, 349, 364
Sheridan, R. B. (1751–1816), 358
Ships sailing apart, 912
Shortcuts for integrals solving $(ax + b)$, 949
Sigma notation, 499
Simplex method
   active variable, 279
   basic solution, 279, 282, 291
   basic variable, 279
   cycling in, 315
   departing variable, 281
   entering variable, 281
   and equality constraints, 315
   exiting variable, 281
   inactive variable, 279
   meaning of slack variables in, 287
   for minimization problems, 310–315
   nonbasic variable, 279
   and nonstandard constraints, 300–309
   for nonstandard problems, 300–315
   selecting the pivot in Phase I, 301, 313–314
   selecting the pivot in Phase II, 280
   slack variable for, 278
   for standard maximization problems, 276–292
   for standard minimization problems, 313–315
   surplus variable for, 300
   tableau, 278
   test ratio, 280
Simpson's rule, 1000–1008
   error in, 1003
   with a computer spreadsheet, A-75–A-76
   with a graphing calculator, A-63–A-65
Sine function, the, 899
Singular maxima and minima, 828
Singular point, 828

Singularity, 12
Skydiving, 960
Slack variable, 278
   meaning of, 287
Sliding ladders. *See* Ladders
Slope, 43–45
   as marginal cost, 59
   as rate of change, 65, 66
   as velocity, 64–66
   formula for, 45
   in demand function, 69
   infinite, 46
   of a curve, 675
   of tangent line, 675
   of secant line, 677–678
   parallel lines, 46–47
   perpendicular lines, 46–47
   undefined, 46
Slope-intercept formula, 51, 55
Slow cars, 966. *See also* Fast cars
Soap bubbles, 777
Soccer, 348
Social Security numbers, 362
Solution, 8
   of equations, numerical, 89–94
Solving equations. *See* Equations
Sound intensity, 641
Speed of light, 41
Sports, 121, 352, 362, 439, 515, 546
   injuries, 471
   team finances, 400
   tournaments, 440
Spreadsheet
   and mean, median, and mode, 497–498
   row-reducing a matrix, 131
   and simplex method, 288
Spreadsheet. *See* Computer spreadsheet
Standard deviation
   of a binomial distribution, 524
   Chebyshev's inequality, 525
   for raw data, 516–520,
   of a random variable, 520
Standardized tests, 473

Stationary maxima and minima, 827
Stationary point, 827
Statistics
   of a collection of data, 495–500, 516–520
   misuses, 548–550
   of a random variable, 501, 520
   using a graphing calculator, A-59–A-61
Steepness of a graph, 676
Stock market, 351, 362, 386, 419–420, 574–575, 597
Stock princes, 5, 15, 651, 762, 749, 1156, 1157
Storage costs, 923
Student evaluations, 472
Student loans, 418, 454
Subset, 332. *See also* Set
   improper, 335
   proper, 332
Subsidies, 319
   clean air, 725–726, 1043, 1044
   medical, 753
   to utilities, 743, 1043, 1044
Substituting, 19
Substitution (integration), 942–951
Sulfur emissions. *See* Pollution
Summation notation, 499
Suntanning, 611
Supply equation/function, 642, 911, 1059. *See also* Modeling
   supply linear supply, 943
Supply, 117–119, 122, 123, 157, 158
Surface area of a cylinder, 39, 94
Surplus
   Consumers.' *See* Consumers' surplus
   Producers.' *See* Producers' surplus
Surplus variable, 300
Surveying, 989
Surveys, 346, 405
Swords and sorcery, 428, 649
Symmetries, 381

System of linear equations
   applications, 150–156
   augmented matrix of, 127, 199
   coefficient matrix of, 198
   consistent, 147,
   general solution of, 144
   in three or more unknowns, 138–147
   in two unknowns, 105–120
   matrix of, 127
   overdetermined, 147
   parametrized solution of, 144
   particular solution of, 144
   possible outcomes, 112
   solving algebraically, 106–120
   solving by common sense, 105
   solving graphically, 106, 108
   solving using matrices, 123–137, 138–147
   solving using matrix inversions, 209–212
   undetermined, 147
   with infinitely many solutions, 111–112, 135–136, 143–146
   with no solutions, 110–111, 136, 146

T

Table of integrals, A-81–A-83
Tan function, the, 899
Tangent line, 675, 679, 715, 805
   and linear approximation, 805
   equation of, 679
   slope of, 675–678
Tangent plane, 1122
Tax bracket, 1079
Tax depreciation, 74
Taxation schemes, 726
Taxes, 22, 74, 1015
   estimating revenue from, 1015, 1078–1083
Telephone numbers, 362, 363
Television ratings, 169, 170
Temperature, 72, 652, 1143, 1171, 1177
Tenure, 161

Test point, 834
Test scores, 863, 880, 899
Tests, 122, 491
Three-dimensional space, 1100–1102
Tides, 1057
Total change, 974
    cost. *See* Cost function demand. *See* Demand, total revenue. *See* Revenue sales. *See* Sales
Toxic waste treatment, 38
Trade deficit, U.S., 15
Traffic flow, 152–154, 159, 160
Transfers, 439
Transportation costs, 726, 812
Transportation, 299, 321, 329
Transposition of a matrix, 177
Trapezoid rule, 997–1008
    error in, 1003
    with a computer spreadsheet, A-71–A-74
    with a graphing calculator, A-63–A-65
Trapezoidal sum, 984
Travel, 456
Traveling salesperson, 365
Tree diagram, 353
    and probability, 446–447
Trial, 405
True/False tests, 436–437
Trust funds, 585
Tuberculosis, 77

U

Undecided voters, 466
Uniform random variable, 489
Union, 337. *See also* Set
    cardinality of, 345
    disjoint, 345
    probability of, 421
    of two events, 396
Unit elasticity, 917
Units of derivative, 665
Units of the integral, 953, 954
Universal set, 341
Using a graphing calculator, A-49–A-68
Utility, 1097, 1111, 1120

V

Variance
    of a binomial distribution, 524
    for raw data, 516–520
    of a random variable, 520
    sample vs. Population, 528
Varying acceleration. *See* Acceleration, varying
Vector. *See also* Matrix
    column, 177
    row, 177
Velocity, 64–66, 669, 715, 748, 957, 958, 965, 966
    average, 669, 670
    instantaneous, 669, 670
    limiting, 867
Venn diagrams, 336
Vertex of parabola, 80, 81
Viewing window, 10, A-50–A-52
Volume, 715
    discount, 951, 962, 1028
    ellipsoid, 715, 1097
    cone, 39, 913, 1098
    cylinder, 913
    maximum, 853, 856
    sphere, 715
Voting, 120, 158

W

Warranty, cost of issuing, 1008–1011
Weapons, 1089
Weather predictions, 451, 457, 465, 476, 580
Weather
    fair, 1055
    foul, 1056
Wheatstone bridge, 160
Window. *See* Viewing window
Wooden beams, 926

X

x-intercept. *See* Intercept
X-scores, 482
xMax, 10, 13
xMin, 10, 13

Y

y-intercept. *See* Intercept
yMax, 10, 13
yMin, 10, 13

Company Index

Abbott Laboratories, Inc., 360, 438, 1044
ABC Network, 171
Aetna Casualty, 158
Airbus, 151, 156
Allstate Insurance Company., 75
American Airlines, 163
American Express, 2198
Ameritech, 360
Amoco, 360
Avis, 466
Batman T-shirts, 1067
Baxter International, 360
Bayshore Trust, 154
Berger Funds, 578, 650
Boeing 747–100, 74
Campbell's, 122, 215
Canadian Western Bank, 154
Carnival Corp., 185, 978
CBS, 298
Chemical Bank, 298
Chicago Bears, 400
Chicago Blackhawks, 400
Chicago Bulls, 400
Chicago Cubs, 400
Chicago White Sox, 400
Chrysler Corp., 601
Coca-Cola, 656
Commonwealth Edison, 360
Delta Airlines, 163
Dial Corporation, 512

Dow Jones Industrial Average, 5, 15, 662, 1157
Eli Lilly Corp., 965
EMC Corp., 978
EMI, 298
Enron Corp., 1017
ESPN, 171
Executive Life, 157
Fidelity Investments, Inc., 121, 511, 608, 704–705
Ford Motor Co., 75, 1099
Genentech Corp., 965
General Mills, Inc., 114
General Motors Corp., 1099
Genzyme, Inc., 965
Gerber, 120, 253, 271–272, 381
GTE Corp., 762
Hertz, 466,
IBM Corp., 714, 988
Japan Air System, 151
Maritime Life, 154
McDonald's, 360, 438, 1053, 1054
McDonnell-Douglas DC-10, 74
Merrill Lynch, 253
Microsoft Corp., 1043
Montreal Trust, 154
Motorola, 360, 438
New York Stock Exchange, 390
Nielsen (ratings), 171
Northrop Grumman, Inc., 626, 652
Northwest Airlines, 163

Old Mutual, Inc., 351
People's Energy Corp., 438
Polygram, 298
Prozac, 1055
Prudential Property and Casualty, 158
Prudential Securities, 253, 272
Radio Shack, 411
Renault, 157
Safety Kleen, 297
Sanlam, Inc., 351
Sara Lee, 360
Sears, Roebuck & Co., 360, 438
Snapple Beverage Corp., 1043
Sony Corporation, 38, 298, 561
TBS Network, 171
Tootsie Roll Industries, Inc., 438
Toronto Dominion Bank, 184
Toys "R" Us, 762
TWA, 163
United Airlines, 974–976, 988
United Pacific, 157
United Parcel Service, 862, 1143
USAir, 163
Vanguard Investments, Inc., 121
Volvo, 157
Walt Disney Co., 1155, 1156
Warner Music, 298
Waste Management, 360
Western Union, 822

Interior design, 476
Interpreting the news, 562
Intramurals, 403
Inventory control, 158, 159, 186
Investments, 38, 94, 121, 154–156, 157, 159, 253,
    260-262, 272, 298, 329, 347, 401–402,
    431–432, 438, 511, 553, 565–569, 572–576,
    578, 598, 613, 614, 624, 643, 650, 651, 657,
    762, 792, 793, 880, 963, 1053, 1068, 1075,
    1085–1121
IQ scores, 546, 547, 551
Isotherms, 1098
Itineraries, 362
Japan, 179, 231–234
Job training, 473
Judges and juries, 428–429
Kenya, 225–226
Kinetic energy, 980
Labor resource allocation, 857
Ladders, 907, 911
Law, 121, 428–429
Learning, 673
License plates, 364, 386
Lie detectors, 420, 457, 461–462
Life expectancy, 549
Life span, 792, 1155
Loan planning, 297
Loans, 418, 454, 526, 542, 556, 559–560, 587–588,
    590–591, 593–596, 597, 599, 601, 624, 1085
Lot size management, 922–925
Lotteries, 439
Lotto, 373, 438–439
Luggage dimensions, 862
Magazine sales, 1012, 1013
Management, 122, 263–265
Manatees, 777
Manufacturing, 158
Marbles, 374–376, 379–380, 387, 403, 418
Market saturation, 1077, 1078
Market share, 1054
Market surveys, 465
Marketing strategy, 861
Marketing, 457, 546
Marriage, 864
Mazes, 364, 386–387
Measurement(s), error, 808
Medical costs, 1072, 1085
medical costs, 1072, 1090
Medical insurance, 391–392, 393, 406–407, 415–417
Medical subsidies, 749
Medicine, 351, 457
Menus, 351, 362

Mergers and acquisitions, 445, 532
Meteor impacts, 1068
Minimizing resources, 854, 863
Minivan sales, 1099, 1119
Money stock, 778, 903
Money supply, 1177
Monkey at the typewriter, the, 439
Monty Hall problem, the, 389, 474–475
Morse code, 380
Mortality rates, 158
Mortgage, 567–590, 593, 601, 813–819
Motion, 88, 99, 941, 951, 956–962, 974, 977, 1024,
    1085
Multiple-choice tests, 362, 369, 379, 440, 506
Municipal bonds, 494–495
Municipal waste, 396
Muscle recovery time, 74
Mutual funds, 608–650, 704–705
Net income, Sony Corporation, 38
Nominal interest rate, 571
Nutrient diffusion, 1121
Nutrition, 114–116, 120, 121, 122, 170, 215, 253,
    265–267, 271, 272, 318, 320, 1144
Oil, 74, 76, 978
Opinion polls, 418–409, 428
Organized crime, 76
Oversupply, 1017
Package dimensions, 862, 1143
Panic sales, 1067
Parking garage fees, 926
Patents, 880, 898
Pedestrian safety, 1085
Pension plans, 157, 579, 592, 593, 599
Pest control, 158
pH, 641
Plantation management, 861, 897
Playing cards, 376, 456
Plutonium-239 dating, 625
Poker hands, 373, 377, 380, 387, 401, 433–434, 438,
    476
Polls, 547
Pollution, 76, 121, 164–167, 454, 659, 725, 741–742,
    777, 819, 862, 989, 1043, 1044, 1154
Popularity, 672, 688
Population, 76, 185, 203, 215, 428, 466, 512, 611, 621,
    625, 649, 651, 753, 792, 818, 1152, 1167, 1171,
    1177
Port exports, 965, 978, 1045, 1046, 1047, 1068
Poultry, price of, 3, 77, 88, 95–97
Presidential pardons, 477
Pricing policy, 761, 916
Principal, 607